ENERGY SCIENCE AND APPLIED TECHNOLOGY ESAT 2016

PROCEEDINGS OF THE INTERNATIONAL CONFERENCE ON ENERGY SCIENCE AND APPLIED TECHNOLOGY (ESAT 2016), WUHAN, CHINA, 25–26 JUNE 2016

Energy Science and Applied Technology ESAT 2016

Editor

Zhigang Fang
Wuhan University of Technology, Wuhan, Hubei, China

CRC Press
Taylor & Francis Group
Boca Raton   London   New York   Leiden

CRC Press is an imprint of the
Taylor & Francis Group, an **informa** business

A BALKEMA BOOK

Published by: CRC Press/Balkema
P.O. Box 11320, 2301 EH Leiden, The Netherlands
e-mail: Pub.NL@taylorandfrancis.com
www.crcpress.com – www.taylorandfrancis.com

First issued in paperback 2020

ISBN 13: 978-0-367-73682-8 (pbk)
ISBN 13: 978-1-138-02973-6 (hbk)

Visit the Taylor & Francis Web site at
http://www.taylorandfrancis.com

and the CRC Press Web site at
http://www.crcpress.com

Typeset by V Publishing Solutions Pvt Ltd., Chennai, India

Table of contents

3 Materials science and chemical engineering

6 Mechanical, manufacturing, process engineering, fluid engineering and numerical simulation

7 Control and automation, system simulation

13 Recognition, video and image processing

14 Civil engineering and geotechnical engineering

Energy Science and Applied Technology ESAT 2016 – Fang (Ed.)
© 2016 Taylor & Francis Group, London, ISBN 978-1-138-02973-6

Preface

The 3rd International Conference on Energy Science and Applied Technology (ESAT 2016: http:// www. ESATconf.org/, Wuhan, China, 25–26 June 2016) aims to provide an interactive platform for the academics, R & D organizations, industries, innovators, entrepreneurs, government agencies and policy-makers from both China and abroad to exchange ideas on recent researches and advances in the domain of energy science and applied technology with its prospective application in the various interdisciplinary domains of engineering.

This two-day conference was organized in technical association with the Asian Union of Information Technology and will include invited keynote talks and oral paper presentations from both academia and industry. The themes covered in this proceedings are: Technologies in Geology, Mining, Oil and Gas Exploration and Exploitation of Deposits; Renewable Energy and Cell Technologies, Energy Recovery, Engines, Generators, Electric Vehicles; Materials Science and Chemical Engineering; Environmental Engineering and Sustainable Development; Electrical and Electronic Technology, Power System Engineering; Mechanical, Manufacturing, Process Engineering, Fluid Engineering and Numerical Simulation; Control and Automation, System Simulation; Communications and Applied Information Technologies; Applied and Computational Mathematics, System Simulation; Methods and Algorithms Optimization; Network Technology and Application; System Test, Diagnosis, Detection and Monitoring; Recognition, Video and Image Processing; Civil Engineering and Geotechnical Engineering; Traffic and Transportation Engineering.

The book is useful to professionals and scientists involved or interested in Energy Science and Applied Technology.

With our warmest regards

Zhigang Fang
Conference Organizing Chair
Wuhan, China

Energy Science and Applied Technology ESAT 2016 – Fang (Ed.)
© 2016 Taylor & Francis Group, London, ISBN 978-1-138-02973-6

Committees

SCIENTIFIC COMMITTEE

Chairman

Dr. Z.G. Fang, *Wuhan University of Technology, China*

Co-Chairmen

Dr. C. Yang, *Wuhan University of Technology, China*
Dr. Y.F. Zhao, *Wuhan University of Technology, China*
Prof. G.J. Fu, *Northeast Petroleum University, China*

Members

Dr. Z.G. Fang, *Wuhan University of Technology, China*
Prof. J.J. Xu, *Northeast Petroleum University, China*
Dr. J. Hu, *Ohio State University, USA*
Dr. W. Zhong, *New York State University at Stony Brook, USA*
Prof. H. Davis, *Boya Century Publishing Ltd., Hong Kong*
Associate Prof. D. Fang, *Wuhan University, China*
Prof. G.J. Fu, *Northeast Petroleum University, China*
Prof. P. Wang, *Guangxi College of Education, China*
Associate Prof. H. Chen, *Shanghai University of Engineering Science, China*
Dr. C. Yang, *Wuhan University of Technology, China*
Dr. Y.F. Zhao, *Wuhan University of Technology, China*
Dr. X.H. Deng, *Wuhan University of Technology, China*
Dr. J. Luo, *Wuhan University of Technology, China*
Dr. G.L. Liu, *Wuhan University of Technology, China*
Dr. X.L. Li, *Peking University, China*
Dr. H.H. You, *Zhicheng Conference Services Ltd., China*

Organizing Committee

Dr. Z.G. Fang, *Wuhan University of Technology, China*
Associate Prof. L. Liu, *Wuhan University of Technology, China*
Dr. C. Yang, *Wuhan University of Technology, China*
Dr. Y.F. Zhao, *Wuhan University of Technology, China*
Dr. J. Luo, *Wuhan University of Technology, China*
Dr. X.H. Deng, *Wuhan University of Technology, China*
Dr. G.L. Liu, *Wuhan University of Technology, China*
Dr. X. Xiao, *Wuhan University of Technology, China*
Dr. C. Zhang, *Wuhan University of Technology, China*
Associate Prof. D. Fang, *Wuhan University, China*
Dr. H.H. You, *Zhicheng Conference Services Ltd., China*
Prof. Y. Ma, *Zhicheng Conference Services Ltd., China*
Dr. J. Shi, *Zhicheng Times Culture Development Co., Ltd., China*
Dr. L. Xu, *Zhicheng Times Culture Development Co., Ltd., China*

Energy Science and Applied Technology ESAT 2016 – Fang (Ed.)
© 2016 Taylor & Francis Group, London, ISBN 978-1-138-02973-6

Sponsors

Northeast Petroleum University
Boya Century Publishing Ltd.
Research Center of Engineering and Science (RCES)
Asian Union of Information Technology
HuBei XinWenSheng Conference Co., Ltd.

1 *Technologies in geology, mining, oil and gas exploration and exploitation of deposits*

Energy Science and Applied Technology ESAT 2016 – Fang (Ed.)
© 2016 Taylor & Francis Group, London, ISBN 978-1-138-02973-6

Numerical simulation of integral fracturing parameters in an ultra-low-permeability oil field

C. Lu, B. Luo, J.C. Guo, Z.H. Zhao, X. Xu & S.W. Xiao
State Key Laboratory of Oil and Gas Reservoir Geology and Exploitation, Southwest Petroleum University, Chengdu, China

Y. Wang
Lusheng Petroleum Development Co. Ltd., Sinopec Shengli Oilfield Company, Dongying, China

B.G. Chen
Luming Oil & Gas Exploration and Development Co. Ltd., Sinopec Shengli Oilfield Company, Dongying, China

ABSTRACT: North block of Zhuang 23 (Z23) belongs to low-porosity and ultra-low-permeability reservoirs and exploitation of integral fracturing is necessary for the efficient development of this block. Appropriate fracture parameters are an immediate requirement when carrying out the hydraulic fracturing stimulations in Z23. In order to select the optimal fracture parameters in Z23, a numerical method was used in this paper. A geological model was developed according to the real physical properties of Z23, and the hydraulic fractures were prefabricated in the five-spot well pattern. The effects of fracture length and fracture conductivity on oil production rate, field recovery ratio, and field water cut were completely studied using the above numerical model. Simulation results showed that the optimal fracture half-length and fracture conductivity in Z23 were from 168 m to 192 m and from 30 D.cm to 35 D.cm, respectively. This paper provided an effective theoretical support to the development of Z23; finally, the overall efficient development of the block was achieved. Meanwhile, the research results of this paper could serve as an important part in the development of this kind of low-permeability reservoirs.

Keywords: hydraulic fracturing; fracture parameter; low permeability; productivity forecast; numerical simulation

1 INTRODUCTION

Hydraulic fracturing is commonly used in the development of ultra-low-permeability reservoirs (Gan et al. 2001 & Zhang Shicheng 2003). Practices and theories show that hydraulic fracture conductivity and fracture length are the essential parameters to ensure the well production performance after fracturing (Zeng et al. 2007 & Osholake 2013).

Hydraulic fracture volume is determined by the volume of the proppant and hydraulic fluid. In the process of fracture propagation, the fracture length competes with that at a given fracture height. Hence, the larger the fracture length, the smaller the fracture width. Different fracture lengths and fracture widths contribute to different reservoir production capacities. Therefore, an optimal fracture length and width is needed to assure the optimal well production performance (Britt LK 1985 & Ma et al. 2011 & Zeng et al. 2012). Fracture conductivity is associated with the fracture width; hence, numerous studies on the effect of fracture conductivity and fracture length were carried out (Cui et al. 2001 & Economides et al. 2002 & Zhang et al. 2004 & Marongiu et al. 2008 & Duan et al. 2013).

In this paper, a five-spot well pattern was established, which completely took into consideration the production characteristics of low-permeability reservoir and fracture flowing characteristics. The numerical simulation method was used to analyze the influence of fracture conductivity and fracture length on well production performance. These parameters, which could guarantee field production rate and field oil recovery to reach the highest value, are selected as the optimal hydraulic fracture parameters. According to the above theory, the fracture parameters of Z23 were optimized, and stimulation results showed that the production performance was largely improved.

2 RESEARCH STRATEGY

2.1 *Governing equations*

The oil reservoir and hydraulic fracture mathematical model was developed based on the production performance of the low-permeability reservoir and the flow characteristics of hydraulic fracture (Meehan D N 1995). Oil reservoir and hydraulic fracture were considered as relatively independent systems while studying their interaction mechanism. The two parts were banded together by the pressure and fluid flowing in their contact surface, and fluid flowing in the reservoir and fracture obey Darcy's law.

2.1.1 *Reservoir model*
We assume that

1. fluid flows in a three-dimensional two-phase oil reservoir;
2. reservoir permeability is anisotropic;
3. the compression coefficient remains constant;
4. the influence of gravity and capillary forces is ignored;
5. the reservoir produces at a constant pressure;
6. The well pattern used is the five-spot well pattern.

Equations of oil and water phases are as follows:

$$\frac{\partial}{\partial x}\left(\frac{\rho_o K K_{ro}}{\mu_o}\cdot\frac{\partial P_o}{\partial x}\right)+\frac{\partial}{\partial y}\left(\frac{\rho_o K K_{ro}}{\mu_o}\cdot\frac{\partial P_o}{\partial y}\right)$$
$$+\frac{\partial}{\partial z}\left(\frac{\rho_o K K_{ro}}{\mu_o}\cdot\frac{\partial P_o}{\partial z}\right)=\frac{\partial}{\partial t}(\varphi\rho_o S_o) \tag{1}$$

$$\frac{\partial}{\partial x}\left(\frac{\rho_w K K_{rw}}{\mu_w}\cdot\frac{\partial P_w}{\partial x}\right)+\frac{\partial}{\partial y}\left(\frac{\rho_w K K_{rw}}{\mu_w}\cdot\frac{\partial P_w}{\partial y}\right)$$
$$+\frac{\partial}{\partial z}\left(\frac{\rho_w K K_{rw}}{\mu_w}\cdot\frac{\partial P_w}{\partial z}\right)=\frac{\partial}{\partial t}(\varphi\rho_w S_w) \tag{2}$$

$S_o + S_w = 1$, $P_o = P_w$, $K_{ro} = K_{ro}(S_o)$, $K_{rw} = K_{rw}(S_w)$, $\rho_o = \rho_o(P_o)$, $\rho_w = \rho_w(P_w)$,

where ρ_o and ρ_w are the oil and water densities, kg/m³; K is the formation permeability; K_{ro} and K_{rw} are the relative permeabilities; μ_o and μ_w are the oil and water viscosities, mPa.s; P_o and P_w are the oil and water pressures, MPa; φ is the porosity,%; S_o and S_w are the oil saturation and water saturation; x, y, and z are the coordinates; and t is the time, s.

2.1.2 *Fracture model*
We assume that

1. hydraulic fracture is a vertical fracture having a rectangular shape;
2. the hydraulic fracture is homogeneous, and the fracture permeability is isotropic;
3. hydraulic fracture is effective in the long term without closure.

Equations of oil and water phases are as follows:

$$\frac{\partial}{\partial x}\left(\frac{K_f K_{rof}}{\mu_o\beta_o}\cdot\frac{\partial P_f}{\partial x}\right)+\frac{\partial}{\partial z}\left(\frac{K_f K_{rof}}{\mu_o\beta_o}\cdot\frac{\partial P_f}{\partial z}\right)$$
$$+q_{ofin}=\frac{\partial}{\partial t}(\varphi\rho_o S_o) \tag{3}$$

$$\frac{\partial}{\partial x}\left(\frac{K_f K_{rwf}}{\mu_w\beta_w}\cdot\frac{\partial P_f}{\partial x}\right)+\frac{\partial}{\partial z}\left(\frac{K_f K_{rwf}}{\mu_w\beta_w}\cdot\frac{\partial P_f}{\partial z}\right)$$
$$+q_{wfin}=\frac{\partial}{\partial t}(\varphi\rho_w S_w) \tag{4}$$

$$S_o + S_w = 1$$

where K_f is the fracture permeability, μm₂; K_{rof} and K_{rwf} are the oil and water relative permeabilities of grid point in fracture; P_f is the pressure of grid point in fracture, MPa; β_o and β_w are the volume factors of oil and water; and q_{ofin} and q_{wfin} are the flow rates of oil and water of grid point in fracture, m³/s.

2.2 *Establishment of optimization model*

Hydraulic fracturing stimulation is necessary in low-permeability reservoirs. However, fracturing may cause the premature wells get flooded; hence, whether the well should be stimulated or how to screen the facture parameters should be taken into consideration. Setting of oil recovery, water cut, and oil production rate as goals to optimize fracture parameters in Z23 is based on black oil model. A 480 m × 150 m five-spot well pattern is used to simulate the production performance of Z23 (Figure 1). Five wells are oil production wells and the extension direction of prefabricated hydraulic fracture is vertical to minimum horizontal principal stress.

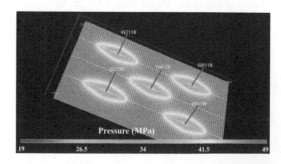

Figure 1. Five-spot well pattern model.

Table 1. Basic simulation parameters.

Parameters	Value
Initial formation pressure (MPa)	49
Depth (m)	3672
Saturation pressure (MPa)	15.0
Formation permeability (mD)	1.0
Formation porosity (%)	14.1
Oil volume factor	1.3
Oil compression factor (MPa^{-1})	1.7×10^{-3}
Surface oil density (g/cm^3)	0.86
Formation oil viscosity (mPa·s)	15
Formation water volume factor	1.1
Formation water compression factor (MPa^{-1})	1.1×10^{-3}
Surface water density (g/cm^3)	1.0
Formation water viscosity (mPa·s)	0.1
Rock compressibility (MPa^{-1})	0.4×10^{-3}

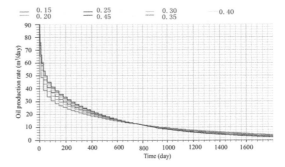

Figure 2. Effects of different RFW on field oil production rate.

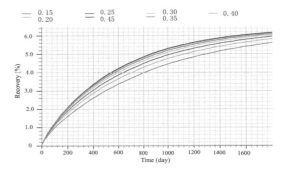

Figure 3. Effects of different RFW on field oil recovery.

According to the exploitation program, the well control area was 960 m × 600 m, the reservoir effective thickness was 12 m. The X direction was divided into 80 girds with a grid size of 12 m, the Y direction was divided into 60 girds with a grid size of 10 m, and the Z direction was divided into 4 girds with a grid size of 3 m. Then, 15 grids were added into the Y direction to set up the hydraulic fracture, and the total grid number is 24000.

Assume that the fracture conductivity is 30 D.cm, and then study the influence of the ratio of the fracture length to the well spacing (RFW) on field recovery, water cut, and oil production rate in 3 years. The RFW range is 0.15, 0.20, 0.25, 0.30, 0.35, 0.40, and 0.45.

2.2.1 Effect of the RFW on oil production performance

Figure 2 shows that hydraulic fractured wells have a high initial production rate, which decreases rapidly and fall into a stable stage with a low production rate finally. A longer fracture length means a greater production rate, but the increase rate decreases gradually. For example, the production rates of fractured wells after the first month are 38.6 m^3/d, 43.5 m^3/d, 46.8 m^3/d, 48.9 m^3/d, 50.1 m^3/d, 53.5 m^3/d, 53.8 m^3/d at the fracture half-length of 72 m, 96 m, 120 m, 144 m, 168 m, 192 m, and 216 m, respectively.

The resistance to fluid flowing near the bottom of the well becomes smaller with an increase in fracture length at the initial production period. Therefore, it is obvious that the bigger the RFW, the greater the production rate. In addition, the well production rate is still related to formation characters, production pressure difference, and so on. When the fracture length increases to some degree, the dimensionless fracture conductivity would decrease. Hence, the fracture length con-

tributed less to the oil production rate. Therefore, from the perspective of dimensionless fracture conductivity, it is better that the fracture length is not longer. However, there is an appropriate proportion between reservoir properties.

Figure 3 shows that the oil recovery increases with the increase in RFW, but the increase rate decreases gradually. Oil recovery reaches the highest value of 6.17% at a RFW of 0.45. It is also observed that the oil recovery difference is 0.42% between a RFW of 0.3 and a RFW of 0.15, while the oil recovery difference is 0.12% between a RFW of 0.45 and a RFW of 0.3.

It is observed from Figure 4 that the water cut increases with the increase in RFW, but the increase rate becomes smaller. Water cut reaches the highest value of 12.86% after 5 years of production at a RFW of 0.45.

2.2.2 Effect of fracture conductivity on oil production performance

Reservoir well spacing is 480 m, row spacing is 150 m, and well control area is 960 × 600 m^2 at the given five-spot well pattern. Field oil recovery, field water cut, and field oil production rate were tested under different fracture conductivities, which ranged from 10 D.cm to 40 D.cm.

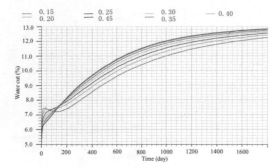

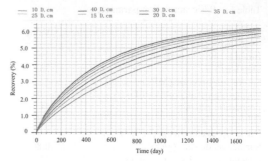

Figure 4. Effects of different RFW on field water cut.

Figure 6. Effects of different fracture permeabilities on field oil recovery.

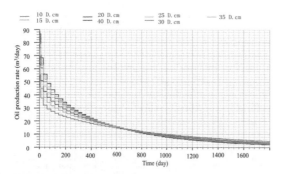

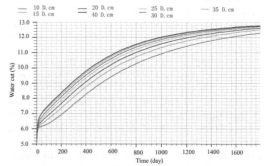

Figure 5. Effects of different fracture permeabilities on field production rate.

Figure 7. Effects of different fracture permeabilities on water cut.

The graph of field oil production rate at varied fracture conductivity (Figure 5) shows that high fracture conductivity is likely to possess a high initial oil production rate. For instance, at day 30 of production, the field oil production rates at the fracture conductivities of 10 D.cm, 15 D.cm, 20 D.cm, 25 D.cm, 30 D.cm, 35 D.cm, and 40 D.cm are $32.2\ m^3/d$, $37.9\ m^3/d$, $42.7\ m^3/d$, $46.7\ m^3/d$, $50.1\ m^3/d$, $56.2\ m^3/d$, and $55.8\ m^3/d$, respectively. Moreover, this figure illustrates that the rate of increase in the oil production rate from 10 D.cm to 25 D.cm is greater than that from 25 D.cm to 40 D.cm.

Figure 6 shows that the field oil recovery increases with the increase in production time; however, the rate of increase gradually becomes constant in the final period. It is clear that the rate of increase in field recovery from 10 D.cm to 20 D.cm is greater than that from 25 D.cm to 40 D.cm. The conclusion can be drawn from Figure 6 that appropriate fracture conductivity should be 30 D.cm.

It is observed from Figure 7 that the water cut increases with the field production time, while the rate of increase gradually becomes stable in the final period. A high fracture conductivity means a high field water cut. It is clear that the rate of increase in field water cut from 20 D.cm to 30 D.cm or from 30 D.cm to 40 D.cm increases only slightly.

3 RESULT ANALYSIS

In this paper, the RFW and fracture conductivity were studied to figure out their influence on field production performance. In the design of this simulation, the above two parameters have seven different variable values. Hence, the total simulation number was 49. In the analysis of a single variable, the influence of RFW and fracture conductivity on oil production rate, water cut, and recovery was studied, respectively.

3.1 *The impact of fracture parameters on oil production rate*

The first month's oil production rate was analyzed after hydraulic stimulation (Figure 8), and it was found that the production rate increases with the increasing RFW and fracture conductivity. Numerical simulation results reveal that it tends to enhance the initial oil production rate at the condition of long hydraulic fracture length and high fracture conductivity.

It could be observed (Figure 8, left) that the production rate increases slowly when the RFW is greater than 0.35, while it increases rapidly when the RFW is less than 0.3. Therefore, simulation

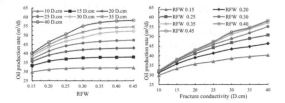

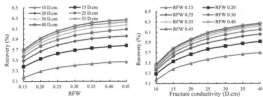

Figure 8. The influence of fracture parameters on oil production rate.

Figure 9. The influence of fracture parameters on recovery.

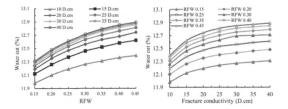

Figure 10. The influence of fracture parameters on water cut.

results confirm that 0.35 is the critical value of RFW. Considering the stimulation cost, the RFW should be confined in the process of hydraulic fracturing. Meanwhile, the simulation results show that the production rate increases linearly with the increase in fracture conductivity (Figure 8 right).

Based on the evaluation index of the first month's production rate after fracturing, the optimal RFW is approximately 0.35 (fracture half-length equaled to 168 m), and a greater fracture conductivity is better.

3.2 The impact of fracture parameters on recovery

The field recovery for 5 years is analyzed after hydraulic stimulation (Figure 9), which indicates that the field recovery increases with the increase in RFW and fracture conductivity. Numerical simulation results reveal that it tends to enhance the recovery at the condition of long hydraulic fracture length and high fracture conductivity.

It could be observed (Figure 9 left) that the increase in recovery is slow when the RFW is greater than 0.35, while it is rapid when the RFW is less than 0.3. Therefore, simulation results argue that 0.35 is the critical value of RFW. Considering the stimulation cost, the RFW should be confined to the process of hydraulic fracturing. Meanwhile, the field recovery increased with the increasing fracture conductivity (Figure 9 right), but it increased slowly when the fracture conductivity is greater than 30 D.cm. Therefore, simulation results argue that 30 D.cm was the critical value of fracture conductivity.

Based on the evaluation index of the recovery rate after fracturing for 5 years, the RFW should be limited to approximately 0.35 (fracture half-length equaled to 168 m), and the fracture conductivity should be set around 30 D.cm.

3.3 The impact of fracture parameters on water cut

The field water cut for 5 years is analyzed after hydraulic stimulation (Figure 10), which shows

that the field water cut increases with the increase in RFW and fracture conductivity.

It is observed (Figure 10 left) that the water cut increases slowly when the RFW is greater than 0.4, while it increases rapidly when the RFW is less than 0.35. Therefore, simulation results argue that 0.35 is the critical value of RFW. Meanwhile, field water increases with the increase in fracture conductivity (Figure 10 right), but, the increase is slow when fracture conductivity is greater than 35 D.cm. Therefore, simulation results verify that 35 D.cm is the critical value of fracture conductivity.

Based on the evaluation index of the water cut after fracturing for 5 years, the RFW should be confined below approximately 0.4 (fracture half-length equal to 192 m), and fracture conductivity should be maintained at 30D.cm

3.4 Optimization results by numerical simulation

The above three evaluation indexes show that the optimal fracture parameters for Z23 should be the following: the RFW was between 0.35 and 0.40, that is, between 168 m and 192 m, and fracture conductivity was between 30 D.cm and 35 D.cm.

4 CONCLUSION

1. The software Eclipse was used in this paper to establish the Z23 simulation model, and the influences of fracture parameters on production rate, field recovery, and water cut were studied.

2. In order to achieve the economic production, it is better that the RFW and fracture conductivity are not greater, but limited in certain range at a given well pattern.
3. Simulation results showed that the optimal fracture parameters for Z23 were the fracture half-length between 168 m and 192 m and fracture conductivity between 30 D.cm and 35 D.cm.

REFERENCES

Britt L K. 1985. Optimization oil well fracture of moderate-permeability reservoir, SPE14371.

Cui Xiantao, Ma Yusheng, Jiao Baofu. 2001. Study on optimum hydraulic parameter in the lowest permeable reservoir, Petroleum Drilling Techniques, 29(5): 61–63.

Duan Weigang. 2013. Optimization design of integral fracturing parameters based on fine 3-D geology model, Journal of Southwest Petroleum University (Science & Technology Edition), 35(5): 109–117.

Economides Michael J, Oligney R E, Valko Peter P. 2002. Unified fracture design. Houston: Orsa Press.

Gan Yunyan et al. 2001. A new method for well pattern optimization and integral fracuring design in low permeability reservoirs, Acta Petroleum Sinica, 32(2): 290–294.

Marongiu P M, Economides Michael J, Holditch S A. 2008. Economic and physical optimization of hydraulic fracturing, SPE111793.

Meehan D N. 1995. Optimization of fracture length and well spacing in heterogeneous reservoirs, SPE Production &Facilities, 10(2): 82–88.

Osholake T, Wang J Y, Ertekin T. 2013. Factors affecting hydraulically fractured well performance in Marcellus shale gas reservoirs, Journal of Energy Resource Technology, 135(1): 1–10.

Xinfang M A. 2011. Analytical method for parameter optimization in hydraulic fracturing, Journal of China University of Petroleum, 35(1): 102–105.

Zeng Fanhui et al. 2007. Factors affecting production capacity of fractured horizontal wells, Petroleum Exploration and Development, 34(4): 474–482.

Zeng Fanhui et al. 2012. Optimization of fracture parameters of fractured horizontal wells in tight sandstone gas reservoirs, Natural Gas Industry, 32(11): 54–58.

Zhang Shicheng et al. 2004. Optimization design of integral fracturing parameters for four-spot well pattern with horizontal fractures, Acta Petroleum Sinica, 25(1): 74–78.

Zhang Shicheng. 2003. The theoretical study and engineering application on integral fracturing in low permeability reservoirs. Beijing: China University of Petroleum.

Energy Science and Applied Technology ESAT 2016 – Fang (Ed.)
© 2016 Taylor & Francis Group, London, ISBN 978-1-138-02973-6

Preparation of porous micro-electrolysis material and treatment of acid mine drainage

Shilin Feng, Xiaojun Xu, Kairui Wang & Han Sun
Faculty of Environmental Science and Technology, Kunming University of Science and Technology, Kunming, China

ABSTRACT: The porous micro-electrolysis material was prepared by direct-reduced iron, also called sponge iron, production process and carbon pellet production technology, and the material was used to treat Acid Mine Drainage (AMD) in the experiment. The influence of coal blending ratio, iron ore particle size, reduction temperature, and reduction time on the metallization rate of the material was investigated, and the influence of metallization rate on the removal effect of copper, lead, and zinc ions in AMD has also been considered. The results showed that the removal efficiency of pollutants was optimum, and the removal efficiencies of Cu^{2+}, Pb^{2+}, and Zn^{2+} were 98.87%, 77.89%, and 47.26%, respectively, when the porous micro-electrolysis material was prepared under the conditions of a coal blending ratio of 35%, a iron ore particle size of 140 mesh, a reduction temperature of 1140°C, and a reduction time of 12 min. At the same time, other pollutants were removed effectively.

Keywords: porous material; micro-electrolysis; acid mine drainage; heavy metals

1 INTRODUCTION

Acid Mine Drainage (AMD) is mainly produced in the mining process of pyrite and polymetallic sulfide ore; the metal sulfide was oxidized to sulfuric acid and metal sulfates, also a variety of metal ions in the ore were dissolved, forming AMD which contains copper, lead, zinc, cadmium, and other heavy metals (X.S. Yang). Therefore, if the AMD is discharged directly into the river and other water sources around the mines without appropriate treatment, it can bring about a change in the pH of the water; meanwhile, the heavy metals can accumulate in secondary mineral phases by precipitation and ion exchange in the water, and then pollute the surface water, groundwater, and soil, and the surrounding environment would be subjected to severe damage (M. Zhou, Sheoran A S).

Research on the treatment of AMD and especially the wastewater that contains lead and zinc has been reported at present; the main treatments were the neutralization precipitation method, the sulfide precipitation method, and the wetland and microbiological method (Z.C. Xu, 2005; Y.L. Li, 2007). But, the most widely used method was neutralization precipitation in the practical application, as this method has advantages of easy operation and low treatment cost; however, this method also generates a large number of hazardous waste sediment, and the treated water is too large to reuse, and hence the resources cannot be recycled (Y. Zhan).

Micro-electrolysis is a simple process that has features of wide application range and good performance in the removal of heavy metals [8]. However, there is less research on AMD treatment by micro-electrolysis. One of the key technologies of the micro-electrolysis method is the preparation of the micro-electrolysis material. In this experiment, we design and prepare a kind of porous micro-electrolysis material, and use this material to treat AMD. We also attempt to provide a new method for the preparation of a porous micro-electrolysis material and AMD treatment in this paper.

2 EXPERIMENT MATERIAL AND METHOD

2.1 Experiment material

The main experiment materials were iron ore and lignite; the main contents of the iron ore elements are shown in Table 1. Analysis results indicate that lignite contains fixed carbon (63.9%), volatiles

Table 1. Content of main elements in iron ore.

Element	Fe	O	Si	C
(%)	57.919	35	2.3712	2.0499
Element	Mg	Ca	Al	Mn
(%)	1.4135	0.4764	0.3044	0.2009

Table 2. Water quality of the AMD.

Index	pH	Pb	Zn	Cu	As	Cd
Concentration (mg/L)	3.5	1.77	25.19	5.86	32.59	1.28

(10.40%, on the dry ash-free basis), and hydrogen (2.94%, on the air-dried basis).

AMD samples were collected from a lead–zinc mine in Yunnan, the water quality of which is shown in Table 2.

2.2 *Experimental method and experimental procedure*

The optimum preparative condition and the treatment conditions of AMD were determined by single factor experiments, and the effects of different preparation conditions on the porous micro-electrolysis material and removal of heavy metals from AMD were studied.

Material preparation: The porous micro-electrolysis material was prepared by direct-reduced iron, also called sponge iron, production process and carbon pellet production technology. The iron ore, coal powder, and starch adhesive were made into a pellet in a certain proportion, and then heated to some uniform temperature in a closed furnace. By making the iron oxides reduced into iron by C, CO, and H_2, which were produced in the weak reductive and reductive environment, many small holes were formed during the volatilization of coal volatiles, and thus the porous micro-electrolysis material needed for the experiment was prepared.

AMD treatment experiment: Take 100 ml of water sample in a 500 ml beaker, and add at different metallization rates different amounts of the porous micro-electrolysis material. This is followed by agitation and aeration for a certain period, and the concentration of heavy metal ions in the supernatant was determined. Then, the separation of solids and liquids is carried out after standing for 1h. Finally, the removal efficiencies of the pollutants were investigated.

2.3 *Analysis method*

This experiment used Germany desktop high-temperature furnace LHT04/17, and the metallic iron in the prepared material was determined by the potassium dichromate titration method after decomposition of the sample by ferric chloride, the iron species, and the content in iron ore was analyzed by X-Ray Diffraction (XRD), and the morphology of the prepared material surface was analyzed by SEM XL30ESEM-TMP (Philips). The contents of Cu, Pb, and Zn were determined using flame atomic absorption spectrophotometer AA-240FS (Varian, Inc.)

3 RESULTS AND DISCUSSION

3.1 *The effect of material preparation process on metallization rate*

1. The effect of coal blending ratio on metallization rate: this is shown in Fig. 1a. With the increase in coal blending ratio, the total iron content of the prepared material decreased; there are two stages of increase in metallization rate with the increase in coal blending ratio; when the coal was insufficient, the metallization rate increased significantly and all depend on the carbon addition, and when the coal blending ratio reached 35%, the metallization rate increased slowly. Therefore, the optimal coal blending ratio chosen was 35%.

2. The effect of iron ore particle size on metallization rate: This is shown in Fig. 1b. The metallization rate increased with the decrease in iron ore particle size until it reached −180 mesh; however, if the raw material used was −180 mesh, the raw material treatment seems to be time— and energy-consuming. We can get −140 mesh iron ore particle after 5 min of crushing, and under the above conditions, the metallization rate of the iron ore particle can reach more than 83%, and the difference with −200 mesh was only 4%. Therefore, in this experiment, the iron ore particle size used is −140 mesh, which is reasonable.

3. The effect of reduction temperature on metallization rate: this is shown in Fig.1c; in the 2000°C reaction temperature, the material is obviously sintered, and the temperature has reached the melting point of coal ash content or other components in the ore. The material at 1160°C and 1180°C reaction temperatures is also mildly sintered, and it is difficult to proceed with crushing. The metallization rate at 1180°C reaction temperature was lower than that at 1160°C. The metallization rate of the material increased with the reaction temperature, between 1060°C and 1180°C; it has been shown that the metallization rate increased rapidly, when the reaction temperature is lower than the melting point of the coal ash content or other components in the ore. In addition, the optimum temperature was selected to be 1140°C, which was the maximum temperature before sintering.

4. The effect of reduction time on metallization rate: this is shown in Fig.1d; the metallization rate increased rapidly, when the reduction time is less than 12 min, and the metallization rate

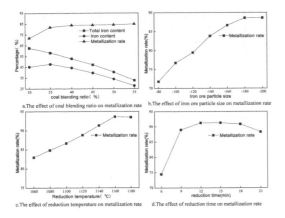

a.The effect of coal blending ratio on metallization rate

b.The effect of iron ore particle size on metallization rate

c.The effect of reduction temperature on metallization rate

d.The effect of reduction time on metallization rate

Figure 1. Influence of material preparation on the metallization rate.

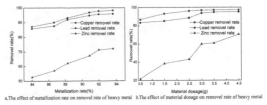

a.The effect of metallization rate on removal rate of heavy metal

b.The effect of material dosage on removal rate of heavy metal

Figure 2. Influence of material metallization on the removal efficiency of heavy metal.

in 15 min of reduction was the same as that in 12 min. When the reduction time was more than 15 min, the metallization rate decreased slightly. As the coal in the reaction material is in excess, the long reaction time would lead to iron carburizing and reduce the metallization rate.

In summary, analysis to determine the optimum preparation conditions were as follows: coal blending ratio, 35%; iron ore particle size, 140 mesh; reduction temperature, 1140°C at which the metallization rate is highest without sintering; and reduction time, 12 min, which was the shortest reaction time possible, without the carburization phenomenon.

3.2 Experimental study on the treatment of AMD by porous micro-electrolysis material

The basic conditions required for the experiment: under the mixing (agitator speed 60R / min) anaerobic condition, the initial pH of the water sample was 3.5, the reaction time was 40 min, the porous micro-electrolysis material particle size was less than 0.3013 mm, and the dosage of the porous micro-electrolysis material was 1.5 g/100 mL.

1. The effect of metallization rate on the removal rate of heavy metal: We can see from Fig2, the removal rate of copper, lead, and zinc increased with the increase in metallization rate. The removal rate of copper and lead was up to 97.71% and 95.61%, respectively, when the metallization rate was 92.01%, and after that, the removal rate changed slightly with the increase in metallization rate. The zinc removal rate increased with the metallization rate rapidly, before the metallization rate reached 92%, and then with the increase in metallization rate, it had little change.

Based on the comprehensive consideration of the content of direct-reduced iron, density, sintering degree, and economic and practical factors in the material preparation process, we selected the best experimental raw materials for which the metallization rate is 92%.

2. The effect of material dosage on the removal rate of heavy metal: in Fig3, the removal rate of copper, lead, and zinc was found to increase with the increase in the porous micro-electrolysis material dosage. In addition, the material dosage had no influence on the removal rate of copper after the dosage of 2.5 g/100 mL. The removal rate of lead had little change when the material dosage was over 3.0 g/100 mL. It can be inferred that the removal of copper and lead in actual wastewater can be divided into two parts: a part is deposited onto the material surface as a simple substance by micro-electrolysis and replacement, and another part is due to the aerated micro-electrolysis reaction, with water pH increased to 4.0, the lead ion and sulfate production, and lead sulfate precipitation and be removed. At the same time, there has been a competitive reaction between copper and lead; according to the Nernst equation, for the replacement reaction, copper has more priority than lead, and as the reaction time increased, lead ion began to replace when the content of copper ion was reduced.

The micro-electrolysis material dosage had a significant effect on the removal rate of zinc, and with the increase in the material dosage, the adsorption capacity of zinc increased and the removal rate also increased gradually. Considering the overall removal efficiency and economy, the material dosage used was 3.0 g/100 mL and the removal rate of copper, lead, and zinc were 96.74%, 93.66%, and 59.86%, respectively. The residual concentrations of copper, lead, and zinc ions in the wastewater were 0.23 mg/L, 0.2 mg/L, and 14.31 mg/L, respectively.

3.3 SEM analysis

The SEM analysis reveals that the material was largely porous and that iron and carbon were well

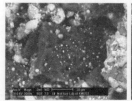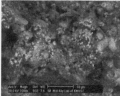

a Surface photograph before reaction
b Surface photograph after reaction

Figure 3. Surface photograph of the prepared material.

distributed in the materials which were prepared under selected preparation conditions. The prepared materials mainly had three kinds of morphology: black porous substance, white smooth substance, and gray loose substance. The binding energy spectrum analysis indicates the black porous substance as carbon, the smooth white substance as elemental iron, and the gray substance as a mixture of impurities such as gangue and iron.

After the micro-electrolysis reaction, the material surface covered a layer of uneven substance, and the original boundaries became unobvious, but still there were certain pores, and the wastewater had no barrier with the inner material, which also proved that the reaction continued until all materials disintegrated.

4 CONCLUSIONS

1. The optimum preparation conditions of the porous micro-electrolysis material were as follows: Coal blending ratio, 35%; iron ore particle size, 140 mesh; reduction temperature, 1140°C; and reduction time, 12 min. Moreover, the products that were prepared in this condition had contents of 60% iron and 13% carbon, and the metallization rate was 92%.

2. The optimum metallization rate is 92% and the optimum material dosage is 30 g/L in the treatment of AMD by the porous micro-electrolysis material. The removal rates of Cu^{2+}, Pb^{2+}, and Zn^{2+} were 98.87%, 77.89%, and 47.26%, respectively. Meanwhile, the other pollutants were removed effectively.

3. Using this method, involving a porous micro-electrolysis material, together with electric flocculation treatment of AMD, the effluent can meet the "integrated wastewater discharge standard" emission requirements, and it provides a new feasible method for AMD treatment, and recyclable use of heavy metals in wastewater.

REFERENCES

Li Y.L., L. Zhao, J. Xu, Progress in Research of Copper Mine Waste Water Treatment, J. Journal of Kunming Metallurgy College. 2007, 23(5) 72–75.
Ma L.X., R.X. Zhao, Study on the Treatment of Wastewater with Iron Internal Electrolysis Process, J. Hebei Journal of Industrial Science & Technology. 20(1) 50–53.
Sheoran A.S., Sheoran V. Heavy metal removal mechanism of acid mine drainage in Wetlands: A critical review, J. Minerals Engineering. 19(2) 105–116.
Xu Z.C., Review on Artificial Wetland for Treatment of Acid Mime Water, J. Mining Safety & Environmental Protection. 2005, 32(2) 40–42.
Yang X.S., L.N. Shao, Development Tendency of Nonferrous Metal Mineral Acidic Wastewater Treatment Technology, J. Nonferrous Metals. 63 (1) 114–117.
Zhan Y., Z. Xiong, Y. Lin, Researches on the pretreatment of ramie wastewater by iron chip inner-electrolysis, J. Industrial Water Treatment. 20(1) 50–53.
Zhou J.M., Z. Dang, M.F. Cai, et al. Speciation Distribution and Transfer of Heavy Metals in Contaminated Stream Waters around Dabaoshan Mine, J. Research of Environmental Sciences.18(5) 5–10.

Energy Science and Applied Technology ESAT 2016 – Fang (Ed.)
© 2016 Taylor & Francis Group, London, ISBN 978-1-138-02973-6

Influence of different cooling techniques on the mineral composition of the zirconia metering nozzle

Xiao Wang
College of Materials and Mineral Resources, Xi'an University of Architecture and Technology, Xi'an, China

Qunhu Xue
College of Materials and Mineral Resources, Xi'an University of Architecture and Technology, Xi'an, China
Shaanxi Techno-Institute of Recycling Economy, Xi'an, China

Liang Zhao & Yonghua Zhou
College of Materials and Mineral Resources, Xi'an University of Architecture and Technology, Xi'an, China

ABSTRACT: The mineral composition and microstructure of the original Mg-PSZ metering nozzle samples were studied in this article to investigate the influence of different cooling techniques. The samples were heated at 1540°C for 5h and then cooled inside the furnace or quenched in water, respectively, simulating the environments of the natural cooling after tilting the tundish and the environment at service temperature. The chemical composition, mineral composition, and microstructure of the samples were analyzed by using modern equipment such as X-ray fluorescence instrument, X-ray diffractometer, and scanning electron microscope. The result showed that the original Mg-PSZ metering nozzle mainly exists as a cubic phase with water quench cooling; the slow cooling rate can cause partial transformation of cubic phase and monoclinic phase, and hence the zirconia metering nozzle mainly exists as a monoclinic phase with natural cooling in the furnace. The granules of the zirconia metering nozzle have a uniform and big particle size (13μm) with water quench cooling, while there may appear many smaller particles (7μm) with natural cooling in the furnace. Zirconium oxide particles in the sample, which was quenched in water, were interconnected to form a staggered submicron columnar structure. This staggered submicron columnar structure could improve the strength of the sample and prevent crack propagation. Therefore, the research results of samples with natural cooling after tilting the tundish cannot represent the actual situation of the zirconia metering nozzle at service environment.

Keywords: metering nozzle; Mg-PSZ; cooling techniques

1 INTRODUCTION

There are many studies on corrosion mechanism of the zirconia metering nozzle at present, mostly got out by research target based on samples with the natural cooling after tilting the tundish (X.W. Li et al, 2006) (X. Li et al, 2012) (H. Zhang et al, 2015). The experimental results indicated that the phase transition occurred in the zirconia metering nozzle after changing the tundish, so researches no longer represent the practical usage status of the nozzle. The binary phase diagram for MgO-ZrO$_2$ shows that MgO that is solid dissolved in ZrO$_2$ and the cubic ZrO$_2$ undergo phase transition at 1400°C; therefore, there is a possibility of phase transition with the zirconia metering nozzle in the process of changing the tundish. To determine the practical usage status of the zirconia metering nozzle, this paper makes a study of the samples that were

heated at 1540°C for 5h. Through the different cooling techniques (water quench cooling or natural cooling in the furnace), we simulate the environment at service temperature and the environment of the natural cooling after tilting the tundish. This paper investigates the mineral composition and microstructure of the zirconia metering nozzle at service temperature using the samples with water quench cooling, which were characterized.

2 EXPERIMENT

2.1 Sampling

The paper makes a study of the original Mg-PSZ metering nozzle. The chemical composition of the metering nozzle is given in Table 1. The residual sample used in this study was obtained from the upper nozzle used in a 5-strand 150 mm × 150 mm

Table 1. Chemical composition of the metering nozzle samples.

Composition	ZrO_2	MgO	Y_2O_3	SiO_2	Al_2O_3	HfO_2	Fe_2O_3	CaO	TiO_2	Total
Wt/%	93.55	2.58	0.21	0.45	0.40	2.82	0.06	0.23	0.11	99.87

Table 2. Heat treatment process parameters.

Temperature/°C	Heating rate/°C · min^{-1}	Time/min
10~1000	10	99
1000~1540	5	108
1540	0	300

billet casting on a 30-ton tundish, whose service life is 40 h. The necessary conditions for the use are as follows: casting steel, with a teeming temperature of 1540°C, at a speed of 1.8-1.9 m/min, lasting for 40 h.

2.2 Simulation procedures

The heat treatment process parameters required to simulate the samples at service temperature by using the electric furnace are given in Tab. 2. Based on the metering nozzle at service temperature in actual situation, for this experiment, the temperature was set at 1540°C, and the holding time at 5 h. Cooling technique included cooling inside the furnace (the sample was cooled inside the furnace after heat treatment) or quenching in water (the sample was removed immediately and let in water in order to maintain the original mineral composition).

3 RESULTS AND DISCUSSION

3.1 XRD analysis

The apparatus used in XRD analysis is the DMAX-2400 X-ray diffractometer produced by Rigaku Corporation, Japan. Tests were conducted on mineral phases of the samples (a. the original sample; b. the sample cooled in the furnace; and c. the sample quenched in water) by different cooling techniques and the results are shown in Fig. 1.

In order to calculate the phase content of the samples with different cooling techniques, the X-ray relative value approximation method was used. The calculation formulas of the monoclinic phase (Vm) and the cubic phase (Vc) are shown as follows (X. Li et al, 2013) (K. Shan et al, 2013):

$$V_m = \frac{I_m(-111) + I_m(111)}{I_m(-111) + I_m(111) + I_c(111)}$$

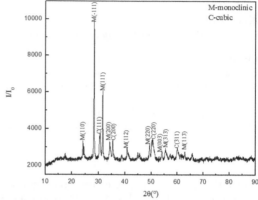

a The original sample

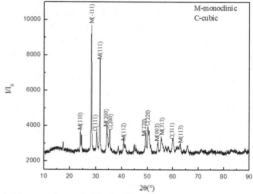

b The sample cooled in the furnace

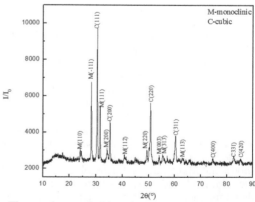

c The sample quenched in water

Figure 1. XRD analysis results of the samples with different cooling techniques.

Table 3. Phase content of the samples with different cooling methods.

Volume fraction/%	The samples		
	Original sample	Cooled in furnace	Quenched in water
V_m	66	64	49
V_c	34	36	51

$$V_c = \frac{I_c(111)}{I_m(-111) + I_m(111) + I_c(111)}$$

Formulas: I_m (−111): the relative peak intensity of the monoclinic zirconia (−111).

I_m (111): the relative peak intensity of the monoclinic zirconia (111).

I_c (111): the relative peak intensity of the cubic zirconia (111).

The phase content of the samples with different cooling techniques in shown in Tab. 3, based on the calculation of the relative peak intensity.

The data based on calculation indicated that the content of monoclinic and cubic zirconia in the original sample and the sample cooled in the furnace were not changed with a proportion of 6:3 and the result showed that the process of simulation in this experiment was similar to the thermal schedule of the practical production process. The content of the cubic zirconia samples increased from 36% to 51%, and the content of the monoclinic zirconia samples decreased from 64% to 49% by quenching in water. The relative peak intensity of the cubic phase on (111), (200), (220) became enhanced and the monoclinic phase on (111) became weakened at a faster cooling rate according to Fig. 1b and c. Therefore, the important phase in the metering nozzle at service temperature was the cubic phase, and a faster cooling rate could inhibit the transition between the monoclinic and cubic phases, so the phase of cubic zirconia was stored in the metering nozzle. The slow cooling rate could lead to the partial transformation from cubic phase to monoclinic phase with natural cooling after changing the tundish, and the content of the cubic phase decreased finally.

From the analysis of experimental data, it has been found that all the samples did not include tetragonal zirconia, and so the metering nozzle could not contain the tetragonal phase in actual environment. There is a critical grain size of the tetragonal phase at room temperature. When the particle size d > dc, the tetragonal zirconia transformed into the monoclinic zirconia; on the contrary, when d < dc, the tetragonal phase was still persistent at room temperature, and the metast-

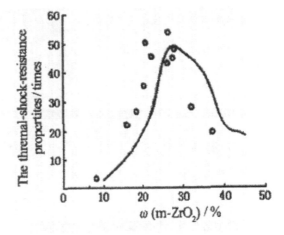

Figure 2. Effect of the phase of ZrO2 on the thermal-shocking-resistance properties.

able tetragonal zirconia plays a significant role in increasing the toughness at room temperature. The particle size was greater than the critical size during the sintering process, and therefore there was absence of tetragonal phase in the zirconia metering nozzle. When the cooling rate was accelerated, the martensite transformation temperature (Ms) decreased, and the martensite transformation finish temperature (Mf) increased, so that the temperature range between the monoclinic phase and the cubic phase narrowed, the content of transformation decreased, and it is in favor of the tetragonal zirconia reserve. Therefore, in order to ensure that the zirconia metering nozzle production does not crack under the premise, it was appropriate to speed up the cooling rate, which was conducive to the retention of tetragonal zirconia, thereby improving the performance of the nozzle. Some studies proved that the thermal-shock resistance was an outstanding property of the metering nozzle with 30% monoclinic phase and 70% cubic phase (Fig. 2).

3.2 SEM analysis

The apparatus used in SEM analysis is the TECAN-VEGA 3MH scanning electron microscope (produced in Czech). The microstructure analysis was performed in a graduated manner along the radical direction. Tests on the microstructure of the samples ((a) the original sample; (b) the sample cooled in the furnace; (c) the sample quenched in water; and (d) the residual sample) by different cooling techniques are shown in Fig. 3.

The results showed that the average particle size of the original zirconia metering nozzle was 8 μm (Fig. 3a), the average particle sizes of the samples

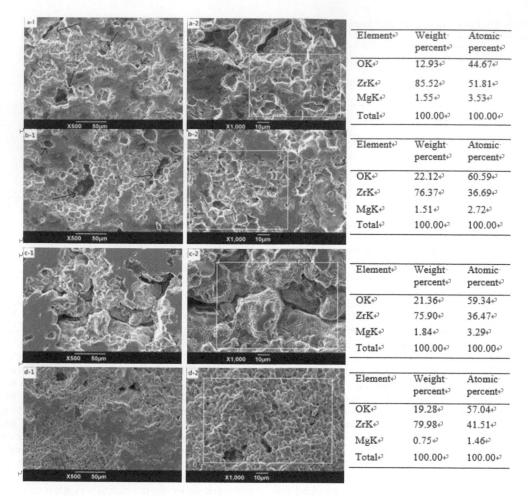

Element	Weight percent	Atomic percent
OK	12.93	44.67
ZrK	85.52	51.81
MgK	1.55	3.53
Total	100.00	100.00

Element	Weight percent	Atomic percent
OK	22.12	60.59
ZrK	76.37	36.69
MgK	1.51	2.72
Total	100.00	100.00

Element	Weight percent	Atomic percent
OK	21.36	59.34
ZrK	75.90	36.47
MgK	1.84	3.29
Total	100.00	100.00

Element	Weight percent	Atomic percent
OK	19.28	57.04
ZrK	79.98	41.51
MgK	0.75	1.46
Total	100.00	100.00

Figure 3.   SEM patterns of samples with different cooling methods. (a, original sample; b, original sample cooled in the furnace; c, original sample cooled in water; d, residual sample after tilting the tundish).

with different cooling techniques were, respectively, 7 μm (Fig. 3b) and 13 μm (Fig. 3c), and the average particle size of the residual sample after natural cooling was only 2 μm (Fig. 3d). The data from the EDS spectrum analysis provided the relationship of the content of Mg element in the samples: $C_{sample\ quenched\ in\ water} > C_{original\ sample} > C_{sample\ cooled\ in\ furnace} > C_{residual\ sample}$.

Figure 3a and b showed that the process of simulation in this experiment was similar to the thermal schedule of the practical production process. Zirconium oxide particles in the sample that was quenched in water were interconnected to form a staggered submicron columnar structure (Fig. 3c). This staggered submicron columnar structure could improve the strength of the sample and prevent crack propagation. According to Fig. 3b and c, the average size of the particles in the sam-

ples quenched in water was greater than that of the particles in the samples cooled in the furnace. This is due to the slow cooling rate that resulted in the stabilizer desolventizing of the zirconia metering nozzle, and partial transformation from cubic phase to monoclinic phase accompanied by volume expansion; therefore, the average particle size of the metering nozzle were smaller. According to Fig. 3c, d, the average size of the particles in the residual sample was significantly smaller than that of the particles in the sample quenched in water, and the stabilizer in the residual sample was at the lowest level, which was due to inclusions in molten steel reacted with the stabilizer (MgO) in the zirconia metering nozzle in actual situation, forming an oxide with the low-melting-point molten steel drain. Instability desolventizing stabilizer caused the cubic-phase zirconia to transform into monoclinic-phase zirconia accompanied by volume expansion, particle breakage, and the cubic phase in the zirconia metering nozzle instability into small particles of monoclinic phase.

4 CONCLUSION

The main phase of the original Mg-PSZ zirconia metering nozzle was the cubic phase during the process of pouring steel at 1540°C, which was due to the slow cooling rate, causing partial transformation of cubic-phase zirconia into the monoclinic-phase zirconia.

There were differences in the microstructure of the samples with different cooling techniques. The zirconium oxide particles were distributed and the average size of the particles in the samples quenched in water was greater than that of the particles in the samples cooled in the furnace.

Zirconium oxide particles in the sample quenched in water were interconnected to form a staggered submicron columnar structure. This staggered submicron columnar structure could improve the strength to the sample and prevent crack propagation.

ACKNOWLEDGMENTS

Foundation item: Project 51372193 was supported by the National Natural Science Foundation of China, and project 2014 JM6224 was supported by the Natural Science Basic Research Plan in Shaanxi Province of China.

REFERENCES

Li, X.W., X.F. Wang, H.B. Zhao. Discussion about Damage Mechanism of Zirconia Nozzle in Operation [J]. HEBEI METALLURGY. 2006, 26(6): 2–4.
Li, X., Q.H. Xue, X.H. Ren. Study on Wearing Mechanism of Zirconia Non-swirl Nozzle [J]. Bulletin of the Chinese Ceramic Society, 2012, 31(6): 1523–1528
Li, X., Q.H. Xue, X.H. Ren. Influence of Al_2O_3-ZrO_2 Composite Powder on Performance of ZrO_2 Metering Nozzle [J]. Bulletin of the Chinese Ceramic Society, 2013, 32(9): 1751–1755.
Shan, K., X.M. Guo. Preparation and characterization of nano-powder zirconia stabilized by magnesia [J]. Materials Science and Engineering of Powder Metallurgy, 2013, 18(3): 430–433.
Zhang, H., Q.H. Xue, L. Zhao. Analysis of the Changes in the Mineral Composition of the Used Zirconia Sizing Nozzle [J]. Bulletin of the Chinese Ceramic Society, 2015, 34(4): 947–950.

Energy Science and Applied Technology ESAT 2016 – Fang (Ed.)

Optimization of horizontal well multistage fracturing in offshore gas reservoirs

Zhenfu Jia
College of Chemistry and Chemical Engineering, Chongqing University of Science and Technology, Chongqing, China

Jianwen Bai
Research Center of Sulige Gas Field, Changqing Oilfield Company, Petrochina Xi'an, Shanxi, China
National Engineering Laboratory of Exploration and Development of Low Permeability Oil/Gas Fields, Xi'an, China

Shilan Chen, Peng Zhang, Chengyu Zhou & Wenzhang Huang
College of Chemistry and Chemical Engineering, Chongqing University of Science and Technology, Chongqing, China

ABSTRACT: Offshore hydraulic fracturing is characterized by limited fracking treatment size and high operation cost, hence reasonable fracture parameters under limited hydraulic fracturing scale play a key role in fracturing design. In this paper, a new method for optimization of the fracture parameters of off-shore gas reservoirs was put forward, and according to this method, the optimum fracture parameters of horizontal well under different fracturing stages of the same sand volume were obtained after assessment of platform deck fracturing capacity. In addition, fracturing stages were optimized according to the prediction of productivity and evaluation of economy; therefore, an integrated optimization method considering fracturing parameters, reservoir characteristics, and economic factors was established. By utilizing this method, the optimum design for gas well P1H in DFF 1-1 Gas field was obtained with five optimum fractures, with an optimum fracture length of 98.6 m and an optimum fracture width of 7.8 mm.

Keywords: offshore horizontal well; multistage fracturing; fracture parameters; optimization design

1 INTRODUCTION

With the increasing offshore prospecting degree and gradually expanding exploration area, more and more low-porosity, low-permeability reservoirs were discovered, and horizontal well-staged fracturing is one of the most important and effective development means (Zhengfu Jia et al, 2015) (Guo Jianchun et al, 2013) (Xin-fang Ma et al, 2005). At present, the principal method of fracturing optimization is productivity simulation by using reservoir numerical simulators; segments of fracture and fracture parameters are optimized based on the maximum productivity or maximum economic profit, but these methods are restricted by the accuracy of reservoir parameters, numerical models, designer's subjective ideas, etc., resulting in poor accuracy of optimized parameters (GuoBo-yun et al, 2008) (Hai-hong Gao et al, 2006) (Fan-hui Zeng et al, 2007) (Yu-guang Zhang, 2010). Supposing that sand amount and reservoir parameters are given, and each fracturing stage was treated as one independent system, there exits an optimum

fracture parameter to obtain the largest productivity index. Besides, offshore fracturing is risky and costly, and fracturing design also needs to limit fracturing size and determine the optimum fracture parameters under economic recovery period.

2 PRINCIPLE OF FRACTURING DESIGN

In terms of a homogeneous rectangular drainage area with a close boundary and several production and injection wells under pseudo-steady state, the pressure drawdown at any point in this multiwell-reservoir system could be calculated by an influence function:

$$\bar{p} - p = \frac{C\mu Bq}{2\pi kh} a[x_{\mathrm{D}}, y_{\mathrm{D}}, x_{\mathrm{wD}}, y_{\mathrm{wD}}, y_{\mathrm{eD}}] \qquad (1)$$

As shown in Fig. 1, fractures are considered to be constituted by n line sources, each stage being treated as an independent line source-reservoir system; the relationship equation between any two

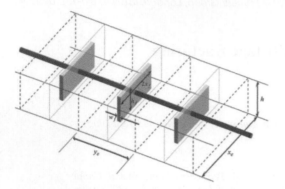

Figure 1.   Sketch map of multistage fracturing.

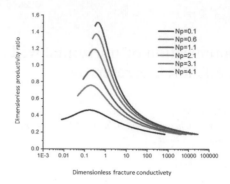

Figure 2. Fracture parameter optimization chart (yeD = 0.75).

line sources can be gained by applying an influence function, and n equations will be derived for n line sources. One-fourth of each fracturing segment was extracted, and the system's overall dimensionless gas productivity can be calculated by equations 2–10:

$$A\bar{q} = \bar{d} \tag{2}$$

$$A = \begin{bmatrix} a_{11} & a_{12} & \cdots & a_{1n} \\ a_{21} & a_{22} & \cdots & a_{2n} \\ \vdots & \vdots & \vdots & \vdots \\ a_{n1} & a_{n2} & \cdots & a_{nn} \end{bmatrix} \tag{3}$$

$$d = \begin{bmatrix} 1 \\ 0 \\ \vdots \\ 0 \end{bmatrix} \tag{4}$$

$$q = \begin{bmatrix} q_{D1} \\ q_{D2} \\ \vdots \\ q_{Dn} \end{bmatrix} \tag{5}$$

$$\begin{aligned} a_{1j} &= a[o_1, w_j] \\ a_{i1} &= a[o_1, w_1] - a[o_i, w_1] \\ a_{ij} &= a[o_1, w_j] - a[o_i, w_j] - \frac{4\pi}{C_{fD}I_x}(x_{Do(min(i,j))} - x_{Do1}) \end{aligned} \tag{6}$$

Dimensionless production rate:

$$q_{Di} = \frac{qB\mu}{2\pi kh(\bar{p} - p_{wf})} \tag{7}$$

Dimensionless distance:

$$x_{Doi} = \frac{x_{oi}}{x_e/2} \tag{8}$$

Fracture penetration ratio:

$$I_x = \frac{x_f}{x_e/2} \tag{9}$$

$$J_D = 4\sum_1^n q_{Di} \tag{10}$$

$$N_p = \frac{2K_{f\text{-eff}}V_p}{KV_r} \tag{11}$$

Given the specific sand volume, reservoir area, and permeability, there exists only one corresponding dimensionless proppant index, which represents the improvement of fracture percolation capacity and the ratio of influence area to the whole reservoir. The relationship chart of three dimensionless parameters in any rectangular drainage area can be plotted based on above analysis, and the optimized chart with length-to-width ratio of 0.75 is illustrated in Fig. 2, from which it is clear that there exists only one optimum dimensionless fracture conductivity at a given proppant index to achieve the largest dimensionless oil productivity.

3   OPTIMIZATION OF FRACTURING STAGES

Dimensionless gas productivity was first computed by applying the above method, and then optimization of fracturing stages was operated by calculating the productivity under different stages first and then screening by economic evaluation. Multistage fracturing horizontal well productivity in a low-permeability gas reservoir can be calculated by equations 12. The economic evaluation model mainly uses the net present value and return on investment function for optimization of fracturing stages, and its expressions are given as follows:

$$q = n_f \frac{Kh(P_e^2 - P_{wf}^2)}{C_2 \overline{\mu Z T}} J_{DTH}$$

$$J_{DTH} = \frac{1}{\dfrac{1}{J_D} + s_c}$$

$$s_c = \frac{Kh}{w_{f\text{-opt}} K_f} \left[\ln\left(\frac{h}{2r_w}\right) - \frac{\pi}{2} \right]$$

$$\frac{\overline{P}}{Z} = \frac{P_i}{Z_i}\left[1 - \frac{G_P}{G}\right]$$

$$NPV = R_F - R_O - C_F = \sum_{j=1}^{n} \frac{(V_F)_j}{(1+i)^j}$$
$$- \sum_{j=1}^{n} \frac{(V_O)_j}{(1+i)^j} - C_F \qquad (13)$$

$$DROI = \left(\sum_{j=1}^{n} \frac{(V_F)_j}{(1+i)^j} - \sum_{j=1}^{n} \frac{(V_O)_j}{(1+i)^j} \right) / C_F \qquad (14)$$

4 OPTIMIZATION PROCEDURES

Optimization procedures are demonstrated in Fig. 3, the fracturing scale is first fixed by platform's operation capacity, and fracturing parameters under

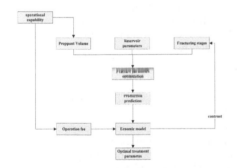

Figure 3. Optimization procedures.

different stages are determined integrated with reservoir characteristics, and then substituted into the productivity model to compute the corresponding productivity. Finally, the ultimate fracturing operation is screened out by the economic evaluation model.

5 FIELD APPLICATION

The second DFF 1-1 gas field formed in Yinggehai Group was buried 1200-1600 m under sea level, with a gas-bearing area of 323 km², porosity of 20.57%, and permeability of 5.1 mD. To improve the recovery of this low-permeability reservoir, a trial of production stimulation was implemented in Well P1H. The horizontal section length of P1H well is 600 m, and the wellbore radius is 108 mm. The effective thickness of the reservoir is 5 m, and the formation pressure is 13.5 MPa. The viscosity of the gas is 0.015 mPa·s, specific gravity 0.8, and gas deviation factor 1.0538. The rectangular drainage area covers 360000 m2. Laboratory test shows that the proppant retains a permeability of 40 D, a porosity of 35%, and a density of 1.65 g/cm3 under reservoir conditions. Combining with the logging data of the prospecting well, the fracture height is approximately 20 m, calculated by fracture height model. As the platform operation space is limited, the maximum sand volume is set at 20 m³. The optimum fracture parameters under different stages are shown in Table 1.

The cumulative gas production volume under different stages is shown in Figure 4, from which it is clear that gas production increases with the increment in fracturing stages, but the increase in production is expected to become smaller when fracture stages reach 5 to 7.

According to the statistical data of offshore fracturing in Bohai Bay, the material cost of one single fracture is 1.29 million yuan and fracturing operating cost is 10.70 million yuan, while

Table 1. The optimum fracture parameters under different fracturing stages.

Fracturing stages	Proppant number	Penetration ratio	Dimensionless fracture conductivity	Length/m	Width/mm	Fracture conductivity/ Dc · cm
2	0.13	0.25	1.06	75.55	10.18	40.73
3	0.20	0.29	0.82	85.74	8.97	35.89
4	0.27	0.31	0.69	93.61	8.22	32.87
5	0.34	0.33	0.62	98.60	7.80	31.21
6	0.40	0.34	0.59	100.99	7.62	30.47
7	0.47	0.34	0.59	101.47	7.58	30.32
8	0.54	0.34	0.60	100.67	7.64	30.56
9	0.60	0.33	0.61	99.06	7.77	31.06
10	0.67	032	0.64	96.97	7.93	31.73

comprehensive daily fee of drilling ship is approximately 1.20 million yuan. The gas price is set at 2.5 yuan per cubic meters, and the discount rate is 15%. Besides, one more fracturing stage increases a day's workload. NPV and DROI curves under different stages in 1-5 years are illustrated in Figs. 5 and 6, respectively. It is apparent that NPV has a larger value with more stages, but the incremental scope becomes smaller. On the other side, the DROI curves demonstrate that different pay back periods have different optimum stages. When investment recovery in 2 years is desired, we can get the following optimum parameters: five stages, a fracture length of 98.6 m, and a width of 7.8 mm.

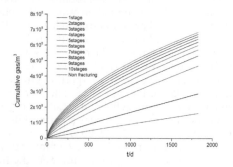

Figure 4. The cumulative gas production volume under different stages.

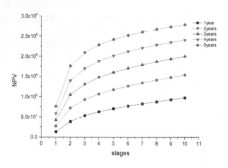

Figure 5. NPV value under different fracturing stages.

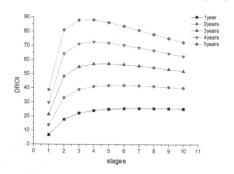

Figure 6. DROI value under different fracturing stages.

6 CONCLUSION

Optimization procedures of fracturing parameters were established and the following optimum parameters were obtained for Well P1H: five stages, a fracture length of 98.6 m, and a width of 7.8 mm. This new approach considers the matching of fracturing parameters between reservoir characteristics and economic conditions, which may provide some guidance for multistage fracturing of offshore horizontal wells.

NOMENCLATURE

μ: oil viscosity, mPa•s
B: oil volume factor, dimensionless
$a[x_D, y_D, x_{wD}, y_{wD}, y_{eD}]$: influence function
C: unit conversion factor
y_{eD}: length-to-width ratio of rectangular drainage area
x_D, y_D: dimensionless coordinates
x_{wD}, y_{wD}: dimensionless coordinates of production well
n_w: number of production wells (line sink)
K: reservoir permeability, mD
h: reservoir thickness, m
Np: proppant index
V_p: proppant volume, m^3
$K_{f\text{-eff}}$: fracture effective permeability, mD
V_f: gas reservoir volume of drainage area, m^3
n_f: fracture number
J_{DTH}: overall dimensionless oil productivity
s_c: skin factor
r_w: well radius
P_i: initial reservoir pressure, MPa
Z_i: gas Z-factor
G_p: Cumulative gas production, m^3
G: original oil in place, m^3
C_F: fracturing operation cost, yuan
N: fixed number of years
I: discount rate
NPV: net present value, yuan
DROI: return on investment
R_F: present value of a fractured well, yuan
R_o: present value of a non-fractured well, yuan
V_F: first j annual revenues of the fracturing well, yuan
V_o: first j annual revenues of the non-fracturing well, yuan

REFERENCES

Fan-hui Zeng, Jian-chun Guo, Yan-bo Xu, etc. Factors affecting production capacity of fractured horizontal wells [J]. Petroleum Exploration and Development, 2007, 34(04): 474–477.

Guo Jianchun, Liang Hao, Zhao zhihong. Optimizing the Fracture Parameters of Low Permeability Gas Reservoirs [J]. Journal of Southwest Petroleum University (Science & Technology Edition), 2013, 35(01): 93–98.

GuoBo-yun, Yu Xian-ce. Simple analytical model for predicting productivity of multifractured horizontal wells [C]. SPE 114452, 2008.

Hai-hong Gao, Lin-song Cheng, Zhan-qing Qu. Optimization of the fracture parameters of fractured horizontal wells [J]. Journal of Xi'an Shiyou University (Natural Science Edition), 2006, 21(02): 29–32.

Xin-fang Ma, Feng-ling Fan, Shou-liang Zhang. Fracture Parameter Optimization of Horizontal Well Fracturing in Low Permeability Gas Reservoir [J]. Natural Gas Industry, 2005, 25(09): 61–63.

Yu-guang Zhang. Optimization Analysis of the Fracture Parameters of Fractured Horizontal Well in Gas Reservoir [J]. Science Technology and Engineering, 2010, 10(12): 2861–2864.

Zhengfu Jia, Jingxia Zhong, Muwang Wu, Hao Liang. Fracturing Optimization Design of Offshore Horizontal Well With Real Field and Economic Constraints [J]. EJGE, Vol. 20 [2015], Bund. 8: 3655–3663.

Energy Science and Applied Technology ESAT 2016 – Fang (Ed.)
© 2016 Taylor & Francis Group, London, ISBN 978-1-138-02973-6

Paleogene strata pathway system and its control on hydrocarbon in Gaoyou Sag

Shuhui Dai & Zhaonian Chen
School of Energy Resources, China University of Geosciences, Beijing, China

Steve Stephen Macateer
National Oilwell Varco, Beijing, China

Yan Liu
School of Energy Resources, China University of Geosciences, Beijing, China

ABSTRACT: As a channel system of hydrocarbon migration, a pathway system is connected to the hydrocarbon source rocks and traps, binding the hydrocarbon migration paths and accumulation places. In recent years, along with the deepening of exploration and development, for conducting research on hydrocarbon migration paths and migration directions in a better way, more in-depth understanding of hydrocarbon accumulation characteristics and rules, and analyzing the system become increasingly important. Gaoyou Paleogene strata, for example, based on existing geological research, proposed different types of possible pathway systems in the study area, such as faults, sand, and unconformity, and other types of complex pathway systems; their characteristics and control on hydrocarbon accumulation are analyzed and summarized.

Keywords: Gaoyou Sag; Paleogene strata; hydrocarbon migration; pathway system

1 INTRODUCTION

The pathway system is a key controlling factor of hydrocarbon accumulation, and its type is controlled by the structural shape of the basin, sedimentary filling characteristics, stratigraphic framework characteristics, fluid activity methods, etc. It is still a weak link in the field of petroleum geology research, because of complex factors affecting the pathway system. The main elements that constitute the hydrocarbon pathway system are fault, sand, and unconformity; various conduction factors can play a conduction role alone, but two or more elements together can play a combined role and compose a compound pathway system, which then controls hydrocarbon accumulation. Owing to the large Paleogene stratigraphic thickness, more reservoir positions, intensive fracture development, and various types of pathway systems in Gaoyou sag, we can improve the recognition accuracy of reservoirs and reduce the exploration risk through research on the pathway system.

2 TYPES OF PATHWAY SYSTEMS

2.1 *Fault pathway system*

This is an important pathway layer having two kinds of vertical and lateral channel modes, and its effec-tiveness depends on the nature, sealing ability, and development scale of fault and space-time configuration relationship with other pathway layers, etc. One of the most important factors is fault sealing (Weihai Zhang et al, 2003); faults did not have the ability to transport if they are completely closed. The fault pathway mode has two types, that is, vertical and lateral. In the vertical pathway mode, an open section connected the strata at different depths, forcing the hydrocarbon do vertical migration from high-potential zone to low-potential zone along the fault under the action of pressure difference. In the lateral pathway mode the hydrocarbon move across the permeable fault to another dish of reservoir by lateral migration, under the condition that the top of fault permeability is low and displacement pressure is greater than the potential gradient vertically (Fig. 1a) (Jinhua Fu et al, 2002) (Changhai Gao et al, 2007). In addition, the fault in space-time configuration relationship is also an important factor affecting the effectiveness of the pathway system, when the fault connected source rocks and reservoirs (Fig. 1b).

2.2 *Sand pathway system*

A connected sand body can be used as a hydrocarbon reservoir space and also a migration channel, and its transportation efficiency depends largely

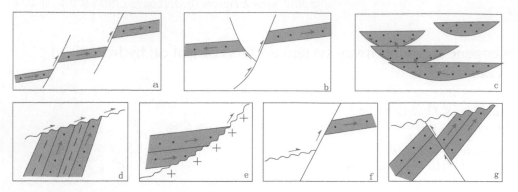

Figure 1. Pathway system elements and their configuration relationship.
(a). fault sand body configuration; (b). fault fault configuration; (c). skeletal sand body configuration; (d). unconformity fault sand body configuration; (e). blocking configuration of sand body and unconformity; (f). overlap configuration of sand body and unconformity; (g). fault wrinkle unconformity configuration of unconformity and fault.

on factors such as the connectivity degree of sand body, space-time configuration relationship, and different properties such as the pore structure, porosity, permeability of sand body, etc. Overall performance is: a higher connectivity degree of sand body in the plane is more favorable for hydrocarbon lateral migration; a higher superposition degree of sand body in the space is more beneficial for hydrocarbon to realize "uphill" movement; better porosity and permeability of sand body help overcome the capillary resistance more easily for realizing hydrocarbon filling during the migration. Thus, we can also consider the connected pathway system largely affected by the sedimentary environment (Fig. 1c).

2.3 *Unconformity pathway system*

This is an effective pathway system when there exists permeable rocks above and below the unconformity surfaces, and is very favorable for hydrocarbon migration and accumulation; then, unconformity becomes an important channel for large long-distance migration, in particular, unconformity overlying of slope and uplift greatly improve the strength of hydrocarbon plane accumulation (Wang Yong et al, 2015). According to the two kinds of contact, overlap and truncation, existing between the unconformity and the reservoir pathway layers, two types of reservoirs may be formed, such as unconformity screening reservoir (Fig. 1d) and formation overlap reservoir (Fig. 1e). Unconformity of truncation contact has high span to different underlying series of strata, in favor of multistrata hydrocarbon gathering to the unconformity. The parallel unconformity developed in the depression area, and its upper and lower layers often developed thick source rocks, a good con-

dition for oil source, but a poor pathway channel (Weihai Zhang et al, 2003). In terms of developmental spatial position, Wupu unconformity widely developed and E_2d_1 unconformity on E_1f_4 source rocks in Gaoyou sag, which is very favorable to channels of hydrocarbon.

2.4 *Fault-Sand body pathway system*

In this class of pathway system, sand bodies are the main migration channel, and there are two types of roles for the fault: ① blocking effect, which causes hydrocarbon accumulation in the condition of better fault sealing area and ② regulating effect, which, if the location of the fault sealing is poorer, changes the series of strata and the direction of hydrocarbon migration with the help of vertical fault transporting ability and lateral permeability. This pathway system mainly formed fault block, fault lithology, fault screened reservoirs, etc. (Fig. 1a) (Carolyn Lampe et al, 2012).

Fault tendency also has certain influence on the pathway system: ① In the consequent fault-sand body mode, hydrocarbon migration occurs stepwise through the combined sand body-fracture-sand body transporting media and the migration distance is larger under an appropriate throw condition (Fig. 2a; Fig. 2b). ② In the reverse fault-sand body mode, the transportation capacity is affected by fault distance and the thickness of sand body, and the hydrocarbon transporting ability is poor, resulting in partial closure of hydrocarbon transport pathway while the sand body thickness is greater than the fault distance (Fig. 2c); docking of plate mudstone layer certainly affects the transport of hydrocarbon, while the sand body thickness is less than the fault distance (Fig. 2d) (Guang Fu et al, 2014).

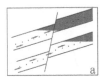

Figure 2.   Connecting relationship between faults and sand body.

2.5   *Unconformity-fault-sand body composite pathway system*

In addition to these common types of pathway systems in the study area, there may be existing the unconformity-sand body-fault composite pathway system and the fault-wrinkle unconformity configuration of unconformity and fault pathway system locally. In the former type, unconformity surface and fault is the transport channel of hydrocarbon migration; the sand body is both a migration and a reservoir channel, and its feature is long hydrocarbon migration distance, E_1d reservoir, and its underlying unconformity surface belongs to the former type in this area (Fig. 1f). In the latter type, three elements of sand body, fault, and unconformity are affected by the spatial configuration relationship, thus the migration distance is not too long, and the reservoir scale is limited (Fig. 1g).

3   PATHWAY SYSTEM CONTROLS HYDROCARBON ACCUMULATION

3.1   *Pathway System controls accumulation time*

Taizhou group accumulated during Neogene Yancheng group depositing in Gaoyou sag, the fault active in that period played a role of migration channel. In Taizhou formation, both break to F-2 oil source fault and connected inner sand body composed the compound hydrocarbon pathway network with, the hydrocarbon migrated to the sand body of Tai-1 member, Fu-1 member along the pathway network and gathered in the sand bodies. The hydrocarbon, which is generated by Fu-2 source rock, migrated upward along long-term active fault and accumulated in the layers of Fu-3 and Dainan, and the second migration formed structural, lithologic-structural reservoirs through sand connectivity on the plane and sand superposition vertically. This shows a fault—and sand-oriented pathway system jointly controlling the accumulation time of a shallow trap.

3.2   *Pathway system controls the part of accumulation*

Taizhou formation thickness is less in Gaoyou sag; Tai-1 member is both the source bed and the cap

rock of the nearby underlying Tai-2 group reservoir. Tai-2 member sand pathway layer has a near-source advantage, which determines that it will inevitably become the preferred place for source rocks expelling hydrocarbon directly and hydrocarbon accumulation nearby, so the Tai-2 sand pathway system has decisive effects on the part of hydrocarbon accumulation. For a long-term active fault, since it is the main channel for vertical hydrocarbon migration and has an effect on adjusting the position and direction of hydrocarbon migration, determining whether the trap can form hydrocarbon accumulation, it determines the distribution of hydrocarbon accumulation zone of each layer series. For example, with the action of the pathway system, the hydrocarbon generated in Fu-2 member source rock can form reservoirs in E_1f_1, E_1f_3, and even a shallower reservoir (Fig. 3).

3.3   *Pathway system controls accumulation scale*

In all kinds of pathway system elements, fault has the strongest variability of transportation ability because of its activity, and thus has the strongest control on hydrocarbon accumulation. If fault scale is a large scale, then accumulation and the capacity of transporting hydrocarbon are strong and form large-scale reservoirs; on the contrary, small dense fault, often resulting in broken fault block, is not conducive to the hydrocarbon flow, transporting and gathering

For the sand pathway system, the greater the scale of its plane distribution, the better the reservoir performance; this will form a larger-scale hydrocarbon reservoir under sufficient oil sources. Dainan formation in study area has developed multi-stage channel deposits, large sand thickness, high degree of spatial superposition, and good horizontal connectivity. Hydrocarbon, through the fracture and unconformity into Dainan formation, does long lateral migration along the skeleton sand bodies, then gathered in the appropriate trap, and formed a lithologic reservoir. Data on well drilling have confirmed that this layer has a higher oil production capacity.

3.4   *Pathway system controls accumulation type*

Pathway system decided the types of oil reservoir trap due to the different characteristics of its

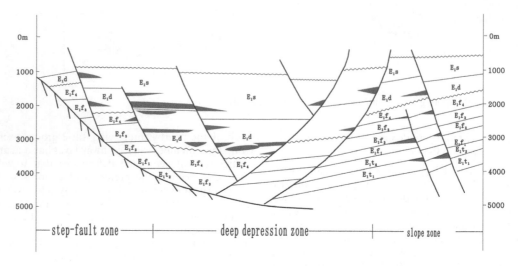

Figure 3. Channel mode of hydrocarbon accumulation in Gaoyou Sag.

geological elements. The main pathway system is a kind of geological factors, and then the trap type is often single, such as stratigraphic trap, fault trap, lithologic trap, etc.; if the pathway system has different types, they tend to form a composite trap, such as lithologic-fault trap. The trap is formed because of a local change in the pathway system or evolution rules; a factor evolution forms a single-factor trap and multi-factors are bound to form a composite trap.

4 SUMMARY

1. The Paleogene strata in Gaoyou sag, all levels of fault, skeletal sand body, and unconformities can be used as the hydrocarbon migration pathway system. Among them, the main factors which control hydrocarbon accumulation are oil source faults and skeletal sand body, and they can form a composite pathway system under certain geological conditions.
2. All kinds of pathway systems have different impacts on hydrocarbon migration, deciding the differences in regional hydrocarbon migration pathway, migration direction, and accumulation part.
3. Pathway system, as a kind of geological factors, has an obvious control effect on hydrocarbon accumulation in accumulation time, hydrocarbon accumulation horizon, trap size, trap type, etc.

REFERENCES

Carolyn Lampe, Guoqi Song and Liangzi Cong, Fault control on hydrocarbon migration and accumulation in the Tertiary Dongying depression, Bohai Basin, China, J. American Association of Petroleum Geologists Bulletin. 96 (2012) 983–1000.

Changhai Gao, Ming Zha and Xinzheng Zhang, Fault Transport System and Hydrocarbon Accumulation Pattern in Chengbei Fault-Ramp Area, J. Xinjiang Petroleum Geology. 28 (2007) 721–724.

Cheng Zhang, Xinong Xie, Xiurong Guo, et al, Pathway System of Large-Scale Petroleum System and Its Controls on Hydrocarbon Accumulation in the Bozhong Sub-Basin, Bohai Bay Basin, J. Earth Science—Journal of China University of Geosciences. 38 (2013) 807–818.

Guang Fu, Wei Wu and Na Li, Controlling effect of three major faults on gas accumulation in the Xujiaweizi faulted depression, Songliao Basin, J. Natural Gas Industry. 1 (2014) 159–164.

Jinhua Fu, Yuliang Liu, Jin Liu, et al, Translocating system of fault block and oil-gas reservoir forming pattern of linnan area, J. Petroleum Geology and Recovery Efficiency. 9 (2002) 55–58.

Qingong Zhuo, Fangxing Ning and Na Rong, Types of Passage Systems and Reservoir-Controlling Mechanisms in Rift Basins, J. Geological Review. 51 (2005) 416–422.

Wang Yong, Yang Daoqing, Guo Juncan, et al, Control Effect of oil/gas Conduction system on Reservoir Forming in Chunguang Area[J]. Special Oil and Gas Reservoirs. 2 (2015) 27–31.

Weihai Zhang, Ming Zha and Jiangxiu Qu, The Type and Configuration of Petroleum Transportation System, J. Xinjiang Petroleum Geology. 24 (2003) 118–120.

Energy Science and Applied Technology ESAT 2016 – Fang (Ed.)
© 2016 Taylor & Francis Group, London, ISBN 978-1-138-02973-6

The characteristics and geological implications of the sporopollen fossils in bauxite in Zunyi, Guizhou

Tao Cui
College of Resources and Environmental Engineering, Guizhou Institute of Technology, Guiyang, China

ABSTRACT: The sporopollen assemblages in the bauxite ore samples taken from Zunyi were analyzed. The research shows that there is only 1 gymnospermae pollen fossil in the ore, and this fossil was formed in Permian. The sporopollen fossils in the ore indicate that the early Permian Epoch was an ore-forming stage for Zunyi bauxite, and its sedimentary environment was mostly a terrestrial one.

Keywords: sporopollen fossils; bauxite; Zunyi

1   INTRODUCTION

Northern Guizhou abounds in bauxite mineral resources, which are concretely distributed in Wuchuan-Zhengan-Daozhen (Wuzhengdao for short) Metallogenic Region and Zunyi Metallogenic Region. There are both connections and differences between the two regions. Most of the previous studies were focused on Wuzhengdao Metallogenic Region (Wu GH et al, 2009) (Du YS et al, 2013) (Du YS et al, 2014) (Wang DH et al, 2013) (Wang Xiaomei et al, 2013) (Gu J et al, 2013) (Yu WC et al, 2014), while there were very few on Zunyi Metallogenic Region (Yin KH et al, 2014) (Liu P et al, 2013) (Li YJ et al, 2014). This paper took samples from Section BT1401 in Zunyi and then tested these samples, identifying and counting the quantity of the sporopollen fossils inside, providing basic data for the research of Zunyi bauxite.

2   GEOLOGICAL BACKGROUND

Zunyi is located in the southwest part of Yangtze Paraplatform, where folds were primarily developed. There are associated faulted structures in the axis and limbs, and that the intra-area structure basically extends in NNE and NE (Li YJ et al, 2014). There are the following exposure stratums and lithological characters in the research area: dolomite in Middle-Upper Cambrian Loushanguan Formation ($\mathcal{C}_{2-3}$ls), whose thickness is greater than 100 m; Lower Ordovician Tongzi Formation (O_1t), with an average thickness of about 40m; Honghuayuan Formation (O_1h), with an average thickness of about 10m; dolomite, shale and bioclastic limestone in Meitan Formation (O_1m); Bioclastic limestone and car-

bonaceous shale in Middle and Upper Ordovician Formation (O_{2-3}); shale and carbonaceous shale in Lower Silurian Longmaxi Formation (S_1l), with an average thickness of about 15 m; Middle Permian Liangshan Formation (P_2l), with a thickness of around 7m; bioclastic limestone and shale in Qixia Formation (P_2q), with a thickness of around 200 m; bioclastic limestone in Middle Permian Maokou Formation (P_2m), with a thickness of around 110m; mudstone, malmstone, silicalite and limestone in Upper Permian Longtan Formation (P_3l) and Changxing Formation (P_3c); mudstone, marlstone, brecciated limestone and dolomite in Lower Triassic Yelang Formation (T_1y) and Maocaopu Formation (T_1m), limestone and dolomite in Middle Triassic Songzikan Formation (T_2s) and Shizishan Formation (T_2sh), and quartz sandstone in Upper Triassic Erqiao Formation (T_3e); quartz sandstone, malmstone and mudstone in Lower Jurassic Ziliujing Formation (J_1zl); mudstone and malmstone in Middle Jurassic Lower Shaximiao Formation (J_2x). The quaternary system contains yellow grey gravelly soil, whose maximum thickness equals to 50 m.

What underlies Zunyi bauxite includes Cambrian Loushanguan Formation and Ordovician Tongzi Formation, while the overlying strata is composed of Permian Liangshan Formation. Different from the bauxite in Wuzhengdao Ore District, Zunyi bauxite is a kind of typical karst-type bauxite. The ores are stacked in a carbonatite ethmolith, showing a disconformable or micro-angle unconformable contact with each other, and a disconformable contact with Upper and Lower Permian Liangshan Formation (P_2l). Middle and upper Silurian formations, Devonian formation as well as middle and Upper Carboniferous Formation are not in sight (Li YJ et al, 2014).

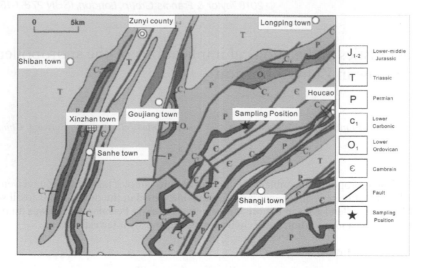

Figure 1.   The geologic diagram of the research area (modified according to [10]).

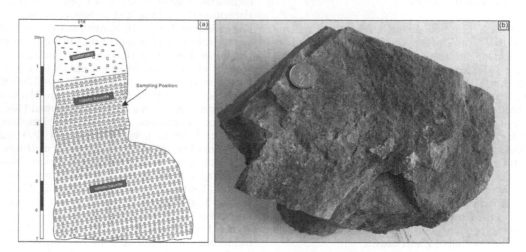

Figure 2.   BT1401 section map and ore samples a. BT1401 section map; b. Black Bauxite ore samples.

3   SAMPLING AND ANALYTICAL TESTING

Samples were taken from typical sections in Zunyi for analytical testing. The sampling position are shown in Fig. 1. The samples were analyzed and handled in Sporopollen Laboratory, Research Center of Palaeontology & Stratigraphy Jilin University, China. The handling process: 100g of screened dry samples were treated in hydrochloric acid, hydrofluoric acid, hydrochloric acid, nitric acid, potassium hydroxide solution and hydrochloric acid; sporopollen was put in a test tube by screening method, and then a stator was placed under a microscope for identification.

4   THE CHARACTERISTICS AND GEOLOGICAL IMPLICATIONS OF SPOROPOLLEN FOSSILS IN THE BAUXITE

The ore samples were taken from BT1402 (Fig. 2, a), known as black bauxite ores (Fig. 2, b). According to analytical processing and identification under microscope, there are very few sporopollens in the ores, but just a sporopollen fossil (Fig. 3). This fossil is a gymnospermae pollen, and there is no pteridophyte spore in this sample. This fossil is a *vestigisporites* sp. (sp. indet.), and formed in Permian.

There are sporopollen fossils that were formed in the Permian in the samples of Zunyi bauxite ore,

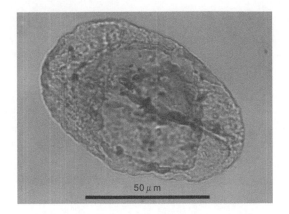

50 µ m

Figure 3.   The sporopollen fossils in Zunyi Bauxite.

suggesting that bauxite ore was still being formed in the early Permian. That ore was formed in the Permian cannot prove that Zunyi bauxite was formed in the early Permian, for the reason that bauxite is a weathered residual-type ore or sedimentary ore, whose formation was a long process (Bárdossy G. et al, 1990) (Bushinsky, 1984). The early Permian is a stage at which bauxite was formed. The early Permian is not the only stage at which metallogenic materials were carried and stacked and Zunyi bauxite was finally formed, but that the formation of Zunyi bauxite underwent several stages. The early Permian might just be an end stage at which Zunyi bauxite was formed, or an important stage at which it was formed. The final ascertainment needs further studies.

5   CONCLUSIONS

After a research of Zunyi bauxite ore samples, we came to the following conclusions: 1) there are very few sporopollen fossils, including only one gymnosperm pollen fossil, in the samples, suggesting that lots of fossils couldn't be preserved in the lager stage of ore formation due to the effect of alteration; 2) the sporopollens are *vestigisporites* formed in Permian, suggesting that the ore-forming process lasted until the Permian; 3) despite the small number of sporopollen fossils in the ores, but there are gymnosperm pollens, suggesting that the ores were mostly formed in a terrestrial environment.

ACKNOWLEDGEMENT

This work was supported by the provincial project of characteristics of REE and its transferring disciplinarian in mineralizing process of the bauxite in northern Guizhou, China (No.LH[2014]7358), the project of analysis on geochemical behaviors of Platinum group elements in Emeishan basalt (No. XJGC 20131204), Guizhou Institute of Technology, Guiyang, 550003, China.

REFERENCES

Bárdossy G. Karst bauxites, bauxite deposits on carbonate rocks. London, Developments in Economic Geology. Vol. 14 (1990) p. 320–332.

Bushinsky. Bauxite geology. Beijing: Geology Press. 1984, p. 129–130.

Du YS, Zhou Q, Jin ZG, Lin WL, Zhang XH. Advances in basic geology and metallogenic regularity study of bauxite in Wuchuan-Zheng, an-Daozhen area, northern Guizhou, China. Geological science and technology information, Vol. 32 (2013) No. 1, p. 1–6.

Du YS, Zhou Q, Jin ZG, Lin WL, Wang XM, Yu WC. Mineralization model for the early permian bauxite deposits in Wuchuan-Zheng, an-Daozhen area, northern Guizhou Province. Journal of palaeogeography, Vol. 16 (2014) No. 1, p. 1–8.

Gu J, Huang ZL, Fan HP, Ye Lin, Jin ZG. Provenance of lateritic bauxite deposits in the Wuchuan-Zheng, an-Daozhen area, northern Guizhou province, China: LA-ICP-MS and sims U-Pb dating of detrital zircons. Journal of Asian earth sciences. Vol. 70 (2013) p. 265–282.

Liu P, Liao YC. The zonation and genetic mechanism of Zunyi high-and low-ferrous bauxites. Geology in China. Vol. 40 (2013) No. 3, p. 949–966.

Li YJ, Zhang ZW, Zhou LJ, Wu CQ. Mineralogical and Geochemistry Feature of Zunyi Bauxite Deposit, Guizhou Province, China. Acta Mineralogica Sinica, Vol. 34 (2014) No. 2, p. 234–246.

Wang DH, Li PG, Qu WJ, Yin LJ, Zhao Z, Lei ZY, Weng SF. Discovery and preliminary study of the high tungsten and lithium contens in the Dazhuyuan bauxite deposit, Guihou, China. Science China: Earth Science. Vol. 56 (2013) p. 145–152.

Wang Xiaomei, Jiao Yangquan, Du Yuanshen. Rare earth element geochemistry of bauxite in Wuchuan—Zheng,an—Daozhen area, northern Guizhou, province. Geological science and technonlogy information. Vol. 32 (2013) p. 27–33.

Wu GH, Zhang BS, Guo CL, Wang CL, Gao H. Detrital zircon U-Pb dating for the Silurian in northern tarin basin and its significance. Geotecton et Metallogenia. Vol. 33 (2009) p. 418–426.

Yin KH, Li KQ, Zhu DB, He XL, Wu Hong. Geological and Geochemical Characteristics of Xianranyan Bauxite Deposit in Zunyi. Guizhou Geology. Vol. 28 (2014) No. 1, p. 24–29.

Yu WC, Du YS, Zhou Q, Jin ZG, Wang XM, Qin YJ, Cui T. Provenance of bauxite beds of the lower Permian in Wuchuan-Zheng, an-Daozhen area, northern Guizhou province: evidence from detrital zircon chronology. Journal of palaeogeography. Vol. 16 (2014) No. 1, p. 19–29.

Energy Science and Applied Technology ESAT 2016 – Fang (Ed.)
© 2016 Taylor & Francis Group, London, ISBN 978-1-138-02973-6

Research on a type of mud pulse communication method in a rotary steering drilling tool

De Shu Lin
Hubei Collaborative Innovation Center of Unconventional Oil and Gas, Yangtze University, Jingzhou, Hubei, China

Ding Feng
School of Mechanical Engineering, Yangtze University, Jingzhou, Hubei, China

ABSTRACT: This paper introduced the development of the communication system between the ground and the underground control system in the rotary steering tool. The transmission mode and transmission principle of the signal were summarized, and the advantages and disadvantages of them were evaluated in terms of reliability, transmission depth, transmission rate, economy, etc. By analyzing the principle of the occurrence and transmission of mud pulse in MWD (Measure While Drilling) system, the characteristics of signal transmission and the requirement of the closed-loop control system of rotary steering tool were analyzed. The ground signal transmission system should be based on the mud negative pulse mode. Based on these summaries and analysis, the design requirements and principles of the ground signal transmission system were put forward, the principle of the negative pulse generation was expounded, and the design scheme of the system was given. In the process of rotary drilling, the transmission of drilling fluid pulse, cable transmission, acoustic transmission, and electromagnetic wave transmission were analyzed. The negative pulse transmission mode of the drilling fluid was chosen to transmit the ground instruction, and the whole scheme of the ground drilling fluid negative pulse signal was designed. The downlink information transmission system had the characteristics of simple structure and high reliability.

Keywords: rotary steering tool; signal transmission technology; drilling fluid pulse

1 INTRODUCTION

In the thirties of the last century, people began the research on application of the measurement technology with drilling. Baker Hughes, Schlumberger, Halliburton, and other petroleum technology companies have developed wireless measurement while drilling tools. Along with the development of the measurement technology, the research and application of underground information real-time to the ground transmission technology has been more mature. At present, due to the emergence of a large number of complex structure wells, conventional drilling tools cannot meet the requirements of the development of the complex reservoir. To accomplish complex well trajectory, the direction of bit drilling must be controlled accurately. The core of the development of the well trajectory control tool is to change the joint force on the drill or to change the eccentric degree of the drilling tool under the working condition (Feng Ding et al, 2011). Well trajectory closed-loop drilling technology is the development of advanced drilling technology at home and abroad, which is the focus and development direction of modern steering drilling. In the case of accurate control of downward drill bit, the problem of drilling guide parameter must be solved, and the drill bit should be drilled along the well trajectory in advance (Feng Ding et al, 2011). The signal transmission system is the key part of the automatic control system of the well trajectory, and it is an important bridge connecting the ground control system and the downward bit guiding mechanism. Into a well which is thousands of meters deep, eye is a channel for the transmission of signals, mainly responsible for the downward condition and parameter monitoring and implementation of downward control system-oriented decision-making, and intervention control function two-way communication tasks. Therefore, the signal transmission system is the key technology to realize the automatic control of well trajectory, which is directly related to the success or failure of the whole control system.

In the downlink communication technology of the well trajectory control system, the drilling direction parameters are mainly transmitted by mud pulse, insulated wire, electromagnetic wave, sound wave, intelligent drill pipe, optical fiber, etc. (Liu Xiu-shan et al, 2000) (Li Fengfei et al, 2012) (Yang Quanjin et al, 2004) (SUN Hao-yu, 2013). As the underground working communication environment is very complex, there are advantages and disadvantages of several transmission media; the transmission characteristics of the above tare shown in Table 1.

Each transmission medium has its own advantages and disadvantages. Now, the application of oil drilling is drilling fluid pulse and electromagnetic wave. The drilling fluid pulse is the most common practical drilling guide parameter of the transmission. The application of high-speed optical fiber communication is more common in the application of coiled tubing drilling.

2 THE BASIC PRINCIPLE OF DOWNWARD TRANSMISSION SYSTEM OF DRILLING GUIDE PARAMETERS

The basic structure of the transmission system of drilling guide parameters is shown in Figure 1. The drilling fluid is more suitable for the actual situation of drilling engineering pulse as the transmission medium; we realized that the drilling parameters are downward transmission oriented.

Drilling fluid pulse can be divided into three types: positive pulse, negative pulse, and continuous wave. In the MWD system, the positive pulse and continuous wave are generally used. Positive pulse mode in that pulse generator is arranged in

Table 1. Comparison of media in rotary steering drilling tool.

Transmission medium	Speed	Reliability	Development cost
Drilling fluid pulse	slow	good	cheap
Insulated wire	fast	good	high
Electromagnetic wave	fast	common	high
Acoustic wave	fast	common	low
Optical fiber	very fast	good	high
Intelligent drill string	fast	good	high

the drill collar. It can instantly impede the flow of the drilling fluid, resulting in a pressure higher than the normal pressure of the pressure wave (Liu Xiushan et al, 2000). In the drilling guide parameter system, the negative pulse form is better. The negative impulse is the pressure wave which is lower than the normal pressure, or the method of judging the flow pulse. The overall scheme of drilling guide parameters based on negative pulse is shown in Figure 2.

A branch pipeline is drawn out on the vertical pipe, and the opening and the closing of the pulse valve are controlled by the pulse valve controller to change the downward displacement of the drilling fluid. When the pulse valve is opened, the drilling fluid is diverted through the one-way valve to a part of the drilling fluid, and the drilling fluid is returned to the mud pool. Owing to the diversion of the drilling fluid, the pressure and flow in the vertical pipe and drill string drop suddenly. When the pulse valve is closed, the pressure of the circulating system returns to the normal value (Li Qi et al, 2007). Therefore, through the opening and closing of the pulse valve, the drilling fluid pressure and the flow rate drop, and the negative pulse waveform signal of the drilling fluid is generated. Through the modulation of the ground control pulse signal, the drilling control parameters can be transmitted down through the pulse code. The overall scheme of the drilling fluid negative pulse guide parameter is shown in Figure 2.

3 PULSE SIGNAL CODING MODE DESIGN

Owing to the high-energy efficiency of Pulse Position Modulation (PPM), it is widely used in the occasions of limited power (Sun Dongkui et al, 2008). PPM was a modulation format that maps message bits to pulse positions.

To achieve the downward transfer of the ground command, the down hole tool must not only need to be able to produce drilling fluid pressure pulse at the bottom, but also be able to identify the drilling fluid pressure pulse from the analysis of the corresponding control instructions. Therefore, the transmission of drilling fluid pressure pulse must be carried out according to a certain rule or time sequence, that is, the signal must be coded accordingly. For the signal transmission system of rotary steering drilling tool, the

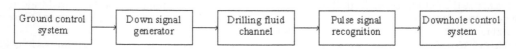

Figure 1. The basic structure of the transmission system of drilling guide parameters.

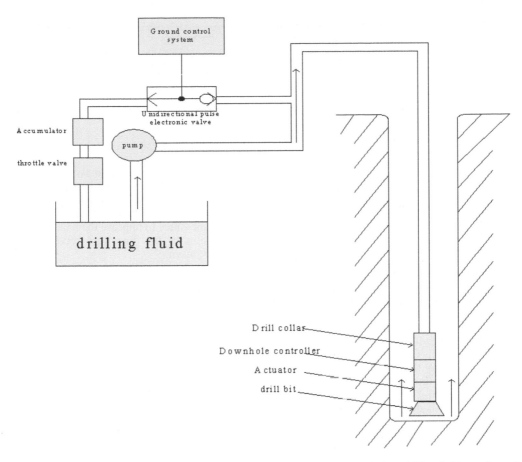

Figure 2. Integral scheme of drilling guide parameters' downward transmission based on drilling fluid negative pulse.

code must follow the principle of accuracy, timeliness, and feasibility.

3.1 *Instruction system design*

The command system of the drilling guidance parameter signal transmission system is mainly used to control the guiding tool to realize the function of stability and orientation. The guiding function comprises four different functions of increasing the slanting, descending the slanting, increasing the position, and lowering the direction. Combined with the characteristics of the guide drilling tool, the design parameters of the system are shown in Figure 3. Corresponding to the cross-section of 0 degrees to 360 degrees, each 30 degrees design one instruction, and so, there must be a total of 12 instructions. At the same time, the design of one stable instruction, one stop instruction, and two spare instructions guided the drilling parameter signal transmission system design under the command of the sixteen.

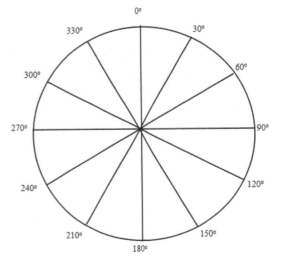

Figure 3. Angle of azimuth.

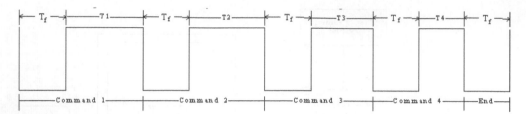

Figure 4. Coding mode of signal transmission system.

3.2 *Signal coding mode design*

For drilling fluid pressure pulse, the falling edge is fast and easy to be recognized. The signal transmission system of the rotary steering drilling tool is set with falling edge as the source of signal coding. In order to receive and recognize the drilling fluid pressure, a pulse signal in each pulse is required after a certain period of recovery. That is to send a pulse after the need to wait for the next recovery cycle in order to carry out the next pulse of the transmission. If the pulse is used to express the digital signal "0" or "1", the transmission process requires frequent switching of the ground leakage device; the system has a great impact on the system. In order to simplify the operation, the system has the ability of anti-interference and fault tolerance of digital coding, and the coding mode of information transmission is designed with five pulses. The whole command is divided into four regions, each of which takes a full pulse as the starting signal. Each region is composed of three parts: pulse fall time, pulse recovery time, and command time (T1, T2, T3, and T4). The coding mode of the signal transmission system is shown in Figure 4.

4 CONCLUSION

Drilling fluid pulse PPM was a simple and feasible method to realize the steering drilling system's downward signal transmission. The control circuit part of the downward instruction steering tool in receiving function can be in the downward tool, for the corresponding hardware and software implementation. By collecting the data of the downward mud motor output voltage, the CPU can detect the change in drilling fluid flow. The key to guide drilling system downward communication software was to achieve input voltage threshold, and the pulse width timing processing. There was more interference in the transmission process in the pulse pressure of fluid drilling site; therefore, effective measures to improve the reliability of the transmission of the downward receiving device need to be taken.

REFERENCES

Feng Ding, Yuan Yongxin, Li Hanxing and Bian Shanyou. The status and development trend of trajectory control tool [J]. China petroleum machinery, (2011), 39(3): 70–73.

Li Fengfei, Jiang Shiquan, Li Hanxing and Gao Deli. Study on downward signaling system of rotary steerable drilling tool [J]. China offshore oil and gas, (2012), 24(6): 42–47.

Li Qi, Peng Yuanchao, Zhang Shaohuai and Liu Zhikun. Study on signal transmission technique in rotary steering drilling [J]. Acta Petrolei Sinica, (2007), 28(4): 108–111.

Liu Xiu-shan and Su Yi-nao. Scheme design of downward signaling system [J]. Acta Petrolei Sinica, (2000), 21(6): 88–92.

Liu Xiushan and Su Yinao. Study on Transmission Velocity of Mud Pulse Signal. Petroleum Drilling Techniques, (2000), 28(5): 24–26.

Sun Dongkui and Dong Shaohua. Time frequency analysis of drilling fluid hydraulic communication channel signal transmission [J]. China Petroleum Machinery, (2008), 36(4): 42–44.

Sun Hao-yu. Technology of magnetic induction transmission of intelligent drill pipe and its channel characteristics. Journal of China University of Petroleum (Edition of Natural Science), (2013), 37(6): 172–176.

Yang Quanjin and Li Jin. A scheme for downward communication in the steerable rotary directional drilling system [J]. Petroleum Geology and Recovery Efficiency, (2004), 11(1): 75–77.

2 Renewable energy and cell technologies, energy recovery, engines, generators, electric vehicles

Energy Science and Applied Technology ESAT 2016 – Fang (Ed.)
© 2016 Taylor & Francis Group, London, ISBN 978-1-138-02973-6

Recovery and utilization process of a vertical tank for sinter waste heat

Junsheng Feng, Hui Dong, Jianye Gao, Jingyu Liu & Hanzhu Li
School of Metallurgy, Northeastern University, Shenyang, China

ABSTRACT: The recovery and utilization of waste heat resources in sintering process is one of the effective ways to reduce the energy consumption of sintering process. In view of the drawbacks of traditional sintering waste heat recovery system, the vertical tank technology for sinter waste heat recovery and utilization is proposed by imitating the structure and technology of a dry quenching furnace, and the two key problems affecting the feasibility of vertical tank technology are elaborated. The research results indicate that the vertical tank for sinter waste heat recovery and utilization is a new way of recycling sinter waste heat resources efficiently, which not only demonstrates the higher recovery rate of sinter waste heat, but also makes the energy level of outlet cooling air become higher. The gas flow resistance characteristics and gas-solid heat transfer characteristics in the sinter bed layer are two key problems in determining the feasibility of vertical tank technology.

Keywords: sinter; waste heat; recovery and utilization; vertical tank; CDQ

1 INTRODUCTION

The waste heat resource carried by the unit mass sinter production (i.e., 1 ton) is approximately 0.94 ~ 1.02 GJ in the large- and medium-sized steel enterprises in our country, which, respectively, accounts for 65% ~ 71% and 11% ~ 12% of the total amount of waste heat resources in the sintering process and steel enterprises (Cai 2009). As the basic research on waste heat resource of the sintering process is relatively backward, the recovery rate of waste heat resource is lower than the average level of foreign major large- and medium-sized iron and steel enterprises, i.e., only 28% ~ 30% (Dong et al. 2011). Therefore, the efficient recovery and utilization of sinter waste heat resource is one of the main ways to reduce the energy consumption of sintering process in the large- and medium-sized iron and steel enterprises in our country (Dong et al. 2014). At present, the sinter waste heat resource is recycled by means of the sinter ring-shaped cooler and straight-line cooler. However, the settings of structure form, structure, and operation parameters of the ring-shaped cooler and straight-line cooler care more about the cooling of sinter particles, rather than the recovery of sinter waste heat resource. In the case of sinter ring-shaped cooler, the sinter cooler system had many disadvantages that are difficult to overcome (Dong et al. 2011, Dong et al. 2012a). Based on this, by imitating the technology of coke dry quenching furnace, the vertical tank for sinter waste heat recovery and utilization is put forward (Cai & Dong 2009). The two key problems

affecting the feasibility of vertical tank for sinter waste heat recovery are further discussed on the basis of existing researches, and the research status of key technological problems are expounded. The present study lays a solid foundation for deepening theoretical research and the subsequent engineering implementation of vertical tank for sinter waste heat recovery and utilization.

2 RECOVERY AND UTILIZATION PROCESS

2.1 Drawbacks of traditional sinter cooler

The drawbacks of sinter ring-shaped cooler are higher leakage rate of the sinter cooler, lower recovery rate of sinter waste heat, lower energy level of air, and higher atmospheric pollutant (Dong et al. 2012a).

1. The air leakage in the sinter ring-shaped cooler is of two types, namely the lower leakage (I) and the upper leakage (II), which are shown in Fig. 1. In general, the upper leakage rate of cooling air is 15% ~ 20%, and the lower leakage rate is 20% ~ 30% for the large- and medium-sized steel plants in China. Owing to the upper leakage, the heat quantity carried by the cooling air is wasted, which results in the reduction of steam quantity and power generation per ton ore. By calculation, the loss of power generation accounts for 27.3% of present power generation. Furthermore, the energy consumption of the

sintering process is increased by 1.25~1.56 kgce. Similarly, the power consumption is increased by 0.25~0.27 kgce due to the lower leakage.

2. The sectional blast along with the cross flow cooling is the main structure form of the large sinter cooling system thus far, which determines that the sinter waste heat is only recycled partly. Taking a typical 360 m² sintering machine as an example, the cooling system is divided into five sections from the beginning to the end, namely first, second, third, fourth, and fifth sections, as shown in Fig. 1. According to the present technology, only the sinter waste heat in the first and second sections have been recycled to generate electricity, and the rest accounting for 36% of overall sinter waste heat resource has not been fully utilized.

3. In the sinter ring-shaped cooler, the heat transfer between the sinter particles and cooling air is the gas-solid heat transfer by cross flow. The height of the sinter bed layer is 1.2~1.5 m, and the flow rate of cooling air is 1.0~1.4 m/s. Therefore, the gas-solid heat transfer time is merely 1.0~1.5 s. For the large sinter ring-shaped cooler in China, the average temperatures of cooling air in the first and second sections are 350~400°C and 300~350°C, respectively. In addition, due to the higher leakage rate of the sinter cooler, the emission of particulate matter is one of the most important pollution sources in the sintering enterprise.

2.2 Determination of vertical tank technology

The sinter cooler could be statically connected with the inlet and outlet devices of cooling air. Changing the "passing through" to "static" type radically avoids the air leakage from the sinter cooler. Changing the "horizontal" type of sinter cooler to "vertical" type radically ensures the energy level of outlet cooling air. As the above types were applied successfully in the dry quenching furnace, the vertical tank technology for sinter waste heat recovery and utilization has been presented by imitating the structure and process of dry quenching furnace (Cai & Dong 2009), as shown in Fig. 2.

The whole process system of vertical tank for sinter waste heat recovery mainly comprises three parts, namely the cooling vertical tank, the waste heat boiler, and the steam turbine generator. The incandescent sinter particles are filled at the top of the vertical tank, then cooled by the recycle cooling air due to the gas-solid heat transfer between the sinter particles and cooling air in the cooling section of the vertical tank, and finally discharged through the rotary discharge valve at the bottom of the vertical tank.

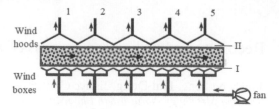

Figure 1. Sketch of a ring-shaped sinter cooler. 1 - Waste heat boiler; 2 - Waste heat boiler; 3 - Drying materials or emission; 4 - Pre-heating ignition or emission; 5 - Emission.

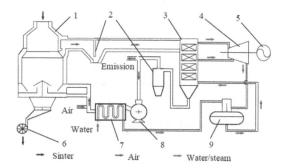

Figure 2. Schematic diagram of vertical tank technology. 1 - Vertical tank; 2 - Dust collector; 3 - Waste heat boiler; 4 - Steam turbine; 5 - Generator; 6 - Discharge valve; 7 - Pre-heater; 8 - Induced draft fan; 9 - Deaerator.

The cooling air discharged from the upper outlet of the vertical tank is cleaned by flowing through the dust collector, and then sent into the dual pressure waste heat boiler. The temperature of cooling air at the inlet of the waste heat boiler is approximately 500°C. The hot air flows through the medium-pressure super-heater, the medium-pressure evaporator, the medium-pressure economizer, and the low-pressure evaporator in turn, and then the air temperature drops to 100~150°C. After flowing through the secondary dust collection device and the feed water pre-heater, the cooling air is sent into the bottom of the vertical tank, as a part of the recycle cooling air.

The temperature of feed water coming out from the feed water pre-heater is 60~90°C, and the feed water is sent into the deaerator with the condensed water out from the waste heat boiler. Through the deoxygenization of the deaerator, a part of the deoxidized water is sent into the low-pressure steam drum. After the vapor-liquid separation, the low-pressure steam as supplementary steam is sent into the steam turbine, and the liquid water sent into the low-pressure evaporator is heated; another part of the deoxidized water is sent into the medium-pressure steam drum through the medium-pressure economizer. After the vapor-liquid separation, the saturated steam heated by the medium-pressure super heater is sent into the turbine to do work, and the separated water sent into the medium-pressure evaporator is heated by vaporization.

2.3 *Advantages of vertical tank technology*

Compared with the technological process of the existing sinter cooler, the vertical tank process for sinter waste heat recovery has many advantages. These advantages are described in detail below.

1. The air leakage rate of cooling equipment is lower, and dust emissions decrease significantly. The vertical tank for sinter waste heat recovery is a sealed device, the top of which is sealed by the water seal tank and the bottom by the rotary sealing valve. Owing to the good air tightness, the air leakage rate of the vertical tank is close to zero. At the same time, as the incandescent sinter particles are cooled by the cooling air in the sealed vertical tank, dust emissions can easily be controlled.
2. The gas-solid heat transfer efficiency, as well as the outlet air quality of the vertical tank and the sinter waste heat utilization, is higher. The reverse heat transfer between the cooling gas and the sinter particles in the vertical tank can make the gas-solid heat transfer more intense, and keep the outlet gas temperature at a higher level. Owing to the existence of the pre-storage section, the heat transfer time between the cooling gas and sinter particles becomes longer, which results in the lower outlet temperature of sinter particles. Therefore, sinter waste heat utilization is greatly improved.
3. The quality of cooled sinter is improved significantly. The existence of a pre-storage section in the vertical tank has the function of heat preservation for the incandescent sinter, and the sinter maturity is improved further. Second, due to the function of mechanical force, the fragile parts and raw ore can be easily sieved out, and the rate of finished products is improved. Meanwhile, the cooling rate of sinter particles becomes slower, and the sinter quality will be more uniform.

By theoretical calculation, the waste heat recovery rate of the vertical tank is 75~80%, and that of the sinter ring-shaped cooler is only 40~45%. Taking the 360 m² sinter machine as an example, the waste heat recovery and utilization of the vertical tank is 121GJ/h higher than that of the sinter ring-shaped cooler, as shown in Table 1.

Table 1. Comparison analysis of sinter waste heat recovery.

Parameters	Ring-shaped cooler	Vertical tank
Sinter inlet temperature/K	923	923
Air outlet temperature /K	623~673	753~803
Cooling system leakage rate /%	15~20	0
Waste heat recovery rate/%	40~45	75–80
Heat recovery quantity/kJ·h⁻¹	1.38×10^8	2.59×10^8

3 MAJOR PROBLEMS OF VERTICAL TANK

As the key equipment of the vertical tank technology for sinter waste heat recovery and utilization, the feasibility of the waste heat recovery and utilization of the vertical tank and how to carry it out must be through the theoretical demonstration and the long-term practice verification. The two problems mentioned below affecting the feasibility of the vertical tank technology are mainly considered.

The first problem is the gas flow resistance characteristic in the sinter bed layer. The cooling air flows into the vertical tank from the bottom and exchanges heat with the sinter particles in the bed layer. A complete contact between the sinter particles and cooling air is the necessary condition to ensure sufficient gas-solid heat exchange. Reducing gas flow resistance loss through the sinter bed layer is one of the important premises to make gas-solid heat transfer fully proceed. Otherwise, it is difficult to ensure the economy of vertical tank technology due to the higher resistance loss of gas flow through the sinter bed layer.

The second problem is the gas-solid heat transfer characteristic in the sinter bed layer, which is closely related to the gas flow resistance characteristics. In essence, the basic rules of gas-solid heat transfer in the sinter bed layer of the vertical tank are discussed through the study of gas-solid heat transfer process, and the relationship among the sinter cooling capacity in the vertical tank, the flow rate and temperature of the outlet cooling air, and the structural parameters of the vertical tank is also studied. In addition, the flow rate and temperature of the outlet cooling air determine the basic situation of follow-up waste heat utilization. Therefore, combined with the subsequent waste heat power generation technology, the ranges of flow rate and temperature of the outlet cooling air, as well as the structural parameters of the vertical tank, can be determined. Meanwhile, the air flow resistance loss during the gas-solid heat transfer process is also considered.

3.1 Gas flow resistance characteristics

In order to obtain the gas flow resistance characteristics in the sinter bed layer, preliminary experiments were carried out, and the research results are shown as follows (Feng et al. 2015a, b, Yang 2013).

1. Based on a homemade gas flow experimental setup, our research team used the dimensionless method of Ergun's correlation to investigate the gas flow pressure drop in the packed beds with sinter particles. The factors affecting the gas flow pressure drop are analyzed in detail (Feng et al. 2015a, b). The results indicate that the main factors affecting the gas flow pressure drop are the gas superficial velocity, sinter particle diameter, and bed layer voidage. Among them, the pressure drop per unit bed layer height increases as a quadratic relationship with the increasing gas superficial velocity and decreases as an exponential relationship with the increasing sinter particle diameter and bed layer voidage. By means of the experimental data fitting, the correlation of gas flow pressure drop is obtained as:

$$\frac{\Delta P}{L} = \left[29.21 + 0.0001056\mathrm{e}^{(31.216\varepsilon)}\right]\frac{\mu(1-\varepsilon)^2}{\varepsilon^3 d_p^{\,2}}u$$
$$+ \left[0.1936 + 0.00007404\mathrm{e}^{(16.162\varepsilon)}\right]\frac{\rho(1-\varepsilon)}{\varepsilon^3 d_p}u^2 \tag{1}$$

where $\Delta P/L$ is the pressure drop per unit bed layer height, μ is the air dynamic viscosity, ε is the bed layer voidage, d_p is the sinter particle diameter, u is the gas superficial velocity, and ρ is the air density.

2. The experimental conditions mentioned above in the literatures (Feng et al. 2015a, b) were further refined on the basis of the previous study of gas flow resistance characteristics in the sinter bed layer, and the non-dimensional correlation of gas flow pressure drop was devised through the method of dimensional analysis. The coefficients and indexes of non-dimensional correlation were determined through the regression analysis of experimental data (Yang 2013). The non-dimensional correlation of gas flow pressure drop is shown as:

$$\frac{\Delta P}{\rho u^2} = 7.448 \times \left(\frac{\rho u D}{\mu}\right)^{-0.319}\left(\frac{d_p}{D}\right)^{-1.56}\left(\frac{L}{D}\right)^{1.068} \tag{2}$$

where ΔP is the gas flow pressure drop through the bed layer, L is the height of the sinter bed layer, and D is the inner diameter of the vertical tank.

3.2 Gas-solid heat transfer characteristics

The methods of experiment and numerical simulation were carried out to obtain the gas-solid heat transfer characteristics in the sinter bed layer, and the results are shown as follows (Feng et al. 2015c, Dong et al. 2012b).

1. The gas-solid heat transfer behavior in a packed bed with sinter particles was experimentally investigated (Feng et al. 2015c). A homemade experimental setup was built to measure the experimental data of sinter and cooling air. The factors affecting the gas-solid heat transfer behavior were analyzed in detail. In addition, the general correlation of gas-solid heat transfer coefficient was derived through the method of dimensional analysis. The results indicate that the gas-solid heat transfer coefficient increases with the increase in gas superficial velocity, and decreases with the increase in sinter particle diameter. A small increase in gas-solid heat transfer coefficient with the sinter temperature is also observed. Finally, the heat transfer Nusselt number is obtained, according to the regression analysis of experimental data, as shown below.

$$Nu = 1.2\varepsilon^{1.38}\,\mathrm{Re_p}^{0.526}\,Pr^{1/3} \tag{3}$$

where Nu is the heat transfer Nusselt number, Re_p is the particle Reynolds number, and Pr is the Prandtl number.

2. The numerical investigation of the gas-solid heat transfer process in a moving bed with sinter particles is conducted using the COMSOL software (Dong et al. 2012b). The temperature distributions of sinter and cooling air were obtained, respectively, and the factors affecting the gas-solid heat transfer process were analyzed in detail. The results indicate that the gas superficial velocity and the bed layer height are the main factors affecting the outlet temperature of the cooling air. For a given gas superficial velocity, the temperature and exergy of the outlet cooling air gradually increase with the increase in bed layer height. For bed layer height, with the increase in gas superficial velocity, both the outlet temperatures of sinter and cooling air decrease, and the outlet exergy of the cooling air increases first and then decreases. The annual output of the 360 m² sintering machine in an iron and steel company is seen as an research object, and the sinter production capacity of

the sintering machine is 547.2 t/h. By numerical calculation, the suitable ratio of the air flow rate to the sinter production capacity for the vertical tank is 1500~1650 m³/t.

4 CONCLUSION AND OUTLOOK

1. The vertical tank technology for sinter waste heat recovery proposed by imitating the structure and technology of a dry quenching furnace is a new way to recycle sinter waste heat resource efficiently, which has many advantages, such as higher sinter waste heat recovery and utilization, lower air leakage rate of cooling system, higher energy level of outlet cooling air, etc.
2. The gas flow resistance characteristics and gas-solid heat transfer characteristics in the sinter bed layer are the two major problems in determining the feasibility of the vertical tank technology for sinter waste heat recovery and utilization. Among them, the pressure drop per unit bed layer height increases as a quadratic relationship with the increasing gas superficial velocity and decreases as an exponential relationship with the increasing sinter particle diameter and bed layer voidage. The gas-solid heat transfer coefficient increases with the increase in gas superficial velocity, and decreases with the increase in sinter particle diameter. Meanwhile, a small increase in gas-solid heat transfer coefficient with the sinter temperature is also observed.
3. The authors, together with the research team, have conducted a lot of theoretical research on the sinter waste heat recovery system of the vertical tank, and some research results have been achieved. Therefore, the technology research and engineering application of the vertical tank for sinter waste heat is a long-term, arduous process, which needs lots of scholars and engineers to work together, and needs strong support from many research institutes and steel enterprises.

ACKNOWLEDGEMENTS

The authors wish to thank the financial support for this work provided by the National Science Foundation of China (No. 51274065) and the Science and Technology Planning Project of Liaoning Province, China (2015020074-201).

REFERENCES

Cai J.J. & Dong H. 2009. Method and device of sinter waste heat recovery and utilization with vertical tank. China Patent 200910187381.8, 2009-01-05.
Cai J.J. 2009. The energy and resources saving technologies employed in Chinese iron and steel industry and their development. World Iron Steel 4: 1–13.
Dong H., Li L., Cai J.J. & Li J. 2012b. Numerical Simulation of Heat Exchange in Vertical Tank of Waste Heat Recovery. Journal of Northeastern University (Natural Science) 33(9): 1299–1302.
Dong H., Lin H.Y., Zhang H.H., Cai J.J., Xu C.B. & Zhou J.W. 2011. Thermal test and analysis of sintering cooling system. Iron and Steel 46 (11): 93–98.
Dong H., Wang A.H., Feng J.S., Zhang Q. & Cai J.J. 2014. Process and prospect in sintering waste heat resource recovery and utilization technology. Iron and Steel 49(9): 1–9.
Dong H., Zhao Y., Cai J.J., Zhou J.W. & Ma G.Y. 2012a. On the air leakage problem in sintering cooling system. Iron and Steel 47(1): 95–99.
Feng J.S., Dong H. & Dong H.D. 2015a. Modification of Ergun's correlation in vertical tank for sinter waste heat recovery. Powder. Technol. 280: 89–93.
Feng J.S., Dong H., Liu J.Y. & Liang K. 2015c. Gas–solid heat transfer characteristics in vertical tank for sinter waste heat recovery. CIESC Journal 66(11): 4418–23.
Feng J.S., Dong H., Liu J.Y., Liang K. & Gao J.Y. 2015b. Experimental study of gas flow characteristics in vertical tank for sinter waste heat recovery. Appl. Therm. Eng. 91: 73–79.
Yang Y.W. 2013. Experimental Study on Resistance Characteristics in Vertical Tank for Recovering Sinter Waste Heat. MS. Northeastern University, China 51–52.

Energy Science and Applied Technology ESAT 2016 – Fang (Ed.)
© 2016 Taylor & Francis Group, London, ISBN 978-1-138-02973-6

A modularized DC-DC converter used in low-power PV-battery power supply applications

Q. Liu
Shenzhen Academy of Aerospace Technology, Shenzhen, China
Harbin Institute of Technology, Harbin, China

T.C. Li & D.L. Zhang
Shenzhen Academy of Aerospace Technology, Shenzhen, China

Z.Z. Zhang
Harbin Institute of Technology, Harbin, China

ABSTRACT: A Photovoltaic (PV)-based stand-alone power system is usually used to interface several PV solar arrays and a battery and deliver a continuous power to the users in an appropriate form. To reduce the cost and make use of the PV energy as much as possible, a modularized DC-DC converter was proposed for low-power PV-battery power supply applications, used as solar Array Power Regulators (APRs) and load Bus Power Regulators (BPRs). The proposed DC-DC converter taking use of buck-boost topology is controlled by four outer loops combining with an inner current loop. With the proposed converter, voltage regulation, taper charge of battery, Maximum Power Point Tracking (MPPT) of solar array, and module parallelization can be analyzed and verified experimentally in a low-power unregulated bus and regulated bus configuration, respectively.

Keywords: PV-battery power supply; voltage regulation; taper charge; MPPT; bus configuration

1 INTRODUCTION

A Photovoltaic (PV)-battery system, interfacing one PV port, one battery port, and one load port, is a good candidate for renewable power system, especially for spacecraft power supply system (Wang & Li 2013). In the PV-battery system, solar array regulators are used to extract the power from the SA in order to recharge the battery and to supply the loads during sunlight. Traditionally, the sequential switching shunt regulator (S3R, introduce by European Space Agency in the seventies) is preferred for the Direct Energy Transfer (DET) configuration (O'Sullivan & Weinberg 1977.). Unfortunately, the DET configuration is not able to make use of all the available solar array power under all the situations. This may penalize the solar array size and mass, in particular for the applications where the available solar array power varies widely, according to the temperature and the solar flux fluctuations (Fatemi et al. 2000).

To overcome that limitation, Maximum Power Point Trackers (MPPT) are proposed. A MPPT system requires a DC-DC converter (usually named Array Power Regulator (APR)) and a MPPT control (Zhu et al. 2015a). The MPPT tracking will be able to extract the maximum solar array power. For different load requirements, regulated bus architecture and unregulated bus architecture are used, respectively. The regulated bus architecture requires at least one APR, one battery charging converter, and one battery discharging converter (as shown in Figure 1), while the unregulated bus architecture just requires an APR (as shown in Figure 2). At system level, the regulated bus architecture (Capel & Perol 2001) brings advantages in terms of mass and simplicity. However, at unit level, the regulated bus architecture is more complex and heavier than the unregulated bus architecture, especially for low-power applications. The inflexibility further affects the costs as a recurrent design becomes difficult. Therefore, the light and simple designs based on unregulated buses are the only choice to reduce the price and mass, despite the advantages at system level of a regulated bus.

Under some special applications, the regulated bus and the unregulated bus may be required at the same time. Usually, a two-stage PV-battery power system is preferred, as shown in Figure 3. The first stage generates an unregulated bus while the second

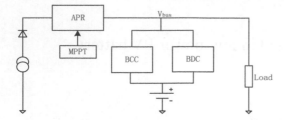

Figure 1. Regulated bus architecture for PV-battery power system.

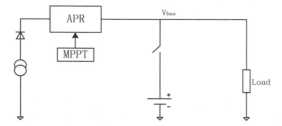

Figure 2. Unregulated bus architecture for PV-battery power system.

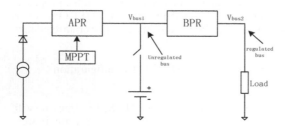

Figure 3. Two-stage PV-battery power system.

stage outputs the required regulated bus using Bus Power Regulators (BPR).

With this bus architecture, the choice of solar arrays and battery becomes easy and flexible. Therefore, the APR with MPPT control (Zhu et al. 2015b) can operate in buck mode (meaning that, in operating mode, the solar array voltage should always be higher than the bus voltage) or in boost mode (meaning that, in operating mode, the solar array voltage should always be lower than the bus voltage). In addition, the BPR may also operate in buck mode or boost mode (Mourra et al. 2010), which depends on the relationship between the required voltages of regulated bus and unregulated bus.

In order to unify the design of APR and BPR, a buck-boost mode DC-DC converter was proposed for the two-stage architecture PV-battery system. Solar array power regulation with MPPT or non-MPPT control, taper charge of battery, and the module parallelization were analyzed and verified experimentally.

2 THE PROPOSED DC-DC CONVERTER

2.1 *Buck-boost topology*

As for APR and BPR, whose input voltage may be below, above, or equal to the required output voltage, a buck-boost topology is chosen. The buck-boost topology is made up of four N-channel MOSFETs and a single inductor, as shown in Figure 4, and can be powered by a solar array panel or a DC voltage source. The resistor Rs is used to sample peak inductor current, which is required for peak-current control of the converter.

2.2 *Control strategy*

The outer voltage (current) control that commands inner current loop is a common technique used for MPPT solar array regulation and battery taper charge (Wu et al. 2009). In this design, a peak current control loop (Karppanen et al. 2008) of inductor current of buck-boost topology is taken as the inner current loop. In order to guarantee the control of APRs and BPRs, four outer control loops are designed in parallel, and connected together with an analog "and"-diode circuit. The four output control loops are designed to regulate or limit the output voltage, output current, input current, and/or input voltage, which are implemented with transconductance amplifiers (EA$_1$~EA$_4$) and using unified PI compensation parameters (Equation (1)), as shown in Figure 5.

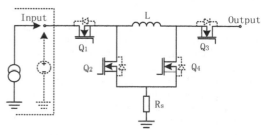

Figure 4. Buck-boost topology.

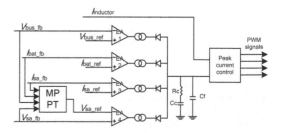

Figure 5. Control scheme of the proposed DC-DC converter.

$$\frac{v_c(s)}{v_d(s)} = \frac{g_m}{(C_C + C_F)s} \cdot \frac{R_C C_C s + 1}{R_C \dfrac{C_F C_C}{C_C + C_F} s + 1} \tag{1}$$

where v_d represents the variation of output voltage, input voltage, output current, or input current.

In a BPR application, the output voltage might be regulated using EA1, while the remaining error amplifiers are monitoring for excessive input or output current or an input under voltage condition.

The input voltage loop is used for MPPT control only when the DC-DC converter is used as an APR. The output voltage loop manages the battery taper charge control. The input and output current loop is used to limit the maximum solar array current and the output current of APRs, respectively. As the AD conversion resolution of input current directly influences MPPT accuracy, the limitation of input current is necessary.

With respect to the proposed control mode, the input voltage loop and input current loop do not work, as the solar array input power is larger than the sum of load power and battery charging power, and the corresponding outputs of EA3 and EA4 are of high level. Under this condition, the operation mode of the DC-DC converter depends on output voltage loop and output current loop. When the unregulated bus voltage (or battery voltage) is increasing and lower than the predetermined voltage, the output voltage loop works, and the voltage of Vc depends on the output of EA$_1$. Usually, the output current loop will not work because the output current will not be larger than the limited maximum output current, with the resulting signal of EA$_2$ remaining at high level. The voltage of Vc after RC compensation will be taken as the reference of inner peak current control loop, which generates the PWM control signal of buck-boost topology.

On the contrary, the input voltage loop and input current loop will take over the outer control loop, as the solar array input power is less than the sum of load power and battery charging power, and the corresponding outputs of EA3 and EA4 determinate the voltage of Vc. No matter whether the bus voltage is smaller than or equal to the predetermined battery voltage, the output voltage loop will not work. The solar array works on its maximum power point determinate by MPPT control algorithm. At this time, the output voltage and current are determined by solar array power and the battery voltage.

2.3 MPPT control algorithm

The MPPT, which is able to set the working point of the solar array along with all its characteristics (Garcia et al. 2013), works with the following concept:

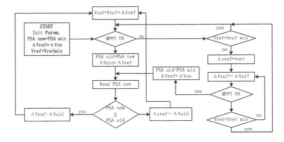

Figure 6.   MPPT control algorithm.

1. applies a small variation (voltage step) to the solar array working voltage (positive or negative);
2. measures the value of the power delivered by the solar array in this new condition;
3. compares this power value with the one relevant to the previous working point;
4. determines the direction of the next voltage step: the same direction if the power is increased and the opposite direction if the power is decreased.

The corresponding algorithm adopted for the MPPT control is shown in Figure 6, and this algorithm is called Perturb and Observe algorithm. This algorithm provides accurate MPPT for slow to moderate changes in panel illumination.

For a correct behavior of the MPPT, it is mandatory to achieve the steady-state condition after perturbation before measuring the new operating point. The time spent to reach the steady-state condition depends on the bandwidth of the proposed DC-DC converter, input filters, and the parasitic capacitance of the solar array.

3 PARALLELIZATION OF DC-DC CONVERTERS

In low-power PV-battery power supply applications, an important concern is the ability to parallelize different modules on a single output bus. The common situation is to have different SA arrays with different operating conditions (for example, irradiance temperature, solar angle, etc.), which result in a very different PV curve for each APRs (Zhu & Zhang 2014). As for the BPRs, the parallelization is usually considered for extending the output capacity. Another typical reason to parallelize different modules comes from reliability and redundancy issues.

Individual DC-DC converter module cannot be parallelized directly to increase the capacity (output current), even if the two modules are completely "the same". The root cause of the problem is the "minute" difference between the two converters. Usually, additional current-sharing control should be adopted. However, the proposed DC-DC converter is controlled with inner current control combined with outer voltage and current control loops. In addition, the transconductance amplifiers are utilized for the outer control loops, which provide the fundamentals for current sharing. Therefore, the proposed DC-DC converters can be parallelized and can connect together directly to build the unregulated bus and regulated bus, as shown in Figure 7.

As for the APRs, MPPT control mode and taper charge control mode are two main working conditions. When APRs are controlled with MPPT, the output voltage depends on the battery voltage, and the output current of each module is controlled by the corresponding solar array power and efficiency of APRs. Therefore, this working condition can be equivalent to input voltage-controlled current source parallelization, and active current sharing is achieved. As for the taper charge control, the output voltage of each module is forced to be equal by the battery voltage. Under this condition, the converters are equivalent to output voltage-controlled current sources, the module parallelization can be directly connected to the unregulated bus, and also active current sharing is achieved.

With respect to BPRs, the output voltage control loop is dominated and the required output is regulated to a constant voltage. Then, the output voltage feedback signal is compared with the internal reference voltage by EA1. Low output voltages would create a higher V_c voltage, and thus more current would flow into the output. Conversely, higher output voltages would cause V_c to drop, thus reducing the current fed into the output. The BPRs can also be taken as voltage-controlled current sources. When the BPRs are parallelized, the BPR with the highest output voltage will provide load current first, the other ones will not work for the parallelized output voltage higher than that of their designed output, no current will feed to the load. When the required load current is more than the limited maximum current of the BPR with the highest output voltage, this BPR will provide maximum current and its output

voltage becomes uncontrollable. In addition, the BPR with the second highest voltage begins to feed load current and its output voltage control loop will take over the control of the parallelized output voltage. Based on this working principle, all of the parallelized BPRs will begin to feed the load current one by one, and the final output voltage is equal to the output voltage of the BPRs with the lowest voltage.

4 EXPERIMENTAL RESULTS

In order to validate the proposed DC-DC converter used as APRs or BPRs and their parallelization, a demo board of a two-stage, low-power, low-voltage PV-battery power system has been designed and built, as shown in Figure 8. In this demo board, two APRs and two BPRs are used.

The power of an APR or a BPR is designed to be 50 W. The limited maximum input and output currents are both 5 A. The APRs are powered by solar array simulator Agilent E4360 A (I_{SC} = 2.2 A, I_{MPP} = 2 A, V_{MPP} = 25V, V_{OC} = 28V). The predetermined voltage of the battery connected to the unregulated bus is set to 13V. also In addition, the required regulated output of BPRs is also set to 13V.

4.1 *Transition between input and output voltage regulation modes*

When the proposed DC-DC converter is used as an APR, the input and output voltage loops will take over the power and voltage regulation alternatively. As the battery charge current is not controlled, the input voltage regulation (MPPT control) dominates until the output voltage reaches the predetermined voltage of battery. When it happens, the output voltage regulation begins to dominate, and the input voltage will move to the point (right side of the MPPT) that equals the power required at the output. The transition between input and output voltage regulation is depicted in Figure 9.

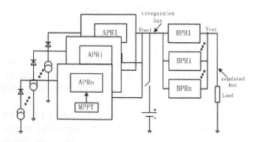

Figure 7. Module parallelization in two-stage systems.

Figure 8. Photo of the designed demo board.

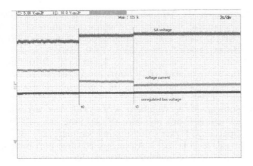

Figure 9.   Transition from input to output regulation.

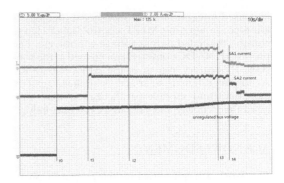

Figure 10.   Parallelization of two APRs.

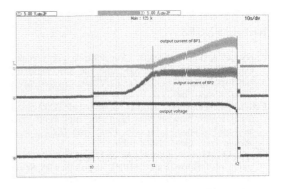

Figure 11.   Transition from input to output regulation.

It can be seen that SA worked on the MPP before t_0, when the input voltage and input current were equal to 25V and 2 A, respectively, which means that the input voltage regulation dominated. The battery voltage increased, such that it reached the predetermined voltage 13V at t0. Then, the output voltage began to take over. At this time, SA voltage increased and the SA current decreased. Correspondingly, the charging current declined. At t_1, the battery was completely charged; the charging current, and the SA output open voltage, became zero.

4.2 Parallelization of APRs

To validate the parallelization of APRs, two APRs were used. Figure 10 shows the working process of the parallelization.

At t_0, the battery began to feed the load, and the unregulated bus voltage becomes equal to the initial voltage of the battery. At t_1, SA_2 was switched on; APR_2 began to charge the battery and feed the load. APR_2 was controlled by MPPT. At t_2, SA_1 was switched on, and APR_1 began to charge the battery and feed the load together with APR_2, which is caused by the charging current of the battery not being controlled. APR_1 also worked with input regulation. At t_3, the battery voltage reached the predetermined voltage of APR_1, approximately 12.96V, and the voltage regulation loop began to take over the control of APR_1. The current fed to the unregulated bus by APR_1 began to decline. However, charging of the battery continued. The battery voltage will have a slight increase. At t_4, the battery voltage reaches the predetermined voltage of APR_2, approximately 13.01V. The voltage regulation loop began to take over the control of APR_2. At this time, the output voltage of APR_1 is bigger than its predetermined voltage, and APR_1 fed less and less current to the unregulated bus.

4.3 Parallelization of BPRs

Two BPRs were also used to validate their parallelization. Figure 11 shows the working process. The output voltages of BPR_1 and BPR_2 were designed to be 12.98V and 13.03V, respectively.

At t0, the battery began to discharge. Owing to the control logic of EA_1, the parallelized output voltage is equal to 13.03V. When the load current is low, BPR_2 provides the load current. The BPR_1 had no output current. At t_1, the load current becomes larger than the limited maximum output current of BPR_2. The BPR_2 output voltage became uncontrolled. The output control loop of BPR_1 began to take over the control of parallelized output. The parallelized output voltage became 12.98V. The output current of BPR_1 is equal to the difference between the load current and maximum current of BPR_2. With the increase in load current required, the output current of BPR_1 will increase until reaching its limited maximum output current of 5 A at t_3.

5 CONCLUSION

This work is aimed to develop a kind of modularized DC-DC converter to be used for low-power PV-battery system. The purpose of this includes cost reduction, short development time and increased integration. The buck-boost topology is adopted, integrating MPPT control, taper charge control, regulated output voltage control, etc. The proposed DC-DC converter can be used in the two-stage PV-battery system as the APRs and BPRs. As for the different solar arrays applied, capacity increment and module parallelization were analyzed theoretically and validated experimentally.

ACKNOWLEDGMENTS

This work was supported by the project of Shenzhen Science and Technology Innovation Plan (JSGG201405191 71924378).

REFERENCES

Capel A. & Perol P. 2001. Comparative performance evaluation between the S4R and the S3R regulated bus topologies. *Proc. 32nd IEEE Power Electron. Spec. Conf.*: 1963–1969.

Fatemi N., Pollard H., Hou H. & Sharps P. 2000. Solar array trades between very high efficiency multi-junction and Si space solar cells. *Proc. 28th IEEE Photovoltaic Spec. Conf.*: 1083–1086.

Garcia O., Alou P., Oliver J.A. & Diaz D. 2013. Comparison of Boost-Based MPPT Topologies for Space Applications. *IEEE Trans. Aero. and Electron.* 49(2): 1091–1107.

Karppanen M., Arminen J., Suntio T. & Savela K. 2008. Dynamical modeling and characterization of peak current mode controlled superbuck converter. *IEEE Trans. Power Electron.* 23(3): 1370–1380.

Mourra O., Fernandez A. & Tonicello F. 2010. Buck boost regulator (B2R) for spacecraft solar array power conversion. *Proc. 25th Annu. Appl. Power Electron. Conf.*: 1313–1319.

O'Sullivan D. & Weinberg A.H. 1977. The sequential switching shunt regulator (S3R). Proc. Rec. ESTEC Spacecraft Power Conditioning Seminar: 123–131.

Wang Z. & Li H. 2013. An integrated three-port bidirectional dc-dc converter for PV application on a dc distribution system. *IEEE Trans. Power Electron.* 28(10): 4612–4624.

Wu C., Ruan X., Zhang R.& Tse C.K. 2009. DC/DC conversion systems consisting of multiple converter modules: Stability, control and experimental verifications. *IEEE Trans. Power Electron.* 24(6): 1463–1474.

Zhu H.Y. & Zhang D.L. 2014 Influence of Multijunction Ga/As Solar Array Parasitic Capacitance in S3R and Solving Methods for High-Power Applications. *IEEE Trans. Power Electron.* 29(1): 179–190

Zhu H., Zhang D., Zhang B. & Zhou Z. 2015a A Nonisolated Three-Port DC–DC Converter and Three-Domain Control Method for PV-Battery Power Systems. IEEE Trans. Ind. Electron. 62(8): 4937–4947.

Zhu H., Zhang D., Qing L. & Zhicheng Z. 2015b Three-Port DC/DC Converter with All Ports Current Ripple Cancellation Using Integrated Magnetic Technique. *IEEE Trans. Power Electron.* pp(99): 1–12.

Energy Science and Applied Technology ESAT 2016 – Fang (Ed.)
© *2016 Taylor & Francis Group, London, ISBN 978-1-138-02973-6*

Comparative study on the dynamic characteristic of an internal combustion compressor

Qingqian Wang & Jizhong Zhang
College of Mechanical and Electronic Engineering, Qingdao University, Qingdao, China

Mei Zhao
School of Foreign Language Education, Qingdao University, Qingdao, China

Xiang Yu & Qun Lu
College of Mechanical and Electronic Engineering, Qingdao University, Qingdao, China

ABSTRACT: In order to reduce the noise and vibration during the work process of piston air compressor driven by the internal-combustion engine, the scheme of horizontally-opposed internal combustion air compressor is proposed. Based on virtual prototype technology, the dynamics simulation model of crankshaft system is built. Using software ADAMS, the kinematics and dynamics analysis in a variety of working conditions are completed, the dynamic load and the change law on the crankshaft are shown, and dynamic characteristic curves in three kinds of work conditions are compared.

Keywords: I. C. engine; compressor; dynamic characteristic analysis; simulation

1 INTRODUCTION

The internal combustion air compressor is new compressor device that can convert heat energy into mechanical energy simple and efficiently. The three cranks and connecting rod mechanism of power piston and compressive piston are horizontally symmetrical, and their sizes and qualities are perfectly symmetrical. The inertia force can be balanced and the noise and vibration during the work process can be reduced (Zhang Jizhong et al, 2014) (Zhang Jizhong et al, 2005).

2 OPERATIONAL PRINCIPLES AND FEATURES

The working principle of horizontally-opposed internal combustion air compressor is shown in Fig. 1. In the course of work the air-fuel mixture combust, push dynamic piston, compressor piston complete breath, compressor piston outputs air pressure energy during the return trip of crank and connecting rod mechanism and the crank and connecting rod mechanism makes subsidiary system of internal combustion engine to work. In the above process, the first studio whose theory is similar to that of four-stroke internal combustion engine translates the heat energy of fuel into mechanical energy of the piston. The second studio translates mechanical energy of the piston into the heat energy.

3 ESTABLISHMENT OF VIRTUAL PROTOTYPE

The basic dimensions of horizontally-opposed internal combustion air compressor which include cylinder diameter, piston stroke, and quality of power piston, length of crank and length of the connecting rod are based on the dimensions of 178F diesel engine, as shown in Table 1.

Based on 3D modeling software Solidworks and dynamic simulation software ADAMS, parts models of crankshaft, the connecting rod,

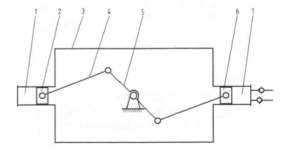

Figure 1. Working principle diagram.

Table 1. The first 4 order natural frequencies of the non zero modes (unit: Hz).

Serial number	Content	Parameters	Unit
1	Cylinder diameter	0.078	m
2	Piston stroke	0.062	m
3	Quality of power piston	0.48	kg
4	Length of crank	0.031	m
5	Length of the connecting rod	0.102	m

Figure 2. 3D model of the crankshaft.

Figure 3. 3D model of the connecting rod.

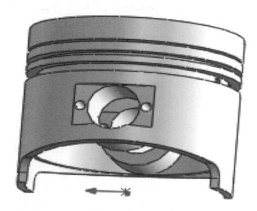

Figure 4. 3D model of dynamic piston.

Figure 5. 3D model of compressor piston.

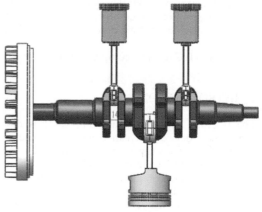

Figure 6. 3D model of the crankshaft system.

dynamic piston, compressor piston and 3D model of the crankshaft system are established, as shown below.

4 MULTI-RIGID-BODY DYNAMICS SIMULATION OF CRANKSHAFT SYSTEM

4.1 Speed of piston

50% throttle position is selected in simulation calculation, and the average speed of crankshaft is 1500r/min, 2000r/min and 2400r/min (Zhang Jizhong et al, 2008) (Zhang Jie, 2010) (Lawrence Kent, 2001). Dynamic piston velocity is shown in Fig. 7.

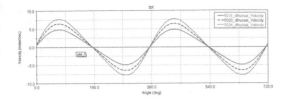

Figure 7.   Dynamic piston speed graph.

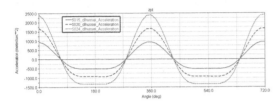

Figure 8.   Dynamic piston acceleration curves.

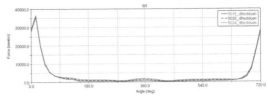

Figure 9.   Force curves on big end of the dynamic piston rod connecting

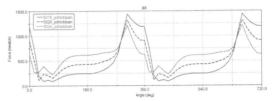

Figure 10.   Force curves on big end of the compressor piston rod connecting

Through the analysis of the result, the trajectories are roughly same. The dynamic piston speed reaches the maximum value at 72 degrees and reaches a negative maximum value at 288 degrees. Results of calculation show

1. When the average speed of the crankshaft is 1500r/min, Vmax = 4.86 m/s and Vmin = –4.84 m/s.
2. When the average speed of the crankshaft is 2000r/min, Vmax = 6.48 m/s and Vmin = –6.45 m/s.
3. When the average speed of the crankshaft is 2400r/min, Vmax = 7.78 m/s and Vmin = –7.74 m/s.

4.2   *Piston acceleration*

The changes of acceleration curves in three kinds of work conditions are as shown in Fig. 8.

1. When the average speed is 1500 r/min, a_{max} = 940.9 m/s² and a_{min} = –524.9 m/s².
2. When the average speed is 2000 r/min, a_{max} = 1672.8 m/s2 and a_{min} = –933.2 m/s².
3. When the average speed is 2400 r/min, a_{max} = 2408.9 m/s2 and a_{min} = –1343.8 m/s².

4.3   *Force on big end of dynamic piston connecting rod*

Through simulation calculation, get the force curves on big end of the dynamic piston connecting rod in three kinds of work conditions, as shown in Fig. 9.

1. When the average speed of the crankshaft is 1500 r/min, F_{max} = 36600.5 N and F_{avq} = 3767.1 N.
2. When the average speed of the crankshaft is 2000 r/min, F_{max} = 36077.9 N and F_{avq} = 3952.2 N.
3. When the average speed of the crankshaft is 2400 r/min, F_{max} = 35552.4 N and F_{avq} = 4141.3 N.

4.4   *Force on big end of Compressor piston connecting rod*

Through simulation calculation, get the force curves on big end of the Compressor piston connecting rod in three kinds of work conditions, as shown in Fig.10.

1. When the average speed of the crankshaft is 1500r/min, F_{max} = 1439.3 N and F_{avq} = 524.1 N.
2. When the average speed of the crankshaft is 2000r/min, F_{max} = 1324.3 N and F_{avq} = 557.7 N.
3. When the average speed of the crankshaft is 2400r/min, F_{max} = 1212.9 N and F_{avq} = 603.2 N.

4.5   *Force on small end of dynamic piston connecting rod*

Through simulation calculation, get the force curves on small end of the dynamic piston connecting rod in three kinds of work conditions, as shown in Fig. 11.

1. When the average speed of the crankshaft is 1500r/min, F_{max} = 36922.8 N and F_{avq} = 3626.2 N.

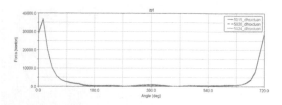

Figure 11. Force curves on small end of the dynamic piston rod connecting.

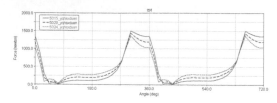

Figure 12. Force curves on small end of the compressor piston rod connecting.

2. When the average speed of the crankshaft is 2000r/min, F_{max} = 36650.9 N and F_{avq} = = 3697.5 N.
3. When the average speed of the crankshaft is 2400r/min, F_{max} = 36377.5 N and F_{avq} = = 3769.3 N.

4.6 *Force on small end of Compressor piston connecting rod*

Through simulation calculation, get the force curves on small end of the Compressor piston connecting rod in three kinds of work conditions, as shown in Fig. 12.

1. When the average speed of the crankshaft is 1500r/min, F_{max} = 1501.7 N and F_{avq} = 496.7 N.

2. When the average speed of the crankshaft is 2000r/min, F_{max} = 1431.7 N and F_{avq} = 501.3 N.
3. When the average speed of the crankshaft is 2400r/min, F_{max} = 1361.5 N and F_{avq} = 507.6 N.

5 CONCLUSION

3D model of the crankshaft system in internal combustion type horizontally opposed compressor is established in 3D software Solidworks. 3D model is imported into dynamic analysis software Adams. Dynamic simulation of internal combustion type horizontally opposed compressor is carried out, providing the feasibility and advantages of internal combustion type horizontally opposed compressor, and providing a basis for the follow-up reliability research and optimization design.

REFERENCES

Lawrence Kent. A Crankshaft System Model for Structural Dynamic Analysis of Internal Combustion Engines. Computers and Structures, 2001, 79 (20).

Zhang Jie, Design and Simulation of the Reciprocal Compressor Dummy Specimen [J]. Coal mine machinery, 2010, 31(03): 232–233

Zhang Jizhong, Shang Pengfei, Zhao Hong, Zhang Yi, Dynamics Analysis of The Internal Combustion Air Compressor [J]. Chinese Mechanism Engineering, 2008, 19(20): 2415–2418.

Zhang Jizhong, Zhang Tiezhu, Zhang Hongxin, Dai Zuoqiang. Conception Design of the Internal Combustion Air Compressor[J]. Modern Manufacture Engineering, 2005, (12): 90–92

Zhang Jizhong, Zhang Tiezhu, Zhang Jipeng, Huo Wei. An Internal Combustion Air Compressor with Dynamic Balance Performance: China, 201210453525.1[P]. 2014-12-31.

Energy Science and Applied Technology ESAT 2016 – Fang (Ed.)
© 2016 Taylor & Francis Group, London, ISBN 978-1-138-02973-6

Diesel common rail pressure control based on feed forward fuzzy PID controller

S. Xiang & Z.M. Ji
School of Automation, Hangzhou Dianzi University, Hangzhou, Zhejiang, China

G.Q. Mo & D.A. Sun
Product Quality Testing Institute, Yulin, Guangxi, China

ABSTRACT: The common rail pressure controls the fuel injection pressure in the common rail fuel injection system, so it will influence the performance, fuel economy and emissions of common-rail diesel engine directly. According to the present situation that the traditional PID control method is difficult to achieve accurate pressure control, the control strategies of common rail pressure under different engine operating conditions are introduced, and the Fuzzy PID control method is used, and it is optimized by feed forward control. The experimental results show that: The control strategies are applicable in starting condition to idle speed condition. The rail pressure control overshoot when used the feed forward fuzzy PID controller is reduced by 75% compared with the traditional PID controller, and the response speed is increased by nearly 70%, which is faster and more stable.

Keywords: common rail diesel engine; pressure control; feed forward fuzzy PID; state of engine

1 INTRODUCTION

In recent years, with the deepening of environmental pollution, automobile emissions regulations become more stringent, and the current situation of energy shortage also becomes increasingly severe, so the fuel injection system of diesel engine will be an important part of innovation and development, it requires not only the performance but also the low emissions and fuel economy. High pressure Common Rail System (HCRS) is now the best fuel injection system, the system has a precise control of fuel injection quantity and rate, and it has advantages such as higher injection pressure, more injection times and more precise injection quality (Catania A.E. et al. 2005). At the same time, HCRS will satisfy the requirements of energy conservation and pollution reduction, representing the development trend of fuel injection technology. In the system, the stability of rail pressure will directly affect the performance of HCRS, and accurate pressure control is one of the many other optimal control prerequisites (Chen H. et al. 2012), so real-time control of common rail pressure is very important.

At present, the research results at home and abroad in the common-rail pressure control are more using the traditional linear PID and its derivative method and some PID parameter tuning methods such as orthogonal experiment or neural networks and so on. But most of these conventional PID control method has some problems that it cannot control precisely or follow the changes with engine operating conditions (Fan G. et al. 2011). Based on the analysis of HCRS's structure and working principle, this paper studies the optimal control strategy of different rail pressure changes under different diesel engine operating conditions; and designs feed forward fuzzy PID controller, compared with the traditional PID control method, it is more flexible, reduces the overshoot and oscillation in the system, and improves the response speed of common rail pressure, so it has certain practical application value

2 PRINCIPLE OF COMMON RAIL PRESSURE CONTROL

The typical common rail system is shown in Figure 1. The high-pressure pump generates fuel in high pressure environment and supplies them to the common rail (Jang J.S. 2007). The task of high pressure common rail is to store fuel in high pressure and unstable condition which is caused by the pump delivery and the injection of fuel stored in common rail volume (Jin J. et al. 2006). A pressure control valve is installed at the inlet part of the high pressure fuel pump, it controls the rate which the fuel flow from the low pressure fuel pump into the high pressure pump (Mallamo F. et al. 2005). Electronically controlled injector can inject the appropriate quantity of fuels into combustor. Electronic Control Unit (ECU) can collect signals from all the sensors on the engine, and precisely control all parameters of controller (Ren W. et al. 2010), such as rail pressure, injection timing and duration. In HCRS, maintaining the stability of the common rail pressure is the premise of ensuring the injector to inject the fuel accurately according to the injection rules. At the same time (Song Q. et al. 2007), rail pressure sensors constantly monitor and send information related to common rail pressure to ECU, Zhang L. et al. 2014as to allow the ECU to adjust the PWM (Wang H. et al. 2008) (Zhang L. et al. 2014), in order to control the rail pressure and fuel injection better.

The common rail pressure depends on the balance between the inlet fuel flow from high pressure pump and outlet flow from the injectors. Oil supply pipelines are containers with limited capacity, so pipe pressure is dynamically affect by the system: when the fuel is injected, pipe pressure drops; and when the fuel flow into the pipe, pipe pressure increases. Because of the changing working conditions of the engine, the set point pressure and pressure control will be very complex. It needs to quickly form pipe pressure, and after the launch, the pressure needs to be able to adapt to changes in operating conditions and remains stable; and under all conditions, the limitation of pressure is required, so is the balance control.

3 DESCRIPTION OF FEED FORWARD FUZZY PID CONTROLLER

Feed forward fuzzy PID controller is shown in figure 2. It consists of feed forward control and fuzzy PID control.

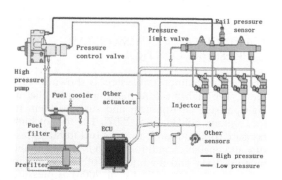

Figure 1. Typical common rail system.

3.1 *Fuzzy PID control*

Traditional PID controls are widely applied in the field of engineering, which can get rid of the steady state error of the system and help the system have a quick response, and track target values accurately. It consists of an integrator. The integrator can reduce static deviation, which can maintain the stability of system. However, the PID parameters not only cannot be altered with the environment changes, but also cannot achieve global optimization, which will lead to the fact that the system is stiff and easily overshoot and oscillate when the PID control was determined.

The common fuzzy control is nonlinear, flexible and robust. But compared with PID control, it has more static deviation. Because its control range is local, both of the deviation and oscillation exist. Even if incremental output is used, which seems to be an integral adding in front of the proportion coefficients, static error and oscillation still are not able to be completely eliminated.

In this thesis, discrete fuzzy PID is used to regulate common rail pressure, because of the combination of strong points of fuzzy control and the PID control, and complementary to its shortcomings, breaking the limitations of the traditional PID control. Absolute pressure deviation |e| and absolute deviation of pressure change rate |Δc| are the input of the fuzzy controller, and K_P, K_I, K_D are the output of fuzzy controller, and PID self-tuning is implemented based on the analysis of e and Δe, as shown in Figure 2.

A fuzzy reasoning process includes five steps, as shown below:

1. The input quantity should be fuzzified in the discourse universe, and the certain input should be transformed into the fuzzy sets described

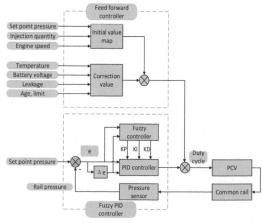

Figure 2. Feed forward fuzzy PID controller.

by membership degree. Which is called fuzzification.

2. The fuzzy operator s such as AND-OR-INVERT should be used in the antecedent of fuzzy rules.
3. The conclusion is inferred from the premise based on fuzzy implication operation.
4. The final conclusion can be derived from the combination of the conclusion of each rule.
5. The output fuzzy quantities should be transformed into the certain output quantities, which is called defuzzification.

According to the actual condition of the diesel engine, the control rules of the fuzzy PID controller are explained as follows: (1) When the deviation is large, K_P should be PB and K_D should be PS, to achieve fast settling; K_I should be ZO, to prevent the emergence of large overshoot and oscillation; (2) When the deviation |e| and rate of pressure deviation |Δe| are PM, K_P and K_D should be PM or PS, to ensure the control error is quickly compensated; K_I should be PS, to decrease overshoot; (3) When |e| is PS, K_P and K_I should be PM or PB, to keep stable; and When |Δe| is bigger, should be smaller, which is to increase anti-jamming ability.

In the thesis, all of the input quantities and output quantities are divided into four fuzzy sets: PB, PM, PS, PO. The computation results of the MATLAB language are determined by the input and output problem domain. The fuzzy PID control parameter table is shown in table 1.

3.2 Feed forward control

Feed forward control should be used in front of the appearance of deviation, which is used to eliminate prediction problem. When the interference can be anticipated, the feed forward control can timely regulate it. Its dynamic characteristics can be well as closed-loop control. In the operation process of the engine, both the rail pressure and injection quantity will change with the rapid change of load, closed-loop control cannot effectively track the set pressure value, because the changes of injection quantity in the closed-loop control is a serious interference. However, in feed forward control, the changes of injection quantity acts as an input, and correction will be done before the injection takes place, and when the errors are checked, the target values will be tracked quickly. In summary, feed forward control is appropriate for transient rail pressure.

The control strategy for the target pressure set value is shown in Figure 2. The initial formation of the feed forward control value is determined according to the basic set point of the rail pressure, the current injection quantity and the engine speed. The initial value is regulated by a variety of possible environmental conditions, such as system temperature, battery voltage, maximal value, injector leakage, pressure control valve leakage, and so on. The corrected values are also multiplied by elements that are affected by environmental hazards and combined with the results obtained before. There is a maximal flow estimated value supplied to systems which have been used for a long time, also the influence of rail transport delay is considered, to obtain the equilibrium value of the quantity of high pressure pump. The initial value plus all the correction values will be the final control value.

The relationship between changes of rail pressure and fuel flow can be seen as follows:

$$\Delta p = (q^P - q^{INJ}) * K_{Bulk} / V_{rail} \tag{1}$$

where:

q^P : High pressure pump fuel flow;
q^{INJ} : Injector fuel flow;
K_{Bulk} : Fuel compression ratio;
V_{rail} : Common rail fuel volume;

The relationship between the coil current of the pressure control valve and the pump fuel flow rate q^P is a non-linear function expression, and high pressure pump fuel which flows into the pipe can be obtained as follows:

$$q^P(t) = \eta(p,n) * f(I)$$

Table 1. Fuzzy PID control parameter table.

| | |e| | | | | | | | | | | |
| | [150,100) | | | [100,60) | | | [60,20) | | | [20,0) | | |
|Δe|	K_P	K_I	K_D	K_P	K_I	K_D	K_P	K_I	K_D	K_P	K_I	K_D
[150,100)	8	0	0.2	5	0.1	0.4	8	0.2	0	8	0.3	0
[100,60)	10	0	0.4	8	0.1	0.4	10	0.3	0.6	10	0.3	0.2
[60,20)	10	0	0.6	8	0	0.6	10	0.3	0.6	10	0.3	0.2
[20,0)	10	0	0.6	8	0	0.6	3	0.3	0	3	0.3	0

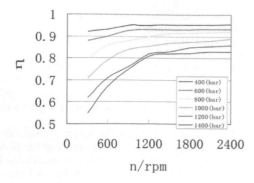

Figure 3. The efficiency value tested by common rail test bench.

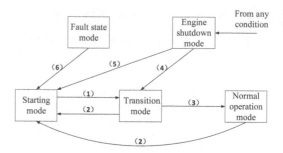

Figure 4. Rail pressure control strategy state machine.

In the expression, coil current I is determined by the resistance and the inductance of the coil, the efficiency $\eta(p,n)$ is determined by rail pressure and engine speed. The efficiency values measured by the common rail test-table are shown in Figure 3.

3.3 Feed forward control

Rail pressure will be regulated by the open loop or closed loop PID based on the current state of the engine. In different engine conditions, the requirements and objectives of the rail pressure control is also different, so different control strategies are necessary. The migration state machine of rail pressure control strategy can be seen in Figure 4.

1. Starting mode

To satisfy the quick start of the diesel engine, open loop control should be applied to regulate the rail pressure, which promotes the rapid establishment of the common rail pressure. PWM of pressure control valve must be ensured in the beginning to deal with different kinds of influence (high throughput when cold-start, fast throughput of the gear pump); when diesel engine speed is bigger than the low voltage starting threshold and rail pressure fluctuation is in a certain range of amplitude, as the condition (1) shown in the Figure 4, the mode changes from the starting mode to transition mode.

2. Transition mode

The quick or transient change of load may cause the consequence that the pressure cannot be controlled, so the open loop control according to MAP is adopted. Feed forward control is supplied by the MAP, and the input of the controller is determined by the interpolation of the current state. The synchronous adjustments of PWM of pressure control valve can avoid unnecessary rail pressure deviations during the period that the load is changing rapidly. When the engine speed is lower than the low voltage starting threshold, or rail pressure fluctuations larger, both of which equates to conditions (2), the mode changes from transition normal-operation; If crankshafts rotate for a period of time in the transition mode, which equates to conditions (3), the engine will be converted from transition mode to normal-operation mode.

3. Normal-operation mode

In the normal-operation mode, the closed loop control and fuzzy PID method are used for different load and speed. Aim value of the rail pressure is different under different work environments. The set value of benchmark fuel injection pressure can be obtained through pulse spectrum (MAP) calibrated by interpolation, and the parameter of MAP is engine speed and current fuel injection quantity. To make a good optimization of the emission performance of engine, benchmark pressure value must regulate in different work environments, like different inlet pressure, different temperature, etc. However, when the condition (2) is met, it will change to starting mode.

4. Engine shutdown mode

When the engine is shut down, it is very necessary for engine to decompress rapidly. In order to ensure the next start of engine can be quickly completed, the system will remain in the lowest pressure environment, and open loop control is adopted. Electrical switch is also necessary, because of two styles of valves: normally open or normally close. In regard to normally open valves, after the shutdown command is received, the fuel injector will avoid injecting fuels, and pressure control valves should be controlled to avoid the fuels entering high pressure pump, which can prevent the damage to HCRS because the rail pressure is too much. If the engine speed is greater than the closed-loop controlled minimal speed, which equates to conditions (4), the engine will be transferred to the transition mode; otherwise, which equates to conditions (5), the engine will be transferred to the starting mode.

Table 2. Key technical data of engine.

Type	Inline/4 Cylinder/16 Valves/DOHC
Cylinder Bore	75 mm
Engine Displacement	1.56 L
Compression Ratio	18:1
Rated Power	80/4000 kW/(r·min)
Max Torque	240/1750 N·m/(r·min)

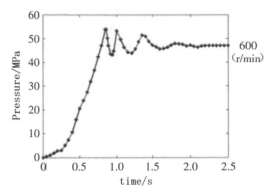

Figure 6. Rail pressure control of starting process.

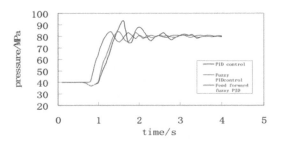

Figure 7. Comparison of experimental results of three control methods.

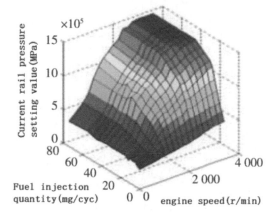

Figure 5. Ideal target injection pressures under different diesel engine conditions.

5. Fault state mode

When the rail pressure sensors receive the error information, which equates to conditions (6), the open loop control method should be used to control rail pressure. Keeping stability of PWM of pressure control valve can retain a part of pressure, which can make engine continue working and "limp home".

4 EXPERIMENTAL RESEARCH

The prototype machine in the research is a 4-cylinder direct injection diesel engine for car with the BOSCH Company's third generation of high pressure common rail fuel injection system—CP3. The fuel injection pressure in this system can reach 160 MPa, and its key technical data can be seen in table 2.

To verify the effects of the control strategies and performance of feed forward fuzzy PID controller, bench tests and simulations are done. Digital experimental test bench is built based on Motorola MPC555 microprocessor, and control algorithms are a combination of the C language and MATLAB/ SIMULINK. In the bench test experiment, electric eddy current dynamometer of Schenck,

transient pressure sensor of Kisler and combustion analyzer of Dewetron are used to get accurate experiment data.

4.1 *Calibration of target rail pressure*

To ensure that the diesel engine has good emissions and power performance, the calibrations of the fuel injection target pressure under different diesel engine conditions are made, the calibration results are shown in Figure 5. It can be found that the fuel injection pressures are different, and it is just a variable pressure process under different diesel engine conditions. With the increasing of fuel injection quantity and engine speed, the ideal target injection pressures are also increasing.

4.2 *Rail pressure control in starting condition*

Because of different kinds of engine conditions, this paper chooses one of these conditions which is starting condition to idle speed condition to verify the control effect of rail pressure control strategies. Parameters are set as follows: idle speed is set to 600 r/min, and aim values of the rail pressure are put to 45 MPa. The calibration procedures are used to obtain data of rail pressure which recorded in Figure 6. As is shown in Figure 6, when it is in the starting mode (0 s–0.7 s), engine is under open loop control, rail pressure increases rapidly, and the fuel injection pressure also rises rapidly; and then, when it is in the transition mode (0.7 s–1.7 s), the rail pressure fluctuates slightly; finally, when it comes in the idle speed mode (1.7 s–2.5 s), feed forward fuzzy PID control plays a role in reducing the fluctuation and making the system stable. So the experiment shows that the control strategy plays a good control effect.

In order to obtain visual comparison, this paper tests different control methods. Simulation experimental tests are done using not only feed forward fuzzy PID control, but also general PID control and fuzzy PID control, which can be seen in the Figure 7. When the loading was changed, the pressure of set points rises from 40 MPa to 80 MPa. In the same period, rail pressure fluctuates when used the general PID control, maximum overshoot is 16%, the response time is 0.5 s, and it costs about 4 s to make the system stable; compared with general PID control, the overshoot of fuzzy PID control is less, which is 4%, and it can stabilize the system faster, which costs 1.7 s; and feed forward fuzzy PID control that this paper mentions can not only reduce the overshoot to 4% but also response very fast, which only costs 0.3 s, what's more, the whole process that uses to stabilize the system costs just 1.2 s. Compared with general PID controllers, the maximum overshoot is reduced by 75%, and the entire response time is reduced by 70%, while compared to the fuzzy PID control, the response time is reduced by 30%.

5 CONCLUSION

1. Common rail pressure control strategies are designed for different engine conditions; the experiment proves that the strategy is adopted to one of engine conditions which is starting condition to idle speed condition, the rail pressure forms rapidly and stabilize rapidly after fluctuating.

2. When feed forward fuzzy PID control method is applied, the rail pressure fluctuates slightly (the overshoot is reduced by 75% compared with general PID control method), and the response speed is faster (the time that stabilize the system is reduced by 70% compared with general PID control method and reduced by 30% compared with fuzzy PID control method).

REFERENCES

Catania A.E. et al. 2005. Development and Application of a Complete Common-Rail Injection System Mathematical Model for Hydrodynamic Analysis and Diagnostics. ASME 2005 Internal Combustion Engine Division Spring Technical Conference. American Society of Mechanical Engineers, 181–192.

Chen H. et al. 2012. Study on combustion and emission characteristics of turbocharged diesel engine with high pressure and high pressure. Chinese Internal Combustion Engine Engineering, 33(6): 39–44.

Fan G. et al. 2011. Diesel engine high-pressure common rail system simulation study. Internal combustion engines, (1): 39–40.

Jang J.S. 2007. Logic Toolbox Fuzzy user's Guide. The Math Work Inc.

Jin J. et al. 2006. Diesel engine with high pressure common rail fuel injection system of the common rail pressure control technology. Diesel engine, 28 (3):5–7.

Mallamo F. et al. 2005. Effect of compression ratio and injection pressure on emissions and fuel consumption of a small displacement common rail diesel engine. Detroit: Society of Automotive Engineers Inc.

Ren W. et al. 2010. Study on the fuzzy adaptive PID control of diesel engine common rail pressure. Computer Engineering and Applications, 46(2): 209–212.

Song Q. et al. 2007. Study on closed loop control algorithm for common rail pressure of high pressure common rail diesel engine. Internal combustion engine, (5): 12–14.

Wang H. et al. 2008. Study on rail pressure control of common rail diesel engine. Journal of Beijing Institute of Technology, 28(9): 778–781.

Zhang L. et al. 2014. The test and performance evaluation method for the transient state of vehicle diesel engine. Journal of Harbin Engineering University, (4): 463–468.

Energy Science and Applied Technology ESAT 2016 – Fang (Ed.)
© *2016 Taylor & Francis Group, London, ISBN 978-1-138-02973-6*

Analysis of the urban electric bus operation parameters under battery replacement

Junxiu Wang
School of Traffic and Transportation, Beijing Jiaotong University, Beijing, China

Xiuyuan Zhang
School of Traffic and Transportation, Beijing Jiaotong University, Beijing, China
Haibin College, Beijing Jiaotong University, Cangzhou, China

Huaiyuan Zhai
Haibin College, Beijing Jiaotong University, Cangzhou, China

ABSTRACT: In recent years, all kinds of transportation structure of Beijing have become more and more perfect, especially the constant improvement of the public transport system. It provides a good example in developing the green public transportation, solving traffic congestion, and reducing energy consumption and carbon emissions. Based on the all-electric battery buses operating technical conditions and the main indexes of the charging station, this paper analyses the key parameters of electric buses operation on the condition of battery replacement and the all-electric spare battery. All those can provide the reference and basic data for making the city electric buses operation plan and scheduling.

Keywords: urban electric buses; battery- replacement; transportation

1 INTRODUCTION

In recent years, Beijing urban traffic is developing rapidly, especially the basic urban public traffic network. Beijing ground public transportation line is widely distributed and this forms intensive passenger transfer services. From the spatial distribution of bus line, Beijing public bus line basically meets different trip purposes. Since the 2008 Beijing Olympic Games, Beijing Public Transport Company actively promotes green public transport system. Urban public transport can be divided into full transportation services based on the rail-based rapid transit and "last kilometer" short-distance travel routine bus system. Analysis was performed using electric vehicles for public transport routes and traffic distribution in order to carry out pure electric buses running exploration, accumulate more experience for the Beijing green, low-carbon, public transportation, and put forward a new idea for sustainable development.

This paper analyzes index calculation from Beijing pure electric bus operators' technical characteristics forming preliminary calibration parameters on how to systematically summarize the technical and economic characteristics of a pure electric bus and construction of Beijing local standards.

2 PURE ELECTRIC BUS LINE-RELATED CONCEPTS

Pure electric bus is a form of electric vehicles. Its structure and related concepts are as follows (Li Jin, 2011) (Xiao Han, 2010) (Haixing Wang, 2007):

1. The electric bus is driven by a motor used on the road. It does not include indoor electric vehicles, rail trolley buses, electric cars, and other vehicles. The motor power is supplied with a

Figure 1. Charging station.

Figure 2.   Replacing spare battery.

Table 1.   Basic parameters of each line.

One-way distance (km)	Peak flow section (person/hour)	Road speed of working condition (km/h)	Peak hour trips (veh/h)
7	325	12	6
8.2	560	17	11
9	410	23	12
11	610	20	8
12.3	1040	18	16
18.08	355	19	6
22.24	1220	16	17
22.35	375	18	7

Note: The table has two symbols, 12 indicates short-distance large passenger flow characteristics and 17 indicates long-distance large passenger flow characteristics.

rechargeable battery or other energy storage devices easy to carry.

2. Battery pack is a single mechanical composition of one or more battery modules.
3. Electric vehicle charging is done directly.
4. Power supply vehicle charging mode does not need the battery to be removed.

Pure electric bus charging for electric work process includes four parts: battery replacement, charging, maintenance, and battery group.

Figure 1 shows the pure electric bus in the charging station replacing the battery by robot. The replaced battery is charging. Figure 2 shows the pure electric bus replacing the spare battery intellectually when operating in a state of emergency problem.

3   PURE ELECTRIC BUS CHARGING MODE ANALYSIS

Depending on the operating mechanism, pure electric bus supply modes can be divided into the following five basic modes (Hongwen He, 2004).

1. "Vehicle stops charging" mode refers to each pure electric bus according to the departure sequence once or several times after the operation, entering the charging stations for power battery charging, and finishing all energy supplements. When charging is over, the pure electric bus re-enters the start sequence, starting the next run.
2. "Vehicle centralized charging" mode, according to the laws of the city bus operators, refers to buses generally running for a certain period during the day and stop running at night. "Vehicle centralized charging" mode refers to every pure electric bus, according to the departure sequence of operation after a certain number of times in the daytime, that stops running and getting into the charging station, waiting for a certain time period in the night, and supple out all the day consumption motive battery energy.
3. "Vehicle stop supplement—concentrated Charge" mode is a combination of the "vehicle stops charging" mode and "vehicle centralized charge" mode. This mode refers to every pure electric bus that first runs a certain number of times during the day and then goes into the charging station to charge the battery power to supplement. But it does not consume the entire energy, only added to the energy needed to run a certain number of times. When charging is over, the vehicle enters the departure sequence, after running a certain number of times, stops, and enters into the charging station, waiting overnight for centralized charging.
4. "Battery replacement—concentrated Charge" mode refers to the pure electric bus running, after a certain number of times, into the charging station. In this station, the robot arm replaces the battery whose power has been consumed by a set of fully charged battery power. Then the pure electric bus enters the start sequence, starting the next round run. All the replaced battery will be charged at night.
5. "Battery replacement—dispersion Charge" mode, which is similar to "Battery change—focused Charge" mode, also needs to replace the battery, but the difference is that all the replaced battery charged dispersion throughout the daytime, and not all centralized charging at night. This charging mode inherits the advantages of "Replacing the battery—concentrated Charge" mode, while making further efficient use of battery power. This mode can reduce the total amount of battery power to complete the same operation tasks. In addition, with scattered charging way, it improves the efficiency of using the charger, and at the same time, reduces the number of charg-

ers. All these reduce the overall capacity of the charging system for the distribution.

Currently, Beijing pure electric bus mainly uses battery replacement mode. In the case of battery replacement, how to determine the vehicle dispatch operation plan and replace the battery case-related technical parameters to determine the scope of indexes and calculation methods of domestic research is still in its preliminary stage.

4 PURE ELECTRIC BUS OPERATING PARAMETER CALCULATION ANALYSIS

Combined with Beijing bus line types, charging station spatial planning, and disposition location, typical bus routes for the basic parameters of simulation calculations were selected. Main line types are shown in table 1.

Forming four main-line features: 1, short-distance and large passenger section; 2, long-distance and large passenger flow section; 3, short-distance and a small section of the passenger; and 4, long distance and a small section of the passenger.

Analysis electric bus operation parameters is mainly based on the following aspects of operating characteristics of ground conventional public and electric vehicles:

1. Vehicle performance statistics. Vehicle number and statistics period, operating time, mileage, maximum speed, average speed, total energy consumption, average energy consumption, one-way time, and SOC.
2. Battery group performance statistics. Operation of the vehicle battery group number, battery case number, statistical period, working

time, total energy consumption, average energy consumption, discharge voltage peak, discharge current peak, the highest monomer voltage position, the lowest monomer voltage position, battery defect information, etc.

3. Run roads and other parameters. Mainly composed of electric bus battery exchange stations including quick-change robot arm, battery charger, the battery pack and battery management system, smoke sensor, DC power supply, battery storage shelf, and quick-change system match. When entering the charging room, vehicles adopt the model of rapid car battery replacement. The replaced battery will recharge in units in a single box.

Electric bus operating performance is divided into three parts: One is the line operation. It refers to the vehicle on the runtime vehicle indicators related to the line. The second is the vehicle performance metrics. It refers to the vehicle in the runtime performance of the vehicle itself. The third is the battery performance. It refers to the discharge of the battery running condition of the vehicle. Key indicators of electric bus operators are shown in Table 2.

Electric bus transport targeting system, based on certain charging station scale, calculates pure electric bus line-operating conditions according to the current conventional bus lines information. First, basic design parameters of pure electric bus are shown in Table 3.

Note 1: Peak departure interval calculated as

$$T = \frac{C_b}{H_m} \qquad (1)$$

where T is the departure interval, H_m is the peak hour section passenger volume, and C_b is the known bus capacity.

Table 2. Main indicators of electric bus operators.

Line operating indices	Vehicle performance index	Battery performance index
one trip time/min	capacity plan/ (person/vehicle)	rated battery capacity/Ah
Parking time/min	operational speed/(km/h)	actual battery capacity/Ah
turnover time and turnover frequency/min	endurance mileage/km	Discharging cycle life/ time
Line number of vehicles/trolley		Rated battery voltage/V
Driving frequency/ (train number/h)		Battery operating voltage/V
Headway/(min/ train number)		SOC/Ah

Table 3. Concrete scheme basic design parameters.

line length (km)	L	charging rate (%)	β
peak departure interval (min)	T	charger power (kw)	P_c
endurance mileage (km)	L_r	car capacity (person)	Cb
peak section flow (person times)	Hm	Battery replacement time (min)	T_c
Number of stations (ind)	N_r sets of quick-change equipment	working hours (h)	T_r
Endurance mileage utilization (%)	μ	spare battery reserve coefficient	γ

63

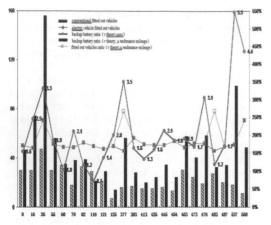

(a) Theoretical speed, driving range utilization (μ\SOC)

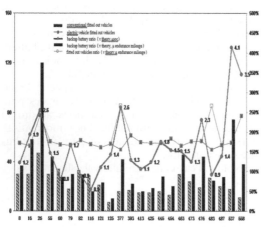

(b) Actual speed, driving range utilization (μ\SOC)

Figure 3. Analysis fitted out vehicle and spare battery coefficients in combination with speed, endurance mileage and soc.
Note: The red line is fitted out pure electric bus coefficient; blue, green are pure electric bus spare battery coefficients.

Whether or not the length of the bus line is reasonable, it affects the utilization of vehicle driving. Bus endurance mileage is an important index that reflects vehicle performance and practicality. This paper also gives the utilization of vehicle endurance mileage based on SOC. This is the accounting unit of electricity in kilometers from each line operating length.

Note 2: Number of spare battery for solving formula

$$N_b = \frac{T_{ch} \times N_r}{T_{ch} \times \gamma} \tag{2}$$

where N_b is the number of spare battery, T_{ch} is the charging time $T_{ch} = 1/\beta$, and; γ is the spare battery reserve coefficient.

This method calculates the maximum amount of backup battery, if the specific operational program designed to optimize the value can also be reduced. Therefore, combined with previous experience, it is generally preferable to a value of 1. Usually, a spare battery has 10% margin.

Note 3: Charging station service ability.

$$N_c = \frac{N_r \times T_r}{T_c} \times C_C \tag{3}$$

where N_c is the most service trains every day and C_c is the reserve coefficient.

Utilization of endurance mileage is the ratio of the length of the line and endurance mileage.

A large number of parameters were estimated as the main technical parameters of the calibration results for electric buses.

1. Fitted out vehicles and spare battery proportion limit when the pure electric bus replaces regular bus are shown in Figure 3.

At 50%, 60%, and 65% different capacity rate conditions, speed, endurance mileage, and SOC are

Table 4. Replaced electric bus parking area after increasing the proportion of the lookup table.

Increased proportion of parking area	Number of regular parking buses							
	<10	10	20	30	40	50	60	……
	Electric bus parking area							
1	<800	800	1600	2400	3200	4000	4800	
1.2	<960	960	1920	2880	3840	4800	5760	
1.25	<1000	1000	2000	3000	4000	5000	6000	
1.3	<1040	1040	2080	3120	4160	5200	6240	
1.4	<1120	1120	2240	3360	4480	5600	6720	
1.45	<1160	1160	2320	3480	4640	5800	6960	
1.5	<1200	1200	2400	3600	4800	6000	7200	
……	……							

estimated to obtain a stable regular bus converted into a pure electric bus with a car ratio in the range of 1.3 to 2.5. The spare battery is used when the replaced battery is not fully changed and the vehicle has a need to replace. For pure electric bus to determine the number of spare batteries, spare batteries give pure electric bus in suitable proportions in the range of 1.4–1.75 in Figure 3 (a), (b) (blue, green lines).

2. Transit depots' parking area calculation, increased correlation

Combining existing Beijing transit depots scale and local standards and new fitted out pure electric buses, the ratio of the parking area grows as shown in Table 4.

5 CONCLUSION

Based on the conventional bus line operating conditions and organization in Beijing, this paper gives the peak passenger flow scheduling scheme corresponding to car and pure electric bus driving range and soc. Measure under the pure electric bus speed, the length of the line of car proportion coefficient. Battery backup scaling factor for the battery under conditions, and current conventional bus terminal stations, major growth in the proportion of intermediate station parking lot, parked vehicle control tables, etc., provide analytical basis for Beijing to vigorously promote the pure electric bus operators, giving a technical reference for Beijing to develop energy saving urban public transport, and realize traffic restructuring.

ACKNOWLEDGEMENTS

This project was supported by Hebei Reform Project 2015GJJG036 (Optimization construction of transportation research and teaching course system Integration) and Beijing Municipal Science and Technology Commission KTL14042530 (Research of key technology of Beijing microcirculation pure electric vehicle demonstration operations).

REFERENCES

Amiri M., M. Esfahanian, M.R. Hairi-Yazdi and V. Esfahanian. Minimization of power losses in hybrid electric vehicles in view of prolonging of battery life. Journal of Power Sources, 2009, 190: 372–379

Chau K.T., Y.S. Wong and C.C. Chan. An overview of energy sources for electric vehicles. Energy Conversion & Management, 1999, 40: 1021–1039.

Haixing Wang. Regional Scheduling Theory and Method Research of Public Transport Vehicles - in the Background of Electric Vehicles [D]. Beijing Jiaotong University, 2007.

Hongwen He, Analysis of Electric Bus BJD 6 100-EV Urban Driving Energy Consumption [J]. Journal of Beijing Institute of Technology, 2004, 3: 222–225.

Li Jin, Research on Matching Relations between Operation and Battery State of Electric Bus [D]. Beijing Jiaotong University, 2011.

Ryu J., Y. Park and M. Sunwoo. Electric power train modeling of a fuel cell hybrid electric vehicle and development of a power distribution algorithm based on driving mode recognition. Journal of Power Sources, 2010, 195: 5735–5748.

Xiao Han, Operating Plan and Simulation of Electric Buses Charge Station [D]. Beijing Jiaotong University, 2010.

Xiuyuan Zhang, Haoming Wang, Fangfang Wang, Hui Zhang. City Electric Bus Using Parameter Estimation Analysis. Technology of Transport, 2013, 2, 225–230.

Energy Science and Applied Technology ESAT 2016 – Fang (Ed.)
© 2016 Taylor & Francis Group, London, ISBN 978-1-138-02973-6

Performance and economic evaluation of the photovoltaic–thermal system for BIPV applications

X.L. Ma, Q. Zhang, X. Zhang & R.J. Hong
Institute for Solar Energy System, Guangdong Provincial Key Laboratory of Photovoltaic Technology,
Sun Yat-sen University, Guangzhou, China

ABSTRACT: As photovoltaic thermal systems become more and more popular and commercial solutions find their way into the market, it is necessary to evaluate both the energetic and economic benefits of such systems at different climatic conditions. This paper describes an experimental study of a centralized PV/T system in South China. The influence of the parameters on its performance including environmental temperature, flow velocity, and solar radiation is investigated. The results showed that the daily thermal efficiency, daily electrical efficiency, and total PV/T efficiency reached 69.24%, 18.96%, and 88.52%, respectively. The system with a PR value of up to 80% and about 6 years payback period was obtained.

Keywords: PV/T; performance; PR value; payback period

1   INTRODUCTION

The Photovoltaic/Thermal (PV/T) collector is an integration of PV cells and solar thermal collector. When part of solar energy is converted into electricity, the rest is converted into heat and leads to an increase in the temperature of solar cells. The PV cells in PV/T collector are cooled down with a coolant (liquid or air) to get a higher electrical efficiency. The heat is collected for water heating in industrial and domestic applications, or for air heating in space heating and agricultural drying. With single area and double output, PV/T collectors have a better performance than conventional solar collectors.

As the concept of PV/T was proposed by Russell and Kern in 1978, plenty of academic and applied research have been performed, which mainly focus on the numerical simulation, system application. Early research refers to PV/T systems performed by refitting it with a heat collector (Huang et al 2001). All kinds of PV/T collectors have considerable technological progress in the past few years (Ibrahim et al 2011). The researcher realized that an economic investigation using actual commercial products is required for the accurate evaluation of a PV/T system's performance (Zhang et al 2011). A few localized studies have been performed. In China, the performance of PV/T systems with natural circulation of water has been investigated based on the climate conditions of Hefei and Hong kong (Chow et al 2005). A flat-box-type thermal absorber was proved to be extremely effective

according to these studies. In Europe and Canada, investigations on PV/T air systems have been primarily conducted in colder climates (Candanedo et al 2010), and the results illustrated that the PV/T collectors with a glass cover exhibited higher thermal efficiencies but lower electrical efficiencies compared with the uncovered collectors. The reduction in electrical efficiency was owing to the glass cover and the higher solar cell temperature. Meanwhile, various PV/T collector designs have been presented with some of the main factors (Chen et al 2010). Detailed studies of a new design of absorber collector under the climate conditions of Malaysia were performed. The parameters like the type of cells, fin efficiency, type of coolant, thermal conductivity, and operating conditions were analyzed. The results illustrated that the thermal efficiency for the amorphous silicon is better than that for the monocrystalline silicon. However, the electrical efficiency is the opposite (Hollick et al 2007). The PV/T systems consisting of thermoelectric generators were investigated. The results indicated that the system shows a linear relationship between the temperature on its plates (Daghigh et al 2011).

As a matter of fact, the high investment, but low energy gain is reminding the primary deficiency of most PV/T systems at present, which is a crucial point for massive commercialization of this technology. Therefore, despite the great potential, it may take decades to payback the initial investment without any government subsidies. The economic cost affects the broader usage of the PV/T system in construction. Without taking into the characteristics of commercial products, applicability of some studies to commercial applications is limited (Dupeyrat et al 2012). In this paper, an experimental PV/T system designed to maximize energy output in the South China climate is presented. Experimental results illustrate the electrical and thermal performance, which aims to offer more references to the practical design and application of PV/T system.

2 EXPERIMENTAL PROCEDURE

The experimental system is located in Guangzhou, Guangdong province, South China. The PV/T system is within the integrated building photovoltaic system. The PV/T system mainly supplies daily hot water and also produces electricity for the DC electrical appliances. The PV components are facing south with the oblique angle of 18°, which is most suitable in Guangzhou, South China according to the calculation.

The system consists of three main components including the photovoltaic one, the photo-thermal one, and system detection and control one. The configuration of the system is shown in Figure 1.

The photovoltaic part was composed of two identical photovoltaic modules. Each module was fabricated by 72 monocrystalline silicon solar cells with a size of 125 mm × 125 mm in series. The parameters of the cell are presented in Table 1.

The photo-thermal part consists of an absorber plate, 150 l water tanks, connecting pipelines, and circulating water pumps. The pumps drove the hot water to the cistern in the house through pipelines. As the core component of the PV/T system, PV/T collector's structure affects the thermoelectric properties of the system directly. The collector we designed is presented in Fig. 2, with its 5 mm cavity separating the blue membrane and the solar cells.

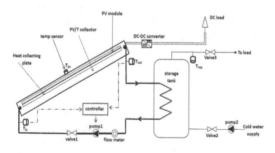

Figure 1. Schematic of the PV/T system configuration.

Table 1. Technical specifications of the PV cell at STC.

P_{max}	V_{mp}	I_{mp}	V_{oc}	I_{sc}
195w	37.2V	5.24 A	45.5V	5.62 A

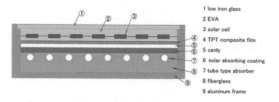

1 low iron glass
2 EVA
3 solar cell
4 TPT composite film
5 cavity
6 solar absorbing coating
7 tube type absorber
8 fiberglass
9 aluminum frame

Figure 2. Structure of PV/T with a cavity.

We adopted the transparent TPT with transmittances higher than 90% and the blue membrane with an absorptivity of 97%, which improves the solar collecting efficiency. At the same time, they keep the cells insulated from the blue membrane. The blue membrane is connected with the collecting pipes by laser welding, and the copper patch agent is filled in the void after welding, which will decrease the thermal contact resistance between the blue membrane and the heat-collecting pipes, and enhance the thermal efficiency as well. The collector uses a centralized passageway design, in which eight branch pipes connect the absorbent plates.

In order to distinguish exactly about their application conditions, this research has collected the electrical performance of different PV components under the same condition since May 2014. We take cooling cycle test for one of the components while giving no further effect on the other one which as a control group. We test the operation of the treatment group under the circumstance of different flowing speeds in various conditions and compare the collected electricity and heat data.

3 RESULTS AND DISCUSSION

3.1 *The research of the concrete output*

The daily thermal efficiency, daily electrical efficiency, and total PV/T efficiency reached 69.24%, 18.96%, and 88.52%, which is a relatively high level.

The real-time power of the components in control and treatment groups under different temperatures of autumn and winter is presented in Fig. 3. By comparing with the relevant results, it can be found that the cooling cycle apparently assists in lowering the temperature on the surface of cells, increasing the output of the power. The power outputting reaches the highest degree in the mid-day, owing to a higher solar elevating angle and external environment temperature. The cooling rate of the components in the experiment group is lower than that of the control group at around 15:30 p.m., as a result of the decreasing temperature and sun radiance. Therefore, the cell temperature of the control group is lower than that of the experiment group, which causes a lower output of the power during the time period.

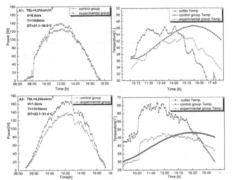

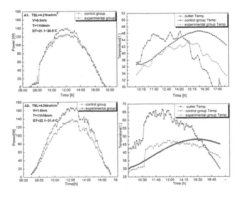

(a) Temperature and power generation contrast in the typical autumn condition (A1 and A2)

(b) Temperature and power generation contrast in the typical winter condition (W1 and W2).

Figure 3. The performance of PV/T system in autumn and winter time.

The environment temperatures for this pair of the groups are close. Comparing the data in these days, it is found that the temperature at the outlet increases and is steadier through an increase in the flowing rate of the fluid, which is around 50°C. However, with the decrease in temperature in the same or even higher radiance level in the winter, the output of the hot water decreased apparently. The average heat goes down to 40°C. It is noteworthy that the power output of the control group surpassed that of the experiment group a lot. It is due to the difference in temperature between morning and evening in the winter time. The temperature changes rapidly in the afternoon, while the circulating water of the experiment group cycles constantly, keeping the temperature of the cell stable, which leads to a relatively lower power output than the control group.

3.2 The comparison of PR value

Besides the cell types and their efficiency, the environment, including temperature and radiance, also affects the electricity output. In order to eliminate the accumulated electricity output differences in different systems caused by various weather conditions. The Performance Ratio (PR) of different systems according to daily output and radiance of different PV components was counted.

The PR value is an important international value that measures the feature and quality of the grid-connected photovoltaic system. Considering the factors such as solar radiance, dust accumulation on the surface of the components, array efficiency, inversion efficiency, and AC distribution efficiency, we can get a better analysis about the overall performance and the electricity output rate of each component. The PR value is obtained by the formula of

$$PR_E = \frac{PRD}{P \times T}, \text{ thereinto } T = \frac{I_i}{I_0} \tag{1}$$

In this formula, PRD is the actual electricity amount in a certain time period Δt, $P \times T$ is the theoretical electricity amount in this time period, P is the standard power of the component under the STC condition. T is the actual efficient time of output in Δt. I_i is the total amount of solar radiance in Δt. I_o is the total amount of solar radiance under the STC condition. The daily efficient PR value of the different components can be calculated with the data collected. The PR value is shown in Fig. 4.

The PR value of both the systems reaches 80%, which is a relatively high level (Reich et al 2012). It proves that the PV/T system functions well in terms of controlling the inversion efficiency and the loss of AC distribution efficiency, and the overall system is also in excellent condition. Comparing the different time periods, the PR value of the PV/T system during June-September is higher than that of during October-December. Both the amount

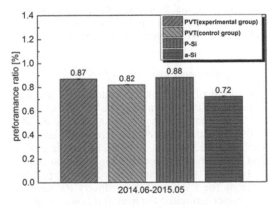

Figure 4. The average annual PR value during the period of 2014.06–2015.05.

69

Table 2. Monthly thermal and electrical energy outputs from PV/T system and four kinds of PV module.

Month	Solar radiation amount [kWh/m²]	PV/T Electrical output [kWh/m²]	PV/T Thermal output [kWh/m²]	P-Si Electrical output [kWh/m²]	a-Si Electrical output [kWh/m²]	UMG Electrical output [kWh/m²]	Double glass Electrical output [kWh/m²]
Jan.	105.56	23.86	45.25	24.75	2.65	12.17	7.11
Feb.	49.84	13.36	26.35	14.41	1.64	7.57	5.18
Mar.	54.56	14.46	27.25	15.69	2.29	7.47	5.12
Apr.	59.58	15.65	36.75	17.75	2.71	8.49	6.95
May	64.50	15.00	39.25	17.23	2.82	8.12	7.12
Jun.	95.02	21.03	37.50	24.77	4.07	12.01	9.04
Jul.	101.81	24.06	46.25	28.49	4.54	16.38	11.36
Aug.	104.15	24.13	49.25	27.79	4.50	15.32	10.63
Sep.	108.96	24.70	42.25	27.93	4.29	14.26	9.89
Oct.	100.91	27.58	44.50	30.15	4.15	14.26	9.89
Nov.	78.24	17.50	30.50	19.01	2.32	9.42	5.80
Dec.	87.06	19.72	30.50	20.71	2.21	10.05	5.81

of the power and the PR value are rising with the increase in the solar radiance, which results from the higher temperature, and solar radiance in more sufficient in South China in summer than in winter.

Comparing the two groups, their working performances are different during the different time periods in one day, even though other factors remain the same. From June to September, the hot and humid climate in South China makes it hard to find out the differences between their PR values. In December, when the weather turns cold, the experimental group begins to function. It is found that the PV/T system has a higher PR value than the control group when comparing the two groups.

To sum up, the solar radiance and the temperature have significant impact on the amount of electricity output in the PV modules. The most sensitive module to the temperature is the a-Si. The a-Si cells do not perform very well even if they have the same nominal power as other polysilicon modules do. Therefore, the polycrystalline silicon solar cells in the PV/T system will be a better choice.

3.3 Performance evaluation of the PV/T system

As a kind of the building attached PV modules, the economic usually affects the broader use of the PV/T components in construction (Beccali et al 2009). In order to assess the economic viability of each installed system, a simple economic analysis has been performed for each kind of solar cell.

In order to compare the economic performance of the PV/T and the ordinary PV modules, this paper estimated the heat and power output of the testing systems first. The climate factor is considered in calculating a more objective estimation.

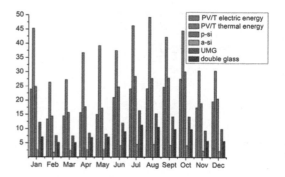

Figure 5. Useful energy output per unit area of the five types of PV construction module.

The energy output of all the modules in the whole year is calculated. All the calculations are made on the basis of the unit collecting area to obtain a more intuitive comparison (Tripanagnostopouloy et al 2005).

As for the estimation of heat output datum, the research presumes the heat of the water that flows out of the PV/T components as useful energy if its temperature is higher than the environment temperature (Tselepis et al 2002). In this way, the estimation of the heat output in this research is the maximum of the system in its environmental condition. Table 2 and Fig. 5 are the energy output data of each PV module collected. According to Fig. 5, the PV/T component heat output is three times as much as its electricity output. The monthly heat output and the energy output of the PV/T components point toward the same trend. From August to October, the PV/T has little heat convection with the environment while getting much more solar radiance at the same time.

Table 3. Capital cost of the system (CNY ¥).

Components	PV/T	P-Si	a-Si	UMG	Double-Glass
PV modules	9000	2000	3900	5600	77457
Support structure		683.2	817.8	2661.9	25257.5
Inverter	1300	1300	650	4000	13000
Other accessories	592	0	0	0	0
Total cost	10892	3983.2	5367.8	12261.9	115714.5
Area	2.8	3.3	3.95	12.9	122
cost per area	3890	1207	1358.9	953.7	948.5

Therefore, it reaches the highest heat output. On the other hand, the heat output reaches the slowest point from February to April owing to the regular rainy weather and a less amount of solar radiance.

3.4 *The payback period of the PV/T system*

Economic analysis is based on the performance results. In order to enrich the types of PV modules and minimize the loss in circuit and mismatch, all the inverters are designed to be micro-inverters in the project. As a most popular inverter, the micro-inverter fits for different kinds of the modules for it is more flexible and more reliable. However, the only flaw is that it costs a bit more than the grouped inverter and the centralized inverter. The micro-inverter in this research costs 2.83 RMB per watt, while the cost of a grouped inverter is only 1.1 RMB. Therefore, the cost calculation in this paper is only about the model project here. It will be lower in mega projects.

The calculation is shown in Table 3; the cost of PV/T system is the highest (3890 RMB/m²), while the cheapest ones are UMG metallurgical silicon PV modules (953.7 RMB/m²) and the double glass components (948.5 RMB/m²), which results from the other fittings of the UMG cost the least among all the systems, and which is 3.5 RMB/m² only. As for the double-glass, it has a larger array size and a lower average inverter cost. However, the PV/T system needs a tank and a bump as well as the expensive components and micro-inverters, which increase the total cost of PV/T system building.

The assumptions in calculating the annual cash flow distribution and the callback period are as follows:

1. The power output all goes through the grid, while the power for own demand is calculated separately. At present, the distributed photovoltaic feed-in tariff in Guangzhou is 0.502 RMB per kilowatt-hour, which is calculated through the coal benchmark price in Guangdong province. The national feed-in tariff subsidy is 0.42 RMB per kilowatt-hour, which means that the PV modules could gain the profit of 0.922 RMB by producing every kilowatt-hour electricity.

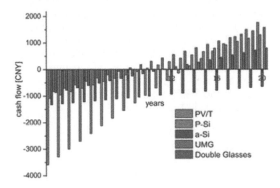

Figure 6. Payback cash flow analysis of the five types of PV construction module.

2. All the heat output is fully used. The user heats the water with the natural gas instead of electric power. For now, the gas price in Guangzhou is 3.45 RMB/m³, while the natural gas heat value is 39.82 MJ/m³, and its burning efficiency is 90%.
3. The annual energy output data are calculated from Fig. 5
4. The initial investment subsidy is 0. The prices of electricity and gas increase by 3.5% every year.
5. The annual efficiency decay rate is 0.4%. The system's service life lasts for 25 years. There is no bank loan.

With the above assumptions taken into comprehensive consideration, the diagram of investment-callback cash flow analysis is drawn in Fig. 6.

According to Fig. 6, the payback period of the PV/T system is around 13 years, which is the longest among all systems. The UMG system has the shortest payback period. However, if we presume that the user uses the same amount of electricity to produce heat, the PV/T payback period will shrink to only about 6 years for the amount of heat energy is tripled than electricity output. Although the PV/T seems more expensive in building at the beginning, it actually profits the most among all the PV modules. In the following 6-year period after the payback period, PV/T system outputs the largest amount of annual energy. Therefore, in the long run, the PV/T's economic performance is better than any other PV module.

4 CONCLUSIONS

The cooling cycle apparently helped in lowering the temperature on the surface of cells, and increasing the output of the power in the components. The solar radiance and the temperature had a significant impact on the PR value and the power output. The PR value of all the systems reached 80%. However, this value may be further improved by using polycrystalline silicon solar cells in the PV/T system. Even though the construction cost of the PV/T system was three to four times higher than the others, it gained much more heat output. Overall, the PV/T system has a better economic performance in the long run.

ACKNOWLEDGEMENTS

This work was financially supported by Guangdong Science and Technology Department (Grant No. 2013 A011402005) and Guangzhou Science and Technology Department (Grant No. 2014Y2–00221).

REFERENCES

Beccali, M., Finocchiaro, P. & Nocke, B., (2009) Energy and economic assessment of desiccant cooling systems coupled with single glazed air and hybrid PV/thermal solar collectors for applications in hot and humid climate, *J. Solar Energy*. 83, 1828–1846.

Candanedo, L., Athienitis, A.K., Candancdo, J., O'Brien, W. & Chen, Y., (2010) Transient and steady state models for open-loop air-based BIPV/T systems. *J ASHRAE Transactions*. 116, 600–612.

Chen, Y., Athienitis, A.K. & Galal, K., (2010) Modeling, design and thermal performance of a BIPV/T system thermally coupled with a ventilated concrete slab in a low energy solar house: Part 1, BIPV/T system and house energy concept. *J Solar Energy*. 84(11), 1892–1907.

Chow, T.T., Ji, J. & He, W., (2005) Photovoltaic–thermal collector system for domestic application, *CProceedings of ISEC 2005, Solar World Congress*, Orlando, USA, (CD ROM).

Daghigh R., Ibrahim, A., Jin, G.L., Ruslan, M.H. & Sopian, K., (2011) Predicting the performance of amorphous and crystalline silicon based photovoltaic solar thermal collectors. *J Energy Conversion Management*. 52(3), 1741–7.

Dupeyrat, P., Ménézo, C. & Fortuin, S., (2012) Study of the thermal and electrical performances of PVT solar hot water system, *J Energy and Buildings*. 68, 751–755.

Huang, B.J., Lin, T.H., Hung, W.C. & Sun, F.S., (2001) Performance evaluation of solar photovoltaic/thermal systems, *J Solar Energy*, 70(5), 443–448.

Ibrahim, A., Othman, M.Y., Ruslan, M.H., Mat, S. & Sopian, K., (2011) Recent advances in flatplate photovoltaic/thermal (PV/T) solar collectors, *J Renewable and Sustainable Energy Reviews*.15(1), pp. 352–365.

Reich, N.H., Mueller, B. & Armbruster, A., (2012) Performance ratio revisited: is PR> 90% realistic, *J. Progress in Photovoltaics: Research and Applications*. 20(6), 717–726.

Tonui, J.K. & Tripanagnostopoulos, Y., (2007) Air-cooled PV/T solar collectors with low cost performance improvements. *J Solar Energy*. 81(4), 498–511.

Tripanagnostopouloy, Y., Souliotis, M. & Battisti, R., (2005) Energy, Cost and LCA Results of PV and Hybrid PV/T Solar Systems, *J. progress in photovoltaics: research and applications*. 13(13), pp. 235–250.

Tselepis, S. & Tripanagnostopoulos, Y., (2002) economic analysis of hybrid photovoltaic/thermal solar systems and comparison with standard PV modules, *J. Solar Energy*. 72(3), 217–234.

Zhang, X., Zhao, X., Smith, S. & Xu, X., (2012) Review of R & D progress and practical application of the solar photovoltaic/thermal (PV/T) technologies, *J Renewable and Sustainable Energy Reviews*. 16(1), 599–617.

Energy Science and Applied Technology ESAT 2016 – Fang (Ed.)
© 2016 Taylor & Francis Group, London, ISBN 978-1-138-02973-6

Effects of the leading and trailing edge shape on the aerodynamic performance of a vertical axis wind turbine

Q. Yao, G.D. Zhao, C. Wang & C.Y. Zhou
Shenzhen Graduate School, Harbin Institute of Technology, Shenzhen, China

ABSTRACT: In this study, the aerodynamic performance of airfoils with changing leading edge radius and trailing edge shape is systematically analyzed using computational fluid dynamic methods. The k-ω SST turbulence model is adopted. According to the tangential force coefficient, the optimal airfoil is selected for the vertical axis wind turbine. It is found that the optimized airfoil could offer 10% more wind energy utilization rate than NACA0015 under the same conditions, and it has a wider range of tip speed ratio of high efficiency, so the vertical axis wind turbine with an installed airfoil has a higher utilization rate of wind energy and a bigger scope of work.

Keywords: vertical axis wind turbine; blade modification; wind energy utilization

1 INTRODUCTION

Wind turbines convert wind energy into electric energy and blades are the core component of the power supply for the wind turbine. Improving the aerodynamic performance of the blades is helpful to improve the wind energy utilization ratio of a wind turbine. Compared with a horizontal axis wind turbine, the vertical axis wind turbine has many advantages, such as always facing the wind, low starting wind speed, easy maintenance, low noise and so on (He, 2006). So in recent years more and more people have become involved in vertical axis wind turbine research and development (Yan, 2012). In order to improve the performance of vertical axis wind turbines, domestic and foreign scholars have carried out a series of research on the vertical axis wind turbine. Numerical simulation by Mohamed (2012) of vertical axis wind turbines found that a symmetrical airfoil has more advantages for vertical axis wind turbines. Xu et al. (2011) studied the effects of various airfoil tail flaps for wind turbines on wind energy utilization effect and Wang (2013) studied the aerodynamic performance of vertical axis wind turbines with the maximum thickness of blade by numerical simulation, finding that the maximum thickness of the airfoil forward can effectively increase the stall angle. In this paper, we mainly numerically study the effects of leading-edge radius and tail shapes on aerodynamic performance of vertical axis wind turbine.

2 PHYSICAL MODEL

There are many methods to evaluate the aerodynamic performance of vertical axis wind turbines, and numerical calculation is one of them. The multiple stream tube model provides one of the more accurate evaluations of the aerodynamic performance of wind turbine; it was proposed by Strickland in 1975. According to the multiple stream tube theory, the wind energy utilization ratio of Π-type vertical axis wind turbines can be expressed as follows:

$$C_P = C_p = \frac{\delta\lambda}{N_t}\sum_{1}^{N_t}\left[C_t\left(\frac{U_R}{U_\infty}\right)^2\right] \tag{1}$$

$$\delta = \frac{Nc}{2R} \tag{2}$$

$$\lambda = \frac{\omega R}{U_\infty} \tag{3}$$

$$C_t = C_l \sin\alpha - C_d \cos\alpha \tag{4}$$

where C_p is the wind energy utilization of wind turbine; δ is the solidity; λ is the tip speed ratio; C_t is the tangential force coefficient; U_R is relative velocity, m/s; U_∞ is the inflow wind speed, m/s; N is the blade number; c is the chord length, m; R is the radius of the wind turbine, m; ω is the rotation angle velocity of the wind turbine, rad/s.

From the above formula, it can be seen that the lift drag characteristics of the airfoil play a decisive role in the value of wind energy

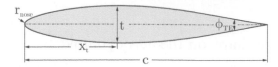

Figure 1. The Roßner airfoil.

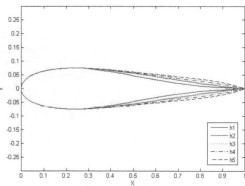

a) Airfoil leading edge modification

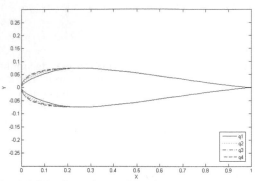

b) Airfoil trailing edge modification

Figure 2. The modification of the leading and trailing edge of the airfoil.

utilization. The leading edge shape and the shape of the airfoil tail directly affect the lift and drag coefficient. The relative thickness of each airfoil is set to 15%. According to the study of Wang (2013), the maximum thickness of airfoil position forward can effectively increase the stall angle and improve the aerodynamic performance; therefore the maximum thickness position of the foil is set to 25%. Blade modification is based on Roßner airfoils (Hepperle, 2011), mainly because of its relatively simple set of airfoil parameters and the number of variables is less as shown in Figure 1. In Figure 1, r_{nose} is the airfoil leading edge radius, t is the airfoil thickness, X_t is the airfoil maximum thickness position, Φ_{TE} is the airfoil trailing edge angle. The airfoil leading

edge shape is determined by r_{nose}, and the tail shape by is determined by r_{nose} and Φ_{TE} joint decision.

Figure 2 shows the schematic diagram of the airfoil shape changed based on Roßner airfoils. The leading edge radius and trailing edge shape of the airfoil changed separately. Firstly, as shown in Figure 2a), keep the airfoil trailing edge while change the leading edge radius, q1, q2, q3 and q4 indicate the leading edge radius are 1%c, 2%c, 3%c and 4%c respectively. Secondly, as shown in Figure 2b), keep the leading edge radius while change the trailing edge shape and several airfoils shape can be obtained. Here, h1 indicates the leading edge radius 1.7%c and trailing edge angle is 10 degrees; h2 indicates the leading edge radius is 2.8%c and trailing edge angle is 10 degrees; h3 indicates the leading edge radius is 2.8%c and trailing edge angle is 20 degrees; h4 indicates the leading edge radius is 2.8%c and trailing edge angle is 40 degrees; h5 indicates the leading edge radius is 2.8%c and trailing edge angle is 60 degrees. In the process of modification of the airfoil, relative thickness and maximum thickness position of the airfoil were not changed.

3 COMPUTATIONAL MODEL

In this study, computational fluid dynamics method was used to calculate the lift and drag characteristics of the blade. Since the blade section of the H-type vertical axis wind turbine is equal, it can be simplified as a two-dimensional model. So in regard to ensuring accuracy it also greatly reduces the calculation time.

The model was built by gambit, the blade chord was set to 1 m, computational domain length was 50 m, width was 60 m, Reynolds number was set to 3.6×10^6, inlet velocity was 5.364 m/s, export design was the outlet pressure, and the airfoil surface underwent boundary layer grid refinement to make sure that Y+ was small enough for the turbulence model, as shown in Figure 3. Using Fluent software for numerical simulation, the k-model of SST was selected. The model has a relatively high accuracy (Wu et al., 2009) and requires the first layer to satisfy Y+<5, using the pressure-based implicit solver, speed, pressure and speed coupled with the SIMPLE algorithm.

In order to verify the accuracy of the turbulence model, the NACA0015 airfoil was selected for comparison and the lift and drag force were analyzed. The simulation results were compared with the experimental data. The experimental results are from the literature (Sheldahl, 1981). Figure 4 shows that the calculated values are in good agreement with the experimental values of the lift drag curve, can better respond to the airfoil stall, and the stall

a) Computational domain mesh b) Blade surface mesh

Figure 3.　The grids of the computational model.

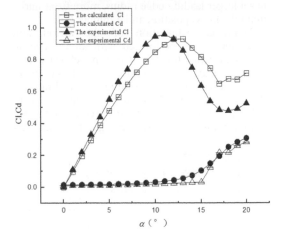

Figure 4.　Comparison of NACA0015 with the experimental results.

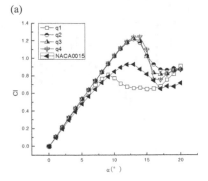

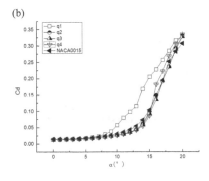

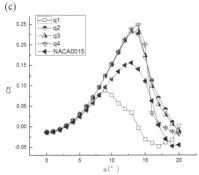

Figure 5.　Aerodynamic performance curve corresponding to the change of the airfoil leading edge. a) C_l, b) C_d, c) C_t.

angle is also close to the experimental results. So, it is feasible and reliable for the following results.

4　RESULTS AND DISCUSSION

4.1　*Effects of the leading edge of the airfoil*

Figure 5 is the aerodynamic performance curve corresponding to the change of the airfoil leading edge. From Figure 5 (a), we can see that the stall angle of the airfoil is very small when the airfoil's leading edge radius is small. The stall angle of the airfoil increases with increasing leading edge radius. At low angle of attack, the large leading edge radius of the airfoil can provide larger lift, but the lift of the airfoil at the leading edge radius after the stall angle of attack decreases sharply. For drag coefficient, in Figure 5 (b), it can be seen that the leading edge radius of the airfoil has a smaller drag coefficient at low angles of attack, but the minimum and maximum of the airfoil has a large drag coefficient after the stall angle. According to Equations (1) and (4), the aerodynamic perform-ance of an airfoil can be judged by the tangen-tial force coefficient C_t. in Figure 5 (c), it is seen that the tangential force coefficient is small when the leading edge radius is small. An airfoil with a

larger leading edge radius has a great advantage when the attack angle is less than 15 degrees, but when the angle of attack is more than 15 degrees, the coefficient of the lift force will decrease and the drag force increases. This means that C_t will decrease sharply when the attack angle is greater than 15 degrees.

4.2　*Effects of the trailing edge of the airfoil*

Figure 6 is the aerodynamic performance curve of airfoils with different tail shapes. As can be seen from Figure 6 (a), an airfoil with a thin tail can generate a larger lift coefficient for the airfoil, but

(a)

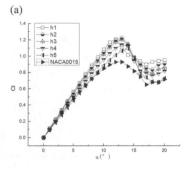

(b)

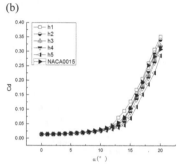

(c)

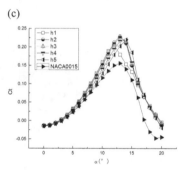

Figure 6. Aerodynamic performance curve corresponding to the change of the airfoil trailing edge. a) C_l, b) C_d, c) C_t.

its stall angle is small. Although the trailing end of the thick airfoil has a larger stall angle, the aerodynamic performance of the stall is poor. From Figure 6 (b), it can be seen that the thinner airfoil has a smaller drag coefficient, but its drag coefficient after the stall sharply increases, which results in higher energy consumption in the process of wind turbine rotation. Figure 6 (c) shows the tangential force coefficient of the shapes of different airfoils. It can be seen that the tangential force coefficient of the thinner airfoil at low angle of attack is higher. After the attack angle is larger than 15 degrees, because of the larger drag coefficient, the tangential force coefficient of the thinner airfoil is smaller than the thicker one.

4.3 Optimal airfoil and comparison

The influence of the airfoil leading edge and the trailing shape may affect the aerodynamic performance of the airfoil. Therefore, the total level of the four kinds of leading edge radius and the five kinds of tail shape with the tangential force coefficient C_t were considered. The results show that the tangential force coefficient curve of q3h2 is best, as shown in Figure 7. Compared with NACA0015, the airfoil has a larger leading edge radius, more front maximum thickness position and a thinner tail.

Numerical simulation of the airfoil and NACA0015 was performed with an installation radius of wind turbine $R = 1$ m, chord length $c = 0.1$ m and blade number $N = 4$. The wind energy utilization ratio with the tip speed ratio of the curve is shown in Figure 8. Compared with NACA0015, under the low tip speed ratio, the airfoil has a high torque, so it has good starting per-

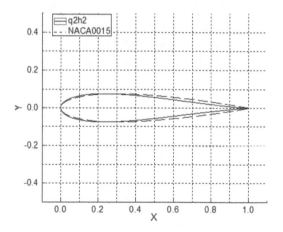

Figure 7. The optimized airfoil shape.

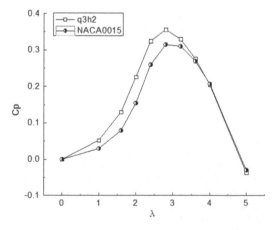

Figure 8. Cp curve of the optimal airfoil and NACA0015.

formance. The maximum wind energy utilization is about 10% higher compared with NACA0015, and its high wind energy utilization ratio of the corresponding range of tip speed ratio increased. This indicates that the vertical axis of the blade of the wind turbine can be effectively operated and the wind turbine work range is larger.

5 CONCLUSION

The airfoil leading edge radius and the shape of the airfoil are changed when the maximum thickness are ensured. Aerodynamic analysis was carried out, and the optimal airfoil was obtained through comparison. Simulation of the whole machine was performed and compared with NACA0015. The conclusions are as follows:

1. An airfoil with a larger leading edge radius will have a larger stall angle and larger lift coefficient at a low angle of attack, but when the leading edge radius is too large, there will be a large drag coefficient after the stall. When the leading edge radius is small the airfoil will stall under a small attack angle. We can see from the tangential force coefficient C_t curve that the leading edge radius should be bigger than 2%c.
2. An airfoil with a smaller trailing edge will have a larger lift coefficient, but its stall angle will be small and its resistance coefficient larger. So an airfoil with a thinner tail shape is more suitable for the H type vertical axis wind turbine, but its thickness should not be too small.
3. The optimal airfoil was obtained by simulation. Compared with NACA0015, the airfoil has a larger leading edge radius, more front maximum thickness position and a thinner tail. The lift coefficient is better than for NACA0015 in the full range of attack, and the resistance coefficient is similar to that of NACA0015; the Tangential force coefficient is better than NACA0015 in the range of 0 degree to 20 degree.
4. Through the simulation of the whole wind turbine, the wind energy utilization ratio of the wind turbine with the new type of airfoil has a bigger wind energy utilization ratio and larger working range.

ACKNOWLEDGEMENTS

The work described in this paper was partly supported by the Shenzhen Science and Technology Plans: Key Laboratory Development Projects (Grant No. ZDSYS201405081615472 9).

REFERENCES

He Dexin. 2006. Wind engineering and Industrial aerodynamics [M]. Beijing: National Defense Industry Press.

Hepperle M. 2011. Javafoil Users Guide [J].

Mohamed M H. 2012. Performance investigation of H-rotor Darrieus turbine with new airfoil shapes [J]. Energy, 2012, 47(1): 522–530.

Sheldahl R E, Klimas P C. 1981. Aerodynamic characteristics of seven symmetrical airfoil sections through a 180-degree angle of attack for use in aerodynamic analysis of vertical axis wind turbines [R]. Sandia National Labs., Albuquerque, NM (USA).

Wang Jianming et al. 2013. Aerodynamic performance of a VAWT with modification of the airfoil in front of a maximum thickness of blade [J]. Renewable Energy Resources, 31(012): 63–67.

Wu Faming et al. 2009. Comparative study of different turbulent models for numerical simulation of wind turbine airfoil friction [J]. Fluid Machinery, 36(12): 11–14.

Xu Zhang et al. 2011. Influence of various flaps on the performance of vertical axis wind turbines [J]. Journal of Chinese Society of Power Engineering, 31(9): 715–719.

Yan Qiang. 2012. The present and future of vertical axis wind turbines [J]. Wind Energy, (5): 50–53.

Energy Science and Applied Technology ESAT 2016 – Fang (Ed.)
© 2016 Taylor & Francis Group, London, ISBN 978-1-138-02973-6

Research on the generation of the bubble and the bubble pump characteristics of different diameters

H.T. Gao, D.J. Kong, C.B. Wang & H.R. Liu
College of Marine Engineering, Dalian Maritime University, Dalian, China

ABSTRACT: In this paper, the effect of the 9.5–6.5 mm different-diameter bubble pump and the 9.5–9.5 mm same-diameter bubble pump on the performance of lithium bromide absorption refrigeration system was experimentally investigated. The contact angle formula was obtained by theoretical derivation. The experimental data showed that: (1) bubble was generated from the core of vaporization in the generator and kept changing its shape from oval to flat until disengagement and (2) the cooling capacity of the former was greater under the same condition, which meant that a secondary lifting pipe with a smaller diameter acted better with the same primary diameter. The increment in cooling capacity was 100–200 W and COP increment was 0.1 under this experimental condition.

Keywords: bubble generation; bubble pump; different diameters; experimental study

1 INTRODUCTION

The combination of declining energy reserves and growing energy consumption calls for energy efficiency improvement and rational use. In the field of ship refrigeration, the researches on waste heat utilization with bubble pump to save the high-grade energy have attracted much attention.

The characteristics of bubble pump have been studied. In 1989, Peng Yichuan et al. (Peng Yichuan, Xiao Zeqiang. 1989.) experimentally studied the effect of bubble pump diameter on phase flow patterns and deduced the bubble pump performance formula with the Nicklin theory. Wang Rujin et al. (Wang Ru-jin et al. 2008.) found a theoretical method to calculate the bubble pump size parameters with a number of experiments and theoretical calculation. As for the structure design, Que Xiongcai et al. (Que Xiong Cai, Li Hong. 1989; Dai Yongqing et al, 2001) got some formulations for diameter options, flow resistance, slug flow pattern judgment, and so on.

In the study of the bubble pump absorption refrigeration system, Gu Xiuya et al. (Gu Yaxiu et al. 2006.) established a pump-free lithium bromide absorption refrigeration system by combining bubble pump and solar pump. The properties of bubble pump in the system had also been studied. Wang Xiaolu et al. (Wang Xiaolu. 2011; Gao Hongtao, Wang Xiaolu. 2013.) experimentally explored the effect of the lifting pipe diameter, concentration, heating power, and immersion height of the single-stage lithium bromide solution bubble pump on the performance of the bubble pump. Ma Guotuan (Ma Guotuan. 2012.) studied the two-stage bubble pump in the pump-free double-effect lithium bromide absorption refrigeration system. The effects of immersion height, concentration, and cooling water flow rate on bubble pump were explored by experiments. Xu Honghao (Xu Honghao. 2013) focused on the effects of two different-diameter bubble pumps using experiments, in which pipe diameters of two stages are the same. Meng Long (Meng Long. 2014.) studied the cooling effect of the bubble pump in two-stage absorption refrigeration system. He mainly explored the effect of condenser temperature and evaporator temperature on the overall cooling performance.

2 FORMATION OF BUBBLES IN THE BUBBLE PUMP

The bubble pump is the power source of pumpless absorption refrigeration system. The heat input boils the solution into bubbles. Then, the bubbles drive the solution to transport. Therefore, the study of bubble generation and motion is very important. As this paper aimed to generate power bubbles with ship waste heat, the research focused on bubble generation in pool boiling.

For this system, the bubbles are generated in an opaque generator. In order to facilitate the observation of bubble generation process, the opaque experimental device was replaced with a glass container as shown in Figure 1b. Figure 1a is the simplified diagram of the experiment device, using a high-speed video camera to shoot the bubble generation process to explore bubble generation principle and its law of motion in a glass container.

Figure 2 shows the bubble generation process when the heat power is 2000 W and the lithium bromide concentration is 50%. As shown in Figure 2a, when the temperature reaches a certain degree, a small bubble first appears in the core parts of vaporization without departing from the bottom. Then, this bubble gradually grows in Figure 2a-c. In Figure 2d, the bubble leaves the bottom with a nearly circular shape. In Figure 2e, it enters the liquid as flat. Then, it grows larger during rising as is shown in Figure 2e-g-h. In Figure 2i, the bubble begins to split into two small bubbles. The whole process describes the generation of bubble on the wall, its departure into the liquid, and final burst. No bubble was observed to be generated inside the liquid. All of them are produced on the high-temperature spot of the bottom.

When the solution temperature keeps rising, the bubble string begins to appear as shown in Figure 3. In a vaporization core, a string of bubbles continues producing, which consists of bubbles generated from the same vaporization core at different times. The generation of the bubble string is similar to the rising process of one bubble. Therefore, the picture of the bubble string can be used to study the rising procedure of a single bubble. It is found that as the bubble grows bigger, its running track makes a curve. In addition, the bubble bursts at the moment it reaches the gas-liquid interface.

Speaking of the theoretical analysis of the generation of bubbles, in 1956, Snyder (Guo Liejin. 2002.) first proposed that there was a very thin liquid layer between the heating surface and bubbles. Moore and Mesler proved its existence. As shown in Figure 4, the bubble radius grows from r_c to r as the time changes from τ_g to τ. In this process, part of the thin film layer evaporates into gas, leading to the increase in bubble volume. If the generator heating wall performed well in heat conduction, the thin film evaporation could be treated as an isothermal process, in which the wall tempera-

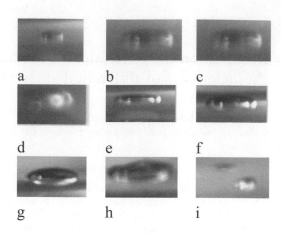

a   b   c

d   e   f

g   h   i

Figure 2.   Process of bubble generation.

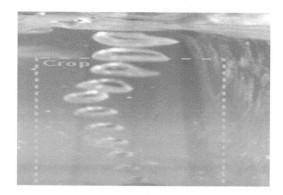

Figure 3.   The figure of a bubble string.

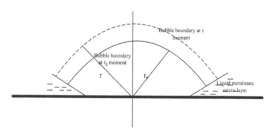

Figure 4.   The figure of bubble growing.

ture is kept as a constant ($T_W = T_{W0}$). Under the assumption that the gas came from the thin film, the bubble radius satisfied the following rule:

$$R = c\tau^n \tag{1}$$

where c is the constant and τ is the time.

The combination of Eq. 1 and the heat balance equation during bubble growth presented by Han and Criffith turns out to be:

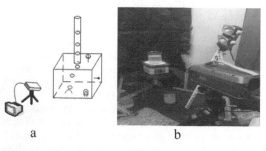

a   b

Figure 1.   Diagram and picture of the bubble generator.

$$\rho_l h_{fg} \frac{d\delta}{d\tau} = -\lambda \frac{(T_W - T_S)}{\delta_l} \qquad (2)$$

where δ is the thickness of the thin film at τ.

Eq. 3 can be obtained by integrating Eq. 2 from τ_g to τ.

$$\delta_0^2 - \delta^2 = 2 \frac{\lambda_l (T_{W0} - T_S)}{\rho_l h_{fg}} (\tau - \tau_g) \qquad (3)$$

Δ_0 is the Initial thickness of the liquid layer and h_{fg} is the gas-liquid enthalpy difference.

Cooper and Lloyd obtained the following Eq. 4 according to micro-layer hydrodynamic.

$$\delta_0 = 0.8(\gamma_l \tau_g)^{1/2} \qquad (4)$$

where γ_l is the kinematic viscosity of the liquid.

During the process of thin film evaporation and gas volume increase, the growing rate of the bubble is assumed to be equal to the evaporating velocity of the thin film layer at τ. Cooper obtained the following formula:

$$R(\tau) \approx 2.5 \frac{j_a}{p r_1^{1/2}} (al\,\tau)^{1/2} \qquad (5)$$

where j_a is the Jacob number. It refers to the ratio of the absorbed heat of the overheated liquid and vaporizing liquid with the same volume, as in Eq. 6:

$$j_a = \frac{c_l \rho_l (T_{W0} - T_S)}{\rho_v h_{fg}} \qquad (6)$$

When the temperature difference between the wall and the liquid increased, the upward force suffered by the gas would be much bigger than the downward force and the bubble would separate itself from the wall and move upwards as in Figure 5. The bubble diameter at the disengaging moment can be got according to liquid-vapor equation 7.

$$P_V - P_l = \sigma\left(\frac{1}{r_2} + \frac{1}{r_1}\right) \qquad (7)$$

At the disengaging moment, there is a relation $r_1 = r_2 = r$. The combination of the relation and Eq. 7 turns out to be radius equation 8.

$$r = \frac{\sigma}{P_V - P_l} \qquad (8)$$

where σ is the surface tension.

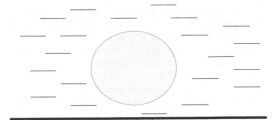

Figure 5.   Figure of a bubble disengagement.

The Fritz formula is often used in bubble disengagement research. For a given contact angle θ, the disengaging radius is shown below:

$$r = 0.0104\theta \sqrt{\frac{\sigma}{g(\rho_l - \rho_v)}} \qquad (9)$$

When the contact angle θ is uneasy to measure, its theoretical expression can be obtained from Eqs. 8 and 9.

$$\theta = \sqrt{\frac{\sigma_g(\rho_l - \rho_v)}{0.0052(P_v - P_l)}} \qquad (10)$$

3   STUDY OF COUPLING CHARACTERISTICS OF DIFFERENT LIFTING PIPE DIAMETERS WITH DOUBLE-STAGE PUMP LIBR-H2O REFRIGERATION SYSTEM

Previous studies focused more on single-stage bubble pump lithium bromide absorption refrigeration and the performance of the bubble pump itself. While researches for two-stage bubble pump in complete absorption refrigeration system were rare. Besides, the two-stage bubble pump mostly used the same diameter for first and second lifting pipes but did not use the different diameters. This chapter focused on exploring the influence of two different-diameter bubble pump lithium bromide absorption refrigeration systems on the cooling performance by experiments.

3.1   3 system introduction

As shown in Figure 6, an experiment table was built. Figures 8 and 9 represent first and second lifting tubes, respectively.

When the system worked, by heating the high pressure generator, the high pressure generator could boil at a lower temperature under the negative pressure. The produced bubbles and steam entered into the first gas-liquid separator 2 through the pump 8 carrying the concentration solution. At the same time, steam and the concentrated solution separated in the first gas-liquid separator 2. Then, the concentrated solution entered into the low-pressure generator 3 after cooling down by pressure reducing valve, in which the solution was called intermediate solution. The separated refrigerant steam entered into the low-pressure generator again through a coiled tube to heat the intermediate solution, making the intermediate solution boiled and pumped again to a higher position at the negative pressure. The vapor carried the highly concentrated solution and entered into the second gas-liquid separator 4. The primary and secondary steam were called first refrigerant vapor and secondary refrigerant vapor, which were all condensed into the condenser 5, becoming refrigerant water. Then, the refrigerant water flowed into the evaporator and absorbed heat to cool the water. The steam generated in the evaporator entered into the absorber and was absorbed by the highly concentrated solution from the secondary gas-liquid separator, producing dilute solution. Dilute solution flowed back into the high-voltage generator with its own potential energy, achieving a refrigerant circulating circuit.

3.2 Effect of heating power on the performance of different-diameter bubble pump absorption refrigeration systems

Two groups of bubble pumps were taken for the experiment, namely a different-diameter group of 9.5–6.5 mm and a same-diameter group of 9.5–9.5 mm. Figures 8 and 9 show the cooling capacity and COP variation of the two groups at heating powers of 1247 W, 1818 W, and 2241 W with 50% concentrated working fluid.

Figure 7 showed that the cooling capacity increased directly with the heating power for the different-diameter groups. Under the same condition,

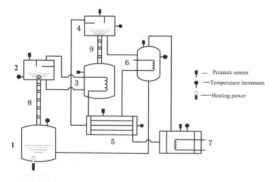

Figure 6. Diagram of double-stage bubble pump LiBr absorption refrigeration system.

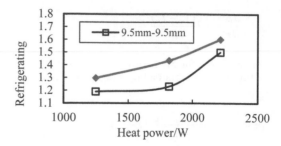

Figure 7. Influence of heat power on cooling capacity under different lifting pipe diameter combination.

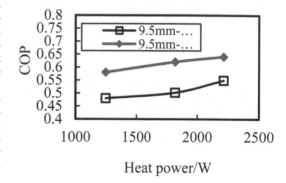

Figure 8. Influence of heat power on COP under different lifting pipe diameters.

the cooling capacity of 9.5–6.5 mm bubble pump is bigger than 9.5–9.5 mm bubble pump, the increment of which was 100–200 W and reached 200 W at 1818 W. This indicated that with the increase in input power, more heat entered into the system during the same period and more primary and secondary cooling capacity would also, which resulted in the evaporation increase. Besides, the reason why different-diameter bubble pump generated greater cooling capacity was that the smaller second-stage diameter bubble pump elevated more rapidly with the same intermediate solution. Thus, the circulation system fasted, achieving better cooling performance.

Figure 8 shows that with the increase in power, COP of the system increases slowly and not obviously. However, the COP of different-diameter groups is 0.1 times greater than the same-diameter group under the same condition. This indicated that power had little effect on COP. While different lifting pipe diameters helped to improve it and created better economy.

3.3 Effect of refrigerant concentration on the performance of different-diameter bubble pump absorption refrigeration system

The initial concentration not only determines the time of the primary pump, temperature, and the crystallization temperature, but also influences the system's cooling effect ultimately. Figures 7 and 8 represent the cooling capacity and COP value variation of 9.5–6.5 mm diameter tubes and 9.5–9.5 mm diameter tubes respectively, with a heating power of 1247 W and the wording fluid concentrations of 50%, 52.6%, and 55%.

Figure 9 shows that, under different diameters, the cooling capacity increases almost linearly with the concentration enhancement and that the cooling capacity of the different-diameter group is higher than the same-diameter one. This means that as the concentration increases, the concentration of the solution into the absorber increases, so does the absorption performance and evaporation, leading to a higher cooling capacity. Besides, the cooling capacity of the different-diameter groups has always been greater than the same one at all of the tested concentrations. The reason could be that the smaller secondary lifting pipe contributed to easily pumping up. In this condition, the intermediate solution inlet valve could open little or the secondary lifting pipe pumped faster, which achieved an increase in the concentration in the absorber, resulting in enhanced cooling capacity.

Figure 10 shows that COP increases with the concentration. As for the different-diameter group,

namely 9.5–6.5 mm, COP increases slowly and linearly. While for the same-diameter group, namely 9.5–9.5 mm, COP increases slowly and then changes at a concentration of 52.6%.

Analyzing the result, it was found that the increasing concentration could improve the absorption performance, resulting in evaporation increase. Therefore, the cooling capacity augmented with the same input heating, achieving a higher COP. Besides, when the concentration was less than 52.6%, the evaporation increase was too little. In addition, a higher concentration means a higher temperature, the loss for which could not be ignored compared with the cooling capacity enhancement.

4 CONCLUSION

In this paper, bubble generation in bubble pump and the effect of different lifting pipe diameters on the cooling performance were studied by experimental methods. Through comparing with the same-diameter group, it was found that the different one performed better. The experimental results are as follows:

1. The bubble was generated from the core of vaporization in the generator and kept changing its shape from oval to flat until disengagement.
2. The cooling capacity of the former was bigger under the same condition, which means that a smaller second-stage diameter pump acts better than the one with the same first-stage diameter. The increment in cooling capacity was 100-200 W and COP increment was 0.1 under this experimental condition.
3. For a given power, the cooling capacity and COP showed an increasing tendency with the concentration.

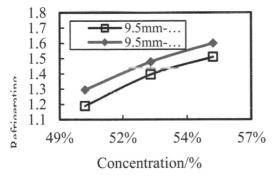

Figure 9. Influence of concentration on cooling capacity under different lifting pipe diameter combinations.

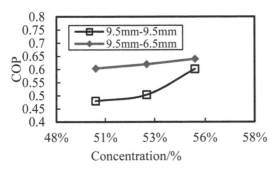

Figure 10. Influence of concentration on COP under different lifting pipe diameter combinations.

ACKNOWLEDGMENTS

This work was financially supported by the National Natural Science Foundation of China (No. 50976015), Liaoning S&T project (No. 2010224002), and the Fundamental Research Funds for the Central Universities (3132014338).

REFERENCES

Dai Yongqing et al. 2001. Review and Prospects of Lithium Bromide Absorption Refrigeration Technology. Refrigeration Technology, 1: 21–24.
Gu Yaxiu et al. 2006. Experimental study on new structure solar pump-free lithium bromide absorption chiller system. SOLAR, 27(7): 473–476.

Gao Hongtao, Wang Xiaolu. 2013.The experimental study on performance of LiBr solution bubble pumps with different concentration of lithium bromide and additive [J]. Journal of engineering thermophysics. 34 (1): 19–21.

Guo Liejin. 2002. Two multi-phase flow dynamics [M]. Xi'an Jiaotong University Press: 520–525.

Ma Guotuan. 2012. Experimental study on lifting characteristic of double-stage lithium bromide bubble pump [D]. Dalian Maritime University.

Meng Long. 2014. Study of coupling characteristics of double-stage bubble pump with lithium bromide refrigeration system [D]. Dalian Maritime University.

Peng Yichuan, Xiao Zeqiang. 1989. Theoretical and experimental study of the phenomenon from the bubble pump [J]. Journal of Northeast University of Technology, 10(2): 112–117.

Que Xiong Cai, Li Hong. 1989. Study on Thermal Siphon Characteristics of Adiabatic Slug Flow in Thermal Siphon Pump of Absorption Refrigerator [J]. Solar, 10(1): 30–38.

Wang Ru-jin et al. 2008. Parameter Design and Determination for Bubble Pump in Single-pressure Einstein Absorption Refrigerator [J]. Fluid Machinery, 36(01): 62–65.

Wang Xiaolu. 2011. Experimental study on Lifting Characteristic of Single Stage Lithium Bromide Bubble Pump [D]. Dalian Maritime University.

Xu Honghao. 2013. Experimental study on lifting characteristic of double-stage lithium bromide bubble pump with different diameters [D]. Dalian Maritime University.

Energy Science and Applied Technology ESAT 2016 – Fang (Ed.)
© 2016 Taylor & Francis Group, London, ISBN 978-1-138-02973-6

Photoelectric properties of sol-gel-derived TiO$_2$ films

G.C. Qi & F.J. Shan
School of Materials Science and Engineering, Liaoning University of Technology, Jinzhou, China

S.L. Zhao & Z.W. Li
Jinzhou Thermal Power Corporation, Jinzhou, China

ABSTRACT: The sol-gel dip-coating method was applied in this study to fabricate TiO$_2$ photoelectrical films. The effects of solvent ratios on condensation property of TiO$_2$ sol were discussed in detail to acquire steady sol. The surface characteristics of the films were analyzed by Scanning Electron Microscopy (SEM), X-Ray Diffraction (XRD), and solar cell IV test system. It shows from the results that approximately 50 μm TiO$_2$ films are fabricated by the sol-gel dip-coating method. After heat treatment at 550°C, strong crystal anatase phase is formed in the TiO$_2$ film, and when used as photo anode, it can improve the photoelectric properties of the dye-sensitized solar cells.

Keywords: Titanium Dioxide (TiO$_2$); sol-gel; Dye-Sensitized Solar Cell (DSSC)

1 INTRODUCTION

Titanium Dioxide (TiO$_2$) thin films are widely used in photoelectric and photocatalysis applications for its wide band gap (3.2 eV) semiconductor properties (Nazeeruddin M.K. et al, 2001). Many ways are applied to make TiO$_2$ films such as magnetron sputtering, thermal spray, screen printing, sol-gel, etc. (Nazeeruddin M.K. et al, 2005). Among them, the sol-gel method is commonly used for its easy preparation on large-area substrates. When used in solar cells, the TiO$_2$ film acted as a photo anode. In order to get excellent performance in the Dye-Sensitized Solar Cell (DSSC) system, the TiO$_2$ film required thicker, strong bonding strength with its substrate and has pure anatase crystal phase for effectively collecting electrons from dye molecules to the anode (Regan O. Grätzel M. 1991) (Wang S.C. et al, 2015).

However, due to the shrinkage and stress, a thicker TiO$_2$ film is easy to have holes, cracks, and coagulation of particles and further brings problems in adhesion properties between the film and its substrate (Yella A. et al, 2011).

Practically, the sol-gel method is widely used to prepare TiO$_2$ nanopowder. When used to prepare TiO$_2$ films, TiO$_2$ sol is very difficult to control the solution properties for its easy condensation and unstable properties. In this study, the sol-gel dip-coating method is applied to prepare TiO$_2$ films and the sol preparation parameters and the characteristics of the acquired films are discussed in detail.

2 EXPERIMENTAL PROCEDURE

2.1 *TiO$_2$ sol preparation*

First, 10 ml tetrabutyl titanate (AR, Kermel, Tianjin) was dissolved in the mixture solution of 20 ml ethanol and 0.5 ml acetylacetone to make solution A. Then, 0.5 ml nitric acid was added in drops to the mixture of 10 ml ethanol and 2.0 ml deionized water to make solution B. To prepare TiO$_2$ sol, solution B was added in drops to solution A, until a desired pH value was achieved. Finally, the mixed solution was magnetic stirred for 3 h and then aged in a sealed beaker kept in a water bath at 40°C for 2 h.

2.2 *Coating preparation*

For coating preparation, a 20 mm × 30 mm Fluorine-Doped Tin Oxide (FTO) glass was cleaned with Deionized (DI) water and anhydrous alcohol. After drying, the substrate was dipped vertically into the sol for 1 min and then withdrawn at a speed of 5 mm/min, dried at 150°C for 10 min. The dipping, drawing, and drying process was repeated four times to get the desired thickness. The films were heat treated at an appropriate temperature for 1 h as TiO$_2$ photo anodes.

2.3 *Solar cell assembly*

The TiO$_2$ film was stained in N719 dye for half an hour and dried at room temperature before cell assembly. Another FTO glass, coated with graphite, was pressed on the anode to make a solar cell.

The electrolyte was added in drops into the cell to fill the middle of the cell surface.

2.4 *Film characterization*

The characteristics of the TiO$_2$ film were analyzed by scanning electron microscopy (SEM, S-3000 N HITACHI) and X-ray diffraction analysis (XRD, Philips, PW1830). The Photovoltaic property test was carried out using an I-V curves tester.

3 RESULTS AND DISCUSSION

3.1 *Effects of solvent ratios on condensation property of TiO$_2$ sol*

3.1.1 *Effect of pH value*
The pH value of TiO$_2$ sol can be modulated by nitric acid in solution B. The effect of pH on gelation time of TiO$_2$ sol is shown in Fig. 1. It shows from the curve that the pH value has an apparent effect on the gelation time. For pH < 1.0, the gelation time is more than 50 h, and if the pH value is in the range of 1.0–2.0, the gelation time decreases quickly, indicating that the gelation time decreases with the increase in pH value. For pH > 2.0, the gelation time becomes steady but takes less than 4 h, which is unsuitable for dip-coating operation. For practical application, according to the film fabrication time, the pH value should be less than 2.0.

The role of pH value in modulating gelation time is through controlling the hydrolysis-condensation reactions in the sol-gel solution. Hydrolysis is a process in which the OC$_4$H$_9$ group in butyl titanate is replaced by the OH group from the solution. This hydrolysis process initially results in the condensation reactions that cross-link Ti (IV) centers together and lead to the precipitation of hydrated TiO$_2$ particles.

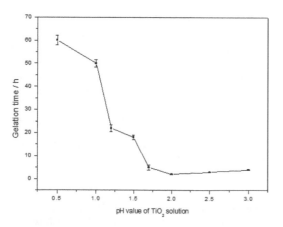

Figure 1. The effect of pH value on the gelation time of TiO$_2$ sol.

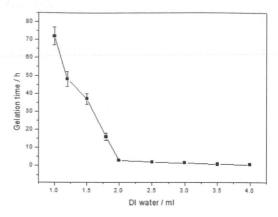

Figure 2. The effect of DI water on the gelation time of TiO2 sol.

As nitric acid ratios in solution B increase, the pH value decreases. H$^+$ from acid can inhibit ionization of water molecules, and decrease the ratio of OH$^-$ group in TiO$_2$ sol. Therefore, the reaction rate can be modulated by controlling the OH$^-$ group in the sol.

3.1.2 *Effect of DI water*
The effect of DI water ratio on the gelation time is shown in Fig. 2.

As shown in Fig. 2, the ratio of DI water has an apparent effect on the gelation time of TiO$_2$ sol. As the amount of DI water increased from 1.0 ml to 2.0 ml, the gelation time decreased from 70 to 3 h, accordingly. When the ratio of DI water further increased, the gelation time becomes steady but less than 3 h, which is too short for the dip-coating operation. Therefore, the ratio of DI water less than 2 ml is necessary for real film fabrication.

3.1.3 *Effect of acetylacetone*
The effect of acetylacetone on the gelation time is shown in Fig. 3. As shown in Fig. 3, acetylacetone can increase the gelation time of TiO$_2$ sol. As the amount of acetylacetone added into the solution increased, the gelation time increased accordingly. As 0.5 ml acetylacetone is added into solution A, after the sol preparation process, the solution state can only be maintained for approximately 3 h. With the increasing ratios of acetylacetone, the gelation time increases. The results of this experiment prove that acetylacetone is an inhibitor to control the condensation reaction of TiO$_2$ in the solution and it improves the stability of the sol.

3.2 *Characteristic analysis of the TiO$_2$ thin films*

3.2.1 *Cross-section morphology analysis*
The typical cross-section morphology of the sol-gel-derived TiO$_2$ thin film is shown in Fig. 4. It is

Figure 3. The effect of acetylacetone on gelation time of TiO$_2$ sol.

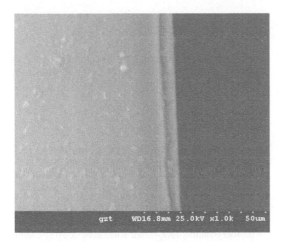

Figure 4. Typical cross-section morphology of the TiO$_2$ film after repeating the sol-gel dip-coating process for four times.

shown in Fig. 4 that the coating surface is continuous and fully dense, and no apparent pores or voids are observable. The cross-section morphology of the TiO$_2$ film shows that after repeating the sol-gel dip-coating process for four times, a film layer which is approximately 50 μm dense is tightly bonded on the FTO substrate.

3.2.2 XRD analysis

The XRD patterns of TiO$_2$ films subjected to the dip-coating process, which was repeated four times, and sintering at 350–550°C are shown in Fig. 5.

There are two crystal phases of TiO$_2$ that can be observed in the patterns, anatase and rutile phase. The films are heat treated at 350 and 400°C, after

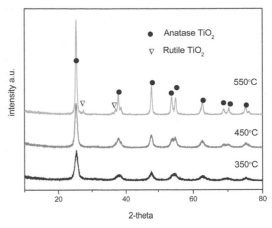

Figure 5. Typical XRD analysis of TiO2 thin films by the sol-gel dip-coating method.

which only the anatase phase can be observed and the strength of the peaks is increased as the temperature increased.

When heated at 550°C, small rutile phase begins to appear. It has been proved that 550°C is the phase-changing temperature, from anatase to rutile phase.

The intensity of TiO$_2$ peaks also show that the sintering temperature has an apparent effect on the crystallization of the sol-gel-derived TiO$_2$ Films.

3.3 Effect of sintering temperature on DSSC photoelectric properties

A series of DSSC was assembled with the sol-gel-derived TiO$_2$ film as photo anodes. The effects of the heat treatment temperature on the open circuit voltage of the DSSC are shown in Fig. 6.

It is shown in Fig. 6 that the heat treatment temperatures have an apparent effect on the photoelectric properties of the sol-gel-derived TiO$_2$ films.

As the heat treatment temperature increases from 350 to 550°C, the open circuit voltages of the DSSC increase accordingly. However, as the heat treatment temperature further increases to 600°C, the open circuit voltage began to decrease.

A similar effect can also be observed in Fig. 7. As the heat treatment temperature increases from 350 to 550°C, the shortcut current of the DSSC increases proportionately and as the heat treatment temperature further increases to 600°C, the current began to decrease too.

Both the open circuit voltage and shortcut current data show that the heat treatment temperature is a key factor to the photoelectric properties of TiO$_2$ films. Owing to the strictly controlled experiment procedure, the effect is from the changed heat treatment temperature.

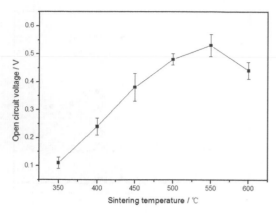

Figure 6. The effect of heat treatment temperature on the open circuit voltage of DSSC.

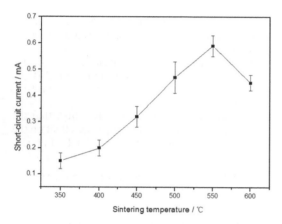

Figure 7. The effect of heat treatment temperature on the shortcut current of DSSC.

As shown in Fig. 5, after heat treatment to a temperature from 350 to 550°C, anatase crystal phase is formed in the TiO_2 film, which is beneficial to the photoelectric properties. As the heat treatment temperature further increases, the intensity of the anatase crystal peaks becomes stronger, which is in accordance with the increase in the open circuit voltage and short current. The results indicate that the crystallization of anatase phase plays an important role in improving DSSC photoelectric

properties. For the real application of TiO_2 films in DSSC, heat treatment at 550°C is suitable.

4 CONCLUSION

TiO_2 films that are approximately 50 μm dense are successfully fabricated by the sol-gel dip-coating method on a conducting (FTO) glass substrate. The gelation time of TiO_2 sol can be modulated by the pH value and other solvents such as deionized water and acetylacetone. The X-ray diffraction results show that in the range of 350–550°C, anatase is the main crystal phase after the sintering process. Only as the temperature increased to 550°C, small crystal peaks from rutile phase appear. When the TiO_2 films are used as photo anodes in dye-sensitized solar cells, the open circuit voltage and short circuit current can be improved by heat treatment and a temperature of 550°C is suitable.

ACKNOWLEDGMENTS

This work was supported by Excellent Talents in Colleges and Universities of Liaoning Province, China (LR2013027).

REFERENCES

Nazeeruddin M.K., Angelis F.D., Fantacci S., et al. 2005, Combined Experimental and DFT-TDDFT Computational Study of Photoelectrochemical Cell Ruthenium Sensitizers. J. Am. Chem. Soc. 127, 16835~16847.

Nazeeruddin M.K., Pechy P., Renouard T., et al. 2001, Engineering of Efficient panch romatic sensitizers for nanocrystalline TiO_2-based solar cells. J. Am. Chem. Soc. 123: 1613~1624.

Regan O., Grätzel M. 1991, A low cost and high efficiency solar cell based on dye-sensitized colloidal TiO_2 films. Nature, 353: 737~740.

Wang S.C., Zhang J., Liu L. 2015. Evaluation of cooling property of high density polyethylene (HDPE) titanium dioxide (TiO_2) composites after accelerated ultraviolet (UV) irradiation. Solar energy materials & solar cells. 143: 120–127.

Yella A., Lee H.W., Tsaok H.N., et al. 2011, Porphyrin-Sensitized Solar Cells with Cobalt(II/III) Based Redox Electrolyte Exceed 12 Percent Efficiency. Science. 334: 629~634.

Energy Science and Applied Technology ESAT 2016 – Fang (Ed.)
© 2016 Taylor & Francis Group, London, ISBN 978-1-138-02973-6

Research on power matching and simulation of a pure electric bus

J.F. Song, W. Xia & H. Chen
Automotive Engineering School, Wuhan University of Technology, Wuhan, China

ABSTRACT: With the intensification of energy and environmental issues, new energy vehicles become the main trend of automotive industry in the future development. Pure electric bus is advantageous to protecting environment and economizing energy as a kind of public transport using clean energy. Based on the vehicle shape and size parameters and design requirements, the required power of electromotor and the theoretical parameters of other units have been analyzed and calculated. The result of simulation showed that the indexes for powerful performance and fuel economy of the pure electric bus can meet the design requirements. The research provided some scientific guidelines and valuable references for the design and development of the pure electric bus.

Keywords: pure electric bus, power system, simulation, parameter matching

1 INTRODUCTION

Nowadays, vehicles have been popularized in human life and they have become an indispensable part of our daily life. People are benefiting from the transportation convenience provided by vehicles, while the negative impacts caused by the automobile industry's high-speed development are also influencing us. The increasing vehicle inventory is the main reason for environmental pollution, fossil energy shortage, traffic accidents, and so on. According to the 2015–2020 China Automotive Manufacturing Market Evaluation and Investment Outlook Report published by chyxx.com, the total car ownership of China has broken through 163 million, which is next only to America. Vehicle fuel consumption accounts for about more than a third of the total oil consumption and the tail emission is one of the biggest air pollution sources. Realizing the urgent situation, the Ministry of Science and Technology indicates that the development of new energy vehicles is the major policy in order to insist the sustainable development of energy and automobile industry. The ministry also drafts detailed goals of new energy vehicle's development.

Researchers conduct real road test or bench test to evaluate the vehicle's economy and power. Given that the variety of power system and the parameters of drive parts, manufacturing vehicles to test, will not only add total costs but also extend the design period. Therefore, computers are used to model each vehicle part and simulate different road conditions in order to find out the preferred plan. Modeling and simulation will simplify the design and shorten the period. Meanwhile, the development costs can be saved and the simulation results can provide reference for the real tests.

2 RESEARCH METHOD OF PARAMETER MATCHING

For satisfying the dynamic property of pure electric bus and increasing the economy, it is important to calculate and choose the parameters. During the initial stage of design, calculation of the part parameters and prediction of the performance of power and economy play an important role in the whole design process. There are three main calculating methods as follows:

1. Automobile theory analysis method
    This method is fundamental. Based on the automobile theory and power balance equations, vehicle parameters can be calculated as the design and power demand. Then, parts and correspondent parameters can be decided. The advantage is that it is easy to understand and operate. However, it is too academic to agree with the actual situation.
2. Vehicle duty cycle method
    This method is based on the real condition test. There are different test conditions and methods according to the vehicle type. Test results are the curves, which represent traveling distance, speed, and acceleration over time. The main vehicle duty cycle methods are USDC, EDC, and JDC. The advantage is that the original parameters can be modified according to the test results, though the process may be repeated until the parameters satisfy the design need.

3. Analysis based on model and simulation

As the technology level develops, the demand of more precise vehicle calculation makes the above two methods incompetent. The computer and simulation system should be combined to model and emulate. The full vehicle model and its performance can be simulated. There are many softwares being used for simulation, such as ADVISOR, EASY5, PAST, V-ELPH, VehProp, Matlab/Simulink, and CRUISE. This paper uses CRUISE to simulate. Using computers and softwares, vehicle's performance can be developed and the research time can be saved a lot.

Comparing the methods above and considering others' research, the method this paper uses can be concluded as:

Based on the dimensional parameters and design requirements of the pure electric bus, calculation of the demand power and the parameters of other main parts, conduction of parameter matching of the power system are performed.

Building model of the bus by CRUISE, setting the calculation task, simulating the power and economy property of the bus, and verifying the matching design are performed.

3 DRIVETRAIN PARAMETER MATCHING

The design of drivetrain parameter matching not only directly decides whether the vehicle power and economy performance can meet the basic requirements, but also provides parameters for the modeling in the environment of CRUISE.

3.1 Vehicle parameter and property demand

This study focuses on pure electric bus. According to the property demand and vehicle parameter, drivetrain parameters are matched. Table 1 shows part of the vehicle parameters.

The matching design has a main impact on vehicle's power and economy property. Table 2 shows

Table 1. Vehicle parameter.

Kerb mass (kg)	Full load mass (kg)	f	Windward area (m²)	C_D	Rolling radius
8700	12000	0.012	6.5	0.65	0.415

Table 2. Design demand.

0–50 km/h acceleration time (s)	Maximum speed (km/h)	Maximum gradability (%)	Driving range (km)
20	70	15	200

the property requirements. The power performance demand consists of three parts: being capable of a 70 km/h top speed, 0–50 km/h in 20 seconds, and the maximum gradability at 15 km/h being 15%. Given that the bus is propelled by electricity, the driving range is the most important sector of economy property.

3.2 Electromotor power matching

Electromotor to pure electric bus is what internal combustion engine is to traditional vehicle. Therefore, the matching method is based on the method used for tradition vehicle matching. First, on the premise satisfying the power demand, the electromotor power can be calculated. Then, there are three evaluation indicators for power performance, and the power provided should meet all the three indicators. Finally, the electromotor power can be decided.

The total motor power demand can be directly calculated according to the vehicle dynamic theory. The basic condition for making a vehicle run is that the propulsion force should overcome all resistance, and the force balance formula can be transferred into a form of power balance:

$$P = P_t \big/ \eta$$
$$= \left(mg\sin\alpha + mgf\cos\alpha + \frac{C_D A}{21.15}v^2 + \delta m\frac{dv}{dt} \right)\frac{v}{3600*\eta}$$

where m is the vehicle mass, f is the coefficient of rolling resistance, v is the velocity, α is the angle of slope, C_D is the coefficient of aerodynamic drag, A is the windward area, δ is the correction coefficient of rotating mass, P_t is the driving wheel power demand, P is the total motor power demand, and η is the transmission system efficiency.

Then, by calculating the three kinds of power demand according to the three power performance indexes, the power can be solved as:

$$P_{max1} = \left(mgf + \frac{C_D A}{21.15}v_{max}^2 \right)\frac{v_{max}}{3600*\eta}$$

$$P_{max2} = \frac{v_m}{3600 t_m \eta}\left(\delta m\frac{v_m^2}{2} + mgf\int_0^{t_m} v_m\left(\frac{t^{0.5}}{t_m^{0.5}}\right)dt \right.$$
$$\left. + \frac{C_D A}{21.15}\int_0^{t_m} v_m^3\left(\frac{t^{1.5}}{t_m^{1.5}}\right)dt \right)$$

$$P_{max3} = \left(mg\sin\alpha_{max} + mgf\cos\alpha_{max} + \frac{C_D A}{21.15}v_i^2 \right)$$
$$\times \frac{v}{3600*\eta}$$

v_{max} is the maximum velocity, δ is the correction coefficient of rotating mass, using $\delta = 1.05$, v_m is

Table 3. Chosen electromotor parameter.

Rated power	Rated rpm	Maximum rpm	Maximum torque	Rated voltage
80 kw	1200	2800	736 N*m	538 V

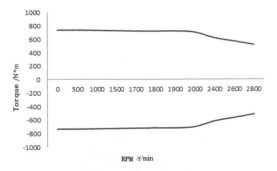

Figure 1. Electromotor characteristic.

the target velocity of the acceleration, t_m is the time needed for acceleration, α_{max} is the maximum angle of the slope, and v_i is the upgrade climbing speed.

P_{maxi} can be calculated by putting the parameters in the equations, and the results are $P_{max1} = 52.3\,kw$, $P_{max2} = 70.7\,kw$, and $P_{max3} = 64.8\,kw$. Given that there are many accessories which need power, the final confirmed motor power should be greater than P_{max2}.

$$P \geq \max\left(P_{\max 1}, P_{\max 2}, P_{\max 3}\right) = 80\ kw$$

Then, we can choose an electromotor from the alternative. The main parameters of the chosen motor are given in Table 3. The characteristic of the electromotor is shown in Figure 1.

3.3 Matching of transmission ratio

Most vehicles have a mounted gearbox and a reducer, which are meant for deceleration and increasing the torque. Drivelines are designed for starting, acceleration, climbing, as well as some conditions, where high output torque is needed. For the purpose of upgrading the whole dynamic performance, a three-speed gearbox is used, and this may offer more choice for picking the electromotor. Meanwhile, the economic performance can be optimized through making the motor work efficiently.

The total drive ration is i_t, $i_t = i_0 i_g$, i_0- is the final drive ratio which is fixed, and i_g- is the gearbox ratio which is based on the gear position. The scope of i_t can be limited by the maximum speed and gradability. Obviously, the top speed occurs

when the ratio is 1. Hence, the final drive ration can be calculated as follows:

$$i_0 \leq 0.377\frac{n_{\max} \cdot r}{v_{\max}}$$

n_{max}–is the maximum speed, r–is the tire rolling radius, and v_{max} is the maximum variation.

In order to make the electromotor work with high efficiency, as the vehicle drives at top speed, the final drive ratio should have a minimum value:

$$i_0 \geq 0.377\frac{n_{ep} \cdot r}{v_{\max}}$$
$$i_{g1} \geq \frac{G(f \cdot \cos\alpha + \sin\alpha) \cdot r}{T_{tq\max} \cdot i_0 \cdot \eta}$$

α is the maximum gradability.

Putting the parameters into the equations, the results are $4.02 \leq i_0 \leq 6.26$. Combining with the driving axle-type selection, i_0 is 4.875. When i_0 is decided, the maximum ratio can be calculated according to the maximum gradability, $i_{g1} \geq 2.41$.

As the gearbox is a three-speed one, the minimum ratio is 1, $i_{g3} = 1$. The median ratio is based on the type of the gearbox. Table 4 shows the detailed driveline ratio.

3.4 Matching of battery parameter

The economic performance index is the driving range when driving on a constant speed of 40 km/h and the battery has discharged 90% energy. The driving range s can be calculated as follows:

$$S = \frac{E_B \times 3600\,\eta_T\,\eta_{mc} \times 0.9\,\eta_q}{F}$$

E_B is the total energy of the battery, η_{mc} is the efficiency of the battery system, and η_q is the efficiency of average discharging.

Then, putting the driving range demand to the equation, the total energy of the battery can be calculated as 107 kw · h. Moreover, taking the electromotor rated voltage into account, the chosen battery parameter is given in Table 5.

Table 4. Driveline ratio.

i_0	i_{g1}	i_{g2}	i_{g3}
4.875	2.8	1.36	1

Table 5. Chosen battery parameter.

Battery capacity (Ah)	Rated voltage (V)
246	560

4 PURE ELECTRIC VEHICLE MODEL ESTABLISHMENT

The pure electric passenger car vehicle model of the traditional power system has the characteristics of dynamic performance and complexity. This also makes the conventional simulation algorithms fail to provide an accurate evaluation of the vehicle's power performance and economy performance. Therefore, we need the help of a more advanced computer simulation software such as CRUISE, for simulation of the power performance and the economic performance of the pure electric bus. The vehicle model built in the CRUISE environment is shown in Fig. 2. According to the structure of the power transmission system and the direction of the power transmission, the battery, the driving motor, the transmission, the main reducer, the moving device, the wheel, and the control unit are included.

After the vehicle model is established, component parameter setting should be done. In the simulation model, according to the vehicle size and design requirements, as well as the selection parameters of the components, the parameters of each component are set up. The specific parameters of the components set are given in Tables 2, 3, and 6, and the motor efficiency, speed, torque, and the corresponding curves of the three three-dimensional views are shown in figs. 3 and 4.

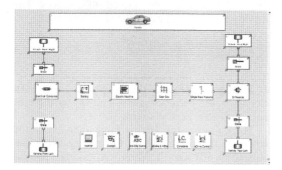

Figure 2. Vehicle model of a pure electric bus.

Table 6. Specific parameters of the battery.

Name	Parameter
Maximum Charge (Ah)	246
Rated Voltage (V)	560
Maximum Voltage (V)	590
Minimum Voltage (V)	490
Initial Charge (%)	90

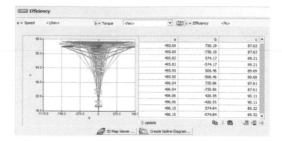

Figure 3. Characteristic curve of motor.

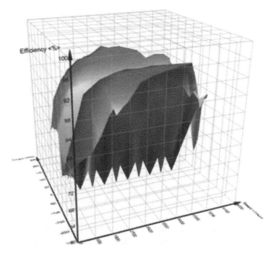

Figure 4. Three-dimensional schematic diagram of the motor.

5 SIMULATION RESULTS ANALYSIS

5.1 *Dynamic performance analysis*

The dynamic performance of the car is one of the decisive factors in the evaluation of vehicle's overall performance. For a pure electric bus as a public transport, the level of its working efficiency and the power performance have a direct relationship.

1. Maximum speed. The results obtained from the simulation calculation of the CRUISE software, including the maximum speed of the vehicle and the speed of the motor at the maximum speed of the three gears, are shown in Fig. 5. The maximum speed of the pure electric bus reached 89 km/h, thereby reaching the design requirements.
2. 0~50 km/h acceleration time. The relationship among acceleration time, speed, distance, and driving time is shown in Fig. 6. From the Figure, we can read out that the 0~50 km/h acceleration time is 13.25 s, meeting the design require-

```
CONSTANT DRIVE - vMax
Theoretical Obtainable Maximum Velocity:   89.00 <km/h>
Real Obtainable Maximum Velocity       :   89.00 <km/h>
Evaluated Gear Step                    :   3
Ratio                                  :   0.500
Single ratio transmission              :   4.875
Maximum velocity in the gears:
    Gear      Velocity       Speed       Measured Speed Ratio
    <->       <km/h>         <1/min>     <->
    1         32.09          2800.00         0.00
    2         66.07          2800.00         0.00
    3         89.00          2773.23         0.00
```

Figure 5.   Simulation results of maximum speed.

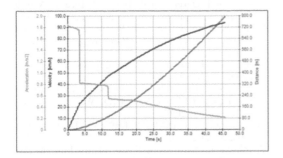

Figure 6.   Simulation results of acceleration time.

ments. For the pure electric bus in the process of starting to accelerate, the speed change is relatively stable. The acceleration decreases with the increase in gear position, and is maximum in first gear, which is also in conformity with the actual gear shift schedule.

3. Maximum gradability of 15 km/h. As shown in Fig. 7, the pure electric bus has the speed of 15 km/h in the first gear, and the maximum can reach the slope of 18.86%, meeting the design requirements. It can also be seen in the figure, that the gradability of second and third gear is significantly lower than that of the first gear, which is in line with the car moving rule.

In summary, the results of the calculation task and the design requirements are compared, the results obtained are given in table 7. From the data in Table 7, we can see that the indexes for powerful performance and fuel economy of the pure electric bus can meet the design requirements.

5.2   *Economic performance analysis*

The economic performance evaluation indexes of pure electric bus are different from those the traditional internal combustion engine, and the economic performance of pure electric bus is evaluated by the driving range of the vehicle when the speed is kept constant at a speed of 40 km/h.

The relationship among the SOC of the battery, travel distance, and the travel time is shown in Fig. 8. When the vehicle starts running, the battery

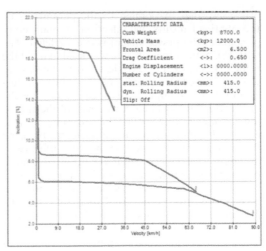

Figure 7.   Simulation results of maximum climbing slope.

Table 7.   Comparison of calculation results and design requirements.

	Maximum speed (km/h)	Acceleration time (s)	Maximum gradability (%)
Design requirements	70	20	15
Calculation results	89	13.25	18.86

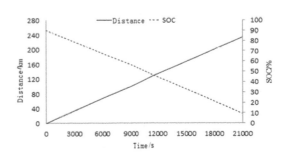

Figure 8.   Curve of state of charge and travel distance.

SOC is 90%, which is 10% at the end of the road, and the travel distance is 231.29 km, which meets the design requirements of 200 km.

6   CONCLUSIONS

According to the relevant requirements of the pure electric vehicle parameters, with the constraint of the vehicle's powerful performance and fuel economy, the whole power system parameter matching was completed. According to the parameters of the obtained results, the selection of main components in the power system was completed. The CRUISE simulation software was used to build the vehicle model of pure electric vehicle. The simulation of the vehicle powerful performance and fuel economy was completed. The calculation results are analyzed to verify the correctness of the matching parameters, and verified the feasibility of modeling and simulation of pure electric bus by using the Cruise software. Based on this, the influence of different transmission ratios on the performance of the whole vehicle is calculated and studied.

In the final optimization scheme, the 0~50 km/h acceleration time is shortened to 13.02 s, and the maximum climb slope of 15 km/h is increased to 19.91%. The above research content provides a basis for the simulation of pure electric bus and the design of power system by using the CRUISE software.

REFERENCES

Chen Qingquan, Sun Fengchun. Hybrid electric vehicle based [M]. Beijing Institute of Technology Press, 2001.

Li Jun. Research on Matching of Powertrain for Coach Based on Cruise [D]. Jilin University, 2013.

Liu Jiang. Optimization and Matching of Bus Power Transmission System [D]. Tongji University, 2007

Ma Youliang, Yan Yunbing. Electric Automobile [M]. China Machine Press, 2012.

Manohar, N. Technologies and Approaches to Reducing the Fuel Consumption of Medium- and Heavy-Duty Vehicles [J]. Washington D.C.: The national academies Press, 2010.

MyUng Seok Lyu. Optimization on vehicle fuel consumption in a highway bus using vehicle Simulation [J]. Automotive Technology, 2006, 7(7):841–846.

Song Yonghua, Yang Yuexi, Hu Zechun. Current Situation and Development Trend of Electric Vehicle Battery [J]. Power System Technology, 2011 (4): 1–7.

Wang Wei. Simulation and Parameter Optimization Design of Automobile Powertrain [D]. Hunan University, 2006.

Wang, L.T. Power Train Matching for Better Fuel Economy [J]. SAE790045.

Yang Chao. The Study on Dynamical Modeling and Simulating of Electric Vehicle [D]. Wuhan University of Technology,2007.

Yu Jin. New energy vehicles [J]. Traffic and Transportation 2008, 24 (2): 28–30.

Yu Zhisheng. Automobile theory (Third Edition) [M]. China Machine Press, 2005.

Zub, R.W., R.G Colello. Effect of Vehicle Design on Top Speed Performance and Fuel Economy [J]. SAE 800215.

3   *Materials science and chemical engineering*

Energy Science and Applied Technology ESAT 2016 – Fang (Ed.)
© *2016 Taylor & Francis Group, London, ISBN 978-1-138-02973-6*

Microstructure and high-damping property of sintered Mn-Cu alloy

F.S. Lu
Functional Materials Research Institute, Central Iron and Steel Research Institute, Beijing, China
Institute for Advanced Materials and Technology, University of Science and Technology Beijing, Beijing, China

B. Wu, J.F. Zhang & D.L. Zhao
Functional Materials Research Institute, Central Iron and Steel Research Institute, Beijing, China

P. Li
Institute for Advanced Materials and Technology, University of Science and Technology Beijing, Beijing, China

ABSTRACT: Mn-Cu alloys exhibit high-damping ability after suitable heat treatment. In order to reduce the density and increase the damping and sound absorption abilities, porous Mn-Cu alloys were prepared by powder metallurgy techniques. The contents of α-Mn phase decrease with the increasing solution temperature after hydrogen sintering at 1233 K for 1.5 h. The diffraction peaks of the fcc-Mn-Cu are split after solid solution treatment at 1173 K for 1.5h and aging treatment at 723 K for 4 h, which implies the occurrence of spinodal decomposition. The porous Mn-Cu alloys possess a density of 5.8 g/cm³ (nominal porosity: 22%), a bending strength of 300 MPa, and a high level of damping ability (tan δ from 0.08 to 0.11). The high damping capacity may be due to the combination of the material itself and the porous structure. The Mn-Cu alloy with porous structure and high damping ability could find wide application in the field of vibration and noise reduction.

Keywords: sinter; Mn-Cu; damping; porous; alloy

1 INTRODUCTION

Mn-Cu alloys are well known as twin-type high damping alloys and have been attracting much interest because of their wide application in industries such as machinery, instruments, and electronics. It has been found in the recent ten years that the doping of Ni and Fe can further improve the properties of Mn-Cu alloys (Zhong, Y. et al. 2007). A representative Mn-Cu-based alloy of high damping capacity is a directionally solidified Mn73-Cu20-Ni5-Fe2 alloy with the highest logarithmic decrement (0.72) (Fukuhara, M. et al. 2006). In recent years, powder metallurgy fabricated porous materials received much attention owing to their advantages such as the energy absorption of impaction, acoustic absorption, low density, low thermal conductivity, high permeability, etc. (Davies, G.J. et al, 1983). The aim of this work is to develop Mn-Cu alloys with low density and high damping capacity by powder metallurgy technology.

2 EXPERIMENTAL

Porous Mn-Cu alloys were prepared with powders of manganese (99.7%), copper (99.7%), nickel (99.5%), and iron (99%). The powders in weight fraction of 73.15% Mn, 20.04% Cu, 4.84% Ni, and 1.97% Fe, were ball-milled in vacuum for 0.5 h and pressurized into green compact with the size of $30 \times 30 \times 100 \, mm^3$. The billets were sintered at 1233 K for 1.5 h in hydrogen atmosphere at 1 atm and then were cut into suitable specimens by wire-electrode cutting. The specimens were solid solution treated at 1123 or 1173 K for 1.5 h and were rapid cooled in a gas quenching furnace, following aging treatment at 723 K for 4 h with furnace cooling. In order to avoid oxidation, the above treatments were carried out in vacuum. The density of sintered porous alloys was determined by mass/volume measurement. The damping capacities of the samples were examined using a Dynamic Mechanical Analyzer (DMA) at frequencies ranging from 0.1 to 100 Hz and the strain amplitude of 5×10^{-5}. The microstructures of the Mn-Cu alloys were investigated by X-Ray Diffraction (XRD) and Optical Microscopy (OM).

3 RESULTS AND DISCUSSION

The chemical compositions of the sintered porous Mn-Cu alloys are listed in table 1. The actual contents of Cu, Ni, and Fe are slightly higher than the nominal compositions due to burning loss of Mn. By comparing the microstructure of investigated alloys between Vacuum Melting (VM) and Powder Metallurgy (PM), it can be seen that the grain size of Mn-Cu alloys by PM is more uniform than that of VM Mn-Cu alloys. The average grain size of PM alloys is approximately 10 μm. Microstructure analysis suggests that some boundaries generated in the grains of PM Mn-Cu alloys during sintering and the formation of twins can be observed. In general, twin boundaries play a key role in the damping capacity for Mn-Cu-based damping materials. When the alloys suffer vibration, the energy dissipates by visco-elasticity effect with twin boundary motion. This could be a predominate reason that porous Mn-Cu alloys possess high damping capacity (Wang, L.T. et al, 1988) (Wang, J.F. et al. 2009).

Table 2 represents the mechanical and physical properties of the porous Mn-Cu alloy, including bending strength, porosity, density, and hardness. In general, porosity could enhance the damping property of Mn-Cu alloys. However, the presence of pores could deteriorate the mechanical properties of the alloys. For example, the bending strength of the porous Mn-Cu alloy is 300 MPa. The existence of pores in the alloy reduces the cross-sectional area, decreasing the critical fracture stress. On the other hand, the pores act as stress release points. When the cracks propagate running into the pores, the pores could make the split end blunt, improving the fracture stress of the alloys. Note that some pores with sharp shape could act as the birthplace for crack, resulting in a lower strength level. Owing to the porosity, the density of Mn-Cu alloys prepared by PM is lower than that of VM Mn-Cu alloys. The porosity of the PM Mn-Cu alloy of the present study is approximately

Table 1. Chemical compositions (mass%).

Elements	Cu	Ni	Fe	Mn
Nominal	20.04	4.84	1.97	balance
Actual	21.94	4.84	2.07	balance

Table 2. Mechanical and physical properties of the PM alloy.

Bending strength (MPa)	Porosity (%)	Density, ρ (g/cm³)	Hardness [HRB] PM	solid solution and aging treatment PM	solid solution and aging treatment VM
300 ± 10	20 ± 3	5.8 ± 0.1	45	59	72

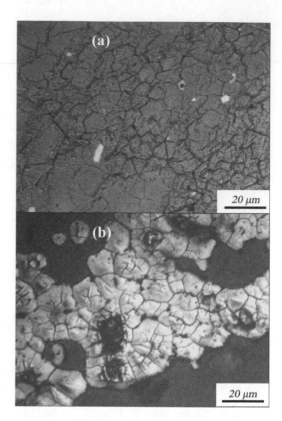

Figure 1. Microstructures of Mn-Cu alloys: (a) VM and (b) PM.

22%. By contrast, the density of the sintered alloy is 5.8 g/cm³ due to the porosity. After solid solution and aging treatment, the Mn-Cu alloy by VM exhibits higher hardness than the Mn-Cu alloy by PM. This is related to the fact that the Mn-Cu alloys prepared by melting have low porosity and high density. The sintered alloys with a porosity of 22% have a lower hardness of approximately 45 HRB after solid solution treatment. Moreover, after aging treatment, the hardness of the sintered alloy increases to 59 HRB seemingly due to the growing of the Mn-rich phase.

Fig. 2 shows the X-ray diffraction patterns of the sintered PM porous Mn-Cu alloy quenched from (a) 1123 K and (b) 1173 K. It can be seen that there are two types of diffraction peaks. In addition to fcc-Mn-Cu phase, there is a large number of bcc-α-Mn phase in Fig. 2(a). Referring to the Mn-Cu binary Phase diagram (Zhang, Z.F. et al, 1991), the residual bcc phase present in the quenched alloy can be attributed to a lower quenching temperature. When the solution temperature was increased to 1173 K, the diffraction peak intensity of α-Mn phase obviously decreases, as shown in Fig. 2 (b).

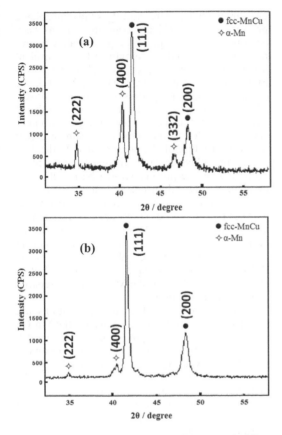

Figure 2. XRD of porous Mn-Cu alloys quenched from (a) 1123 K and (b) 1173 K.

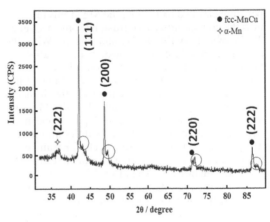

Figure 3. XRD of the Mn-Cu alloy. Alloy state: 1173 K × 1.5h water-cooled, 723 K × 4h aging treatment.

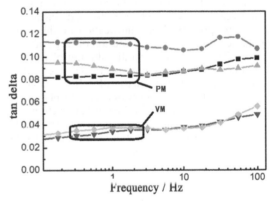

Figure 4. Damping capacity of Mn-Cu alloys by VM and PM.

Figure 3 shows the X-ray diffraction pattern of the Mn-Cu alloy after quenching followed by aging treatment. According to the figure, the main component of the alloy is the fcc-Mn-rich phase. Note that a weak peak can be observed on the right of the main peaks, suggesting that the γ-Mn-Cu phase decomposes into a Mn-rich region and a Cu-rich region with the same crystal structures during quenching followed by aging treatment. Owing to higher Mn content than Cu content in the alloy and larger lattice constant of Mn than that of Cu, the stronger peak of the fcc structure represents Mn-rich region, and weak diffraction peaks for Cu-rich region. Therefore, the spinodal decomposition could result in a large number of Mn-rich regions, which causes microscopic inhomogeneity. During the cooling process after aging treatment, the antiferromagnetic transition and fcc-fct martensitic transformation occur in these Mn-rich microdomains (Deng, H.M. et al, 2002) (Lu, W.L. et al, 2002).

In order to characterize the damping properties of the PM Mn-Cu-Ni-Fe alloys, DMA was applied to examine their damping capacity. It was found that the damping capacity is almost kept constant as the frequency increases from 0.1 to 100 Hz. By contrast, the damping capacity of the VM alloy increases with the increasing frequency in the same test conditions, as shown in Fig. 4. PM Mn-Cu alloys exhibit an excellent damping capacity at different frequencies, with an average damping capacity beyond 0.08, and the highest value of 0.12.

The alloys of the present studies were prepared by sintering reaction of pure Mn, Cu, Fe, and Ni powders, in which Mn and Cu occupied a predominant fraction in weight. Therefore, the sintering reaction process could be analyzed on the basis of Mn-Cu binary alloy phase diagram (Golovin, I.S. et al. 2004). Referring to the diagram, a continuous Mn-Cu solid solution with fcc crystal structure exists below 1144 K. Above 1144 K, solidus and liquidus lines exist at Cu-rich and Mn-rich sides. When the pure Mn and Cu powders were mixed and the temperature elevated to 1223 K, the Mn atoms resolved into the copper lattice by diffusion (Song, Y.Q. et al, 2001) owing to the Kirkendall effects. With the increasing Mn in copper powders, the solidus and liquidus lines in Cu-rich sides could be crossed in the phase diagram. In the case, a few volume of liquid phase could be formed, enhancing the diffusion and sintering processes. Finally, when the Mn content in the liquid phase reach the liquidus line in the Mn-rich side, the liquid phase solidifies around the Mn powders, and the sintering necks form and grow with residual pure Mn cores. This can be certified by X-ray diffractions shown in Fig. 2.

As for the high damping capacity of the sintering alloy as shown in Fig. 4, the porous structure, the intrinsic dislocations as well as the sub-structured twinning and grain boundaries must play key roles. For porous materials, the elastic modulus of the pores becomes close to zero. The internal friction results from the contributions from pores and matrices. The liquid phase is formed during the sintering process. The solidification of the liquid phase could lead to the formation of large thermal stress. Thus, non-uniform distribution of pores in the micro-scale could result in stress concentration areas or thermal cracks in the alloy. It is reasonable to believe that these stress fields could play a key role in the mechanism for the internal friction.

The energy absorption capacity refers to the absorption of energy per unit volume, denoted by C (Gent, A.N. et al, 1966).

$$C = \int_0^\varepsilon \sigma \cdot d\varepsilon \tag{1}$$

The second parameter for the energy absorption is the energy absorption rate expressed by E (Miltz, J, et al, 1981).

$$E = \int_0^\varepsilon \sigma \cdot d\varepsilon / (\sigma\varepsilon) \tag{2}$$

Σ is the yield stress of the material and ε is the strain. When plastic materials begin yielding, pores in the materials could collapse. According to the Gibson-Ashby theory, the theoretical formula for the equivalent yield stress of the porous material can be expressed as follows:

$$\sigma_{p1}^* = \sigma_{ys} \left[C_2 \left(\phi \frac{\rho^*}{\rho_s} \right)^{3/2} + C_2'(1-\phi)\frac{\rho^*}{\rho_s} + \frac{P_0 - P_{at}}{\sigma_{ys}} \right] \tag{3}$$

Σ_{ys} indicates the yield stress of the matrix material and ϕ is the void fraction of pore edge. ρ^* is the equivalent density of porous materials, and it is inversely proportional to the porosity. ρs indicates the density of the base material. P0 is the fluid pressure including pores and P_{at} is the atmospheric pressure. C1 and C2 are constants. It has been demonstrated that σpl of the porous alloy is proportional to the matrix strength and the equivalent density, and inversely proportional to the porosity of the material. From equations 2 and 3, the energy absorption properties of the porous Mn-Cu alloy are closely related to the yield strength (σ) and the strain (ε). As is known, the yield strength is closely related to the density, the pore size, and the pore structure of the porous alloy. As the density of the alloy increases, the energy absorption properties of the porous Mn-Cu-Ni-Fe alloy can be significantly enhanced. Actually, the average pore size is an important factor to influence the performance of porous alloys, such as the strength and damping capacity. In general, the smaller the pore size, the higher the absorption energy. Owing to high porosity of Mn-Cu alloys prepared by the PM method, the damping properties of PM alloy are higher than those of the Mn-Cu alloy prepared by the melting method.

It should be noted that the dislocations are another important damping source for the Mn-Cu alloys. Dislocation damping, i.e. pinning model, was known as the Granato-Lücke model (Granato, A.V, 1985). When external vibration takes place, the movement of the dislocations will occur in the material, making it unpin with a type of avalanche at the point where the vacancy and the solute atoms existed. In strong pinning points such as dislocation mesh nodes and the dislocation loops around precipitates, the movements of dislocation cause stress relaxation and mechanical energy consumption.

It has been recognized that the high damping property of Mn-Cu alloys arise from the paramagnetic-antiferromagnetic transition and the movement of twin boundaries in martensites. The α-Mn precipitates formed in the twin and/or grain boundaries could hinder the movement of interface and twins in martensite, decreasing the damping properties of the Mn-Cu alloys. However, with the increase in these α-Mn precipitates, new Mn-rich regions form, which could produce new martensite and enhance the damping properties of the Mn-Cu alloys. Semi-coherency between the α-Mn precipitates and the surrounding matrix could cause inelastic strain, and the stress changes in the sample will change the local balance around the pellet, causing the pellet to be dissolved in some regions' local produce. Accompanying the growth of α-Mn precipitates and the formation of new Mn-rich regions, new martensites could form in the Mn-rich regions. This twin-structured martensite could improve the damping properties of Mn-Cu alloys. This suggests that the solid solution and aging treatment exerts complex effect on the damping properties of the Mn-Cu alloys.

4 CONCLUSION

The Mn-Cu-Ni-Fe alloy sintered at 1233 K for 1.5 h in hydrogen atmosphere at ambient pressure possesses a porosity of approximately 22% and a density of 5.8 g/cm³ due to pores by the powder sintering technology. There are intensive diffraction peaks of fcc-Mn-Cu phase and some weak diffraction peaks of α-Mn phase detected by XRD, when quenched from 1173 K. Aging treatment is a key factor for damping capacity of the alloys which could cause the germination of spinodal decomposition. After aging at 723 K for 4 h, the average value (tan δ) of the damping capacity of PM Mn-Cu alloy is over 0.08, obviously higher than that of vacuum melting Mn-Cu alloy. Meanwhile, the PM Mn-Cu alloy has excellent mechanical properties with a bending strength of 300 MPa.

REFERENCES

Davies, G.J. & Zhen, S. 1983. Metallic foams: Their production, properties and applications, Mater. Sci., 18: 1899–1911.

Deng, H.M. & Zhong, Z.Y. 2002. Antiferromagnetic Distortion and Formation of High Damping Twins in γ Mn-Based Alloys, J. Shanghai Jiaotong Univ., 36: 28–32.

Fukuhara, M. et al. 2006. High-damping properties of Mn-Cu sintered alloys, Mater. Sci. Eng. A, 442: 439–443.

Gent, A.N. & Rusch K.C. 1966. Permeability of open-cell foamed materials. Journal Cellular Plastics, 2(1): 46–51.

Granato, A.V, 1985. Dislocation and Properties of Real Materials, London: The Institute of Metals, 266.

Golovin, I.S. et al. 2004. Damping in some cellular metalic materials due to microplasticity, Mater. Sci. Eng. A, 370: 531–536.

Lu, W.L. & Jiang, Y.F. 2002. Effects of heat treatment on the damping properties of the extruded Mn-Cu damping alloy, Spec. Cast. Nonferr. Alloys, 6: 10–11.

Miltz, J, & Gruenbaum, G. 1981. Evaluation of cushioning properties of plastic foams from compressive measurements, Polymer Engineering and Science, 21(15): 1010–1013.

Song, Y.Q. & Li, S.C. 2001. The sintering performances of copper and manganese powders, Journal of the university of petroleum, 25(5): 76–78.

Wang, L.T. & Ge, T.S. 1988. Internal friction in martensite and martensitic transformation in Mn-Cu alloys, Acta Metall. Sin., 24: A147–154.

Wang, J.F. et al. 2009. New development and prospect of research on metallic damping materials, Mater. Rev., 23: 15–19.

Zhang, Z.F. & Zhang, Z.D. 1991. Internal friction of martensitic transformation and precipitation in Mn-Cu alloys, Acta Metall. Sin., 27: 7–10.

Zhong, Y. et al. 2007. Dislocation structure evolution and characterization in the compression deformed Mn-Cu alloy, Acta Materialia, 55: 2747–2756.

Energy Science and Applied Technology ESAT 2016 – Fang (Ed.)
© *2016 Taylor & Francis Group, London, ISBN 978-1-138-02973-6*

Methanol aromatization reaction on ZSM-5 zeolites at different conditions

J.Y. Xun, K.M. Ji, J.M. Deng, H.P. Deng, K. Zhang & P. Liu
State Key Laboratory of Coal Conversion, Chinese Academy of Sciences, Institute of Coal Chemistry, Taiyuan, China

ABSTRACT: HZSM-5 zeolite was modified by the impregnation method to prepare Zn/HZSM-5 catalyst in this research. The performances of Zn/HZSM-5 and HZSM-5 catalysts on methanol aromatization reaction were tested in a fixed bed reactor, and the reaction temperature, space velocity, and reaction time were taken into consideration. The result indicates that the yield of oil and selectivity of Benzene, Toluene, and Xylene (BTX) on Zn/HZSM-5 are highest at 430°C and weight hourly space velocity of 1 h^{-1}. With the increase in reaction time, the methanol conversion remained to be 100%, the selectivity of aromatics slightly decreased by 6%, and the selectivity change of BTX is less than 2.5%.

Keywords: aromatization reaction; HZSM-5; methanol conversion rate.

1 INTRODUCTION

Benzene, Toluene, and Xylene (BTX), known as important aromatic species, are widely used in the field of energy, traffic, material, daily, and farm chemical. The industrialized process to produce BTX is based on petrochemical route and aromatic combination unit. Developing the technology of Methanol to Aromatics (MTA), which uses coal as a raw material, can effectively reduce the dependence on crude oil resource and has good economic value, is becoming highly attractive in the research of coal chemical industry [Bi, Y. 2014].

Bi [Kitagawa, H, 1986] and coworkers researched the MTA reaction on different Zn salt-modified HZSM-5 zeolite, they found that the distribution of acid sites and Zn species in the HZSM-5 catalyst modified with zinc sulfate effectively improved the MTA performance, and the total aromatic selectivity can reach 77.9 wt%. Niu [Namuangruk, S. 2011] and coworkers prepared Zn-containing HZSM-5 by four methods including impregnation, ion exchange, physical mixing, and direct synthesis to study the influence of preparation method on the catalytic performance of methanol aromatization. They found that Lewis acid sites of zinc species (ZnOH$^+$) are formed by introducing zinc into HZSM-5, and there is a linear correlation between the amount of surface ZnOH$^+$ species and the selectivity to aromatics for MTA. The ZnOH$^+$ species act as the active sites for the dehydrogenation of light hydrocarbons, and existence of Zn cations can reduce the Brønsted acidity, which is helpful to suppress the formation of alkanes. The Zn/HZSM-5 prepared by ion exchange has the highest fraction of surface ZnOH$^+$ species and gives the highest selectivity to aromatics.

In this article, we used the impregnation method to modify HZSM-5 zeolite with Zn species and studied the MTA performance between Zn/HZSM-5 and HZSM-5, and also discussed the influence of reaction temperature, space velocity, and reaction time, which provide foundation to the catalyst modification and process conditions optimization.

2 EXPERIMENTAL

An HZSM-5 zeolite catalyst (State Key Laboratory of Coal Conversion, Institute of Coal Chemistry, Chinese Academy of Sciences, GY1) was modified with Zn(NO$_3$)$_2$ (AR, Tianjin Berchenfangzheng Chemical Reagent). The HZSM-5 catalysts were crushed to 20–40 mesh and then added into Zn(NO$_3$)$_2$ solutions using an isometric impregnation method for 16 h, after the catalyst was dried at 120°C for 3 h and calcined at 550°C for 5 h; the samples with 3 wt% Zn loadings obtained were denoted by Zn/HZSM-5. The HZSM-5 catalyst was also treated by wetting out, drying, and calcining.

The performance of the two catalysts was investigated using a laboratory-scale fixed-bed reactor. In each run, the catalyst (3 g) was loaded into the middle part of the reactor with an inner diameter of 10 mm, and then heated at 10°C/min rate to the

Table 1. Phase distribution of products.

Yield	Hydrocarbon in gas phase/wt.%			Oil phase/ wt.%			Aqueous phase/ wt.%		
	400°C	430°C	460°C	400°C	430°C	460°C	400°C	430°C	460°C
HZSM-5	22.1	17.3	21.5	15.7	22.6	16.2	52.8	53.5	50.7
Zn/HZSM-5	19.4	16.9	17.4	19.1	25.4	21.3	51.7	51.5	53.6

* Reaction conditions: 400 to 460°C, 1 h⁻¹.

reaction temperature. The methanol feedstock was fed into the reactor by a liquid pump (Dalian Jiangshen LC-05P), the product was cooled by a congealer which was filled with 2°C ethylene glycol solution. The quantity of tail gas was recorded by the wet-type gas flow meter and the reaction product was analyzed in a gas chromatograph. We used a separating funnel to separate the liquid products into oil product and aqueous product, and the oil product was analyzed by gas chromatography (Beifen SP-2100 A, HP-INNOWAX column, FID detector) to detect the BTX, aromatic components, and other oil-phase products. The aqueous product distribution was obtained by gas chromatography (Beifen SP-2100 A, Porapak-Q column, TCD detector) to detect methanol and water components. The gaseous products were analyzed by gas chromatography (Beifen SP-2100 A, Al$_2$O$_3$ column, FID detector) to detect hydrogen and gas-phase hydrocarbon.

3 RESULTS AND DISCUSSION

3.1 *Effects of reaction temperature*

Catalytic MTA performance of HZSM-5 and Zn/HZSM-5 was investigated and we also analyzed the product distribution of different phases, and the results are shown in Table 1. From Table 1, we can see that the gas-phase hydrocarbon products are less, while the oil-phase products are more at a reaction temperature of 430°C. The aqueous product yield is close to the theoretical yield at all reaction temperatures. Compared with product composition at 430°C on modified HZSM-5 catalyst, the yields of gas hydrocarbon and oil-phase products on untreated HZSM-5 catalyst are 17.3% and 22.6%, respectively. At a reaction temperature of 430°C, Zn/HZSM-5 and HZSM-5 catalysts showed the highest oil-phase component mass composition, and 430°C is the optimum temperature for the MTA reaction.

Currently, the most widely accepted methanol aromatization reaction mechanism holds the view that methanol first occurred methanol conversion reaction (1), the acid catalytic dehydration, and then polymerization to form light olefins. Later, the cyclization of olefins (2) occurred and cycloolefin

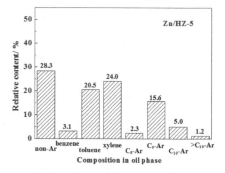

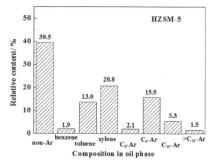

Figure 1. Product distribution in oil phase *Reaction conditions: 430°C, 1 or 2 h⁻¹.

was formed. After that, part of cycloolefin occurred in the dehydrogenation reaction (3) to form aromatics and H$_2$, and the remaining parts of cycloolefin and olefin occurred in the hydrogen transfer reaction (4) and formed aromatic and alkane byproducts. The whole process was acid catalytic reaction. The total acid content of Zn-modified zeolite catalyst increased and is helpful to the aromatization reaction, while the yield of oil-phase products increased.

3.2 *Effects of supported metal*

The oil-phase products contain many different hydrocarbon compositions in both HZSM-5 and Zn/HZSM-5 catalysts, and the analysis results are shown in Fig. 1. From Fig. 1, we can see that the content of non-aromatic hydrocarbon is 39.5% on HZSM-5 catalyst, the benzene, toluene, and xylene content increased, in turn, reached 1.9%, 13.0%, and 20.8%, respectively. The content of C_9, C_{10}, and heavyweight aromatic hydrocarbon decreased, in turn, and reached 15.9%, 5.3%, and 1.5%, respectively. Compared with the non-modified HZSM-5 catalyst, the non-aromatic hydrocarbon content on Zn/HZSM-5 catalyst is significantly lower than that on HZSM-5 catalyst, i.e., only 28.3%. The benzene, toluene, and xylene content is higher, which reached 3.1%, 20.5%, and 24.0% respectively. The content of C_9 and heavier aromatic hydrocarbon is slightly lower than the non-modified catalyst.

3.3 Effects of reaction space velocity

The process of aromatization on HZSM-5 and Zn/HZSM-5 catalysts at a space velocity 1 h^{-1} and 2 h^{-1} was tested, the yield of oil phase, total aromatics, and BTX is shown in Fig. 2. It is obvious in Fig. 2 that the yield of oil phase, total aromatics, and BTX

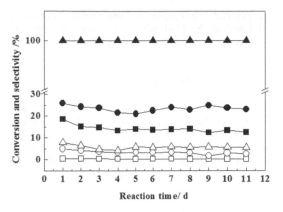

Figure 3. Methanol conversion and product selectivity of Zn/HZ-5 catalyst ▲ is conversion of methanol; ●, ■, □, ○, and △ are selectivities of oil, aromatics, benzene, toluene, and xylene. * Reaction conditions: 430°C, 1.5 h^{-1}.

decreased when the space velocity increased; levels of yield dropped by an average of 4%, 7%, and 4%, respectively. When the space velocity increased, the yield of oil-phase products decreased by 4% on average, similar to HZSM-5. The yield of total aromatics and BTX decreased by 5% and 2%, and the degree of reduction is less than that of HZSM-5. Owing to the Zn-modified catalyst, more acid sites and more active centers exist; it exhibits excellent selectivity at a higher space velocity and a longer reaction time.

3.4 Effects of reaction time

A long evaluation of Zn/HZSM-5 catalyst for 11 days has been made to record the methanol conversion rate and selectivity of each component, and the results are shown in Fig. 3. In full reaction time, the methanol conversion rate remained at 100%, while the selectivity of oil-phase product, aromatics, and BTX appears to decline. The selectivity of aromatics decreased by 6%, followed by BTX whose decline is only less than 2.5%.

4 CONCLUSIONS

The Zn-modified HZSM-5 shows a higher yield of oil-phase product, aromatics, and BTX than the untreated HZSM-5. We investigate the reaction condition of this process, and the results indicated that the yield of oil-phase products is highest at 430°C; the yield of oil-phase products, total aromatics, and BTX decreases when the space velocity increases; Zn-modified catalyst has good stability and selectivity, and benzene, toluene, and xylene contents reached 3.1%, 20.5%, and 24.0%,

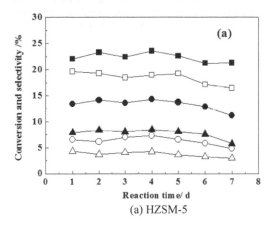

(a) HZSM-5

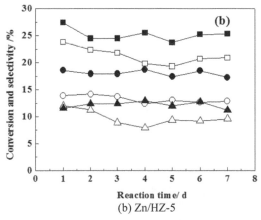

(b) Zn/HZ-5

Figure 2. Yield of oil, aromatics, and BTX. ■, ● and ▲ are yield of oil, aromatics, and BTX at 1 h^{-1}; □, ○, and △ are yield of oil, aromatics, and BTX at 2 h^{-1}. * Reaction conditions: 430°C, 1 or 2 h^{-1}.

respectively; the methanol conversion rate remained at 100% in the 11-day-long reaction time; and the selectivity of aromatics changed slightly.

ACKNOWLEDGMENT

This work was supported by Key Technology and Demonstration on Low Rank Coal Clean, Efficient and Cascade Application Project, Synthesis Technology of Coal-based Bulk Chemical and Fuel, and Research on Key Technology of Methanol to Aromatics (MTA) (XDA 07070800).

REFERENCES

Bi, Y. 2014. Methanol aromatization over HZSM-5 catalysts modified with different zinc salts. Chinese Journal of Catalysis 35(10): p. 1740–1751.

Conte, M. 2012. Modified zeolite ZSM-5 for the methanol to aromatics reaction. Catalysis Science & Technology 2(1): p. 105–112.

Choi, M. 2009. Stable single-unit-cell nanosheets of zeolite MFI as active and long-lived catalysts. Nature 461(7261): p. 246–249.

Groen, C. 2004. Optimal Aluminum-Assisted Mesoporosity Development in MFI Zeolites by Desilication. The Journal of Physical Chemistry B 108(35): p. 13062–13065.

Kitagawa, H, Y. Sendoda, and Y. Ono. 1986. Transformation of propane into aromatic hydrocarbons over ZSM-5 zeolites. Journal of Catalysis 101(1): p. 12–18.

Namuangruk, S. 2011. A Combined Experimental and Theoretical Study on the Hydrolysis of Dimethyl Ether over H-ZSM-5. The Journal of Physical Chemistry C 115(23): p. 11649–11656.

Ni, Y. 2011.Aromatization of Methanol over La/Zn/HZSM-5 Catalysts. Chinese Journal of Chemical Engineering 19(3): p. 439–445.

Niu, X. 2014. Influence of preparation method on the performance of Zn-containing HZSM-5 catalysts in methanol-to-aromatics. Microporous and Mesoporous Materials 197(0): p. 252–261.

Song, Y, Q. 2006. An effective method to enhance the stability on-stream of butene aromatization: Post-treatment of ZSM-5 by alkali solution of sodium hydroxide. Applied Catalysis A: General 302(1):p. 69–77.

Energy Science and Applied Technology ESAT 2016 – Fang (Ed.)
© 2016 Taylor & Francis Group, London, ISBN 978-1-138-02973-6

Study on the migration amount of Bisphenol A and the thermal stability of polycarbonate with photo-oxidation aging

Yongxian Zhao, Yangyang Zhao, Wei Xu, Congheng Zhou & Yanhong Huang
Key Laboratory of Rubber-Plastics of Ministry of Education/Shandong Provincial Key Laboratory of Rubber-Plastics, Qingdao University of Science and Technology, Qingdao, China

Jianguo Gao
Qingdao Exit_&_Entry Inspection and Quarantine Bureau, Qingdao, China

ABSTRACT: The present work was concerned with the study on relationship between the degree of photo-oxidation aging and the amount of Bisphenol A (BPA) precipitating from Polycarbonate (PC) used as food containers and packaging materials. And the effects of photo-oxidation aging time on the precipitation amount of BPA and on the structure and performance of PC were observed. The results showed that: in 0–192 hours the precipitation amount of BPA was increasing with the increase of aging time, but the increase was small, the precipitation amount of BPA was in the range of 3.0–6.4 mg/kg. After photo-oxidation aging, PC Yellowness Index (YI) and color difference (ΔC) were increased significantly, the maximum increase was respectively up to 6 and 8 times than before; after PC photo-oxidation aging, the initial weight loss temperature, the maximum weight loss temperature and the activation energy presented a degradation trend, and the thermal stability also declined.

Keywords: polycarbonate; photo-oxidation aging; bisphenol A; thermal stability

1 INTRODUCTION

Polycarbonate (PC) is a kind of versatile engineering thermoplastic because of its excellent performance and widely used in food containers and packaging materials. However, the Bisphenol A (BPA) will precipitate from PC in the process of processing and using, which impacts the application of PC. BPA is an endocrine disrupting chemical and its disturbance has been confirmed (R. F. Chen et al, 2009).

In this paper, xenon lamp irradiation aging methods were taken in PC to implement the artificial accelerated aging. Measured the actual precipitation amount of BPA from PC at different aging times by Gas Chromatography-Mass Spectrometry (GC-MS) and analyzed its changes of structure by means of infrared spectroscopy (FTIR) to establish the correlation between the actual precipitation amount of BPA and the characteristic absorption peak intensity of BPA in FTIR spectra.

2 EXPERIMENTAL

2.1 Sample preparation and aging test

Polycarbonate (grade 110) was purchased from Taiwan Chi Mei Industrial Co., Ltd. Put PC pellets in a vacuum oven to dry for 24h at 120°C to ensure PC pellets not contain moisture. (A) Powdered sample preparation. After frozen in liquid nitrogen refrigeration system, the 30 g PC samples were crushed into a powder sample whose size was less than 1 mm³ with a grinder. (B) Film sample preparation: Took PC particle for hot forming with vulcanizing machine at 200°C, and the pressure was 10 MPa, the pressing time was 5 min.

Conducting Xenon artificial accelerated aging test in according to GB/T16422.2–1999, placed the specimen into anti-yellowing aging box (set the working temperature of 65°C, xenon lamp irradiation intensity of 25 W/m², relative humidity of 65%–5%) experiencing the photo-oxidation oxygen aging. Set the aging sampling time of 0h, 96h, 120h, 168h and 192h.

2.2 Characterization

The actual precipitation amount of BPA was tested by using Agilent 5975C gas chromatograph mass spectrometer (GC-MS) according to SN/T 2379–2009 standards.

The German company Bruker's Vertex 70 type full digital FT-IR spectrometer (FTIR) was used to characterize the relative precipitation amount of BPA on PC film samples. When quantitatively analyzing, the absorption peak 2969 cm^{-1} (Y. H. Pan et al, 2012) was taken as a reference. The peak area ratio (A3515/A2969) was the absorbance ratio (Ra) of two groups, which could characterize the relative precipitation amount of BPA after PC aging degradation.

The PC film samples before and after aging was tested by using the HITACHI U-4100 UV/spectro-photometer. The °C and YI was calculated through the formula (W. F. Chen, 1994–2012).

The German NETZSCH company's TG-20 type thermogravimetric analyzer (TG) was used to test PC granular samples. Under the air atmosphere at a heating rate of 10°C/min, temperature ranged from 30~800°C.

3 RESULTS AND DISCUSSION

It can be seen from the Fig. 1 that the precipitation amount of BPA shows an increasing trend in PC after aging. Within the 0–120h, the precipitation amount of BPA is increasing with aging time; aging after 120h, the precipitation amount of BPA reduces. Overall, the photo-oxidation has little effect on the precipitate of BPA in the aging PC. PC products usually would not cause dramatic increase in precipitation amount of BPA under normal sunlight conditions.

From the Fig. 2 (a) and (b) can be seen, the intensity of BPA absorption peak at 3515 cm^{-1} in PC experienced optical oxygen aging is markedly enhanced; and the precipitation amount of BPA is increasing with aging time that is in the range of 0–96h; after 96h, the precipitation amount of BPA decreases slightly with aging time. That is consistent with the law of the precipitation amount of BPA that directly measured by GC-MS.

Fig. 3 shows that with the aging time, the 71 nm absorption peak belonging to B absorption band increased and the red shift occurred. This indicated that PC took on the fracture of ester bond under the action of light radiation and Fries rearrangement occurred, to generate BPA (maximum absorption wavelength was in 280–340 nm) reacting with photo-oxidation reaction and salicylic acid phenyl ester (maximum absorption wavelength was in 290–330 nm), which increased the levels of aromatic compounds (W.B. Gao et al, 2008) (G.F. Tjandraatmadja et al, 2002).

Fig. 4 (a) is the UV spectra of PC aging 0h, 96h, 120h, 168h and 192h. Fig. 4 (b) is the influence of aging time on the color difference and yellow index of PC. It can be seen from Fig. 4 (a) and (b) that with the increasing aging time, the uv transmittance decreases until reached a minimum at 168h, then it basically remains unchanged after 168h; and there is a significant increase in the yellow index and the color of the PC which the maximum value increase is 6 times and 8 times than aging before.

Table 1 is the characteristic parameter valued in the first phase of thermal weight loss in PC. From Tab.1 can obviously see that, under air atmosphere, along with the aging time, the initial thermal weightlessness temperature, maximum thermo-

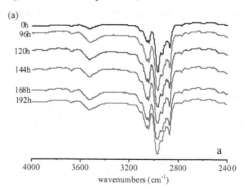

(a)

wavenumbers (cm^{-1})

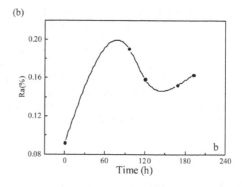

(b)

Time (h)

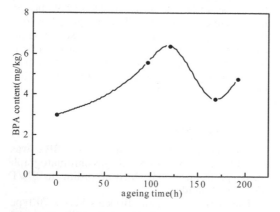

Figure 1. The relationship between photo-oxidation aging time and the the migration amount of BPA.

Figure 2. IR spectra of PC under different photo-oxidation aging time (a) and the influence of photo-oxidation aging time on relative precipitation amount of BPA(b).

gravimetric degradation temperature and activation energy of PC are declining, which shows that the longer optical oxygen aging time is, the more significantly thermal stability of PC reduced.

As can be seen from Fig. 5 (a), before 340°C, the PC presents a weight-gaining progress, and the weight increase of PC is different at different aging time. Further explored the impact of light oxygen

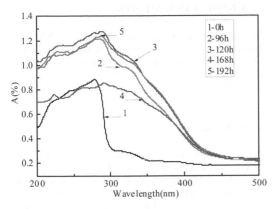

Figure 3.   UV-VIS spectra of PC at different aging time.

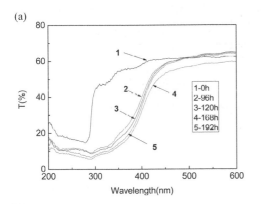

(a)

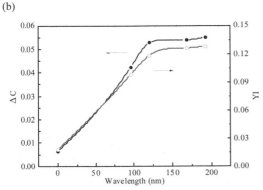

(b)

Figure 4.   Influence of aging time on light transmittance, chromatism and yellowness index.

aging time on the phenomenon would do further analysis the oxidation induction period of PC. The aging degree of polymer can be evaluated through the speed of oxygen or oxygen amount within a certain time (F. Kurihara et al, 1977) (B. N. Jang et al, 2005). From Fig. 5 (b) shows, the oxygen amount is reduced with the increase of aging time, and the quantity of oxygen is significantly reduced when the aging time greater than 144h.

As can be seen from the Fig. 6, in the first stage of weight loss, with the increase of the aging time, the TG rate first increases and then decreases; reasons were speculated as follows: in the first stage of weight loss, during the aging process of 0h–96h, the degradation product of PC is diphenyl carbon-

Table 1.   Pyrolysis characteristic parameter values of PC in photo-oxidation aging method.

Aging time/h	Temperature of initial spot/°C	Temperature of biggest loss rate/°C	E/kJ·mol⁻¹
0h	435.4	467.0	248.0
96h	436.6	466.9	229.9
144h	435.4	466.9	219.4
192h	429.4	466.1	172.7

(a)

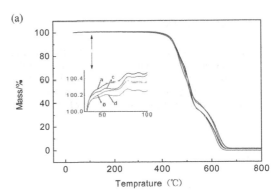

(b)

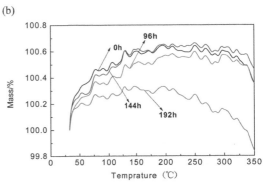

Figure 5.   Effect of aging time on TG spectra and oxygen absorbed.

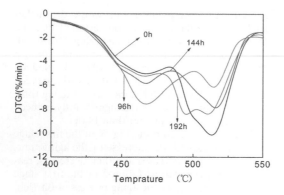

Figure 6. Influence of aging time on DTG.

ate and ethers, which are still reactive substance, increased the aging rate; when the aging time up to a certain extent, to the reaction of the material already in the process of aging the reaction is completed, the resulting product relatively stable, and because of the aging PC unaged sample material margin decrease heat loss rate of the PC play a "dilution effect", which in the aging 120h–168h TG sample rate decreases.

4 CONCLUSION

With the extension of the photo-oxidation aging time, BPA precipitation levels will increase, but the rate of increase was small as 2–4 mg/kg. PC optical properties change significantly with photo-oxidation aging time. Yellowness Index (YI) and color difference (ΔC) were increased, the maximum was increased respectively by 6 times and 8 times; meanwhile the transparency (T) decreased. After photo-oxidation aging, PC initial weight loss temperature, the maximum weight loss temperature and the activation energy were decreased degradation and the thermal stability decreased.

ACKNOWLEDGEMENTS

This work was financially supported by the National Natural Science Foundation of China (51373084, 51243002), Natural Science Foundation of Shandong province of China (ZR2010EM025), and Public Welfare Project of AQSIQ (201410083).

REFERENCES

Chen, R.F., Z.H. Gu: Chinese Journal of Food Hygiene, Vol. 21 (2009) No. 6, p. 529–533.

Chen, W.F.: China Academic Journal Electronic Publishing House (All Right Reserved, 1994–2012).

Gao, W.B., S.C. Han, M.J. Yang, et al: Polymer Materials Science and Engineering, Vol. 24 (2008) No. 10, p. 67–70.

Jang, B.N., C.A. Wilkie: Thermochimica Acta, Vol. 433 (2005) No. 2, p. 1–12.

Kurihara, F., S.S. Wu (translation): Plastics Aging (Shanghai: National Defense Industry Press, China 1977, p. 61).

Pan, Y.H., M.J. Yang, S.M. Han, et al: Journal of Applied Polymer Science, Vol. 125 (2012) No. 3, p. 1–9.

Tjandraatmadja, G.F., S.L. Burn, M.C. Jollands: Polymer Degradation and Stability, Vol. 78 (2002) No 3, p. 435–448.

Energy Science and Applied Technology ESAT 2016 – Fang (Ed.)
© 2016 Taylor & Francis Group, London, ISBN 978-1-138-02973-6

Influence of sintering temperature on microstructure and magnetic properties of $Ni_{0.26}Cu_{0.16}Zn_{0.58}O(Fe_2O_3)_{0.985}$ ferrite

Conggen Dong, Difei Liang & Weijia Li
State Key Laboratory of Electronic Thin Films and Integrated Devices, National Engineering Research Center of Electromagnetic Radiation Control Materials, University of Electronic Science and Technology of China, Chengdu, China

ABSTRACT: $Ni_{0.26}Cu_{0.16}Zn_{0.58}O(Fe_2O_3)_{0.985}$ ferrite was prepared by the oxide-ceramic method. Microstructure and magnetic properties of NiCuZn ferrite with different values of sintering temperature (900–1100°C) were investigated. X-ray diffraction, a scanning electron microscope, a B-H analyzer, agilent-E4991 A Impedance Analyzer and the Archimedes method were used to characterize NiCuZn ferrite samples. It is found that, with the increase in sintering temperature, the grain size and sintering density of NiCuZn ferrite first increase and then decrease; meanwhile, the peaks of the imaginary part of magnetic permeability (μ'') transfer to the lower frequency in 3–5 MHz. The saturation flux density (Bs), remanent magnetization (Br), the coercivity (Hc), and the power loss (Pcv) were also studied. The values of Bs increase first and then decrease with the increase in sintering temperature, while the values of Br, Hc, and Pcv change in the opposite trend. Additionally, the temperature stability of NiZnCu ferrite becomes poor with the increase in sintering temperature. It has been observed that, when the sintering temperature is 1050°C, the NiCuZn ferrite has the optimum performance.

Keywords: ferrite; sintering temperature; permeability; saturation flux density; power loss

1 INTRODUCTION

NiCuZn ferrite materials, as having good electromagnetic properties, have been used in many fields such as inductors, electromagnetic interference filters, and microwave devices. However, the research about NiZnCu ferrite being applied to wireless charging is rarely reported. In recent years, plenty of scientific studies focus on the effect of sintering temperature on magnetic properties of NiZnCu ferrite (K. Kawano et al, 2009) (C.Y. Liu et al, 2008) (X.H. He et al, 2008) (H. Su et al, 2013) (H. Su et al, 2011). Changing sintering temperature to improve properties of NiZnCu ferrite is a common and effective method. In this paper, the microstructure and magnetic properties of $Ni_{0.26}Cu_{0.16}Zn_{0.58}O(Fe_2O_3)_{0.985}$ ferrite sintered at different sintering temperatures were investigated systematically, the optimal sintering temperature was chosen, and NiCuZn ferrite materials with excellent performance that is applied to wireless charging were prepared.

2 EXPERIMENTAL

In this paper, NiCuZn ferrites with the composition of $Ni_{0.26}Cu_{0.16}Zn_{0.58}O(Fe_2O_3)_{0.985}$ were prepared by the conventional ceramic techniques, with Fe_2O_3

(99.4wt%), NiO (99.45wt%), ZnO (98wt%), and CuO (98.45wt%). The raw materials were milled in an attritor with deionized water at the speed of 250 rad/min for 2.5 h. After drying, the mixture of raw materials were homogenized and calcined at 800°C for 3 h in high-temperature furnace, with a heating rate of 2°C/min. Then, pre-sintered materials were homogenized and wet-milled for 3 h. After drying, the powder was granulated with 8–9% polyvinyl alcohol (PVA) and pressed into toroid shape with a dimension of o.d. = 16 mm, i.d. = 8 mm, and h = 4 mm at 10 MPa. The toroidal samples were sintered at 900°C, 950°C, 1000°C, 1050°C, and 1100°C for 4 h. The surface morphology was observed using a Scanning Electron Microscope (SEM) (JSM-7600F). Agilent-E4991 A Impedance Analyzer was used to measure complex permeability in the frequency from 1 MHz to 1 GHz. Sintering densities were measured by the Archimedes method. A B-H analyzer (SY-8232) was employed to test the value of Bs, Br, Hc, and Pcv. The TH-2828 LCR was used to measure the temperature property of permeability.

3 RESULTS AND DISCUSSION

The X-Ray Diffraction (XRD) patterns and the sintering density of $Ni_{0.26}Cu_{0.16}Zn_{0.58}O(Fe_2O_3)_{0.985}$

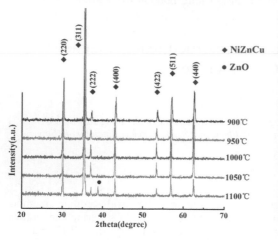

Figure 1. XRD patterns for $Ni_{0.26}Cu_{0.16}Zn_{0.58}$ $O(Fe_2O_3)_{0.985}$ ferrite sintered at 900°C, 950°C, 1000°C, 1050°C and 1100°C.

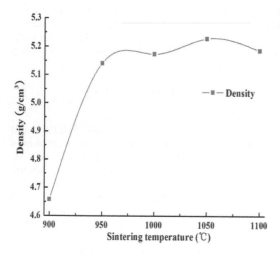

Figure 2. Variation in density of $Ni_{0.26}Cu_{0.16}Zn_{0.58}$ $O(Fe_2O_3)_{0.985}$ ferrite with the sintering temperature.

Figure 3. SEM micrographs for $Ni_{0.26}Cu_{0.16}Zn_{0.58}O$ $(Fe_2O_3)_{0.985}$ ferrite sintered at: (a) 900°C, (b) 950°C, (c) 1000°C, (d) 1050°C, and (e) 1100°C.

ferrite sintered at 900°C, 950°C, 1000°C, 1050°C, and 1100°C are shown in Figs. 1 and 3, respectively. Seven diffraction peaks in all the samples are associated with the (220), (311), (222), (400), (422), (511), and (440) planes of NiZnCu. All samples generate spinel phase. When the sintering temperature rises to 1100°C, ZnO hexagonal crystal phase appears at 38.5 degree. The sintering densities, which were measured by the Archimedes method, as shown in Fig. 3, increase first and then decrease with the increase in sintering temperature and reach the maximum value, which is 5.228 g/cm³, at 1050°C.

Figure 3 shows the SEM micrographs of $Ni_{0.26}Cu_{0.16}Zn_{0.58}O(Fe_2O_3)_{0.985}$ ferrite with the different values of sintering temperature. At 900°C, the grain boundary is fuzzy, there are many voids on the crystal surface and the grain growth is not enough. The reason is that the solid-phase reaction is not complete at low temperatures. At 950°C, the grain size increases slightly and the voids on the crystal surface decrease obviously. At 1000°C, the grain size increases obviously and the size is uniform, which indicate that the solid-phase reaction is almost complete, but the grain boundary is thick. At 1050°C, the structure of NiZnCu ferrite becomes uniform, the grain boundary becomes thin, and the grain size reaches the maximum value. At 1100°C, the grain size becomes uneven. An explanation might be as follows. The sintering temperature is too high, grains grow too fast, and then uneven grain growth resulted. The average grain sizes of ferrite sintered at 900°C, 950°C, 1000°C, 1050°C, and 1100°C (0.7 μm, 1.6 μm, 4.8 μm, 5.6 μm, and 4.5 μm) were obtained by calculation, which is consistent with Fig. 3.

Figure 4 illustrates the permeability changes of $Ni_{0.26}Cu_{0.16}Zn_{0.58}O(Fe_2O_3)_{0.985}$ ferrite with the sintering temperature. The complex permeability increases first and then decreases with the increase in sintering temperature at a frequency of 1 MHz

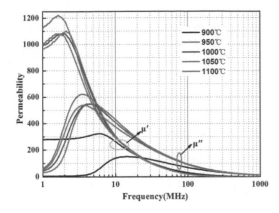

Figure 4. Variation in permeability spectra with the sintering temperature.

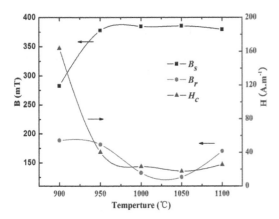

Figure 5. Variation in Bs,Br,Hc with the sintering temperature.

and the peaks of the imaginary part of μ″ transfer to a lower frequency of 3–5 MHz, which is consistent with Snoek's limit formula (T. Nakamura, 2000). The changes in μ′ and μ″ could be attributed to the variations in the grain size, as shown in Fig. 3, and the value of Bs and Hc, as shown in Fig. 5 and Table 1. According to the following relation:

$$\mu i \propto \frac{DMs}{K_1}$$

where μi is the original permeability, D is the grain size, Ms(Bs) is the saturation flux density, and K_1 is the magnetocrystalline anisotropy constant. μ′ has the same change trend with μi at low frequency, so the permeability has changed, as shown in Fig. 4.

Variations in Bs, Br, Hc, and Pcv of $Ni_{0.26}Cu_{0.16}Zn_{0.58}O(Fe_2O_3)_{0.985}$ ferrite are shown in

Table 1. Variation in Bs, Br, and Hc with the sintering temperature.

Sintering temperature (°C)	B_s (mT)	B_r (mT)	H_c (A/m)
900	282.7	188.6	163.5
950	377.3	181.6	40.2
1000	384.6	132.5	23.0
1050	385.4	125.3	17.6
1100	379.4	170.7	25.5

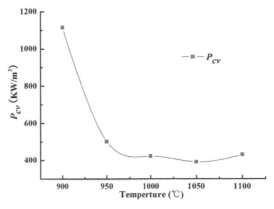

Figure 6. Variation in Pcv with the sintering temperature.

Fig. 5, Table 1, and Fig. 6. It has been found that, the value of Bs first increases and then decreases with the increase in sintering temperature and Bs has a maximum value, which is 385.4 mT. The values of Br, Hc, and Pcv have the opposite trend. Br, Hc, and Pcv all have the optimum values, when the sintering temperature is 1050°C, as shown in Table 1 and Fig. 6.

From Fig. 7, it can be seen that the permeability of NiCuZn ferrite in 100 KHz increases first and then decreases at normal atmospheric temperature and, at 1050°C, μ′ has a maximum value, which is consistent with that in 1 MHz. Before 150°C, the permeability increases gradually with the increase in the test temperature, and then decreases sharply. The Curie temperature of $Ni_{0.26}Cu_{0.16}Zn_{0.58}O(Fe_2O_3)_{0.985}$ ferrite with different sintering temperatures is almost equal, as shown in Table 2. Curie temperature is the macro performance of electrostatic exchange of iron (K. Sun, 2009). The formula of NiCuZn ferrite is constant, the strength of exchange is unchanged, and the energy required to overcome the ability of the parallel orientation of the spins is also constant, so Curie temperature remains constant (R. Islam et al, 2013). The loss coefficient in 100 KHz decreases first and then increases gradually. In addition,

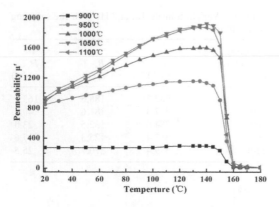

Figure 7. Temperature characteristics of the permeability of $Ni_{0.26}Cu_{0.16}Zn_{0.58}O(Fe_2O_3)_{0.985}$ ferrite with the sintering temperature at 100 KHz.

Table 2. Curie temperature, temperature coefficient, and loss of coefficient of $Ni_{0.26}Cu_{0.16}Zn_{0.58}O(Fe_2O_3)_{0.985}$ ferrite with the sintering temperature.

Sintering temperature (°C)	$\tan \zeta/\mu' \times 10^{-6}$ (100 KHz)	T_c (°C)	α_u (%/°C) (20°C–70°C)
900	53.7	150	0.01
950	11.6	150	0.44
1000	12.1	150	0.87
1050	14.5	150	1.06
1100	15.9	150	1.15

it has the maximum value at 900°C, as shown in Table 2, which can be caused by incomplete crystallization and many voids on crystal surface, as shown in Fig. 3. From Table 2, it is observed that the temperature coefficient increases gradually with the increase in sintering temperature between 20°C and 70°C, indicating that the temperature stability of $Ni_{0.26}Cu_{0.16}Zn_{0.58}O(Fe_2O_3)_{0.985}$ ferrite becomes worse with the increase in sintering temperature.

4 CONCLUSION

In this paper, $Ni_{0.26}Cu_{0.16}Zn_{0.58}O(Fe_2O_3)_{0.985}$ ferrite samples were prepared by the conventional ceramic techniques. The microstructure and magnetic properties of $Ni_{0.26}Cu_{0.16}Zn_{0.58}O(Fe_2O_3)_{0.985}$ ferrite with different sintering temperatures were investigated. It has been concluded that, with the increase in sintering temperature, the grain size and sintering density of NiCuZn ferrite first increase and then decrease; meanwhile, the peaks of the imaginary part of magnetic permeability transfer to a lower frequency in 3–5 MHz. Bs, Br, Hc, and Pcv were also studied. The values of Bs increase first and then decrease with the increase in sintering temperature, while the values of Br, Hc, and Pcv change in the opposite trend. Additionally, the temperature stability of NiZnCu ferrite becomes worse with the increase in sintering temperature. $Ni_{0.26}Cu_{0.16}Zn_{0.58}O(Fe_2O_3)_{0.985}$ ferrite materials with optimal performance used in wireless charging were obtained at 1050°C.

ACKNOWLEDGMENTS

This work was supported by the Application Foundation Research of Sichuan Science and Technology Agency under Grant 2015 JY0179.

REFERENCES

He, X.H., Z.D. Zhang and Z.Y. Ling: Ceramics International, Vol. 34 (2008) No. 7, p. 1409–1412.
Islam, R., M.A. Hakim, M.O. Rahman, H.N. Das and M.A. Mamun: J. Alloys. Compounds, Vol. 559 (2013) p. 174–180.
Kawano, K., M. Hachiya, Y. Iijima, N. Sato and Y. Mizuno: J. Magn. Magn. Mater, Vol. 321 (2009) No. 16, p. 2488–2493.
Liu, C.Y., Z.W. Lan, X.N. Jiang, Z. Yu, K. Sun, L.Z. Li and P.Y. Liu: J. Magn. Magn. Mater. Vol. 320 (2008) No. 7, p. 1335–1339.
Nakamura, T.: J. Appl. Phys, Vol. 88 (2000) No. 1, p. 348–353.
Su, H., X.L. Tang, H.W. Zhang, Y.L. Jing, F.M. Bai and Z.Y. Zhong: J. Appl. Phys, Vol. 113 (2013) No. 17, p. 17B301.
Su, H., X.L. Tang, H.W. Zhang, Z.Y. Zhong and J. Shen: J. Appl. Phys, Vol. 109 (2011) No. 7, p. 07A501.
Sun, K.: Study on Two Soft Magnetic Ferrite Materials for High Frequency Power Conversion (Ph.D., University of Electronic Science and Technology of China, China 2009), p. 35.

Energy Science and Applied Technology ESAT 2016 – Fang (Ed.)
© 2016 Taylor & Francis Group, London, ISBN 978-1-138-02973-6

Effect of a stabilizer on the phase composition and mechanical properties of ZrO$_2$ materials

Yonghua Zhou
College of Materials and Mineral Resources, Xi'an University of Architecture and Technology, Xi'an, China

Qunhu Xue
College of Materials and Mineral Resources, Xi'an University of Architecture and Technology, Xi'an, China
Shaanxi Techno-Institute of Recycling Economy, Xi'an, China

Baikuan Liu
Puyang Refractories Group Co. Ltd., Puyang, Henan, China

ABSTRACT: This paper researched the effect of compound stabilization, which used MgO, TiO$_2$, Y$_2$O$_3$, and CeO$_2$ as stabilizer on ZrO$_2$ materials by co-stabilizing the monoclinic zirconium fine powder in an orthogonal experiment. The effects of stabilizer on the phase composition, compressive strength, and thermal shock resistance of ZrO$_2$ materials were studied by XRD and analyzed by extremum difference analysis. Results indicate that all samples generated cubic zirconium after sintering at 1710 °C and the variation of cubic and monoclinic phases of each sample was different because of the different kinds and amounts of stabilizers in raw materials. For the samples with 30% (in volume fraction) cubic phase, the values of compressive strength and thermal shock resistance were 1 and 41 times, respectively, higher than those of the samples with too much or too little cubic phase. The optimum feed ratio of the stabilizer should be 3.0% (in mass fraction) CeO$_2$, 3.0% MgO, 0 Y$_2$O$_3$, and 5.0% TiO$_2$.

Keywords: compound stabilization; zirconia; phase composition; properties

1 INTRODUCTION

Zirconia would shift in the three, monoclinic, tetragonal, and cubic, phases with a certain volume change as the temperature changes (X.M. Yu et al., 2007) (Q. Liu et al., 2013), so the application of pure zirconia is limited. By adding MgO, Y$_2$O$_3$, CeO$_2$, and other rare earth oxides as stabilizer to zirconia, these oxides and zirconia form a solid solution to make the zirconia exit as the metastable state of tetragonal and cubic phases and phase transformation not to occur, as the cation radius of these oxides (Mg^{2+}, Y^{3+}, Ce^{4+}, etc.) is close to Zr^{4+}, and the solubility in the zirconia is very big (B.Y. Dai et al., 2007) (T. Onda et al., 2010). The partially stabilized zirconia formed by this way is widely used in the preparation of the functional device sizing nozzle in iron and steel casting (H. Zhang et al., 2014) (X.F. Wang et al., 2003).

This paper co-stabilized the monoclinic zirconium fine powder with four different stabilizers (MgO, Y$_2$O$_3$, CeO$_2$, TiO$_2$) using an orthogonal experiment, and sintering at 1710°C for 2 hours after being pressed into samples. In addition, the compressive strength and thermal shock resistance of all samples after sintering were measured, and the effects of different stabilizers on the phase composition and mechanical properties of zirconia material were studied and analyzed by XRD and extremum difference analysis to provide important basis for selecting raw materials of sizing nozzle.

2 EXPERIMENT

According to the orthogonal experiment table, different types and contents of the stabilizers were added to the monoclinic zirconia powder, the types of stabilizers were MgO, Y$_2$O$_3$, CeO$_2$, and TiO$_2$, and their contents and the orthogonal experiment table are shown in Table 1. Table 2 shows the contents of the stabilizer of the sample for each group.

The samples were pressed in Φ50 × 10 mm cylindrical specimen by molding on the hydraulic press at a pressure of 500 MPa. The samples after pressing were dried at 110°C for 24 hours using a constant-temperature drying oven. Finally, the specimens after drying were burned at 1710°C for

Table 1. Orthogonal experiment of *L9* (3⁴).

Level	Factor			
	$w(CeO_2)/\%$	$w(MgO)/\%$	$w(Y_2O_3)/\%)$	$w(TiO_2)/\%$
1	3.0	2.6	0.0	3.0
2	5.0	2.8	1.0	5.0
3	7.0	3.0	2.0	7.0

W: Mass fraction.

Table 2. Content of stabilizer added in each sample.

Sample no.	$w(CeO_2)/\%$	$w(MgO)/\%$	$w(Y_2O_3)/\%$	$w(TiO_2)/\%$
1	3.0	2.6	0.0	3.0
2	3.0	2.8	1.0	5.0
3	3.0	3.0	2.0	7.0
4	5.0	2.6	1.0	7.0
5	5.0	2.8	2.0	3.0
6	5.0	3.0	0.0	5.0
7	7.0	2.6	2.0	5.0
8	7.0	2.8	0.0	7.0
9	7.0	3.0	1.0	3.0

2 hours in a high-temperature furnace. The phase analysis was performed using a X-ray diffractometer produced by Rigaku of Japan.

3   RESULTS AND DISCUSSION

3.1   *XRD analysis*

As the XRD spectra of all the samples were similar and only the contents of the cubic and monoclinic phases were different, the spectra of samples 1 and 2 were listed in this paper, as shown in figures 1 and 2, in which a and b are the spectra before and after burning, respectively. The phase composition of each sample before sintering was mainly monoclinic phase without cubic and tetragonal phases because the raw materials for the samples were monoclinic zirconia powder and various oxide stabilizers.

The phase composition of the samples after firing at 1710°C was mainly cubic and monoclinic phases. Each sample generated cubic phase after sintering, and the content of monoclinic phase decreased, that is, a part of monoclinic zirconia transformed into cubic zirconia. For the samples with different types and contents of stabilizers, the amount of phase transformation of monoclinic was different; thus, the relative content of cubic and monoclinic phases of all samples was also different.

In this paper, the following formula (J. SUN et al, 2005) can be used to calculate the volume

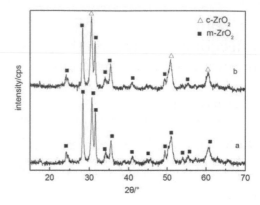

Figure 1.   XRD spectra of sample 1.

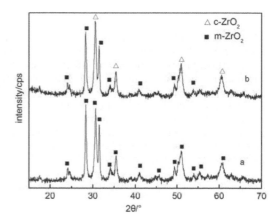

Figure 2.   XRD spectra of sample 2.

fraction of monoclinic and cubic phases of zirconia after stabilizing:

$$V_m = \frac{1.6031 I_m(111)}{1.6031 I_m(111) + I_c(111)}$$

(*I*m and *I*c refer to the diffraction intensity of monoclinic and cubic phases.)

The results of the phase composition and properties of all samples are shown in table 3.

We can see that the content of cubic and monoclinic phases of all the samples were different from table 3; also, there were differences in the compressive strength and thermal shock resistance. The compressive strength and the times of thermal shock resistance of the sample were high when the content of cubic phase was approximately 30 vol%. This is mainly because monoclinic zirconia would undergo martensitic transformation, and it is easily cracked in the firing process, and the strength of products is low, but the stabilized zirconia has volume sta-

Table 3. Results of phase composition and properties of each sample.

Sample no.	$\phi(V_m)$/%	$\phi(V_c)$/%	Compressive strength/MPa	Thermal shock resistance/time
1	62.5	37.5	299.20	9
2	60.3	39.7	313.15	10
3	64.6	35.4	258.60	42
4	73.6	26.4	204.62	1
5	97.9	2.1	136.81	1
6	65.5	34.5	349.63	33
7	9.4	90.6	172.46	1
8	14.9	85.1	190.21	1
9	4.8	95.2	154.69	1

ϕ—Volume fraction.

bility in the sintering process, and thermal shock resistance of the product is low. Partially stabilized zirconia is used as a raw material in sizing nozzle preparation industry, firing crack is inhibited, and a certain amount of martensite phase transformation occurs at the same time. It could improve the thermal shock resistance of sizing nozzle through the phase transformation toughening, and solve the crack problem in the process of using.

The thermal expansion coefficient of the material is directly related to thermal shock stability; also, the smaller the thermal expansion coefficient, the better the thermal shock resistance. The thermal expansion coefficient is also an important factor that affects the compressive strength of the stabilized zirconia (J. Juwoong et al, 2004). When the content of cubic phase was too large, the thermal expansion coefficient of the sample was big and this is against the compressive strength and thermal shock resistance because the thermal expansion coefficient of cubic zirconia in three crystal zirconia was the biggest and increased with the temperature (X. H. LI et al, 1998). The content of cubic phase was too small, as the thermal expansion coefficient of the monoclinic zirconia was smaller than that of the cubic and tetragonal zirconia (J. Juwoong et al, 2004) and it had obvious anisotropy, and there was phase transition between tetragonal and monoclinic phases with a certain volume effect, which will make the sample easy to crack and was adverse to the improvement of thermal shock resistance and compressive strength. Therefore, the appropriate content of monoclinic and cubic phases and the appropriate amount of phase shift was an important process parameter of zirconia sizing nozzle with good thermal shock stability.

3.2 Performance analysis

Extremum difference analysis was used to analyze the effects of various factors on the compression

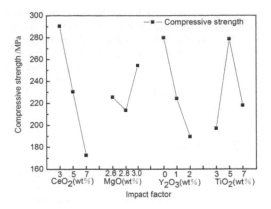

Figure 3. The relationship between the compressive strength and impact factor.

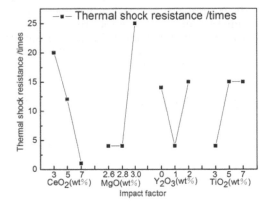

Figure 4. The relationship between the thermal shock resistance and impact factor.

strength and thermal shock resistance and the results are shown in Figures 3 and 4, respectively. The higher range analysis results indicate the greater pressure impact on strength and thermal shock resistance.

It can be seen from Figures 3 and 4 that, for the compression strength and thermal shock resistance of all samples, the poor values of R of each factors were $R(CeO_2) > R(Y_2O_3) > R(TiO_2) > R(MgO)$ and $R(MgO) > R(CeO_2) > R(Y_2O_3) = R(TiO_2)$, respectively. Then, the most important influence factor on the compression strength and thermal shock resistance of the sample was the content of CeO_2 and MgO, respectively. In the factor of CeO_2, the compressive strength and the times of thermal shock resistance were the biggest when the content of CeO_2 was 3% (in mass fraction). In the factor of MgO, the compressive strength and the times of thermal shock resistance were the biggest when the content of MgO was 3%. In the factor of Y_2O_3, the compressive strength and the times of thermal shock resistance were the biggest when the contents of Y_2O_3 were 0% and 1%, respectively. In the factor of TiO_2, the compressive strength and the times of thermal shock resistance were the biggest when the content of TiO_2 was 5%. Therefore, the optimum feed ratio of stabilizer should be 3.0% CeO_2, 3.0% MgO, 0 Y_2O_3, and 5.0% TiO_2.

4 CONCLUSION

1. Samples mixed with stabilizers generate the cubic phase after sintering at 1710°C and the different kinds and amounts of stabilizers in raw materials made cubic and monoclinic phase variation of each sample different.
2. For the samples with 30% cubic phase, the values of compressive strength and times of thermal shock resistance were higher than those of the samples with too much or too little cubic phase.
3. The optimum feed ratio of the stabilizer should be 3.0% (in mass fraction) CeO_2, 3.0% MgO, 0 Y_2O_3, and 5.0% TiO_2.

ACKNOWLEDGMENTS

This work was supported by the funds from the National Natural Science Foundation of China (51372193) and the Natural Science Basic Research Plan in Shaanxi Province of China (2014 JM6224).

REFERENCES

Dai, B.Y., T.C. Chen and J.L. Shang, et al. Chin Ceram Soc. Vol. 35(2007) No. 3, p. 193–196.

Juwoong, J., K. Hakkwan and L. Deukyong. The effect of tetravalent dopants on the unit cell volume of 2Y-TZP and 8Y-SZ. Mater Lett. Vol.58 (2004), p. 1160–1163.

Li, X.H., T. Zhao and W.G. Zhao, et al. J Baotou Univ Iron Steel Tech. Vol. 17(1998) No. 2, p. 91–93.

Liu, Q., Q.H. Yang and G.G. Zhao et al. Chin J Inorg Chem. Vol. 29(2013) No. 4, p. 798–802.

Onda, T., J. Hara and C.C. Zhong, et al. Isothermal and a thermal martensitic transformations of ceria doped zirconia. J Mater Met. Vol. 9(2010) No. 3, p. 211–216.

Sun, J., C.Z. Huang and H.L. Liu, et al. Mater Mech Eng. Vol. 29(2005) No. 8, p. 1–3.

Wang, X.F., H.X. Li and B. Yang. Continuous Casting. Vol. (2003) No. 3, p. 37–39.

Yu, X.M., B.K. Xu and F.D. Yuan. Stabilizing and Applications of Zirconia. Rare Metals Letters. Vol. 26 (2007) No. 1, p. 28–32.

Zhang, H., Q.H. Xue and W. Liu. Bull Chin Ceram Soc. Vol. 33(2014) No. 7, p. 1762–1768.

Energy Science and Applied Technology ESAT 2016 – Fang (Ed.)
© 2016 Taylor & Francis Group, London, ISBN 978-1-138-02973-6

The effect of heat treatment on the microstructure and properties of M40/ZM6 composites

Meihui Song, Yu Zhang & Xiaochen Zhang
Institute of Advanced Technology, Heilongjiang Academy of Sciences, Harbin, China

Longtao Jiang & Gaohui Wu
School of Materials Science and Engineering, Harbin Institute of Technology, Harbin, China

ABSTRACT: In this work, M40/ZM6 composites had been successfully fabricated by pressure infiltration method. The effect of heat treatment process on microstructure and mechanical properties of the composites was studied by means of Scanning Electron Microscope (SEM), Transmission Electron Microscope (TEM) and universal electronic tensile testing machine. The results revealed that the mechanical properties of composites were decreased after heat treatment, which was attributed to increase the number and the size of precipitates at the interface. However, heat treatment had little effect on elasticity modulus.

Keywords: Cf/Mg composites; heat treatment; interface; mechanical properties

1 INTRODUCTION

Magnesium alloys are the lightest structural metallic materials (R. Zhen et al, 2012). Carbon fiber reinforced Mg composites (Cf/Mg) are widely applied in propellant bottle, missile instrument bay, satellite antenna, subpanel and mirror substrates or its support (K. Sapozhnikov et al, 2009) of Space Station since Cf/Mg composites offer the highest specific strength and stiffness of all structural materials (H. Imamura et al, 2002). Unfortunately, the poor interface bonding between carbon fiber and magnesium alloys is unfavorable for the fabricating process and transverse strength (F. Wu et al, 1997). In order to improve the interface problems mentioned above, recent researches on Cf/Mg composites have focused on fiber surface modification. For example, Zhang et al. prepare coatings of SiC or ZrO_2 (P. Zhang et al, 2011) (F. Reischer et al, 2007) (W.G. Wang et al, 2007) or plate metals, such as Ni or Cu (X.P. Luo et al, 2012) (F. Wu et al, 2000) on the surface of the carbon fiber. Moreover, mixing other particles with the matrix is also an efficient method, including SiC particle (C. Körner et al, 2000) and cenosphere (Z.L. Pei et al, 2009). Compared with the two ways mentioned before, matrix alloying is a promising method because of convenience and effectiveness (H.W. Wang et al, 1992) (S.F. Zhang et al, 2014).

In terms of M40/ZM6 composites, element Nd will significantly enrich on the surface of ZM6 matrix, which will form the discontinuous distribution of massive precipitated phase $Mg_{12}Nd$ (G.H. Wu et al, 2009). Based on the above research, the effect of different heat treatment processes on the microstructure and mechanical properties was further studied in this paper.

2 MATERIALS AND METHODS

2.1 Materials

In present study, unidirectional fibre M40 reinforced Mg matrix composites were fabricated by pressure infiltration method. M40 carbon fibre (60vol%) was used to reinforce the matrix Mg-Re-Zr alloy, which contained Re (2.0wt%) and Zr (0.2wt%). Annealing process was 285°C/3h and then furnace cooling, while T6 process was 530°C/12h and then air cooling, followed by artificial aging 205°C/15h. Before heat treatment, put the sample into quartz glass tube, whose vacuum degree was 10-5 pa for preventing the oxidation of the sample. During heat treatment, the temperature control precision was ±5°C.

2.2 Methods

The Bending fracture morphology of the composites was observed by S-570 Scanning Electron

Microscopy (SEM). The interface characteristics of the composites were investigated by Philips CM-12 Transmission Electron Microscopy (TEM) with acceleration voltage 100kV~120kV. Bending tests were conducted by using an Instron-5569 universal electronic tensile testing machine with a cross-head speed of 2 mm/min and data acquisition of 10 points/min. The sample size was 60 mm × 10 mm × 2 mm. BE120-3 AA strain gage was pasted on the center of the tension face, whose effective area was 2 mm × 3 mm. In the test, the upper ram's diameter was 10 mm and the lower ram's diameter was 4 mm, while the span was 40 mm.

3 RESULTS AND DISCUSSION

3.1 The effect of heat treatment on interface

Figure 1 shows the interface of M40/ZM6 composites of different heat treatment processes. The interface area was widened and larger due to forming precipitates with the heat treatment, as shown in Fig. 1. Element Nd in the as-cast matrix alloy was segregated to form small Mg_{12} Nd precipitates (Fig. 1a) on the surface during preparation. The analyses of precipitates are reported in reference (G.H. Wu et al., 2009). Annealing process led to the results that a large amount of Mg_{12} Nd phase was deposited on the interface and the area of interface was widened. Because of recovery and recrystallization, sub-grains of magnesium took shape in ZM6 matrix (Fig. 1b). The diffusion of solute atoms got energy from high temperature, which led element Nd in the alloy to diffuse to the interface and gather together during T6 process. In the aging process, element Nd nucleared grew on the surface and ultimatcly formed large stable bulks of Mg_{12} Nd phase (Fig. 1c).

3.2 The effect of heat treatment on mechanical properties

Bending property of M40/ZM6 composites after different heat treatment processes is shown in

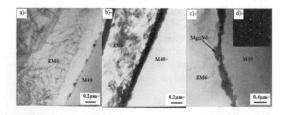

Figure 1. The effect of heat treatment on the interface of M40/ZM6 composites: a) As-cast; b) Annealed; c) T6 treated; d) Diffraction pattern of Mg12 Nd.

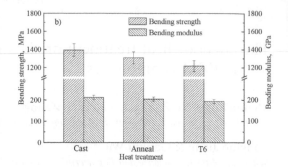

Figure 2. Effect of heat treatment on the bending properties of M40/ZM6 composites.

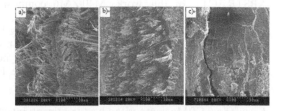

Figure 3. Effect of heat treatment on the bending fractographs of Cf/Mg composites: a) As-cast; b) Annealed; c) T6 treated.

Fig. 2. It can be seen from Fig. 2 that strength of composites decreased after heat treatment and especially that bending property of M40/ZM6 dipped approximately 13% after T6 process. However, heat treatment had little effect on elasticity modulus of composites. As different heat treatment processes are applied, the interface of the composites is widened and more larger precipitates formed on the interface. It needs to emphasize is that a large number of precipitates grow on the interface, which enhance interface bonding strength of the material. As a result, the interface cannot offer channel to microcracks for their proliferation and expansion, which is the reason for stress concentration and prevention of stress releasing when the composites undertook load. Therefore, the strength of composites is lower than that before annealing process. Because microstructure has little effect on elasticity modulus, there is no obvious changing of elasticity modulus of the composites after heat treatment. Compared with Fig. 3, bending fracture of materials with different state, it can be seen clearly that with the change of the heat treatment, the fiber pulling out of the fracture surface of the composites decreased obviously. In the material, the elastic modulus is not affected by the change of microstructure, so the influence of the heat treatment on the elastic modulus of the composites is not obvious.

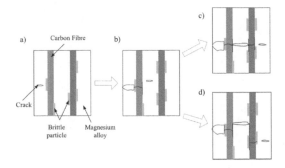

Figure 4. Schematic diagram of crack propagation in the Cf/Mg composites: a) Crack initiation at the magnesium matrix alloy; b) Crack propagation into the magnesium matrix and interface and; c) Fracture of strong interface composites, d) Fracture of weak-bonding interface composites.

3.3 *Effect of interfacial reaction on mechanical properties*

Figure 4 reveals the influencing mechanism of interface on Cf/Mg composites. Because the strength of carbon fiber is much higher than that of the matrix, cracks of the composites firstly formed in magnesium alloy, as shown in Fig. 4a). Cracks in the matrix proliferated and expanded to the interfaces with the increasing of stress (Fig. 4b). Because of the brittle materials which led to stress concentration on the surfaces, cracks cannot expand to other areas through interfaces to release stress. The composites would be destroyed with low stress, and the fracture surface is relatively flat (Fig. 4c). In contrast, if there is less interfacial brittle phase in the composites, then stress around the fracture of the carbon fiber could be transferred to other complete fiber through the interface and caused fiber fracture. This process would constantly repeat until the sample completely fracture with stress transmission and releasing,. The composites possess high strength as the fracture obviously presented carbon fiber pulled out (Fig. 4d). When T6 heat treatment is carried out, larger $Mg_{12}Nd$ form at the interface, which enhance the interfacial bonding strength between carbon and the matrix. However, this would also lead to stress concentration, which is the reason for decreasing stress, straight fracture and less carbon fiber pulled out.

4 CONCLUSION

1. Through heat treatment process, the area of the interface was widened and precipitates at the interface increased. In addition, as a result of annealing, sub-grains formed in the matrix alloy compared with which the effect of T6 process on interface was more obvious as precipitates grew up.
2. Annealing and T6 process could decrease the strength of the composites, and T6 process can result in a larger decline. However, heat treatment had little effect on elastic modulus.
3. Types of fracture failure of Cf/Mg composites were determined by interfacial bonding strength. If interfacial bonding strength was large, the fracture of the composites would be straight, and less carbon fiber would be pulled out. In contrast, the fracture of the composites whose interfacial bonding strength was suitable obviously exhibited pulled-out carbon fiber.

ACKNOWLEDGEMENTS

The research is supported by the national natural science foundation of China (Grant No. 51401079).

REFERENCES

Imamura, H., S. Tabata, N. Shigetomi: J Alloy Comp d, Vol. 330–332 (2002) No. 1, p. 579.
Körner, C., W. Schäff, M. Ottmüller, R. Singer: Adv Eng Mater, Vol. 2 (2000) No. 6, p. 327.
Luo, X.P., M.G. Zhang, C.X. Lv, X.X. Lv. Rare Met Mater Eng, Vol. 41 (2012) No.4, p. 743.
Pei, Z.L., K. Li, J. Gong, N.L. Shi: J Mater Sci, Vol. 44 (2009) No. 15, p. 4124.
Reischer, F., E. Pippel, J. Woltersdorf, S. Stockel, G. Marx: Mater Chem Phys, Vol. 104 (2007), p. 83.
Sapozhnikov, K., S. Golyandin, S. Kustov: Compos Part A, Vol. 40 (2009) No. 2, p. 105.
Wang, H.W., B.L. Shang, L.S. Zheng: Mater Sci prog, Vol. 6 (1992) No. 5, p. 445.
Wang, W.G., B.L. Xiao, Z.Y. Ma: Compos Sci Techno, Vol. 72 (2007) 152.
Wu, F., J. Zhu, Y. Chen, G.D. Zhang: Mater Sci Eng A, Vol. 277 (2000) No. 1–2, p. 143.
Wu, F., J. Zhu: Compos Sci Technol, Vol. 57 (1997) No. 6, p. 661.
Wu, G.H., M.H. Song, Z.Y. Xiu, N. Wang, W.S. Yang: J Mater Sci and Technol, Vol. 25 (2009) No. 3, p. 423.
Zhang, P., Y.Z. Zhang, F.Z. Yin: Nonferrous Met, Vol. 63 (2011) No. 1, p. 19.
Zhang, S.F., G.Q. Chen, R.S. Pei: Materials Science & Engineering A, Vol. 613 (2014) No. 4, p. 111.
Zhen, R., Y.S. Sun, J. Bai: Acta Metallurgica Sinica, Vol. 48 (2012) No. 6, p. 733.

Energy Science and Applied Technology ESAT 2016 – Fang (Ed.)
© *2016 Taylor & Francis Group, London, ISBN 978-1-138-02973-6*

Fabrication of Cu(In,Ga)Se2 films with the magnetron sputtering method

G.C. Qi, Q. Wu & J. Xiang
School of Materials Science and Engineering, Liaoning University of Technology, Jinzhou, China

S.L. Zhao & Z.W. Li
Jinzhou Thermal Power Corporation, Jinzhou, China

ABSTRACT: Cu(In,Ga)Se2 (CIGS) thin films were deposited with a single target by the magnetron sputtering method. The effects of sputtering parameters such as base temperature, heat treatment temperature, and sputtering power on the properties of CIGS films were analyzed. The surface morphology, composition, and crystal structural property of the films were characterized by Scanning Electron Microscopy (SEM) and X-Ray Diffraction (XRD). The results indicate that CIGS films fabricate without substrate heating that can lead to Se coagulation bubbles on the surface after heat treatment process. Substrate heating at 200°C can effectively prevent such drawbacks. Heat treatment temperature and sputtering power have apparent effect on the crystallization of CIGS phase. The optimized parameters of the sputtering process are heat treatment at 400°C and a sputtering power of 165W.

Keywords: Cu(In,Ga)Se2 (CIGS); magnetron sputtering; solar cell (DSSC)

1 INTRODUCTION

Solar cells made from Cu(In,Ga)Se2 (CIGS) have been extensively studied for their high optical absorption coefficient and tunable band gap (Huang, C. Cheng, 2012, Kaelin, M, 2004). Recently, much efforts on the CIGS solar cells are directly paid to thinner films for the limited natural reserves of In and Ga (Liu, C. Chuang, C. 2012, Ramanathan, K. Contreras, M. Perkins, C. Asher, S. 2003, Saji, V. Choi, I. Lee C. 2011). Two-step selenization process of metal precursors was commonly applied to achieve CIGS absorber films. First, Cu, In, and Ga metallic precursors are deposited by a multi-source sputtering process, and post-selenization is then followed using H₂Se or Se vapor. However, this two-step process has many technical drawbacks, such as high selenization temperature, rough surface of films, and agglomeration of Ga on the film–Mo layer interface, etc. (Tian, J. Peng, 2014]. In order to simplify the fabrication process, one-step technique for directly sputtering CIGS thin films has attracted much attention among researchers (Vasekar, P. Dhere, N. 2009). In this process, a single quaternary target containing In, Ga, Se, and Cu is hot compressed to become a dense target first, and coevaporated at the same time to achieve a CIGS film. This one-step coevaporation process is considered as an effective way to simplify the fabrication of large-scale CIGS films.

For further improving the performance of the one-step sputtered CIGS films, more efforts should be put on the fabrication process. In the present work, a specially designed CIGS target was employed to directly coevaporate CIGS films on a conductive glass substrate by magnetron sputtering. The influence of sputtering parameters on the properties of the deposited CIGS films was studied in detail.

2 EXPERIMENTAL PROCEDURE

2.1 Film deposition process

Glass substrates (20mm × 30mm) were cleaned with deionized (DI) water, and anhydrous alcohol. CIGS films were deposited in a RF magnetron sputtering system (JGP-450a, Shenyang branch, Chinese academy of sciences) with a chamber size of Φ450mm × 350mm and reachable maximum vacuum is 6.67×10^{-5} Pa.

The atom ratio of sputtering target is Cu:In:Ga:Se = 1:0.7:0.3:2. The size of the target is Φ60 mm × 3 mm, and it is fixed on a copper-based plate before deposition process.

2.2 Film characterization

Scanning Electron Microscopy (SEM, S-3000N, HITACHI) was applied to observe the surface morphology of the films. The crystal phase of the

CIGS films were detected by X-Ray Diffraction Analysis (XRD, Philips, PW1830) and analyzed using the Jade software.

3 RESULTS AND DISCUSSION

3.1 *Surface morphology*

Fig. 1 shows the typical surface morphologies of deposited CIGS films before heat treatment. The films are uniform and crack-free, with only some small pinholes on the surface; no much difference can be observed. However, after heat treatment, the surface morphologies show apparent difference as shown in Fig. 2. Some bubbles with cracks can be observed clearly on the surface.

The SEM morphology shows that heat treatment has an apparent effect on the quality of the CIGS films. When the films are fabricated, the excited Cu, In, Ga, and Se atoms coevaporated on the substrate,

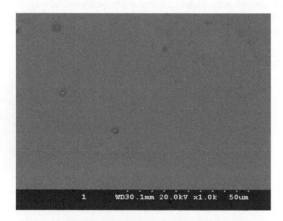

Figure 1. Surface morphology of CIGS films deposited by magnetron sputtering.

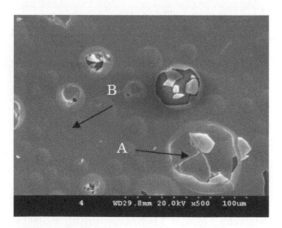

Figure 2. Surface morphology of CIGS films after heat treatment at 400°C.

the newly formed CIGS layer is cooled by the substrate and contracted quickly. When this contracted layer is heat treated, the contracted bonding layer becomes swelled and forms bubbles on the film.

To overcome the drawback from the contraction, the substrate heated over 200°C is proved to be useful during the sputtering process in our experiment. When the excited Cu, In, Ga, and Se atoms coevaporated on the heated substrate, the temperature difference between atom vapor and the substrate is less and the thermal stress is reduced. When reheated, the bonding layer has no deformation and no bubbles can be observed.

In order to further analyze the forming reasons of the bubbles, Energy Display Analysis (EDS) experiments are carried out in (point A) and out (B) of the bubbles in Fig. 2. The results show that the atom ratio of Se element is ~64.6% for point A on bubbles which are much higher than those on even area (point B), ~43.4%. The EDS data indicate that the bubbles are mainly formed by the coagulation of Se from the CIGS film.

3.2 *Crystal characteristics of CIGS films*

Fig. 3 shows the effects of substrate temperatures on the crystal characteristics of the deposited CIGS films. It is clear from the patterns that the CIGS films are amorphous when the substrate temperatures are below 200°C. However, as the temperature increased, the peaks at ~27° from CIGS crystal phase appear and increase accordingly.

To further analyze the effect of heat treatment temperature on the crystal characteristics, the CIGS films fabricated at room temperature are heat treated at different temperatures and the XRD analysis results are shown in Fig. 4.

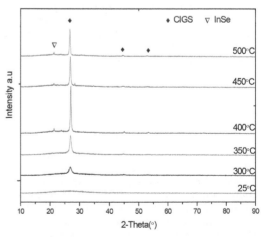

Figure 3. XRD analysis of CIGS films under different sputtering base treatments.

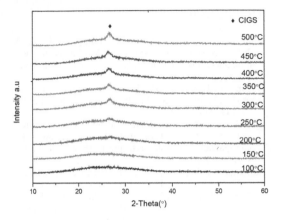

Figure 4. XRD analysis of CIGS films heat treated at different temperatures.

It is shown in Fig. 4 that the intensities of CIGS peaks of heat-treated samples are much stronger than those of untreated samples. The results indicate that heat treatment is very important for CIGS crystallization. The peak at 26.78° is from the CuGa0.3In0.7Se2 phase (112) direction. The intensity of the samples represents the crystal degree of CIGS phase. For the samples with heat treating temperature lower than 400°C, the intensity of (112) peaks increases as the temperature increase, which indicates that the increasing heat treatment temperature is helpful to the CIGS crystallization process. However, as the heat treating temperature is higher than 400°C, the intensity of (112) peaks decreases as the temperature further increases accordingly. Y. Lin et al. studied CIGS films which were prepared at different annealing temperatures and they found similar changes. They summarized that the increasing heat treatment temperature can cause the film composition to deviate from the designed element ratios, and finally lead to poor film structure characteristics.

In Fig. 4, as heat treating temperature is over 400°C, the peaks from InSe phase can be observed. The InSe peaks prove the coagulation of Se as shown in Fig. 2. Therefore, in order to acquire well-crystallized CIGS phase and decrease other unwanted phases, heat treatment at ~400°C is suitable.

The CIGS films fabricated at different substrate temperatures are heat treated at 400°C and the XRD analysis results are shown in Fig. 5. For CIGS film deposited at 100°C and 200°C, only one crystal phase at 26.78° from the CuGa0.3In0.7Se2 phase (112) can be observed. As the substrate temperature increased to above 300°C, other crystal phases from InSe phase appear. The XRD analysis indicates that a higher substrate temperature can lead to an increase in InSe phase coagulation,

which may be harmful to the properties of CIGS films in solar cells. It is also shown in Fig. 5 that the crystal peak at 26.78°from CIGS film formed at 200°C base temperature has the strongest intensity. Both lower and higher than this temperature, the intensity of crystal peak will decrease. It indicates that 200°C is the optimum substrate temperature for the one-step CIGS film deposition.

3.3 Effect of sputtering power

The sputtering power has an apparent effect on the crystal properties of CIGS films as shown in Fig. 6. The films are deposited at 200°C substrate temperature and then heat treated at 400°C for 2 h.

Sputtering power can have effects on the crystal strength of CIGS films. As the sputtering powers increase from 85 w to 165 w, the intensities of CIGS peaks increase. The results indicate that higher

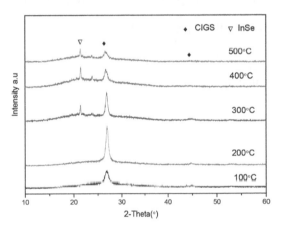

Figure 5. XRD analysis of different sputtering base temperatures with 400°C heat treatment.

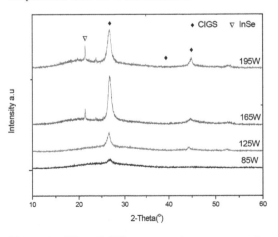

Figure 6. Effect of different sputtering powers on the crystal properties of CIGS films.

sputtering power is helpful to the crystallization process of the CIGS phase. However, as the sputtering power further increase to 195 w, the intensity of the CIGS peak is decreased. InSe phase can also be observed as a sputtering power higher than 165 W and it increased as the power increased. It is evident from the experiment that sputtering CIGS films at 165 W is suitable.

4 CONCLUSION

Cu(In, Ga)Se2(CIGS) films are deposited with a single target by the magnetron sputtering process. CIGS film fabricate without substrate heating can lead to formation of Se-coagulated bubbles on the surface after heat treatment process. The optimized parameters of the sputtering process are: substrate temperature of approximately 200°C, and the sputtering power of 165 W. Further heat treatment at 400°C will improve the crystallinity of the films.

ACKNOWLEDGMENTS

This work was supported by excellent talents in colleges and universities of Liaoning Province, China (LR2013027).

REFERENCES

Huang, C., Cheng, H., Chang, W., Huang, M., Chien, Y. 2012. Investigation of sputtered Mo layers on soda-lime glass substrates for CIGS solar cells. Semicond. Sci. Technol, 27, 1–9.
Kaelin, M., Rudmann, D., Tiwari, A.N. 2004. Low cost processing of CIGS thin film solar cells. Sol. Energy, 77, 749–756.
Liu, C., Chuang, C., 2012. Fabrication of copper indium gallium diselenide absorber layer by quaternary-alloy nanoparticles for solar cell applications, Sol Energy, 86, 2795–2801.
Ramanathan, K., Contreras, M., Perkins, C. Asher, S. 2003. Properties of 19.2% Efficiency ZnO/CdS/CuInGaSe2 Thin film Solar Cells, Res. Appl. 11, 225–230.
Saji, V., Choi, I., Lee C. 2011. Progress in electro deposited absorber layer for CuIn(1–x)GaxSe2 (CIGS) solar cells, Sol Energy, 85, 2666–2678.
Tian, J., Peng, L., Chen, J., Wang, G., Wang, X., Kang, H., Wang, R. 2014. Comparison of Cu(In, Ga)Se2 thin films deposited on different preferred oriented Mo back contact by RF sputtering from a quaternary target, Appl. Phys. A., 10, 1007–1015.
Vasekar, P., Dhere, N. 2009, Effect of sodium addition on Cu-deficient CuInGaSe2 thin film solar cells, Sol Energy Mat Sol C., 93, 69–73.

Energy Science and Applied Technology ESAT 2016 – Fang (Ed.)
© 2016 Taylor & Francis Group, London, ISBN 978-1-138-02973-6

Study on the electrochemical property of a pure titanium plate in a corrosive medium after anodic oxidation treatment

Zhefeng Xu
PanGang Group Research Institute Co. Ltd., State Key Laboratory of Vanadium and Titanium Resources Comprehensive Utilization, Panzhihua, Sichuan, China

Shan Liu
Department of Biological and Chemical Engineering, Panzhihua University, Panzhihua, P.R. China

Haoqing Zheng
PanGang Group Research Institute Co. Ltd., State Key Laboratory of Vanadium and Titanium Resources Comprehensive Utilization, Panzhihua, Sichuan, China

ABSTRACT: Anodic oxidation treatment was carried on commercial pure titanium with voltages of 2V, 3V, 5V respectively in a mixed solution containing 40 g/L sodium gluconate and 10 ml/L phosphoric. Electrochemical Properties of the passive film was studied by Electrochemical Impedance Spectroscopy (EIS) measurements. The electrochemical impedance spectroscopy was analyzed by equivalent circuit. The obtained results show that passive film obtained in 3 V has the best corrosion resistance.

Keywords: commercial pure titanium; anodize; impedance spectroscopy; corrosion resistance

1 INTRODUCTION

Titanium and its alloys have many advantages such as strong corrosion resistance, high strength to weight ratio, good toughness in high/low temperature. They are widely used in nuclear power industry, chemical industry, aerospace industry and marine industry. The studies found that titanium and its alloys have good corrosion resistance due to the passive film of titanium dioxide formed on the surface is highly anticorrosive. However, the passive film becomes unstable in the solution of hydrochloric acid or containing chlorine ions, which limits the application of titanium alloy to certain extents. For instance, in marine industry and petrochemical industry, that surface protection only relies on TiO_2 passive film is not enough. In order to overcome the shortage of titanium and its alloys in the field of anticorrosion, researchers proposed some surface strengthening methods, such as laser melting deposition, plasma penetrating and physical vapor deposition, but these methods are too complicated and expensive to be applied. Therefore, it is significant to improve a surface treatment process of titanium and its alloys with less cost and complexity. The anodic oxidation process is an optimal choice for a simple surface treatment. Anodic oxidation technology (a technology which uses external current to form a layer of oxide film on the metal surface (anode) in a corresponding electrolyte and specific conditions) can generate a more compact and tight layer of TiO_2 passive film without changing substrate composition. This method is simple and easy for operating, the passive film of titanium dioxide is stability, and the anodic oxidation film shows gorgeous color so getting aesthetic. Therefore, this paper applied anodic oxidation treatment on commercial pure titanium in sodium gluconate and phosphoric acid mixed solution system to generate a layer of anodic oxidation coloring film. Then studied the coloring rules and characteristics of anodic oxidation film under different voltage conditions, and applied electrochemical polarization curves and EIS to measure the corrosion resistance of commercial pure titanium in 3.5% NaCl solution after anodic oxidation, thereby providing a reference of surface protection of titanium and its alloys.

2 MATERIALS AND METHODS

2.1 *Physical model*

Commercial pure titanium plate (produced by Panzhihua Iron and Steel Group, China) was used as substrate in this study, the mass fraction of component is O< 0.10%, N< 0.05%, C< 0.05%, H< 0.015%, Si< 0.15%, Fe< 0.30%, the rest of Ti. The titanium plate was cut into $10 \times 10 \times 1$ mm plate as specimens. Specimens were degreased in anhydrous ethanol as cathode, then polished the surface to a mirror by shot blasting machine. After cleaned in acetone, alcohol and deionized water, specimens were dried in 75°C oven. Electrolyte solution of anodic oxidation is a mixed solution of 40 g/L sodium gluconate (Zhengzhou Chemical Industry Co. Ltd., Zhengzhou, China) and 10 ml/L phosphate (Shanghai Shicheng Co., Ltd., Shanghai, China). The anodic oxidation treatment was proceed in the potentiostat (IPD-20001SLU, Shenzhen Huaqing Instrument Co. Ltd., Shenzhen, China) with a dual electrodes system. Specimens worked as anode and 304# stainless steel worked as counter electrode; temperature was 55°C, the applied voltages were 2V, 3V and 5V respectively. The surface morphology and composition of specimens surface were observed by electron microscopy (Quanta650, the Fei company, USA) and EDAX.

2.2 Methods

The electrochemical polarization curves and EIS measured by PMC1000 electrochemical workstation (U.S.A Mei Turk, PAR, USA). Corrosive medium was 3.5% NaCl solution, a triple electrodes system was applied, the specimen worked as working electrode, saturated calomel (SCE) worked as reference electrode, and a platinum sheet worked as auxiliary electrode. Impedance analysis (measuring range: $10^4 \sim 10^{-2}$ Hz, voltage amplitude of $\pm$ 5 mV) was applied. Analysis of EIS was proceed by ZsimpWin software. Electrochemical test was repeated for three times to ensure the stability of obtained experimental data.

3 ANALYSIS AND DISCUSSION

3.1 SEM micro zone morphology observation

The surface of specimen before and after anodizing was observed by Quanta650 electron microscope (FEI, America). The micro zone morphology of the surface of specimen after shot blasting and polishing is shown in Figure 1 a, and the micro zone morphology of the specimen after anodizing is shown in Figure b. The micro zone morphology indicates that, compared with the specimen after shot blasting, the film on specimen's surface after anodizing became very dense, the film completely covered the hole and pit defects of specimens. EDAX analysis (shown in Figure 2a, b) also verifies this conclusion. The energy spectrum curves indicate that the composition of the surface of

Figure 1. Micro zone morphology of the surface of commercial pure titanium before and after the anodic oxidation (a) after shot blasting pretreatment; (b) after anodic oxidation treatment.

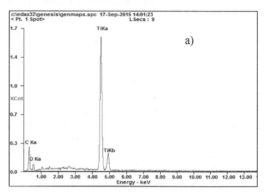

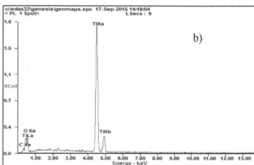

Figure 2. The EDAX energy spectrum analysis of commercial pure titanium before and after anodic oxidation (a) after shot blasting pretreatment (b) after anodic oxidation treatment.

commercial pure titanium before anodizing was mainly Ti, and a little TiO_2, but the surface was completely covered by TiO_2 film after anodizing, which also indirectly proves that a dense TiO_2 film formed on the surface of commercial pure titanium through anodizing.

3.2 Polarization curve

Figure 3 shows the polarization curve of commercial pure titanium in the mixed solution of 40 g/L sodium gluconate and 10 ml/L phosphoric

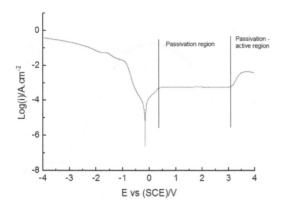

Figure 3. Polarization curve of commercial pure titanium in the mixed solution of 40 g/L sodium gluconate and 10 ml/L phosphoric acid at 55°C.

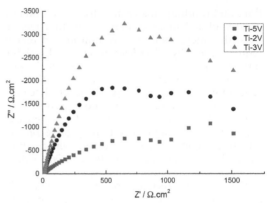

Figure 4. Nyquest diagram of specimens in 3.5% NaCl solution at 25°C.

acid. As shown, 0.5–3.2 VSCE is passive region, after 3.2 VSCE is a passive-active area. 2, 3, 5 V were selected as anodic oxidation potential. Under these potentials, the anodic oxidation process is a dynamic process, on the one hand, the generation of passive film relies on the reaction of anodic oxidation, on the other hand, passive film dissolved in interaction of H^+, Cl^- ions and the increase of current density. The current density kept constant in passive region, which indicates that the growth of passive film is relatively stable, but after 3.2 VSCE, the current density increased, which probably means that the passive film was partial collapsed, and the growth rate of passive film decreased while dissolution rate increased.

3.3 Electrochemical Impedance Spectroscopy

In order to study the effect of anodic oxidation treatment on the passive film on the surface of commercial pure titanium under different potentials, the method of EIS was selected to do the analysis. Figure 4 shows the Nyquest chart of specimens measured in 3.5% NaCl solution, at 25 °C, under open circuit potential. Obviously, two capacitive arcs appeared in high and low frequency region of Nyquest respectively. The capacitive arc of Ti-5V is very obvious, Ti-2V follows, second capacitive arc of Ti-3V is about to disappear. The phenomenon that the same Nyquest shows two capacitive arcs indicates that two electrode process occurred in the interface of passive film / electrolyte at least. The first capacitive arc in high frequency region is likely to be related to the space charge, while the second capacitive arc is likely to be related to double layer structure. In general, the passive film formed on the surface of Ti possesses some semiconductor properties (J. Jayaraj et al, 2012) (D.G Li et al, 2015), when the passive film

contacts with electrolyte, a charge transfer occurs on the interface, which leads to the appearance of space charge layer; due to the effect of metal surface tension, the surface of passive film adhered some components of the solution, such as water molecules, and due to the effect of static electricity, some ions were gathered at the interface, thus, a double electric layer structure formed between the passive film and the electrolyte, and a thermodynamic equilibrium at the both sides of the interface was immediately established.

The unit energy in the electric double layer of electrolyte was much greater than the energy in the side of the space charge layer near passive film, therefore the thickness of space charge layer is much larger than the thickness of double electric layer, which also indicates that in the Nyquest curve, semicircular radian of high frequency region is slightly larger than that of the low frequency region. By contrasting each specimen in high frequency region, it is easy to find that capacitive arc size of Ti-3V is maximum, followed by Ti-2V, the capacitance arc of Ti-5V is minimum. Capacitive arc in high frequency region is mainly resulted from the space charge layer, so it primary reflects the charge transfer on the metal / solution interface, thus, there is a close relationship between the capacitive arc size in high frequency region and corrosion resistance, larger capacitance arc means better corrosion resistance (A. Fattah-alhosseini, 2012) (Z. Feng et al, 2010).

High corrosion resistance of Ti and its alloys is mainly due to the passive film, however, the passive film can be easily corroded in the presence of halogen elements, especially the solution containing fluorine ion and chlorine ion, thereby leading to a decline in corrosion resistance. Because there are many Cl^- ions, which adhere to the passive film thereby promoting the dissolution of the passive film, and resulting in a greater dissolution rate of passive film than generation rate in that region, thus, the thickness of passive film in that region reduces gradually, which leads to a faster dissolving rate of metal ions. On the one hand, the hydrolysis of passive film resulting in the increase of the concentration of H^+ ions in the part of electrolyte near the passive film, on the other hand, the concentration of Cl^- ions also increase, the both increases accelerate the dissolution rate of passive film in return, thereby forming a so called "self catalysis Effect", eventually leading to the consistent thinning of the passive film in that region.

During this process, two time constants will appear in the impedance diagram. The capacitive arc in region of low frequency probably related to Cl^- diffusion, the capacitive arc in Ti-5V low frequency region was smallest and most obvious, which indicates that the diffusion of Cl^- on the surface of Ti-5V specimen is relatively difficult, and Cl^- is easy to gather at the passive film/electrolyte interface. The increase of capacitive arc in low frequency region of Ti-2V indicates that the diffusion of Cl^- on the surface of Ti-2V specimen became easier. The phenomenon that capacitive arc in low frequency region of Ti-3V almost disappeared indicates that the Cl^- hardly gathered at the interface. And the disappearance of capacitive arc in low frequency region is probably due to the high density of specimen's surface, less pits provide less vacancy for Cl^- ions gatering.

Figure 5 (a) shows the Bode (Frequency-Phase) measured in 3.5% NaCl solution, at 25°C, under open circuit potential. As shown, the curve of Ti-5V has two obvious phase angle peak, which represents two time constants. The first peak appeared in the low frequency region, around 10–2 Hz, the phase angle is about –40 degrees. The second peak appeared on one side of the high frequency region, around 10^{-1}~10^2 Hz, and the phase angle is close to –60°. The curve of Ti-2V in the low frequency region (10^{-2} Hz) also has a similar phase angle peak, the peak value is around –23 degrees; on one side of the high frequency region (10^{-1}~10^1 Hz) shows the second peak, the phase angle is around –73°. The curve of Ti-3V specimen almost has no phase angle peak in low frequency region, but there is an around –76° angle peak in the high frequency (10^{-1}~10^1 Hz) region. Comparing with Ti-5V specimen, phase angle peaks of Ti-2V, Ti-3V specimens moved to the high frequency region. Phase angle peak of Ti-5V moved to low frequency region, which indicates that on the surface of Ti- 5V specimen, charge transferred easier (Z.F Xu et al, 2015) (A. Fattah-alhosseini, 2012) (Z. Feng et al, 2010) (M.M Verdian et al, 2014); by contrasting the values of phase angle peaks, it can be found that the phase

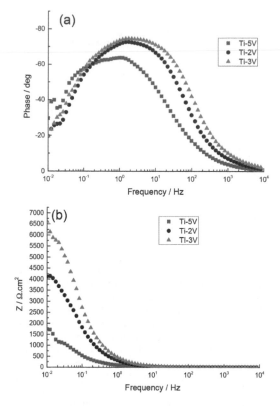

Figure 5. Bode diagrams of specimens in NaCl solution at 25 °C.

angle peaks of Ti-2V, Ti-3V specimens are higher, and that of Ti-5V specimens are slightly lower, this phenomenon mainly relates to the physical characters of the passive film formed on the surface of specimens (M. Gannouni et al, 2015). Figure 5 (b) shows the Frequency-Z of Bode diagram, the Z value of Ti-2V specimen reached a larger number ($\approx$6750 Ω.cm^2) finally, and the minimum Z value is around 2000 Ω.cm^2.

In order to determine specific electrochemical parameters of each layer, it is necessary to establish an equivalent circuit (R (Q (R (QR)))) to fit the EIS spectrum. In the equivalent circuit (Figure 6), an equivalent component was introduced to simulate the electrochemical impedance spectroscopy (CPE) better. The value of CPE reflects the deviating degree from the real capacitance forming on the passive film to ideal capacitance, n value should remain in the range between 0 and 1, when n = 1, CPE represents an ideal capacitance; when n = 0, CPE represents an ideal resistance; when n = 0.5, CPE represents the Warburg impedance with diffusion characteristic. Some factors such as surface roughness, not uniform surface grains and adsorbed ions will lead to not ideal capacitance (R. Leiva-García et al, 2015).

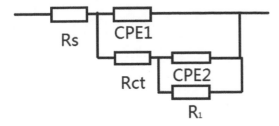

Figure 6. The equivalent circuit of the EIS spectra in the NaCl solution at 25 °C.

Rs is the electrolyte resistance; impedance was measured under open circuit potential, after the passive film partial dissolved, the surface of passive film consisted of two parts: one part is the complete region, the Faraday impedance of this part was represented by CPE1 in equivalent circuit, the value of CPE1 relates to the density and roughness of the passive film surface; the other part is the region where the passive film was penetrated, the process of metal anodic dissolution mainly occurred there and generated a higher anodic current density, thereby leading to a rapid depression of this region, and the gather of Cl⁻ to this region under the effect of anodic current density to form a region where mass transfer was difficult and ion concentration was high. The equivalent circuit of the anodic dissolution process is the series connection of the charge transfer resistance during anodic dissolution process Rct and, the parallel connection of the charge transfer resistance in the pit and crack of passive film R1 and the equivalent element CPE2.

4 CONCLUSION

A stable passive film was formed on the surface of commercial pure titanium in mixed solution of 40 g/L sodium phosphate ml/L +10 phosphoric acid through anodic oxidation treatment with using 2~5V potentials. The passive film obtained with 3V potential showed beat corrosion resistance, thus 3V is the best treatment potential. Passive films treated with different potentials varied greatly in electrochemical characteristics. The passive film of specimen with anodic oxidation voltage of 5V has smaller charge transfer resistance $(114.45\Omega.cm^2)$ during anodic dissolution process. In contrast, the passive film of specimen with anodic oxidation voltage of 2 and 3 V have greater charge transfer resistance (173.63, $185.79\Omega.cm^2$ respectively), this phenomenon most likely resulted from that the passive film formed with 5 V anodic oxidation voltage was not compact, leading to the gather of Cl⁻ thereby accelerating the dissolution of the passive film.

REFERENCES

Chen, S., Xiong, W. & Yao, Z. et al: International Journal of Refractory Metals and Hard Materials, vol.47 (2014) no.12, p. 139–144.
Fattah-alhosseini, A: Arabian Journal of Chemistry, vol. 49(2012) no.7, p. 572–573.
Feng, Z., Cheng, X. & Dong, C. et al: Corrosion Science, vol. 52(2010) no.11, p. 3646–3653.
Gannouni, M. Assaker, I.B. Chtourou, R. l: Materials Research Bulletin, vol. 61(2015) no.7, p. 519–527.
Geetha, M., K. Singh, A. & Asokamani, R. et al: Progress in Materials Science, vol. 54 (2009) no.3, p. 397–425.
Jayaraj, J. Ravi Shankar, A: U. Kamachi Mudali. Electrochimica Acta, vol. 85(2012) p. 210–219.
Leiva-García, R., Fernandes, J.C.S. & Muñoz-Portero, M.J. et al: Corrosion Science, vol. 94(2015) p. 327–341.
Li, D.G., Wang, J.D. & Chen, D.R. et al. Ultrasonics Sonochemistry, vol. 26(2015) p. 99–110.
Verdian, M.M., Raeissi, K. & Salehi, M. l: Surface and Coatings Technology, vol. 240 (2014) no.7, p. 70–75.
Xu, Z.F., Liu, S. & Gan, G.Y. et al: Optoelectron and Mat, vol. 9(2015) no.11–12 p. 1487–1490.
Xu, Z.F., Liu, S. & Gan, G.Y. et al: Optoelectron and Mat, vol. 9(2015) no.1–2 p. 260–265.

Energy Science and Applied Technology ESAT 2016 – Fang (Ed.)
© 2016 Taylor & Francis Group, London, ISBN 978-1-138-02973-6

Research on photo-degradation of typical sulfonylurea herbicides-nicosulfuron

Fanli Chen & Wenqiang Jiang

School of Environmental Science and Engineering, Qilu University of Technology, Jinan, China

ABSTRACT: The photo-degradation of nicosulfuron in aqueous solution was investigated. The effects of light source, initial concentration of the nicosulfuron, initial pH value of the solution were studied in this experiment. Results show that the photo-degradation of the nicosulfuron under the mercury lamp was better than xenon lamp. The best photo-degradation condition proved by experiment was upon mercury lamp irradiation when initial pH is 7. Meanwhile, with increasing of initial concentration of nicosulfuron, the half-life period of the photo-degradation becomes longer.

Keywords: nicosulfuron; photo-degradation; pesticide

1 INTRODUCTION

As for ensuring the high yield of the crops and human demand for food, pesticides has become an essential material in agricultural production (D.S. Chen et al, 2008) (S.L. Chu, 2014). However, problems of environmental pollution and food safety brought by the extensive use of pesticides received people's attention. Traditional pesticides (eg. DDT, BHC) with high toxicity and high residue not only pollute the ecological environment, but also damage human's health (Meyer et al, 2010). Pesticides have become one of the important sources of water pollution.

Nicosulfuron, as an effective herbicide, was widely used for removing weeds in Maize Field (D.S. Chen et al, 2008), which is discovered by Ishihara Industrial Co (C.H. Lv et al, 2010). The molecular formula of nicosulfuron is $C_{15}H_{18}N_6O_6S$ and under the standard condition is white powder.

In the past twenty years, due to the degradation process with clean and efficient, research on photodegradation has been favored by many scholars. At present, the photochemical degradation of pesticides has become a hot field at home and abroad. Through the photochemical degradation of pesticides, human can master the way of degradation of pesticides in the natural environment, to verify the safety of pesticides use and to prevent environmental pollution (J. Song et al, 2013). Therefore, the research on the photo-degradation of pesticides is of great significance.

2 EXPERIMENT

Weigh 100 mg nicosulfuron (93.5%) by using the electronic balance, dilute it with deionized water in a 100 mL volumetric flask and obtain standard solution which concentration is 100 mg/L. Dilute standard solution to 100 mg/L, 50 mg/L, 10 mg/L, and 1 mg/L, then through the liquid phase analysis to draw the standard curve (C.H. Jia et al, 2014). During the experiment, we put simulated wastewater which concentration is 100 mg/L and volume is 200 mL in reactor, turn on the light of photocatalytic reactor which cooling water was also used for reactor. Sampling analysis at intervals.

3 RESULTS AND DISCUSSION

3.1 *Effects of different light sources on photo-degradation of nicosulfuron*

From Figure 1, we can tell that the photo-degradation of nicosulfuron works best under the irradiation of high pressure mercury lamp. After 30 minutes' reaction, the photo-degradation rate is almost 100%. The photo-degradation under the irradiation of the xenon lamp is less effective. The photo-degradation rate is only 5% after 30 minutes' reaction. That is because the optical wavelength produced by the high pressure mercury lamp is 365 nm, which is in the UV range. Due to the high energy of the UV, the nicosulfuron is motivated after absorbing the light, and then gets oxidative degradation. The light of the xenon lamp is in the visible region and can not motivate the e nicosulfuron molecules.

3.2 *Effects of pH on nicosulfuron photo-degradation*

From Figure 2, we can tell that the initial pH affects the photo-degradation of nicosulfuron. The photo-

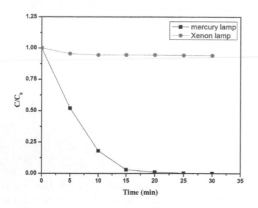

Figure 1. Effects of different light sources on photo-degradation of nicosulfuron.

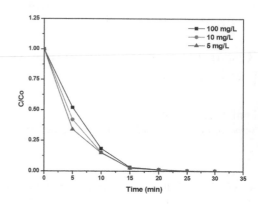

Figure 3. Effects of initial concentration on nicosulfuron photo-degradation..

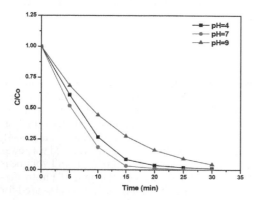

Figure 2. Effects of pH on nicosulfuron photo-degradation.

degradation rate varies with the different pH conditions. When the pH is 9, the photo-degradation rate of nicosulfuron is slow and its half-life is 8 min; when the pH is 4, its photo-degradation rate is also slow with half-life time as 6 min. When the pH is 7, the photo-degradation rate is the highest with half-life time as 4 min. It may be because of the different stabilities of nicosulfuron at different pH conditions.

3.3 Effects of initial concentration on nicosulfuron Photo-degradation

From Figure 3, we can tell that the photo-degradation rate of nicosulfuron decreases with the increase of the initial concentration. When the initial concentration is 100 mg/L, the half-life of photo-degradation is about 5 min; when the initial concentration is 10 mg/L, the half-life of photo-degradation is about 4 min. That is because, with the increase of the concentration, the light intensity remains unchanged and produces the same energy, which leads to the motivated nicosulfuron molecules becoming fewer and fewer. This phenomenon also follows the first-order degradation kinetics.

4 CONCLUSION

This paper has studied the photo-degradation of nicosulfuron in water environment. The experiment results show that the photo-degradation of the nicosulfuron under the UV produced by the high pressure mercury lamp is better than that of the visible light produced by the xenon lamp. Under the mercury lamp, the 100 mg/L nicosulfuron is almost entirely degraded within 30 min. The pH of the solution has an effect on the photo-degradation rate. When the pH is 7, the photo-degradation rate is the highest. With the increase of the initial concentration, the photo-degradation rate of the nicosulfuron decreases, which follows the first-order degradation kinetics (D.S. Chen et al, 2008).

REFERENCES

Chen, D.S., G.B. Gong and Y.D. Wang: Research on the Effect and Safety of Controlling Weed in Corn Field with 19% Nicosulfuron Bromoxynil Octanoate Suspension Concentrate, Anhui Agricultural Science, (2008)No. 36, p. 617–618.

Chu, S.L: Analysis on the Cause of Phytotoxicity of the Nicosulfuron on the Corn and the Remedial Measures, Henan Agriculture, (2014) No. 3, p. 36–38.

Jia, C.H., X.D. Zhu and E.C. Zhao: Determination of the Atrazine in Corn and the Nicosulfuron Residues by Liquid Chromatography-Mass Spectrometry, Jiangsu Agricultural Scien, (2014), No. 42(2),p. 253–255.

Lv, C.H., J.J. Wen and Q. Dai: The Make of the Nicosulfuron Microemulsion with the Method of Ternary Phase Diagram, Anhui Chemical Industry, (2010), No. 36, p. 6–8.

Meyer, D. Michael and K. Pataky: Genetic Factors Influencing Adverse Effects of Mesotrione and Nicosulfuron on Sweet Corn Yield, Agronomy Journal, (2010), No. 102 (4), p. 1138–1144.

Song, J., J. Gu and Y. Zhai: Biodegradation of nicosulfuron by a TalaRomyces Flavus LZM1, Bioresource Technology, (2013), No. 140, p. 243–248.

Energy Science and Applied Technology ESAT 2016 – Fang (Ed.)
© 2016 Taylor & Francis Group, London, ISBN 978-1-138-02973-6

Electrical conductivity of carbon nanotube fibers synthesized by chemical vapor deposition

Jingdong Zhu

Key Laboratory of Advanced Ceramics and Machining Technology, Ministry of Education, School of Material Science and Engineering, Tianjin University, Tianjin, China

ABSTRACT: Carbon nanotube fibers synthesized by chemical vapor deposition can be used for a variety of applications, due to their facile fabrication. In this paper, their electrical property was studied. We measured their room-temperature electrical conductivity, giving the value of 5×10^5 S/m for the typical fiber, and revealed their conduct transport according to the temperature dependence of the conductance, indicating Mott's variable range hopping mechanism. Moreover, the measurement of thermoelectrical power was achieved, suggesting that the carrier within the fiber is a hole.

Keywords: carbon materials, nanomaterials, electrical conductivity

1 INTRODUCTION

Carbon Nanotubes (CNTs) with unique one-dimensional nanostructures possess super strength (Yu MF, 2000), high electrical (Collins PG, 2000) and thermal conductivities (Balandin AAm, 2011), and multifunctional properties. CNT fiber, an oriented assembly of CNTs, has several potential applications such as artificial muscle (Lima MD, 2012), super capacitor (Dalton AB, 2003), ultra-high-conductivity fiber (Behabtu N, 2013), stretchable electrical conductor (Zu M, 2013), and high-performance structural fibers (Koziol K, 2007). Currently, there are four main routes to synthesize CNT fibers: 1) by spinning from a CNT solution (Ericson LM, 2004), 2) by spinning from an aligned CNT array (Zhang M, 2004), 3) by spinning from a CNT aerogel synthesized by chemical vapor deposition (CVD) (Li YL, 2004), and 4) by twisting the CNT film (Ma WJ, 2009). CNT fibers fabricated by CVD process can be commercially available on a large scale, because of facile synthesis and a stable continuously spinning process. Currently, the fabrication technology of CVD CNT fibers has matured; however, the properties of CNT fibers should be evaluated for their applications, particularly the electrical property of CNT fibers. In this study, the electrical property of CNT fibers was investigated.

2 EXPERIMENTAL METHODOLOGY

CNT fibers were fabricated through the CVD process. A mixture of catalysts and carbon source was put into a high-temperature reactor along with H_2 stream, affording a continuous stream of CNT fiber at the bottom of the reactor. The CNT fiber was removed from the reactor after online fiber densification with acetone. A detailed description of the synthesis process was mentioned elsewhere (Zhong XH, 2010). Scanning electron microscope (JSM-6700F), tunneling electron microscope (Tecnai-G20 F20), and Raman spectrometer (Renishaw) were used to characterize the cross-section and surface morphologies of carbon nanotube fibers.

Electrical measurement was performed through the four-probe method in the range of 2 to 300 K. The fiber was subjected to a constant current, which was set at 20 µA. To ensure a good contact between the electrode and the fiber, a pre-patterned glass substrate with four Al electrode strips was prepared by sputtering through a shadow mask. The four Al strips were 1 mm wide, 5 mm long, and separated by 1 mm. Then, the fiber was transferred onto the pre-patterned substrate as shown in Fig. 1 and covered with a thin layer of

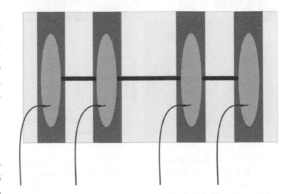

Figure 1. Schematic illustration of four-probe electrical measurement.

silver paste at each Al electrode. For the Thermo-Electric Power (TEP) measurement, the two ends of a strand (length 3 cm) were attached to hot and cold copper plates using the silver paste. Then, the hot plate was heated, and a constant temperature gradient ($\Delta T = T_{high} - T_{low} = 5$ K) was established along the direction of the length of the fiber. The voltage difference (ΔV) between the two ends of the fiber was measured simultaneously using the Keithley 2400 nanovoltmeter with two copper wires welded to the hot and cold copper plates. Therefore, the TEP can be obtained as follows: $S = \Delta V / \Delta T$.

3 RESULTS AND DISCUSSION

3.1 Structure of CNT fibers

The CNT fibers were continuous and soft as cotton yarns. Three hierarchy structures were observed in the Scanning Electron Microscope (SEM) and Transmission Electron Microscope (HRTEM). At the macroscopic scale, the fiber is hollow, with a fiber wall of 1–10 µm (Fig. 2a). The fiber wall consists of a large quantity of CNT bundles with a diameter of 10–100 nm, slightly aligned with the fiber axis (Fig. 2b). At the microscopic scale, the bundles comprise large-diameter double-wall CNT stacks (Fig. 2c). The crystallization of the fiber was characterized by Raman spectroscopy, and I_G/I_D value of 6.3 was obtained (Fig. 2d), indicating that the CNT fibers have less defect than those obtained from array (Li QW, 2007).

3.2 Electrical property of CNT fibers

To determine the electrical conductivity of the CNT fibers using the formula $\rho = (RS)/L$, cross-sectional area of the fiber should be determined first. However, it is difficult to measure the cross-sectional area directly because of their irregularity. Therefore, the cross-sectional area was obtained by an indirect method—by measuring the linear density through the vibration principle—as typically employed in the spinning industry. The cross-sectional area was calculated by dividing the linear density with the volume density of the graphite (2.1 g/cm³).

The electrical resistivity of a typical CNT fiber is 2×10^{-6} Ωm at room temperature, and the corresponding conductivity is 5×10^{5} S/m, which is higher than that of the CNT fiber from array, but lower than that of the CNT fiber from wetting spinning. To understand the electrical conductivity transport mechanism within the CNT fiber, the electrical resistivity dependent on the temperature of the CNT fiber was measured. The temperature dependence of the resistance from 2 to 300 K was plotted in Fig. 3 The resistance decreased smoothly and monotonically with the increase in temperature, indicating the semiconducting behavior. The temperature dependence of resistance is helpful to understand the intrinsic conduction. The plot of resistance versus temperature was found to be consistent with Mott's three-dimensional variable range hopping mechanism, which are expressed as follows: $R(T)/R(300\ K) = A_0 \exp((T_0/T)^{1/4})$, where A_0 and T_0 are constants. This is probably because of the contact between CNT bundles. Furthermore, to determine the type of electrical carrier within the fiber, the TEP of the fiber was measured

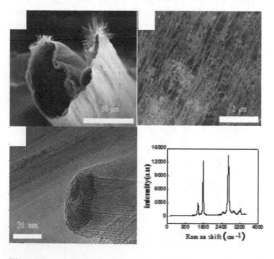

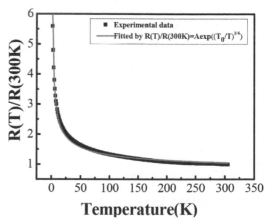

Figure 2. a) Low-magnification SEM image of a typical CNT fiber. b) SEM image at high magnification. c) Cross-section TEM image of the CNT at high resolution. d) Raman spectrum of the fiber.

Figure 3. Temperature dependence of fiber resistance R(T), normalized with respect to the R (300 K) and the fitting of resistance data with the VRH mechanism.

to be +65 μV/K. The positive sign of the TEP value indicates that the electrical carrier is a hole, similar to the P-type semiconductor.

4 CONCLUSION

The CNT fibers from CVD process were found to comprise CNT bundles with an electrical conductivity of 5×10^5 S/m at room temperature. The temperature dependence of the resistance of the CNT fiber indicates that the 3D variable-range hopping mechanism operates. The TEP measurement reveals that the electrical carriers within the fiber are holes.

ACKNOWLEDGMENT

The author is grateful to professor Ya-Li Li at Tianjin University for providing CNT fibers from CVD and Professor Li Zhiqing at Tianjin University for his help with electrical measurements.

REFERENCES

Balandin AA, Thermal properties of graphene and nanostructured carbon materials. Nat Mater 2011; 10(8): 569-81.

Behabtu N, Young CC, Tsentalovich DE, et al. Strong, light, multifunctional fibers of carbon nanotubes with ultrahigh conductivity. Science 2013; 339: 182-6.

Collins PG and Avouris P, Nanotubes for electronics. Scientific American 2000; 283(6): 62.

Dalton AB, Collins S, Munoz E, et al. Super-tough carbon-nanotube fibres. Nature 2003; 423(6941): 703.

Ericson LM, Fan H, Peng HQ, et al. Macroscopic, neat, single-walled carbon nanotube fibers. Science 2004; 305(5689): 1447–50.

Koziol K, Vilatela J, Moisala A, et al. High-Performance carbon nanotube fiber. Science 2007; 318: 1892–95.

Lima MD, li N, Endrade MJ, et al. Electrically, chemically, and photonically powered torsional and tensile actuation of hybrid carbon nanotube yarn muscles. Science 2012; 338: 928–32.

Li YL, Kinloch IA, and Windle AH, Direct spinning of carbon nanotube fibers from chemical vapor deposition synthesis. Science 2004; 304(5668): 276–8.

Li QW, Li Y, Zhang XF, Chikkannaavar SB et al. Structure-dependent electrical properties of carbon nanotube fibers. Adv Mater 2007; 19(20): 3358–63.

Ma WJ, Liu LQ, Yang R, et al. Monitoring a micromechanical process in macroscale carbon nanotube films and fibers. Adv Mater 2009; 21(5): 603–8.

Yu MF, Files BS, Arepalli S, Ruoff RS. Tensile loading of ropes of single wall carbon nanotubes and their mechanical properties. Phys Rev Lett 2000; 84(24): 5552–55.

Zhong XH, Li YL, Liu YK, et al. Continuous multilayered carbon nanotube yarns. Adv Mater 2010; 22(6): 692–6.

Zhang M, Atkinson KR, Baughman RH, Multifunctional carbon nanotube yarns by downsizing an ancient technology. Science 2004; 306(5700): 1358–61.

Zu M, Li QW, Wang GJ, et al. Carbon nanotube fiber based stretchable conductor. Adv Func Mater 2013; 23: 789–93.

Energy Science and Applied Technology ESAT 2016 – Fang (Ed.)
© 2016 Taylor & Francis Group, London, ISBN 978-1-138-02973-6

Preparation and properties of dredged sediment baking-free bricks

R. Jia, P.Q. Yang, H. Liu, X. Peng & Y. Wu
*College of Chemical Engineering and Materials Science, Tianjin University of Science and Technology,
Tianjin, P.R. China*

ABSTRACT: In this paper, a method of utilizing dredged sediment to prepare a waterproof aggregate was developed, and then vibrating molding technology was used to fabricate the Dredged Sediment Aggregate Baking-free Bricks (DSABBs). In order to compare with the bricks made by directly adding sediment (DSBBs), two kinds of bricks were characterized, that includes pressure, water absorption rate, and aggregate brick acid and salt frost resistance by contrast. The compressive strength of DSABBs cured for 28 d could reach 4.8 MPa, while the strength of DSBBs is lower than 2.7 MPa. Water absorption of DSABBs was approximately 4%, which was lower than that of DSBBs under the same condition. The value of water absorption meets the requirements of JC /T422-1991 standard, which provide that the water absorption rate should be less than 20% after immersing in water for 24 h.

Keywords: dredged sediment; baking-free aggregates; waterproof aggregates; baking-free bricks

1 INTRODUCTION

At present, the sediments are produced by dredging engineering at the rate of billion cubic meters per year and these data would continue increasing rapidly in the following years. Generally, dredged sediment is a type of very soft soil with notably low mechanical strength and extremely high moisture content. Taking into account the resource conservation and environmental protection, many countries have restricted, or even forbidden, to make use of natural resources to produce any construction materials. Therefore, scientific and circulatory utilized dredged sediment as secondary resources would solve the pollutant treatment problem, and it has important practical significance.

Baking-free bricks, which have the advantage of reducing fossil energy consumption, would meet the requirements of the building materials, including safe and durable new solid or hollow brick. Main constituents contained in dredged sediment was silicon oxide, which would act as a main component of baking-free bricks. Thus, preparation of baking-free bricks with dredged sediment will not only achieve sediment recycling utilization and energy conservation, but also avoid the gas generated from burning.

Lightweight aggregates, which were made from mineral resources such as clay, shale, and fly ash in common, were a kind of building materials, which can be used as the skeleton of concrete with cements. In this study, waterproof lightweight aggregates were prepared by utilizing dredged sediment as main raw materials, and then the baking-free bricks were fabricated with these aggregates with cement. Moreover, compressive strength, degree of water absorption, acid and alkali resistance, and salt resistance to freezing of baking-free bricks of DSABBs and DSBBs were both characterized, in order to compare and improve the utility of dredged baking-free bricks, and provide some scientific basis for the further study.

2 MATERIALS AND METHODOLOGY

2.1 Properties of materials used in aggregates and baking-free brick production

The dredged sediments utilized in this study were collected from Dianchi Lake, and their basic properties are listed in Table 1. Fly ash was obtained from Tianjin Huadian Nanjiang thermal power plant. Other chemicals used in the experiment are the following: Portland cement (42.5R, Tianjin Kun Bo Yi Co., Ltd, China), calcium oxide (Hua Sheng Tian He Chemical Co., Ltd, China), phosphogypsum (Jin Hui TaiYa Chemical reagent Co., Ltd, China), water glass (Le Tai Chemical Co., Ltd, China), white glue (Xin Shuang Ying Building Materials

Table 1. Basic properties of sediments.

Water content /%	Organic content/%	pH	Unit weight /g/cm^3	Porosity /%
48.2	9.1	6.8	1.4	28.9

Table 2. Chemical composition of dredged sediments, cement, fly ash, calcium oxide, and phosphogypsum (% by weight).

Oxide (%)	Dredged sediments	CEMI 42.5R	Fly ash	Calcium oxide	Phosphogypsum
SiO_2	70.11	20.51	54.00	3.18	3.08
Al_2O_3	11.76	4.82	21.58	2.10	0.41
Fe_2O_3	4.89	3.44	6.30	0.30	0.21
CaO	0.91	63.74	6.54	85.76	29.82
MgO	0.89	1.90	1.55	0.59	0.11
SO_3	–	2.77	0.65	–	41.86
Na_2O	1.26	0.35	0.86	–	0.04
K_2O	1.76	0.66	0.85	0.12	0.04
TiO_2	0.87	–	–	–	—

Co., Ltd, China), and a waterproofing agent (Kun Nai Building Materials Co., Ltd, China). Chemical composition of some raw materials, such as dredged sediments, Portland cement, fly ash, calcium oxide, and phosphogypsum, are given in Table 2.

2.2 Preparation of baking-free lightweight aggregates

In this study, the aggregates were prepared by the disc granulation machine. While the pelletizer operated, the disc worked in a condition of the angle 45° and the speed is 45 r/min based on the previous study. According to our previous work, the optimum compositions of raw materials were recommended as the following: dredged sediment, 80%; cement, 3%; quicklime, 3%; phosphogypsum, 3%; fly ash, 5%; and binder, 6%. After the dry raw materials were mixed uniformly, the mixture was poured into the disc pelletizer. Then, the binder aqueous solution was sprayed onto the powder mixtures for the purpose of forming spherical pellets. After the disc pelletizer rolling for 0.5 h, the obtained pellets were naturally cured for 7 days and labeled as Baking-Free Lightweight Aggregates (BLAs). Waterproof Lightweight Aggregates (WLAs) were prepared by utilizing BLAs sprayed waterproofing agent (10wt%) on their surface in the weight ratio of 1:5. The flowchart of preparing baking-free aggregates is shown in Fig. 1.

2.3 Preparation of baking-free brick production

In the laboratory, waterproof aggregates for preparing baking-free bricks were prepared by a shock forming method, the model of shock block forming machine was QMR 2-45, and the size of the obtained brick was 100 mm × 50 mm × 50 mm.

In our previous study, the optimum compositions of molding the baking-free bricks were recommended as the following: waterproof lightweight aggregates, 54.16%; cement, 12.5%; quicklime, 6.97%; phosphogypsum, 1.99%; and fly ash,

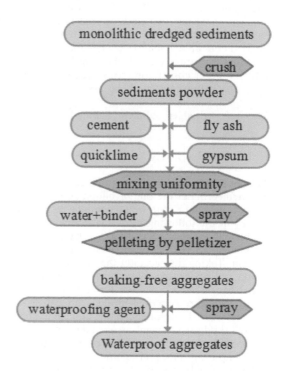

Figure 1. Flow chart of the manufacturing process for Baking-free lightweight aggregates and Waterproof lightweight aggregates.

24.38%. The photographs of waterproof aggregates, raw materials, equipment, and obtained baking-free bricks are shown in Fig. 2. The waterproof lightweight aggregates and cement, fly ash, gypsum, lime, and other cementitious materials were mixed evenly, a certain amount of water was added and mixed, and then transferred into the brick machine after aging for 10 min, the baking-free bricks appeared after the brick machine shocking for 1 min. The dredged sediment aggregate baking-free bricks were obtained after natural curing for 28 days.

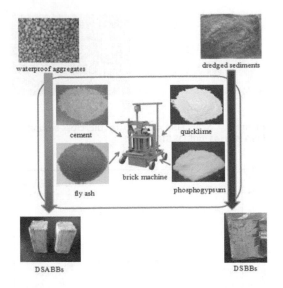

Figure 2. Production of dredged sediment baking-free bricks.

DSBBs were fabricated with similar methods of DSABBs, according to the correlative reference, and the addition ratio of sediment is the same as WLAs. The optimum compositions of raw materials were recommended as follows: dredged sediment, 43.2%; cement, 14.12%; quicklime, 8.62%; phosphogypsum, 3.62%; and fly ash, 30.44%. For testing and comparing purposes, DSBBs and DSABBs were formed in the same shock forming method.

2.4 Baking-free brick performance characterization

Compressive strength
The experimental instrument for measuring the compressive strength is electronic universal testing machine GNT200. In addition, for the experiment using the displacement method, the press speed is 4 mm/min. The specimens were placed in the middle of the bearing plate, pressing the loading with a uniform speed of 2–6 kN/s until the blocks are destroyed, and the maximum failure load P was recorded. According to Eq. 1, the compressive strength of the specimen (Rp) is calculated:

$$Rp = \frac{P}{LB} \qquad (1)$$

Rp—compressive strength, MPa; P—maximum failure load, N; L—length of compression surface (junction), mm; and B—width of compression surface (junction), mm.

Water absorption
The bricks were dried in the oven at 105°C until a constant weight, and then their value of weight was recorded after removing the dust from the surface. Then, the dry bricks were soaked into water (10–30°C) for a certain while. After removing excess water, weigh the bricks immediately. The wet mass of soaking for 24 h was labeled as G24, and the water absorption for 24 h at room temperature was W24, which was calculated based on Eq. 2:

$$W_{24} = \frac{G_{24} - G_0}{G_0} \times 100\% \qquad (2)$$

G_0—Quality of sample under dry condition, g;

Resistance to acid and alkali
In the resistant acid-base test for bricks, pieces in the oven were dried at 105°C first, then soaked into three solutions within different pH values (pH = 2, 7 and 12), and the whole specimens was kept immersed. The weight of pieces after natural air drying was measured, and the mass loss rate at different corrosion times was calculated as Eq. 3.

$$\Delta m = \frac{m - m_d}{m} \times 100\% \qquad (3)$$

Δm—mass loss rate at different corrosion times,%; m—original weight of the brick, g; and m_d—average quality of samples after corrosion, g.

Frost resistance
Specimens were placed in the oven for drying at 105°C, until the time interval of 1 h mass reduction is less than 0.1 g. Specimens were placed in different concentrations of $CaCl_2$ solution (5%, 10%, and 15%;) and the temperature was adjusted from −18°C to 18°C. Finally, the specimens were laid under room temperature for 2 h. All these above operations were termed as a cycle. After certain cycles of freezing and thawing, the samples were dried naturally.

The mass loss rate of freezing and thawing is calculated as Eq. 4, and their changes in compressive strength rate were calculated as Eq. 5:

$$\Delta M = \frac{M - M_D}{M} \times 100\% \qquad (4)$$

ΔM—weight loss rate of brick specimens after freeze-thaw cycles,%; M—average quality of pre-salt samples, g; and M_D—average quality of samples after salt freeze, g.

$$\Delta R = \frac{R - R_D}{R} \times 100\% \qquad (5)$$

ΔR—brick compressive strength loss rate in freeze-thaw cycles,%; R—average value of compressive strength of pre-salt samples, MPa; RD-average value of compressive strength of specimens after salt freezing, MPa.

3 RESULTS AND DISCUSSION

3.1 *Compressive strength*

The compressive strength of bricks is an important factor, which would affect their engineering and practical characteristics, and the values of DSABBs and DSBBs at different curing ages were obtained as shown in Fig. 3. The average compressive strength of DSABBs at the age of 7 d, 14 d, 21 d, and 28 d were 4.3 MPa, 4.50 MPa, 4.7 MPa, and 4.8 MPa, respectively, while the maximum compressive strength of DSBBs was below 2.7 MPa. It could be clearly seen that the compressive strength value of DSABBs is much larger than DSBBs. In addition, the result shows that the method of preparing baking-free bricks in dredged aggregates can greatly enhance the mechanical properties of baking-free bricks, and this strategy of utilizing sediment in construction material is feasible. The reinforcement of DSABBs probably lies in continuous CMs. While the sediment is prepared as aggregates, CMs in baking-free bricks could form a continuous phase of cement, and these CMs would work as a skeleton to present a strong binding force of baking-free bricks. In the case of sediment directly mixing with CMs, all the raw materials are completely uniform, and the presence of sediment will limit the pozzolanic reaction, which was reducing the strength of baking-free bricks. Based on the above-mentioned factor, high-strength continuous phase, such as CMs, is the guarantee of building material strength. The schematic diagrams of the cross-section of the bricks are shown in Fig. 4.

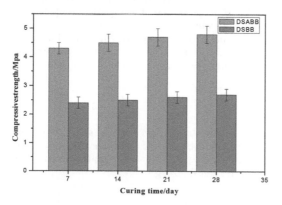

Figure 3. Compressive strength test of DSABBs and DSBBs.

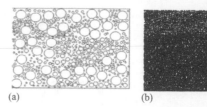

(a)   (b)

Figure 4. Schematic diagram of the cross-section of DSABBs (a) and DSBBs (b).

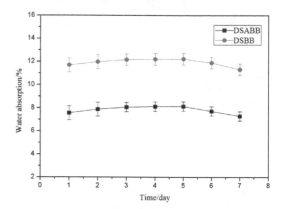

Figure 5. Water absorption of Dredged Sediment Baking-free Bricks (DSBBs) and Dredged Sediment Aggregate Baking-free Bricks (DSABBs).

3.2 *Water absorption*

Ability to resist water and rain is the necessary property of most building materials, and the baking-free bricks are not excluded. The water absorption measurement of DSABBs and DSBBs was carried out in immersing the pieces, and the obtained water absorption rate in different times (1 d, 2 d, 3 d, 5 d, 7 d) is shown in Fig 5. It can be seen from the figure that the changes of two kinds of baking-free brick immersion in water for 5 days are not oblivious, which could indicate that the baking-free brick soaking reached saturation after 1d, and the water absorption decreased 5 days later. Analysis of water absorption test results shows that DSBBs and DSABBs immersed in water at different times are satisfied with the requirements of JC/T422-1991 standard, which is the water absorption rate less than 20% for 24 h.

With the advantage of aggregates and CMs in DSABBs compacted, the surface of the brick is smooth and without any cracks. In the process of absorption, the brick surface was first wetted with water, which then penetrates into their inner. In particular, the aggregates themselves have a waterproof surface, which prevent the sediment core from touching the water, and both the CMs and aggregate would not be damaged due to water immersion. In conclusion, making maximum degree of efforts to prevent the most hydrated part of building materials from contacting moisture is an important mean to improve the practical performance of building materials.

3.3 Resistance to acid and alkali

Baking-free bricks in the use of the process will encounter with the corrosion of acid or alkali, such as acid rain. Therefore, the resistance to acid and alkali is one of the most important indexes of their practical application. The photographs of the baking-free bricks before and after acid and alkali corrosion are shown in Fig. 6, respectively, from left to right; immersion corrosion solution followed by hydrochloric acid solution (pH = 2), water (pH = 7), and dilute sodium hydroxide (pH = 12). Comparing specimens before and after immersion, it could hardly observe any change. It was especially noteworthy that DSBBs collapsed after immersion for 5 d in water, and thus the measurement of water or acid and alkali resistance could not be performed at DSBBs.

By studying the internal structure of DSABBs, binding of aggregates and CM was very tight, and CM was still a continuous phase, which ensures that the brick maintains a hard skeleton and smooth surface without any cracks, so that DSABBs have a good waterproof effect, even for acid or alkali solution. With the improvement of waterproof effect of bricks and the aggregates, the resistance to acid or alkali would also be promoted. It is obvious that brick corrosion resistance mainly depends on the properties of cementitious phase, rather than inner aggregates.

3.4 Frost resistance

Qualified building materials should be applied to different temperatures, regardless of the high or low temperature; therefore, the test of frost resistance has the significance of practical application of baking-free bricks. The performance was determined at a certain concentration of $CaCl_2$ solutions for 5, 10, or 15 cycles, and their mass loss rate is shown

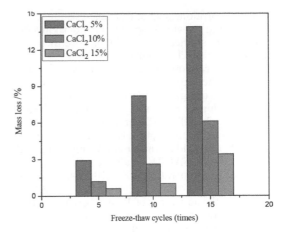

Figure 7. Variation of weight loss as a function of freeze-thaw cycles for DSABBs.

in Fig. 7. In the same concentration of $CaCl_2$ solution, the mass loss rate of DSABBs was increasing with multiple freeze-thaw cycles, and the mass loss rate was less than 6% after 15 cycles. It shows that DSABBs has good frost resistance even in salt solution. It is worth noting that, after 15 cycles, the mass loss rate of DSABBs in 5% $CaCl_2$ solution was as high as 13.9%. Obviously, application of aggregate baking-free brick should avoid suffering low-concentration salt solution, and their appearance changes consistent with this conclusion.

For DSABBs, the continuous phase is the cementing material, which grated tightly, and so the aggregate is. on the one hand, the infiltration of water is reduced, the volume of expansion force decreases. The specimens have to bear hydraulic pressure and osmotic pressure when it was subjected to freeze-thaw cycles, and DSABBs were able to withstand these pressures because of their strong water resistance. On the other hand, the strength of DSABBs is originally higher, and it is the reason of strong ability to resist frost force.

In the freezeing-thawing process, the diffusion of chloride ion was according to Fick's second law, which suggests that the concentration difference between outside and inside of the brick determined the extent of its damage. Therefore, the brick soaking in a solution of low concentration was most likely to be damaged, because of their high internal concentration.

4 CONCLUSIONS

Dredged sediment was employed as a raw material for preparing DSABBs and DSBBs, and the preparation process and the raw materials ratio were explored. Furthermore, compressive strength,

(a)   (b)

Figure 6. Exterior morphology of DSABBs before and after acid and alkali corrosion.

water absorption, acid and alkali resistance, and freezing and thawing properties of the baking-free bricks were studied in order to improve their performance in practical application.

Mechanical strength of DSABBs was increasing with the natural curing time, from the first 7 days of curing 4.3 MPa to curing for 24 days, while reaching 4.8 MPa with the value of DSBBs lower than 2.7MP, which demonstrated that a continuous high-strength phase is the guarantee of the strength of building materials. Comparing with DSBBs and DSABBs, in any soaking time, the water absorption rate of DSABBs was less than DSBBs, due to the waterproof shell of WLAs. Owing to the advantage of DSABBs, acid or alkali corrosion resistance and freeze-thaw performance were better than for DSBBs. Moreover, the difference in concentration between the outside and the inside of the brick would determine the extent of its damage in the process of frost resistance, thus reducing the inner ion concentration of the brick, which is the direction of our future study.

ACKNOWLEDGMENTS

We gratefully acknowledge our colleagues for their assistance with some of the experimental works. The work was financially supported by the fund of CCCC Tianjin Port & Waterway Prospection & Design Research Institute Co. Ltd (1500030023).

REFERENCES

Baruzzo D. et al. 2005. Possible production of ceramic tiles from marine dredging spoils alone and mixed with other waste materials. J Hazard Mater; 134(1) 202–210.

Huang Y.H. et al. 2011. Change of mechanical behavior between solidified and remolded solidified dredged materials. Eng Geol; 119(3): 112–119.

Bing J.I. et al. 2010. Dredged mud solidification disposal techniques and resource, J. Safety and Environmental Engineering, 17(2): 54–56.

Kornmann M. Clay bricks and roof tiles: manufacturing and properties. Paris: Sociétéde l'industrie minérale; 2007.

Liu Z. et al. 2011. Utilization of the sludge derived from dyestuff-making waste water coagulation for unfired bricks, J. Construction & Building Materials; 25(4): 1699–1706.

Mizuriaev S.A. et al. 2015. Production Technology of Waterproof Porous Aggregates Based on Alkali Silicate and Non-bloating Clay for Concrete of General Usage, J. Procedia Engineering; 111: 540–544.

Niyazi U.K., Turan O. 2010. Effects of lightweight fly ash aggregate properties on the behavior of lightweight concretes, J. Hazard Mater; 179 (1–3): 954–965.

Naganathan S et al. 2015. Performance of bricks made using fly ash and bottom ash, J. Construction & Building Materials; 96: 576–580.

Saeed A., Lianyang Z. 2012. Production of eco-friendly bricks from copper mine tailings through geopolymerization. Constr Build Mater; 29: 323–331.

Xu G.R. et al. Effect of sintering temperature on the characteristics of sludge ceramsite. J Hazard Mater 2007; 150 (2): 394–400.

Wang Fazhou et al,. 2014. Influence of Chloride Solution Concentration Difference on Salt Frost Scaling of Concrete, J. Journal of building materials; 17(1): 138–142.

Zhang W. 2015. Low-velocity flexural impact response of steel fiber reinforced concrete subjected to freeze-thaw cycles in NaCl solution, J. Construction & Building Materials; 101: 522–526.

Zou J.L. et al. 2008. Ceramsite obtained from water and waste water sludge and its characteristics affected by Fe_2O_3, CaO and MgO. J. Hazard Mater; 165(1): 995–1001.

Energy Science and Applied Technology ESAT 2016 – Fang (Ed.)
© 2016 Taylor & Francis Group, London, ISBN 978-1-138-02973-6

Baotou fluorite component analysis and the possibility of market application in Banyan Obo, China

Zihuan Wang & Hao Ren
Shandong Provincial Key Laboratory of Fluorine Chemistry and Chemical Materials, School of Chemistry and Chemical Engineering, University of Jinan, Shandong, China

Xiaogang Li, Wenxiang Meng & Guoying Yan
Baotou Steel Group Mining Research Institute (Limited Liability Company), Inner Mongolia, China

Yufen Zhang & Taizhong Huang
Shandong Provincial Key Laboratory of Fluorine Chemistry and Chemical Materials, School of Chemistry and Chemical Engineering, University of Jinan, Shandong, China

ABSTRACT: The Bayan Obo ore system in northern China is the world's largest comprehensive deposit, and also contains considerable amounts of iron, REE, niobium metals and fluorite. Although there are abundant fluorite resources, it has been no effective recovery and utilization. In this study we report the detailed mineral chemistry of fluorite, supporting data were collected with Thermogravimetry (TG), grain size analysis and X-Ray Diffraction (XRD) compared with commercially available fluorite (JiangXi fluorite, JX). It was found that the average particle size of Baotou fluorite (BG) concentration to 23 μm compared with the average particle size 42 μm JiangXi fluorite concentrate granularity, Baotou fluorite concentration significantly more fine granularity. And prove superior purity of Baotou fluorite (BG) at the same time, with commercial acid powder—JiangXi fluorite (JX) almost no difference. By contrast, Baotou fluorite has much better application in the market, has ambitious prospects. In addition, the component analysis shows that fluorite constituents contain calcium fluoride, can be transformed into raw materials of fluorine chemical industry and fluorine-related products, and can make the maximum benefits of comprehensive utilization of resources.

Keywords: fluorite, mineral chemistry, average particle size, purity, component analysis

1 INTRODUCTION

There are almost 1 billion tons total reserves of fluorite in the world (LID Li-xia et al. 2011, Hu Yuehua et al. 2003), China is one of the most populous country of fluorite mineral in the world, which has 35% of the world's reserves. Fluorite is an important industrial raw material, because it contains the element fluorine and has a low melting point, widely used in chemical industry (Qi Li et al. 2014), metallurgy (E. Aghion & Y. Perez 2014), building material industry and etc. With the development of economy, science and technology, fluorite has become infrequent mineral raw materials in the modern industrial and many developed countries regard it as important strategic material reserves. Bayan Obo ore belongs to polymetallic intergrowth fluorite ore (Xiao-Wen Huang et al. 2015), Bayan Obo integrated utilization of mineral resources research for decades, but so far, really for

industrial use only iron and rare earth two useful ingredients, fluorite part is still in the laboratory research stage (Hongen Gu et al. 2011).Fluorite is a natural calcium fluoride mineral with general chemical formula CaF_2 (Shoeleh Assemi et al. 2006, J. A. Mielczarski et al. 1999). Fluorite with 51.1% calcium and 48.9% fluoride, in theory, has the highest content of fluoride among the minerals in the crust (Sahar Afzal et al. 2010). Tailings dam (Yang Yi et al. 2011) was established in 1965 in Baotou iron and steel company is dedicated to the accumulation of Baotou dressing up the rest of the slag (F. Engström et al. 2013), mineral powder, and slag powder in circulating water cinder flushing way through open chute into the dam. After 36 years accumulate over a long period, quantity of tailings dam has amounted to 170 million tons, 67% of the quantity of mining, rare earth 8.5 million tons and tailings dam covers an area of 11 square kilometers, is China's largest tailings dam. Herein, fluorite in tailings dam recycling cannot only mitigate the pollution to the environment, alleviate steel a year into the overhead expenses of tailings dam, and the protection of limited mineral resources.

2 EXPERIMENTAL

BG fluorite is purple powder, and commercial acid powder—JX fluorite, is white powder. Natural pure fluorite is colorless and transparent. In general, natural fluorite exhibits various colors due to different color centers or impurities (Hongen Gu et al. 2011). The precursor of fluorite powder has been prepared by a simple test sieve screening method, since it is an informal, fast and effective method of analysis and separation.

In order to check the quality of the samples X-ray diffraction powder patterns were collected using an X'Pert PRO X-ray diffraction system equipped with a PIXcel ultrafast line detector, focusing mirror and Soller slits for CuKα1 radiation (λ = 1.54056 Å). Four-hour scans were applied in order to attain the best accuracy of results.

In order to check the flow of components of the samples TG-DTG patterns were collected at discrete temperature using Perkin-Elmer Diamond TG/DTG equipped. Temperature accretion was 10°C/min in the air flow when measuring the samples. The main characteristic of thermogravimetric method is strongly quantification, can accurately measure the change in the material and the rate of change.

3 RESULTS AND DISCUSSION

3.1 *The grain size analysis of the sample*

Component contents of experimental samples are shown in Table 1. It can be proved that CaF2 make up most of BG fluorite and initial calcium fluoride content is 93.94%. This also proves that the fluorite can promote the development of fluorine chemical industry. The development and comprehensive utilization of BG fluorite resources reflects an effective way of improving the economic benefits, social benefits and environmental benefits. Herein, to build a world-class large-scale fluorine chemical material production and research and development base will have crucial significance in Baotou, China.

Figure 1 shows the BG fluorite concentrate granularity accumulation of −500 meshes

accounted for about 50%, to −200 mesh accumulation accounted for about 98%. The average particle size is 23 microns. And Fig. 2 shows the JX fluorite particle size accumulation of −500 meshes accounted for about 27%, covering around 80.4% to −200 meshes. To sum up, the BG fluorite concentration granularity more granular compared with JX fluorite significantly.

3.2 *X-Ray Diffraction study*

Figure 3 shows XRD images for the fluorite sample and Table 2 show the diffraction angle of the main peak position. This figure confirms that the BG fluorite (blue curve) closely related to the JX fluorite (black curve) and their five peak positions dovetail beautifully with reference material (red short line). Prove lofty purity of BG fluorite at the same time, with commercial grade acid powder—JX fluorite concentrate almost no difference.

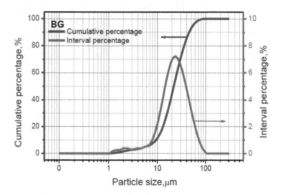

Figure 1.   BG fluorite granularity analysis curve.

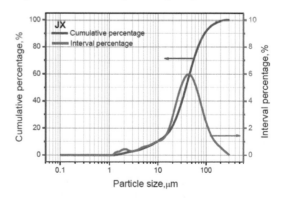

Figure 2.   JX fluorite granularity analysis curve.

Table 1.   BG fluorite composition (mass percent,%).

Component	CaF$_2$	P	S	REE	Fe$_2$O$_3$	K$_2$O	Na$_2$O	SiO$_2$	CaCO$_3$	BaO	MgO
	93.94	0.16	0.083	1.53	1.00	0.008	0.12	0.72	1.35	0.35	0.17

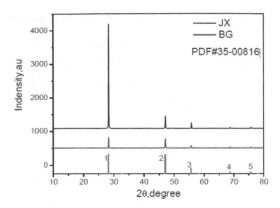

Figure 3. XRD analysis of BG fluorite and JX fluorite.

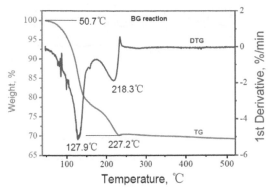

Figure 4. TG-DTG profiles of the reaction what BG fluorite and sulfuric acid.

Table 2. The main peak position corresponding to the diffraction angle (2θ).

Ore sample and standard sample	Diffraction angle (2θ)				
	1	2	3	4	5
JX fluorite	28.280	47.001	55.763	68.672	75.859
BG fluorite	28.270	47.002	55.757	68.633	75.823
Standard specimen	28.266	47.004	55.765	68.672	75.848

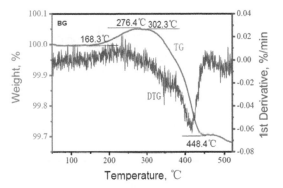

Figure 5. TG-DTG profiles of BG fluorite.

3.3 TG-DTG under fluorite and sulfuric acid reaction and only fluorite

Hydrogen fluoride can be obtained by contacting fluorspar with concentrated sulfuric acid. The reaction is endothermic. Fig. 4 TG-DTG curves showed that, first generation intermediate compound dissolved in liquid of HF, and calcium sulfate crystallization precipitation from solution, hydrogen fluoride vapor emissions in twice. Under the condition of low temperature, hydrogen fluoride may be dissolved in HSO3F, when the temperature is higher than 100°C, HSO3F hydrolysis and reduction hydrogen fluoride quickly.

The first stage, it reacts in 50.7°C~127.9°C, heated to 127.9°C, weightlessness rate of the reaction system was 13.8%; next stage, it reacts in 127.9°C~227.2°C, weightlessness rate in the process of the reaction was 14.0%. For the TG curves of the whole process, the weightlessness rate reached 27.8%, however, 22.72% for calcium fluoride and sulfate reaction, and at the same time has accounted for about 5.08% of the CO2, SiF4, H2O gas emission, etc. And Table 1 illustrates the BG fluorite contains a variety of impurities, Fig. 5 shows that the sample weight within the range of 168.3°C~276.4°C with the rise of small change, can infer sample contains reducible components, such as P, S and bivalent

iron ion absorbed oxygen or water in the air, which show the weight, the weight gain rate was +0.05%. In addition, 276.4°C~302.3°C with tiny weightlessness, as temperatures continue to rise what weightlessness rate is about 0.35%, to 448.4°C turning points, the situation illustrates the decomposition of mineral composition happened, according to the known mineral component analysis estimates for magnesium carbonate in weightlessness.

Figure 6 shows that the reaction of JX fluorite and equimolar sulfuric acid also could be divided into two main stages. The first stage, it reacts in 66.3°C~122.8°C, heated to 122.8°C, weightlessness rate of the reaction system was 9.8%; next stage, it reacts in 122.8°C~268.6°C, weightlessness rate in the process of the reaction was 32.8%. For the TG curves of the whole process, the weightlessness rate reached 42.6%, however, 22.72% for calcium fluoride and sulfate reaction, and at the same time has accounted for about 19.88% of the CO2, SiF4, H2O gas emission, etc. Fig. 7 (TG curves) shows that the specimen more apparent weightlessness in326°C, more significant compared with the Fig.5, the highest weight loss rate of 4.4%, it

147

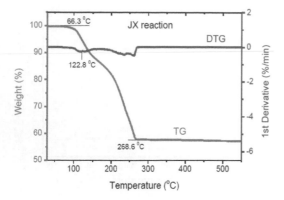

Figure 6. TG-DTG profiles of the reaction what JXfluorite and sulfuric acid.

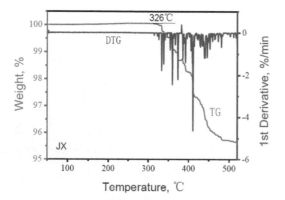

Figure 7. TG-DTG profiles of JX fluorite.

indicates weightlessness is due to test the decomposition reaction, such as magnesium carbonate decomposition, the volatilization of S and P, etc.

4 CONCLUSIONS

China is the biggest fluorine chemical industry production capacity in the world, yet as a whole laying the bottom of the global industrial chain, low value-added products and high-end products rely on imports. In order to improve the situation of fluorine chemical industry is big but not strong, the national policy also began gradually preference to the fluorine chemical industry in China, continue to regulate and support to the fluorine chemical industry development. However, fluorine chemical industry development cannot leave the fluorite mineral. m Compared to BG fluorite and JX fluorite, not only the grain size is fine, but also the content of calcium fluoride is higher. Herein, BG fluorite can replace JX fluorite in the market. And more importantly, with the development of

the fluorine industry and the increasing demands from other industries, the current reserves of fluorite could not meet the demands, but BG fluorite resource is abundant. More importantly, the development and comprehensive utilization of BG fluorite resources reflects an effective way of improving the economic benefit, social benefit and environmental benefit.

ACKNOWLEDGMENTS

This work was financially supported by the Baotou Steel Group Mining Research Institute (limited liability company) (2014034). Special thanks are given to the research team for providing support in this study.

REFERENCES

Aghion, E., Y. Perez. 2014 Effects of porosity on corrosion resistance of Mg alloy foam produced by powder metallurgy technology. Materials Characterization 96 78–83.

Engström, F., D. Adolfsson, C. Samuelsson, Å. Sandström, B. Björkman. 2013 A study of the solubility of pure slag minerals. Minerals Engineering 41 46–52.

Hongen Gu, Dongliang Ma, Weiwei Chen, Rui Zhu, Yutong Li, Yang Li. 2011Electrolytic coloration and spectral properties of natural fluorite crystals containing oxygen impurities. Spectrochimica Acta Part A 82 327–331.

Hu Yuehua, Ruan Chi, and Zhenghe Xu. 2003 Solution Chemistry Study of Salt-type Mineral Flotatio n Systems: Role of Inorganic Dispersants. Ind. Eng. Chem. Res. 42, 1641–1647.

Lid Li-xia, Wang Shi-jun, Dong Yuan-chi, Jia X iao-hui. 2011. Performance of Low Fluoride Depho sphorization Slag of Hot Metal. Journal of I Ron and Steel Research, Internxfion AL. 18(1):11–15

Mielczarski, J.A., E. Mielczarski, and J.M. Cases.1999 Dynamics of Fluorite-Oleate Interactions. Lan gmuir 15, 500–508.

Qi Li, Ya-Ni Wei, Guizhen Liu, Hui Shi. 2014 CO 2-EWR: a cleaner solution for coal chemical ind ustry in China. Journal of Cleaner Production xx x1–8.

Sahar Afzal, Amir Rahimi, Mohammad Reza Ehsani, Hossein Tavakoli. 2010 Experimental study of hydrogen fluoride adsorption on sodium fluoride. Journal of Industrial and Engineering Chemistry 16 147–151.

Shoeleh Assemi, Jakub Nalaskowski, Jan D. Miller, and William P. Johnson. 2006 Isoelectric Point of Fluorite by Direct Force Measurements Using Atomic Force Microscopy. Langmuir, 22, 1403–1405.

Xiao-Wen Huang, Mei-Fu Zhou, Yu-Zhuo Qiu, Liang Qi.2015 In-situ LA-ICP-MS trace elemental analyses of magnetite: The Bayan Obo Fe-REE-Nb deposit, North China. Ore Geology Reviews 65 884–899.

Yang Yi, Sun Wei, Li Shufen. 2011Tailings dam stability analysis of the process of recovery. Procedia Engineering 26 1782–1787.

Energy Science and Applied Technology ESAT 2016 – Fang (Ed.)
© *2016 Taylor & Francis Group, London, ISBN 978-1-138-02973-6*

Effect of stabilizers on the electrical conductivity of zirconia materials

L.P. Tian, Q.H. Xue & L. Zhao
College of Materials and Mineral Resources, Xi'an University of Architecture and Technology, Xi'an, China

ABSTRACT: In this paper, the zirconia materials were prepared by using desilication zirconium power as raw materials, and adding various stabilizers of 2.1% MgO, 2.6% MgO and 2.1% MgO + 1% Y_2O_3 respectively. After sintering at 1710°C for 2 h. The phase composition and microstructure of the obtained samples were characterized through X-ray diffraction and scanning electron microscopy, and then the conductivity of zirconia materials was tested from ambient temperature (300 K) to 1073 K. The results show that the addition of stabilizers played an important role in the contents of cubic and monoclinic phases. The content of cubic zirconia increases, and the content of the monoclinic phase decreases obviously with the increasing addition of stabilizers in all sintered samples. The conductivity increases with the increasing content of MgO and testing temperature, the conductivity of the sample combining 2.1% MgO + 1% Y_2O_3 resches the maximum value.

Keywords: zirconia materials; stabilization; electrical conductivity; cubic zirconia

1 INTRODUCTION

Zirconia materials are widely used in the high-temperature field of metallurgy and chemical industry, because of its outstanding properties, such as high temperature resistance, corrosion resistance, wear resistance and so on. The conductivity of pure zirconia is very low at room temperature, which is a favorable insulator. The conductivity increases obviously and increases with the increasing testing temperature after adding various stabilizers. It is known that the conductive mechanism of ZrO_2 is due to the addition of some metal ion (Mg^{2+}, Ca^{2+}, Y^{3+}), whose ionic radius are more similar with Zr^{4+} in size, when the Zr^{4+} ions are partially replaced by dopant ions having a valence less than 4^+, result in anionic vacancy for keeping electric neutrality. Then the zirconia materials would have the electrical behavior because of the migration of O^{2-} ions through vacant ion sites at high temperature.

The ionic conductivity of zirconia materials reflects the difference between oxygen ions vacancy concentration and activation energy of migration at high temperature. On the basis of the effect of the variety and the addition of stabilizers on electrical conductivity of zirconia materials, we need further investigate the correlativity between oxygen ions vacancy concentration, activation energy of migration and the diffusion speed of slag and alloy element in liquid steel when at the erosion process.

The purpose of this paper is, to mainly investigate the differences of electrical conductivity, the relation between electrical conductivity and the phase composition and microstructure of zirconia materials by adding 2.1% MgO, 2.6% MgO and 2.1% MgO +1% Y_2O_3 respectively, in order to provide basis for studying the relation between electrical conductivity and corrosion resistance to slag.

2 EXPERIMENT

2.1 Raw materials

The raw materials for the experiment were desilication zirconium powers (94.9% pure) with a mean particle size of 7.51 um (Yunjia Refractories, Ltd, Luoyang) which were prepared by plasma method, and MgO powers (99.9% pure), Y_2O_3 powers (99.9% pure) for the additions were purchased from Tianli Chemical Reagents, Ltd, Tianjin.

The addition of stabilizers were 2.1% MgO, 2.6% MgO and 2.1% MgO + 1% Y_2O_3, then were marked with A,B,C, respectively, and their compositions are shown in Table 1.

2.2 Sample preparation

Three batch compositions were prepared by varying the kind and amount of stabilizer. All the batches were individually attrition milled using alumina pot for 24 h. Powdered samples were uniformly mixed with 10% Polyvinyl Alcohol (PVA) solution (wt5%) as green binder and subsequently granulated for 0–2 mm and uniaxially pressed into pellets (dia 20 mm × 5 mm) under the specific

Table 1. Chemical composition of samples.

Code	ZrO$_2$	MgO	Y$_2$O$_3$	CaO	SiO$_2$	Al$_2$O$_3$	Fe$_2$O$_3$
A	95.19	2.10	/	0.07	0.46	0.15	0.08
B	94.53	2.60	/	0.08	0.53	0.13	0.04
C	94.04	2.10	1.00	0.06	0.44	0.17	0.08

pressure of 300 MPa. The pellets were kept over-night for natural drying followed by 24 h oven drying at 110°C. Dried samples were sintered at 1710°C with 2 h of soaking at the peak tempera-ture in a zirconia high-temperature furnace. The crystalline phases and microstructure of sintered samples were identified by x-ray diffraction tech-nique and scanning electron microscopy.

The polished samples were prepared with a ultra-sonic cleaner in ethanol medium and dried in an oven for testing the conductivity. The conductivity of zirconia materials was tested from ambient tem-perature (300 K) to 1073 K in a program controlled electric furnace and the heating rate was maintain at 8 k/min from room temperature to 473 K and then from 473 K to peak temperature at a rate of 5 K/min. The resistance of samples were recorded by a data collector (Key sight 34465 A, America).

3 RESULT AND DISCUSSION

3.1 Crystalline phase

The crystalline phases of sintered samples were identified by x-ray diffraction technique. The x-ray diffraction patterns of the samples were obtained in a Rigaku x-ray diffractometer (D/MAX200) operating at 45 kV and 80 mA, using Cu-Kα radi-ation. The XRD datas were recorded with a step size of 0.02°(2θ) and a scanning rate 10°(2θ) per minute from 10° to 90°.

The x-ray diffraction patterns of the samples sintered at 1710°C are shown in Fig. 1. In all cases, Cubic zirconia is the major crystalline phase along with monoclinic zirconia as minor phase. Tetrago-nal zirconia is not detected in all samples. It can be observed that an increase in stabilizer (MgO) content from 2.1% to 2.6% increases the cubic phase content of sintered samples in A and B. It means that the stabilization degree of sample B is higher. The cubic phase content of sample C is the highest in all sam-ples. Due to the ionic radius of Y^{3+} (0.093 nm) and Zr^{4+} (0.082 nm) is relatively similar, it's easier to form solid solution with adding Y$_2$O$_3$ to ZrO$_2$. It makes the ZrO$_2$–MgO materials which is doped Y$_2$O$_3$ better chemical stability. By controlling the kind or content of stabilizer and the radius or valence of stabilizer ionic, the stability and phase structure of the zirco-nia materials changes dramatically.

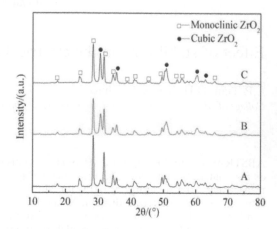

Figure 1. X-ray diffraction of the samples.

3.2 Densification

The bulk density and apparent porosity were meas-ured by standard liquid displacement method using Archimedes principle in water medium accord-ing to GB/T2997-2000 bulk density, porosity test method of dense shaped refractory products.

Variation of bulk density and apparent porosity of the different samples have been shown in Fig. 2. With increase the content of stabilizer (MgO) from 2.1% to 2.6%, the bulk density of samples increases and the apparent porosity decreases. It can be seen that sample with additive (2.1% MgO and 1% Y$_2$O$_3$) has achieved maximum bulk density and minimum apparent porosity.

It shows that the stabilization degree of dif-ferent samples vary with the kind and content of stabilizer. The type and content of stabilizer to sin-tering densification degree also have a significant impact. As we known, ZrO$_2$ is a type of high tem-perature structural material. If it has high density, the shorter the diffusion distance is, the easier the diffusion of oxygen air becomes.

3.3 Microstructure

Microstructure evaluation of the sintered sam-ples was done by scanning electron microscopy (Quanta200, Make FEI, America). The scanning electron photomicrographs of the samples with different stabilizer sintered at 1710°C are shown in Fig. 3 (a)-(f).

SEM photomicrographs of the samples also reveal that as the amount of stabilizer (MgO) has increased from 2.1% to 2.6%, the uniformity of grain size is increasing, the densification of grain is enhancing, the amount of porosity and grain which was growing up abnormal are decreasing. Compared with the former two, the uniformity of

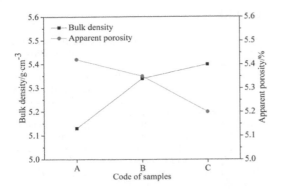

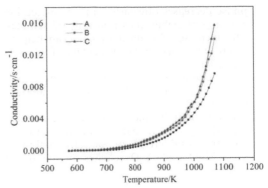

Figure 2. Bulk density and apparent porosity of different samples.

Figure 4. Variation of electrical conductivity of the samples with temperature.

Figure 3. SEM photomicrograph of samples sintered at 1720°C with different stabilizer content (a) (d)2.1% MgO, (b)(e) 2.6% MgO, (c)(f) 2.1% MgO +1%Y$_2$O$_3$.

grain size and the combination between different grains of sample which is added composite stabilizer (2.1% MgO and 1% Y$_2$O$_3$) are all better.

3.4 Conductivity measurements

To measure the conductivity of the sintered samples, the samples were heated in the furnace which is made by Sinosteel Luoyang Institute of Refractories Research Co., LTD and simultaneously the resistance of samples were recorded by a data collector (Keysight 34465 A, America). The conductivity was calculated by Eq. (1).

$$\alpha = \frac{L}{R \cdot A} \qquad (1)$$

where α = conductivity; and R = resistance; A = superficial area of sintered sample; L = electrode spacing on both ends.

Fig. 4 shows the change of the electrical conductivity with different sintering temperature. With increasing the sintering temperature, both electrical conductivity of the samples have increased. Under the same temperature, the electrical conductivity increases with the increase of the content of stabilizer (MgO).

As we can see from the XRD analysis results, the cubic phase content of three samples increases in turn. The stable cubic phase structure makes the

increase of electrical conductivity with admixing the low valence kation (Mg^{2+}, Ca^{2+}, Y^{3+}) which has the similar ionic radius compared with Zr^{4+}, in zirconia materials. In cubic phase structure, Zr^{4+} lies in the body center position of the simple cubic which is composed of O^{2-}, the coordination number is eight. And O^{2-} lies in the center position of the tetrahedron which is composed of Zr^{4+}, the coordination number is four. In the simple cubic structural composed of O^{2-}, only half of body center positions are dominated by Zr^{4+}. In consequence, there is a big gap in the zirconia unit cell center. From the point of view of conductive, the existence of this gap contribute to ionic conduction.

On the other hand, it considers that it's easy to form solid solution with adding Y$_2$O$_3$ to ZrO$_2$, the amount of oxygen vacancy involved in the transition increase. Comprehensiving the density and microstructure results of the samples, with the increase of sintering temperature, the average grain size increases, the amount of grain boundary and the porosity of samples decrease, the densification of materials increase. If the presence of porosity in the sample, it will change the migration path of oxygen ions and hinder the motion, resulting in low conductivity. In all samples, the uniformity of grain size and the combination between different grains of sample C are better, so it has the highest conductivity. Due to the worse uniformity of grain size and the higher porosity of sample A, it has a bad effect on oxygen ionic conduction. Therefore, the conductivity is relatively low.

4 CONCLUSION

1. The addition of stabilizers played an important role in the contents of cubic and monoclinic phases. The content of cubic zirconia increases, and the content of the monoclinic

phase decreases obviously with the increasing addition of stabilizers for all sintered samples. The increase of content of cubiccan help to improve the electrical conductivity of zirconia materials.

2. It is helpful to optimize the sintering conditions of samples with the doping amount increasing.

3. The conductivity increases with the increasing MgO content and testing temperature, and the conductivity of sample combining 2.1% MgO + 1% Y_2O_3 reaches the maximum value.

ACKNOWLEDGEMENTS

The present work is supported by the funds of the National Natural Science Foundation of China (51372193) and the Natural Science Basic Research Plan in Shaanxi Province of China (2014 JM6224).

REFERENCES

Chen, L. Jia, C. Influence of additive on the sintering of zirconia[J]. *Powder metallurgy technology*, 2008(2): 138–144.

Liu, J.J. He, L.P. Chen, Z.Z. Study on the electrical properties of Yttria-doped tetragonal zirconia[J], *Journal of the chinese ceramic society*, 2003, 31(3): 241–245.

Li, P. Luo, F. Xu, J. Mechanical and dielectric properties of partially stabilized zirconia ceramic[J], *Journal of the chinese ceramic society*, 2008, 36(3): 306–310.

Li, Y.M. Electrical properties and defect structure of ZrO_2[J]. *Journal of ceramics*, 1999(4): 215–219.

Wang, W.B. Refractory materials technology[M]. *Beijing, Metallurgy industry press*, 2007, 174.

Xiong, B.K. Lin. Z.H. Zirconium dioxide preparation technology and application[M]. *Metallurgy industry press*, 2008.

Zhao, W.G. An, S.L. Song, X.W. Electrical behavior study of ZrO_2 materials doped with MgO and Y_2O_3[J]. *Chinese rare earths*, 2006, 27(4):59–62.

Zhang, Q. Lin, Z.H. Tang, H. Ionic Conductivity and research progress of yttria-stabilized zirconia electrolyte[J]. *Rare metals letters*, 2008, 27(3): 1–5.

Energy Science and Applied Technology ESAT 2016 – Fang (Ed.)
© 2016 Taylor & Francis Group, London, ISBN 978-1-138-02973-6

Research of the Cu/Ti overlapping region produced via friction stir lap welding of dissimilar copper and titanium

Bo Li
Shanghai Institute of Special Equipment Inspection and Technical Research, Shanghai, P.R. China

Yifu Shen & Lei Yao
College of Materials Science and Technology, Nanjing University of Aeronautics and Astronautics, Nanjing, P.R. China

ABSTRACT: The technological innovation for joining of dissimilar titanium (Ti) and copper (Cu) still stays in demand. The promising joining method of friction stir lap welding has been successfully introduced to produce the defect-free and high-performance joints of dissimilar Cu and Ti sheets. A special cutting pin was utilized for promoting Cu/Ti mixing and transferring, and limiting the evident formation of Cu-Ti intermetallic compound phases in the dissimilar material overlapping region. After the process parameter optimization, the macro- and microstructures of the lap joints were carefully analyzed. The formation mechanisms of the Cu/Ti alternate band structure were discussed in detail. In addition, the composite-like structure had a mechanical locking effect on the Cu/Ti overlapping region. The strengthening effect was also provided by the metallurgical bonding at Cu/Ti band interfaces without cracking in micro-scale. The tensile strength of the joint reached 95% of that of the parent copper.

Keywords: dissimilar; friction stir lap welding; overlapping region; microstructure

1 INTRODUCTION

The implementation of hybrid structures or components through joining methods of different materials is tailored to local needs (Çam, G. et al, 1998). A dissimilar joint structure of pure Ti to Cu is significantly of technological importance, as this combination of materials finds a widespread application for the heat exchanger components in power generation industries (Lee, M. et al, 2008) (Dehghani, M. et al, 2013). However, a reliable and strong joining of Ti to Cu is difficult owing to a lack of the metallurgical compatibility between these two materials, such as different melting temperatures, thermal expansion mismatch, and structural incoherency (Dehghani, M. et al, 2013) (Wei, Y. et al, 2013).

Friction Stir Welding (FSW), patented by TWI in the UK, has emerged as a promising welding process. FSW principle is based on extreme plastic deformation in the solid state, where no associated bulk melting is involved for metallic materials. Till date, the FSW technique can confirm a reliable joining of Al alloys, Mg alloys, Cu-based alloys, Ti alloys, and steels. At present, to our knowledge, investigation on FSW of dissimilar Cu and Ti has not yet been reported. In this research, a combined-type tool with a hard alloy cutting pin was designed and utilized for the Friction Stir Lap Welding (FSLW) of dissimilar Cu and Ti, aiming to achieve a reliable joining and effectively control the IMC formation at the lap Cu/Ti dissimilar welded interfaces.

2 EXPERIMENTAL DETAILS

Commercially available pure titanium sheets of 2 mm thickness and copper sheets of 2 mm thickness were used as parent materials to be welded via FSLW. As shown in Fig. 1, the FSLW tool was composed of a holder, a shoulder, and a cutting pin. As illustrated in Fig. 2, a shielding atmos-

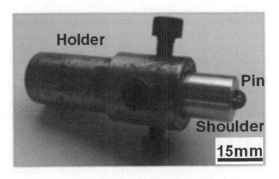

Figure 1. Appearance of the FSLW-tool.

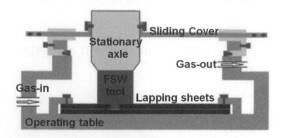

Figure 2. Illustration of shielding atmosphere device utilized for Cu/Ti FSLW procedure.

phere device was designed and utilized for the Cu/Ti FSLW procedure. In the present work for Cu/Ti FSLW, the following process conditions were tailored and changeless after the previous FSLW parameter optimizations. The tool rotation speed (n, r/min) was set at 800 r/min, and the tool travel speed (v, mm/min) was set at 40 mm/min. The macro- and microstructures of the overlapping region were carefully analyzed using an Optical Microscope (OM), X-ray Diffractometer (XRD), and Scanning Electron Microscope (SEM) with an Energy-Dispersive X-ray Spectroscope (EDS) to detect the related chemical compositions.

3 RESULTS AND DISCUSSION

3.1 *Macroscopic formability and cross-sectional microstructure*

The macroscopic view of a defect-free joint morphology, with a typical key-hole structure of dissimilar lap joint, is shown in Fig. 3. Although the joint external ring-like texture was inevitable according to the technical feature of FSW, the engineering requirements of the lap weld surface roughness could still be ensured using mechanical milling processing. When v was 40 mm/min, the overlapping region material density was further improved, so that no obvious void was generated, as shown in Fig. 4. More band-like flow patterns of the dissimilar materials presented in the overlapping region, due to which the lower pin-travel value led to more pin-rotation behaviors provided in unit time. Thus, it has been indicated that the adequate material plasticized deformations, with a higher friction heat quantity generated between tool pin and its adjacent metals, benefited the dissimilar material mixing and transferring at the Cu/Ti lap weld interface. In addition, as shown in Fig. 4, the macrostructure characteristics at AS (the advancing side) and RS (the retreating side) were differentiated that more dramatic material mixing and transferring behaviors presented at RS.

Fig. 5 is the magnification of the overlapping region microstructure in Fig. 4 by OM. It exhibited the dissimilar lap weld interfacial microstructure produced via FSLW. More significant behaviors of the material mixing and transferring generated the complex turbulent-flow patterns. As indicated in these OM images in micro-scale, the typical Cu/Ti alternate band structure was formed. The minimum value of the bandwidth was no more than 1 μm. In addition, the micro-interfaces between the adjacent bands of dissimilar metals were soundly bonded, so that no evident cracking existed. It indicated that the bonding interfaces were good between the zones of Cu/Ti alternate band structure and their adjacent bulky Ti sheet or Cu sheet. Although the trend or flow direction of the Cu/Ti alternate bands was not stable, neither consistent everywhere, the microstructure characteristics of some local banding patterns reflected the degree of Cu/Ti dissimilar material mixing and transferring behaviors there. Moreover, micro-sized Ti bands with whirling features were embedded in their adjacent Cu matrix. It has been believed that the complex morphology of the Cu/Ti interface was in favor of strengthening for the joining. Fig. 6 gives the EDS detection results on the Cu/Ti alternate band structure in the overlapping region. As indi-

Figure 3. Macroscopic appearance of the surface.

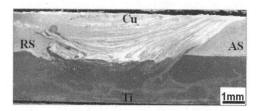

Figure 4. Macrostructure of the cross-section.

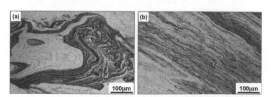

Figure 5. OM images of the overlapping region cross-sectional microstructures produced via FSLW.

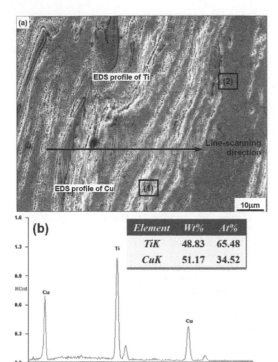

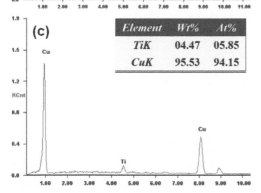

Figure 6. EDS detection results on the Cu/Ti alternate band structure in the overlapping region: (a) EDS line scanning with main element content profiles; (b) EDS scanning of Region (1); and (c) EDS scanning of Region (2).

cated in Fig. 6a, along the EDS scanning direction line, the profiles of Ti and Cu element contents were exhibited. Many peak alternations and gradient changes on the contents of the dissimilar elements formed the whole EDS profiles. It has been suggested that the dissimilar mixing degree of the Cu/Ti alternate band structure was high. The EDS results of Region (1) and Region (2) marked in Fig. 6a are shown as Fig. 6b and Fig. 6c, respectively. It showed that the amount of Ti or Cu was in not homogeneously distributed. Meanwhile, it indicated the local differences in Ti and Cu with changing location in the mixing zone.

The phase determination results by XRD on the joint sample, which was produced using a v value of 40 min/mm and n value of 800 r/min. Two faces or positions of the sample were separately detected. The lap-welded interface overlapping region beneath the upper Cu sheet was obtained by cutting and grinding, aiming to be detected by XRD. While the overlapping region above the lap-welded interface, within the upper copper sheet, was also extracted for its XRD detection. The two XRD spectrums indicated that no significant Cu-Ti IMC phases were formed at the Cu/Ti lap welding interface or in the overlapping region, except for the parent Ti and Cu phases. Although the two chosen local faces or regions for XRD contained multiple Cu/Ti dissimilar material mixing structures, no large-sized, bulky IMC phase was evident enough to be detected out by XRD. However, tiny Cu-Ti IMC particles, in micro-scale, were also formed and observed by SEM.

3.2 Mechanical tensile properties

Based on the fracture position, the cracking path view, and the specimen geometry, the highest anti-fracture strength value of overlapping joint was calculated to be 279.7 MPa. The experimentally tested tensile strength of the used parent copper was 295 MPa. Thus, the anti-fracture strength of the produced joint was ~95% of that of the parent copper. It has been believed that the high joint strength was mainly contributed by the Cu-grain refinement effect provided by the FSLW process, and the reinforcement effect of the embedded fine Ti-particles and Ti-bands within the Cu-matrix in the overlapping region.

3.3 Formation and mechanical locking effect of Cu/Ti alternate structure

The formation of the Cu/Ti alternate band structure in the overlapping region at the lapping interface location was strongly related to the tool pin mechanical behaviors and the heating effect provided by FSLW procedure. The heat generation during Cu/Ti FSLW was provided by both tool/material friction heating and material plastic-deformation heating. Owing to that the overlapping region of Cu in the upper sheet occupied a large part of the weld, the heat input from the upper sheet was more than that of the lower Ti sheet. In addition, the more outstanding thermal conductivity of Cu benefited the heat transport toward the lapping interface location and the lower Ti sheet. On the other hand, the sustaining rotation behavior of tool pin under the relatively lower travel speed promoted the dissimilar metal flow and mixing in their plasticized state. The cutting pin, which had milling cutters on the surface, could make the harder Ti in the lower lapping sheet to be fractions, stripes, or particles. Thus, they were migrated and mixed with the softer, the plastic-deformed Cu in the upper sheet. And the fine harder Ti particles or bands then embedded within the softer Cu-matrix in the overlapping region behind the traveling tool, after the cooling procedure. Meanwhile, during the heating procedure during and after FSLW, the Cu/Ti inter-diffusion could exist at the interfaces between the adjacent Cu/Ti bands in the overlapping region, to form soundly bonded interfaces in micro-scale. The micro-metallurgical bonding effect also benefited the mechanical locking effect on joint strength by the Cu/Ti alternate band structure.

ACKNOWLEDGMENTS

The research work was sponsored by the Shanghai Rising-Star Program (Grant No. 16QB1403200) and supported by the China Postdoctoral Science Foundation-funded project (Grant No. 2015M580342) and the National Natural Science Foundation of China (Grant No. 51505293).

REFERENCES

Çam, G. and Koçak, M. Progress in joining of advanced materials. Int. Mater. Rev., 1998; 43:1–44.

Dehghani, M., Amadeh, A. and Akbari Mousavi, S.A.A. Investigations on the effects of friction stir welding parameters on intermetallic and defect formation in joining aluminum alloy to mild steel. Mater. Des. 2013; 49:433–41.

Dressler, U., Biallas, G. and Alfaro Mercado, U. Friction stir welding of titanium alloy TiAl6V4 to aluminium alloy AA2024-T3. Mater. Sci. Eng. A. 2009; 526:113–7.

Lee, M., Lee, J., Lee, J., Park, J., Lee, G. and Uhm, Y. Strong bonding of titanium to copper through the elimination of the brittle interfacial intermetallics. J. Mater. Res. 2008; 23:2254–63.

Wei, Y., Li, J., Xiong, J. and Zhang, F. Effect of Tool Pin Insertion Depth on Friction Stir Lap Welding of Aluminum to Stainless Steel. J. Mater. Eng. Perf. 2013:1–9.

Energy Science and Applied Technology ESAT 2016 – Fang (Ed.)
© 2016 Taylor & Francis Group, London, ISBN 978-1-138-02973-6

Research on the prediction of the storage life of HTPB propellant

Yongqiang Du, Jian Zheng, Wei Peng, Xiao Zhang & Zhixu Gu
Ordnance Engineering College, Shijiazhuang, China

ABSTRACT: The high temperature accelerated life test was carried out on the HTPB propellant, and the maximum elongation rate was selected to characterize the performance. The dualistic regression model was put forward based on the linear model and the logarithmic model. The storage life of HTPB propellant at 25°C is predicted for 13.19 years by using this aging model, and the relative error is 6.45%, which is lower than the normal temperature extrapolation test results for 14.1 years. The aging model can meet the actual need on the consideration of security.

Keywords: HTPB propellant, aging model, life prediction, correlation analysis

1 INTRODUCTION

HTPB composite solid propellant is a kind of energetic composite material with high polymer as matrix, metal powder and oxidant as filler (Hou L F, 1994). HTPB propellant is the important source of the rocket engine, which performance will have direct impact on the operational capability of the rocket. Under the influence of the load / environment spectrum during storage, the physical and chemical properties will be changed, which leads to the deterioration of the mechanical properties of the propellant with the increase of storage time, especially when the rocket propellant is whole adherent casting, the charge cannot be replaced during the service (Zhang W, 2014). Therefore, in order to avoid the waste caused by early retirement and the harm caused by expired service, the storage life of solid propellant of rocket motor needs to be predicted accurately. At present, the commonly used research methods at home and abroad are the combination of high temperature accelerated life test and Arrhenius equation, and establish the aging model for the change of propellant properties to predict the storage life of propellant at room temperature (Pang A M, 2014). Celina. M et al. (Celina M, 2005) estimated the propellant storage life through the combination of Arrhenius equation and aging test method. At present, the aging models often used mainly include linear model and nonlinear model, and the selection of specific models should be determined according to the changes of the aging performance evaluation parameters in the aging process. However, it is known from the relationship between propellant and storage time that the single aging model is difficult to accurately describe the properties of the propellant. In order to match with the actual storage situation, it is needed to research more accurate aging model.

Accelerated life test of HTPB propellant was carried out in this paper, and the dualistic regression model was proposed by combination with the characteristics of the logarithmic model and linear model. The proportion of the logarithmic model and the linear model in the model was fitted by influencing factor, and the correlation degree of the model under different influence factors was analyzed. And the storage life of HTPB propellant at room temperature was estimated by this aging model.

2 TEST

2.1 *Aging and testing*

A certain type of dumbbell shaped HTPB propellant specimen was selected to carry out the high temperature accelerated life test for one year according to the methods of GJB 770B-2005 506.1. The specimen was sealed packing bag and put in the DU288 type electric heating oil bath constant temperature box. The aging temperature was set at 50°C, 60°C and 70°C respectively, and control the temperature fluctuations in the range of + 1°C. The relative humidity in the constant temperature box was adjusted less than 30%. A quantitative test specimen was taken out every month, and natural cooling in a closed dryer for 24 hours. Then used the Instron 5982 type material stretching machine to test the maximum elongation according to the method of GJB770B-2005 413.1. Each test point of the aging test piece was not less than five groups, and recorded the test data, then draw the curve of the maximum elongation of HTPB propellant with storage time. There were inevitably subjective and objective factors in the testing process, which makes the measurement data may contain abnormal data and may have an impact on the accuracy of the test data. Therefore, using Romanov criteria (t test criteria) to remove abnormal data, and using the average value to solve the other normal test data.

2.2 Test results

The variation curves of maximum elongation and storage time of HTPB propellant under high temperature accelerated life test are shown in Figure 1. Aging of HTPB propellant is a complicated physical and chemical process, during the aging process, the interaction of the post curing, the oxidative crosslinking and the degradation of the broken chain is accompanied. In addition, the process of dehydration, crack, hole growth and growth is exist (Zhang XG, 2009). It can be concluded from Figure 1 that the maximum elongation of HTPB propellant has the same trend with the change of storage time under three aging temperatures, which shows that the propellant under three temperatures follow the same aging mechanism. In addition, with the increase of temperature, the reaction rate of propellant elongation increases obviously, which is due to the acceleration of the aging process under high temperature conditions. Overall, the maximum elongation of HTPB propellant showed a decreasing trend with the increase of storage time, but in term of the reaction rate, which decline quickly at first, and then the rate becomes slow, and finally the reaction rate tends to zero.

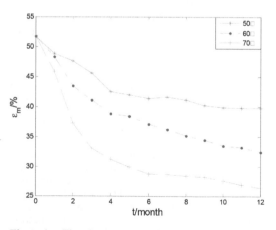

Figure 1. The change curve of the maximum elongation of HTPB propellant with storage time.

3 AGING MODELING AND SOLUTION

3.1 Modeling analysis

The research methods at home and abroad are the combination of the high temperature accelerated life test and the Arrhenius equation (Gao DY, 2006; Gao J, 2011).

High temperature accelerated aging method assumes that the aging reaction rate of solid propellant obeys the Arrhenius equation:

$$k = Ae^{-E_a/RT} \qquad (1)$$

In the formula: A is the pre exponential factor; E_a is the apparent activation energy; R is the universal gas constant; T is the thermodynamic temperature.

In order to predict the storage life of propellant, it is needed to establish the aging model of propellant performance change. The high temperature accelerated life test data of HTPB propellant were analyzed by using the logarithmic model and linear model, and then the service life of the propellant at room temperature was estimated (Tang GJ, 2012; Wang GQ, 2015).

Logarithmic model:

$$\varepsilon_m = \varepsilon_{m0} + k \lg t \qquad (2)$$

Linear model:

$$\varepsilon_m = \varepsilon_{m0} + kt \qquad (3)$$

In the formula: ε_m is the HTPB propellant aging at a time of performance parameters, here refers to the maximum elongation rate; k is the aging reaction rate, which complies with the Arrhenius equation; t is the aging reaction time; ε_{m0} is the initial value of the performance.

The results of accelerated life tests show that, the maximum extension rate of the propellant with the change of storage time contains two parts, linear and nonlinear. The single linear or nonlinear model cannot describe the change process of the maximum elongation of the propellant accurately, so it can be considered that the combination of the linear model and the nonlinear model to obtain the dualistic regression model:

$$\varepsilon_m = \varepsilon_{m0} + \theta_1 k_1 t + \theta_2 k_2 \lg t \qquad (4)$$

In the formula: k_1, k_2 are the aging reaction rate of the linear model and the logarithmic model respectively; θ_1, θ_2 are the impact factor, which used to fit the proportion of the logarithmic model and the linear model, and there are:

$$\begin{cases} \theta_1 + \theta_2 = 1 \\ 0 \le \theta_1 \le 1 \\ 0 \le \theta_2 \le 1 \end{cases} \qquad (5)$$

When $\theta_1 = 0$, the model is transformed into the logarithmic model (2); when $\theta_2 = 0$, the model is transformed into the linear model (3).

3.2 Model solution

In this paper, the parameters of the aging model were solved by using the method of stepwise regression and the least square method.

According to the relationship between the mechanical properties and storage time of the propellant, the formula (2) and (3) were fitted, and the correlation analysis of the test data at different aging temperatures was carried out. The results are shown in Table 1. In Table 1, n is the number for regression analysis, R is the correlation coefficient, $R_{0.01}$ indicates the critical correlation coefficient when the significant level is 0.01.

The formula (1) logarithm available:

$$\ln k = \ln A - E_a / RT \qquad (6)$$

Assuming that $y = \ln k, x = -1/RT, b = \ln A$, the formula (6) is transformed into the form of linear equation $y = E_a x + b$. According to the relationship between the aging reaction rate and storage temperature in Table 1, the parameters E_a and $\ln A$ can be obtained by means of the least square method, and the aging reaction rate $k_{25°C}$ of propellant is also introduced. As shown in Table 2.

It can be drawn from Table 1 that the linear model and exponential model are best fitting under at 60°C. Therefore, the experimental data of this condition was used to fit the influence factors. Assuming $\theta_1 = 0, 0.1, 0.2, ..., 1.0$, the correlation degree of the model under different influence factors were analyzed, which is shown in Table 3.

It can be seen from the Table 3 that the correlation coefficient of the aging model combined with the linear model and the logarithmic model is larger than that of the single model. Different combination of affecting factors show different correlation coefficients, and with the increase of the proportion of the logarithmic model, the relevance of the model improves gradually.

Take the parameters into formula (4), the storage life of HTPB propellant at 25°C is predicted for 158.32 months, which is 13.19 years, with the propellant maximum extension rate decreased by 15% as the failure criterion. Compared with the normal temperature extrapolation test results for 14.1 years, the relative error is 6.45%, which is lower than the normal temperature extrapolation test result, and is considered to meet the actual demand from the point of view of security.

Table 1. Data processing results.

Models	T	k	F_{m0}	n	R	$R_{0.01}$
Linear model	50°C	−0.7801	47.5955	12	0.9053	0.708
	60°C	−1.2364	45.7030	12	0.9436	0.708
	70°C	−1.2710	39.2530	12	0.8268	0.708
Logarithmic model	50°C	−9.2412	49.2098	12	0.9765	0.708
	60°C	−14.3667	48.0590	12	0.9833	0.708
	70°C	−16.2219	42.7260	12	0.9608	0.708

Table 2. Regression results of Arrhenius equation.

Models	E_a	$\ln A$	$k_{25°C}$
Linear model	2.2663×10^4	8	−0.4110
Logarithmic model	2.6042×10^4	11.9747	−4.3003

Table 3. Correlation analysis of different influence factors.

$\theta 1: \theta 2$	0:1.0	0.1:0.9	0.2:0.8	0.3:0.7	0.4:0.6	0.5:0.5	0.6:0.4	0.7:0.3	0.8:0.2	0.9:0.1	1.0:0
R	0.9833	0.9977	0.9962	0.9936	0.9899	0.9851	0.9791	0.9720	0.9637	0.9542	0.9436

4 SUMMARY

1. The dualistic regression model was put forward by combining the linear model and the logarithmic model. The results show that the combination of the logarithm model and the linear model can improve the correlation coefficient of the aging model, and the higher the proportion of the logarithm model, the greater the correlation coefficient of the model;

2. The storage life of HTPB propellant at 25°C is predicted for 13.19 years by using the dualistic regression model with the propellant maximum extension rate decreased by 15% as the failure criterion. Compared with the normal temperature extrapolation test results for 14.1 years, the relative error is 6.45%, which is lower than the normal temperature extrapolation test result, and is considered to meet the actual demand from the point of view of security.

REFERENCES

Celina M, Gillen KT, Assink RA. Accelerated aging and lifetime prediction: review of non-Arrhenius behaviour due to two competing processes, J. Polym Degrad Stab. 90 (2005) 395.

Celina M, Graham AC, Gillen KT, Assink RA, Minier LM. Thermal degradation studies of a polyurethane propellant binder, J. Rubber Chem Technol. 73 (2000) 678.

Celina M, Wise J, Ottesen DK, Gillen KT, Clough RL. Correlation of chemical and mechanical property changes during oxidative degradation of neoprene, J. Polym Degrad Stab. 68 (2000) 171.

Gillen KT, Celina M, Bernstein R. Review of the ultra-sensitive oxygen consumption method for making more reliable extrapolated predictions of polymer lifetimes, J. Ann Tech Conf Soc Plast Eng. 62 (2004) 2289.

Bernstein R, Derzon DK, Gillen KT. Nylon 6.6 accelerated aging studies: thermal-oxidative degradation and its interaction with hydrolysis, J. Polym Degrad Stab. 88 (2005) 480.

Fu HM, Yang LB, Lin FC, et al. Integral method for storage life prediction of solid propellant, J. Journal of Mechanical Strength, 29 (2007) 754–759.

GJB 770B-2005, S. Test method of propellant. Christiansen AG, Layton LH, Carpenter RL. HTPB propellant aging, J. J.Spacecraft. 18 (1981) 211–215.

Gao DY, He B, He SW, et al. Discussion on limitations of the Arrhenius Methodology, J. Chinese Journal of Energetic Materials. 14 (2006) 132–135.

Gao J, Wang X M, Song QQ. Combined life evaluation method for solid propellant grain, J. Tactical Missile Technology. 3 (2011) 43–47.

Hou LF. Composite solid propellant, first ed., China Aerospace Publishing House, Beijing, 1994.

Pang AM. Theory and engineering of solid rocket propellant, first ed., China Aerospace Publishing House, Beijing, 2014.

Shen TF, Zhao BH. Analyses of theory and experiment for void damage of solid propellant, J. Journal of Propulsion Technology. 23(2002) 71–73.

Tang GJ, Shen ZB, Tian SP, et al. Probabilistic storage life prediction of solid rocket motor grain, J. Acta Armamentarii. 33 (2012) 301–306.

Wang GQ, Shi AJ, Ding L, et al. Mechanical properties of HTPB propellant after thermal accelerated aging and its life prediction, J. Chinese Journal of Explosives & Propellants. 38 (2015) 47–50.

Zhang W. The rocket propellant, first ed., National Defence Industry Press, Beijing, 2014.

Zhang XG. Study on the aging properties and storage life prediction of HTPB propellant, D. National University of Defense Technology, Changsha 2009.

Energy Science and Applied Technology ESAT 2016 – Fang (Ed.)
© 2016 Taylor & Francis Group, London, ISBN 978-1-138-02973-6

Numerical study of complex rheological behaviors of polymer melt past a cylinder

Wei Wang & Wenwen Li
Key Laboratory of Rubber-Plastics, Ministry of Education/Shandong Provincial Key Laboratory of Rubber-Plastics, Qingdao University of Science and Technology, Qingdao, China

ABSTRACT: The viscoelastic flow past a cylinder is a benchmark problem for studying the rheological behaviors of polymer melt by the numerical method. The Giesekus and S-MDCPP models were applied to characterize the rheological behaviors of low-density polyethylene melt, respectively. The discrete elastic-viscous stress split method with streamline-upwind/Petrov–Galerkin or Streamline Upwind (SU) techniques and the modified Finite Increment Calculus (FIC) were employed to ensure the computational stability. Results indicate that the velocity and stress predicted by S-MDCPP and Giesekus models agree well with the experimental data. In addition, the predictability of the S-MDCPP model has much better consistency with the experiment than the Giesekus model.

Keywords: polymer melts, viscoelastic flow, numerical simulation, rheological behavior

1 INTRODUCTION

The viscoelastic flow past a cylinder is not only employed to study the rheological behavior of polymer melts or solutions by rheologists, but also common in the practical processing of polymer fluids, such as extrusion, injection molding, and fiber spinning. Thus far, various investigators have reported the rheological behaviors of polymer melt around a cylinder (Y.J. Choi, 2012). In this study, our purpose is to verify the numerical algorithm for viscoelastic flow, and to evaluate the predictable ability of the viscoelastic constitutive equations.

During the computation of complex viscoelastic flow problem, a suitable constitutive equation needs to be chosen to describe the rheological behaviors of polymer melt or solution, because this greatly affects the calculation results. In the present study, the commercial LDPE melt (Stamylan LD 2008 XC43, DSM) is described by the S-MDCPP (W. Wang, 2010) and Giesekus models in this investigation, respectively. The fitting curve of simple rheology experimental data with the S-MDCPP model may be found in W. Wang (2014). The constitutive parameters of S-MDCPP and Giesekus models are shown in Table 1. The viscoelastic flow past a cylinder is simulated by the iterative fractional step algorithm, along with the Discrete Elastic-Viscous Stress Split (DEVSS) technique and the SU or the Streamline-Upwind/Petrov–Galerkin (SUPG) method to ensure the stability of solving the con-vection-dominated problem in this work. Mean-

Table 1. Constitutive parameters of S-MDCPP and Giesekus models (T=170°C).

Maxwell parameters		S-MDCPP			Giesekus
G(Pa)	λ_0(s)	q	r	ξ	α
2.3832E+3	1.7415	4	1.2	0.02	0.25

while, the modified Finite Incremental Calculus (FIC) method is applied to reconstruct the mass conservation equation to achieve a stable pressure field, in order to solve the velocity, pressure, and stress fields with the equal low-order interpolation elements. The same numerical schemes have been employed to predict the 4:1 contraction and cross-slot flows (W. Wang, 2016).

2 DESCRIPTION OF THE VISCOELASTIC FLOW PAST A CYLINDER

The planar geometry of viscoelastic flow past a cylinder problem is shown in Fig. 1. The static cylinder is located in the middle of two parallel plates. The polymer melt enters the inlet and pass around the cylinder. The radius R of the cylinder is 1.1875 mm, and the distance H of the two plates is 4.95 mm, i.e. $H = 4R$. The total length of the channel L is 40 mm, in order to ensure the full developed flow. The origin of coordinates is located in the center of this cylinder. The no-

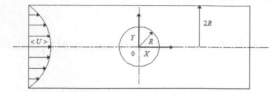

Figure 1. Schematic of cylinder flow problem.

Figure 2. Computational mesh around the cylinder.

slip boundary is assumed to be the wall, and the exit pressure of zero is applied to the symmetric point of the outlet. The initial velocity and stress are applied in the similar way reported in reference. According to the symmetry of geometry and boundaries, we only calculate the half channel in order to reduce the calculation cost. The mesh of the half channel is displayed in Fig. 2. Owing to the large stress gradient in the vicinity of the cylinder surface, the number of meshes is increased.

3 RESULTS AND DISCUSSION

Comparison of results calculated by different methods for the S-MDCPP model

We calculated the viscoelastic flow of the S-MDCPP model around a cylinder by the above-mentioned numerical scheme. In order to deal with the convection-dominated problem and discuss the reliability of numerical results, two streamline-upwinding methods, i.e. SU and SUPG methods, are used, respectively. The predicted velocity and Principle Stress Difference (PSD) of the S-MD-CPP model along the channel centerline, are displayed in Fig. 3, where the experimental results of Verbeeten (W.M.H, 2001) were also given in Fig. 3. We observed that the results calculated by the SU and SUPG methods agree well with the experimental data. The results predicted by two numerical schemes have slight difference except for the velocity in the rear of the cylinder. Consequently, the results indicate that the SU and SUPG methods

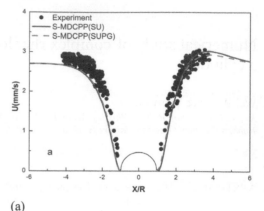

(a)

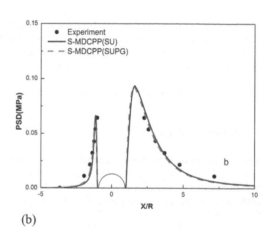

(b)

Figure 3. Comparison of velocity (a) and PSD (b) profiles between experimental results and numerical results along the centerline of the channel with the SU and SUPG methods for the S-MDCPP model.

can do a good job for the viscoelastic flow around the cylinder. Meanwhile, the predictability of the S-MDCPP model is verified by comparison with the experimental results.

Comparison of results calculated by different methods for the Giesekus model

Fig. 4 displays the distributions of horizontal velocity and PSD along the channel centerline predicted by the Giesekus model. To verify the reliability of the SUPG method, the experimental data and the calculated results of Verbeeten using the discontinuous-Galerkin (DG) method are shown in Fig. 4. We may observe that the results calculated by the SUPG method have better agreement with the experiment than those of the DG method, especially for the distribution of PSD. However, the velocity predicted by the DG

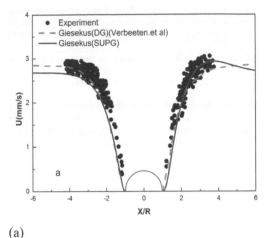

(a)

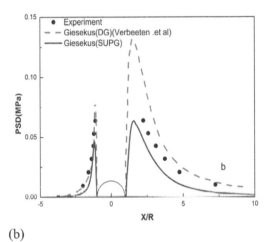

(b)

Figure 4. Comparison of velocity (a) and PSD (b) profiles between experimental results and numerical results along the centerline of the channel with the DG and SUPG methods for the Giesekus model.

(a)

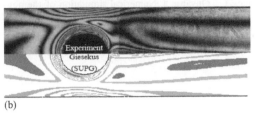

(b)

Figure 5. Comparison of the isochromatic fringe patterns between experimental results and numerical results with the DG and SUPG methods for the Giesekus model.

method in the upstream channel is slightly better than that of the SUPG method. Moreover, the isochromatic fringe patterns of principle stress difference of experiment and numerical simulation are shown in Fig. 5. We may observe that the results of the four-mode Giesekus model reported by Verbeeten using the DG method have better agreement than our results of the single-mode Giesekus model calculated by the SUPG method, which displays that the multimode viscoelastic constitutive model has much better predictability, due to the polydispersity of the polymer. Although we applied only the single-mode Giesekus model

to predict the complex rheological behaviors of viscoelastic flow, the calculated results agree well with the experiment. More importantly, the computational cost can be reduced.

4 SUMMARY

The LDPE melts were described by the S-MDCPP and Giesekus models, and the viscoelastic flows past the cylinder were calculated by our numerical scheme. The SU and SUPG methods were used to cope with the convection term of the viscoelastic constitutive models, respectively. The velocity and the PSD calculated by the S-MDCPP and Giesekus models agreed well with the experimental results. Hence, the two models can capture and reproduce the rheological behaviors of the polymer melt. Results indicate that both of the streamline-upwinding methods are stability and reliability for predicting the rheological behaviors of polymer melt past a cylinder. In addition, by comparison with the experimental results, the predictability of S-MDCPP model has much better than Giesekus model.

ACKNOWLEDGEMENT

This work was funded by the National Natural Science Foundation of China (Grant number 2127 4072). The corresponding author of this work is Wei Wang.

REFERENCES

Choi, Y.J., Hulsen, M.A., Meijer, H.E. Simulation of the flow of a viscoelastic fluid around a stationary cylinder using an extended finite element method, Comput. Fluids, 57 (2012) 183–194.

Claus, S., Phillips, T. Viscoelastic flow around a confined cylinder using spectral/hp element methods, J. Non-Newtonian Fluid Mech., 200 (2013) 131–146.

Ruan, C., Ouyang, J. Numerical simulation of the non-isothermal viscoelastic flow past a confined cylinder, Chin. J. Chem. Eng., 18 (2010) 177–184.

Verbeeten, W.M.H. Computational polymer melt rheology, Ph.D. Thesis, Eindhoven University of Technology. (2001)

Wang, W., Hu, C.X., Li, W.W. Time-dependent rheological behavior of branched polymer melts in extensional flows, Mech. Time-Depend. Mater. 20 (2016) 123–137.

Wang, W., Li, X.K., Han, X.H. A numerical study of constitutive models endowed with Pom-Pom molecular attributes, J. Non-Newton Fluid Mech. 165 (2010) 1480–1493.

Wang, W., Wang, X.P., Hu, C.X. A comparative study of viscoelastic planar contraction flow for polymer melts using molecular constitutive models, Korea-Aust. Rheol. J. 29 (2014) 365–375.

Energy Science and Applied Technology ESAT 2016 – Fang (Ed.)
© *2016 Taylor & Francis Group, London, ISBN 978-1-138-02973-6*

Effect of stacking faults on twinning in hexagonal close-packed crystals

Shan Jiang

Research Institute for New Materials and Technology, Chongqing University of Arts and Sciences, Yongchuan, Chongqing, China

ABSTRACT: In the light of the Giant Atomic Clusters (GAC) model, we have established elsewhere, here we focus on exploring the effect of stacking fault on twinning behaviors. Based on the analysis in the law of atomic migration in the twinning process, we speculate that the presence of stacking-fault will disorder the arrangement of the GAC units and create a new type of GAC units named R-GAC, which have a chiral relation to the normal GAC units. Further investigation indicates that the R-GAC units move in an absolutely different way from that of the GAC units by decomposing themselves to form a distorted lattice region in grains, which is likely to supply nucleation sites for recrystallization, when conditions allow for, due to their high-energy state.

Keywords: stacking fault, twinning, magnesium; hexagonal close-packed; slip

1 INTRODUCTION

Magnesium alloys are the lightest available metallic structural materials, increasingly used in a variety of engineering applications instead of conventional materials, such as auto, aircraft, and electronic industry, giving rise to essential weight savings (Y. N. Wang, 2007). However, the poor plasticity and limited ductility of the material have severely impeded the practical processing and the widespread applications further (M. A, 1960), in spite of their superplasticity exhibited at high temperatures (C. N. Tomé, 1991), which has become one of the most urgent issues to be solved. For this reason, quite a few investigations have been performed to explore the origin of their poor mechanical properties at low temperatures. Theoretically, an inevitable problem is the limited number of independent slip systems of the materials due to their special Hexagonal Close-Packed (HCP) crystal structure, in which most of the slip systems except basal slip system, such as {$10\bar{1}0$} <$11\bar{2}0$> prismatic slip system and {$10\bar{1}1$} <$11\bar{2}0$> pyramidal slip system, have a much large critically Resolved Shear Stress (CRSS), that prevents them from being activated at room temperature, and therefore, from satisfying the requirement of the five independent slip systems in terms of the Von Mises law (R. Z. von Mises, 1928). However, the deformation of magnesium alloys at low temperature can probably produce available high yield strength and anisotropy in practical uses. Considering the controversies in the deformation mechanism at low temperature and the increasing needs for a better application of magnesium alloys, a comprehensive investigation on the correlation between microstructure and mechanical properties in the materials is necessary.

Although, a number of intriguing phenomena concerning twins/twinning in HCP polycrystal materials including magnesium alloys, such as twin growth, detwinning, twin interception, and twinning induced dynamic recrystallization nucleation, have been discovered thus far (T. Al-Samman, 2008), the deformation mechanism triggered by twins in magnesium alloys at room temperature has not been well clarified because of the complications in a direct observation of twin evolution in the experiment, together with the lack of theoretical models. Moreover, it is still unknown what kind of atomic motion (e.g., microstructure change) emerges as twins behave and how twinning affects the tensile plastic deformation and texture formation. These critical issues remain unclear because of the following reasons: (i) Twinning occurs at intermediate states, but current studies on the deformation of magnesium alloys are generally restricted to initial and final states only. (ii) A few fundamental issues are still unknown, including which is easier to occur between basal slip and twinning, whether it is likely to initiate non-basal and twinning slips, and whether twinning is beneficial to deformation or harmful for introducing cracks.

Here, following the previously proposed concepts on twins, we then constructed the GAC model to describe atomic motion during twinning, and found that the twin boundaries can be reasonably addressed by the GAC units at the equilibrium state in strain and become mobile once that equilibrium is unbalanced, resulting microscopically in the important twinning and detwinning phenomena. Based on the above suggestion, the underlying influence of stacking faults on twinning is discussed in a new way.

2 MODELING

Jaswon and Dove had proposed that some atoms assume shear motion while the others reconstruct during twinning, and that both the shearing amplitude and the proportion of shearing atoms play a key role in accomplish twinning. However, it is still difficult to figure out how reconstructed atoms displace based on this theory. In addition, how the atoms belonging to different subareas are connected and interacted during twinning remains unclear. In a separate paper, we have developed the GAC rotation model to uncover atomic motions in the $\{10\bar{1}2\}$ extension twinning and $\{10\bar{1}1\}$ contraction twinning. Following this model, the two twinning modes can be realized by rotating the GAC units as a whole. The location of the twinning atoms before and after the $\{10\bar{1}2\}$ twinning is illustrated in Fig. 1a, where the displacement vectors for mobile atoms are labeled. Note that, for a better description, we have classified the crystal planes in magnesium alloys into two species according to the stacking sequence ABAB... in the HCP structure: (i) main planes (labeled M) and (ii) subplanes (S), which are defined as the planes passing through A- and B-layer atoms, respectively.

Here, we systematically clarify the transfer mechanism between GAC units and expand the GAC units to twin boundaries, growth, and detwinning, through which the influence of stacking faults in the matrix on twining is revealed. Figure 1 shows an arrangement of the GAC units before and after the $\{10\bar{1}2\}$ twinning, where it can be seen that the GAC units are arranged periodically. In addition, rotations of GAC units are transferable and take place synchronously, i.e., with the same angle at an identical time. These therefore suggest that the

occurrence of twinning is transient once stress is sufficiently large, which is quite different from the case of basal slip.

3 DISCUSSION

As aforementioned, it is usual to see both twinning and detwinning in magnesium alloys, which can be attributed to the motion of twin boundaries. In general, twinning occurs when twin boundaries move outward from twins, while detwinning is a result of inward movement of twin boundaries. In addition, motion of twin boundaries differs completely from that of GB sliding because the latter is along GB while the former is along the normal of twin boundaries. Since twining is able to be realized via rotation of GAC units and the misorientation between matrix and twin is minor, we speculate that twin boundaries are likely to be composed of the GAC units as well (Fig. 2), with the orientations of units being in between those of matrix and twin. In addition, misorientation of the GAC units in twin boundaries is gradual. As seen in Fig. 2, the GAC units around twin boundary are repeated periodically both along (T_1 to T_8) and across (U_0 to U_3) the twinning direction. In this sense, despite that the GAC units are distorted at the twin boundary, they can still satisfy the transferability required by a GAC unit. As a twin boundary is in equilibrium in stress, the GAC units do not rotate at all and the twin boundary stands still. As soon as the stress is unbalanced, the GAC units around twin boundary start to rotate, resulting in the motion of boundary. Specifically, the twin boundary shifts to the right (twin growth) when the GAC units rotate counterclockwise, while to the left (detwinning) when rotating clockwise.

Figure 3 shows schematically how stacking fault, a kind of partial dislocation, affects the $\{10\bar{1}2\}$ twinning, where the stacking-fault atoms are highlighted in yellow. The stacking-fault atoms can misarrange the GAC units, resulting in the formation of a new

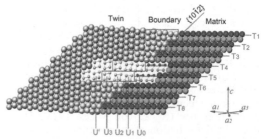

Figure 2. Schematic illustration of atomic arrangements around $\{10\bar{1}2\}$ twin boundary. The GAC units are labeled T along twinning and U across twining direction.

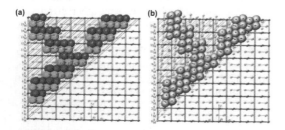

Figure 1. Model illustration of transfer of GAC units in the $\{10\bar{1}2\}$ twinning: (a) before and (b) after twinning.

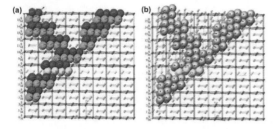

Figure 3. Illustration of the influence of stacking faults on the {10$\bar{1}$2} twinning process: (a) before and (b) after twinning. Atoms in the stacking fault are highlighted in yellow.

type of GAC units having a chiral relation to the normal GAC units. Since the twinning requires a symmetric relation between twins and matrixes, the presence of stacking fault in the matrix disorders the normal GAC units, so that there emerges a row of disturbed atoms in between matrix and twinning planes in twins. In this sense, the distorted crystal lattice, which is induced by the stacking faults, is considered to serve as nucleation sites for recrystallization, when conditions allow for.

4 SUMMARY

It is assumed that the twin boundaries in HCP crystals can be composed of transitional GAC units with their orientation aligned between those of matrix and twin. In addition, the GAC units may be transferable making occurrence of twinning instantaneous. The presence of stacking-fault will disorganize the appearance of GAC units, resulting in a formation of a new type of GAC units having a chiral relation to the normal GAC units, which rotate in a different way from the normal units and finally leave a distorted lattice region in the twin. In this sense, this distorted crystal lattice, induced by the stacking faults is believed to be able to serve as nucleation sites for recrystallization in twins.

ACKNOWLEDGMENTS

This work was supported in part by the National Natural Science Foundation of China (No. 51301215), the Talent project of Chongqing University of Arts and Sciences (R2014CJ04; Y2014CJ29), the Natural Science Foundation Project of Chongqing (No. cstc2012jjA50016), and the Chongqing Post-doctoral Support Program (Rc 201334).

REFERENCES

Al-Samman, T., Gottstein, G. Dynamic recrystallization during high temperature deformation of magnesium. Mater. Sci. Eng. A. 490 (2008) 1-2, 411–420.

Jaswon, M.A., Dove, D.B. The crystallography of deformation twinning. Acta Cryst. 13 (1960) 232–240.

Jiang, S., Liu, T., Lu, L. et al. Atomic motion in Mg-3-Al-1Zn during twinning deformation. Scripta Mater. 62 (2010) 556–559.

Serra, A., Bacon, D.J. Computer simulation of screw dislocation interactions with twin boundaries in H.C.P. metals. Acta Metall. Mater. 43 (1995) 12, 4465–4481.

Tomé, C.N., Lebensohn, R.A., Kocks, U.F. A model for texture development dominated by deformation twinning: Application to zirconium alloys. Acta Metall. Mater. 39 (1991) 11, 2667–2680.

Wang, Y.N., Huang, J.C. The role of twinning and untwinning in yielding behavior in hot-extruded Mg-Al-Zn alloy. Acta Mater. 55 (2007) 3, 897–905.

Yamashita, A., Horita, Z., Langdon, T.G. Improving the mechanical properties of magnesium and a magnesium alloy through severe plastic deformation. Mater. Sci. Eng. A. 300 (2001) 1-2, 142–147.

von Mises, R.Z. Angew. Math. Mech. 8 (1928) 161.

Energy Science and Applied Technology ESAT 2016 – Fang (Ed.)
© 2016 Taylor & Francis Group, London, ISBN 978-1-138-02973-6

GGA+U Study of the elastic, electronic, and optical properties of Ceb$_6$

L.H. Xiao

School of Materials and Metallurgical Engineering, Guizhou Institute of Technology, Guiyang, China
School of Materials Science and Engineering, Central South University, Changsha, Hunan, China

Y.C. Su

School of Materials Science and Engineering, Central South University, Changsha, Hunan, China

Fang Shao

School of Mechanical Engineering, Guizhou Institute of Technology, Guiyang, China

Ping Peng

School of Materials Science and Engineering, Hunan University, Changsha, Hunan, China

ABSTRACT: The optical and electronic properties of CeB$_6$ have been explored in detail based on a density functional theory and a GGA+U approach. It was found that the calculated structural parameters were consistent with previously reported experimental and theoretical data. At $U' = 3$ eV, the polycrystalline values of the elastic constants and bulk modulus are consistent with values determined experimentally. Furthermore, the optical properties of CeB$_6$ including the absorption spectrum, dielectric function,, extinction coefficient, refractive index, energy loss spectrum, and reflectivity, were also calculated and analyzed in detail. The optical spectra were assigned primarily to interband contributions from B $2p$ valence bands to Ce $5d$ conduction bands. These results show that this GGA+U' approach is an efficient and useful method for analysis of rare-earth elements that contain $4f$ electrons, which are critical in determination of elastic constants and bulk modulus.

Keywords: CeB$_6$; first-principles calculations; electronic structures; elastic constants; optical properties

1 INTRODUCTION

As a group, the rare-earth hexaborides (RB_6) not only have similar crystal structures (Bai,L. & Ma, N. 2010), but also show many similar properties such as efficient field emission (Balakrishnan, G. et al. 2003) (Givord, F. et al. 2003) (Goto, T. et al. 1983), high melting point, good thermal stability (Jang, H. et al. 2014), and hardness. Previous research has so far investigated the electric (Jie, D. et al. 2015), thermal (Kimura, et al. 1992), and magnetic properties (Kimura, S. et al. 1990) (Leger, J. et al. 1985) (Loschen, C. et al. 2007) of RB_6. Interest in these compounds is high due to the $4f$ electrons present in the rare earth elements, the presence of which play a chief role in determining the structural, electronic, elastic, and magnetic properties of the compounds. Among RB_6 compounds, CeB$_6$ is particularly interesting. CeB$_6$ is a dense Kondo material (Lüthi, B. et al. 1984) and exhibits heavy fermion behavior (Magnuson, M. et al. 2001). In addition, CeB$_6$ has been widely used in metallurgy, wave-absorbing materials, cath-ode emission materials, nuclear industry, and other applications (Nakamura, S. et al. 1994) (Pickard, C. J. et al. 2000).

Due to its scientific importance and potential applications, the structural, elastic, vibrational, thermodynamical, magnetic, and optical properties of CeB$_6$ have been investigated experimentally (Ripplinger, H. et al. 1997) (Plata, J. J. et al. 2012) (Samwer, K. et al. 1976) and theoretically by several groups. Magnuson *et al.* used an experimental approach to study the hybridization between the localized $4f$ orbitals and the delocalized valence-band states by identifying the various spectral contributions due to inelastic Raman scattering and normal fluorescence. Significantly different values have been reported for The measured elastic compliance constants of CeB$_6$ showed significant differences, particularly C_{12}. Within the literature values range from $C_{12} = -93$ GPa ((Plata, J. J. et al. 2012)), to 53 GPa reported by Nakamura *et al.* for the same constant. There are similar differences, though less drastic, among the various experimental reports on the bulk modulus of

CeB$_6$. It is clear that current experimental results are sometimes contradictory and are insufficient overall. On the theoretical side, the Fermi surface of CeB$_6$ has been studied by Suvasini *et al* using the fully relativistic spinpolarized mean muffin-tin orbital method within the local density approximation. The electronic structures of CeB$_6$ have also been studied by Ripplinger *et al.* using the full-potential Linearized-Augmented Plane-Wave (LAPW) method within density functional theory. The structural and chemical bond properties of CeB$_6$ have been calculated by Bai *et al.* using an all-electron Full-Potential Linearized Augmented Plane Wave (FPLAPW) and a so-called GGA+U method, which were implemented within the EXCITING code. Singh *et al.* used the full potential linearized augmented plane wave method to study the electronic and optical properties of CeB$_6$. To better explain the on-site *f*-electron correlation, Singh *et al* applied the Coulomb corrected Local Spin Density Approximation (LSDA + U) to the exchange correlation functional in the calculations. Gürel *et al* have completed a comprehensive *ab initio* investigation of the elastic, lattice-dynamic, and thermodynamic properties of CeB$_6$ within the density-functional and density-functional perturbation-theory frameworks by using quasiharmonic approximation to account for the anharmonicity of the vibrations. However, this work claimed that the plasma energy ($\hbar\omega_p$) for CeB$_6$ is 4.90 eV, [21] in apparent contrast to relevant experimental values ($\hbar\omega_p$ = 2.0 eV and 1.96 eV). Therefore, it is important to investigate the electronic structures used in this work, as well as the corresponding phases in order to clarify the differences in optical properties and elastic constants based on the results of first-principles calculations.

One alternative that could be used to integrate correlation effects while also requiring less computational effort is the LDA+U or GGA+U method, where correlation effects are incorporated through the on-site Coulomb interaction, U. Recently, calculations using a GGA+U approach have been performed for CeO$_2$, where not only the on-site U for Ce 4f electrons, but also for O 2p electrons, were taken into account. However, the choice of U is ambiguous. Though there have been attempts to obtain this value from first-principles calculations, it has proven crucial to confirm its value a priori. Therefore, U is often determined through fitting certain sets of experimental data such as optical properties and elastic constants.

Here, we use GGA and GGA+U schemes formulated by Loschen *et al.* to calculate the electronic structures, structural parameters, optical properties, and elastic constants of CeB$_6$. It was determined that after taking into account the on-side Coulomb interactions of the 4f orbitals on the Ce atom (U^f)

and of the 2p orbitals on the B atom, the elastic constants, bulk modulus, and optical properties obtained using this method are consistent with previously reported data, and are more accurate than those obtained using scissor operations.

2 COMPUTATIONAL METHODOLOGY

First-principles calculations were carried out with plane wave ultrasoft pseudopotential using GGA with a Perdew–Burke–Ernzerhof (PBE) functional, and GGA+U approach was used in the CASTEP code. The ionic cores of the Ce and B atoms were represented by ultrasoft pseudopotentials. For the Ce atom, the configuration was [Xe]4f^1, where the 4f^1, 5s^2, 5p^6, 5d^1, and 6s^2 electrons were treated as valence electrons. In the case of the B atom, the configuration was [He]2$s^2$2p^4, with the 2s^2 and 2p^1 electrons treated as valence electrons. The detailed experimental conditions are similar with those described previously (Ref 13, 9).

As is well known, magnetic systems can be effectively studied using spin-polarized DFT analysis with the CASTEP code. The initial spin moment value is crucial for calculation of spin-polarizeds, and should be very close to the expected value. In the following sections, the geometric structure of CeB$_6$ is initially optimized using the GGA method. Using the optimized structure, U^f values for Ce 4f orbitals are then introduced. The formal charge and initial spin values were set as +3 and seven respectively for each Ce atom. Through comparison of the bulk modulus and the elastic constants with experimental values, the optimal values of U^f were determined. The resulting optical properties and electronic structures of CeB$_6$ obtained by the GGA + U^f calculations are then presented and compared with experimental data.

3 RESULTS AND DISCUSSION

3.1 *Structural properties and elastic constants*

At ambient pressure, CeB$_6$ crystallizes as a CsCl-type structure (cerium and the hexaboride cluster taking the place of cesium and chlorine, respectively) and has a *Pm3m* space group. Ce is located at the Wyckoff position 1a (0, 0, 0) and B is at the 6f (1/2, 1/2, z) site. Z is a positional parameter that determines the ratio of inter- and intra-octahedron B–B distances. The optimized lattice constants, positional parameters of the boron atoms, calculated spin magnetic moment m_s, and the plasma energy ($\hbar\omega_p$) for CeB$_6$, both experimental and theoretical values are listed in Table 1 and are based on the initial crystal structure. These values clearly reveal that for CeB$_6$, the positional parameters of

Table 1. The optimized lattice constant a (Å), positional parameters z, spin magnetic moment ms (μ_B) and the plasma energy $\hbar\omega_p$ (eV) for CeB_6.

	U	a (Å)	z	ms (μ_B)	$\hbar\omega_p$ (eV)
GGA		4.1415	0.2001	1.48	1.99
GGA+U	0	4.1415	0.2001	1.48	1.99
	1	4.1415	0.2001	1.48	1.99
	3	4.1415	0.2001	1.48	1.99
	5	4.1415	0.2001	1.48	1.99
Expt.		4.025[27]	0.1996[27]	1.00[20]	1.96[23]
		4.1397[28]	0.19923[28]		2.0[22]
Calc.		4.1542[7]	0.2002[7]	1.03[21]	
		4.098[29]	0.200[29]		
		4.1585[30]	0.2001[30]		

Table 2. Elastic constants C_{ij}(GPa), bulk modulus B(GPa) of CeB_6.

	U	C_{11}	C_{12}	C_{44}	B
GGA		459.40	26.64	187.26	170.89
GGA+U	0	452.56	16.60	86.11	161.92
	1	456.73	57.48	147.77	190.56
	3	451.53	16.36	84.86	161.42
	5	451.49	12.63	−19.97	158.91
Calc.		452[7]	34[7]	98[7]	173[7]
		460[30]	31[30]	82[30]	173.5[30]
Expt.		473[31]	16[31]	81[31]	168[31]
Expt.		508[31]	19[31]	79[31]	182[31]
Expt.					166[32]

the boron atoms and lattice constant are consistent with the corresponding experimental results for various effective parameter U_{eff} values. However the elastic constants and bulk modulus are in better line with the experimental results for an effective parameter $U_{eff} = 3$ eV, as shown in Table 2. Therefore, the following date reported for density of states, Mulliken charge population, band structure, and optical properties were obtained using $U_{eff} = 3$ eV.

3.2 Electronic structures

The analysis results of band structure and Density of States (DOS) of CeB_6 are plotted and compared in Fig. 1 and Fig. 2, respectively. The split in the electronic energy band corresponding to spin-up and down is attributed to the exchange coupling between electrons. The calculated spin-polarized electronic band structures of CeB_6 are displayed in Fig. 1. Together Fig. 1 and Fig. 2 show the Ce 4f energy level curves for majority spin (spin up) states and minority spin (spin down) states cross the Fermi energy (E_F) level, indicating that the electronic energy band of CeB_6 has metallic character-

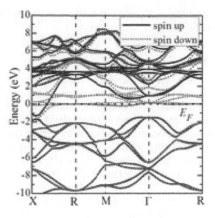

Figure 1. The spin energy band structure of CeB_6.

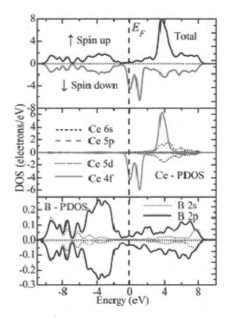

Figure 2. Spin-projected densities of states of CeB_6.

istics. Fig. 2 demonstrates that the the bottommost Conduction Bands (CBs) and upmost Valence Bands (VBs) of CeB_6 are mainly composed of either B 2p or Ce 4f states, respectively. The conduction band can be separated into two intervals. The upper valence band is dominated by the Ce 4f and 5d states, while the lower one originates mainly from Ce 4f states. Similarly, the valence band has two parts (upper and lower valence bands). The upper valence band (−10 to −7 eV) originates from nearly equal contributions of the B 2s and 2p states, while the lower one (−7 to −1 eV) is dominated by the B 2p states. Furthermore, the B 2p states extend into and above the vicinity of the Fermi energy level and show a strong hybridization with the Ce $4f$ and $5d$ states.

3.3 *Optical properties*

The frequency-dependent dielectric function $\varepsilon(\omega) = \varepsilon_1(\omega) + i\varepsilon_2(\omega)$, which is related to the electronic structures of compound was used to characterize the optical properties of CeB_6. The imaginary portion, $\varepsilon_2(\omega)$ can be analyzed based on the band structure, during which, the inter-band transitions were taken into account. Kramers–Kronig relations was used to determine the real part, $\varepsilon_1(\omega)$. As shown in Fig. 3, the complex dielectric function, as a function of photon energy, was obtained using GGA + U^f calculations at $U^f = 3$ eV. The calculated results of the complex dielectric function of CeB_6 are quite similar to those for LaB_6. The static dielectric constant $\varepsilon_1(0)$ was calculated to be 61.92. From Fig. 3 and in combination with Ref., the critical peaks in the optical spectra of $\varepsilon_2(\omega)$ were assigned as peak A (0.53 eV), peak B (4.08 eV), and peak C (6.66 eV), which correspond primarily to transitions from B $2p$ states in the valence bands to Ce $5d$ states in the conduction bands.

All other optical constants including the extinction coefficient $k(\omega)$, the refractive index $n(\omega)$, the electron energy-loss spectroscopy $L(\omega)$, the absorption spectrum $\alpha(\omega)$ and the reflectivity $R(\omega)$, are shown in Figs. 4(a)–4(e). Gaussian smearing at 0.5 eV was used in the calculation. The static refractive index was calculated to be 7.92. In the range from 0 to 1.60 eV, the reflectivity shows a sharp dip of more than 35%, indicating that the CeB_6 material is reflective primarily in the infrared wavelength range. The absorption coefficient is very large (above 10^5 cm^{-1}) in the ultraviolet to infrared wavelength range, showing that it is a solar radiation/heat-absorbing material. The energy-loss spectrum reflects the energy loss during a fast traversing of electron through the CeB_6 material, which is usually high at the plasma energy. Three

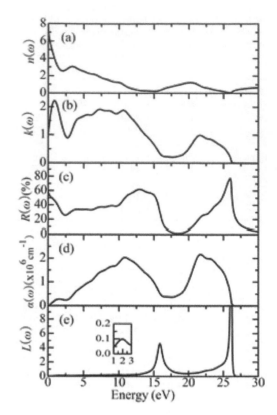

Figure 4. Optical constants of CeB_6. (a) refractive index, (b) extinction coefficient, (c) reflectivity, (d) absorption spectrum, and (e) energy-loss spectrum.

prominent peaks were observed in the energy-loss spectrum, and are consistent with the roots of $\varepsilon_2(\omega)$. In combination with Ref., the plasma energy ($\hbar\omega_p$) of CeB_6 was calculated to be 1.99 eV, which is well consistent with available experimental values: $\hbar\omega_p = 2.0$ eV and 1.96 eV.

3.4 *Conclusion*

In summary, the electronic structures, structural parameters, optical properties, and elastic constants of CeB_6 were obtained using first-principles GGA and GGA+U^f methods. The on-site coulomb interactions within the $4f$ orbital of the Ce atom (U^f) and the $2p$ orbital of the B atom were taken into account. Optimal values of U^f were obtained and were able to reproduce the experimental values of elastic constants and bulk modulus for CeB_6. These results show that the GGA+U^f approach used herein is a convenient and powerful method for investigation of rare-earth elements that include $4f$ electrons as they play a primary role in determination of elastic constants and bulk modulus.

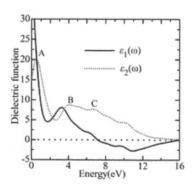

Figure 3. Dielectric function of CeB_6.

ACKNOWLEDGMENTS

This research was financially supported by Guizhou Province Science and Technology Department-Guizhou Institute of Technology Joint Fund (Guizhou Science and Technology Agency LH [2014] 7357), the Natural Science Foundation of Guizhou Province (Guizhou Science and Technology Agency J [2014] 2086, [2014] 2085), the National Natural Science Fund (No. 21361007, and 51465009), the Talents of high level scientific research fund in Guizhou Insititute of Technology (No. XJGC20140604, XJGC20131201) and the Scientific &Technological Innovation Talent Team of Guizhou Province (Qian Ke He Talent Team [2015] 4008). This manuscript has been thoroughly edited by a native English speaker from an editing company. Editing Certificate will be provided upon request.

REFERENCES

Bai,L. & Ma, N. 2010. GGA+ U method investigating structural and chemical bond properties of CeB6 and EuB6, Physica B, 405 (22): 4634–4637.

Balakrishnan, G. et al. 2003. Growth of large single crystals of rare earth hexaborides, J. Cryst. Growth, 256 (1):206–209.

Givord, F. et al. 2003. Non-anomalous magnetization density distribution in CeB6, J. Phys.: Condens. Matter. 15 (19) (2003) 3095–3106.

Goto, T. et al. 1983. Elastic properties in CeB6, J. Magn. Magn. Mater. 31–34: 419–420.Gürel, T. et al. 2010. Ab initio lattice dynamics and thermodynamics of rare-earth hexaborides LaB6 and CeB6, Phys. Rev. B 82 (10): 104302.

Jang, H. et al. 2014. Intense low energy ferromagnetic fluctuations in the antiferromagnetic heavy-fermion metal CeB6, Nat. Mater. 13 (7): 682–687.

Jie, D. et al. 2015. Elastic properties and electronic structures of lanthanide hexaborides, Chin. Phys. B, 24(9):096201.

Kimura, et al. 1992. Electronic structure of rare-earth hexaborides, Phys. Rev. B 46 (19):12196–12204.

Kimura, S. et al. 1990. Anomalous infrared absorption in rare-earth hexaborides, Solid State Commu. 75 (9): 717–720.

Leger, J. et al. 1985. High pressure compression of CeB6 up to 20 GPa, Solid State Commun. 54 (11):995–997.

Loschen, C. et al. 2007. First-principles LDA+ U and GGA+ U study of cerium oxides: Dependence on the effective U parameter, Phys. Rev. B 75 (3): 035115.

Lüthi, B. et al. 1984. Elastic and magnetoelastic effects in CeB6, Z. Phys. B: Condens. Matter 58 (1):31–38.

Magnuson, M. et al. 2001. Electronic-structure investigation of CeB6 by means of soft-x-ray scattering, Phys. Rev. B 63(7): 075101.

Nakamura, S. et al. 1994. Quadrupole-strain interaction in rare earth hexaborides, J. Phys. Soc. Jpn. 63 (2): 623–636.

Pickard, C. J. et al. 2000. Structural Properties of Lanthanide and Actinide Compounds within the Plane Wave Pseudopotential Approach, Phys. Rev. Lett. 25(84):5122–5125.

Plata, J. J. et al. 2012. Communication: Improving the density functional theory+U description of CeO2 by including the contribution of the O 2p electrons, J. Chem. Phys. 136 (4): 041101.

Ripplinger, H. et al. 1997. Electronic structure calculations of hexaborides and boron carbide, J. Solid State Chem. 133 (1): 51–54.

Samwer, K. et al. 1976. Magnetoresistivity of the Kondo-system (La, Ce) B6, Z. Phys. B 25 (3): 269–274.

Sato, S. et al. 1985. J. Magn. Magn. Mater. Elastic anomalies in intermediate valence rare-earth intermetallics, 52(1): 310–312.

Segall, M. D. et al. 2002. First-principles simulation: ideas, illustrations and the CASTEP code, J. Phys.: Condens. Matter 14 (11):2717.

Shannon, J. R. et al. 1972. Long-range magnetic interactions in europium hexaborides, Inorg. Chem. 11 (4):904–906.

Singh, N. et al. 2007. Electronic structure and optical properties of rare earth hexaborides RB6 (R = La, Ce, Pr, Nd, Sm, Eu, Gd), J. Phys.: Condens. Matter. 19 (34): 346226.

Suvasini, M. B. et al. 1996. The Fermi surface of, J. Phys.: Condens. Matter. 8 (38):7105.

Takahashi, K. et al. 1997. Single Crystal Growth and Properties of Incongruently Melting TbB6, DyB6, HoB6, and YB6, J. Solid State Chem. 133 (1): 198–200.

Tarascon, J. M. et al. 1981 Magnetic structures determined by neutron diffraction in the EuB∈-xCx system, Solid State Commu. 37 (2):133–137.

Van der Heide, et al. 1986. Differences between LaB6 and CeB6 by means of spectroscopic ellipsometry, J. Phys. F, 16 (10): 1617.

Xiao, L. et al. 2012. Origins of high visible light transparency and solar heat-shielding performance in LaB6, Appl. Phys. Lett. 101 (4):041913.

Xu, J. et al. 2013a. Fabrication of vertically aligned single-crystalline lanthanum hexaboride nanowire arrays and investigation of their field emission, NPG Asia Mater. 5 (7):e53.

Xu, J. et al. 2013b, Excellent Field - Emission Performances of Neodymium Hexaboride (NdB6) Nanoneedles with Ultra-Low Work Functions, Adv. Funct. Mater., 23 (40):5038–5048.

Zhang, H. et al. 2005. Single-crystalline GdB6 nanowire field emitters, J. Am. Chem. Soc., 127 (38):13120–13121.

Zhang, H. et al. 2006. Field emission of electrons from single LaB6 nanowires, Adv. Mater. 18 (1):87–91.

Zhang, T. A. & J. C. He. 1999. Ceramic Micro-Powders of TiB2 and LaB6 by SHS Metallurgy, Shenyang: Northeastern University Press, 69.

Zou, C. Y. et al. 2006. Synthesis of single-crystalline CeB6 nanowires, J. Cryst. Growth, 291 (1): 112–116.

173

Energy Science and Applied Technology ESAT 2016 – Fang (Ed.)
© *2016 Taylor & Francis Group, London, ISBN 978-1-138-02973-6*

CuSO$_4$ · 5H$_2$O catalyzed N-H homocoupling of N-monoalkylanilines in air: Facile construction of 1, 2-diphenylhydrazines and o-semidines

Hongying Zhang
College of Chemistry and Chemical Engineering, University of South China, Hengyang, China
Hengyang Finance Economics and Industry Polytechnic, Hengyang, China

Xueming Yan
College of Chemistry and Chemical Engineering, University of South China, Hengyang, China

ABSTRACT: A copper (II)-catalyzed N-N bond forming reaction was developed by the direct dehydrogenative homocoupling of N-alkylanilines, affording N, N'-dialkyl or N, N'-diphenylhydrazines in 52–82% yields. This new method has advantages of using inexpensive CuSO$_4$ · 5H$_2$O as a catalyst and CuO as an oxidant, under air condition, direct synthesis from N-alkylanilines, simple manipulations, mild reaction conditions, and satisfactory yields. The possible mechanism through ligand exchange and reductive elimination was proposed.

Keywords: N-H homocoupling; oxidative dehydrogenation; hydrazine; copper catalyst

1 INTRODUCTION

Many substituted hydrazines are widely used in photo- and thermo-chemical studies (Ragnarsson, U. 2001) (A.R. Katritzky et al, 2004); moreover, they show significant biological activity and are key components of biologically active azapeptides (Han, H. et al, 1996) (Zhang, R. et al, 2002) (Powers, J. C. et al, 2005), and this has prompted the continued development of new and efficient methods for obtaining substituted hydrazines (Han, H. et al, 1996) (Zhang, R. et al, 2002) (Powers, J. C. et al, 2005) (Bredihhin, A. 2007). Nevertheless, these methods still need using substituted hydrazines as starting material, and thus suffer from harsh conditions such as multiplicity of steps (protection and deprotection), use of dangerous organolithium agent, and low temperature (−78°C).

Coupling reactions catalyzed by transition metals are useful methods for constructing C-C and C-heteroatom bonds (A. de Meijere, 2004) (Alberico, D. 2007). In particular, transition metal-catalyzed oxidative dehydrogenative coupling is the even more attractive methodology because of the numerous advantages such as high efficiency, atom economy, and minimal environment impact. In recent years, considerable efforts have been made toward oxidative Cross-Dehydrogenative Coupling (CDC) and dehydrogenative homocoupling (scheme 1) (Li, C.-J. 2009) (Truong, T. 2010) (Collins, J.C. et al, 2010) (Wang, G. 2009), However,

the application for N-N bond formation via N-H bond direct oxidative dehydrogenation coupling has rarely been reported (Zhang, C. 2010) (Ragnarsson, U. 2001) (H. Han and K. D. Janda 1996) (Bredihhin, A. et al, 2007). Herein, we would like to report an efficient copper-catalyzed N-H homocoupling of alkylanilines under air condition.

2 SCHEME 1: VARIOUS CROSS-DEHYDROGENATIVE COUPLING AND DEHYDROGENATIVE HOMOCOUPLING

Previous work:

This work: homocoupling

Initially, we examined the dehydrogenative homocoupling reactions of N-methylaniline using $CuSO_4 \cdot 5H_2O$ (20%) as a catalyst in the presence of TMEDA (2.0 equivalent) with air as the oxidant at 120°C in DMF for 24 h; 45% yield was obtained (Table 1, entry 2).The reactivity of this homocoupling was found to depend significantly on the ligands of TMEDA. In comparison with the other popular ligands, such as L-proline, EDA, or 2,2′-bipyridine, TMEDA was found to be the best (Table 1, entries 2, 3, and 4). Only trace product was obtained while the reaction proceeded under N_2, indicating that oxygen is an indispensable oxidant for this homocoupling. Further investigation revealed that addition of inorganic metaloxides such as CuO, SnO_2, and SnO_2 could significantly improve the yield (Table 1, entries 5, 6, and 7). while organic peroxides t-BuOOH and t-BuOOt-Bu, which are commonly used for Cross-Dehydrogenative Coupling (CDC), failed (Table 1, entries 8 and 9). Attempts of using copper salt such as $CuCl_2 \cdot 2H_2O$, $CuBr_2$, and $Cu(OAc)_2$ also gave moderate to good yields (Table 1, entries

Table 1. Optimization of the dehydrogenative homocoupling conditions[a].

Entry	Cu Catalysis	Additive	ligand	Yield [%]
1	$CuSO_4 \cdot 5H_2O$	-	-	ND
2	$CuSO_4 \cdot 5H_2O$	-	TMEDA	45
3	$CuSO_4 \cdot 5H_2O$	-	L-proline,	ND
4	$CuSO_4 \cdot 5H_2O$	-	EDA	ND
5	$CuSO_4 \cdot 5H_2O$	SnO_2	TMEDA	68
6	$CuSO_4 \cdot 5H_2O$	SeO_2	TMEDA	70
7	$CuSO_4 \cdot 5H_2O$	CuO	TMEDA	82
8	$CuSO_4 \cdot 5H_2O$	t-BuOOH	TMEDA	0
9	$CuSO_4 \cdot 5H_2O$	t-BuOOt-Bn	TMEDA	0
10	$CuCl_2 \cdot 2H_2O$	CuO	TMEDA	35
11	$CuBr_2$	CuO	TMEDA	42
12	$Cu(OAc)_2$	CuO	TMEDA	15
13	-	CuO	TMEDA	0

[a] General conditions: N-methylaniline (0.1 mmol), Cu salt (20 mol%), K_2CO_3 (2.0 equivalent), DMF (2 mL), TMEDA (2.0 equivalent), additive (0.5 equivalent) under air.

Table 2. Direct synthesis of hydrazines via the N-H Homocoupling reaction[a].

entry	Alkylaniline	Yield[b][%]
1	R_1=Et	81
2	R_1= n-Pro	80
3	R_1= n-Bu	78
4	R_1= Me,R_2=4-Cl	52
5	R_1= Me,R_2=4-Cl	57
6	R_1= Me,R_2=4-Me	82
7	R_1= Me,R_2=3-Me	71
8	R_1=Et ,R_2=3-Me	75
9	R_1= Me,R_2=3-OMe	78

10,11,12), but $Cu(OAc)_2 \cdot H_2O$ were less effective. No expected products were obtained at all in the absence of copper catalysis (Table 1, entry 13). Solvents were also screened, and DMF was found to be the best choice.

Subsequently, the generality and substrate scope of this homocoupling reaction were investigated under above optimal conditions. As indicated in Table 2, Et, n-Pro, and n-Bu substitutions at N atom were tolerated (Table 2, entries 1, 2, and 3) and good yields were obtained. However, the electronic nature and positions of substituents at the phenyl moiety were found to exert a significant influence on the efficiency, electronic-rich alkylanilines (1f~1i) were exceptionally effective (Table 2, entries 6–9), and reactions can be performed smoothly at a lower temperature and with shorter reaction time. Mostly because of the steric hindrance, 2-methyl and 2-methoxy-substituted alkylanilines proved unreactive. Substrates with electron-withdrawing groups at the 4-position were operative, but provided slightly lower yields (Table 2, entries 4 and 5); 3-halo-substituted alkylanilines and nitro-substituted alkylanilines were not capable of undergoing this oxidation presumably due to the low electron density on the N atom.

Most interestingly, with a strong electron-donating group of methoxy or ethoxy on the para-position of the alkylanilines, no desired homocoupling product was obtain at all under above optimal condition. To our knowledge, lowering the reaction temperature to RT led to excellent yields of o-semidines (Table 3), whereas

Table 3. The direct formation of o-semidines from the alkylanilines[a].

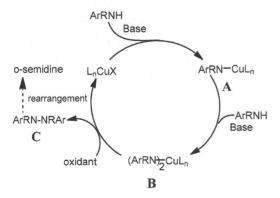

entry	Alkylaniline	Yield[b][%]
1	R₁= Me, R₂=4-OMe	82
2	R₁= Et, R₂=4-OMe	81
3	R₁=-Pro, R₂=4-OMe	74
4	R₁=Me, R₂=4-OEt	72
5	R₁=Et , R₂=4-OE	78

3-methoxy-substituted alkylanilines still gave the N-N coupling product (Table 2, entry 9).

Although the exact mechanism of this reaction is indistinct, the most possible approach would be similar to oxidative C-H /N-H cross-coupling and the oxidative homocoupling of terminal alkynes. As shown in Scheme 2, base-assisted cupration of the alkylaniline forms organocopper **A** and **B**. The subsequent step of reductive elimination led to the synthesis of the homocoupling product **C**, and a lower oxidation state of copper was also attained, which was easily converted into Cu(II) oxidized by oxygen, and thus the catalytic cycle was completed perfectly. Homocoupling product bearing strong electron donating group (Table 3, entries 1 5) immediately rearranges to o-semidines, which is in accordance with the previous studies.

3   SCHEME 2: THE PROPOSED MECHANISM OF THE COPPER-CATALYZED HOMO-COUPLING OF N-ALKYLANILINES

In conclusion, we have developed a novel copper-catalyzed N-N bond-forming method through direct dehydrogenative homocoupling. This new protocol requires inexpensive $CuSO_4·5H_2O$ as a catalyst, TMEDA as a ligand, and CuO as an additive. It is the most direct and economical strategy for efficient access to tetrasubstituted hydrazines and o-semidines. Investigation of the scope and mechanism is under study. We are also applying this method to construct complex molecules.

REFERENCES

de Meijere, A., Diederich, F. Metal-catalyzed crosscoupling reactions, 2nd edn., Wiley, New York, 2004. (b) Handbook of C-H Transformations (Eds.: G. Dyker), Wiley-VCH, Weinheim, 2005. (c) Transition Metals for Organic Synthesis (Eds.: M. Beller, C, Bolm), Wiley-VCH, Weinheim, 2004. (d) E. Negishi, Handbook of Organopalladium Chemistry for Organic Synthesis Wiley, New York, 2002.

For a review see: Ragnarsson, U. Synthetic methodology for alkyl substituted hydrazine [J]. Chem. Soc. Rev. 2001, 30, 205–213. (b) Hydrazine and its Derivatives, in Kirk-Othmer Encyclopedia of Chemical Technology, Wiley, New York, 4th, 1995, vol. 13. (c) A.R. Katritzky; R.J.K. Taylor; Comprehensive Organic Functional Group Transformations II, Elsevier, 2004; 345–353.

C-N bond formation through cross-dehydrogenative-coupling (CDC) see: (a) Armstrong, A.; Collins, J.C. Gold-Catalyzed Halogenation of Aromatics by N-Halosuccinimides [J]. Angew. Chem. Int. Ed. 2010, 49, 2282 (b) Wang, Q.; Schreiber, S.L. Copper-Mediated Amidation of Heterocyclic and Aromatic C–H Bonds[J].Org. Lett. 2009, 11, 5178. (c) Monguchi, D.; Fujiwara, T.; Furukawa, H.; Mori, A. Direct Amination of Azoles via Catalytic C–H, N–H Coupling[J]. Org. Lett. 2009, 11, 1607. (d) Hamada, T.; Ye, X.; Stahl, S.S. Copper-Catalyzed Aerobic Oxidative Amidation of Terminal Alkynes: Efficient Synthesis of Ynamides[J]. J. Am. Chem. Soc. 2008, 130, 833. (e) Reed, S.A.; Mazzotti, A.R.; White, M.C. A Catalytic, Brønsted Base Strategy for Intermolecular Allylic C–H Amination[J]. J. Am. Chem. Soc. 2009, 131, 11701. (f) Sherman, E.S.; Chemler, S.R.; Tan, T.B.; Gerlits, O. Copper(II) Acetate Promoted Oxidative Cyclization of Arylsulfonyl-o-allylanilines[J]. Org. Lett. 2004, 6, 1573.

C-Si, C-P and C-O bond formation through cross-dehydrogenative-coupling (CDC) see: (k) Tsukada, N.; Hartwig, J.F. Intermolecular and Intramolecular, Platinum-Catalyzed, Acceptorless Dehydrogenative Coupling of Hydrosilanes with Aryl and Aliphatic Methyl C–H Bonds[J]. J. Am. Chem.Soc. 2005, 127, 5022. (l) Gao, Y.X.; Wang, G.; Chen, L.; Xu, P.X.; Zhao, Y.F.; Zhou, Y.b.; Han, L.B. Copper-Catalyzed Aerobic Oxidative Coupling of Terminal Alkynes withH-Phosphonates Leading to Alkynylphosphonates[J]. J. Am. Chem. Soc. 2009, 131, 7956. (m) King, A.E.; Huffman, L.M.; Casitas, A.; Costas, M.; Ribas, X.; Stahl, S.S. Copper-Catalyzed Aerobic Oxidative Functionalization of an Arene C–H Bond: Evidence for an Aryl-Copper(III) Intermediate[J]. J. Am. Chem.Soc. 2010, 132, 12068.

For recent reviews see: (a) Alberico, D.; Scott, M. E.; Lautens, M. Aryl–Aryl Bond Formation by Transition-Metal-Catalyzed Direct Arylation[J]. Chem. Rev. 2007, 107,174–238. (b) Diaz-Requejo, M. M.; Perez, P. J. Coinage Metal Catalyzed C–H Bond Functionalization of Hydrocarbons[J]. Chem. Rev. 2008.108, 3379–3394. (c) Kakiuchi, F.; Murai, S. Catalytic C–H/Olefin Coupling [J]. Acc. Chem. Res. 2002, 35, 826–834.

Han, H. and Janda, K. D. Azatides: Solution and Liquid Phase Syntheses of a New Peptidomimetic[J]. J. Am. Chem. Soc. 1996, 118, 2539. (b) Zhang, R.; Durkin, J. P.; Windsor, W. T. Azapeptides as inhibitors of the hepatitis C virus NS3 serine protease[J]. Bioorg. Med. Chem. Lett. 2002, 12, 1005.

Han, H.; Janda, K. D. Azatides: Solution and Liquid Phase Syntheses of a New Peptidomimetic[J]. J. Am. Chem. Soc., 1996, 118, 2539. (b) Zhang, R.; Durkin, J. P.; Windsor, W. T. Azapeptides as inhibitors of the hepatitis C virus NS3 serine protease[J]. Bioorg. Med. Chem. Lett. 2002, 12, 1005–1008. (c) Lee, T.-W.; Cherney, M. M.; Huitema, C.; Liu, J.; James, K. E.;Powers, J. C.; Eltis, L. D.; James, M. N. G. Crystal structures of the main peptidase from the SARS coronavirus inhibited by a substrate-like aza-peptide epoxide.[J]. J. Mol. Biol. 2005, 353, 1137–1151.

Hydrazine and Its Derivatives, in Kirk-Othmer Encyclopedia of Chemical Technology; Wiley-VCH: New York, 4th edn., 1995, vol. 13. (b) Encyclopedia of Reagents for Organic Synthesis; ed. L. A. Paquette, Wiley, Chichester, 1995. (c) Ragnarsson, U. Synthetic methodology for alkyl substituted hydrazines[J]. Chem. Soc. Rev. 2001, 30, 205.

Oxidative cross-dehydrogenative coupling (CDC), For some reviews see: (a) Li, C.-J. Cross-Dehydrogenative Coupling (CDC): Exploring C–C Bond Formations beyond Functional Group Transformations[J]. Acc. Chem. Res. 2009, 42, 335. (b) Wei, Y.; Zhao, H.-Q.; Kan, J.; Su, W.-P. Hong, M.-C. Copper-Catalyzed Direct Alkynylation of Electron-Deficient Polyfluoroarenes with Terminal Alkynes Using O2 as an Oxidant [J]. J. Am. Chem. Soc. 2010, 132, 2522. (c) Matsuyama, N.; Kitahara, M.; Hirano, K.; Satoh, T.; Miura, M. Nickel- and Copper-Catalyzed Direct Alkynylation of Azoles and Polyfluoroarenes with Terminal Alkynes under O2 or Atmospheric Conditions [J]. Org. Lett.2010, 12, 2358. (d) Lin, S.; Song, C.-X.; Cai, G.-X.; Wang, W.-H.; Shi, Z.-J. Intra/Intermolecular Direct Allylic Alkylation via Pd(II)-Catalyzed Allylic C–H Activation[J]. J. Am. Chem. Soc. 2008, 130, 12901. (e) Li, Z.-P.; Li, C.-J. Catalytic Allylic Alkylation via the Cross-Dehydrogenative-Coupling Reaction between Allylic sp3 C–H and Methylenic sp3 C–H Bonds[J]. J. Am. Chem. Soc. 2006, 128, 56.

Oxidative dehydrogenative homocoupling, For representative references, see: (a) Truong, T.; Alvarado, J.; Tran, L.D.; Daugulis, O. Nickel, Manganese, Cobalt, and Iron-Catalyzed Deprotonative Arene Dimerization[J]. Org. Lett.2010, 12, 1200. (b) Chen, Q.-A.;Dong, X.;Chen, M.-W.; Wang, D.S.; Zhou, Y.G.; Li, Y.X. Highly Effective and Diastereoselective Synthesis of Axially Chiral Bis-sulfoxide Ligands via Oxidative Aryl Coupling[J]. Org. Lett. 2010, 12, 1928. (c) Egami, H.; Katsuki, T. Iron-Catalyzed Asymmetric Aerobic Oxidation: Oxidative Coupling of 2-Naphthols[J]. J. Am. Chem. Soc. 2009, 131, 6082. (d) Hull, K.L.; Lanni, E.L.; Sanford, M.S. Highly Regioselective Catalytic Oxidative Coupling Reactions: Synthetic and Mechanistic Investigations[J]. J. Am. Chem. Soc. 2006, 128, 14047. (e) Matsushita, M.; Kamata, K.; Yamaguchi, K.; Mizuno, N. Heterogeneously Catalyzed Aerobic Oxidative Biaryl Coupling of 2-Naphthols and Substituted Phenols in Water[J]. J. Am. Chem. Soc. 2005, 127, 6632. (f) Masui, K.; Ikegami, H.; Mori, A. Palladium-Catalyzed C–H Homocoupling of Thiophenes: Facile Construction of Bithiophene Structure[J]. J. Am. Chem. Soc. 2004, 126, 5074.

Representative papers for the synthesis of hydrazines from the compounds containing N-N or N = N bonds: (a) Bredihhin, A.; Groth, U. M.; Maeorg, U. Efficient Methodology for Selective Alkylation of Hydrazine Derivatives[J]. Org. Lett. 2007, 9, 1097. (b) Kisseljova, K.; Tsubrik, O. Addition of Arylboronic Acids to Symmetrical and Unsymmetrical Azo Compounds[J]. Org. Lett. 2006, 8, 43. (c) Bredihhin, A.; Maeorg, U. Use of Polyanions for Alkylation of Hydrazine Derivatives[J]. Org. Lett. 2007, 9, 4975. (d) Tsubrik, O.; Maeorg, U. Combination of tert-Butoxycarbonyl and Triphenylphosphonium Protecting Groups in the Synthesis of Substituted Hydrazines[J]. Org. Lett. 2001, 3, 2297.

Selected examples: (a) Bredihhin, A.; Groth, U.M.; Mäeorg, U. Efficient methodology for selective alkylation of hydrazine derivatives[J]. Org. Lett. 2007, 9, 1097–1099. (b) Kisseljova, K.; Tsubrik, O. Addition of Arylboronic Acids to Symmetrical and Unsymmetrical Azo Compounds[J]. Org. Lett. 2006, 8, 43–45. (c) Bredihhin, A.; Maleorg, U. Use of Polyanions for Alkylation of Hydrazine Derivatives [J]. Org. Lett. 2007, 9, 4975–4977. (d) Tsubrik, O.; Mäeorg, U. [J]. Org. Lett. 2001, 3, 2297–2299.

Zhang, C.; Jiao, N. Copper-Catalyzed Aerobic Oxidative Dehydrogenative Coupling of Anilines Leading to Aromatic Azo Compounds using Dioxygen as an Oxidant [J]. Angew. Chem. Int. Ed. 2010, 49, 6174. (b) Grirrane, A.; Corma, A.; Garcia, H. Gold-catalyzed synthesis of aromatic azo compounds from anilines and nitroaromatics[J]. Science 2008, 322, 1661.

178

Energy Science and Applied Technology ESAT 2016 – Fang (Ed.)
© 2016 Taylor & Francis Group, London, ISBN 978-1-138-02973-6

The influence factors of fluorite for making hydrogen fluoride and its additional value

Zihuan Wang & Hao Ren
Shandong Provincial Key Laboratory of Fluorine Chemistry and Chemical Materials,
School of Chemistry and Chemical Engineering, University of Jinan, Shandong, China

Xiaogang Li, Guoying Yan & Wenxiang Meng
Baotou Steel Group Mining Research Institute (Limited Liability Company), Inner Mongolia, China

Yufen Zhang & Taizhong Huang
Shandong Provincial Key Laboratory of Fluorine Chemistry and Chemical Materials,
School of Chemistry and Chemical Engineering, University of Jinan, Shandong, China

ABSTRACT: Fluorite resources are enriched in Baiyun Obo Region of China, but have not been effectively recycled and processed. Results of hydrogen fluoride preparation technology under the different influence factors are reported. In this study, we mainly study the influence of the different feeding methods, different sulfuric acid and fluorite powder molar ratio, different reaction temperatures, different reaction times and different granularity on the reaction between fluorite powder and sulfuric acid. We can conclude that the higher the temperature, the faster the reaction and the reaction time are longer, the reaction is more thorough. Using the smaller particle size product experiment can greatly improve the reaction efficiency. In addition, the impurity iron and rare earth in the fluorite generated to mostly waters soluble sulfate, easy to wash leaching, and it shows that the rare earth can be recycled so that further improve the additional value of comprehensive utilization of fluorite mineral. More importantly, through this experiment, we determined the best conditions what the reaction of calcium fluoride (CaF_2) powder with sulfuric acid which can solve the problem of baotou steel company's tailings.

Keywords: fluorite, hydrogen fluoride, influence factors, iron and rare earth, additional value

1 INTRODUCTION

Fluorine chemistry is an important part of chemical industry. A process for the preparation of HF and anhydrite from reaction of calcium fluoride (CaF_2) powder with sulfuric acid (H. Grass 2011) and in the form of anhydrous hydrogen fluoride or hydrofluoric acid products are the basic materials of all kinds of inorganic and organic fluoride production. Fluorite resource is widely distributed and can be found in continent worldwide; there are more than 40 countries having explored the fluorite's reserves (S. Chen & S. Zhang 2012). A rude ore of fluorite may contain, in addition to calcium fluoride (CaF_2) as a main component, impurities such as silicon dioxide (SiO_2), calcium carbonate ($CaCO_3$), ferric oxide (Fe_2O_3) and Rare-Earth Oxide (REO) (R. Haglwara & K. Matsumoto 2011, Y.Hu et al. 2003). Baotou fluorite ore concentrate is a by-product of baiyun obo mineral resources comprehensive utilization, which is to choose the iron, rare earth, niobium and scandium of tailings

and then to choose fluorite (D. Candido & G. P. Mathur 1974, L. Li et al. 2011, W. L. Augenstein et al. 1991).

A method of fluorite-sulfuric acid procedure is utilized for production of anhydrous hydrogen fluoride; it mainly comes from the production practice experience and the introduction of technology.

2 EXPERIMENTAL

This experiment uses the mixture mixing process, –200 mesh particle size for the materials making up 98%, –500 mesh accounting for 55%, CaF_2 content is 93.94% of fluorite ore concentrate. And sulfuric acid use analytical pure what it's the concentration of 98% (E. A. Naumova et al. 2010).

First of all, the mole ratio of sulfuric acid and fluorite powder is 0.96, 1.00, 1.04, 3, 5; and the reaction time is 4 hours. In addition, we need to study the effect of different reaction temperature with the reaction result. In the process of experiment, others things being equal, reaction temperature set in 90, 120, 150, 190, 230°C respectively. Then, we continue to study the influence of reaction time on the reaction result when the reaction temperature at 150°C (S. Ying & Z. Zhou 2013, P. M. Arnow et al. 1994). When the acid powder mole ratio is 1:1, the reaction time is 2, 5, 10, 20, 30, 60, 120, 240, 120, 240 min. Finally, we study the effect of fluorite granularity on the reaction result when the reaction temperature at 150°C. Grinding raw fluorite powder first, and then respectively take screen mesh + 25, −25 ~ + 13, −13 ~ + 6.5, −6.5 μm fluorite powder of 3 parts. However, the acid powder mole ratio is 1:1; the reaction time is 1, 2, 4 hours of the experiment (L. Wang & T. Liang 2014).

3 RESULTS AND DISCUSSION

3.1 Different mole ratio of acid powder

Table 1 and Figure 1 show that the reaction result what the ratio is 0.96:1, 1, 1.04:1 is slightly different but no significant difference, CaF_2 conversion rate is around 91%. However, from the reaction result what the molar ratio is 3:1, 5:1 can see CaF_2 conversion rate is around 98% from sulfuric acid and fluorite powder. This suggests that the larger the molar ratio of concentrated sulfuric acid and fluorite powder, the more complete the reaction.

The experimental results of different mole ratio of acid powder show that the content of Fe_2O_3 is between 0.44%~0.84% and REO is between 1.13%~1.36% in gypsum residue, iron and rare earth generated mostly water soluble sulfate after the reaction, easy to wash leaching, which from another aspect shows that rare earth can be further recovery, further improve the additional value of comprehensive utilization of fluorite mineral.

3.2 Different reaction temperature

Can be seen from Table 2 and Figure 2 is that the reaction temperature had great effect on the

result of the reaction, the higher reaction temperature more completely. It also shows that the yield of anhydrous hydrogen fluoride is constantly improving.

The experimental results of different reaction temperature show that the Fe_2O_3 content in the fluorgypsum residue after washing is between 0.56%~0.99% and REO between 0.12%~0.12%, the iron and rare earth generated mostly water soluble sulfate after the reaction, easy to wash leaching, also shows that rare earth can recycling, further improve the additional value of comprehensive utilization of fluorite mineral.

3.3 Different reaction time

What can be seen from Table 3 and Figure 3 is that the reaction rate of 2~60 minutes is the fastest stage, reaction leveling off after 60 minutes, but the reaction continues, and 5.8% the content of CaF_2 didn't respond when the reaction time to 720 minutes.

The experimental results of different reaction time show that the Fe_2O_3 content in the fluorgyp-

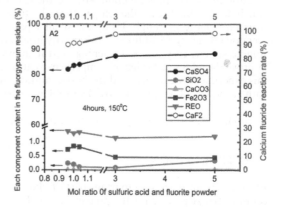

Figure 1. The molar ratio of sulfuric acid and fluorite powder, the conversion of CaF_2 and the relationship between various component content in the fluorgypsum residue.

Table 1. The conversion of CaF_2 under different mole ratio of acid powder and the ratio (K) of the input of fluorite powder and the output of anhydrous hydrogen fluoride.

Molar ratio	$CaSO_4$/%	SiO_2/%	$CaCO_3$/%	$Fe2O_3$/%	REO/%	CaF_2 Percentage/%	Conversion rate/%	K
0.96:1	82.1	0.237	0	0.72	1.36	9.0	90.4	2.29
1:1	83.3	0.191	0	0.84	1.29	8.1	91.4	2.27
1.04:1	84.0	0.107	0	0.81	1.33	8.2	91.3	2.28
3:1	87.4	0.085	0	0.45	1.13	1.95	97.9	2.12
5:1	88.3	0.325	0	0.44	1.18	1.37	98.5	2.11

Table 2. The conversion of CaF_2 under different reaction temperature and the ratio (K) of the input of fluorite powder and the output of anhydrous hydrogen fluoride.

Temperature/°C	$CaSO_4$/%	SiO_2/%	$CaCO_3$/%	Fe_2O_3/%	REO/%	CaF_2 Percentage/%	Conversion rate/%	K
90	73.8	0.129	0	0.99	1.18	14.4	84.7	2.45
123	79.4	0.259	0	0.79	0.87	11.0	88.3	2.35
150	82.4	0.259	0	0.80	0.12	8.2	91.3	2.28
190	84.9	0.259	0	0.56	1.39	5.9	93.7	2.22
230	86.6	0.259	0	0.63	0.91	4.8	94.9	2.19

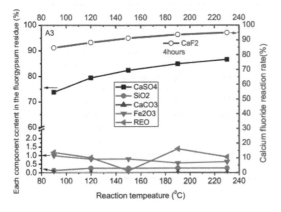

Figure 2. Reaction temperature, the conversion of CaF_2 and the relationship between various component content in the fluorgypsum residue.

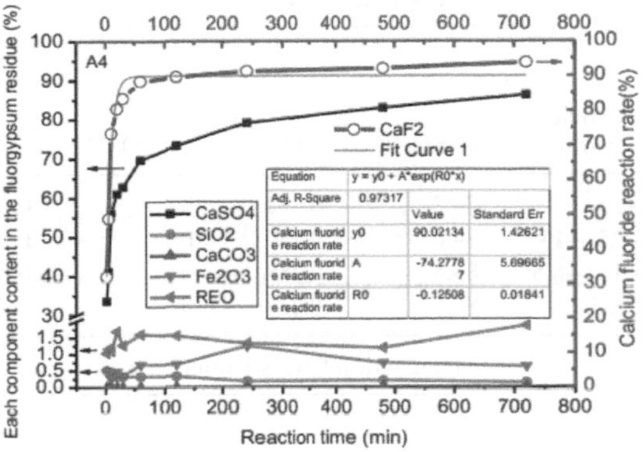

Figure 3. Reaction time, the conversion of CaF_2 and the relationship between various component content in the fluorgypsum residue.

Table 3. The conversion of CaF_2 under different reaction time and the ratio (K) of the input of fluorite powder and the output of anhydrous hydrogen fluoride.

Reaction time/min	$CaSO_4$/%	SiO_2/%	$CaCO_3$/%	Fe_2O_3/%	REO/%	CaF_2 Percentage/%	Conversion rate/%	K
2	33.65	0.512	0	0.46	1.03	63.9	32.0	6.49
5	41.1	0.477	0	0.50	1.16	48.1	48.8	4.26
10	56.3	0.325	0	0.28	1.10	25.0	73.4	2.83
20	61.1	0.325	0	0.47	1.69	18.3	80.5	2.58
30	62.8	0.325	0	0.35	1.27	15.5	83.5	2.49
60	69.7	0.325	0	0.68	1.60	10.9	88.4	2.35
120	73.4	0.325	0	0.68	1.55	9.7	89.7	2.32
240	79.2	0.172	0	1.04	1.32	8.2	91.3	2.28
480	83.0	0.194	0	0.73	1.18	7.4	92.1	2.26
720	86.3	0.129	0	0.62	1.86	5.8	93.8	2.22

sum residue after washing is between 0.28%~1.04% and REO between 1.03%~1.86%, the iron and rare earth generated mostly water soluble sulfate after the reaction, easy to wash leaching, also shows that rare earth can recycling, further improve the additional value of comprehensive utilization of fluorite mineral.

3.4 Different particle size

Table 4 and Figure 4a show that the conversion of calcium fluoride gradual growth with the increase of reaction time. Until 4 hours later, the conversion of calcium fluoride is between 87%~95%, and the ratio of the input of fluorite powder and the

Table 4. The conversion of CaF_2 under different particle size and the ratio (K) of the input of fluorite powder and the output of anhydrous hydrogen fluoride.

The reaction time is 4 hours

Particle size/μm	$CaSO_4$/%	SiO_2/%	$CaCO_3$/%	Fe_2O_3/%	REO/%	CaF_2 Percentage/%	CaF_2 Conversion rate/%	K
+25	80.4	0.259	0	1.01	1.25	6.0	93.6	2.22
−25~+13	80.5	0.172	0	1.05	1.31	6.1	93.5	2.22
−13~+6.5	87.0	0.194	0	0.53	0.89	4.7	95.0	2.19
−6.5	87.5	0.041	0	0.38	0.79	4.8	94.9	2.19

The reaction time is 2 hours

Particle size/μm	$CaSO_4$/%	SiO_2/%	$CaCO_3$/%	Fe_2O_3/%	REO/%	CaF_2 Percentage/%	CaF_2 Conversion rate/%	K
+25	74.2	0.342	0	0.47	1.32	10.4	88.9	2.34
−25~+13	74.6	0.236	0	0.42	1.30	10.5	88.8	2.34
−13~+6.5	83.8	0.288	0	0.39	1.12	9.3	90.1	2.31
−6.5	86.3	0.217	0	0.40	0.97	8.8	90.6	2.29

The reaction time is 1 hours

Particle size/μm	$CaSO_4$/%	SiO_2/%	$CaCO_3$/%	Fe_2O_3/%	REO/%	CaF_2 Percentage/%	CaF_2 Conversion rate/%	K
+25	69.7	0.291	0	0.46	1.45	11.6	87.7	2.37
−25~+13	70.2	0.335	0	0.36	1.38	11.3	88.0	2.37
−13~+6.5	79.8	0.286	0	0.35	1.21	15.7	88.7	2.35
−6.5	85.6	0.278	0	0.33	1.09	15.3	89.1	2.34

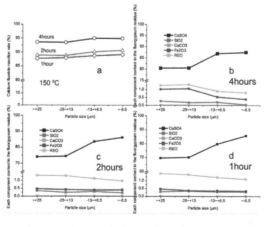

Figure 4. The particle size of fluorite, the conversion of CaF_2 and the relationship between various component content in the fluorgypsum residue.

output of anhydrous hydrogen fluoride can control below 2.30.

Figure 4b indicates the content of Fe_2O_3 is between 0.38%~1.01% and REO is between 0.79%~1.31% under the condition of the reaction time is 4 hours. Figure 4c indicates the content of Fe_2O_3 is between 0.39%~0.47% and REO is between 0.97%~1.32% under the condition of the reaction time is 2 hours. Figure 4d indicates the content of Fe_2O_3 is between 0.33% ~ 0.46% and REO is between 1.09%~1.45% under the condition of the reaction time is 1 hour. Different granularity and reaction time to the results of Fe_2O_3 and REO without certain regularity, but each is lower than the corresponding component content of raw materials.

The experimental results of different particle size show that the fluorgypsum residue after washing the content of Fe_2O_3 and REO both in the scope, the iron and rare earth generated mostly water soluble sulfate after the reaction, easy to wash leaching, also shows that rare earth can recycling, further improve the additional value of comprehensive utilization of fluorite mineral.

4 CONCLUSIONS

From the series of the experimental results can be reference to the daily production, namely, the content of $CaCO_3$, SiO_2 of fluorite powder, sulfuric acid concentration, the weight ratio of acid and fluorite powder, the reaction heat and reaction time are the important factors that affect the production, these factors should be well controlled and deal with the relationship between them in order to achieve the best reaction condition. More importantly, through this experiment, we determined the best conditions about the reaction of calcium fluoride (CaF_2) powder with sulfuric acid which can solve the problem of baotou steel company's tailings. Through the study:

1. The higher the temperature, the faster the reaction, so improving the temperature of reaction device on the reaction is favorable.
2. The longer the reaction time, the more thoroughly the reaction goes on.
3. With the product testing of smaller particle size, can greatly improve the efficiency of reaction, the finer the fluorite powder, the faster the reaction speed, the shorter the time is needed to fully reaction.
4. From the experimental results it can be seen the iron and rare earth generated mostly water soluble sulfate after the reaction, easy to wash leaching, and the rare earth can recycle, further improve the additional value of comprehensive utilization of fluorite mineral.

5. Through the experiment, we can not only solve the problem of long-term legacy of tailings pollution, also can achieve the goal of recycling the waste resource and make the benefit maximization.

ACKNOWLEDGMENTS

This work was financially supported by the Baotou Steel Group Mining Research Institute (limited liability company) (2014034). Special thanks are given to the research team for providing support in this study.

REFERENCES

Arnow, P.M., Bland, L.A. and Garcia, H.S. 1994, ANN INTERN MED. 121: 339–344.
Augenstein, W.L., Spoerke, D.G., Kulig, K.W. 1991, PEDIA TR RES. 88: 907–912.
Candido, D. and Mathur, G.P. 1974, Ind. Eng. Chem., Process Des. Develop, Vol. 13, No. 1.E.A. Naumova, P. Gaengler, S. Zimmer and W.H. Arnold. 2010, Open Dent J, 4, 185–190.
Grass, H., Hengst, M., Eicher, J. 2011. Pub. No.: US 2011/02065 98 A1. Pub. Date: Aug. 25, Patent Application Publication.
Haglwara, R., Matsumoto, K. 2011. Pub. No.: US 2011/0286911 A1. Pub. Date: Nov. 24, Patent Application Publication. S. Chen, S. Zhang. 2012, Advanced Materials Research, 10, 142–145.
Hu, Y., Ruan, C. and Xu, Z. 2003, Ind. Eng. Chem. Res. 42, 1641–1647.
Li, L., Wang, S., Dong, Y., Jia, X. 2011, J IRON STEEL RES INT. 18(1): 11–15.
Wang, L., Liang, T. 2014, ATMOS ENVIRON. 88: 23–29.
Ying, S., Zhou, Z. 2013, Chemical Production and Technology. 20(1): 6–8.

Energy Science and Applied Technology ESAT 2016 – Fang (Ed.)
© 2016 Taylor & Francis Group, London, ISBN 978-1-138-02973-6

Surface modification and mechanism of a calcium sulfate whisker treated by calcium stearate surfactant

Huaiyou Wang, Min Wang, Youjing Zhao & Jinlin Li
Qinghai Institute of Salt Lakes, Chinese Academy of Sciences, Xining, China
Key Laboratory of Comprehensive and Highly Effect Utilization of Salt Lake Resources, Chinese Academy of Sciences, Qinghai, China

ABSTRACT: Hydrophobic-lipophilic Calcium Sulfate Whiskers (CSW) was synthesized by wet modificaton with Calcium Stearate (CaSt). The hydrophobicity of CSW and modified CSW (CSW-CaSt) were determined by Water Contact Angle (WCA) instrument and the optimum surface modification conditions of CSW are obtained. The surface morphology of CSW and CSW-CaSt was measured with Field Emission Scanning Electron Microscope (FESEM) and the phase of CSW and CSW-CaSt were determined by X-Ray powder Diffraction (XRD). It is shown that modification occurred on the surface of CSW exclusively. The existence of physisorbed CaSt was proved by Fouier Transform Infrared (FT-IR) spectroscopy and thermogravimetric analyses. X-ray Photoelectron Spectroscopy (XPS) was used to study the varieties and states of different elements on the surface of CSW and CSW-CaSt. The result also indicates that CaSt exists on the surface modification of CSW. Hence, it is confirmed that the type of surface modification of CSW with CaSt is physical adsorption.

Keywords: Calcium Sulfate Whiskers; surface modification; surface morphology

1 INTRODUCTION

In recent years, CSW with high aspect ratios has been recognized as promising reinforcing materials widely used in rubber, ceramics, paper making and plastics because of its some properties such as, low cost, non-toxic, chemical resistance, high strength and high modulus. (S.C. Hou, 2014). But CSW is in a state of thermodynamic instability due to large surface area and high surface energy, which leads to CSW reunite easily. Furthermore, there is poor compatibility between CSW and organic matrix because of the strong polarity of CSW surface. When CSW was directly filled in organic matrix, it is hard to disperse in the organic matrix and the adhesion strength of the matrix is very weak. Two phase interface defects between CSW and organic matrix result in a decline in the performance of the composite materials. Therefore it is necessary that CSW is modified in order to adjust the hydrophobicity, to decrease surface energy and enhance the compatibility with organic matrix. Furthermore modified CSW can improve and enhance the performance of the composite materials. (W.L. Hao, 2013).

Wang et al. (J.C. Wang, 2011) Studied modification effects of two kinds of organic silicon surface modification agent, hydrogenated silicon oil (OCSW-H) and silane coupling agent KH-560 (OCSW-K) on CSW. CSW was added to the rubber to prepare LSR/OCSW complex. The tensile strength, thermal stability of the two different modified CSW were measured. The results show that the tensile strength of LSR/OCSW-K raises 60%. Yin et al (W.Z. Yin, 2007). studied influences of the factors such as surfactant dosage, modification temperature/ time and stirring speed on the surface modification of CSW with borate surfactant SBW-181. The results indicated that the borate surfactant SBW-181 has good effect on surface modification of CSW. The activation index of modified CSW is 0.996 with a contact angle 103°. A reaction model was developed between CSW and surfactant. Wang et al. (X.L. Wang, 2006) chose several kinds of surfactants to perform wet modification experiments of CSW. The results indicate that stearic acid has good effect on surface modification of calcium CSW. Under the optimum condition, active exponent of modified product is 1.00. They also used these surface modification of calcium CSW to synthesized polypropylene composites, and the mechanical properties were determined. Shen et al (H.L. Shen, 2011). reported two different kinds of maleated polypropylene (MPP), as surfactant, were fixed respectively with PP/ $CaSO_4$ composite. The results show that the tensile strength had a peak at CSW concentration 30%. In order to solve the high solubility and low retention problems used as paper filler, Liu et al had characterized and modified the calcium sulfate by dissolution-inhabiting method. The results showed that stearic acid treatment can obviously decrease solubility of calcium sulfate. Liao focused on the use of chitosan coating method in modifying CSW to reduce its water solubility and the application of modified CSW in paper making.

However, surface modification of CSW with CaSt is barely reported. In this paper, a facile method was developed to prepare Hydrophobic-lipophilic CSW) by surface modification with Calcium stearate(CaSt). The hydrophobicity of CSW and CSW-CaSt were determined. The surface morphology of CSW and CSW-CaSt was examined with FESEM and the corresponding mechanisms were studied.

2 EXPERIMENTAL

Materials. Calcium sulfate whiskers with a diameter of 1~4 μm and length of 20–200 μm was synthesized by hydrothermal synthesis, using the natural gypsum as raw material. Calcium stearate $(CH_3(CH_2)_{16}COO_2Ca)$ (chemical pure 99%, Shanghai Sailing Reagent Factory, China) and ethanol absolute (C_2H_5OH) (Aldrich 99%, Tianjin Yongda Chemical Reactant Company Limite, China) were used as received.

Surface modification of CSW. The CSW was dispersed in ethanol absolute in the proportion of 5.1285 g to 50 ml ethanol absolute, and then immersed in water bath at 223.15 K~363.15 K. Subsequently, the slurrier of CaSt was dropped to the slurrier containing CSW at stirring rate 600 rpm~1400 rpm for 5 min~25 min. The mixture slurrier was filtered the liquid phase at a high temperature. The solid phase was dried at 373.15 K for 3h.

Measurement. The hydrophobicity of CSW and CSW-CaSt were determined by water contact angle apparatus (DSA30, Kruss, Germany). The final value was the average of three measurements. The morphology of CSW and CSW-CaSt was measured with FESEM (SU8010/Aztec(X-MaxN), Hitachi, Japan). XRD pattern was carried out by a X'Pert X-ray spectrometer (Philips, Holland). FT-IR analysis was determined by a Nicolet Nexus 670 FTIR Spectrophotometer (Thermo Nicolet Corporation, Madison, USA) in liquid films using KBr salt tablets and the wavenumber range was from 4000 cm^{-1} to 400 cm^{-1} at room temperature. Thermal analysis is measured by a thermal analysis instrument (STA449F3, NETZSCH, Germany) under 5 °C/min heating rate. XPS measurement of CSW and CSW-CaSt was carried out using X-ray photoelectron spectrometer (ESCALAB 250Xi, American Thermo Scientific Physical Electronics, USA) equipped with an Al Kα X-ray source.

3 RESULTS AND DISCUSSION

Surface modification of CSW. The influence factors of CaSt on surface modification of CSW include surfactant concentration, stirring rate, modified temperature, modified time and so on. The wetting behavior and surface hydrophobicity of the CSW and The CSW coated CaSt were characterized by WCA measurements.

Figure 1 shows the WCA of CSW-CaSt as a function of modifying agent concentration. It is noticeable that the contact angel of CSW-CaSt increases with the increasing mass fraction of CaSt. It indicates the surface hydrophobic of calcium sulphate whisker increases with the increasing of dosage of modifying agent, thus leading to the increasing of the contact angel of CSW-CaSt. The WCA of CSW-CaSt changes slightly when the mass fraction of CaSt increases from 14% to 16%. which indicates that the surface modification of calcium sulphate whisker is completely. Considering the cost of the CaSt, the optimum mass fraction of the modifying agent to CSW is 14%.

The WCA of CSW-CaSt as a function of temperature is exhibited in Fig. 2. It reveals that the WCA of CSW-CaSt increases firstly and decreases with temperature. It indicates that the activity of surfactant enhance firstly and decreases with temperature increaseing. Therefore, the optimum modifying temperature is 353.15 K.

Fig. 3 shows that the WCA of CSW-CaSt as a function of stirring rate (A) and time (B). It is noticeable that the WCA of CSW-CaSt increases firstly and decreases with stirring rate (A) and time (B).

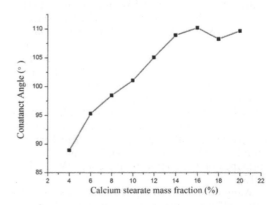

Figure 1.   The WCA of CSW-CaSt as a function of surfactant concentration.

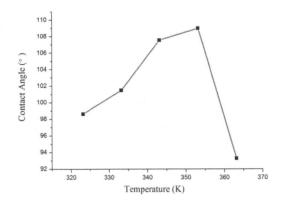

Figure 2. The WCA of CSW-CaSt as a function of temperature.

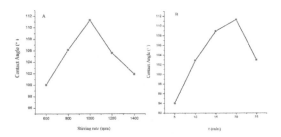

Figure 3. The WCA of CSW-CaSt as a function of stirring rate (A) and time (B).

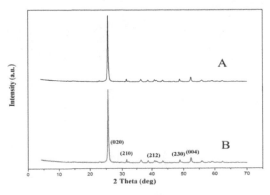

Figure 4. XRD patterns of CSW (A) and modified $CaSO_4$ (B) whiskers.

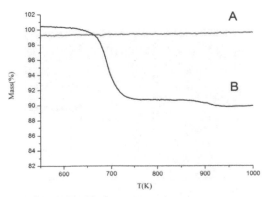

Figure 5. TG curves of CSW (A) and modified $CaSO_4$ (B) whiskers.

XRD of CSW-CaSt. The identity of a solid material is confirmed by Powder X-Ray Diffraction (XRD). A crystallographic study for CSW (A) and modified CaSO4 (B) whiskers were confirmed using XRD with a tube voltage and current of 40 kV and 30 mA, respectively. The scanning position 2θ is from 5.0013° to 69.9915°. Fig. 4 shows XRD patterns of the CSW (A) and modified $CaSO_4$ (B) whiskers. According to the powder diffraction file (PDF#01-074-2124), these two patterns present uniform diffraction peaks at 2θ around 25 °, 31 °, 40 °, 48 ° and 52 °, assigned to (020), (210), (212), (230) and (004) crystal faces, respectively. So, we conclude that the modification occurs only on the surface of CSW without changing its crystal structure.

TG curves of CSW-CaSt. Fig. 5 shows the TG curves of CSW (A) and modified CaSO4 (B) whiskers. It is noticeable that there is almost no mass loss below 1000 K in Fig. 5A, which indicates the CSW (A) is stable from room temperature to 1000 K. But in Fig. 5B, there is about 9.72% mass loss from 669.85 K to 711.75 K, which reavals the CaSt adsorbed on $CaSO_4$ (B) whiskers starts to break down at about 669.85 K.

FTIR spectroscopy of CSW-CaSt. The interaction between different groups and structures can be analyzed and identified by FTIR. The FTIR spectroscopy of CSW (A), CSW-CaSt (B) whiskers and CaSt (C) are presented in Fig. 3. The vC-H, vC-H and δsC-H at 2919.19 cm^{-1}, 2850.19 cm^{-1}, 1473.91 and 2917.46 cm^{-1}, 2849.70 cm^{-1}, 1470.20 cm^{-1} in Fig. 6(B) and (C) indicate that CaSt exists on the surface modification of CSW. In the Fig. 6(A) and (B), the frequency at 1155.77 cm^{-1}, 1153.96 cm^{-1} can be associated to asymmetric stretching vibration of (SO_4^{2-}), the frequency at 674.24 cm^{-1}, 595.85 cm^{-1}, 673.27 cm^{-1}, 595.85 cm^{-1} was associated to bending vibration of (SO_4^{2-}). The absorption peak of v_{asCOO} is changed from 1575.78 cm^{-1} in Fig. 7(C) to 1584.05 cm^{-1} in Fig. 7(B), which indicates the physisorbed occurs between CSW and CaSt in the process of modification. The δ_{O-H} at 1624.09 cm^{-1} and 1626.09 cm^{-1} in the Fig. 6(A) and Fig. 6(B) were resulted from the water adorbed on the surface of CSW (A) and CSW-CaSt(B).

Morphology of CSW-CaSt. The morphology of CSW and CSW-CaSt with a diameter of 1~5 μm and a length of 20~200 μm was exhibited in

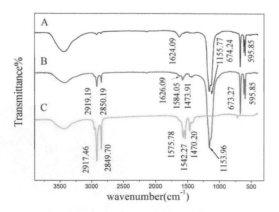

Figure 6. The FTIR spectroscopy of CSW (A), modified CaSO4 (B) and CaSt (C).

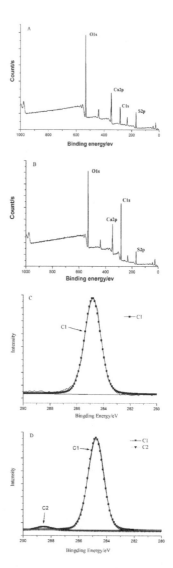

Figure 8. XPS spectra of CSW (A) and CSW-CaSt (B) and high-resolution spectra of C1 s peak: CSW (C) and CSW-CaSt (D).

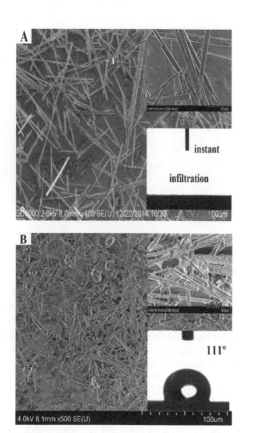

Figure 7. FESEM and WCA images of CSW (A) and CSW-CaSt(B).

Fig. 7(A). The surface of fibrous CSW is smooth. CSW-CaSt, by contrast, possess a lower transparency than CSW and a white coating layer, which indicates that the surface of CSW was effectively coated with CaSt. In the Fig. 7(A) and (B), it can

be noted that the WCA of CSW and CSW-CaSt is 0° and 111° separately, we conclude that the modification occurs on the surface of CSW.

XPS was used to study the varieties and states of different elements on the surface of CSW and CSW-CaSt. Fig. 8A to B show the full scan spectra and high-resolution C1 s spectra of CSW and CSW-CaSt. It can be seen from Fig. 8A to B, the main character peaks located at 284 eV, 532 eV, 348 eV and 169 eV can be assigned to C1 s, O1 s, Ca2p and S2p signals, respectively. The C1 s signal in Fig. 8A indicates that the surface of CSW adsorbs organic chemicals during analysis test.

Table 1. The O/C and O/Ca ratios and the fraction of the decomposed C1 signal of CSW, CSW-CaSt and CaSt.

Samples	O/C	O/Ca	Bending energy (eV)	
			C1	C2
			C-C C-H	O-C=O
CSW	1.76[1]	4.25	100[1]	–
CSW-CaSt	0.55	4.17	96.69	3.31
CaSt	0.11[2]	4[2]	94.44[2]	5.56[2]

[1] The surface contamination of organic chemicals during analysis test.
[2] The theoretical values.

Figure 8C to D show the curve-fitting analysis of high-resolution C1 s spectra of CSW, which indicates the signal located at 284.75 eV is assigned to C1 (C-C, C-H). However, there is a new signal appeared at 288.51 eV for CSW-CaSt associated to C2 (O-C=O) and it demonstrates there is CaSt on the surface of CSW. Table 1 shows that the O/C and O/Ca ratios and the fraction of the decomposed C1 s signal. The ratios of O/C and O/ Ca are calculated by Equation (1) as follows:

$$\frac{\rho_a}{\rho_b} = \frac{I_a/S_a}{I_b/S_b} \qquad (1)$$

where a, b are two different elements, ρ is the atom ratio (molar ratio), I is the intensity, S is the atomic sensitivity factor of element. The O/C ratios of CSW and CSW-CaSt measured by XPS are 1.76 and 0.55. The O/C ratio of CSW-CaSt decreases compare to CSW. This is because the O/C ratio of CaSt is low, only 0.11 theoretically. Similarly O/Ca ratios of CSW-CaSt decrease from 4.25 to 1.17 after modification, since the O/Ca ratios of CaSt was lower. The fraction of C2 (O-C=O) of CSW-CaSt increases from 0 to 3.31%, but still is less than the theoretical value (5.56) of CaSt, which indicates tha there is the presence of alkyl groups of impurities.

4 CONCLUSIONS

CSW-CaSt was synthesized by wet modifying. The CSW-CaSt has obvious hydrophobic with the WCA of 111° when the modification conditions are 14 wt% of CaSt, 353.15 K, 1000 rpm and 20 min. FESEM and XRD indicate that the surface of CSW was effectively coated with CaSt and there is no changes of crystal structure of CSW after modification. FTIR illustrates the presence of COO⁻ group on the surface of CSW-CaSt. It is confirmed that the CaSt exists on the surface modification of CSW and the type of surface modification between CSW and CaSt was physical adsorption based on the results of FESEM and FTIR. The varieties and states of different elements on the surface of CSW and CSW-CaSt was determined by XPS. This paper provides the possibility for the preparation and Application of whisker/polymer matrix composites.

REFERENCES

Dang, L., Nai, X.Y., Zhu, D.H., Jing, Y.W., Liu, X., Dong, Y.P., Li, W.: Applied Surface Science, (2014), no.317, p. 325–331.

Guan, B., Ma, X., Wu, Z., Yang, L., Shen, Z.: Journal of Chemical and Engineering Data, (2009), no. 54, p. 719–725.

Hao, W.L., J.C. Wang, X.Q. Chen, Y.C. Du, D. Wu, Z. Xi: Industrial minerals and processing, (2013), p. 19–21.

Hou, S.C., Wang, J., Wang, X.X., Chen, H.Y. and Xiang, L.: Langmuir, (2014) no. 30, p. 9804–9810.

Jiang, Q.: The Surface Modification Research of Gypsum Whisker. (Ph.D., East China University of Science and Technology, China 2014).

Li, L., Zhu, Y.J., Ma, M.G.: Materials Letters, (2008), no. 62, p. 4552–4554.

Liao, X.L., X.R. Qian, B.H. He: Paper Science and Technology, vol. 29 (2010), no.6, p. 82–87.

Liu, F.F., Wang, Y.L., Qin, S.T., Liu, Y.X.: Hunan Papermakin, (2012), no.1, p. 21–23.

Nissinen, T., M. Li, N. Brielles, S. Mann: Cryst Eng Comm, (2013), no. 15, p. 3793–3798.

Ru, X., B. Ma, J. Huang: Journal of the American Ceramic Society, (2012), no. 95, p. 3478–3482.

Shen, H.L., Z.N. LQ. Zhao: Plastics, vol. 30 (2001), no. 4, p. 59–62.

Song, X., L. Zhang, J. Zhao, Y. Xu, Z. Sun, P. Li, J. Yu: Crystal Research and Technology, (2011), no. 46, p. 166–172.

Wang, J.C., Yang, K., Lu, S.: High Performance Polymers, vol. 23 (2011), no. 2, p. 141–150.

Wang, X.L., Y.M. Zhu and Y.X. Han: Mining Metall, (2006), no.15, p. 30–37.

Wang, X.L., Y.M. Zhu, Y.X. Han, Z.T. Yuan and W.Z. Yin: Advanced Materials Research, (2009), no.58, p. 225–229.

Wang, X.L., Y.M. Zhu, Y.X. Han, Z.T.Yuan: Journal of Northeastern University(Natural Science), vol. 129 (2008), no.10, p. 1494–1497.

Wang, X.L.:Research on surface modification of calcium sulphate whiskers (Ph.D., Northeastern University, China 2005).

Wang, Y.L., S.T. Qin, H.Y. Zhan, Y.X. Liu: Non-Metallic Mines, vol. 36 (2013), no.1, p. 42–45.

Yin, W.Z., X.L. Wang, Y.X. Han and Z.T. Yuan: Journal of Northeastern University (Natural Science), (2007), no. 28, p. 580–583.

Zhang, L.H.: Research Progress on Preparation of Calcium Sulfate Whiskers and its Application. (Ph.D., Northeastern University, China 2010).

Zhang, L.Q., Wu, S.M., Feng, Y.X. and Feng, W.: China Synthetic Rubber Industry, vol. 21 (1998), no. 5, p. 261–265

Zhu, Y.M., Y. Zhang, X.L. Wang, Y.X. Han: China Powder Science and Technology. vol. 21 (2015), no. 2, p. 35–37.

Energy Science and Applied Technology ESAT 2016 – Fang (Ed.)
© *2016 Taylor & Francis Group, London, ISBN 978-1-138-02973-6*

Influence of unidirectional hot-stretching on the mechanical and dielectric properties of polymer-based composites

Zejun Pu, Renbo Wei & Xiaobo Liu
School of Microelectronic and Solid-State Electronic, University of Electronic Science and Technology of China (UESTC), Chengdu, China

ABSTRACT: In this work, novel polymer-matrix composites with tunable dielectric property and crystalline degree were fabricated via unidirectional hot-stretching of semicrystalline Polyarylene Ether Nitriles (PEN) in the presence of different contents of tetra-amino-phthalocyanine copper (CuPc-NH$_2$). It is found that the crystallinity, thermostability and mechanical properties of samples were improved via unidirectional hot-stretching. Most importantly, the result of microstructure analysis pointed out the orientation of the PEN polymer chain in crystalline regions was changed during the unidirectional hot-stretching process and the agglomeration of CuPc-NH$_2$ particles in polymer was reduced. Thus, the dielectric properties and breakdown strength was also ultimately improved. After the hot-stretching, the energy density of the polymer-based composites containing 30 wt% CuPc-NH$_2$ increased by 74%, from 1.8 to 3.13 J/cm^3.

Keywords: phthalocyanine copper; polyarylene ether nitriles; composites; hot-stretching; dielectric

1 INTRODUCTION

Materials with excellent dielectric properties had been widely studied in recent years for their applications in gate dielectric materials and high energy storage material, etc (Z. M, 2014). Among those dielectric materials, polyme-based composites have attracted considerable attention because of their mechanical flexibility, good processability, high breakdown strength and low cost. But, conventional polymers generally possess a low permittivity. Moreover, those polymers are usually used at conventional environment due to their low glass transition temperatures. Thus, the key approach is to improve dielectric constant while retaining other outstanding performance.

Phthalocyanine (Pc) oligomer is a semi-conductor with excellent chemical stability and high thermostability. Moreover, it has an extended π-conjugation and thus shows excellent electrical performance (N. Kobayashi, 2002). Hence, it is an effective approach to improve the dielectric properties of polymer by introducing phthalocyanine copper (CuPc) with high dielectric constant into polymers (M. Guo, 2006). While, polymer/CuPc composites usually suffer high dielectric loss (tan δ) and low breakdown strength (E_b) owing to insulation–conduction transformation. Thus, decreasing the tan δ and maintaining high E_b is still a challenge in the development of high-k materials. As a

typical semi-crystalline polymer, Polyarylene Ether Nitriles (PEN) possesses the advantage of hot-stretching and unique properties such as excellent thermostability, high breakdown strength, outstanding mechanical strength and chemical inertia, etc(Y. You, 2015). Thus, PEN was selected as an ideal polymer matrix for preparing high dielectric polymer-based materials.

In this work, PEN/CuPc-NH2 composites were fabricated by solution-casting method. After that, the hot-stretching process was carried out in a high temperature oven by the previously reported method. The stretching ratio was fixed at 150%. Then, the influence of unidirectional hot-stretching on mechanical and dielectric properties of polymer-based composites was systematically studied.

2 EXPERIMENTAL

PEN and CuPc-NH$_2$ were synthesized according to the previous reports (D. F. Ren, 2012). Meanwhile, PEN/CuPc-NH$_2$ composites with different CuPc-NH$_2$ loading (0%, 5%, 10%, 20%, 30%) were obtained by solution-casting method, then hot-stretched in an oven. The fracture morphology of samples were investigated using scanning electron microscopy (SEM, JSM-6490LV). The thermal transition and crystalline structures of samples were characterized by differential scanning calor-

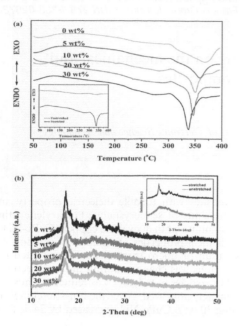

Figure 1. (a) DSC curves and (b) XRD patterns of hot-stretching PEN/CuPc-NH2 composites.

imetry (DSC, Q100, TA) and X-ray diffraction (XRD, Rigaku RINT2400), respectively. Besides, the mechanical and dielectric properties of PEN/CuPc-NH$_2$ composites are also being systematically investigated.

3 RESULTS AND DISCUSSION

Fig. 1(a) presents the DSC curves of hot-stretched PEN/CuPc-NH$_2$ composites. As the CuPc-NH$_2$ content increases, the melting temperature (T_m) increases from 337 °C to 369 °C. Moreover, the melting peak intensity decreases while the peak width increases. The changes in T_m indicated that the crystallinity of PEN decreases in PEN/CuPc-NH$_2$ system, owing to the effect of the restriction of the excess CuPc-NH$_2$ on the crystallization behavior. XRD was used to further investigate the effect of CuPc-NH$_2$ on the crystallization behavior of the hot-stretched samples. As shown in Fig. 1(b), there are two obvious diffraction peaks at 17.2° and 23.1°, and the percent of the crystallinity of samples are calculated from the deconvoluted reflections of the XRD profiles by using Jade 6 software. The results reveal that the crystallinity of these films slightly decreases from 17.3% to 13.8% as the CuPc-NH$_2$ content increases. Moreover, the inset of Fig. 1(a) presents the DSC curves of pure PEN before and after hot-stretching. The melting curves show obvious increase after hot-stretching,

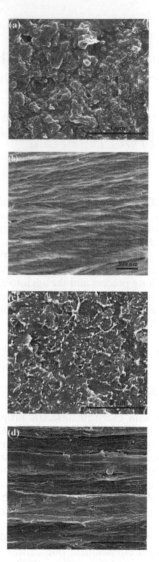

Figure 2. Microstructure of pure PEN and 20 wt% PEN/NH2-CuPc composites before and after hot-stretching.

indicating the crystallinity of PEN increases. The result was further proved by the XRD analysis from the inset of Fig. 1(b). The crystallinity of hot-stretched pure PEN increases from 4.6% to 17.3%, which is in good agreement with DSC results.

Fig. 2 shows the microstructure evolution of pure PEN and 20 wt% PEN/CuPc-NH$_2$ composites before and after hot-stretching. Comparing with the un-stretched film (Fig. 2a and c), it is worth noticing that many fiber bundles are arranged in rows and polymer chain in crystalline regions also become highly orientated along the stretch direction after hot-stretching (Fig. 2b and d). Most importantly, the size agglomeration of CuPc-NH$_2$ particles was improved in polymer after hot-stretching, thereby effective blocking off the filler-filer contacts and providing a chance to rebuild the conductive network. Therefore, these highly orientated crystals will contribute a lot to the dielectric and mechanical properties of the stretched PEN.

As expected, the mechanical properties of composites were tremendously enhanced by hot-stretching. For the unstretched composites, the tensile strength of PEN composites with NH$_2$-CuPc content of 0 to 5, 10, 20 and 30 wt% correspond to 124, 122, 117, 112 and 108 MPa, respectively. After hot-stretching, the tensile strength increased to 330, 328, 305, 300 and 281 MPa, respectively. Meanwhile, the tensile modulus are as high as 4129, 4412, 4150, 3897 and 3461 MPa, respectively, with an increment rate of 113, 117, 106, 103 and 98% compared with the un-stretched films. The result is attributed to that the highly orientation of PEN polymer chains and the formation of irregular crystals.

Fig. 4(a) and Fig. 4(b) show the dielectric constant (ε) and loss tangent ($tan\delta$) for unstretched composites. it is worth noticing that the ε of composites are strongly enhanced by the introduction of CuPc-NH$_2$. For instance, at 1 kHz, the ε of 30 wt% PEN/CuPc-NH$_2$ composite shows a 13-fold increase (up to 51.6 from 4.0 of pure PEN),

whereas the $tan\delta$ is up to 0.61. But, the ε and $tan\delta$ of hot-stretched composites decreases obviously, compared with the un-stretched one. It is due to the effective blocking off the Dc leakage conductance by the insulating layers (pere PEN), as a result, the ε and $tan\delta$ decreased. Importantly, after hot-stretching, the ε of pure PEN film increases from 4.0 to 7.1, which was in consistent with our former reports.

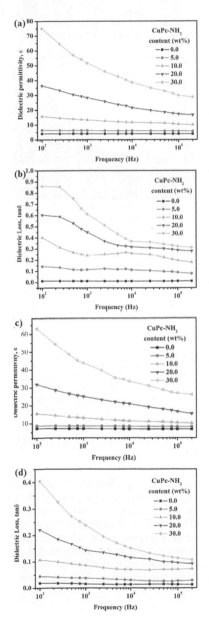

Figure 4. Dielectric properties of the PEN/NH2-CuPc composites.

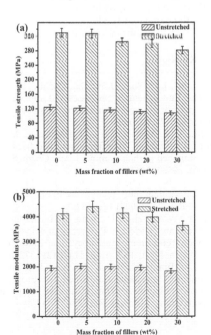

Figure 3. (a) Tensile strength and (b) tensile modulus of the PEN/NH2-CuPccomposites.

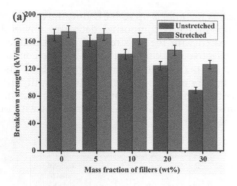

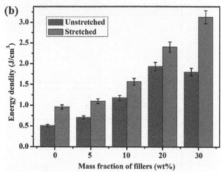

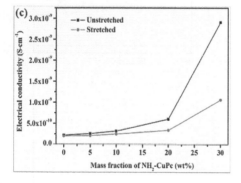

Figure 5. (a) Breakdown strength, (b) energy density and (c) electrical conductivity of PEN/NH2-CuPc composites.

Fig. 5(a) and Fig. 5(b) show the breakdown strength (E_b) and energy density (U) of PEN/CuPc-NH$_2$ composites. With the increasing content of CuPc-NH$_2$, the E_b of all samples decreased gradually, while the decrement of E_b of un-stretched samples is more obvious, compared with the stretched one. It is well known that E_b has a stronger effect on the U of the specimen. The calculated results

proved that the U of the stretched composites increased from 0.96 to 3.13 J/cm^3, with an increment rate of 74% compared with the un-stretched film. This was the result of obvious increase in E_b and slight decrease in ε. Besides, the electrical conductivity of composites decreases after hot-stretching. The decrement in electrical conductivity is possibly due to the effective blocking off the Dc leakage conductance by PEN insulating layers.

4 CONCLUSIONS

In summary, unidirectional hot-stretching is a useful method to improve the crystallization behaviors and mechanical as well as thermal and dielectric properties of PEN/CuPc-NH$_2$ composites. For the stretched film, the T_m is increased by 34 °C. Compared with un-stretched film with 30 wt% NH$_2$-CuPc loading, the tensile strength and modulus of the hot-stretched composites increase by 160% and 98%, respectively. After hot-stretching, the ε of 30 wt% PEN/CuPc-NH$_2$ composites reached to 43.9 at 1 kHz, while the $tan\delta$ is as low as 0.23. In addition, the breakdown strength of composites has also been obviously enhanced. As a result, the U of the composites with 30 wt% NH$_2$-CuPc loading increases from 0.96 to 3.13 J/cm^3, with an increment rate of 74% than the one without hot-stretching. Thus, the unique features will make the PEN/CuPc-NH$_2$ composites attractive for applications in the capacitor fields.

REFERENCES

Dang, Z. M., J. K. Yuan, J. W. Zha, T. Zhou, S. T. Li, G. H. Hu, Fundamentals, processes and applications of high-permittivity polymer–matrix composites, Prog Mater Sci 57 (2014) 660–723.

Guo, M., X. Z. Yan, Y. Kwon, T. Hayakawa, M. Kakimoto, T. Goodson, High frequency dielectric response in a branched phthalocyanine, J. Am. Chem. Soc., 128 (2006), 14820–14821.

Kobayashi, N., Dimers, trimers and oligomers of phthalocyanines and related compounds, Coordin. Chem. Rev. 227 (2002) 129–152.

Ren, D. F., W.W. Wei, Z. H. Jiang, Y. H. Zhang, Microstructure and Dielectric Properties of Poly(aryl ether ketone)s/Tetra-amino-phthalocyanine Zinc Composites, Polym, Plast. Tech. Eng. 51 (2012) 1372–6.

You, Y., X. Huang, Z. J. Pu, K. Jia, X. B. Liu, Enhanced crystallinity, mechanical and dielectric properties of biphenyl polyarylene ether nitriles by unidirectional hot-stretching, J. Polym. Res. 22 (2015) 1–9.

Energy Science and Applied Technology ESAT 2016 – Fang (Ed.)
© *2016 Taylor & Francis Group, London, ISBN 978-1-138-02973-6*

Influence of an electron-donating active group on the properties of organic nonlinear optical chromophores containing multiple cyano-based acceptors

Xiang Tang, Kun Jia, Xianzhong Tang & Lei Chen

School of Microelectronics and Solid State Electronics, State Key Laboratory of Electronic Thin Films and Integrated Devices, University of Electronic Science and Technology of China (UESTC), Chengdu, P.R. China

ABSTRACT: A range of organic Nonlinear Optical (NLO) chromophores containing a hydroxyl-modified electron donor moiety, different π electron-conjugated bridges of vinyl, furfuran, or azobenzene, and tricyanofuran as an electron acceptor have been synthesized. In addition, their chemical structures, and optical and thermal properties have been thoroughly characterized with a variety of techniques such as FTIR, NMR, UV-Vis absorption spectra, and TGA. The solvatochromism method was explored to calculate the microscopic second-order molecular polarizability of chromophores based on their absorption spectra recorded in several solvents of different refractive indexes. We discovered that the hydroxyl modification to electron donor group significantly improved the thermal stability (increase of 5% decomposition temperature around 10–20°C) of the synthesized chromophores. Moreover, we found that a blue-shift of maximum absorption wavelength up to 80 nm as well as the 150% enhancement of molecular microscopic NLO coefficient was observed after attaching an active hydroxyl group to the internal electron donor moiety of chromophores containing a longer π-conjugated bridge.

Keywords: nonlinear optics, chromophore, active group, solvatochromism, hydroxyl

1 INTRODUCTION

Present days, organic Nonlinear Optical (NLO) materials have attracted increasing research interests and have been considered as a hot topic in high-tech field worldwide (F. Liu, 2015). Generally, the NLO chromophores with good optical properties have been synthesized mainly on the basis of the charge transfer theory that normally contains electron donor (D)-π conjugate bridge-acceptor (A) molecular structure (J. Liu, 2012). However, the performance of electric-optical (E-O) devices fabricated from classical D-π-A chromophores, especially for the macroscopic optical properties, would be largely limited because of strong aggregation tendency of planar chromophores. An effective strategy to overcome the aggregation disadvantage is to link these conventional chromophores into larger entity, such as NLO chromophore grafting onto macromolecular backbone/side chains, or combining several NLO chromophores to construct two-dimensional (2D) planar complexes with multiple charge transfer tunnels (K. Y. Choe, 2014). On the other hand, it has been reported recently that introducing active groups with electron donating ability to the electron donor

moiety of NLO chromophores (auxiliary donor groups) would be a facile way to both overcome the inherent aggregation tendency and enhance the molecular optical properties of NLO chromophores (H. Wang, 2015).

Hydroxyl is an active group with electron donating ability, thus NLO chromophores modified with a hydroxyl group not only exhibit improved microscopic NLO properties due to enhanced electron donating, but also can be linked with many functional groups of other molecules. Therefore, both the preparation of precursor and clarification of structure-properties relationship for NLO chromophores can be obtained by a systematic study of hydroxyl-modified chromophores.

Herein, a range of different NLO active chromophores bearing vinyl, furfuran, or azobenzene as π-electron conjugate bridge and electron-donating moieties modified by hydroxyl at different positions have been synthesized as displayed in Figure 1. By combining optical spectroscopy and thermal analysis, the influence of hydroxyl group on the molecular nonlinear optical properties of chromophores was investigated, which would be of great importance for novel NLO chromophore design and synthesis.

2 RESULTS AND DISCUSSION

Synthesis. Chromophores A, B, C, and II_A were synthesized according to the previously published work of L. Han, Y (2008), Tang, X (2012), Wang, Y (2011) and You, Y (2013), respectively, while the chemical synthesis route for chromophores of I_A, I_B, and II_A is shown in Figure 2. Specifically, compounds A, B, and C are previously synthesized chromophores without any hydroxyl modification and they exhibited good microscopic NLO properties according to our previous work. In order to determine the influence of hydroxyl group, the chromophores A and B were substituted with hydroxyl group at R1 position to obtain I_A and I_B, while chromophores A and C were modified with hydroxyl group at R2 position to obtain II_A and II_C, respectively. Moreover, the synthesis of I_A, I_B and II_A, II_C was basically based on the synthesis of A, B, and C, which indicated that all these chromophores can be obtained in less than three steps with a high yield. It has been found that the reaction process was virtually unchanged during hydroxyl substitution, while the chromophore solubility was obviously alternated after hydroxyl modification, which would have a potential influence on the

R=C₂H₅, R₁=H, R₂=H

A: X= C=C
B: X=
C: X= N:N

I_X: R=CH₃, R₁=OH, R₂=H II_X: R=C₂H₅, R₁=H, R₂=OH

I_A: X= C=C I_B: X= II_A: X= C=C II_C: X= N:N

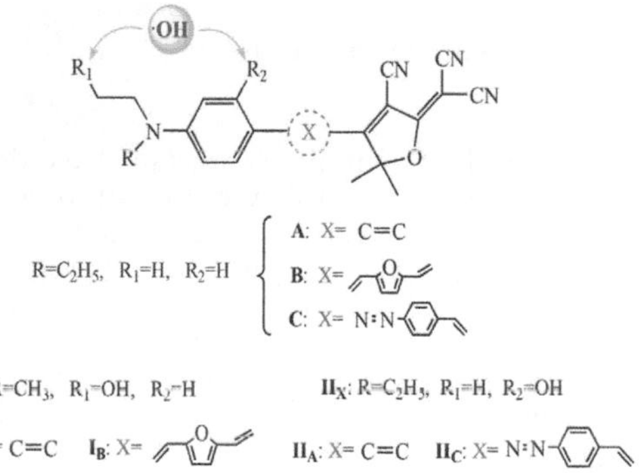

Figure 1. Chemical structures of seven chromophores involved in this work.

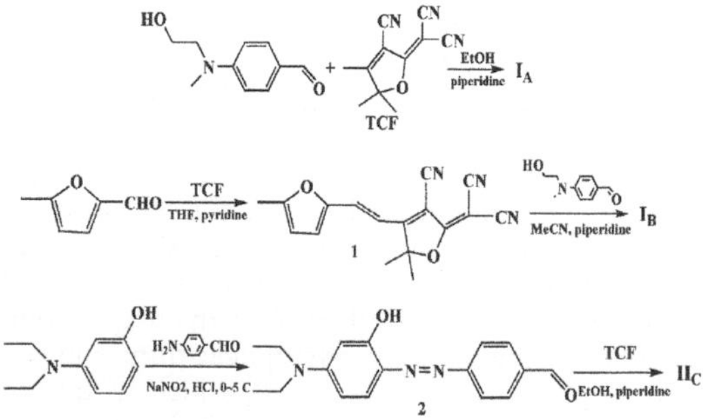

Figure 2. The chemical synthesis route for chromophores IA, IB, and IIC.

chromophore preparation and optical spectroscopy testing. The detailed synthesis and spectral characterizations of synthesized compounds were shown in supporting materials.

Linear optical properties. The UV-Vis absorption spectroscopy was used to evaluate the linear optical properties of the synthesized chromophores. More specifically, the obtained chromophores were Dissolved in Dimethyl Sulfoxide (DMSO) for UV-Vis absorption test, and the absorption spectra along with maximum absorption wavelength were demonstrated in Figure 3 and Table 1, respectively. Based on these results, the maximum absorption wavelength of chromophores with short conjugated bridge (A and I_A) was nearly unchanged after modification with the hydroxyl group. For the chromophores containing longer π-conjugated bridge (I_B and II_C), their maximum absorption wavelength was obviously blue-shifted when compared with the corresponding chromophores (B and C) without any hydroxyl modification. For instance, the maximum absorption wavelength of II_C was blue shifted around 80 nm when compared with that of the C chromophore, which indicated that the visible light transmittance of chromophore was obviously enhanced after hydroxyl modification in the internal electron donor.

Microscopic nonlinear optical properties. The solvatochromism method is a simple and effective protocol to evaluate the microscopic nonlinear optical properties of NLO chromophores. In this method, chromophores of closely related molecular structures were dissolved in several organic solvents with different refractive indexes for UV-Vis absorption spectra test, as the maximum absorption wavelength of the same chromophore is highly dependent on the refractive index of the solvent, thus the second-order molecular polarizability, a representative

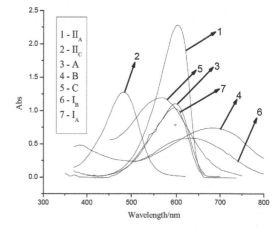

Figure 3. The UV-Vis absorption spectra of different chromophores solubilized in DMSO solvent.

Table 1. Characterization data of the chromophores.

Chromophore	λ_{max} (nm)	$\mu_g\beta_{(1064\,nm)}$ (Relatively to A)	Td (°C)	Chromophore	λ_{max} (nm)	$\mu_g\beta_{(1064\,nm)}$ (Relatively to A)	T_d (°C)
A	591	1	262	B	676	4.90	250
I_A	595	2.95	281	I_B	647	2.17	–
–	–	–	–	C	567	9.16	254

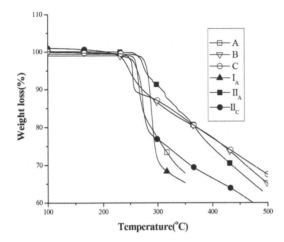

Figure 4. The thermal analysis of synthesized chromophores.

parameter for the microscopic NLO properties of chromophores, can be easily calculated on the basis of the recorded UV-Vis absorption spectra. In our experiment, typical organic solvents of ethyl acetate, acetone, chloroform, and DMF were used to dissolve synthesized chromophores for UV-Vis absorption measurement, and the calculated second-order molecular polarity ($\mu_g\beta$) are summarized in Table 1 as well. According to the calculated data, for the chromophores with a shorter conjugation bridge, a threefold enhancement of second-order molecular polarity ($\mu_g\beta$) was observed, no matter which position of electron donor moiety was substituted by the hydroxyl group. However, the different conclusions were obtained for chromophores containing a longer π-conjugated bridge. Specifically, the molecular hyperpolarizability of II_C chromophores was only slightly larger than that of chromophore C. For the chromophores with a shorter conjugation bridge, the enhancement of molecular hyperpolarizability can be rationalized according to the fact that the hydroxyl group was an active group with electron donating ability, thus the overall charge transfer efficiency would be improved, no matter which position of the electron donor moiety was modified. However, when the conjugate bridge became longer, the additional electron pushing ability from the hydroxyl substitute would be declined in the overall molecular hyperpolarizability, and thus we

assumed that the introduction of the other group with a stronger electron donating ability would be more suitable for the chromophores containing a longer conjugation bridge.

Finally, we found that the molecular hyperpolarizability of chromophore I_B was even lower than its unmodified counterpart of chromophore B. Theoretically, the presence of an auxiliary electron-donating group (i.e. hydroxyl) is supposed to enhance the microscopic NLO properties of chromophores, which is in contrast to the measurement results. This unusual spectroscopy behavior would be attributed to the currently unknown influence (e.g. distortion of molecular configuration) derived from the hydroxyl modification at the end of chromophores, thus we assumed that the chromophore I_B would be more suitable to be served as a precursor for constructing advanced NLO chromophores.

Thermal analysis. As shown in Figure 4 A, a thermogravimetric analyzer (TGA) was explored to evaluate the thermal stability of all the chromophores in nitrogen atmosphere, and the decomposition temperatures corresponding to 5% weight loss (T5%) from TGA curves are listed in Table 1. It is clear that all the hydroxyl modified chromophores exhibit better thermal stability when compared with their corresponding unmodified chromophores, which should be attributed to the two reasons. First, the molecular mass was increased after the incorporation of the hydroxyl group. Second, the formation of inter-molecular hydrogen bond would be enhanced after hydroxyl substitution. Moreover, the hydroxyl substitution position played an important role in determining the thermal stability of chromophores. For instance, a 20% increase in Td was detected from Ix chromophores, while the Td increased only by 10% for IIx chromophores. Therefore, we conclude that the more efficient method to enhance the thermal stability of chromophores would be attaching hydroxyl on the far end of molecules or incorporation of larger aromatic substitutes to the internal position of the electron donor moiety of NLO chromophores.

3 CONCLUSION

In this report, a group of organic nonlinear optical chromophores with similar backbone structures containing a hydroxyl substituent at different positions have been synthesized. Furthermore, the effects of the hydroxyl substituent on the molecular optical properties of the obtained chromophores were investigated using a range of different techniques including UV-Vis absorption spectroscopy, solvatochromism, and thermal analysis. First, we found that all the synthesized chromophores demonstrated good thermal stability as confirmed by the high decomposition temperature over 250 °C. Second, the chromophores with hydroxyl substitution at internal position of electron donor moiety possess better transparency (*i.e.* shorter absorption wavelength). Finally, the enhancement of molecular microscopic NLO properties (e.g. molecular hyper-polarizability) was observed from the chromophores containing a shorter conjugation bridge, no matter which position was substituted with hydroxyl, but for the chromophores containing longer conjugation bridge, the auxiliary hydroxyl group should be incorporated into the internal position of the electron donor of chromophores to enhance their nonlinear optical properties. Therefore, the present work will be of great importance for the design of novel high-performance NLO chromophores.

ACKNOWLEDGMENTS

The financial supports from the National Natural Science Foundation of China (No. 51403029) and the Scientific Research Foundation for the Returned Overseas Chinese Scholars from State Education Ministry for this work are highly appreciated.

REFERENCES

Choe, K.Y., J. Lee, J.Y. Lee. Polymer Bull, vol. 71 (2014) no. 9, p. 2369–2382.
Gao, J., Y. Cui, J. Yu, et al. Dyes and Pigments, vol. 87 (2010) p. 204–208.
Gong, W., Q. Li, S. Li, et al. Materials Letters, vol. 61 (2007) no. 4, p. 1151–1153.
Guo, L., Z. Guo, H. Chen. J Mater Sci: Mater Electron, vol. 25 (2014) p. 3633–3638.
Han, L., Y. Jiang, W. Li, et al. Materials Letters, vol. 62 (2008) p. 1495–1498.
Liang, T., Y.J. Cui, J. C, Yu. Dyes and Pigments. vol. 98 (2013) p. 377–383.
Liu, F., H. Wang, Y. Yang, et al. Dyes and Pigments, vol. 114 (2015) p. 196–203.
Liu, J., A.V. Franiv, X. Liu, Zhen Zhen. J Mater Sci: Mater Electron, vol. 71 (2013) p. 2701–2705.
Liu, J., L. Wang, Z. Zhen, et al. Colloid Polym Sci, vol. 290 (2012) p. 1215–1220.
Liu, L., H. Li, S. Yao. J Mater Sci, vol. 50 (2015) p. 57–65.
Morales, A.R., A. Frazer, A.W. Woodward, et al. J. Org. Chem. vol. 78 (2013) p. 1014–1025.
Piao, X., X. Zhang, S. Inoue, et al. Organic Electronics, vol. 12 (2011)p. 1093–1097
Tang, X.; Tang, X.Z.; You, Y.C, et al. Acta Chim. Sinica, vol.70 (2012) p. 1565.
Wang, H., F. Liu, Y. Yang. Dyes and Pigments, vol. 112 (2015) p. 42–49.
Wang, Y.; Tang, X.Z.; Tang, X. Chem. J Chin. Univ, vol. 32 (2011) p. 2327.
Wu, J., C. Peng, H. Xiao, et al. Dyes and Pigments, vol. 104 (2014) p. 15–23.
Wu, J., J. Liu, T. Zhou, et al. Rsc Advances, vol. 71 (2012) p. 1416–1423.
Xu, H., M. Zhang, A. Zhang, et al. Dyes and Pigments, vol. 71 (2014) p. 142–149.
You, Y. C; Tang, X. Z, Tang, X. Wang, Y. Org. Chem, vol. 33 (2013) p. 530.

Energy Science and Applied Technology ESAT 2016 – Fang (Ed.)
© 2016 Taylor & Francis Group, London, ISBN 978-1-138-02973-6

Polyarylene ether nitrile/zinc tetraamino-phthalocyanine nanofibrous mats with high visible-light photodegradation performance

Xu Huang, Kun Jia & Xiaobo Liu

Institute of Microelectronic and Solid State Electronic, University of Electronic Science and Technology of China, Chengdu, P.R. China

ABSTRACT: A new photocatalyst of nanosized zinc tetra-amino-phthalocyanine (ZnPc) immobilized on the Polyarylene Ether Nitrile (PEN) nanofibers have been successfully fabricated by an electrospinning method. Scanning electron microscopy images indicated that ZnPc were homogeneously distributed on the exterior of nanofiber. The PEN@ZnPc non-woven mat ca be used to decolor the organic wastewater. It is found that the nanofiber mats (PEN@ZnPc) have high-efficiency ability to the degradation of rhodamine B (RhB), more than 90% of rhodamine B can be eliminated by PEN@ZnPc in 2.5 h. Moreover, the as-prepared PEN@ZnPc can be parted easily for repeated usage.

Keywords: zinc tetra-amino-phthalocyanine, electrospinning, degradation

1 INTRODUCTION

As we all know, metallophthalocyanines (MPcs) and their derivatives have attracted considerable attention because of their structural composed by a neutral metallic atom bound to a π-conjugated ligand, suggesting their potential as catalysts for oxidation (V. Iliev,1999). While, photocatalysis of Pcs in this system has been limited by the aggregation of the Pcs and the separation of the photocatalyst from the products (J. Cen, 2006). To overcome these hurdles, MPc are often anchored onto solid substrates (L. U. Hua, 2007). Besides, heterogeneous catalyst provides many advantages, such as, separation of catalysts from products easily, repeated usage of catalysts.

The structure of the substrates has huge influence on the properties of the immobilized MPcs. The nano-structured frames stand out of other frames due to the high specific surface area. Hence, nano-structured frames has been exploited to support fine fibers in electrospinning fields, with diameters in the nano-micro meter range (J. N. Young, 2010).

Photocatalyst of nanostructured ZnPc immobilized on the PEN nanofibers were successfully prepared via electrospinning technique. Photocatalytic performnces of the as-prepared PEN@ZnPc composite nanofibers mats to degrade RhB under visible-light irradiation were studied in detail. The catalytic oxidation products were analyzed by ultraviolet-visible absorption spectrum.

2 EXPERIMENTAL

Firstly, 2.0 g PEN gradually dissolved in 10.0 mL DMF solvent within 2 hours to form a 16.6 wt% solution (precursor). Then, 1 g of ZnPc was added to the PEN solution. Subsequently, the above mixture solution was stirred for 1 hour to produce a thick solution. Then the solution was carefully added into a 10 mL syringe, which is has a metallic needle (0.5 mm in diameter); besides, a piece of Al foil overlapped by a sheet of wires workd as the collector. Electrospinning has been carried out at room temperature with a relative humidity (RH) of 60%. The obtained nanofibrous mats were soaked in ethanol for 8 hours. Then, the obtained mat was placed in an oven overnight at about 80 °C. Schematic diagram to prepare porous nanofibrous mats was shown in Scheme 1.

3 CHARACTERIZATION

The final concentration (C), the initial concentration (C_0), and the degradation (D%) is calculated via following formula:

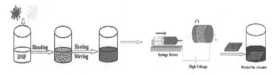

Scheme 1. Schematic diagram of the fbrication procedure of nanofibrous film.

$$D\% = \frac{C_0 - C}{C} \times 100\% \tag{1}$$

4   RESULTS AND DISCUSSION

Fig. 1a and b present SEM images of the as-prepared PEN@ZnPc. The PEN@ZnPc nanofibrous mat did not show obvious cracks and presented an almost similar morphology as that of nanofibers. An enlarged SEM image has been shown in Fig. 1b, indicating that the PEN@ZnPc fibers exhibit uniform sizes with an average diameter of 1.08 μm. The PEN/FePc fibers show some particles on the fiber surface, it indicated that the ZnPc were uniformly distributed on the nanofiber. Fig. 1c and d show the water contact angles images of untreated PEN@ZnPc and treated PEN@ZnPc, respectively. It should be noted that the initially gave a water droplet with measurable contact angles 126°. The droplet could remain unchanged over a period of 2 h on the PEN@ZnPc. However, the treated PEN@ZnPc absorbed water quickly. Hence the

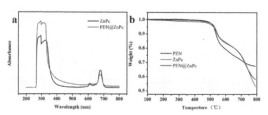

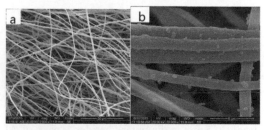

Figure 1.   (a)and (b): SEM images of PEN@ZnPc; (c) and(d): water contact angles images of PEN@ZnPc and treated PEN@ZnPc.

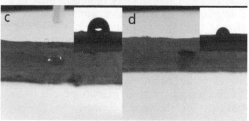

Figure 2.   (a) UV–vis absorption spectra of the samples, (b) TGA curve of samples.

treated PEN@ZnPc allowed dye water diffusion through nanofibrous mat and accelerated decoloration of dye wastewater.

The UV–vis absorption spectrum of PEN@ZnPc nanofibers and neat ZnPc were shown in Fig. 2a. All of the samples were dissolved in DMF solvent. As can be seen in Fig. 2a, the UV–vis absorption spectra of PEN@ZnPc and neat ZnPc showed two absorption regions, which were the characteristic B and Q bands of Pcs. The absorption at 550–750 nm should be attributed to the π–π* transition of monomers from the HOMO state to the LUMO state of the Pc^{-2} ring, which was the characteristic Q-band electronic spectra of ZnPc. The absorption from 300 to 450 nm is attributed to B-band absorption of Pcs. Moreover, the thermal properties of the samples is characterized by the initial decomposition temperature (T_{id}). Fig. 2b shows TGA curves of all samples, all the samples exhibit almost the same decomposition behavior. It can be inferred that the T_{id} values of the samples are in the range of 505°C–520°C, suggesting that these samples possess outstanding thermal properties.

The prepared PEN@ZnPc was then used as a catalyst for the degradation of organic wastewater. Fig. 3a is the absorption spectrum of an aqueous solution of RhB in the presence of PEN@ZnPc irradiated by visible light at different periods of time. The intensity of absorption peak decreases rapidly with prolonging reaction time. In order to further investigate the photocatalytic efficiency of the sample, further experiments were carried out in the presence of PEN@ZnPc under the same experimental conditions. As shown in Fig. 3b, the change of absorption spectra of RhB aqueous solution represented the change of its concentra-

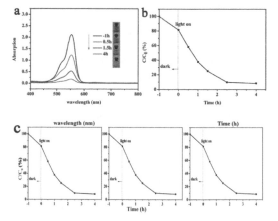

Figure 3.   (a) Adsorption spectra of RhB solution at different periods of time; (b) the degradation rate of RhB solution at different periods of time. (c) The photocatalytic degradation of RhB for 3 cycles.

tion. 1 hour later, 19.6% of adsorption has been achieved under the catalysis activity of PEN@ZnPc. The high adsorption capacity of PEN@ZnPc can be attributed to the strong electrostatic interactions between PEN@ZnPc and dye molecules. This experiment was conducted for 4 hours and a maximum degradation of 91.2% of the dye was achieved in the presence of PEN@ZnPc under visible light irradiation, implying that the PEN@ZnPc nanofibers possess excellent photocatalytic performance. Two major factors, the retrievability and stability, are of great importance when applied in environmental technology. The stability of the PEN@ZnPc nanofibers mats is investigated for degradation of RhB by a three cycle tests. As shown in Fig. 3c, the degradation efficiency of PEN@ZnPc remained almost unchanged during a three cycle experiment. It was indicated that that PEN@ZnPc can be recycled for adsorption and catalytic degradation of RhB in water.

5 SUMMARY

PEN@ZnPc nanofibers mat is prepared by the electrospinning method. PEN@ZnPc demonstrated high photocatalytic performance for the degradation of RhB under visible light, providing another photocatalyst to deal with organic pollutants. Repetitive measurements show that this PEN@ZnPc is quite stable and remains efficient with no obvious decrease of catalytic activity. It is believed that the obtained PEN@ZnPc nanofibers mat has huge potential to be used for eliminating the dye pollutants in the wastewater.

ACKNOWLEDGMENTS

This work was financially supported by the National NSF (No.51173021).

REFERENCES

Cen, J., X. Li, M. He, et al, The effect of background irradiation on photocatalytic efficiencies of TiO2 thin films (J). Chemosphere, 62(2006)810–816.

Hua, L.U., Q. U. Yun-He, T. S. Zhou, et al, Determination of Chemical Oxygen Demand Values by Photocatalytic Oxidation Using Nano ZnO Film on Quartz. Journal of East China Normal University, 21(2007) 55–62.

Iliev, V., V. Alexiev, L. Bilyarska, Effect of metal phthalocyanine complex aggregation on the catalytic and photocatalytic oxidation of sulfur containing compounds. Journal of Molecular Catalysis A Chemical, 137(1999)15–22.

Young, J.N., T. C. Chang, S.C. Tsai, et al, Preparation of a nonleaching, recoverable and recyclable palladium-complex catalyst for Heck coupling reactions by immobilization on Au nanoparticles. Journal of Catalysis, 272(2010)253–261.

Energy Science and Applied Technology ESAT 2016 – Fang (Ed.)
© *2016 Taylor & Francis Group, London, ISBN 978-1-138-02973-6*

Enhanced drug loading and encapsulation efficiency of microencapsulated herbal medicine via an optimization process

Wenyi Wang, Patrick C.L. Hui, Frency S.F. Ng & Chi-Wai Kan
Institute of Textiles and Clothing, Hong Kong Polytechnic University, Hong Kong

Clara B.S. Lau & Pingchung Leung
Institute of Chinese Medicine, Chinese University of Hong Kong, Hong Kong
State Key Laboratory of Phytochemistry and Plant Resources in West China, Chinese University of Hong Kong, Hong Kong

ABSTRACT: This study aims to increase the Drug Loading (DL) and Encapsulation Efficiency (EE) of Chitosan/Sodium Alginate (CSA) microcapsule loading PentaHerbs (PHF) via optimization of the fabrication variables. Effects of three fabrication variables, namely, mass ratio of sodium alginate to chitosan, ratio of water to oil and ratio of core to shell material, on the DL and EE, were fully investigated. Moreover, the fabricated microcapsule was also examined in terms of physicochemical properties such as surface morphology, particle size distribution, and drug release behavior.

Keywords: microcapsules, herbal medicine, encapsulation efficiency, drug loading

1 INTRODUCTION

Nanotechnology has been extensively studied and applied in the development of novel drug delivery system for herbal medicines in the past few decades (B.V. Bonifacio, 2014). Several dosage forms, such as liposome, nanoemulsion, nanoparticles, microcapsules, etc. (D.-C. Li, 2009), have been successfully fabricated with advantageous characteristics for herbal drugs, e.g., enhancement of solubility and bioavailability, protection from toxicity, enhancement of pharmacological activity, sustained delivery, etc. (P.L. Lam, R, 2014).

Herbal medicines have been widely used for thousands of years around the world due to its fewer adverse effects compared with modern medicines (B.C. Chan, et al, 2008). Modern analytical technique has confirmed that herbal medicines contain a large variety of active constituents and each component plays a significant role in determining the overall function of herbal medicines. Moreover, each active constituent provides synergistic action in the therapeutic values of herbal medicines (R. Yuan, 2000). Hence, traditional herbal medicines are routinely used as an entirety without isolating its each active component. This medical approach has proven to be much more appropriate to prevention of diseases and the treatment of chronic diseases without an unacceptably high level of collateral damage (C.S., et al, 2014).

For the treatment of Atopic Dermatitis (AD), a functionalized fabric coated with PHF-containing CSA microcapsules has been reported (P.C. Hui, et al, 2013). PHF is a traditional Chinese prescription of five herbs, i.e., *Cortex Moutan, Cortex Phellodendri, Flos Lonicerae, Herba Menthae,* and *Rhizoma Atractylodis*, in the proportion of 2:2:2:1:2. However, low DL and EE of the fabricated CSA microcapsules increased the quantity of carrier materials for the microcapsule fabrication and the amount of microcapsules for the coating on the functional fabric. In view of the loss of the content of active components during microcapsule fabrication, this may thus lower and even eliminate the clinical efficacy of PHF. Therefore, it is of high importance to enhance the DL and EE of the microcapsules for the development of herbal medicine functionalized garments. The DL and EE normally depend on the adopted fabrication approaches and variables. For the fabrication of CSA microcapsules by emulsion-chemically cross-linking approach, the major fabrication variables consist of the weight ratio of sodium alginate and chitosan, the volume ratio of water, and paraffin oil, and core/shell ratio, which are further studied. Therefore, the present study is aimed at enhancing the DL and EE of PHF-containing CSA microcapsules by optimizing the fabrication variables.

2 EXPERIMENTAL METHODOLOGY

Materials. PHF was a gift from the Institute of Chinese Medicine (ICM), the Chinese University of Hong Kong. Chitosan (Mw = 2.5×10^5, DD = 90%) was obtained from Sigma-Aldrich Co. LLC. Alginate sodium was purchased from International Laboratory (U.S.A.). All other chemicals and reagents were of analytical grade.

Fabrication and characterization of CSA microcapsule emulsion. The chemical cross-linking method was employed to fabricate CSA microcapsules, as described in our previous work (P.C. Hui, et al, 2013). A scanning electron microscope (SEM, JEOL 6490 Hitachi, Japan) was used to observe the surface morphological structure of CSA microcapsules. The particle size was measured by a laser particle size analyzer (Brookhaven Instrument Corporation, USA).

Drug Loading (DL) and Encapsulation Efficiency (EE) Analysis. This study selected gallic acid (GA), one of the major active components of PHF, to determine DL and EE of the prepared microcapsules. The analytical tool was High-Performance Liquid Chromatography (HPLC) system (Hewlett Packard Agilent 1000 series). DL and EE were calculated using Eqs. (1) and (2), respectively.

DL/% = GA within microcapsule/weight of dried microcapsule × 100% (1)

EE/% = GA within microcapsule/GA within PHF × 100% (2)

Controlled Release Property. Two pieces of dried microcapsules (5 mg) were dispersed into 5 ml of buffered saline (PBS, pH = 5.4 and 5.0). The temperature of release medium was maintained at 37°C and the shaking rate of water bath was 100 rpm. Periodically, the suspension was withdrawn and centrifuged at 4000 rpm to collect the supernatant for GA determination by the HPLC analytical method.

3 RESULTS AND DISCUSSION

Effects of mass ratio of sodium alginate to chitosan. The sodium alginate-to-chitosan (SA/CH) ratio was set at 0:1, 1:2, 1:1, 2:1, and 3:1, while other conditions remained unchanged. As shown in Fig. 1(a), a higher proportion of sodium alginate tends to result in a higher DL and EE on the whole. However, a decrease was seen in both the DL and EE when the proportion of sodium alginate was increased to 75%. The highest DL and EE (around 17% and 34%, respectively) were

obtained at SA/CH ratio of 2:1, indicating the optimum ratio. The reason lies in the high degree of deacetylation (90%) of chitosan, which results in tremendous amine groups in acidic solution. More sodium alginate is thus required to form polyelectrolyte with chitosan, exactly between the carboxyl group of alginate and amine groups of chitosan, and the optimum mixing ratio of sodium alginate and chitosan is 2:1.

Effects of water-to-oil ratio. The ratio of water to oil has a significant impact on the properties of the microcapsules, primarily because the properties of emulsion that are greatly influenced by the ratio of each phase. In this study, the volume ratio of water to oil (liquid paraffin) was kept at 1:1, 1:3, 1:5, and 1:7, and other fabrication conditions remained unchanged. As shown in Fig. 1(b), CSA microcapsule prepared with a water-to-oil ratio of 1:5 had the highest DL and EE, around 19% and 35%, respectively. Thus, the optimum ratio of water to oil (liquid paraffin) was 1:5. This was primarily caused by the emulsifier (Span-80) used in the formation of water-in-oil emulsion. The dosage of emulsifier greatly influenced the stability of emulsion. A small amount of emulsifier may lead to unstable emulsion and the emulsion generally tends to break, whereas high dosage of emulsifier usually produces stable emulsion as well as more foam. This study employed the same amount of surfactant for fabrication of microcapsules under different conditions and the final concentration of surfactant was maintained at 2% in emulsion. Therefore, for the water-to-oil ratio of 1:7, the dosage of emulsifier was relatively low and thus unstable emulsion was formed. This provides an explanation for the low DL and EE of microcapsules at this ratio while the ratio of 1:5 of water and oil resulted in the highest DL and EE for the microcapsules.

Effects of ratio of core to shell. The core-to-shell ratio is generally regarded as one of the most important influential factors in determining DL and EE. A higher ratio of core to shell material is more likely to lead to a higher DL but a lower EE due to a higher loss of drug during the microencapsulation process, especially for the hydrophilic drugs. This means that DL cannot be unlimitedly enhanced by increasing the ratio of core to

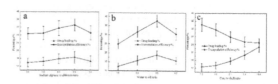

Figure 1. DL and EE of CSA microcapsules prepared with different ratios of sodium alginate to chitosan (n = 3); (b) ratios of water to oil (n = 3); and (c) ratios of core to shell material (n = 3).

Table 1. DL and EE of PHF-containing CSA microcapsules prepared under various conditions (n = 3).

	2:1 of SA/CH	1:5 of water/oil	10:1 of core/shell	Optimal condition
DL/%	16.6 ± 1.89	18.82 ± 2.21	22.92 ± 2.45	24.49 ± 3.01
EE/%	33.21 ± 2.77	37.64 ± 3.13	25.07 ± 2.11.	36.77 ± 3.24

Figure 2. Micrograph (a) ($\times$ 400) and SEM (b) ($\times$ 10000) of CSA microcapsules and size distribution (c).

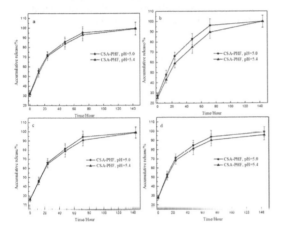

Figure 3 Release properties of CSA microcapsules in PBS of pH = 5.0 and pH = 5.4 (n = 3) (a: sample under optimum condition; b: sample with 2:1 ratio of SA/CH; c: sample with 1:5 ratio of water to oil; d: sample with 10:1 ratio of core to shell).

shell material. On the other hand, the increase in the ratio of core to shell material reveals that the proportion of core material becomes higher, which may be limited by the solubility of core materials in water. Therefore, the solubility of core material, DL and EE should be taken into consideration when optimizing the ratio of core and shell material.

In this study, the core-to-shell material ratio was set as 1:2, 1:1, 2:1, 5:1, and 10:1 while other fabrication conditions remained unchanged. From Fig. 1(c), it can be clearly seen that EE was inversely proportional to the ratio of core to shell material, while DL was in the same (positive) proportion to the ratios. The reason may be that high proportion of PHF results in higher loss of the drug in the fabrication of microcapsules, and thus low EE was obtained. By contrast, in spite of

higher loss of PHF, the microcapsules had more drugs per unit mass when core-to-shell material ratio was increased. This means that high proportion of PHF tends to produce high drug loading in microcapsules, per unit mass, and likewise, low core-to-shell material ratio is more likely to lead to low drug loading. The highest DL was seen when core-to-shell material ratio was 10:1; EE at this ratio was the lowest, however, being around 23% and 25%. Therefore, the ratio of 10:1 was regarded as the optimum ratio of core to shell material and selected for further experiments.

CSA microcapsules fabricated under optimum conditions. Based on the above analysis, the optimum conditions for CSA microcapsules are as follows: (i) 2:1 ratio of sodium alginate to chitosan; (ii) 1:5 ratio of water to liquid paraffin; and (iii) 10:1 ratio of core to shell material. For the purpose of the optimum condition verification, an extra experiment was conducted. The result is shown in Table 1, together with the highest result under each condition in Fig. 13. It can be clearly seen that there is an increase in the DL under optimum condition, at 24.49%, while EE is at 36.77%. This indicates that DL and EE could be improved by optimizing the fabrication condition of CSA microcapsules.

Morphological structure and particle size Fig. 2(a) shows that the particle size of CSA microcapsules was uniform in shape. The spherical CSA microcapsules with a relatively smooth surface were clearly observed (Fig. 2(b)). The particle size ranged from 0.9µm to 5µm in diameter, averaged 2.2µm, and approximately followed a normal distribution (Fig. 2(c)). This is in accordance with the observations in Fig. 2(a) and Fig. 2(b).

In vitro controlled release. The release properties of PHF-containing CSA microcapsules under optimum conditions were investigated by monitoring the release behavior of GA in PBS with a pH of 5.4 and 5.0 over 144 hours (6 days) (Fig. 3(a)). Fig. 3 indicates that the release of GA from CSA microcapsules saw a faster rate during the first 24 hours, and then the rate dropped steadily and reached a constant value. For all the samples, over 60% of GA was released during the first 24 hours and eventually achieved a constant level of around 97%.

Fig. 3 also reveals that the release rate of GA in pH = 5.0 PBS was markedly higher than that in pH = 5.4 PBS, within the same period, indicating that the release kinetics are negatively influenced by pH values of PBS. The release of GA reached equilibrium after around 72 hours in both pH = 5.0 and 5.4 PBS, for all the four samples. Moreover, there was no significant difference among four samples in terms of the release rate and accumulative release amount of GA. This means that the release of GA may depend on the release medium, rather than the microcapsule matrix.

205

4 CONCLUSION

This study investigated the fabrication variables of CSA microcapsules containing PHF to enhance the DL and EE. The optimized DL and EE were 24.49% and 38.77%, respectively.

ACKNOWLEDGMENTS

We gratefully acknowledge Mr. So Tung Lam, Mr. Henry Ng, and Prof. T.K. Tsui for their financial support.

REFERENCES

Bonifacio, B.V., P.B.d. Silva, S. Ramos, K.M. Silveira Negri, T.M. Bauab, M. Chorilli, Nanotechnology-based drug delivery systems and herbal medicines: a review, Int. J. nanomed., (2014) 1–15.

C.S., et al., Nano Carriers of Novel Drug Delivery System for "Ayurveda Herbal Remedies" Need of Hour– A Bird's Eye View, Am J Pharm Tech Res, 4 (2014).

Chan, B.C. et al., Traditional Chinese medicine for atopic eczema: PentaHerbs formula suppresses inflammatory mediators release from mast cells, J Ethnopharmacol, 120 (2008) 85–91.

Hui, P.C. et al., Microencapsulation of Traditional Chinese Herbs-PentaHerbs extracts and potential application in healthcare textiles, Colloid Surface B, Biointerfaces, 111 (2013) 156–161.

Lam, P.L., R. Gambari, Advanced progress of microencapsulation technologies: in vivo and in vitro models for studying oral and transdermal drug deliveries, J Control Release, 178 (2014) 25–45.

Li, D.-C. et al., Application of targeted drug delivery system in Chinese medicine, J Control Release, 138 (2009) 103–112.

Yuan, R., Y. Lin, Traditional Chinese medicine: an approach to scientific proof and clinical validation, Pharmacol Therapeutic, 86 (2000) 191–198.

Energy Science and Applied Technology ESAT 2016 – Fang (Ed.)
© *2016 Taylor & Francis Group, London, ISBN 978-1-138-02973-6*

Electrochemical deposition of Fe_2O_3/graphene composites for energy storage

Ke Ma, Daoming Zhang & Quansheng Zhang
School of Chemical and Environmental Engineering, Shanghai Institute of Technology, Shanghai, China

ABSTRACT: Fe_2O_3/graphene composites were successfully fabricated by the electrochemical deposition method. The obtained samples were characterized by X-Ray Diffraction (XRD) and Scanning Electron Microscopy (SEM). The Fe_2O_3/graphene nanocomposites are uniformly loaded on the electrode. To assess the electrochemical capacity of composites, cyclic voltammetry and galvanostatic charging–discharging measurements were performed in 1 mol/L Na_2SO_3. The result demonstrates that the composite electrode exhibits an excellent specific capacitance of 194 F/g at 1 A/g and the charge-discharge stability measurements indicate a capacity retention of 87.8% after 1000 cycles. The outstanding capacitance and good cycling stability demonstrate that the nanocomposite is a quite promising electrode material for supercapacitors.

Keywords: supercapacitors; graphene; electrochemical deposition

1 INTRODUCTION

Energy is one of the most pressing problems of the human kind. Considerable efforts have been expended on energy storage applications for the purpose of overcoming the challenge of fossil fuel consumption. The desire for clean and secure energy demands a worldwide scale to search for advanced electrical energy storage (Chan P et al., 2014) (Kim H et al., 2014). Supercapacitors, with important features such as long life expectancy, high efficiency, and safety, constantly draw increasing attention for energy storage researches (Zhang H et al., 2012) (He Y et al., 2012). There are two kinds of supercapacitors, which depend on their charge-storage mechanism. The first category includes the Electrical Double Layer Capacitors (EDLCs). Carbonaceous materials are widely used in EDLCs, such as graphene, carbon nanotubes, activated carbon, and carbon aerogels (Wu Z-S et al., 2012). Another kind is redox electrochemical capacitor, including conducting polymers and metal oxides (Shakir I et al., 2011). Many metal oxides have been widely investigated for charge storage applications in supercapacitors, such as NiO (Yan X et al., 2014), MnO_2 (Yu G et al., 2011), etc. Effort to search for more alternative, inexpensive electrode materials with good capacitive characteristics is now quite active. Fe_2O_3 show a large potential for supercapacitor electrode materials (Wang D et al., 2012), considering its significant advantage of inexpensive, environmental friendly,

and high redox activity. However, the pristine Fe_2O_3 suffers from the drawback of insufficient high-rate charge/discharge, fast capacity degradation, and poor cycling stability due to its low electrical conductivity and volume change during repeated redox processes (Yang S et al., 2014). As a rising star, graphene has drawn considerable attention in various research fields, due to its large specific surface area, excellent electrical conductivity, high specific surface area, and chemical stability (Xia X et al., 2011). Hence, one of the most effective approaches to solve these problems is to combine the pseudo active material Fe_2O_3 with highly conductive graphene nanosheets. The nanocomposite electrode shows a great potential as an effective way to improve the conductivity and increase the capacitance value. Further, intercalating metal oxide nanoparticles into graphene nanosheets may also avoid agglomeration and restacking graphene simultaneously. Electrochemical deposition is a fast and green approach and has been widely used to prepare films and nanocomposites (Liu X et al., 2014). What is more, compared with other methods, electrochemical deposition eliminates the fabrication process of films or hybrid structures and the as-prepared nanomaterials are directly immobilized on the conducting substrate, facilitating further applications.

In the present report, Fe_2O_3/graphene composites were synthesized by cathodic electrochemical deposition using $Fe(NO_3)_3$ and Graphene Oxide (GO) as precursors. The electrochemical performance of Fe_2O_3/graphene composites as a supercapacitor electrode material was investigated in detail.

2 EXPERIMENTAL METHODOLOGY

2.1 Synthesis of the Fe_2O_3/graphene composites

For synthesis of the Fe_2O_3/graphene composite, Graphene Oxide (GO) was first synthesized by modified Hummer's method (Qi X et al., 2013) and 45 ml 0.1mol/L GO suspension was prepared by dispersing GO into N-Methyl-2-Pyrrolidone (NMP) solution. Subsequently, $Fe(NO_3)_3·9H_2O$ (0.018 g) was added to the above mixture solution. The obtained dispersion was sonicated for 30 min to yield a stable suspension. The electrochemical synthesis was performed at 60 V voltage direct current at 40°C for 5 min. The cathode and anode were spaced 5 mm apart and immersed in the deposition bath. The Fe_2O_3/graphene was obtained by heating the prepared cathode at 300°C for 4 h under ambient environment. For comparison, pure Fe_2O_3 were obtained via the same electrodeposition process without adding GO. The payload mass of Fe_2O_3/graphene in the electrode was evaluated by a microbalance with an accuracy of 0.01 mg.

2.2 Characterization

The morphology of products was examined using the Hitachi S-3400N Scanning Electron Microscope (SEM) at 15 kV. The crystal structure was determined by X-Ray Diffraction (XRD) analysis using the Bruker AXS D8 Advance diffractometer in the 2θ range of 10–80° at a scanning rate of 5°/min, and an accelerating voltage of 40 KV and emission current of 40 mA.

2.3 Electrochemical test

The electrochemical capacity of Fe_2O_3/graphene electrode was examined by using the CHI660C electrochemical workstation. Cyclic Voltammetry (CV) and Galvanostatic Charge-Discharge (GCD) were conducted in a three-electrode system arrangement in 1 M Na_2SO_3 electrolyte. The Pt electrode and saturated calomel electrode (SCE) were utilized as the counter electrode and reference electrode, respectively. Before testing, the electrode was steeped in 1 M Na_2SO_3 solution for 12 h with high purity nitrogen in the solution for about 30 minutes to exclude oxygen. The potential was scanned between −1 and 0 V at different scanning rates. The specific capacitance ($F g^{-1}$) of the electrode material was evaluated by GCD curves according to the following equation (1):

$$Cm = I\Delta t/m\Delta V \qquad (1)$$

where Cm ($F g^{-1}$) is the specific capacitance, I(A) is the current of the charge-discharge, Δt (s) and ΔV (V) represent the total discharge time and the potential window (vs. SCE), respectively, and m (g) is the mass load of the active material in the electrode.

3 RESULT AND DISCUSSION

The schematic diagram of the synthesis procedure of Fe_2O_3/graphene is illustrated in Scheme 1. First, GO was dispersed in the NMP solution with $Fe(NO_3)_3·9H_2O$ salt added, which has been formulated as the electrochemical deposition bath. Then, the Fe^{3+} ions get adsorbed on the surface of GO, imparting Fe^{3+} the positive charges and head toward the cathode under the influence of an external electric field. The Fe^{3+} can be deposited on the conduct substrate SS and the graphene oxide was electrochemically reduced to graphene simultaneously. Finally, the Fe_2O_3/graphene electrode was obtained via thermal treatment of the prepared cathode.

Figure 1 represents the XRD patterns of Fe_2O_3/graphene composites. According to Figure 1, no

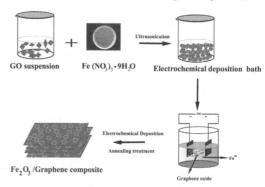

Scheme 1. Schematic of the synthesizing procedure of Fe_2O_3/graphene composite via electrodeposition.

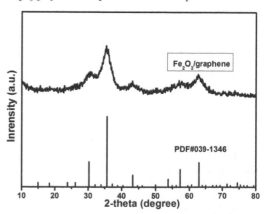

Figure 1. XRD pattern of Fe_2O_3/graphene composites.

208

apparent diffraction peaks, at 26° corresponding to graphene, were observed. All the diffraction peaks can be easily indexed to Fe_2O_3 (JCPDS No. 039–1346). The Fe_2O_3/graphene has no other diffraction peaks, which suggest its high purity and crystallinity.

To further confirm the surface composition and the pressure of Fe_2O_3, the Fe_2O_3/graphene nanocomposites were further studied by XPS analysis. A wide survey scan of XPS spectra is displayed in Figure 2. The survey of XPS spectrum indicates that F_2O_3/graphene nanocomposites consist of Fe, O, and C elements. The inset shows the high-resolution XPS spectrum for the Fe 2p. The Fe 2p spectrum shows two distinct peaks at binding energies of 710.5 eV and 724.5 eV, which are ascribed to Fe $2p_{3/2}$ and Fe $2p_{1/2}$. Moreover, a shake-up satellite at nearly 718.9 eV is a characteristic of Fe^{3+} (Mu J et al., 2011). The presence of a satellite peak excluded the existence of Fe_3O_4 phase in the Fe_2O_3/graphene nanocomposites.

The morphological structure and particle size of the sample were characterized by using SEM. Figure 3 shows the surface morphology of the Stainless Steel (SS) mesh, and pure Fe_2O_3 and Fe_2O_3/graphene composite deposited on the SS mesh, respectively. The three-dimensional network architecture of SS is shown in Figure 3(a). The special 3D network porous structure makes it an appropriate conduct current collector for supercapacitor electrode. Pure Fe_2O_3 particles without GO were directly deposited on the SS mesh via electrochemical deposition. The Fe_2O_3 products displayed in Figure 3(b) have a massive structure and lose its uniformity, showing a serious agglomeration among the Fe_2O_3 particles. Figure 3(c) shows a low-magnification SEM image of the Fe_2O_3/graphene composite. Fe_2O_3 nanoparticles are heterogene-ously anchored on the surface of graphene. Fe_2O_3/graphene composites show a different morphology from pure Fe_2O_3. Figure 3(d) further demonstrates a high-magnification SEM image of the Fe_2O_3/graphene composite. The SEM image indicates that the Fe_2O_3/graphene composites are composed of intertwined graphene and Fe_2O_3 nanoparticles which are approximately 400 nm in diameter. The agglomeration among Fe_2O_3 nanoparticles is effectively avoided by introducing GO and they form a perfect multilayer structure. The special 3D nanoporous network structure provides a high contact area between the electrode and the electrolyte and short ion diffusion path for faster charge transfer reaction at the interface between the the electrode and the electrolyte.

The electrochemical behavior and capacitor performance of the Fe_2O_3/graphene electrode were evaluated by CV and GCD measurements. CV was carried out on nanocomposites over various sweeping rates of 0.005, 0.01, 0.02, 0.05, and 0.1 Vs^{-1} within a potential window of $-1\sim0$ V. Figure 4(a) presents the CV curves for the graphene/Fe_2O_3 composites, which exhibit a nearly rectangular shape at a scanning rate of 5 mV s^{-1}. A pair of cathodic and anodic peaks are identified at a low scanning rate at a potential of approximately -0.4 and -0.3 V, showing the contribution of pseudocapacitances from Fe_2O_3 to the total capacitances. The anodic peaks shift positively and the cathodic peaks shift negatively with increased scanning rates. As the scanning rates increased, the current density of the Fe_2O_3/graphene composite electrodes increased. The CV curves exhibit a gradual deviation from the symmetry shape when the applied scanning rates increase to 0.1 Vs^{-1}. This is due to the kinetics of electron transfer in the elec-

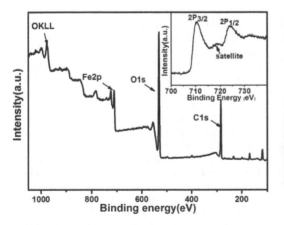

Figure 2. XPS fully scanned spectra of the Fe_2O_3/graphene nanocomposites. The inset shows the Fe 2p scan.

Figure 3. SEM image of (a) 3D network architecture of SS, (b) pure Fe_2O_3 deposited on a stainless steel mesh, and (c) and (d) Fe_2O_3/graphene composite films deposited on a stainless steel mesh at different magnifications.

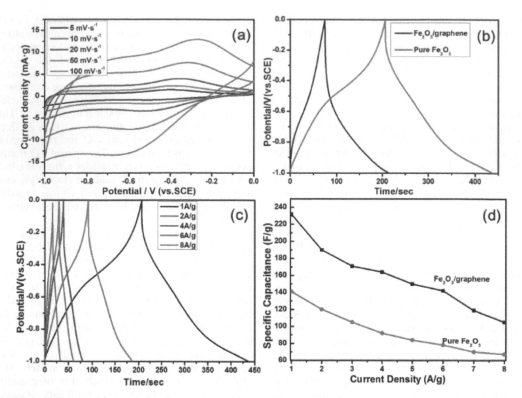

Figure 4. Electrochemical performance of Fe_2O_3/graphene composite: (a) CV curves at different scanning rates, (b) galvanostatic charge-discharge curves of Fe_2O_3/graphene composite and pure Fe_2O_3 at 1 A g^{-1}, (c) galvanostatic charge-discharge curves of Fe_2O_3/graphene composite, and (d) rate capacities of pure Fe_2O_3 capacitance and Fe_2O_3/graphene composite.

trode materials and the limited ion adsorption–desorption at the electrode and electrolyte interface, causing the active materials not to be fully utilized. The GCD curves of bare Fe_2O_3 and Fe_2O_3/graphene nanocomposite are shown in Figure 4(b) at current densities of 1A/g. The composite yield specific capacitance is 231 F g^{-1}, which is 1.6 times higher than the pure Fe_2O_3 electrode (141 F g^{-1}) in capacity. The outstanding electrochemical performance suggested that poorly conducting Fe_2O_3 is capable of storing high charge when highly dispersed and nanosized by being decorated with GO. The GCD curves of the Fe_2O_3/graphene nanocomposite at various current densities are presented in Figure 4(c). The IR drops on all curves that are negligible, suggesting little overall resistance of this hybrid material. The specific capacitance of pure Fe_2O_3 and Fe_2O_3/graphene nanocomposite at different current densities is shown in Figure 4(d). It could be seen that the specific capacitance of the composites decreases with the decrease in the discharge current density. The pseudocapacitance values of composites are 232, 190, 171, 164, and

105 F g^{-1} at current densities of 1, 2, 3, 4, and 8 A/g, respectively. The rate of performance of Fe_2O_3/graphene hybrids is excellent. However, the specific capacitance for the pure Fe_2O_3 electrode is 141, 120, 105, 92, and 67 F g^{-1} at 1, 2, 3, 4, and 8 A g^{-1}, respectively.

The cyclability of Fe_2O_3/graphene electrode was also examined by conducting continuous charge-discharge cycles at a constant discharge current density of 1A/g, as shown in Figure 5. The Fe_2O_3/graphene composite electrode retains 87.8% of its initial value after 1000 cycles. The excellent cyclability can be contributed to the cohesion between nanocomposites and substrates. The cohesion is greatly improved and it efficiently avoided the aggregation of Fe_2O_3 particles by introducing GO. However, the specific capacitance of pure Fe_2O_3 decreases from 140 F g^{-1} to 82 F g^{-1}, corresponding to 64.4% capacity retention of the initial capacitance. It might be due to the utilization of Fe_2O_3, which is limited and parts of Fe_2O_3 active materials fall off from the electrode during the charge-discharge process.

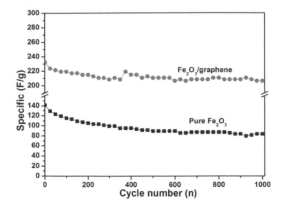

Figure 5. Cycle stability comparison of Fe_2O_3/graphene composite and Fe_2O_3.

4 CONCLUSION

Fe_2O_3/graphene composites were synthesized by the electrochemical deposition method and obtained layered Fe_2O_3/graphene structures under a rather low applied voltage. Electrochemical deposition technology eliminates the fabricating electrode process. Fe_2O_3/graphene nanocomposites can combine the advantages of Fe_2O_3 and graphene, which show good electrochemical capacitive behavior. The evaluation about the specific capacitance of the composite is 231 F g^{-1}, which is 1.6 times higher than the bare Fe_2O_3 electrode in capacity at 1A/g. The charge-discharge stability measurements suggest a retention of specific capacitance of approximately 87.8% after 1000 cycles at a current density of 1A/g. The fine specific capacitance and good cycling stability demonstrate that the Fe_2O_3/graphene electrode has the potential application in supercapacitors.

REFERENCES

Chan P, Majid S. RGO-wrapped MnO2 composite electrode for supercapacitor application. Solid State Ionics 2014, 262:226–229.

He Y, Chen W, Li X, et al. Freestanding three-dimensional graphene/MnO2 composite networks as ultra-light and flexible supercapacitor electrodes. ACS nano 2012, 7:174–182.

Kim H, Park K-Y, Hong J, et al. All-graphene-battery: bridging the gap between supercapacitors and lithium ion batteries. Scientific reports 2014, 4.

Liu X, Qi X, Zhang Z, et al. One-step electrochemical deposition of nickel sulfide/graphene and its use for supercapacitors. Ceramics International 2014, 40:8189–8193.

Mu J, Chen B, Guo Z, et al. Highly dispersed Fe3O4 nanosheets on one-dimensional carbon nanofibers: synthesis, formation mechanism, and electrochemical performance as supercapacitor electrode materials[J]. Nanoscale, 2011, 3(12): 5034–5040.

Qi X, Zou X, Huang Z, et al. Ultraviolet, visible, and near infrared photoresponse properties of solution processed graphene oxide. Applied Surface Science 2013, 266:332–336.

Shakir I, Shahid M, Cherevko S, et al. Ultrahigh-energy and stable supercapacitors based on intertwined porous MoO3–MWCNT nanocomposites. Electrochimica Acta 2011, 58:76–80.

Wang D, Li Y, Wang Q, et al. Nanostructured Fe2O3–graphene composite as a novel electrode material for supercapacitors[J]. Journal of Solid State Electrochemistry, 2012, 16(6): 2095–2102.

Wu Z-S, Zhou G, Yin L-C, et al. Graphene/metal oxide composite electrode materials for energy storage. Nano Energy 2012, 1:107–131.

Xia X, Tu J, Mai Y, Chen R, Wang X, Gu C, Zhao X: Graphene sheet/porous NiO hybrid film for supercapacitor applications. Chemistry 2011, 17:10898–10905.

Yan X, Tong X, Wang J, et al. Synthesis of mesoporous NiO nanoflake array and its enhanced electrochemical performance for supercapacitor application. Journal of Alloys and Compounds 2014, 593:184–189.

Yang S, Song X, Zhang P, et al. Self_Assembled α_Fe2O3 Mesocrystals/Graphene Nanohybrid for Enhanced Electrochemical Capacitors[J]. small, 2014, 10(11): 2270–2279.

Yu G, Hu L, Vosgueritchian M, et al. Solution-processed graphene/MnO2 nanostructured textiles for high-performance electrochemical capacitors. Nano Lett 2011, 11:2905–2911.

Zhang H, Zhang X, Zhang D, et al. One-step electrophoretic deposition of reduced graphene oxide and Ni(OH)2 composite films for controlled syntheses supercapacitor electrodes. The Journal of Physical Chemistry B 2012, 117:1616–1627.

Energy Science and Applied Technology ESAT 2016 – Fang (Ed.)
© *2016 Taylor & Francis Group, London, ISBN 978-1-138-02973-6*

A facile one-step solvothermal method to prepare morphology-controlled Fe_3O_4@SiO_2 core-shell nanocomposites

Maosheng Fu

Department of Environmental and Chemical Engineering, Nanchang Hangkong University, Nanchang, China

ABSTRACT: Fe_3O_4@SiO_2 core-shell nanocomposite was successfully prepared by a new facile one-step solvothermal synthesis method using $FeCl_3 \cdot 6H_2O$ as the iron resource, $NaHCO_3$ and $NaSiO_3 \cdot 9H_2O$ as mixed precipitants, glucose as the assistant reducing agent, and ethylene glycol as the reaction solvent, which was characterized by X-ray diffraction, Fourier transform infrared spectroscopy, transmission electron microscopy, and vibrating sample magnetometer methods. The mass ratio of $NaSiO_3 \cdot 9H_2O$ to $NaHCO_3$ played an important role to phase composition and particle morphology of the as-prepared nanocomposite. Controllable synthesis of fine fibrous one-dimensional Fe_3O_4@SiO_2 core-shell nanocomposite and spherical-like Fe_3O_4@SiO_2 core-shell nanocomposite was implemented, and its formation mechanism of morphology-controlled nanocomposite via one-step solvothermal synthesis was specially proposed.

Keywords: solvothermal synthesis; Fe_3O_4@SiO_2 core-shell nanocomposite; morphology control

1 INTRODUCTION

In recent years, silica-coated Fe_3O_4 nanoparticles have attracted increasing interest especially in biomedical applications like targeted drug delivery and magnetic resonance imaging enhancement (M. Abbas et al, 2014) (C. Hui et al, 2011), because the use of silica as a coating layer to Fe_3O_4 nanoparticles not only helps in enhancing the advantages of their high biocompatibility, hydrophilicity, dielectric property, and stability against degradation, but also facilitates easy surface modification due to the availability of abundant silanol groups (–SiOH) on its surface (L. Caruana et al, 2012). To date, many procedure strategies have been developed to fabricate silica-coated Fe_3O_4 hybrid nanocomposite (B. Mojic et al, 2012). Among these procedure strategies, they mostly require multi-step procedures, that is, the first step is for preparation of magnetic nanoparticles and the second step is for coating. However, the multi-step procedure strategies involve the intrinsic agglomeration phenomenon and phase transformation of the as-prepared Fe_3O_4 nanoparticles, long reaction time, and higher costs. To the best of our knowledge, there are few reports concerning preparation of silica-coated Fe_3O_4 hybrid nanocomposite by one-step procedure strategy (M. Abbas et al, 2014).

In the present study, we developed a new one-step solvothermal synthesis method to prepare morphology-controlled Fe_3O_4@SiO_2 core-shell nanocomposite in the presence of $FeCl_3 \cdot 6H_2O$, ethylene glycol, glucose, $NaHCO_3$, and $NaSiO_3 \cdot 9H_2O$. Effect of the mass ratio of $NaSiO_3 \cdot 9H_2O$ to $NaHCO_3$ on the phase composition and particle morphology of silica-coated Fe_3O_4 hybrid nanocomposite was focused on investigation, and its synthesis mechanism was specially proposed.

2 EXPERIMENTS

2.1 Chemicals

$FeCl_3 \cdot 6H_2O$, $NaHCO_3$, $NaSiO_3 \cdot 9H_2O$, ethylene glycol (EG), glucose, and Polyvinyl Pyrrolidone (PVP, K30) were purchased from Shanghai Chemical Reagents Company (Beijing China). All reagents were of analytical grade and were used as received without further purification.

2.2 Procedures

Fe_3O_4@SiO_2 nanocomposite was prepared in an analogous manner as described in our previous work (M.S. Fu et al, 2014). In a typical synthesis, first, 1.6 g of $NaHCO_3$, 1.6 g of $NaSiO_3 \cdot 9H_2O$, 0.05 g of glucose, and 0.05 g of PVP were dissolved in 70 mL of EG to obtain the mixed solution by ultrasonication. Then, 3.6 g of $FeCl_3 \cdot 6H_2O$ was slowly added to the mixed solution under ultrasonication to get a yellow and brown even solution after CO_2

bubbles nearly gave off. Finally, the resulting mixture was transferred into a 100 mL Teflon-lined stainless steel autoclave, and heated at a temperature of 200°C for 10 hours. The products were obtained by magnetic separation, and sequentially washed with DI water and ethanol several times and then dried in a vacuum oven at 60°C for 12 hours. Based on the above-mentioned procedure scheme, the samples with different mass ratios of $NaSiO_3 \cdot 9H_2O$ to $NaHCO_3$ were prepared.

2.3 *Characterization*

All samples were examined by XRD on a Bruker Advance D8 X-ray powder diffractometer equipped. FTIR spectra were recorded on Nicolet Avatar 370 Fourier Transform Infrared Spectrometer using the KBr discs. TEM images were taken using a transmission electron microscope (JEOL JEM-2010, JAPAN) at an accelerating voltage of 200 kV. Magnetic properties were investigated on a Vibrating Sample Magnetometer (VSM, JDAW-2000D) at 300K.

3 RESULTS AND DISCUSSION

Fig. 1 shows the XRD patterns of the samples prepared at 200°C for 10 h by different mass ratios of $NaSiO_3 \cdot 9H_2O$ to $NaHCO_3$. As was observed from Fig. 1, the sharp and strong diffraction peaks for the sample with a mass ratio of 1:1 appeared at $2\theta = 30.09°, 35.42°, 43.06°, 53.46°, 57.06°,$ and $62.52°$, which was corresponding to the standard cubic-structured magnetite (Fe_3O_4) (JCPDS No.: 19-0629) (J. C. Zhan et al, 2014), and they were also quite similar to that of 2:1, which indicated that

the as-obtained samples with the mass ratios of 1:1 and 2:1 could be both assigned to Fe_3O_4 crystals. For the sample with a mass ratio of 1:2, in addition to the standard diffraction peaks of Fe_3O_4, the standard diffraction peaks of siderite ($FeCO_3$) (JCPDS No.: 29-0696) were also observed, which indicated that the as-prepared sample was made up of Fe_3O_4 and $FeCO_3$. In addition, for the sample prepared in the presence of only $NaSiO_3 \cdot 9H_2O$, the standard diffraction peaks of maghemite-C (γ-Fe_2O_3) (JCPDS No.: 39-1346) and the standard diffraction peaks of NaCl (JCPDS No.: 05-0628) as well as a typical hump diffraction peak of amorphous SiO_2 at around $2\theta = 24.0°$ could all be observed, which indicated that the as-obtained sample consists of γ-Fe_2O_3, SiO_2 as well as NaCl. It has been explained that NaCl was the residue resulting from incomplete washing during the magnetic separation of the product. Moreover, a weak similar hump diffraction peak of amorphous SiO_2 at around $2\theta = 22.0°$ could also be observed in the XRD patterns of the sample with a mass ratio of 2:1.

Fig. 2 exhibits the FTIR spectra of the as-prepared samples with different mass ratios of $NaSiO_3 \cdot 9H_2O$ to $NaHCO_3$. For the FTIR spectra of the as-prepared sample with a mass ratio of 2:1 in Fig. 2b, the peaks at 578 cm^{-1} were ascribed to the Fe-O stretching vibration, and the strong bands at 1031 cm^{-1} were assigned to the stretching vibration of Si-O (M. Abbas et al, 2014). Based on the results of XRD and FTIR, it is certain that the sample with the mass ratio of 2:1 was made up of Fe_3O_4 and amorphous SiO_2. Furthermore, the FTIR spectra in Fig. 2a was very similar to the FTIR spectra in Fig. 2b, so it was convinced that the as-prepared sample with the mass ratio of 1:1 was also made up of Fe_3O_4 and amorphous SiO_2.

The morphology of the samples was investigated by TEM, as shown in Fig. 3. For the sam-

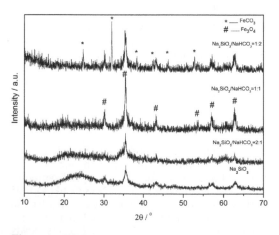

Figure 1. XRD patterns of the samples prepared at 200°C for 10h with different mass ratios of $NaSiO_3 \cdot 9H_2O$ to $NaHCO_3$.

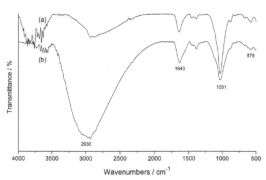

Figure 2. FTIR spectra of the as-prepared samples with different mass ratios of $NaSiO_3 \cdot 9H_2O$ to $NaHCO_3$: (a) 1:1 and (b) 2:1.

ple with the mass ratio (1:1) of $NaSiO_3·9H_2O$ to $NaHCO_3$, it could be clearly observed in Fig. 3a that the surface of the as-prepared Fe_3O_4 nanoparticles was coated by amorphous silica, Fe_3O_4 nanoparticles were generally spherical-like in shape and their average diameter was ca. 90nm, and the core-shell structured $Fe_3O_4@SiO_2$ nanocomposites were slightly agglomerated together by means of combination of the respective coated amorphous silica. However, for the sample of 2:1, besides a few spherical core-shell structured $Fe_3O_4@SiO_2$ nanocomposite particles with an average diameter of *ca.* 30 nm, many one-dimensional fibrous Fe_3O_4 nanoparticles with a fiber length of ca. 30–60nm were mostly observed in Fig. 3b, and they were uniformly embedded in amorphous silica matrix, that is, fine fibrous $Fe_3O_4@SiO_2$ core-shell nanocomposite was mainly formed.

The magnetic properties of the as-prepared samples with different mass ratios (that is, 1:1 and 2:1) of $NaSiO_3·9H_2O$ to $NaHCO_3$ were measured at room temperature using VSM, as shown in Fig. 4. The saturation magnetization (M_s) of the as-prepared samples decreased dramatically with the increase in the mass ratio of $NaSiO_3·9H_2O$ to $NaHCO_3$, and their M_s for the samples with the mass ratio of 1:1 and 2:1 at 300 K were 35.4 emu/g

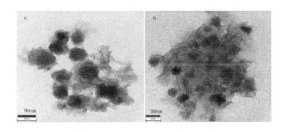

Figure 3. TEM micrographs of the as-prepared samples with different mass ratios of $NaSiO_3·9H_2O$ to $NaHCO_3$: (a) 1:1 and (b) 2:1.

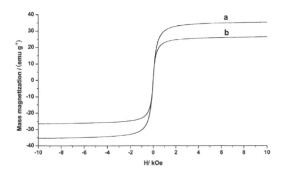

Figure 4. Magnetic curves of the as-prepared samples with different mass ratios of $NaSiO_3·9H_2O$ to $NaHCO_3$: (a) 1:1 and (b) 2:1.

and 26.5 emu/g, respectively, which may be attributed to spherical-like morphology and higher crystallinity degree of Fe_3O_4 nanoparticles (R.F. Chen et al, 2012).

Based on the above-mentioned observations and analysis, the synthesis mechanism of morphology-controlled $Fe_3O_4@SiO_2$ core-shell nanocomposite *via* one-step solvothermal synthesis was proposed, and the detailed synthesis mechanism is shown in Fig. 5. In our synthesis process, the trace amount of water originated from the crystal water of $FeCl_3·6H_2O$ and $NaSiO_3·9H_2O$. First, the double hydrolysis reaction occurred between Fe^{3+} and HCO_3^- as well as SiO_3^{2-} in EG, when the reaction raw materials were mixed at room temperature, which instantaneously resulted in emission of CO_2 bubbles and formation of $Fe(OH)_3$ and H_4SiO_4 precursors. When the semitransparent colloidal solution of $Fe(OH)_3$ and H_4SiO_4 precursors was dealt with by solvothermal treatment, dissolution-precipitation equilibrium of $Fe(OH)_3$ precursors occurred at high temperature and high pressure, and $Fe(OH)_3$ precursors could be transformed to crystalline nuclei of Fe_3O_4 nanoparticles by the reduction effect of EG and glucose at high temperature (S. W. Cao et al, 2008); meanwhile, H_4SiO_4 precursors were also transformed into amorphous SiO_2 in the solvothermal process. The mass ratio of $NaSiO_3·9H_2O$ to $NaHCO_3$ could affect the distribution between $Fe(OH)_3$ and H_4SiO_4 precursors, so as to play an important role in the particle morphology of the as-prepared composite. That is to say, when the mass ratio of $NaSiO_3·9H_2O$ to $NaHCO_3$ was 2:1, the as-prepared $Fe(OH)_3$ precursors could arrange in a linear array, and were separated and coated by H_4SiO_4 precursors, so that one-dimensional core-shell structured $Fe_3O_4@SiO_2$ fine fibrous nanocomposite was prepared by the following solvothermal treatment. However, when the mass ratio of $NaSiO_3·9H_2O$ to $NaHCO_3$ was 1:1, the as-prepared $Fe(OH)_3$ precursors could agglomerate together in twos and threes, and were also separated and coated by H_4SiO_4 precursors, so that core-shell structured $Fe_3O_4@SiO_2$ spherical-

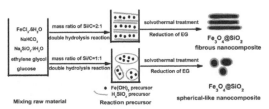

Figure 5. Schematic illustration of the mechanism of synthesis of morphology-controlled $Fe_3O_4@SiO_2$ core-shell nanocomposite.

like nanocomposite was prepared by the following solvothermal treatment.

4 CONCLUSION

In summary, in the new facile one-step solvothermal synthesis of $Fe_3O_4@SiO_2$ core-shell nanocomposite, the mass ratio of $NaSiO_3 \cdot 9H_2O$ to $NaHCO_3$ played an important role in the phase composition, crystallinity degree, and particle morphology of the as-prepared composite. Moreover, controllable synthesis of an one-dimensional core-shell structured $Fe_3O_4@SiO_2$ fine fibrous nanocomposite and a core-shell-structured $Fe_3O_4@SiO_2$ spherical-like nanocomposite was successfully implemented, and the formation mechanism of the synthesis method was also proposed.

REFERENCES

Abbas, M., B.P. Rao, M.N. Islam, S.M. Naga, M. Takahashi, Ceram. Int. 40(2014)1379–1385.

Cao, S.W., Y.J. Zhu, J. Chang, New J. Chem., 32(2008)1526–1530.

Caruana, L., A.L. Costa, M.C. Cassani, E. Rampazzo, L. Peodi, N. Zacceroni, Colloid Surface A., 410(2012) 111–118.

Chen, R.F., S.S. Song, Y. Wei, Colloid Surface A., 395(2012)137–144.

Fu, M.S., J.G. Li, Nanosci. Nanotechnol. Lett., 6(2014)1116–1122.

Hui, C., C. Shen, J. Tian, L. Bao, H. Ding, C. Li, Y. Tian, X. Shi, J. Nanoscale. 3(2011)701–705.

Mojic, B., K.P. Giannakopoulos, Z. Cvejic, V.V. Srdic, Ceram. Int. 38(2012)6635–6641.

Zhan, J.C., H. Zhang, G.Q. Zhu, Ceram. Int. 40(2014) 8547–8559.

Energy Science and Applied Technology ESAT 2016 – Fang (Ed.)
© 2016 Taylor & Francis Group, London, ISBN 978-1-138-02973-6

Synthesis of alginate bonded with zinc (II) phthalocyanine and their fluorescent property

Mingzhen Xu
School of Microelectronics and Solid-state Electronics, University of Electronic Science and Technology of China, Chengdu, China

Andrew Clive Grimsdale
School of Materials Science and Engineering, Nanyang Technological University, Singapore

Kun Jia & Xiaobo Liu
School of Microelectronics and Solid-state Electronics, University of Electronic Science and Technology of China, Chengdu, China

ABSTRACT: Zinc (II) phthalocyanine (ZnPc) has been employed as fluorophore in the synthesis of water-soluble alginate-phthalocyanine zinc (alginate-ZnPc) via aqueous-phase carbodiimide activation chemistry. The structures of the alginate-ZnPc macromolecular were characterized by FTIR and UV-vis. Also, the extent of modification of alginates by ZnPc was calculated as 4.8%. Fluorescent properties of ZnPc and alginate-ZnPc were also characterized.

Keywords: Zinc (II) phthalocyanine; alginate; fluorescence properties

1 INTRODUCTION

Liposomal zinc (II) phthalocyanine (ZnPc) has been proved to be very useful for photokilling of HeLa cells and tumor locatization in vivo and PDT (Photodynamic Therapy) of murine tumors, due to their strong absorption in the red (Q-band) which overlaps with the region of maximum light penetration in tissues (Gürol I et al, 2007) (Juan-José C et al, 2009). However, attributing to their unique π electronic structures which resulted to serious aggregation effects, the excellent biological functions of MPcs will not be functional. Additionally, it cannot be acceptable that the effects of MPcs in organisms are hard to predict during and after the aggregation. Lots of researchers have tried to distribute the MPcs in the silica and sol-gels, hydrophilic/hydropholic organic polymers, semiconducting organic polymer dots and so on (İlke Gürol et al, 2010). Hydrogel alginates, which belong to a family of copolymers containing 1, 4-linked β-D-mannuronic acid (M units) and α-L-guluronic acid (G units) that vary in both proportion and sequential arrangement, have been widely used as hydrogel synthetic Extracellular Matrices (ECMs) (Abu-Rabeah K et al, 2005). Considering of the fair biocompatibility of hydrogel alginate and the outstanding fluorescent property of zinc phthalocyanine, herein, we proposed a new hybrid molecule hydrogel where a new zinc phthalocyanine monosubstituted with 4-amino group (ZnAPc, Fig. 1) is linked to the alginate matrix. The structures of the new macromolecules were investigated, as well as their fluorescence properties.

2 EXPERIMENTS PART

Chemicals obtained from commercial sources were of analytical grade and used without further purification. Sodium alginate (from brown algae, viscosity of the 1% (w/v) solution is 20cp), 1-ethyl-[3-(dimethylamino)propyl]-3-ethyl-carbodiimide HCl (EDAC; E-1769), 2-[N-morpholino]-ethanesulfonic acid (MES buffer, M-8250), N-Hydroxysulfosuccinimide (NHSS, 24510) were acquired from Sigma-Aldrich Chemical company (Singapore). All other starting materials, reagents and solvents used in the project were purchased from commercial sources (e.g. Sigma/Aldrich, Alfa Aesar, and so on), and used as obtained. 4-(p-tolyloxy) phthalonitrile and [2-amino-10, 16, 24-tri-4-(p-tolyoxy) phthalonitrile phthalocyaninato]

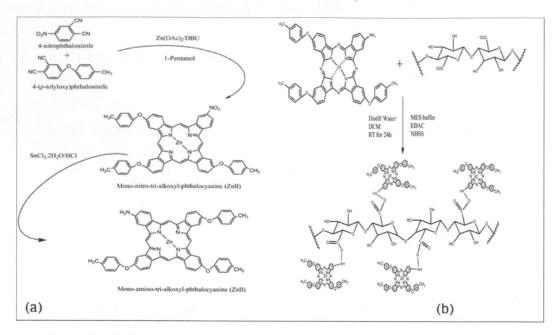

Figure 1. Synthetic route of (a) nitro- and amino-substituted phthalocyanine zinc (II) and alginate—phthalocyanine zinc.

zinc(II) were synthesized in our lab. Alginate bonded with zinc (II) phthalocyanine was achieved by aqueous-phase carbodiimide activation chemistry. The structures of the new macromolecules were investigated by FTIR, UV-vis and their fluorescence were characterized by spectrofluorophotometer (Shimadzu RF-5301PC).

3 RESULTS AND DISCUSSION

The chemical composition of the ZnAPc, alginate and ZnAPc-alginate product were characterized by FTIR, shown in Fig. 2. It is possible to observe that ZnAPc (shown in Fig. 2 (a)) presents characteristic absorption bands at 3433 cm^{-1} and 2929 cm^{-1} attributed to the NH$_2$ and CH$_3$ groups, respectively. The characteristic absorption bands assigned to the phthalocyanine ring frame C-N and Zn-N vibration are observed at 1088 cm^{-1} and 737 cm^{-1}, respectively (Juan-José C et al, 2009). Origin alginate (shown in Fig. 2(b)) shows two characteristic absorption bands around 1616 and 1421 cm^{-1}, corresponding to the asymmetric and symmetric stretching vibration of the COO$^-$ groups, respectively. The band assigned to C-O-C stretching occurs around 1031 cm^{-1} and a large absorption band at 3398 cm^{-1}, attributed to the stretching vibration of OH groups. For the spectrum of ZnAPc-alginate product shown in Fig. 2(c), the characteristic features are strong

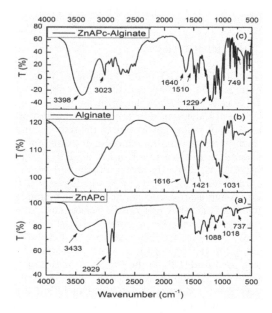

Figure 2. FTIR spectra of (a) ZnAPc, (b) origin sodium alginate and (c) ZnAPc-alginate conjugate.

sharp peaks at 1640 cm^{-1} and 1510 cm^{-1}, which corresponded to the amide I band, C = O stretching vibration, and amide II band, N-H bending vibrations, respectively. A strong-sharp band at 1229 cm^{-1} (amide II band, interaction between the

Table 1. Relative date collected from the UV measurements.

Item	Samples	3.15E-6M	6.3E-6M	9.45E-6M	1.26E-5M	1.575E-5M	ZnAPc-Alginate
Absorption	611 nm	0.0105	0.0145	0.0175	0.02057	0.02457	0.0474
Intensity	676 nm	0.01025	0.0145	0.0183	0.0209	0.02454	0.0449
Equation					$y = a + b*x$		
Adj. R-square					0.91585		
Slope					1426.35		

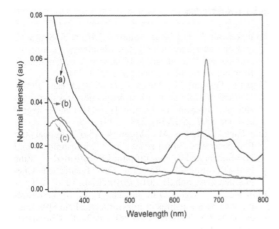

Figure 3. UV-vis bsorption spectra of (a) ZnAPc-Alginate in distill water (0.1 w/v%), (b) origin sodium alginate dissolved in distill water (0.1 w/v%) and (c) ZnAPc dissolved in DMF (10^{-9}M).

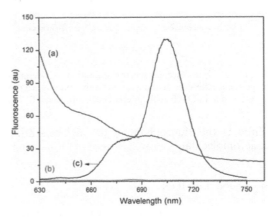

Figure 4. Fluorescence spectra of (a) ZnAPc-Alginate dissolved in distill water (0.1 w/v%), (b) origin sodium alginate in distill water (0.1 w/v%) and (c) ZnAPc dissolved in DMF (10^{-9}M), λ_{exc} = 590 nm.

N-H bending and C-N stretching vibrations) is also observed (Abu-Rabcah K et al, 2005).

The absorption spectrum of ZnAPc-alginate product was performed to compare with that of ZnAPc monomers shown in Fig. 3. It can be observed the typical Soret (320–400 nm) and Q-bands (600–750 nm) bands characteristic of ZnPc in both Fig. 3(a) and (c), which ascribed to ZnAPc-alginate and ZnAPc (İlke Gürol et al, 2010). In comparison with that of pure alginates (Fig. 3(b)), the Q band at 674 nm of ZnAPc-alginate shown in Fig. 3(a) is obvious and wider than that of ZnAPc, indicating the unsymmetrical bond of ZnAPC and the alginates.

The extent of modification of alginate by ZnAPc was evaluated by UV-vis absorption spectroscopy, which is a useful technique for detection and quantitative measurement of chromophores that undergo n→π* or π→π* transitions. The characteristic π→π* transition occurs in ZnAPc ring present in the ZnAPc-alginate polymer, allowing quantification of the amount of ZnAPc by UV-vis absorption spectroscopy at 611 nm. Several samples within a concentration range (50–250μM) of ZnAPc dissolved in the alginate solution (0.1 w/v%)

was measured by UV absorption at 611 nm and 676 nm, respectively. The data collected from the UV absorption measurements (shown in Table 1) had a very good linear relationship at the function of the studied concentration (R^2 = 0.91585). The extinction coefficient of 1426.35M^{-1} cm^{-1} for ZnAPc-alginate conjugate was calculated from the slope of the calibration curve. The average value of the degree of alginate modification obtained was about 4.8% of weight modification at the reaction condition. The relative low bonding yield may be ascribed to the poor solubility of ZnAPc in water.

The photoluminescence properties of obtained ZnAPc monomer and ZnAPc-alginate conjugate in solution were studied and presented in Fig. 4 and 5. As shown, the maximum excitation wavelength of ZnPc is detected at 590–700 nm, while the alginates-ZnPc shows the maximum fluorescent intensity in the range from 560 to 750 nm. When these two solutions are excited at 590 nm, the maximum emission wavelength of ZnPc and alginates-ZnPc is 708 nm and 695 nm, respectively (Fig. 4). Fig. 5 shows the dependency of the fluorescence properties of alginates-ZnPc on the excitation wavelength, indicating the fact that

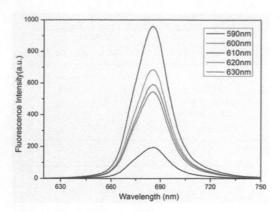

Figure 5. Fluorescence spectra: excitation wavelength-dependent emission of mono-amino substituted phthalocyanine zinc in DMF with different concentration: 10^{-6}M.

ZnPc is the fluorophore in the alginates-ZnPc macromolecular.

4 CONCLUSION

Water-soluble alginate bonded with phthalonitrile zinc (alginate-ZnPc) was achieved via the aqueous-phase carbodiimide activation chemistry. Results of UV-vis tests show the characteristic absorption of ZnPc at 600–750 nm for alginate-ZnPc macromolecular, and the extent of modification of alginate by ZnAPc was evaluated. The average value of alginate modification obtained was about 4.8% of weight modification at the reaction condition and the alginate-ZnPc shows expected photoluminescence properties, due to the fact that ZnPc acts as the fluorophore.

REFERENCES

Abu-Rabeah K., Polyak B., Ionescu R.E., et al. Synthesis and characterization of a pyrrole-alginate conjugate and its application in a biosensor construction[J]. Biomacromolecules, 2005, 6(6): 3313–3318.

Gürol I., Durmuş M., Ahsen V., et al. Synthesis, photophysical and photochemical properties of substituted zinc phthalocyanines. Dalton Trans[J]. Dalton Transactions, 2007, 2007(34): 3782–3791.

İlke Gürol, Durmuş M., Ahsen V. Photophysical and Photochemical Properties of Fluorinated and Nonfluorinated n—Propanol-Substituted Zinc Phthalocyanines[J]. Berichte Der Deutschen Chemischen Gesellschaft, 2010, 2010(2010): 1220–1230.

Juan-José C., Miguel G.I., Jun-Ho Y., et al. Structure-function relationships in unsymmetrical zinc phthalocyanines for dye-sensitized solar cells.[J]. Chemistry, 2009, 15(15): 5130–7.

4 Environmental engineering and sustainable development

Energy Science and Applied Technology ESAT 2016 – Fang (Ed.)
© 2016 Taylor & Francis Group, London, ISBN 978-1-138-02973-6

Evaluation method of the fast-charging infrastructure service capability based on the demand of electric vehicles

L. Xia, C. Liu, B. Li & K. Li
Beijing Electric Vehicle Charging/Battery Swap Engineering and Technology Research Center, China Electric Power Research Institute, Beijing, China

L. Yu
School of Control and Computer Engineering, North China Electric Power University, Beijing, China

ABSTRACT: Considering the problem of whether service capability of urban public fast-charging infrastructure can meet the demand of Electric Vehicle (EV) charging, a comprehensive evaluation method is proposed in this paper. First, the demand of EV charging in public places is analyzed. Second, according to the demand of EV charging and distribution of charging infrastructure, a service capability evaluation model for EV fast-charging infrastructure is built. Finally, the service capability of EV fast-charging infrastructure of each hour in a day is obtained based on the data of EV and fast-charging infrastructure. Based on this evaluation method, Beijing acts a sample city. According to Beijing's planning number of public charging infrastructure and EVs in 2017, the current service capability of public fast-charging infrastructure in Beijing is calculated and acquired.

Keywords: electric vehicle; fast charging; service capability

1 INTRODUCTION

Recently, the government issued a series of supporting policies to promote the development of the electric vehicle industry. The electric vehicle ownership maintains a high growth (Z.R, 2015, C.W Gao, 2011). Electric vehicle charging network as a supporting infrastructure is also rapidly developing (W. Su, 2012, K. Sun, 2013, T.T. Lu, 2014). Yet, the battery life has not achieved a substantial breakthrough, to promote the development of electric vehicles, and a larger number of charging infrastructure need to be relied on (X.N. Xiao, 2014, J.B. Xin, 2010, X.Y. Zhao, 2012). On the one hand, under the requirement that the battery lifecycle cannot be affected significantly and the distribution network capacity would meet power demand, public fast-charging mode can provide temporary charging service for electric vehicles, and significantly improve charging network service capability. On the other hand, a research has to be conducted on the evaluation method and simulation technology of fast-charging service capability of electric vehicles, the results of which will provide feedback to the existing fast-charging infrastructure and services. It is particularly important as guidance for the development of electric vehicle public fast-charging infrastructure.

A related research on electric vehicle charging network service capability has been carried out at home and abroad. In a paper by W.C. Liu (2014), by analyzing and defining the mechanisms and associations of each component, from the perspective of the overall evaluation, 7 evaluation systems of electric vehicle demonstration are built, which comprised a supporting infrastructure system, operating data acquisition system, etc. However, it only mentions the evaluation framework, and further in-depth discussion is not proposed. In a paper by B.W. Li & C.L (2009), using the Delphi and Analytic Hierarchy Process (AHP) method, the feasibility evaluation index system of the electric vehicle project is constructed based on the geographical location, economic effects, and so on, yet without any actual data to estimate. A paper by S.X. Wang (2013) combines the balanced score card theory and enterprise ecology to build the evaluation index system for the business mode of electric vehicle charging services, yet without considering the real charging demand on the premise of saving the cost of the construction of infrastructure.

Considering the fast-charging demand, the capacity of charging infrastructure configuration, and other factors, this paper proposes an evaluation method of service capability of urban public fast charging for electric vehicles. The service capability of urban public fast-charging infrastructure is obtained by simulation and calculation. It is a quantitative assessment method, which can be used for the promotion of the service capability of urban public fast-charging network, ensuring working efficiency, and accelerate the development of electric vehicle industry.

2 CHARGING DEMAND ANALYSIS OF ELECTRIC VEHICLES

By estimating the index allocation of Beijing new energy vehicles, the total number of Beijing electric passenger vehicles is expected to reach 170000 in 2017. Electric passenger vehicles are mainly composed of commercial electric vehicles and private electric vehicles. Take electric passenger vehicles as the research object, with the commercial electric vehicles and private electric vehicles ratio of 1:1, i.e. 85000 as the number of each vehicle type, to build a model and calculate, respectively.

2.1 Charging demand analysis of commercial electric vehicles

The property of the commercial electric vehicle determined that it could be used at any moment during the working time, and a part of the commercial electric vehicles will also be driven during non-working time. For commercial electric vehicles, the charging probability is constant in every moment of working time, and the charging probability will be declined during non-working time. Thus, the charging probability coincides with a trapezoidal distribution within 24 hours a day. $P_{comm}(t)$ expresses the probabilities to start charging in each hour, as is shown in Table 1. The probability distribution is shown in Figure 1.

Table 1. The probability of commercial vehicles starts charging in public facilities per hour.

Time	0	1	2	3	4	5
P_{comm}	0.0125	0.011	0.011	0.011	0.0125	0.026
Time	6	7	8	9	10	11
P_{comm}	0.04	0.056	0.056	0.056	0.056	0.056
Time	12	13	14	15	16	17
P_{comm}	0.056	0.056	0.056	0.056	0.056	0.056
Time	18	19	20	21	22	23
P_{comm}	0.056	0.056	0.056	0.04	0.026	0.0125

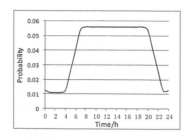

Figure 1. The charging start probability distribution of a commercial electric vehicle.

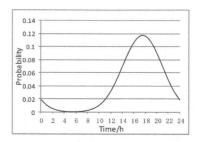

Figure 2. The charging start probability distribution of a private electric vehicle.

2.2 Charging demand analysis of private electric vehicles

For private electric vehicles, it is assumed that the user takes an electric vehicle to start charging right after work at about 17:30. The charging start probability in each hour coincided with the normal distribution, where the mean value $\mu = 17.5$, and the standard deviation $\sigma = 3.4$. The charging probability distribution of private electric vehicle is expressed as $P_{priv}(t)$, and is shown in Figure 2:

$$P_{priv}(t) = \begin{cases} \dfrac{1}{\sigma\sqrt{2\pi}} e^{-\frac{(t-\mu)^2}{2\sigma^2}} & \mu - 12 < t \le 24 \\ \dfrac{1}{\sigma\sqrt{2\pi}} e^{-\frac{(t+24-\mu)^2}{2\sigma^2}} & 0 < t \le \mu - 12 \end{cases} \quad (1)$$

2.3 Total charging demand analysis

Assuming that commercial vehicles and private vehicles accept charging service in the same charging system, the probability of vehicles starting charging per hour is expressed as:

$$P(t) = P_{comm}(t)\frac{n_{comm}}{n_{comm} + n_{priv}} + P_{prvi}(t)\frac{n_{priv}}{n_{priv} + n_{comm}} \quad (2)$$

According to the central limit theorem, the number of vehicles starting charging per hour is given in (3), and as is shown in Figure 3:

$$n_{car}(t) = P(t)(n_{comm} + n_{priv}) \quad (3)$$

After combining the number distribution of commercial vehicles and private vehicles, the number of vehicles that accept service from charging infrastructure is peaking at 18:00, i.e. 6618/h, the service capability of charging infrastructure is minimum at this time. The number of vehicles that accept service from charging infrastructure is bottoming at 3:00, i.e. 514/h, and the service capability of charging infrastructure is maximum at this time.

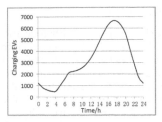

Figure 3. The number of all the vehicles reaching the charging station per hour.

3 SERVICE CAPABILITY EVALUATION MODEL OF FAST CHARGING BASED ON CHARGING DEMAND

3.1 *The influence factor of the service capability of charging*

The service capability of EV public charging infrastructure is defined as the capability of public charging infrastructure configuration meeting EV public charging demand. According to the characteristics of EV charging demand, the influence factors of the service capability of EV public charging infrastructure for passenger vehicle are the two aspects of EV and charging infrastructure, as shown in Table 2.

3.2 *Service capability evaluation model of fast charging*

The electric vehicle charging service capability model can be divided into the charging infrastructure supply model and charging demand model. In addition, the charging infrastructure supply model can be divided into slow charging and fast charging.

The charging infrastructure supply model

The charging infrastructure supply model is expressed as:

$$C = C_n / C_t \qquad (4)$$

where C is the allowed maximum charging times per hour for public charging infrastructure, C_n is the number of charging spots, and C_t is the average charging time length.

For equation (4), when the charging infrastructure is AC charger (for slow charging), C_{nA} is defined as the corresponding number of AC charger, C_{tA} is defined as the average charging time, and C_A is defined as the allowed maximum charging times per hour for AC charging. When the charging infrastructure is DC charger (for fast charging), the corresponding symbolic representations are C_{nD}, C_{tD}, and C_D, respectively. Assuming that both kinds of charging spots have the same serviceability rate, α:

$$C = \alpha \frac{C_{nA}}{C_{tA}} + \alpha \frac{C_{nD}}{C_{tD}} = C_A + C_D \qquad (5)$$

Table 2. The influence factors of the service capability of EV charging infrastructure.

Aspects	Factors	Influence
EV	The number of EV	The larger the number, the larger the charging demand
	The driving mileage	The larger the driving mileage on a single charge, the lower the charging demand
Infrastructure	The number of charging infrastructure	The larger the number, the better the service capability
	The charging time	The shorter the charging time, the better the service capability

Table 3. Coefficients of charging possibility function.

Area	a	b	c	d
Office	0.007 483	1.289	0.019 61	3.95
Commercial	0.012 720	2.474	1.528(10⁵	10.95
Recreational	0.025 760	2.566	2.244(10⁷	14.86

The charging demand *model*

The average charging frequency per day for an electric vehicle is given by:

$$V_t = V_{kday} / (V_{kmax} - V_{dump}) \qquad (6)$$

where V_t is the average number of times that an electric vehicle charges in a day, V_{kmax} is the average driving mileage with full energy support, V_{dump} is the average driving mileage with dump energy support before recharging, and V_{kday} is the daily average driving distance of each electric vehicle.

A paper by (J. Smart & S. Schey. 2012) fitted the electric vehicles in urban different areas (office, commercial, recreational) to obtain the charging demand probability p_c:

$$p_c(x) = ae^{bx} + ce^{dx} \qquad (7)$$

where x is the SOC when a user needs charging.

From equation (7), the fitting coefficients a, b, c, and d are obtained, as shown in Table 3.

$$V_{tp} = (p_{c\text{-}w} + p_{c\text{-}s} + p_{c\text{-}r})V_t \qquad (8)$$

where V_{tp} is the average charging number of using public charging infrastructure of each vehicle every day, $p_{c\text{-}w}$, $p_{c\text{-}s}$, $p_{c\text{-}r}$ is the probability of using public charging infrastructure in the office, commercial, and recreational areas, respectively.

The users usually charge at home during 22:00–07:00 (next day), and charge at public infrastructures during 07:00–22:00. Assume that the ratio of the number of charging at time t and the number of total charging is defined as $\varphi(t)$, where the proportions of fast and slow charging are defined as $\varphi(t)_D, \varphi(t)_A$ respectively. In addition, $\varphi(t) = \varphi(t)_D + \varphi(t)_A$.

Since the number of fast charging is C_{tD}/C_{tA} times the number of slow charging within the same time period, then

$$\varphi(t)_D = \frac{C_{tA}}{C_{tA} + C_{tD}} \varphi(t) \qquad (9)$$

The total number of electric vehicle charging is defined as V_n, the number of fast-charging demand at time t is obtained:

$$V(t) = \varphi(t)_D \cdot V_{tp} \cdot V_n \qquad (10)$$

The service capability of fast-charging model

$$E(t) = \frac{C_D}{V(t)} \qquad (11)$$

where $E(t)$ is the service capability of fast charging at time t.

4 CALCULATION AND ANALYSIS

According to "Beijing traffic development annual report in 2013", the annual average driving distance of electric passenger vehicle is 11, 803 km, thus the daily average driving distance $V_{kday} = 32.3$ km. According to the main electric vehicle performance index at the current domestic market, the average driving mileage $V_{kmax} = 158$ km. Charging is started at the average SOC of 0.55, and the corresponding average driving mileage with dump energy support before recharging $V_{dump} = 86.6$ km. The average time of slow charging $C_{tA} = 3.14$h, and the average time of slow charging $C_{tD} = 1$h. The Beijing government is planning to build a fast-charging pile of about 10000 in 2017, and the ratio of DC and AC charging pile is about 1:1, i.e., $C_{nA} = C_{nD} = 10000$. Based on the above data, the result of charging service capability in Beijing is given in Table 4.

The minimum value of fast-charging service capabilities within a day reflects how the charging demand is met. The minimum value of fast-charging service capability in Beijing is 5.75 within a day, which indicates that even there is a peaking time, and the public fast-charging infrastructure can supply charging service 5.75 times the current demand from electric passenger vehicles.

Table 4. Fast-charging service capability in Beijing.

Equipment	Service capability C	Service capability $E_r(t)$		
		Minimum	Average	Maximum
Fast charger	12876.56	5.74	41.89	175.45

5 CONCLUSION

Taking the characteristics of urban electric passenger vehicles and the actual public charging infrastructure into account, from the perspective of the balance of supply and demand, and considering various factors, this paper proposes an evaluation and simulation method of service capability of public fast-charging infrastructure for urban electric vehicles. From the evaluated and simulated results, Beijing's planning on charging infrastructure will provide adequate public fast-charging service in 2017 to meet the demand from electric passenger vehicles.

ACKNOWLEDGMENT

This work was financially supported by the Research Project of SGCC (YD71-15-032).

REFERENCES

Gao C.W. & L. Zhang. 2011. A Survey of Influence of Electrics Vehicle Charging on Power Grid (J). Power System Technology, 35(2):127–131.

Li B.W. & C.L. Yang. 2009. Research on Evaluation Method of Extent of Feasibility of Electric Vehicle Project (J). Technology Economics, 2009(9):78–82.

Liu W.C., & M.J. Zhou. 2014. Electric Vehicle Regional Demonstration Evaluation System (J). Automobile Applied Technology, 2014(3):18–21.

Liu, Z.R., J. Chen, K. Lin, Y.J. Zhao & H.P. Xu. 2015. Domestic and Foreign Present Situation and the Tendency of Electric Vehicle (J). Electric Power Construction, 36(7):25–32.

Lu T.T. & C.W. Gao. 2014. A General Model for Optimal Scheduling of Battery Charging and Renewal Network (J). Power System Technology, 38(10):2700–2707.

Smart J. & S. Schey. 2012. Battery Electric Vehicle Driving and Charging Behavior Observed Early in the EV Project (J). SAE International Journal of Alternative Powertrains, 1(1):27–33.

Su, W., E.H Rahimi & W. Zeng. 2012. A Survey on the Electrification of Transportation in a Smart Grid Environment (J). IEEE Transactions on Industrial Informatics, 8(1):1–10.

Sun, K., M.W. Peng & C.Y Song. 2013. A Requirement Analysis Model of Charge and Battery Replacement Service Network for Electric Vehicles (J). Power and Energy, 34(6):650–654.

Wang. S.X., R. Rao & Y.H. Song. 2013. Research on Comprehensive Evaluation of Business Mode of Electric Vehicle Charging Service (J). Modern Electric Power, 30(2):89–94.

Xiao, X.N., J.F. Wen, S. Tao & Q.S. Li. 2014. Study and Recommendations of the Key Issues in Planning of Electric Vehicles' Charging Facilities. Transactions of China Electrotechnical Society, 29(8):1–10.

Xin, J.B., Y.B. Wen & R. Li. 2010. Discussion on Demand Forecast Method for Vehicle Charging Facilities. Jiangxi Electric Power, 34(5):1–5.

Zhao X.Y. & Y.Q. Zhao. 2012. Research on Charging Infrastructure Layout for Electric Vehicles under Smart Grid. Power System and Clean Energy. 28(11):61–64.

Energy Science and Applied Technology ESAT 2016 – Fang (Ed.)
© *2016 Taylor & Francis Group, London, ISBN 978-1-138-02973-6*

Study of the influence of coal exploitation on groundwater resources and the environment in the Hanglaiwan mine of Northern Shaanxi

Zhenyu Dong & Shuangming Wang

College of Environmental Science and Engineering, Chang'an University, Xi'an, China

ABSTRACT: The study of influence of coal exploitation on groundwater and ecological environment is the focusing subject in the process of energy development, especially for the region of Northern Shaanxi. Using the coal exploiting area in Hanglaiwan coal mine as an example, based on analysis of hydrological geology condition, this paper ascertains the aquifer that fracture gets through by calculating the height of caving and fissure zone. This paper analyzed the influence of coal mining on the groundwater resources and environment quantitatively and qualitatively through calculation and test of water sample. The calculation results of static reserve and dynamic reserve show that the loss of groundwater is great and the aquifer is gradually drained out. The results of test show that, at present, coal mining has little impact on water quality.

Keywords: coal mining; groundwater; Hanglaiwan coal mine; Northern Shaanxi; static reserve; dynamic reserve

1 INTRODUCTION

In China, there are abundant coal resources and coal is the main energy source and a very important industrial raw material. In China's primary energy structure, oil accounts for 20.4%, natural gas accounts for 3%, hydropower and renewable energy accounts for 7.2%, and coal accounts for 76.7% (Shuang-ming Wang et al, 2010). The proportion of coal is far greater than other energies. With the rapid development of China's economy and society, production of coal resources increased year by year. In 2006, coal production was 2.37 billion tons, with year-on-year growth of 8.0%; in 2007 coal production was 2.53 billion tons, with year-on-year growth of 6.9%. In 2015, coal production reached 3.5 billion tons (Zhong Huang et al, 2010). "The 13th Five-year Period" is a critical stage to build a well-off society in China. Therefore, as the driving force for economic development, energy consumption is still to maintain sustained and rapid growth trend. Under the premise to achieve the stated objectives of economic development, it is estimated that the energy gap will reach more than 660 million tons of coal in 2020 (Yan-de Dai et al, 2015). Coal plays an important role in economic and social development, but at the same time, coal mining had a negative impact which cannot be ignored to natural environment. For example, in Northern Shaanxi, the coal industry had developed rapidly since 1989, especially after 2000, raw coal production increased from less than 50 million tons to 338 million tons in 2013

(Shuang-ming Wang et al, 2010). However, the large-scale, high-intensity mining project resulted in a significant decline of the groundwater level, in coal mining area and the surrounding area, and so many springs and rivers dried up, and there are many other problems happening such as geological environmental variation, ecological environmental pollution and destruction, and other things (De-xin Zhang, 2000) (Li-min Fan, 2015).

2 THE RESEARCH STATUS OF COAL MINING IMPACTS ON GROUNDWATER

Under conditions of coal mining, a study of the influence on groundwater has made many achievements. Han boping systematically analyzed the relationship between influencing factors, phenomena, and harmfulness of the hydrogeological problems in the coal mining process. He established a hydrogeological conceptual model under the disturbance of coal mining, which had a great significance of explaining the effect of hydrogeology related to this and the way of evolution and strength under the condition of coal mining (Bao-Ping Han et al, 1994). Cao Zhiguo revealed the groundwater migration law of coal mining in Shendong mining area by using geophysical exploration and drilling observation, model experiments, numerical simulation, and other means (Zhi-guo Cao et al, 2014). Under the condition of coal mining, Zhang Shukui made a regionalization of the influence of groundwater on the coal mining area in the middle reaches of the Yellow River

(Shu-kui Zhang et al, 2015). Li Zhenshuan and SHI Hong calculated the destruction of groundwater reserves by using the evaluation method of dynamic and static reserve, and put forward the rationalization proposals and countermeasures for the rational utilization of water resources in Shanxi Province (Zhen-shuan Li, 2007) (Hong Shi et al, 2011). Szzeppansk J pointed out that the groundwater pollution of coal mining is the main aspect of groundwater destruction, and in a very long period of time, it is difficult to eliminate the hazards (Szzeppansk J et al, 1999). Wu Qiang, Dong Donglin studied the effects of mining on river runoff and water, and analyzed the formation mechanism of acidic water in Xishan coal mine (Qiang Wu et al, 2002). Deng Qiang wei discussed the influence of coal mining on groundwater level and water by using simulation of mathematical models in Daheng coal mine (Qiang-wei Deng et al, 2014).

3 HYDROGEOLOGICAL CHARACTERISTICS OF HANGLAIWAN COAL MINE

According to storage conditions and hydraulic characteristics, the Hanglaiwan mine groundwater aquifer can be divided into two types: pore and pore fissure water in Quaternary loose rock, fissure and water in clastic rock.

3.1 Pore and pore fissure phreatic water in quaternary loose rock

Holocene valley alluvium pore water. It is first distributed in terraces and overbank of river in the southern minefield. Valley sides for poor water Liang Mao loess area have poor lateral supply condition. It is water-poor loess hilly region on both sides of the valley; therefore, the conditions of lateral recharge are poor. Stacking thickness of sand and gravel layer is thin, usually 1.50–7.85 m. The lithology of the upper aquifer is silty sand. The lower part is sand, gravel, and fine sand. The aquifer has good permeability and is weakly water rich.

Pleistocene series in the lacustrine formation pore water (aquifer in Sala Wusu Group). Aquifer has planar continuous distribution in desert region. In northern and southern minefields, the terrain is conducive to surface water supplies, and the aquifer is composed of loose silty sand and coarse sand, and has larger thickness, generally 20–50 m. It is medium-strong and water rich.

Quaternary Pleistocene loess pore fissure water. It is distributed all over the minefield, mostly in the local area in the north and south of minefield buried beneath Salawusu Formation; thickness is 9.80–120.49 m, generally 40–80 m. Aquifer lithology is mainly for the powder soil loess, local central location with silty sand layer, and the thickness is 9.80–90.00 m in general. Loess aquifer is weakly water rich.

3.2 Fissure water in clastic rock

The Jurassic clastic rock weathering crust fracture confined water. It is distributed all over the minefield and concealed under the loess of quaternary pleistocene. The aquifer is in the bedrock weathering fracture belt, its thickness is approximately 20 m. As bedrock weathering belt is covered by large thick loess aquifer, so the belt of weathering fissure water is confined and weakly water rich.

Clastic rock fracture-confined water. It is divided into two water-bearing sections by No. 3 coal seam. The upper is distributed from the No. 3 coal seam to the belt or weathering fractured bedrock, including each lithologic section in J_2z, J_2y^4, and its thickness increases from east to west. The aquifer is mainly composed of "Seven Mile town sandstone" in J_2z and "Zhenwu hole sandstone" in J_2y^4, rock fractures are not developed, and it is weakly water rich. According to the pumping test, the water depth is 6.20–66.20 m, and the thickness of aquifer is 15.04–69.10 m, when the drawdown is 31.25–50.88 m, water inflow is 1.607–32.31 m³/d, and the permeability coefficient is 0.0013–0.014 m/d. The lower portion is distributed between the No. 3 coal seam and the lower boundary of Yanan; it is composed of light gray siltstone, fine sandstone, and dark gray mudstone. Aquifer with complete rock is thin and buried deep, its cracks are not developed, and it is weakly water rich.

3.3 Aquiclude

Aquiclude is located in the coal between two fracture confined water, alternates with the aquifer, and has continuous distribution. Aquiclude is mainly composed of mudstone, silty mudstone, siltstone and argillaceous siltstone, and its thickness is 10–40 m generally.

4 CALCULATION OF CAVING AND FISSURE ZONE IN HANGLAIWAN COAL MINE

In Hanglaiwan coal mine, No.3 coal seam is the main mining coal, the thickness of coal seam is 8.27–10.66 m, and the average mining height is 9.13 m. Immediate roof is composed of siltstone, mudstone, and a small amount of fine-coarse grain arkoses, it is medium-hard rock. According to "Regulation of buildings, water bodies, railways and coal pillar retention of the main roadway and coal mining", the formula of medium-hard rock

Table 1. Relationship between aquifers and caving and fissure zone.

Region	Drills number	Coal seam Thickness/m	Coal floor depth/m	Height of caving and fissure zone/m	Depth of clasolite fractured zone/m	Distance from caving and fissure zone to clasolite fractured zone/m	Distance from caving and fissure zone to weathering bedrock fractured zone/m
First mining area	26	6.72~11.90	200.85~241.00	67.90~76.24	112.18–151.74	–14.05~–1.12	47.64~113.45

which is to calculate the height of caving and fissure zone is as follows:

$$H_c = \frac{100\sum m}{4.7\sum m + 19} \pm 2.2 \qquad (1)$$

$$H_f = \frac{100\sum m}{1.6\sum m + 3.6} \pm 5.6 \qquad (2)$$

where H_c is the height of caving zone, m; H_f is the height of fissure zone, m; $\sum m$ is the cumulative mining height, m.

In first mining area of Hanglaiwan minefield, the heights of caving and fissure zone of 26 drillings is calculated, and the calculation results are shown in Table 1. From Table 1, it can be seen that the height of caving and fissure zone is bigger than the distance between caving and fissure zone and clasolite fractured zone, but does not reach the height of weathering bedrock fractured zone. So, it is concluded that the caving and fissure zone which is formed by roof strata collapsed has broken through clastic rock fracture confined aquifer and becomes the mine water filling channel. The clastic rock fracture-confined aquifer is the main aquifer that is broken by coal mining.

5 THE INFLUENCE OF COAL EXPLOITATION ON GROUNDWATER AND ENVIRONMENT IN HANGLAIWAN COAL MINE

5.1 Calculation of the amount of loss of groundwater resources

Computing method of static reserve of groundwater. After coal mining, the collapse of roof would undermine the aquifer overlying the seam. Then, groundwater of aquifer is emptied along the cracks and this part is called static reserve of groundwater. The destruction of static reserves is permanent, and its quantity is a fixed amount related to the character of the aquifer.

Static reserve calculation formula:

$$Q_J = \sum_{i=1}^{n} H_i \cdot S_i \cdot \mu_i \qquad (3)$$

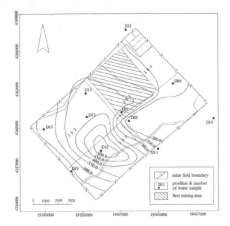

Figure 1. The position of first mining area and water samples.

Q_J is the static reserve of groundwater after coal mining, 10^4 m^3; H_i is the thickness of aquifer, m; S_i is the acreage of worked-out section, 10^1 m^2; and μ_i is the specific water yield.

Computing method of dynamic reserve of groundwater. Under the conditions that the aquifer is damaged, the mine water inflow increases rapidly first, then it tends to be stable gradually over time, and finally reaches a relatively stable quantity, it is called dynamic reserve. The destruction of dynamic reserves is also permanent, and its quantity is variable and affected by topography, structure and precipitation, depth of coal seam and coal mining methods, and other factors.

Dynamic reserve calculation formula:

$$Q_D = \sum_{i=1}^{n} S_i \cdot M_i \qquad (4)$$

Q_D: dynamic reserve of groundwater after coal mining, 10^4 m^3/h.

M_i: modulus of groundwater runoff after coal mining, m^3/h·m^2.

In Hanglaiwan mine field, the length and width of the first mining area are 5.6 km and 5.2 km. Its position is shown in Fig. 1. The thickness of the fissure-confined aquifer of the clasolite is approximately

Table 2. Calculation parameters and result of underground water loss.

Region	Length/km	Width/km	$S_i/10^4\,m^2$	H_i/m	μ_i	$M/\,m^3/h\cdot m^2$	$Q_J/10^4\,m^3$	$Q_D/10^4\,m^3/h$
First mining area	5.6	5.2	2912	42.06	0.12	0.464×10^{-7}	14697.45	1.35

Table 3. Result of chemical analysis (mg/L).

Number	Hg	As	Cr^{6+}	F^-	NO_3^-	NO_2^-
D01	0.000049	0.00606	/	0.11	19.11	/
D02	0.000023	0.00919	/	0.38	3.45	/
D06	/	0.00167	/	0.1	1.83	/
D10	0.000054	0.00392	0.004	0.12	8.6	/
D11	0.00019	0.00242	/	0.21	12.7	/
D12	0.000055	0.00063	/	0.23	2.58	/
D13	0.000027	0.00081	0.004	0.06	14.57	/
D14	0.00011	0.0022	0.006	0.22	3.44	/
D15	0.000022	0.00016	/	0.06	39.26	0.91
D16	0.000022	0.00417	0.004	1.53	2.65	/
D17	0.00001	0.00543	0.004	1.54	1.86	/
D18	0.00012	0.00328	0.004	0.12	5.44	/
D19	0.000012	0.00182	0.006	0.08	2.03	/
limits	≤0.001	≤0.01	≤0.05	≤1.00	≤20	≤0.02

42.06 m. According to the test data, the other relevant parameters and the calculation results of static reserve and dynamic reserve are shown in Table 2.

5.2 The influence of coal exploitation on groundwater quality

In Hanglaiwan minefield, 19 water samples were collected, and they had a full analysis of water quality. The location and number of sampling points is shown in Figure 1 and the main results of water sample analysis are shown in Table 3.

According to the results from the analysis of water samples, it is found that, in Hanglaiwan mine, the cations of shallow groundwater is mainly Ca^{2+}, followed by Na^+ and Mg^{2+}, anion is mainly HCO_3^-, and the main water chemistry type is HCO3-Ca and widely distributed in the region. The test results of groundwater samples of 13 monitoring indicators (PH; Fe; Mn^{2+}; Cl^-; SO_4^{2-}; TDS; $CaCO_3$; COD_{Mn}; Hg; As; Cr^{6+}; F^-; NO_3^-; NO_2^-) show that the total number of bacteria in water samples generally exceed the limit, especially the coli of D10 and D14. Compared with "groundwater quality standards" (GB /T14848-93), the monitoring indicators of most water samples collected meet class III water standard; therefore, water quality is generally good. But some indicators of three water samples exceed the limit, as follows: Fe and fluoride of D17, nitrate and nitrite of D15, fluoride of D16; they does not meet the Class III water standard, so the groundwater cannot be directly used as drinking water and it needs appropriate treatment.

6 CONCLUSION

In the first mining area of Hanglaiwan mine, the caving and fissure zone will communicate clastic rock fracture-confined water after mining and aquifer of Sarah Wusu group, which is strongly water rich, is not communicated. Groundwater damage of exploitation is large, including static reserve of $14697.45 \times 10^4\,m^3$ and dynamic reserve of $1.35 \times 10^4\,m^3/\,h$. The destruction of static and dynamic reserves is permanent and unrecoverable, which has a huge adverse effect on the groundwater system and the environment. Groundwater of Hanglaiwan is less contaminated, it can meet the national class III water standard, but the total number of bacteria exceeds the limit generally and toxicological indicator of local samples also exceeds the limit; therefore, the groundwater cannot be consumed directly.

REFERENCES

Bao-Ping Han, Shi-shu Zheng: Journal of China University of Mining & Technology, vol. 23 (1994) No.3, p.70–77.

De-xin Zhang: Geology in China, (2000) No.3, p.12–13.

Hong Shi, Yong-bo Zhang: Shanxi Science and Technology, vol. 26 (2011) No.1, p.36–37.

Li min Fan: Hydrogeology & Engineering Geology, vol. 42 (2015) No.1, p.65.

Qiang-wei Deng, Yong-bo Zhang: Bulletin of Soil and Water Conservation, vol. 34 (2014) No.6, p.123–125.

Qiang Wu, Dong-lin Dong and Yao-jun Fu: Journal of China University of Mining & Technology, vol. 31 (2002) No.1, p.19–22.

Shuang-ming Wang, Li-min Fan and Xiong-de Ma: China Minging Magazine, vol. 19 (2010) p.212–216.

Shu-kui Zhang, Nan Zhang and Hong Jiang: Water Resources and Hydropower Engineering, vol. 46 (2015) No.10, p.12–16.

Szzeppansk J. Twardowska L: Environmental Geology, vol. 38 (1999) No.3, p.249–258.

Yan-de Dai, Bin Lv and Chao Feng: Journal of Beijing Institute of Technology (Social Sciences Edition), vol. 17 (2015) No.1, p. 1–7.

Zhen-shuan Li: Coal Geology of China, vol. 19 (2007) No.5, p.35–37.

Zhi-guo Cao, Rui-min He and Xing-feng Wang: Coal Science and Technology, (2014) No.12, p.113–116.

Zhong Huang, Rui-feng Li:China Coal, vol. 36 (2010) No.8, p.27–29.

Energy Science and Applied Technology ESAT 2016 – Fang (Ed.)
© *2016 Taylor & Francis Group, London, ISBN 978-1-138-02973-6*

Sorption characteristics of soil absorbing β-HCH

Hong Li
College of Environmental Science and Engineering, Guilin University of Technology, Guilin, China

Honghu Zeng & Yanpeng Liang
College of Environmental Science and Engineering, Guilin University of Technology, Guilin, China
Guangxi Collaborative Innovation Center for Water Pollution Control and Water Safety in Karst Area,
Guilin, China

ABSTRACT: Adsorption experiments are making use of the studies on the soil adsorbing β-HCH at a temperature of 25°C, under different initial concentrations, oscillation time of adsorption. In addition, it is analyzed by GC-ECD to determine the adsorption equilibrium time and adsorption isotherm. Experimental results show that: the adsorption of soil of initial concentration of 100 μg/L β-HCH can reach high peak at 8.7833 μg/g after 5 hours, can reach adsorption equilibrium after 12 hours, and adsorption is in the range of 5.0488 ~5.9486 μg/g. The sorption amount of β-HCH increased with the increasing amount of initial concentration at 25°C; the Freundlich model and the Langmuir model are well modeled with adsorption isotherms of β-HCH, and they show that sorption of β-HCH is a nonlinear behavior.

Keywords: Mine ventilation gas; orthogonal test; operating parameter; reactor performance

1 INTRODUCTION

Organochlorine pesticide is a kind of Persistent Organic Pollutant (POP), which has been widely used in many countries around the world. Its use was prohibited in our country since 1983, because of its high residual quantity, bio-concentration and difficulty in degradation. It can be still founded from the research in recent years that pesticide residues can be detected in the surface water, bottom mud of the soil, sediment, groundwater, and the foreign water environment. HCHs is a kind of commonly used organochlorine pesticide, which has been widely used in the farmland, and β-HCH is one of the five stable isomeric compounds of HCHs, which was listed in the POPs which should be controlled in *Stockholm Convention* in 2005.μg/g

β-HCH cannot be degraded through photolysis and hydrolysis, whose degradation rate of microorganisms is also very slow. In the investigation of organochlorine pesticides in water environment in Beijing, Liu (Chen Ming et al, 2007) found that β-HCH in the water was of the highest concentration among the detected pollutants; Zhong (Gong Zhongming et al, 2003) found there were comparatively high isomeric compound residues of HCHs in the topsoil of Tianjin, among which β-HCH is the main pollutant accounting for more than 50%. In the survey of organic chlorine pesticide residue on drinking water reservoir, β-HCH was detected in every water sample and soil sample obtained from Qingshitan reservoir in Guilin and its surrounding farmland soil, and what is worse, the residues account for a higher proportion.

There are also some studies conducted at home which are about the adsorption properties of β-HCH or other Pops and its influencing factors, the adsorption properties of HCHs in soil (Wu Haibing et al, 2014), cinder (Yang Shenke et al, 2007), and modified material (Yang Shenke et al, 2007) (Yuan Xiaoling et al, 2004). However, the studies only on adsorption properties of β-HCH have not been found yet.

The study on the adsorption properties of β-HCH in soil is a new topic, which contributes to their search on the main reasons for the secondary pollution of the β-HCH under natural conditions, and is of great significance in controlling and coping with organic chlorine pesticide pollution.

2 EXPERIMENTAL MATERIAL AND METHODS

2.1 *The adsorbent and the adsorption solution*

First, get some soil from the flower beds and seal it into the polyethylene plastic bags, wash the soil with deionized water after removing sundries in the laboratory, dry it in ventilate cupboard, then sieve the soil with mesh sieve of 100 holes (150 um) after grinding with a ceramic mortar, and finally put the adsorbent in ventilated and dry place. Dilute 200 ug/ml β-HCH with acetone to obtain the β-HCH standard solution, then further dilute the β-HCH standard solution with ultrapure water, to make up the volume to a constant level in a volumetric flask, and finally make the adsorption solution of different concentrations.

2.2 *Experimental methods*

As instructed, 3.0 g of adsorption agent was taken in a centrifugal tube, with a PVC plug, to this added 60 ml of adsorption solution, and placed in a steam bath thermostat oscillator spring table at 45 degree slanting position. Avoid light shock for sometime, centrifuge at a speed of 4000 r/min for 20 min, and remove the supernatant; 20 ml of the the extract was used as the sample. The sample was analyzed by gas chromatography, equipped with an electron capture detector.

2.3 *The adsorption kinetics experiment*

The concentration of the adsorption solution is 100 μg/L, and the temperature is 25°C, vibrate the adsorption solution, respectively, for 0.5 h, 1.0 h, 1.5 h, 2.0 h, 3.0 h, 4.0 h, 8.0 h, 12 h, 16 h, and 24 h.

2.4 *Adsorption isotherm test*

The initial concentration of the adsorption solution is 10, 20, 30, 40, 80, 100, and 200 μg/L, and the temperature is 25°C; vibrate the adsorption solution for 18 h.

2.5 *Blank test and parallel sample*

In order to ensure that the experiment data are reliable, it is necessary to eliminate the test error and interference caused by human factors. Therefore, there are three parallel samples and a blank control (only add the adsorption solution, without the soil matrix) in the experiment. The min using will be adopted to calculate the adsorption quantity of β-HCH on the adsorbent.

3 DETERMINATION OF THE ADSORPTION TIME OF β-HCH IN THE SOIL

According to the measured data, draw the changing curve of the adsorption quantity qe (μg/g) of β-HCH in the soil, which will change with the time t(h), as shown in Figure 1.

As shown in Figure 1, when the soil concentration is 100 μg/L, the aqueous solution of β-HCH

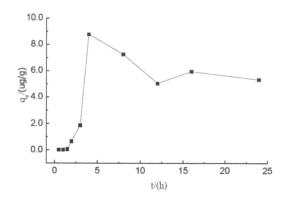

Figure 1. The curve graph of the adsorption kinetics of β-HCH in the soil (25°C).

is at the initial stage, it reaches maximum at 4 h and begins to decline at 8 h. Although there are fluctuations after 12 h, compared with the previous 5 h, it remained constant (5~6 μg/g). In order to achieve the adsorption equilibrium of the soil to β-HCH as far as possible, the vibration time of the following adsorption test is 18 h.

β-HCH is a kind of hydrophobic organic pollutant, which reacts intensively with the soil in the vibration process. In the early stage of the vibration, it adsorbs quickly and soon reaches the peak, a part of the adsorbed β-HCH would be slowly released in soil with the increase in vibrating time, and the rest is steadily desorbed in the soil and eventually reaches the equilibrium state.

4 THE DISCUSSION ON THE ISOTHERMAL ADSORPTION OF SOIL TO β-HCH

As can be seen from the slope of isothermal adsorption curve in Figure 2, when the temperature is 25°C, the tendency is divided into two sections: when the mass concentration of β-HCH aqueous solution is 10~80 μg/L, the adsorption quantity of soil to β-HCH rises rapidly; when the mass concentration of β-HCH aqueous solution is100~200 μg/L, compared with the previous section, the adsorption quantity of soil to β-HCH rises slowly. The slope of the isotherms decreases with the increase in the concentration of the aqueous solution, which indicates that there are more than two interactions between the β-HCH and the adsorbent. This phenomenon also shows that the adsorption process of organochlorine pesticides is complex. With the increase in the concentration of β-HCH in the adsorption solution, the soil adsorption quantity also increases and reaches the saturation.

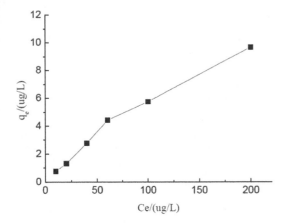

Figure 2. The isothermal adsorption of soil to β-HCH (25°C).

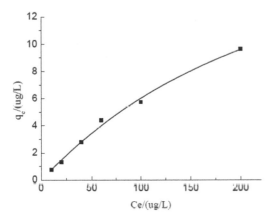

Figure 3. Langmuir adsorption isotherm graph (25°C).

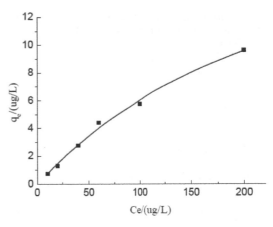

Figure 4. Freundlich adsorption isotherm graph (25°C).

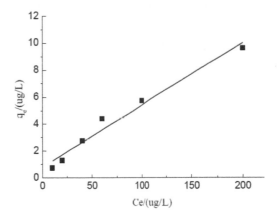

Figure 5. The slope graph of the adsorption isotherm (25°C).

Match the three kinds of adsorption model with the isothermal adsorption data of soil to β-HCH, and the results are shown in Figures 3~5.

As can be seen from Figures 3~5, when the temperature is 25°C, the imitative effect between the adsorption thermodynamic data of soil to β-HCH and linear adsorption isothermal equation, Freundlich isotherm equation and Langmuir isotherm equation is good. Three adsorption isotherms rise with the increase in the initial concentration of β-HCH aqueous solution, indicating that the test data are consistent with model data. Among them, Freundlich isotherm equation and Langmuir isotherm equation are consistent with the experimental data trend, when the concentration rises from 100 μg/L to 200 μg/L, the adsorption rate decreases, Freundlich model curve and Langmuir model curve also show their decreasing trend in the adsorption rate.

The imitative effect results from Figs. 3~5 show that the three models all match the thermody-namic data of soil well. In the Freundlich model, the correlation coefficient R^2 is 0.9720, and the n is greater than 1, which shows that the imitative effect of Freundlich adsorption isotherm is good.

In the related papers, the value of n is less than 1, and the closer to 1 the n is, the better the adsorption of line is. It is generally believed that the n value is between 0.1 and 0.5 when it is easy to absorb, and it is difficult to absorb with n > 2 to adsorption. The value of the Table is 1.226, which indicates that it is a nonlinear adsorption. In addition, the adsorption capacity of the soil to β-HCH is moderate, which can also be proved by the adsorption strength kf. In terms of the correlation coefficient R2, the imitative effect of the Langmuir model is better; the Langmuir model is used to describe nonlinear adsorption model, and its correlation coefficient R^2 value is 0.9910, which is the highest among the three.

As shown in Figs. 3~5, the imitative effects of the adsorption pattern of soil to β-HCH with Freundlich model and Langmuir model are better than that of linear model. Generally, the Freundlich model is used to describe the nonlinear adsorption phenomena, the Langmuir model is usually used to describe the monolayer absorption process of the adsorbate on the surface layer.

5 CONCLUSION

When the initial concentration of β-HCH aqueous solution is 100 μg/L and the temperature is 25°C, the adsorption of soil to β-HCH reaches the highest value 8.7833 ug/after having been vibrated for 5 h, and reaches a steady state after having been vibrated for 12 h. When the adsorption reaches the equilibrium, the soil adsorption quantity is 5.0488~5.9486 9.56 μg/g.

When the temperature is 25°C, the adsorption concentration of soil to β-HCH increases with the increase in the initial concentration of β-HCH; the adsorption of soil to β-HCH is comparatively consistent with the Freundlich model and the Langmuir model, and the fitting correlation coefficient $R^2 > 0.9$. The adsorption of β-HCH in the soil is a kind of nonlinear adsorption.

REFERENCES

Chen Ming, Ren Ren, Wang Zijian, et al. The Analysis of Organic Chlorine Pesticide Residue of Industrial Waste in Beijing City Sewage [J].China Environmental Monitoring 2007, 23 (4): 29–32.

Gong Zhongming, Cao Jun, Li Bengang, et al. The Residues of HCH in Soil in Tianjin Area (HCH) Of and Distribution Characteristics [J]. Chinese Environmental Science, 2003, 23 (3): 311–314.

HuanJu Yao, Yang Zhiqun, Zhou Gaizhou etc. On BHC Adsorption Desorption in Lanzhou Section of the Yellow River Sediment [J]. People of the Yellow River, 2009, 31 (5): 54–56.

Wu Haibing, Ceng Honghu, Liang Yanpeng et al. Integrated Solid Phase Extraction-GC/ECD the Measurement of Organochlorine Pesticides In Water [J]. Industrial Safety and Environmental Protection, 2014, (12): 4–7.

Xu Min, Zhang Gangya, Chao Shengbo. Cetyl Trimethyl Ammonium Bromide Modified Clinoptilolite and Its Adsorption of BHC in Water [J]. Soil, 2009, 41 (1): 116–120. (1) 133–138.

Yang Shenke, Liu Xiaoni, Deng Xiufang etc. Organic Modified Sepiolite's Adsorption to BHC [J]. Journal of Safety and Environment, 2007, 7 (3) 54–57.

Yuan Xiaoling, Zhang Lanying, Liuna etc. Experimental Study on the Cinder Adsorption of BHC (HCH) [J]. Jilin University Journal (Earth Science Edition), 2004, 34 (2).

Energy Science and Applied Technology ESAT 2016 – Fang (Ed.)
© 2016 Taylor & Francis Group, London, ISBN 978-1-138-02973-6

The economic feasibility analysis of an integration air source heat pump system based on an energy management contract

Shuwei Li
Department of Power Engineering, School of Energy, Power and Mechanical Engineering,
North China Electric Power University, Baoding, Hebei, China

ABSTRACT: This paper analyzes some problems of the existing heat pump system, and puts forward the concept of integration air source heat pump system introduced and the advantages of energy management contract. In addition, it can be applied to the BOT operation mode.

Keywords: integration; heat pump; energy management contract; advantage; economic feasibility

1   INTRODUCTION

The existing heat pump technology has been very mature, and it has also experienced numerous engineering practice tests and scholars' continuous optimization. However, with the progress of the society, the continuous development of economy, the constant change of architectural form, and lifestyle factors, there are still some problems which need to be solved. The problems are as follows: First, it will make the user spend a huge sum of money to purchase and maintain equipment. Second, the installation construction cycle is long, and it is not convenient to maintain and replace equipment. Third, there is a hidden trouble whether the indoor environment can achieve the desired design requirements. If some problems appear, it is difficult for the party A, the designer and the contractor to solve. Fourth, ordinary central air conditioning will cause great waste of energy because of the unit, the design, and maintenance, in the building energy consumption, the energy consumption used for HVAC accounts for 30–50% of building energy consumption. It will also cause great damage to the environment. The energy of HVAC system is basically the high grade of non-renewable energy resources, where electric power accounts for absolute proportion. For the use of energy, it makes the earth's resources increasingly scarce, and at the same time, it also brings serious environmental problems, such as acid rain in some areas of our country, fly ash, a trend growing of haze weather, and great influence on the ecological environment and sustainable development.

1.1   *New type of integration air source heat pump system*

Based on the above defects of the common cooling system, we put forward a new type of modular inverter cooling station. This new type of refrigeration station consists of integration air source heat pump system (described by the new cold standing in the following), it has good energy saving and environmental protection, advanced automatic control system, and a high degree of integration.

1.2   *The introduction of the new system*

The cooling system is a modular cold station, which is easy to transport and install. Energy consumption of the whole refrigeration station will be reduced by more than 40–60% when compared with the traditional constant-speed refrigeration station and the station will run steadily after the whole frequency conversion refrigeration station being controlled by using Hartmann control algorithm and the professional control software which are implanted into this control algorithm (WD J, 2008). At present, many refrigeration stations are artificial or half-artificial operation, in order to change this situation, the professional manufacturer has provided a control box product which contains some functions such as simple operation, observation, automatic alarm, and the online help, which can help them to monitor the unit energy consumption of refrigeration station dynamically and other operating conditions.

2 COMPOSITION OF THE NEW SYSTEM

2.1 *Hardware devices of the new system*

The main components of the new system are given in Table 1 (the new integration cold station configuration list), the stand structure is shown in Figure 1 (the cold station structure simulation diagram), and the cold station transportation simulation diagram is shown in Figure 2.

Table 1. The new integration cold station configuration list.

The new integration cold station configuration list

Centralized control system	Arctic Energy Loop control system
Refrigeration host	McGwire maglev chiller
Frozen water pump	GRUNDFOS (Denmark) VFD
Cooling water pump	GRUNDFOS (Denmark) VFD
Inverter	VACON (Finland)
Cooling tower	KINGSUN (Shanghai)

Figure 1. The cold station structure simulation diagram.

Figure 2. The cold station transportation simulation diagram.

We can have a standing to cold hardware equipment and layout idea through the chart.

2.2 *Advantages of the new system*

1. It can reduce the running cost and save space
The physical space that this system's computer room accounts for is the smallest of all the complete set of systems on the market, and its infrastructure space reduces more than 40%. It will save 10–20% in the total operating costs of the existing buildings.
2. Energy conservation and efficiency
The operation efficiency of the cooling room is high, and the unit cold power consumption of this system is less than 0.5 KW or the COP which is based on the annual running is 7.3, which can save more than 50% in comparison with traditional cooling room. It can implement the remote monitoring and control by using the Internet. There is no lubricating oil in the compressor running (saving maintenance time and cost of lubricant). The operation is ultra-quiet of the compressor (magnetic bearing), which is less than 73 db. Vertical water pumps which have low vibration and magnetic levitation compressor are present. Compressor, water pump, and pipeline are connected by using the design, which is easy to maintain. Through energy loop man-machine interface, we can get the system's parameters of running and debugging.
3. The system's cost of maintenance is low
The system adopts the integrated control strategy, and super-efficient integrated refrigeration station makes devices running speed lower and reduces the wear and tear. This leads to the fact that failure time is reduced, and the maintenance cost was reduced by 50%. All parts are easy to maintain, and it reduces the time of failure and the maintenance cost. Using closed-type cooling tower, it greatly reduces the water consumption, and the cooling tower is used for cleaning and maintenance work.
4. It extends the service life of equipment
The continuous operation speed of all frequency conversion motors is low, it reduces the number of on/off, and the use of magnetic suspension bearing makes the compressor without any abrasion. The factory renovation in the controlled environment improves the quality of the product. Durable heavy components and the structure of high-quality guarantee the reliability of the system.

3 NEW OPERATION MODE OF COLD STANDING: CONTRACT ENERGY MANAGEMENT

3.1 *Introduction of contract energy management*

Energy management contract (EPC) is the contract between the professional energy conservation service company and energy-using units, and it provides energy-using units with energy saving diagnosis, financing, design, renovation, purchase, equipment installation, debugging, and other services, and it can recycle the investment fund and obtain the corresponding reasonable profit by sharing the way of energy saving efficiency, thus it greatly saves the energy saving renovation of energy-using units and reduces the technical risk of energy-using units, and it fully arouses the enthusiasm of energy-saving reconstruct energy-using units. Energy management contract is a new kind of market-oriented energy-saving mechanism. The connotation of energy management contract is a kind way of energy saving which reduces the energy consumption cost to paying full cost of energy saving project. The practice of developed countries proves that the energy management contract is the energy-saving measure which is effective and should be popularized (XM S, 2012).

4 PROJECT CASES

4.1 Project profile

Qingdao international shipping center (as shown in Figure 4.1) is located in ShiBei district, Qingdao city. The main building height is 249.95 meters, and the total construction area is 239700square meters. The description of air conditioning project of the main building is: the main building construction area of 160300 square meters. The building elevation is 250 meters. Five layers are underground, and 56 layers on the ground. The building is divided into two parts, shopping malls and office. The mall part is divided into two parts, underground and on the ground; the cooling load is 1474 KW and 2847 KW, respectively. All the offices are located on the ground, and the total load is 8518 KW.

4.2 Investment income comparison of two schemes

In this project, the new program saves 9.3 million Yuan investment. It also saves about 90 million Yuan of operation and maintenance cost com-pared with the original program, and shorten the payback period of about seven years. In addition, the new program can also save party distribution capacity and space of distribution room.

5 CONCLUSION

Integrated conventional heat pump systems have more advantages. They are more in line with contemporary buildings, user groups, mode of operation, and other factors. By carrying through the economic comparison of the best saving ice storage systems available and the new integrated air-source heat pump system (new freezing station), we draw a conclusion that the latter has a good economic feasibility. Combining the contract energy management with the new full-frequency integrated combination of cold station, the prospects are bright, and the benefits created for social and enterprises are substantial.

REFERENCES

Green Building Becoming Mainstream [J]. Energy Policy, 1998(16):297–301.
Hartman, T. Designing efficient systems with the equal marginal performance principle. ASHRAE Journal.2005, 47:64–70.
LY F. Hvac energy saving technology analysis [J]. China's residential facilities, 2011 (3): 37–38.
WD J. Frequency-converting refrigeration station and its control strategy Hartman Loop is introduced [J]. Journal of technology platform, 2008 (90): 45–47.
XM S. Energy contract management in the field of building energy conservation research [D]. University of finance and economics, 2012:1–55.
ZL H. HVAC energy saving analysis [J]. Energy and environment. 2010, 02.

Energy Science and Applied Technology ESAT 2016 – Fang (Ed.)
© *2016 Taylor & Francis Group, London, ISBN 978-1-138-02973-6*

Purification efficiency of Cr(VI) by constructed wetlands using six different substrates

Yonghui Bu, Jie Liu, Shaohong You & Qingxin Wu
Collaborative Innovation Center for Water Pollution Control and Water Safety in Karst Area,
Guilin University of Technology, Guilin, China

ABSTRACT: To screen suitable substrate of Constructed Wetlands (CWs) for Cr(VI) removal, six different kinds of substrate (paddy soil, peat, red loam, sand, diatomite and fly ash) were tested using a microcosm experiment. The Cr removal efficiency, together with pH, Eh, organic content and plant biomass, was analyzed. The results showed that Cr removal efficiency in the CWs using peat and paddy soil as substrate were significantly higher than those in other CWs. The Cr(VI) concentration reached 0.003 mg/L and 0.023 mg/L in the CWs with peat and paddy soil substrates when influent Cr(VI) was 5 mg/L. The six kinds of substrate had different pH values. In general, diatomite and fly ash with high pH values were unfavorable for Cr(VI) removal. The CWs with peat and paddy soil substrates had higher organic content than the CWs with other substrates, thereby enhancing the Cr reduction and removal. Moreover, the plant biomass was significantly higher in the CWs with peat and paddy soil substrates than those in the CWs with other substrates ($p < 0.05$). This finding indicated that peat and paddy soil were favorable for plant growth and enhanced plant action for Cr(VI) purification. Therefore, peat and paddy soil are proposed to be used as substrates in CWs for Cr(VI) removal.

Keywords: Cr(VI); substrates; constructed wetland; purification

1   INTRODUCTION

Chromium is a kind of heavy metal element which is widely used and is harmful to the human body. It is one of the necessary trace elements in the process of human and animal metabolism, but excessive intake of chromium can cause poisoning (Guo S. L. 1992). Cr(VI) is acknowledged as the teratogenic carcinogens (Shanker A. K. 2005). Therefore, Cr(VI) is one of the implementation of total amount control of pollutants in our country.

Previous studies have shown that the substrate is an important part of the wetland system, not only directly participate in the removal of pollutants, also provides the survival of live environment for wetland plant and microbes (Wan J. J. et al, 2009). On the one hand, the substrate provides the stable surface of the attachment for the growth of microorganisms, but also provides the carrier and nutrients for plant, and provides most of interface reaction for the physical, chemical and biological reactions in wetlands (Wu H. et al, 2015). Therefore, screening of substrate of artificial wetland, not only has to consider the physical and chemical properties of the substrate itself, also consider the synergistic effect of substrate, wetland plants and microorganisms.

To screen suitable substrate of constructed wetlands for Cr(VI) removal, six different kinds of substrates(paddy soil, peat, red loam, sand, diatomite and fly ash) were tested using a microcosm experiment, and analyzed the plant growth in the different substrates of wetland system, whice provides a scientific basis for screening the ideal matrix material, optimizing the design of the wetland and improving the chromium purification performance of the wetland system.

2   MATERIALS AND METHODS

2.1   *Device and method*

Experiments were conducted in the greenhouses in guilin university of technology, based on vertical flow constructed wetland of PVC sheet bonded high growth x width x = 20 cm × 20 cm × 40 cm. The design surface load rate was 0.0125 m³ / (m²,d) and the hydraulic retention time of the constructed wetlands were 12 h, 24 h and 36 h, 48 h and 60 h. The depth of substrates was 0.3 m. The lower level was 0.05 m of gravel whice was 1 cm of the diameter, and the upper respectively were paddy soil, red soil, peat, sand, fly ash and diatomite whice were 1~2 mm of the diameter. Each matrix wetland sets two repeats, a total of 12 groups of wetland system. Wetland plants were collected from guilin suburbs, and each wetland planting density was 125 plants/m^2.

Leersia hexandra Swartz were planted in late May 2014, growing to late July 2014, the plant were basically fbestrid the full of the wetland. Wetland ran since early August 2014 with the method of continuous water, and ended until early January 2015. The Cr(VI) polluted water was compounded by $K_2Cr_2O_7$ and water, and the Cr(VI) concentration was 5 mg/L. After a sampling period changed a hydraulic retention time.

2.2 Sampling and analysis

The 50 mL tubes were used to sample from the outlet in 12 groups of wetlands at the same time. In addition the frequency was once every 3 days and cycle was 13 times. After collecting water samples, immediately conducted 3 times repetition measurement, whice monitoring indicators included Cr(VI) and total Cr. Determinated the organic matter using potassium dichromate capacity— spectrophotometry. Matrix pH and Eh were measured with pH meter and ORP meter.

3 RESULTS AND ANALYSIS

3.1 Removal efficiency of Cr(VI) and total Cr in constructed wetlands with different substrates

Cr(VI) concentration of peat was the lowest, reaching 0.003 mg/L, and the concentration of paddy soil timed, reaching 0.023 mg/L, but the concentration of fly ash was the hightest, and the concentration of Cr (VI) was only 2.757 mg/L. The concentration on various substrates of Cr(VI) were the size of the order: peat, paddy soil, red soil, sand, diatomite, fly ash. The concentration of total Cr and Cr (VI) was significantly different between different substrates ($p<0.05$).

Results are means ± standard deviation, n = 5. Different letters indicate statistically significant differences (LSD, $p< 0.05$).

Table 1. Concentration of Cr(VI) and total Cr in constructed wetlands with different substrates.

Substrate	Cr(VI)(mg/L)	total Cr(mg/L)
paddy soil	0.023 ± 0.03a	0.150 ± 0.11a
red soil	0.290 ± 0.04a	0.440 ± 0.07b
sand	0.832 ± 0.03b	0.128 ± 0.10b
peat	0.003 ± 0.01a	0.031 ± 0.07a
fly ash	2.757 ± 0.05d	2.929 ± 0.23d
diatomite	1.674 ± 0.09c	1.732 ± 0.33c

Table 2. Removal rates of Cr(VI) and total Cr in constructed wetlands with different substrates at hydraulic retention time of 12h.

Substrate	Cr(VI)(mg/L)	total Cr(mg/L)
paddy soil	98.7%~99.5%	98.4%~99.4%
red soil	90.7%~94.2%	89.8%~91.6%
sand	90.7%~94.2%	86.8%~89.3%
peat	90.7%~94.2%	99.7%~99.9%
fly ash	46.2%~57.0%	33.4%~44.8%
diatomite	46.2%~57.0%	51.5%~55.6%

As is shown in Table 2, when the hydraulic retention time was 12 h, Cr(VI) and total Cr removal rates in effluent in wetlands with different substrates remained stable. The removal rates of Cr(VI) in paddy soil, red soil, sand, peat, fly ash and diatomite wetland respectively were 98.7%~99.5%, 90.7%~94.2%, 90.7%~94.2%, 90.7%~94.2%, 46.2%~57.0%, 46.2%~57.0%, total Cr removal rates respectively were 98.4%~99.4%, 89.8%~91.6%, 86.8%~89.3%, 99.7%~99.9%, 33.4%~44.8%, 51.5%~55.6%. The removal rates of Cr(VI) and total Cr in fly ash and diatomite wetlands whice Cr(VI) total Cr treatment effect were the worst, which were inferior to those of the other four substrates.

3.2 The consumption of organic matter, pH and Eh of different wetland substrates after the running of the wetlands

After the running of the wetlands, redox potential difference of six kinds of matrices was not large, between 141 mV and 221 mV. Studies had shown that Cr(VI) at Eh <+300 mV in wetlands could be reduced to Cr(III) (Masscheleyn P. H. et al, 1992). Therefore, six matrices Eh values all met the redox conditions of Cr(VI) reduction. However, the pH value of six kinds of matrices obviously differenced. The pH of peat was the lowest, whereas fly ash and diatomite were the highest, all more than 8. The Cr(VI) was not easily adsorbed and reducted with strong mobility in alkaline conditions (Selvia K. et al, 2001). Therefore, fly ash and diatomite wetlands were not conducive to Cr(VI) removal. The chromium removal effect also had certain correlation with organic matter content in matrices. The organic matter content in peat and soil wetlands were the highest, correspondingly total Cr and Cr(VI) removal rates were the highest. The organic matter content in diatomite wetlands was the lowest, correspondingly total Cr and Cr(VI) removal rates were the lowest.

3.3 The growth of wetland plants in the wetlands of different substrates

Based on the substrates to be fixed, *Leersia hexandra* Swartz in the growth was relatively stable in the environment. Found in the process of constructed wetland running, the *Leersia hexandra* Swartz of paddy soil, clay and peat wetlands all grew well, and its biomass and plant height all had varying degrees of increase. The *Leersia hexandra* Swartz of sand, fly ash and diatomite wetlands grew worsely, and some leaves became yellow and withered. The biomass of *Leersia hexandra* Swartz of paddy soil, peat wetland wetlands increased significantly compared with the rest of several ($p<0.05$). The plant height and root length in paddy soil and peat wetlands were significantly higher than those in the rest of several matrix wetlands ($p<0.05$).

Results are means ± standard deviation, n = 3. Different letters indicate statistically significant differences (LSD, $p< 0.05$)

4 DISCUSSIONS

The removal efficiency of total Cr and Cr(VI) were significantly different between different substrates ($p<0.05$). The chromium removal rate of peat is 2.4 times of fly ash (Table 1). This suggested that choosing excellent substrates to improve the processing efficiency of chromium in *Leersia hexandra* Swartz wetland systems was crucial. Substrate played an important role in fixing and removing chromium (Dotro G. et al, 2009). Reduction, adsorption, sedimentation, filtration and biological function of purification process associated with chromium mostly occurred in the substrates (Sultana M. Y. et al, 2014).

The pH and redox potential (Eh) of wetland substrates were the key factors to determine the valence state and morphology of the chromium. In alkaline conditions (pH > 8.0), being a relatively stable anion form, Cr(VI) was not easy to be adsorbed by minerals and organic matter, so it is not easy to be removed from water (Kotaś J. et al, 2000). In the present study, it was found that the pH value of fly ash and diatomite were higher than 8.0 (Table 3), therefore the removal rates of Cr were poor. The pH value of peat was the lowest, which was favorable for Cr(VI) reduction to Cr(III). Cr(III) was easy to be adsorbed and formed hydroxide precipitation, and eventually intercepted in the substrates. Therefore, the removal rate of Cr(VI) was higher than others in the wetland with peat as substrate. Although the paddy soil pH value was neutral and weak alkaline, but the paddy soil wetland system still had the very high chromium removal efficiency. This showed that the pH and Eh is not the only factor to chromium purification performance

and chromium removal efficiency of *Leersia hexandra* Swartz wetland systems but also other factors affected the removal efficiency.

The organic matter content of the substrate was likely to be another important factor that influenced efficiency of chromium removal. The organic matter in the substrates played two roles in the process of the purification of chromium. On the one hand, organic matter in the substrates might directly involved in the reduction of Cr(VI) as an electron donor (Liu J. et al, 2014). On the other hand, the organic matter in the substrates had a large number of functional groups which could adsorb and fix chromium. In this study, organic matter content of peat and paddy soil were high than others, so the removal rates of chromium of two substrates wetland systems were relatively higher than others. With low organic matter content of diatomite and sand, the removal efficiency of chromium were relatively weak.

The ideal substrates could form a healthy ecosystem with the wetland plants and microbes, and improved the function of the systems. In this study, *Leersia hexandra* Swartz biomass in the wetland systems of peat and paddy soil were significantly higher than others (Table 4). This showed that the two kinds of substrates were more suitable for the growth of *Leersia hexandra* Swartz. Root of *Leersia hexandra* Swartz had the strong ability

Table 3. After the running of the wetlands, Eh, and pH in different substrates used in constructed wetlands.

Substrate	Organic matter content (%)	Consumption of organic matter(%)	pH	Eh(mV)
paddy soil	4.6	0.46	7.28	151.5
red soil	1.19	0.92	7.42	165
sand	0.78	1.23	7.93	151
peat	14.18	0.43	5.45	221
fly ash	1.64	3.96	8.44	176.5
diatomite	0.37	3.17	9.39	141

Table 4. Biomass, plant height and root length of *Leersia hexandra* Swartz in wetlands of different substrates.

Substrate	Biomass/ (g/per plant)	Plant height/cm	Root length/ cm
paddy soil	1.14 ± 0.13a	50.78 ± 5.39b	9.66 ± 2.37a
red soil	0.89 ± 0.10b	43.90 ± 5.21b	6.98 ± 2.22b
sand	0.58 ± 0.08c	33.49 ± 4.16d	6.53 ± 2.16bc
peat	1.63 ± 0.20a	60.92 ± 5.97a	10.49 ± 2.89a
fly ash	0.35 ± 0.07d	38.04 ± 1.21c	6.35 ± 1.09c
diatomite	0.39 ± 0.07d	40.27 ± 1.77bc	7.16 ± 2.24b

to absorption chromium, and the root exudates of *Leersia hexandra* Swartz could also participate in the chromium reduction and immobilization (Liu J. et al, 2015). Therefore, the high biological amount of *Leersia hexandra* Swartz might be the reason for good purifying effect of chromium in peat and paddy soil wetland systems.

5 CONCLUSION

There were significant differences in the purification effect of Cr(VI)-polluted water in *Leersia hexandra* Swartz wetlands with the six kinds substrates. The chromium purification effect of peat and paddy soil wetlands were the best, red soil and sand wetlands ranked in the second place, and the chromium purification effect of the fly ash and the diatomite wetlands were the worst. Peat and paddy soil were ideal substrates of *Leersia hexandra* Swartz Cr(VI)-contaminated wetlands treatment systems. The main reasons included: (1) An enabling environment of the pH value and Ph for Cr(VI) reduction; (2) The content of organic matter were high, which provided the reducing agent and the adsorption site for the purification of chromium; (3) There were suitable for the growth of *Leersia hexandra* Swartz, which were beneficial to plants chromium purification function into full play.

ACKNOWLEDGEMENTS

The corresponding author of this paper is Liu Jie. This paper is supported by the Natural Science Foundation of China (41273142, 41471270) and High Level Innovation Team and Outstanding Scholar Program in Guangxi colleges and universities (002401013001).

REFERENCES

Beijing municipal environmental protection bureau. A selection of environmental protection standard[M]. Standards Press of China. 2002.

Dotro G., Palazolo P., Larsen D. Chromium fate in constructed wetlands treating tannery wastewaters[J]. Water Environment Research, 2009, 81:617–625.

Guo S.L., The properties and application of chromium and manganese[M]. Higher Education Press, 1992.

Shanker A.K., Cervantes C., Loza A.H., et al. Chromium toxicity in plants[J]. Environment International, 2005, 31(7):739–753.

Kotaś J., Stasicka Z. Chromium occurrence in the environment and methods of its speciation[J]. Environmental Pollution, 2000, 107(3):263–283.

Liu J., Zhang X.H., You S.H., et al. Cr(VI) removal and detoxification in constructed wetlands planted with Leersia hexandra Swartz[J]. Ecological Engineering, 2014, 71:36–40.

Liu J., Zhang X.H., You S.H., et al. Function of Leersia hexandra Swartz in constructed wetlands for Cr(VI) decontamination: A comparative study of planted and unplanted mesocosms[J]. Ecological Engineering, 2015, 81:70–75.

Masscheleyn P.H., Pardue J.H., DeLaune R.D., et al. Chromium redox chemistry in a lower Mississippi Valley bottomland hardwood wetland[J]. Environmental Science and Technology, 1992, 26: 1217–1226.

Selvia K., Pattabhia S., Kadirvelu K. Removal of Cr(VI) from aqueous solution by adsorption onto activated carbon[J]. Bioresource Technology, 2001, 80(1):87–89

Sultana M.Y., Akratos C.S., Pavlou S., et al. Chromium removal in constructed wetlands: A review[J]. International Biodeterioration & Biodegradation, 2014, 96:181–190.

Wan J.J., Wang Z., Li J., et al. Function of the substrates in the constructed wetlands[J]. Science of Environmental Protection, 2009, 35(3):16–19.

Wu H., Zhang J., Ngo H.H., et al. A review on the sustainability of constructed wetlands for wastewater treatment: Design and operation[J]. Bioresource Technology, 2015, 175:594–601.

Energy Science and Applied Technology ESAT 2016 – Fang (Ed.)
© 2016 Taylor & Francis Group, London, ISBN 978-1-138-02973-6

Research on the operation modes of the microporous aerator for efficient wastewater aerobic biological treatment

Fei Li, Jingdong Zhang, Chaoyang Liu & Zhiguang Qu
Research Center for Environment and Health, Zhongnan University of Economics and Law, Wuhan, China

Gaoqi Duan
Zhongkeluyu Water Ecological Investment Co. Ltd., Beijing, China

Xue Li
Department of Biological and Environmental Engineering, Changsha University, Changsha, China

ABSTRACT: Aeration was the main cost process in whole aerobic biological treatment. Transferring coefficient of oxygenic aeration (K_{La}) plays a key role in affecting the energy cost, which also makes a decisive effect on the normal metabolism of microorganism. With the background of sustainable development and energy saving posed by Chinese government, different setting ways of the microporous aerator were studied in order to improve the efficiency of energy consumption based on the medium-sized aeration tank and microporous aerator. According to the theoretical dynamic efficiency (E_p) of different experimental designs, the results indicated that (1) at fixed ventilatory capacity (Q), E_p of four different experimental designs were in descending order: aerator oblique upward, aerator upward, aerator inclined down, and aerator downward. (2) In present experiments, at fixed setting ways of the microporous aerator, the more the Q, the higher the efficiency, and tends to be constant finally. (3) At fixed Qs and fixed setting ways of the microporous aerator, ascending aerator height could not make its E_p increase simply. (4) At the same amount of the microporous aerators, E_p were improved greatly when relatively scattered arrangement was replaced by the relatively centralized layout.

Keywords: aerators; aerobic treatment; efficiency of energy use; wastewater

1 INTRODUCTION

Wastewater aerobic biological treatment is the technology to take certain measures to create a controllable, suitable temperature, pH, dissolved oxygen, nutrients, and other environmental conditions for feeding lots of microorganisms which have good ability of the oxidative decomposition of organic matters (Eddy & Tchobanoglous 1979). With the fast development of Chinese urbanization, the urban environment in lots of Chinese cities, including soil environment, aquatic environment, atmospheric environment, etc., has been contaminated to different extents (Chen et al. 2015; Huang et al. 2016a,b; Li et al. 2013; Li et al. 2015a,b; Li et al. 2016; Liu et al. 2015; Yuan et al. 2015; Zeng et al. 2015). The need for urban sewage treatment is increasing unprecedentedly. In China, urban sewage treatment is one of the most energy-intensive industries (Li et al. 2012) and the sewage treatment rate has been more than 45% throughout the whole country. In the recent years, a large number of sewage treatment plants have been built and put into operation. Therefore, how to save energy and reduce consumption in a better operation mode during the sewage treatment process is of significance for being explored not only from the viewpoint of scientific research but also from the viewpoint of cost-benefit analysis.

Wastewater aerobic treatment need to have enough oxygen through sufficient aeration to ensure normal microbial growth and metabolism (Gao et al. 2007; Ren & Ma 2003). Gu *et al.* showed that the rank of energy consumption steps during Chinese typical secondary urban sewage treatment was: oxygen aeration equipment (57%), input water pump (22%), sludge treatment (7.9%), return sludge pump (7.5%), other ancillary equipment (2.6%) and grille, grit chamber, settling basin, and enrichment pool (2.4%) (Jiang et al. 2011). Obviously, the blast aeration stage is the most energy-consuming step. Therefore, under the enthusiastic promotion of saving energy and reducing consumption by Chinese government, this study aims at the optimization of the aeration process and tries to explore the relationship between different microporous aerator arrangement methods and the corresponding theoretical dynamic efficiency (E_p) in a medium-sized aeration tank in order to make a valuable theoretical basis for saving energy and reducing consumption in process of wastewater aerobic biological treatment.

2 MATERIALS AND METHODS

2.1 Experimental materials and equipment

The self-built medium-sized aeration tank was adopted in the present study. In addition, the design parameter of the aeration tank and overflow dam parameters are shown in Fig. 1. Two parallel aerator pipes were designed at position of 500 mm and 1000 mm in the width line. Moreover, each exposure tracheal path can be most installed in 12 equidistant microporous aerators.

2.2 Reagent and instrument

The main reagent and instrument used in this study were as follows: the three-blade root blower (CSR50, Zhangqiuhengfeng Machinery Co., LTD.), the Dissolved Oxygen (DO) tester (INESA,

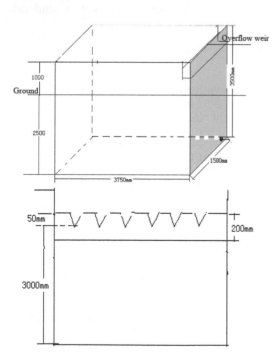

Figure 1. Design parameters of the medium-sized aeration tank and overflow dam parameters.

Shanghai), the second chronograph, the tape, the three-phase asynchronous motor, the microporous aerators (PWX-250), and the glass rotameter. Moreover, the technical parameters of the selected microporous aerator were: air volume of 5–8 m^3/h, oxygen utilization rate of 25–30%, diameter of the bubble of 1–3 mm, the dynamic efficiency ≥ 5 kgO_2/kW·h, the proper applied water depth of 1–6 m, and each service area of 0.35–0.6 m^2.

2.3 Principle of the Experiment

Based on ideal conditions, the oxygen transfer efficiency analysis method is adopted under the unstable state. Namely, with sodium sulfite deoxidization, it makes the dissolved oxygen in the water to zero first, and then the aeration is started with recording a dissolved oxygen value every 1 minute until dissolved oxygen value is close to the saturation level.

Assume that the process is in completely mixed liquid state, in line with the first-order kinetics reaction, the changes of dissolved oxygen in the water can be used under the expression (Gao et al. 2007; Zhao et al. 2006):

$$\frac{dc}{dt} = K_{La}(C_s - C) \tag{1}$$

$$\ln(C_s - C) = -K_{La} \cdot t + k \tag{2}$$

where dc/dt is the oxygen transfer velocity (mg/L·h), C_s is the saturated dissolved oxygen concentration under certain experiment conditions (mg/L), C is the dissolved oxygen concentration (mg/L) at aeration time t, K_{La} is the oxygen transfer coefficient of aerator under certain experiment conditions (min^{-1}), and k is a constant. In addition, equation 2 was obtained through integral calculation of equation 1. According to equation 2, K_{La} will be obtained in the relationship curve graph between In (C_s–C) and t.

The basis of evaluation of aeration equipment performance indicators includes: K_{La} (oxygen transfer coefficient, min^{-1}), q_c (oxygenation capacity, kg/h), E_A (oxygen utilization coefficient, %), and E_p (theoretical dynamic efficiency, kg/kW·h). In the present study, E_p was selected as the main assessment criteria. Furthermore, the corresponding computational formula of E_p was (Gao et al. 2007):

$$E_p = q_c / N_T \tag{3}$$

where E_p is the theoretical dynamic efficiency under the standard conditions, kg/kW·h, q_c is the oxygenation capacity under the standard conditions, kg/h, and N_T is the theoretical power of aerator oxygen filling, kW.

2.4 Technical route

The technical route of the present study included six steps as follows: (1) To determine the monitoring points throughout the aeration tank, 12 monitoring points were selected to fully understand the spatial distribution of DO in the present study. (2) Infusion of the water aeration tank and the determination of the volume of aeration in the inland waters, and the determination of dissolved oxygen in water. (3) Through the chemical equations $Na_2SO_3 + 1/2O_2 \rightarrow Na_2SO_4$, to determine the demand quantity of $CoCl_2$ and Na_2SO_3 in order to reduce the DO to zero mg/L. (4) To add an overdose of $CoCl_2$ and Na_2SO_3 into the aeration tank and stir it uniformly. (5) When the DO reaches zero mg/L in the 12 monitoring points, the regular aeration started and the DO in the 12 monitoring points was measured every 1 min until DO was saturated. (6) To improve the quality of testing, repeat twice in each group and finally determine the variation tendency of DO under different aeration patterns. Notes: to ensure the ventilator capacity constant during the experiment.

3 RESULTS AND DISCUSSION

3.1 Influence of different aerator directions on the dynamic efficiency

To study the effect of aerator direction on the dynamic efficiency, the different aeration patterns, including the direction sets of the aerator in upward, oblique upward (45°), inclined downward (45°), and downward, were explored under the fixed air volume (24 m³/h, 32 m³/h, 36 m³/h, and 38 m³/h), respectively. According to the technical route in the previous section, the dynamic efficiency along with the changes of direction sets of the microporous aerators is shown in Fig. 2.

Figure 2 shows that, at fixed ventilatory capacity (Q), E_p of four different experimental designs were in descending order: aerator oblique upward, aerator upward, aerator inclined down, and aerator downward. The reasons of the obtained results probably were the relative increase in the air-liquid contact distance and time when aerators were set oblique upward. Through the relative increase in the air-liquid contact distance and time when aerators were set adown, the obvious bubble coalescence gave rise to the decrease in the theoretical dynamic efficiency. In practice, conventional wastewater treatment plants take aerator upward and their theoretical dynamic efficiency is generally 2–5 kgO₂/kW·h. Therefore, the mode of oblique upward aerators set was recommended.

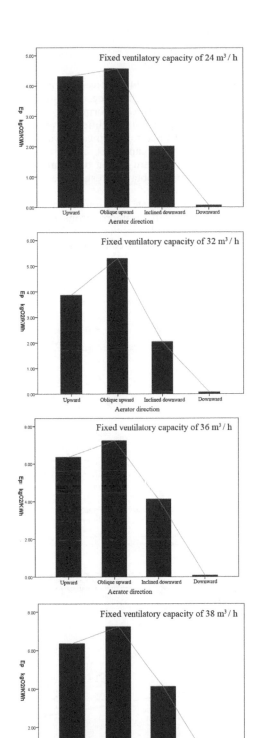

Figure 2. Influence of different aerator directions on the dynamic efficiency.

245

3.2 Influence of different ventilator capacities on the dynamic efficiency

To study the effect of different ventilatory capacities on the dynamic efficiency, this study conducted a series of three experiments, including (1) 12 aerator upward under different ventilatory capacities, (2) 12 aerator inclined upward under different ventilatory capacities, and (3) 12 aerator inclined downward under different ventilatory capacities; the obtained results are shown in Fig. 3a.

Figure 3a shows that, in present experiments, at fixed setting ways of the microporous aerator, the more the Q was, the higher the theoretical dynamic efficiency, and tends to be constant finally.

The reasons of the obtained results probably were: the higher the ventilatory capacity, the higher the concentration of DO. Though the dynamic efficiency was increasing with the increasing ventilatory capacity to some extent, it tended to be constant while the the power consumption was increasing simultaneously.

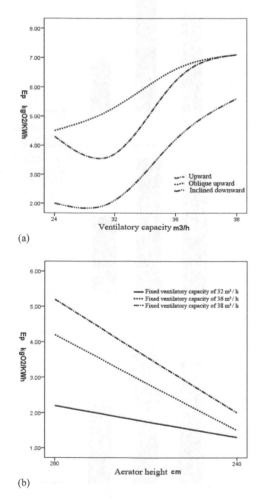

(a)

(b)

Figure 3. (a) Influence of different ventilator capacities on the dynamic efficiency. (b) Influence of different aerator heights on the dynamic efficiency.

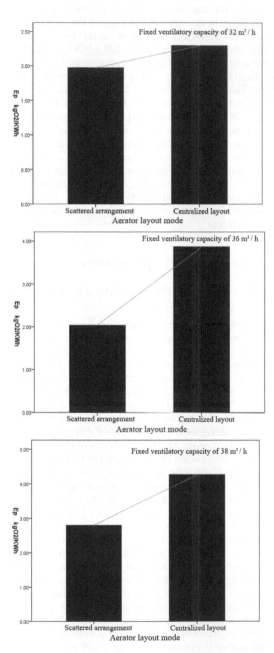

Figure 4. Influence of different aerator layout modes on the dynamic efficiency.

3.3 Influence of different aerator heights on the dynamic efficiency

In order to further study the influence of the aerator height (distance from the surface of 280 cm and 240 cm) on the theoretical dynamic efficiency, the background condition of the 12 aerator inclined downward and fixed Q were set. Furthermore, this study conducted a series of three experiments, including (1) fixed Q at 32 m^3/h, (2) fixed Q at 36 m^3/h, and (3) fixed Q at 38 m^3/h. The obtained results are shown in Fig. 3b.

Figure 3b shows that at fixed Qs and setting ways of the microporous aerator, ascending aerator height could not make its E_p increase simply. Furthermore, unlike our assumption, the ventilator capacity was not increasing with the decreasing aerator height.

3.4 Influence of aerator with different dense dispersions on dynamic efficiency

In order to further explore the effect of different aeration density on dynamic efficiency, this study conducted a series of three experiments, including (1) fixed Q was 30 m^3/h, six-aerator centralized arrangement, and dispersion, respectively, (2) fixed Q was 36 m^3/h, six-aerator centralized arrangement and dispersion, respectively, (3) fixed Q was 38 m^3/h, six aerator centralized arrangement, and dispersion, respectively. The obtained results are shown in Fig. 4.

Figure 4 shows that at the same amount of the microporous aerators, E_p were improved greatly when relatively scattered arrangement was replaced by relatively centralized layout. The relatively centralized layout aerators probably make the obvious high-dissolved oxygen region and low-dissolved oxygen region, which improves the oxygen transfer rate by the drive of water flow.

4 CONCLUSION

(1) With fixed ventilatory capacity (Q), E_p of four different experimental designs were in descending order: aerator oblique upward, aerator upward, aerator inclined down, and aerator downward. (2) In present experiments, at fixed setting ways of the microporous aerator, the more the quantitative value of Q, the higher the efficiency. It finally tended to be constant finally. (3) At fixed Qs and setting ways of the microporous aerator, ascending aerator height could not make its E_p increase simply. (4) At the same amount of the microporous aerators, E_p were improved greatly when relatively scattered arrangement was replaced by a relatively centralized layout.

ACKNOWLEDGMENTS

This study was financially supported by the Graduate Innovative Education Program of Zhongnan University of Economics and Law (2013 JY07) and the National Natural Science Foundation of China (51308076; 71503268).

REFERENCES

Chen, H., Teng, Y., Lu, S., et al. 2015. Contamination features and health risk of soil heavy metals in China. Science of The Total Environment 512–513: 143–153.

Eddy, M., Tchobanoglous, G. 1979. Wastewater Engineering Treatment Disposal Reuse, second ed. Columbus: Mc Graw-Hill Book Company.

Gao, Y., Gu, G., Zhou, Q. 2007. Water pollution control engineering. Beijing: Higher Education Press.

Huang, J.H., Liu, W.C., Zeng, G.M., et al. 2016a. An exploration of spatial human health risk assessment of soil toxic metals under different land uses using sequential indicator simulation. Ecotoxicology and Environmental Safety 129: 199–209.

Huang, J.H., Li, F., Zeng, G.M., et al. 2016b. Incorporating hierarchical bioavailability and possible receptor distribution into potential eco-risk assessment of heavy metals in soils: A case study in Xiandao District, Changsha city, China. Science of the Total Environment 541: 969–976.

Jiang, C., Yang, A., Gan, Y., 2011. Energy consumption analysis and energy saving solutions in WWTP. China Water & Wastewater 27: 33–36.

Li, F., Huang, J.H., Zeng, G.M., et al. 2013. Spatial risk assessment and sources identification of heavy metals in surface sediments from the Dongting Lake, Middle China. Journal of Geochemical Exploration 132: 75–83.

Li, F., Huang, J.H., Zeng, G.M., et al. 2015a. Toxic metals in topsoil under different land uses from Xiandao District, middle China: Distribution, relationship with soil characteristics and health risk assessment. Environmental Science and Pollution Research 22: 12261–12275.

Li, F., Huang, J.H., Zeng, G.M., et al. 2015b. Spatial distributions and health risk assessment of heavy metals associated with receptor population density in street dust: A case study of Xiandao District, Middle China. Environmental Science and Pollution Research 22: 6732–6742.

Li, F., Zhang, J.D., Huang, J.H., et al. 2016. Heavy metals in road dust from Xiandao District, Changsha city, China: Characteristics, health risk assessment and integrated source identification. Environmental Science and Pollution Research, DOI: 10.1007/s11356-016-6458-y.

Li, P., Zheng, X., Sun, Y., et al. 2012. Study on Identification of Main Energy Consumption Points and Energy-saving Methods for WWTP Using A2/O Process. China Water & Wastewater 28: 6–10.

Liu, J., Liang, J., Yuan, X.Z., et al. 2015. An integrated risk model for assessing heavy metal exposure to migratory birds in wetland ecosystem: a case study in Dongting Lake wetland, China. Chemosphere 135: 14–19.

Ren, N., Ma, F. 2003. Microbiology of Environmental Engineering: Principle and Application. Beijing: Chemical Industry Press.

Yuan, Y.J., Zeng, G.M., Liang, J., et al. 2015. Variation of water level in Dongting Lake over a 50-year period: Implications for the impacts of anthropogenic activity and climate change. Journal of Hydrology 525:450–456.

Zeng, X.X., Liu, Y.G., You, S.H., 2015. Spatial distribution, health risk assessment and statistical source identification of the trace elements in surface water from the Xiangjiang River, China. Environmental Science and Pollution Research 22(12): 9400–9412.

Zhao, J., Zheng, X., Gao, J., 2006. Research on Transfer Coefficient of Oxygenic Aeration, J Beijing Institute of Civil Engineering and Architecture 22: 11–17.

Energy Science and Applied Technology ESAT 2016 – Fang (Ed.)
© 2016 Taylor & Francis Group, London, ISBN 978-1-138-02973-6

Simulation of the multimedia fate of persistent environmental organic pollutants based on the fugacity theory

Fei Li, Jingdong Zhang, Chaoyang Liu & Jun Yang
Research Center for Environment and Health, Zhongnan University of Economics and Law, Wuhan, China

Xue Li
Department of Biological and Environmental Engineering, Changsha University, Changsha, China

Xufeng Cui
School of Business Management, Zhongnan University of Economics and Law, Wuhan, China

ABSTRACT: Multimedia model, including air, water, soil, and sediment, for persistent environmental organic pollutants was developed based on steady-state assumption of the fugacity theory proposed by Mackay. Afterwards, the multimedia model was utilized to simulate the concentration distribution of PAHs in air, water, sediment, and soil in a practical study area in order to test and verify the model based on available data collection. The results showed that the direct discharge and advection input were the main sources of PAHs, and advection output was the main loss pathway. Soil was the highest accumulation environmental medium for PAHs. Besides, with the increasing amount of benzene rings, the percentage of soil degradation loss for PAHs increased gradually and degradation loss ratio gradually reduced in the atmosphere degradation. Finally, the present study made comparative analyses between the simulative concentrations and the measured concentrations, and the results indicated the developed multimedia model fitted well in simulation of multimedia fate of persistent environmental organic pollutants.

Keywords: organic pollutants; fugacity; multimedia fate; steady state; model development

1 INTRODUCTION

With the fast development of Chinese urbanization, the urban environment in lots of Chinese cities, including soil environment, aquatic environment, atmospheric environment, etc., has been contaminated to different extents (Huang et al. 2016a, b; Li et al. 2013; Li et al. 2015a,b; Li et al. 2016; Liu et al. 2015; Yuan et al. 2015; Zeng et al. 2015). Persistent Organic Pollutant (POP) is a kind of common organic toxic pollutants in our living environment. Most of the POPs have the teratogenic, carcinogenic, mutagenic effects and they mainly derive from incomplete combustion of fossil fuels and biomass (Jones & Voogt 1999). In lots of literatures, POPs, including PAHs, DDTs, and PCB, can be in the form of gas and particles into the atmosphere and cause direct and indirect adverse influences on human health through wet sedimentation and diffusion into the water, soil, and sediment and other environmental media (Douben 2003). Therefore, it is of significance to explore POPs pollution mechanism in multi-environment migration and transformation.

At present, a lot of studies are conducted on Persistent Organic Pollutants (POPs) for their field monitoring, evaluation, and analysis in a single environment medium (Jiao et al. 2010; Ma et al. 2009). However, the local emissions, geographic conditions, and climate conditions can make influence on their migration and transformation. Therefore, with budget constraints and the human limit, to establish an effective pollutant tending model in multi-environmental media is necessary and required. The fugacity theory is of capacity to quantitatively characterize POPs' sources, distribution, migration, transformation, etc. Furthermore, the fugacity theory has been widely applied in the environmental behaviors of POPs simulation and prediction (Wang et al. 2011; Dong et al. 2008). Based on the classic fugacity theory developed by Mackay (Mackay 2001), this study tries to establish the migration and transformation model in environmental multi-media (including atmosphere, water, sediment and soil) under the hypothesis of steady state. Moreover, to test and verify the developed model, the five PAHs from the multi-environment in the Chinese Dalian city and

1 km waters close to the city shore were used as a real study case.

2 MATERIALS AND METHODS

2.1 Model framework

Based on the classic fugacity theory developed by Mackay (Mackay 2001), Level 3 steady-state models were developed combined with the geographical and meteorological field information in Dalian city and its nearshore 1 km waters. This model is divided into four major media, mainly including atmosphere (1), water (2), sediment (3), and soil (4). Based on POPs behaviors in multi-environment following the law of conservation of mass, the mass balance equations for each environment medium are shown below.

Atmosphere equation (subscript 1):

$$E_1 + G_{A1}c_{B1} + f_2 D_{21} + f_3 D_{31} = \atop f_1\left(D_{12} + D_{13} + D_{R1} + D_{A1}\right) = f_1 D_{T1} \tag{1}$$

Water equation (subscript 2):

$$E_2 + G_{A2}c_{B2} + f_1 D_{12} + f_3 D_{32} + f_4 D_{42} = \atop f_2\left(D_{21} + D_{24} + D_{R2} + D_{A2}\right) = f_2 D_{T2} \tag{2}$$

Soil equation (subscript 3):

$$E_3 + f_1 D_{13} = f_3\left(D_{31} + D_{32} + D_{R3}\right) = f_3 D_{T3} \tag{3}$$

Sediment equation (subscript 4):

$$E_4 + f_2 D_{24} = f_4\left(D_{42} + D_{R4} + D_{44}\right) = f_4 D_{T4} \tag{4}$$

where f and D represent the fugacity and the migration flux; E_i means the rate of emission, *mol/h*; G_A means the advective flow rate, m^3/h; c_{Bi} is the advective flow concentration, mol/m^3; D_{Ri} is the reaction rate D value; D_{Ai} is the horizontal flow rate D value; and D_{Ti} means the sum value of D value decreasing from environment medium i.

Figure 1 shows the POPs behaviors in the multi-environment migration and transformation. For a detailed description of the process, read the reference developed by Mackey.

2.2 Case region PAH process and parameters

PAHs in the migration and transformation process of the study area mainly include: (1) the input process of PAHs in the study area include fuel motor vehicle exhaust emission, heating coal emissions, etc., (2) the exchange process between different environmental media including the settlement

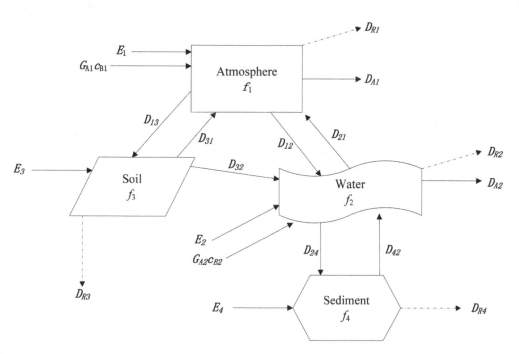

Figure 1. Model frame of the study area.

of exchange between gas and water, the water particle/resuspension, etc., (3) output process including PAHs degradation in multi-media, buried in the sediment, gas-water advection output, etc.

Model parameters are classified into the input parameters and the certifiable parameter. Input parameters including the nature of PAHs (Table 1 and Table 2), the environmental parameter value in the study area (Table 3) and environmental migration parameters (Table 4) (Liaoning province bureau of statistics 2003; Wania & Mackay 1999). The certifiable parameter for each material is in the measured average concentration in the medium.

Furthermore, through solving these equations in section 2.1, the fugacity capacity of each medium and the migration of energy between different media

Table 1. Degradation half-life in various media of PAHs/h (Sun 2005).

PAHs	Atmosphere	Water	Soil	Sediment
Phenanthrene (PHE)	55	550	5500	17000
Pyrene (PYR)	170	1700	17000	55000
Benz[a]anthracene (BaA)	170	1700	17000	55000
Chrysene (CHR)	170	1700	17000	55000
Benzo[a]pyrene (BaP)	170	1700	17000	55000

Table 2. Physical and chemical properties of PAHs.

PAHs	Vapour pressure /Pa	Fusing point /°C	Octanol-water partition coefficient	Henry's constant/ Pa·(m^3·mol)$^{-1}$
PHE	0.02	101	4.57	3.24
PYR	0.0006	156	5.18	0.92
BaA	0.000606	162	5.91	0.581
CHR	0.00000057	255	5.61	5.86
BaP	0.0000007	175	5.94	4.6×10^{-2}

Table 3. Environmental parameters value.

Main medium	Atmos.	Coastal seawater	Soil	Sedi.
Area/m^2	3.0×10^8	1.1×10^8	7.7×10^7	1.1×10^8
Depth/m	1000	3.78	0.05	0.02
Volume/m^3	2.3×10^{11}	4.2×10^8	3.8×10^6	2.2×10^6
Density/kg/m^3	1.19	1000	1578	1420
Organic carbon content	0.2	5.7×10^{-3}	5.7×10^{-3}	1.1×10^{-2}
Meteorological volume ratio	1	NA	0.2	0
Water phase volume ratio	NA	1	0.3	0.7
Particle volume ratio	9.4×10^{-11}	2.23×10^{-4}	0.5	0.3

Table 4. Input parameters and coefficients.

Parameters	Symbols	Value
Gas lateral mass transfer coefficient of air/water interface	k_{VA}	3
Gas/water interface lateral mass transfer coefficient	k_{VW}	0.03
Precipitation rate/m/h	U_R	6.47×10^{-5}
Clearance rate	Q	200000
dry deposition velocity/m/h	U_P	10.8
Gas lateral mass transfer coefficient of air/soil interface	k_{AS}	1
Runoff rate/m/h	U_{WW}	2.59×10^{-5}
Loss rate/m/h	U_{SW}	2.30×10^{-8}
Water/sediment interface lateral mass transfer coefficient	k_{SedW}	0.01
Sediment deposition rate/m/h	U_{BS}	1.37×10^{-7}
Sediment resuspension rate/m/h	U_{RSed}	9.13×10^{-8}
Sediment burial rate/m/h	U_{BS}	1.37×10^{-7}

were calculated. In addition, the studied 5 PAHs concentrations in different environmental media and the migration flux between different media were finally calculated by the equation of $c = Zf$.

3 RESULTS AND DISCUSSION

3.1 5 PAHs mass distribution percents in the major media

Mass distribution percents of 5PAHs in the major media are shown in Figure 2. Figure 2 shows that PHE is the highest level in the atmosphere, as high as 70%. PYR and BaA had similar contents of the atmosphere, accounting for 4.6% and 6.4%, respectively. In addition, there are little CHR and BaP in the atmosphere. Furthermore, except for PHE, other studied PAHs had higher concentration in the soil than in other environmental media. Also, a certain amount of PAHs had been in sediment, but the content is all below 1%. In short, for most PAHs, the soil is their main enrichment environment medium and the atmosphere is the main enrichment medium for the PAHs, which is composed of a number of low rings. The cause of this difference may be due to their different physical and chemical properties.

3.2 Simulation of PAHs migration in the multi-media

Based on the Eqs. (1)~(4), the PAHs migrations in the multi-media were simulating and the results

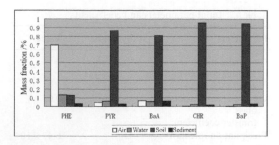

Figure 2. Mass distribution percents of five PAHs in multimedia.

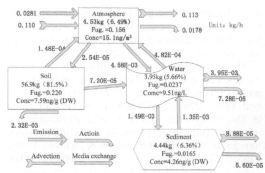

Figure 5. Estimated rates of chemical movement and transformation of BaA in multimedia.

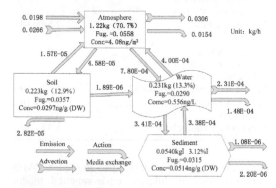

Figure 3. Estimated rates of chemical movement and transformation of PHE in multimedia.

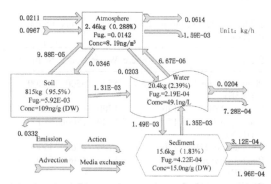

Figure 6. Estimated rates of chemical movement and transformation of CHR in multimedia.

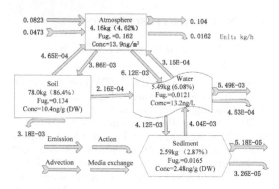

Figure 4. Estimated rates of chemical movement and transformation of PYR in multimedia.

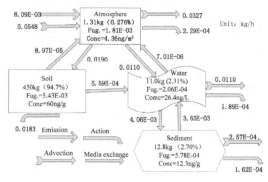

Figure 7. Estimated rates of chemical movement and transformation of BaP in multimedia.

are shown in Figs. 3~7. Figs. 3~7 show that PHE mainly derive from the direct discharge of the atmosphere and the atmospheric advection input. In addition, the atmospheric advection output and degradation are the main way to decrease. BaA and PYR mainly derive from the atmospheric advection input, and the atmospheric advection output and degradation are also its main way to decrease (Fig. 4 and Fig. 5). Degradation in

the soil loss rate increased and degradation loss decreased in the atmosphere obviously for BaP and CHR (Figs. 6 and 7).

In conclusion, for all of the studied PAHs, the direct emissions and atmospheric advection input are the main sources of PAHs, and the advection output loss is their main loss way. Along with the

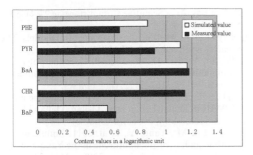

Figure 8. Comparison between calculated and measured concentrations of PAHs in air.

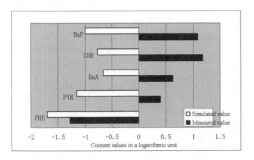

Figure 9. Comparison between calculated and measured concentrations of PAHs in Sediment.

increasing ring number, the soil degradation loss ratios of those PAHs increase gradually

3.3 *Model validation and analysis*

Really, five PAH concentrations in the atmosphere and sediment concentration were used to test and verify the model simulation results. The model validates data from the following two reports: (1) Wan XL's report on the Dalian city PAH concentrations in the atmosphere (Wan & Yang 2003); (2) Liu XM's study on PAH concentration in sediment in the Dalian bay (Liu et al. 2001). The comparison results are shown in Figs. 8 and 9.

Based on Figs. 8 and 9, five PAHs show good consistency between the simulation values and the measured values. The difference between the measured values and predicted values are within 0.4 logarithmic unit. The results verified the validity of the model. Furthermore, model calculation value is on the high side in the sediment except PHE. The reason may be that the study area includes 1 km offshore waters and, key hydrological parameters in the waters such as flow rate, flow velocity, water sediment volume change due to seasonal variation, influence of the tides, etc.

4 CONCLUSION

1. Based on the classic fugacity theory developed by Mackay, the migration model based on Level 3 steady state theory in environmental multimedia were developed combined with the geographical and meteorological field information in case city.
2. Five studied PAHs discharged into the environment system through the advective input and direct atmospheric emission, and those low ring components mainly existed in atmosphere, whereas moderate and high ring components mainly existed in the sediment and soil.
3. For all of the studied PAHs, the direct emissions and atmospheric advection input are the main sources of PAHs, and the advection output loss is their main loss way. Along with the increasing ring number, the soil degradation loss ratios of those PAHs increase gradually.
4. Difference between the measured values and predicted values are within 0.4 logarithmic unit and therefore the results verified the validity of the model.

ACKNOWLEDGMENTS

This study was financially supported by the Graduate Innovative Education Program of Zhongnan University of Economics and Law (2013 JY07) and the National Natural Science Foundation of China (51308076; 71503268).

REFERENCES

Dong, J., Wang, S., Gao, H., et al. 2008. Simulation of multi-media transfer and fate of benzo[a]pyrene in Lanzhou Region. Ecology and Environment 17(6): 2150–2153. (in Chinese).

Douben P.E.T. 2003. PAHs: an ecotoxicological perspective. New York: Wiley.

Huang, J.H., Liu, W.C., Zeng, G.M., et al. 2016a. An exploration of spatial human health risk assessment of soil toxic metals under different land uses using sequential indicator simulation. Ecotoxicology and Environmental Safety 129: 199–209.

Huang, J.H., Li, F., Zeng, G.M., et al. 2016b. Incorporating hierarchical bioavailability and possible receptor distribution into potential eco-risk assessment of heavy metals in soils: A case study in Xiandao District, Changsha city, China. Science of the Total Environment 541: 969–976.

Jiao, L., Meng, W., Zheng, B., et al. 2010. Distribution of polycyclic aromatic hydrocarbons (PAHs) in different size fractions of sediments from intertidal zone of Bohai Bay, China. China Environmental Science 30(9): 1241–1248. (in Chinese).

Jones, K.C., Voogt, P.D. 1999. Persistent organic pollutants (POPs): State of the Science. Environment Pollution 100(1–3): 209–221.

Li, F., Huang, J.H., Zeng, G.M., et al. 2013. Spatial risk assessment and sources identification of heavy metals in surface sediments from the Dongting Lake, Middle China. Journal of Geochemical Exploration 132: 75–83.

Li, F., Huang, J.H., Zeng, G.M., et al. 2015a. Toxic metals in topsoil under different land uses from Xiandao District, middle China: Distribution, relationship with soil characteristics and health risk assessment. Environmental Science and Pollution Research 22: 12261–12275.

Li, F., Huang, J.H., Zeng, G.M., et al. 2015b. Spatial distributions and health risk assessment of heavy metals associated with receptor population density in street dust: A case study of Xiandao District, Middle China. Environmental Science and Pollution Research 22: 6732–6742.

Li, F., Zhang, J.D., Huang, J.H., et al. 2016. Heavy metals in road dust from Xiandao District, Changsha city, China: Characteristics, health risk assessment and integrated source identification. Environmental Science and Pollution Research. DOI: 10.1007/s11356–016–6458-y.

Liaoning province bureau of statistics. Liaoning statistical yearbook 2003. Peking: China Statistic Press, 2003. (in Chinese).

Liu, J., Liang, J., Yuan, X.Z., et al. 2015. An integrated risk model for assessing heavy metal exposure to migratory birds in wetland ecosystem: a case study in Dongting Lake wetland, China. Chemosphere 135: 14–19.

Liu, X., Xu, X., Zhang, X., et al. 2001. A preliminary study on PAHs in the surface sediment samples from Dalian Bay. Acta Scientiae Circumstantiae 21(4): 507–509. (in Chinese).

Mackay D. 2001. Multimedia environmental models: The fugacity approach, 2ed. FL, USA: CRC Press.

Ma W., Li, Y., Suo D., et al. 2009. Gaseous polycyclic aromatic hydrocarbons in Harbin air. Environmental Science 30(11): 3167–3172. (in Chinese).

Sun, H., Source and multimedia environmental behavior of PAHs in Dalian urban area. Dalian: Dalian Unversity of Technology, 2005. (in Chinese).

Wang Z., Liu M., Yang, Y., et al. 2011. Simulation of multimedia fate of phenanthrene in the Yangtze Estuary. Environmental Science 32(8): 2444–2449. (in Chinese).

Wania, F., Mackay, D. 1999. The evoluation of mass balance models of persistent organic pollutant fate in the environment. Environmental Pollution 100(3): 223–240.

Wan, X., Yang, F. 2003. Distribution and seasonal changing and source identification of polycyclic aromatic hydrocarbons in atmosphere of Dalian. Journal of Dalian University of Technology (43): 170–173. (in Chinese).

Yuan, Y.J., Zeng, G.M., Liang, J., et al. 2015. Variation of water level in Dongting Lake over a 50-year period: Implications for the impacts of anthropogenic activity and climate change. Journal of Hydrology 525:450–456.

Zeng, X.X., Liu, Y.G., You, S.H., 2015. Spatial distribution, health risk assessment and statistical source identification of the trace elements in surface water from the Xiangjiang River, China. Environmental Science and Pollution Research 22(12): 9400–9412.

254

Energy Science and Applied Technology ESAT 2016 – Fang (Ed.)
© *2016 Taylor & Francis Group, London, ISBN 978-1-138-02973-6*

Evaluation of water resources utilization based on AHP and the fuzzy comprehensive evaluation method

Y. Kang, L.X. Ren, Y.Z. Chen, H.H. Zhao & H.W. Lu
North China Electric Power University, Beijing, China

ABSTRACT: This study applied the AHP (Analytic Hierarchy Process) and fuzzy comprehensive evaluation method to evaluate the exploitation and utilization of water resources. The AHP can calculate the weight of various indicators in the evaluation system, and fuzzy comprehensive evaluation method can evaluate the exploitation and utilization degree of water resources. The approach is applied to the Yancheng coastal reclamation area in China, and the results showed that the exploitation and utilization degree of water resources in Yancheng is unevenly distributed in different counties. This study have great significance to the large-scale land reclamation in Jiangsu Province.

Keywords: water resources; evaluation of exploitation and utilization; analytic hierarchy process; fuzzy comprehensive evaluation

1 INTRODUCTION

Water is one of the most important natural resources. With the swift development of economy and the increasing human's activities at present, water scarcity is considered as one of the most important threats for human societies and a constraint for sustainable development. Therefore, it is urgent to exploit and utilize water resource scientifically and reasonably, which could be helpful for the sustainable exploitation and utilization of water resources, the national economy and human development.

Many optimization methods have been proposed to deal with water resources exploitation and utilization evaluation problems. For example, Yang et al. (2003) introduced a Genetic Projection Pursuit Interpolation Model (GPPIM) to evaluate water resources exploitation and utilization; Zang et al. (2016) used the Three Red Lines method to make quantitative characterization and comprehensive evaluation of regional water resources. However, due to the inherently complex and uncertain conditions, this paper uses the fuzzy comprehensive evaluation mathematical model to evaluate the exploitation and utilization of water resources.

In this paper, a comprehensive evaluation system of Yancheng coastal reclamation area of water resources exploitation and utilization is established. After determining the evaluation level of the indicators, AHP is used to calculate the weight of each index. Then fuzzy comprehensive evaluation method is used to evaluate the exploitation and utilization of water resources in Yancheng coastal reclamation area.

2 METHODOLOGY

2.1 *Analytic Hierarchy Process*

AHP is one of the most popular Multi-Criteria Decision-Making (MCDM) tools which has been widely applied in many decision-making situations (Wan et al. 2015; Gangadharan, R. et al. 2016). The basic steps of this method can be described as following:

1. Establish the judgment matrix. According to the affiliation between the upper and lower-level indicators which determined by the level indicator system, the same level indicators are pairwise compared and the judgment result can be showed as T.L. Saaty 1–9 Scaling method.
2. Use the sum-product method to calculate the relative weight of each indicator. First, normalize each column of the judgment matrix.

$$\overline{u_{ij}} = \frac{u_{ij}}{\sum_{k=1}^{n} u_{kj}} (i, j = 1, 2, \cdots, n) \qquad (1)$$

Add together all elements of each row of the judgement matrix, and then normalize each column of the judgment matrix.

$$\overline{\omega}_i = \sum_{j=1}^{n} \overline{u}_{ij} \, (i=1,2,\cdots,n) \qquad (2)$$

$$\omega_i = \frac{\overline{\omega}_i}{\sum_{j=1}^{n} \overline{\omega}_j} \, (i=1,2,\cdots,n) \qquad (3)$$

Next calculate the maximum eigenvalue.

$$\lambda_{\max} = \frac{1}{n} \sum_{i=1}^{n} \frac{(P\omega^T)_i}{\omega_i} \qquad (4)$$

3. Put the judgement matrix into consistency check. Define the consistency Index CI as follows:

$$C I = \frac{1}{n-1}(\lambda_{\max} - n) \qquad (5)$$

$$C I / C R \begin{cases} =0 & \text{if consistent} \\ >0 & \text{otherwise} \end{cases} \qquad (6)$$

$$C R = C I / R I \qquad (7)$$

where CR is the consistency ratio and RI is the random consistency index.

When $CI/CR<0.1$, which holds that the consistency of the judgment matrix is satisfied and the weight distribution is reasonable; Otherwise, adjust the judgment matrix until a satisfactory consistency.

2.2 *Fuzzy comprehensive evaluation method*

Based on the membership degree theory of fuzzy Mathematics, fuzzy comprehensive evaluation is an application of fuzzy mathematics which is used to evaluate all relevant factors to make a comprehensive evaluation. Because of the multi-classification and large number of evaluation factors, this study uses multi-level comprehensive evaluation model. The basic steps are shown in Figure 1.

1. Determine the evaluation object to make clear the target of the evaluation and the intended target or desired results of the analysis.
2. Determine the factors set of the evaluation object. Factors are used to characterize the various properties or performances of evaluation object. On different occasions, it was also known as parameters or quality indicators, which synthetically reflects the quality of the object.
3. Determine the comments set of evaluation object. Comments set $V = \{v_1, v_2, \dots, v_m\}$ is a collection of all the evaluation results that the evaluators make.
4. Establish the weight set. Assuming that $A = (a_1, a_2, \dots a_n)$ is the fuzzy vector for weight distribution, where ai (ai > 0) is the weight of factor i, reflecting the importance of each factor.

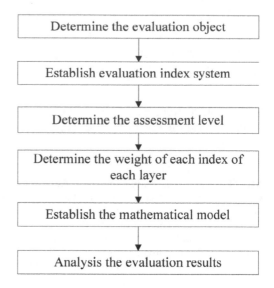

Figure 1. Fuzzy comprehensive evaluation flowchart.

5. Single factor fuzzy evaluation, namely to establish a fuzzy mapping from U to $F = (V)$..

$$f : U \to F(V), \forall u_i \in U, u_i| \to f(u_i)$$
$$= \frac{r_{i1}}{v_1} + \frac{r_{i2}}{v_2} + \dots + \frac{r_{im}}{v_m} \qquad (8)$$

where rij represents the membership that ui belongs to vj.

By calculating $f(\mu_i)$, single factor evaluation set can be obtained. The matrix composed of the single factor evaluation set is called single factor evaluation matrix.

$$R_i = (r_{i1}, r_{i2}, \dots, r_{im}), i=1,2,\dots,n \qquad (9)$$

$$R_i = \begin{bmatrix} r_{11} & r_{12} & \dots & r_{1m} \\ r_{21} & r_{22} & \dots & r_{2m} \\ \dots & \dots & \dots & \dots \\ r_{n1} & r_{n2} & \dots & r_{nm} \end{bmatrix} \qquad (10)$$

6. Conduct fuzzy comprehensive evaluation. The corresponding weight of the factors can reasonably reflect the fuzzy comprehensive effect of all factors.

$$B = A \bullet R = (a_1, a_2, \dots, a_n) \begin{bmatrix} r_{11} & r_{12} & \dots & r_{1m} \\ r_{21} & r_{22} & \dots & r_{2m} \\ \dots & \dots & \dots & \dots \\ r_{n1} & r_{n2} & \dots & r_{nm} \end{bmatrix} \qquad (11)$$
$$= (b_1, b_2, \dots, b_m)$$

$$b_j = \overset{n}{\underset{i=1}{V}}(a_i \wedge r_{ij}) \qquad (12)$$

where a_i is the weight of the index and b_j represents fuzzy comprehensive evaluation factor.

7. Deal with the evaluation factor. This study chooses multiplication addition method to deal with the evaluation results.

$$B = A \bullet R = d_j \ (j = 1,2,3,\dots,n) \qquad (13)$$

$$d_j = \sum (a_j \wedge r_{ij}) \qquad (14)$$

where $\wedge$ means that the minimum value of the composite operation should be chose, d_j is the membership that sample for standard.

3 RESULTS

According to the actual situation of Yancheng coastal reclamation area, this study chooses social subsystem, economic subsystem, water resources subsystems and ecological subsystems as the basic layer. The weight of all indexes of Yancheng coastal reclamation area are listed in Table 1.

The results of the level eigenvalues of Yancheng coastal reclamation area can be obtained by using the multi-level fuzzy comprehensive evaluation model, which are list in Table 2. The evaluation results of the counties are showed in Figure 2.

Table 2 and Figure 2 show that the exploitation and utilization of water resources in Xiangshui are in the lower middle level and keep a rising momentum, the economic development and ecological protection are relatively backward. The situation of Binhai is similar to Xiangshui. The exploitation and utilization degree of water resources of Sheyang is close to the moderate level but the economic development is in a poor level, which indicates that there are large waste and low efficiency of the use of water resources. Tinghu's water resources utilization degree is on the middle level and of which the economic development is the best among all counties in Yancheng. In the future it should pay more attention on the social and ecological development. The water resources exploitation degree of Dafeng is close to the middle level.

Table 2. The level eigenvalues of each county of Yancheng coastal reclamation area.

County	Social	Economic	Water resources	Ecological	Total
Xiangshui	2.5408	2.3581	3.1454	2.3852	2.6191
Binhai	2.3495	2.3089	3.0861	2.3378	2.5538
Sheyang	3.0683	2.4752	3.1998	2.3251	2.7299
Tinghu	1.9772	3.6101	3.0413	1.9042	3.0397
Dafeng	3.3785	2.6728	3.3629	2.4165	2.9123
Dongtai	2.7131	2.8324	3.4961	2.0687	2.9141

Table 1. The weight of all indexes of Yancheng coastal reclamation area.

Target layer	Rule layer and weight	Index layer	Weight
Water Resources Exploitation and Utilization Evaluation	Social subsystem (0.0933)	Population density	0.2375
		Urbanization rate	0.1362
		Arable land per capita	0.6263
	Economic subsystem (0.4584)	GDP per capita	0.1102
		Unilateral water GDP	0.4039
		Water consumption per 10,000 GDP value	0.0488
		Effective utilization coefficient of irrigation water	0.2697
		Water consumption per 10,000 yuan of industrial added value	0.1674
	Water resources subsystem (0.3049)	Water resources per capita	0.1354
		Water production modulus	0.3656
		Water supply per capita	0.0801
		Groundwater Water ratio	0.0519
		Water supply modulus	0.1354
		Water resources exploitation rate	0.2316
	Ecological subsystem (0.1434)	Sewage runoff ratio	0.2503
		Ecological and environmental water use rate	0.7497

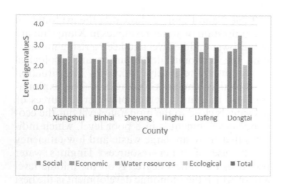

Figure 2. Evaluation results of each counties of Yancheng coastal reclamation area.

The score of social subsystem is the highest indicates that the social factors are very stable, while the lower score of ecological subsystem indicates that the amount of water used on ecological and environmental protection needs to be improved. Dongtai's water resources utilization degree is close to the middle level. The highest score of all subsystems is water resources subsystem so it should pay attention to rational exploitation and utilization of water resources and prevent water waste.

4 CONCLUSIONS

In this paper, AHP and fuzzy comprehensive evaluation method are applied to evaluate the exploitation and utilization of water resources of Yancheng coastal reclamation area. Through the analysis about the present situation of the exploitation and utilization of water resources in Yancheng, the main factors that affect the degree of exploitation and utilization of water resources were concluded. Then according to the evaluation results and combined with the actual situation of Yancheng, the problems existing in the exploitation of various counties were analysed.

ACKNOWLEDGEMENTS

This research was supported by the China National Funds for Excellent Young Scientists (51422903), National Natural Science Foundation of China (41271540), Program for New Century Excellent Talents in University of China (NCET-13-0791), and Fundamental Research Funds for the Central Universities.

REFERENCES

Gangadharan, R et al. 2016. Assessment of groundwater vulnerability mapping using AHP method in coastal watershed of shrimp farming area. *Arabian Journal of Geosciences.* 9(2): 1–14.

Wan, K.Y. & Hu, Q.C. 2013. Water Resources Safety Status Evaluation and Trends Prediction in Hangzhou City Based on AHP Method. *Water Resources & Power.* 2013.

Yang, X.H et al. 2003. A genetic projection pursuit interpolation model for comprehensive assessment of development level of regional water resources. *Journal of Natural Resources.* 18(6): 760–765.

Zang, Z et al. 2016. Quantitative characterization and comprehensive evaluation of regional water resources using the Three Red Lines method. *Journal of Geographical Sciences.* 26(4): 397–414.

Energy Science and Applied Technology ESAT 2016 – Fang (Ed.)
© 2016 Taylor & Francis Group, London, ISBN 978-1-138-02973-6

Multimedia exposure risk assessment for lead in parts of the Xiandao district based on the IEUBK model

Liyun Zhu, Jingdong Zhang, Fei Li & Chaoyang Liu
Research Center for Environment and Health, Zhongnan University of Economics and Law, Wuhan, China

ABSTRACT: The harmfulness of lead in environmental multimedia is latent and cumulative, which is especially harmful for children and pregnant women. There will be irreversible damages to the nervous system and brain development when children's blood lead (PbB) level exceeds 10 ug/dL. However, considering the cost and infectious blood diseases, children's PbB levels are difficult to be widely monitored by medical methods. The Integrated Exposure Uptake Biokinetic (IEUBK) Model is used to estimate the PbB levels of children (0–7 years old) who are exposed to multimedia environmental lead. This article selected seven sampling sites to estimate the local children's PbB levels. All the selected parameters are based on EPA, but most of them are adjusted to be suitable for the actual situation of the sampling sites and Chinese children's behaviors through investigations and experiments. Fortunately, according to the model, the calculated results of samples from 0–7-year-old children's PbB did not exceed the social interference value (10 ug/dL). However, the PbB concentration level geometric mean value of Xiazepu town is 2.526 ug/dL, and the proportion of over 10 ug/dL PbB concentration level is 17.1%. Both of the values are higher than the other sampling sites in the Xiandao district. Therefore, the soil Pb concentration in Xiazepu town needs to be seriously detected and controlled.

Keywords: blood lead; multimedia environment; heavy metal; risk assessment; IEUBK

1 INTRODUCTION

With the rapid development of urbanization and industrialization all over the world, the heavy metal pollution especially in the soil was already considered as one of the most important environmental problems. Human behavior is the chief cause of heavy metal pollution in the soil, such as polluted water irrigation, improper stacking of industrial solid waste, and extensive use of agricultural and chemical fertilizers (Chen et al. 2015; Huang et al. 2016a,b; Li et al. 2015a,b; Li et al. 2016; Zeng et al. 2015).

Lead is a widely existing heavy metal that can be exposed to the crowd through the soil, air, dust, drinking water and so on. The harmfulness of lead in environmental multimedia is latent and cumulative, which is especially harmful for children and pregnant women. There will be irreversible damages to the nervous system and brain development when children's blood lead (PbB) level exceeds 10 ug/dL. However, considering the cost and infectious blood diseases, children's PbB levels are difficult to be widely monitored by medical methods. The Integrated Exposure Uptake Biokinetic (IEUBK) Model is used to estimate the PbB levels

of children (0–7 years old) who are exposed to the multimedia environmental Pb. Xiandao district is an advanced demonstration area of developing the resource conservation and environment friendly society approved by the Chinese State Council. This article selected seven sampling sites to assess the local children's PbB levels based on IEUBK. The objectives of this study were to (1) detect sampling sites' soil lead concentration, (2) introduce the model of IEUBK and adjust the parameters to suit the actual situation, and (3) predict 0–7-year-old children's PbB levels and analyze the results.

2 MATERIALS AND METHODS

2.1 *Study area*

Changsha is located in the northeastern part of Hunan province. The area of Changsha is approximately 11,819 square kilometers, and the urban area is nearly 1,910 square kilometers. Xiandao district is the main pilot district of Changsha. Nearly 1.2 million permanent residents live here, and the area of Xiandao district is approximately 1,200 square kilometers. Moreover, the land uses in the Xiandao district include traditional agriculture, forest, industry, and resident uses. Xiandao district is a national pilot district of constructing the resource conservation and environment friendly society approved by the Chinese State Council. Therefore, Xiandao district was selected as the study area, which has been experiencing the results of environmental pollution due to the rapid development of economy.

2.2 Sample collection

Seven residential areas in Xiandao district were selected as the main investigation areas, respectively, named as Leifeng town, Hanpu town, Lianhua town, Pingtang town, Yuchangping town, Xiazepu town, and Huangjin town. According to the area characteristics and the rules of pollutant migration and diffusion, the radial distribution sampling method was selected. Pollution monitoring points' selection and layout were based on the main investigation areas, beyond the leading and dominant wind leeward. With reference to soil environmental monitoring technical specifications of the HJ/166–2004, nine sampling points were confirmed in every investigation area, and each point has one parallel sample. The samples were vertically collected downward 0–30 cm from the topsoil. Samples in each sampling point were totally mixed by quartering, and were well preserved in polyethylene bags with label for details.

2.3 Laboratory analysis

Soil samples were collected in a freeze dryer drying to a constant weight. In the first step, the obvious stones and dead organisms were picked out and the soil samples were repeatedly crushed and grinded by wooden sticks until they can be smoothly filtered through the 0.2 mm nylon sieve. Then, 0.1000 g soil samples were settled in a 50 mL Teflon crucible. Then digested and heated with HCl, HNO_3, HF, and $HClO_4$. Subsequently, atomic absorption spectrometry was used to analyze the total concentrations of lead based on GB/T 17141–1997.

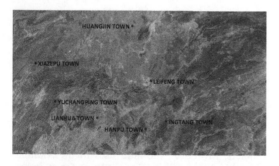

Figure 1. Topsoil sampling sites in Xiandao district.

3 THE INTEGRATED EXPOSURE UPTAKE BIOKINETIC MODEL (IEUBK)

The Integrated Exposure Uptake Biokinetic Model (referred to in this article simply as IEUBK) is a model design for children (under the age of seven) who are exposed to environmental Pb from a lot of sources and used to estimate the PbB concentration level geometric mean values and the proportions of PbB concentration level over 10 ug/dL. There are four modules of the IEUBK, exposure, uptake, biokinetic, and probability distribution, respectively. (James & Michael 1992; Xu 2010; Wang et al. 2011; Zhang et al. 2013).

3.1 Expose module

This module is used to calculate the intake rates of Pb that is based on different media Pb concentrations and different media intake rates. Pb is often exposed in air (indoor or outdoor), dust, water, soil, diet, etc. Children can take in Pb from all the listed sources through touching, breathing, and ingesting. The exposure module calculates how much Pb may enter children' body of one confirmed polluted media by the following equation:

Pb Intake Rate = Media Pb Concentration *
Media Intake Rate   (1)

Some of the values of media Pb concentrations are based on experimental results. EPA defined the other values of media lead concentrations and all the media intake rates' suggested default values.

3.2 Uptake module

This module is used to predict the uptake rates of Pb based on the intake rates calculated by expose module. Uptake means the intake environmental Pb that through the lungs or stomachs finally transform to bloodstream's lead. The uptake module calculates how much intake lead can finally transform into the bloodstream's Pb by using the following equation:

Pb Uptake Rate
= Pb Intake Rate * Absorption Factor   (2)

The Pb intake rates can be calculated through expose model and the factors of absorption can consult the default values suggested by EPA. Both of the Pb intake rates and Pb uptake rates will be different depending on the varied environmental situations and residents' behaviors. Therefore, they need to be adjusted into the local level.

3.3 Biokinetic module

This module is used to describe the Pb transfer between the bloodstream and body tissues or organs. The Pb uptake in our bloodstream can also be excreted through urine, stools, and so on. However, the other PbB will be stored in our body tissues or organs, like liver and kidney. Both of the excreted and stored PbB can be calculated by the total Pb uptake rates, which are calculated by uptake module. This need a series of complex equations to obtain the results, so they will not be listed in this article. However, each equations and parameters can be obtained through the EPA web site. After all, the geometric mean of PbB concentration can be calculated.

3.4 Probability distribution module

This module is used to predict the probability distribution of PbB concentration in 0–7-year-old children based on the geometric mean PbB concentrations described by the biokinetic module. This module can also predict the quantity of the PbB concentration exceeding the standard level in children, approximately 10 ug/dL. This module also needs to use a lot of complex equations, so the equations will not be listed either.

4 SELECTED PARAMETERS

4.1 Air Pb concentration

The default value of air Pb concentration recommended by EPA is 0.1 ug/m³; however, China's urban air is severely polluted by traffic emissions, industrial dust, and exhaust gas. Therefore, according to the monitoring results of recent years, the average value of the city's air Pb concentration is 0.38 ug/m³. The ratio of indoor to outdoor air Pb concentration recommended by EPA is 30%. The default air intake rates of 0–7-year-old children recommended by EPA are 2 m³/d (0–1 year), 3 m³/d (1–2 years), 5 m³/d (2–5 years), and 7 m³/d (5–7 years), respectively. The times children spend outdoor are 1 hour/d (0–1 year), 2 hours/d (1–2 years), 3 hours/d (2–3 years), and 4 hours/d (3–7 years), respectively. The ventilation rates are 2 m³/day (0–1 year), 3 m³/day (1–2 years), 5 m³/day (2–5 years), and 7 m³/day (5–7 years), respectively. The ratio of lung absorption is 32% (Khoury & Diamond 2003; Diamond et al. 2006).

4.2 Diet Pb concentration

The default diet intake rates of 0–7-year-old children recommended by EPA are 5.53 ugPb/d (0–1 year), 5.78 ugPb/d (1–2 years), 6.49 ugPb/d

(2–3 years), 6.24 ugPb/d (3–4 years), 6.01 ugPb/d (4–5 years), 6.34 ugPb/d (5–6 years), and 7.00 ugPb/d (6–7 years), respectively. Considering the significant differences between Chinese and Western children's diet, after consulting the relevant literatures, the default values are adjusted to 2.26 ugPb/d (0–1 year), 1.96 ugPb/d (1–2 years), 2.13 ugPb/d (2–3 years), 2.04 ugPb/d, (3–4 years), 1.95 ugPb/d (4–5 years), 2.05 ugPb/d, (5–6 years), and 2.22 ugPb/d (6–7 years), respectively (Lanphear et al. 2005; Yang et al. 2016).

4.3 Drinking water Pb concentration

The default value of water Pb concentration recommended by EPA is 4 ug/L, the maximum permissible content of Pb in the < sanitary standard for drinking water > (GB5749–2006) of china is 10 ug/L, and the maximum permissible content of drinking water Pb concentration in the < underground water quality standard > (GB/T-14848–93) of china is 50 ug/L. Considering China's situation, the default value of water Pb concentration is set at 6 ug/L. The default drinking water intake rates of 0–7-year-old children recommended by EPA are 0.2 L/d (0–1 year), 0.5 L/d (1–2 years), 0.52 L/d (2–3 years), 0.53 L/d (3–4 years), 0.55 L/d (4–5 years), 0.58 L/d (5–6 years), and 0.59 L/d (6–7 years), respectively.

4.4 Dust and soil Pb concentration

Soil Pb concentration has already been measured. The indoor dust Pb concentration is nearly 150 ug/g. The default dust and soil Pb intake rates of 0–7-year-old children recommended by EPA are 0.85 g/d (0–1 year), 0.135 g/d (1–4 years), 0.100 g/d (4–5 years), 0.090 g/d (5–6 years), and 0.085 g/d (6–7 years).

5 RESULTS AND DISCUSSION

5.1 Soil lead concentrations

Figure 2 reveals that all the sampling areas' Pb concentration is stable at 20–40 mg/kg except Xiazepu town's Pb concentration. According to soil environmental quality standards in China (GB15618–1995), all the sampling areas' Pb concentrations are under the secondary standard (350 mg/kg).

5.2 Calculated progress

Figure 3 shows the detailed steps of IEUBK model, including inputs of all of the parameters and experimental results. There are three things valued to be output, PbB geometric mean, probability distribution, and proportion of over 10 ug/dL.

mg/kg

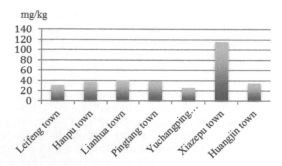

Figure 2. Lead concentration in 7 sampling areas.

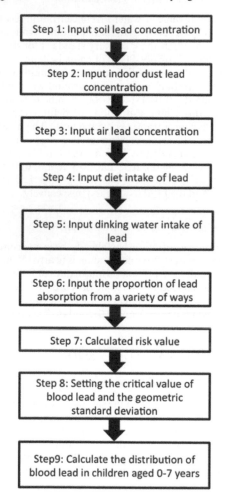

Figure 3. Flow chart of calculated progress.

5.3 *Results discussion*

The PbB concentration level geometric mean values of Leifeng town, Hanpu town, Lianhua town, Pingtang town, Yuchangping town, Xiazepu town, and Huangjin town are 2.101 ug/dL, 2.136 ug/

Table 1. Calculated results in Xiandao district.

Sampling site	Lead concentration mg/kg	PbB geometric mean ug/dL	Proportion of over 10 ug/dL
Hanpu town	38	2.136	5.1%
Hanpu town	38	2.151	5.4%
Pingtang town	39	2.141	5.2%
Yuchangping town	26	2.075	4.1%
Xiazepu town	116	2.526	17.1%
Huangjin town	35	2.121	4.8%
Leifeng town	31	2.101	4.5%

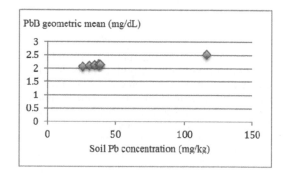

Figure 4. The relationship of Pb concentration and PbB geometric mean.

dL, 2.151 ug/dL, 2.141 ug/dL, 2.075 ug/dL, and 2.121 ug/dL, respectively. The proportions of PbB concentration level over 10 ug/dL are 4.5%, 5.1%, 5.4%, 5.2%, 4.1%, and 4.8%, respectively. Therefore, based on the set parameters, 0–7-year-old children's PbB levels in Xiandao district were at a lower risk.

When the concentrations and intake rates of dinking water, air, and diet are constant values, the prediction of children's blood lead level is positively correlated with the soil lead content.

The PbB concentration level geometric mean value of Xiazepu town is 2.526 ug/dL, and the proportion of over 10 ug/dL PbB concentration level is 17.1%. Both of the values are higher than the other sampling sites in Xiandao district. Therefore, the soil Pb concentration in Xiazepu town needs to be seriously detected and controlled.

6 CONCLUSION

1. The soil Pb concentration in Xiazepu town needs to be seriously detected and controlled.
2. Some selected parameters are uncertain and subjective. If time and budget allows, comparing PbB parts of concentration results that

are calculated by IEUBK model and detected by blood samples is beneficial for correcting parameters.

3. Soil Pb concentration standard setting needs to consider the quantitative effects of lead on human health based on the IEUBK model.

ACKNOWLEDGEMENTS

The financial support of this article was received from the National Natural Science Foundation of China (51308076; 71503268).

REFERENCES

Cornelis, C., Berghmans, P., Van, S.M., et al. 2006. Use of the IEUBK Model for Determination of Exposure Routes in View of Site Remediation. Hum Ecol Risk Assess 12(5): 963–982.

Chen, H., Teng, Y., Lu, S., et al. 2015. Contamination features and health risk of soil heavy metals in China. Science of The Total Environment 512–513: 143–153.

Duan, X.L. 2012. Research Methods of Exposure Parameters and its Application in Environmental Health Risk Assessment. Beijing: Science Press 2012: 122–141.

Huang, J.H., Liu, W.C., Zeng, G.M., et al. 2016a. An exploration of spatial human health risk assessment of soil toxic metals under different land uses using sequential indicator simulation. Ecotoxicology and Environmental Safety 129: 199–209.

Huang, J.H., Li, F., Zeng, G., et al. 2016b. Incorporating hierarchical bioavailability and possible receptor distribution into potential eco-risk assessment of heavy metals in soils: A case study in Xiandao District, Changsha city, China. Science of the Total Environment 541: 969–976.

Khoury, G.A. & Diamond, G.L. 2003. Risks to Children from Exposure to Lead in Air during Remedial or Removal Activities at Superfund Sites: A Case Study of the RSR Lead Smelter Superfund Site. Journal of Exposure Analysis and Environmental Epidemiology 13(1): 51–65.

Li, F., Huang, J.H., Zeng, G.M., et al. 2015a. Toxic metals in topsoil under different land uses from Xiandao District, middle China: Distribution, relationship with soil characteristics and health risk assessment. Environmental Science and Pollution Research 22: 12261–12275.

Li, F., Huang, J.H., Zeng, G.M., et al. 2015b. Spatial distributions and health risk assessment of heavy metals associated with receptor population density in street dust: A case study of Xiandao District, Middle China. Environmental Science and Pollution Research 22: 6732–6742.

Liu J.Y, Liu J., Yuan X.Z., et al. 2015. An integrated risk model for assessing heavy metal exposure to migratory birds in wetland ecosystem: a case study in Dongting Lake wetland, China. Chemosphere135: 14–19.

Lanphear, B.P., Hornung, R., Khoury, J., et al. 2005. Low-level Environmental Lead Exposure and Children's Intellectual Function: An International Pooled Analysis. Environmental Health Perspectives 113(7): 894–899.

James, C.C. & Michael, J.W. 1992. Predicting Blood Lead Concentrations from Environmental Concentrations [J]. Regulatory Toxicology and Pharmacology 16(3): 280–289.

Wang, B., Shao, D.C., Xiang, Z.H., et al. 2011. Contribution of Environmental Lead Exposure to Blood Lead Level among Infants Based on IEUBK Model. Journal of Hygiene Research 40(4): 478–480.

Xu, S. 2010. Health Risk Assessment via IEUBK model and Epidemiological Investigation for Environmental Lead Exposed Children. Wuhan: Huazhong University of science and technology, 2010: 11–21.

Yang, K.L., Zhang, H.Z, Zhang, Z.G, et al. 2016. Localization Study of Environmental Health Risk Assessment Model for Lead Exposure. China Population, resources and environment 2: 163–169.

YuanY.J., Zeng G.M., Liang J., et al. 2015. Variation of water level in Dongting Lake over a 50-year period: Implications for the impacts of anthropogenic activity and climate change. Journal of Hydrology 525: 450–456.

Zeng, X.X., Liu, Y.G., You, S.H., et al. 2015. Spatial distribution, health risk assessment and statistical source identification of the trace elements in surface water from the Xiangjiang River, China. Environmental Science and Pollution Research 22(12): 9400–9412.

Zhang, H.Z., Yan, P.S., Yang, K.L., et al. 2013. The Prediction Model of Air Pollution and Blood Lead of Children. Wuhan: Huazhong University of Science and Technology Press 2013: 94–110.

Energy Science and Applied Technology ESAT 2016 – Fang (Ed.)
© *2016 Taylor & Francis Group, London, ISBN 978-1-138-02973-6*

Variation of stable isotopes of the precipitation at Kun Ming station in China

Haiying Hu & Wuhua Li

School of Civil and Transportation Engineering, South China University of Technology, Guangzhou, China

ABSTRACT: Based on the stable isotopic composition in precipitation and meteorological data at Kun Ming station during 1986–2003, the regression analysis, Mann-Kendall, and correlation analysis methods are used to find out the variation law of stable isotopic composition in precipitation and its influencing factors. The results show that the oxygen isotopic composition of precipitation in the rainy season (May-October) is low, while the value in the dry season (November-April) is high. The highest frequency of the maximum values occurs in April, and the highest frequency of minimum values appears in July. It also shows that the correlation coefficients of stable isotopic composition in monthly precipitation and monthly average rainfall amount, air temperature, and water vapor pressure at Kun Ming station are all negative.

Keywords: precipitation; hydrogen and oxygen isotope; variation characteristic; Kun Ming station

1 INTRODUCTION

The isotope hydrology has been an important tool in studding the origin and dynamics of surface and ground water (Harrington G A et al, 2002), soil water infiltration (Kwang-Sik Lee et al, 2007), atmospheric circulation, and paleoclimatic investigations (Tsutomu Yamanaka et al, 2004) (Araguas-Araguas L et al, 2000). The International Atomic Energy Agency (IAEA), in co-operation with the World Meteorological Organization (WMO), is conducting a worldwide survey of oxygen and hydrogen isotope content in precipitation. The program was initiated in 1958 and the large-scale and organized collection of precipitation samples was initiated in 1961. Precipitation samples for the environmental isotopic compositions in precipitation and meteorological elements are collected monthly. The isotopic compositions in precipitation and meteorological data are regularly published by IAEA.

Like other hydrological variables, the time series of stable isotopes in precipitation affected by the comprehensive effects of meteorological factors and geographic parameters is a complex dynamic process. To analyze the change characteristic and impact mechanism of hydrogen and oxygen isotopes in precipitation, the stable isotope data of precipitation in Kun Ming station are selected for statistical analysis of monthly scale. The main objective of this study is to analyze the change trends of hydrogen and oxygen isotopes in precipitation on monthly scale, and to identify the possible influencing factors driving the observed changes of the hydrogen and oxygen isotope time-series.

2 DATA AND METHODS

2.1 Data

In this paper, the Kun Ming station was selected as the study region. Kun Ming station is located in the southwest of China (25°01'00" N, 112°41'01" E, 1892 m). It is in a typical monsoon climate zone. The data used in this paper are from the Global Network for Isotopes in Precipitation (GNIP) project. Kun Ming station are early established in China by the IAEA and the WMO for stable isotopic surveys of precipitation. The data of monthly average oxygen isotope (denoted by M_{18O}), monthly average air temperature (denoted by $M_{tempera}$), monthly average rainfall amount (denoted by $M_{rainfall}$), and monthly average water vapor pressure (denoted by $M_{pressure}$) from January 1986 to December 2003(Kun Ming station) can be obtained from the database, although little data were missed.

2.2 Methodology

A simple method to detect long-term trends in time series is linear regression analysis. Another highly recommended method for long-term trend detection is the Mann-Kendall test, which is originally used by Mann (Mann HB, 1945) and subsequently developed by Kendall (Kendall MG, 1975).

For a time series x_1, x_2, ..., x_n, where $n > 10$, the standard normal statistic, Zc, is given as

$$Z_c = \begin{cases} (S-1)\big/\sqrt{\mathrm{var}(S)}, & S > 0 \\ 0, & S = 0 \\ (S+1)\big/\sqrt{\mathrm{var}(S)}, & S < 0 \end{cases} \qquad (1)$$

where

$$S = \sum_{i=1}^{n-1} \sum_{k=i+1}^{n} \mathrm{sgn}(x_k - x_i) \qquad (2)$$

$$\mathrm{sgn}(\theta) = \begin{cases} 1, & \theta > 0 \\ 0, & \theta = 0 \\ -1, & \theta < 0 \end{cases} \qquad (3)$$

where n is the length of the data set, and x_i and x_k are the i-th and k-th element in the sequential time series, respectively. Under the null hypothesis of no trend, and the assumption that the series are independent and identically distributed, the variance of S, which is denoted by var (S), is described as

$$\mathrm{var}(S) = \frac{n(n-1)(2n+5)}{18} \qquad (4)$$

The estimated values of Z_c from equation (1) represent the trend of the time series. In the case of $Z_c > 0$, the time series show an increasing tendency, while the case of $Z_c < 0$ indicates a decreasing sequence. On condition that $|Z_c| > 1.96$, the trend is significant at the 95% confidence level. In addition, if $|Z_c| \leq 1.96$, the null hypothesis is accepted, which indicates that there is no obvious variation trend in the series.

3 RESULTS AND DISCUSSION

3.1 Trend analysis of monthly average oxygen isotopes

The M_{18O} in the period from January 1986 to December 2003 at Kun Ming station are shown in Figure 1. The mean value of M_{18O} was also plotted for reference. It can be seen that these values of M_{18O} fluctuate around the mean value, but are not symmetrically distributed. It can be seen that the monthly variations of δ_{18O} in precipitation are with markedly lower values in the rainy season and higher values in the dry season.

The occurrence times of the maximum value of M_{18O} max(M_{18O}), and the minimum values of M_{18O}, min(M_{18O}), in each calendar year were counted. All occurrence numbers of every month for max(M_{18O}) and min(M_{18O}) at Kun Ming station are shown in Table 1. It is shown that max(M_{18O}) occurred in April with a relatively high occurrence frequency and the occurrence number is 6 and takes 40% in all statistical years. On the other hand, the July has the highest occurrence frequency for min(M_{18O}) than other months and the occurrence number is 7 and it takes 46.7% in all years. It is also shown that a high occurrence frequency for max(M_{18O}) generally appears from February to April in these years at Kun Ming station, while that for min(M_{18O}) mostly emerges in the months from July to August.

3.2 Trend analysis of maximum and minimum monthly average oxygen isotopes

Figure 2 shows max(M_{18O}) and min(M_{18O}) in the years from 1986 to 2003 at Kun Ming station. It is shown that the max(M_{18O}) time series are scattered over the whole range of statistical years, and so as the

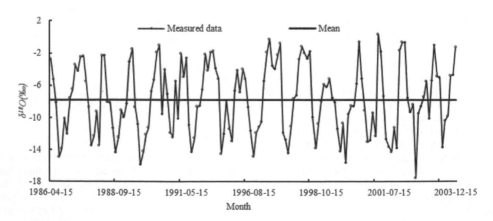

Figure 1. Monthly average oxygen isotopes from April 1986 to December 2003 at Kun Ming station.

Table 1. Occurrence frequency of every month for the maximum and minimum values of monthly average oxygen isotopes in a calendar year.

	Month	1	2	3	4	5	6	7	8	9	10	11	12	Total
max (M_{18O})	Number	0	5	3	6	1	0	0	0	0	0	0	0	15
	Frequency (%)	0	33.3	20	40	6.7	0	0	0	0	0	0	0	100
min (M_{18O})	Number	0	0	0	0	0	0	7	6	1	1	0	0	15
	Frequency (%)	0	0	0	0	0	0	46.7	40	6.7	6.7	0	0	100

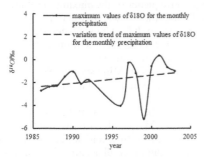

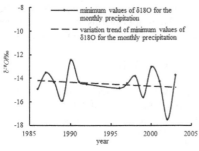

Figure 2. The maximum and minimum values of monthly average oxygen isotopes in every year from 1986 to 2003 at Kun Ming station.

min(M_{18O}) time series. According to the Mann-Kendall method, the value of Z_c for the max(M_{18O}) was calculated as 1.97. This positive Z_c indicates an increasing sequence, while the value of Z_c for the min (M_{18O}) was calculated as −0.22, which indicates a decreasing sequence. Therefore, max(M_{18O}) time series become more enriched along with the time, but min(M_{18O}) time series become more depleted. This change may result from the extreme weather. From the linear trend regression lines and the calculated results of the Mann-Kendall method, the long-term inter-annual variability is not significant.

3.3 Influence factors of oxygen isotopes in precipitation

By using Pearson correlation analysis, the correlation coefficients for M_{18O} time series and the meteorological elements at Kun Ming station were calculated and listed in Table 2, respectively. To study seasonal variations of M_{18O} time series

Table 2. Pearson correlation coefficients for the monthly average oxygen isotopes and meteorological elements at Kun Ming station.

Monthly average value	Rain amount	Air temperature	Water vapor pressure
correlation coefficient**	−0.609	−0.440	−0.665
significant probability	0.000	0.000	0.000

** Correlation is significant at the level of 0.01.

and the three meteorological elements, the mean monthly values of these parameters at Kun Ming station are plotted in Figure 3. As an example, the mean monthly value of M_{18O} is defined as the mean value of M_{18O} in the same months during the statistical period, and denoted as mean(M_{18O}).

It is seen from Table 2 that the correlation coefficient of M_{18O} and monthly average rainfall amount $M_{rainfall}$ at Kun Ming station, is −0.609, and the significant probability is 0 at the significance level of 0.01. It indicates that the variation trend between mean(M_{18O}) and mean($M_{rainfall}$) time series at Kun Ming station is also opposite. Moreover, it can be seen from Figure 3(a) that the oxygen isotopic composition is low in summer seasons with large rainfall amount, while the oxygen isotopic composition is high in winter seasons with small rainfall amount. The relationship between mean(M_{18O}) and mean($M_{rainfall}$) time series is significant negative correlation either in wet or dry seasons, which is called the amount effect (Zhang Xinping et al, 1998).

Table 2 shows that the correlation coefficient of M_{18O} and monthly average air temperature, $M_{tempera}$, is −0.440 at Kun Ming station. At the significance level of 0.01, the significant probability is 0, which indicates that the negative correlation between these two sets of data is significant. This indicates that the effect of air temperature is obvious for the long-term variation of M_{18O} at Kun Ming station. The mean(M_{18O}) and mean($M_{tempera}$) time series are plotted in Figure 3(b). It can be seen that the oxygen isotopic composition is low in the warm half of the year, while the oxygen isotopic composition is high in the cold half at Kun Ming station. It shows that there are marked positive correlations between mean (M_{18O}) in precipitation and mean monthly temperature. Since the sampling station Kun Ming station belong to low-latitude typical monsoonal climates, the temperature effect (Dansgaard W, 1964) is not notable.

As can be seen from Table 2, the correlation coefficient of the δ_{18O} in the average monthly precipitation and vapor pressure is -0.665 at Kun Ming station. At the significance level of 0.01, the significant probability is 0, which indicate that the negative correlation between these two sets of data is significant. Therefore, there was also significant negative correlation between the stable isotopic composition of precipitation and vapor pressure under the scale of month. The relationship between the seasonal variation of δ_{18O} in precipitation and vapor pressure is shown in Figure 3(c). As can be seen from the fig-

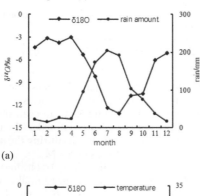

(a)

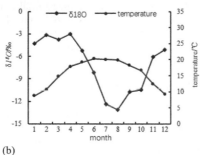

(b)

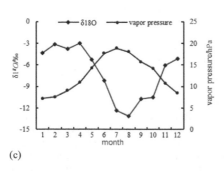

(c)

Figure 3. The mean values of monthly average oxygen isotopes, (a) monthly average air temperature, (b) monthly average rainfall amount, and (c) monthly average water vapor pressure.

ure, seasonal changes of δ_{18O} in the precipitation are inversely related to the vapor pressure.

4 CONCLUSION

In this paper, the variation characteristic and impact mechanism of oxygen isotope in precipitation under monthly scale at Kun Ming Station are analyzed in detail. Some conclusions are obtained as follows: The highest frequency of the maximum values occurs in April, and the highest frequency of minimum values appears in July. The statistics show that the max(M_{18O}) time series become more enriched along with the time, but min(M_{18O}) time series become more depleted at Kun Ming station. The oxygen isotopic composition of precipitation in the rainy season (May-October) is low, while the value in the dry season (November-April) is high. The correlation coefficients of M_{18O} and monthly average rainfall amount, air temperature, and water vapor pressure at Kun Ming station are all negative. The results indicate that the amount effect is notable, and the temperature effect is not obvious. It is mainly affected by the low-latitude typical monsoonal climates.

ACKNOWLEDGMENT

This study was supported by the National Natural Sciences Foundation of China (grant number: 51209096) and the Fundamental Research Funds for the Central Universities, SCUT (grant number: 2015ZM110).

REFERENCES

Araguas-Araguas L, Froehlich K, Rozanski K. Deuterium and oxygen18 isotope composition of precipitation and atmospheric moisture. Hydrological Processes, 2000, 14: 1341–1355.

Dansgaard W. Stable isotopes in precipitation. Tellus, 1964, 16 (4): 436–468.

Harrington G A, Cook P G, Herczeg A L. Spatial and Temporal Variability of Ground Water Recharge in Central Australia: A Tracer Approach. Ground Water, 2002, 40(5): 518–528.

Kendall MG. Rank correlation Measures. Charles Griffin: London. 1975.

Kwang-Sik Lee, Jun-Mo Kim, Dong-Rim Lee, et al. Analysis of water movement through an unsaturated soil zone in Jeju Island, Korea using stable oxygen and hydrogen isotopes. Journal of Hydrology. 2007, 345, 199–211.

Mann HB. Non-parametric tests again trend. Econometrica, 1945, 13: 245–259.

Tsutomu Yamanaka, Jun Shimada, Yohhei Hamada, et al. Hydrogen and oxygen isotopes in precipitation in the northern part of the North China Plain: Climatology and inter-storm variability. Hydrological Processes, 2004, 18: 2211–2222.

Zhang Xinping, Yao Tandong. Distributional features of $\delta 18O$ in precipitation in China. Acta Geographica Sinica, 1998, 53 (4): 356–363.

Energy Science and Applied Technology ESAT 2016 – Fang (Ed.)
© 2016 Taylor & Francis Group, London, ISBN 978-1-138-02973-6

Application of an adaptive neuro-fuzzy inference system for the performance analysis of an air-cooled chiller

Chengwen Lee, Yungchung Chang & Yunting Lu
Department of Energy and Refrigerating Air Conditioning Engineering, National Taipei University of Technology, Taipei, Taiwan

ABSTRACT: Old machines that use the R-22 refrigerant face a refrigerant replacement problem. This study replaced the R-22 refrigerant in an air-cooled chiller unit with R-290 and compared the different Coefficients Of Performance (COPs) to assess the respective energy efficiencies. Data from the R-290 refrigerant were used to develop an Adaptive Neuro-Fuzzy Inference System (ANFIS) model, and a patented process was used to select data within the overlapping range. R-22 refrigerant operating data were then input into the ANFIS model to obtain the COP for R-290 under the same conditions, to compare the power consumption. The ANFIS model developed in this study obtained an R2 greater than 0.999 and an average modeling error of 0.06%, which was more accurate than the commonly used backpropagation neural network model. Recent studies have shown that substituting hydrocarbon refrigerants into existing refrigeration units can improve the performance efficiency. In this study, chiller units using R-22 consumed an average of 1.44 kW of power, whereas the same chiller units using R-290 only consumed an average of 1.15 kW, for a total energy savings of 20%. Using hydrocarbons in old chiller units reduces refrigerant refill requirements, decreases power consumption, prolongs the life of the machine, and improves performance.

Keywords: hydrocarbon refrigerant; adaptive neuro-fuzzy inference system; coefficient of performance; patented process-air conditioning energy authentication method

1 INTRODUCTION

With the rapid development of high-technology industries in Taiwan, power consumption has grown drastically. A statistical analysis of power consumption using government offices to represent typical buildings shows that air-conditioning equipment consumes 56.29% of electricity, lighting 17.59%, other equipment 19.06%, and utility equipment 6.7%. As shown in Fig. 1, air-conditioning equipment consumes more than 50% of the power used in an entire building. Reducing the energy consumed in air-conditioning equipment alone can lead to a more efficient power consumption for the entire building, decreased costs, and reduced greenhouse gas emissions to mitigate the global warming crisis. Most commercial chiller units traditionally used the refrigerants R-22 and R-134a. R-22 has an atmospheric lifetime of 11.9 years, an ozone depletion potential (ODP) of 0.040, and a global warming potential (GWP) of 1790. In comparison, the hydrocarbon refrigerant R-290 has a shorter atmospheric lifetime, a smaller ODP, and a smaller GWP, and is therefore more environmentally friendly (Table 1) (James M. Calm et al, 2011).

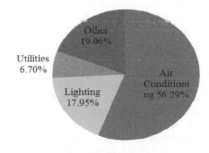

Figure 1. Office power consumption analysis.

Table 1. Environmental impact of various refrigerants (James M. Calm et al, 2011).

Refrigerant	R-22	R-134a	R-600a	R-290
Type	HCFC	HFC	HC	HC
Atmospheric lifetime (years)	11.9	13.4	<1	<1
Ozone depletion potential	0.040	0.000	0.000	0.000
Global warming potential (100 years)	1,790	1,370	~20	~20

With the rapid development of the high-technology industry, air conditioning has become an indispensable piece of equipment both at work and at home. The summers in Taiwan are hot and humid. As summer temperatures continue to rise year after year, air-conditioning units have become increasingly vital to the quality of life and work. In central air-conditioning units, chillers are responsible for approximately 60% of power consumption. Improving the energy efficiency of chillers could reduce the overall power consumption, save energy, and decrease costs. Relevant research in this area includes the following:

1. Chan et al. (Chan KT et al, 2002) proposed an energy saving method to control the condensing temperature in current machines. Unlike previous water-pressure control methods, this method can maintain the energy consumption to under 2 kW/refrigeration ton. After conducting research on the two types of machines, the results showed that this method had an energy saving potential of 18.2 and could reduce the annual power consumption by 29%.
2. Palm (Palm B. 2008) compared the features and performance of hydrocarbon refrigerants in small heat-pump and refrigeration systems (< 20 kW) and cited the commercially feasible systems. The study presented systems that increased or maintained their Coefficient Of Performance (COP) after switching from hydrofluorocarbon refrigerants to hydrocarbon refrigerants. Palm also referenced ways to mitigate the risks associated with using hydrocarbon refrigerants, such as carefully considering ventilation or installing part of the system outside.
3. Yang (T. Yang, 2012) demonstrated the crucial role refrigerants play in affecting the global climate and ecological environment, and how refrigerant features directly affect refrigeration system efficiencies. Units that use R-22 and R-134a face a refrigerant replacement problem. After directly changing to hydrocarbon refrigerants without altering the original system, this study compared and analyzed R-22, R-134a, and hydrocarbon refrigerants. The results showed that replacing R-134a with R-610a increased the COP by 16%, and replacing R-22 with HyChill Minus 50 increased the COP by 1.9%.
4. Chau et al. (K.T.Chau et al, 2003) used an adaptive neuro-fuzzy inference system (ANFIS) on the temperature, current, and capacity of nickel-metal hybrid batteries to construct a new nickel-metal hybrid battery model and to indicate the remaining battery capacity. The results showed an average error of less than 2.67%.
5. Zaheeruddin et al. (Zaheeruddin et al, 2006) created a work-efficiency model for noisy environments using an ANFIS with critical parameters such as the noise level, work type, and time in the noisy environment to predict the work efficiency of people under noisy conditions. The results showed a high level of accuracy.
6. Entchev et al. (Evgueniy Entchev et al, 2007) used artificial neural networks and ANFISs to construct a performance prediction model for residential solid-oxide fuel cells. The results showed that for incomplete data, the ANFIS predictions were more accurate than those obtained using the artificial neural networks technique.
7. Chan et al. (K.T. Chan et al, 2004) developed an energy consumption model for air-cooled chillers using TRNSYS programming. They analyzed chiller parameters such as the chilled water supply flow rate, chilled water supply temperature, outdoor temperature, set condensing temperature, evaporator overheating, condenser overcooling, and partial load rate, to algorithmically calculate chiller energy consumption.

2 ADAPTIVE NEURO-FUZZY INFERENCE SYSTEM

The ANFIS combines artificial neural networks with fuzzy logic principles. First presented in (J.-S. Roger Jang, 1993), ANFIS solves expert-defined fuzzy-rule-based problems in traditional fuzzy systems by using hybrid learning procedures to construct a set of if-then rules that gradually adjust to a proper membership function that satisfies the desired fuzzy inference input–output relationship. ANFIS also solves the problem in artificial neural network systems of determining the initial number of hidden layers. Too many or too few hidden layers will inhibit network convergence. Furthermore, the number of hidden layers required depends on the situation. As changes in the number of hidden layers cannot be traced, determining the optimal number of hidden layers is entirely based on trial and error.

To simplify the ANFIS, assume that the fuzzy inference system in question only has two input variables (x, y) and one output variable (z). For a first-order Sugeno fuzzy model, the resulting four fuzzy if-then rules would be as follows:

Fuzzy Rule R_1: If x is A_1 and y is B_1, then $z = f_1 = p_1 x + q_1 y + r_1$

Fuzzy Rule R_2: If x is A_1 and y is B_2, then $z = f_2 = p_2 x + q_2 y + r_2$

Fuzzy Rule R_3: If x is A_2 and y is B_1, then $z = f_3 = p_3 x + q_3 y + r_3$

Fuzzy Rule R_4: If x is A_2 and y is B_2, then $z = f_4 = p_4 x + q_4 y + r_4$

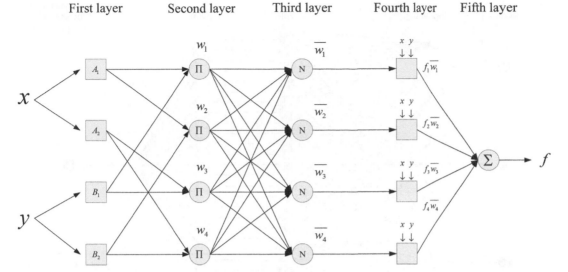

Figure 2. ANFIS structure corresponding to a first-order Sugeno fuzzy model.

The ANFIS structure corresponding to the fuzzy inference in this first-order Sugeno fuzzy model is shown in Fig. 2. Nodes in the same layer have the same function. The features and operations for each layer are described below.

1. **First layer**: Every node i in this layer is a square node with a node function:

$$O_{1,i} = \mu_{A_j}(x), \text{ for } i = 1, 2, \quad j = 1, 2$$
$$O_{1,i} = \mu_{B_j}(y), \text{ for } i = 3, 4, \quad j = 1, 2 \tag{1}$$

where x or y is the input to node i and A_i or B_{i-2} is a fuzzy set (such as "small" or "large") associated with node i. In other words, $O_{1,i}$ is the degree of membership for the input variable x or y to fuzzy set $A(= A_1, A_2, B_1, B_2)$. The membership function for fuzzy set A can be any appropriate parameterized membership function, such as the bell-shaped membership function:

$$\mu_A(x) = \frac{1}{1 + \left| \dfrac{x - c_i}{a_i} \right|^{2b_i}} \tag{2}$$

where $\{a_i, b_i, c_i\}$ is the parameter set for the membership function $\mu_A(x)$. As the values of these parameters change, the bell-shaped function varies accordingly. Parameters in this layer are referred to as premise parameters.

2. **Second layer**: Every node in this layer is a circle node labeled II, whose output is the product of all input signals:

$$O_{2,i} = w_i = \mu_{A_j}(x) \cdot \mu_{B_j}(y), \ i = 1, 2, 3, 4, \ j = 1, 2. \tag{3}$$

Each node output represents a rule's firing strength. Generally, the nodes in this layer can be the fuzzy intersection operator for any triangular-norm operator.

3. **Third layer**: Every node in this layer is a circle node labeled N. The ith node calculates the ratio of the ith rule's firing strength to the sum of all rules' firing strengths:

$$O_{3,i} = \overline{w_i} = \frac{w_i}{w_1 + w_2 + w_3 + w_4}, \ i = 1, 2, 3, 4. \tag{4}$$

Outputs in this layer are called normalized firing strengths.

4. **Fourth layer**: Every node i is a square node with a node function:

$$O_{4,i} = \overline{w_i}, f_i = \overline{w_i}(p_i x + q_i y + r_i), \ i = 1, 2, 3, 4 \tag{5}$$

where $\overline{w_i}$ is the normalized firing strength from layer 3 and $\{p_i, q_i, r_i\}$ is the parameter set for nodes in this layer. Parameters in this layer are referred to as consequential parameters.

5. **Fifth layer**: The single node in this layer is labeled Σ and calculates the overall output for the previous layer as the final output value:

$$O_{5,1} = \sum \overline{w_i} f_i = \frac{\sum_i w_i f_i}{\sum_i w_i}, \ i = 1, 2, 3, 4. \tag{6}$$

271

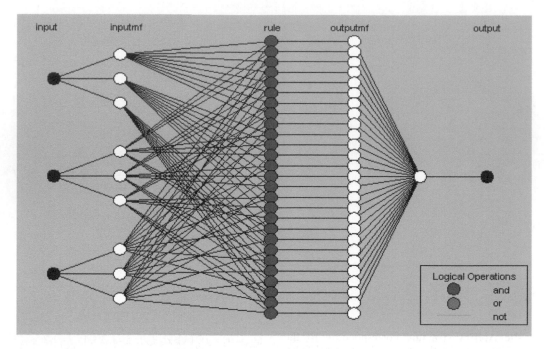

Figure 3.   Basic structure for the ANFIS model (C.-H. Hsu, 2013).

This study developed an ANFIS performance model for air-cooled chiller units with three input parameters (chilled water supply temperature, wet bulb temperature, and refrigerating capacity) and one output parameter (COP). Following the sigmoidal membership function, each input variable had three membership functions, producing $3^3 = 27$ rules (more membership functions, and therefore more rules, not only improve the accuracy, but also prolongs the training time). The basic structure of this model is depicted in Fig. 3 (C.-H. Hsu, 2013). This study used the Fuzzy Logic Toolbox in the MATLAB software along with tailored programming to complete the model. The procedure is outlined in Fig. 4.

Step 1: Collect the operational information for the air-cooled chiller unit.
Step 2: Determine the input and output variables.
Step 3: Set the ANFIS parameters, such as the type of membership function, the number of membership functions, and the number of iterations.
Step 4: Input the above parameters into the ANFIS and initiate training.
Step 5: Complete the ANFIS training once the results achieve the preset iterations.

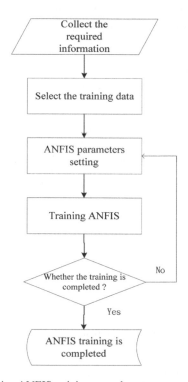

Figure 4.   ANFIS training procedure.

3 AIR-CONDITIONING ENERGY AUTHENTICATION PROCESS (Y. ZHANG, 2008)

This study used patent publication number 200839204 (air-conditioning energy authentication method) to verify the improved COP for a single chiller unit. The independent variables were the wet bulb temperature, the chilled water supply temperature, and the refrigerating capacity. The dependent variable was the COP. The process is outlined in Fig. 5.

1. Collect the independent and dependent variables for R-290;
2. Develop a performance model for the air-cooled chiller unit with R-290;

3. Select the performance model with the minimum error;
4. Collect the independent and dependent variables for R-22;
5. Enter the R-290 and R-22 independent variables into a scatter diagram, with each independent variable as the coordinate axis;
6. Select the R-22 independent variables that overlap with those of R-290;
7. Input the selected R-22 independent variables into the R-290 chiller unit performance model and solve for the R-290 COP; and
8. Compare and analyze the performance of R-290 and R-22 according to the COP from Step 4.

4 RESULTS AND DISCUSSION

Between May 14 and June 3, 2014, this study measured the chilled water supply temperature (Tchws), chilled water return temperature (Tchrs), wet bulb temperature (WB), and total power consumption (kW) for a scroll chiller with a refrigerating capacity of 10 (Q kW) using two different types of refrigerant: R-22 and R–290. The data collected were verified through the patented process air-conditioning energy authentication method (patent publication number 200839204) and unreasonable data were removed. A chiller unit COP model was then developed using an ANFIS. Under the same basis, the COP was simulated and converted into energy consumption for comparison.

To determine the model's accuracy, R-290 refrigerant data from May 14 to May 31 were simulated using a backpropagation neural network model and an ANFIS simulation. The errors between the actual and simulated COP for both models were then calculated. The ANFIS simulation was more accurate than the commonly used backpropagation neural network model. Fig. 6 shows the

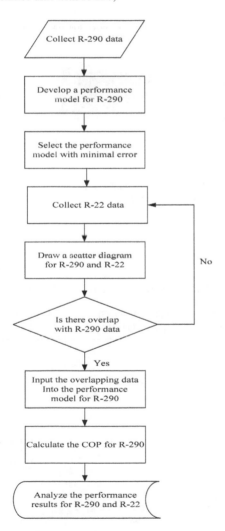

Figure 5. Air-conditioning energy authentication process (Y. Zhang, 2008).

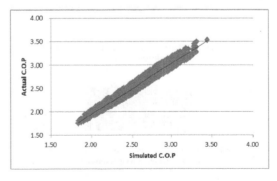

Figure 6. Regression analysis for the backpropagation neural network model (R2 = 0.9806).

regression analysis results for the backpropagation neural network and Fig. 7 depicts the regression analysis results for the ANFIS. The error analysis for the two models is outlined in Table 2.

After the models were completed, patent publication number 200839204 was used to filter R-22 refrigerant data from June 1–3. The overlapping parts were selected as the filter range. Figs. 8–10 illustrate the overlapping data-sets for each variable, and Table 3 outlines the filter range. The filtered data were then fed into the model to obtain the R-290 data under an identical load.

Figs. 11 and 12 illustrate the change in power consumption between using R-290 and R-22 as

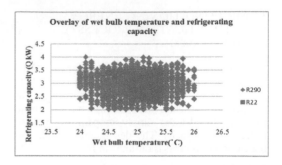

Figure 9.   Overlay of wet bulb temperature with refrigerating capacity.

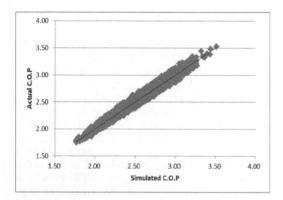

Figure 7.   Regression analysis for the ANFIS (R2 = 0.999).

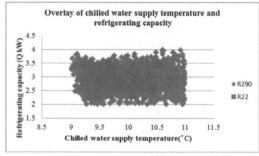

Figure 10.   Overlay of chilled water supply temperature with refrigerating capacity.

Table 2.   Error analysis for backpropagation neural network and ANFIS modelling.

Research method	R2	Average error
Backpropagation neural network	0.9806	0.19%
ANFIS	0.999	0.06%

Table 3.   Variable selection range.

Variable	Selection range
Chilled water supply temperature T_{chws} (°C)	9.2–10.8
Wet bulb temperature WB (°C)	24.5–25.4
Refrigerating capacity Q (kW)	2.25–3.5

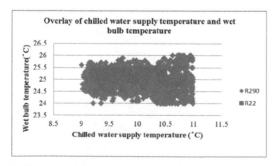

Figure 8.   Overlay of chilled water supply temperature with wet bulb temperature.

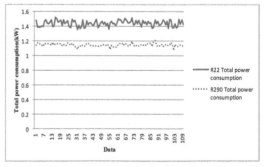

Figure 11.   Comparison of power consumption between R-22 and R-290 refrigerants total power consumption.

274

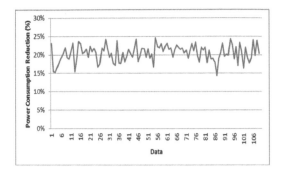

Figure 12. Power consumption reduction after switching to R-290.

refrigerants. The power consumption range for R-290 was between 1.08 and 1.21 kW with an average consumption of 1.15 kW. The power consumption range for R-22 was between 1.35 and 1.51 kW with an average consumption of 1.44 kW. This difference clearly demonstrates that using the R-290 refrigerant can reduce power consumption by 20%.

5 CONCLUSION

This study replaced the R-22 refrigerant with R-290 in an air-cooled chiller unit and compared the COPs to understand their energy efficiencies. Operational data were collected from the chiller in two phases: when the R-290 refrigerant was used between May 14 and 31, 2014, and when the R-22 refrigerant was used between June 1 and 3, 2014. Data from R-290 refrigerant operations were used to develop an artificial neural network model, and patent publication number 200839204 was used to select the overlapping data. R-22 refrigerant operating data were then fed into the model to obtain the COP for the chiller when the R-290 refrigerant was used under the same conditions, subsequently comparing the power consumption. The ANFIS model developed in this study obtained an R2 greater than 0.999 and an average error of 0.06%, proving the accuracy of the model. Recent research results on hydrocarbon refrigerants have shown that replacing R-22 with R-290 could increase the energy efficiency by 20%–25%. These results confirm that using hydrocarbon refrigerants in existing refrigeration units can significantly improve the performance efficiency. Chiller units using the R-22 refrigerant consume an average of 1.44 kW of power, whereas units using the R-290

refrigerant only consume an average of 1.15 kW of power, for a total energy savings of 20%. Using hydrocarbons in old chiller units reduces the amount of refrigerant needed for refills, decreases the power consumption, prolongs the lifespan of the machine, and improves performance. Despite its flammable nature and the need to practice caution when handling it, using hydrocarbons can contribute to ensuring environmental sustainability.

REFERENCES

Chan, K.T., F. W.Yu, "Optimum Setpoint of Condensing Temperature for Air-Cooled Chillers", HVAC&R RESEARCH, vol. 10, no. 2, 2004, pp. 113–128.

Chan K.T., Yu F.W., "Applying condensing-temperature control in air-cooled reciprocating water chillers for energy efficiency," Appl Energy, 2002, 72(3–4): 565–581.

Chau, K.T., K.C. Wu, C.C. Chan and W. X. Shen, "A new battery capacity indicator for nickel–metal hydride battery powered electric vehicles using adaptive neuro-fuzzy inference system", Energy Conversion and Management 44: 2059–2071, 2003.

"Energy-Saving Techniques Manual for Government Agency Offices," Bureau of Energy, Min. of Econ. Affairs, Taipei, Republic of China, 2007.

Evgueniy Entchev and Libing Yang," Application of adaptive neuro-fuzzy inference system techniques and artificial neural networks to predict solid oxide fuel cell performance in residential microgeneration installation", Journal of Power Sources 170: 122–129, 2007.

Hsu, C.-H. "Application of Adaptive Neuro-Fuzzy Inference System for the Performance Analysis of an Air-Cooled Scroll Chiller with Variable Speed," M.S. thesis, Dept. of Energy and Refrig. Air-Cond. Eng., Nat. Taipei Univ. Tech., Taipei, 2013.

James M. Calm, Glenn C. Hourahan, "Physical, Safety, and Environmental Data for Current and Alternativke Refrigerants," Proceedings of the 23rd IIR International Congress of Refrigeration, 2011, pp. 4–10.

Palm B., "Hydrocarbons as refrigerants in small heat pump and refrigeration systems - A review," Int J Refrig., 2008, pp. 552–563.

Roger Jang, J.-S. ANFIS: Adaptive-Network-based Fuzzy Inference Systems. IEEE Transactions on Systems, Man, and Cybernetics, 23(03): 665–685, May 1993.

Yang, T. "Investigation of Air-Cooled Conditioner Performance by Retrofitting to Hydrocarbon Refrigerants," M.S. thesis, Dept. of Energy and Refrig. Air-Cond. Eng., Nat. Taipei Univ. Tech., Taipei, 2012.

Zhang, Y. "Air-Conditioning Energy Authentication Method," R.O.C. Patent 200839204.

Zaheeruddin and Garima,"A neuro-fuzzy approach for prediction of human work efficiency in noisy environment", Applied Soft Computing 6:283–294, 2006.

Energy Science and Applied Technology ESAT 2016 – Fang (Ed.)
© *2016 Taylor & Francis Group, London, ISBN 978-1-138-02973-6*

Kinetics of As desorption from soil aggregates by KH_2PO_4-aided soil washing

Ranran Zhao
College of Ecological Technique and Engineering, Shanghai Institute of Technology, Shanghai, China

Xiaojun Li
Key Laboratory of Pollution Ecology and Environmental Engineering, Institute of Applied Ecology, Chinese Academy of Sciences, Shenyang, China

Zhiguo Zhang
College of Ecological Technique and Engineering, Shanghai Institute of Technology, Shanghai, China

ABSTRACT: Desorption kinetics of As from soil aggregates (>2.0, 2.0-0.25, 0.25-0.053, <0.053 mm) by soil washing with KH_2PO_4 and the changes of As forms in soil aggregates and bulk soil after soil washing were studied. The As contents in soil aggregates decreased with the following sequence: (0.25–0.053 mm) > bulk soil > (2.0–0.25 mm) > (<0.053 mm) > (>2.0 mm). The optimum washing time for As removal was determined to be 360 min. Soil washing with KH_2PO_4 was effective in extracting specially sorbed As and As associated with oxides of Fe and Al both in bulk soil and soil aggregates. Elovich model was the best to describe the kinetic of As desorption both from bulk soil and soil aggregates, the mechanisms of As removed from soil aggregates might be the same as that from bulk soil.

Keywords: arsenic (As); soil aggregates; potassium dihydrogen phosphate (KH_2PO_4); soil washing

1 INTRODUCTION

The arsenic (As) contamination of soil was becoming an important environmental problem throughout the world (H.D. Moon et al, 2004), and given that soil played an irreplaceable role in the environment, if it was polluted by As, it could bring huge damage to environmental security and human health (G.P. Warren et al, 2003) (W. Guo et al, 2007). Therefore, many researches had been done on how to relieve the risks of As polluted soils (A. Giacomino et al, 2010) (D.H. Moon et al, 2004) (P. Bhattacharya et al, 2002) (R. Shi et al, 2009).

Soil aggregates were the basic units of soil constructer (H.R. Schulten et al, 2000). The physicochemical properties and contents of heavy metals were distinct among soil aggregates of different sizes separated from contaminated soil (K. Ilg et al, 2004) (F. Ajmone-Marsan et al, 2008) (G.N. Fedotov et al, 2008). Furthermore, sorption/desorption of As by soil aggregates of different sizes was also different (C.X. Jiang et al, 2015). Some studies had been done on the removal of As from contaminated soils by soil washing with phosphate solutions (M. Zeng et al, 2008) (M.G.M. Alam et al, 2015) (C. Chen et al, 2015) (E.H. Jho et al, 2015). However, researches on the desorption of As from soil aggregates of contaminated soil by soil washing was rare.

The present study was therefore conducted to analyze the effects of KH_2PO_4 on release of As from soil aggregates collected at a farmland near to a smelter in Northeast China. The kinetics of As desorption efficiency with KH_2PO_4 from different soil aggregates was performed. Fractions of As in samples after washing was also discussed to find the changes of the contaminated soil after washing.

2 MATERIALS AND METHODS

2.1 *Soil sample preparation*

Surface soil samples (0–20 cm) were collected from a farmland near to a smelter in Liaonin Province, Northeast China. Soil samples were avoided extrusion to maintain their structure and transported back to the laboratory in 2 days. Part of the samples was taken out and air-dried at room temperature and subsequently sieved through a 2-mm polyethylene sieve as the bulk soil (S0), the rest was kept fresh in the refrigerator (4°C) before separated into soil aggregates with different sizes.

2.2 Separation of soil aggregates

Fraction of soil aggregates was achieved by using a wet-sieving procedure (C.A. Cambardella et al, 1994). Four sizes of aggregates of >2.0 mm(S1), 2.0–0.25 mm(S2), 0.25–0.053 mm(S3) and <0.053 mm(S4) were finally collected.

2.3 Desorption of As from soil aggregates with different washing time

Twenty four of centrifugal tubes were added with 1.0 g of bulk soil or soil aggregates for each one, and then 10 mL of KH_2PO_4 solution (in 0.1 mol/L) was added to the tube. The suspension was shaken in a thermostat at $25 \pm 1°C$. At the time of 20, 40, 60, 120, 240, 360, 720, 1440 min, these tubes were taken out from the thermostat, centrifuged and then filtered through a 0.45 μm membrane.

2.4 As fractionations in soil aggregates after KH_2PO_4 soil washing

10 mL KH_2PO_4 (in 0.1 mol/L) solution was added into 50 mL centrifugal tubes contained 1.0 g soil or soil aggregate sample. The suspension was shaken for 360 min in a thermostat at 25 ± 1 °C and then centrifuged to discard the supernatant. The soil pellets were rinsed twice with deionized water by shaking on a thermostat and then dried after discarding the supernatant.

2.5 Chemical analysis

The pH of soil and aggregates was measured using a pH electrode (pHs-3C, Shanghai Jing Ke Test Equipment Institute, China) (soil: water = 1:2.5). Organic Matter (OM) content and Free iron oxide was measured by the method described by Lu (R.K. Lu, 2000). Total Phosphorus (TP) of soil samples was measured according to the method introduced by Bao (S.D. Bao, 2000). Arsenic content in the soil aggregates was detected using the atomic fluorescence spectrometry (AFS-9700A) after aqua regia extraction. The properties of the soil and soil aggregates was shown in Table 1.

The As fractions in soil aggregates and bulk soil before and after washing were extracted following the method described by Wenzel et al. (E. Lombi et al, 2001), which divided As in soils into five fractions (1) non-specifically sorbed As (F1); (2) specifically sorbed As (F2); (3) As associated with amorphous hydrous oxides of Fe and Al (F3); (4) As associated with crystalline hydrous oxides of Fe and Al (F4) and (5) residual As fraction (F5).

2.6 Data analysis

The experimental results were subjected to a one-way analysis of variance and least significance difference using SPSS 16.0 statistical software at 95% confidence, SigmaPlot 10.0 was used to generate the graphs.

3 RESULTS AND DISCUSSION

3.1 Physicochemical properties of soil

The properties of the soil examined were presented in Table 1. The percentages of < 0.25 mm aggregates (S3 and S4) in bulk soils were the highest (76.2%), while > 2 mm soil aggregates was the lowest (2.36%). The contents of As in bulk soil and soil aggregates were in the order: S2 (146.78 mg/kg) > S0(126.07 mg/kg) > S3(120.32 mg/kg) > S4(107.16 mg/kg) > S1(44.72 mg/kg). The distribution of As in soil aggregates of different particle sizes was distinct (C. Gong et al, 2012) (C. Gong et al, 2014) (W.W. Wenzel et al, 2001).Gong et al. (C. Gong et al, 2012) reported that the contents of As in 0.053~0.25 mm and >1 mm soil aggregates was higher than other soil aggregates in black soils.

3.2 Desorption of As from soil aggregates at different time

Kinetic study was undertaken by washing bulk soil and soil aggregates with 0.1 mol/L KH_2PO_4. Result was presented in Fig. 1.

The kinetic experiment indicated that As removal from bulk soil was the same with that from soil

Table 1. The properties of the soil and soil aggregates.

Size/mm	pH/$H_2$0	OM/(g·kg⁻¹)	TP/(g·kg⁻¹)	Free iron oxide / (g·kg⁻¹)	As/ (mg·kg⁻¹)	Mass/ (%)
Bulk soil (S0)	6.79±0.02	19.08±0.22	0.93±0.05	14.54±0.06	126.07±6.88	
S1	6.00±0.14c	26.04±0.06a	1.58±0.05a	12.97±0.18d	44.72±0.67d	2.36
S2	6.28±0.03b	25.06±0.31b	1.22±0.10b	15.94±0.16a	146.78±0.98a	21.36
S3	6.36±0.04b	17.18±0.31c	0.78±0.03c	15.04±0.16b	120.32±3.21b	42.91
S4	6.91±0.04a	14.17±0.08d	0.75±0.04c	13.70±0.07c	107.16±1.7c	33.37

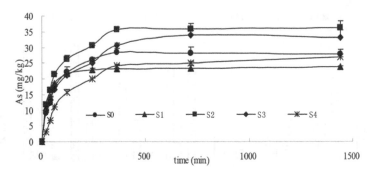

Figure 1.　Kinetics of As removed from bulk soil and soil aggregates.

Table 2.　Parameters of kinetic models for As extraction.

Kinetic models	Parameter	Bulk soil (S0)	S1	S2	S3	S4
Parabolic	r^2	0.65	0.66	0.78	0.79	0.81
diffusion	a	12.86	14.59	15.28	11.58	5.05
$q_t = a + bt^{1/2}$	b	0.55	0.58	0.75	0.71	0.72
Elovich	r^2	0.92	0.91	0.95	0.94	0.97
$q_t = a + bInt$	a	−3.72	−2.70	−5.59	−7.95	−14.85
	b	4.99	5.21	6.45	6.07	6.19
Two-constant	r^2	0.89	0.90	0.89	0.95	0.95
$Inq_t = a + bInt$	a	1.87	1.97	1.32	1.65	1.82
	b	0.21	0.20	0.28	0.25	0.23

aggregates. In the first 360 min, As desorption was rapid, after which As desorbed slowly. This result was similar with the study of Xu et al. (C. Xu et al, 2009) who documented that removal of heavy metals from contaminated soil could be resolved into two distinct phases: a fast removal and a relatively slower removal. The optimum washing time for As removal was 360 min in our study. For soil aggregates and bulk soil, the removal of As increased significantly with the increase of washing time from 0 to 360 min (p<0.05), after which only 3.7–9.1% of the As was removed during the 1440 min washing process. A longer washing period meant a higher operating cost (J.S. Hwang, 2001). Hence, 360 min washing could be chosen as the optimum washing period to remediate As-contaminated soil. This result was similar with the study of Zeng et al. (M. Zeng et al, 2008).

The parameters of kinetic models for As extraction from soil aggregates were given in Table 2. According to the correlation coefficients obtained from the kinetic models, it was easily to find that Elovich model was the best to describe the process of As removal while parabolic diffusion model was the worst. Elovich model was also the best to describe the kinetic of desorption of Pb and Cd by organic acids washing (D.P. Xu et al, 2015). In addition, the release of As both from bulk soil and soil aggregates could be best described by the same model, which might concluded that the mechanisms of As removed from soil aggregates might be the same as that from bulk soil.

3.3　Effects of soil washing on As fractionations

Fig. 2 compared the distribution of As fractions in bulk soil and soil aggregates before and after washing with KH_2PO_4. The results showed that soil washing with KH_2PO_4 resulted in the significant reduction of specially sorbed As and As associated with oxides of Fe and Al (F3 and F4), while made slight increase of non-specially sorbed As in bulk soil and soil aggregates. The removal of specially sorbed As was probably due to that PO_4^{3-} replaced AsO_4^{3-} in the adsorption sites (M.G.M. Alam et al, 2015). While the release of As associated with oxides of Fe and Al might resulted from the dissolution of oxides of Fe and Al led by the acid-based KH_2PO_4 solution (S. Tokunaga et al, 2002). The non-specially sorbed As increased slightly might be due to that the acid-based KH_2PO_4 solutions dissolved soil fractions during washing, which made some As sorpted in the sites easier to be extracted by $(NH_4)_2SO_4$ than before.

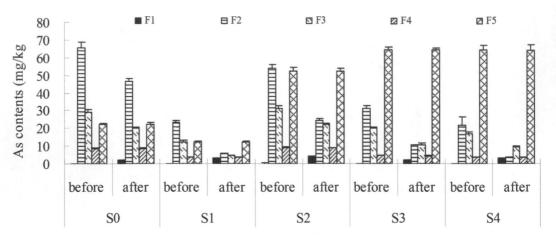

Figure 2. Arsenic fractions before and after soil washing. F1 non-specially sorbed As; F2 specially sorbed As; F3 As associated with amorphous hydrous oxides of Fe and Al; F4 As associated with crystalline hydrous oxides of Fe and Al; F5 residual As.

4 CONCLUSION

The contents of As in soil aggregates of different particle sizes was distinct and was in the order: S2 (146.78 mg/kg) > S0(126.07 mg/kg) > S3(120.32 mg/kg) > S4(107.16 mg/kg) > S1(44.72 mg/kg). The optimum washing time for As removal was determined to be 360 min. Soil washing with KH_2PO_4 was effective in extracting specially sorbed As and As associated with oxides of Fe and Al both in bulk soil and soil aggregates. Elovich model was the best to describe the kinetic of desorption of As both from bulk soil and soil aggregates, the mechanisms of As removed from soil aggregates might be the same as that from bulk soil.

ACKNOWLEDGMENTS

The research was financially supported by National Science Foundation of China (Nos. 41101295, 41271336, 41201310).

REFERENCES

Ajmone-Marsan, F., M. Biasioli and T. Kral: Environmental Pollution, Vol. 152 (2008), p. 73–81.
Alam, M.G.M., S. Tokunaga and T. Maekawa: Chemosphere, Vol. 43 (2015) No. 8, p. 1035–1041.
Bao, S.D.: Soil and Agriculture Chemistry Analysis. 3rd ed. (Chinese Agricultural Press, China 2000).
Bhattacharya, P., A.B. Mukherjee and G. Jacks: The Science of the Total Environment, Vol. 290 (2002), p. 165–180.
Cambardella, C.A., E.T. Elliott: Soil Sci. Soc. Am., Vol. 58 (1994), No. 1 p. 123–130.
Chen, C., X.F. Chen and X.M. L: Acta Sci. Circumst, Vol. 35 (2015) No. 8, p. 2582–2588.
Fedotov, G.N., G.G. Omel'yanyuk and O.N. Bystrova: Doklady Akademii Nauk, Vol. 420 (2008) No. 3, p. 346–350.
Giacomino, A., M. Malandrino and O. Abollino: Environmental Pollution, Vol. 158 (2010), p. 416–423.
Gong, C., L. Ma and H.X. Cheng: Ecology and Environmental Sciences, Vol. 21 (2012) No. 9, p. 1635–1639.
Gong, C., L. Ma and H.X. Cheng: Journal of Geochemical Exploration, Vol. 139 (2014), p. 109–114.
Guo, W., Y.G. Zhu and W.J. Liu: Environmental Pollution, Vol. 14 (2007) No. 1, p. 251–257.
Hwang, J.S.: Soil Ground Water Environ., Vol. 9 (2004), p. 104–111.
Ilg, K., W. Wilcke and G. Safronov: Geoderma, Vol. 123 (2004), p. 153–162.
Jho, E.H., J. Im and K. Yang: Chemosphere, Vol. 119 (2015), p. 1399–1405.
Jiang, C.X., J.G. Xia and W.L. He: Acta Sci. Circumst, Vol. 29 (2015) No. 5, p. 212–219.
Lombi, E., R.S. Sletten and W.W. Wenzel: Water Air and Soil Pollution, Vol. 124 (2001) No. 3–4, p. 319–332.
Lu, R.K.: Analytical methods for soil and agrochemistry (Chinese agricultural Science and Technology Press, China 2000).
Moon, D.H., D. Dermatas and N. Menounou: Science of the Total Environment Vol. 330 (2004), p. 17–185.
Moon, H.D., D. Dermatas and N. Menounou: Sci. Total Environ. Vol. 330 (2004) No. 1, p. 171–185.
Schulten, H.R., P. Leinweber: Biol. Fertil. Soils, Vol. 30 (2000) No. 5–6, p. 399–432.
Shi, R., Y.F. JIA and CH.Z. Wang: Journal of Environmental Sciences, Vol. 21 (2009), p. 106–112.
Tokunaga, S. and T. Hakuta: Chemosphere, Vol. 46 (2002), p. 3–38.
Warren, G.P., B.J. Alloway and N.W. Lepp: Sci. Total Environ. Vol. 311 (2003) No. 1, p. 19–33.
Wenzel, W.W., R.S. Sletten and E. Lombi: Analytica Chimica Acta., Vol. 436 (2001), p. 309–323.
Xu, C., B.C. and Y. Lin: Ecology and Environmental Sciences, Vol. 18 (2009) No. 2, p. 507–510.
Xu, D.P., X.B. Li and L. Sun: Journal of safety and environment, Vol. 15(2015) No. 3, p. 261–266.
Zeng, M., B.H. Liao and M. Lei: J. Environ. Sci. Vol. 20 (2008) No. 1, p. 75–79.

5 Electrical and electronic technology, power system engineering

Energy Science and Applied Technology ESAT 2016 – Fang (Ed.)
© 2016 Taylor & Francis Group, London, ISBN 978-1-138-02973-6

Impact of the effective assets on the costs of power transmission and distribution of grid enterprises

Liping Guo, Hengjie Ren & Jie Zhou
State Grid Zhejiang Yuyao Power Supply Company, Yuyao, P.R. China

Jianjun Xu
College of Science and Technology, Ningbo University, Ningbo, P.R. China

ABSTRACT: The inherent relationship between the effective assets and the cost of transmission and distribution of power grid enterprises lack empirical evidence. In order to reveal the interrelationship between them, this paper constructs the panel data model including six cities in Zhejiang Province and use the pooled EGLS approach to explore the impact of effective assets on the cost of transmission and distribution. The results indicate that the effective assets are positive to the unit of transport and distribution cost, and if the effective assets increased by 1%, the unit transmission and distribution cost would increase by 0.23%. The estimation results based on the varying coefficient panel data model showed that effective assets have different elasticities on the transmission and distribution cost of gird enterprises in six different cities and the elasticity range is within [0.52–0.62]. In addition, it can be found that the factors reflecting grid structure, the characteristics of power load, and the power consumer size are positively correlated to the unit cost of power distribution, respectively. The factors reflecting the economic and geographical characteristics except for the level of economic development are positively related to the unit cost of power supply.

Keywords: effective assets; power transmission and distribution; cost; grid enterprises

1 INTRODUCTION

State Electricity Regulatory Commission of China (2005) clearly pointed out that the transmission and distribution cost refers to the cost of "power grid enterprises for transporting and providing with power in transmission and distribution sectors" in a file named "the power transmission and distribution cost accounting approach". A lot of literatures had explained the compositions of the power transmission and distribution cost. Chi (2004) divided the transmission and distribution cost into fixed cost and variable cost in accordance with the rule of cost variation. Li & Zhou (2008) divided the costs of transmission and distribution into five components of fixed assets costs, power grid operation costs, the congestion costs, ancillary service costs, and reliability costs. According to the different types of concrete production and management activities, Liu (2013) regarded that the transmission and distribution costs could be divided into transmission costs, wages, depreciation, proprietary material costs, outsourcing material costs, outsourcing maintenance fee, commissioned operation maintenance fees, social insurance premiums, low-value consumable amortization, property insurance, research and development expenses, and other

operating expenses. In fact, the State Electricity Regulatory Committee (2005) had made it clear that power transmission and distribution costs belong to the category of production cost, and could be subdivided into material costs, wages, welfare costs, depreciation costs, repair costs, and other expenses in a file named "the transmission and distribution cost accounting approach". The State Electricity Regulatory Commission (2011) had once again made it clear that the transmission and distribution costs of power grid enterprises include material costs, workers compensation, depreciation, repairs, transmission charges, entrusted operation, and maintenance fees in a file named "interim measures for the supervision and control of power transmission and distribution costs".

The total cost law has been used to account for the cost of power transmission and distribution in China. In spite of China's accounting system reform since 1993, it demands that the manufacturing cost accounting method should substitute for the full cost method in the process of costs accounting for enterprises, and management costs, marketing costs, and other expenses should single out from the product cost. However, considering the particularity of power products, grid enterprises did not change the full cost accounting method into the manufacturing cost accounting method. Liu (2007) contrasted the advantages and disadvantages of manufacturing cost method and full cost method for cost accounting of the power grid enterprises, and the results showed that the manufacturing cost would increase the management cost of the grid enterprises, while the full cost method could more realistically reflect the actual operation situation of grid enterprises.

There are many factors that will affect the cost of power transmission and distribution of grid enterprises. Zhao & Li (2010) found that the level of economic development, grid network structure, and load characteristics, and geographical conditions are mainly aspects of the cost of grid enterprises. Zhang et al (2014) believed that the factors influencing power grid operation and maintenance costs mainly include the network scale and the load characteristics, regional economic development level, geographical and natural conditions, and so on. Han (2012) summarized many factors influencing the operation and maintenance costs, such as economic development level, line length, power capacity, the maximum load, the volume of sales electricity, reliability, the number of users, the power supply area, power supply density, topography, climate, etc. At the same time, he also considered the impact of asset depreciation on the costs of power transmission and distribution.

The above literatures neglect one of the most important factors, that is, the effective asset. Xiao (2004) regarded that the effective asset refers to the necessary assets for the power grid to run. Fan (2012) believed that the effective assets of power grid enterprises are the necessary economic resources to maintain the normal production and operation of the regulated business. Effective assets of the power grid enterprises can also be defined as the assets that are operating or not in operation but have the potential to operate, and they can contribute to the profitability of the grid enterprises. Hu (2009) and Feng & Zhang (2011) both discussed the definition of assets from the utility and benefit of the assets, and they believed that the effective assets are in the amount of assets which meet the needs of the normal production and business operation and the excess assets for the normal production and operation are invalid assets. Zhang & Ding (2006) regarded that the effective assets of the power grid enterprise including transmission and distribution lines, electrical equipment, and utilities. Utilities include electric metering equipment, communication lines and equipment, automation control equipment and instrumentation, manufacturing and maintenance of equipment, production and labor apparatus, transportation equipment, non-production equipment and furniture, houses, buildings, and land. However, few people quantify

the impact of effective assets on the cost of power transmission and distribution enterprises.

In a word, although the above literatures discussed and summarized the many factors having influence on the cost of transmission and distribution, literatures rarely paid attention to the impact of effective assets on the cost of the power transmission and distribution. Thus far, the quantitative relationship between the effective assets and the cost of power transmission and distribution is not clear. In this paper, we collected the basic data related to the power grid enterprises in six cities of Zhejiang Province and constructed panel data models to empirically analyze the impact of the effective assets on the cost of power transmission and distribution and further disclosed the regional differences in six cities.

2 MODEL, VARIABLES, AND DATA

In order to quantitatively analyze the impact of the effective assets on the costs of power transmission and distribution, we explored the cost of power transmission and distribution index as explanatory variables, and applied the effective assets and the other relevant influencing factors to explanatory variables, and constructed the constant coefficient panel data model (1) and the varying coefficient panel data model (2) as follows:

$$y_{it} = \alpha + \beta' X_{it} + u_{it} \qquad (1)$$

$$y_{it} = \alpha + \beta'_i X_{it} + u_{it} \qquad (2)$$

Among them, y_{it} is explained as a variable; $X'_{it} = (x_{1it}, x_{2it}, L, x_{kit})$ is the matrix vector composed of explanatory variables; $\beta'_{it} = (\beta_{1it}, \beta_{2it}, L, \beta_{kit})$ is a vector parameter; K is the number of the explained variables; T is the periods numbers, which ranges from 2011 to 2014; N is the number of cross-sections, which includes six cities named Yinzhou, Cixi, Yuyao, Fenghua, Ninghai, and Xiangshan; and u_{it} is a stochastic term.

The concrete indexes for the regression model are shown in Table 1. The basic data of the six cities come from the "Ningbo electric power statistical data" and "Ningbo Statistical Yearbook" during the year 2011–2014.

3 THE ESTIMATION RESULTS BASED ON THE PANEL DATA MODEL

3.1 The estimation results based on the constant coefficient panel data model

First of all, we construct the constant panel data model (1), and apply the pooled EGLS regression approach to fit the coefficients and the results

Table 1. Variable names and related instructions.

Sign	Variable name	Measuring method	Unit
Y	Cost of transmission and distribution	Total distribution cost or total distribution cost/LC3	100 million Yuan or 100 million Yuan/10000kwh
EA	Effective assets	Total fixed assets or total fixed assets/LC3	100 million Yuan or 100 million Yuan/10000kwh
WS1	Line length	Addition of loop length of different voltage lines	Kilometer
WS2	Electric capacity	Varying electric capacity	10 thousand KV
WS3	Loss rate	Line loss load/power supply load	%
LC1	Maximum load	The maximum load of the power consumers	10 thousand kilowatt per hour
LC2	Number of users	Electricity customers	Household
LC3	Sales volume of electric consumer	Total electricity consumption in the area	10 thousands kilowatt per hour
EG1	Area of power supply	Area of power supply	Square kilometer
EG2	Number of employees	Population at the end of the year	Population
EG3	Degree of economic development	GDP/population	Ten thousand Yuan/population
EG4	Annual wage per capita	Total wages divided by the number of employees	Ten thousand Yuan/population

are shown in the left half of Table 2. Taking into account the statistical significance of estimation coefficients, we screen out the variables LNWS1 and LNWS3 from model 1 and re-estimate model (1), the results are shown in the right half part of Table 2. As is shown in the right part of Table 2, the adjusted R-squared is 0.98, which illustrates that the simulated effect of the model is very well, so we can further analyze the meaning of the estimated parameters of each variable. The coefficient of the variable LNEA indicating the impact of the effective assets on the cost per unit of power transmission and distribution is 0.23, which means that if effective assets per unit increased by 1%, the cost per unit of power transmission and distribution would increase by 0.22%. Then, we analyze the grid structure indicators, including variables LNWS2 reflecting power capacity and LNWS3 reflecting line loss rate, on the cost per unit of power transmission and distribution, and the estimated parameters of them are 0.23 and 0.10, respectively. Next, let us investigate the grid load characteristic index. The estimated parameters of the variable LNLC1, LNLC2, and LNLC3 are 0.03, 0.11, and −0.77 respectively, which indicate that the highest load characteristic and the power consumer size are positively related to the cost per unit of power transmission and distribution, while the power sale scales are negatively related to the cost per unit of power transmission and distribution. Finally, let us take into account the effects of the economic geography condition indexes on the cost per unit of power transmission and distribution. The estimated parameters of the indica-

Table 2. The estimation results based on the constant coefficient panel model.

Variable	Coeffi.	t-Stat.	Prob.	Coeffi.	t-Stat.	Prob.
LNEA	0.23	11.17	0.00	0.22	13.96	0.00
LNWS1	−0.14	−1.61	0.13			
LNWS2	0.28	2.43	0.03	0.23	2.94	0.01
LNWS3	0.16	1.91	0.08	0.10	1.77	0.10
LNLC1	0.04	3.29	0.01	0.03	3.20	0.01
LNLC2	0.11	7.99	0.00	0.11	7.90	0.00
LNLC3	−0.74	−56.7	0.00	−0.77	−28.5	0.00
LNEG1	0.10	2.83	0.01	0.13	6.38	0.00
LNEG2	0.39	2.38	0.03	0.37	2.26	0.04
LNEG3	−0.09	−6.78	0.00	−0.10	−5.94	0.00
LNEG4	0.03	12.81	0.00	0.03	15.58	0.00
R-squared			0.99			0.99
Adjusted R-squared			0.98			0.98
Sum squared residual			1.22			1.24

Method: Pooled EGLS (Cross-section weights)

tor LNEG1 reflecting the area power supply, the indicator LNEG2 reflecting the number of workers, and the indicator LNEG4 reflecting workers per capita wage of the grid enterprises on the cost per unit of power transmission and distribution are 0.13, 0.37, and 0.03, respectively, which suggest that the area of power supply, the employees of the power grid enterprise, and per capita wage of workers are positively related to the cost per unit of power transmission and distribution. That is to say, if we increase the area of power supply, increase the employees of the power grid enterprise, and improve per capita wage of workers in

Table 3. The estimation results based on the varying coefficient panel model.

Variable	Coefficient	t-Statistic	Prob.
LNEA_Yinzhou	0.59	52.84	0.00
LNEA_Cixi	0.60	103.94	0.00
LNEA_Yuyao	0.62	67.95	0.00
LNEA_Fenghua	0.57	25.86	0.00
LNEA_Ninghai	0.54	35.52	0.00
LNEA_Xiangshan	0.52	37.12	0.00
R-squared			0.96
Adjusted R-squared			0.95
Sum squared residual			0.10

Method: Pooled EGLS (Cross-section weights)

the power grid enterprises, it will further increase the cost per unit of power transmission and distribution. In particular, it needs to be pointed out that the variable LNEG3 reflecting the regional economic development level on the cost per unit of power transmission and distribution is –0.10, which means that the high level of economic development in this region will decrease the cost per unit of power transmission and distribution.

3.2 *The estimation results based on the varying coefficient panel data model*

In order to compare different impacts of effective assets on the cost of power transmission and distribution, we construct the varying coefficient panel data model (2); fitting results are shown in Table 3. It is clear that the adjusted goodness of fit of the model in Table 3 is 0.95, which indicates that the parameters fit well. Then, we compare the regional difference of estimated parameters of effective assets among Yinzhou, Cixi, Yuyao, Fenghua, Ninghai, and Xiangshan on the total cost of power transmission and distribution, the influencing parameters of effective assets of six cities on the total cost of power transmission and distribution ranging from 0.522 to 0.622. Among them, the elasticities of effective assets on the total cost of power transmission and distribution very from large to small, followed by Yuyao (0.622), Cixi (0.601), Yinzhou (0.590), Fenghua (0.566), Ninghai (0.540), and Xiangshan (0.522).

4 CONCLUSIONS AND DISCUSSIONS

This paper constructs the panel data model using the pooled EGLS approach to explore the impact of effective assets on the cost of power transmission and distribution. Following are the results:

1. The effective assets are positively related to the cost per unit of power transmission and distribution.

If the effective assets per unit increase by 1%, the unit cost of power transmission and distribution would increase by 0.23%. The indexes reflecting grid structure, and power system load characteristics are positively related to the cost per unit of power transmission and distribution. Economic geographical factors such as the index for the power supply area, the employees of grid enterprises, the level of economic development, and average wages of staff in the power supply enterprises are positively related to total cost of the power transmission and distribution, respectively; while the index reflecting the level of economic development is negatively related to the cost per unit of power transmission and distribution.

2. The impact of the effective assets on the total cost of power transmission and distribution are different in Yinzhou, Cixi, Yuyao, Fenghua, Ninghai, and Xiangshan, and the estimated elasticity of the effective assets on the total cost of transmission and distribution in the six cities ranges from 0.52 to 0.62. Among them, the biggest elasticity is in Yuyao, and the smallest elasticity is in Xiangshan.

Although this paper employs the different econometric models to compare the impact of effective assets on the costs of power transmission and distribution in the six cities of Zhejiang Province, we cannot confirm the robustness of the above conclusions getting by the other samples. In addition, the inter mechanism between effective assets on the costs of power transmission and distribution needs to be discussed in the future.

ACKNOWLEDGEMENT

This work was financially supported by the project of State Gird Zhejiang Yuyao Power Supply Company (No.: HS2015000149).

REFERENCES

Chi Y.Y. 2004. Empirical study and pricing methods of power market in the northeast China. Master's Degree Thesis of Jilin University.

Fan X.W. 2012. Study on the particularity of electricity regulatory accounting. Master's Degree Thesis of Changsha University of Science and Technology.

Feng L.X & Zhang Y.Y. 2011. Research on the cost standard of transmission and distribution in China. Price Theory and Practice (12): 73–74.

Han Y. 2012. Transmission and distribution price formation mechanism and application study under the government regulation. Doctor's Degree Dissertation of North China Electric Power University.

Hu C. 2009. The objectives, factors and report analysis of accounting regulation. China Township Enterprises Accounting (5): 87–89

Li F. & Zhou Y.P. 2008. Study on the electricity pricing method and the division of transmission and distribution cost in the power market [J]. Modern Management Science (7): 42–44.

Liu Y. 2013. Research on the supervision cost accounting and information disclosure of the grid enterprises. Master's Degree Thesis of North China Electric Power University.

Liu J. 2007. Analysis and suggestion on cost accounting of the grid enterprises. East China Electric Power, (1): 92–93.

State Electricity Regulatory Commission of China. 2005. The transmission and distribution cost accounting method (Trial Implementation). China Electric Power Newspaper, July 28.

State Electricity Regulatory Commission of China. 2011. Interim measures for the supervision and control of power transmission and distribution costs. China Electric Power Newspaper, November 8.

Xiao J. 2004. Study on the financial problems of transmission and distribution electricity price reform and the operation of the power grid. Modern Electric Power (6): 56–59.

Zhao Q & Li C.R. 2010. Study on the approved method of power grid operation and maintenance fee in China Based on DEA. Chinese Journal of Management Science (18): 607–611.

Zhang L.Z, Fu N.N, Wang C.L, Wang Tao & Zhou Xianfei. 2014. The reasonable evaluation on the grid operation and maintenance costs based on regression analysis. Automation of Electric Power Systems (13):140–144.

Zhang J.L & Ding S.L. 2006. Study on the price calculation method of the transmission and distribution power. China Price (8): 18–23.

Energy Science and Applied Technology ESAT 2016 – Fang (Ed.)
© *2016 Taylor & Francis Group, London, ISBN 978-1-138-02973-6*

A novel power quality data reconstruction algorithm using compressed sensing

Y.T. Jia
Shenzhen Academy of Aerospace Technology, Shenzhen, China
Harbin Institute of Technology, Harbin, China

T.C. Li & D.L. Zhang
Shenzhen Academy of Aerospace Technology, Shenzhen, China

Z.C. Wang
Harbin Institute of Technology, Harbin, China

ABSTRACT: In this paper, power quality disturbances are sampled by an under-sampling Compressed Sensing (CS) system. Breaking through the limit of Shannon sampling theorem, CS could reconstruct the original signal with fewer sampling data. A conjugated gradient method using optimal weighted Total Variation Minimization (TVM) is proposed for data reconstruction. The simulation and experimental data verified the effectiveness of the pro-posed method. Higher compression ratio and better reconstruction accuracy can be guaranteed by the pro-posed method. Under the noisy system environment, even when the compression sensing ratio is 1/1500, the reconstruction mean square error of the proposed method is just 3.13%.

Keywords: power quality disturbances; fast measurement matrix optimization; Reweighted Total Variation Minimization (WTVM); conjugate gradient method.

1 INTRODUCTION

When large-scale new energy power is merging with the existing power grid, and large number of nonlinear loads are connected to the power systems, they cause problems like deteriorated Power Quality (PQ), complicated and various fault modes, and decreased predictability in power grid (Lin et al. 2013). High sampling frequency is required to monitor, locate, or estimate transient disturbances in power system (Lin 2007), due to the high fluctuation and random occur time of PQ disturbances.

The structure of the traditional PQ monitoring equipment (Muscas 2010) is shown in Figure 1. Complex frequency tracking algorithm, equal phase sampling, and other synchronous acquisition calibration structure are equipped in the process of sampling (Testa et al. 2007). High-frequency PQ data is sampled by high-cost acquisition system, and then compressed by high-quality central processing units. Extensive storage media resources and high bandwidth networks (Zhu et al. 2014) are used to store and transmit the compressed data, respectively.

Figure 1. Structure of the traditional PQ monitors equipment.

Sub-Nyquist sampling structure with high precision and low noise is one of emphases to improve compression ratio. Compressed Sensing is an innovative and revolutionary structure that samples and compresses sparse signals simultaneously, and reconstructs the original signal with few sample points (Candès & Wakin 2008). It has become a hot topic in signal processing and has improved markedly in image acquisition and processing, video signals and multi-sensor network field.

CS theory includes three aspects, as sparse representation basis, incoherent observation, and non-linear optimal reconstruction of the sparse signal. The sparser the original signal is, or the less the number of nonzero elements of the sparse representation matrix is, the less observation elements are needed for high probability accurate reconstruction. Standard orthogonal basis, including Fourier transform (Bertocco et al. 2014), Gabor basis (Matusiak & Wakin 2012) and Curvelet transform (Jianwei 2011) are used as sparse representation basis. Random matrix is widely used in measurement matrix. Some researchers have been working on optimizing the measurement matrix Φ to increase the reconstructing probability and take fewer measurements at the same time (Schnass & Vandergheynst 2008). Shrinkage-based alternating projection algorithm is considered for determining measurement matrix Φ according to incoherency theory (Wenjie et al. 2014). High-resolution signal can be recovered from low-resolution and sparse measurement data by CS theory, only if the sensing matrix satisfies Restricted Isometry Property (RIP). Verify the RIP of a fixed Φ is a NP-hard problem that only a few random matrices which have particular distributions could satisfy the so-called RIP with a constant parameter. To solve these problems, researchers construct non-RIP and non-convex CS frame based on l_0-minimization (Colbourn et al. 2011). Then l_1-optimization is used to take place of l_0-minimization and construct convex optimization and relaxation CS frame based on RIP theory (Candes et al. 2011). Approximation techniques, which are known as greedy pursuit algorithms, such as Matching Pursuit algorithm (MP) (Mallat & Zhang 1993), Orthogonal Matching Pursuit algorithm (OMP) (Tropp & Gilbert 2007), and Regularized Orthogonal Matching Pursuit (ROMP) (Needell & Vershynin 2007), et al., emerged for such sparse signals in recent years. Total Variation Minimization (TVM) algorithm is presented and applied for compressed sensing image restoration. It is similar to the l_1-optimization method. The difference is that the former controls the discrete gradient of the image and the latter controls the l_1-norm of the sparse signal (Chambolle 2010).

The characteristic of CS makes itself suitable for data acquiring and compressing in power quality monitoring system which has extremely information redundancy and heavy transmission burden. Fast Fourier transform basis is used as a fixed Ψ in this paper due to its essential characteristics. A new Gradient method using optimal Weighted Total Variation Minimization (WTVMG) is proposed to reconstruct the original signal. The proposed algorithm has been verified by frequency and time-domain simulation and experimental analysis.

2 COMPRESSED SENSING PRINCIPLE

Compressed sensing is a particular under-sampling system, which can be equivalently described by (1).

$$y = \Phi x + \varepsilon = \Phi\Psi\eta + \varepsilon = \Theta\eta + \varepsilon \qquad (1)$$

where the original signal x. Φ is a measurement matrix whose dimension belongs to $R^{M \times N}$. Ψ is the sparse matrix whose dimension belongs to $R^{N \times P}$. Θ is the sensing matrix. The measurement error including system noise is denoted by ε. CS reconstruction is a NP-hard problem and is simplified as

$$\min_{\eta}\left\{\|y - \Phi\Psi\eta\|_2 + \lambda\|\eta\|_1\right\} \qquad (2)$$

The measurement matrix Φ should satisfy the RIP theory. There exists a constant δ ($0<\delta<1$), which makes all the K sparse vectors of x satisfying (3).

$$(1-\delta)\|x\|_2^2 \leq \|\Phi x\|_2^2 \leq (1+\delta)\|x\|_2^2 \qquad (3)$$

System noise ε is the white Gaussian noise with a variance of σ^2 and zero mean. If the measurement matrix Φ satisfies the RIP theory with $0 < \delta < \sqrt{2} - 1$ and $\lambda = \sqrt{2(1+a)(1+\delta)}\lg N$ ($a \geq 0$), then (2) can be expressed as

$$\hat{x} = \min\|x\|_1 \, subjuct \, to \left\|\Phi^T(y - \Phi x)\right\|_\infty \leq \delta\lambda. \qquad (4)$$

The core issues in this paper focus on the reconstruction method.

3 PQ DATA RECONSTRUCTION ALGORITHM AND OPTIMIZATION BASED ON CS

3.1 WTVMG Analysis in reconstruction procedure

One-dimensional PQ signals are segmented and reshaped to two-dimensional image which has linear singularity properties in horizontal, vertical, and diagonal directions, similar to a gradient structure. It is suitable to use total variation function to reconstruct the signal based on direction singularity. An optimal weighted total variation minimization gradient model, which denotes as WTVMG, is proposed in this paper to solve the reconstruction problem of two-dimensional image for PQ signal.

The total variation minimization model of the two-dimensional image of the PQ signal which is sparse is represented by (5).

$$\arg\min_{\eta}\|\Phi\Psi\eta - Y\|_2^2 + \lambda\|\eta\|_1 + WO_{\text{TV}}(\Psi\eta) \qquad (5)$$

where the sparse representation of the original signal is denoted by η. Y is the vector that projects on Φ to get the original signal. λ is the regularization parameter. O_{TV} is the total variation operator, which is the sum of discrete gradient in an image. $WO_{TV}(\Psi\eta)$ is stated as (6).

$$\begin{aligned}\underset{\eta}{\text{argmin}}(WO_{TV}(\Psi\eta)) &= \underset{X}{\text{argmin}}(WO_{TV}(X))\\ &= \underset{x_{i,j}}{\text{argmin}}(\sum_{i,j}W_{i,j}|Q(x)_{i,j}|)\end{aligned} \quad (6)$$

where $|Q(x)_{i,j}| = \sqrt{\left(X_{i+1,j}-X_{i,j}\right)^2+\left(X_{i,j+1}-X_{i,j}\right)^2}$.

The weight coefficients for each pixel $X_{i,j}$ is denoted as $W_{i,j}$, which can be written as a function of $|Q(x)_{i,j}|$ and denoted by $W = g(|Q(x)|)$. The weighted total variation minimization model which is described as an enhanced sparse representation and can improve the quality of reconstructed image partly is stated in (6). The key issue is how to find the applicable weight coefficients W to represent the innate characteristics and enhance sparsity for the image itself. Weight coefficients for the smoothed areas should be larger than the strong mutated areas. When solving the weighted total variation model, solution is associated with the history one each iteration.

Assume that $W = W^k$, then we get

$$W^k = \underset{W}{\arg\min}\|W\|_{TV} + \frac{k}{2}\left\|W-\hat{W}\right\|_2^2 \quad (7)$$

where $\|W\|_{TV}$ is the regularization item which decreases the value of the total variation; $\frac{k}{2}\|W-\hat{W}\|_2^2$ is the truth-preserving item that makes the value of the optimized weight being consistent with the pre-estimated one.

Then equation (5) can be rewritten to the weighted total variation minimization model as (8).

$$\underset{\eta}{\text{argmin}}\|\Phi\Psi\eta-Y\|_2^2 + \lambda\|\eta\|_1 + \sum_{i,j}W_{i,j}^k|Q(x)_{i,j}| \quad (8)$$

Using Chambolle projection algorithm [17], with a fixed step $\tau\in(0,1/8)$ and initial value of $p^{(0)} = 0$, the weight is obtained by (9).

$$W = \hat{W} - \frac{\text{div}(p^*)}{k} \quad (9)$$

where p^* is the convergence value of (10).

$$p_{i,j}^{(n+1)} = \frac{p_{i,j}^{(n)} + \tau(\nabla(\text{div}(p^{(n)})-k\hat{W})_{i,j})}{1+\tau\left|(\nabla(\text{div}(p^{(n)})-k\hat{W})_{i,j}\right|} \quad (10)$$

To solve the optimal weighted total variation minimization model in (8), a non-linear conjugate gradient descent algorithm with backtracking line search is proposed in this paper. Hestenes-Stiefel (HS), Fletcher-Reeves (FR), and Polak-Ribi'ere-Polyak (PRP) conjugate gradient are in common usage. FR has a fast convergence speed but poor numeric performance, and may bring about small step size for a continuous time, which makes it difficult to guarantee the global convergence. PRP has been proven to be one of the best conjugate gradient methods, but it is still probably to generate non-global convergence sequences (Powell 1984). A modified PRP conjugate gradient method is adopted in this paper to solve the $\min\{f(x)|x\in R^n\}$, and its iterative formula is shown in (11).

$$x_{k+1} = x_k + t_k d_k \quad (11)$$

where the step size is t_k, and the searching direction is d_k which is determined by (12).

$$d_k = \begin{cases} -g_k, & k=1, \\ -g_k + \beta_k d_{k-1}, & k\geq 2, \end{cases} \quad (12)$$

Generalized formula of PRP gradient algorithm is

$$\beta_k^{PRP} = \frac{g_k^T(g_k-g_{k-1})}{g_{k-1}^T g_{k-1}} \quad (13)$$

β_k is calculated by the modified PRP conjugate gradient method as shown in (14) and (15). It will avoid generating small step size for a continuous time, effectively improve the efficiency of numerical calculation, and possess global convergence.

$$\beta_k = \max\left\{0, \beta_k^{PRP}+\min(0,\gamma_k)\right\} \quad (14)$$

$$\gamma_k = \frac{g_k^T(g_k-g_{k-1})}{\|g_k\|^2}\frac{g_k^T g_{k-1}}{\|g_{k-1}\|^2} \quad (15)$$

By solving the weighted total variation minimization model in (8), two-dimensional reconstructed image X for PQ signal can be obtained

$$X = \Psi\eta \quad (16)$$

The proposed reconstruction algorithm is shown in detail as follows

3.2 Evaluation index

Three indexes are used to evaluate the results of PQ signal based on CS theory. Signal Noise Ratio (SNR) is used to evaluate the distortion of the reconstructed signal while Mean Square Error (MSE) evaluates the error between the original signals and reconstructed one. Compressed sensing ratio is defined in this paper to represent the ratio between the length of CS sampling points and Nyquist sampling points.

$$R_{\text{SNR}} = 10 \log_{10}\left(\sum_{i=1}^{N} \left\| x(i) \right\|^2 \Big/ \sum_{i=1}^{N} \left\| x(i) - \hat{x}(i) \right\|^2 \right) \quad (17)$$

$$R_{\text{MSE}} = \left(\sum_{i=1}^{N} \left\| x(i) - \hat{x}(i) \right\|^2 \Big/ \sum_{i=1}^{N} \left\| x(i) \right\|^2 \right)^{1/2} \times 100\% \quad (18)$$

$$C_{\text{R}} = M/N \quad (19)$$

where R_{SNR} stands for signal noise ratio, unit dB; R_{MSE} is the mean square error; $x(i)$ is the original signal; $\hat{x}(i)$ stands for the reconstructed signal after CS; C_{R} is the compressed sensing ratio; N is the length of the original signal; M is the length of the reconstructed signal after CS.

4 SIMULATION AND EXPERIMENTAL RESULTS

The proposed WTVMG algorithm has been tested on various noisy PQ signals. Signals denoted as A, B, C, D, and E are used for simulation. White Gaussian noise with SNR of 40 dB is added to the aforementioned signals A-E. Signal F is a measured PQ disturbance. An analytical comparison has been made between the normal OMP, ROMP, and the proposed WTVMG algorithm. For the simulation signals A-E, compressed sensing ratio C_{R} is fixed to be 1/2. The simulated signals A-E contain harmonics no larger than 32th, assumed that the Nyquist frequency is 1.6 kHz and the total original data has a dimension belongs to $R^{4096 \times 1}$. When the one-dimensional power quality signal is mapped to the two-dimensional image whose dimension belongs to $R^{64 \times 64}$, the dimension of compressed sensing observation matrix is $R^{32 \times 64}$, which means the sampling rate of CS is only 1.6 kHz, half of Shannon sampling frequency.

a. Signal A is a signal with fundamental frequency and inter-harmonics.

$$x(t) = \begin{cases} \sin(2\pi 50 t_a) + 0.33\sin(2\pi 1550.5 t_a), & 0 \leq t_a < 0.64s \\ \sin(2\pi 50 t_b) + 0.2\sin(2\pi 1570.5 t_b), & 0.64 \leq t_b < 1.28s \end{cases}$$

b. Signal B is a signal with fundamental frequency, harmonics, and sags.

$$x(t) = \begin{cases} \sin(2\pi 50 t_a) + 0.2\sin(2\pi 250 t_a) \\ \quad + 0.1\sin(2\pi 1550 t_a), & 0 \leq t_a < 0.32s \\ 0.9\sin(2\pi 50 t_b) + 0.18\sin(2\pi 250 t_b) \\ \quad + 0.08\sin(2\pi 1550 t_b), & 0.32 \leq t_b < 0.96s \\ \sin(2\pi 50 t_c) + 0.2\sin(2\pi 250 t_c) \\ \quad + 0.1\sin(2\pi 1550 t_c), & 0.96 \leq t_c < 1.28s \end{cases}$$

Algorithm Iterative algorithm for l_1-penalized reconstruction

Objective: Find the suitable η to make the given expression $\|\Phi\Psi\eta - Y\|_2^2 + \lambda\|\eta\|_1 + WO_{\text{TV}}(\Psi\eta)$ attained its minimum value.

Input
2: • k-space measurement Y.
   • Random measurement matrix Φ.
4: • Sparsifying transform operator Ψ.
   • Regularization parameter, a data consistency tuning constant λ.
6: • Iteration counter $k = 0$, and parameter $t = 1$.

Optional Parameters:
8: • TolGrad - stopping criteria by gradient magnitude (default 10^{-5}).
   • MaxIter - stopping criteria by number of iterations (default 100).
10: • α, β - line search parameters (defaults $\alpha = 0.01$, $\beta = 0.6$)

Output
   • η - the numerical approximation in (5).
12: ***Algorithm begin:***
   Initialization:
14:   $k = 0$; $\eta_0 = Y$; $g_0 = \nabla f(\eta_0)$; $\Delta\eta_0 = -g_0$.
   Loop
16: 1) Update parameter t as follows:
      while ($f(\eta_k + t\Delta\eta_k) > f(\eta_k) + \alpha t \cdot Real(g_k \Delta\eta_k)$)
        $t = \beta t$;
      end while
18: 2) Update η: $\eta_{k+1} = \eta_k + t\Delta\eta_k$
    3)* Conjugate gradient calculation: $g_{k+1} = \nabla f(\eta_{k+1})$
20: 4) Modified PRP-Conjugate Gradient Method
    $\sigma_{k+1}^{PRP} = g_{k+1}^T(g_{k+1} - g_k)/(g_k^T g_k)$
    $\xi_{k+1} = g_{k+1}^T(g_{k+1} - g_k)g_{k+1}^T g_k/(\| g_{k+1} \|^2 \| g_k \|^2)$
      $\gamma = \max\{0, \sigma_{k+1}^{PRP} + \min(0, \xi_{k+1})\}$
    $\Delta\eta_{k+1} = -g_{k+1} + \gamma\Delta\eta_k$
22: **if** $\|g_k\|_2 < \text{TolGrad}$ or $k > \text{maxIter}$ **then**
    stop the iterations and terminate the algorithm.
24: **else**
    $k = k + 1$ and go to 1);
26: **end if**
    end loop.
28: **Update** $X = \Psi\eta$.

In this algorithm line 19 which labled *:
The conjugate gradient requires the computation of $\nabla f(m)$
      $\nabla f(\eta) = 2(\Phi\Psi)^*(\Phi\Psi\eta - Y) + \lambda\nabla\|\eta\|_1 + \nabla WO_{\text{TV}}(\Psi\eta)$
   The ℓ_1 norm is the sum of absolute values. The absolute value function however, is not a smooth function and as a result it is not well defined for all values of η. Instead, we approximate the absolute value with a smooth function by using the relation $|x| \approx (x^*x + \mu)^{1/2}$, where μ is a positive smoothing parameter. With this approximation, $d|x| / dx \approx x(x^*x + \mu)^{-1/2}$.
   It can be approximated by,
      $\nabla f(\eta) \approx 2(\Phi\Psi)^*(\Phi\Psi\eta - Y) + \lambda\eta(\eta^*\eta + \mu)^{-1/2} + \nabla WO_{\text{TV}}(\Psi\eta)$
   Update $\nabla WO_{\text{TV}}(\Psi\eta)$ by (8)~(11).
   Update g_{k+1}.

c. Signal *C* is a signal with fundamental frequency, harmonics, and swells.

$$x(t) = \begin{cases} \sin(2\pi 50 t_a) + 0.2\sin(2\pi 250 t_a) \\ \quad + 0.1\sin(2\pi 1550 t_a), & 0 \le t_a < 0.32s \\ 1.1\sin(2\pi 50 t_b) + 0.22\sin(2\pi 250 t_b) \\ \quad + 0.12\sin(2\pi 1550 t_b), & 0.32 \le t_b < 0.96s \\ \sin(2\pi 50 t_c) + 0.2\sin(2\pi 250 t_c) \\ \quad + 0.1\sin(2\pi 1550 t_c), & 0.96 \le t_c < 1.28s \end{cases}$$

d. Signal *D* is a signal with fundamental frequency, harmonics, and transient.

$$x(t) = \begin{cases} \sin(2\pi 50 t_a) + 0.1\sin(2\pi 1550 t_a), & 0 \le t_a < 0.32s \\ \sin(2\pi 50 t_b) + 0.1\sin(2\pi 1550 t_b) \\ \quad + \sin(2\pi 500 t_b)e^{(-90 t_b - 0.32)}, & 0.32 \le t_b < 0.96s \\ \sin(2\pi 50 t_c) + 0.1\sin(2\pi 1550 t_c), & 0.96 \le t_c < 1.28s \end{cases}$$

e. Signal *E* is a signal with fundamental frequency, harmonics, and flickers.

$$x(t) = \sin(2\pi 50 t_a)(1 + 0.1\sin(2\pi 2 t_a)) + 0.1\sin(2\pi 1550 t_a),$$
$$0 \le t_a < 1.28s$$

The signal noise ratio R_{SNR} results of simulated signal *A-E* are shown in Table 1. It confirms that the proposed WTVMG algorithm is robust to various power quality disturbances.

Measured voltage signal *F* of North China Electricity Board is illustrated here to test the proposed algorithm

The fundamental frequency of signal *F* is varied randomly within ± 3 Hz, and there contains voltage

sags and swells, transients, and any other PQ disturbances. The sampling frequency is 10 kHz, data during 4 s are obtained and shown in Figure 2(a). After normalizing the sampled data, and doing FFT transform, the spectrogram is shown in Figure 3(a). It is obviously observed that the measured signal *F* contains fundamental component, 2th-100th harmonics, and some interharmonics.

Sampled by the proposed WTVMG method based on CS theory, with a fixed compressed sensing ratio C_R of 1/2, the obtained data are shown in Figure 2(b). After normalizing the reconstructed data, and doing FFT transform, the spectrogram is shown in Figure 3(b). It is impossible to obtain the real value of the measured data, while the signal noise ratio is beyond computation, and the mean square error R_{MSE} is used to describe the signal reconstruction accuracy. The mean square error R_{MSE} is calculated with 0.0018%. Comparing Figure 3(a) and (b), it is evident that the proposed method gives a reliable result by the sampling frequency of 5 kHz only. The mean square error

Table 1. The signal noise ratio R_{SNR} results of simulated signal *A-E*.

Simulated signal	Algorithms	R_{SNR} /dB
A	Normal OMP	19.0502
	Normal ROMP	21.0475
	WTVMG	36.5657
B	Normal OMP	37.0733
	Normal ROMP	36.8393
	WTVMG	38.8632
C	Normal OMP	36.8393
	Normal ROMP	37.0577
	WTVMG	37.7637
D	Normal OMP	29.6843
	Normal ROMP	32.1286
	WTVMG	36.5587
E	Normal OMP	37.0078
	Normal ROMP	38.4147
	WTVMG	39.0348

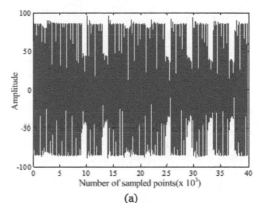

(a)

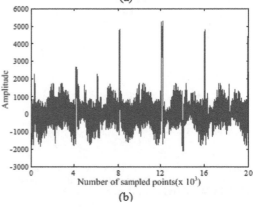

(b)

Figure 2. (a) Measured signal *F* by the sampling frequency of 10 kHz based on Nyquist sampling algorithm. (b) Measured signal *F* by the sampling frequency of 5 kHz based on WTVMG.

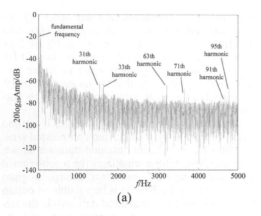

(a)

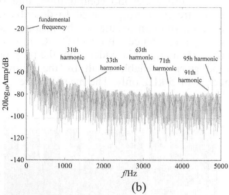

(b)

Figure 3. (a) Spectrogram of the measured signal F by the sampling frequency of 10 kHz based on Nyquist sampling algorithm. (b) Spectrogram of the measured signal F by the sampling frequency of 5 kHz based on WTVMG after reconstruction.

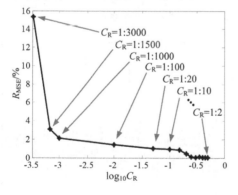

Figure 4. Values of R_{MSE} when the compressed sensing ratio C_R changes for measured signal F based on FMMCA-WTVMG algorithm.

R_{MSE} decreases while compressed sensing ratio C_R increases as shown in Figure 4. The proposed algorithm achieves extremely small mean square error R_{MSE} (that is, 0.001%) when compressed

sensing ratio C_R is 1/2. Due to the complexity of measured signal F, the mean square error R_{MSE} by the proposed WTVMG algorithm is 3.13% while compressed sensing ratio C_R is 1/1500. It maintains precision of the reconstruction performance while achieves high compression ratio. It is obviously observed in Figure 4 that the compressed sensing ratio C_R can be further decreased and good reconstruction performance can be achieved at the same time with wider the allowed tolerance.

5 CONCLUSION

Breaking the limitations of the traditional Shannon-Nyquist sampling theorem, a reconstruction method based on CS theory is proposed in this paper to allow power quality disturbances to be sampled at sub-Nyquist rates which saves sampling hardware cost, minimizes hardware complexity, and occupies less storage media and network bandwidth resources with high reconstruction accuracy. Power quality disturbances signals are in analogy with two-dimensional image to achieve faster processing speed. Based on gradient method which is suitable for two-dimensional signal, WTVMG reconstruction algorithm is proposed which obtained high compression ratio and guaranteed reconstruction accuracy, even when the compression sensing ratio is 1/1500. The proposed method maintains the reconstruction mean square error to 3.13%.

ACKNOWLEDGEMENTS

Project Supported by National Natural Science Foundation of China (51247009, 50977016); International Science & Technology Cooperation Program of China (2010DFB63050); Key Laboratory of Wind Power and Smart Grid (CXB20100 52500 25 A).

REFERENCES

Bertocco, M. & Frigo, G. & Narduzzi, C. & Tramarin, F. 2014. Resolution Enhancement by Compressive Sensing in Power Quality and Phasor Measurement. IEEE Transactions on Instrumentation and measurement 63(10): 2358–2367.

Chambolle, A. 2004. An algorithm for total variation minimization and applications. Journal of Mathematical Imaging and Vision 20(1–2): 89–97.

Colbourn, C.J. & Horsley, D. & McLean, C. 2011. Compressive sensing matrices and hash families. IEEE Transactions on Communication 59(7): 1840–1845.

Candes, E.J. & Romberg, J. & Tao, T. 2006. Robust uncertainty principles: Exact signal reconstruction from highly incomplete frequency information. IEEE Transactions on Information Theory 52(2): 489–509.

Candès, E.J. & Wakin, M.B. 2008. An introduction to Compressive Sampling. IEEE Signal Processing Magazine 25(2): 21–30.

Lin, W. & Wen, J. & Liang, J. 2013. A three-terminal HVDC system to bundle wind farms with conventional power plants. IEEE Transactions on Power System 28(3): 2292–2300.

Lin, H.C. 2007. Fast tracking of time-varying power system frequency and harmonics using iterative-loop approaching algorithm. IEEE Transactions on Industrial Electronics 54(2): 974–983.

Muscas, C. 2010. Power quality monitoring in modern electric distribution systems. IEEE Instrumentation and measurement Magazine 13(5): 19–27.

Testa, A. & Akram, M.F. & Burch, R. 2007. Interharmonics: theory and modeling. IEEE Transactions on Power Delivery 22(4): 2335–2348.

Mallat, S.G. & Zhang, Z. 1993. Matching pursuit in a time-frequency dictionary. IEEE Transactions on Signal Processing 41(12): 3397–3415.

Matusiak, E. & Eldar, Y.C. 2012. Sub-Nyquist sampling of short pulses. IEEE Transactions on Signal Processing 60(3): 1134–1148.

Ma, J.W. 2011. Improved iterative curvelet thresholding for compressed sensing and measurement. IEEE Transactions on Instrumentation and measurement 60(1): 126–136.

Needell, D. & Vershynin, R. 2010. Signal recovery from incomplete and inaccurate measurements via regularized orthogonal matching pursuit. IEEE Journal of Selected Topics in Signal Processing 4(2): 310–316.

Powell, M.J.D. 1984. Nonconvex minimization calculations and the conjugate gradient method. Lecture notes in Mathematics 1066:122–141.

Schnass, K. & Vandergheynst, P. 2008. Dictionary preconding for greedy algorithms. IEEE Transactions on Signal Processing 56(5): 1994–2002.

Tropp, J.A. & Gilbert, A.C. 2007. Signal recovery from random measurements via orthogonal matching pursuit. IEEE Transactions on Information Theory 53(12): 4655–4666.

Yan, W.J. & Wang, Q. & Shen, Y. 2014. Shrinkage-based alternating projection algorithm for efficient measurement matrix construction in compressive sensing. IEEE Transactions on Instrumentation and measurement 63(5): 1073–1084.

Zhu, K. & Chenine, M. & Nordström, L. & Holmström, S. & Ericsson, G. 2014. Design requirements of wide-area damping systems—using empirical data from a utility IP network. IEEE Transactions on Smart Grid 5(2): 829–838.

Energy Science and Applied Technology ESAT 2016 – Fang (Ed.)
© 2016 Taylor & Francis Group, London, ISBN 978-1-138-02973-6

'Oblique pulling method' for live replacing V-type insulator string on UHVDC lines

Longfei He, Yang Long, Hao Guo & Hui Li
Live Operations Centre in Hunan, Changsha, China

ABSTRACT: This paper first describes the limitations of the existing methods for live replacing insulator of the 800 kV UHVDC transmission lines, then it puts forward to the improvement of the "oblique pulling" method and the verification of its feasibility. Accordingly, we developed hydraulic tensioners with low-density/high-strength aluminum alloy, reducing the self-weight of the tight line tools in order to meet the UHV operation requirements. By structural re-design, force analysis, and testing, the supporting tools have finally met the standards for live working regulations and can be successfully applied to the actual line. As a result, this optimized method is able to reduce the physical intensity of workers and improve the safety of operation.

Keywords: 800 kV UHVDC line; live replacement insulator string; the oblique pulling method; tight line tools

1   INTRODUCTION

UHVDC transmission is characterized as high voltage level, long delivery distance, high delivery capacity, and non-simultaneous networking (GAO Jian et al., 2015). Facilitating the construction of UHVDC networks is a significant strategy which can secure the overall situation of national energy (HU YI et al., 2011), satisfy the demands of the economic development, and maintain the balance of electricity supply (HU YI et al., 2010), so as to achieve the sustainable development of the electrical industry. In recent decades, China is the only country worldwide building this 800 kV UHVDC transmission network (LI Qingfeng et al., 2009). However, the maintenance or reparation of this network with power cut will lead to tremendous financial loss as it delivers 620~640 wkwh electricity for domestic and industries (LI Ruhu, 2009). Therefore, live operation is a critical method maintaining and repairing these UHVDC transmission lines, and it is extremely important to ensure the safety, reliability, and stable working of the network (LUO RI Cheng et al., 2015).

During the operation, insulator strings are easily affected by internal voltage and threatening by external factors, such as working voltage, operating voltage and thunderstorm voltage changing, natural/man-made damages, environmental pollution, and inevitable extreme weather conditions (XIAO Yong et al., 2010). Accordingly, when broken and zero-value happens, insulator strings may endanger the stable operation of the network and need to be replaced immediately (Figure 1-A). However, recent methods for live replacing insulator strings are still underdeveloped and heavy tools make it more difficult to be completed (Xu XIAO Yi et al., 2012). Therefore, there is an urgent demand for specific research of live replacing insulator strings on 800 kV UHVDC transmission lines.

V-type insulator strings are the widest applied install shape among diverse types, consisted of a quarter of overall numbers. The project of UHVDC transmission on vertical lines generally employs large-tonnage complex strings, which have over 300 kN mechanical strength level (Figure 1-B).

In this article, the ± 800 kV UHVDC transmission V-type insulator strings are researched, in order to find suitable replace methods and develop re-designed tight line tools. Meanwhile, the improved method is illustrated and the application of tool sets is displayed using ± 800 kV UHVDC transmission electrical tower BinJ in line, part of network in Fenghuang town, Xiangxi Autonomous district, as a typical example.

A Damage of insulator   B V-type insulator string

Figure 1.   Damage and assembly of insulator.

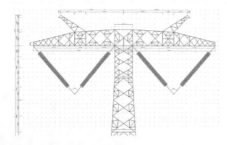

Figure 2. BinJin line diagram (± 800 kV).

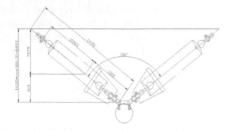

Figure 3. BinJin line fittings assembly (± 800 kV).

2 RESEARCH OF PRESENT SITUATION

2.1 Tower-shaped structure and hardware connection

The UHVDC lines have considerably high electricity poles, long insulator strings, and heavy tools. Taking No. 1115 linear tower, for instance, the pole type is ZC27203 with 60 m height and overall 73.5 m height (Figure 2). The wire type is JL/G2 A-900/75; the insulator type is FXBZ-±800/420 with 10600 mm diameter; the string type is XS10 with duplex V shape and included angle 106° (Figure 3).

2.2 Research on recent methods

The existing method for c tightens strings as a whole and using the professional tool, which is installed on the cross arm to implement reparation (Figure 4).

However, it has drawbacks shown below:

1. V-type insulator strings display 53° angle, which leads to dis-separation between up and down suspension points, due to the self-weight of insulators causing arc-shaped load transfer during the tightening.
2. Tightening strings entirely requires professional tools installed on the right top of suspension points, while there is no specific position on the cross arms for installation.
3. The safety requirement of the distance between workers' positions on the cross arm and the working spots can hardly be satisfied during the tool installation period.

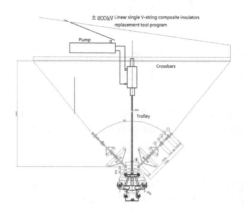

Figure 4. A whole way to tight line.

4. Professional tools are excessively heavy and hard to operate, which cost a great deal of physical strength.

For above reasons, the existing method of replacing insulators has considerable potential problems in safe manufacture and restricts the development of UHVDC live operation to a certain degree. Therefore, it is reasonable and necessary to study the method of live replacing V-type insulator string on 800kv UHVDC transmission lines.

3 RESEARCH ON 'OBLIQUE PULLING METHOD'

3.1 Introduction of 'oblique pulling method'

The 'oblique pulling method' is a different line lifting method compared with the entire pulling method mentioned above. By installing tighten tools alongside the V-type insulator string so as to disperse loads, tightened strings are relaxed which makes it easier for the workers to change the socket of the insulator. This method is widely applied in live replacing V-type insulator string on 500 kV UHVDC transmission lines, but it has not been utilized in live replacing 800 kV UHVDC transmission lines yet. Owing the particular characteristics of 800 kV UHVDC transmission lines, the Strong Electric Field and labor protection have to be carefully analyzed and verified, and the safe distance and the combined gap have to be checked and tested, in order to ensure the practicability of the 'oblique pulling method'.

3.2 The verification of the 'oblique pulling method'

3.2.1 The surface field strength of operators and protective measures

By calculating and analyzing different operating positions (Figure 5), results for the specific distri-

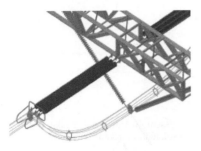

(a) Worker at the wires above the cross arm

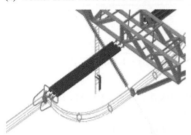

(b) Worker in the middle position of insulation string

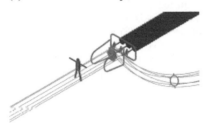

(c) Worker in the wire

Figure 5. Typical position used in calculation of field strength.

(a) Worker at the wires above the cross arm

(b) Worker in the middle position of insulation string

(c) Worker in the wire

Figure 6. Distribution of electric field around the human body.

bution of the body surface field strength are displayed in Figure 6. The surface field strengths of the ground potential and during entering are both lower than 100 kV/m, while it rises to 195.3 kV/m on the head or tip parts. After entering the equal potential, the electrical field distortion of tip parts will lead to a dramatic increase in the intensity, which is maximum 1706 kV/m and far above the 240 kV/m perception level.

According to these results, it is clear that tip parts of human body, like head and hands, have higher surface field strength, while that is lower on chest and back of the body. For operators working on ±800 kV wires, the maximum surface field strength is 1706 kV/m, which can be protected by wearing 60dB shield effectiveness clothes specially designed for UHVDC operation. Moreover, considering the time during entering the equal potential, the arc discharge may occur between the operator's face and live wire, so it is necessary to wear the shield mask over 20 dB effectiveness.

3.2.2 The verification of the safe distance in live operations

The safe distance is one of protective barriers for operators, which is too dangerous to be crossed. Taking 800 kV UHVDC live operation as an example, the safe distance between the body and live parts should be no less than 6.8 m. Operators who entering the field by "lifting basket method" should keep a minimum distance of 6.6 m against the combination of the ground and live parts.

• Verification of horizontal distance

The horizontal distance between the side of the cross arm and the centre of a tower is 25000 mm, the distance from the side of the cross arm to the centre of a bundled conductor is 11325 mm, and the horizontal distance from the centre of a tower to the side of it is 2350 mm. The short circuit gap between foot nails is neglected.

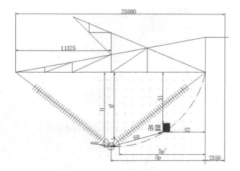

Figure 7. Diagram of safety distance and combined gap.

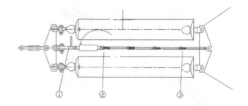

Figure 8. Tool assembly drawing 1, insulated pull rod; 2, wire rod; 3, insulated pull rod of wire rod.

$$\begin{cases} S_0 = \sqrt{8480^2 + (8480 - 1200 - 1000)^2} \\ \quad = 10500 \ mm \\ S_1 = 1000 \ mm \\ S_2 = 11325 - \sqrt{8480^2 - 1000^2} = 2900 \ mm \end{cases} \quad (7)$$

$$\begin{cases} S_a = S_0 + S_1 - r \\ \quad = 10500 + 1000 - 1000 = 10500 \ mm \\ S_b = S_0 + S_2 - r \\ \quad = 10500 + 2900 = 13400 \ mm \end{cases} \quad (8)$$

Thus, the horizontal distance between the centre of a bundled conductor and the side of a tower is:

$$SP = 25000 - 11325 - 2350 = 11325 \ mm \quad (1)$$

Considering the air gap of equal voltage circle connection, short circuit r equals to 1000 mm, so the distance from the live part to tower itself is:

$$SP' = SP - 1000 = 10325 \ mm \quad (2)$$

Therefore, the minimum safe distance 6.8 m is verified.

• Verification of vertical distance
According to relevant data of insulator strings, the overall length can be calculated to be 14105 mm, and the included angle is 106°. The air gap r of equal voltage circle connection is equal to 1000 mm, so the vertical distance and safe distance are:

$$H = L \times Cos \ (106 / 2) = 8480 \ mm \quad (3)$$

$$H' = H - r = 7480 \ mm \quad (4)$$

Therefore, the minimum safe distance 6.8 m is verified.

3.2.3 *Verification of the gap between live parts*
As shown in Figure 7, the main gaps are S0 + S1 and S1 + S2. Using the process of "lifting basket method", the width of a basket is 600 mm and the height is 1200 mm. Two combinations of gaps are:

$$S_a = S_0 + S_1 - r \quad (5)$$

$$S_b = S_0 + S_2 - r \quad (6)$$

where r is the air gap of equal voltage circle connection, which is equal to 1000 mm.

• when the basket is 1 m away from the cross arm, track rope is 8480 mm, then:

The combination gap between Sa and Sb meets the 6.6 m requirement.
• when the basket is 1 m away from the wire, based on formulas (7) and (8), the gap is calculated that Sa is 9725 mm and Sb is 8280 mm. This combination gap between Sa and Sb is dissatisfied with the 6.6 m requirement.

4 RESEARCH ON TIGHT LINE TOOLS

When applying the 'oblique pulling method' for live replacing insulator strings on 800 kV UHVDC transmission lines, and the pulling tools, such as tight line tools and insulated pull rods, are installed on the prepared holes and loads will therefore transfer alongside insulator strings. However, as the length of UHVDC insulator strings is over 10 m and loads of it are over 10 tonnes, the existing tools are unable to complete the operation. Thus, it is urgent to redesign and improve existing tight line tools, which can achieve UHVDC operation without having problems like lack of the tonnage.

4.1 *Improved hydraulic tighter*

Two main types of existing tight line tools are hydraulic tighter and wire-rod tighter and the hydraulic one consisted of hydraulic pump and power system.

According to previous analysis of the insulator's structure and hanging method, 40 mm is the required tension distance in order to replace insulators and transfer loads. In comparison of stress conditions during tightening up tools, it shows that

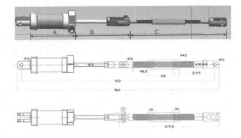

Figure 9. Schematic diagram of the improved hydraulic line A, hydraulic cylinder; B, piston rod; C, rod.

tighter experiences less than 3 tonnes force in the first 30 mm but it will suddenly soar in the following 10 mm. Based on operational experiences, workers can swing the wire-rod manually and transfer no more than 3 tonnes force. Therefore, the hydraulic tighter is improved and shown in Figure 9.

The improved hydraulic tighter has a more reasonable structure, which reduces the length of the cylinder. The mechanical wire-rod part is used to tighten first 300 mm and the minimum distance of the piston inside the cylinder is reduced from 400 mm to 100 mm, and these improvements can decline the hydraulic tighter's self-weight to a great degree.

The improved hydraulic tighter is made up of several materials. The cylinder is made by low-density/high-strength aluminum alloy, so as to cause further reduction in self-weight, while the piston rod and wire-rod are still made of high-strength steel alloy, which can undertake more force. This new tighter not only can satisfy the requirement of live operation, but also can make it easier to be completed.

4.2 The analysis of V-type insulator strings' loads

According to the format of wires, the unit weight is 3.0472 kg/m and the cross-sectional area is 973.16 mm².

- self-weight/load ratio g1 and vertical load G:

$$g_1 = 9.807 g_2 \times 10^{-3} / A \tag{9}$$

$$G = n g_1 A L_v + G_j \tag{10}$$

where n is the number of bundle conductors, which is equal to 6 in this example, and the vertical distance Lv of the tower is 743 m; and G is the total weight of insulator strings, which is equal to 500 kg in this example, and the vertical load G is 139.4 kN.

- pulling force F2 of insulator strings:

$$F_2 = (G / 2) / \mathrm{Cos}(106 / 2) = 116.7 \ kN \tag{11}$$

Figure 10. Physical diagram of the improved hydraulic line A, wire rod tight line part; B, hydraulic tight line part.

4.3 The force analysis of the improved tighter

Improved tighter consists of two parts, namely wire-rod tighter and hydraulic tighter, and the designed structure as well as the selected materials have to be calculated and verified separately so as to ensure that the tool satisfies the strength requirement during the load transfer period.

4.3.1 Calculations for hydraulic tighter part
① verification of the piston rod's strength:

$$P = K_n F_2 \tag{12}$$

$$\sigma = \frac{P}{S} = \frac{4P}{\pi d^2} \leq [\sigma] \tag{13}$$

P is the undertaken force, Kn is the safety coefficient, S is the rod's cross-sectional area, and $[\sigma]$ is the material's yield stress.

Where K_n is 3, d is 35 mm,
$F_2 = 116.7 \ kN [\sigma] = 463, \sigma = 364.1, N / mm^2 \leq [\sigma]$

Therefore, the design of the structure of the rod and material meets the requirement.
② verification of the piston barrels' strength:

$$\delta = \frac{D_1 - D_2}{2} \tag{14}$$

$$\sigma = \frac{P}{S} = \frac{4P}{\pi(D_1^2 - D_2^2)} \leq [\sigma] \tag{15}$$

where $D_1 = 53 \ mm$, $\delta = 10 \ mm$, $D_2 = 43 \ mm$, $[\sigma] = 510 \ N / mm^2$, $\sigma = 464.6 \ N / mm^2 \leq [\sigma]$

Therefore, the design of barrel's structure and material meets the requirement.

4.3.2 Calculations for wire-rod tighter part
- verification of the spiral rod's strength:
  Because of (12) and (13), when $K_n = 3$, d = 30 mm, material 40Cr:
  $[\sigma] = 550 \ N / mm^2 \ \sigma = 495.6 \ N / mm^2 \leq [\sigma]$.

Therefore, the design of structure and material meets the requirement.
- verification of the socket's strength:
  Because of (14) and (15)

A Into the potential field   B Transfer load spot

Figure 11. The worker into the potential field.

$D_1 = 43\ mm,\ \delta = 10\ mm,\ D_2 = 30\ mm$

material: 40Cr

$[\sigma] = 550\ N\,/\,mm^2\ \sigma = 470.0\ N\,/\,mm^2 \le [\sigma]$.

Therefore, the design of rod's structure and material meets the requirement.

• verification of the dowel's strength:

$$Q = \frac{P}{2} \qquad\qquad (16)$$

$$S \ge \frac{Q}{[\tau]} \text{ then } \frac{\pi d^2}{4} \ge \frac{Q}{[\tau]} \qquad\qquad (17)$$

$$d \ge \sqrt{\frac{4Q}{\pi[\tau]}} \qquad\qquad (18)$$

Q represents the stress of cutting, $[\tau] = 300\ N/mm^2$

$Q = 58.38\ kN$, due to (18) $d \ge 15.7\ mm$, $\phi = 16\ mm$ meets the requirement.

4.4 The trail of tools

Improved tools have already been texted and confirmed at electricity-technology research institution. Therefore, the theoretical analysis and practical trail both show that the capability of improved tools satisfies the safe requirement.

5 THE FIELD APPLICATION OF THE METHOD AND TOOLS

In May of 2015, the field team has applied the 'oblique pulling method' for live operation at Tongmu village, Qiangongpin town, Fenghuang County, and Xiangxi autonomous district. The replacement of damaged insulators and the reparation of deficient wires have been completed, which ensured the safe operation of the electricity network.

The process of 'oblique pulling method' was implemented successfully and the safe distance between operators and wires was kept strictly. Meanwhile, the set tools were conveniently used and the physical intensity of operators reduced significantly.

6 CONCLUSION

1. In view of the safety risk of the existing methods for live replacing V-type insulator string on 800 kV UHVDC transmission lines, the 'oblique pulling method' was advanced. By conducting electrical field simulation analysis and safe distance/combination gap verification, the results provide the practical implementation with theoretical basis and the feasibility of this method is accordingly confirmed.

2. The low-density but high-strength aluminum alloy material is applied to produce the improved hydraulic tight line tool. By doing mechanical strength analysis and laboratory tests, the safety and reliability of this tool are fully verified, which provides the implementation of the method with a supportive tool.

3. The 'oblique pulling method' is successfully employed to live replacing V-type insulator string on 800 kV UHVDC transmission lines. The successful application of the improved tight line tool in real lines has accumulated valuable experiences for the future operation.

REFERENCES

DL/T 1242–2013. Technical specifications of ±800 kV DC live line job, [S].

Gao Jian, Zhong Zhuo Yin. Study on adhesion mechanism and protective measures of 220 kV vertical double split wire [J]. Journal of electric power science and technology 2015, 11 (2): 1–3.

Hu Yi et al. Field Strength of Body Surface During the Live Working on the UHV AC and DC Transmission Lines [J]. High Voltage Engineering, 2010, 26(1):13–18. 74–77.

Hu Yi, Liu Kai. Live operation of ultra / extra high voltage AC / DC transmission line [M]. Beijing: China Electric Power Press, 2011.

Li Qing-feng et al. Shielding Protection for Live Working on ±800 kV DC Transmission Line [J]. Proceedings of the CSEE, 2009, 29(34): 96–101.

Li Ru-hu. Comments on the Enter Ways into Equal Potential in Electrification Operation [J]. Southern Power System Technology, 2009, 3(2):77–80.

Luo Ri Cheng, Li Zhi. Qian Analysis of induced voltage and current of four circuit transmission lines on the same tower [J]. Journal of electric power science and technology 2015, 04(1): 2–4.

Q/GDW 1799.2–2013. Line portion of State Grid Corporation of electrical safety engineering disciplines [S], 2013.

Xiao Yong, Fan Ling-meng. Analysis of live working on Yun-Guang ±800kV DC Transmission Line [J]. High Voltage Engineering, 2010, 36(9): 2206–2211.

Xu Xiao Yi et al. Design of reactive power voltage characteristic and voltage control coordination system for extra high voltage AC / DC hybrid power grid [J]. Journal of electric power science and technology 2012, 2(4): 1–5.

Energy Science and Applied Technology ESAT 2016 – Fang (Ed.)
© 2016 Taylor & Francis Group, London, ISBN 978-1-138-02973-6

Design of a temperature control circuit for a semiconductor laser

Li Zhen
Changchun University of Science and Technology, Changchun, China

ABSTRACT: Based on the deep research on the temperature control theory, a new type of temperature control circuit was designed. The circuit uses ADN8831 as the core device, combined with the PID control method. The temperature control precision can reach 0.01°C. The circuit has the advantages of small size, high efficiency and reliability, high performance, and strong driving capability, and can provide constant temperature control for a semiconductor laser.

Keywords: semiconductor Laser; temperature control; PID control

1 INTRODUCTION

Through the study of the characteristics of the semiconductor laser, it is known that the temperature has an important influence on the normal operation of the laser (Zheng Xing et al, 2013). The temperature will directly affect the working parameters of the semiconductor laser, including (Lin Zhiqi et al, 2014): threshold current, V-I relation, output wavelength, P-I relation, etc. At the same time, a high temperature will have a great influence (Zhou Qi, 2012). The influence of its service life and efficiency is seriously affected. In this paper, the ADN8831 temperature control chip is used to provide a constant and adjustable operating temperature to ensure the high efficiency of the laser (Feng Guangzhou et al, 2009). The chip can make the set temperature error control at about 0.01°C.

2 TEMPERATURE CONTROL CIRCUIT DESIGN

The system diagram of the temperature method of the TEC stabilized laser is shown in Figure 1.

A thermistor is used to measure the temperature of laser. Expectations of the temperature of the laser using fixed voltage values, and thermistor voltage values were compared using a high-precision operational amplifier (Zhu Hongbo, et al, 2011). After comparing, the error voltage is passed through a high gain amplifier amplification, also network compensation of laser hot and cold side is caused by the phase delay compensation, the compensation after driving in the H-bridge output. H-bridge not only controls TEC (Gu Yuanyuan, et al, 2009) current size but also controls the direction of the TEC current.

2.1 Input circuit design

Bridge is composed of R_1, R_2, R_{TH}, and the R_{TH} thermistor with a negative temperature coefficient. The value of R_1 can be calculated by formula (1):

$$R_1 = \frac{R_L R_M + R_H R_M - 2R_L R_H}{R_L + R_H - 2R_M} \qquad (1)$$

where R_L is the resistance at the lowest temperature, R_H is the resistance at the highest temperature, and R_M is the resistance of average temperature. The input voltage in ADN8831 is:

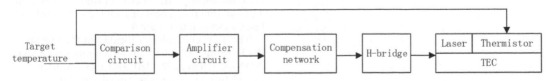

Figure 1. TEC control circuit.

$$V_{IN} = V_{REF} \frac{R_{TH}}{R_1 + R_{TH}} \qquad (2)$$

where R_{TH} is the resistance of the thermistor. After testing, the room temperature is 25°C, and the resistance of the NTC thermistor is about 10k Ω; then, select $R_2 = 10k\ \Omega$, $R_3 = 10k\ \Omega$. The induction temperature proportional to the output voltage is:

$$V_{TEMPOUT} = 0.5 \times V_{REF} \times \left(\frac{1}{R_3} - \frac{1}{R_1} + \frac{1}{R_2 + R_{TH}} \right) \qquad (3)$$

The temperature voltage conversion circuit is shown in Figure 2.

2.2 Compensation circuit design

PID compensation network is the key part of TEC temperature control circuit, which determines the response speed and temperature stability of TEC controller. PID equivalent magnification adjustable amplifier, using proportional and integral arithmetic operations, is used to improve the accuracy of regulation, accelerate the transition process with differential operation, and solve the problem of adjusting speed and precision. PID mathematical model can be shown as:

$$U = K_p \left(E + \frac{1}{T_i} \int_0^Y edt + T_d \frac{de}{dt} \right) \qquad (4)$$

where K_p is the proportion coefficient, T_i is the integral time constant, and T_d is the differential time constant. In the correction, response time and accuracy by adjusting the compensation circuit parameters to the TEC control system become optimal. In the circuit design, the output of the error amplifier is connected to the temperature compensation circuit of input pins, thus completing the temperature compensation circuit design

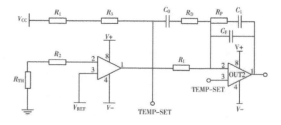

Figure 2. Temperature-voltage conversion circuit.

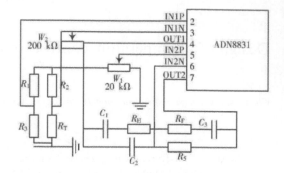

Figure 3. Differential amplification and PID compensating circuit.

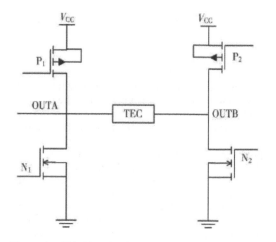

Figure 4. H-bridge circuit.

and the specific circuit connection diagram as shown in Figure 3.

Since the temperature targets for laser in this paper, according to the requirements of the design and calculation, the parameters of the system are usually selected such that: $R_5 = 100\Omega$, $R_H = 1M\Omega$, $R_F = 200k\Omega$, C1 = 1uF, and $C_2 = 10uF$ feedback capacitor and 330pF.

2.3 Output circuit design

ADN8831 is a differential output mode TEC controller. A peripheral H bridge circuit is built to generate an appropriate electric current to drive the TEC, so that it can heat or cool the semiconductor laser, as shown in Figure 4.

The TEC controller is located in the middle of the H-bridge, which forms an asymmetric bridge. ADN8831 drives the left and right branches of the H-bridge, when the transistor N1 is open, transistor P1 is closed, P2 is open, and N2 is

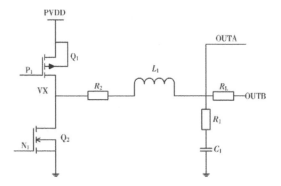

Figure 5.   Low-pass filter circuit.

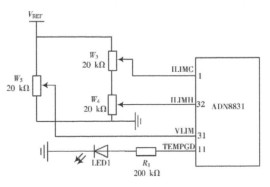

Figure 6.   Protection and detection circuit.

closed, current flows from TEC OUTB to TEC OUTA, which is the cooling state; when the transistor N1 is closed, transistor P1 is open, transistor P2 is closed, and N2 is open, current flows from TEC OUTA to TEC OUTB, which is the heating state. The flexible and convenient external H-bridge can improve the power efficiency, reduce the ripple current, and increase the heat dissipation path.

3 FILTER CIRCUIT

In order to make the ADN8831 effectively drive the TEC, the voltage must be stable. We need to design a filter circuit to achieve the purpose of driving. The design uses the RLC low-pass filter network; the equivalent circuit is shown in Figure 5.

R_L is the TEC resistor, R_1 is the series resistance of C_1, R_2 is equal to the parasitic resistance of the L_1 and Q_1 or Q_2 conduction resistance, and R_1 and R_2 are far less than the R_L, VX is varied between the PVDD and PGND pulse width modulation voltage, and the circuit constitutes a second-order low-pass filter network.

4 PROTECTION AND DETECTION CIRCUIT

ADN8831 provides the protection circuit, so as to protect the TEC laser to prevent damage due to overheating. As the TEC current would be greater than the rated voltage sometimes, it will burn Tec and semiconductor laser, resulting in economic losses. Figure 6 is the protection and detection circuit.

5 CONCLUSION

Through the experiment and analysis, the temperature control deviation is ± 0.01°C. The temperature control precision of the system depends on the temperature sampling value and the characteristic of the temperature setting value, the precision of the sensor itself is higher, the sensitivity of the sensor depends on its own characteristics. High stability, high precision, and low temperature drift are needed if the voltage setting value of high stability is to be obtained. In addition, the system circuit also uses low temperature drift and high stability of the device.

REFERENCES

Feng Guangzhou, Gu Yuanyuan, Shan Xiaonan. 808 nm high power diode laser tack with polarization [J]. Chinese Journal of Luminescence, 2009, 29(4): 695–700(in Chinese).

Gu Yuan-yuan, FENG Guangzhou, Deng Xin-li. 808 nm and 980 nm high power laser diode stack with wavelength [J]. Optics and Precision Engineering, 2009, 17 (1): 8–13(in Chinese).

Lin Zhiqi, Zhang Yang. The semiconductor laser with PN junction temperature characteristics of semiconductor laser temperature control [J]. Journal of luminescence, 2014, 30 (2): 223–225.

Zheng Xing, Jiang Yadong. Design of ADN8830 based non refrigeration infrared focal plane temperature control circuit [J]. Modern electronic technology, 2013, 32(24): 154–156.

Zhou Qi. Low power semiconductor laser driver design method of [J]. Infrared and laser engineering, 2012, 41(10): 2689–2693.

Zhu Hongbo, Liu Yun, Hao Mingming, et al. High efficiency module of fiber coupled diode laser [J]. Chinese Journal of Luminescence, 2011, 32(11): 1147–1151(in Chinese).

Energy Science and Applied Technology ESAT 2016 – Fang (Ed.)
© 2016 Taylor & Francis Group, London, ISBN 978-1-138-02973-6

Calculation of the electromagnetic parameters in a dry-type air-core reactor based on FEM

Liming Bo, Huiping Zheng, Jie Hao & Xinjie Hao
Shanxi Electric Power Research Institute, Taiyuan, China

ABSTRACT: The calculation of a reactor's electromagnetic parameters, such as its capacitance, inductance, resistance, and conductance is fundamental for predicting the frequency behavior or internal overvoltage of the device. In this paper, the author calculates a dry-type air-core reactor's electromagnetic parameters based on Finite Element Method (FEM). By comparing the different frequency characteristics between the experimental result and the simulation result, we get the conclusion that the method of calculating parameters is accurate. The conclusion could be the foundation for modeling technology of air-core reactor.

Keywords: electromagnetic parameters, air-core reactor, FEM

1 INTRODUCTION

The evaluation of transient voltage distributions in windings of electrical equipment is important for electrical insulation sizing a design optimization (R. Rüdenberg, 1940). Historically, electromagnetic parameters, such as capacitance, inductance, resistance, and conductance, have been used primarily in power engineering applications, while little attention is given to air-core reactor. Yu Q. (2001) proposes a method for modeling the self-capacitance of rod inductors, and calculainge the stray capacitance of the inductor, but there are not so many inductors to be calculated. Li Y. (2011) computes the equivalent series capacitance and inductance of a unit coil for the transient analysis in large transformers based on FEM, but it models the wingding as a calculating unit, not the conductor, and there are not so many inductors too. This paper presents a method of calculating electromagnetic parameters of an air-core reactor based on FEM. The result calculated by the method of finite element method compared with the result of experiment is discussed in this paper.

2 CALCULATION OF THE PARAMETERS

2.1 *Capacitance*

FEMM is a finite element analysis software, which is compatible with MATLAB, then the capacitance parameter can be automatically calculated through MATLAB programming.

During the calculation, give the incentive conductor the voltage of 1V and other conductor the voltage of 0V. As a result, the capacitance could be calculated by

$$\sum_{i=1}^{N} \frac{1}{2} C_{ij} \Delta u_{ij}^2 = W_j \qquad (1)$$

where W_j is the electric energy generated by the jth incentive conductor, and $\Delta u_{ij} = 1$ is the voltage between ith conductor and jth conductor.

While the number of wingding conductors of air-core reactor is large, it brings a large amount of calculation methods to calculate the capacitance directly using the FEM software. In order to make the calculation easier, we will calculate from three aspects based on the capacitance distribution regularities.

- capacitance between conductors
- capacitance between packages
- direct earth capacitance.

The capacitance unit is adopted by the Pf.

Capacitance between conductors
There are three questions to be solved: first, different calculated models; second, different calculated layers; and the last one, different positions.

Different calculated model
As shown in Figure 1, the capacitances between conductors are calculated by four layers of wingdings in different positions. Only the capacitance between the selected conductor and its

surrounding conductors should be calculated and others could be neglected.

Figure 2 shows the calculated model and conductor numbers. Table 1 lists the capacitances between conductors calculated by FEMM. From the data, the capacitances between adjacent conductors are far larger than the non-adjacent conductors. Hence, the capacitance between non-adjacent conductors could be neglected and the capacitance matrix becomes sparse.

Different calculated layers

Through the comparison in Table 2, the difference is quite small. That is to say, there is little influence in capacitance between conductors by adding the layer. As a result, a three-layer construct is accurate enough to calculate.

Different position

Just as shown in Figure 1, we also compare the difference in the position. In addition, Table 3 gives the differences of three positions. From the result, we can make the conclusion that the capacitance between inject conductors is slightly affected by the position. Then, we can get the capacitance between inject conductors by modeling conductors in any position of the reactor.

Capacitance between packages

The capacitances between the package and near conducts of its adjacent package are all within the range of 2 to 4 pF. The capacitances are far smaller than the capacitance between conductors. Therefore, this capacitance could be neglected.

Direct earth capacitance

The direct earth capacitance is calculated by modeling as shown in Figure 3: (a) the electric field intensity and (b) the construct in detail.

2.2 Inductance

According to the Neumann formula, the mutual inductance between two circular coils whose radii are a and b, respectively, can expressed as

$$M = \mu_0 (ab)^{1/2} \frac{(2-k^2)K(k) - 2E(k)}{k} \qquad (2)$$

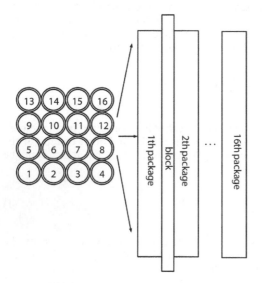

Figure 1. Position of conductors.

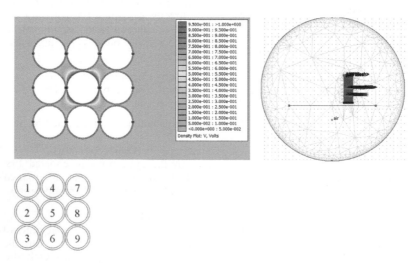

Figure 2. Calculated model and conductor numbers.

Table 1. Capacitances between nine conductors (pF).

1	2	3	4	5
134.791	64.044	1.050	63.967	1.213
6	7	8	9	
0.100	0.892	0.069	0.153	

Table 2. comparison of capacitance on different models.

Model	Direction	Max (%)	Min (%)
3 and 4 layers	radial	1.086	0.012
4 and 5 layers	(like C_{1-2})	0.216	0.010
3 and 4 layers	length wise	1.025	0.014
4 and 5 layers	(like C_{1-5})	1.610	0.014

Table 3. Comparison of capacitance in different positions.

Direction	Max (%)	Min (%)
Radial	0.28	0.01
Length wise	0.27	0

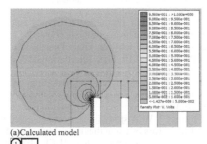

(a)Calculated model

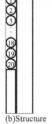

(b)Structure

Figure 3. Calculation of direct earth capacitance.

where, $k = \dfrac{2(ab)^{1/2}}{[(a+b)^2+d^2]^{1/2}}$, $K(k) = \int_0^{\pi/2} \dfrac{d\beta}{(1-k^2\sin^2\beta)^{1/2}}$,

$E(k) = \int_0^{\pi/2} (1-k^2\sin^2\beta)\, d\beta$.

The difference in result between formula and FEM is shown in Table 4. From the table, the formula is accurate enough to calculate the inductance.

Table 4. Difference between formula and FEM.

Inductance	Difference
L_{11} (μH)	3.47%
L_{22} (μH)	2.34%
M_{12} (μH)	1.69%

2.3 Resistance

At very low frequency or dc condition, the resistance of conductors assumes (Liang Y, 2010),

$$R_{dc} = \frac{\rho}{\pi r^2} \tag{3}$$

For high frequencies, when alternating current flows through a conductor, the current density is not uniform across the conductor, it tends to concentrate on the conductor surface and continuously increase from the conductor center to its surface, and thus, makes the resistance frequency dependent.

Define $\delta(f) = \sqrt{\rho / \pi f \mu}$ as the skin depth, which is assumed to be a dividing line for current density. In addition, $f_\delta = \frac{\rho}{\mu \pi r^2}$ is defined as the frequency when $\delta(f) = r$. When $\delta(f) < r$, the resistance of the conductor approximates to equation (4):

$$R(f) = \mathrm{Re}(Z(f)) = \frac{\rho}{\pi[r^2 - (r - \delta(f))^2]} \approx \frac{1}{2r}\sqrt{\frac{\rho f}{\pi \mu}} \tag{4}$$

where ρ is the conductivity, f is the corresponding frequency, and μ is the magnetic permeability. Therefore, the resistance value of a conductor can be expressed as

$$R = \begin{cases} R_{dc} & f \le f_\delta \\ R(f) & f > f_\delta \end{cases} \tag{5}$$

2.4 Conductance

The conductance of a conductor can be expressed as $G = \frac{\sigma}{\varepsilon} C$, where ρ is the conductivity, ε is the dielectric permittivity, and C is the capacitance. If the frequency is lower than 1GHz, the conductance could be neglected.

3 COMPARISON OF RESULTS

As the electromagnetic parameters have been calculated, the frequency characteristic of a simple wingding, which is shown in Figure 4, is tested. The comparison of frequency characteristics between

Figure 4. Tested by Agilent4395 A network analyzer.

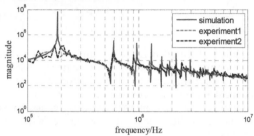

(a) Magnitude-frequency characteristics

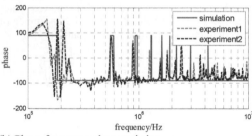

(b) Phase-frequency characteristics

Figure 5. Comparison between experiment and simulation.

experiment and simulation is shown in Figure 5: (a) shows the magnitude-frequency characteristics and (b) shows the phase-frequency characteristics. The experiment results are tested using the Agilent4395 A network analyzer.

From the two figures, we can learn that the simulation result approximately agrees with the experimental result. That is to say, the electromagnetic parameters calculated by this method are reliable.

4 SUMMARY

In this paper, the reactor's electromagnetic parameters, such as capacitance, inductance, resistance, and conductance, have been calculated based on FEM. As it is very complex to calculate capacitance directly, we have to separate into several parts to calculate the capacitance, respectively. Then, we get the conclusion that the method of calculating parameters is accurate, through comparing the different frequency characteristics between the experimental result and the simulation result. The conclusion could be the foundation for further modeling technology of air-core reactor.

REFERENCES

Li Y, Du J, Li X, et al. Calculation of capacitance and inductance parameters based on FEM in high-voltage transformer winding[C]//Electrical Machines and Systems (ICEMS), 2011 International Conference on. IEEE, 2011: 1–4.
Liang Y, Li Y. High frequency resistance calculation and modeling of Through Silicon Vias[C]//Electronic Packaging Technology & High Density Packaging (ICEPT-HDP), 2010 11th International Conference on. IEEE, 2010: 563–566.
Rüdenberg, R. "Performance of Travelling Waves in coils and Winding", AIEE Trans., Vol. 59,1940, pp. 1031–1040
Yu Q, Holmes T W. A study on stray capacitance modeling of inductors by using the finite element method[J]. Electromagnetic Compatibility, IEEE Transactions on, 2001, 43(1): 88–93.

Energy Science and Applied Technology ESAT 2016 – Fang (Ed.)
© *2016 Taylor & Francis Group, London, ISBN 978-1-138-02973-6*

Shunt active power filter model predictive current control

Suxia Jiang, Hao Chen, Nan Jin & Guangzhao Cui
College of Electric and Information Engineering, Zhengzhou University of Light Industry, Zhengzhou, China

Yang Zhang
School of Electric Power of North China University of Water Resources and Electric Power, Zhengzhou, China

Tao Tao
State Grid Xuchang Power Supply Company, China

ABSTRACT: Shunt active power filter is a new power electronic device which is an effective compensation of the harmonic and reactive current. The current detection and control of active power filter directly affect the compensation performance. The prediction model of inverter output current ab coordinate is constructed by coordinate transformation. Based on the cost function, different voltage vector of inverter output is evaluated to predict all possible output current values. The active power filter circuit and current control circuit are built by using MATLAB software as the simulation platform. The simulation results show that the predictive current control strategy has good dynamic and static performance and compensation function for the finite state model.

Keywords: shunt active power filter; instantaneous reactive power; model predictive current control; cost function

1 INTRODUCTION

Electronic devices, which are significant to improve the efficiency and automation of modern industry, can cause some serious harmonic problems in power system. With the wide spread of these devices among the modern industry, the harmonic problems are getting severer. Harmonic problems, which can result in network voltage fluctuation and current waveform distortion, are becoming a key factor that affects the power quality of power grid. Therefore, active power filter has been widely studied.

At present, the technology of controlling active power filter compensating has been a hot spot in the power electronics field. The paper adopts harmonic detection method based on instantaneous reactive power theory to detect harmonic currents. Carrier Control and Hysteresis Current Control are both traditional control methods, the former is simple in circuit implementation, but is limited in large-power application. The latter has defects that its controlled quantity usually fails to be effectively controlled in great varying range of switching frequency and load short circuit.

With the improvement of the performance of digital processor. The new control method has been put forward more, including fuzzy control, sliding mode control and predictive control.

Predictive control adopts multi-step test, rolling optimization, feedback correction and some control strategies, mainly including model predictive control, deadbeat control and tracking control. Among these control strategies, model predictive control has the ability of explicit constraints and handling constraints, through the establishment of system prediction model, it can choose the control mode that makes the minimal cost function, so as to achieve the best control.

2 OPERATING PRINCIPLE OF SHUNT ACTIVE POWER FILTER

Shut Active Power Filter (SAPF) comprises command current detection circuit and comp- ensating current generating circuit. As known in Fig. 1, three-phase load current: i_{sa}, i_{sb} and i_{sc} are three-phase power supply output currents; i_{ca}, i_{cb} and i_{cc} are three-phase SAPF output currents; V_a, V_b and V_c are three-phase output voltages in SAPF alternating current side. Supposing that the three-phase circuits in Fig. 1 are symmetric, then a-phase current is $i_{la} = i_{sa} + i_{ca}$, if the a-phase output current i_{ac} of SAPF is equal to harmonic current component in load current i_{la}, then power output current i_{sa} only contains fundamental currents.

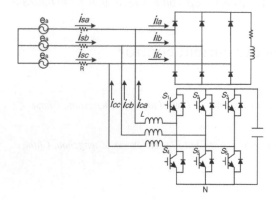

Figure 1. The main structure of shunt active power filter.

2.1 Predictive current model of three-phase active power filter

Control signals Sa, Sb and Sc determine the switching state of inverter-bridge silicon-controlled rectifier:

$$S_a = \begin{cases} 1 & S_1 \text{ on}, S_4 \text{ off} \\ 0 & S_1 \text{ off}, S_4 \text{ on} \end{cases} \quad (1)$$

$$S_b = \begin{cases} 1 & S_2 \text{ on}, S_5 \text{ off} \\ 0 & S_2 \text{ off}, S_5 \text{ on} \end{cases} \quad (2)$$

$$S_c = \begin{cases} 1 & S_3 \text{ on}, S_6 \text{ off} \\ 0 & S_3 \text{ off}, S_6 \text{ on} \end{cases} \quad (3)$$

Supposing that switching state vector is S, then:

$$S = \frac{2}{3}\left(S_a + aS_b + a^2 S_c\right) \quad (4)$$

Among them, $a = e^{j2\pi/3}$, inverter-bridge output voltage vector is V, then:

$$V = \frac{2}{3}\left(V_{aN} + aV_{bN} + a^2 V_{cN}\right) \quad (5)$$

By Fig. 1, the vector equation of SAPF switching in grid current state can be written:

$$V = Ri + L\frac{di}{dt} + e \quad (6)$$

$$\begin{bmatrix} V_{aN} \\ V_{bN} \\ V_{cN} \end{bmatrix} = R\begin{bmatrix} i_{sa} \\ i_{sb} \\ i_{sc} \end{bmatrix} + L\frac{d}{dt}\begin{bmatrix} i_{ca} \\ i_{cb} \\ i_{cb} \end{bmatrix} + \begin{bmatrix} e_a \\ e_b \\ e_c \end{bmatrix} \quad (7)$$

The formula (7) is the matrix form of (6).
Expression of obtained equation (6) under $\alpha\,\beta$ coordinate is:

$$\begin{bmatrix} V_\alpha \\ V_\beta \end{bmatrix} = R\begin{bmatrix} i_\alpha \\ i_\beta \end{bmatrix} + L\frac{d}{dt}\begin{bmatrix} i_\alpha \\ i_\beta \end{bmatrix} + \begin{bmatrix} e_\alpha \\ e_\beta \end{bmatrix} \quad (8)$$

Regarding signal sampling period as Ts, then

$$\frac{d}{dt}\begin{bmatrix} i_\alpha \\ i_\beta \end{bmatrix} = \frac{1}{T_s}\begin{bmatrix} i_\alpha(k+1) - i_\alpha(k) \\ i_\beta(k+1) - i_\beta(k) \end{bmatrix} \quad (9)$$

Substitute equation (9) into equation (8), disperse and simplify as

$$\begin{bmatrix} i_\alpha(k+1) \\ i_\beta(k+1) \end{bmatrix} = \frac{T_s}{L}\begin{bmatrix} V_\alpha(k) - e_\alpha(k) \\ V_\beta(k) - e_\beta(k) \end{bmatrix} + \left(1 - \frac{RT_s}{L}\right)\begin{bmatrix} i_\alpha(k) \\ i_\beta(k) \end{bmatrix} \quad (10)$$

Among them, $i_\alpha(k)$ and $i_\beta(k)$ are obtained from grid connected current in the k sampling period through coordinate transformation; $V_\alpha(k)$ and $V_\beta(k)$ are the components of different voltage vectors in the k sampling period under $\alpha\beta$ coordinate; $e_\alpha(k)$ and $e_\beta(k)$ are obtained from network voltage in the k sampling period through coordinate transformation; $i_\alpha(k+1)$ and $i_\beta(k+1)$ are the predictive current values in the $k+1$ sampling period under $\alpha\beta$ coordinate.

2.2 Model predictive current control principles and algorithm design

The model predictive control is based on the discrete time model, and the control principle diagram is shown in Fig. 2, through selecting control signal Sn, so that the control system output $x(t)$

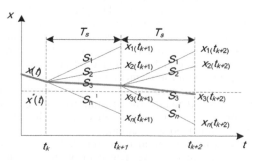

Figure 2. Principle of model predictive control.

can be quickly approximate the reference value $x^*(t)$. If the system has n kind of control signals, according to $x(t)$ at the moment of t_k and prediction function $f(x)$, all the possible predictive values at the t_{k+1} moment are obtained, and prediction function $f(x)$ is usually obtained through system model. In order to select the optimal control signal, cost function $g(x^*(t), x(t))$ is defined based on the absolute value of difference value between reference value and predicted value, then when the value is minimal, control signal is optimal. $x(t)$ corresponding to S_3 at the t_k moment in the figure is closest to the reference value $x^*(t)$, and the value of cost function is minimal, then control signal corresponding to S_3 at the t_{k+1} moment is chosen and the downward selection and so on.

Reference currents $i_\alpha^*(k + 1)$ and $i_\beta^*(k + 1)$ are obtained by three-phase reference currents i_a^*, i_b^* and i_c^* through coordinate transformation under α-β coordinate. Three-phase network voltages e_a, e_b and e_c, currents i_a, i_b and i_c as well as DC side voltage V_{dc} are gathered, and various parameters e_α, e_β, i_α, i_β, v_α and v_β required by prediction model under α-β coordinate are obtained through coordinate transformation, and all the predicted values $i_\alpha(k+1)$ and $i_\beta(k+1)$ corresponding to different voltage vectors of prediction function output at the next moment are also obtained. Comparing according to cost function, voltage vector that makes cost function take the minimum value is selected, and its corresponding switching state S_a, S_b and S_c are applied.

According to different switching states of SAPF compensation module, the corresponding 8 kinds of voltage vectors have different output compensating currents. In order to make output current at the t_{k+1} moment close to reference current, optimal switching state is chosen, it is necessary to predict different current values at the t_{k+1} moment, and the function of prediction function in Fig. 3 is to

predict output value under the action of different voltage vectors at the next moment. According to inverter mathematical model, the paper selects the expression (10) of current values $i_\alpha(k + 1)$ and i_β $(k +1)$ at the t_{k+1} moment under $\alpha\beta$ coordinate as prediction function.

In Fig. 3, the function of cost function is to select the optimal set of switch variables from different predictive values, so that the output currents $i_\alpha(k+1)$ and $i_\beta(k+1)$ at the t_{k+1} moment have minimum variance with reference currents $i_\alpha^*(k+1)$ and $i_\beta^*(k+1)$. The sum of absolute value of the difference value between predictive current and reference current is selected as cost function:

$$ g = \left| i_\alpha^*(k+1) - i_\alpha(k+1) \right| + \left| i_\beta^*(k+1) - i_\beta(k+1) \right| \quad (11) $$

Among them, $i_\alpha^*(k+1)$ and $i_\beta^*(k+1)$ are components of reference current at t_{k+1} moment under $\alpha\beta$ coordinate; $i_\alpha(k+1)$ and $i_\beta(k+1)$ are predicted current values at the t_{k+1} moment under α-β coordinate, the process of control algorithm is shown in Fig. 4. In the figure, $S(n)$ is the switching state of compensating circuit, m and j are variable parameters, j is respectively corresponding to 8 voltage vectors of inverter-bridge output from 0 to 7, and $g(j)$ is the value of cost function in the j switching state.

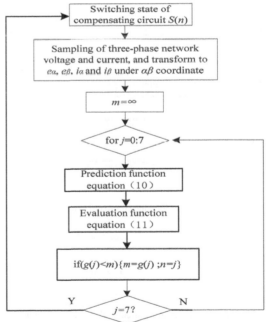

Figure 4. Flow diagram of model predictive control algorithm.

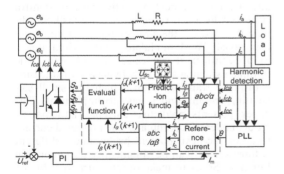

Figure 3. Model predictive control of active power filter structure.

313

3 MATLAB SIMULATION AND EXPERIMENT OF SHUNT ACTIVE POWER FILTER

Based on the theory of instantaneous reactive power, Simulink is adopted to build ip-iq harmonic current detection module and compensating current generation module of model predictive current control. The cut-off frequency of low pass filter is 50 Hz in detection part, and the order is 2. Simulation parameters are shown in Table 1, when in the 0.2S of simulation process, load side resistance variation is 40Ω.

In order to more intuitively analyze harmonic part of the grid, FFT analysis has been conducted on grid current and the spectrogram is obtained as shown in Fig. 6(a). As seen from the figure, after the intervention of non-linear load, grid current has distortion and current distortion rate reaches to 27.78%, which is that current contains fundamental wave, 5 times, 7 times, 11 times, 17 times and other harmonic waves.

According the analysis of Fig. 6 (c), it can be seen that current distortion rate of grid current after compensation is 2.5%, which is in line with standard of grid phase current harmonic rate. The current loop of active power filter by is controlled

Table 1. Simulation and experimental parameters.

Parameters	Value
DC side capacitance $C/\mu F$	5000
Filter inductance L/mH	20
Network voltage e/V	380
Load side resistance R/Ω	40/50
Load side electrical inductance L/mH	5

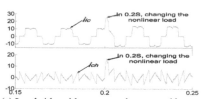

(a) Load side grid current and command harmonic current

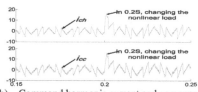

(b) Command harmonic current and compensating current

Figure 5. Output side and input side current waveform of SAPF.

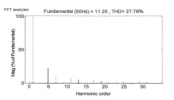

(a) Load side harmonic analysis

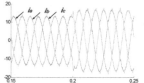

(b) Grid current after compensation of SAPF

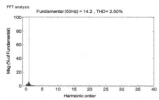

(c) Grid current harmonic analysis after compensation of SAPF

Figure 6. Grid current waveform and harmonic analysis before and after compensation of SAPF compensation.

through predictive control, which can effectively compensate harmonic currents in grid and reduce higher harmonics in grid current. Therefore, SAPF can use the control strategies of model predictive current control.

4 CONCLUSION

After simplifying model predictive control according to the mathematical model of the active filter, studies on control of active power filter are made and a kind of model predictive current control strategy which can be achieved under $\alpha\beta$ coordinate is proposed, the strategy has no need to use traditional PWM modulator and current inner-loop controller, it can calculate and obtain the optimal switching state only by means of one time coordinate transformation, which has small calculating amount and is easy to implement. This paper presents predictive current mathematical model under $\alpha\beta$ coordinate. The predicted current value of the next moment bridge under different voltage vectors can be obtained by the prediction function. So as to make output current value of SAPF quickly approximate given reference value and achieve compensation effect.

ACKNOWLEDGEMENTS

This work is supported by the key scientific research of colleges and universities in Henan Province (15 A120021).

REFERENCES

Cortes P, Rodriguez J, Silva C, et al. Delay Compensation in Model Predictive Current Control of a Three-Phase Inverter [J]. IEEE Transactions on Industrial Electronics, 2012, 59(2): 1323–1325.

Curkovic M, Jezernik K, Horvat R. FPGA-based Predictive Sliding Mode Controller of a Three-Phase Inverter [J]. IEEE Transactions on Industrial Electronics, 2013, 60(2): 637–644.

Curkovic M, Jezernik K, Horvat R. FPGA-Based Predictive Sliding Mode Controller of a Three-phase Inverter [J]. IEEE Transactions on Industrial Electronics, 2013, 60(2): 637–644.

Muyeen S M, Al-Durra A. Modeling and Control Strategies of Fuzzy Logic Controlled Inverter System for Grid Interconnected Variable Speed Wind Generator [J]. IEEE Systems Journal, 2013, 7(4): 817–824.

Preindl M, Schaltz E, Thogersen P. Switching Frequency Reduction Using Model Predictive Direct Current Control for High-power Voltage Source Inverters [J]. IEEE Transactions on Industrial Electronics, 2011, 58(7): 2826–2835.

Rahmani S, Mendalek N, Al-Haddad K. Experimental Design of a Nonlinear Control Technique for Three-phase Shunt Active Power Filter [J]. IEEE Transactions on Industrial Electronics, 2010, 57(10): 3364–3375.

Tang Y, Loh P C, Wang P, et al. Generalized Design of High Performance Shunt Active Power Filter with Output LCL Filter [J]. IEEE Transactions on Industrial Electronics, 2012, 59(3): 1443–1452.

Trinh Q N, Lee H H. An Advanced Current Control Strategy for Three-phase Shunt Active Power Filters [J]. IEEE Transactions on Industrial Electronics, 2013, 60(12): 5400–5410.

Wang Zhaoan, Yang Jun, Liu Jinjun. Harmonic Suppression and Reactive Power Compensation [M]. Beijing: Mechanical Industry Press, 1998.

Wang Zhaoan, Li Min, Zhuo Fang. Researches on Three-phase Circuit Instantaneous Reactive Power Theory [J]. Transactions of China Electrotechnical Society, 1992, 03: 55–59.

Zhou Ke, Wang Kai, Liu Lu, Qin Qi, Chu Hongbo. A Kind of Improved Shunt Hybrid Active Power Filter and its Control [J]. Proceedings of the CSEE, 2012, 30: 67–72.

Energy Science and Applied Technology ESAT 2016 – Fang (Ed.)
© *2016 Taylor & Francis Group, London, ISBN 978-1-138-02973-6*

Study of the evaluation index system of fault-tolerant technology

Weifa Peng
School of Mechatronics Engineering and Automation, Shanghai University, Shanghai, China
College of Electrical Engineering, East China Jiaotong University, Nanchang, China

Surong Huang, Haishan Liu & Xianglin Wei
School of Mechatronics Engineering and Automation, Shanghai University, Shanghai, China

ABSTRACT: Reasonable fault tolerant technology index system can lead and promote the development of fault tolerant technology for electric drive system. At the same time, the development of fault tolerance will also promote the research on the index system. In this paper, a series of new fault tolerance indexes are proposed, and the evaluation index system of fault tolerant technology for electric drive system is first established. Based on the analytic hierarchy process, the evaluation system model is built up and the three kinds of fault tolerant electric drive system are evaluated. The results show that the fault-tolerant technology evaluation system and the evaluation model of the electric drive system have good applicability and operability.

Keywords: fault tolerant; electric drive system; index system; Analytic Hierarchy Process (AHP)

1 INTRODUCTION

The concept of fault tolerant first appeared in the computer field, and later on introduced into the electrical industry.

This paper will cover two definitions of 'fault-tolerance'. The first definition is that the machine needs to maintain the same or comparable performance under fault to that when the machine was healthy. The second definition (which is more important in safety-critical applications, where there is a lot of redundancy) is that in case of fault (especially turn-to-turn fault which is most severe), the machine needs to fail safely without any catastrophic damage (A.M. El-Refaie, 2011).

For electric vehicles, electric propulsion ship, rail transit, and aerospace vehicles which are related to the safety of life, the fault-tolerant performance of the electric drive system is crucial. Electric drive system generally consists of motor, power converters, sensors, and controllers; each part of the fault are likely to affect the normal operation of the entire system, and even cause the system to crash.

Research on fault tolerant motor and fault tolerant inverter and its control algorithm is now a hot topic, and research on fault tolerant indexes also appeared in succession. Research team of Thomas A. Lipo has a comparison of the features, cost, and limitations of fault-tolerant three-phase AC motor drive topologies and first proposed two fault-tolerant quantitative indicators in 2004 (Brian A. Welchko et al, 2004). One of two technical indicators is FPRF,

and the other is SOCF. The Fault Power Rating Factor (FPRF) is proposed to quantify the output capacity of each of the systems in the presence of a fault. In addition, Silicon Overrating Cost Factor (SOCF) is proposed to compare the cost associated with the extra silicon devices in the fault-tolerant topologies. In 2010, the research team of Thomas W. Nehl proposed two technical indicators for fault-tolerant inverter: the Post-Fault Performance Factor (PFPF) and the inverse of the Cost Factor (CF) (Malakondaiah Naidu, 2010). Two papers on the part of quantitative indicators do a very detailed study, but do not involve qualitative indicators. It is not comprehensive enough, nor establishes an index system. At present, there is an urgent need for systematic and comprehensive fault-tolerant technology index system in the industry and academia. It can provide reasonable performance evaluation and data support for fault-tolerant high-quality electric drive system, then put forward the research goal, to lead and promote the development of fault-tolerant technology.

2 EVALUATION INDEX SYSTEM OF FAULT TOLERANT TECHNOLOGY FOR ELECTRIC DRIVE SYSTEM

2.1 *Overall framework*

In the establishment of evaluation index system, on the one hand, to ensure the full performance of the index system, it is not only to reflect the excellent work performance, but also to reflect the fault tolerance performance. On the other hand, it needs to achieve the independence between the various indicators, to avoid duplication of indicators. Based on the above principles, combined with the characteristics of the fault-tolerant technology, this paper establishes the evaluation index system of fault-tolerant technology from three aspects: performance testing, motor body design, and fault-tolerant features. Performance testing indicators include torque response time, speed response time, torque control precision, and high-efficiency motor range. Motor body design indicators contain motor power density, peak torque and peak power, overspeed performance, and cost. Fault-tolerant feature index contains fault-tolerant response time, system failure rate, the number of fault-tolerant species, single-phase bridge open output power ratio, single-phase winding open output power ratio, single-phase bridge open torque ripple, and single-phase winding open torque ripple.

Based on the idea of AHP, the evaluation index system of fault tolerant technology for electric drive system is established, and the overall framework is shown in Figure 1. From top to bottom, it is divided into three levels, followed by the object layer, criterion layer, and the index layers.

2.2 Evaluation index

The fault tolerant system of the multiphase motor is more freedom and fault type when compared with the three-phase system, and it is the trend and hotspot of fault tolerance research at present. Therefore, evaluation index system of fault-tolerant technology must be evaluated very well for the three-phase system and multiphase system.

To evaluate the electric drive system's ability of fault tolerance, the system cost, the volume, the response time of the fault, and the output power are all considered. In order to establish the evaluation index system of fault tolerant technology, this paper puts forward a series of new technical indexes. Details are explained below.

1. FTRT (Fault-Tolerant Response Time): FTRT is defined as the time of the fault tolerant electric drive system from failure to fault tolerant control of stable operation. The shorter the fault-tolerance response time, the smaller the loss caused by the system.

2. NFS (the Number of Fault-tolerant Species): NFS is defined as the number of faults for the system can achieve fault tolerance. These fault types include a single-phase open circuit inverter, a two-phase open circuit inverter, a single-phase open circuit of stator winding, a two-phase open circuit of stator winding, stator winding single-phase end short circuit, and stator winding two-phase turn-to-turn short circuit.

3. PFOPR (Post-Fault Output Power Ratio): it is the ratio of the maximum power in fault-tolerant operating state to the rated output power.

$$PFOPR = \frac{the\ maximum\ power\ of\ fault\ tolerant\ operating\ state}{the\ rated\ output\ power}$$

Different fault, power landing of fault-tolerant operating state is different.

4. PFTR (post-fault torque ripple): the ratio of the maximum torque plus the minimum torque to the maximum torque minus the minimum torque in the fault tolerant operation state for a period of time.

$$PFTR = \frac{T_{max} - T_{min}}{T_{max} + T_{min}} \times 100\%$$

Different fault, waveform quality is also different in fault-tolerant operation state.

3 ESTABLISHING COMPREHENSIVE EVALUATION MODEL BY AHP

Evaluation system for the determination of index weights largely determines the credibility of the comprehensive evaluation scheme. In this paper, the AHP will be applied to obtain weights of 15 fault-tolerant technology evaluation indexes of the electric drive system for objectivity.

3.1 Establishing hierarchical structure

Establishing a hierarchical model is the first and most critical step of AHP. Hierarchical model established in this paper is shown in Figure 1.

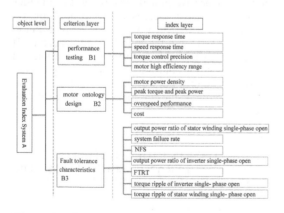

Figure 1. Overall framework of evaluation index system.

3.2 Constructing judgment matrix and consistency test

In the process of obtaining the indexes weight by AHP, 1 to 9 scale method proposed by Professor Saaty (Thomas L. Saaty, 1978) is introduced to make quantitative decision. In this way, to measure the relative importance of a series of pairs of elements, the qualitative evaluation is transformed into quantitative evaluation. Opinions of 15 experts are collected to establish pairwise comparison matrices (and judgment matrices) as follows:

$$A = \begin{bmatrix} 1 & 1 & 1/3 \\ 1 & 1 & 1/3 \\ 3 & 3 & 1 \end{bmatrix},$$

$$B_1 = \begin{bmatrix} 1 & 1 & 1/2 & 1/3 \\ 1 & 1 & 1/2 & 1/3 \\ 2 & 2 & 1 & 1/2 \\ 3 & 3 & 2 & 1 \end{bmatrix},$$

$$B_2 = \begin{bmatrix} 1 & 2 & 3 & 2 \\ 1/2 & 1 & 2 & 1 \\ 1/3 & 1/2 & 1 & 1/2 \\ 1/2 & 1 & 2 & 1 \end{bmatrix},$$

$$B_3 = \begin{bmatrix} 1 & 1/2 & 1/2 & 1 & 1/2 & 1 & 1 \\ 2 & 1 & 1 & 2 & 1 & 2 & 2 \\ 2 & 1 & 1 & 2 & 1 & 2 & 2 \\ 1 & 1/2 & 1/2 & 1 & 1/2 & 1 & 1 \\ 2 & 1 & 1 & 2 & 1 & 2 & 2 \\ 1 & 1/2 & 1/2 & 1 & 1/2 & 1 & 1 \\ 1 & 1/2 & 1/2 & 1 & 1/2 & 1 & 1 \end{bmatrix}.$$

The eigenvalues and λ_{max}, CI, and CR calculated using the MATLAB software are given in Table 1.

The values CR of all judgment matrices are less than 0.1. This satisfies the requirement of consistency checking of hierarchy single sort.

3.3 Total level sorting and consistency checking

Total sort results are checked for consistency, and the value of the comprehensive test indicator CR is equal to 0.010, so the consistency of comprehensive ranking is acceptable. Each index weight is shown in Table 2.

4 EXAMPLE ANALYSIS

In this paper, three sets of the electric drive system provided by the manufacturers A, B, and C are selected for evaluation using the evaluation index system for its comprehensive evaluation.

There are three sets of the electric drive system: A system is three-phase four-bridge arm PMSM drive system from company A, B system from company B is five-phase induction motor drive system, and C system from company C is double three-phase PMSM drive system. Test data of electric drive system bench is processed by quantitative measurement. The dimensionless data are unified, and the corresponding score ranges from 0 to 1.

For efficiency index data (the bigger the indicator value, the better, such as high-efficiency motor range, the motor power density, etc.), the higher bench test data correspond to the higher value. For cost-type data (the smaller, the index value, the better, such as the cost of the motor, the torque response time, etc.), the lower bench test data correspond to the higher score. The data of three sets of electric drive system after quantification process are given in Table 3.

The three-phase system gets the highest score in motor ontology design. In performance testing, it gets flat score compared with the double three-phase system. While it gets the lowest score in fault tolerance characteristics. For three-phase system, the three phases can operate normally while inverter single-phase is open, in which the output power ratio is 1, and the torque ripple is also small. It cannot run when stator winding single-phase is open, in which the single-phase output power ratio is 0. Therefore, the value of its NFS is 1.

The five-phase system gets a higher score in fault tolerance characteristics than the three-phase system. It can run normally while the inverter single-phase is open or the stator winding single-phase is

Table 1. Eigenvalues of each matrix, λ_{max}, CI, and CR.

	w_1	w_2	w_3	w_4	w_5	w_6	w_7	λ_{max}	CI	CR
A	0.2	0.2	0.6					3	0	0
B_1	0.141	0.141	0.263	0.455				4.014	0.005	0.006
B_2	0.423	0.227	0.123	0.227				4.012	0.004	0.004
B_3	0.1	0.2	0.2	0.1	0.2	0.1	0.1	7	0	0

Table 2. Each index weight in evaluation index system.

Project	Major index	Specific indexes	Weight
Evaluation Index System	performance testing	torque response time	0.0282
		speed response time	0.0282
		torque control precision	0.0526
		motor high efficiency range	0.091
	motor ontology design	motor power density	0.0846
		peak torque and peak power	0.0454
		over speed performance	0.0246
		Cost	0.0454
	fault tolerance characteristics	output power ratio of stator winding single-phase open	0.06
		system failure rate	0.12
		NFS	0.12
		output power ratio of inverter single-phase open	0.06
		FTRT	0.12
		torque ripple of inverter single-phase open	0.06
		torque ripple of stator winding single-phase open	0.06

Table 3. Comprehensive score of three different products.

Project	Major index	Specific indexes	Weight	A	B	C
Evaluation Index System	performance testing	torque response time	0.0282	0.7	0.8	0.9
		speed response time	0.0282	0.8	0.8	0.9
		torque control precision	0.0526	0.8	0.85	0.9
		motor high-efficiency range	0.091	0.9	0.7	0.75
	subtotal		0.2	0.1663	0.1535	0.1664
	motor ontology design	motor power density	0.0846	0.9	0.7	0.75
		peak torque and peak power	0.0454	0.75	0.9	0.9
		overspeed performance	0.0246	0.7	0.75	0.8
		cost	0.0454	0.8	0.7	0.75
	subtotal		0.2	0.1637	0.1503	0.1580
	fault tolerance characteristics	output power ratio of stator winding single-phase open	0.06	0	0.8	0.85
		system failure rate	0.12	0.9	0.7	0.6
		NFS	0.12	0.2	0.7	0.7
		output power ratio of inverter single-phase open	0.06	1	0.8	0.85
		FTRT	0.12	0.9	0.9	0.9
		torque ripple of inverter single-phase open	0.06	1	0.7	0.9
		torque ripple of stator winding single-phase open	0.06	0	0.7	0.9
	subtotal		0.60	0.36	0.456	0.474
Total			1	0.6900	0.7598	0.7984

open, in which the output power ratio is 0.8 with a large torque ripple. It can also operate in fault tolerance state, while inverter or stator winding two-phase is open, so the value of its NFS is 4.

The double three-phase system is better than the five-phase system in performance testing, motor ontology design, and fault tolerance character-istics. It can also run normally while the inverter single-phase is open or the stator winding single-phase is open, in which the output power ratio is 0.83 with a smaller torque ripple. With the increase in the phase number, the component is increased, and the system failure rate is also increased. It can also operate in fault tolerance, while the inverter or stator winding two-phase is open, so the value of its NFS is also 4.

A final calculation showed that the score of system A is 0.6900, that of system B is 0.7598, and that of system C is 0.7984, that is C > B > A. The data may reflect that the three-phase system got the highest score in motor ontology design, and the lowest score in fault tolerance characteristics. Technology of the three-phase system is mature, with high-power density, small devices, and low cost, but fault tolerance performance is poor. Technology of multi-phase system is relatively immature, with multi-device, high cost, but better control performance. Fault-tolerant performance of multi-phase system is also greatly improved.

5 CONCLUSION

In this paper, the evaluation index system framework is established in performance testing, motor ontology design, and fault tolerance characteristics. A number of new indicators are proposed. The AHP evaluation system of the weight of each index is manifested. Examples show that the evaluation index system and the comprehensive evaluation model are both scientific in theory, meeting the needs of the practice of electric drive system fault-tolerant technology. It is possible to measure the overall level of fault tolerance. Comparing three kinds of fault-tolerant solutions is significant for the development of fault-tolerant technology.

ACKNOWLEDGMENTS

This work was supported by the School Scientific Research Fund of East China Jiaotong University (No. 24441012).

REFERENCES

Brian A. Welchko, Thomas A. Lipo, Thomas M. Jahns, and Steven E. Schul, Fault Tolerant Three-Phase AC Motor Drive Topologies: A Comparison of Features, Cost, and Limitations, J. IEEE Transactions On Power Electronics, Vol. 19 (2004) No.4, p. 1108.

El-Refaie, A.M. Fault-tolerant permanent magnet machines a review, J. IET Electr. Power Appl., Vol. 5 (2011) No.1, p. 59.

Malakondaiah Naidu, Suresh Gopalakrishnan, and Thomas W. Nehl, Fault-Tolerant Permanent Magnet Motor Drive Topologies for Automotive X-By-Wire Systems, J. IEEE Transactions on Industry Applications, Vol. 46 (2010) No.2, p. 841.

Thomas L. Saaty, Modeling Unstructured Decision Problems: The Theory of Analytical Hierarchies, J. Mathematics and computers in Simulation, Vol. 20 (1978) No.3, p.147.

Energy Science and Applied Technology ESAT 2016 – Fang (Ed.)
© *2016 Taylor & Francis Group, London, ISBN 978-1-138-02973-6*

City-type analysis of the electricity supply and demand matching degree

Donghui Wang
North China Electric Power University, Baoding, China

ABSTRACT: In the new smart grid, demand response system has been improved gradually. There are many smart power terminal design ideas having been proposed. In this paper, I analyze the general urban intelligent power terminal design ideas. I select the appropriate methods from the perspective of supply and demand to investigate the extent of both supply and demand match degree. The results can effectively guide the design of intelligent power terminal. And it helps to regulate the behavior of users of electricity and the further development of intelligent grid.

Keywords: smart power terminal design; Matching Supply and Demand Degree (MSADD); City type

1 INTRODUCTION

Smart power terminal generally consists of two parts (Z.X. Li et al, 2013): Smart power terminal system and mobile smart power terminal. The former is connected with the grid information platform. It exchanges information with the latter through sensor technology. On one hand, users can access information to facilitate decision-making power dispatch; On the other hand, it provides important information via mobile intelligent electricity power for the user terminal. It guides the user uses electricity flexibility.

The degree of matching supply and demand is the practice guidelines for evaluation of urban intelligent power terminal design. We combine the indicators proposed later. Through a series of mathematical methods, we finally get the effective electricity of user side and effective power supplied by the grid. Then make the ratio of the two and the result is regarded as the quantitative value of matching supply and demand. The greater value is, the more difficult to meet the needs of users. The smaller value is, the lower power efficiency of the grid is. Through above, we analyze the MSADD between different regions. It has some significance for the electricity plan.

2 MATHEMATICAL ANALYSIS

2.1 *Parameter selection*

2.1.1 *From the perspective of demand*
Through analysis of the user side, I decided to choose the total urban population, per capita electricity consumption times, the power company's

share rate, Per capita household duration electricity and the power company's effective supply time as the parameters. These factors are directly related to the user side of the effective power consumption. The type of the urban population also has a great impact on the demand for electricity. Floating population's demand for electricity is not as permanent as the residents of the city. Especially some cities have more floating population. Therefore, when estimate the user's electricity demand, it is necessary to consider the floating population and permanent residents of the city separately.

2.1.2 *From the perspective of supply*
When consider the model parameters selected from the supply, supply of grid is affected by many factors. Especially those power supply control policies from the government's decisions. But these factors are difficult to quantify. So I consider the impact of power company's operating costs for supply. Due to the power company's profitability, it is different from institutions. Power grid company's operating costs can be divided into two parts: change costs and fixed costs.

Change costs can be expressed as:

$$F_1 = C_1 \times L_1 = C_1 \times v \times T \qquad (1)$$

Or

$$F_1 = C_1 \times L_1 = C_1 \times L_2 / (1-k) \qquad (2)$$

From equations above, we know change costs is related to the average operational efficiency, the average operating time, effective supply time, and air ratio. Fixed costs can express as:

$$F_2 = C_2 \times T \qquad (3)$$

F_1 – Change costs of the power company in the statistical period
F_2 – Fixed costs of the power company in the statistical period
C_1 – Change costs per unit time
C_2 – Fixed costs per unit time
T – Statistics time
L_1 – Effective power supply of the power company in the statistical period
L_2 – Power supply of the power company in the statistical period
From equation above we know fixed costs is related to average operating time.

2.2 Model set

Through the above analysis, we know the total effective power supply time is related to the demand for electricity and supply of power. Therefore, I choose the company's total power supply time as an effective model variable to do the further analysis.

2.2.1 The effective power supply time L on the supply side

Urban residents and migrants' power turnover can be respectively expressed by equations (4) and (5) as:

$$Q_1 = P_1 \times \lambda_1 \times \varepsilon_1 D_1 \cdot \qquad (4)$$

$$Q_2 = P_2 \times \lambda_2 \times \varepsilon_2 D_2 \cdot \qquad (5)$$

In the process of the power gird company's operations, power supply is not the same for each segment. The total effective power supply time is the suppliers of the power gird company's power supply turnover and average effective power supply number from demand perspective (S.K. Si, 2007). So the power gird company's the total effective power supply time can be expressed by equations (6) as:

$$L = \frac{Q_1}{S_1} + \frac{Q_2}{S_2} \cdot \qquad (6)$$

It is noted that the total effective power supply time is not a fixed value. It increases as the urban economic growth, consumption growth and people income's increase.

2.2.2 The effective power supply time L' on the demand side

Air ratio is an important parameter which the governments use to regulate the power supply company. And effective power supply time is directly related to air ratio (Y.G. Gan et al, 2005). The

air ratio K of the power supply company can be expressed by equation (7).

$$K = 1 - \frac{L'}{T \cdot v \cdot n} \cdot \qquad (7)$$

Deform the equation (8) and we can get the effective power supply time's formula:

$$L' = n \cdot (1 - K) \cdot T \cdot v \cdots \qquad (8)$$

Through the above analysis, we know that L/L' reflects the degree of matching supply and demand for electricity. When L/L' is closer to 1, MSADD is higher. Otherwise when L/L' is away from 1, MSADD is low.
$Q_1 Q_2$ – Power supply provided for urban residents and migrants
$P_1 P_2$ – The total urban population and migrants
$\lambda_1 \lambda_2$ – The daily per capita electricity consumption frequency of the urban and migrants
$\varepsilon_1 \varepsilon_2$ – The household electricity proportion of the urban and migrants
$D_1 D_2$ – The average effective power used per time of the urban and migrants
$S_1 S_2$ – The average effective power used times of the urban and migrants
n – City stored electricity
v – Power Supply Company time loss

3 COMPREHENSIVE ANALYSIS

3.1 The results of the MSADD

According to previous research, we divide the city into different zones on the basis of electricity consumption: Central Business District, Residential, Wholesale business district, Peri-urban areas. Based on the above model, we get the result:

3.2 Theoretical analysis

By the above results, it can obtain the degree of matching between supply and demand for electric-

Table 1. Different areas of supply and demand for electricity match.

Zone	CBD	Residential	WB district	Peri-urban areas
F1(10^4 yuan)	15876	12443	9745	6872
F2(10^4 yuan)	16274	12039	9302	5936
Q1(neh)	213.6	172.6	153.2	80.5
Q2(neh)	197.5	164.2	147.8	92.1
L(10^4 Kwh)	334.77	250.61	181.78	136.63
L'(10^4 Kwh)	535.83	296.42	193.39	125.94
L/L'	0.6246	0.8455	0.9399	1.0849

ity in different zones. From the data of the table above, we know that MSADD is low in the region where electricity consumption is large. Power is easily wasted in such places. Otherwise the MSADD is high in region where electricity consumption is not that large.

4 CONCLUSION

Through the above research, I firstly establish the factors that affect the degree of matching supply and demand of electricity for the city. And I find out that the main aspect among these factors is the power company's effective power supply time. Based on this I build up the mathematical model to solve out the expression of the effective supply time respectively from the electricity company and user's point. So I get the conclusion that different electricity consumption zones in the city have the order of MSADD as below: Peri-urban areas> Wholesale business district>Residential>Central Business District.

Based on the indicators we proposed above, we can carry out a quantitative analysis of the smart power terminal design. We can evaluate the feasibility of the program from the view of the degree of matching supply and demand. It is helpful to improve the design. Overall it provides a new perspective for the research of the grid scheduling and resource allocation.

REFERENCES

Gan, Y.G. and Tian, F. *Operations Research* (Tsinghua University Press, China 2005), p. 102.
Li, Z.X. and Li H.H. Technology Innovation and Application, Vol. 1(2013) No. 3 p. 144.
Si, S.K. *Mathematical modeling algorithms and procedures* (National Defense Industry Press, China 2007), p. 185.

Energy Science and Applied Technology ESAT 2016 – Fang (Ed.)
© *2016 Taylor & Francis Group, London, ISBN 978-1-138-02973-6*

An application of the Adomian decomposition method for analysis of the SMIB power system

Z.H. Liang & J.F. Gao
School of Electrical Engineering, Zhengzhou University, Zhengzhou, China

ABSTRACT: We use the Adomian decomposition method to analyze the dynamics of a Single-Machine Infinite-Bus (SMIB) power system. This method provides series solution of the system equation which usually converges very rapidly. Using the series solution, the dynamics are analyzed numerically. Period motions, period-doubling bifurcations and chaotic attractors are observed by varying the system parameter. The positive Lya-punov exponents confirm the existence of chaotic attractors.

Keywords: Adomian decomposition method; power system; chaos; bifurcation

1 INTRODUCTION

Power systems are highly nonlinear dynamical systems which can demonstrate complex dynamical phenomena as chaos under certain conditions. Undesirable chaotic behaviors can bring about serious consequences such as power blackout. Thus, it is of considerable importance to study chaotic phenomena in power systems from the point of view of preventing the damage generated by chaotic behaviors (Grillo et al. 2008). Kopell & Washburn 1982 studied chaotic motions in a three-machine swing equations using Menikov's method. Lee & Ajjarapu 1993 observed a period-doubling bifurcation route to chaos in an electrical power system. Chiang et al. 1995 found that a three-bus power system and a nine-bus power system both exhibited a period-doubling transition to chaos. Ji & Venkatasubramanian 1996 investigated chaotic behavior in a SMIB power system. Chen et al. 2005 studied the stability of a SMIB power system employing the Lyapunov direct method and observed the period doubling bifurcation and chaotic phenomena.

The Adomian decomposition method is an effective method for solving nonlinear continuous-time dynamical systems without transformation, linearization, perturbation or discretization (Adomian 1990). This method provides analytical solutions in terms of infinite series that converge very rapidly towards accurate solutions (Abbaoui & Cherruault 1994). The nonlinear terms in the system equations are decomposed into a special series of polynomials in which each term is the so-called Adomian polynomial (González-Parra et al. 2009). The Adomian decomposition method has been applied to various chaotic systems and its accuracy

has been tested with Runge-Kutta method (or its variants). Guellal et al. 1997 proved that the Adomian decomposition method converges faster than Runge-Kutta method when studying the extended Lorenz equation. Vadasz & Olek 2000 observed that the Adomian decomposition method are generally more accurate than Runge-Kutta-Verner method when solving the classical Lorenz equation. Noorani et al. 2007 found that the Adomian decomposition method achieved more accurate results compared with the fourth-order Runge-Kutta method for the chaotic and non-chaotic Chen systems.

In this paper, we use the Adomian decomposition method for analyzing the dynamic behaviors of a Single-Machine Infinite-Bus (SMIB) power system. The rest of the paper is organized as follows. In Section 2, the Adomian decomposition method is briefly introduced. In Section 3, we employ the Adomian decomposition method to solve the SMIB power system and obtain an truncated series solution. In section 4, the chaotic dynamics of the SMIB power system are analyzed numerically using the series solution. Complex dynamical behaviors are observed, such as period motions, period-doubling bifurcations and chaotic attractors.

2 ADOMIAN DECOMPOSITION METHOD

Consider the differential equation

$$F(x(t)) = g(t), \qquad (1)$$

where $x(t) = (x_1(t), x_2(t), \cdots, x_n(t))^T$, $g(t) = (g_1(t), g_2(t), \cdots, g_n(t))^T$ are real functions and F is a general nonlinear differential operator. Assume F can be divided into three terms $F = D + L + N$, where the differential operator D is always invertible, L is the linear term and N is the nonlinear term (Adomian 1990). Thus, the following initial value problem is obtained

$$Dx(t) = g(t) - Lx(t) - Nx(t), \qquad (2)$$

with $x(t_0) = x_0$.

If D is a first-order operator, the integral operator D^{-1} is a definite integral from t_0 to t. Applying the integral operator D^{-1} to both sides of (2), we can obtain the solution $x(t)$ is obtained

$$x(t) = x_0 + D^{-1}g(t) - D^{-1}Lx(t) - D^{-1}Nx(t), \qquad (3)$$

According to Adomian decomposition method, the solution $x(t)$ and the nonlinear term $Nx(t)$ can be written in the series forms

$$x(t) = \sum_{i=0}^{\infty} x^i = \sum_{i=0}^{\infty} \begin{pmatrix} x_1^i \\ x_2^i \\ \vdots \\ x_n^i \end{pmatrix}, Nx(t) = \sum_{i=0}^{\infty} A^i = \sum_{i=0}^{\infty} \begin{pmatrix} A_1^i \\ A_2^i \\ \vdots \\ A_n^i \end{pmatrix}, \qquad (4)$$

where the vector A^i are the so-called Adomian polynomials given by (Adomian 1986)

$$A^i = \frac{1}{i!}\left[\frac{d^i}{d\lambda^i}N\left(\sum_{k=0}^{\infty}\lambda^k x^k\right)\right]_{\lambda=0}, \qquad (5)$$

where λ is a parameter.

Considering a nonlinear term $Nx = f(x)$, Adomian 1986 introduced convenient computational forms for Adomian polynomials

$$A^0 = f(x^0),$$
$$A^1 = x^1 f'(x^0),$$
$$A^2 = x^2 f'(x^0) + \frac{1}{2!}(x^1)^2 f''(x^0),$$
$$A^3 = x^3 f'(x^0) + x^1 x^2 f''(x^0) + \frac{1}{3!}(x^1)^3 f'''(x^0), \qquad (6)$$
$$A^4 = x^4 f'(x^0) + \left(x^1 x^3 + \frac{1}{2!}(x^2)^2\right)f''(x^0)$$
$$+ \frac{1}{2!}(x^1)^2 x^2 f'''(x^0) + \frac{1}{4!}(x^1)^4 f^{(4)}(x^0),$$
$$\vdots$$

Substituting equation (4) into equation (3) yields the infinite series solution of equation (1)

$$\sum_{i=0}^{\infty} x^i = x_0 + D^{-1}g(t) - D^{-1}L\left(\sum_{i=0}^{\infty} x^i\right) - D^{-1}N\left(\sum_{i=0}^{\infty} x^i\right), \qquad (7)$$

Each component x^i can be determined through the following recursive equations

$$x^0 = x_0 + D^{-1}g(t),$$
$$x^1 = -D^{-1}Lx^0 - D^{-1}A^0,$$
$$x^2 = -D^{-1}Lx^1 - D^{-1}A^1,$$
$$\vdots \qquad (8)$$
$$x^i = -D^{-1}Lx^{i-1} - D^{-1}A^{i-1},$$
$$\vdots$$

However, the exact solution (7) is an infinite series, in practice, an approximate solution can be used from the truncated series

$$\tilde{x}(t) = \sum_{i=0}^{m} x^i, \text{ with } \lim_{m\to\infty}\tilde{x}(t) = x(t). \qquad (9)$$

3 ADOMIAN DECOMPOSITION FOR THE SMIB POWER SYSTEM

The Single-Machine Infinite-Bus (SMIB) power system is described as (Chen et al. 2005)

$$M\ddot{\theta} + D\dot{\theta} + P_{max}\sin\theta = P_m, \qquad (10)$$

where θ is the rotor angle of the generator, M is the moment of inertia, D is the damping constant, P_{max} is the maximum power of the generator, and $P_m = A \cdot \sin\omega t$ is the power of machine.

Assuming $x_1 = \theta$, $x_2 = \dot{\theta}$, $x_3 = \sin\omega t$ and $x_4 = \cos\omega t$, equation (10) can be rewritten as

$$\begin{cases} \dot{x}_1 = x_2 \\ \dot{x}_2 = -cx_2 + fx_3 - \beta\sin x_1 \\ \dot{x}_3 = \omega x_4 \\ \dot{x}_4 = -\omega x_3 \end{cases}, \qquad (11)$$

where

$$c = \frac{D}{M}, f = \frac{A}{M}, \beta = \frac{P_{max}}{M}. \qquad (12)$$

In order to solve the SMIB power system (11) via Adomian decomposition method, define the following identities

$$\begin{pmatrix} Dx_1 \\ Dx_2 \\ Dx_3 \\ Dx_4 \end{pmatrix} = \begin{pmatrix} \dot{x}_1 \\ \dot{x}_2 \\ \dot{x}_3 \\ \dot{x}_4 \end{pmatrix}, -\begin{pmatrix} Lx_1 \\ Lx_2 \\ Lx_3 \\ Lx_4 \end{pmatrix} = \begin{pmatrix} x_2 \\ -cx_2 + fx_3 \\ \omega x_4 \\ -\omega x_3 \end{pmatrix},$$
$$-\begin{pmatrix} Nx_1 \\ Nx_2 \\ Nx_3 \\ Nx_4 \end{pmatrix} = \begin{pmatrix} 0 \\ -\beta\sin x_1 \\ 0 \\ 0 \end{pmatrix}, \begin{pmatrix} g_1 \\ g_2 \\ g_3 \\ g_4 \end{pmatrix} = \begin{pmatrix} 0 \\ 0 \\ 0 \\ 0 \end{pmatrix}, \qquad (13)$$

Hence, by doing integrals from t_0 to t on the both sides of (11), the solution is obtained

$$\begin{pmatrix} x_1(t) \\ x_2(t) \\ x_3(t) \\ x_4(t) \end{pmatrix} = \begin{pmatrix} x_1(t_0) \\ x_2(t_0) \\ x_3(t_0) \\ x_4(t_0) \end{pmatrix} + \int_{t_0}^{t} \begin{pmatrix} x_2 \\ -cx_2 + fx_3 \\ \omega x_4 \\ -\omega x_3 \end{pmatrix} d\tau$$

$$+ \int_{t_0}^{t} \begin{pmatrix} 0 \\ -\beta \sin x_1 \\ 0 \\ 0 \end{pmatrix} d\tau, \tag{14}$$

According to the algorithm (6), the first five terms of A^i are calculated as

$$A^0 = -\beta \sin x_1^0,$$
$$A^1 = -\beta x_1^1 \cos x_1^0,$$
$$A^2 = -\beta x_1^2 \cos x_1^0 + \frac{1}{2}(x_1^1)^2 \sin x_1^0,$$
$$A^3 = -\beta x_1^3 \cos x_1^0 + \beta x_1^1 x_1^2 \sin x_1^0 + \frac{1}{6}(x_1^1)^3 \cos x_1^0, \tag{15}$$
$$A^4 = -\beta x_1^4 \cos x_1^0 + \beta \left(x_1^1 x_1^3 + \frac{1}{2}(x_1^2)^2 \right) \sin x_1^0$$
$$+ \frac{\beta}{2}(x_1^1)^2 x_1^2 \cos x_1^0 - \frac{\beta}{24}(x_1^1)^4 \sin x_1^0.$$

Now, by using the first equation $x^0 = x_0 + D^{-1}g(t)$ in (8), the component x^0 is

$$x_j^0 = d_j^0, j = 1,2,3,4, \tag{16}$$

where the coefficients d_j^0 are

$$d_j^0 = x_j(t_0), j = 1,2,3,4. \tag{17}$$

Successively, by utilizing the equation $x^1 = -D^{-1}Lx^0 - D^{-1}A^0$, the coefficients (17) and the polynomial A^0 in (15), the component x^1 is

$$x_j^1 = d_j^1(t - t_0), j = 1,2,3,4, \tag{18}$$

where the coefficients d_j^1 are

$$\begin{cases} d_1^1 = d_2^0 \\ d_2^1 = -cd_2^0 + fd_3^0 - \beta \sin d_1^0 \\ d_3^1 = \omega d_4^0 \\ d_4^1 = -\omega d_3^0 \end{cases}. \tag{19}$$

Similarly, by exploiting the equation $x^2 = -D^{-1}Lx^1 - D^{-1}A^1$, the coefficients (19) and the polynomial A^1 in (15), the component x^2 is

$$x_j^2 = \frac{1}{2}d_j^2(t - t_0)^2, j = 1,2,3,4, \tag{20}$$

where the coefficients d_j^2 are

$$\begin{cases} d_1^2 = d_2^1 \\ d_2^2 = -cd_2^1 + fd_3^1 - \beta d_1^1 \cos d_1^0 \\ d_3^2 = \omega d_4^1 \\ d_4^2 = -\omega d_3^1 \end{cases}. \tag{21}$$

In the same way, the components x^3, x^4 and x^5 can be obtained, respectively

$$x_j^3 = \frac{1}{3}d_j^3(t - t_0)^3, j = 1,2,3,4, \tag{22}$$

$$x_j^4 = \frac{1}{4}d_j^4(t - t_0)^4, j = 1,2,3,4, \tag{23}$$

$$x_j^5 = \frac{1}{5}d_j^5(t - t_0)^5, j = 1,2,3,4, \tag{24}$$

where the coefficients d_j^3, d_j^4 and d_j^5 are, respectively

$$\begin{cases} d_1^3 = d_2^2 \\ d_2^3 = -cd_2^2 + fd_3^2 - \beta d_1^2 \cos d_1^0 + \frac{1}{2}\beta(d_1^1)^2 \sin d_1^0 \\ d_3^3 = \omega d_4^2 \\ d_4^3 = -\omega d_3^2 \end{cases}, \tag{25}$$

$$\begin{cases} d_1^4 = d_2^3 \\ d_2^4 = -cd_2^3 + fd_3^3 - \beta d_1^3 \cos d_1^0 + \beta d_1^1 d_1^2 \sin d_1^0 \\ \qquad + \frac{1}{6}\beta(d_1^1)^3 \cos d_1^0 \\ d_3^4 = \omega d_4^3 \\ d_4^4 = -\omega d_3^3 \end{cases}, \tag{26}$$

$$\begin{cases} d_1^5 = d_2^4 \\ d_2^5 = -cd_2^4 + fd_3^4 - \beta d_1^4 \cos d_1^0 + \beta \left(d_1^1 d_1^3 + \frac{1}{2}(d_1^2)^2 \right) \\ \qquad \cdot \sin d_1^0 + \frac{1}{2}\beta(d_1^1)^2 d_1^2 \cos d_1^0 - \frac{1}{24}\beta(d_1^1)^4 \sin d_1^0 \\ d_3^5 = \omega d_4^4 \\ d_4^5 = -\omega d_3^4 \end{cases}. \tag{27}$$

Therefore, by taking the first six terms in the series solution (7), the approximate solution $\tilde{x}(t) = \sum_{i=0}^{5} x^i$ of the SMIB power system in the interval $[t_0, t]$ is obtained

$$\tilde{x}_j(t) = d_j^0 + d_j^1(t - t_0) + \frac{1}{2}d_j^2(t - t_0)^2 + \frac{1}{3}d_j^3(t - t_0)^3$$
$$+ \frac{1}{4}d_j^4(t - t_0)^4 + \frac{1}{5}d_j^5(t - t_0)^5, j = 1,2,3,4. \tag{28}$$

Note that the approximate solution $\tilde{x}(t)$ is in accordance with the exact solution $x(t)$ only in the neighborhood of $t = t_0$ (Cafagna & Grassi 2009). As a result, for large values of time t, the interval $[t_0, t]$ must be separated into N subintervals $[t_0, t_1], [t_1, t_2], \ldots, [t_{k-1}, t_k = t], k = 1,2, \ldots, N$. Each subinterval has the same step size $h = t/N$. The initial conditions in

329

the subinterval $[t_{k-1}, t_k]$ are taken as $\tilde{x}_j(t_0) = \tilde{x}_j(t_{k-1})$ $(j = 1,2,3,4)$ with $t_0 = t_{k-1}$.

4 NUMERICAL SIMULATIONS

The dynamics of the SMIB power system are analyzed numerically using the series solution (28) obtained via Adomian decomposition method. The initial conditions are chosen as $x_1(0) = 1$, $x_2(0) = -0.3$, $x_3(0) = 0$ and $x_4(0) = 1$. The parameters are taken as $c = 0.5$, $\beta = 1$ and $\omega = 1$.

Let f vary from 2.2 to 2.5. The bifurcation diagram is plotted in Figure 1, from which a period-doubling bifurcation route to chaos can be found in the system. To observe the period-doubling cascade, the phase-space portraits for various f values are plotted in Figure 2. At first, taking $f = 2.28$, a period-1 limit cycle is obtained (Fig. 2a). Then, along with the increase of f, a period-2 limit cycle (Fig. 2b) and a period-4 limit cycle (Fig. 2c) are obtained via period-doubling bifurcations at $f = 2.32$ and $f = 2.38$, respectively. Finally, after a cascade of period-doubling bifurcations, two chaotic attractors are obtained at $f = 2.41$ (Fig. 2d) and $f = 2.45$ (Fig. 2e).

With the increase of f, the behaviors of the system become more and more complicated. As f is larger than 2.4, the system becomes chaotic. Figure 3 shows the time domain diagrams of the generator angle x_1. The sustained irregular chaotic oscillations (Fig. 3a,b) are harmful to the secure and stable operation of the power system. When f increases further, for example $f = 2.471$ (Fig. 3c), the angle is diverging away and the system becomes unstable. By using the method proposed independently by Shimada & Nagashima 1979 and by Benettin et al. 1980, the largest Lyapunov exponents for $f = 2.41$ and $f = 2.45$ are 0.0497 and 0.1023 respectively, which demonstrate the existence of chaos in the SMIB power system.

5 CONCLUSIONS

The Adomian decomposition method has been applied to analyze the dynamics of the SMIB power system. The method provides series solution by decomposing the sine nonlinear term into a series of polynomials. It has been proved that the series solutions generally converge very rapidly towards accurate solutions. The simulation results demonstrate complex dynamic behaviors, such as

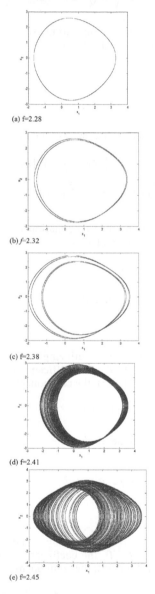

(a) f=2.28

(b) f=2.32

(c) f=2.38

(d) f=2.41

(e) f=2.45

Figure 2. Period-doubling bifurcation route to chaos in the SMIB power system: (a) period-1 cycle; (b) period-2 cycle; (c) period-4 cycle; (d) chaotic attractor; (e) chaotic attractor.

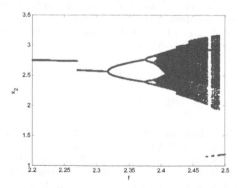

Figure 1. Bifurcation diagram of the SMIB power system.

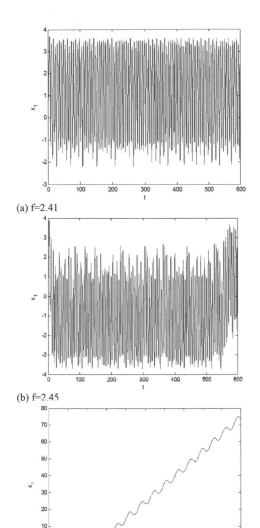

(a) f=2.41

(b) f=2.45

(c) f=2.471

Figure 3. Time domain diagrams for different values of f: (a) chaotic behavior; (b) chaotic behavior; (c) phase divergence.

period orbits, period-doubling bifurcations and chaotic attractors. The existence of chaos in the SMIB power system is verified by the positive Lyapunov exponents.

ACKNOWLEDGEMENTS

Project supported by the Natural Science Foundation of Henan Province, China (Grant No.14A120005) and Excellent Young Scientist Development Foundation of Zhengzhou University, China (Grant No. 142131 9086).

REFERENCES

Abbaoui, K. & Cherruault, Y. 1994. Convergence of Adomian's method applied to differential equations. Comput. Math. Appl. 28(5): 103–109.

Adomian, G. 1986. Nonlinear Stochastic Operator Equations. Orlando: Academic Press.

Adomian, G. 1990. A review of the decomposition method and some recent results for nonlinear equations. Mathl Comput. Modelling 13(7): 17–43.

Benettin, G., Galgani, L., Giorgilli, A & Strelcyn, J.-M. 1980. Lyapunov Characteristic Exponents for smooth dynamical systems and for hamiltonian systems; a method for computing all of them. Meccanica 15(1): 9–20.

Cafagna, D. & Grassi, G. 2009. Hyperchaos in the Fractional-Order Rossler System with Lowest-Order. Int. J. Bifurcation and Chaos 19(1): 339–347.

Chen, H.K., Lin, T.N. & Chen, J.H. 2005. Dynamic analysis, controlling chaos and chaotification of a SMIB power system. Chaos, Solitons and Fractals 24(5): 1307–1315.

Chiang, H.D., Conneen, T.P. & Flueck, A.J. 1995. Bifurcations and chaos in electric power systems: Numerical studies, J. Frankl. Inst. 331(6): 1001–1036.

González-Parra, G., Arenas, A.J. & Jódar, L. 2009. Piecewise finite series solutions of seasonal diseases models using multistage Adomian method. Commun. Nonlinear Sci. Numer. Simulat.14(11): 3967–3977.

Grillo, S., Massucco, S., Morini, A., Pitto, A. & Silvestro, F. 2008. Bifurcation analysis and Chaos detection in power systems. 2008 43rd International Universities Power Engineering Conference (UPEC), Padova, Italy: 1–6.

Guellal, S., Grimalt, P. & Cherruault, Y. 1997. Numerical study of Lorenz's equation by the Adomian method. Comput. Math. Appl. 33(3): 25–29.

Ji, W. & Venkatasubramanian, V. 1996. Hard-limit induced chaos in a fundamental power system model. Electr. Power Energy Syst. 18(5): 279–295.

Kopell, N. & Washburn, R.B. 1982. Chaotic motions in the two-degree-of-freedom swing equations. IEEE Trans. Circ. Syst. 29(11): 738–746.

Lee, B. & Ajjarapu, V. 1993. Period-doubling route to chaos in an electrical power system. IEE Proceedings C (Generation, Transmission and Distribution) 140(6): 490–496.

Noorani, M.S.M., Hashim, I., Ahmad, R., Bakar, S.A., Ismail, E.S. & Zakaria, A.M. 2007. Comparing numerical methods for the solutions of the Chen system. Chaos, Solitons and Fractals 32(4): 1296–1304.

Shimada, I. & Nagashima, T. 1979. A numerical approach to ergodic problem of dissipative dynamical systems. Progress of Theoretical Physics 61(6): 1605–1616.

Vadasz, P. & Olek, S. 2000. Convergence and accuracy of Adomian's decomposition method for the solution of Lorenz equation. Int. J. Heat Mass Transfer 43(10): 1715–1734.

Energy Science and Applied Technology ESAT 2016 – Fang (Ed.)
© 2016 Taylor & Francis Group, London, ISBN 978-1-138-02973-6

Peripheral circuit design and electromagnetic shielding of the PLC control system

Jingwei Liu
Northeastern University, Boston, USA

ABSTRACT: In this paper, based on the possible phenomenon in the practical work of the PLC control system, the electro-magnetic interference problems of PLC control system are analyzed and discussed. According to the characteristics of PLC control system, the electromagnetic compatibility design of the PLC control system is analyzed which is composed of the circuit design and the system shielding. Through the design of the actual PLC control circuit, the feasibility of the electromagnetic compatibility design is verified.

Keywords: PLC control system, circuit design, system shield, electromagnetic compatibility

1 INTRODUCTION

With the development of the automation control technology, PLC has been developed rapidly on the basis of the continuous development of micro-electronic technology and computer technology (Norashikin M, 2011). PLC is widely used in the control of industrial equipment. Therefore, the reliability of PLC control system is more important (Nishant Kumar, 2013). In recent years, the application of PLC control system is more extensive. Most of the time, its working environment is high voltage circuit and strong electromagnetic environment (SHEN Ling-yun, 2011). In the work, PLC will encounter a lot of electromagnetic interference problems. So the electromagnetic compatibility design is very necessary when the circuit of PLC control system is designed.

PLC is provided by the manufacturer and the manufacturer has carried out a series of electromagnetic compatibility experiment at the beginning of the design (Xiaoqun Liu, 2011, V. Prasad Kodali, 2006). So we only need to consider the electromagnetic compatibility design of PLC peripheral circuit when we design the PLC control system (Mark I, 2008). There are three parts which are connected to the peripheral circuit, including the power circuit, the input circuit and the output circuit (Evgeni Genender, 2014). The electromagnetic compatibility design of the PLC control system is considered mainly from the three aspects (Richard Hoad, 2013, Bruce Archambeault, 2013). In addition, in order to ensure the operation of the PLC control system in the strong electromagnetic interference, so the PLC control system must be required to do some shielding measures (Milad Mehri, 2015). PLC control system in the electromagnetic compatibility design carries out the following five steps: demonstration stage, project stage, engineering development stage, setting stage, generation and use stage, which is shown in Fig. 1.

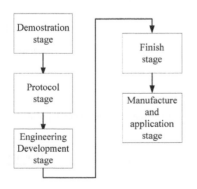

Figure 1. Design flow of PLC control system electromagnetic compatibility.

2 EMC DESIGN OF PERIPHERAL POWER SUPPLY CIRCUIT OF PLC CONTROL SYSTEM

Reliable power supply system can not only guarantee the normal operation of the PLC control system, but also can prevent the external interference signal from entering the system through the power supply circuit and affects the normal operation of the system. Therefore, the design of power supply system plays an important role in the PLC control system. The PLC power supply is used to provide the working power supply for the integrated circuit of each module of the PLC, wherein the input type of the power supply is provided with an Alternating Current (AC) power source and a Direct Current (DC) power supply. In China, the AC is mainly supplied by State Grid, which is the 220V/380V voltage and 50 Hz power frequency. Through the AC to DC converter, the DC power can be converted through the AD power, which is needed by the PLC circuit. Due to the power grid in the process of providing power supply will be affected by the external environment, including wind, snow, lightning strike, and so on. The DC power supply for the PLC control system produces fluctuations. In order to make sure the circuit of the PLC system is not affected by the fluctuation of the DC power supply, the power supply circuit for preventing surge current must be designed.

In order to ensure the normal operation of the DC power supply circuit, the electromagnetic compatibility of peripheral power supply circuit in PLC control system must be analyzed. Power supply electromagnetic compatibility work is mainly concentrated in the power circuit Anti-Surge design, power circuit isolation design, power circuit filter design and power circuit design.

2.1 Anti Surge design of power supply circuit

In Fig. 2, Anti Surge protection circuit ensures that the transient overvoltage of PLC control system is less than the insulation resistance value of the electronic equipment. The protection ability of the selected device is 1.5~2 times of the rated current when the over current protection requires the selection of components. The response time of surge protection device of PLC control system is less than that of surge transient response time of equipment. He can avoid the device is too late to respond, making the surge through the power circuit into the work circuit, and the circuit components in a permanent damage. Grounding pro-

tection requires PLC control system to achieve a good grounding in the installation. In particular, the power circuit part should be able to make good contact with the electronic equipment. Otherwise, surge energy generated by lightning and static electricity cannot effectively release and damages the components in the circuit.

Surge suppression devices are divided into two categories, including: crowbar device and clamp protector. After the breakdown of the device, the voltage surge rapidly discharges by crowbar device, for example Gas Discharge Tube (GDT), bidirectional silicone tube, etc. Clamp protection device is characterized in breakdown device can maintain breakdown voltage does not rise, such as ZnO varistor, Transient Voltage Suppressor (TVS). Due to the characteristics of these protective devices, a combination of surge protection system will be chosen, which is shown in Fig. 3. The circuit can use the advantages of various surge protection devices to achieve the maximum possible protection of the PLC control system.

2.2 Isolation design of power supply circuit

Through the use of isolation transformer, circuit isolation can filter harmonic noise, surge current and interference noise, etc. The isolation transformer is used in the power circuit and can only filter the partial interference, which is shown in Fig. 4. In order to better suppress all kinds of noise interference, the filter circuit is needed to filter the

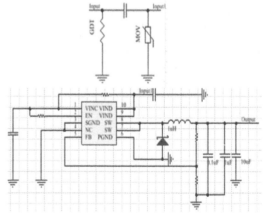

Figure 3.   Two stage combined surge protection circuit.

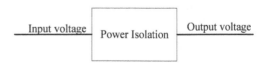

Figure 4.   Power isolation.

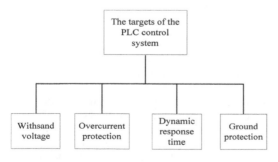

Figure 2.   The targets of the PLC control system.

electromagnetic interference noise in the power supply circuit.

There are two main types of noise, common mode and differential mode noise in the power circuit. Common mode noise is the main part of the circuit in the whole frequency domain, especially in the high frequency domain. In the design of power filter, the main design is the common mode noise filter circuit. Ring, E- and U-shaped ferrite cores are used for the power filter of PLC control system. Metal powder pressing magnetic core to filter out differential mode noise, which the filter circuit is shown in Fig. 5.

2.3 *Grounding design of power supply circuit*

The purpose of grounding design of power circuit is the introduction of transient surge voltage into the earth. Another purpose of grounding design is that preventing interferences into the power supply and then interferes with other load circuit. In order to conduct the transient surge through the case to avoid the transient surge of the components in the load circuit, the ground in the power circuit is connected with the ground of the equipment case. The power circuit and the load circuit are carried out in many places. After that, the connect between the ferrite beads and the PLC control system can

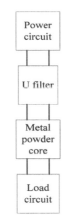

Figure 5.   Filter circuit design.

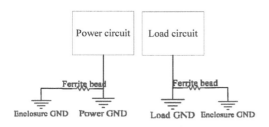

Figure 6.   PLC control system GND design.

suppress electromagnetic interference noise. The grounding design of PLC control system is shown in Fig. 6.

3   PERIPHERAL INPUT / OUTPUT CIRCUIT EMC DESIGN OF PLC CONTROL SYSTEM

The input circuit of PLC control system is the port of PLC receiving analog input signal, the quality of its components and connection mode are the important factors that affect the PLC control system. In the PLC input circuit, the electromagnetic interference is directly from the field of lead into the PLC internal, so loop signal should be as short as possible, and cannot cross with each other. In the circuit design, the filter circuit should be added in the signal front end. The signal output end and the A/D converter also should add the filter or the isolation circuit. The differential amplifier circuit is used to reduce the influence of common mode interference in the signal amplification circuit of the input circuit. In the process of circuit design, the input and output ports which are not used can take short or ground.

In order to prevent the interference of mixed noise from the signal output, the isolation protection measures are added to the output circuit. The main purpose of the isolation is to suppress noise interference into the output circuit. Output circuit of the isolation method mainly includes transformer isolation method, pulse transformer isolation method, relay isolation method, and so on.

4   SHIELDING METHOD FOR PLC CONTROL SYSTEM

Shielding is an important measure to improve the electronic system and electronic equipment EMC, It can effectively prevent or reduce the radiation electromagnetic energy which is not needed and make sure that the equipment can work in the electromagnetic environment. Because the PLC control system is more likely to be disturbed by electromagnetic noise, it is particularly important to effectively shield the PLC control system. The factors of the shielding design should be taken into consideration in Fig. 7.

In Fig. 7, the electromagnetic environment of the PLC control system, including the type of electromagnetic field, the intensity of the field, frequency and the distance between the control system and the source of the interference, should be determined at first. Secondly, according to electromagnetic environment, electromagnetic shielding requirements and the nature of the electromagnetic

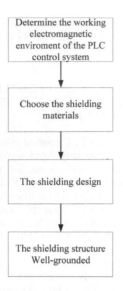

```
Determine the working
electromagnetic
enviroment of the PLC
control system
        ↓
Choose the shielding
materials
        ↓
The shielding design
        ↓
The shielding structure
Well-grounded
```

Figure 7. The shielding design of the PLC control system.

field, the appropriate material should be chosen. Thirdly, after the determination of shielding materials, the shielding structure is designed, when the strong magnetic field is shielded, the multi-layer shield structure can be used. The material inner layer is made of low permeability material, the material middle layer is made of medium permeability material, and the material outer layer is made of high permeability material. Finally, the shield of the PLC control system should be well grounded.

5 CONCLUSION

In this paper, the PLC control system power supply circuit, PLC control system circuit peripheral input/output circuit, and the PLC control system shielding are analyzed and discussed. After verification, the EMC design of the PLC control system can ensure that PLC can work normally in the complex electromagnetic environment.

REFERENCES

American National Standards Institute, American National Standard Dictionary of Electromagnetic Compatibility (EMC) including Electromagnetic Environmental Effects (E3)// ANSI C63.14–2014 (Revision of ANSI C63.14–2009), IEEE Standards, 2014:1–79.

Bruce Archambeault, Sam Connor, Matthew S. Halligan, James L. Drewniak, Albert E. Ruehli, Electromagnetic Radiation Resulting From PCB/High-Density Connector Interfaces//IEEE Transactions on Electromagnetic Compatibility, IEEE Journals & Magazines, 2013: 614–623

Evgeni Genender, Heyno Garbe, Frank Sabath, Probabilistic Risk Analysis Technique of Intentional Electromagnetic Interference at System Level//IEEE Transactions on Electromagnetic Compatibility, IEEE Journals & Magazines, 2014: 200–207

Mark I. Montrose. Printed Circuit Board Design Techniques For EMC Compliance (Section Edition). Beijing, China Machine Press, 2008.

Milad Mehri, Nasser Masoumi, Salomeh Heidari, Electromagnetic susceptibility analysis of PCBs using predictive method//Synthesis, Modeling, Analysis and Simulation Methods and Applications to Circuit Design (SMACD), 2015 International Conference on, IEEE Conference Publications, 2015: 1–4

Nishant Kumar, Tanmoy Mulo, Vikash Prakash Verma, Application of computer and modern automation system for protection and optimum use of High voltage power transformer//Computer Communication and Informatics (ICCCI), 2013 International Conference on, 2013: 1–5

Norashikin M. Thamrin, Mohd. Mukhlis Ismail, Development of virtual machine for Programmable Logic Controller (PLC) by using STEPS™ programming method//System Engineering and Technology (ICSET), 2011 IEEE International Conference on, IEEE Conference Publications, 2011: 138–142

Richard Hoad, William A. Radasky, Progress in High-Altitude Electromagnetic Pulse (HEMP) Standardization//IEEE Transactions on Electromagnetic Compatibility, IEEE Journals & Magazines, 2013: 532–538

SHEN Ling-yun, GUO Xiao-lan, LI Yun-e. Anti-interference Problems in Design and Application of PLC Control System// Automation Application, 2011:48–50

V. Prasad Kodali. Engineering Electromagnetic Compatibility Principles, Measurements, Technologies, and Computer Models (Second Edition). beijing, POSTS&TELECOM PRESS Copyright, 2006.

Xiaoqun Liu, Qingbo Geng, Xiaosong Huang, The EMC analysis and design of the PCB//Mechanic Automation and Control Engineering (MACE), 2011 Second International Conference on, IEEE Conference Publications, 2011: 5383–5386

Energy Science and Applied Technology ESAT 2016 – Fang (Ed.)
© *2016 Taylor & Francis Group, London, ISBN 978-1-138-02973-6*

Research and application of the structure and model of the CCHP energy management system

Yijie Li, Zhen Wang & Li Huang
NARI Group Corporation (State Grid Electric Power Research Institute), Nanjing, China

Jiancheng Yu & Yingqiu Wang
State Grid Tianjin Electric Power Company, Hebei District, Tianjin, China

ABSTRACT: With higher energy efficiency and flexibility requirements of modern power systems, CCHP (Combined Cooling, Heating and Power) has been highly valued, because it can achieve energy efficient conversion and comprehensive utilization and reduce the pollution. Compared with other power supply system, the CCHP system can overcome the weakness of conventional centralized power supply system and is being developed into important trend of energy utilization. This paper studies a variety of data structure of CCHP for information transfer and develop a general data information model. Firstly, the data information characteristics of operation state of CCHP system are studied and the corresponding data information model is established. Then, this paper studies multi-core parallel data processing technology based on the thread structure and develops database technology of management platform. Finally, this paper studies the data information model of CCHP energy management system.

Keywords: CCHP (Combined Cooling, Heating and Power); energy management system; data information model

1   INTRODUCTION

With the state attaching great importance to energy saving in recent years, DG (distributed generation, DG) rapidly develops and scholars research the DG based environmental benefit more and more. In (Liao G C, 2010), the optimization model of electricity generation costs and emissions cost is established based on environmental benefits of DG. Due to the further increase of environmental benefits with energy efficiency and higher flexibility requirements, CCHP (El-Khattam W, 2004) (CCHP, Combined Cooling, Heating and Power) has been highly valued.

CCHP system is the main form of distributed energy systems. In CCHP system, the heat energy (900~1200°C) released by chemical energy is converted to electric energy by advanced micro or small generator and the efficiency is up to 30%~42%. At the same time, the waste heat between 250°C and 450°C can be further utilized and converted by heat pump or absorption refrigeration. The waste heat below 250°C can be used for hot water supply system. Therefore, the energy utilization efficiency of CCHP system is high. In addition, the performance of CCHP system can be further improved by the complementary of renewable energy and the adjusting methods of energy storage. Thus, CCHP

system has good social benefit and economic benefit and is being developed into important trend of energy utilization (Wu D W, 2006).

2   THE OVERALL INFRASTRUCTURE DESIGN OF CCHP ENERGY MANAGEMENT SYSTEM

2.1   *Overall structure*

The overall architecture design of hot-electric hybrid integrated energy management system is divided into five layers (perception layer, transport layer, data layer, application layer, and visual presentation layer) proposed by this paper, as shown in Figure 1.

Perception layer: collecting data from the data acquisition terminal of user, including the user's energy and operational data.

Transport layer: transporting energy data of user into data layer by the Internet, wireless network, 2G/3G/4G Mobile Internet, and optical fiber communication network.

Data layer: storing, processing, analyzing energy data, and storing real-time data, historical data, spatial data, and application data of the end users. At the same time, it supports the data mining of industrial data and its spatial data stores spatial

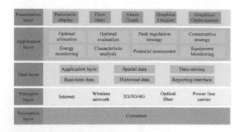

Figure 1. Integrated energy management system of CCHP.

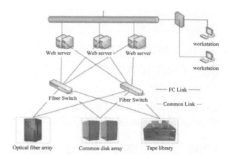

Figure 3. Network structure of storage area.

1. Data storage management

In view of the huge data in integrated energy management of CCHP, this paper uses SAN storage structure of file sharing system, and all server (client) directly access the disk array on the shared file system in a Fiber Channel (FC), as shown in Figure 3.

Data is shared in the storage and data can be accessed directly from the FC link on any one server (client) without considering the operation platform of the server (client), and Storage Area Network (SAN) avoids dependence and influence on the traditional LAN bandwidth. Storage structure of SAN can facilitate achievement of the purpose of expanding the storage capacity by expensing the number of disk arrays and does not affect the efficiency of data sharing. According to the need, it is divided into three levels of storage: online, near-line, and offline. Online devices use a higher performance of the high-end disk array (such as optical fiber array). Near-line devices use a large array of high-capacity disk arrays (such as SATA disk array) in general. Offline devices generally use the tape library equipment. When offline data is needed, the data in the tape library can be restored to near line (or online) devices. Hierarchical storage technology ensures the high availability of important data and reduces the cost of the entire data storage system maximally.

For the massive data of CCHP (especially large file data), it does not take all the way into the relational database, which can lead the database to expand rapidly. Archiving operations complete data files shifting from the reception area to data storage sharing, and they extract metadata information for management and query from the file name, file header, or specialized metadata files. At the same time, metadata information is inserted into a relational database because mature optimization performance of relational database facilitates the data query and the data management.

2. Buffer management

In the database system, transactions need to access the data blocks on the disk and then make an access

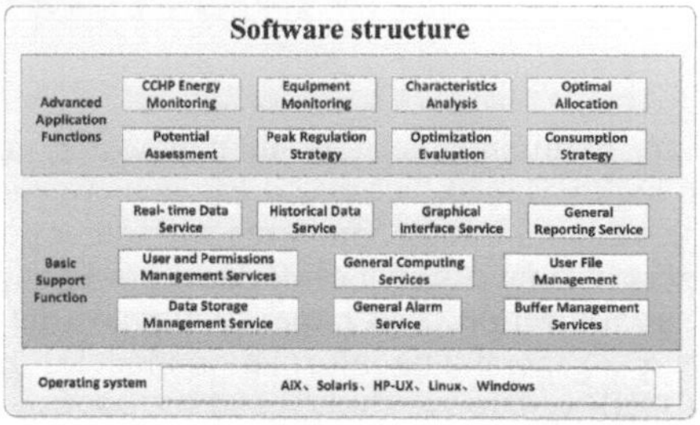

Figure 2. Software structure of integrated energy management system of CCHP.

data of GIS (Geographic Information System), which can support GIS to monitor and display the energy use. The application data stores platform application analysis report and other data and the platform also left reported interactive interface between the interfaces and other systems (Alan Shalloway, 2004).

Application layer: energy integrated management function includes CCHP energy monitoring, energy optimal allocation, energy optimal evaluation, peak load and storage regulation strategy, photovoltaic consumption strategy, analysis of energy characteristics, adjusting potential evaluation, equipment testing, and so on.

Visual presentation layer: based on GIS technology, the energy information visualization show includes flow chart, alarm figure, graphic report, energy optimization and display, and so on. It can display the process of monitoring, alarm, analysis, optimization in the platform using visualization technology to achieve visual display, which enhance the entire platform interaction and presentation, and make the operation and display friendlier.

2.2 Software structure

The software structure of the integrated management system of CCPH is shown in Figure 2 and is described in detail in functional data storage management services of the system (DING Jie, 2012), buffer management services of users and recovery mechanisms.

request to the buffer manager. If the required data is already in the buffer, the buffer manager passes the address of the data block to the transaction. If the required data is not in the buffer, buffer manager first allocates a certain memory space in the buffer of the memory space for the data block. If the buffer does not have the available space, the buffer manager obtains some available memory space through replacing some old data block.

Then, the buffer manager reads the requested data block from the disk to the available memory space of buffer, and passes the address of the data block in memory to the transaction. If the data block needed to be replaced is modified in the last save to the disk, the data block will be written to the disk. Thus, the buffer manager is similar to the virtual memory manager, but the latter is located in the operating system. The buffer manager uses different techniques (such as buffer replacement policy, pinning block, buffer cache, etc.) to provide efficient service for database systems.

3. Recovery mechanism

When the system restarts every time, the log block is read first to find the largest recorded of the file size. Then, through the file name and cluster number of file directory, the file information is found in the FDT (if the card is not the file FDT item, the file is created according to the log records). If the relevant data is consistent, the restore operation is not needed.

Otherwise, according to the contents of the log file in accordance with the size of the increasing order of the file, the FAT data and FDT data are modified from the file record of FDT start, which make the data clusters of the file in the CF card linked to the FAT table and ensure the consistency of the file. The specific process is shown in Figure 4.

3 DATABASE STRUCTURE DESIGN OF CCHP ENERGY MANAGEMENT SYSTEM

3.1 Ideas of database design

Database is used to record data collected from cold thermal energy equipment, to save the processing results of the control system, and to integrate the logical relationship between each module. Therefore, the database is the core of CCHP energy systems.

The data designed by CCHP has the following characteristics:

1. Large amount of data are collected.
2. A complex relationship exists between different data.
3. The database must be able to adapt to unknown loads and energy sources.
4. The database should be stored in a unified way for transfer and control of upper application.
5. Each module of the database should be loose coupling, which is easy to adjust the system upgrade.

In order to solve these problems in the CCHP, the following points are needed:

1. researching multi-core parallel data processing technology based on thread structure in order to deal with the data quantity, data association, and so on;
2. researching the aggregation service of massive data tags in order to solve the problem of multi-load and multi-energy;
3. researching scheduling algorithm of distributed data processing to solve the problem of system development.

The above can be summarized as the following ideas:

Based on this, the design of the simulation database structure should be implemented to meet the demand of massive real-time energy supply and demand data processing.

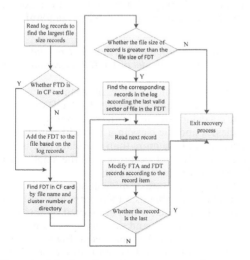

Figure 4. The flowchart of recovery processing.

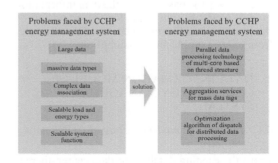

Figure 5. Ideas of database design.

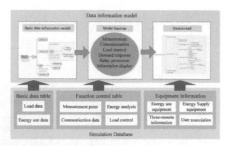

Figure 6. Simulation data structure of the CCHP energy management system.

3.2 Structure design of database

In order to realize simulation database of CCHP with real-time energy data of supply and demand, this paper designs the basic data information, the model function, and the way to read data of equipment as shown in Figure 6. It realizes the construction of the simulation data structure model of the CCHP energy management system. Based on the simulation data structure model, the basic data table, the function control table, and the equipment information table are established to support the data model according to the inherent characteristics of the CCHP system.

The basic data information table includes two database tables of load data and energy data. Load data includes database tables of the cold, heat, and electricity. Energy data includes database tables of electricity, gas, and oil. These tables and data models are corresponding, which is conducive to excavate the relationship between them. The functional control table includes measurement point, energy analysis, load control, and data communication. This table covers the basic functions of simulation database and is easy to enhance the efficiency of simulation database. The simulation database includes equipment information table, covering functional equipment, power supply equipment, remote information, association information of the user, and so on (DING Ming, 2008).

In general, the design of CCHP simulation database realizes the basic idea of the data information model, which is convenient for the expansion of the function and the improvement of the efficiency from energy supply and consumption.

4 DATA INFORMATION MODEL OF CCHP ENERGY MANAGEMENT SYSTEM

4.1 Information model structure of CCHP data platform

From the existing database or data center, there is no standard information model building and most

of the construction is to consider their own data center or platform application scenarios. Overall, the general problem is that the model is not complete and the standard is not unified. Multiple data models lead to many sources needed to maintain, the lack of authority, duplication of investment, and waste of resources.

In view of the problems, in order to avoid repeated model maintenance, to reduce the waste of resources (human resources and equipment resources), and the cost of investment (construction costs, management costs, and maintenance costs), uniform standards (coding standards and data exchange standards) need to be established according to the energy network model to achieve the control of the whole network energy data, to improve the overall sharing of data and to form external service standards and global coding standards.

4.2 Representation and structure of the data information model of CCHP operation state

1. Information model representation
The data information model of CCHP operation state is a demonstration of the relationship between all classes and the class in a bag with graphics. A package can have more than one class that is analyzed and defined according to the relationship between attributes and classes.
2. Package structure of information model
The following table lists the components of data information model, which are sorted by the package's hierarchy path and the package's name:

The package can be expressed in Figure 7, which is involved in data information model of CCHP operation state.

Table 1. The package of information model in the energy management system of CCHP.

Package	Path
AMS	
Cunsumers	61968
ERP_Supports	61968
AssetBasics	61968::Assets
PointAssetHierarchy	61968::Assets
Location	61968::Core2
Toplevel	61968::Core2
DocumentInheritance	61968::Document
Core	61970
Domain	61970
Meas	61970
SCADA	61970
Topoly	61970
Wires	61970
Production	61970::Generation

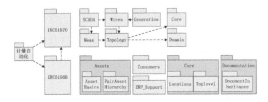

Figure 7. The package structure of information model of CCHP energy management system.

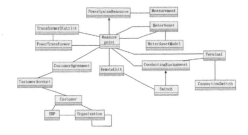

Figure 8. Data information overall model of operation state.

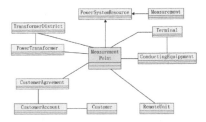

Figure 9. Measurement point.

CCHP energy management system quotes the package of Core, Domain, Topology, Meas, Wires, Generation, and SCADA of IEC 61970 and the package of AssetBasics, PointAssetHierarychy, Consumers, Assets, ERP_support, DocumentInheritance, Core2, Locations, Toplevel, and Documentation of IEC 61968, and placed the extension part of classes, attributes, and associations in the AMS package.

The expansion package AMS of CCHP energy management system is an extended model of the package based on IEC 61968 and 61970CIM to adapt to the information description of CCHP energy management system.

4.3 Description of the data information model of CCHP operation state

1. Overall description

Figure 8 illustrates the panorama of the data information model of operation state in general. In order to make the data of operation state adapt to the electric data analysis, power quality analysis, real-time line loss analysis and other advanced application analysis, topological relation, and meters are associated to ensure good connection between measurement and model.

2. Measurement point model

The measurement point is the core of the information model of CCHP energy management system. The design of measurement points directly determines that the whole model is good or bad.

When designing the measurement point model, it is needed to consider the measurement point of the charging mode adopted by a common resident through the contract, the measurement point of transformer and line used to charge and check. In addition, as a sub-class of the power system resource package, the relationship between the meter and the CT/PT need to be established when describing multiple features of measurement, which is used to support the operation of equipment replacement.

In the measurement configuration, the relationship between measurement point and the conductive equipment specifies measurement point

measuring a certain conductive equipment. If the measurement point set needs to be specified at a point of the conductive equipment that is a terminal, the association of MeasurePoint and Terminal is established. The whole measurement of special transformer or public transformer is embodied by the correlation of measurement point and PowerTransformer. When multiple stations are measured in a certain district, the relationship between measurement point and TransformerDistrict is established.

3. Transformer model

In Figure 10, the measurement point is related to the transformer as a conductive device (the subclass of power system resource), each of which is associated with a CT and a PT.

Generally, CT assets and PT assets, respectively, play PT and CT role at operation time. If CombinedTAsset is adopted, then the asset plays CT and PT roles at the same time.

4. Measurement model

The core of CCHP energy management system is the measurement point. It contains multiple measurement, such as power, voltage, current, and operation information of measurement points, as a subclass of PowerSystemResource. The measurement model is shown in Figure 11.

In order to facilitate the realization, various data types of the measurement and the corresponding measurement values are independently built as sub-categories by the measurement system, including four series: Analog and Analog Value,

341

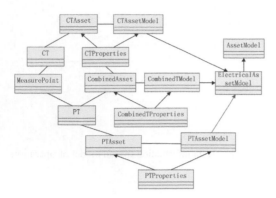

Figure 10.   Transformer model.

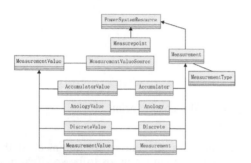

Figure 11.   Measurement model.

Accumulator and Accumulator Value, Discrete and Discrete Value, and String Measurement and String Measurement Value.

5   CONCLUSION

This paper studies a variety of data structures of CCHP for information transfer and develops a general data information model. First, the data information characteristics of the operation state of CCHP system are studied and the corresponding data information model is established. Then, this paper studies multi-core parallel data processing technology based on the thread structure, introduces the scheduling algorithm of distributed data processing, develops database technology of management platform based on distributed data processing and scheduling optimization algorithm, and forms the simulation data structure that is applicable to massive real-time energy supply and demand data. Finally, the data information model of CCHP energy management system is studied, including the overall structure, data model structure, real-time transaction model and processing design, buffer management, data storage management, recovery mechanism, cluster server structure, and so on.

ACKNOWLEDGMENT

The support by the project of SGTJDK00D-WJS1500100 is gratefully acknowledged.

REFERENCES

Al-Sulaiman F A, Hamdullahpur F, Dincer I. Trigeneration: a comprehensive review based on prime movers(J). International journal of energy research, 2011, 35(3): 233–258.

Alan Shalloway, James. R. Trott. Design Patterns Explained: A New Perspective on Object-Oriented Design (M). Tsinghua University Press, 2004.

Cui Wei, Shi Yong, Sun Bing. Construction and integration of grid model based on IEC61970/61968 (J). Power System Protection and Control, 2011, 39(17): 60–63.

Ding Jie, Xi Houwei, Han Haiyun, Zhou Aihua, etc. A Smart Grid-oriented Data Placement Strategy for Data-intensive Cloud Environment (J). Automation of Electric Power Systems, 2012, 36(12): 66–70.

Ding Ming, Yang Wei, Zhang Yingyuan, etc. IEC61970 based MicroGrid energy management system (J). Electric Power Automation Equipment, 2009, 29(10).

Ding Ming, Zhang Zhengkai, Bi Rui, etc. Distributed Generation System Oriented CIM Extension (J). Automation of Electric Power Systems, 2008, 32(20): 83–87.

El-Khattam W, Salama M M A. Distributed generation technologies, definitions and benefits(J). Electric power systems research, 2004, 71(2): 119–128.

Gu Qiang, Wang Shou-xiang, Li Xiao-hui, etc. CIM extension to distribution system components (J). Electric Power Automation Equipment, 2007, 27(10): 91–95.

Jin Hongguang, Zheng Danxing, Xu Jianzhong. Device and application of distributed cold combined heat and power system (M). China Electric Power Press, 2010.

Liao G C. Using chaotic quantum genetic algorithm solving environmental economic dispatch of smart microgrid containing distributed generation system problems(C)//Power System Technology (POWERCON), 2010 International Conference on. IEEE, 2010: 1–7.

Qi Y D, Liu Z G, Song G M. CCHP and Reliability of Electricity Supply(C)//Advanced Materials Research. 2011, 250: 3173–3176.

Wu D W, Wang R Z. Combined cooling, heating and power: a review(J). progress in energy and combustion science, 2006, 32(5): 459–495.

Zhang Shenming, LIU Guoding. Introduction of Standard IEC61970 (J). Automation of Electric Power Systems, 2002, 26(14): 1–6.

Zhang J, Wang Y, Wang R, et al. A new particle swarm optimization solution to nonconvex economic dispatch problem(M)//Advances in Swarm Intelligence. Springer Berlin Heidelberg, 2010: 191–197.

Energy Science and Applied Technology ESAT 2016 – Fang (Ed.)
© 2016 Taylor & Francis Group, London, ISBN 978-1-138-02973-6

Smart grid load forecasting of gray model with optimization of weight function

Wenhao Zhu
Department of Electrical Engineering, Tongji University, Shanghai, China
Schneider Shanghai Low Voltage Terminal Apparatus Co. Ltd., Shanghai, China

Qiyi Guo
Department of Electrical Engineering, Tongji University, Shanghai, China

ABSTRACT: Smart grid has advantages such as strong, interactive, economic, compatible, self-healing etc., and short-term load forecasting has very good security for the normal operation of the smart grid system. According to the demand of the smart grid load forecasting, a gray forecasting model is put forward based on the weight function optimization. First, advantages and disadvantages of four gray whitenization weight functions of the gray forecasting model are analyzed, and then using the cut-off points of index value range, the membership degree of subclass is stipulated. In addition, comprehensive clustering weights are distributed. Finally, the smart grid load forecasting model of the gray system is built based on weighting function optimization. Simulations show that, compared with the traditional gray prediction model, the proposed smart grid load forecasting model has higher accuracy.

Keywords: whiten function; gray prediction; weight function optimization; smart grid; load forecasting

1 INTRODUCTION

The construction of smart grid has higher demand for the accuracy of power resources deployment, and the short-term power load forecasting (generally refers to power consumption forecast) is the focus of the competitive electricity market. How to predict the power consumption is an important task of all market participants (Hu, et al., 2015). As a result, accurately predicting the smart grid power consumption is particularly important. As power consumption changes constantly, peak appeared frequently, forecasting that low electricity consumption causes power outages due to the lack of electrical distribution, while forecasting high electricity consumption brings unnecessary cost and wastes energy (Chen, et al., 2012).

As the electric power system show up, the researchers in the field of electric power for the research of load forecasting technology never stopped. Chen Changsong et al. obtained the main factors influencing the photovoltaic power generation by photovoltaic array analysis of historical data and meteorological data, output sequence of photovoltaic array and day-type index and temperature are used to build a neural network prediction model which considers meteorological factors (Chen, et al., 2009). Cao et al. fully considered the influencing factors such as time-of-use power price of decline index trend electricity and load

status of obeying the probability distribution, and an optimized charging model is established, which can greatly reduce the charging cost and the pressure of the grid at peak demand (Cao, et al., 2012). Stanton has analyzed the charging load of a variety of electric vehicles under the different battery capacities and different charging loads (Stanton, 2007). According to the different types of plug-in hybrid electric car, Liao Feng et al. put forward a charging load model based on the needs of charging, calculate the actual charging time by charging method, the initial charging status, and charge needs, and finally, greatly improve the prediction accuracy based on the probability distribution of the initial charging time and charging by the Monte Carlo method to determine the starting point (Liao, et al., 2011). A. Barbato et al. put forward a model based on the wireless sensor monitoring domestic electrical installation done automatically to achieve the next January's households (Barbato, et al., 2011). J. Matsumoto et al. put forward a method that smart sensors are connected to the solar panels, values from multiple spots, and a forecasting server can be used to predict the intensity of the recent period of time. This method replaces the original provided by the meteorological monitoring light intensity data, and provides a solution to predict light intensity (Matsumoto, et al., 2011). Y. Depeng et al. propose a short-term load forecasting model of the sparse Bayesian learning method based on the load of the geographical similarities and historical parallels, by collecting load data of multiple users in the smart grid (Liang, et al., 2011). Chan, et al. propose a load forecasting model based on the multiple classifier of the micro-network environment, the historical load data of last 24 hours, three days and a week and one month before prediction for classification training, respectively, and finally, for weighted combination (Chan, et al., 2011). N. Amjady et al. propose two levels of load forecasting model which is applied to network based on neural network and improved differential evolution algorithm, and its essence is using differential evolution algorithm for parameter optimization of the neural network (Amjady, et al., 2010).

According to the needs of the smart grid load forecasting, a gray forecasting model is put forward based on the weight function optimization, and simulations show its performance.

2 FORECASTING MODEL BASED ON THE GRAY THEORY

The gray forecasting model (Peng & Han, 2016) is a method that defines categories for several observation indexes or observation object clustering based on gray correlation matrix or whitenization weight function.

The common gray whitenization weight function has the following four types:

Assume that there are n clustering objects, m clustering indexes, and s different gray classes, according to the $i(i=1,2,...,n)$ object observations x_{ij}, regarding the indicators $j(j=1,2,...,m)$, then the i object in the $k(k \in \{1,2,...,s\})$ gray class becomes gray clustering.

The value of indicators j about n objects can be divided into s gray classes, the so-called subclass of indicators j. Whitenization weight function k subclass of indicator j index is marked as $f_j^k(\cdot)$. Figure 1 is the whitenization weight function.

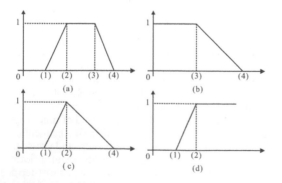

Figure 1.   Whitenization weight function graph.

1. The typical whitenization weight function is shown in Figure 1 (a). $x_j^k(1), x_j^k(2), x_j^k(3), x_j^k(4)$ is the turning point of function $f_j^k(\cdot)$; the whitenization weight function gives equation (1):

$$f_j^k(x) = \begin{cases} 0, x \notin [x_j^k(1), x_j^k(4)] \\ \dfrac{x - x_j^k(1)}{x_j^k(2) - x_j^k(1)}, x \notin [x_j^k(1), x_j^k(2)] \\ 1, x \notin [x_j^k(2), x_j^k(3)] \\ \dfrac{x_j^k(4) - x}{x_j^k(4) - x_j^k(3)}, x \notin [x_j^k(3), x_j^k(4)] \end{cases} \quad (1)$$

2. The lower measure whitenization weight function is shown in figure 1 (b); this function has no first and second turning points $x_j^k(1)$ and $x_j^k(2)$, the whitenization weight function is given in equation (2):

$$f_j^k(x) = \begin{cases} 0, x \notin [0, x_j^k(4)] \\ 1, x \notin [0, x_j^k(3)] \\ \dfrac{x_j^k(4) - x}{x_j^k(4) - x_j^k(3)}, x \notin [x_j^k(3), x_j^k(4)] \end{cases} \quad (2)$$

3. The moderate measure whitenization weight function is shown in figure 1 (c); the second and third turning points $x_j^k(2)$ and $x_j^k(3)$ overlap, and the whitenization weight function is given as equation (3):

$$f_j^k(x) = \begin{cases} 0, x \notin [x_j^k(1), x_j^k(4)] \\ \dfrac{x - x_j^k(1)}{x_j^k(2) - x_j^k(1)}, x \notin [x_j^k(1), x_j^k(2)] \\ \dfrac{x_j^k(4) - x}{x_j^k(4) - x_j^k(2)}, x \notin [x_j^k(2), x_j^k(4)] \end{cases} \quad (3)$$

4. The limit measure whitenization weight function is shown in Figure 1 (d); the functions without the third and fourth turning points are $x_j^k(3)$ and $x_j^k(4)$, and the whitenization weight function is given in equation (4):

$$f_j^k(x) = \begin{cases} 0, x < x_j^k(1) \\ \dfrac{x - x_j^k(1)}{x_j^k(2) - x_j^k(1)}, x \notin [x_j^k(1), x_j^k(2)] \\ 1, x \geq x_j^k(2) \end{cases} \quad (4)$$

In addition, the smart grid load forecasting is a gray system with certificate. Therefore, we can fully use the basic theory of gray system and the method to solve.

3 SMART GRID LOAD FORECASTING OF WEIGHT FUNCTION OPTIMIZED GRAY MODEL

There are n objects, m indicator values, and s gray classes of j indicator. Cluster objects are in accordance with the observed value x_{ij}. The gray forecasting model with optimization of the weight function can be described as follows:

1. In the determination of the number s of subclasses, through actual situation or qualitative research, results indicate the cut-off point $a_k(k = 1,2,...,s,s+1)$ of the j indicator values range, divided into s ranges.

$$[a_1,a_2],[a_2,a_3],...,[a_{k-1},a_k],...,[a_s,a_{s+1}] \quad (5)$$

2. In the interval $[a_k,a_{k+1}]$, the value of whitenization weight function of $\lambda = (a_k + a_{k+1})/2$ to k subclass is 1. With the join point $(\lambda_k,1)$, the $k-1$ subclass starting point $(a_{k-1},0)$, and the $k+1$ subclass ending point $(a_{k+2},0)$, the triangular whitenization weight function $f_j^k(\cdot)$ of k subclass j indicator is obtained. $f_j^1(\cdot)$ and $f_j^s(\cdot)$ work with the j indicator value of left and right to continuation, respectively, and each subclass whitenization weight function is shown in Figure 2.
Observation x of index j is, according to,

$$f_j^k(x) = \begin{cases} 0, x \notin [a_{k-1}, a_{k+2}] \\ \dfrac{x - a_{k-1}}{\lambda_k - a_{k-1}}, x \notin [a_{k-1}, \lambda_k] \\ \dfrac{a_{k+2} - x}{a_{k+2} - \lambda_k}, x \notin [\lambda_k, a_{k+2}] \end{cases} \quad (6)$$

The obtained observed value belongs to the membership degree of k subclass.

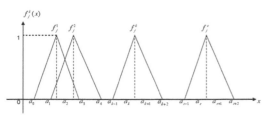

Figure 2.   Whitenization weight function.

3. According to the observed value of all indicators of the object i belonging to the membership degree of k subclass, calculate the comprehensive clustering coefficient σ_i^k of the object i about the k subclass:

$$\sigma_i^k = \sum_{j=1}^{m} f_j^k(x_{ij}) \cdot \eta_j \quad (7)$$

η_j is the weight of the j indicator in the comprehensive clustering.

4. According to $\max_{1 \le k \le s}\{\sigma_i^k\} = \sigma_i^{k*}$, get the object i which belongs to a subclass. In addition, according to the size of σ_i^k, the quality and order of clustering object of the same subclass can be determined.

4   PERFORMANCE SIMULATION

In order to verify the effectiveness of the proposed smart grid load forecasting model, we carry on the simulation test of the model. With smart grid, power consumption of data provided by a company, they offer the consumption data on 25–30 August,

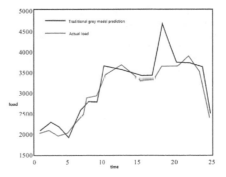

Figure 3.   Power load forecasting results of traditional gray model.

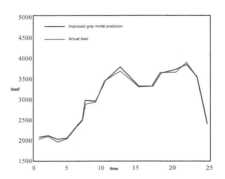

Figure 4.   Power load forecasting results of improved gray model.

which is used for training the gray forecasting model. After obtaining the forecasting model, forecast the power consumption of 30 August; the results are shown in Figures 3 and 4.

Seen from the simulation results, the proposed gray forecasting model has smaller error and higher accuracy for judgment of the smart grid load.

5 CONCLUSIONS

Short-term load forecasting is very good for the normal operation of the smart grid system security, and electricity load forecasting influences the precision of resources deployment of smart grid, and can help develop a plan of power, and optimize the power resources. It has the very high decision-making. According to the demand of smart grid load forecasting, a gray forecasting model is proposed based on the weight function optimization. Simulations show that compared with the traditional gray system, the proposed gray system has smaller error.

REFERENCES

Amjady, N. et al. 2010. Short-Term Load Forecast of Microgrids by a New Bilevel Prediction Strategy. *Smart Grid, IEEE Transactions on* (1):286–294.

Barbato, A. et al. 2011. Forecasting The Usage of Household Appliances Through Power Meter Sensors for Demand Management in The Smart Grid. *Smart Grid Communications* 4(13): 404–409.

Cao, Yijia et al. 2012. An Optimized EV Charging Model Considering TOU Price and SOC Curve. *IEEE Transactions on Smart Grid* 3(1): 388–393.

Chan, P.P.K. et al. 2011. Multiple Classifier System for Short Term Load Forecast of Microgrid. *Machine Learning and Cybernetics* (1):1268–1273.

Chen, Changsong et al. 2009. Design of the neural network predictive model Based on the PV array power generation. *Journal of electrotechnics* 24(9): 153–158.

Chen, Minyou et al. 2012. Micro grid remaining ultra short-term load prediction based on the hybrid intelligent technology of. *Electric power automation equipment* 32 (5): 13–18.

Hu, Shiyu et al. 2015. Based on the multivariable LS-SVM and fuzzy reasoning system load forecasting. *Journal of computer applications* 35 (2): 595–600.

Liang, X. et al. 2011. Joint Electrical Load Modeling and Forecasting Based on Sparse Bayesian Learning for The Smart Grid. *Information Sciences and Systems* 4(7): 1–6.

Liao, Feng et al. 2011. Centralized land-saving integration bus load forecasting system. *Electric power demand side management* 13(3):54–58.

Matsumoto, J. et al. 2011. Highly-accurate Short-term Forecasting Photovoltaic Output Power Architecture without Meteorological Observations in Smart Grid. *Access Spaces* (5):186–190.

Peng, Huaiwu & Han, Jiatong. 2016. A Research on Wind Power Ultra-Short Term Forecasting Based on Trend Analysis. *Acta Scientiarum Naturalium Universitatis Neimongol* 47(1): 96–101.

Stanton W. 2007. Evaluating the Impact of Plug-in Hybrid Electric Vehicles on Regional Electricity Supplies. *Symposium-Bulk Power System Dynamics and Contro* (3):1–12.

Energy Science and Applied Technology ESAT 2016 – Fang (Ed.)
© 2016 Taylor & Francis Group, London, ISBN 978-1-138-02973-6

Research on relay protection device life cycle based on the gray Markov model

Yongtian Jia, Liming Ying, Jinwei Wang, Yulei Wang & Lei Yang
School of Electrical Engineering, Wuhan University, Wuhan, China

ABSTRACT: Predicting the life cycle of relay protection device accurately is beneficial to eliminating security risks and enduring power systems. This paper analyzes the main factors affecting the life cycle of the relay protection device, and determines their weight with AHP. Combined with example analysis, a relay protection devices' life cycle model based on gray Markov chain model is established. Finally, the validity and accuracy of the model are verified.

Keywords: relay protection device; life cycle; ahp; gray markov model; average failure rate; weight

1 INTRODUCTION

Relay protection device is a significant defense line to guarantee the safe operation of power system, and its life cycle has a direct effect on power quality of consumers (Tan J C. 2001). Though relay protection device is composed of a software part and a hardware part, the software part generally will not degrade during operation. Therefore, when it comes to researching on relay protection device life cycle, the hardware part of relay protection equipment is mainly studied (Yu Xingbin, Singh. 2004).

MTBF (Mean Time Between Failures) is a vital factor for predicting equipment life cycle. It means the average working time between two adjacent faults. However, it is not accurate to predict the life cycle of equipment with MTBF only. Therefore, the average failure rate is used as one of the important factors in this paper, either. In general, the higher the average failure rate is, the greater the probability of stop running will be. The life cycle of relay protection is generally not less than 12 years. Owing to unstable working environment, devices' practical life cycle commonly tends to be 10–12 years (Chen J. 2005).

In this paper, the main factors that affect equipment life cycle and their weight are determined with AHP.

2 WEIGHT DETERMINATION OF MULTIPLE FACTORS

2.1 *Hierarchical analysis diagram*

Form a multi-level analysis structure model by gathering them on the basis of their subordinate relationship (Yingchen W. 2008). For this model, the hierarchical analysis diagram is shown in Figure 1.

In Figure 1, the study of equipment life cycle is located in the top layer. The rule layer is composed of humidity, temperature, electromagnetism, dust, and other factors. Electronic component's faults, software bug, and faults of output device make up the lowest layer.

2.2 *Judgment matrix*

To obtain objective weight of each factor, after finishing pairwise comparison of factors, a nine-level judgment table shown in Table 1 is designed.

In Table 1, the numbers 2,4,6,8 represent the median of two adjacent states. The judgment matrix of the rule layer can be obtained by using the nine-level judgment table:

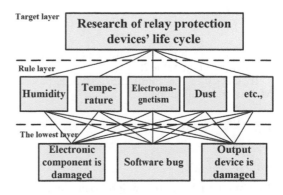

Figure 1. Hierarchical analysis diagram.

Table 1. Nine-level judgment table.				
Implication	Paramount	Very important	Quite important	Important
Degree	9	7	5	3
Equal	Unimportant	Not quite important	Not very important	Least
1	1/3	1/5	1/7	1/9

Table 2. Satty random index.

N	1	2	3	4
0	0	0.58	0.90	1.12
5	6	7	8	...
1.12	1.24	1.32	1.14	...

Table 3. Data of each layer.

$w(2)$	0.263	0.475	0.055	0.090	0.110
$w(3)$	0.592	0.082	0.429	0.633	0.166
	0.277	0.236	0.429	0.193	0.166
	0.129	0.682	0.142	0.175	0.668
λ_k	3.005	3.002	3	3.009	3
CI_k	0.003	0.001	0	0.005	0

$$A = \begin{bmatrix} 1 & \dfrac{1}{2} & 4 & 3 & 3 \\ 2 & 1 & 7 & 5 & 5 \\ \dfrac{1}{4} & \dfrac{1}{7} & 1 & \dfrac{1}{2} & \dfrac{1}{3} \\ \dfrac{1}{3} & \dfrac{1}{5} & 2 & 1 & 1 \\ \dfrac{1}{3} & \dfrac{1}{5} & 3 & 1 & 1 \end{bmatrix} \tag{1}$$

2.3 Consistency check

Consistency check refers to the logical consistency of the judgment theory. When confirming a factor's influence on each object of high level, namely single-level sequencing, the nine-level judgment index being introduced makes it possible for quantitative research. Only getting through consistency check, can weight value obtained by AHP be considered to be correct.

Consistency index:

$$CI = \frac{\lambda - N}{N - 1} \tag{2}$$

λ is the greatest characteristic root of n-order positive and negative matrix, and N is the non-zero characteristic root of the n-order consistent matrix.

Consistency ratio:

$$CR = \frac{CI}{RI} \tag{3}$$

RI is the random index, and the Satty random index is shown in Table 2:

If $CR < 0.1$, the consistency check is passed. Moreover, in this model:

$$CR = \frac{0.018}{1.12} = 0.016 < 0.1 \tag{4}$$

2.4 Weight value

To obtain each factor's weight value, total-level sequencing is figured out after finishing single-level sequencing from target layer to the lowest layer in turn.

Consistency ratio of total-level sequencing:

$$CR(m) = \frac{a_1 CI_1 + a_2 CI_2 + \ldots + a_m CI_m}{a_1 RI_1 + a_2 RI_2 + \ldots + a_m RI_n} \tag{5}$$

$w(t)$ is defined as the weight value of the t th layer, and the results are listed in Table 3.

The lowest layer contains three elements. By using methods of tale look-up, RI can be got with Table 2, and when put it into formula (4), RI and CR can be obtained as follows.

$$\begin{cases} RI = 0.58 \\ CR(5) = 0.00296 < 0.1 \end{cases} \tag{6}$$

Consistency check is passed, and weight vector of the lowest layer is $(0.300, 0.246, 0.454)^T$.

3 RELAY PROTECTION DEVICE LIFE CYCLE BASED ON GRAY MARKOV MODEL

The gray Markov model is adopted to predict the life cycle of relay protection. In addition, the model is proved to be correct with example analysis.

3.1 Determine influencing factors and weight value

Since there are so many factors that can have a negative effect on the equipment life cycle in operation, for the purpose of predicting, more accurately, the main causes which lead to equipment failures including electronic component's faults, software bug and faults of output device are found out, and their weight values are 30%, 24.6%, and 45.4%.

3.2 GM(1,1) prediction model

Original data sequence $X^{(0)}$ can be turned to be AGO sequence $X^{(1)}$.

The corresponding linear differential equation is set up as follows:

$$\frac{dx^{(1)}}{dt} + ax^{(1)} = u \tag{7}$$

Parameters a and u are calculated by using the least squares method.

$$\begin{pmatrix} a \\ u \end{pmatrix} = (BB^T)^{-1}B^T Y \tag{8}$$

Forecasting model of original data:

$$\hat{x}^{(0)}(t) = \hat{x}^{(1)}(t) - \hat{x}^{(1)}(t-1) \tag{9}$$

Then, the residual absolute value sequence can be expressed as $\varepsilon^{(0)}(t) = \left| x^{(0)}(t) - \hat{x}^{(0)}(t) \right|$, $(t = 1,2,\cdots,n)$. Next, put the residual absolute value AGO sequence into GM(1,1) model. The improved forecasting model of original data is shown as:

$$\hat{x}^{(0)}(t+1) = (1-e^a)[x^{(0)}(1) - \frac{u}{a}]e^{at}$$
$$+ \operatorname{sgn}(t+1)(1-e^{a_1})[\varepsilon^{(0)}(1) - \frac{u_1}{a_1}]e^{-a_1 t} \tag{10}$$

3.3 State partition

Fitted value of original data sequence $\hat{y}(t)$ can be figured out on the basis of multivariate gray forecasting model. So as to reduce the error of each state data, we divide the state after ranking the ratio of actual and virtual value in light of increasing sequence.

3.4 Transition probability matrix

According to statistical data of discrete-time state information, transition probability matrix based on Markov chain is as follows:

$$P = \begin{bmatrix} p_{11} & p_{12} & \cdots & p_{1n} \\ p_{21} & p_{22} & \cdots & p_{2n} \\ \cdots & \cdots & \cdots & \cdots \\ p_{n1} & p_{n2} & \cdots & p_{nn} \end{bmatrix} \tag{11}$$

p_{ij} represents the probability from state i to state j, and $p_{ij} \geq 0 \cap \sum_{j=1}^{n} p_{ij} = 1$.

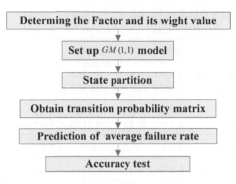

Figure 2. Procedure of this model.

3.5 Prediction of average failure rate

In this paper, the average failure rate is used to measure the relay protection device life cycle. Putting historical data into the program based on Matlab, we can get the data of predicted average failure rate in the coming year.

3.6 Accuracy test

With the purpose of verifying accuracy of this model, there is a function to analyze:

$$E(n) = \frac{\sum_{i=1}^{n} \left| \frac{y_i^{(\blacklozenge\blacklozenge)} - y_i^{(\blacklozenge\blacklozenge)}}{y_i^{(\blacklozenge\blacklozenge)}} \right| \times 100\%}{n} \tag{12}$$

$E(n)$ reflects the curve deviation. If $E(n) < 5\%$, the accuracy test is passed.

In conclusion, model's flow path is shown in Figure 2.

4 EXAMPLE ANALYSIS

As shown in Table 4, it contains past 8-year average failure rate data in a region.

The number in Table 4 means the failure frequency of a set of equipment in the whole year. Fault factors statistical line chart is shown in Figure 3.

In Figure 3, relay protection equipment failure rate caused by software bug has changed little as time continues. However, the failure rate resulted from damage of electronic component and output of the device increased obviously in the past 8 years. After inputting the weight value and parameter, fitted curve of the average failure rate can be obtained as Figure 4 with AHP.

Through this model, it is indicated that the predicted value of the average failure rate in 2016 is 9.279; and $E(8) = 1.3177\% < 5\%$. The accuracy test is passed, which means that the predicted curve is valid.

Table 4. Historical data of the average failure rate.

Year	2008	2009	2010	2011	
Component	0.0371	0.0514	0.1569	0.2986	
Software bug	0.0448	0.0349	0.0547	0.0583	Output
Divicel	0.0422	0.0865	0.3105	0.6372	
Total	0.1241	0.1728	0.5221	0.9941	
Year	2012	2013	2014	2015	
Component	0.3708	0.5292	0.9733	1.4494	
Software bug	0.0667	0.0625	0.0795	0.0868	Output
Divicel	0.7974	1.1724	2.1916	3.2957	
Total	1.2349	1.7641	3.2444	4.8319	

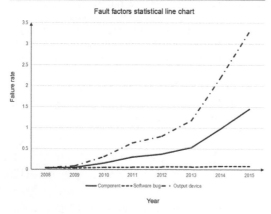

Figure 3. Fault factors statistical line chart.

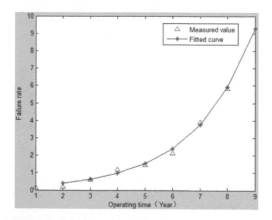

Figure 4. Fitted curve of average failure rate.

5 CONCLUSION

Main factors affecting the life cycle of relay protection device are determined. The weight value of these factors is calculated by using AHP. In addition, the gray Markov model aimed at studying the relay protection device life cycle is established to obtain the fitted curve of the average failure rate and predict the average failure rate in the coming year. We can draw the following conclusions with the example analysis. The following work has been done in this paper.

1. Allowing for there are so many factors associated with each other, which will have an effect on the equipment life cycle, to improve the accuracy of this model, the research takes multiple factors into consideration, and the weight value of each factor is determined with AHP.

2. Through the example analysis of the relay protection device of the power system based on the gray Markov chain model with weight value, the predicted average failure rate of the next year is obtained, which makes it possible to do quantitative research. Meanwhile, the validity and accuracy of the model are verified.

REFERENCES

Chen J. 2005. Cascading dynamics and mitigation assessment in power system disturbances via a hidden failure model(J), Electrical Power and Energy Systems,27(4):318–326.

Ohnishi S, Saito T, Yamanoi T, et al. 2011. A weights representation for absolute measurement AHP using fuzzy sets theory(C). Floriana.

Tan J C. 2001. Intelligent wide area back-up protection and its role in enhancing transmission network reliability(C). Seventh International Conference on Developments in Power System Protection (IEE):446–449.

Xiao X P, Li F Q. 2006.The theoretical study on the stability of the grey forecasting model(J). Dynamics of Continuous Discrete and Implusive Systems—Series B—Applications & Algorithms.13:570–573.

Yu Xingbin, Singh. 2004. Practical approach for integrated power system vulnerability analysis with protection failures(J). IEEE Transactions on Power Systems, 19(4): 1811–1820.

Yingchen W.2008. Research of AHP Model on Combination between the Social Responsibility and the Economic Benefits in Corporation(C). Wuhan.

Energy Science and Applied Technology ESAT 2016 – Fang (Ed.)
© *2016 Taylor & Francis Group, London, ISBN 978-1-138-02973-6*

Research of the maintenance cost allocation of secondary electrical equipment based on nucleolus solution

L. Yang, L.M. Ying, Y.L. Wang, J.W. Wang & Y.T. Jia
School of Electrical Engineering, Wuhan University, Wuhan, China

ABSTRACT: The maintenance cost allocation has been the key and difficult point for a grid company to carry out the life cycle cost accounting and management of secondary equipment. On the basis of proper selection of indexes for cost allocation, a model for maintenance cost allocation of secondary equipment based on nucleolus solution in DEA game is proposed in this paper. The linear programming algorithm and genetic Particle Swarm Optimization (PSO) of nucleolus solution are also put forward according to the number of equipment in the allocation, and thus conducting analysis with relevant examples. The results show that the allocation model is a relative scientific, reasonable, and accurate allocation strategy, and that it has certain reference values for grid company to promote the life cycle cost management of secondary equipment.

Keywords: secondary equipment; cost allocation; nucleolus solution; genetic particle swarm optimization

1 INTRODUCTION

With the further development of China's electric power reform, the economic and social comprehensive benefits are gradually put in the same important position in the power system. Secondary equipment plays a decisive role in the stable operation of power system, and its Life cycle Cost Management (LCM) has become one of the focuses of grid company. In the whole life cycle of secondary equipment, maintenance costs account for a relatively large proportion (approximately 45-90% (Zhen Z.M., 2012)). Unlike primary equipment, the maintenance of secondary equipment is often conducted with station or bay as the unit, and the cost obtained after statistic accounting by actual activity-based costing method is binding. How to allocate the cost to relevant secondary equipment in a scientific, fair and reasonable method has become the key and difficult work for a grid company to implement the cost accounting of maintenance of secondary equipment under LCM.

The original value of equipment is generally used as an indicator to allocate the maintenance cost, which cannot fully reflect the consumption of labor, material and per machine-team in the actual maintenance of equipment and it becomes less accurate under the differentiated maintenance mode widely carried out by the grid company, which takes equipment status and importance as basics. Therefore, scientific allocation indicators must be multi-dimensional to reflect the consumption of labor, material and per machine-team, and the differentiated maintenance strategy.

DEA game is a multi-index allocation strategy combining Data Envelopment Analysis (DEA) with the game theoretical model (Nakabayashi K., 2006), and nucleolus solution is a widely used calculation method with reasonable apportion result. Based on the nucleolus solution of DEA game, this paper conducts allocation of the binding maintenance cost of the secondary equipment, and verifies the feasibility and rationality of the method by related examples.

2 THE BASIC THEORY OF MAINTENANCE COST ALLOCATION

2.1 *Acquisition of maintenance cost of secondary equipment*

To improve economic benefits and realize the scientific and effective management of activity cost, power enterprises generally introduce ABC (activity-based cost) (Li L. 2012), which completes the maintenance cost accounting based on the concept of "resources-activity-cost object". Maintenance of secondary equipment is abundant in category, easy to be affected by external environment and different in task difficulty, so the actual cost method is suitable for the acquisition of its maintenance costs. The actual cost method is a method where the system automatically calculates the activity cost according to the consumption of labor, per machine-team and materials recorded in the operation sheets in actual maintenance operations and the standard rate of labor, materials, and per machine-team when the operation is completed.

Most of the maintenance items of primary equipment are regulated by more perfect quotas, and often conducted with independent equipment as the unit, which makes the maintenance cost unbinding. The maintenance of secondary equipment is different from that of primary equipment with station or bay as the unit, besides it is also difficult to achieve respective statistics on labor, material, and per machine-team consumption of maintenance of all secondary equipment actually, so the result obtained by the actual cost method is often binding.

2.2 Selection of cost allocation index

According to the cost driver theory of ABC, we need to follow the principle of causality, gained benefits, rationality, and cost effectiveness when choosing the cost allocation index (GU J. et al. 2006). Currently, the grid company generally takes the differentiated maintenance strategy based on equipment status and importance (Wang Q.W. et al. 2013). For equipment running well with less importance, there is no need to spend too much effort and cost, thus saving resources and ensuring the safe operation of equipment; otherwise, attention should be focused on its maintenance. Therefore, the cost allocation indexes selected in this paper are as follows:

1. Labor intensity
   Labor intensity reflects human resource consumption in the maintenance of secondary equipment. Not only the quantity of maintenance personnel but also the time needed is to be considered.
2. Material consumption
   Material consumption reflects materials consumed in secondary equipment maintenance, which considers not only the type of the materials but also the quantity and price of materials consumed.
3. Mechanical consumption
   Mechanical consumption mainly reflects the amount of per machine-team in the maintenance, considering not only the type and price of per machine-team but also time of application.
4. Running state
   The running state is mainly obtained through the state evaluation of secondary equipment carried out by the power company, and its running state is generally divided into normal state, attention state, abnormal state, and serious state. The state evaluation generally considers the situations before operation, history condition, maintenance condition, real-time running status, and equipment operation environment (Li Y.L. 2008). The state parameters are selected through classification

to obtain the state score of secondary equipment, thus determining the running state.
5. Importance of equipment
   The quantization of equipment importance is similar to the state evaluation process, generally considering the voltage level and the impact degree on primary equipment and users.

3 THE COST ALLOCATION METHOD OF SECONDARY EQUIPMENT MAINTENANCE BASED ON NUCLEOLUS SOLUTION

3.1 Nucleolus theory of DEA game

DEA game is a theory integrating data envelopment analysis with the game theory model to solve the cost allocation problem with multiple indicators. Its basic model is as follows:

Provided coalition S is a subset of the players set, and its input and output are, respectively, recorded as:

$$x_i(S) = \sum_{j \in S} x_{ij}, y_i(S) = \sum_{j \in S} y_{ij} \tag{1}$$

In game theory, all participants and coalitions are assumed selfish so that they take the strategy minimizing allocated costs. The allocation characteristic value $C(S)$ of coalition S can be expressed as the following linear programming problem:

$$C(S) = \min_{u_r, v_i} \left[\sum_{r=1}^{s} u_r y_r(S) - \sum_{i=1}^{m} v_i x_i(S) \right] \tag{2}$$

$$s.t. \sum_{j=1}^{n} R_j = R$$

$$R_j = \sum_{r=1}^{s} u_r y_r(S) - \sum_{i=1}^{m} v_i x_i(S) \geq 0$$

$$u_r, v_i \geq 0, \forall r, j$$

The constraint conditions of model (2) are the scheme set of cost allocation, and the methods to solve and analyze game problems include bargaining set, stable set, core, nucleolus, and the Shapley value, among which the nucleolus solution, has one and only one, is an optimal cost allocation strategy meeting individual and overall rationality.

Define the excess value of coalition S on allocation x as $\varepsilon(S, x) = \sum_{j \in S} R_j - C(S), \forall S \subset N, 1 \leq |S| < n$, among which $|S|$ is the number of coalition S. The value of $\varepsilon(S, x)$ reflects satisfaction of coalition S about x, and the greater the value, the more unsatisfied with the allocation strategy. All the excess values are listed in descending order and expressed as vector $\theta(x)$:

$$\theta(x) = \begin{pmatrix} \theta_1(x) \\ \theta_2(x) \\ \vdots \\ \theta_{2^{n-1}}(x) \end{pmatrix} = \begin{pmatrix} e(S_1, x) \\ e(S_2, x) \\ \vdots \\ e(S_{2^{n-1}} x) \end{pmatrix} \quad (3)$$

where $S_1, S_2, ... S_{2^{n-1}}$ is an array of all coalitions of N and satisfies:

$$e(S_1, x) \ge e(S_2, x) \ge ... \ge e(S_{2^{n-1}}, x) \quad (4)$$

For $\theta(x) = (\theta_1(x), \theta_2(x), ... \theta_{2^{n-1}}(x))$

$\theta(y) = (\theta_1(y), \theta_2(y), ... \theta_{2^{n-1}}(y))$, if there exists $1 \le h \le 2^{n-1}$ to make

$$\theta_i(x) = \theta_i(y) \quad (i = 1, 2... h-1) \quad (5)$$

$$\theta_h(x) < \theta_h(y) \quad (6)$$

We can get vector $\theta(x)$ smaller than vector $\theta(y)$ in lexicographical order, written as $\theta(x) \overset{L}{<} \theta(y)$. Nucleolus is the allocation entirety making $\theta(x)$ the smallest in lexicographical order; that is:

$$\left\{ x \in E(v) \middle| \forall y \in E(v), \theta(x) \overset{L}{<} \theta(y) \right\} \quad (7)$$

3.2 Solution algorithm for nucleolus allocation scheme

The solution method of nucleolus in DEA game is: according to the order of dissatisfaction degree, let the most dissatisfied coalition select allocation first to minimize its dissatisfaction degree; and then consider the second unsatisfactory coalitions; cycle this process until the dissatisfaction degree of all coalitions is minimized. Therefore, two kinds of solving methods are given.

1. Linear programming solution
When the number of secondary equipment in the maintenance cost allocation is small, the nucleolus solution can be transformed into the following formula:

$Min\ \alpha$

$s.t. \sum_{j \in S} R_j - C(S) \le \alpha,\ 1 \le |S| < n$

$$R_j = \sum_{r=1}^{s} u_r y_r(S) - \sum_{i=1}^{m} v_i x_i(S) \ge 0 \quad (8)$$

$\sum_{j=1}^{n} R_j = R$

$u_r, v_i, R_j \ge 0,\ \forall r, i, j$

where α is an arbitrary real number and S is all the non-empty sets of N.

2. GA-PSO optimization algorithm
When the number of equipment in the cost allocation is relatively large, the complexity of linear programming solution is $O(4^n)$, which means spending much time on solution. As an efficient global optimization technology combining GA with PSO, GA-PSO can have it well solved. However, we should notice that the solution of the GA-PSO is only an approximate solution, not the optimal.

GA is a global optimization search algorithm based on the theory of natural selection and genetics. It first initializes the population to randomly generate a certain scale of the chromosome X, and then calculate the individual fitness to conduct selection, crossover, and mutation according to selection operator T_s, crossover operator T_c, and mutation operator T_m; finally, determine whether the termination conditions are satisfied according to cyclic algebra or iterative precision, thus outputting the optimal solution.

PSO is a kind of global optimization technique based on swarm intelligence proposed by Kennedy and Eberhart after inspiration by birds' predation. There is a location parameter x_i and a velocity parameter v_i reflecting the position changes for every particle in this algorithm. The particle moves in the space and iterates the velocity and position according to the individual optimal solution $Pbest$ and group optimal solution $Pgest$; its expression is:

$$v_{id}^{(k+1)} = \omega v_{id}^{(k)} + c_1 r_1^{(k)} \left(Pbest_{id}^{(k)} - x_{id}^{(k)} \right) + c_2 r_2^{(k)} \left(Pgbest_{id}^{(k)} - x_{id}^{(k)} \right) \quad (9)$$

$$x_{id}^{(k+1)} = x_{id}^{(k)} + v_{id}^{(k+1)} \quad (10)$$

where $i = 1, 2... n$ and n is the total number of particles; k denotes times of the current iteration; $x_{id}^{(k)}$ represents the location of d-dimensional component in the iteration of particle i; c_1, c_2 are the acceleration coefficients, indicating the maximum step of flying particles; $r_1^{(k)}$ and $r_2^{(k)}$ are random numbers between 0 and 1; and ω is the inertia weight.

PSO, with its memory function, can realize internal information sharing and ensure the population moves toward the optimal solution. Compared with GA, PSO is simple and faster in the rate of convergence, but the convergence efficiency is not high in the later period of calculation and becomes easily trapped in local extremum; GA improves the diversity of solution through evolution patterns of crossover and mutation. Therefore, we can consider combining the two methods, retaining the strong global search ability of GA and integrating the memory and position shifts of PSO so as to improve the accuracy and efficiency of the algorithm. The general steps of GA-PSO algorithm are shown in Figure 1.

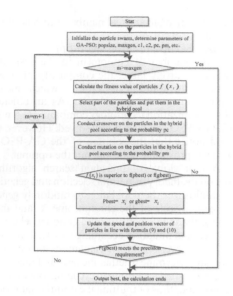

Figure 1. General steps of GA-PSO algorithm.

According to the basic concept of nucleolus solution, it can be transformed into the optimization problem of multivariate function:

$$\begin{cases} \min_{x}\left\{\omega_1\theta_1(x)+\omega_2\theta_2(x)+...+\omega_{2^{n-1}}\theta_{2^{n-1}}(x)\right\} \\ x_1+x_2+...+x_{2^{n-1}}=C \end{cases} \quad (11)$$

And satisfies
$$\begin{cases} \theta_1(x)>\theta_2(x)>...>\theta_{2^{n-1}}(x) \\ \omega_1>\omega_2>...>\omega_{2^{n-1}} \end{cases} \quad (12)$$

where ω_i is the weight coefficient. Generally, for optimization problems of multidimensional continuous function, the convergence of model (11) with constraint conditions cannot be guaranteed when using GA-PSO algorithm. Therefore, it is modified as an unconstrained model through adding penalty terms to the objective function. The penalty term v_p is designed as Figure 1.

$$v_p(x)=\left|\omega_0\left(C-\sum_{i=1}^{n}x_i\right)\right| \quad (13)$$

To ensure the complete allocation of the total cost by participants, the weight coefficient of the penalty term is required as $\omega_0>>\omega_1$, and the final fitness function $F(x)$ can be expressed as:

$$F(x)=-\sum_{i=1}^{2^{n-1}}\omega_i\theta_i(x)-v_p(x) \quad (14)$$

The corresponding $F(x)$ of the cost allocation scheme selected by nucleolus solution is the largest.

3.3 The application of nucleolus solution in cost allocation of secondary equipment maintenance

Under the LCM concept, the manufacturers of secondary equipment are unwilling to be allocated more cost, avoiding higher cost accounting of products. Therefore, the independent secondary equipment and its coalition can be seen as selfish. When allocating the cost with the nucleolus in DEA game, the cost allocation indexes are listed as i, $i = 1,2,3...m$, and the score x_{ij} (indicating the score of the jth secondary equipment under the ith indicator, $j = 1,2,3...n$) is determined. If x_{ij} is greater, the equipment j is allocated more cost under the index i. Form a matrix X with all the index scores as $X=\left\{x_{ij}\right\}_{m\times n}$, and row standard processing on it, then $\sum_{j=1}^{n}x_{ij}=1(i=1,2,...,m)$.

The methods of determining the index score of all equipment are mainly analytic hierarchy process, the entropy weight method, the Delphi method, survey and statistics, weighted statistical method, and so on. In the above five parameters, labor intensity, material consumption, mechanical consumption, and equipment importance are positive indexes (i.e. the larger the number, the more the cost allocated), and the running state is an inverse indicator. To conduct unified evaluation on cost allocation of secondary equipment maintenance, the inverse index needs to be converted into a positive index. The formula is:

$$x_{ij}^*=\frac{1}{x_{ij}} \quad i=1,2...n; j=1,2,...m \quad (15)$$

$X_i(S)$ represents the score of coalition S under index i. The formula is:

$$x_i(S)=\sum_{j\in S}x_{ij}(1,2,...,m) \quad (16)$$

$C(S)$ is the characteristic function of coalition $S(C(\theta)=0)$, indicating the minimum maintenance cost of it. It can be expressed with linear programming:

$$C(S)=\min\sum_{i=1}^{m}w_ix_i(S) \quad (17)$$

s.t.

$$\begin{cases} \sum_{i=1}^{m}w_i=1 \\ w_i\geq 0 \quad (i=1,2,...m) \end{cases}$$

where ω_i is the weight value of the ith index. The nucleolus method of DEA game in 3.2 is used to get the optimal cost allocation of relevant secondary equipment and its coalition.

4 EXAMPLE ANALYSIS

4.1 *Example 1: Maintenance cost allocation with less number of equipment*

The secondary equipment maintenance team of substation carries out a maintenance operation on a bay, involving three secondary equipment a, b, c ($N = 3$). After the operation, the maintenance cost obtained by the actual cost method according to the operation sheets totals 5800 yuan.

First, score the five cost allocation indexes of secondary equipment by AHP, and conduct standardized treatment of the matrix X (see Table 1).

The characteristic function value $C(S)$ of different coalitions can be obtained through formula (17) (see Table 2):

Substitute the characteristic function values in Table 3 into equation (5), and z_a, z_b, and z_c are respectively the allocated cost ratio of secondary equipment a, b, and c.

$$\min \varepsilon st. \begin{cases} z_a - \varepsilon \leq 0.2 \\ z_b - \varepsilon \leq 0.15 \\ z_c - \varepsilon \leq 0.05 \\ z_a + z_b - \varepsilon \leq 0.7 \\ z_a + z_c - \varepsilon \leq 0.5 \\ z_b + z_c - \varepsilon \leq 0.2 \\ z_a + z_b + z_c = 1 \end{cases}$$

The solutions obtained using Matlab 7.0 are $z_a = 0.5$, $z_b = 0.3179$, and $z_c = 0.1821$, and the corresponding allocated maintenance cost of secondary equipment a, b, and c are 2900 yuan, 1843.82 yuan, and 1056.18 yuan.

2. GA-PSO optimization algorithm

The fitness function of secondary equipment maintenance cost allocation using GA-PSO optimization algorithm is as shown in formula (14). This paper determines $\omega_0 = 10000$, $\omega_i = 1/i$, $i = 1, 2, ... 7$, and according to the process of GA-PSO, we get

Table 1. The evaluation score of each cost driver.

	a	b	c
Labor intensity	0.5	0.4	0.1
Material consumption	0.7	0.2	0.1
Mechanical consumption	0.8	0.15	0.05
Running status	0.2	0.5	0.3
Equipment importance	0.4	0.3	0.3

Table 2. Coalition mode and characteristic function value.

Coalition mode		Characteristic function
Independent	a	0.2
	b	0.15
	c	0.05
Coalition of two equipment	ab	0.7
	ac	0.5
	bc	0,2
Coalition of three equipment	abc	1

Table 3. Evaluation score of each cost driver.

	Labor intensity	Material consumption	Machinery consumption	Runing state	Equipment importance
a	0.23	0.31	0.30	0.11	0.36
b	0.31	0.26	0.33	0.26	0.27
c	0.16	0.18	0.15	0.15	0.18
d	0.16	0.18	0.15	0.32	0.14
e	0.14	0.07	0.07	0.16	0.05

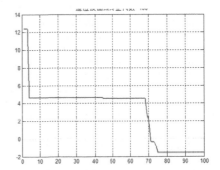

Figure 2. Fitness curve (example 1).

the cost ratio of a, b, and c: $z_a = 0.5283$, $z_b = 0.3290$, $z_c = 0.1428$. The fitness curve is as shown in Figure 2.

The allocated maintenance costs of a, b, and c are, respectively, 3064.14, 1908.2, and 828.24 yuan.

Thus, there is no big difference in the cost allocation scheme of the nucleolus solution obtained by GA-PSO optimization algorithm and linear programming algorithm. It can be used to allocate maintenance cost with a large number of equipment.

4.2 *Example two: Maintenance cost allocation with large number of equipment*

When the number of secondary equipment involved in the maintenance cost allocation is relatively large, the linear programming solution is rather complicated and time-consuming, while GA-PSO optimization algorithm has more advantages for this.

The secondary equipment maintenance team of substation carries out a maintenance operation on an bay, involving five equipment ($n = 5$) of a 220 kV line protection and control device (a), a transformer protection and control device (b), two sets of 110 kV line protection and control devices (c and d) and a 110 kV buscouple protection and control device (e). After the operation, the maintenance cost obtained with actual cost method totals 8600 yuan. The relevant index score of cost allocation is as shown in Table 3.

According to the fitness function (14) of secondary equipment maintenance cost allocation and the process of GA-PSO, the parameters are set as $\omega_0 = 10000$, $\omega_i = 1/i$, $i = 1, 2, \ldots 2^n - 1$, to obtain the corresponding nucleolus solution allocation scheme and the allocated cost of all equipment (see Table 4).

The fitness curve is shown in Figure 3:

Table 4 shows that compared with other three equipment involved in the maintenance, the related index values of a and b are relatively high, correspondingly bearing more cost. The cost allocation index values of e are relatively low, bearing less cost. Although the two sets of 110 kV line protection and control devices c and d have the same value in labor intensity, material consumption, and mechanical consumption index, the device d operates worse than device c, taking more effort and cost for maintenance. According to the qualitative analysis, the allocated maintenance costs of relevant secondary equipment are as follows in the descending order: b, a, d, c, and e. Therefore, the cost allocation scheme of the secondary equipment

maintenance based on nucleolus solution of DEA game by GA-PSO method has certain rationality.

5 CONCLUSIONS

Based on the reasonable selection of cost allocation indexes of maintenance, a cost allocation model of secondary equipment maintenance based on nucleolus solution in DEA game is proposed in this paper. According to the number of devices in the allocation, proposing the linear programming algorithm and GA-PSO of nucleolus solution, the latter of which integrates the crossover and mutation of GA and memory and position transfer concept of PSO, being more suitable for maintenance cost allocation with a large number of equipment with a better convergence rate and accuracy. The example analysis shows that the model is a scientific, reasonable, and accurate allocation strategy, providing useful technical support for power companies to promote LCM of secondary equipment.

REFERENCES

Biljana S. & Vlandimir P. 2004. AHP method application in evaluation of plant genetic resources conservation strategy. Neural, Paraller & Scientific Computations 12(1):53–72.

Gu J. et al. 2006. The derivation of cost allocation and its defects under the cost driver theory. Coal Economic Research (6):56–57.

Holland J.H. 1992. Adaptation in natural and artificial systems: an introductory analysis with applications to biology, control, and artificial intelligence (2nd ed). Cambridge: MIT Press.

Kennedy J. & Eberhart R.C. 1997. A discrete binary version of the particle swarmoptimization algorithm. International Conference on Systems, Man, and Cybernetics. New York, USA. IEEE Service Center(5): 4104–4108.

Li L. 2012. The application of activity-based method in the cost accounting system of power engineering enterprises. Baoding: North China Electric Power University.

Li Y.L. 2008. Research on fixed cost allocation based on DEA theory. Hefei: University of Science and Technology of China.

Nakabayashi K. & Tone K. 2006. Egoist's dilemma: a DEA game. Omega 34(2): 135–148.

Qiu Y.H. 2002. Management decision and entropy application. Beijing: China Machine Press: 101–127.

Wang Q.W. et al. 2013. Condition based maintenance of secondary equipment based on analytic hierarchy model. China Southern Power Grid 7(4): 97–102.

Wu J. & Wang Y. 2013. The design of risk quantification evaluation system of electric secondary equipment. China Power 46(1):75–79.

Zhen Z.M. & Niu S.L. 2012. The application of LCC management in secondary equipment of intelligent substation. Comprehensive Research (33):150–151.

Table 4. Allocation results of the nucleolus solution in DEA.

Equipment	a	b	c	d	e
Allocation ratio	0.2436	0.2639	0.1765	0.2008	0.1142
Allocated cost (yuan)	2094.96	2269.54	1517.9	1726.88	982.12

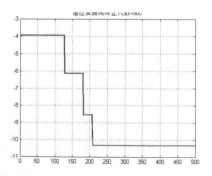

Figure 3. Fitness curve (example 2).

Energy Science and Applied Technology ESAT 2016 – Fang (Ed.)
© *2016 Taylor & Francis Group, London, ISBN 978-1-138-02973-6*

Integrated custom-tailored model for RF power amplifiers

Leyu Zhai
Xi'an Communications Institution, Xi'an, China
School of Electronic Science and Engineering, National University of Defense Technology, Changsha, China

Yi He & Qiang Liu
Xi'an Communications Institution, Xi'an, China

ABSTRACT: An Integrated Custom-Tailored (ICT) model for RF Power Amplifiers (PAs) is proposed in this paper. Referring to the Vector-Switched (VS) model, the ICT model is the result by implementing the Recursive Optimum-term Selecting (ROS) approach in every sub-region. This is an offline system identification process. The ICT models vary with the PAs from individual to individual, and also are different with the PAs' working status, so they are private. Experiments and simulations show that the ICT models characterize the PAs more effectively than those common ones.

Keywords: behavioral model; Digital Predistorter (DPD); Power Amplifiers (PAS); Integrated Customer-Tailored (ICT) model

1 INTRODUCTION

BEHAVIORAL model is a general way to characterize nonlinear systems. When operating near the saturation point, Power Amplifiers (PAs) behave nonlinearly with memory effects. In order to linearize PAs, Digital Pre-Distortion (D-PD) technique is usually implemented. On this occasion, the inverse behavioral model of PAs is critical, since it directly affects the linearization. A right behavioral model could make linearization easy to actualize, but a bad one will lead to a result which is not satisfactory. A Recursive Optimum-term Selecting (ROS) approach (L. Zhai et al, 2014) is implemented to develop a Custom-Tailored (CT) model. The CT model could be used as an effective inverse behavioral model of PAs, because it selects out the dominant terms from the general Volterra series. When considering more complex PAs, a Suboptimum Custom-Tailored (SCT) model is proposed (L. Zhai et al, 2014). The SCT model is based on the pruned Volterra series by compensating accuracy, so it is still available in the conditions of high-order nonlinearity and long-term memory effects.

To further improve the accuracy of behavioral model, a Vector-Switched (VS) model (S. Afsardoost et al, 2012) classifies the input space into a number of regions and implements different models in every sub-region. Considering that different behavioral models will lead to wastage of the computer resource, such as the Logical Cell (LC) in Field-Programmable Gate Arrays (FPGA), an

Integrated Customer-Tailored (ICT) model is presented in this paper. By using the ROS approach, the terms in every sub-region can be sorted according to the extent of the dominance. Then, an ICT model can be reconstructed with some rules. Since the ICT model is developed by an offline identification process with the measurement data, its terms vary with the individual PA and operating status. Therefore, the ICT model is private.

This paper is organized as follows. In Section II, we introduce a procedure for constructing the ICT model. Then, we evaluate the model by experiment and simulation in Section III. At last, we draw a conclusion in Section IV.

2 THE PROPOSED MODEL

Referring to the Vector-Switched (VS) model, the input space of PAs could be partitioned into several sub-regions. Each sub-region is independent relatively. Then, the ROS algorithm would be implemented in every sub-region. Based on the base model, such as the general Volterra series, a regressor matrix is also constructed in every sub-region. Here, for the convenience of application, the ROS algorithm (L. Zhai et al, 2014) is amended. First, at the seventh step of the ROS algorithm, the remainder terms are updated by signal decomposition of qr method to cancel the influence of the selected optimum term. Second, the terms selecting orders by implementing the ROS approach are recorded.

Therefore, we have a whole and clear knowledge about the terms' dominance. The results of sorting may be different in every sub-region, because the terms act differently on the condition of the working point. Subsequently, we can identify the terms in total by a rule. An ICT model would be obtained by selecting the front terms according to the result of identification. When implementing the ICT model, each region has its private parameters, which are extracted in every sub-region. In summary, the ICT model construction procedure can be written as follows:

Procedure: The ICT model construct process

1. Select a base model.
2. Build a regressor matrix Φ_x based on the selected model.
3. Divide the input space into K sub-regions, correspondingly, the matrix Φ_x is partitioned into K sub-matrixes.
4. Implement ROS process in every sub-region, recording the terms' selecting order.
5. Employing some rules, sort the terms of the base model.
6. Construct an ICT model based on the result of the sort.

For example, given that the general Volterra is set as the base model, the regressor matrix Φ_x of the base model is similar to that in L. Zhai et al. (2014). The terms in the matrix Φ_x are all odd-order products and the number of $\tilde{x}$ is more than that of its complex conjugate $\tilde{x}^*$ by exactly one (D. R. Morgan et al, 2006); this makes the

output of the PAs near the fundament of the carrier frequency. Given that the input space is partitioned into K sub-regions, which are denoted by (1, 2, ..., k, ..., K), the regressor matrix Φ_x would be written as (1), where $()^{(k)}$ indicates that the element belongs to the kth sub-region. As the number of the sub-region is smaller than that of the input data, it is normal that some elements are denoted by the same sub-region index. Here, the first and the second rows of the regressor matrix are denoted by '(1)', which means that they belong to the same sub-region. Clustering the element with the same sub-region index into a new matrix, a sub-matrix of the sub-region can be denoted as $\Phi_x^{(k)}$. Based on this sub-matrix, we can implement the amended ROS procedure in every sub-region.

The procedure of the ICT model construction is listed in Table 1. When setting nonlinearity order P = 5 and setting memory length M = 1, there are 20 terms, which are indexed from 1 to 20. By implementing the amended ROS algorithm, the sorted result of terms in every sub-region is listed in the column from 'Region1' to 'Region4'. Comparing with the common models, such as the Memory Polynomial (MP) model (L. Ding et al, 2004) and the Generalized Memory Polynomial (GMP) model (D. R. Morgan et al, 2006), we find that the sorted results are accordant with them. The terms with the marker ① belong to the MP model, and the terms with the marker ② belong to the GMP model. In general, those terms selected by the common models are sorted in the front of the list. This means that those terms are dominant

Table 1. The ICT model (M = 1, P = 5).

Index	The Volterra Terms	Region 1	Region 2	Region 3	Region 4	Weight	Sort	MP	GMP(G=1)	ICT
1	$\tilde{x}_{n-0}$	1	1	1	1	4	1	①	②	✓
2	$\tilde{x}_{n-1}$	2	3	3	14	22	3	①	②	✓
3	$\tilde{x}_{n-0}\tilde{x}_{n-0}\tilde{x}^*_{n-0}$	13	2	2	4	21	2	①	②	✓
4	$\tilde{x}_{n-0}\tilde{x}_{n-0}\tilde{x}^*_{n-1}$	17	12	10	18	57	17			
5	$\tilde{x}_{n-0}\tilde{x}_{n-1}\tilde{x}^*_{n-0}$	20	11	18	20	69	20		②	
6	$\tilde{x}_{n-0}\tilde{x}_{n-1}\tilde{x}^*_{n-1}$	18	9	17	8	52	14		②	
7	$\tilde{x}_{n-1}\tilde{x}_{n-1}\tilde{x}^*_{n-0}$	19	6	12	19	56	15			
8	$\tilde{x}_{n-1}\tilde{x}_{n-1}\tilde{x}^*_{n-1}$	10	19	16	15	60	19	①	②	
9	$\tilde{x}_{n-0}\tilde{x}_{n-0}\tilde{x}_{n-0}\tilde{x}^*_{n-0}\tilde{x}^*_{n-0}$	14	10	5	2	31	5	①	②	✓
10	$\tilde{x}_{n-0}\tilde{x}_{n-0}\tilde{x}_{n-0}\tilde{x}^*_{n-0}\tilde{x}^*_{n-1}$	5	13	6	5	29	4			✓
11	$\tilde{x}_{n-0}\tilde{x}_{n-0}\tilde{x}_{n-0}\tilde{x}^*_{n-1}\tilde{x}^*_{n-1}$	16	5	11	10	42	9			
12	$\tilde{x}_{n-0}\tilde{x}_{n-0}\tilde{x}_{n-1}\tilde{x}^*_{n-0}\tilde{x}^*_{n-0}$	12	4	8	12	36	7		②	
13	$\tilde{x}_{n-0}\tilde{x}_{n-0}\tilde{x}_{n-1}\tilde{x}^*_{n-0}\tilde{x}^*_{n-1}$	3	17	13	11	44	11			
14	$\tilde{x}_{n-0}\tilde{x}_{n-0}\tilde{x}_{n-1}\tilde{x}^*_{n-1}\tilde{x}^*_{n-1}$	4	18	7	16	45	12			
15	$\tilde{x}_{n-0}\tilde{x}_{n-1}\tilde{x}_{n-1}\tilde{x}^*_{n-0}\tilde{x}^*_{n-0}$	7	8	14	7	36	8			
16	$\tilde{x}_{n-0}\tilde{x}_{n-1}\tilde{x}_{n-1}\tilde{x}^*_{n-0}\tilde{x}^*_{n-1}$	6	14	4	9	33	6			✓
17	$\tilde{x}_{n-0}\tilde{x}_{n-1}\tilde{x}_{n-1}\tilde{x}^*_{n-1}\tilde{x}^*_{n-1}$	8	15	20	13	56	16		②	
18	$\tilde{x}_{n-1}\tilde{x}_{n-1}\tilde{x}_{n-1}\tilde{x}^*_{n-0}\tilde{x}^*_{n-0}$	9	16	15	3	43	10			
19	$\tilde{x}_{n-1}\tilde{x}_{n-1}\tilde{x}_{n-1}\tilde{x}^*_{n-0}\tilde{x}^*_{n-1}$	15	7	19	17	58	18			
20	$\tilde{x}_{n-1}\tilde{x}_{n-1}\tilde{x}_{n-1}\tilde{x}^*_{n-1}\tilde{x}^*_{n-1}$	11	20	9	6	46	13	①	②	

and effective. On the other hand, the conformance between the sorted result and the common models proves that the amended ROS algorithm is sensible.

In order to reconstruct an efficient ICT model, it is necessary to take the sorted results in every sub-region into account. In this study, for the sake of simplification, we add up the sorted number of terms and make the result as its weight, which is listed in the column 'Weight'. Then, the terms are sorted again according to the weight, and the result is listed in the column 'Sort'. Therefore, the degree of the terms' dominance is clear, and the terms in the front are more effective than the latter in the sequence. Here, we select six terms as an ICT model to compare with the MP model. The terms of the ICT model are denoted by √ and listed in the column 'ICT'. It is clear that most of the terms are similar to that of the MP model, but there are also some different terms. It will be illustrated that

3 MODEL EVALUATION

A test bench was set up to evaluate the proposed model, which is similar to that in L. Zhai et al. (2014) (L. Zhai et al, 2014). The Device Under Test (DUT) was a solid wideband PA employed in the fourth-generation (4G) base station. The baseband I/Q source data was generated in MATLAB and downloaded into a Vector Signal Generator (VSG), where the baseband I/Q signal was modulated at the carrier centered at a frequency of 2 GHz. In succession, the RF signal was amplified by the DUT and filtered by a vacuum filter, then went into an attenuator (ATT), which weakened the amplified signal by 30 dB. A Digital Oscillograph (DSO) sampled and digitalized the analog RF signal. At last, the data was uploaded into PC via Ethernet for analysis by a Vector Signal Analysis (VSA) software.

$$
\Phi_x = \begin{pmatrix}
(\tilde{x}_1)^{(1)} & \cdots & (\tilde{x}_1 - M)^{(1)} & (\tilde{x}_1 \tilde{x}_1 \tilde{x}_1^*)^{(1)} & \cdots & (\tilde{x}_1 - M \cdots \tilde{x}_1 - M \tilde{x}_{1-M}^* \cdots \tilde{x}_{1-M}^*)^{(1)} \\
(\tilde{x}_2)^{(1)} & \cdots & (\tilde{x}_2 - M)^{(1)} & (\tilde{x}_2 \tilde{x}_2 \tilde{x}_2^*)^{(1)} & \cdots & (\tilde{x}_2 - M \cdots \tilde{x}_2 - M \tilde{x}_{2-M}^* \cdots \tilde{x}_{2-M}^*)^{(1)} \\
\vdots & \vdots & \vdots & \vdots & \vdots & \vdots \\
(\tilde{x}_n)^{(k)} & \cdots & (\tilde{x}_n - M)^{(k)} & (\tilde{x}_n \tilde{x}_n \tilde{x}_n^*)^{(k)} & \cdots & (\tilde{x}_n - M \cdots \tilde{x}_n - M \tilde{x}_{n-M}^* \cdots \tilde{x}_{n-M}^*)^{(k)} \\
\vdots & \vdots & \vdots & \vdots & \vdots & \vdots \\
(\tilde{x}_N)^{(K)} & \cdots & (\tilde{x}_N - M)^{(K)} & (\tilde{x}_N \tilde{x}_N \tilde{x}_N^*)^{(K)} & \cdots & (\tilde{x}_N - M \cdots \tilde{x}_N - M \tilde{x}_{N-M}^* \cdots \tilde{x}_{N-M}^*)^{(K)}
\end{pmatrix} \tag{1}
$$

the different terms are more efficient than those in the MP model according to the experiments and simulations in section III.

After reconstructing the ICT model, the parameters of every sub-region can be calculated. Define the output vector $Y = (\tilde{y}1, \tilde{y}2, ..., \tilde{y}n, ..., \tilde{y}N)^T$, and define model parameters vector $B = (\tilde{b}1, \tilde{b}2, ..., \tilde{b}s, ..., \tilde{b}S)^T$, where $\tilde{b}s$ represents the sth parameter and S is the total number of the parameters, then each sub-region has a matrix equation

$$Y^{(k)} = \Phi_x^{(k)} B^{(k)} \tag{2}$$

Using the LS method, the linear matrix equation (2) can be solved into the form of

$$B^{(k)} = (\Phi_x^{(k)})^\dagger Y^{(k)} \tag{3}$$

where $(\Phi_x^{(k)})$ is the Moore-Penrose pseudo-inverse of $\Phi_x^{(k)}$ with $()^H$ indicating the Hermitian transpose.

$$(\Phi_x^{(k)})^\dagger = ((\Phi_x^{(k)}))^H ((\Phi_x^{(k)}))^{-1} (\Phi_x^{(k)})^H \tag{4}$$

In this study, a 20M-bandwidth Orthogonal Frequency-Division Multiple-Access (OFDM) signal was adopted as the probing signal, which was a Long-Term Evolution (LTE) like signal with 9.5-dB Peak-to-Average Ratio (PAR). A total of 8,192 baseband samples were captured by DSO. Of these, 4,096 samples were used for model extraction after alignment, and the remainder of the data was used for validation separately by simulation. Normalized Mean Square Error (NMSE) was used to evaluate the performance of the proposed model, which represents the difference between the real signal and the expected signal in the time domain. In order to evaluate the performance of the behavioral model when employed into DPD application, the inverse model is considered in this paper. The normalized output data is worked as the input of the behavioral model, and the normalized input data is worked as the output. As a consequence, when setting the nonlinearity order P = 5 and varying the memory delay, the NMSEs were plotted in Fig. 1.

As shown in Fig. 1, there are four curves denoting different models' NMSEs. It is evident that the VS-ICT model performs well than others, as its

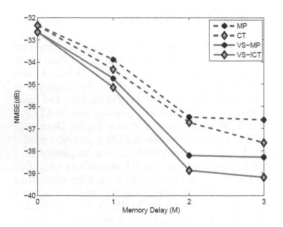

Figure 1. Effectiveness Comparison in terms of NMSE.

Table 2. The NMSE comparison (M = 3, P = 5).

Model	MP	CT	VS-MP	VS-ICT
NMSE	−36.6	−37.7	−38.3	−39.2

NMSE is lower than the others. Especially, according to the data in Table 2, when M = 3, the VS-ICT model has the advantage of about 0.9-dB than the VS-MP model in terms of NMSE, and has an advantage of about 1.5-dB/2.6-dB than the CT/MP model. It is distinct that the proposed model is effective.

4 CONCLUSION

This paper proposes an ICT model for RF PAs' linearization. The ICT model is efficient when actualized in FPGA or DSP, so it is suitable to be implemented in the digital pre-distortion application as inverse model. Experiments and simulations show that the ICT model is more effective than those common ones.

REFERENCES

Afsardoost, S., T. Eriksson, and C. Fager, "Digital Predistortion Using a Vector-Switched Model" *IEEE Trans. Microwave Theory Tech.*, vol. 60, pp. 1166–1174, Apr. 2012.

Ding, L., G.T. Zhou, D.R. Morgan, Z. Ma, J.S. Kenney, J. Kim, and C.R. Giardina, "A robust digital baseband predistorter constructed using memory polynomials, *IEEE Trans. Commun.*, vol. 52, pp. 159–165, Jan. 2004.

Morgan, D. R., Z. Ma, J. Kim, M.G. Zierdt, and J. Pastalan, "A Generalized Memory Polynomial Model for Digital Predistortion of RF Power Amplifiers," *IEEE Trans. Signal Processing.*, vol. 54, pp. 3852–3860, Oct. 2006.

Zhai, L., H. Zhai, Z. Zhou, E. Zhang, and K. Gao, "A novel approach to pruning the general Volterra series for modeling power amplifiers, *IEICE Electron. Express.* vol. 11, pp. 20131030, Feb. 2014.

Zhai, L., H. Zhai, Z. Zhou, E. Zhang, and R. Zhang, "Suboptimum custom-tailored model based on the pruned Volterra series for power amplifiers, *IEICE Electron. Express.* vol. 11, pp. 20140693, Sep. 2014.

Energy Science and Applied Technology ESAT 2016 – Fang (Ed.)
© 2016 Taylor & Francis Group, London, ISBN 978-1-138-02973-6

Electronic and magnetic properties of Ni$_n$Al (n = 2–13) clusters

Wei Song & Bin Wang
Department of Physics and Electronic Engineering, Xinxiang University, Xinxiang, China

Kai Guo
The Shale Oil Plant Fushun Mining Group Co. Ltd., Fushun, China

Chaozheng He
Physics and Electronic Engineering College, Nanyang Normal University, Nanyang, China

ABSTRACT: Using the density functional theory calculations with the PBE exchange-correlation energy function, we have studied the electronic properties including binding energy, magnetic property, ionization potential, and electron affinity for adsorbing an aluminum atom into Ni$_n$ (n = 2–13) neutral and ionic clusters.

Keywords: first-principle; magnetic properties; ionization potentials; electron affinities

1 INTRODUCTION

The clusters of atoms or molecules are referred to as clusters; they are made up of several or even thousands of atoms, molecules, or ions through physical or chemical binding force, to form relatively stable microscopic or submicroscopic aggregates. With further research of clusters, metal clusters research has paid more and more attention. They are used widely in many material movement processes and phenomena such as catalysis, combustion, crystal growth, nucleation and solidification, phase change, sol, film formation, sputtering, and so on. Currently, more and more attention has been paid to the study of the physical and chemical properties of bimetallic Transition Metal (TM) clusters. Of particular interest are the binary clusters, which may vary with the composition and atomic ordering, as well as the size of the clusters. Nickel clusters are largely concerned because of their extensive catalytic and important magnetic properties. By adsorption, different elements make nickel clusters show different characteristics. They show superiority in catalysis and magnetic aspects. The double metal clusters are more stable than single nickel clusters. The practical application of nickel clusters constrained severely because of the expensive of nickel clusters, but the nickel-based alloy could alleviate this problem largely. Therefore, Ni-alloy clusters research has paid more and more attention. With further research of transition-metal alloy clusters, the bimetallic systems of many metals such as Al (Z.W. Ma et al, 2015) (H.Q. Sun et al, 2013) (C. Woodward et al, 2014), Fe (T. Mohri et al, 2015), Rh (J. Teeriniemi et al, 2015), Pd (J.X. Zhu et al, 2015), Ti (N.S. Venkataramanan et al, 2010), Cu (J.A. Mary et al, 2014), Mn (B.R. Wang et al, 2014), or others adsorbed Ni clusters, as has been reported. Among many Ni-alloy clusters, nickel-aluminum alloy clusters are the primary target because of their wide applications in advanced material technology.

Owing to the complicated electronic ground-state structures and delocalized d electrons, the studies on Ni-Al alloy clusters are limited. Therefore, in this study, we have investigated adsorbing aluminum to the Ni$_n$ (n = 2–13) clusters theoretically by applying the DFT method implemented in the VASP code with the spin polarized PBE (Perdew, Burke, and Ernzerhof) gradient-corrected exchange-correlation functional method. Based on the neutral and ionic cluster structures, the binding energy, magnetic property, ionization potential, and electron affinity of the clusters, as the cluster size increases, have been systematically analyzed.

2 COMPUTATIONAL METHODS

Density functional theory is a quantum mechanical method, which is used to study the multi-electron system of the electronic structure. It has wide applications in physics and chemistry, in particular, it is used to study the properties of molecules and condensed matter and computational chemistry. With the rapid development of computer technology in recent years, the density functional theory has been widely used in materials, emerging interdisciplinary life, and laid the foundation for the manufacture of new functional materials. Therefore, our study is based on DFT using the VASP code with the Perdew, Burke, and Ernzerhof (PBE) gradient-corrected exchange-correlation energy functional. Parameters for the calculation of 269.5eV for cutoff energy of Plane Wave (PW) and 10^{-5}eV are converged to an accuracy till the energy. The simple cubic supercell is used 20Å.

3 RESULTS AND DISCUSSION

3.1 Structures of the Ni_nAl and $Ni_nAl^{+/-}$ clusters

At the DFT-PBE level, the lowest energy of the neutral Ni_nAl (n = 2–13) clusters are displayed in Fig. 1. The stable structures of $Ni_{2-5}Al$ are triangle, tetrahedron, tetragonal pyramid, and square bipyramid (or octahedron), respectively. For Ni_{7-9} structures, the growth pattern is mainly based on the mechanism of capping extra atoms on the Ni_6 octahedron (W. Song et al, 2011). In addition, the $Ni_{6-9}Al$ structures are formed by capping one atom to the corresponding Ni_{6-9} structure. The structures of $Ni_{10}Al$ and $Ni_{11}Al$ are found to be an Al atom located at the hollow triangle site of the Ni_{10} tetrahedron and Ni_{11} tetracapped pentagonal bipyramid, respectively. The ground-state structure of $Ni_{12}Al$ is a perfect icosahedron. Moreover, the $Ni_{13}Al$ is viewed as attaching to an Al atom at the hollow triangle site of the icosahedral structure. The ionic structures similar to the neutral structures are not repeated in Fig. 1.

3.2 Relative stabilities

Binding energy indicated that several particles from the free state combined into a composite particle to release energy. The greater the binding energy, the more stable the structure of clusters. Therefore, we have analyzed the binding energy (E_b) for Ni_nAl compared with Ni_n including neutral and ionic clusters. The calculated results are shown in Fig. 2(a–c). The E_b of Ni_n, $Ni_n^{+/-}$ and Ni_nAl, $Ni_{n-1}Al^{+/-}$

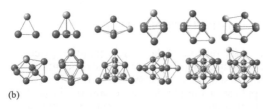

(b)

Figure 1. Lowest-energy structures of the Ni_nAl (n = 2 – 13) clusters calculated.

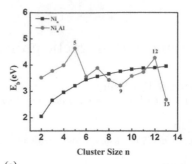

(a)

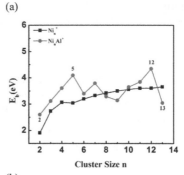

(b)

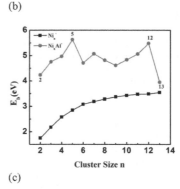

(c)

Figure 2. Average binding energy of Ni_n and Ni_nAl (n = 2 – 13) for (a) neutral, (b) cationic, and (c) anionic clusters.

clusters are calculated according to the following definition:

$$E_b(Ni_n) = [E_{total}(Ni_n) - nE(Ni)]/n;$$

$$E_b(Ni_nAl) = E_{total}(Ni_n) + E(Al) - E_{total}(Ni_nAl)$$

$$E_b(Ni_n^{+/-}) = [(n-1)E(Ni) + E(Ni^{+/-}) - E_{total}(Ni_n^{+/-})]/n;$$

$$E_b(Ni_nAl^{+/-}) = E_{total}(Ni_n) + E(Al^{+/-}) - E_{total}(Ni_nAl^{+/-})$$

in which $E_{total}(Ni_n)$, $E_{total}(Ni_n^{+/-})$ and $E_{total}(Ni_nAl)$, $E_{total}(Ni_nAl^{+/-})$ are the total energy of the Ni_n, $Ni_n^{+/-}$ and Ni_nAl, $Ni_nAl^{+/-}$ clusters, E (Ni) and Ni (Al) are the energy of a free Ni and Al atom; and E ($Ni^{+/-}$) and

E ($Al^{+/-}$) are the energy of a free $Ni^{+/-}$ and $Al^{+/-}$ ion, respectively. It is found from observing Fig. 2 that the E_b of Ni_n and $Ni_n^{+/-}$ clusters increases monotonically with the increase in cluster scale. However, the curve of binding energy for Ni_nAl and $Ni_{n-1}Al^{+/-}$ exhibits an oscillating behavior, and there are several peaks on the binding energy curve at n = 5 and 12, showing that these cluster ions are relatively more stable. In addition, the binding energy of anionic clusters is more than that of neutral and cationic clusters.

3.3 Magnetic properties

The magnetic material is not only widespread, but also varied, which has been widely studied and applied. Near from our body and the surrounding material, as far as the various stars and interstellar matter, the microscopic world of atoms, atomic nuclei and elementary particles, various materials of macroscopic world, have a kind of magnetism. The magnetic material can be divided into diamagnetism, paramagnetism, ferromagnetism, antiferromagnetism, and ferrimagnetism. Nickel clusters are typically ferromagnetic transition-metal clusters, which have large magnetic moment. As shown in Fig. 3(a–c), the total magnetic moments obtained for Ni_nAl and $Ni_nAl^{1/}$ clusters from our calculation have the same trend as

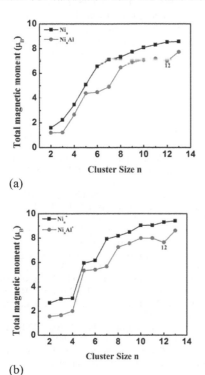

(a)

(b)

Figure 3. (*Continued*)

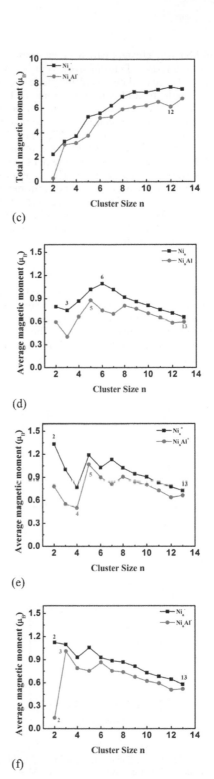

(c)

(d)

(e)

(f)

Figure 3. The total and average magnetic moment (in units of μ_B) of Ni_n and Ni_nAl (n = 2 – 13) clusters for (a, d) neutral, (b, e) cationic, and (c, f) anionic clusters.

Ni_n and $Ni_nAl^{+/-}$ clusters, which increases monotonically as a function of cluster size except $Ni_{12}Al$ and $Ni_{12}Al^{+/-}$. The calculated total magnetic moments decrease with the addition of Al atom. To further understand the impact of cluster size on the magnetic moment, we can compare the average magnetic moment per atom of Ni_n, $Ni_n^{+/-}$ and Ni_nAl, $Ni_nAl^{+/-}$ clusters in Fig. 3(d–f). As a result, it is found that the average magnetic moment per atom exhibits an oscillating behavior. The local maxima and minima of the average magnetic moment per atom exhibit at n = 6, 3 (Ni_n), n = 2, 13 ($Ni_n^{+/-}$) and n = 5, 13 (Ni_nAl), n = 5, 4 (Ni_nAl^+), n = 3, 2 (Ni_nAl^-). The results obtained by the present study help us choose magnetic materials.

3.4 *Ionization potential and electron affinity*

The Adiabatic Ionization Potential (AIP) is the ground state of the gaseous atoms or gaseous ions that lose an electron from a neutral cluster requiring the minimum ionization energy. The Adiabatic Electron Affinity (AEA) is the difference of the total energies between the neutral and anionic clusters. The AIP and AEA of Ni_n and Ni_nAl (n = 2–13) clusters are presented in Fig. 4. The AIP and AEA are the total energy difference of neutral and ionic clusters, which can be defined by

$$AIP\,(Ni_n) = E_T\,(Ni_n^+) - E_T\,(Ni_n);\ AIP\,(Ni_nAl) = E_T\,(Ni_nAl^+) - E_T\,(Ni_nAl)$$

$$AEA\,(Ni_n) = E_T\,(Ni_n) - E_T\,(Ni_n^-);\ AEA\,(Ni_nAl) = E_T\,(Ni_nAl) - E_T\,(Ni_nAl^-)$$

where E_T is the total energy. The Adiabatic Ionization Potentials (AIP) and Adiabatic Electron Affinities (AEA) of Ni_n and Ni_nAl clusters with n = 2–13 are investigated with the DFT-PBE calculations. It can be seen that the overall trend curve of AIP decreases and AEA increases with the increase in cluster scale.

4 CONCLUSION

The electronic properties of Ni_nAl (n = 2–13) clusters were located by first-principle calculations at the DFT-PBE level. The binding energy of Ni_n and $Ni_n^{+/-}$ clusters increases monotonically with the increase in cluster scale. However, the curve of binding energy exhibits an oscillating behavior for Ni_nAl and $Ni_nAl^{+/-}$ clusters. From the magnetic moment analyses, it is found that the addition of the Al atom can decrease the magnetic moments of Ni_n clusters. Finally, the overall trend curve of AIP decreases and AEA increases as the cluster size increases.

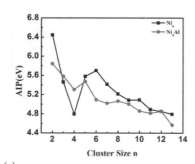

(a)

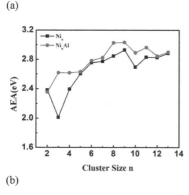

(b)

Figure 4.   (a) The Adiabatic Ionization Potential (AIP) and (b) the Adiabatic Electron Affinitiy (AEA) of Ni_n and Ni_nAl (n = 2 – 13) clusters.

REFERENCES

Ma, Z.W., B.X. Li, Comput. Theor. Chem., vol. 1068 (2015) p. 88–96.
Mary, J.A., A. Manikandan, L.J. Kennedy, M. Bououdina, R. Sundaram, J.J. Vijaya, T. Nonferr. Metal. Soc., vol. 24 (2014) p. 1467–1473.
Mohri, T., J. Mater. Sci., vol. 50 (2015) p. 7705–7712.
Song, W., W.C. Lu, C.Z. Wang, K.M. Ho, Int. J. of Quant. Chem., vol. 978 (2011) p.41–46.
Sun, H.Q., Y.J. Bai, H.Y. Cheng, B.L. Wang, G.H. Wang, J. Mol. Struct., vol. 1031 (2013) p. 22–29.
Teeriniemi, J., J. Huisman, P. Taskinen, K. Laasonen, J. Alloy. Compd., vol. 652 (2015) p. 371–378.
Venkataramanan, N.S., R. Sahara, H. Mizuseki, Y. Kawazoe, J. Phys. Chem. A, vol. 114 (2010) p. 5049–5057.
Wang, B.R., H.Y. Han, Z. Xie, J. Mol. Struct., vol. 1062 (2014) p. 174–178.
Woodward, C., A. van de Walle, M. Asta, D.R. Trinkle, Acta Materialia, vol. 75 (2014) p. 60–70
Zhu, J.X., P. Cheng, N. Wang, S.P. Huang, Comput. Theor. Chem., vol. 1071 (2015) p. 9–17.

6 Mechanical, manufacturing, process engineering, fluid engineering and numerical simulation

Energy Science and Applied Technology ESAT 2016 – Fang (Ed.)
© 2016 Taylor & Francis Group, London, ISBN 978-1-138-02973-6

Research of the cloud manufacturing resources based on the supply sub-chain

Quping Qiu, Jianfei Tu & Lingling Yi
School of Mechanical Engineering and Mechanics, Ningbo University, Zhejiang, China

ABSTRACT: Because of the network and multi-agent features of cloud manufacturing resources, it is necessary to define and classify the cloud manufacturing resources based on supply chain. Through the features and composition of the cloud sub-chain manufacturing resources, this thesis puts forward the definition and classification of the cloud sub-chain, and finally gives an instance to demonstrate the cloud sub-chain.

Keywords: cloud manufacturing; supply sub-chain; cloud sub-chain manufacturing resources

1 INTRODUCTION

With the rise of cloud computing, a newly service-oriented networked manufacturing mode—cloud manufacturing based on the service philosophy of cloud computing has been given great concern gradually. How to organize multiple and complex manufacturing resources rationally and how to optimize the resource allocation and achieve resource sharing under the cloud manufacturing environment have become the important problems confronted by its current advanced manufacturing mode development.

The previous scholars have conducted extensive researches on the heterogeneous platform's manufacturing resource integration, optimal configuration, integrated service mode and virtualization etc. For example, in order to integrate a multiple manufacturing resource heterogeneous platforms, Fan Wenhui et al put forward an integrated scheme concerning the cloud manufacturing integration framework and execution supporting environment based on federation mode; Huang Gang et al put forward a unique cloud manufacturing service framework, whose core is to separate the basic data with the application cloud. Aiming at the design collaboration of complex product, He Dongjing et al designed a method for distributed collaboration and resource sharing among designers at different design links in the complex product design process; Zhan Dechen et al explored the concepts of cloud manufacturing service oriented by the enterprise cloud manufacturing service platform and achieved service innovation in the manufacturing and management process based on resource integration, demand-supply integration and optimal configuration. Yi Sheng et al proposed and built an integrated service mode that can promote the rapid sharing and high-efficient application of cooperative manufacturing resources under the cloud manufacturing environment with wide integration and distributed service.

However, under the networked and cloud manufacturing environment, the manufacturing resources have been distributed at different physical positions and heterogeneous platforms. The cloud manufacturing resource is characterized by multilayer, multi-agent and network chain structure. Those manufacturing resources are related and interacted closely. When a manufacturing resource is affected by the market with fluctuation, the other related manufacturing resources will also be affected accordingly. This incidence relation among manufacturing resources is reflected on the supply-demand relationship in enterprises, whose nature lies in the supply chain relationship.

Therefore, to explore and analyze the cloud manufacturing resource from the perspective of supply chain has important practical significance in reflecting the cloud manufacturing resource's nature and characteristics. This paper introduced the concept of supply sub-chain to analyze the manufacturing resource in the cloud manufacturing environment aiming to provide a new method for the cloud manufacturing resource analysis from the perspective of supply chain.

2 DEFINITION AND CHARACTERISTICS OF CLOUD SUB-CHAIN MANUFACTURING RESOURCE

2.1 *Definition of cloud sub-chain manufacturing resource*

Some supply and demand relationship is existed among cloud manufacturing resources in the supply chain environment, which is a reflection of supply chain relationship among real enterprises in the cloud manufacturing. In order to define the supply chain relationship in the cloud manufacturing resource, this paper introduced the cloud sub-chain manufacturing resources. The cloud sub-chain manufacturing resource is based on the cloud manufacturing environment of supply sub-chain, which refers to several chain cloud manufacturing resources integrated by node cloud manufacturing resources and existed in some certain business connection or relations. The cloud sub-chain manufacturing resource is presented as a whole with certain manufacturing capacity externally.

The cloud manufacturing pool based on supply sub-chain, see Figure 1. Two different types of manufacturing resources are existed in the pool. Namely, the simple cloud manufacturing resource with single nodes and the cloud sub-chain manufacturing resource existed via sub-chain. Compared with the simple cloud manufacturing resource of single enterprise, the cloud sub-chain manufacturing resource with wide coverage can better reflect the practical capacity of component manufacturing and raw material supply, which can help to consider the influence of manufacturing resource fluctuation in the cloud production plan comprehensively so as to provide a more effective manufacturing resource integration method applicable to the manufacturing environment.

2.2 *Characteristics of cloud sub-chain manufacturing resource*

1. Network chain possibility

The manufacturing resource is mutually related and interacted with the manufacturing capacity under the cloud manufacturing environment. When a manufacturing resource is affected by the market or society with certain fluctuation, the other related manufacturing resource and manufacturing capacity will be affected accordingly. Meanwhile, various cloud sub-chains consist of a whole supply chain. Therefore, the cloud sub-chain manufacturing resource is characterized by the network chain possibility.

2. Hierarchy

The cloud sub-chain is composed by various nodes as well as its neighboring child node enterprise with supply-demand relationship; meanwhile, the nodes of cloud sub-chain also have certain matching or subcontracting relationship with the nodes at the upstream of cloud sub-chain, which then consists of

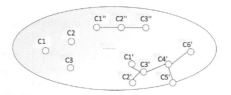

Figure 1. The cloud manufacturing platform based on the supply chain.

another sub-chain. This reflects the hierarchy of cloud sub-chain manufacturing resource longitudinally.

3. Independence and recombination

When some node enterprise quits from the cloud sub-chain for some reason, this cloud sub-chain can find new manufacturing resource or manufacturing capacity on the cloud manufacturing resource platform to replace the outgoing node enterprise. This reflects certain independence and recombination of cloud sub-chain manufacturing resource.

3 CLASSIFICATION OF CLOUD SUB-CHAIN MANUFACTURING RESOURCE

The cloud sub-chain manufacturing resource can be classified from various perspectives according to the form structure and application mode etc.

3.1 *Classification according to the form structure of cloud sub-chain*

The cloud sub-chain can be divided into simple cloud sub-chain and complex sub-chain in the cloud manufacturing environment according to the structure. The simple cloud sub-chain means that the main body only has one root node, which is a member of cloud manufacturing service platform. That is to say, the whole cloud sub-chain manufacturing resource is existed in the cloud manufacturing service platform via a single manufacturing resource for user's search, acquisition and application. The complex cloud sub-chain means that the main body has several root nodes, which are the manufacturing resource of the cloud manufacturing service platform, as shown in Figure 2.

The manufacturing resources provided by the cloud sub-chain include simple single cloud sub-chain manufacturing resource. As shown in Figure 2, the main body is S_1. The manufacturing capacity required by S_1 can only be provided by S_2 and S_3. Meanwhile, the manufacturing resource provided by complex could sub-chain include complex cloud sub-chain manufacturing resource. As showed in Figure 3, the main body is S_2 and S_3. The manufacturing capacity required by S_2 can not be met by only relying on S_{21} and S_{22}. S_2 still needs S_{22} to provide manufacturing capacity service.

Obviously, the manufacturing resource of simple cloud sub-chain is rather simple, but the complex sub-chain reflects the practical complicated production manufacturing process.

3.2 *Classification according to the longitudinal relation of cloud sub-chain*

There existed two relations between two neighboring nodes longitudinally on the cloud sub-chain: matching and subcontracting. According to those two relations, the cloud sub-chain can be divided into matching cloud sub-chain and subcontracting cloud sub-chain.

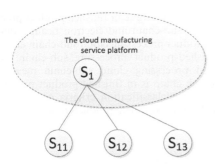

Figure 2.   The simple cloud sub-chain.

The matching cloud sub-chain, as shown in Figure 4; A_1 is the cloud node; A_{12} and A_{13} are child nodes; the child node A_{13} is the manufacturing resource or manufacturing service provider of A_{12}. From the perspective of material, A_{13} is the raw material or semi-finished product supplier of A_{12}.

The subcontracting sub-chain, as shown in Figure 5; the subcontracting means the cloud node enterprise cannot complete the task independently and has to subcontract partial production task to other node enterprises. For example, S2 and S3 in the sub-chain belong to cooperative relation, namely subcontracting, which is different from the above mentioned matching. The subcontract-

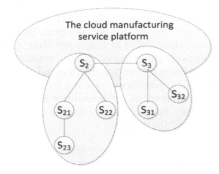

Figure 3.   The complex cloud sub-chain.

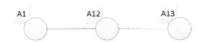

Figure 4.   The matching cloud sub-chain.

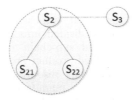

Figure 5.   The subcontracting cloud sub-chain.

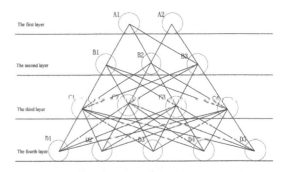

Figure 6.   The layer of cloud sub-chain.

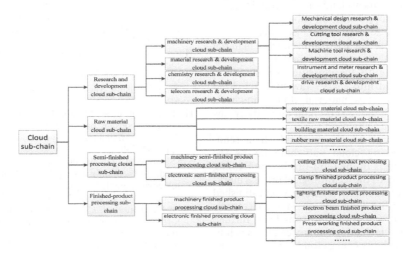

Figure 7.   The classification instance of cloud sub-chain.

ing cloud sub-chain is often a complex cloud sub-chain.

3.3 Classification according to the layer of cloud sub-chain

According to the layer of cloud sub-chain, as shown in Figure 6, it can be divided into as follows:

The first layer, finished product processing cloud sub-chain;
The second layer, semi-finished product processing cloud sub-chain;
The third layer, raw material cloud sub-chain;
The fourth layer, research and development cloud sub-chain.

Therefore, the cloud sub-chain manufacturing resource can be divided into finished product processing cloud sub-chain manufacturing resource, semi-finished product processing cloud sub-chain manufacturing resource, raw material supply cloud sub-chain manufacturing resource and research & development cloud sub-chain manufacturing resource.

4 EXAMPLES OF CLOUD SUB-CHAIN MANUFACTURING RESOURCE

According to the layer of cloud sub-chain, this paper conducted detailed classification on the cloud sub-chain manufacturing resource based on the supply sub-chain, as shown in Figure 7.

Research and development cloud sub-chain: research and development cloud sub-chain refers to the activities under the cloud manufacturing environment such as new product design, stimulation and testing ctc, which can further be divided into machinery research & development cloud sub-chain, chemistry research & development cloud sub-chain and telecom research & development cloud sub-chain according to the research & development subjects etc.

Raw material cloud sub-chain: the raw material cloud sub-chain refers to the material procurement. According to the classification standard, it can be divided into energy raw material cloud sub-chain, textile raw material cloud sub-chain, building material cloud sub-chain and rubber raw material cloud sub-chain etc.

Semi-finished processing cloud sub-chain: semi-finished processing cloud sub-chain means the cloud sub-chain is in the semi-finished processing layer. According to the technological process marking, it can be divided into machinery semi-finished product processing cloud sub-chain and electronic semi-finished processing cloud sub-chain; further, according to the technological process, the machinery semi-finished processing cloud sub-chain can be divided into cutting semi-finished product processing cloud sub-chain and electron beam semi-finished product processing cloud sub-chain etc.

Finished-product processing sub-chain: semi-finished processing cloud sub-chain means the cloud sub-chain is in finished-product processing layer, whose classification criteria are consistent with the semi-finished product processing cloud sub-chain.

5 SUMMARY

The cloud manufacturing is a new network manufacturing mode. The key to achieve its real "practice" is to integrate the complex manufacturing resources. In order to fully reflect the relations among cloud manufacturing resources, this paper introduced the concept of supply sub-chain to integrate the node cloud manufacturing resource with certain business connection into a cloud sub-chain with certain manufacturing capacity, which can avoid the influence of single resource on the whole cloud supply chain effectively so as to improve the robustness of the system.

ACKNOWLEDGEMENTS

This work has been supported by the Star Program of the State Ministry of Science and Technology (2014GA701032) and Zhejiang Provincial Natural Science Foundation (Y14G010012), and also sponsored by K.C. Wong Magna Fund in Ningbo University.

REFERENCES

Fan Wen-hui, et al. Integrated architecture of cloud manufacturing based on federation mode. *Computer Integrated Manufacturing Systems*. Vol. 17 (2011) No. 3, p. 469–476.

Huang Gang, et al. Manufacturing resources cloud locating services platform based on separated data cloud and application cloud. *Computer Integrated Manufacturing Systems*. Vol. 17 (2011) No. 3, p. 519–524.

He Dong-jing, et al. Method for complex product collaborative design based on cloud service. *Computer Integrated Manufacturing Systems*. Vol. 17 (2011) No. 3, p. 533–539.

Yin Sheng, et al. Outsourcing resources integration service mode and semantic description in cloud manufacturing environment. *Computer Integrated Manufacturing Systems*. Vol. 17 (2011) No. 3, p. 525–532.

Zhan De-chen et al. Cloud manufacturing service platform for group enterprises oriented to manufacturing and management. *Computer Integrated Manufacturing Systems*. Vol. 17 (2011) No. 3, p. 487–494.

Energy Science and Applied Technology ESAT 2016 – Fang (Ed.)
© *2016 Taylor & Francis Group, London, ISBN 978-1-138-02973-6*

A method of part lightweight based on filling lattice structure

E.C. Liu, L.Z. Wan, G.X. Li, J.Z. Gong & B.Z. Wu
College of Mechatronic Engineer and Automation, National University of Defense Technology, Changsha, China

ABSTRACT: Filling with lattice structure is an effective lightweight method, which can obtain larger lightweight ratio, meanwhile guaranteeing the strength, stiffness, and other related properties. A lightweight method presented in this paper is oriented to fill lattice structure and solve the problems of mechanical properties assessment and simulation analysis. By experimental analysis of the mechanical properties and failure modes of the lattice structure, lattice structure with better mechanical properties is obtained and the homogenization equivalent model of the lattice structure is established. In the overall simulation analysis of the part, the lattice structure is filled into the part in entity with its equivalent performance to reduce the difficulty and workload of the simulation analysis. The method can guide lattice structure filling process of parts and reduce the costs of time.

Keywords: lattice structure; mechanical properties; homogenization equivalent model

1 INTRODUCTION

The design of lattice structure conforms to the idea of coordinate optimization design, which integrates design materials, structural design, and functional design. Lattice structure is a porous ordered microstructure, simulating the configuration of molecules lattice which consists of nodes and connecting rod unit between nodes. The meso configurations of this structure are two- and three-dimensional grid systems with no fillers in the gap of the grid to carry. The structure has the properties of ultra-light, high specific stiffness, high specific strength, impact resistance and so on. It also has the properties of heat distribution, heat isolation, sound absorption, electromagnetic absorption, and multi-functional integration in functionality. Using a lattice structure instead of solid structure in the interior of the part, the weight of the part can be lost obviously, but ensuring the overall performance.

It has become a hot research of academic and industrial sectors gradually.

In recent years, studies of the mechanical properties of the lattice structure mainly focus on a specific lattice structure: Dede introduced a technology about designing a single-layer or layers periodic lattice structure, and analyzed the mechanical properties of the single-layer lattice structure. Zhang Qiancheng analyzed the mechanical properties of the lightweight lattice structure according to its cell structure, and the method to enhance the mechanical properties of the lattice structure. Chen Liming established the strength model and stiffness model of the cylindrical shell of lattice structure, through the study of the mechanical properties of lightweight lattice sandwich. Tekoglu studied the relationship between variation on the cell size of the lattice structure and its mechanical properties, based on the theoretic and simulated analysis of compression, bending and shear conditions. Fan put forward the theoretical model of mechanical properties of lightweight lattice structure and experiment it.

For now, however, we do not have a complete and effective solution to the problem of analysis and evaluation of mechanical properties after filling lattice structure. Additionally, the existing simulation software is difficult to analyze and evaluate the part filling lattice structure and requires a lot of experiments to verify whether the part filling lattice structure meets the requirements, which is a waste of time and money.

Regarding those problems, this paper researches the problem of lattice structure filling of the connecting rod, by three typical lattice structures: tetrahedral lattice structure, pyramid lattice structure, and 3D kagome lattice structure.

The study of the mechanical properties and failure modes of these three lattice structures arranged in different ways, and the mechanical properties and failure modes of lattice structures fill in the connecting rod in different ways, and guide the selection of the type, arrangement mode, and filling mode of the lattice structure. Finally, we can solve the problem of lattice structure filling of the connecting rod, and then solve the problem of lattice structure filling of more parts.

2 OVERALL TECHNICAL ROUTE

The basic problem to be solved in this paper is how to achieve the requiring lightweight, high rigidity, and impact resistance properties of the connecting rod. The overall technical route is shown in Fig. 1.

As shown in Fig. 1, optimization of the connecting rod's topology and analysis of stress are first processed to determine the filling position, the density, and the stress form of the lattice structure. Then, the lattice structure is made into a standard specimen, to determine whether the lattice structure meets the requirements, meanwhile to obtain the homogenization equivalent model of the lattice structure. Finally, the lattice structure is filled into rod, to determine whether the filling results meet the requirement.

3 OPTIMIZATION ANALYSIS OF THE PART

We need to determine the load type and working conditions of the part before topology optimization. Connecting rod appears frequently in the mechanical products, and it usually bears uniformly distributed tensile load and compressive load at its two ends. However, it does not bear the torsional load. This paper treats the connecting rod as a reference part in filling the lattice structure, and considers only the optimized result under tensile load in order to simplify the analysis process. In this paper, we topologically optimize the part, to find the optimal material distribution in the given design space. We use US Altair's HyperWorks for finite element mesh and structural optimization. HyperMesh is used as CAE pre-processing tools, and OptiStruct as the structural optimization solver.

First, we mesh the connecting rod by tetrahedral mesh with the scale of 0.4 mm which obtains finite element analysis structure containing 143,488 nodes. The connecting rod's load restraint and finite element structure are shown in Fig. 2.

Data initialization is followed after meshing which includes definition of the material properties, operating conditions of the part, optimization problems and optimization variables.

We use the method of equivalent static load to build the optimization model. The defined attributes and its values are shown in Table 1:

After the definition, if the setting conditions are correct, we can obtain the result which is shown in Fig. 3.

As shown in Fig. 3, the relative density of junction is largest because of the maximum stress. Meanwhile, the relative density of its upper and lower surfaces is relatively large. However, the relative density of the center is relatively small, so the center of the connecting rod has the space of optimization. Meanwhile, the relative density of the material has small changes in the Y-axis direction, due to the symmetry of the connecting rod's load.

Figure 2. Connecting rod's load restraint and finite element structure.

Table 1. Defined attributes and its values of optimization.

Attributes	Definition	Remark
Material	AL6061	Nu = 0.3; E = 2.1e+05
Load	Equivalent stretched uniform load at both ends of hinge hole	F = 1000 N
Constraint	Both ends of the hinge hole, limiting the displacement of three directions	dof1 = dof2 = dof3 = 0
Topology design variables	Add minimum size control	Mindim = 0.05 mm
Response	Response of strain energy (target) Response of volume fraction (constraint)	
Reference target	Range of strain energy	(−1.0,1.0)
The constraint of volume fraction	Definite the upper border	0.5

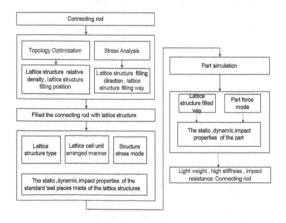

Figure 1. Overall technical route.

Figure 3. Relative density after optimization.

Figure 4. Relative density is less than the threshold value.

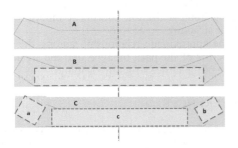

Figure 5. Two programs.

In summary, the junction, upper surface, and the lower surface of the connecting rod are set to the non-design area in which we will not fill the lattice structure in the next step. Dividing the part is based on the lower density of the non-design area according to the grading and the relative density value 0.60514497. The units less than the value are shown in Fig. 4.

We use a cuboid to split the part into three-dimensional structures. For the rest of the structures, the problem can be degraded to a two-dimensional problem, then we can use a rectangle to split the part. We proposed two segmentation schemes as shown in Fig. 5:

Scheme C has a larger fitting area compared with Scheme B; therefore, Scheme C is used.

4 STUDY ON THE MECHANICAL PROPERTIES AND FAILURE MODES OF THE LATTICE STRUCTURE.

The mechanical properties and failure modes of the lattice structure produced by 3D printing technology are unknown. The existing researches on mechanical properties and failure modes of the lattice structure are mostly for traditional manufacturing methods. Therefore, we need to study the mechanical properties and failure modes of 3D printing manufacturing lattice structure. We can choose a number of types and arranged manner of lattice structure to meet the requirements by researching the mechanical properties and failure modes of the lattice structure; we can also obtain the homogenization equivalent model of the lattice structure.

I Choose arranged manner of the lattice structure with better performance: the three lattice structures are made to standard samples according to different arrangement manners. Tests include tensile test, compression test, buckling test, and impact test, according to the working conditions of the connecting rod. One or several arranging modes of the lattice structure that meet the requirements are chosen to do more analysis according to the test results.

II Obtain the homogenized equivalent model: Lattice structures used in this paper are uniform lattice structures, which have a periodic arrangement. Therefore, we can analyze the mechanical properties by theoretical methods. According to the stress map of the part after topology optimization, stress in the boundary of the filling lattice structure position is known, that is the border force of the filling lattice structure is known. The mechanical properties and failure modes of the material can be obtained after the simulation analysis in which

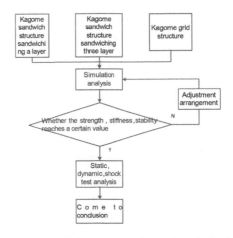

Figure 6. Simulation and experimental analysis about the arranging mode of lattice structure.

the force of the border is known. We can get the equivalent mechanical properties and homogenization equivalent model of the lattice structure by the results of theoretical analysis, simulation, and experimental analysis. Kagome lattice structure is an example in Fig. 6.

5 SIMULATION OF THE LATTICE STRUCTURE FILLED PARTS

Simulation analysis of lattice structure filled part is difficult, due to the difference in the lattice structure meshing, and the computation is huge after meshing. To solve this problem, the lattice structure is treated as an entity material filled into connecting rod in the simulation, according to their equivalent mechanical properties. When setting material parameters, material parameters of the lattice structure are the same as the material parameters of the homogeneous entity material. The contact between the two materials is surface contact.

The simulation can be divided into three parts: establish the finite element model of the connecting rod; mesh the model; add loads, restraints, and other conditions according to the actual working conditions; and simulate the model.

6 DESTRUCTIVE TEST OF THE CONNECTING ROD

According to the simulation results, the selected lattice structure is filled into the connecting rod and manufactured by the method of 3D printing. Then,

tests are progressed to analyze whether the lattice structure meets the requirements according the results. If the selected lattice structure meets requirements, the optimization model of the connecting rod filling lattice structure is obtained. If not, we should adjust the arrangement manner and filling patterns of the lattice structure until the results can meet the requirements. The process is shown in Fig. 7.

7 CONCLUSION

We take the connecting rod as an example to research the lattice structure filling method of part and get the optimization model of the connecting rod filled with a lattice structure, which can reduce the weight of the connecting rod and meet the requirements of strength, stiffness, and stability at the same time. It solves some problems about the process of the lattice structure filling. The method presented in the paper helps to achieve the following objectives:

The amount of the tests during the process of the lattice structure filling is reduced and the time and money is saved.

The CAE software can be used to analyze and evaluate the mechanical properties of the part filled with lattice structure by the method of establishing homogenization equivalent model of the lattice structure.

The approximate relationship among the final properties of the part, the type, and the arranging mode of the lattice structures is found.

REFERENCES

Chen Liming, Dai Zheng, Fan Hualin, et al. Design and Analysis of Light-weight Lattice Sandwich Cylinder [J]. j. Tsinghua Univ. (Sci & Tech.), 2012, 52(4): 489–493.

Dede E.M., Hulbert G.M. Computational Analysis and Design of Lattice Structures with Integral Compliant Mechanisms [J]. Finite Elem. Anal. Des., 2008, 44: 819–830.

Fan H.L., Jin F.N., Fang D.N. Mechanical Properties of Hierarchical Cellular Materials. Part I: Analysis [J]. Compos. Sci. Technol., 2008, 68: 3380–3387.

Tekoglu C., Gibson L.J, Pardoen T., et al. Size Effects in Foams: Experiments and Modeling [J]. Prog. Mater. Sci., 2011, 56(2):109–138.

Yang Yazheng, Yang Jialing, Zeng Tao, Fang Daiyu. Progress in Research Work of Light Material[J] Chinese Quarterly of Mechanics, 200728(4): 489–493.

Zeng song, Zhu Rong, Jiang Wei, Cai Xiaotian, Liu Jinqiang. Research Progress of Matal Lattice Materials [J], Material Review, 201203(3): 18–23.

Zhang Qiancheng, Liu Tianjian, Wen Ting. Processes in the Study on Enhanced Mechanical Properties of high-Performance Lightweight Lattice Metallic Materials [J]. Advances in Mechanics, 2010, 40(2): 157–169.

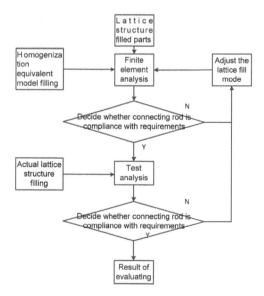

Figure 7. Simulation and experiment about the connecting rod.

Energy Science and Applied Technology ESAT 2016 – Fang (Ed.)
© 2016 Taylor & Francis Group, London, ISBN 978-1-138-02973-6

A comparative study of bearing radial stiffness with the analytical method and the finite element method

J. Li
University of Chinese Academy of Sciences, Beijing, China
Institute of Engineering Thermophysics, Chinese Academy of Sciences, Beijing, China

Y.S. Tian, H.L. Zhang, Q. Gao & C.Q. Tan
Institute of Engineering Thermophysics, Chinese Academy of Sciences, Beijing, China

ABSTRACT: Rolling bearing is widely used as support in rotary machine, and its stiffness play an important role in the dynamic performances of the rotor system. The key of studying rolling element bearing-rotor system lies in calculating bearing stiffness accurately. In consideration of oil film lubrication, this paper is based on traditional Hertz theory and self-programming while the analytical solution of the radial stiffness of cylindrical roller bearing is proposed. Meanwhile, the effects of film lubrication, radial load, roller number and rotation velocity on bearing stiffness are studied. Finally, the finite element method is also used to analyze the stiffness of cylindrical roller bearing. Through the contact analysis, the contact stress between the rings and rollers, strain and the friction stress distribution can be calculated. And the bearing radial stiffness can be extracted by post-processing method. This work can provide good theoretical basis for the optimization design of cylindrical rolling bearings.

Keywords: stiffness, cylindrical roller bearing, hertz theory, lubrication, finite element method

1 INTRODUCTION

Rolling bearing is widely used as support in rotary machine, and its dynamic mechanical performances have important influence on the dynamic performances of the rotor system (Bugra H, 2004; Deng Chi, 2013).The radial stiffness of cylindrical roller bearing is the basis of the vibration analysis for a shafting. Rolling bearing stiffness has many complicated factors, so it has not yet set up the proper calculation model (Harris, 2001; H Prashad, 2004; P. K. Gupta, 1979). Under static loading, we apply Hertz theory to the contact between the rolling bearing roller and race. In fact, for the elastic hydrodynamic lubrication will be formed in the bearing, the purely Hertz contact between roller and race is not existed (P Ehret, 1997). In this paper, a more scientific and reasonable roller bearing radial stiffness calculation model will be deduced based on Hertz theory, elastic fluid dynamic pressure lubrication theory and relevant stiffness calculation method.

2 MECHANICAL MODEL

2.1 Force analysis

Take the cylindrical bearing with twelve rolling elements as the analysis object, the structure steel for the bearing is selected and geometric parameters of a cylindrical bearing are as follow:

We can see from Fig. 1, for the center of inner ring moved from O to O_1, the load of any other roller is different from the radial load Fr and the load on the lowest roller is the largest.

Due to the force balance, we can get this:

$$F_r = F_0 + 2\sum F_\varphi \cos\varphi \ (\varphi \le \pi/2) \tag{1}$$

Table 1. Geometric parameters of a cylindrical bearing.

Parameters	Values	Parameters	Values
Outer diameter	90 mm	Outer ring groove bottom diameter	83.5 mm
Inside diameter	65 mm	Radial load	5000 N
Width	24 mm	Modulus of elasticity E	2E11 Pa
Roller number	12	Poisson's ratio μ	0.3
Roller diameter	12 mm	coefficient of friction f	0.2
Inner ring groove bottom diameter	71.5 mm		

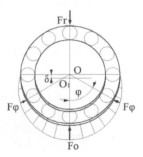

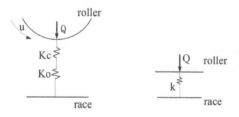

Figure 3. Simplified model of single contact stiffness.

Figure 1. Load distribution.

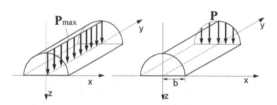

Figure 2. Pressure distribution.

φ is the angle between the center line of the bearing and the rolling element, F_φ is the contact load of the roller which the included angle is φ, F_0 is the load of the roller in the F_r direction. The deformation relationship of the bearing is:

$$\delta_\varphi = \varphi_{max}\cos\varphi \tag{2}$$

while δ_φ is the total elastic deformation of the roller which the included angle is φ.

2.2 Hertz contact analysis

Rolling bearing contact stress and strain can be successfully solved by Hertz elastic contact theory (Sharad Shekhar Palariya, 2013). We adopt the following assumptions when solve the elastic solid deformation: (1) only elastic deformation exists between the contact objects and obeys Hooke's law. (2) The contact surface is smooth. (3) Body size is far less than contact surface curvature radius (Xia Sheng, 2014). According to the Hertz theory, contact surface is a line, and the surface pressure is linear distributed, as show in Figure 2. Line contact computation formula is as follows:
Total curvature function:

$$\sum\rho = \rho_{11} + \rho_{12} + \rho_{21} + \rho_{22} \tag{3}$$

Curvature difference function:

$$F(\rho) = \frac{(\rho_{11} - \rho_{12}) + (\rho_{21} - \rho_{22})}{\sum\rho} \tag{4}$$

Half of the contact width:

$$b = \left[\frac{4F}{\pi l \sum\rho}\left(\frac{(1 - \upsilon_1^2)}{E_1} + \frac{(1 - \upsilon_2^2)}{E_2}\right)\right]^{\frac{1}{3}} \tag{5}$$

Empirical formula of contact elastic deformation:

$$\begin{aligned}\delta &= \frac{2F(1 - \nu^2)}{\pi El}\ln\left[\frac{\pi El^2}{F(1 - \nu^2)(1 \mp \gamma)}\right]\\ &= 3.84 \times 10^{-5}\frac{F^{0.9}}{l^{0.8}}\end{aligned} \tag{6}$$

Empirical formula of normal contact load:

$$F = \frac{4.08F_r}{Z} \tag{7}$$

3 STIFFNESS ANALYTICAL MODEL

According to the definition of contact stiffness:

$$K = \frac{\partial F}{\partial\delta} = 2.894 \times 10^7 l^{0.8}F^{0.1} \tag{8}$$

In 1979, Dowson put forward the minimum oil film thickness formula of line contact under the isothermal condition:

$$h_c = 3.533\frac{\alpha^{0.54}(\eta_0 u)^{0.7}R^{0.43}l^{0.13}}{E^{0.03}F^{0.13}} \tag{9}$$

Then, we can get the oil film stiffness:

$$K = \frac{\partial F}{\partial h} = 459.3\frac{E^{0.03}F^{1.13}}{\alpha^{0.54}(\eta_0 u)^{0.7}R^{0.43}l^{0.13}} \tag{10}$$

3.1 Contact analysis of kinematic pair

Oil film stiffness (Ko) will exist when the bearing rotates as shown in Fig. 3, there are contact stiffness (Kc) and oil film stiffness (Ko) between roller and race and the two are in series. The equivalent stiffness (K) is used to replace Kc and Ko. The relationship between K, K_O, Kc is:

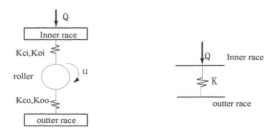

Figure 4. A simplified model of double contact stiffness.

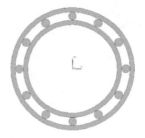

Figure 5. UG model of bearing.

$$\frac{1}{K} = \frac{1}{K_o} + \frac{1}{K_c} \qquad (11)$$

Considering the effect of the roller and centrifugal force, the load of inner ring-roller and that of outer ring-roller are different when the bearing rotates. As shown in Fig. 4, it can be assumend that the load of roller in the normal direction of the inner race is Q. Then the contact force between roller and outer race is:

Figure 6. Mesh of the bearing.

$$Q_o = Q + mg + mR_w \left(\frac{2\pi n_w}{60}\right)^2 \qquad (12)$$

while m is roller's weight, R_w is the revolution radius of the roller, n_w is the revolution velocity of the roller. Finally,, it can be obtained the models of contact stiffness and the oil film stiffness by putting Q_o, Q into Eq.(8) and Eq. (10). The stiffness of a roller:

$$K_i = \frac{1}{K_{ci}} + \frac{1}{K_{co}} + \frac{1}{K_{oi}} + \frac{1}{K_{oo}} \qquad (13)$$

The stiffness of a bearing:

$$K = \sum K_i \qquad (14)$$

4 FINITE ELEMENT MODEL

4.1 Finite element bearing model

First of all, three-dimensional modeling software UG is used to establish the bearing model with twelve rollers (Fig. 5). Secondly, input the model into the finite element software ANSYS with the bearing's geometric parameter, then define material properties, determine constraint position and apply the corresponding load. In fact, to simplify the computation, the small force distribution and deformation for bearing chamfer, edge and dust cover are ignored in the analytical model. It can

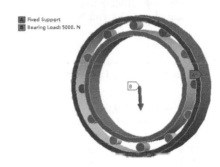

Figure 7. Boundary constraint and load model.

be seen from the analytical solutions that the effect of oil film lubrication on bearing stiffness is small enough. So the effect of oil film lubrication is out of consideration by ANSYS analysis.

To make the calculation convergence, the mesh of model is generated regularly. Mapped face meshing is used to generate in the inner and outer race. Multi zone is used to generate mesh of the twelve roller (Fig. 6).

4.2 Boundary conditions, loading and set solution options

The degrees of freedom of the side surface and outer surface for outer ring in Fig. 7 are constrained.

In order to calculate the contact problem with ANSYS, we usually define the contact pair to realize it. The contact pair consists of contact surface and target surface. ANSYS offers a variety of algorithms: pure lagrange multiplier method, augmented lagrange method, pure penalty method, penalty function method, augmented-lagrange contact algorithm. For the accuracy and convergence of results, corresponding contact algorithms are applied to different contact models.

5 DISCUSSION AND RESULTS

The bearing stiffness changes with a series of factors. In this section, the effects of lubrication, radial load, velocity and number of rollers on stiffness are studied.

5.1 Radial load

The load of inner and outer ring is increasing along with the increasing of the external radial load (Fig. 8).

5.2 Oil lubrication stiffness

Lubrication conditions have little influence on the stiffness of bearing because of the static state. There is one order of magnitude difference between oil lubricated stiffness and bearing contact stiffness. When two types of stiffness are series, the bigger one, oil lubricated stiffness, can be ignored (Fig. 9).

5.3 Roller number

The influence of roller number to radial stiffness can be shown in Fig. 10. It can be seen that the

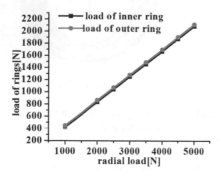

Figure 8.   Load of inner/outer ring.

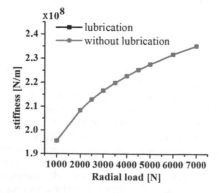

Figure 9.   Lubrication effects on stiffness.

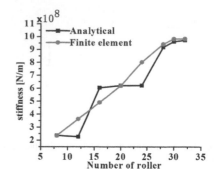

Figure 10.   Effects of rollers on stiffness.

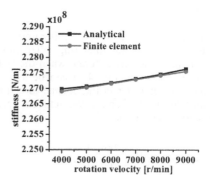

Figure 11.   Effects of rotation velocity on stiffness.

stiffness increase along with the increase of roller number. For the discontinuity of the number of loaded rollers, the solution of stiffness have a leap expression. Finally, the theoretical solution have a good agreement with the numerical solution except the number of roller is too few.

5.4 Rotation velocity

The stiffness is increasing along with the increasing of the rotation velocity, but the growth is slow. Based on this result, the effect of centrifugal force on the bearing stiffness is small (Fig. 11).

6 CONCLUSIONS

In this paper, the radial stiffness of a cylindrical rolling bearing is calculated by analytical method and finite element method, and the effects of lubrication, roller number and rotation velocity on it are studied. With the increase of radial loads, the radial stiffness will be linearly increased and the oil lubrication effects can be ignored because of the static state. Furthermore, the influence of roller number and the inner race rotation speed to the radial stiffness also remarkable. Finally, the analytical method is good agreeable with the finite element method except the fewer roller number cases.

However, the stiffness results via analytical method and finite element method in this paper are just about the analysis of static stiffness of the rolling bearing. Effects of external unbalance force, gyroscopic torque, oil film force and so on are out of consideration. The study of dynamic stiffness will be carried out in the future work.

ACKNOWLEDGMENT

The work described in this paper was supported by the Science Foundation of The Chinese Academy of Sciences (No.Y61H0112T1).

REFERENCES

Bugra H. Ertas, John M. Vance. Effect of static and dynamic misalignment on ball bearing radial stiffness. Journal of propulsion and power, 2004, 20(4):634–647.

Deng Chi, Yang Guang-hui. Finite element analysis of 6300 deep groove ball bearing. Computer Aided Drafting, Design and Manufacturing, September 2013:41–45.

Ehret, P., D. Dowson, C.M. Taylor and D. Wang. Analysis of isothermal elasto hydrodynamic point contacts lubricated by Newtonian fluids using multigrid methods. Proceedings of the Institution of Mechanical Engineers,1997, 211:493–508.

Gupta, P, K. Dynamics of rolling-element bearings—Part I Cylindrical roller bearing analysis. Journal of lubrication technology.1979,101:293–302.

Harris. Rolling bearing analysis[M]. 4th ed. New York: John Wiley and Sons Inc, 2001.

Hernot, X, M. Sartor and J. Guillot. Calculation of the stiffness matrix of angular contact ball bearings by using the analytical approach. Journal of Mechanical Design. 2000, 122:83–90.

Prashad, V. Determination of stiffness of roller bearings an alternative approach. Journal of mechanical design. 2004, 84:186–192.

Sharad Shekhar Palariya, M. Rajasekhar and J. Srinivas. Experimental identification of bearing stiffness in a rotor bearing system, National Symposium on Rotor Dynamics, 2013, 102:359–366.

Xia Sheng, Beizhi Li, Zhouping Wu and Huyan Li. Calculation of ball bearing speed-varying stiffness. Mechanism and Machine Theory. 2014, 81:166–180.

Yang Jing, Yang Liang. Stiffness calculation cylindrical roller bearing. China sciencepaper, 2014, 9(8):897–901.

Yi Guo, Robert G. Parker. Stiffness matrix calculation of rolling element bearings using a finite element/contact mechanics model. Mechanism and Machine Theory, 2012:3 2–45.

Zhang Yu-yan, Wang Xiao-li, Zhang Xiao-qing. Dynamic analysis of a high-speed rotor-ball bearing system under elasto hydrodynamic lubrication. Journal of Vibration and Acoustics, 2014:1–11.

Energy Science and Applied Technology ESAT 2016 – Fang (Ed.)
© *2016 Taylor & Francis Group, London, ISBN 978-1-138-02973-6*

The design and simulation of a fatigue wear testing system of the flexspline in harmonic drive

Zelin Wu & Lijie Chen
School of Aerospace Engineering, Xiamen University, Xiamen, China

ABSTRACT: The major failure mode of a harmonic drive is the fatigue wear of the flexspline. In order to investigate the fatigue wear strength of the flexspline during transmission, this paper design a fatigue wear testing system for the harmonic drive to simulate the meshing process of the flexspline and the circular spline. The finite element model of the testing system is built by ABAQUS and the dynamic simulation is conducted to analyze the deformation and stress for the critical positions of the model. The simulation results will be used for the further failure analysis and parameter optimization of the flexspline.

Keywords: flexspline, fatigue wear, testing system, ABAQUS simulation

1 INTRODUCTION

The harmonic drive is made up of flexspline, circular spline and wave generator, which is shown in Fig. 1. The power transmission is achieved by the elastic deformation of the flexspline. Compared with traditional driving methods, the harmonic drive has smaller size, lighter weight, higher reduction ratio and torsional stiffness (Dong H M, 2011, Folega P, 2013). Hence it is widely used in industrial fields. The flexspline usually experiences a periodic deformation under dynamic load as an elastic thin-walled structure (Yan F 2013). Because of the complex stress state, the flexpline particularly suffers from fatigue failure, which is a main failure mode for the harmonic drive. The fatigue life and dynamic features of the flexspline are the main factors for the general performance of the harmonic drive. Therefore the fatigue wear analysis for the flexspline is significant for improving the performance of the harmonic drive.

During the transmission of harmonic drive, some flexspline teeth are coming into the circular splines, namely meshing-in, while some teeth are coming out of contact, namely meshing-out. There are many teeth on flexspline, and each tooth experiences the contact and non-contact with a tooth of circular spline. By this cyclic motion, cyclic stress is subjected to the root of a tooth of flexspline, and the fatigue strength is a serious problem (Kikuchi M, 2003). Currently, there are few researches about the experimental device for investigating the fatigue of the flexspline in harmonic drive. The current fatigue wear testing system is unsuitable for the working environment of the flexspline. In this paper, a new fatigue wear testing system is designed to simulate the process of the flexspline meshing with circular spline. In the meanwhile, a finite element model of the testing system is also built in ABAQUS to analyze the fatigue properties of contact positions. Based on the simulation analysis, the working performances of both flexspline and circular spline are predicted to obtain the optimal design.

2 THE DESIGN OF FATIGUE WEAR TESTING SYSTEM

The fatigue wear testing system is based on the QBG-100 high frequency fatigue testing machine, as shown in Fig. 2. The advantages of the high frequency fatigue testing machine are simple structures, easy operation, high efficiency and low power consumption. It is widely used in aeronautics, aerospace, defense, education and industrial

Figure 1. Main components of harmonic drive [6].

production fields. The design of new fatigue wear testing system is shown in Fig. 3. The main part of the testing system is made up of the base, hydraulic cylinder, tie rod, flange plate, piston rod, tooth holder and flexspline tooth. There are two hydraulic cylinders sitting symmetrically on both sides of the testing system to provide stable and large-range normal forces, while the axial force is provided by high frequency fatigue testing machine. When starts the experiment, the testing specimen is clamped by the fixture of the high frequency fatigue testing machine. Then the piston rod is pushed by the hydraulic cylinders and starts moving inward. When the flexspline tooth contacts with the specimen, they show a minute misplace and the groove in specimen is slightly higher than the flexspline tooth. Then the normal and axial forces are applied together to simulate the meshing process of the flexspline and circular spline until the flexspline tooth reaches the maximum depth of meshing and fully meshes with the specimen.

The assembly drawing of specimen and flexspline tooth is shown in Fig. 4. The thicknesses of the flexspline tooth and specimen are both 4 mm. During the assembly, the flexspline tooth should offset downward to make the top of the groove in specimen slightly higher than the flexspline tooth. A small gap is supposed to be existed between them. In order to increase the contact area, the

Figure 2. QBG-100 high frequency fatigue testing machine.

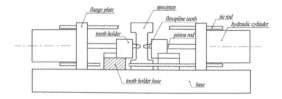

Figure 3. Design of new fatigue wear testing system.

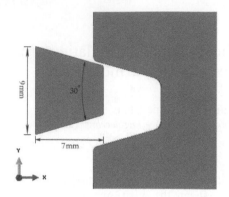

Figure 4. Assembly drawing of specimen.

trapezoidal tooth is used and the dimensions are shown in the assembly drawing. The material of the flexspline tooth is 40CrNiMoA, which has density of 7850 kg/m^3, elasticity modulus of 209GPa and Poisson's ratio of 0.295. The plate specimen is used for the testing system, which is widely used in the fatigue wear experiments. However, compared with the normal fatigue wear testing specimen, there are two grooves on both sides right in the middle section of the specimen, which have the same dimensions with the circular spline. The height of the specimen is 100 mm. In order to avoid concentration of stress, the corners of the grooves in specimen are rounded off to 0.5 mm radius, and the top clearance of the flexspline tooth is 0.25. The material of the circular spline is 2Cr13, which has density of 7750 kg/m^3, elasticity modulus of 229GPa and Poisson's ratio of 0.3.

3 THE SIMULATION ANALYSIS OF FATIGUE WEAR TESTING SYSTEM

Before producing the actual fatigue wear testing system, the simulation analysis is conducted by ABAQUS to predict the performance of the testing system. ABAQUS is capable to analyze the complex solid mechanics and structure mechanics, as well as the highly nonlinear problems The implicit method in ABAQUS/Standard is applied to the simplified model of the fatigue wear testing system to simulate the meshing process of the flexspline. The displacement and stress of the critical contact positions of the model are calculated during the simulation.

To simplify the simulation analysis, we take the flexspline tooth and specimen to simulate with finite element method. Due to the symmetry of the fatigue wear testing system, a quarter of the assembly model is used for the analysis, which is separated from y-direction symmetry axis and z-direction symmetry axis of the original model. The analysis is a rigid-flexible contact problem where the flexspline tooth is flexible and the specimen is a rigid-body. The surfaces on the flexspline tooth and the groove in specimen are defined as contact surfaces. Finite element mesh of the model is shown in Fig. 5.

Symetric boundary conditions are also applied, in which the bottom surface of the specimen is fixed and the right section of the specimen is restricted by a DOF of x-direction. The specimen is given a surface load on the top surface. The flexspline tooth is given a displacement load along its top surface moving inward and outward. When starts the simulation, the specimen deforms under the surface load and starts to contact with the moving flexspline tooth. The flexspline tooth and specimen meshes with each other till the maximum depth of meshing. Then the flexspline tooth returns by opposite path moving outward and the simulation ends as soon as the flexspline tooth move back to the origin position. The surface load and the initial size of the gap between the flexspline tooth and the specimen are adjusted to optimize the output of stress. The result of stress has a linear approximation with both the surface load and the initial size of the gap. The surface load can't be too large or the specimen will fail quickly. The initial size of the gap can't be too large either, otherwise the flexspline tooth won't contact with the specimen. To obtain the most adequate result, the surface load is defined by 43 MPa and the initial size of the gap is defined by 0.02 mm. The deformation and stress distribution of the model is solved throughout the simulation.

According to the maximum principal shear stress theory, the distribution of the Tresca stress in contact regions of the model are shown in Fig. 6, and the distribution of the Tresca stress in the flexspline tooth are shown Fig. 7. The Tresca stress reaches the maximum value as soon as the flexspline tooth

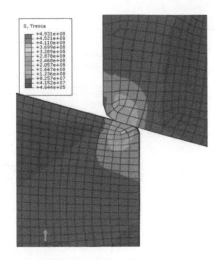

Figure 6.　The distribution of the Tresca stress in contact regions.

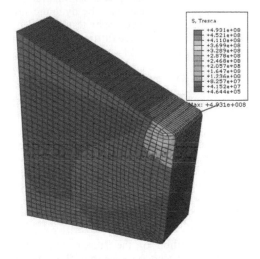

Figure 7.　The distribution of the Tresca stress in the flexspline tooth.

and the specimen come into contact and start to mesh with each other. The maximum Tresca stress is right at the subsurface of the flexspline tooth, and decreases gradually with the distance from the contact positions. The first row in Table 1 lists each stress component of the maximum Tresca stress point of the model, with the second row lists three known maximum stress components of the maximum Tresca stress point of flexspline in harmonic drive carried out by Patran, in which the maximum Tresca stress, the maximum Von Mises stress and the minimum principle stress of the flexspline structure are at the same point. From Table 1, for the simulation specimen we design, the maximum Tresca stress is 493 MPa, the maximum Von Mises

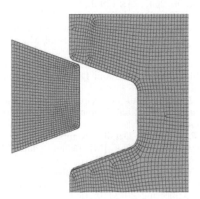

Figure 5.　Finite element mesh of the model.

Table 1. Stress components of the maximum Tresca stress point.

	Maximum Tresca Stress (MPa)	Node Number	Von Mises Stress (MPa)	Principal Stress Components (MPa)		
				First Principal	Second Principal	Third Principal
Simulation specimen	493	2290	443	82.6	−110	−470
Flexspline structure [8, 9]	499	620048	462	–	–	−464

stress is 443 MPa and the minimum third principle stress is −470 MPa. They are all numerically large, which is caused by meshing of the flexspline tooth with the rigid specimen. However, the model keeps elastic during the entire process of simulation. Compared with the result from stress analysis of flexspline in harmonic drive, they are all in proportion with their respective stress component of the flexspline structure. More specifically, they are approximately equal, which demonstrates the rationality of the simulation. The dynamic simulation of the whole fatigue wear testing system is expected to carry out in the future.

4 SUMMARY

In this paper, a new fatigue wear testing system is designed to simulate the working process of flexspline and circular spline in harmonic drive. A model of the fatigue wear testing system is built and the simulation of the model is conducted with finite element method by ABAQUS. The results obtained from the simulation analysis can be used to provide certain reference significance for further failure analysis of the flexspline. The main conclusions are as follows:

1. The distribution of the Tresca stress of the simulation specimen shows that: the maximum Tresca stress is at the subsurface of the flexspline tooth specimen. It reaches the maximum value as soon as the flexspline tooth and the specimen in the fatigue testing system come into contact and mesh with each other. At that moment, the Von Mises stress in the flexspline tooth reaches the maximum value and the third principal stress in the flexspline tooth reaches the minimum value. The result is linearly changing with the load and the initial gap size between the flexspline tooth and the specimen in the fatigue testing system.
2. The stress components of the maximum Tresca stress point of the simulation specimen shows that: the maximum Tresca stress, the maximum Von Mises stress and the minimum principle

stress are all numerically large while the model is elastic during the entire simulation. However, compared with the result of stress analysis of flexspline structure carried out by Patran, they are approximately equal to the respective maximum stress components of the flexspline structure. It tells that the result of the simulation analysis is reasonable and the design of the new fatigue wear testing system can be used for fatigue wear experiments.

ACKNOWLEDGEMENTS

The authors would like to thank the support of the National Basic Research Program of China (No. 2013CB733004) and the National Natural Science Foundation of China (No. 51475396).

REFERENCES

Dong H M , Zhu Z D, Lu Y. Dynamic Simulation of Harmonic Drives Based on FEM[C]//Advanced Materials Research. 2011, 323: 28–33.

Folega P. The study of the dynamic properties of some structural components of harmonic drive [J]. Journal of Vibroengineering, 2013, 15(4): 2096–2102.

Information on http://www.wogo58.com/article/show–163.aspx.

Kikuchi M, Nitta R, Kiyosawa Y, et al. Stress analysis of cup type strain wave gearing[C]//Key Engineering Materials. 2003, 243: 129–134.

Li Y G, Ji Z G. Technical Status and development of the domestic high frequency fatigue testing machine[J]. Experimental Technology and Testing Machine, 2006, 46(1): 1–4. (in Chinese).

Liu C, Chen L, Wei C. Deformation and Stress Analysis of Flexspline in Harmonic Drive based on Finite Element Method[J].

Liu, C.J., Behavior of a Flexspline in Harmonic Drive, Master Dissertation, Xiamen University, 2015. (In Chinese).

Shi Y P, Zhou Y R. Detailed examples of finite element analysis of ABAQUS [M]. Beijing: China Machine Press, 2006. (in Chinese).

Yan F, Yang W, Duan C C. The analysis of fatigue life of flexspline in harmonic drive[J]. Modern Manufacturing Engineering, 2013 (10). (in Chinese).

Energy Science and Applied Technology ESAT 2016 – Fang (Ed.)
© 2016 Taylor & Francis Group, London, ISBN 978-1-138-02973-6

The dynamic simulation of harmonic drive considering flexible ball bearing based on ABAQUS

Zhihua Fu, Tieqiang Gang & Sen Liu
School of Aerospace Engineering, Xiamen University, Xiamen, China

ABSTRACT: In this paper, a new dynamic simulation model of Harmonic Drive (HD) considering flexible ball bearing is presented. The finite element method is used to analyze the deformation and stress as well as movement characteristics of the HD in a slow speed of rotation based on ABAQUS. The results show an accurate match between simulation and experiment, which indicates the reliability of the proposed model.

Keywords: harmonic drive, dynamic simulation, finite element method, flexible ball bearing, flexspline

1 INTRODUCTION

Harmonic drive is a new transmission technology with the development of aerospace technology and robotics since 1950 s. As shown in Fig. 1, it contains three main components, the Flexspline (FS) composed of a thin-walled flexible cup with small external gear, the Circular Spline (CS) which is a rigid ring with internal teeth machined along a slightly larger pitch diameter than that of the flexspline and the Wave Generator (WG), a ball bearing assembly with a rigid elliptical cam (T. Tjahjowidodo, 2013). The rotation of the elliptical WG causes a periodic clastic deformation of the FS. Then external teeth of the FS mesh in and out internal teeth of the CS alternately resulting in a relative rotation between the FS and the CS. The HD has been widely used in industrial fields due to its small size, high precision, light weight, compactness properties and high reduction ratio and torsional stiffness characteristics. Many studies have been conducted on

HD tooth profiles, kinematics, deformation, stress, position accuracy and dynamic behavior (S.H. Oh, 1997, H.S. Jeong, 1999) to improve HD performance using theoretical calculation and experiment methods. Compared with the traditional methods which is high cost and long periods, the finite element method can not only provides HD research with a cost-less, efficient and convenient approach but also be conducive to the parameter optimization (Huimin Dong, 2012).

However, the WG is simplified considering the complexity of flexible bearing, that is to say, only using a rigid elliptical cam instead of a real WG composed of a flexible ball-bearing and a rigid elliptical cam in most of HD simulations literature. It does not conform with reality and can not explain some experiment phenomena well, such as wear of middle part of gear along the axial direction of the FS. So, considering a deep groove flexible ball bearing with 23 balls, one establishes a new HD finite element simulation model based on ABAQUS in this paper. Then, finite element stress analysis results of the FS and the deep groove flexible ball bearing is presented. Finally, the output kinetic characteristics of the HD under a slow speed of input rotating speed is studied.

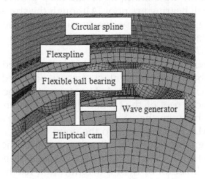

Figure 1. Finite element model.

2 FINITE ELEMENT MODEL

This paper establish a harmonic drive model with involute tooth profile whose speed reduction ratio is 100. The WG contains a elliptical cam and a deep groove flexible ball bearing with 23 balls. The key meshing parameters of HD model are shown in Table 1. The theoretical conjugate profile of circu-

Table 1. Meshing parameters.

Part	Tooth number	Module	Addendum coefficient	Pressure angle
Flexspline	200	0.55	0.35	20°
Circular spline	202			

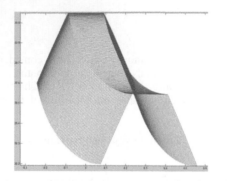

Figure 2. The theoretical conjugate profile of circular spline.

Table 2. Material parameters.

Part	Material type	Density (kg/m^3)	Elastic modulus (Gpa)	Poisson ratio
Flexspline Circular spline	40CrNiMoA	6691	204	0.29
Elliptical cam	C45E4	7830	210	0.29
Cage	PA66+25%GF	1180	140	0.24
Ball bearing Bearing enclose	GCr15	7850	210	0.29

lar spline is shown in Fig. 2. The material parameters without considering the plastic are shown in Table 2. The finite element model is meshed using C3D8R hexahedral reduction element as shown in Fig. 1. The number of element is 1038114, and number of nodes is 1331027.

The contact of harmonic drive simulation is complicated because of the deep groove flexible ball bearing. It contains three kinds of motion including sliding friction motion, rolling friction motion and frictionless motion. Thus, we set up three different friction coefficient as shown in Table 3, to simulate different frictional characteristics.

The process of solving the dynamic simulation of the HD is divided into three steps in ABAQUS/Explicit. The first step is simulating the assembling

Table 3. Friction parameters.

Master surface	Slave surface	Friction coefficient	Master surface	Slave surface	Friction coefficient
Elliptical cam	Bearing Inner Ring	0.15	Bearing roller	Bearing Inner Ring	0.02
Bearing outer Ring	Flexspline		Bearing roller	Bearing outer Ring	0.02
Internal teeth of the CS	External gear of the FS		Bearing roller	cage	0

process. The elliptical cam is installed into the deep groove flexible ball bearing, and then, the wave generator containing elliptical cam and flexible ball bearing is installed into the flexspline. The second step is loading a slow rotating speed of input smoothly. In this step, the minimum specified increment should be less than time increment required to ensure the convergence in ABAQUS on account of sophisticated nonlinear contact. The third step is stable running step process with rotating speed n = 120 rpm. It is better to use restart-method to solve this complicated problem which can not only make sure every step controllable but also save time. Finally, wc compile directives to make every step result together through ABAQUS Command. All calculation are solved by HP-Z800 workstation (8CPU-24G).

3 RESULTS ANALYSIS

The stress and kinetic characteristics of the HD have changed a lot because of the deep groove flexible ball bearing.

Firstly, the stress of the deep groove flexible ball bearing has been investigated. It presents a gradual step-up trend from minor axis to major axis nearly to 1.1GPa in Fig. 3, which indicates that working conditions of the deep groove flexible ball bearing are poor. Although the stress may be bigger than reality due to without considering the plastic, it is still below the yield strength which is 1667~1814 MPa [8].

Then the stress of the FS has been studied. The distribution of the contact stress is obtained during the meshing in Fig. 4. The stress main focus on the tooth root of the FS and the region of engagement between the FS and the CS nearly to about 500 MP. However, due to the point contact of the deep groove flexible ball bearing, it causes the contact stress of the middle part of gear along the axial direction of the FS deterioration reaching to nearly to 800 MP as shown in Fig. 5. The deterioration of the contact stress exacerbates wear of the

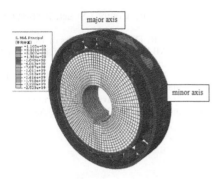

Figure 3. The stress of the deep groove flexible ball bearing.

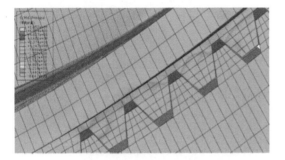

Figure 4. The contact stress of gears.

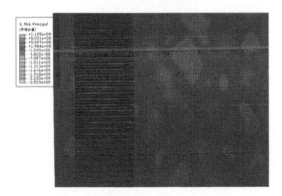

Figure 5. The stress of the FS.

gear of the FS, which coincides well with experimental result in Fig. 6.

Finally, the kinetic characteristics of the HD is analyzed. The input angular velocity of the elliptical cam is $120/60 \times 360° = 720°/s$. So the theory stability output angular velocity of the FS is $720/100 = 7.2°/s$. The simulation result indicates a unsteadiness harmonic signal about $7.18°/s$ shown in Fig. 7, which matches the theory approximately. The date of fluctuations are caused by incoherent motion of the deep groove flexible ball bearing with 23 balls.

Figure 6. Experimental result.

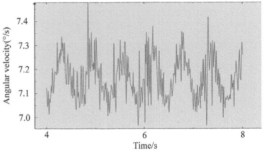

Figure 7. Output angular velocity.

4 SUMMARY

The above simulation results show that it is the deep groove flexible ball bearing that make not only the stress of the FS deteriorates, especially the middle part of gear along the axial direction, which causes gear wear severely, but also and the output of the HD vibrates which brings to inaccuracy position control. So it is better to establish a finite element model of the HD with flexible ball bearing for further research. The main work of this paper is worth considering.

ACKNOWLEDGEMENTS

The authors would like to thank the support of the National Basic Research Program of China (No. 2013CB733004) and the National Natural Science Foundation of China (No. 51475396).

REFERENCES

Ghorbel, F.H., P.S. Gandhi, F. Alpeter, On the kinematic error in harmonic drive gears, J. Mech. Des. 123 (2001) 90–97.

Handbook of engineering materials, Vol 1[M], Beijing: China Standards Press, 2001, pp. 380–386. (in Chinese).

Huimin Dong, Zhengdu Zhu, Weidong Zhou, Dynamic Simulation of Harmonic Gear Drives Considering Tooth Profiles Parameters Optimization*, Journal of computers, Vol. 7, No. 6, June 2012, pp.1429–1436.

Jeong, H.S., S.H. Oh, A study on stress and vibration analysis of a steel and hybrid flexspline for harmonic drive, Compos. Struct. 47 (1–4) (1999) 827–833.

Oh, S.H., S.H. Chang, D.G. Lee, Improvement of the dynamic properties of a steel-composite hybrid flexspline of a harmonic, Compos. Struct. 38 (1–4) (1997), 251–260.

Shi Y P., Zhou Y R. Detailed examples of finite element analysis of ABAQUS[M]. Beijing: China Machine Press, 2006. (in Chinese).

Tjahjowidodo, T., F. Al-Bender, H. Van Brussel, Theoretical modelling and experimental identification of nonlinear torsional behaviour in harmonic drives, Mechatronics, Volume 23, Issue 5, August 2013, 497–504.

Energy Science and Applied Technology ESAT 2016 – Fang (Ed.)
© *2016 Taylor & Francis Group, London, ISBN 978-1-138-02973-6*

Analysis of the influence of groove size on the sealing property of hydraulically expanded joints

Tao Li & Chenghong Duan
School of Beijing University of Chemical Technology, Beijing, China

ABSTRACT: In this paper, the hydraulic expanding process of heat exchange is simulated by the finite element method. The residual contact pressure is calculated so as to figure out the influence of groove size and expending pressure on the sealing property of expanded joint. The relation between optimal groove width and expending pressure is revealed, and the conclusion that the sealing property reaches the best when the ratio of groove depth to width equals 0.05. The seal test is conducted to prove the conclusion.

Keywords: residual contact pressure, groove width, grooves depth, expanding pressure

1 INTRODUCTION

There are several joint forms between tube and tube-sheet. Hydraulic expansion has become the most universal technical means applied to the connecting of tube and tube-sheet with the advantages of few mechanical damage to material, accurate control of depth of the expanded joint and less time-consuming (H.F. Li, 2010, H.F. Wang, 2005). The sealing property of hydraulically expanded joint is one of the main performance indexes. In order to enhance the sealing performance usually there need to be grooves in the tube-sheet. The groove width and depth is always the research hotspot (A. RAkisanya, 2011, Z. Wei, 2012).

2 NUMERICAL MODEL

The material of the tube-sheet is a kind of stainless steel, the tube material is 0Cr18 Ni10Ti and its specification is $\Phi25 \times 2.5$ mm. The tube-hole diameter is $\Phi25.2$ mm and the center distance of tube holes is 32 mm, the axial thickness of the tube-sheet is 100 mm. The tube holes are in triangular pattern, shown as Fig. 1. The tube-sheet structure around each hole is diverse, the minimum and maximum thicknesses are D_1 and D_2 respectively. It should be noticed that the thickness here is the radial thickness between two holes (i.e., the bridge thickness).

Plastic deformation usually occurs near the groove edge in the tube-sheet during hydraulic expansion, and a seal annular region with high residual contact pressure will appear here after unloading. The expansion is carried out more dif-

ficultly to the position with larger radial thickness, so the minimum residual contact pressure appears at the annular position with radial thickness D_2, and the radial thickness is relatively even here; therefore, a planar axisymmetric model can be used to calculate the minimum residual contact pressure (H.F. WANG, 2003, W. Zhou, 2010). The model is shown in Fig. 2. Fig. 3 shows the true stress-strain curves of the materials that are obtained from tensile tests.

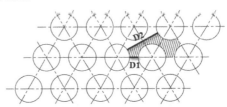

Figure 1.   Arrangement pattern of tube holes.

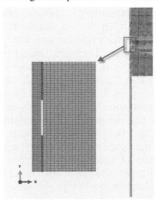

Figure 2.   Numerical finite element model.

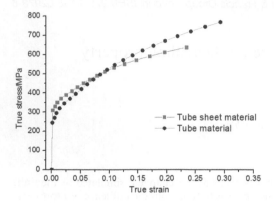

Figure 3.　True stress-strain curve.

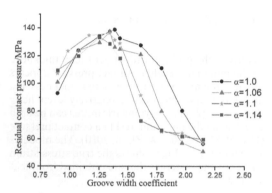

Figure 4.　Residual contact pressure ($\alpha = 1.0\text{~}1.15$).

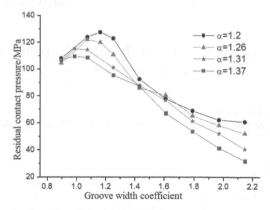

Figure 5.　Residual contact pressure ($\alpha = 1.0\text{~}1.15$).

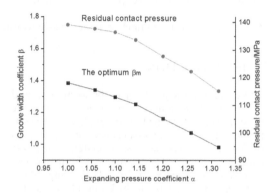

Figure 6.　Relation between βm and α.

3　GROOVE WIDTH

In order to study the impact of the groove width w, set the groove depth large enough, to ensure that the outer surface of the tube cannot contact the groove bottom during expanding, to eliminate the impact of the groove depth. In the real situation, there needs to be residual contact pressure to some extent in the non-grooved region. Therefore, the expanding pressure larger than P_0 is needed. Consider the expanding pressure P in the range of 175~230 MPa, and the groove width w in the range of 5~12 mm. P and w are non-dimensionalized in analysis according to Eq. (1) and (2), define α as the expanding pressure coefficient and β as the groove width coefficient, where r_0 is the outer radius of the tube, and t is the wall thickness of the tube. Finally, 8×11 sets of data are obtained, the curves with $\alpha = 1.0\text{~}1.14$ and $\alpha = 1.2\text{~}1.37$ are shown in Figs. 4 and 5, respectively.

$$\alpha = P/P_0 \tag{1}$$

$$\beta = w/\sqrt{r_0 t} \tag{2}$$

From Figs. 5 and 6, it can be seen that when α is constant, the residual contact pressure rise and then decrease as β increases, different maximum residual contact pressures correspond to different αs.

4　GROOVE DEPTH

If the expanding pressure is high enough or a softer tube material is used, the maximum displacement of the outer surface becomes greater than the groove depth, and then contact happens. The maximum displacements of the outer surface under expanding pressure 220 MPa are listed in Table 1.

The maximum displacements of the outer surface under expanding pressure 220 MPa are listed in Table 1.

If the groove depth is smaller than the displacements in Table 1, the outer surface contacts with the groove bottom during expanding. As a result

Table 1. Maximum displacement of the outer face of tube.

Groove width /mm	5	6	7	8	9	10	11	12
Displacement /mm	0.2	0.312	0.476	0.678	0.92	1.182	1.467	1.753

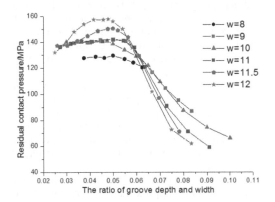

Figure 7. Residual contact pressure and h/w.

of the compression in the contact region, expansion results at the edges of the groove vary with the changing of contact region. Generally, shallow groove leads to lager contact region, while deep groove leads to smaller contact region, so the degree of influence varies.

As showed in Fig. 7, if the value of h/w is less than 0.05 the residual contact pressures remain unchanged or keep increasing. The larger the groove width, the more the residual contact pressures, and it reaches a peak when the value of h/w is about 0.05. If the value of h/w is greater than 0.5, the residual contact pressures keep falling.

Simultaneously, the rising and falling step of residual contact pressures changes gently when w = 8 as shown in Fig. 7. This is because the maximum displacement is small and the changing of groove depth has little impact on the contact region. It can be seen that the curves change more gently if groove widths equal 7 mm or 6 mm. Therefore, when the groove width is less than 6 mm, groove depth value equals 0.3 mm. When the groove width is larger than 6 mm, the groove depth value equals 0.05 times groove width.

5 EXPANDING PRESSURE

Define the residual contact pressure when h/w = 0.05 as a peak value of residual contact pressure. The larger the groove width (less than

12 mm), the higher the peak value. In order to explore the influences of expanding pressure on the peak value of residual contact pressure, the peak values are calculated when expanding pressure P = 200~240 MPa, groove width w = 8~14 mm, and groove depth h = 0.05 w. Finally, the relation between peak values and groove width values is shown in Fig. 8.

As is shown in Fig. 8, the peak values of residual contact pressure increase first and then decrease with the increase in groove width. As is shown in Fig. 9:, (1) the expanding pressures have little impact on the residual contact pressure when expanding pressure coefficients vary from 1.14 to 1.37; and (2) there is a inflection point corresponding to expanding pressure 220 MPa when the groove width is greater than 12 mm. If the expanding pressures are greater than 220 MPa, the residual contact pressures decrease. Moreover, the laws show that the expanding pressures should not be too large.

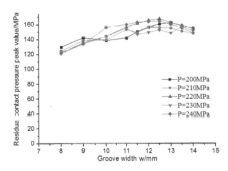

Figure 8. Residual contact pressure peak value and groove width.

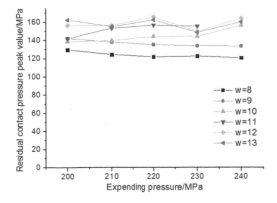

Figure 9. Residual contact pressure peak value and expanding pressure.

6 SEAL TEST

To verify whether the residual contact pressure meets the seal requirement when h = 0.05 w, a seal test is designed (Y. Lu, H.G, 2003). The tube holes distribution of test model are shown in Fig.10, and each region contains three holes. The groove width and groove depth of each hole are shown in Table 2. The center hole is set as a comparison that does not have a groove. The test model is shown in Fig. 11.

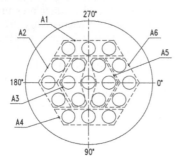

Figure 10.   Tube hole distribution.

Table 2.   Groove width and depth.

Region	A1	A2	A3	A4	A5	A6
Groove width/mm	8	9	10	11	12	13
Groove depth/mm	0.4	0.45	0.5	0.55	0.6	0.65

Figure 11.   Test model.

Finally, the test results show that a good sealing is achieved when groove width values are 8 ~ 13 mm and the groove depth h = 0.05 w. The residual contact pressures corresponding to the groove size shown in Fig. 9 are about 120~165 MPa.

7 SUMMARY

1. When the groove depth is large and the tubes do not contact with the groove bottoms, if the expanding pressure is larger than the initial expanding pressure, the optimal groove width decreases with the increase in expanding pressure.
2. When the tube contacts with the groove bottom during the expanding process, the sealing property of expanded joint achieves the best if the groove depth and groove width satisfy the condition h = 0.05 w, and the residual contact pressures are really high when the width is larger than 8 mm.
3. Once the ratio of the groove depth to groove width reaches approximately 0.05, the more appropriate expanding pressure coefficient is between 1.2 and 1.3.

REFERENCES

Akisanya, A. R., F.U. Khan, W.F. Dean, Cold hydraulic expansion of oil well tubular, International Journal of Pressure Vessel and Piping. 88(2011) 465–472.

Li, H.F., C.F. Qian, W. Pan, Study of the residual stresses in a hydraulically expanded thick tube-sheet, Journal of Beijing University of Chemical Technology (Natural Science Edition). 37(2010) 121–125.

Lu, Y., H.G. Yan, Seal pressure of hydraulically expanded tube to tube sheet joints, Chemical Engineering & Machinery. 03(2003) 156–159.

Wang, H.F., Z.F. Sang, Effect of geometry of grooves on connection strength of hydraulically expanded tube-to-tube-sheet Joints, ASME J. Pressure Vessel Techno. 127(2005) 430–435.

Wang, H.F., Z.F. Sang, Residual contact pressure in hydraulically expanded tube-tube-sheet joints of heat exchangers, Nanjing University of Chemical Technology (Natural Science Edition). 52(2003) 52–56.

Wei, Z., Analysis of residual contact pressure during hydraulic expanded tube-to-sheet joint, Modern Manufacturing Engineering. 01(2012) 73–76.

Zhou, W., Y.C. Zhang, Y. Lu, Relation between material mechanics performance and hydraulic expansion performance, Chemical Engineering & Machinery. 04(2010) 422–424.

Energy Science and Applied Technology ESAT 2016 – Fang (Ed.)
© *2016 Taylor & Francis Group, London, ISBN 978-1-138-02973-6*

Computational study regarding the evolution of volute noise radiation in a centrifugal pump under variable medium temperature conditions

Peixin Dong, Ming Gao, Dongyue Lu & Chang Guo
School of Energy and Power Engineering, Shandong University, Ji'nan, China

ABSTRACT: A three dimensional (3D) unsteady hydro/aero acoustic computational model is developed to explore the 3D flow field in a centrifugal pump based on a Reynolds time-averaged k-epsilon two-equation model, with a view to investigate principally, the acoustic signature of variable medium temperature conditions. In order to obtain the total sound pressure level, a frequency-domain analysis is conducted utilizing the practice of acoustically monitoring points outside the centrifugal pump, to arrive at a realistic model to elucidate the evolution of the radiation of volute noise under variable medium temperature conditions. The simulation results demonstrate that noise radiation is not influenced significantly due to different medium temperatures. These conclusions lay a theoretical foundation for further research on the noise radiation behavior and control in a typical centrifugal pump and provide guidance for 3D design of the low-noise impeller and flow delivery system.

Keywords: centrifugal pump; 3D flow-field; total sound pressure level; variable medium temperature conditions

1 INTRODUCTION

The centrifugal pump is often accompanied by a wide frequency range of noise radiation (F.G. Johann, 2008). The radiation of noise in a centrifugal pump typically degrades the pump performance, resulting in deviation from the optimal working point (S. Benedek, 1994). Thus, it is extremely important to study the hydro/aero acoustic behavior of centrifugal pumps. The early research methods concerning the radiation of noise in the centrifugal pump depended on experimental studies (J. Wang et al, 2007) (J.S. Choi et al, 2003). Because of the shortcomings of experimental research methods, Computational Fluid Dynamics (CFD) methods are now widely used in pump analysis including the hydro/aero acoustic emissions. (M. Wang et al, 2006) (J. Parrondo et al, 2011) (W.H. Jeon et al, 2003).

Hence, it is extremely important to study the flow-induced noise inside the volute. As a consequence, in this paper studies are conducted involving 3D unsteady flow fields in centrifugal pumps via CFD to reveal the evolution of volute noise radiation under scenarios pertaining to variable medium temperature conditions.

2 THEORETICAL MODEL AND CALCULATION METHODS

The solutions for the acoustic and the flow field should be integrated, via the FW-H acoustic model (S.B. Kenneth et al, 1997), which is usually based on the inhomogeneous wave operator derived by manipulating the Navier-Stokes equations to result in, viz:

$$\frac{1}{a_0^2}\frac{\partial^2 p'}{\partial t^2} - \nabla^2 p' = \frac{\partial}{\partial t}\{[\rho_0 v_n + \rho(u_n - v_n)]\delta(f)\}$$
$$- \frac{\partial}{\partial x_i}\{[p_{ij}n_j + \rho u_i(u_n - v_n)]\delta(f)\} \quad (1)$$
$$+ \frac{\partial^2}{\partial x_i \partial x_j}\{T_{ij}H(f)\}$$

where, a_0 is the far-field acoustic source and p' is far-field acoustic pressure. u_i is a typical component of flow velocity at the direction x_i and u_n is one normal component of flow velocity at the control plane f_0. v_i is component of control plane velocity at the direction x_i and v_n is component of control plane velocity at the normal direction. n_j is unit normal vector. $\delta(f)$ is Dirac delta function and $H(f)$ is Heaviside function. P_{ij} is stress tensor and T_{ij} is Lighthill stress tensor.

3 NUMERICAL COMPUTATIONS

3.1 *Flow field computation*

1. Geometric model
A single-stage, single-suction centrifugal pump was set up with structural and performance parameters as shown in Table 1. The computational domain for the model pump is shown as Fig. 1.

Table 1. Structural and performance parameters of the model pump.

Parameters	Value
Impeller inlet diameter D_1/mm	130
Impeller outlet diameter D_2/mm	270
Impeller outlet width D/mm	30
Design flow rate $Q/m^3 \cdot h^{-1}$	80
Design head H/m	28
Design rotational speed $n/(r \cdot min^{-1})$	1450
Blade number Z	6
Shaft frequency f_a/Hz	24.2
Blade passing frequency f_b/Hz	145

Figure 2. Mesh of global computational domain and volute tongue of the centrifugal pump.

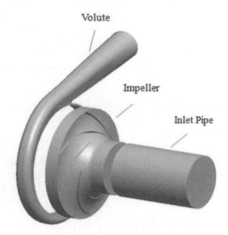

Figure 1. Computational domain of the centrifugal pump.

2. Mesh generation

Fluent associated Gambit software is adopted for the mesh generation exercise, as shown in Fig. 2.

3.2 *Computation of the radiated acoustic field*

The symmetry face of the volute has been selected as the monitoring surface for volute noise radiation, (the monitoring surface is a circular surface, with radius as 1000 *mm*). Monitoring point P_1 faces exactly the volute tongue, and eight monitoring points $(P_1 \sim P_8)$ have been set up at intervals of 45 degree, as shown in Fig. 3. Additionally, the Sound Pressure Level (SPL) for each of the monitoring point can be obtained to yield the SPL as defined by,

$$SPL = 20\log_{10}\frac{P_e}{P_{ref}}(dB) \qquad (2)$$

where, P_e is effective value of acoustic pressure, P_{ref} is reference sound pressure.

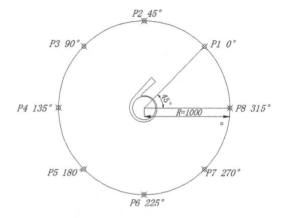

Figure 3. External monitoring points around the centrifugal pump.

On the other hand, the Total Sound Pressure Level (TSPL) can be obtained by Eq. 3,

$$TSPL = 10\lg\sum_{i=1}^{n}10^{SPL_i/10} \qquad (3)$$

4 RESULTS AND ANALYSIS

As widely used power convertors, the typical medium temperature in a centrifugal pump changes according to different operational requirements; thus, it is a practical requirement to study the impact of different medium temperatures on the volute noise radiation. The fluid used in this study is water, and the corresponding relationship between densities, viscosity and other parameters are provided by compiling a table via the User Defined Function (UDF) attribute of Fluent. Subject to the premise regarding neglect of cavitation concerns, the numerical simulation regarding volute noise radiation is conducted

394

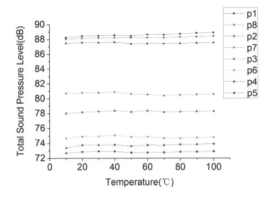

Figure 4. Dependency curves between the total sound pressure level and medium temperature.

under different medium temperatures, including 10°C, 20°C, 30°C, 40°C, 50°C, 60°C, 70°C, 80°C, 90°C, and 100°C, respectively. The functional dependency curves between the total sound pressure level and medium temperature at every monitoring point are shown in Fig. 4.

It can be observed from Fig. 9 that with the increasing of the medium temperature, the total sound pressure level of volute noise radiation keeps essentially constant at each monitoring point, and fluctuates in a small range. The conclusion has consequently been arrived at via the numerical simulations reported, that the total sound pressure level for each of the monitoring points will not change substantially with the variation of water temperature. It can be inferred that water temperature is henceforth not a primary factor influencing radiation of volute noise.

5 CONCLUSION

Subject to the premise regarding the neglect of cavitation, the radiation of noise will not change acutely with the variation of water temperature and water temperature is thus not a primary factor influencing radiation of volute noise.

In summary, a representative hydro/aero acoustic module has been applied to integrate to provide the sound field based on the evolution of the pressure fluctuations on volute wall, a sensible and fundamentally formulated way to research the evolution of radiation of volute noise in a typical centrifugal pump under variable medium temperature conditions. This research has provided some ideas for the establishment of a design oriented model regarding the abatement of radiation of noise in a future study on variable operating conditions and laid a the theoretical foundation for further control concerning the radiation of noise from a typical centrifugal pump.

REFERENCES

Benedek, S. Fluid vibration induced by a pump. Journal of sound and vibration 177(1994) 337–348.
Choi, J.S., D.K. McLaughlin, D.E. Thompson. Experiments on the unsteady flow field and noise generation in a centrifugal pump impeller. Journal of Sound and Vibration 263(2003) 493–514.
Jeon, W.H., D.J. Lee. A numerical study on the flow and sound fields of centrifugal impeller located near a wedge. Journal of sound and vibration 266(2003) 785–804.
Johann, F.G. Centrifugal pumps[M]. New York; Springer Berlin Heidelberg, 2008
Kenneth, S.B., F. Farassat. An analytical comparison of the acoustic analogy and Kirchhoff formulation for moving surfaces. AIAA Journal 36(1997) 1379–1386.
Parrondo, J., J. Pérez, R. Barrio, et al. A simple acoustic model to characterize the internal low frequency sound field in centrifugal pumps. Applied Acoustics 72(2011) 59–64.
Wang, J., T. Feng, K. Liu, et al. Experimental research on the relationship between the flow-induced noise and the hydraulic parameters in centrifugal pump. Fluid Machinery 35(2007) 8–11.
Wang, M., J.B. Freund, S.K. Lele. Computational prediction of flow-generated sound. Annu. Rev. Fluid Mech 38(2006) 483–512.

Energy Science and Applied Technology ESAT 2016 – Fang (Ed.)
© 2016 Taylor & Francis Group, London, ISBN 978-1-138-02973-6

Research on the structure optimization of air-cooled condensers

Yu Zhou, Xiaorui Dong, Youliang Cheng & Hongkuan Hu
*Key Laboratory of Ministry of Education of Condition Monitoring and Control for Power Plant Equipment,
North China Electric Power University, Baoding, Hebei, China*

ABSTRACT: Direct air-cooled condenser is used as cold-end equipment in power plants, and its operation condition affects the safe and economic operation of the whole unit directly. This paper presents an optimized design of the air-cooled unit for the direct air-cooled condenser. As the triangular side of the air-cooling unit has not yet been used for cooling the steam from the condenser, a heat sink was arranged on the triangular side of the air-cooling unit to increase the heat exchange area of the air-cooling unit. In this design, there are three aspects that need to be considered, including the arrangement mode, the thickness of the sink, and the fine porosity of the triangular radiator. As an example, a 600 MW direct air-cooling unit was studied, the best arrangement for the heat transfer effect was obtained, and the heat transfer characteristics of the air-cooling unit was tested. The results indicate that this design of the air-cooling unit can improve the effect of heat transfer of the direct air-cooling condenser, and hence, can contribute to the energy saving and emission reduction effectively.

Keywords: air-cooling unit; optimized design; triangle radiator; heat transfer; condenser

1 INTRODUCTION

As one type of the air-cooling system (Li Yuning, et al., 2008; Wen Gao, 2008), direct air-cooling system condenses steam turbine exhaust directly using ambient air and heat exchanges between air and exhaust gas through the metal wall. Its effect of heat transfer directly affects the safety and economy of the whole unit.

Different structure and shape of the diversion device were installed inside the air-cooling unit by Yang Lijun, Du Xiaoze (Yang Lijun, et al., 2009; Yang Lijun, et al., 2011) et al., to make the air velocity of the whole radiator more uniform, so as to improve the heat transfer performance of air-cooling condenser. Zhou Lanxin (Zhou Lanxin, et al., 2011) et al. proposed the measures for installing arc deflector to reduce the flow rotation inside the air-cooling unit, formed a relatively uniform rising gas from the axial flow fan. Numerical simulation of a single row of fins with cylindrical trapezoidal vortex generators was carried out by Yang Laishun (Zhou Guobing, Yang Laishun, 2011) et al, and the results show that the average heat transfer Nu number of the fin tube had been improved after the vortex generator was added. In the study by Jin Youyin (Jing Youyin, et al., 2008), the measures of installing the grid with the effect of reducing flow rotation were proposed to reduce the nonuniformity, in order to achieve the effect of uniform heat transfer. Sohal and O'Brien

(Sohal M S, O'Brien J E, 2001) proposed the use of elliptical tube instead of circular tube in the air-cooled condenser in geothermal power station, so that the heat transfer performance of the tube bundle was increased (Torii K, et al., 2002; Joardar A, Jacobl A M, 2007). Hutchkiss et al. (P. J. Hotchkiss, et al., 2006; J. A.Van. Rooyen, et al., 2008) used commercial software CFD carrying out simulation analysis, and the results show that the performance of the axial flow fan is affected by the transverse wind. In the research by Owen and Kroger (M. T. F. Owen, et al., 2010), porous shields were installed in the bottom of the air-cooling island platform.

Recently, Cheng Youliang et al. (Cheng Youliang, et al., 2014) proposed an air-cooling heat sink for the direct air-cooling unit from the view of increasing the cooling area of the air-cooling unit, namely arranging radiator at the triangle side of "A" type of air-cooling unit for increasing the heat exchange area of the air-cooling unit in order to improve the effect of heat transfer.

Based on the actual data of the air-cooling unit of a 600 MW unit in a power plant, a 3D model is established. The influence of the arrangement of the triangular radiator, the thickness, and the fin porosity on the heat transfer efficiency of the air-cooling unit with triangular radiator was simulated and analyzed, the optimal arrangement of the heat transfer effect was obtained, and the heat transfer characteristics of the air-cooling unit were improved.

2 MODEL ESTABLISHMENT AND CALCULATION METHOD

2.1 Physical model

In this paper, with a single air-cooling unit as the research object, air-cooling island is constructed on the platform of 45 m away from the ground, the local average elevation is 505, the atmospheric pressure value of this height is 94.6 kPa and the height of geometric model is 18.27 m, and the size of the radiator with "A"-type layout is for 11.44 m × 0.219 m × 9.567 m.

2.2 The air-cooling units for increasing heat transfer area

The traditional air-cooling unit only possessed two radiators for "A" shaped arrangement used for cooling the turbine exhaust steam, and between the two adjacent air-cooling units with only a plate to separate, yet not the arrangement of radiator, caused the loss of cooling space utilization at last.

The idea of the novel air-cooling unit is: turbine exhaust steam is assigned by the steam distribution pipe at the upper side into the "A"-shaped radiator for cooling, meanwhile part of the steam is assigned by the steam distribution pipe at the edge side into the triangle radiator to conduct heat transfer with the cold air blew from the axial flow fan, and this increased the radiating area of air-cooling unit at last.

2.3 Boundary conditions and layout model

The fan of air-cooling unit model is regarded as a thin surface, whose inlet is set as velocity inlet, then the air flow rate is 5.3 m/s. The porous medium model is adopted in the finned tube bundle, which simplified the problem. The top of the model is set as the pressure outlet. The model is surrounded by a symmetrical boundary. The bottom of the unit, the cylinder wall of the wind turbine, and the wall of the steam pipe are all set as thermal insulation wall. The heat load of the air condenser unit is 14.98 MW, and the ambient temperature is 306 K.

The triangular radiator is arranged between two adjacent air condenser units, the design and layout of the triangular radiator had the following two kinds:

Scheme 1: Dismantle the area near the triangle side of the A-type radiator, making the triangular radiator to be arranged at intervals between two adjacent air-cooling units.
Scheme 2: Without changing the original A-type radiator, a certain distance is reserved between two adjacent air-cooled condenser units for the layout of the triangle radiator, and no change is observed in such A-type radiator.

3 RESULTS AND DISCUSSION

3.1 The influence of the arrangement mode of the triangular radiator

Numerical simulation and analysis of the traditional air-cooling units

First, the traditional air-cooling unit was simulated, and the results were processed using the Tecplot software to obtain the temperature distribution of the outer surface of the traditional air-cooling condenser unit radiator, as shown in Fig. 2.

From the diagram of temperature distribution, it can be seen that there is a local high-temperature region at the bottom of the radiator, the temperature is even higher than the saturated steam temperature value, forming a local "heat transfer dead zone".

Comparison of different arrangement schemes of triangle heat transfer

Numerical simulation is carried out according to the design of the upper section of the two triangular radiator layout schemes. The simulation results of the heat sink surface temperature distribution are obtained by the Tecplot treatment, as shown in Figs. 3 and 4.

The temperature distribution of the cooling surface of the air-cooling heat transfer unit had a local change, in which trends are basically consistent for schemes 1 and 2, under the working condition of constant heat load of air-cooled condense (discussions in this paper are all in this condition if no special instructions are given) and two triangle radiators mounted to the traditional air-cooled condenser unit. The general temperature distribution of the entire cooling surface of the triangular radiator is the low temperature at the bottom, and increased gradually along the vapor in a counter flow direction; the temperature at the bottom corners was the highest; and there is a small area of high temperature on the top.

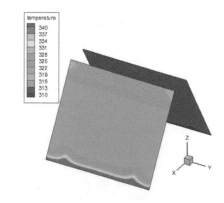

Figure 2. Surface temperature distribution of the traditional air-cooling unit radiator.

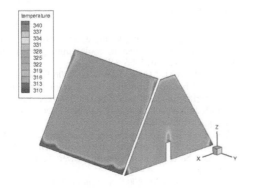

Figure 3. Temperature distribution of the heat sink of scheme 1.

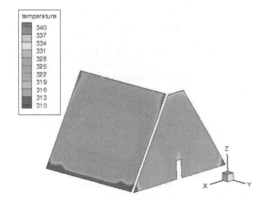

Figure 4. Temperature distribution of the heat sink of scheme 2.

Table 1. The outlet temperature of air-cooling unit radiator.

	Outlet temperature/K		Temperature decrease range/K	
	A-type radiator	Triangular radiator	A-type radiator	Triangular radiator
Conventional air-cooling unit	333.37	-	-	-
Scheme 1	333.88	334.29	−0.51	-
Scheme 2	332.37	333.97	1.00	-

Simulation results of the specific data are shown in Table 1, it can be seen clearly from the table that the outlet temperature of A-type radiator of scheme 1 is increased compared with the traditional air-cooling unit, yet the outlet temperature of A-type radiator of scheme 2 is more reduced than the conventional air-cooling unit. At the same time, the outlet temperature of the triangular radiator

of scheme 1 is higher than scheme 2. Based on the above factors, scheme 2 is better than scheme 1, the layout scheme should be adopted in the program 2.

The influence of the thickness of the triangular radiator is studied. By conclusion in the last section, the arrangement of scheme 2 is better, the layout thickness of triangle radiator is optimized in this section with the other conditions remain unchanged. Specific program is that thicknesses of triangular radiator: 0.219 m, 0.27375 m, 0.3285 m, 0.38325 m, and 0.438 m are arranged, respectively, between the two air-cooled condenser unit, and half of the given thicknesses of triangular radiator are arranged for each air-cooling unit. The specific data of the radiator outlet temperature are shown in Table 2 after using the software FLUENT to simulate the calculation.

It can be seen from the data of above table that the effect of the layout thickness of the triangular radiator on the air-cooled condenser is very large. The outlet temperature of A-type reduce first with triangular radiator thickness increases, then gradually smooth and almost no change; the triangle radiator outlet temperature decreased first and then increased with the triangular radiator thickness increases, when the thickness of the layout is 0.3285 m, outlet temperature of the triangle radiator is the lowest, 332.66 K. Therefore, from the point of the outlet temperature of the radiator, the thickness of the triangular radiator is best for the 0.3285 m.

3.2 The influence of fin porosity of triangular radiator

The porous medium model is adopted in the FLUENT software to reflect the different porosity. The simulation of different porosity is set up to analyze the influence of different porosity on the whole air condenser unit. The simulation results show that the heat transfer effect is best when the thickness of the triangular radiator is 0.3285 m. Based on this, numerical simulation for the triangular radiator is carried out, different porosity are set to reflect the

Table 2. The outlet temperature of air-cooling unit radiator.

Thickness of the triangle radiator/m	Outlet temperature/K		Temperature decrease range/K	
	A-type radiator	Triangular radiator	A-type radiator	Triangular radiator
0.219	332.37	333.97	0	0
0.27375	332.07	333.88	0.3	0.09
0.3285	332.05	332.66	0.32	1.31
0.38325	332.06	334.71	0.31	−0.74
0.438	332.07	340.88	0.3	−6.91

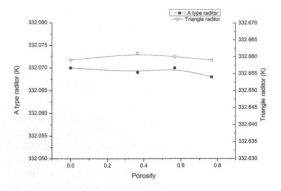

Figure 5. The influence of the porosity of triangular radiator on the outlet temperature of the condenser.

different fin arrangement according to the porous medium model used by the heat sink. The specific porosity is set as follows: 0,0. 367, 0.567, and 0.767. The FLUENT software is used for simulation. The outlet temperature of the condenser after a change in the parameter porosity is shown in Fig. 9.

Fig. 5 indicates the influence of the triangular radiator on the outlet temperature of the condenser. It can be seen from the above that the porosity of triangle radiator has almost no effect for the heat transfer of whole unit. Its value is set to 0.567 for facilitating the subsequent simulation, and the following sections are all the use of this value without a special note.

4 CONCLUSION

1. It can be seen that thermal effect of scheme 2 is obviously superior to the traditional air condenser unit and scheme 1, the layout scheme of scheme 2 is adopted, and the scheme is optimal.
2. Based on scheme 2, the results can be obtained according to the comparison that when the thickness of triangle radiator is 0.3285 m, outlet temperature and pressure of the air-cooled condenser is the best.
3. Changes of porosity almost have no effect on the air-cooled condenser unit, and triangular radiator fin arrangement cannot be with many restrictions, mainly limited to the manufacturing process design.

ACKNOWLEDGMENTS

This project was funded by Fundamental Research Funds for the Central Universities (2016 MS154).

REFERENCES

Cheng Youliang, Hu Hongkuan, et al. 2014. An air cooled heat sink for direct air cooled units [P]. China: 201320679596.3, 2014–06–18.

Hotchkiss, P.J., C.J. Meyer, T.W. Von. 2006.Backstrom. Numerical investigation into the effect of corss-flow on the performance of axial flow fans inforced draught air-cooled heat exchangers [J]. Applied Thermal Engineering, 26(8):200–208.

Jing Youyin, Gui Yanding, Wang Changhai. 2008. Air cooling island airflow distribution uneven cooling effect [J]. Northeastern Electric Power Technology, (11): 6–7.

Joardar A, Jacobi A M. 2007. A numerical study of flow and heat transfer enhancement using an array of Delta-Winglet Vortex Generators in a fin-and-tube heat exchanger [J]. ASME Journal of Heat Transfer, 129:1156–1167.

Li Yuning, Jiang Wenjun, Wu Dong. 2008.The development of air cooling technology in power plant [J]. China Hi tech enterprises, (11); 80–81.

Owen, M.T.F., D.G. Kroger. 2010. The effect of screens on air-cooled steam condenser performance under windy conditions[J]. Applied Thermal Engineering, 30(16):2610–2615.

Sohal M S, O'Brien J E. 2001. Improving air-cooled condenser performance using winglets and oval tubes in a geothermal power plant [J]. Geothermal Resources Council Transactions, 25, 1–7.

Torii K, Kwak K M, Nishino K. 2002. Heat transfer enhancement accompanying pressure-loss reduction with winglet-type vortex generators for fin-tube heat exchangers [J]. International Journal of Heat and Mass Transfer, 45(18):3795–3801.

Van. Rooyen, J.A., D.G. Kroger. 2008. Performance trends of an air-cooled steam condenser under windy conditions [J]. Journal of Engineering for Gas Turbines and Power, 130(2):1–7.

Wen Gao. 2008.Air cooling technology in power plant [M]. Beijing: China Electric Power Press, 9.

Yang Lijun, Du Xiaoze, Yang Yongping. 2011. A canopy type cooling air flow guide of direct air cooling unit [P]. China: 201010551093.9, 2011–04–13.

Yang Lijun, Du Xiaoze, Yang Yongping. 2009. Direct air cooling unit cooling air flow guiding device [P]. China: 200810227060.1, 2009–04–08.

Zhou Lanxin, Li Haihong, Zhang Shuxia, et al. 2011. Numerical simulation of the installation of a meso baffle in the direct air cooled condenser unit [J]. proceedings of the Chinese Journal of electrical engineering, 31 (8): 7–12.

Zhou Guobing, Yang Laishun. 2011. Numerical simulation of heat transfer enhancement of direct air cooled single row tube in vortex generator [J]. modern electric power, 28 (2): 53–57.

Energy Science and Applied Technology ESAT 2016 – Fang (Ed.)
© 2016 Taylor & Francis Group, London, ISBN 978-1-138-02973-6

Influence of vortex shedding of H-vertical axis wind turbines on the structural dynamics of rotors

Q. Yao, J.Y. Wang, C. Wang & C.Y. Zhou
Shenzhen Graduate School, Harbin Institute of Technology, Shenzhen, China

Q.L. Xie
Liugong Forklift Co. Ltd., China

ABSTRACT: In this paper, the vortex shedding effect was studied using the Computational Fluid Dynamics (CFD) method and the Finite Element Method (FEM) for the 5 KW H-vertical axis wind turbine rotor. Lift and drag coefficient of the vortex shedding of a rotating shaft was studied, and also the relationship between the frequency and the inlet velocity. The ANSYS finite element method was used to analyze the harmonic response of a rotating shaft under vortex-induced vibration. And then the safety of the structure under the exciting resonance force was checked to provide a certain theoretical basis for the design of vertical axis wind turbines.

Keywords: vortex shedding; vortex-induced vibration; harmonic response; exciting vibration

1 INTRODUCTION

H-vertical axis wind turbines are booming since the advantages of easy maintenance, manufacturing, and processing (Jha, 2013). However the flow field structure is more complex than that of horizontal axis wind turbines due to the typical large separation unsteady flow. The research of vortex vibration interference between wind and rotation has important significance in revealing the steadiness, pneumatic elasticity and vibration of H-vertical axis wind turbine systems. As a result, domestic and foreign scholars have done a lot of research regarding this. Loth (1985) analyzed double-bladed, three-bladed and four-bladed vertical axis wind turbine tower stress by introducing an appropriate wind interference model to the closed-form solution for the tower force vectors. It shows that the magnitude of the largest tower shake force vector, using a three-bladed rotor, is four times smaller than for a two-bladed rotor. Dong et al. (2012) analyzed the vortex-induced transverse vibration of the 2 MW VAWT unit tower and checked the structural security of the tower when the exciting force causes tower resonance, providing a theoretical basis for the hoisting. Wang (1985) analyzed the natural frequency, vibration mode and other modal parameters of the tower as well as under the action of wind load displacement along the direction of the wind speed and displacement produced by the blade rotation. Dou (1995) established a relatively perfect rotor/tower system model. Through

the matrix iteration method that gives the method for calculating the natural frequencies and natural modes, he expounds the torsional vibration of the tower, bending and torsional coupled vibration. But there are few articles which study rotor vortex induced vibration for the VAWT. But as the wind machine runs within a certain wind speed range, rotation induced by vortex shedding vortex-induced resonance is sometimes unavoidable. If this question is not considered at the beginning of the design, some security problems will exist. So this paper mainly studies vortex-induced vibration to explore its effect on the stability of the vortex-induced vibration of the rotor structure.

2 COMPUTATIONAL FLUID DYNAMICS CALCULATION

2.1 Computational model

In this paper, we select k-omega SST turbulence model to solve the two-dimensional incompressible Navier-Stokes equations. The time derivative is discretized by a second-order implicit scheme and convective term by a second-order upwind differencing scheme and a second-order central differencing scheme for the diffusive term. In solving the final resulting algebraic equation system the SIMPLER algorithm is applied.

Because of the existence of a blunt body in the flow field, it is very difficult to generate a single domain of computational grid. Even if grids are force-generated, grid quality cannot be guaranteed and this will affect the precision of the flow field. So grid design is adopted in this paper, based on the partition, namely according to the outline of the flow field characteristics of the whole flow field which is divided into several sub domains and for each subdomain a structured grid is established. The grid system partition is shown in Figure 1.

Uniform flow velocity is prescribed at the upstream boundary and outlet is set as free export. The sliding grid technology is used to simulate a vertical axis wind turbine rotor and the rotation of the airfoil problem. Rotor-wall and airfoil-walls are set as no-slip wall conditions. On both up and down side-walls of the parallelogram, symmetry conditions are applied to simulate infinite space.

2.2 CFD results

The curves of lift coefficient C_L and drag coefficient C_D changing with the time are shown in Figure 2 when only the rotor uniformly rotates. C_L and C_D coefficients periodically change with the time. And as time elapses, the periodic law is more obvious. The time averaged C_L and C_D are very close to experimental data in the literature (Swanson, 1961) which is shown in Table 1. C_{Lm} and C_{Dm} represent the time averaged C_L and C_D, respectively.

Wind turbines work within a range of wind speed between the cut-in wind speed (3 m/s) and extreme wind speed (18 m/s). The calculation results of lift,

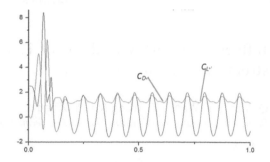

(a) α=0.2

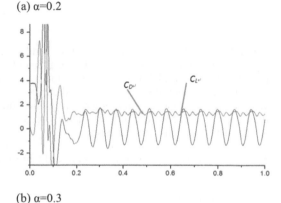

(b) α=0.3

Figure 2. Curves of lift and drag coefficient changing with time.

Table 1. The comparison of the calculated results with the experimental results reported in the literature (Swanson, 1961).

α	The Calculated C_{Lm} and C_{Dm}	The experimental C_{Lm} and C_{Dm}
0.2	(0.16, 1.387)	(0.15, 1.32)
0.3	(0.24, 1.31)	(0.225, 1.26)
0.4	(0.30, 1.25)	(0.30, 1.22)

drag coefficient and vortex shedding frequency are shown in Table 2 when the H-VAWT rotor runs at different wind speeds and its corresponding rotational speed. Here, U_∞ is the inlet wind speed (m/s), n is the rotational speed (rpm), f is the vortex shedding frequency (Hz).

In the reference (Swanson, 1961), lift and drag coefficients basically remain at a fixed value under constant speed ratio α. As the wind turbine is designed according to the set tip speed ratio which is equal to the set speed ratio α and the disturbance of the existing of airfoils, the lift coefficient of calculation results remains at around 0.007, and the drag coefficients remain at around 0.16, which are reasonable.

(a) Whole grid display

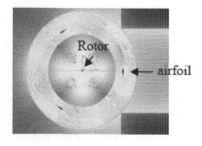

(b) Local grid display

Figure 1. Computational grids.

Table 2. Lift and drag coefficients and vortex shedding frequency at different wind speeds.

$U\infty$ (m/s)	n (rpm)	C_{Ln}	C_{Dm}	f (Hz)
8	80	0.0085	0.1684	5.78
9	92	0.0074	0.1691	6.86
10	115	0.0082	0.1632	7.56
11	126	0.0081	0.1689	8.82
12	137.5	0.0067	0.1675	9.78
13	149	0.0076	0.1658	10.68
13.5	154.7	0.0073	0.1632	10.95
13.6	156	0.0070	0.1636	11.20
13.8	158	0.0065	0.1657	11.43
14	160	0.0081	0.1649	11.69

Table 3. Basic parameters and physical parameters of the 5 kW H-vertical axis wind turbine rotor model.

Title	Value
Inside diameter (m)	0.08
Outside diameter (m)	0.12
Length (m)	3
Material	C_r
Density (kg/m³)	7870
Poisson's ratio	0.277
Elasticity modulus (GPa)	211
Maximum permissible stress (MPa)	800

Table 4. The first six-order modal calculation results of the modal of the rotor.

Order	Frequency (Hz)
1	11.208
2	11.210
3	69.557
4	69.564
5	191.84
6	191.86

3 MODAL ANALYSIS

In order to analyze the security of H-vertical axis wind turbine rotors under air vortex-induced vibration, we first provide the modal analysis.

3.1 Rotor modal analysis

The rotor is modelled according to the design. The rigid beam element is used to establish rotor torque at the top of the rotating shaft. Lumped mass point is used to simulate the quality of the wheel hub and blades whose coordinates are (0, 1500 mm, 0) and total quality is 295 kg. Moment of inertia in the Y direction $I_y = 1843.75$ kg·m². Basic parameters and physical parameters of the 5 kW H-vertical axis wind turbine rotor model are shown in Table 3. The finite element model of the rotor uses a solid 185 hexahedral grid. Model unit is 27119 and the total number of nodes is 36119. UX, UY, UZ, ROTX and ROTZ five direction constraints are exerted at the bottom of the rotor. The first six-order modal calculation results of the modal of the rotor are shown in Table 4.

3.2 Preliminary safety check

As we know, the rotor's rotational speed range is 25 ~ 194 rpm. We can obtain the rotation frequency of the 1P which is 1 times the rotor vibration frequency of 0.42 ~ 3.23 Hz, and 3P is 1.26 ~ 9.7 Hz. From the modal analysis of the rotor, the first-order modal frequency of 11.208 Hz is far away from wind turbines' 3p oar frequency of 9.7 Hz.

$$\frac{11.208 - 9.7}{9.7} = 15.5\%$$

Self-excitation frequency of the rotor is far away from more than 10% of the rotor's rotating frequency and the oar frequency meets the engineering requirements (Zhu, 2008) and satisfies the requirement of GL specification (2003) which require the two frequencies to avoid 5%.

4 EFFECTS OF VORTEX SHEDDING

Airflow of the rotating shaft can produce periodic vortex shedding which forms vertical to the periodic force on it. This force will excite transverse wind vibration which is the vortex-induced vibration. When the frequency of the periodic force is close to or equal to the natural frequency of the rotor, it will be possible for the rotor to vibrate significantly and generate the locking phenomenon. All this will result in vibration being very strong and having the risk to damage the rotor. Vortex shedding frequency follows the change of wind speed leading to vortex shedding frequency which can reach the natural frequencies of the rotor. When this happens, it will cause vortex resonance. In order to analyze the rotor vibration of the H-vertical axis wind turbine, the harmonic response analysis is conducted.

4.1 Rotor harmonic response analysis

The vortex shedding vortex frequency of 11.200 Hz is close to the first-order natural frequency of the rotor of 11.208 Hz at the Reynolds number

Re $= 10.95 \times 10^4$. In this situation the wind speed is 13.6 m/s. Uniformly distributed load is 5.72 N/m (according to the GL specification, the load is multiplied by the safety coefficient of 1.35 Wang (1985)). Assuming that the flow direction of wind is along the Z axis, the direction of the transverse vibration force produced by the wind flow around the tower is along the X axis. Using theoretical mechanics the simplified load amplitude can be obtained as $F_x = 17.16$ N, $M_z = 25.74$ N·m whose acting point is at the top of the tower drum. Assume that the damping coefficient is 0.005 and the phase angle values are zero. We take the vibration frequency range as 11–11.4 Hz and with eight calculation steps, the full solution method and the output format are the amplitudes and phase to conduct the harmonic response analysis. Take the X direction amplitude of the top node of the rotor as the observation point. The amplitude changes subject to the vibration frequency.

Figure 3 shows that when the excitation frequency is 11.2 Hz, the node displacement of the top of the rotor is maximal. The maximal displacement is 29.9822 mm. Further observation under rotor resonance integral displacement and stress distribution is shown in Figure 4 and Figure 5.

The real component calculation results: the maximum stress is 99.7409 MPa, and the maximal displacement is 19.9125 mm. Imaginary part calculation results: the maximum stress is 116.792 MPa,

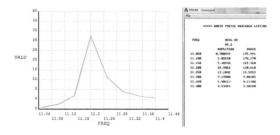

Figure 3. X-direction amplitude of the top point of the rotor changing with the vibration frequency.

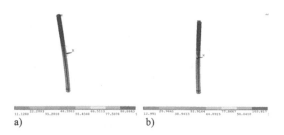

a)   b)

Figure 4. The stress distribution considering rotor resonance. a) stress of real component; b) stress of imaginary component.

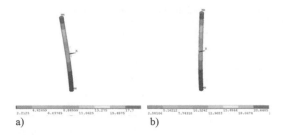

a)   b)

Figure 5. The strain distribution considering rotor resonance. a) displacement of real component; b) displacement of imaginary component.

Table 5. The strain and stress comparison between vortex-induced resonance analysis and static analysis.

Analysis type	Strain (mm)	Stress (MPa)
Static analysis	0.066	3.20
Vortex-induced resonance analysis	30.5960	153.5859

and the maximal displacement is 23.2295 mm. Comprehensive real part and imaginary part calculation results: the maximum stress of the rotor is 153.5859 MPa and the maximal displacement is 30.5960 mm, meeting GB50135–2006 tower structure design specifications, that is

$$\frac{\Delta l}{H} = \frac{30.5960}{3000} < \frac{1}{75}.$$

The result conforms to the safety requirements. Stress is less than the allowable stress $[\sigma] = 800/3 = 266.7$ MPa. The influence generated by the rotation of the lateral vibration of the rotor is small and cannot cause serious damage to the rotor.

The strain and stress comparison between vortex-induced resonance analysis and static analysis are shown in Table 5. The comparison shows that the strain vortex-induced resonance is 464 times that of the conventional strain and stress is 48 times that of the conventional stress. The harm caused by vortex-induced resonance cannot be ignored.

5 CONCLUSION

The analysis of vortex shedding frequency, lift and drag coefficient, inherent frequency and harmonic response of a 5 kW H-vertical axis wind turbine rotor resulted in the following conclusions:

1. H-vertical axis wind turbine rotor vortex shedding frequency, and lift and drag coefficients are related to the speed and Reynolds number, largely in keeping with the Magnus effect;
2. H-vertical axis wind turbines operate within a certain wind speed range and there is difficulty to avoid the vortex-induced resonance. As stress and strain of vortex-induced resonance is greater than the static stress and strain analysis, the vortex-induced resonance safety problem should be checked at the beginning of the design, providing the basis for the design of the wind turbine.

ACKNOWLEDGEMENTS

The work described in this paper was partly supported by the Shenzhen Science and Technology Plans: Key Laboratory Development Projects (Grant No. ZDSYS20140508161547729).

REFERENCES

Dong Zhanzhuo, Liao Hui. 2012. The Research for Vortex-induced Transverse Vibration of the Wind Turbine Tower [J]. Journal of Oriental steam turbine, (2):4–11.

Dou Xiurong. 1995. Horizontal axis wind turbine aerodynamic performance and structure dynamic characteristics research [D]. Jinan: Shandong Industrial University Ph.D. Thesis, 74–79.

GL specification.2003. GL Wind.

Jha, AR. 2013. Wind machine technology [M]. Yue Dawei, Li Jie et al. translate, Beijing: mechanical industry press.

Loth JL. 1985. Aerodynamic Tower Shake Force Analysis for VAWT[J]. Journal of Solar Energy Engineering, (107): 45–49.

Swanson WM. 1961. The Magnus Effect: A Summary of Investigations to Date[J]. Journal of Fluids Engineering, 83(3).

Wang Yongzhi, Tao Qibin. 1995. Wind turbine tower structure dynamic analysis [J]. Journal of solar energy, (2): 162–169.

Zhu Zhenmin. 2008. Megawatt wind turbine vibration analysis and protection [D]. Lanzhou: Lanzhou University of Technology, 2008.

Energy Science and Applied Technology ESAT 2016 – Fang (Ed.)
© 2016 Taylor & Francis Group, London, ISBN 978-1-138-02973-6

Product integration process of the CMMI grade 3 model

C.M. Su, Y.G. Li & L.B. Guo
China Satellite Maritime Tracking and Controlling Department, Jiangyin, China

ABSTRACT: The purpose and requirements of CMMI grade 3 product integration process are analyzed. The purpose of product integration is made clear to assemble product components into product, ensure a product integrated function accommodation, and consign a product. Based on accumulation of practical knowledge, a specific product integration process of implement and management solutions is proposed, and an application case is given. The research in this paper has some reference value for enterprise software process improvement.

Keywords: capability maturity model integration; product integration; interface

1 INTRODUCTION

Product Integration (PI) process area is the process area of CMMI maturity grade 3. The purpose of PI is to assemble the product from the product components, ensure that the product, as integrated, behaves properly, and deliver the product. Therefore, the effective overlay implementation of PI will contribute to raising software engineering level and software product quality, lowering software product cost, and promising to consign a qualified software product according to the date.

Although CMMI standard medium explicit PI process area carries out a request, the operability is not very strong, for attaining CMMI model PI process area-specific practice request, carrying out specific goal, and the writer summarizes the concrete practice and success experience of department implementation CMMI3 certification evaluation process, puts forward a set of reasonable and feasible product integration process, and combines an application for presentation, and expects to provide useful reference for the implementation of CMMI process improvement of relevant units.

2 METHOD OF PI CARRYING OUT

PI is a process activity that it assembles Technical Solution (TS) process area output of product component into product. The scope of PI is gradually construction and evaluation product component at a stage or some incremental order according to defined integration sequence and regulation, until the completion of the complete PI.

For attaining PI process area-specific practice of request, carry out a specific goal; this paper makes detailed analysis and research from integration plan, PI, and interface management.

2.1 Integration plan

Product integration is not complete, therefore we need to draw up an integration plan and make of the file turn. The main purpose of integration plan is: describe an integrating, carry out strategy by support; describe each integrated step that needs to be done what toward the participant; and identify the resources of demand and where, when, and what location is needed. The degree of plan is different for the different scale and application-type project product.

Integration plan include following contents:

1. Certain integration order: a) marking integrated item (software unit, component, configuration, system etc.); b) marking verifying activity; and c) choosing integrated sequence and reason.
2. Certain integration environment: a) marking integration environment and b) marking the identification standard of integration environment.
3. Certain integration regulations and standard: a) marking integration test detail and b) certain integration standard.

2.2 Product integration

According to the grain level that needs to be integrated as product component, PI is divided into component level integration, configuration level integration, and system level integration and cut implementation according to the project actual scale and characteristics in the project development. After component (configuration, system)-level integration after completion unit (component, configuration) test, carry out integration according to a component (configuration, system) level integration request and integration order in the PI plan.

PI is divided into three steps: preparing and confirming integrated item and integration environment, implementing product integration, and product packaging and delivery.

Confirming integrated product is to confirm that integrated items have already passed test, and version is controlled, according to product integration plan we check against to treat integration parts pursue item reviewing, and confirm that interface and design of integrated item are consistent.

Confirming integration environment is to build integration environment first. Developer and tester according to the environment request in the product integration plan build the integrated soft-hardware environment, like test equipment and tool which ensures integrated reliability and usefulness. The general recommendation builds good integration environment before coding work begins, easy to develop and test. Then, press acceptance standard for integration of the environment carries on confirmation, and record check confirms result in "Product integration report".

Implementing PI step is performed as follows:

1. Integrating director must obtain exactitude treating and integrating document and version from the configuration database, and install appointed machine. Integrating director uses the related tool to carry on editing and translating, integrating, and check whether compilation and integration pass according to integration verification regulations.
2. If integration passes and then integrating director deploy integration passed package to appointed test machine, hand it to configuration database in the meantime, and integration be over.
3. If integration impasses and then integrating director analytical and check to seek the reason of integrating failure, adopt a different countermeasure according to the different reason. If source code problem then notice developer modify a source code or supply to hand in a source code.

Product packaging and delivery: 1) If integration is completed, integrator carry on check toward a product handed in. 2) According to the document contract request, integrator will pack text needed, executable code, configuration data, etc., to create install file, and hand over configuration manager upload to configuration database of appointed catalogue, and provide product consignation.

2.3 *Interface management*

To ensure interface compatibility is an all-important link in the product integration process area. Interface includes interior interface and exterior interface, interior interface is the interface of the product module piece, and then exterior interface is interface with external system to being integrated. However, regardless of interior or exterior interface, all need to carry on the necessity management, such as the control and review of interface of change etc.

When Coders is carrying on interface realization, if the interface cause changed because of the requirement changes, the design blemish etc. factor, then coders should put forward interface change apply in time, and hand in to designer and project manager carry on a review. If review conclusion is impassability, then coders according to review opinion continue to modify, or re-adjust change claim, re-review after negotiating. If review conclusion is passed, then designer completes the interface design document after interface change, and hand in to coder's modification to interface procedure after redesigning interface.

3 APPLICATION

Emulate Simulate Control Center software system (ESCC) was the CMMI3 evaluated project for my department. After the project starts up, project manager chooses a life cycle model according to CMMI3 systematic document and the project characteristics, and plans integrating activity in the software develop plan, explicit product integrated person and integration process, establishment of integrated plan and resolve the integration activity to the life cycle model different software develop stage in the work breakdown structure. According to the CMMI3 systematic cuts guidebook, the C class software needs to carry out a configuration

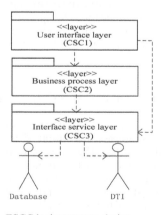

Figure 1.   ESCC logic structure design.

integration and the component to independently circulate needs to carry out component integration. This system wraps to independently circulate for each component of the C classes software and demarcation, therefore certain integration activity carries out configuration integration after unit test completion. The continuation carries on a description to part of integration activities.

1. Integration strategy

ESCC is divided into three layers: user interface layer, business process layer, and interface service layer. Each layer is an integrated software component. The ESCC logic structure is shown in Figure 1.

We can see that interface service layer is the lowest layer, the upper layer is business process layer, and topside layer is user interface layer from the Figure 1, and three layers exist a dependence relation with each other. Comprehensive consider an above factor, we select from bottom to upward integrating method and follow to transfer relation order implementation repetitious integration. The concrete step is as follows.

a. According to "ESCC software design description", to be clear to the logic relationship between the integrated unit and layer, then develop integrated sequence. Developing integration sequence of principle for layer with layer of integrate first not to need exterior to dependent

layer, layer inside integrate first not to need external dependent unit, there is no unit and layer of mutual dependence relation integrated the in proper order does not matter, there are the unit and layer to depend on a relation integrating unit and layer that is depended on first successively.

b. According to step, an explicit of integration in proper order, software unit is integrated layer first, again layer is integrated configuration, and to test and verify the accuracy of interface and function of integrated component.

c. Testing a software configuration whether can work as usual. It is necessary that tester develop some drive code to drive tested configuration.

2. Integration entry and exit standard

Entry standard: a) the classes in the layer have already passed unit test; b) the classes in the layer

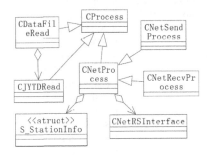

Figure 3. Business process layer relation.

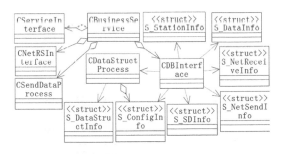

Figure 2. Interface service layer relation.

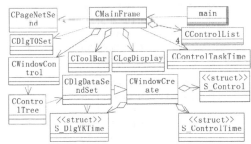

Figure 4. User interface layer relation.

Table 1. PI environment and check (configuration level).

Integrated environment name	Acquire way and description	Use/type	Result
NeoKylin	From organize wealth library. Domestic OS: Kylin 3.2–4(64).		√
QT Creator/ QT	From organize wealth library. QT Creator2.7/ QT 4.8.	Developing-testing equipment	√
Computer	CPU Core i 3, RAM 2 GBs, hard disk 500 GBs.		√
Database	From organize wealth library. Shentong databaseV7.0.		√
DTI	The engine room environment provides,to provice time and interruption service.		√
ESCC	From controlled library. Version V0.9 after unit test, it must match the integration entry standard.	Software sourcecode integrated	√
Replay Software	From organize wealth library. Version V1.00.	Test and verify	√

Table 2. PI interface list and check (configuration level).

Interface pack name	Interface name	Pack depends on relation	Interface description
CSC1←→CSC2	CServiceInterface	CSC1 depends on CSC2	To check that exterior interface definition and each interface unit description should be consistent with all operate in the PI plan and software design
CSC1←→CSC3	CDBInterface, CDataStructProcess, CNetRSInterface, CServiceInterface	CSC1 depends on CSC3	
CSC3←→CSC2	CDBInterface, CDataStructProcess, CNetRSInterface	CSC2 depends on CSC3	

Table 3. PI report (configuration level).

No.	Integrated pack name	Integrated unit name	Dependent unit No.	Function/Interface verify	Complete date	Conclusion
1	Interface service layer (CSC3)	S_DataInfo	/	√	2015.9.23	pass
2		S_NetReceiveInfo	/	√	2015.9.23	pass
3		S_NetSendInfo	/	√	2015.9.23	pass
4		S_SDInfo	/	√	2015.9.23	pass
5		S_StationInfo	/	√	2015.9.23	pass
6		S_DataStructInfo	/	√	2015.9.23	pass
7		CDataStructProcess	6	√	2015.9.23	pass
8		CDBInterface	1,2,3,4,5,7	√	2015.9.23	pass
9		CNetRSInterface	/	√	2015.9.23	pass
10		CServiceInterface	/	√	2015.9.23	pass
11		CSendDataProcess	/	√	2015.9.23	pass
12		CBusinessService	8,9,10,11	√	2015.9.23	pass
13	Business process layer (CSC2)	CProcess	/	√	2015.9.23	pass
14		CJYTDRead	13	√	2015.9.23	pass
15		CDataFileRead	13,14	√	2015.9.23	pass
16		CNetProcess	5,9,13	√	2015.9.23	pass
17		CNetSendProcess	16	√	2015.9.23	pass
18		CNetRecvProcess	16	√	2015.9.23	pass
19	User interface layer (CSC1)	S_DlgYKTime	/	√	2015.9.23	pass
20		S_ControlTime	/	√	2015.9.23	pass
21		S_Control	/	√	2015.9.23	pass
22		CPageNetSend	/	√	2015.9.23	pass
23		CDlgT0SetT0	/	√	2015.9.23	pass
24		CWindowControl	/	√	2015.9.23	pass
25		CLogDisplay	/	√	2015.9.23	pass
26		CControlList	/	√	2015.9.23	pass
27		CControlTaskTime	/	√	2015.9.23	pass
28		CToolBar	/	√	2015.9.23	pass
29		CDlgStationInfoSet	/	√	2015.9.23	pass
30		CDlgDataSendSet	31,19	√	2015.9.23	pass
31		CControlTree	29,28	√	2015.9.23	pass
32		CWindowCreate	20,21,19	√	2015.9.23	pass
33		CMainFrame	22,23,24,25,26,27,28	√	2015.9.23	pass
34		main	33	√	2015.9.23	pass

have already marked, function, and description agree with; c) all interfaces have matched the functions and capability requests of design; d) configuration integrated environment has to be prepared to complete and got a verification.

Exit standard: a) Various problem in the integration process have got effective solution. b) Integration test detection of blemish have satisfied quality target request. c) Software has already integrated an configuration, and it can carry out functions of design and satisfy a capability request.

3. Integrated object and sequence

As shown in Figure 1, the upper layer depends on the lower layer realization, selection from bottom to upward of integrated method is clearer. First to integrate interface service layer being never needed exterior dependence; again to integrate business process layer who depend on interface service layer; end to integrate user interface layer. Each layer inner relation is shown in Figures 2–4. Software integrated order is shown in Table 3.

4. Integrating environment and validation

Integration of environment verification verifies whether soft hardware and equipment's model type exactitude, whether function can use as usual. Then constitute and check it as shown in Table 1.

5. Integrating interface and validating

Checking and verifying of interface accuracy is point of software configuration level integration. Interface list and check records are given in Table 2.

6. Product integration result

Software integrating process: first to verify whether integrated environment satisfies integration request, then to verify integrated software (or unit) to being ready a circumstance according to integrating order, and to check each unit function and interface whether is consistent with design document, to verify integrated unit transfer depended unit of function or interface whether can use as usual if software unit have a dependence relation, product integration result, see Tables 1, 2, and 3.

To carry out PI according to CMMI3 Request, by product of strict norm integration process management, effectively reduced a judge blemish and test BUG, raised the implementation efficiency of verification and confirmation activity, ensure thus according to the date guarantee both quality and quantity to consign a qualified product.

4 CONCLUSION

This paper introduced PI process area specific goal and specific practice in the CMMI's model, and PI process putted forward nicely overlaid PI process area all specific goal and specific practice and satisfied PI process area request in the CMMI's model. This paper from integration plan, product integration and interface management three concrete analyzed to study a product integration process implementation methods, and gave applied case. This paper research aim is improving the software process for software enterprise, and providing reference for software project manager carrying on PI.

REFERENCES

Gao, Y. et al. 2011. Application of Software Verification and Validation in MIS Development, Computer Engineering, 37(1): 84–86.

Li, X.M. et al. 2013. Research on the Construction of GJB 5000 A—2008 Grade-3 System Based on Software Engineering Process. Journal of Telemetry, Tracking and Command, 34(2): 72–76.

NASA Headquarters. 2007. NASA Systems Engineering Handbook. USA: National Aeronautics and Space Administration, 78–81.

Ni, T. Promote 2013. Gjb5000a in Software Development and Testing. Software, 34(2): 31–35.

Watts Humphrey 2011. CMMI for Development Guidelines for Process Integration and Product Improvement for Third Edition. USA: Addison-Wesley.

Ye, Z.L & Wang, X.H. 2013. Research of Testing Technology about Embedded Software of Satellites Based in CMMI. Computer Measurement & Control, 21(4): 72–76.

Energy Science and Applied Technology ESAT 2016 – Fang (Ed.)
© *2016 Taylor & Francis Group, London, ISBN 978-1-138-02973-6*

A novel cohesive zone model for the finite thickness of adhesive layers

Lili Chen, Boqin Gu & Xiaoming Yu
School of Mechanical and Power Engineering, Nanjing Tech University, Nanjing, P.R. China

Bin Zhang
School of Mechanical Engineering, Changshu Institute of Technology, Changshu, P.R. China

Jiahui Tao
School of Mechanical and Power Engineering, Nanjing Tech University, Nanjing, P.R. China

ABSTRACT: A novel Cohesive Zone Model (CZM) for a finite thickness adhesive layer was proposed in this paper. The damage constitutive model of the new CZM is derived according to a continuum damage mechanics formulation with two important parameters. The smaller the parameter β, the more the energy needed for the failure of the adhesive material. The parameter ε0 has a significant influence on the shape of the constitutive curve. The finite element simulation results based on the User Material Subroutine (UMAT) developed in this paper agree very well with the theoretical values.

Keywords: CZM, finite thickness, damage constitutive model

1 INTRODUCTION

The Cohesive Zone Model (CZM) was developed by Dugdale (1960) and Barenblatt (1962). Its fundamental idea is concerned mainly with the whole process of fracture to a thin region called cohesive zone (M. Elices, 2002). The cohesive zone is a fracture process zone around the crack tip, where the material still transfer stresses and strains after being damaged. When the crack tip opens, the stress would not fall to zero immediately, but would decrease gently when the crack width increases. The interaction vanishes when a critical displacement is reached (Q. Ye, 2011).

Thus far, most of the available CZMs in the literature are integrated with zero-thickness interfaces (S. T, 2006). However, practical adhesive materials have finite thickness and different elastic properties than those of the materials neighboring. The use of a CZM that accounts for the finite thickness would avoid a redundant construction and a finer finite element mesh of adhesive material layer. Some studies have been conducted on the CZMs with finite thickness. Zhang et al. (P. Zhang, 2012) used finite-thickness cohesive zones, which is automatically generated, to model the grain boundaries and junctions. However, the constitutive models of the CZMs used in above researches still follow the traditional empirical ones of the CZMs with zero thickness. A complex process of trial and error is necessary to get the appropriate parameters when

the empirical constitutive models are used. An effective way to obtain physically meaningful damage constitutive model is the theoretical derivation based on classical damage mechanics.

In the present work, a novel CZM for finite thickness adhesive layer was presented. The degradation of the mechanical properties in the adhesive layer is modeled based on a damage propagation formulation. The shape of the CZM comes directly from the damage evolution law. The User Material Subroutine (UMAT) for ABAQUS based on the proposed model was developed, and the explicit expression of the Jacobian matrix of the UMAT was given. The parameter studies of the proposed model and comparisons between the finite element results with the theoretical results were carried out.

2 PROPOSED FINITE THICKNESS COHESIVE ZONE MODEL

A layer of adhesive material with Young's modulus E_a and finite thickness l_a embedded between two layers of matrix (Fig.1a) is considered in this paper. The two material interfaces are considered to be bonded perfectly. The specimen is subjected to a normal tension σ (Mode I crack). The behavior of the materials is linear elastic initially.

This elastic relationship is assumed to hold until the normal strain of adhesive material ε_a reaches a

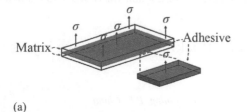

(a)

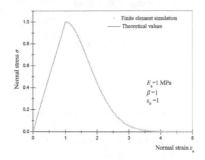

(b)

Figure 1. (a) Adhesive layer embedded in matrix. (b) Mechanical behaviors obtained from simulation and theoretical calculation.

certain threshold value, say ε_0. Then, for a result of complex dissipative mechanisms, such as interface debonding, interatomic bond breaking, and formation of micro voids, progressive damage develops. By introducing a damage variable, D, ranging from zero ($\varepsilon_a = \varepsilon_0$) to unity (finial failure), the dissipative mechanisms at lower length scale can be taken into account. The distributed defects induce progressive material deterioration. Hence, Young's modulus of the damaged adhesive material is $E_a(1-D)$, and the stress-strain relationship becomes:

$$\varepsilon_a = \frac{\sigma}{E_a(1-D)} \tag{1}$$

The damage evolution law has to be specified. Thus far, different formulations based on damage mechanics could be found, as discussed in the review paper by Krajcinovic (D. Krajcinovic, 1981). In this paper, damage is assumed to be related to the level of strain. An ideal quantity is the normal strain of the adhesive material. Then the D is given by a differential equations as below:

$$\beta E_a \varepsilon_a (1-D) = \frac{\Delta D}{\Delta \varepsilon_a} \tag{2}$$

where β is a material constant related to the damage mechanisms of the adhesive material existing at a low scale.

Integrating equation (2) and substituting the integral boundary condition ($D = 0$ when $\varepsilon_a = \varepsilon_0$) yields the explicit expression of damage variable D:

$$\begin{cases} D = 0 & \varepsilon_a \leq \varepsilon_0 \\ D = 1 - e^{-\left[\frac{\beta E_a}{2}\left(\varepsilon_a^2 - \varepsilon_0^2\right)\right]} & \varepsilon_a > \varepsilon_0 \end{cases} \tag{3}$$

The relationship of normal stress-strain of the adhesive material can be obtained by inserting equation (3) into equation (1):

$$\begin{cases} \sigma = \varepsilon_a E_a & \varepsilon_a \leq \varepsilon_0 \\ \sigma = \varepsilon_a E_a e^{-\left[\frac{\beta E_a}{2}\left(\varepsilon_a^2 - \varepsilon_0^2\right)\right]} & \varepsilon_a > \varepsilon_0 \end{cases} \tag{4}$$

Finally, for $D = 1$, a macroscopic crack is generated and the adhesive layer can no longer be able to endure external load (which means $\sigma = 0$).

The behavior of the cohesive zone model is illustrated in Figs. 2 and 3 by means of two examples. The driving parameter is the strain of the adhesive layer, ε_a. For $E_a = 1$ MPa, $\varepsilon_0 = 0.1$, and there are five different values of $\beta = 0.02, 0.1, 0.5, 1$ and 10. The normal stress-strain relationship of the adhesive material is depicted in Fig. 2a. The damage evolution curves are also shown in Fig. 2b.

It can be found that parameter β does not affect the shape of the constitutive curve. An initial hardening after the onset of damage is observed in all the curves (Fig. 2a). What is more, the smaller the β, the larger the area enveloped by the constitutive curve. This means the toughness of the adhesive layer increases with the decrease in β. It is found from Fig. 2b that the damage evolution curves with different β are similar in shape. The speed of damage evolution increases first and then decreases. Moreover, the smaller the β, the slower the damage evolution.

The parameter ε_0 also has an important effect on the mechanical behavior of the adhesive material. This effect is studied by considering a β to be equal to 1 and five values of ε_0 to be 0, 0.5, 1, 2 and 4 respectively. All the other parameters are taken as the same with the example before.

It can be seen from Fig. 3a that ε_0 has a significant influence on the shape of the constitutive curves. When ε_0 is larger than approximately 2, the mechanical response is brittle, with a sudden drop of the stress after the onset of damage ($\varepsilon_a = \varepsilon_0$). Otherwise, the mechanical response is relatively ductile, with an initial hardening followed by a smooth softening branch after the onset of damage. This is because the damage evolutions are different with the decrease of ε_0, as shown in Fig. 3b.

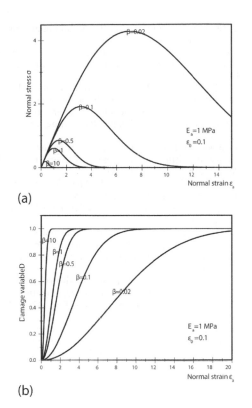

(a)

(b)

Figure 2. (a) Normal stress vs. normal strain of adhesive layer. (b) The effect of parameter β on the damage evolution.

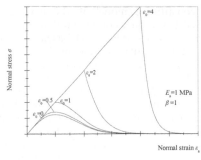

(a)

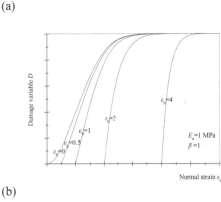

(b)

Figure 3. (a) Normal stress vs. normal strain of adhesive layer; (b) The effect of parameter ε_0 on the damage evolution.

$$\dot{\sigma} = E_n \dot{\varepsilon}_n \qquad\qquad \varepsilon_a < \varepsilon_0$$
$$\dot{\sigma} = (E_a + \beta E_a^2 \varepsilon_a^2)(1-d)\dot{\varepsilon}_a \qquad \varepsilon_a > \varepsilon_0 \qquad (5)$$

where '·' above the variable represents the first time derivative.

The case of plane strain is considered in the simulation. The adhesive layer is modeled by 4-node two-dimensional cohesive elements (COH2D4) in ABAQUS. The displacement load along the thickness is applied on the top surface of the adhesive layer. Moreover, the lower surface is fixed. The parameters of the damage constitutive model used in the simulation are listed in Fig. 1b.

The mechanical behaviors of the adhesive layer obtained by the finite element simulation and the theoretical calculation (equation (4)) are shown in Fig. 1b. It can be seen that the finite element simulation results agree very well with the theoretical values, which verifies the effectiveness of the UMAT developed in this paper.

It can be seen that the larger the ε_0 is, the more rapid the damage evolution is. It should be noted that when the value of ε_0 is larger than about 1.5, the speed of damage evolution keeps decreasing after the onset of damage.

3 NUMERICAL APPLICATIONS

The User Material Subroutine (UMAT) for ABAQUS Standard finite element code based on the damage constitutive model described in section 2 was developed. Then the UMAT was combined with the cohesive element of ABAQUS to model the damage process of the adhesive material.

The Jacobian matrix has to be given for the UMAT in ABAQUS. As membrane stresses are not included in the cohesive element and only Mode I crack is considered in this paper. Therefore, only the ratio of the normal stress increment and the normal strain increment (equation (5)) is needed in the Jacobian matrix.

4 CONCLUSION

In the present work, a novel CZM for the finite thickness adhesive layer was presented. The damage constitutive model of the new CZM was derived based on a continuum damage mechanics formulation, and the UMAT based on the damage constitutive model was developed.

The β of the damage constitutive model do not affect the shape of the constitutive curve. The smaller the β is, the more the energy is needed for the failure of the adhesive material. When ε_0 is larger than about 2 the mechanical response is brittle, with a sudden drop of the stress after the onset of damage ($\varepsilon_a = \varepsilon_0$). Otherwise, the mechanical response is relatively ductile, with an initial hardening followed by a smooth post-peak softening branch after the onset of damage. The finite element simulation results based on the UMAT agree very well with the theoretical values. The method of finite element can be used to model the complex damage problems of adhesive layer in the future studies.

REFERENCES

Barenblatt, G.I. The mathematical theory of equilibrium cracks in brittle fracture, Adv Appl Mech. 7 (1962) 55–129.

Dugdale, D.S. Yielding of steel sheets containing slits, J Mech Phys Solids. 8 (1960) 100–104.

Elices, M. Guinea, G.V. Gómez, J. Planas, J. The cohesive zone model: Advantages, limitations and challenges, Eng Fract Mech. 69 (2002) 137–163.

Floros, I.S. Tserpes, K.I. Loebel, T. Mode-I, mode-II and mixed-mode I plus II fracture behavior of composite bonded joints: Experimental characterization and numerical simulation, Compos Part B-Eng. 78 (2015) 459–468.

Krajcinovic, D. Fonseka, G.U. The continuous damage theory of brittle materials, part 1: General theory, J Appl Mech. 48 (1981) 809–815.

Lei, H. Wang, Z. Zhou, B. Tong, L. Wang, X. Simulation and analysis of shape memory alloy fiber reinforced composite based on cohesive zone model, Mater Design. 40 (2012) 138–147.

Pinho, S.T. Iannucci, L. Robinson, P. Formulation and implementation of decohesion elements in an explicit finite element code, Compos Part A-Appl S. 37 (2006) 778–789.

Ye, Q. Chen, P. Prediction of the cohesive strength for numerically simulating composite delamination via CZM-based FEM, Compos Part B-Eng. 42 (2011) 1076–1083.

Zhang, P. Karimpour, M. Balint, D. Lin, J. Cohesive zone representation and junction partitioning for crystal plasticity analyses, Int J Numer Meth Eng. 92 (2012) 715–733.

Energy Science and Applied Technology ESAT 2016 – Fang (Ed.)
© 2016 Taylor & Francis Group, London, ISBN 978-1-138-02973-6

Stamp forming of commingled GF/PP fabric with vacuum technology

Zhou Zhou, Bingyan Jiang & Wangqing Wu
School of Mechanical and Electrical Engineering, Changsha, China

ABSTRACT: Stamp forming of unconsolidated commingled GF/PP fabric composite has been studied for plate geometry. The effect of the process parameters (stamping temperature, mold temperature and holding time) on the flexural properties and void content are determined. Vacuum technology is introduced to evacuate the air from the fabrics. Flexural strength and modulus have a good correspondence with the void content, which could predict approximately the flexural properties by only measuring the void content.

Keywords: commingled fabrics; vacuum system; stamp forming; mechanical properties

1 INTRODUCTION

Continuous Fiber Reinforced Thermoplastics (CFRT) have taken the place of thermosets composite as structural materials in the field of aircraft industry (A.R. Offringa, 1996; M. Hou, 1997). Recently, numerous works have been performed to introduce the CFRT in the automotive industries (H. Kuch, F, 2000; M.D. Wakeman, 2000; H. Kuch, 2000). Since thermoplastic composites offer new and automated manufacturing process with low cycle time, the development of superior manufacturing techniques determines the extensive utilization of of such materials in the future (A.R. Chambers, 2006).

In the past 20 years, numerous manufacturing methods were developed whereas stamp forming has been considered as one of the most attractive processing methods to process the CFRT, due to the short molding cycle (M. Hou, 1997). And the content of void is identified as the most important parameter affecting the mechanical strength (A.R. Chambers, 2006).

Experimental investigations ware conducted in this paper, which presents the manufacturing process of stamp forming unconsolidated continuous fiber composite fabrics. The research object of this work was to characterize the thermally consolidated fabrics plies to be stamped. To achieve this objective, processing parameters such as stamping temperature, mold temperature and holding time were determined. In order to eliminate entrapped air, vacuum technology is introduced to improve the mechanical properties of CFRT. Further, isothermal annealing is also introduced to increase the degree of crystallinity of the matrix resulting from the great temperature gradient between the mold and the fabrics during the process of molding.

2 EXPERIMENTAL

Materials and Equipment. The CFRT studied in this work were supplied by Zhenshi Group Hengshi Fiberglass Fabrics Co. Ltd., which is consisted of two unconsolidated continuous fiber fabrics woven into balanced 2-2 cross-plied fiber cloth. The weaves were made up of Compofil™ yarns. The melting temperature of polypropylene was 165°C. Fabrics nominal density was calculated to be 980 g/m². The fiber weight fraction was 60%, which means that the final consolidated fiber volume fraction is 35%. Four layers of fabrics were superimposed, whose thickness is approximate 8 mm and 2.5 mm after consolidation. The unconsolidated fabrics were heated in a convection oven before stamp forming, which was performed using a plate metal mold installed on a 300kN hydraulic press (Delishi DSB-30A). The working area of mold is a square plate of 280×280 mm² with a circular groove which was placed by sealed silicone rubber. Dies could be internally heated by means of electrical cartridges. The entrapped air was eliminated using a vacuum pump and mechanical properties were performed using a SANS mechanical testing machine (CMT 4204) with ISO 14125 standard.

Processing procedure. There are five steps of the stamp forming stage: (i) supplying the vacuum, (ii) closing of the mold, (iii) application of the pressure, (iv) holding of the pressure, and (v) open the mold to reduce/drop pressure. In order to create a sealed environment, the upper die was lowered quickly onto the sealing silicon rubber (Fig. 1(a)). After applying vacuum, the upper die was further lowered onto the fabrics until reaching the setting pressure using the pressure gauge (Fig. 1(b)).

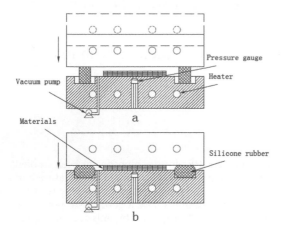

Figure 1. The procedure of stamp forming with vacuum.

Table 1. The values of stamping parameters.

Stamping temperature (°C)	175, 185, 195, 205, 215
Molding temperature (°C)	25, 50, 75, 100, 130
Holding time(s)	15, 30, 45, 60, 75
Vacuum degree (bar)	0, 0.4, 0.6, 0.8, 0.95

Four stamping parameters, such as the stamping temperature, mold temperature, holding time and vacuum pressure were studied. 4 MPa and 8 mm/s were chosen for stamping pressure and a closing speed respectively, which brings the upper mold onto the fabric. Every parameter was studied respectively, that is to say one parameter was studies when the others were kept constant. The stamping parameters values are listed in Table 1. The underlined values of parameters were constant when the other parameters were studied.

3   RESULTS AND DISCUSSION

Effect of stamping parameters. Stamping temperature is a vital processing parameter for stamp forming. Figures 2 show the variation of the flexural strength with respect to the stamping temperature, respectively. A great improvement of the flexural strength is observed with increasing the stamping temperature from 175 to 185°C. It is evident that higher stamping temperature is required to acquire the final stabilization. Void content suddenly decrease at a temperature of 185°C. A possible explanation for the higher temperature needed to obtain the highest mechanical properties may be related to the difficulty of evacuation the air from the unconsolidated layers. At the temperature of 185 to 195°C, the void content amount remains almost unchanged. Beyond 195°C or higher, the

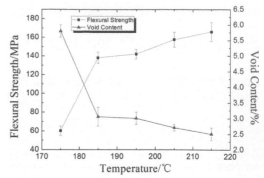

Figure 2. Influence of the stamping temperature on the flexural strength and void content.

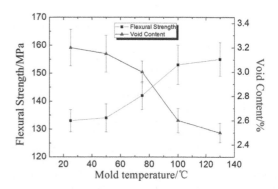

Figure 3. Influence of the mold temperature on the flexural strength and void content.

void content gradually decreases with increasing stamping temperature, which suggests that the air is evacuated a higher stamping temperature.

The flexural properties are found to be poor at relatively low temperature, such as 25 and 50°C (Figure 3). There are two plateaus: one for the initial two temperature and other for the final two mold temperature. Consequently, 75°C is found to be the minimum mold temperature to obtain better flexural properties with lower stamping temperature. The variation of the void content with respect to the mold temperature resembles to be the flexural properties, i.e. Void amount is decreased and flexural properties is increased simultaneously with increasing mold temperature.

Hold time is defined as the compression time allowed to condense the materials with applying stamping pressure. Figure 4 show that flexural properties obtained using a short holding time (15 s) is insufficient and considered to be very low in amount. This could be explained that the holding time of 15 s is slightly shorter than the solidification time of the polymer matrix. However, flexural properties increase slowly with increasing

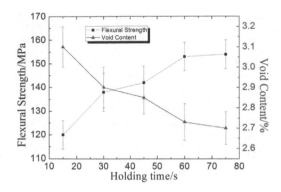

Figure 4. Influence of the holding time on the flexural strength and void content.

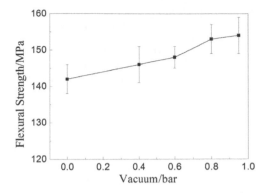

Figure 5. Influence of the vacuum on flexural strength with variation stamping temperature.

the holding time from 15 to 60 s. The flexural properties and the void content have a very good correspondence. The abrupt reduction in void content corresponds to increase in mechanical properties during holding time of 15 to 30 s.

The vacuum pumping technology is generally used for processing of thermoset composite molding or CFRT autoclave molding. Zhongyou (Wu Zhongyou, 2010) indicated that vacuum could increase the driving force of thermoset matrix to impregnate the fibers. In the case of CFRT moiling vacuum benefited for excluding the air as the viscosity and stamping pressure were high enough. Thus, vacuum was applied before the step of application of the pressure in the study to study the effect of vacuum on flexural strength. Vacuum applied before applying the pressure was investigated. Figure 5 shows the effect of the degree of vacuum on the flexural strength. The flexural strength increased slightly with increasing the vacuum up to 0.6 bars. However, the flexural strength increased rapidly and reached to 154 MPa at 0.8 bars. It is interesting to note that the increment rate of the flexural strength was not very pronounced beyond 0.8 bars. It can be summarized that vacuum imposed before applying the pressure could exclude the entrapped air and increased the flexural properties.

Effect of isothermal Annealing. Stamp forming products are annealed isothermally to improve the mechanical properties of semi-crystalline polymers. In order to reveal the effect of isothermal annealing on flexural strength, the flexural strength was measured after annealing using various mold temperature. Based on the aforementioned investigation, annealing temperature was selected to be 160°C and the annealing time was set to be 30 min. Figure 6 shows the relationship between the flexural strength and molding temperature before and after annealing. It is evident that flexural strength

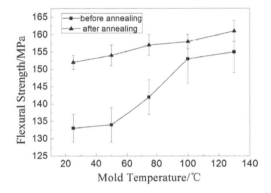

Figure 6. Effect of mold temperature on flexural strength before and after annealing.

increased after isothermal annealing with increasing the mold temperature from 25 to 75°C. If the mold temperature is reached at 100°C and above, the improvement of flexural strength was not significant. It can be concluded that low mold temperature reduces the crystallization of PP with higher cooling rate and isothermal annealing could improve the flexural strength without increasing the mold temperature.

4 CONCLUSION

In this paper, stamp forming process of unconsolidated PP/GF fabric was investigated using a sheet molding. The effect of the stamping temperature, mold temperature, holding time and the degree of vacuum on the void content and flexural mechanical properties have been determined. Results indicate that vacuum before stamping affected positively in eliminating the air from fabric, which ultimately reduced the void content and increased the flexural properties, which obtained using stamp forming process of unconsolidated fabrics were compared to the properties obtained by the annealing treated one, and the mechanical properties increased obviously after isothermal annealing at low mold temperature. Therefore, stamp forming process with vacuum pumping is considered to be very promising to process the studied materials.

REFERENCES

Chambers, A.R. Earl, J.S. Squires, C.A. International Journal of Fatigue, Vol. 28 (2006), p.1389–1398.

Hedesiu, C. Demo, D.E. Kleppinger, R. Buda, A.A. Blumich, B. Remerie, K. Polymer, Vol. 48 (2007), p.763–777.

Hou, M. Composites: Part A, Vol. 87 (1997), p.695–702.

Kuch, H. Henning, F.: 15th Annual Technical Conference of the American Society for Composites Proceeding (Dallas, TX, USA, 2000). Vol. 1, p.256–259.

Nowacki, J. Neitzel, M. Kunstoffe plast Europe, Vol. 87 (1997), p.1154–1156.

Offringa, A.R. Composites: Part A, Vol. 27 (1996), p.329–336.

Wakeman, M.D. Rudd, C.D. Cain, T.A. Brooks, R. Long, A.C. Composite Science Technology, Vol. 60 (2000), p.1901–1918.

Wu, Zhongyou., Zuli, Sun., Nian, Li. Fiber Reinforced plastics/Composites, Vol. 4 (2010), p.62–66.

Energy Science and Applied Technology ESAT 2016 – Fang (Ed.)
© 2016 Taylor & Francis Group, London, ISBN 978-1-138-02973-6

Characteristics research of a low-specific-speed centrifugal pump with splitter blades and a vaned diffuser

Cheng Qiu & Xiangjun Fang
School of Energy and Power Engineering, Beihang University, Beijing, China

ABSTRACT: Numerical simulation and experiment are carried out to study the characteristics of a low-specific-speed centrifugal pump with splitter blades and a vaned diffuser. Reynolds average Navier-Stokes equation and the standard k-ε turbulence model are used in numerical simulation. Head curves obtained by numerical simulation at different speeds are compared with the experimental values, and the average error is about 5%. Centrifugal pump without splitter blades is also calculated. The results show that splitter blades increase the head and efficiency by 5.3% and 1.6% respectively. Comparative analysis of the distribution figures of streamlines, pressure field and velocity vector of two pumps imply that, splitter blades diminish the vortex near the pressure surface of blades, reducing the losses; the fluids into the diffuser become more uniform, increasing the static pressure rise in diffuser. The consistency of the head coefficient at different speeds implies that the similarity principle is effective in this situation. Data obtained from numerical simulation is fitted to get a quick prediction of pump head in different conditions. This article provides reference to splitter blades design and application in low-specific-speed centrifugal pump with a vane diffuser.

Keywords: low-specific-speed centrifugal pump; splitter blades; vaned diffuser; numerical simulation; similarity

1 INTRODUCTION

Low-specific-speed centrifugal pumps generally refer to the pumps whose specific speeds are between 30 to 80 (min⁻¹, m³/h, m). Because of the small flow rate and high head, low-specific-speed centrifugal pumps are widely used in agricultural irrigation, urban water supply, mining, petrochemical industry, and aerospace industry (Yuan Shouqi, 1997). Due to the larger impeller diameter, the narrower export width, and the longer flow channel, the disc friction loss and hydraulic loss are bigger, resulting in lower efficiency. In order to improve the efficiency of the low-specific-speed centrifugal pumps, adding splitter blades is a common method. Numerical simulation can help to have insight into the complex inner flow, and to calculate the characteristics in a significant lower cost of time and money than constructing and testing physical prototypes. CFD analysis plays a more and more important part to optimize the design and to shorten the design process.

Adding splitter blades in the passage of adjacent main blades can improve the performance of centrifugal pump. Li (Li Yibin, 2008) and Shehata (Shehata M H, 2012) analyzed the influence of the number, the radial and circumferential positions of splitter blades on the performance and flow field

of centrifugal pump. Three-dimensional turbulent numerical simulation and PIV test were mutual authenticated by He (He Youshi, 2004), revealed that the splitter blades improved the "jet-wake" flow structure in the flow field of centrifugal pump. Ye et al (Ye Liting, 2012) studied the effects of the splitter blades on the performance and unsteady characteristics of centrifugal pump. Zhang et al (Zhang Yunlei, 2015) studied the influence of splitter blades inlet diameter on the cavitation performance of centrifugal pump.

The research work of centrifugal pump is mainly based on the volute diffuser, and there are few literatures about the research of the centrifugal pump with vaned diffuser. Arndt et al (Arndt N, 1990) studied interference between rotor and stator of the centrifugal pump with a vaned diffuser. Pavesi et al (Pavesi G, 2004) researched the influence of the diffuser on the instability flow and noise in the centrifugal pump with a vaned diffuser. Segala (Segala W, 2011) and Stel et al (Stel H, 2013) studied internal flow of centrifugal pump with a vaned diffuser, the numerical calculation results showed that, at low load the liquidity was bad, turbulence intensity and instability were high, especially in the diffuser. The similarity principle was studied, and used to get an expression of the pump in the whole operating range.

In this paper, the effect of splitter blades on the performance and the inner flow of low-specific-speed centrifugal pump with a vaned diffuser are studied. The parameters of the centrifugal pump are acquired through the previous design work. CFX software is adopted to simulate low-specific-speed centrifugal pumps with and without splitter blades for 11 operating point, and the results are compared with experimental data. Influence of splitter blades on internal flow and characteristics of low-specific-speed centrifugal pump with a vaned diffuser is analyzed. According to the similarity principle, formula of the pump head for different working condition is given.

2 NUMERICAL SIMULATION

2.1 Governing equation

Water flow in Centrifugal pump follows the mass and momentum conservation equations. The flow in the impeller at a constant speed is considered to be steady, and the relative motion is described by the rotation of the relative coordinate system. The water is considered incompressible, and the continuous equation and the mean momentum Reynolds equation for the turbulent flow are listed as follows:

$$u_x + u_y + u_z = 0$$

$$\frac{\partial(\rho u_x)}{\partial t} + \nabla(\rho u_x \bar{u}) = -\frac{\partial p}{\partial x} + \frac{\partial \tau_{xx}}{\partial x} + \frac{\partial \tau_{yx}}{\partial y} + \frac{\partial \tau_{zx}}{\partial z} + F_x$$

$$\frac{\partial(\rho u_y)}{\partial t} + \nabla(\rho u_y \bar{u}) = -\frac{\partial p}{\partial y} + \frac{\partial \tau_{xy}}{\partial x} + \frac{\partial \tau_{yy}}{\partial y} + \frac{\partial \tau_{zy}}{\partial z} + F_y$$

$$\frac{\partial(\rho u_z)}{\partial t} + \nabla(\rho u_z \bar{u}) = -\frac{\partial p}{\partial z} + \frac{\partial \tau_{xz}}{\partial x} + \frac{\partial \tau_{yz}}{\partial y} + \frac{\partial \tau_{zz}}{\partial z} + F_z$$

The standard k-ε turbulence model is adopted, the equations are:

$$S_k = \frac{\partial(\rho u_j k)}{\partial x_j} - \frac{\partial}{\partial x_j}\left[\frac{\mu_e}{\sigma_k}\frac{\partial k}{\partial x_j}\right]$$

$$= \mu_t \left[\frac{\partial u_i}{\partial x_j} + \frac{\partial u_j}{\partial x_i}\right]\frac{\partial u_i}{\partial x_j} - \rho\varepsilon$$

$$S_\varepsilon = \frac{\partial(\rho u_j \varepsilon)}{\partial x_j} - \frac{\partial}{\partial x_j}\left[\frac{\mu_e}{\sigma_\varepsilon}\frac{\partial \varepsilon}{\partial x_j}\right]$$

$$= \frac{\varepsilon}{k}\left[C_1\mu_t\left[\frac{\partial u_i}{\partial x_j} + \frac{\partial u_j}{\partial x_i}\right]\frac{\partial u_i}{\partial x_j} - C_2\rho\varepsilon\right]$$

in which

$$u_c = u + u_t = u + \rho C_u \frac{k^2}{\varepsilon}, C_u = 0.09, C_1 = 1.44,$$
$$C_2 = 1.92, \sigma_k = 1.0, \sigma_\varepsilon = 1.3$$

2.2 Model and grid

The design parameters of the impeller are obtained by the preliminary work, and the three-dimensional model of the centrifugal pump is carried out by using Solidworks software. Figure 1 shows the flow channel of two impellers, and they are divided into three parts: inlet, impeller and vaned diffuser. The specific parameters of impeller and diffuser of centrifugal pump are shown in table 1. The design parameters of centrifugal pump are: flow rate, Q = 40 m³/h; head, H = 160 m; rotate speed, n = 5510 r/min; the specific speed, n_s = 47.12.

ICEM software is used to mesh the three-dimensional model. Unstructured grid is generated, because of it is considered to be more adaptable to complex model. The model is divided into three parts, which are inlet, impeller and diffuser, and the total number of grid nodes is about 520,000. The boundary layer is encrypted, and the influence of the number of grid is less than 0.5%, which is considered as the influence of computer resources.

2.3 Flow solver and boundary conditions

ANSYS-CFX software is adopted, and standard k-ε turbulence model is chosen for steady cal-

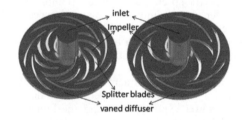

Figure 1. Flow channel of centrifugal pumps with and without splitter blades.

Table 1. Parameters of the impeller.

Parameters	Symbol	Value
blade inlet diameter	D_1	51.5 mm
impeller outlet diameter	D_2	202 mm
blade inlet angle	β_1	10°
blade outlet angle	β_2	23°
diffuser blade inlet diameter	D_3	209 mm
diffuser blade outlet diameter	D_4	273 mm
diffuser blade inlet angle	β_3	2.5°
diffuser blade outlet angle	β_4	19.5°
splitter blade inlet diameter	D_5	110 mm

culation. water at 25°C and water vapor at 25°C with the saturation pressure of 3169 Pa are set as working fluid. Plesset Rayleigh equation is used to describe the formation and collapse of the bubble, the average diameter of the bubble is set to 2 mm. the volume fraction of the inlet cavity was set to 0, the volume fraction of liquid water was set to 1.

3 TEST RIG

The experiment is carried out on the test rig of pump microcomputer test system, and the schematic diagram of the test rig is shown in figure 2. During the experiment, the input frequency of the motor is changed through the frequency converter to obtain different rotate speed. The throttle valve is used to control the flow rate. Experiments are carried out at room temperature, and the clean water is used as the test medium. The test execution standard is MT/T671-2005. The flow rate is measured by the LWGY turbine flow meter, and the outlet pressure of the pump is measured by the pressure transmitter. The accuracy level of turbine flow meter is 1.0, and the accuracy of the pressure transmitter is 0.5. Test rig of pump microcomputer test system use computer to record flow rate, head, power, current, voltage, and rotate speed of pumps. During the test, manual control of the throttle valve to obtain the desired flow rate, the system automatic collect, analyze and process data, giving out the head, efficiency and power curve.

(a) Impeller and diffuser (b) Console and computerer

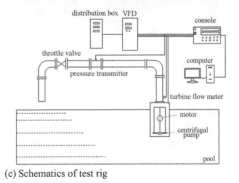

(c) Schematics of test rig

Figure 2. Impeller and diffuser, console and computer, schematics of test rig.

4 NUMERICAL RESULTS ANALYSIS

4.1 Head curve

Cases at different rotate speeds are calculated, and the head performance at different rotate speed (flow rate ranges from 10 m³/h to 60 m³/h) is compared with the experimental results, as it is shown in figure 3. It can be seen from the figure that the numerical simulation can accurately predict the head of the centrifugal pump. The average error is about 5%. The results of low rotate speeds cases are more accurate, and the error is smaller at design point. For larger rotating speed, the difference between the numerical simulation and the experimental results is larger. The numerical simulation does not take into account the leakage loss and the friction of the outer surface of front and back cover. Furthermore, at high speed rotate and large flow rate, the load of the pump is larger, thus results in motor speed decreased. So compared with the numerical simulation using a constant rotate speed, the pump head will be lower. In summary, the experimental result has verified the accuracy of numerical method, so the results of numerical simulation can be used for further analysis.

4.2 Effects of the splitter blades on the external performance of pump

To study the influence of splitter blade on internal flow and external performance of low-specific-speed centrifugal pump, pumps with and without splitter blades are calculated, head and efficiency curves are shown in Figure 4. Pump 1 and pump 2 refer to the pump with and without splitter blade respectively.

It can be seen from the comparison of head and efficiency curves of two cases, that the head and efficiency of pump 1 are both higher than that of pump 2. At the design point, the head is increased by 5.3%, about 8.7 meters by adding splitter blades. The efficiency at the design points is increased from 60.7% to 62.3%. The application of splitter blades

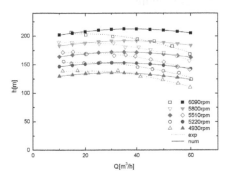

Figure 3. Comparison of numerical and experimental head curves for different rotor speeds.

in low-specific-speed centrifugal pump improves the head and efficiency, but also does not decrease the throat of the pump, so its ability of solid particle passing through is not affected.

4.3 Internal flow fields of low-specific-speed centrifugal pump with a vaned diffuser

The interference between two components in relative motion has great influence on the flow fields and external performance. In Figure 5, the static pressure distribution of 2 pumps on middle section (a plane located in the middle height of the impeller-diffuser interface) for design condition is given. It can be seen that the splitter blades improve the pressure near suction surface of main blades. The pressure rise in the blade passages is increased, and more uniform. High pressure and low velocity of the flow out of impeller make the loss smaller. From the point of view of minimum pressure, the minimum pressure of the impeller with splitter blades is smaller, that is, the suction ability is stronger, which is an important ability of centrifugal pump.

Streamline on the cross section at design flow rate is showed in figure 6. It's clear that the splitter blades make the recirculation zone near pressure surface of main blades disappear. The secondary flow loss is reduced in the blade passage. At the outlet of the impeller, the wake flow is washed out because the existence of splitter blades. Therefore, the splitter blades improve the flowing characteristics before water goes into the diffuser, that is, flow separation in the high speed rotating impeller is suppressed. Better flowing characteristics inside the impeller can be beneficial to the external characteristics, namely, higher head and higher efficiency.

Figure 7 shows the two-dimensional velocity vector distribution of two pumps on the middle section at design rotate speed for different flow rates. It is clear that splitter blades make the vector distribution more uniform at all three flow rates, and low-velocity zone near pressure surface of main blades is diminished.

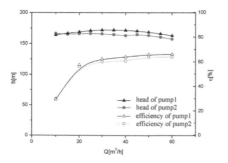

Figure 4. Comparison of head and efficiency curves for two pumps.

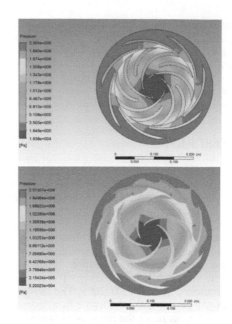

Figure 5. Pressure distribution of two pumps on middle section for design condition.

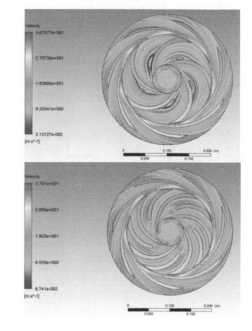

Figure 6. Streamlines of two pumps at middle section for design condition.

Splitter blades also improve the diffusion capacity of the diffuser. Figure 8 shows static pressure rise from diffuser inlet to outlet and the ratio of outlet static pressure and inlet static pressure, 1 for pump with splitter blades, 2 for pump without splitter blades. The figure shows that, whether static or

424

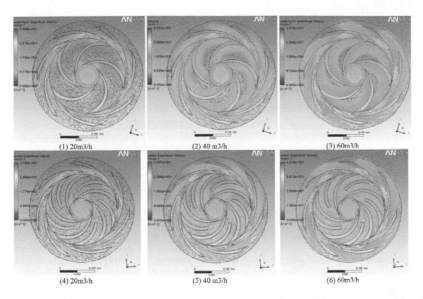

Figure 7. Velocity vector distribution of two pumps on the middle section at design rotate speed for different flow rates.

static pressure ratio, the pump with splitter blades is higher than the pump without splitter blades (except for the first flow rate point). This is because the splitter blades make the liquid more uniform before flow into the diffuser, thereby improving the capacity of the diffuser, so a higher static pressure and static pressure ratio are obtained.

4.4 Similarity and head prediction

Similarity principle indicates that the flow pattern of the same centrifugal pump should be similar at different speeds, if the flow coefficient is the same. The flow coefficient ϕ and head coefficient ψ is defined below:

$$\phi = \frac{Q}{2\pi R_2^2 b_2 \omega}, \psi = \frac{Hg}{(\omega D_2)^2}$$

As seen in Figure 9, the head coefficient curves collapsed to each other. The head coefficient is approximately equal for different rotate speeds, as long as the flow coefficient is the same. As expected, the results match each other confirming that similarity holds for head coefficient.

According to the similarity principle, the numerical head at different speeds can be expressed by a second-order polynomial equation with the form $H(Q, n) = AQ^2 + BnQ + Cn^2$, Where Q is the volumetric flow rate (m³/h), n is the rotate speed (rpm). The constants A, B, C that provides the best fit for the numerical values are obtained according to numerical database, the fitting formula is

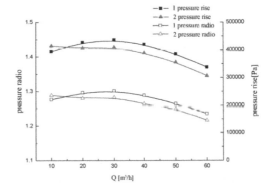

Figure 8. Pressure rise and pressure rise radio of diffuser in two pumps.

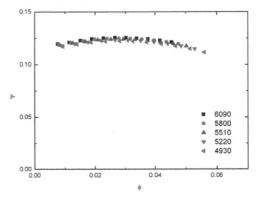

Figure 9. Relationship between head coefficient and flow coefficient for different rotate speeds.

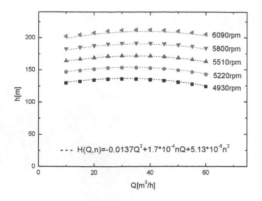

Figure 10. Comparison of numerical head curves and fitting formula in different rotate speeds.

$H(Q, n) = 0.0137Q^2 + 1.7 \times 10^{-4} nQ + 5.13 \times 10^{-6} n^2$. The fitting formula is compared with numerical values in figure 10, and they are well matched. In engineering application, it can predict the pump head of certain speed and flow rate; or when the pump needs a specified flow and head, through the fitting formula, rotate speed can be quickly obtained, and then adjust the converter to get the corresponding rotate speed. That provides convenience for engineering application.

5 CONCLUSIONS

Optimized blade profile and splitter blades method have been adopted in the preliminary design work, in order to improve the efficiency of the low-specific-speed centrifugal pump. The numerical simulation and experiment are carried out to research the characteristics of the centrifugal pump with splitter blades and a vaned diffuser. Pressure distribution, streamlines and velocity vector of two pumps are analyzed, it is found that splitter blades diminish the generation of backflow in the blade passages, making the flow field in impeller better. The wake flow is washed out, making the flow field in impeller exit more uniform, thus improving the diffusion capacity of diffuser. The centrifugal pump head and efficiency are improved by adding splitter blades, according to both numerical and experimental methods. Because of the simplification of the model, the numerical simulation average error is about 5%. The similarity of the head of the centrifugal pump is studied, and it is found that the head coefficient is consistent when the flow coefficient is the same. The fit formula of the head pump is obtained, making the engineering application more convenient. The research provides reference to the research of low-specific-speed centrifugal pump with a vaned diffuser and splitter blades.

REFERENCES

Arndt N, Acosta A J, Brennen C E, et al. Experimental investigation of rotor-stator interaction in a centrifugal pump with several vaned diffusers. ASME Journal of Turbomachinery. Journal of Turbomachinery, 1990, 112, 98–108.

He Youshi, Yuan Shouqi, Guo Xiaomei, et al. Numerical simulation of three dimensional incompressible turbulent flow field in the impeller of centrifugal pump with splitter blades. Chinese Journal of Mechanical Engineering, 2004, 40(11):153–157. "in Chinese"

Li Yibin, Zhang Desheng, Zhao Weiguo, et al. Influence of blade number and splitter-blade position on performance of centrifugal pumps. Journal of Lanzhou University of Technology, 2008, 34(2):45–48. "in Chinese"

Pavesi G, Ardizzon G, Cavazzini G. Numerical and Experimental Investigations on a Centrifugal Pump With and Without a Vaned Diffuser. Asme International Mechanical Engineering Congress & Exposition, 2004:485–493.

Segala W, Stel H, Hungria V, et al. Numerical simulation of the flow in a centrifugal pump with a vaned diffuser// ASME-JSME-KSME 2011 Joint Fluids Engineering Conference. American Society of Mechanical Engineers, 2011:1791–1800.

Shehata M H, Abd Elganny M E, Abd ELhafez A S, et al. Effect of shorted blade circumferential positions on centrifugal pump characteristics// 10th International Energy Conversion Engineering Conference. American Institute of Aeronautics and Astronautics. 2012–4098.

Stel H, Amaral G D L, Negrão C O R, et al. Numerical analysis of the fluid flow in the first stage of a two-stage centrifugal pump with a vaned diffuser. Journal of Fluids Engineering, 2013, 135: págs. 235–244.

Ye Liting, Yuan Shouqi, Zhang Jinfeng, et al. Effects of splitter blades on the unsteady flow of a centrifugal pump. American Society of Mechanical Engineers, 2012:435–441.

Yuan Shouqi. Theory and design of low specific speed centrifugal pump. Beijing, China Machine Press, 1997: 16–19. "in Chinese"

Zhang Yunlei, Yuan Shouqi, Zhang Jinfeng, et al. Numerical analysis on effects of splitter blades on cavitation performance in a centrifugal pump. Journal of Drainage and Irrigation Machinery Engineering, 2015(10). "in Chinese"

Energy Science and Applied Technology ESAT 2016 – Fang (Ed.)
© *2016 Taylor & Francis Group, London, ISBN 978-1-138-02973-6*

Research on the microstructure of WC-Ni-Fe-cemented carbides

Jun Wang & Xianquan Jiang
Chongqing Academy of Science and Technology, Chongqing, China

ABSTRACT: WC-Ni-Fe-cemented carbides with different Fe/Ni ratios were fabricated by tumbling ball milling together with vacuum sintering. When investigated using a scanning electron microscope and EDS, it was found that doping small amounts of Fe-powder had inhibited the growth of WC grain during the sinter process, and had decreased the dissolubility of W in the binding phase. Additionally, the number of "the binding phase pool" was seen to rise with the increment in the Fe content. The porosity of the sintered body also rose with the increment in the Fe content for values higher than 1%.

Keywords: WC-Ni-cemented carbide; Fe powder; Microstructure; mechanical properties

1 INTRODUCTION

WC–Co-cemented carbides are widely used in industries (Zhang L, 2012). However, Co is an expensive and rare metal. Therefore, Co powder is often replaced by cheap Ni powder. On the upside, WC-Ni-cemented carbides have high anti-corrosion properties (Huiyong Rong, 2012).

At present, comprehensive mechanical properties of WC-Ni-cemented carbides still have a certain gap as compared with Co-based cemented carbides (K. Bonny, 2010). Therefore, the main research focus in the field of WC-Ni-cemented carbides is to improve their performance. Research shows that solid solution strengthening performed by adding Cr and Mo can improve the performance of WC-Ni-cemented carbides (N. Lin, 2012). Both Cr and Mo are rare precious metals. The aim is to replace these with cheaper metals. Therefore, in this paper, we report a detailed study on the influences of Fe powder addition on the microstructures of WC-Ni-cemented carbides.

2 EXPERIMENTAL

2.1 Raw and supplemental materials

The specifications of the powders used for the fabrication of the WC-Ni-cemented carbides are listed in Table 1. Dehydrated alcohol (analytical pure) was used as the milling medium. All powders were commercially available (Table 1).

2.2 Sample preparation

Materials were obtained using a mixture of WC, Ni and Fe powders by powder metallurgy methods. The amount of the hard phases (in wt.%) was kept constant and only the binder composition was changed.

The mixtures were mixed in a tumbling ball mill in anhydrous alcohol for 48 h. The paraffin wax was used as the formative agent. The green compacts were prepared by the way of the double-action pressure at 100 MPa. These were then sintered in vacuum at 1480°C for 60 min (Table 2).

2.3 Material characterization

A scanning electron microscope was used to examine the microstructure of the materials. The phases in the sintered samples were identified using X-Ray Diffraction (XRD). The porosity was detected by the NIKON ECLIPSE 801 optical microscope according to the international standards (ISO/BIS4505).

Table 1. Specifications of raw materials.

Powder	Average particle size (μm)	Purity (%)	O content (wt.%)	Total carbon (wt%)	Free carbon (wt%)
WC	1.0 (Fsss)	99.78	0.21	6.14	0.08
Ni	2.45 (Fsss)	99.75	0.078	*	*
Fe	3.0(Fsss)	99.6	0.38	*	*

Table 2. Composition of 92 WC-(8-x)Ni–xFe-cemented carbides (wt.%).

Specimen	WC	Ni	Fe	Fe/Ni
A	92	8	0	0
B	92	7.5	0.5	0.07
C	92	7	1	0.14
D	92	6	2	0.33
E	92	4	4	1

3 RESULTS AND DISCUSSION

3.1 *Microstructure*

Figure 1 shows the BSE pictures of samples A, B, C, and D, in which the polygonal grayish white part is the hard tungsten carbide phase and the irregularly shaped black part is the binding phase. The mean particle size of the tungsten carbide powder used in this experiment was 1 μm. Sample A has WC-Ni-cemented carbides without any Fe powder. The mean grain size of tungsten carbide in sample A is clearly larger than 1 μm, indicating that, in this sample, tungsten carbide grew significantly during the sintering process. The solubility of W in Ni is about 10–30%, which is higher than that in Co (2–10%). Hence, the nickel-based cemented carbides are of strong dissolution-precipitation in the liquid-phase sintering and of strong diffusive mass transfer in the solid-phase sintering, thus leading to the overgrowth of WC-Ni-cemented carbides.

Figure 1 presents the grain sizes of the hard phase in samples B, C, and D, which clearly are lower than those of sample A. The grain size of tungsten carbide gradually decreases with the increase in Fe content. The binding phase components of samples A, B, C, and D were analyzed by an EDS. Changes in the W content of the binding phase are shown in figure 2. It is clear that the W content decreases with an increase in the Fe content, which leads to the decrease in liquid phase number and diffusive mass transfer speed of W in the sintering, thus playing a role in restraining the grain growth of tungsten carbide.

From Figure 1, it is estimated that the number of "the binding phase pool" increases with the increase in the amount of Fe content. This is due to the increase in large iron particles in the powder.

In Figure 3, specimen C contains only two phases: the hard tungsten carbide phase and the binding phase. Compared with the nickel powder, the diffraction peaks of the binding phase are similar, but the positions of the peaks have shifted to the left, suggesting that the lattice parameter of nickel has increased. Results of the EDX analysis for the binding phase in sample C are shown in Figure 4. The binding phase contains four kinds of

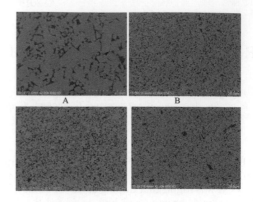

Figure 1. SEM-BSE micrographs of samples A, B, C, and D.

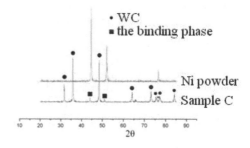

Figure 2. The effect of Fe content on the solid solution content of W in the binding phase.

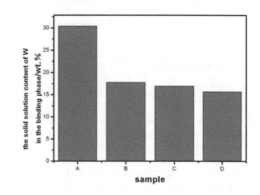

Figure 3. Typical XRD patterns of the sample C and Ni powder.

elements, namely, Ni, W, Fe, and C. Therefore, the binding phase is a quaternary solid solution, with Ni base. The solid solution effect leads to widened interatomic spacing between the Ni atoms; therefore, diffraction peaks of the binding phase were shifted toward the left with respect to the diffraction peak of Ni.

Table 3 presents the values for porosities for the five groups of specimens. The porosity of a sintered sample rises with the Fe content when

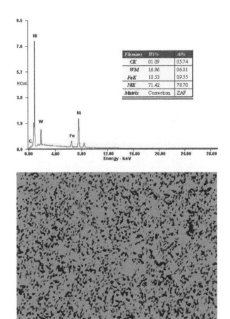

Element	W%	A%
CK	01.09	05.74
WM	16.96	06.01
FeK	10.53	09.55
NiK	71.42	78.70
Matrix	Correction	ZAF

Figure 4.   EDX results of simple C.

Table 3.   The porosities of cemented carbide.

Specimen	A	B	C	D	E
Porosity	A02	A02	A02	A04	A04B02

the addition of iron is more than 1%. Fe has a higher Melting point than Ni. Therefore, the viscosity increases and fluidity of the binding phase decreases with the increase in Fe content at a given sintering temperature. This is only one of the reasons that results in an increase in the porosity of the sintered sample. But the porosity can be influenced by many other factors. The true important reasons for the increase in porosity are not established yet.

4   CONCLUSIONS

The effect of Fe powder content on 92 WC–(8-x) Ni–xFe system-cemented carbides was studied. The results show that:

1. Fe doping can inhibit the growth of WC grains during sintering as in the case of WC-Ni-cemented carbides.
2. Distributions of the mixtures get worse, and the number of "the binding phase pool" increases with the increase in the amount of Fe content.
3. The porosity of the sintered body increases with the increase in the amount of Fe content when the addition of iron is more than 1%.
4. For WC-Ni-Fe-cemented carbides, the binding phase is quaternary solid solution, which contains Ni, W, C, and Fe. Additionally, the W content decreases with the increase in Fe content.

ACKNOWLEDGMENTS

This work was financially supported by the Chongqing General Research Program of Advanced Technology and Application Foundation (Contract Grant NO. cstc2014 jcyjA50010).

REFERENCES

Bonny, K., P. De Baets, J. Van Wittenberghe, Y. Perez Delgado, J. Vleugels, O. Vander Biest, B. Lauwers. Influence of electrical discharge machining on sliding friction and wear of WC–Ni cemented carbide [J]. Tribology International, 2010, 43: 2333–2344.

Fernandes, C.M., A.M.R. Senos. Cemented carbide phase diagrams: A review [J]. Int. Journal of Refractory Metals and Hard Materials, 2011, 29: 405–418.

Hu Hai-bo. Effects of Additives on Microstructure and Mechanical Properties of WC-Ni Cemented Carbide [J]. Rare Metals and Cemented Carbides, 2013, 41(1): 55–59.

Huiyong Rong, Zhijian Peng, Xiaoyong Ren, Ying Peng, Chengbiao Wang,

Jiang Yuan-yuan, Yi Dan-qing, Li Jian, Zhang Lu-huai, SU Hua, Huang Liang, Peng Zhen-wen. Corrosion Characteristic of WC-9 Ni-0.57Cr Cemented Carbide in Artifical Seawater Environment [J]. Journal of Materials Science and Engineering, 2008, 26(5):750–753.

Leech, Patrick W., Li, Xing S., Alam, Nazmul. Comparison of abrasive wear of a complex high alloy hardfacing deposit and WC-Ni based metal matrix composite [J]. WEAR, 2012, 294:380–386.

Lin, N., C.H. Wu, Y.H. He, D.F. Zhang. Effect of Mo and Co additions on the microstructure and properties of WC-TiC-Ni cemented carbides [J]. Int. Journal of Refractory Metals and Hard Materials, 2012, 30: 107–113.

Shi Kai-hua, Zhou Ke-chao, Li Zhi-you, Zan Xiu-qi, Xu Shang-zhi, Min Zhao-yu. Effect of adding method of Cr on microstructure and properties of WC-9 Ni-2Cr cemented carbides [J]. Int. Journal of Refractory Metals and Hard Materials, 2013,38:1–6.

Zhang L. Chen S., Shan C., Cheng X., Ma Y., Xiong X.-J. Effects of cobalt additions on WC grain growth [J]. Powder Metallurgy, 2012, 55(3): 200–205.

Zhiqiang Fu, Longhao Qi, Hezhuo Miao. Ultrafine WC–Ni cemented carbides fabricated by spark plasma sintering [J]. Materials Science and Engineering A, 2012, 532:543–547.

Energy Science and Applied Technology ESAT 2016 – Fang (Ed.)
© *2016 Taylor & Francis Group, London, ISBN 978-1-138-02973-6*

Numerical simulation of the force by particles acting on the blades of a rotary valve

X.S. Zhu & L.W. He
School of Mechanical and Power Engineering, East China University of Science and Technology, Shanghai, China

ABSTRACT: The rotary valve is one of the transportation equipment of particles. First, the discharging process of the rotary valve is simulated with the discrete element method to obtain the force by particles acting on the blades. Then, the load data is imported into ANSYS workbench to calculate the stress. The result can provide a reference for the rotary valve design.

Keywords: rotary valve; particles; DEM; ANSYS workbench; stress

1 INTRODUCTION

Many scholars have conducted research on granular flow in the rolling container; for example, Sergio M. Savaresi (Savaresi, M. et al, 2001) and Sheehan M.E. et al (Britton, P.F. et al, 2006). Researched the rolling granular desiccant flow with the signal plate. M.Silvina. Tomassone's (Chaudhuri, B. et al, 2006) experiment reflected the flow status of dispersion in rotary drying container. However, studies of the force by particles acting on a rotary machine have been relatively less so far. Based on the discrete element method, the particles' acting force on the blades of the rotary valve is obtained more accurately in this paper. The load data can be imported into the ANSYS workbench to calculate the stress and deformation of the blades. The result can provide a referencefor the rotary valve design.

2 HERTZ-MINDLIN CONTACT MODEL

2.1 *Hertz-Mindlin contact model*

The Hertz-Mindlin contact model (Wang, G.J. et al, 2010) is a general contact model in Discrete Element Method (DEM) (Cundall, P.A. et al, 1979). Normal force in the model is based on the Hertz contact theory, while the tangential force is based on the research of Mindlin-Deresiewicz, which makes the force calculation more accurately and efficiently.

2.2 *Time step*

Selecting an appropriate time step is the key to ensure the accuracy and efficiency for simulation.

When the particles contact, surfaces are influenced by variable stress, generating a polarized wave propagating along the surface of particles, which is called the Rayleigh wave. The experiment showed that 70% of the total energy consumption is consumed by Rayleigh wave when particles collide. So, the critical time step should be determined by the Rayleigh wave velocity spreading along a solid spherical particle surface. The time step of the system with different particles is given by

$$\Delta t = \pi \left[\frac{R}{0.163\nu + 0.877} \sqrt{\frac{\rho}{G}} \right]_{\min} \quad (1)$$

The time step Δt determined can guarantee the calculation stability of the particle system. An appropriate time step should be chosen according to the particles' movement intensity to ensure the stability of the calculation; generally the time step is 10%~40% of Δt.

3 SIMULATION OF THE DISCHARGING PROCESS OF THE ROTARY VALVE

The discharging process of the rotary valve is simulated in EDEM. The particles' acting force on the blades of the rotary valve can be extracted by post-processing function of EDEM, where the load file can be imported into the ANSYS workbench to calculate the equipment of stress and deformation of rotary blades.

A virtual panel is built above the entrance as a particle factory, which is used to generate particles. PE particles are adopted as the simulated media, whose basic parameters are given in Table 1.

Table 1. DEM model parameters.

Parameters	Value
Poisson's ratio of granular	0.203
Shear modulus of granular [pa]	8.31e8
Density of granular [kg/m³]	945
Coefficient of static friction between granular and steel	0.5
Coefficient of rolling friction between granular and steel	0.01
Radius of spherical particles [mm]	3

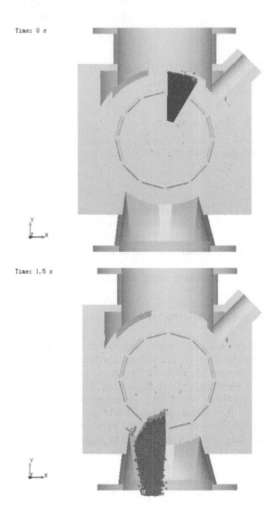

Figure 1. The granular distribution of the rotary valve at 0s and 1.5s.

The time step is $1e^{-5}s$, which is 25% of Δt. The rotor speed is 20 rpm. The rotor has twelve blades. The degree between two adjacent blades is 30°. Considering that particles' movements in each sector are the same as others during the process of discharging, then the model can be simplified into one sector. The discharging model is shown in Fig. 1. The particles are filled in one sector, where particles are supported by two blades on the left and right. Then, the rotor rotates in the counter-clockwise direction to simulate for $3s$ (turning a circle). The total number of particles is about 30000, whose total mass is 3 kg.

Fig. 2 to Fig. 5, respectively, expresses the force on the blades in the X, Y direction. The horizontal axis represents the distance from the top to the root of the blade while the vertical axis is the value of the total force. The forces in different time are represented with different curves.

Fig. 2 shows the forces on the left blade in the X direction. At the beginning of discharging, the force values in all distance are negative, whose maximum force is at the root. The forces decrease with time, reducing nearly to zero at 0.75s. Then, the value increases gradually along X positive direction, reaching the maximum at 1.25s. Finally, the forces reduce to zero at 1.5s. Fig. 3 represents the forces on the right blade in the X component. The forces in different distance are positive. The maximum occurs in the distance of 0.14 m. The values decrease with valve rotating, even rapidly closing to zero after 0.75s. We can see that the value in the upper half of the left blade is less than the lower part before 0.75s. After that the forces mainly act on the upper half.

Fig. 4 indicates the forces on the left blade in the Y component. All forces on the blade are negative during the discharging process. The forces are near to zero at the beginning of discharging. Then, they increase along the Y negative direction, running up to a maximum at 1s. Then, the forces decrease rapidly, reaching the minimum at 1.5s. We can see that the forces mainly act on root before 0.75s. After that the forces mainly act on the upper half. Fig. 5 shows the forces on the right blade in the Y

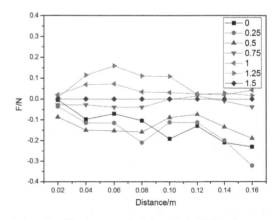

Figure 2. The force on the left blade in X direction.

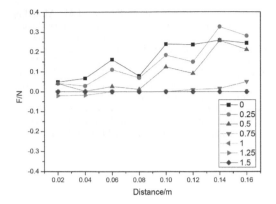

Figure 3.　The force on the right blade in X direction.

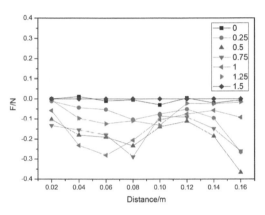

Figure 4.　The force on the left blade in the Y direction.

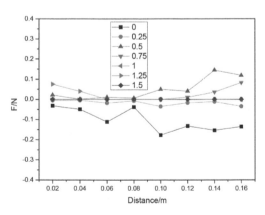

Figure 5.　The force on the right blade in the Y direction.

component. The forces are negative at the beginning. They reduce to zero at 0.25s. After that, the forces increase along the Y positive direction, running up to maximum at 0.5s. Then the value decrease to zero gradually once more.

4　SIMULATION OF ROTOR'S STRESS

The load data from EDEM are loaded into the ANSYS workbench. The load distribution at 0s in the ANSYS workbench is shown in Fig. 6.

From the calculated stress nephogram of Fig. 7, we can see that the maximum stress occurs at the junction of the blade and reinforcement. The maximum value is 0.16345 MPa. The stress value at the root of blade is far less than the maximum.

5　SUMMARY

1. Particles' acting force on the blades of the rotary valve is obtained with the Hertz-Mindlin model. The results illustrate that the forces of the left and right blade in the X direction decreases with time; however, the left one increases along the opposite direction after rotating 90°. The

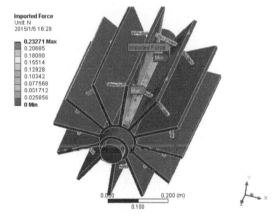

Figure 6.　Load distribution of the blades.

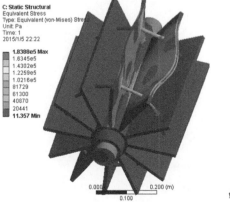

Figure 7.　Stress nephogram of the blades.

433

value in the Y direction mainly acts on the left blade, adding up to a maximum when rotated to the position of 75°.

2. The load data from DEM are imported into the ANSYS workbench to calculate the rotor's stress the maximum stress occurs at the junction of the blade and reinforcement.

3. Combining DEM and FEA, we can simulate the force by particles acting on the rotary machine easily.

REFERENCES

Britton, P.F. Sheehan, M.E. & Schneider, P.A. A physical description of solid transport in flighted rotary dryers. Powder Technology, 165(2006)153–160.

Chaudhuri, B. Muzzio, J. & Silvina. T.M. Modeling of heat transfer in granular flow in rotary vessels. Chemical Engineering Science, 61(2006) 6348–6360.

Cundall, P.A. & Strack, O.L. A discrete numerical model for granular assembles. Geotechnique, 29(1979):47–65.

Savaresi, M. Bitmead, R. & Peirce, R. On modeling and control of a rotary sugar dryer. Control Engineering Practice, 9 (2001)249–266.

Wang, G.J. Hao, W.J. & Wang, W.X. Discrete element method and its application in EDEM. Northwestern Polytechnical University press, 2010, pp.16–18.

Energy Science and Applied Technology ESAT 2016 – Fang (Ed.)
© 2016 Taylor & Francis Group, London, ISBN 978-1-138-02973-6

Numerical simulation of the elastic stability of cracked plates under uniaxial compression

Cheng Geng

School of Business Administration, Xinyang Agriculture and Forestry University, Xinyang, China

ABSTRACT: Buckling failure is a common occurrence in plates under compression. In this paper, compressive buckling of thin cracked plates are investigated by the finite element method. Effects of some factors such as length and direction of crack on the compressive buckling loads are studied. Some interesting results obtained and useful conclusions are drawn to understand the buckling failure of cracked compressed plates.

Keywords: buckling, plate, cracked, finite element method

1 INTRODUCTION

Thin walled structures is widely used in different branches of engineering, such as ship grillages and hulls, dock gates, plate and box girders of bridges. The most typical problem of thin steel structures is the stability. In many cases, cracks in the thin structures can reduce significantly the ultimate bearing capacity and strength. So the elastic buckling of such thin cracked plates has received the attention of many researchers recently.

Carlson et al. (Carlson RL et al, 1970) conducted an experimental study of the buckling of thin cracked plates subjected tensile loads and concluded an empirical formula for tensile critical buckling stress of cracked plates. Markstrom and Strorakers (Markstrom K et al, 1980) studied the buckling load of the cracked plates under uniaxial tensile loads with the finite element method based on the linear bifurcation theory. Shaw and Huang (Shaw D et al, 1990) researched the buckling of tensioned cracked plates with the numerical method based on the Von Karman's linearized theory. The local tensile buckling and fracture response of a thin flat composite plate with an inclined interior crack was numerically studied by Barut (Barut A et al, 1997). Finite element models was employed to study the buckling of cylindrical shells and panels with through cracks (Vafai A et al, 1999) (Brighenti R, 2005). Brighenti (Brighenti R, 2005) used the numerical method to study the the effects of various geometrical and mechanical properties and boundary conditions on the buckling of cracked plates under tension and compression. TPaik (Paik JK et al, 2005) conducted an experimental and numerical research on the ultimate

strength of cracked plate under axial compression and tension. Jahromi and Vaziri (Jahromi BH et al, 2012) (Kim YT et al, 2013) numerically studied the compressive buckling cracked cylindrical shells.

In this study, the finite element method is used to analyze the elastic buckling of cracked plates subjected to uniaxial compression. The effects of the length and direction of crack on the compressive buckling loads on are studied. The results can make it possible to decrease the buckling loads.

2 GEOMETRICAL MODEL

A rectangular steel plate (height H, width W) containing an inclined crack with the length of *2a*, is subjected to a uniform compression stress in the y derection, as shown in Fig. 1. The inclined angle of the crack is θ. The steel plate has the

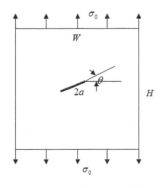

Figure 1. Geometrical model of the plate with a crack.

following elastic properties: Young's modulus $E_p = 2100\,Mpa$; Poisson ratio $\nu = 0.33$.

3 FINITE ELEMENT MODELING

The commercial software package ABAQUS was used to conduct the finite element analysis for the problem. The shell element S4R with reduced integration was adopted for modeling and analyzing the cracked plates. The material of the plates was assumed to be homogeneous, isotropic and elastic. The material properties for Young's modulus E = 210 GPa and Poision's ratio = 0.3 were selected. The dimensions of the square plate were fixed at 200 × 200 and the thickness of plates was taken as 1 mm In the simulation. Fig. 2 shows the typical mesh model of the whole structure and the mesh near the crack tip.

4 ELASTIC BUCKLING ANALYSIS

In this research, the lowest eigenvalue obtained in the eigenvalue analysis is defined as the buckling load σ_{crit}. A typical buckling mode of the plate is shown in Fig. 3.

A parametric study was conducted here to get a good understanding of the buckling of the cracked plate. Effects of the length and direction of the crack on the compressive buckling loads are studied as follows:

1. Effect of crack length on the buckling load.
In this study, the inclined angle of the crack is selected to be 0. Seven cases of the crack length are considered: 20 mm, 40 mm, 60 mm, 80 mm. to 140 mm. And the buckling load-crack length curve obtained are shown in Fig. 4.

As we can see from Fig. 4, with the increase of the length of the crack, the buckling load of the plate decrease obviously. This means cracks in the thin structures can reduce significantly the ultimate

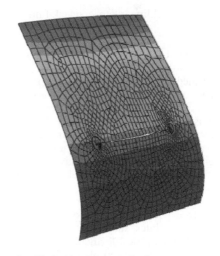

Figure 3. Typical buckling mode plot.

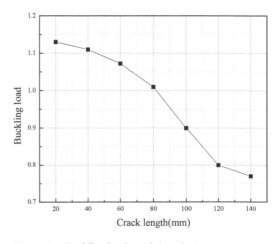

Figure 4. Buckling load-crack length curve.

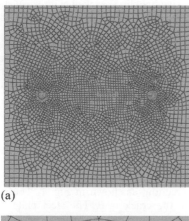

(a)

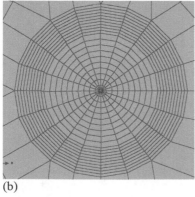

(b)

Figure 2. The finite element model: (a) the mesh of the whole model; (b) the mesh near the crack tip.

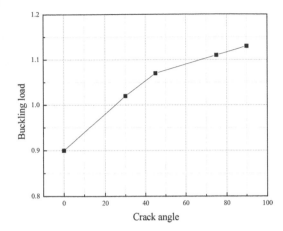

Figure 5.   Buckling load-crack angle curve.

bearing capacity and strength with the increase of the crack length.

2. Effect of the inclined angle on the buckling load.

In this study, the length of the crack is fixed at 60 mm. and the angle of the crack varies from 0 to 90. The buckling load-crack angle curve obtained is shown in Fig. 5.

As we can see from Fig. 5, in this case, the buckling load of the plate also increase dramatically with the increase of the angle of the crack. The plate has the highest bearing capability when the crack is perpendicular to the applied stress, while it has the lowest bearing capability when the crack is parallel to the applied stress.

5   DISCUSSION AND CONCLUSIONS

In this paper, the finite element method is used to analyze the elastic stability of cracked plate subjected to uniaxial compression. Effects of the length and direction of the crack on the compressive buckling loads are studied. The obtained results allow us to deduce the following conclusions:

1. With the increase of the length of the crack, the buckling load of the plate decrease obviously.
2. The plate has the highest bearing capability when the crack is perpendicular to the applied stress, while it has the lowest bearing capability when the crack is parallel to the applied stress.

REFERENCES

Barut A, Madensi E, Britt VO, Starnes JH. Buckling of a thin, tension-loaded, composite plate with an inclined crack.Eng Fract Mech. 58 (1997) 233–48.

Brighenti R. Buckling of cracked thin-plates under tension or compression Thin-walled Struct. 43 (2005) 209–24.

Brighenti R. Numerical buckling analysis of compressed or tensioned cracked thin plates. Eng Struct. 27 (2005) 265–76.

Carlson RL, Zielsdorff GF, Harrison JC. Buckling in thin cracked sheets. In: Proceedings of the Air Force Conference on Fatigue and Fracture of Aircraft Structures and Materials. Miami Beach, Florida, USA; 1970. p. 193–205.

Jahromi BH, Vaziri A. Instability of cylindrical shells with single and multiple cracks under axial compression.Thin-Walled Struct. 54 (2012) 35–43.

Kim YT, Haghpanah B, Ghosh R, Ali H, Hamouda AMS, Vaziri A. Instability of a cracked cylindrical shell reinforced by an elastic liner. Thin-Walled Struct. 70 (2013) 39–48.

Markstrom K, Stroralero B. Buckling of cracked members under tension. Int J Solids Struct. 16 (1980) 217–29.

Paik JK, Kumar YVS, Lee JM. Ultimate strength of cracked plate elements under axial compression or tension.Thin-Walled Struct. 43 (2005) 237–72.

Shaw D, Huang YH. Buckling behavior of a central cracked thin plate under tension. Eng Fract Mech. 35 (1990) 1019–27.

Vafai A, Estek a chiHE. Aparametric finite element study of cracked plates and shells. Thin-Walled Struct. 3 (1999) 211–29.

Energy Science and Applied Technology ESAT 2016 – Fang (Ed.)
© *2016 Taylor & Francis Group, London, ISBN 978-1-138-02973-6*

Numerical simulation research on the natural convection heat transfer of diamond-water nanofluid in a rectangular enclosure

Yongliang Wan, Cong Qi, Dongtai Han & Chunyang Li
School of Electric Power Engineering, China University of Mining and Technology, Xuzhou, China

ABSTRACT: Natural convection heat transfer characteristics of diamond-water nanofluid in rectangular enclosure are investigated by a single-phase model in this paper. The effects of temperature difference, nanoparticle volume fraction, and heated areas on the natural convection heat transfer characteristics are discussed. It is found that diamond-water nanofluid can enhance the heat transfer by 10.25%.

Keywords: natural convection, enhanced heat transfer, nanofluid, numerical simulation

1 INTRODUCTION

Nanofluid has been applied in many fields due to its high heat conductivity coefficient, and some researchers have investigated the natural convection heat transfer of nanofluid by simulations or experiments (M. Sheikholeslami et al, 2016) (M. Sheikholeslami et al, 2014) (M. Bouhalleb et al, 2014) (I. Rashidi et al, 2014). In addition, some researchers have also investigated the natural convection heat transfer characteristics of nanofluid by experimental methods (A. Zamzamian et al, 2011) (R. Mansour et al, 2011) (S. Heris et al, 2007) (Y. Yang et al, 2005).

Most of above investigations are on the heat transfer of metal or metal oxide nanoparticle-based fluid. It is well known that different nanoparticles have different heat transfer characteristics. It is found that the chemical property of diamond is quite steady, and the thermal conductivity is even higher than some metal. While the information on the natural convection of diamond-water nanofluid is quite little. In addition, the influences of heated areas on the natural heat transfer are less investigated. Hence, the effects of temperature difference, nanoparticle volume fraction, and heated areas on the natural convection heat transfer characteristics of diamond-water nanofluid are discussed in this paper.

2 METHODS

In order to simplify the natural convection heat transfer model, the Boussinesq hypothesis is adopted. Moreover, the diamond-water nanofluid can be seen as a single-phase fluid. The basic control equations are as follows:

Continuity equation:

$$\frac{\partial \rho_l}{\partial t} + \nabla \cdot (\rho_l \mathbf{v}_l) = 0. \tag{1}$$

Momentum equation:

$$\frac{\partial \rho_l \mathbf{v}_l}{\partial t} + \nabla \cdot (\rho_l \mathbf{v}_l \mathbf{v}_l) = -\nabla p + \nabla \cdot \mathbf{T}_l + \rho_l \mathbf{g}. \tag{2}$$

where p is the pressure of nanofluid, and the stress tensor $\mathbf{T}_l$ is defined as follows:

$$\mathbf{T}_l = \mu_l [\nabla \mathbf{v}_l + \nabla \mathbf{v}_l{}^T] - \frac{2}{3} \mu_l \nabla \cdot \mathbf{v}_l \mathbf{I}. \tag{3}$$

where μ_l is the dynamic viscosity of nanofluid, and $\mathbf{I}$ is the unit tensor.

Energy equation:

$$\rho_l c \left[\frac{\partial T}{\partial t} + v_l \cdot \nabla T \right] = \nabla \cdot [k \nabla T]. \tag{4}$$

where c, k, and T are, respectively, the specific heat, thermal conductivity, and temperature of the mixed nanofluid.

3 RESULTS AND DISCUSSIONS

The schematic of the rectangular enclosure is shown in Figure 1. The ratio between width (W) and height (H) is 0.5. The top wall is adiabatic, and the right wall is maintained at $T_L = 328$ K/338 K. There are two conditions in this paper, one is that the left wall is maintained at $T_H = 339$ K, and the

Table 1. Comparison of Nusselt numbers with different grids (Ra = 3.3×10^6).

Physical property	96×192	128×256	150×300	170×340	Literature (C. Ho et al, 2010)
Nu_{avg}	11.828	11.935	12.149	12.151	12.154

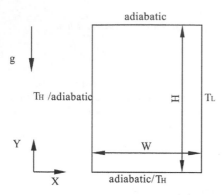

Figure 1. Schematic of the rectangular enclosure.

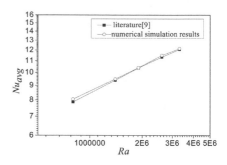

Figure 2. Comparison between present work and other published experimental data.

bottom wall is adiabatic. The other is that the left wall is adiabatic, and the bottom wall is maintained at T_H = 339 K. The grid independence test is performed, and details are shown in Table 1. Compared with the results from the literature (C. Ho et al, 2010), a grid size of 150×300 was chosen in this paper. Initial simulations are carried out on the natural convection heat transfer of Al_2O_3-water nanofluid (ϕ = 0.01) at different Rayleigh numbers ($Ra = 8 \times 10^5$, $Ra = 1.4 \times 10^6$, $Ra = 1.9 \times 10^6$, $Ra = 2.6 \times 10^6$, and $Ra = 3.3 \times 10^6$) to check the reliability and accuracy. The simulation results have a good agreement with the results of the literature (C. Ho et al, 2010), as shown in Figure 2, which proves that the model has a high accuracy.

After the verification of the model, natural convection heat transfer characteristics of diamond-water nanofluid are investigated. Figure 3 shows the temperature distribution of diamond-water nanofluid at ϕ = 0.01. It can be seen that the isotherm becomes more crooked at ΔT = 11 K compared with that at ΔT = 1 K, this is because that heat conduction plays a major role at ΔT = 1 K, while convection plays a major role at ΔT = 11 K and have a stronger disturbance on laminar boundary layer than heat conduction. It can be obtained that high temperature difference can enhance the natural convection heat transfer of the fluid in the enclosure.

All above discussions are not comprehensive, and the heat transfer characteristics of water and diamond-water nanofluid will be quantitatively investigated in the following section. Figure 4 shows the Nusselt number distribution of fluid with vol-

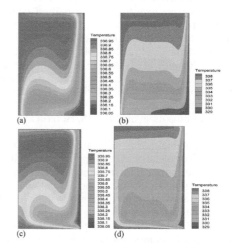

Figure 3. The temperature distribution of diamond-water nanofluid in the rectangular enclosure (a) the left wall heated and the right wall cooled, ΔT = 1 K; (b) the left wall heated and the right wall cooled, ΔT = 11 K; (c) the bottom wall heated and the right wall cooled, ΔT = 1 K; (d) the bottom wall heated and the right wall cooled, ΔT = 11 K.

ume fractions (ϕ = 0.01, ϕ = 0.03 and ϕ = 0.05) along with the heated wall at different temperature differences (ΔT = 1 K and ΔT = 11 K). It can be seen that the addition of nanoparticles causes the Nusselt number distribution along the heated wall to increase basically. Moreover, the phenomenon

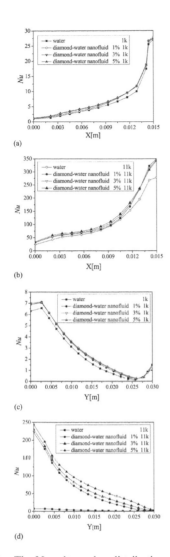

Figure 4. The Nusselt number distribution along with the heated wall (a) The bottom heated and the right wall cooled, $\Delta T = 1$ K; (b) The bottom heated and the right wall cooled, $\Delta T = 11$ K; (c) The left wall heated and the right wall cooled, $\Delta T = 1$ K; and (d) The left wall heated and the right wall cooled, $\Delta T = 11$ K.

becomes more and more obvious with the increase of temperature difference. In addition, when the heat conductivity coefficient plays a major role in the heat transfer, addition of nanoparticles will cause the Nusselt number distribution along the heated surface to increase. Conversely, when the viscosity plays a major role in the heat transfer, addition of nanoparticles will cause the Nusselt number distribution along the heated surface to decrease.

4 CONCLUSIONS

Natural convection heat transfer characteristics of diamond-water nanofluid in rectangular enclose are investigated in this paper. Following are some of the main results:

1. With the increase in temperature difference, the way of heat transfer changed from heat conduction to natural convection in rectangular enclosure.
2. Addition of nanoparticles causes the Nusselt numbers increase basically. In addition, the phenomenon becomes more and more obvious with the increase in temperature difference. The Nusselt numbers along the bottom surface are larger than that along the left surface, and it can enhance the heat transfer by 9.62% at large.
3. Compared with water, diamond-water nanofluid shows larger enhancement. The diamond-water nanofluid can enhance the heat transfer by 10.25%. In addition, the enhancement ratio at $\Delta T = 1$ K is larger than that at $\Delta T = 11$ K.

ACKNOWLEDGMENTS

This work was financially supported by the Fundamental Research Funds for the Central Universities (NO. 2015XKMS063).

REFERENCES

Zamzamian, A., S. Oskouie, A. Doosthoseini, A. Joneidi and M. Pazouki: Experimental Thermal and Fluid Science, Vol. 35 (2011) No. 3, p. 495.

Ho, C., W. Liu, Y. Chang and C. Lin: International Journal of Thermal Sciences, Vol. 49 (2010) No. 8, p. 1345.

Rashidi, I., O. Mahian, G. Lorenzini, C. Biserni and S. Wongwises: International Journal of Heat and Mass Transfer, Vol. 74 (2014) No. 3, p. 391.

Bouhalleb, M. and H. Abbassi: International Journal of Hydrogen Energy, Vol. 39 (2014) No. 27, p. 15275.

Sheikholeslami, M., M. Gorji-Bandpy, D. Ganji and S. Soleimani: Journal of the Taiwan Institute of Chemical Engineering, Vol. 45 (2014) No. 1, p. 40.

Sheikholeslami, M., T. Hayat, A. Alsaedi: International Journal of Heat and Mass Transfer, Vol. 96 (2016) No. 1, p. 513.

Mansour, R., N. Galanis and C. Nguyen C: International Journal of Thermal Sciences, Vol. 50 (2011) No. 3, p. 403.

Heris, S., M. Esfahany and S. Etemad: International Journal of Heat and Fluid Flow, Vol. 28 (2007) No. 2, p. 203.

Yang, Y., Z. Zhang, E. Grulke, W. Anderson and G. Wu: International Journal of Heat and Mass Transfer, Vol. 48 (2005) No. 6, p. 1107.

7 Control and automation, system simulation

Control and optimization. System simulation

Energy Science and Applied Technology ESAT 2016 – Fang (Ed.)
© 2016 Taylor & Francis Group, London, ISBN 978-1-138-02973-6

Current sensorless control system of PMSM based on a hybrid system

Weifa Peng
School of Mechatronics Engineering and Automation, Shanghai University, Shanghai, China
College of Electrical Engineering, East China Jiaotong University, Nanchang, China

Surong Huang, Haishan Liu & Xianglin Wei
School of Mechatronics Engineering and Automation, Shanghai University, Shanghai, China

ABSTRACT: The PMSM drive system contains the continuous state variables and the discrete variables, which is a typical hybrid system. In this paper, the hybrid system theory is introduced into the analysis process, and the discrete voltage model of the inverter is embedded into the continuous state equation of the motor, and the hybrid logic dynamic model of the motor drive system is established. Based on this, a new control method of permanent magnet synchronous motor without current sensor is proposed, and the validity and reliability of the proposed control method are proved by MATLAB/Simulink simulation.

Keywords: PMSM, hybrid system, current sensorless, mixed logic dynamic model

1   INTRODUCTION

In order to realize high precision and high dynamic performance of PMSM, various control methods are emerging in endlessly, among which the vector control (Amit Vilas Sant et al, 2015) and direct torque control (Shashidhar Mathapati et al, 2013) are the most typical two kinds of methods. Vector control requires high-performance current, speed, and position feedback control. At present, the research of less sensor and no sensor in the speed control system becomes a hot spot, and it is more typical does not use the position sensor, but in the positioning system the position sensor cannot be missing. The main methods of current reconstruction are the modulation method and the model method. The model method is based on a precise mathematical model to calculate the three-phase current relying on other physical quantity. In the literature (Shigeo Morimoto et al, 2003), the three-phase current of PMSM is estimated by the precise mathematical motor model and only low-resolution position sensor, and finally realizes the current sensorless control. In the literature (Manuele Bertoluzzo et al, 2006), the three-phase current of induction motor is calculated by the precise mathematical motor model, and it is modified by the DC current. The PMSM drive system is a typical hybrid system, which contains the discrete event switch on-off form, including specific switch mode by continuous state equation constraint evolution consisting of continuous events (ZHANG Zhi-xue et al, 2005). In this paper, from the perspective of hybrid system, a unified mathematical model of the motor drive system is built for a high-performance servo control system without any current sensor.

2   MIXED LOGIC DYNAMIC MODEL OF MOTOR DRIVE SYSTEM OF PMSM

The system composed of PMSM and inverter is depicted in Figure 1, and its DC bus voltage is V_{dc}. The electric motor can be equivalent to a circuit composed of resistance, inductance and back EMF. The continuous model of the motor can be described as:

$$\begin{cases} u_{an} = Ri_a + L\dfrac{di_a}{dt} + e_a \\[2mm] u_{bn} = Ri_b + L\dfrac{di_b}{dt} + e_b \\[2mm] u_{cn} = Ri_c + L\dfrac{di_c}{dt} + e_c \end{cases} \tag{1}$$

where u_{an}, u_{bn}, and u_{cn} are the stator phase voltages, i_a, i_b, and i_c are the stator phase currents, e_a, e_b, and e_c are the stator phase back-EMFs, R is the stator resistance, and L is the stator inductance.

$S_1 \sim s_6$ stand for the switching signals of IGBT $V_1 \sim V_6$, respectively, and $s_i = 1 \Rightarrow V_i = ON$, $s_i = 0 \Rightarrow V_i = OFF$; $\forall i \in \{1,2,3,4,5,6\}$.

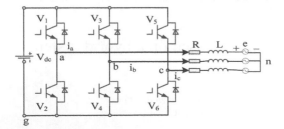

Figure 1. Equivalent circuit diagram of inverter and motor system.

According to the system topology constraint:

$$\begin{cases} u_{an} = u_{ag} - u_{ng} \\ u_{bn} = u_{bg} - u_{ng} \\ u_{cn} = u_{cg} - u_{ng} \end{cases} \tag{2}$$

$$i_a + i_b + i_c = 0 \tag{3}$$

Generally, the three-phase back EMFs of the PMSM are balanced, and so $e_a + e_b + e_c = 0$, Consequently, the following equation can be achieved:

$$\begin{cases} u_{an} = \dfrac{1}{3}(2u_{ag} - u_{bg} - u_{cg}) \\ u_{bn} = \dfrac{1}{3}(-u_{ag} + 2u_{bg} - u_{cg}) \\ u_{cn} = \dfrac{1}{3}(-u_{ag} - u_{bg} + 2u_{cg}) \end{cases} \tag{4}$$

Assuming that the current is continuous, define auxiliary logical variables δ_a, δ_b and δ_c as:

$$\begin{aligned} \left[\delta_j = 1\right] &\leftrightarrow \left[i_j > 0\right], j = a,b,c \\ \left[\delta_j = 0\right] &\leftrightarrow \left[i_j < 0\right], j = a,b,c \end{aligned} \tag{5}$$

According to the various states of the inverter, the MLD model of the inverter can be obtained:

$$\begin{cases} u_{ag} = V_{dc}\overline{s_2}(s_1\delta_a + \overline{\delta_a}) \\ u_{bg} = V_{dc}\overline{s_4}(s_3\delta_b + \overline{\delta_b}) \\ u_{cg} = V_{dc}\overline{s_6}(s_5\delta_c + \overline{\delta_c}) \end{cases} \tag{6}$$

Substituting (6) into (4), the phase voltages can be obtained:

$$\begin{bmatrix} u_{an} \\ u_{bn} \\ u_{cn} \end{bmatrix} = \frac{V_{dc}}{3}\begin{bmatrix} 2 & -1 & -1 \\ -1 & 2 & -1 \\ -1 & -1 & 2 \end{bmatrix}\begin{bmatrix} \overline{s_2}(s_1\delta_a + \overline{\delta_a}) \\ \overline{s_4}(s_3\delta_b + \overline{\delta_b}) \\ \overline{s_6}(s_5\delta_c + \overline{\delta_c}) \end{bmatrix} \tag{7}$$

Substituting (7) into (1), the MLD model description of the voltage equation of the motor drive system is obtained:

$$\begin{bmatrix} u_{an} \\ u_{bn} \\ u_{cn} \end{bmatrix} = \frac{V_{dc}}{3}\begin{bmatrix} 2 & -1 & -1 \\ -1 & 2 & -1 \\ -1 & -1 & 2 \end{bmatrix}\begin{bmatrix} \overline{s_2}(s_1\delta_a + \overline{\delta_a}) \\ \overline{s_4}(s_3\delta_b + \overline{\delta_b}) \\ \overline{s_6}(s_5\delta_c + \overline{\delta_c}) \end{bmatrix}$$

$$= \begin{bmatrix} R & 0 & 0 \\ 0 & R & 0 \\ 0 & 0 & R \end{bmatrix}\begin{bmatrix} i_a \\ i_b \\ i_c \end{bmatrix} + \frac{d}{dt}\begin{bmatrix} L & 0 & 0 \\ 0 & L & 0 \\ 0 & 0 & L \end{bmatrix}\begin{bmatrix} i_a \\ i_b \\ i_c \end{bmatrix} + \begin{bmatrix} e_a \\ e_b \\ e_c \end{bmatrix} \tag{8}$$

The above formula can be transformed into the system state equation with the winding current as the state variable:

$$\begin{bmatrix} \dot{i}_a \\ \dot{i}_b \\ \dot{i}_c \end{bmatrix} = \begin{bmatrix} -\dfrac{R}{L} & 0 & 0 \\ 0 & -\dfrac{R}{L} & 0 \\ 0 & 0 & -\dfrac{R}{L} \end{bmatrix}\begin{bmatrix} i_a \\ i_b \\ i_c \end{bmatrix} + \begin{bmatrix} -\dfrac{1}{L} & 0 & 0 \\ 0 & -\dfrac{1}{L} & 0 \\ 0 & 0 & -\dfrac{1}{L} \end{bmatrix}\begin{bmatrix} e_a \\ e_b \\ e_c \end{bmatrix}$$

$$+ \frac{V_{dc}}{3L}\begin{bmatrix} 2 & -1 & -1 \\ -1 & 2 & -1 \\ -1 & -1 & 2 \end{bmatrix}\begin{bmatrix} \overline{s_2}(s_1\delta_a + \overline{\delta_a}) \\ \overline{s_4}(s_3\delta_b + \overline{\delta_b}) \\ \overline{s_6}(s_5\delta_c + \overline{\delta_c}) \end{bmatrix} \tag{9}$$

In order to simplify the model, define a new auxiliary logical variable vector $\delta = \begin{bmatrix} \delta_1 & \delta_2 & \delta_3 \end{bmatrix}^T$, and let:

$$\begin{cases} \delta_1 = \overline{s_2}(s_1\delta_a + \overline{\delta_a}) \\ \delta_2 = \overline{s_4}(s_3\delta_b + \overline{\delta_b}) \\ \delta_3 = \overline{s_6}(s_5\delta_c + \overline{\delta_c}) \end{cases} \tag{10}$$

So, the MLD model of the motor drive system including continuous variables and discrete variables is obtained:

$$\begin{bmatrix} \dot{i}_a \\ \dot{i}_b \\ \dot{i}_c \end{bmatrix} = \begin{bmatrix} -\dfrac{R}{L} & 0 & 0 \\ 0 & -\dfrac{R}{L} & 0 \\ 0 & 0 & -\dfrac{R}{L} \end{bmatrix}\begin{bmatrix} i_a \\ i_b \\ i_c \end{bmatrix} + \begin{bmatrix} -\dfrac{1}{L} & 0 & 0 \\ 0 & -\dfrac{1}{L} & 0 \\ 0 & 0 & -\dfrac{1}{L} \end{bmatrix}$$

$$\begin{bmatrix} e_a \\ e_b \\ e_c \end{bmatrix} + \frac{V_{dc}}{3L}\begin{bmatrix} 2 & -1 & -1 \\ -1 & 2 & -1 \\ -1 & -1 & 2 \end{bmatrix}\begin{bmatrix} \delta_1 \\ \delta_2 \\ \delta_3 \end{bmatrix} \tag{11}$$

The system state function can be achieved by the MLD model as follows:

$$i = Ai + B_1e + B_2\delta \qquad (12)$$

where $i = \begin{bmatrix} i_a & i_b & i_c \end{bmatrix}^T$ is the system state vector. The auxiliary logic vector δ is the discrete input vector of the system. $e = \begin{bmatrix} e_a & e_b & e_c \end{bmatrix}^T$ is the back EMF vector, and can be regarded as the continuous input vector. The state coefficient matrix is,

$$A = \begin{bmatrix} -\dfrac{R}{L} & 0 & 0 \\ 0 & -\dfrac{R}{L} & 0 \\ 0 & 0 & -\dfrac{R}{L} \end{bmatrix}, \quad B_1 = \begin{bmatrix} -\dfrac{1}{L} & 0 & 0 \\ 0 & -\dfrac{1}{L} & 0 \\ 0 & 0 & -\dfrac{1}{L} \end{bmatrix} \text{ and}$$

$$B_2 = \frac{V_{dc}}{3L}\begin{bmatrix} 2 & -1 & -1 \\ -1 & 2 & -1 \\ -1 & -1 & 2 \end{bmatrix}.$$

For a PMSM system, if the electric angle for rotor d axis rotating over a axis is set to θ, the three-phase back EMFs are expressed as:

$$e = -\sqrt{\frac{2}{3}}\omega\psi_f\begin{bmatrix} \sin\theta \\ \sin(\theta - 2\pi/3) \\ \sin(\theta + 2\pi/3) \end{bmatrix} \qquad (13)$$

where ω is the electric angular speed of the rotor and ψ_f is the rotor's permanent magnet flux.

The state estimator of the motor drive system by means of its MLD is built as

$$\dot{\hat{i}} = A\hat{i} + B_1e + B_2\delta \qquad (14)$$

3 REALIZATION OF CURRENT SENSORLESS CONTROL SYSTEM

As the PMSM rotor flux is constant, this control system adopts a vector control method of rotor field oriented. The PMSM is modeled in the d-q reference frame fixed to the rotor. The d axis is oriented along the permanent magnet flux, whose angle in the stator reference frame is θ in electrical radians. The d-axis current is controlled to zero, so stator current contains only the q-axis component. Through the photoelectric encoder, the system can get the position of the rotor, and then calculate the speed of the motor. As there is no current sensor, the current module in Figure 2 can only be obtained by the current state estimator. The current state estimator is based on formula (14). The vector control system shown in Figure 2 includes a block diagram of the cascade speed and current control loop.

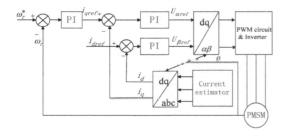

Figure 2. Current sensorless control of PMSM drive.

4 SIMULATION RESULTS

Parameters of the considered motor are listed as: rating speed, 3000 rpm; number of pole-pairs, 4;

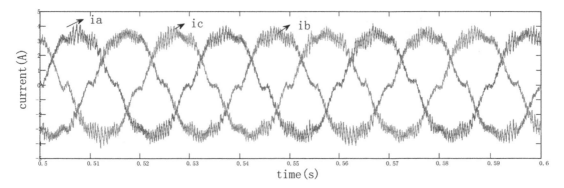

Figure 3. Experimental three-phase current waveform.

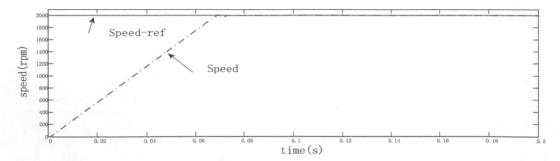

Figure 4.   Speed input signal "speed-ref" and speed response curve "speed".

rating torque, 3N·m; stator resistance, 2.875 Ω; d- and q-axis inductances, $L_d = L_q = 0.0085$ H; permanent magnet flux linkage, 0.174 Wb; and rotor inertia, 0.008 kg.m². Simulation experiment of motor speed control is carried out based on the simulation model. In addition, the given speed is 2000 r/min with 2 N·m load. Figure 3 shows the experimental three-phase current waveform. Figure 4 shows the speed input signal "speed-ref" and speed response curve "speed". Speed response in 0~0.07 s quickly increased to 2000 rpm, then PMSM enters steady-state operation stage.

5   CONCLUSIONS

In this paper, a PMSM AC servo system based on the mixed logic dynamic model for the current sensorless is presented. The system is constructed to estimate the current module to obtain the voltage reference value. The simulation results show that the system is simple and feasible, can achieve the system control requirements, save system costs, and has a good application prospect.

ACKNOWLEDGMENT

This work was supported by School scientific research fund of East China Jiaotong University (No. 24441012).

REFERENCES

Amit Vilas Sant, Vinod Khadkikar, Weidong Xiao, and H.H. Zeineldin, Four-axis vector-controlled dual-rotor PMSM for plug-in electric vehicles, J. IEEE Transactions on Industrial Electronics, Vol. 62 (2015) No.5, p. 3202.

Manuele Bertoluzzo, Giuseppe Buja, Roberto Menis, Direct torque control of an induction motor using a single current sensor, J. IEEE Transactions on Industrial Electronics, Vol. 53 (2006) No.3, p. 778.

Shashidhar Mathapati, and Joachim Böcker, Analytical and Offline Approach to Select Optimal Hysteresis Bands of DTC for PMSM, J. IEEE Transactions on Industrial Electronics, Vol. 60 (2013) No.3, p. 885.

Shigeo Morimoto, Masayuki Sanada, Yoji Takeda, High-performance current-sensorless drive for PMSM and SynRM with only low-resolution position sensor, J. IEEE Transactions on Industry Applications, Vol. 39 (2003) No.3, p. 792.

ZHANG Zhi-xue, MA Hao, MAO Xing-yun, Fault diagnosis for power electronic circuits based on hybrid system theory and event identification, J. Proceedings of the CSEE, Vol. 25 (2005) No.3, p. 49.

Energy Science and Applied Technology ESAT 2016 – Fang (Ed.)
© *2016 Taylor & Francis Group, London, ISBN 978-1-138-02973-6*

Hierarchical modeling method for intelligent plant simulation

Zhen Zhang & Xujie Hu
China Aerospace Construction Group Co. Ltd., Beijing, China

ABSTRACT: Conventionally, Plant simulation model is classified into macroscopic and microscopic. An Enterprise Resources Planning (ERP)-level model and a Process Controlling System (PCS)-level model belong to macro and micro simulation model respectively, while an Manufacturing Executive System (MES)-level model lies in between. Simulation models of certain level established by former researchers usually have poor reusability and scalability. Therefore, a new four-layered approach, which is composed of model entity structure/model base, model abstraction, plant modeling and Intelligent Plant System (IPS) simulation layer, is proposed for plant modeling and simulation. Configurable and reusable simulation model can be easily obtained by adopting the abstraction modeling and simulation method with respect to plant network. In this paper, a multi-level intelligent plant simulation system is constructed with the proposed method.

Keywords: intelligent plant, modeling method, computer simulation

1 INTRODUCTION

Intelligent Plants include many kinds of technology and application such as computer, communication, controlling and other digital technology. Workshop layout, logistics system, and production decision making such as production scheduling, supply chain optimization should be accomplished with simulation method. The national institute of standards and technology of US (NIST) initiated the plan to support modeling and simulation technology of intelligent plant (F. Proctor, 2013). For the enterprise level optimization, in order to optimize the whole supply chain—planning—scheduling—controlling process, a hierarchical modeling method and a universal modeling and simulation platform oriented enterprise level should be proposed (I. Grossmann, 2005) (S.M.L. Coalistion, 2011).

Conventionally, Plant simulation model is classified into macroscopic or microscopic. For example, a supply chain model is on the Enterprise Resources Planning (ERP) level and a produce process model is on Process Controlling System (PCS) level. The extraction of the model on PCS level, MES level, and ERP level is very important to analyze and optimize the manufacturing situation of the plant (B. Scholten, 2007). Hierarchical modeling method can simplify the model complexity and at the same time can describe the object and its environment correctly. However, researchers need to provide new independent models for each plant simulation problem solving. This can result in long developing time. Furthermore, the universality of the model is too weak to support model's reuse and expansibility (C. Herrmann et al, 2011).

To solve mentioned problems, this paper proposes a four-layered framework for simulation modeling, composed of model entity structure/model base, model abstraction, plant modeling and Intelligent Plant System (IPS) simulation layer. A hierarchical modeling method is developed based on this framework. This method can supply plant manager with reusable and expansible simulation model to adjust for better simulation result.

2 PLANT MODELING AND SIMULATION METHODS

Until now, there are many kinds of technology and methods for simulation systems. Simulation technology is divided into continuous time simulation and discrete event simulation. The discrete event simulation always has difficulty to solve hierarchical mapping of enterprises-level models, because different models on different level have their own time granularity, which need consider equipment character, material character, and their relationships.

Based on discrete time simulation methods, this paper proposes a four-layered framework for discrete production modeling and simulation. As shown in Fig. 1, layer 1 provides an assemble simulation model base, ensured the system's configurability and expansibility. Models of layer 1 can be

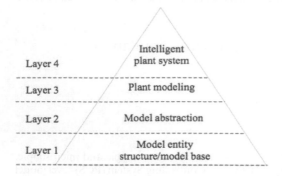

Figure 1. Layered approach for designing plant simulation systems.

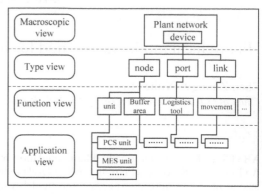

Figure 2. Model entity structure of plant system.

abstracted on layer 2, and the mapping between the abstract model and the initiative model is constructed. Plant simulation modeling of different layers can be accomplished on layer 3. Finally, the intelligent plant simulation system is achieved on layer 4.

3 LAYER 1: MODEL ENTITY STRUCTURE/MODEL BASE

Model entity structure/model base is the base of intelligent plant hierarchical modeling and simulation environment. It includes entity structure and model base.

Entity structure (as shown in Fig. 2) describes a series of decomposition rules, combination rules, and restriction conditions for the real environment. A certain model is the mapping of a real entity. An entity can have different kinds of description, with each description representing certain variant of the entity. Therefore, a model instance can be obtained from its inheritance relationships to the entity structure, and an intelligent plant hierarchical simulation model can be obtained by splicing model instances. The entity structure can be described in four perspectives.

A. From macroscopic perspective, a plant can be considered as a combination of huge numbers of devices with certain characters, which is the plant network configured by devices.
B. From type perspective, a plant device can be divided into node model, port model, and link model. The node represents manufacturing device, buffer area, conveyance. The port model represents the entrance or exit of the node model. The link node is used to link to two ports. The topological structure of a plant can be represented clearly and completely by node, port, and link models.

C. From function perspective, taking the node model as an example, considering its different character and attributes, the node model can be instantiated into manufacturing unit model, buffer area model, or logistics equipment.
D. From application perspective, a same model may have different role when it is applied on different layer. Taking the manufacturing unit as an example, it can be instantiated into PCS unit and Manufacturing Executive System (MES) unit.

The model base includes huge numbers of models. The model is described by computer with high configurability and reusability. The topological structure of the plant is composed of devices, and devices derive node model, port model, and link model. The three kinds of model can describe complete plant topological structure. Simulation module can search and traverse every node according to the topological structure and simulate the node's behavior.

Component libraries can be constructed according to different function of models. Component libraries can be divided into four types: calculator library, static library, dynamic input library, and dynamic output library. The detail of the model component is described in layer 3, plant modeling.

4 LAYER 2: MODEL ABSTRACTION

Abstraction is the key to simulation modeling. The information of the model is decided by the model scale and model resolution. The model scale means how large environment size the model should include. The model resolution means the granularity of model's information. The scale and the resolution should be balanced in modeling. Therefore, the abstract process is important for reducing model's complexity and ensuring its reliability in

450

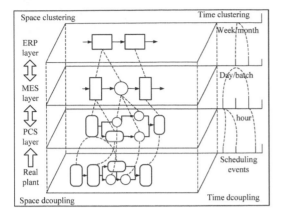

Space clustering   Time clustering

ERP layer

MES layer

PCS layer

Real plant

Week/month

Day/batch

hour

Scheduling events

Space dcoupling   Time dcoupling

Figure 3. Abstraction process for plant modelling.

modeling objective and simulation process. During the abstract process, the clustering role is bellow (as shown in Fig. 3): ① Buffer area and logistic equipment on PCS layer are abstracted to logic buffer on MES layer. ② Logic buffers on MES layer are abstracted to logic buffer on ERP layer. ③ Logic buffer area on MES layer is abstracted to a point.

In model's abstract process, models on different layer have its own application. Such as PCS layer is manufacturing control layer, and it is used to optimize manufacturing process, manage manufacturing operation, and monitor workshop environment. MES layer is a medial scheduling layer which is used to execute the implementation plans, track logistics, and manage product quality. ERP layer is the highest management layer which is used to optimize enterprise planning and manage key performance indicator. Each layer has its different time and space information granularity, and clustering and decoupling on different layers can be accomplished by model abstraction.

5 LAYER 3: PLANT MODELING

The plant model is composed of models from component library. First, the model component should be constructed and registered in the relative component library according to its function. If the model is constructed before, the existed model can be used directly. By choosing and composing the component models, the plant model can be configured and constructed.

The component librarie has four types: calculator library, static library, dynamic input library, and dynamic output library. Static library is used

to define plant static attributes information. Such information is important to plant device model. It may influence some important parameters during the simulation process. For example, StaticModel PCSUnit describes manufacturing device information on PCS layer, such as device name, device area, device type, production capacity. Dynamic input library and dynamic output library emphasize on the behavior of device, such as the change of buffer capacity and conveyance capacity. Calculator library is the core of the simulation. It describes the behavior of plant devices, such as production process, material input or output, conveyance operation. Calculator model is the necessity of a node model.

6 LAYER 4: INTELLIGENT PLANT SIMULATION SYSTEM

Based on the proposed hierarchical abstract modeling method and simulation framework, a modeling and simulation platform can be constructed. The platform provides intelligent plant simulation with expansible environment for composing component models into different entity model.

7 CONCLUSION

Aiming at some problems in plant simulation, the paper proposes a four-layer framework for modeling and simulation. It includes model entity structure/ model base, model abstraction, plant modeling and IPS simulation layer. A hierarchical modeling method is developed based on this framework. Based on the framework, a modeling and simulation system is constructed to efficiently reduce the difficulty in simulation model reuse and configuration.

REFERENCES

Coalistion, S.M.L. Implementing 21st Centrury Smart Manufacturing: Workshop Report, SMLC and US DOE, 2011.
Grossmann, I. AIChE Journal, Vol. 51(2005) No. 7, p.1846–1857.
Herrmann, C., S. Thiede: CIRP Annals-Manufacturing Technology, Vol. 60 (2011) No. 1, p. 45–48.
Proctor, F. Smart manufacturing and construction control systems, information on http://www.nist.gov/el/isd/cs/upload/SMCSprogram2013.
Scholten, B. The Road to Integration: A Guide to Applying the ISA-95 Standard in Manufacturing. (Isa 2007).

8 *Communications and applied information technologies*

Energy Science and Applied Technology ESAT 2016 – Fang (Ed.)
© *2016 Taylor & Francis Group, London, ISBN 978-1-138-02973-6*

A GDOP-weighted AOA localization method based on ray-trace

F.Z. Kong, X.K. Ren & N.E. Zheng
Zhengzhou Institute of Information Science and Technology, Zhengzhou, China

ABSTRACT: For the existing AOA localization method based on the weighted intersection, which caused the shortage of positioning accuracy due to undesirable intersection of bearing lines, meanwhile taking into consideration the introduction of ray tracing to eliminate the effect of NLOS noise, a GDOP-weighted intersection method for ray-trace-based target localization using AOA Measurements (GWIAOA-RT) is proposed in this paper. This method first transforms the NLOS path into LOS path based on the building Mirror Base Station (M-BS) in terms of the track results of multipath positioning signals; then obtain all the bearing lines and their intersection by using the AOA measurements of LOS paths; furthermore, determine the effective intersection according to the CEP and give corresponding weight according to the GDOP of effective intersection, and make weighted summation of effective intersection as target position estimation. The method makes use of multipath positioning signals, and eliminates the effect of NLOS noises on the positioning accuracy. Meanwhile, take into consideration the position error of M-BS in the calculation of GDOP, with strong robustness. The simulation result shows that the GWIAOA-RT method has higher positioning accuracy and better adaptability in different application scenarios compared with the existing method.

Keywords: localization; Angle of Arrival (AOA); Geometric Dilution of Precision (GDOP); Circular Error Probable (CEP); Ray Trace

1 INTRODUCTION

The Location-Based Service (LBS) can provide significant convenience to people's life, and has become an important part of our modern life. Accurate location information is the base of LBS, so localization technology is always the research hotspot. Among many localization technologies, localization technologies based on ray-tracing (Berman, Z. 1997) can effectively solve such problems as multipath propagation, hearability (Yuan, Z. W. 2007), and Non-Line-of-Sight (NLOS) noise, which can seriously influence the localization effect of the range-based localization method, thus being widely concerned by people.

The positioning technique based on ray tracing realizes the tracking of the signal propagation path by using the measurements of various positioning signals, constructs a virtual positioning station with a direct path, and calculates the position of the target by using the range-based location method. A number of studies have been carried out on ray tracing for localization. In Lui, K. (2010), the multi-sensor based method is proposed to trace the direct propagation path and the once-reflected propagation path of the multipath signal under the condition of only one reflecting surface. In Liu, D. L. (2014), the multi-sensor-based method

is adopted to establish virtual sensors through the measured TOA and AOA as well as geometric constraint relationship and accordingly complete the ray tracing process for perpendicular incidence, reflection, and diffraction of the first signal arrival path. A single sensor is used to realize the tracking of multi-path positioning signal based on indoor structure combined in Setlur, P. G. et al. (2012). The ray tracing method is the basis of this kind of localization method, and will be discussed in other papers. It will not be discussed in this paper.

The typical range-based localization methods mainly include Time of Arrival (TOA) localization method, Time Difference of Arrival (TDOA) (Compagnoni, M. & Notari, R. 2014) localization method, Angle of Arrival (AOA) (Shao, H. J. et al. 2014) localization method, etc. AOA-based localization method is an important branch of research. This paper is the study of ray-trace-based target localization method using AOA measurements and its CRLB.

The scholars have made many researches in the AOA-based localization method. Stansfield proposed the Weighted Least Square (WLS) method. In (Pages-Zamora, A. et al. 2002), a Pseudo Linear-Squares (PLS) method, was proposed and gave the closed-form solution of the target localization. The two methods do not need the initial estimation of target localization and are easily realized, but both are biased, and adding the anchor node cannot totally eliminate the bias. In order to eliminate the estimation error, the Constrained Least Squares (CLS) and the Total Least Squares (TLS) method are proposed. These methods are iterative methods, and the computation is large. CLS and TLS method are proved to be asymptotically Maximum Likelihood Estimation (MLE) in (Gu, G. 2010).

The method in Wang, Z. et al. (2012) is the Maximum Likelihood Estimation (MLE) method. Under the condition of the Gaussian hypothesis, MLE can minimize the total AOA measuring error, realize the asymptotic unbiased estimation, and the method performance is close to the Cramer-Rao Low Bound (CRLB). However, all of them are iterative methods and need an initial guess, which approaches the true position of the target; an irrational initial guess might cause the method convergence to be with locally optimal solution or failure of convergence.

The literature (Zhou, Q. & Duan, Z. 2014) conducts the weighted summation to estimate the target position by using the intersections between the base stations and the target's bearing line, and the method can be called the weighted intersection localization method using AOA measurements. Complex AOA (CAOA) considers the mean of all Intersections of Bearing Lines (IBL) as the estimation of the target position. The literature proposed the grid-based and vector-based approximation methods. The two methods are iterative methods; they have large calculation workload, but their positioning accuracy is lower than that of PLS. The literature proposed the Sensitivity Analysis (SA) localization method; it defines the sensitivity of every intersection by using the first or second derivative of the closed-form solution of the IBLs, generates the corresponding weights and then conducts the weighted summation. The literature, based on the positioning error analysis, proposed the weighted IBL localization method. The weighted intersection AOA (WIAOA) localization method in (Zhou, Q. & Duan, Z. 2014) uses the distance between the anchor nodes, the AOA measurement of anchor nodes and the targets, and the standard deviation to construct the weight of the IBLs. The CRLB-weighted intersection AOA (CWIAOA) localization method in Duan, Z. & Zhou, Q. (2015) uses the CRLB value of IBL to get the intersection weight. The positioning accuracy of the WIAOA method and CWIAOA method are both superior to that of PLS; but they do not consider about the following situation, that is, some intersections have larger weights, but its deviation with the true position of the target is larger than that between the intersections with smaller weights

and the true position of the target. Using this kind of intersection in the target position estimation will seriously influence the estimation accuracy, and can be called the Undesirable Intersection of Bearing Lines (UIBL).

In this paper, first based on microcell where Base Station (BS) and Mobile Station (MS) are located, the AOA-based localization model has been established, including the signal propagation environmental model and localization model, and the formula of Mirror Base Station (M-BS) position is given based on the mirror-image principle. Second, the effect of different bearing lines intersection on positioning accuracy is analyzed by using the CRLB theory. Based on this point, a GDOP-weighted intersection method for ray-trace-based target localization using AOA measurements (GWIAOA-RT) is proposed in this paper. GWIAOA-RT method can obtain all the IBLs by using the M-BSs and AOA measurements of M-BSs based on ray trace, and then eliminate UIBLs according to CEP, regard the rest IBLs as effective IBLs, give corresponding normalized weight and make weight summation for the effective IBLs as the estimation of MS position. The method fully uses the multipath propagation of positioning signal to realize the single-station positioning, and just make weighted summation on effective IBLs. Considering about the position error of BS in the calculation of weight will make the method obtain higher accuracy and better robustness in applied environments.

2 AOA-BASED LOCALIZATION MODEL OF MICROCELL

The model of the microcell in this paper mainly includes two parts, namely, the signal propagation environment model and the localization model.

2.1 Signal propagation environment model

For the ray tracing process for the received signal, it is necessary to establish the wireless signal propagation environment model according to the geographic information of the mobile station area. The model can be obtained through field measurement or GIS system. The proposed method is mainly used for microcell, so the ground of the target area can be assumed as the horizontal flat ground. For model simplification, the geometric model of the wireless signal propagation environment is established in a two-dimensional plane, and the geometric building model is the polygon composed of the projections of the buildings on the ground. The edges of the polygon are the building surfaces, which are smooth, and rays can be reflected on these surfaces. The ith surface can be expressed in 2D model by the following formula:

$$F_i : a_i x + b_i y + c_i = 0 \tag{1}$$

The intersection point of two edges is the intersecting array of the building surfaces and rays can be diffracted when arriving at the arrays.

If a signal only includes perpendicular incidence or reflection during the propagation process, the signal propagation path formed thereby is called as Position Orientating Path (POP) and the corresponding signal is called the Position Orientating Signal (POS). POS that occurs N times the reflection in the propagation process, is called N-order POS, and its propagation path is called N-order POP. The ith N_i-order POP is expressed by Eq.(2), where $F_{i_1}, \ldots, F_{i_{Ni}}$ are build surfaces that the POP is passing through.

$$PL_i^{N_i} : \left\{ F_{i_1}, \ldots, F_{i_{Ni}} \right\} \tag{2}$$

POP can be uniquely identified by the ray-trace method, the both ends of the path are Base Station (BS) and Mobile Station (MS). In addition to the direct path, using the principle of specular reflection, we can get the mirror-image BSs (M-BS) according to BS and POPs. NLOS path could be converted into LOS path to M-BS. In order to describe the convenience, if the POP is direct path, it could be called 1-order POP, the surface which it passes through is F_0, corresponding signal could be known as 1-order POS, BS could be also known as M-BS.

2.2 Localization model

The AOA-based localization method uses the AOA measurements of the MS signals from BSs to establish the bearing line between the BSs and MS, and the IBL is the rough estimate of the MS position. The BS receives the positioning signals through array antenna, and realizes the AOA estimation by using the Beam-forming method, Multiple Signal Classification (MUSIC) algorithm or Estimation of Signal Parameters via Rotational Invariance Techniques (ESPRIT) algorithm.

Let us assume that there are N POPs between BS and MS, namely, it could get N M-BSs. $\tilde{X}_0 = [\tilde{x}_0, \tilde{y}_0]^T$ and $X_0 = [x_0, y_0]^T$, respectively represent the known position and true position of the BS, n_{x0} is the error between x_0 and $\tilde{x}_0$, n_{y0} is the error between y_0 and $\tilde{y}_0$, n_{x0} and n_{y0} are the zero mean Gaussian noise, their standard variance are σ_{x0} and σ_{y0}, respectively. n_{x0} and n_{y0} are irrelevant, $N_0 = [n_{x0}, n_{y0}]^T$. the relationship between $\tilde{X}_0$ and X_0 is represented by the following formula:

$$\tilde{X}_0 = X_0 + N_0 \tag{3}$$

$\tilde{X}_i = [\tilde{x}_i, \tilde{y}_i]^T$ and $X_i = [x_i, y_i]^T$, respectively, represent the known position and true position of the i M-BS, and the position of M-BSs can be got according to position of BS and POPs. According to Eq. (1) (2) by using the mirror-image principle, the position of the ith N_i-order M-BS of BS can be got by the following formula:

$$\begin{cases} \tilde{X}_i = A_i \tilde{X}_0 + B_i \\ X_i = A_i X_0 + B_i \end{cases} \tag{4}$$

where,

$$A_i = \prod_{n=N_i}^{1} A_{i,n} = A_{i,N_i} A_{i,N_i-1} \cdots A_{i,1},$$

$$B_i = \sum_{n=1}^{N_i} A_i^n B_{i,n}, \quad A_i^m = \begin{cases} \prod_{n=N_i}^{m+1} A_{i,n} & , m < N_i, \\ I & , m = N_i \end{cases}$$

$$A_{i,n} = \begin{bmatrix} \dfrac{b_{i_n}^2 - a_{i_n}^2}{a_{i_n}^2 + b_{i_n}^2} & \dfrac{-2a_{i_n} b_{i_n}}{a_{i_n}^2 + b_{i_n}^2} \\ \dfrac{-2a_{i_n} b_{i_n}}{a_{i_n}^2 + b_{i_n}^2} & \dfrac{a_{i_n}^2 - b_{i_n}^2}{a_{i_n}^2 + b_{i_n}^2} \end{bmatrix}, \quad B_{i,n} = \begin{bmatrix} \dfrac{-2a_{i_n} c_{i_n}}{a_{i_n}^2 + b_{i_n}^2} \\ \dfrac{-2b_{i_n} c_{i_n}}{a_{i_n}^2 + b_{i_n}^2} \end{bmatrix}.$$

If the ith POP is direct path, then $A_i = I$ and $B_i = 0$.

Let us set $\tilde{X}_s = \left[\tilde{X}_1^T, \cdots, \tilde{X}_N^T \right]^T$, $X_s = \left[X_1^T, \cdots, X_N^T \right]$, $A = \left[A_1^T, \cdots, A_N^T \right]^T$, $B = \left[B_1^T, \cdots, B_N^T \right]^T$, according to Eqs.(3) and (4), we can get

$$\tilde{X}_s = A\tilde{X}_0 + B = AX_0 + B + AN_0 = X_s + AN_0 \tag{5}$$

$\tilde{\theta}_i$ represents the AOA measurement of the ith M-BS, θ_i represents the AOA true value of the ith M-BS. n_i represents the AOA measurement error of the ith M-BS and is the zero mean Gaussian noise with the standard variance σ_i. The position of the MS is unknown, and $p = (x_T, y_T)$ indicates its true position. $\hat{p} = (\hat{x}_T, \hat{y}_T)$ indicates the estimation result of the MS position from the method. The localization model can be represented by the following formula:

$$\tilde{\Theta} = \Theta + N_\theta \tag{6}$$

$$\theta_i = f_i(p_T) = \arctan\left(\frac{y_T - y_i}{x_T - x_i} \right) \tag{7}$$

where, $\tilde{\Theta} = [\tilde{\theta}_1, \cdots, \tilde{\theta}_N]^T$, $\Theta = [\theta_1, \cdots, \theta_N]^T$, $N_\theta = [n_1, \ldots, n_N]^T$.

3 ANALYSIS OF THE WEIGHTED INTERSECTION LOCALIZATION METHOD USING AOA MEASUREMENTS

The main idea of the weighted intersection localization method using AOA measurements is to conduct the weighted summation as the target position estimation based on IBLs of all the BSs and MS, and its general formula is:

$$\begin{bmatrix} \hat{x}_T \\ \hat{y}_T \end{bmatrix} = \begin{bmatrix} \sum_{i=1}^{N} \sum_{j=i+1}^{N} \omega_{i,j}^x \hat{x}_T^{i,j} \\ \sum_{i=1}^{N} \sum_{j=i+1}^{N} \omega_{i,j}^y \hat{y}_T^{i,j} \end{bmatrix} \qquad (8)$$

where N is the number of M-BSs, $\omega_{i,j}^x$ and $\omega_{i,j}^y$ are the normalized weights of the IBL $(\hat{x}_T^{i,j}, \hat{y}_T^{i,j})$ on the x-axis and y-axis, and the difference among different method is the deterministic process of $\omega_{i,j}^x$ and $\omega_{i,j}^y$.

As to the CAOA method, $\omega_{i,j}^x = \omega_{i,j}^y = 2 / N(N-1)$ in Eq. (8), estimation of MS position is the mean of all IBLs. The disadvantage of the CAOA method is that it averages all the IBLs undiscriminatingly for MS position estimation; in Figure 1, we can see that the IBLs represented by hexagon and square deviate a lot from the target's true position, which will definitely influence the positioning accuracy.

The SA method, based on the sensitivity of the IBLs, assigns the corresponding weights to the IBLs. The weights of IBLs in Eq. (4) can be obtained from the following formula:

$$\omega_{i,j}^x = \left(\sqrt{ \left(\frac{\partial x_{i,j}}{\partial \theta_i} \right)^2 + \left(\frac{\partial x_{i,j}}{\partial \theta_j} \right)^2 } \right)^{-1},$$

$$\omega_{i,j}^y = \left(\sqrt{ \left(\frac{\partial y_{i,j}}{\partial \theta_i} \right)^2 + \left(\frac{\partial y_{i,j}}{\partial \theta_j} \right)^2 } \right)^{-1}$$

where $x_{i,j}$ and $y_{i,j}$ are the closed-form solutions of IBL.

The CWIAOA method thinks the non-normalized weights $\tilde{\omega}_{i,j}^x$ and $\tilde{\omega}_{i,j}^y$ of the IBL. $(\hat{x}_T^{i,j}, \hat{y}_T^{i,j})$ is inversely proportional to the CRLB, that is:

$$\tilde{\omega}_{i,j}^x \propto \left(C_{i,j}(1,1) \right)^{-1}, \tilde{\omega}_{i,j}^y \propto \left(C_{i,j}(2,2) \right)^{-1}$$

where $C_{i,j} = J_{i,j}^{-1}, J_{i,j}$ is the Fisher information matrix of the ith M-BS and the jth M-BS. Normalize, respectively, $\tilde{\omega}_{i,j}^x$ and $\tilde{\omega}_{i,j}^y$ to get $\omega_{i,j}^x$ and $\omega_{i,j}^y$ in Eq. (8).

The simulation result in Duan, Z. & Zhou, Q. (2015) shows that the performance of the CWIAOA method is superior to that of the CAOA method and the SA method. However, the existing weighted intersection AOA localization method still has two problems.

Problem I: Let us assume that the positions of two BSs are (0,50) and (0,150), and the standard deviation of the AOA measurement is $\sigma = 2°$, then the CRLB of the target localization within the range of 200×200 is shown in Figure 2. From the Figure, it can be discovered that the farther the target is from the BS baseline, the larger the CRLB is. All the marks in Figure 2 represent the possible IBL of two BSs, and we can see that the CRLB where the hexagrams are is smaller than that where other marks are; according to the CWIAOA method, it should have larger weights and larger contribution to the target position estimate. However, the circular symbols approach true position of the target much closer than the hexagrams, and their weights are relatively smaller. The CWIAOA method does not remove the IBL, which are represented by the hexagrams in the process of the target position estimation, so the positioning accuracy is finally influenced.

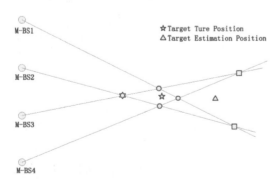

Figure 1. The diagram of the CAOA method.

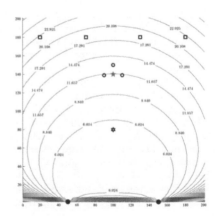

Figure 2. CRLB contour distribution of the target.

Problem II: the existing method generally assumes that the position of BS is accurate and has no error, but in the practical application, the BS position does have error. If we do not consider about the position error of the BS, it will definitely influence the practical application of the method.

To overcome the aforesaid problems, this paper proposes GDOP-weighted intersections for ray-tracing-based target localization using AOA measurement.

4 GDOP-WEIGHTED INTERSECTIONS FOR RAY-TRACING-BASED TARGET LOCALIZATION USING AOA MEASUREMENTS

To increase the positioning accuracy of the weighted intersection AOA localization method, it is necessary to remove those IBLs which have larger weights but deviate from the true position of the MS much more on the basis of ray tracing of the positioning signal. To solve this problem, Circular Error Probable (CEP) can be introduced. In the 2D environment, CEP is defined as the probability that the positioning result falls into the circular area C, which centers on the MS true position. The circular area C is called circle of error probability, and its radius is defined as the n times of that of GDOP, and confirmed by Eq. (9), and R_{CEP-P} indicates the radius of the circular area C into which the positioning result falls with the probability P. In accordance with the literature, the relation between P and n in Eq. (9) can be confirmed by Eq. (10), and Eq. (11) can be obtained through deduction.

$$R_{CEP-P} = n \cdot GDOP \qquad (9)$$

$$P = \int_0^n 2\rho e^{-\rho^2}\, d\rho = 1 - e^{-n^2} \qquad (10)$$

$$n = \sqrt{-\ln(1-P)} \qquad (11)$$

The center of the circular area C is the true position of the MS, and the MS position for localization problem is unknown. Here, we use the IBL $\left(\hat{x}_T^{i,j}, \hat{y}_T^{i,j}\right)$ to replace the true position of the MS (x_T, y_T), and the radius of the circle of error probability can be reached by

$$R_{CEP-P}^{i,j} \approx R_{CEP-P}\big|_{x_T = \hat{x}_T^{i,j},\, y_T = \hat{y}_T^{i,j}} \qquad (12)$$

where $R_{CEP-P}^{i,j}$ represents the radius of the circle of error probability for the intersection of the ith bearing line and jth bearing line. There are C_n^2 intersections for N bearing lines; if the distance $d_{i1,j1,i2,j2}$ between the IBL $\hat{p}_{i1,j1}$ and IBL $\hat{p}_{i2,j2}$ meets

$$d_{i1,j1,i2,j2} < R_{CEP-P}^{i1,j1} \quad or \quad d_{i1,j1,i2,j2} < R_{CEP-P}^{i2,j2} \qquad (13)$$

Then, $\hat{p}_{i1,j1}$ and $\hat{p}_{i2,j2}$ are considered as the valid IBLs. Under the condition when the probability is P, discovering the valid IBLs and using them for the position estimation of MS will reduce the influences on the positioning accuracy from the IBL represented by the hexagrams in Figure 1.

Below we will discuss the computing problem of GDOP. Differentiate θ_i according to Eq. (7) in the condition that the BS position has error, then we can get

$$d\theta_i = \frac{\partial f_i(p)}{\partial x_T} dx_T + \frac{\partial f_i(p)}{\partial y_T} dy_T + \frac{\partial f_i(p)}{\partial x_i} dx_i + \frac{\partial f_i(p)}{\partial y_i} dy_i \qquad (14)$$

where

$$\frac{\partial f_i(p)}{\partial x_T} = -\frac{\partial f_i(p)}{\partial x_i} = \frac{-u}{1+u^2}\frac{1}{x_T - x_i};$$
$$\frac{\partial f_i(p)}{\partial y_T} = -\frac{\partial f_i(p)}{\partial y_i} = \frac{1}{1+u^2}\frac{1}{x_T - x_i}; u = \frac{y_T - y_i}{x_T - x_i}.$$

Let us set $d\mathbf{X} = [dx_T \; dy_T]^T$, $d\Theta = [d\theta_1 \; d\theta_2 \; \ldots \; d\theta_N]^T$, $d\mathbf{X}_s = [dx_1 \; dy_1 \; dx_2 \; dy_2 \; \ldots \; dx_N \; dy_N]^T$,

$$\mathbf{C}_1 = \begin{bmatrix} \frac{\partial f_1(p)}{\partial x} & \frac{\partial f_2(p)}{\partial x} & \cdots & \frac{\partial f_N(p)}{\partial x} \\ \frac{\partial f_1(p)}{\partial y} & \frac{\partial f_2(p)}{\partial y} & \cdots & \frac{\partial f_N(p)}{\partial y} \end{bmatrix}^T,$$

$$\mathbf{C}_2 = \begin{bmatrix} \frac{\partial f_1(p)}{\partial x_1} & \frac{\partial f_1(p)}{\partial y_1} & & & & \\ & & \frac{\partial f_2(p)}{\partial x_2} & \frac{\partial f_2(p)}{\partial y_2} & & \\ & & & & \ddots & \\ & & & & \frac{\partial f_N(p)}{\partial x_N} & \frac{\partial f_N(p)}{\partial y_N} \end{bmatrix} \qquad (15)$$

Moreover, we can get $\mathbf{C}_1 d\mathbf{X} = d\Theta - \mathbf{C}_2 d\mathbf{X}_s$

Use pseudo-inverse to solve the error estimate of the MS positioning as

$$d\hat{\mathbf{X}} = \mathbf{D}(d\Theta - \mathbf{C}_2 dX_s) \qquad (16)$$

$$\mathbf{P}_{d\hat{X}} = E\left[d\hat{\mathbf{X}} d\hat{\mathbf{X}}^T\right]$$
$$= \mathbf{D}\left\{E\left[d\Theta d\Theta^T\right] + \mathbf{C}_2 E\left[dX_s dX_s^T\right]\mathbf{C}_2^T\right\}\mathbf{D}^T \qquad (17)$$

where $\mathbf{D} = \left(\mathbf{C}_1^T \mathbf{C}_1\right)^{-1}\mathbf{C}_1^T$. Assuming the AOA measurement errors of each M-BS are irrelevant, we can get

$$E\left[d\Theta d\Theta^T\right] = diag(\sigma_1^2, \sigma_2^2, \cdots, \sigma_N^2) \qquad (18)$$

459

According to Eq. (5), the position of the M-BSs are determined by the position of BS and the model of the microcell, and all the factors for the position errors of BS are irrelevant, so

$$E\left[d\mathbf{X}_s d\mathbf{X}_s^T\right] = \mathbf{A}\begin{pmatrix} \sigma_{x0}^2 & 0 \\ 0 & \sigma_{y0}^2 \end{pmatrix}\mathbf{A}^T \qquad (19)$$

Let us set $\mathbf{P}_{d\hat{x}} = \begin{bmatrix} \sigma_{\hat{x}}^2 & \sigma_{\hat{x}\hat{y}} \\ \sigma_{\hat{y}\hat{x}} & \sigma_{\hat{y}}^2 \end{bmatrix}$, then $GDOP = \sqrt{\sigma_{\hat{x}}^2 + \sigma_{\hat{y}}^2}$.

The above computing method of GDOP not only considers about the AOA measurement errors of the M-BSs, but also takes the BS position errors into account, which is strongly adaptable. Here, we still use the intersection of the ith bearing line and jth bearing line to replace the true position of the MS, we can get

$$\mathbf{P}_{d\hat{X}}^{i,j} = \mathbf{P}_{d\hat{X}}\big|x_T = \hat{x}_T^{i,j}, y_T = \hat{y}_T^{i,j} \qquad (20)$$

$$GDOP_{i,j} = GDOP\big|x_T = \hat{x}_T^{i,j}, y_T = \hat{y}_T^{i,j} \qquad (21)$$

Based on the aforesaid analysis, the intersection of the ith bearing line and jth bearing line is $\hat{p}_{i,j} = \left(\hat{x}_T^{i,j}, \hat{y}_T^{i,j}\right)$, and its non-normalized weight in the x-axis and y-axis is:

$$\tilde{\omega}_{i,j}^x = \left(\mathbf{P}_{d\hat{X}}^{i,j}(1,1)\right)^{-1}, \tilde{\omega}_{i,j}^y = \left(\mathbf{P}_{d\hat{X}}^{i,j}(2,2)\right)^{-1} \qquad (22)$$

All the valid IBLs form the set $\hat{\mathbf{P}}$, let us set

$$\mathbb{I}\left(\hat{p}_{i,j}\right) = \begin{cases} 1 & \hat{p}_{i,j} \in \hat{\mathbf{P}} \\ 0 & \hat{p}_{i,j} \notin \hat{\mathbf{P}} \end{cases} \qquad (23)$$

Normalize the weight of the valid IBL ($\hat{p}_{i,j} \in \hat{\mathbf{P}}$)

$$\begin{cases} \omega_{i,j}^x = \mathbb{I}\left(\hat{p}_{i,j}\right)\tilde{\omega}_{i,j}^x\left(\sum_{m=1}^{N-1}\sum_{n=m+1}^{N}\mathbb{I}\left(\hat{p}_{m,n}\right)\tilde{\omega}_{m,n}^x\right)^{-1} \\ \omega_{i,j}^y = \mathbb{I}\left(\hat{p}_{i,j}\right)\tilde{\omega}_{i,j}^y\left(\sum_{m=1}^{N-1}\sum_{n=m+1}^{N}\mathbb{I}\left(\hat{p}_{m,n}\right)\tilde{\omega}_{m,n}^y\right)^{-1} \end{cases} \qquad (24)$$

If the IBL $\hat{p}_{i,j}$ does not belong to $\hat{\mathbf{P}}$, then

$$\omega_{i,j}^x = 0, \omega_{i,j}^y = 0 \quad \hat{p}_{i,j} \notin \hat{\mathbf{P}} \qquad (25)$$

In accordance with Eqs. (24) and (25), conduct the weighted summation for every IBLs, and the estimation of the MS position is

$$\begin{bmatrix} \hat{x}_T \\ \hat{y}_T \end{bmatrix} = \begin{bmatrix} \sum_{i=1}^{N-1}\sum_{j=i+1}^{N}\omega_{i,j}^x\hat{x}_T^{i,j} \\ \sum_{i=1}^{N-1}\sum_{j=i+1}^{N}\omega_{i,j}^x\hat{y}_T^{i,j} \end{bmatrix} \qquad (26)$$

GWIAOA-RT method

Initialization:

BS $\tilde{\mathbf{x}}_0$; according to ray-tracing result $PL_i^{N_i} \left(i = 1, \cdots, N\right)$ and Eq.(4)(5), get $\tilde{\mathbf{X}}_s$; the AOA measurements from M-BSs, $\tilde{\Theta} = \left[\tilde{\theta}_1, \cdots, \tilde{\theta}_N\right]$; CEP $P = 0.99$.

The method process: position the MS $\left(x_T, y_T\right)$

STEP 1: get all the IBLs, $\hat{p}_{i,j} = \left(\hat{x}_T^{i,j}, \hat{y}_T^{i,j}\right)$;

STEP 2: calculate $GDOP_{i,j}$ according to Eq. (21);

STEP 3: calculate $R_{CEP-P}^{i,j}$ according to Eq. (12);

STEP 4: calculate the distance of IBLs, and confirm the set $\hat{\mathbf{P}}$ of valid IBLs according to Eq.(13);

STEP 5: calculate the weights of IBLs according to Eqs. (20-24);

STEP 6: estimate the position of the MS according to Eq.(26).

To sum up, the GWIAOA-RT method proposed in this paper mainly includes the following three steps: 1) get the M-BSs and the AOA measurements from the M-BSs; 2) acquire the IBLs, and confirm the valid IBLs; and 3) conduct the weighted summation for the valid IBLs, and obtain the MS position estimation. The concrete steps of the GWIAOA-RT method are shown as follows:

5   CRLB

Assume that, based on the signal propagation model, there are N POPs between BS (x_0, y_0) and MS (x_T, y_T) in the microcell, namely $PL_i^{N_i}(i = 1, \cdots, N)$. According to Eqs. (4) and (5), we can get N M-BSs and their true position $\mathbf{X}_s$. In accordance with the literature, the Fisher information matrix of the MS is

$$\begin{cases} J = \sum_{i=1}^{N}\mathbf{M}_i \\ \mathbf{M}_i = \begin{bmatrix} \dfrac{(y_T - y_i)^2}{\sigma_i^2((x_T - x_i)^2 + (y_T - y_i)^2)^2} & \dfrac{-(y_T - y_i)(x_T - x_i)}{\sigma_i^2((x_T - x_i)^2 + (y_T - y_i)^2)^2} \\ \dfrac{-(y_T - y_i)(x_T - x_i)}{\sigma_i^2((x_T - x_i)^2 + (y_T - y_i)^2)^2} & \dfrac{(x_T - x_i)^2}{\sigma_i^2((x_T - x_i)^2 + (y_T - y_i)^2)^2} \end{bmatrix} \end{cases}$$

The inverse of Fisher information matrix gives the CRLB matrix $\mathbf{C} = \mathbf{J}^{-1}$, then CRLB of unbiased estimator of MS (x_T, y_T) using AOA measurements is

$$CRLB = \sqrt{Trace(\mathbf{C})} \qquad (27)$$

6   SIMULATION RESULT AND ANALYSIS

To verify the performances of the GWIAOA-RT method proposed in this paper, this section shall compare the positioning performances of the

GWIAOA-RT method, the CAOA method, the WIAOA method, the CWIAOA method, and the SA method in two typical scenarios. In each scenario, each method conducts 1000 times of Monte-Carlo simulation, makes statistics of the Root Mean Square Error (RMSE), and the positions error cumulative distribution of the method.

In scenario 1 (Figure 3), there are first-order POPs only between MS and BS, and M-BSs are located on one side of the MS. In scenario 2 (Figure 4), there are first-order and second-order POPs between MS and BS, and M-BSs are located on both sides of the MS. In these two scenarios, we can analyze the influence of different M-BS distribution on the position accuracy of the methods.

Scenario 1: There are three buildings in microcell 1, the projection coordinates of the buildings in the horizontal plane are shown in Table 1. The actual coordinate of BS is (0,45), the actual coordinate of MS is (240,40). n_{x0} and n_{y0} are errors of the known position of the BS along x-axes and y-axes, $n_{x0} \sim N(0, \sigma_{x0}^2)$, $n_{y0} \sim N(0, \sigma_{y0}^2)$, $\sigma_{x0} = \sigma_{y0} = \sigma_s = 3$, and n_{x0} and n_{y0} are irrelevant. There are four first-order POPs between MS and BS, which mean that there are 4 M-BSs in Table 2, as shown in Figure 3. Let us assume that the ray-tracing method can correctly track all the POPs, which mean the building surfaces tracking-paths passing through are consistent with the building surfaces actual-paths passing through. n_i which is the AOA measurement error of the *ith*

Table 1.   Coordinates of building in microcell 1 and microcell 2.

	X1	Y1	X2	Y2	X3	Y3	X4	Y4
Building1	100	80	100	95	180	95	180	80
Building2	100	10	100	25	150	25	150	10
Building3	100	–25	100	–5	235	–5	235	–25
Building4	255	0	255	80	270	80	270	0

Table 2.   Coordinates of M-BS in microcell 1 and microcell 2.

	M-BS1	M-BS2	M-BS3	M-BS4	M-BS5	M-BS6	M-BS7
X	0	0	0	0	510	510	510
Y	45	115	5	–55	45	115	–55

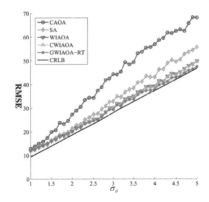

Figure 5.   The RMSE of all the methods in Scenario 1.

M-BS, $n_i \sim N(0, \sigma_{x0}^2) \, i = 1, \cdots, 4$, $\sigma_1 = \cdots = \sigma_4 = \sigma_\theta$, the value range of σ_θ is $[1°, 5°]$, and all the AOA measurement errors are irrelevant.

Figure 5 shows the RMSE curves of all the methods with the changes in σ_θ in Scenario 1, and Figure 6 shows the cumulative distribution of positioning error of all the methods when $\sigma_\theta = 5°$ in Scenario 1. From Figure 5, we can see that in Scenario 1, the CAOA method has the worst performance, the general performance of the GWIAOA-RT method is better than the other methods, RMSE of CAOA and SA method deviates from the CRLB further as the σ_θ increases. However, as the known position of the BS have errors and are used for the position calculation, the RMSE of the aforesaid five methods have certain deviation compared with the CRLB. From Figure 6, we can see that the positioning accuracy of GWIAOA-RT method is superior to that of the other methods.

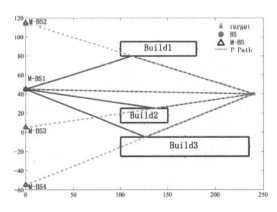

Figure 3.   Microcell 1.

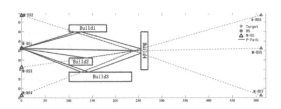

Figure 4.   Microcell 2.

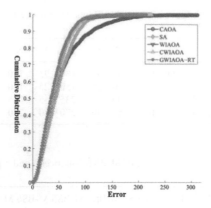

Figure 6. The error cumulative distribution of all the methods in Scenario 1 ($\sigma_\theta = 5$).

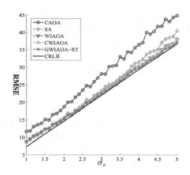

Figure 7. The RMSE of all the methods in Scenario 2.

Scenario 2: There are four buildings in microcell 2, and the projection coordinates of the buildings in the horizontal plane are shown in Table 1. The actual coordinate of BS is (0,45), and the actual coordinate of MS is (240,40). n_{x0} and n_{y0} are errors of the known position of the BS along x-axes and y-axes, $n_{x0} \sim N(0, \sigma_{x0}^2)$, $n_{y0} \sim N(0, \sigma_{y0}^2)$, $\sigma_{x0} = \sigma_{y0} = \sigma_s = 3$, and n_{x0} and n_{y0} are irrelevant. There are five first-order POPs and two second-order POPs between MS and BS, which mean that there are 7 M-BSs in Table 2, as shown in Figure 4. Let us assume that the ray-tracing method can correctly track all the POPs. n_i is the AOA measurement error of the *ith* M-BS, $n_i \sim N(0, \sigma_i^2)$ $i = 1, \cdots, 6$, $\sigma_1 = \cdots = \sigma_6 = \sigma_\theta$, the value range of σ_θ is [1°, 5°], and all the AOA measurement errors are irrelevant.

Figure 7 shows the RMSE curves of all the method with the changes in σ_θ in Scenario 2, and Figure 8 shows the cumulative distribution of positioning error of all the methods when $\sigma_\theta = 5°$ in Scenario 2. We can see from Figure 7 that in Scenario 2, the performance of CAOA method is the worst, and the performance of SA method is better than the CAOA method. When $\sigma_\theta < 4°$, performance of GWIAOA-RT, WIAOA and CWIAOA method is approached. When $\sigma_\theta > 4°$, the performance of GWIAOA-RT method is better than the WIAOA and CWIAOA methods. Owing to the error of the known position of BS, the RMSE of CAOA and SA methods obviously deviates from CRLB, while the RMSE and CRLB have a certain deviation for GWIAOA-RT, WIAOA and the CWIAOA method is with relatively smaller AOA measurement error. However, from Figure 8, it is found that the whole performance of the GWIAOA-RT method is still better than other methods.

By comparing Scenario 1 with Scenario 2, we can discover that the aforesaid methods have better positioning performance in Scenario 2 than they

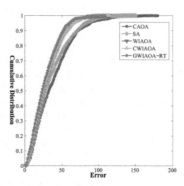

Figure 8. The error cumulative distribution of all the methods in Scenario 2 ($\sigma_\theta = 5°$).

do in Scenario 1, that is, when the M-BSs are distributed around the MS, the methods have better positioning performance. Meanwhile, in Scenario 1 and Scenario 2, as the known positions of the M-BSs have errors, the RMSEs of the aforesaid five methods all have certain biases compared with CRLB. From Figures 5 and Figure 7, it is also discovered that when σ_s is at a fixed value, the influence from σ_s on the method performance weakens as σ_θ increases.

The simulation result shows that the GWIAOA-RT method has better adaptivity than the other methods, and its positioning accuracy is superior to that of the other methods as a whole.

7 CONCLUSION

In this paper, the GDOP-weighted intersection method for ray-trace-based target localization using AOA measurements is proposed. The main idea of the GWIAOA-RT method is that BS realizes the ray trace and building of M-BS according to the related measurements of multipath positioning signal, transfers the NLOS path to LOS path, generates the corresponding bearing lines and obtains the IBLs, confirms the effective IBLs by using the CEP of all the IBLs, and gives corresponding weights according to the GDOP of effective IBLs, and makes weighted summation on effective IBLs and obtains the position estimation of MS. The GWIAOA-RT method makes full use of the multipath propagation of positioning signal and takes consideration of the effect of the position error of BS on the results in the calculation of GDOP, and also it does not need initial guess for iteration. The simulation result shows that the whole positioning accuracy of the GWIAOA-RT method is much higher and the adaptability is much better in different application scenarios for the GWIAOA-RT method compared with the WIAOA method, the CWIAOA method, the SA method, and the CAOA method.

ACKNOWLEDGMENTS

This project was supported by the National Natural Science Foundation of China (61401513).

REFERENCES

Berman, Z. 1997. A reliable maximum likelihood algorithm for bearing-only target motion analysis. in Proceedings of the 36th IEEE Conference on Decision and Control: 5012–5017.

Brida, P. et al. 2010. A new complex angle of arrival location method for ad hoc networks. Proceedings of the 7th Workshop on Positioning Navigation and Communication: 284–290.

Compagnoni, M. & Notari, R. 2014. TDOA-based localization in two dimensions: the bifurcation curve. Fundamenta Informaticae, 135: 199–210.

Duan, Z. & Zhou, Q. 2015. CRLB-weighted intersection method for target localization using AOA measurements. Computational Intelligence and Virtual Environments for Measurement Systems and Applications (CIVEMSA).

Foy, W. 1976. Position-location solutions by Taylor-series estimation. IEEE Transactions on Aerospace and Electronic Systems, 12(2): 187–194.

Gu, G. 2010. A novel power-bearing approach and asymptotically optimum estimator for target motion analysis. in Proceedings of the IEEE Conference on Decision and Control: 5013–5018.

Ho, K. & Chan, Y. 2006. An asymptotically unbiased estimator for bearings only and doppler-bearing target motion analysis. IEEE Transactions on Signal Processing, 54(3): 809–822.

Ikegami, F. & Yoshida, S. 1983. Feasibility of Predicting Mean Field Strength for Urban Mobile Radio by Aid of Building Data Base. Proc. IEEE Int'l. Conf. on Commun., ICC'83: 68–71.

Jiang, J.R. et al. 2010. Localization with rotatable directional antennas for wireless sensor networks. Proceedings of the 39th International Conference on Parallel Processing Workshops: 542–548.

Le Cadre, J.P. & Jaetffret, C. 1999. On the convergence of iterative methods for bearings-only tracking. IEEE Transactions on Aerospace and Electronic Systems, 35(3): 801–818.

Liu, D.L. 2014. Joint TOA and DOA Localization in Indoor Environment Using Virtual Station. IEEE Communications Letters, 18(8): 1423–1426.

Lui, K.W.K. & So, H.C. 2010. Range-based Source Localization with Pure Reflector in Presence of Multipath Propagation. Electronics Letters, 46(8), 957–958.

Pages-Zamora, A. et al. 2002. Closed-form Solution for Positioning Based on Angle of Arrival Measurements. in Proceedings of the 13th IEEE International Symposium on Personal, Indoor and Mobile Radio Communications: 1522–1526.

Pezeshk, A.A.M. & Dallai, M. 2010. A novel method for position finding of stationary targets using bearing measurements. Proceedings of the C4I Conference at Sharif University of Technology.

Schmitz, A. et al. 2011. Efficient Rasterization for Outdoor Radio Wave Propagation. IEEE Transactions on Visualization & Computer Graphics, 17(2), 2011, 159–170.

Setlur, P.G. et al. 2012. Target Localization with a Single Sensor via Multipath Exploitation. IEEE Transactions on Aerospace and Electronic Systems, 48(3): 1996–2014.

Shao, H.J. et al. 2014. Efficient Closed-Form Algorithms for AOA Based Self-Localization of Sensor Nodes Using Auxiliary Variables. IEEE Transactions on Signal Processing, 62(10): 2580–2594.

Soltanian, M. et al. 2011. A new iterative position finding algorithm based on taylor series expansion. Proceedings of the 19th Iranian Conference on Electrical Engineering: 1–4.

Stansfield, R. 1947. Statistical theory of d.f. fixing. Journal of the Institution of Electrical Engineers-Part IIIA: Radiocommunication, 94(15): 762–770.

Torrieri, D. 1984. Statistical theory of passive location systems. IEEE Transactions on Aerospace and Electronic Systems, 20(2): 183–198.

Wang, Z. et al. 2012. A novel location-penalized maximum likelihood estimator for bearing-only target localization. IEEE Transactions on Signal Processing, 60(12): 6166–6181.

Yu, K. et al. 2009. Ground-Based Wireless Positioning, Wiley-IEEE Press, Chichester.

Yuan, Z.W. 2007. Terminal ray tracing theory and methods in mobile communications, 109–140. Beijing: Publishing House of Electronics Industry.

Zhou, Q. & Duan, Z. 2014. Weighted intersections of bearing lines for AOA based localization. Information Fusion (FUSION), 2014 17th International Conference on. IEEE: 1–8.

Zhou, X.P. et al. 2013. An improved algorithm on field strength calculation in ray tracing technology. Chinese Journal of Radio Science, 28(4): 669–675.

Zhu, G.H. et al. 2015. TOA localization algorithm using the linear-correction technique. Journal of Xidian University (Natural Science), (3): 22–25.

Energy Science and Applied Technology ESAT 2016 – Fang (Ed.)
© *2016 Taylor & Francis Group, London, ISBN 978-1-138-02973-6*

Military agent framework with multi-layered intelligence

Jiang Zhu, Yingchun Wang, Jin Feng & Chuanhua Wen
Nanjing Army Command College/Combat Experiment Department, Nanjing, China

ABSTRACT: In C2 (Command & Control) domain, there are lots of complexities from different aspects in solving military decision problems. Current C2 model, due to short of adaptability and lack of intelligence, has become a weak link in the chain of enhancing decision quality. After analyzing the military needs, MAFMI (A military agent framework with multi-layered intelligence), an integrated and flexible framework is put forward to bring more bounce and intelligence to the current C2 model. The framework integrates the advantages of BDI and reactive agent architecture, and consists of five ladders indicating varying intelligence abilities such as thought, estimation, and cognition. It had been applied in military training case, and proved to have good significance and applied value.

Keywords: C2; agent; architecture; intelligence

1 INTRODUCTION

In recent years, the world goes pace toward the "intelligent world". The latest event that AlaphGO play I-go with Lee Sedol seemed to be a landmark of the "singularity coming" (Ray K, 1988). As to military domain, whether in reality or in simulation, intelligent army is more likely to defeat the ordinary ones.

C2 (command and control) models, considered as a brain and engine of military system, are concerning emerging group behavior from bottom behavior, so as to get better synchronization and coordination. In the 1980s, the US army began to study c2 modeling. In 1998, existing models are surveyed and, a "human behavior and C2 modeling" committee is founded to do some advanced research. Up to now, a bunch of models such as RPD (Klein G, 2008), CECA (Maria Strofky et al, 2003), and PMF (Barry G et al, 2006) having been put out as theoretic stuff. As to the implement technology, the agent method is suitable for constructing simulated force with intelligent C2 decision ability (D Barteneva et al, 2008) (C Becker et al, 2006).

In the next section, a mixed and flexible agent framework is put forward to bring more intelligence to the current C2 model. Afterward, five layers of agent architecture and its concrete implementation are designed and illustrated in section 3. Lastly, this framework proved to have good applied value.

2 AGENT ARCHITECTURE

Architecture decides the whole structure and operational style of the agent model. There are several typical classes such as reactive architecture and BDI agent architecture.

Table 1. Comparison of architecture.

Architecture	Decision model	Knowledge base	Simple question	Complex question
Reaction	Simulate-Respond	Rule-action	Good	Poor
BDI	Deliberate thinking	Domain knowledge	Poor	Good

As can be seen from Table 1, the reaction agent is too simple to handle complex combat events, while BDI agent think too much but act slowly. Herein, the hybrid architecture is provided in this paper. As Fig. 1 shows deliberate agent as the brain and reaction agent as the local nerve center, they both corporate to take responsibility to translate the reactive knowledge and decision knowledge into execution module for conducting the physical action of agent. Hybrid architecture not only has the reasoning ability, but also responses quickly. Therefore, it becomes a mainstream architecture (John A Sokolowski, 2007) (Ostantini S et al, 2007).

3 DECISION MAKING

The combat system is an open, dynamic, and complex system, evolving various C2 units, big or small, wise or stupid. The commanders' task and the corresponding decision keep changing. Military decision always took place in different levels and involved various entities. It has the functionality of mission analysis, plan making, communica-

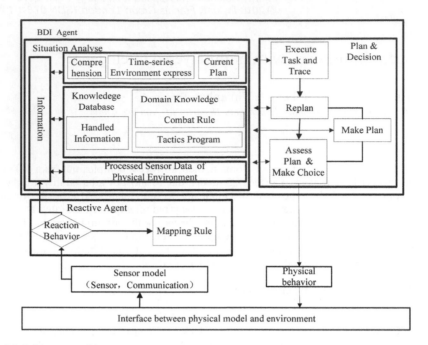

Figure 1. A hybrid agent architecture.

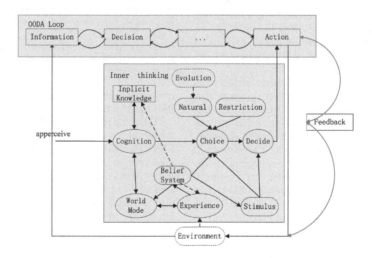

Figure 2. C2 process.

tion coordination, battlefield supervision, mission management, route choosing, collaboration, task allocation, and so on.

C2 process is shown in Fig. 2; a commander gets information and apperceives it from the environment, then through a inner-thinking module, facilities provided by the brain such as cognition, reaction, evolution, and leaning work together, going with the knowledge from implicit to explicit,

intelligence from low to high, a most appropriate plan or behavior is chosen. Finally, after exerting an influence on physical world, people get the feedback of environment, and then began another cycle. Seen from the outside, it is an OODA loop of decision.

Physical action of C2 equipment always needs a quick OODA loop, where decision is taken more simply but promptly. Command center is of high hierarchy, and decision is more complex but needs more time, always after a serious task planning under time and resource restriction according to the combat mission. Intelligence is a mixture of conditioned reflex, learning, reasoning, synthesizing. Different decision needs different intelligence ability, knowledge base, and modeling technology, convenient for solving the problem and fulfilling the different needs of C2.

Five level of intelligence (Levels 1–5) is divided and implementing technology is given in MAFMI, as Table 2 shows.

A single-agent architecture and the corresponding method is incapable of balancing between speed and performance. Therefore, different architectures could be appropriate for specific problems in whole decision space. As Figure 3 shows, problems are classified into two classes:

1. Indirect-to-handle problems always occurred in high hierarchy of decision space, such as main-attack-selection, plan or predict, and operation-planning. Intelligence in level 3 to level 5 is preferable. By a word, deliberate architecture with four-level intelligence with full consideration is a good solution for specific planning problem

Table 2. Five intelligence level.

Level-Name	Description
① R ule-based stimulus-response	A quite straightforward decision strategy, reacts to environmental signal, and is controlled by the so-called "IF-Then" rule. The agent may change his behavior when environment status, his own status, or other agents' status came up changing. The IF-Then Rule is expressed by BNF norm.
② Knowledge-based stimulus-response	A strategy combining a set of simple responses puts emphasis on linking relative rules into a knowledge framework.
③ Learning	Leaning strategy works as follows: First, the agent set the propensity to calculate the probability of each action in selectable list. Then, the agent chose action with impact on environment. Finally, it gets feedback as reward or punishment to evaluate his done. After continually trying that routine and adjusting propensity, a lot of experience is acquired and a best way is formed.
④ Deliberate thinking	Concerns on the comprehensive process of people thinking, described as a mental-driven deliberate thinking. Deliberate agent chose behavior by the mental status and domain knowledge. It is good at handling information in a high abstract form. BDI architecture, composed of formalized symbols and inference engine, is implemented under this guide
⑤ integration	Integration in a way to get high level intelligence by combining levels 1–4.

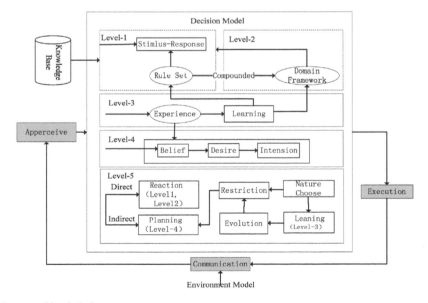

Figure 3.   Integrated levels 1–5.

2. Direct-to-handle problems, occurred in low hierarchy, needs to handle promptly, and is fit for if-then architecture with 1-level intelligence. Make decisions like how to induct the weapon, fire or move in times of emergency, and intelligence in level 1 or level 2 is preferable. The deliberate agent can provide plan for a tough maneuver. Under a "war dense frog". Level 5 can solve both the indirect-to-handle problems and direct-to-handle problems using level 3 and level 1 intelligence, respectively. What is more, in view of a long-term decision, a self-learning and an adapt model inherited from Level 4 are utilized to optimize the behavior here.

4 SUMMARY

Those architecture and key technologies have been utilized in combat experiment of warfare rehearsals, especially for headquarters to improve the decision skills; it has been proved to have the following advantages:

1. The architecture is an open one, accompanying with the development of artificial intelligent technology, and could be expanded step by step.
2. By dividing into several levels, some mature and typical module can be packaged independently, that enhances the reusability and easy to deploy.
3. Architecture in this study can handle most kinds of problem in decision, and thus, it enhances the intelligence and flexibility of agent and simulates a lively commander.

ACKNOWLEDGMENT

This work was supported by NSFC (No. 71401177).

REFERENCES

Barry G. Silverman, Michael Johns, Jason Cornwell, and Kevin O'Brien, "Human Behavior Models for Agents in Simulators and Games: Part I: Enabling Science with PMFserv". April 2006.
Becker, C., N Lessmann, S Kopp. I Wachsmuth. Connecting feelings and thoughts-modeling the interaction of emotion and cognition in embodied agents [C]//Proceedings of Seventh International Conference on Cognitive Modeling (ICCM-06). Ann Arbor, Michigan, USA, 2006:32–37.
Barteneva, D., N Lau, L P Reis. A computational study on emotions and temperament in Multi-agent system [DB/OL]2008, http://arxiv.org/abs/0809.4784.
John A Sokolowski, Enhanced Military Modeling Using a Multi-Agent System Approach//Proceeding of the 12th ICCRTS, 2007
Klein, G. (2008). Naturalistic decision making. Human Factors, 50, 456–460.
Maria Strofky, Cmy Hogan, John Prince. A Common Architecture for Behavior and Cognitive Modeling[C]//2003 conference on Behavior Representation in Modeling and Simulation. 2003.
Ostantini S, Tocchio A, Toni F, et al. A multi-layered general agent model[C]//Artificial Intelligence and Human-Oriented Computing. Italy Rome: The 10th Congress of the Italian Association for Artificial Intelligence on AI*IA, 2007:121–132.
Ray K., The Singularity Is Near: Wen Humans Transcend Biology[M], Penguin Books, 1988.

Energy Science and Applied Technology ESAT 2016 – Fang (Ed.)
© *2016 Taylor & Francis Group, London, ISBN 978-1-138-02973-6*

Comparison of performances of variational mode decomposition and empirical mode decomposition

Yingjuan Yue, Gang Sun, Yanping Cai, Ru Chen, Xu Wang & Shixiong Zhang
School of Xi' an Institute of High Technology, Xi'an, China

ABSTRACT: Variational Mode Decomposition (VMD) is a non-recursive adaptive signal processing method which is completely different from the EMD structure model. In order to analyze the performances of EMD and VMD, the two signal processing methods, the Hilbert-based VMD is proposed. The EMD and VMD are analyzed in noise effect, mode mixing, pseudo-components, energy leaking, and other problems by constructing multi-frequency mixed signal, am-fm signal, and noised-signal. Simulation results show that: the VMD decomposition performance is superior to EMD and it can achieve a signal in the frequency domain of each component of the adaptive mesh with no pseudo-components; the Hilbert-based VMD has a better signal characterization.

Keywords: empirical mode decomposition, variational mode decomposition, pseudo-components

1 INTRODUCTION

Empirical Mode Decomposition (EMD) (Huang N E et al, 1998) is an adaptive signal analysis method, for the first time proposed by Norden E. Huang NASA in 1998. The EMD decomposition method not only is used for linear stationary signal analysis, but also carries on the analysis to the non-steady signal. EMD has been widely used in medicine, geology, oceanography, meteorology, mechanical equipment, and fault diagnosis signal analysis (LIU Yang et al, 2015) (Wang Xuehuan et al, 2014) (Hao Guocheng et al, 2015). However, there are many problems in the EMD algorithm itself, such as the end effect problem (Xu Zhuofei et al, 2015), the envelope fitting problem (Chen Q H et al, 2006), the mode mixing problem (Wu Z H et al, 2009), and so on.

VMD (Dragomiretskiy, K et al, 2014) is a new adaptive signal processing analysis method proposed by Dragomiretskiy et al. in 2014, which effectively overcomes the shortcomings of the EMD algorithm. The essence of the VMD algorithm is a set of adaptive Wiener filter banks with good noise robustness.

This paper makes a comparative analysis of EMD and VMD algorithm by constructing different types of simulation signals, and proposes a Hilbert time-frequency distribution based on VMD calculation method. Simulation analysis result shows that the VMD algorithm can achieve signal adaptive decomposition in frequency domain for various components, and effectively overcome

the EMD algorithm in decomposition of modal aliasing and pseudo components, and compared with EMD has strong robust to noise and smaller energy leakage; Hilbert time-frequency distribution based on VMD calculation method has a very good effect of the signal characteristics.

2 VMD ALGORITHM AND HILBERT TIME-FREQUENCY ANALYSIS BASED ON VMD

2.1 *VMD algorithm*

VMD is a new method of adaptive signal decomposition. The VMD decomposition method determines each Intrinsic Mode Component (IMF) of center frequency and bandwidth through an iterative search variable model of optimal solution. It is a completely non-recursive signal decomposition method, implementing the signal frequency domain adaptive subdivision of each component. It is a completely non-recursive signal decomposition method.

The signal is decomposed into a series of intrinsic mode components (IMF) by VMD, and each IMF can be expressed as an amplitude modulation signal $u_k(t)$:

$$u_k(t) = A_k(t)\cos(\phi_k(t)) \qquad (1)$$

where $A_k(t)$ is the instantaneous amplitude for $u_k(t)$, and $A_k(t) \geq 0$. $\omega_k(t)$ is the instantaneous frequency for the $u_k(t)$, $\omega_k(t) = \phi_k'(t)$, $\omega_k(t) \geq 0$. In addition, the change in $A_k(t)$ and $\omega_k(t)$ is slow with respect to the phase $\phi_k(t)$. In the time range $[t - \delta, t + \delta](\delta \approx 2\pi / \phi_k'(t))$, $u_k(t)$ can be seen as the harmonic signal with amplitude $A_k(t)$, $\omega_k(t)$ frequency.

Supposing that the signal is decomposed into IMF K components by VMD, the variational constraint model can be obtained as:

$$\begin{cases} \min_{\{u_k\},\{\omega_k\}} \left\{ \sum_k \left\| \partial_t \left[\left(\delta(t) + \frac{j}{\pi t} \right) * u_k(t) \right] e^{-jw_k t} \right\|_2^2 \right\} \\ s.t. \sum_k u_k = f \end{cases} \quad (2)$$

where: $\delta(t)$ is Dirac distribution, * denotes convolution. $\{u_k\}$ on behalf of the signal after the VMD decomposition of K IMF component; $\{u_k\} = \{u_1, \ldots, u_k\}$; $\{\omega_k\}$ is each center frequency for IMF component, $\{\omega_k\} = \{\omega_1, \ldots, \omega_k\}$.

In order to obtain the optimal solution of the variational constraint model, the two penalty function term and Lagrange multiplier are bring in:

$$L(\{u_k\}, \{\omega_k\}, \lambda) = \alpha \sum_k \left\| \partial_t \left[\left(\delta(t) + \frac{j}{\pi t} \right) * u_k(t) \right] e^{-jw_k t} \right\|_2^2 \\ + \left\| f(t) - \sum_k u_k(t) \right\|_2^2 + \left\langle \lambda(t), f(t) - \sum_k u_k(t) \right\rangle \quad (3)$$

where α is the punish parameter and λ is the Lagrange multiplier.

The VMD method uses multiplication operator alternating to calculate the variational constraint model, and the optimal solution will become a component of narrowband signal decomposition IMF.

2.2 Hilbert time-frequency analysis based on VMD

Hilbert spectrum can clearly reflect the signal frequency and amplitude varying with time. The time series $X(t)$ can be decomposed by VMD, using Hilbert to each IMF and can get Hilbert spectrum transform. According to the principle of Hilbert transform, the VMD-Hilbert process is as follows:

VMD-Hilbert algorithm:

Step 1: The VMD decomposition is carried out to $X(t)$ obtain a set of IMF components;

Step 2: Hilbert transform for each component;

Step 3: Based on the Hilbert transform of the IMF component, the analytic form of the component is constructed.

Step 4: For each calculate component Instantaneous amplitude and instantaneous phase;

Step 5: By using the instantaneous phase to obtain the derivative, the instantaneous frequency of the component is obtained;

Step 6: By using the instantaneous phase to obtain the derivative, the instantaneous frequency of the component is obtained;

Step 7: According to the Hilbert spectrum of each component, calculate $X(t)$ HHT time spectrum.

3 SIMULATION ANALYSIS

3.1 Comparison of multi frequency synthetic signals' decomposition

Simulation signal 1 using the multi frequency cosine and sine signal as its component. The simulation signal 1:

$$\begin{aligned} X(t) = & \sin(2\pi \times 3t) + 0.3\sin(2\pi \times 11t) \\ & + 0.8\sin(2\pi \times 25t) + \sin(2\pi \times 40t) \\ & + 0.6\cos(2\pi \times 100t) \end{aligned} \quad (4)$$

The EMD decomposition method is used to decompose the simulation signal 1 as shown in Figure 1:

You can see from Figure 1 that, after the signal decomposition, there are 10 components, which are a large number of pseudo components. The pseudo component enhanced the difficulty of analysis of signal, to facilitate analysis of the contrast of algorithm, and the removal of pseudo component method based on mutual information (Zhang Zhigang et al, 2013).

$$MI = I(X | Y) = -\sum_{-i,j} P(x_i, y_i) \log \frac{P(x_i, y_i)}{P(x_i)P(y_i)} \quad (5)$$

where $P(x_i, y_i)$ is the joint probability of x_i and y_i.

Based on mutual information to remove the pseudo component, it can be divided into the following two steps: 1) calculation of the breakdown of the i-th item of IMF and the original signal mutual information MI_i; 2) the definition of false judgment threshold δ, which is that the $MI_i > \delta$, the component for the real component to be retained, otherwise false component to be removed, $\delta = 0.5 \, mean(MI_1, MI_2, \ldots, MI_n)$.

Using EMD decomposition to the simulation signal 1, and using the method based on mutual information to remove the pseudo component in the decomposition results. The decomposition results are shown in figure 2:

From Figure 2, you see, signal simulation 1 by EMD decomposition and, after eliminating pseudo component, the remaining five IMF components,

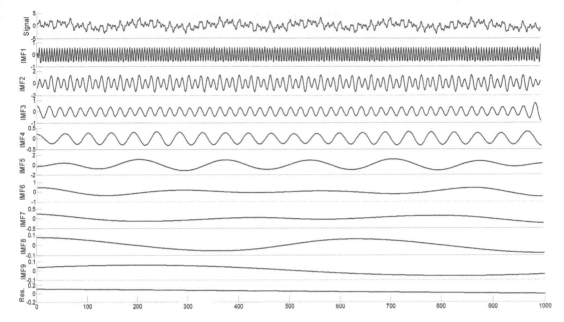

Figure 1.　EMD decomposition of signal 1.

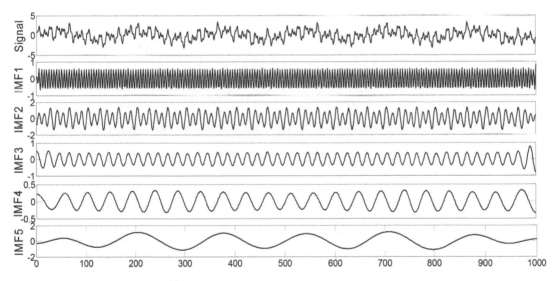

Figure 2.　The signal without pseudo-components.

but from the time domain graph of each IMF component that remains difficult to each component, are what frequency components, so using EMD-Hilbert to the simulation signal 1, as shown inFigure 3:

From the original signal, it can be known that the signal contains the 3Hz, 11Hz, 25Hz, 40Hz, and 100Hz frequency components. Figure 8 shows the EMD-Hilbert spectrum decomposition due

to deviation in EMD three spline envelope fitting result, resulting in EMD decomposition results has many pseudo components, this can be seen from the 100Hz frequency.

The VMD is used to simulate signal 1 decomposition, and the results are shown in Figure 4; EMD-Hilbert spectrum is shown in Figure 5:

It can be seen from Figure 4, EMD decomposition in the case of the removal of the pseudo

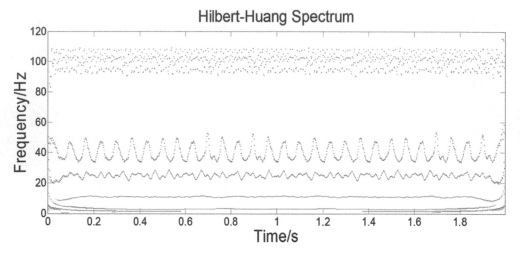

Figure 3. EMD-Hilbert image of the signal 1.

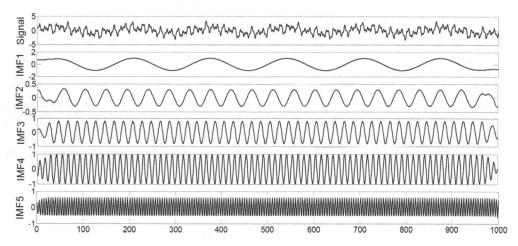

Figure 4. VMD decomposition of the signal 1.

component decomposition effect is still not as good as VMD. The simulation signal is decomposed by VMD, each IMF component has a certain range of scale, and there is no mode mixing phenomenon, which can realize the multi-scale representation of the simulation signal. Figure 5 shows the VMD-Hilbert time spectrum, and also a good description of the original signal characteristics, reflecting the different components of the signal with the frequency and amplitude changing with time.

The simulation results show that VMD is better than EMD in the decomposition of multi-component synthetic simulation signals, and there is no mode mixing and pseudo component phenomenon. The VMD-Hilbert spectrum calculation method for characterization of the signal is also more accurate.

3.2 Decomposition and comparison of modulation signals

The simulation signal 2 is generated by the superposition of two FM components:

$$\begin{aligned}
X(t) = &[1+0.5\cos(2\pi\times 4t)]\\
&\times\cos[2\pi\times 100t + 2\cos(2\pi\times 5t)]\\
&+[1+\sin(2\pi\times 5t)]\\
&\times\sin[2\pi\times 50t + 0.5\cos(2\pi\times 5t)]
\end{aligned} \quad (6)$$

In order to facilitate the observation and comparative analysis, the simulated signal 2 is decomposed by EMD, and used to remove the pseudo component decomposition results based on mutual information method. The decomposition results are shown in Figure 6, and the EMD-Hilbert spectrum is shown in Figure 7.

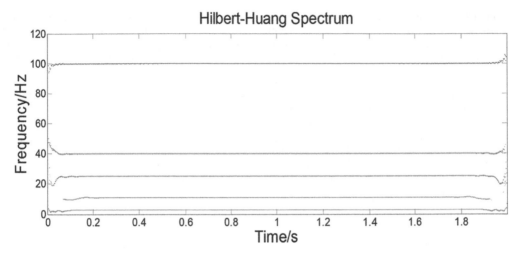

Figure 5. VMD-Hilbert image of the signal 1.

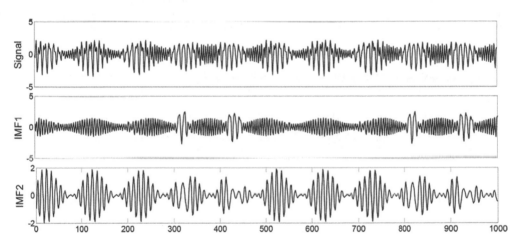

Figure 6. EMD decomposition of the signal 2.

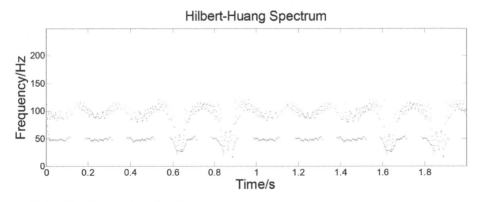

Figure 7. EMD-Hilbert image of the signal 2.

From Figure 6, it can be seen that the main component of the IMF1 for a given simulation signal is 100Hz frequency and amplitude modulation components, but mixed in with the part of 50Hz frequency and amplitude modulation components; IMF2, the main ingredient for a given signal of 50Hz frequency and amplitude modulation components. Figure 7 shows the EMD-Hilbert spectrum, due to the presence of EMD decomposition of the more serious mode aliasing, resulting in Hilbert based on the EMD spectrum of the original signal analysis which is weak.

Using VMD decomposition to the simulation signal 2, the results are shown in Figure 8, and after the decomposition, VMD-Hilbert analysis is used to signal 2, as shown in figure 9:

The simulation signal 2 is decomposed into two given frequency and amplitude components by VMD, and the decomposition effect is good. From simulation signal 2, the VMD-Hilbert spectrum can be very clear to reflect the original signal in the 100Hz and 50Hz frequency and amplitude modulation components with change in time.

Decomposition of synthesis modulation signal shows that: for synthesis of frequency and amplitude signal decomposition effect, VMD decomposition method is better than the EMD decomposition method and based on VMD Hilbert spectrum calculation method for the characterization of the effect is superior to that based on EMD Hilbert spectrum calculation method.

3.3 *Effect of noise on EMD and VMD decomposition*

Simulation signal 4 is composed of signal in Section 3.3 with standard deviation of 1 random noise, and the time domain waveform is shown in figure 17.

Using EMD decomposition of the signal with noise, but because the decomposition results contained in the pseudo component are too much, it is not easy to observe, analyze, and compare, and so the use of the mutual information method to remove pseudo components then draw the EMD-Hilbert spectrum as shown in Figure 12:

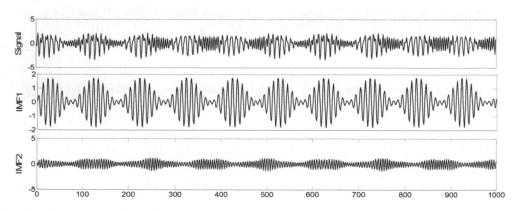

Figure 8. VMD decomposition of the signal 2.

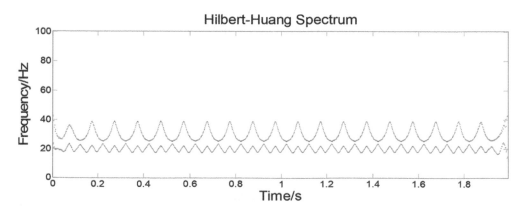

Figure 9. VMD-Hilbert image of the signal 2.

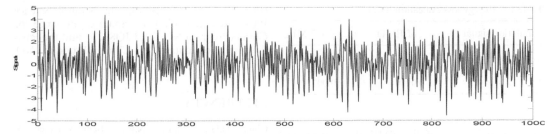

Figure 10.   Time domain image of the simulation signal 3.

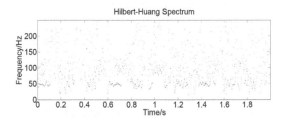

Figure 11.   EMD-Hilbert image of the signal 4.

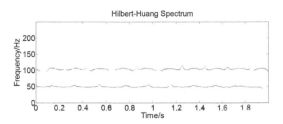

Figure 12.   VMD-Hilbert image of the signal 4.

Figure 11, due to the decomposition process envelope fitting deviation caused by mode mixing and the phenomenon of overlapping and pseudo component, the EMD-Hilbert spectrum signal simulation for the four analysis has lost its meaning.

Using VMD decomposition to the simulation signal 3, draw the VMD-Hilbert spectrum as shown in figure 12:

In Figure 12, Hilbert spectra show very good simulation signal components in the vicinity of frequency and amplitude modulation component existing near 50Hz and 100Hz, which is accurate analysis for signal. Comparison of Figures 11 and 12 found that the noise effect on the VMD decomposition is small.

The comparison of the noise effects on EMD and VMD decomposition shows that the method of VMD decomposition and the Hilbert spectral calculation method based on VMD have better robustness in the noisy environment.

3.4   Energy leakage contrast between EMD and VMD decomposition

This section analyzes two methods, EMD and VMD, before and after the decomposition of energy values of the changes to compare the two methods of energy leakage (Xiang Ling et al, 2014).

1. Calculate energy of each IMF component generated by the given signal $X(t)$ after EMD and VMD decomposition.

$$E = \sqrt{\frac{\sum_{i=1}^{n} x^2(i)}{n}} \tag{7}$$

where E is the given signal $X(t)$ or the energy of each IMF, $x(i)$ is the signal sequence, and n is the signal sampling points.

2. By comparing the difference between the energy of each IMF component and the given signal energy value, we get the evaluation index ξ:

$$\xi = \frac{\left| \sqrt{\sum_{p=1}^{k} E_p^2} - E_x \right|}{E_x'} \tag{8}$$

where E_x is the energy of a given signal; E_p is the energy of the P-th IMF component; and K is the total number of the IMF component, including the residual term of the decomposition.

By definition, the greater the value ξ, the greater the energy leakage of EMD and VMD decomposition, $\xi = 0$ for the end effect of EMD and VMD decomposition has no energy leakage.

The simulation signal 1, signal 2, and signal 3 is the simulation signal. When the EMD and VMD are used, there is no image extension. The simulated signals are decomposed by standard EMD and VMD algorithms, and their values are obtained, as shown in Table 1:

The data in Table 1 show that the energy leakage of the EMD decomposition method is larger, and

Table 1. Energy loss evaluation index.

Simulation signal	ξ of EMD	ξ of VMD
1	0.1277	0.0066
2	0.0589	0.0088
3	0.0186	0.0075

the energy leakage of the VMD decomposition method is not obvious, and the VMD algorithm is better than the EMD algorithm.

4 CONCLUSION

1. To compare the effects of the EMD and VMD decomposition advantages and disadvantages, VMD-Hilbert is proposed, constructs frequency mixed signal and frequency and amplitude modulation signal and with noise signal, using the EMD and VMD method, respectively, to characterize the signal. Results show that the two methods all can characterize signals at different scales, but when the EMD decomposition method is used in the decomposition and FM frequency closes, multi-component composite signal amplitude modulation, multi-component signal synthesis, modal aliasing, and pseudo components will appear easily, while using VMD decomposition method can avoid the modal aliasing and the generation of pseudo components, and can represent the original signal in different scales, and the VMD-Hilbert spectrum calculation method can better depict the characteristic of signal.
2. The effects of noise in EMD and VMD methods are compared and analyzed, and leakage of energy before and after the signal decomposition is analyzed, respectively. The results show that the noise robustness of the VMD method is better than that of the EMD method, and the energy leakage is smaller than that of the EMD decomposition method.

ACKNOWLEDGMENTS

This research was financially supported by the National Natural Science Foundation of China (Program NO. 51405498), the Natural Science Basic Research Plan in Shaanxi Province of China (Program NO. 2013 JQ8023), and the Postdoctoral Foundation of China (Program NO. 2015M582642).

REFERENCES

Chen Q.H., Huang N.E., Riemenscheider S. et al. A B-spline approach for empirical mode decomposition[J].Advances in Computational Mathematics, 2006, 24: 171–195.

Dragomiretskiy, K., et al. Variational Mode Decomposition [J], Ieee transactions on signal processing, VOL. 62, NO. 3, February 1, 2014, pp. 531–544.

Hao Guocheng, Gong Ting, et al. Time-frequency characteristics and energy analysis of the Earth's natural pulse electromagnetic fields based on ensemble empirical mode decomposition: The Lushan MS7.0 earthquake as an example[J]. Earth Science Frontiers, 2015, 22(5): 231–238.

Huang N.E., Shen Z., et al. The Empirical Mode Decomposition and Hilbert Spectrum for Nonlinear and Non-stationary Time Series Analysis[J]. Proceedings of the Royal Society of London. London A, 1998, 454: 903–993.

Liu Yang, Cao Yundong, et al. The Cable Two-terminal Fault Location Algorithm Based on EMD and WVD[J]. Proceedings of the CSEE, 2015, 35(16): 4086–4093.

Wang Xuehuan, Liu Jianguo, et al. An Empirical Mode Decomposition Alogorithm Based on Cross Validation and Its Application to Lidar Return Signal Denoising[J]. Chinese Journal of Lasers, 2014, 41(10): 11–18.

Wu Z.H., Huang N.E. Ensemble Empirical Mode Decomposition: A Nosie Assisted Data Analysis Method [J], Advances in Adaptive Data Analysis, 2009(1): 1–41.

Xiang Ling, Yan Xiaoan. Performance Contrast Between LMD and EMD in Fault Diagnosis of Turbine Rotors[J]. Journal of Chinese Society of Power Engineering, 2014, 34(12): 945–951.

Xu Zhuofei, Liu Kai. Method of Empirical Mode Decomposition End Effect Based on Analysis of Extreme Value Symbol Sequence[J]. Journal of Vibration, Measurement & Diagnosis, 2015, 35(2): 309–315.

Zhang Zhigang, Shi Xiaohui, et al. Fault feature Extraction of Rolling Element Bearing Based on Improved EMD and Spectral Kurtosis[J]. Journal of Vibration, Measurement & Diagnosis, 2013, 33(3): 478–482.

Energy Science and Applied Technology ESAT 2016 – Fang (Ed.)
© *2016 Taylor & Francis Group, London, ISBN 978-1-138-02973-6*

Costas loop effect from the demodulation on space optical communication of the BPSK scheme with atmospheric turbulence

Zhaoqi Wang
School of Software Engineering, Huazhong University of Science and Technology, Wuhan, China

Moyu Liu
School of Optoelectronic Engineering, Changchun University of Science and Technology, Changchun, China

Shuhao Ye
Guangdong University of Technology, Guangzhou, China

ABSTRACT: The communication quality of space optical communication with effect of Costas loop effect is investigated. Based on analyses, BER performances hardly change when phase deviation is under $\pi/16$. BER definitely increases when phase deviation caused by Costas loop enhances. In addition, the effect of phase deviation is sensitive about BER in lower level of divergence angle, and it has optimal gain factor for practical system. Besides, phase deviation hardly influences the optimal point of gain factor. These works are beneficial for researching Costas loop effect on BER performance and designing the practical communication system.

Keywords: costas loop, space optical communication, BER, atmospheric turbulence

1 INTRODUCTION

Nowadays, people pay much attention to space optical communication because of numerous advantages such as high data rate, and great bandwidth (C. C. Wei et al, 2009) (L. Andrews et al, 2001) (S. T. Le et al, 2014) (L. Yang et al, 2014). As we know, the atmospheric turbulence is usually considered as the essential factor that can deteriorate the communication quality (J. Ma et al, 2008) (S. Y. Lin et al, 2015) (M. Li et al, 2014). The atmospheric turbulence will lead to the intensity scintillation effect for optical signal in optical communication system.

As the data rate increases to the high level, coherent communication becomes the important technique. Thus, Costas loop, as the significant demodulation system in coherent communication, has been utilized for a long time (X. R. Ma et al, 2015) (J. C. Ding et al, 2013) (Ivan. B. Djordjevic et al, 1999). When it comes to the demodulation principle of Costas loop, it will create the reproduced carrier wave and this carrier wave will work with the received signal. And then, the original information can be demodulated finally. However, the reproduced carrier wave is not totally the same as the former one. It has the inevitable

phase deviation which can decrease the accuracy of demodulated signal and increase the Bit Error Rate (BER). For the phase modulation system, the digital information is modulated in the phase of carrier wave and this phase deviation will definitely affect the demodulation quality. As we know, some people pay much attention to Costas loop and atmospheric turbulence. The application of Costas loop in fiber has been already researched (K. Masa et al, 2014) and the effect of atmospheric turbulence is also investigated in reality (J. F. Campbell et al, 2014). However, to the best of our knowledge, no one has researched the Costas loop effect combined with atmospheric turbulence.

Up to now, Binary Phase Shift Keying (BPSK) has been used in the practical communication system (Y. Li et al, 2014) (Z. Shu et al, 2015). For BPSK, digital information is modulated in the phase of carrier wave and digital information '0' and '1' have different phase information. In this paper, the phase deviation effect by Costas loop with consideration of atmospheric turbulence effect on BER performance with BPSK scheme is investigated. We research the relationship between BER and practical system parameters in different Costas loop effects. Based on numerical simulation results, the BER as the function of transmission power and receiver diameter in different Costas loop effects is discussed. The environment factor, named zenith angle, is also analyzed for discovering the BER performance. Besides, for Avalanche Photodiode (APD), the relationship between BER and gain factor is further discussed. The BER as the function of divergence angle is also analyzed. These works are beneficial for researching the Costas loop effect on BER performance and designing the practical communication system.

2 THEORY

Speaking of the atmospheric turbulence effect, it will lead to the intensity scintillation in space optical communication system. The Probability Density Function (PDF) of receiving light can be shown as (L. C. Andrews et al, 1998):

$$P_I(r,I) = \frac{1}{\sqrt{2\pi\sigma_I^2(r,L)}}$$

$$\frac{1}{I}\exp\left(-\frac{\left(\ln\frac{I}{\langle I(0,L)\rangle} + \frac{2r^2}{W^2} + \frac{\sigma_I^2(r,L)}{2}\right)^2}{[2\sigma_I^2(r,L)]}\right). \quad (1)$$

where $\langle I(0,L)\rangle = \alpha P_T D_r^2 / 2W^2$ is the mean intensify, D_r is the receiving diameter, P_T is the transmission power, $W = \theta L/2$ is the radius of beam at the receiving plane, α is the energy loss of the link, r is the distance deviation between the beam center and receiving point, θ is divergence angle, $L = (H - h)\sec(\zeta)$ is the length of the laser link, ζ is the zenith angle, H and h_0 are the heights of the receiver and the emitter, and $\sigma_I^2(z,L)$ is the variance (L. C. Andrews et al, 1998).

Apart from the atmospheric turbulence, we can consider the optical signal modulated by BPSK scheme. The original optical signal of BPSK scheme affected by Gaussian noise is (Jones et al, 1967)

$$s(t) = a\cos(w_c t + \varphi) + n(t). \quad (2)$$

where $n(t)$ is the Gaussian noise from channel, a is the intensity of receiving signal, φ is the phase information, and ω_c is the frequency of carrier wave. As it is of BPSK scheme, when the digital information '1' is transmitted, the phase of carrier wave φ turns to be 0. When the digital information '0' is transmitted, the phase of carrier wave φ turns to be π (Y. C. Chi et al, 2014).

Besides, when it comes to Costas loop, it can produce relative carrier wave for demodulation.

The output voltage shows the carrier wave, and it is (R. M. Gagliardi et al, 1998)

$$v_g = \cos(\omega_c t + \varphi + \Delta\varphi). \quad (3)$$

where ω_c is carrier wave frequency, φ is original signal phase, $\Delta\varphi$ is the phase deviation. The $\Delta\varphi$ is created in the circuit in Costas loop, which is shown as (R. M. Gagliardi et al, 1998):

$$\Delta\varphi = \Delta f / K_r. \quad (4)$$

where Δf is the frequency difference and K_r is the direct current gain.

As we know, BER performance is often associated with the Signal-to-Noise Rate (SNR). Based on the research (M. Li et al, 2015), we can know how the phase deviation from Costas loop works with the BER performance. The overall BER of BPSK can be calculated as (M. Li et al, 2015):

$$BER_{BPSK} = \frac{1}{2}P(0/1) + \frac{1}{2}P(1/0) = \frac{1}{2}erfc(\frac{a\cos\Delta\varphi}{\sqrt{2}\sigma_n}). \quad (5)$$

For space optical communication, the transmission distance is so far that the received signal is too weak to detect. Thus, we use the avalanche photodiode (APD) to amplify the receiver signal. Thus, the mean value a and variance of noise σ_n^2 can be expressed as (R. M. Gagliardi et al, 1998):

$$a = G \cdot e \cdot \left(K_s(I) + K_b\right) + I_{dc}T_s. \quad (6)$$

$$\sigma_n^2 = (G \cdot e)^2 \cdot F \cdot \left(K_s(I) + K_b\right) + \sigma_T^2. \quad (7)$$

where e is the electron quantity, G is the gain factor, K_b is the photon count of the background light, T_s is the bit time, K_s is the photon count, $\sigma_T^2 = 2\kappa_c T T_s/R_L$ is thermal noise, T is the temperature, k_c is the Boltzmann constant, and R_L is the load resistance.

Thus, if considering the intensity scintillation, the BER performance can be calculated by

$$BER = \int_0^{+\infty} BER_{BPSK}P_I(r,I)dI. \quad (8)$$

3 NUMERICAL SIMULATIONS

Simulation parameters are as follows: zenith angle $\zeta = 0°$, the divergence angle $\theta = 30\ \mu rad$, wavelength $\lambda = 850$ nm, the altitude of the ground station $h_0 = 100$ m, the altitude of the satellite H = 36,000 km, quantum efficiency of APD $\eta = 0.75$, load resistance $R_L = 50\Omega$, additional noise factor $F = G^{0.5}$, photomultiplier gain factor $G = 100$, the receiving aperture $D_r = 0.4$ m, temperature $T = 300$ K, the dark current $I_{dc} = 1$nA, the time duration per slot $T_s = 10$ ns, spectral density $I_b = 10$ nW/m², the energy loss $\alpha = 1$, P_T is 1 W, and the atmospheric refractive index $C_n^2 = 10^{-16} m^{-2/3}$.

It is true that zenith angle is also an essential parameter in space optical communication practical system. Fig. 1 indicates the relationship between BER and zenith angle in different phase deviations. We can see from the figure that BER gradually increases when the zenith angle increases from 0° to 60°. It means that BER is sensitive to zenith angle with effect of $\Delta\varphi$. When $\Delta\varphi$ increases from 0 to $\pi/16$ $2\pi/16$, $3\pi/16$, $4\pi/16$, $5\pi/16$, BER significantly increases. It indicates that the effect of phase deviation $\Delta\varphi$ plays an important part on BER. This work can help investigate the performance of communication system with relationship with the zenith angle in the Costas loop effect.

In Fig. 2, the relationship between BER and transmission power in different phase deviations $\Delta\varphi$ is shown. We can see that BER are the same when $\Delta\varphi$ is 0 and $\pi/16$. However, with the increasing transmission power, BER in other $\Delta\varphi$

$(2\pi/16, 3\pi/16, 4\pi/16, 5\pi/16)$ are different. Thus, it is reasonable that BER cannot be influenced by phase deviation from the Costas loop when $\Delta\varphi$ is under $\pi/16$. Besides, the figure also shows that when the transmission power increases from 0W to 4W, BER gradually decreases. The effect of $\Delta\varphi$ also gradually increases when the transmission power increases from 0W to 4W. These analyses are helpful to research the BER performance with the effect of phase deviation from Costas loop.

Fig. 3 shows the relationship between BER and divergence angle in different $\Delta\varphi$. We can see that BER increases in different phase deviations when the divergence angle increases from 10 μrad to 70 μrad. As we know, enhancing the divergence angle will decrease the receiving power at the receiver point and this will lead to the worse BER performance. Besides, when the phase deviation is considered, BER increases when the divergence angle is at low level. It has about 6dB, 11dB, 22dB, 29dB BER increasing in different $\Delta\varphi$ ($2\pi/16$, $3\pi/16$, $4\pi/16$, $5\pi/16$), if the zenith angle is 20°.

Fig. 4 shows the BER as the function of receiver diameter in different phase deviations. We can see that BER gradually decreases in different $\Delta\varphi$, when

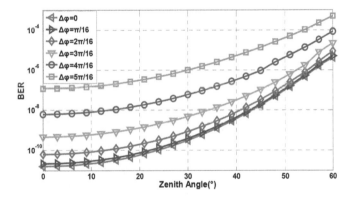

Figure 1. BER versus zenith angle in different phase deviations.

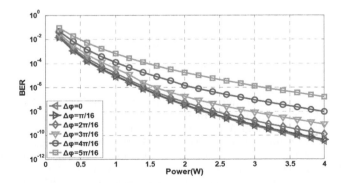

Figure 2. BER versus transmission power in different phase deviations.

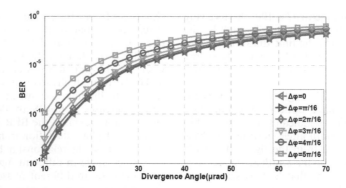

Figure 3. BER versus divergence angle in different phase deviations.

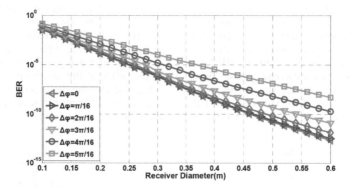

Figure 4. BER versus receiver diameter in different phase deviations.

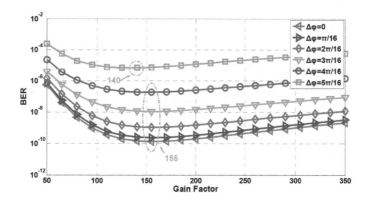

Figure 5. BER versus gain factor in different phase deviations.

the receiver diameter increases from 0.1m to 0.6m. It has nearly 35dB, 40dB, 45dB, 55dB, 65dB BER increasing in different phase deviations $\Delta\varphi$ ($2\pi/16$, $3\pi/16$, $4\pi/16$, $5\pi/16$). We can enlarge the receiver diameter to improve BER performance to much extent. Although enlarging receiver diameter can enhance communication quality, this parameter cannot be too large because this will increase the

cost of practical system. Thus, we should carefully design the receiver diameter in the communication system.

When it comes to the parameters in the Avalanche Photodiode Detector (APD), the gain factor is an essential parameter. In Fig. 5, it indicates the relationship between BER and gain factor. We can see from curves that there exists the optimal gain factor with effect of $\Delta\varphi$ and at the optimal point of gain factor, the BER is the minimum. Specifically, the optimal gain factor is 155 in different $\Delta\varphi$ (0, $\pi/16$, $2\pi/16$, $3\pi/16$, $4\pi/16$) and it becomes to be 140 when the phase deviation grows to $5\pi/16$. The optimal point of gain factor does not change a lot in different $\Delta\varphi$. Thus, we must carefully design the gain factor and we can get the best BER performance in practical system.

4 SUMMARY

In this paper, the communication quality of space optical communication with effect of Costas loop effect is investigated. During research process, the intensity scintillation from atmospheric turbulence is considered. Based on analysis on numerical simulation, we can see that BER performance hardly change when phase deviation is under $\pi/16$. BER definitely increase when phase deviation caused by Costas loop enhances at the same time. Besides, BER is sensitive to zenith angle and receiver diameter, and we must carefully consider the design of these parameters. The effect of phase deviation is obvious about BER in lower level of divergence angle, and it has an optimal gain factor for practical system. In addition, phase deviation hardly influences the optimal point of gain factor. We can adjust the APD for better BER performance in Costas loop effects. These works are beneficial for researching Costas loop effect on BER performance and designing the practical communication system.

ACKNOWLEDGMENTS

This work was supported by the National Natural Science Foundation of China.

REFERENCES

Andrews L.C. and R.L. Phillips, Laser beam propagation through random media, SPIE Optical Engineering Press, Bellingham, 1998.

Andrews, L., R. Phillips, & C. Hopen, Laser beam scintillation with applications, New York, SPIE Press, 2001.

Campbell, J.F., B. Lin, A.R. Nehrir, F.W. Harrison, & M.D. Obland, Binary phase shift keying on orthogonal carriers for multichannel CO2 absorption measurements in presence of thin clouds, Opt. Express 22 (2014) A1634–A1640.

Chi, Y.C. & G.R. Lin, Self optical pulsation based RZ-BPSK and reused RZ-OOK bi-directional oc-768 transmission, J. Lightw. Technol. 32 (2014) 372–83734.

Ding, J.C., M. Li, M.H. Tang, Yan. Li, & J.Y. Song, BER performance of MSK in ground-to-satellite uplink optical communication under the influence of atmospheric turbulence and detector noise, Opt. Lett. 38 (2013) 3488–3491.

Gagliardi, R.M. & S. Karp, Optical telecommunications, Publishing House of Electronics Industry, 1998.

Ivan. B. Djordjevic, Mihajlo. & C. Stefanovic, Performance of optical heterodyne PSK systems with Costas loop in multichannel environment for nonlinear second-order PLL model, J. Lightw. Technol. 17 (1999) 2470–2480.

Jones, & J. Jay, Modern Communication Principle with Application to Digital Signaling, New York, McGraw Hill, 1967.

Le, S.T., K.J. Blow, V.K. Mezentsev, & S.K. Turitsyn, Bit error rate estimation methods for QPSK co-OFDM transmission, J. Lightw. Technol. 32 (2014) 2951–2959.

Li, M., B.W. Li, Y.J. Song, X.P. Zhang, L.Q. Chang, & J. Liu, Investigation of Costas loop synchronization effect on BER performance of space uplink optical communication system with BPSK scheme, IEEE Photon. J. (2015) 790239.

Li, M., W.X. Jiao, Y.J. Song, X.P. Zhang, S.D. Dong, & Y. Poo, Investigation of the EDFA effect on the BER performance in space uplink optical communication under the atmospheric turbulence, Opt. Express 22 (2014) 25354–25361.

Li, Y. M. Li, Y. Po, J.C. Ding, M.H. Tang, & Y.G. Lu, Performance analysis of OOK, BPSK, QPSK modulation schemes in uplink of ground-to-satellite laser communication system under atmospheric fluctuation, Opt. Commun. 317 (2014) 57–61.

Lin, S.Y., Y.C. Chi, Y.C. Su, Y.C. Li, & G.R. Lin, An injection-locked weak resonant cavity laser diode for beyond bandwidth encoded 10-Gb/s OOK transmission, IEEE Photonic. J. 7 (2015) 1109–1112.

Ma, J., Y.J. Jiang, L.Y. Tan, S.Y. Yu, & W.H. Du, Influence of beam wander on bit-error rate in a ground-to-satellite laser uplink communication system, Opt. Lett. 33 (2008) 2611–2613.

Ma, X.R., Y.J. Xu, X. Wang, & Z.C. Ding, A novel high precision adaptive equalizer in digital coherent optical receiver, Opt. Commun. 351 (2015) 63–65.

Masa, K. & M. Akira, Decision-Directed Costas loop stable homodyne detection for 10Gb/s BPSK signal transmission, IEEE Photon. Technol. Lett. 26 (2014) 319–323.

Papoulis, A. & S.U. Pillai, Probability random variables and stochastic processes, New York, Mc GrawHill, 1984.

Shu, Z. & C.C. John, A SSP-Based control method for a nonlinear Mach-Zehnder interferometer DPSK regenerator, J. Lightw. Technol. 33 (2015) 3788–3795.

Wei, C.C., W. Astar, J. Chen, Y.J. Chen, and G.M. Carter, Theoretical investigation of polarization insensitive data format conversion of RZ-OOK to RZ-BPSK in a nonlinear birefringent fiber, Opt. Express 17 (2009) 4306–4316.

Yang, L., X.Q. Gao, & M.S. Alouini, Performance analysis of relay-assisted all-optical FSO networks over the strong atmospheric turbulence channels with pointing errors, J. Lightw. Technol. 32 (2014) 4011–4018.

Energy Science and Applied Technology ESAT 2016 – Fang (Ed.)
© *2016 Taylor & Francis Group, London, ISBN 978-1-138-02973-6*

Design and implementation of key modules of the OFDM receiving system based on FPGA

H.Y. Li, G.H. Cai & J. Yang
School of Information Science and Engineering, Yunnan University, Kunming, China

ABSTRACT: With the development of communication technology, the traditional channel multiplexing technology has been difficult to meet the higher needs of users. Orthogonal Frequency Division Multiplexing (OFDM) is a hot technology in the field of wireless communication, which can effectively eliminate the intersymbol interference and the inter-carrier interference. Starting the analysis from the model of the wireless communication system, this paper designs and realizes the key module of the OFDM receiver system based on FPGA, which includes interleaving and de-interleaving module, RS encoding and decoding module, QPSK modulation and demodulation modules, and some auxiliary modules. Besides, it uses the reconstruction and operation principle of FPGA to make the top document composed of each module complete the control and the realization of specific functions through written drivers and software platform. Through the performance test and analysis, the system has the advantages of less hardware resources, fast speed, good stability, etc., which can be widely used in the field of wireless communication.

Keywords: OFDM; interleaving and de-interleaving; RS encoding and decoding; QPSK modulation and demodulation; FPGA

1 INTROUCTION

In recent years, the communication technology develops rapidly. The key technology of the fourth-generation wireless communication (4G) is Orthogonal Frequency Division Multiplexing (OFDM) (Zhen L et al., 2002), which is a heated topic in the field of wireless communication. Wireless networks share the same wireless communicational channel in a special space and time environment, so the multiplexing technology of the channel is the key technology of the wireless communication (Mastronarde N & van der Schaar M, 2013). Most traditional channel multiplexing technology including FDMA, TDMA, and CDMA is flawed (Park J et al., 2013), for example, their frequency utilization rate is low, the cost is high, and they are unable to meet the higher needs of users. OFDM technology distributes high-speed data signals to several orthogonal sub-channels, and each sub data flow is distributed to each channel for transmission, then realizes the diversity reception at receiver, which can effectively eliminate intersymbol interference and reduce the inter carrier interference (Sangeetha M et al., 2014; Mao X, 2008).

OFDM technology has become an important part of wireless communication because of its unique advantages. But now, the understanding of OFDM technology is stuck in the simulation stage, where is no in-depth discussion on the modules. The paper adopts the FPGA technology to design an OFDM receiving systems including interleaving and de-interleaving module, RS encoding and decoding module, modulation and demodulation module, etc. Finally, it carries out the simulation test by ModelSim software to obtain the analysis of efficiency, cost and bit error rate. This provides a further reference for the study of OFDM system based on FPGA.

2 PRINCIPLE

2.1 *Principle of OFDM*

An OFDM signal is composed of many sub-carrier signals, which can be modulated by PSK or QAM (Petig T, et al, 2014). We can determine the orthogonality of complex exponential signals by calculating whether the integral quantity is zero in the interval. That is,

$$\frac{1}{T_0}\int_0^{T_0} e^{j2\pi} f_k^t dt = \frac{1}{T_0}\int_0^{T_0} e^{j2\pi\frac{k-i}{T_0}t} dt = \begin{cases} 1, & k=i \\ 0, & others \end{cases} \quad (1)$$

Then to introduce the discrete samples $t = nTs = n\frac{T_0}{N}, n = 0,1,2..., N-1$, the above equation can be written in the discrete time domain:

$$\frac{1}{N}\sum_{n=0}^{N-1}e^{j2\pi\frac{k}{T_0}nTs} - e^{j2\pi\frac{i}{T_0}nTs} = \frac{1}{N}\sum_{n=0}^{N-1}e^{j2\pi\frac{k-i}{N}n} = \begin{cases} 1, & k=i \\ 0, & others \end{cases} \quad (2)$$

By the formula, we can draw that the OFDM technology can save about 50% of the bandwidth. In practice, we often select of the enough number of subcarriers for the sub-channel bandwidth less than the channel coherence bandwidth, in order to effectively eliminate the frequency selective fading (Bindhaiq S, 2014).

2.2 OFDM communication model

The whole OFDM system is divided into two parts: the transmission system and the receiving system. In the system, there are several techniques, such as modulation and demodulation, source channel coding and decoding, serial and parallel conversion, digital and analog conversion, and so on. We briefly introduce the system process.

Error source and channel coding part: In the process of digital communication, what we firstly need to consider is the problem of the safety and reliability of the information. The information error in the process of transmission channel may occur, due to road noise and the channel characteristics are not ideal.

Modulation and demodulation part: after the source channel coding, the data are modulated by the interleaving technique. The process of modulation is that the frequency spectrum of the digital baseband signal modulates to the transmitted frequency band through the carriers, which is more advantageous to the transmission and can use the frequency band resources more efficiently.

As the sending and receiving process of OFDM system is reversible, this paper focuses on the OFDM system in the receiving process, and uses FPGA technology to realize source channels decoding, de-interleaving and interleaving, and modulation and demodulation module, then designs an OFDM system receiving model.

3 THE DESIGN

As each module in the system is independent, we adopt a "bottom-up" approach based on the modules. We design and optimize each module in separate prior to the completion of the top-level design, and then integrate the system.

3.1 Interleaving technology and interleaver

In communication system, in order to deal with the transmission information, the interleaving device transforms the original structure of the information, but does not change the information content.

First, we send a group of information $w = (w_1, w_2..., w_{36})$, and put the information into the 6×6 array storage interleaver. In the interleaver, the output information gets into the error channel, after de-interleaver, the interleaver output is read by columns, thus it becomes independent random error. Interleaver matrix is as follows:

$$H_\alpha = \begin{bmatrix} w_1 & w_7 & w_{13} & w_{19} & w_{25} & w_{31} \\ w_2 & w_8 & w_{14} & w_{20} & w_{26} & w_{32} \\ w_3 & w_9 & w_{15} & w_{21} & w_{27} & w_{33} \\ w_4 & w_{10} & w_{16} & w_{22} & w_{28} & w_{34} \\ w_5 & w_{11} & w_{17} & w_{23} & w_{29} & w_{35} \\ w_6 & w_{12} & w_{18} & w_{24} & w_{30} & w_{36} \end{bmatrix} \quad (3)$$

Therefore, we can get the signal $w = (w_1\ w_7\ w_{13}\ w_{19}\ w_{25} ... \ w_6\ w_{12}\ w_{18}\ w_{24}\ w_{30}\ w_{36})$. Assume that the channel generates burst errors, "$w_{20}\ w_{26}\ w_{32}$" and "$w_1\ w_{14}\ w_{28}\ w_{30}$". The two paragraphs are wrong, we have received the signal by using interleaving matrix according to column output, and row read:

$$H_\alpha = \begin{bmatrix} w_1 & w_2 & w_3 & w_4 & w_5 & w_6 \\ w_7 & w_8 & w_9 & w_{10} & w_{11} & w_{12} \\ w_{13} & w_{14} & w_{15} & w_{16} & w_{17} & w_{18} \\ w_{19} & w_{20} & w_{21} & w_{22} & w_{23} & w_{24} \\ w_{25} & w_{26} & w_{27} & w_{28} & w_{29} & w_{30} \\ w_{31} & w_{32} & w_{33} & w_{34} & w_{35} & w_{36} \end{bmatrix} \quad (4)$$

From the above formula, we can find that the burst error in the original channel changes into unrelated independence error, namely, the original continuous 4 error codes are independent of each other. From this, we can see that it is possible to realize complex weaving technique by using more order matrixes.

When implementing the system, it outputs a 6×6 matrix by row, while reads by column, and selects dual port RAM to realize the buffer register. Thus, the original continuous data can be transmitted distributedly, and the continuous errors can be uniformly distributed to a plurality of data packets by interleaving.

De-interleaver can recompose the scattered data into the original sequence. The de-interleaving is the inverse process of interleaving, in which the input data is read by row and output by column.

3.2 RS encoding and decoding

Error correction encoding is in order to achieve the reliability of communication system in the transmission process. The RS code has the same encoding efficiency, its error correction ability is particularly strong, and the performance is closer to the theoretical value, which is widely used in the field of wireless communication and data storage units (Natori T et al., 2014). In the OFDM system, the error correction ability must be high, and the data transmission speed needs to meet certain requirements. We need not only a high encoding efficiency, at the same time to consider from the encoding algorithm, but also a circuit structure, therefore it is suitable to choose RS codes.

In the (n, k) RS code, there are k, m, n three parameters, where n is the length of the codeword, k is the length of the segment information, and m is the domain value over GF (2^m) of the symbols. The symbol length needs to meet the $n=2^m-1$ symbols, $d=2t+1$ is the minimum code length of a symbol, and segment information is set as k; the supervisory position is $2t=n-k$ symbols. The main idea of RS code is to find $k(x)$ polynomial, and make the $k(x)$ is an integer multiple of segment information. Then, the receiver receives the codeword polynomial of type $k(x)$ is not divisible, so that we can know there is an error in the information. Generator polynomial is:

$$k(x) = (x - \theta)(x - \theta^2)... \approx \prod_{i=1}^{2t}(x - \theta^i) \qquad (5)$$

where θ^i is an element of GF (2^m). We use $a(x)$ to represent segment polynomials and $b(x)$ to represent code polynomials, if the remaining type is not 0, it means the code polynomial $b(x)$ is not divisible by generated polynomial $k(x)$, and there is an error in the received codes. RS codes can correct t m-hex codes, which is suitable for the burst errors. In addition, for RS codes, RS decoding circuit is more complex than coding circuit; in brief, the main purpose of decoding is according to the received code words and their relationship, to determine whether the received codeword is wrong.

In the OFDM system, it is also important to consider the hardware resource consumption of the decoding circuit by FPGA, so the choice of the circuit must be as efficient and simple as possible. The decoding circuit is composed of an accompanying calculation circuit, an error position polynomial circuit, a wrong position circuit, an error value circuit, and a decoding output circuit.

The relations between the error location polynomial $\sigma(x)$, the error value polynomial $\omega(x)$, and Omega with polynomial $S(x)$ are:

$$\sigma(x)S(x) = \omega(x) \bmod x^{2t} \qquad (6)$$

The above formula called the key equation, and the key of RS decoding is solving the key equation. There are a lot of decoding algorithms, and the BM algorithm is a fast recursive method with easy hardware implementation, so the paper uses this method to solving the key equation.

From the view of algebra, we analyze RS encoding and decoding in the time domain. First, we calculate the adjoint formula from the received information. Then, the error location and error values are calculated by the adjoint formula, and then their values can be solved. Finally, the error correction can be conducted. For the finite field multiplier in RS encoder, we can use the look-up table method. For the RS codes, each element occupied space of the 8 bit, and we need to use the look-up table method to query 16 tables where each has 16 inputs, so each of the look-up tables requires 64KB memory space. In order to reduce the occupied resources, we can fix the multiplier which designs the multiplier, which also satisfies the fixed coefficient characteristics of generated polynomial $g(x)$.

3.3 QPSK modulation and demodulation

Quadrature Phase Shift Keying (QPSK) is a kind of digital modulation method with excellent performance, which is widely used in digital signal modulation. Compared with ASK, PSK, and FSK, it has a high spectrum utilization and strong anti-interference ability (Hung Y & Tsai S L, 2014). QPSK modulation uses the random processing of energy diffusion to deal with data stream. QPSK can also handle the RS coding, convolution interleaving, shrinkage, and so on, which ensures data transmission performance. QPSK modulation has a strong anti-interference, with good spectrum utilization and good cost performance, besides a large number of devices that can be compatible.

Each sine wave carrier phase of the QPSK signal is provided with two binary symbols and four possible discrete phase states. The orthogonal modulation principle of QPSK signal is shown in Figure 1.

Where $\{b_i\}$ means binary non-return to zero sequence, T_S means the period of symbol, and T_b is the bit cycle. The power spectrum consists of an orthogonal branch and a branch circuit. These two orthogonal carriers, 2PSK linear, superpose to generate QPSK signal, of which the power spectrum is also superimposed.

As for the demodulation of QPSK, we use the coherent demodulation. In the process of coherent demodulation, the receiver provides a local carrier with the same frequency and direction, and the synchronization signal of the carrier can be extracted directly.

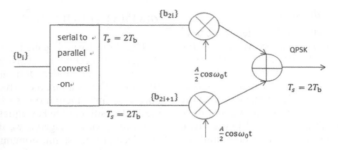

Figure 1.　The orthogonal modulation principle of QPSK signal.

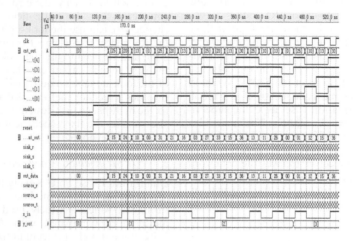

Figure 2.　Simulation waveforms of system.

4　SYSTEM SIMULATION AND PERFORMANCE ANALYSIS

4.1　System simulation

The following figure is the waveform of system simulation through the QuartusII, where clk is the clock signal, reset is the system reset signal, and enable is the enable signal, and the output is shown in Figure 2.

From the graph, in the 170ns, the signal is located in the phase {20}, the register value is 00101, and the system receives the match range 3, so the output results are basically the same with the input value. We can draw that the logic functions of the receiving system are in conformity with the expected design, and the design is correct.

4.2　Performance analysis

In the traditional design, the Phase Locked Loop (PLL) is usually used in the design process (Pandey K K et al., 2012). In this paper, there is no use of

Table 1.　Comparison of the comprehensive results of system resources.

Comparison	Applied PPL	This design
Total logic elements	14132/33216(43%)	13008/33216(39%)
Total pins	24/475(5%)	21/475(4%)
Total memory bits	45331/483840(9%)	40215/483840(8%)
Total PLL	1/4(25%)	0/4(0%)

embedded phase-locked loop, because the frequency of the system in the MHz level, and the synchronization requirements are not particularly high, which also ensures the stability of the system. Table 1 gives the system resources comparison between PPL and this design.

5　CONCLUSION

OFDM is the key technique in the field of the fourth-generation communication (4G), this paper realizes OFDM receiving system modules based on FPGA technology. The system models consists of interleaving and de-interleaving module, RS encoding and decoding module, QPSK modulation and demodulation module, and some auxiliary modules. Compared with the traditional method, the system has obvious advantages in system performance and efficiency, and it can be widely used in the field of wireless communication.

ACKNOWLEDGMENTS

This work was financially supported by the Yunnan Science Research Fund "Research on the key technology of OFDM system based on FPGA" (2014Y020).

REFERENCES

Bindhaiq S., Syed-Yusof S.K., Hosseini H. (2014). Performance analysis of Doppler shift effects on OFDM-based and MC-CDMA-based cognitive radios[J]. International Journal of Communication Systems, 27(11): 2658–2669.

Hung Y., Tsai S.L. (2014). PAPR Analysis and Mitigation Algorithms for Beamforming mimo ofdm Systems[J]. IEEE transactions on wireless communications, 13(5): 2588–2600.

Mao X., Maaref A., Teo K.H. (2008). Adaptive soft frequency reuse for inter-cell interference coordination in SC-FDMA based 3GPP LTE uplinks[C]//Global Telecommunications Conference. IEEE GLOBECOM. IEEE. IEEE, (1–6).

Mastronarde N, van der Schaar M (2013). Joint Physical-Layer and System-Level Power Management for Delay-Sensitive Wireless Communications[J]. Mobile Computing IEEE Transactions on, 12(4): 694–709.

Natori T, Fanabe N, Furukawa T (2014). A MIMO-OFDM channel estimation algorithm for high-speed movement environments[J]. Communications, Control and Signal Processing (ISCCSP), 6th International Symposium on, 2014: 348–351.

Pandey K K, Bhatt U R, Upadhyay R. (2012) Investigation effect of phase noise of OFDM system and realization LO with and without phase locked loop (PLL) [C]//Wireless and Optical Communications Networks (WOCN), 2012 Ninth International Conference on. IEEE, 2012: 1–5.

Park J, Chun J, Jeong B J (2013). Cross-relation-based frequency-domain blind channel estimation with multiple antennas in orthogonal frequency division multiplexing systems[J]. Iet Communications, 7(16): 1753–1768.

Petig T, et al (2014). Self-stabilizing TDMA Algorithms for Wireless Ad-hoc Networks without External Reference[J]. Ad Hoc Networking Workshop, 2014:87–94.

Sangeetha M, Bhaskar V, Cyriac A R (2014). Performance Analysis of Downlink W-CDMA Systems in Weibull and Lognormal Fading Channels Using Chaotic Codes[J]. Wireless Personal Communications, 74(2): 259–283.

Zhen L et al. (2002). Consideration and research issues for the future generation of mobile communication[J]. IEEE CCECE. Canadian Conference on, 3: 1276–1281.

Energy Science and Applied Technology ESAT 2016 – Fang (Ed.)
© 2016 Taylor & Francis Group, London, ISBN 978-1-138-02973-6

Blind estimation of satellite wideband background noise based on statistics

Boyu Yu, Canhui Liao & Ling You
National Key Laboratory on Blind Signals Processing, Chengdu, China

ABSTRACT: In this paper, a method was proposed to estimate the wideband background noise. In order to accurately estimate the background noise, we extract the statistical feature of the wideband spectrum, and estimate dynamic threshold to separate signal and noise. A large number of actual test results show that this method is suitable for most of the wideband satellite signals. Moreover, it is simple and easy to realize.

Keywords: noise estimate, wide-band, statistics, threshold

1 INTRODUCTION

As the communication system is influenced by the environment electromagnetic interference, its own heating, and so on, there is always noise in the receiver passband, which is affecting the quality of signal reception. Wideband noise estimation is an important part of the power spectrum detection of satellite signals. In a non-cooperative satellite communication system, to monitor the target satellite signal and to carry out the detection of carrier wave, a common method is to observe whether the signal energy at a certain frequency is higher than the expected threshold value. In addition, the accuracy of noise energy estimation will directly affect the accuracy of the test results.

Currently, the method of noise estimation is mainly aimed at some kinds of signal. Such as frequency hopping signals noise estimation, speech signal noise estimation, and narrow band signal noise estimation. There are some limitations and particularities in these signal noise estimation methods. Moreover, most of these methods are not suitable for wideband noise estimation. The hopping noise estimation method utilizes the frequency's sparsely of hopping signal (Xin J, 2014). The speech signal noise estimation method requires accurate speech frame/non speech frame detection (Martin R, 2001). Some simple background noise estimation for wideband signals such as energy average method and word statistics method (Xu Qi-hua) do not consider that the wideband noise may not be smooth and the wideband signals are dense, which may cause an inaccurate estimation result.

This paper presents a wideband noise estimation method based on power spectrum statistics, which can estimate the wideband noise accurately. A large number of actual signal test results show that this method is suitable for most of wideband signals, and it is simple and easy to realize.

2 ALGORITHM THEORY

2.1 *Problem statement*

Wideband spectrum background noise estimation means using the received actual wideband waveform and power spectrum to estimate the passband noise spectrum, which can be used for signal detection and signal-to-noise ratio estimation.

For the narrowband signal, as the noise is very smooth in the passband, we always use the average noise which is in the passband instead of that at the signal frequency, as shown in Figure 1.

However, the wideband signal may appear in the situation that signals are intensive and even overlapping. In this case, we cannot filter each signal to estimate its noise. In addition, due to a wide range of spectrum, electromagnetic wave propagation and front-end equipment will introduce the passband response uneven whose degree cannot be ignored, resulting in the traditional noise estimation method for failure.

2.2 *The theory of wideband background noise estimation based on statistics*

The method of noise estimation presented in this paper can be divided into three steps: spectrum histogram statistics, threshold estimation; and noise estimation.

2.2.1 *Spectrum histogram statistics*
As the signal energy is generally higher than the noise for wideband signals, the logarithmic spectrum histogram has some regularity. In Figure 2, a satellite signal power spectrum have been shown, and in Figure 3, it is the histogram of this signal. The horizontal axis represents energy histogram, dB, the vertical axis represents the frequency, and each peak's meaning is indicated in this figure.

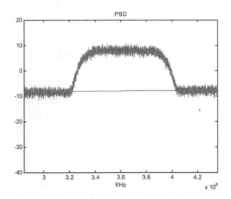

Figure 1. Narrow band signal noise estimation.

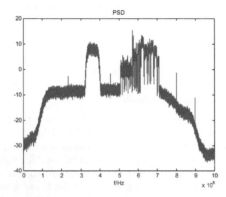

Figure 2. Power spectrum.

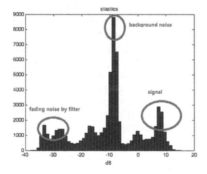

Figure 3. The peak's meaning of histogram.

2.2.2 Threshold estimation

How to separate the noise and the signal accurately is the key of the wideband background noise estimation algorithm based on statistics. Therefore, the accuracy of threshold estimation will directly affect the result of separation. The traditional separation methods, logarithmic average value method, and week energy preservation method make a simple energy average or short-time filter to estimate the threshold of the signal spectrum. The accuracy and adaptability of threshold estimation are poor,

and the signal\noise can be separated in the case of relatively smooth background noise, as shown in Figure 4. For non-smooth noise floor of the signal, these methods cannot separate the signal and noise effectively. In this case, some signals would be considered as noise, as shown in Figure 5. In this paper, a method of dynamic threshold estimation based on twice statistics is proposed, which can obtain a more accurate dynamic threshold, as in Figure 6.

The dynamic threshold estimation method can be divided into two steps:

a. The global threshold estimation.

  a.1. Power spectrum histogram statistics and the average value of the power spectrum are calculated, as shown in the following equation:

$$xSpec_{ave} = \frac{1}{N}\sum_{i=1}^{N} xSpec(i) \qquad (1)$$

  a.2. Find the location of $xSpec_{ave}$ in the histogram, and find the first maximum (or

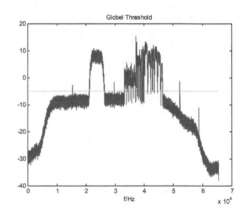

Figure 4. Globel threshold.

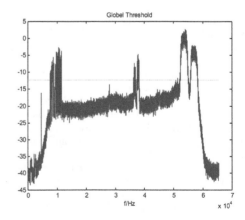

Figure 5. Globel threshold (noise uneven).

490

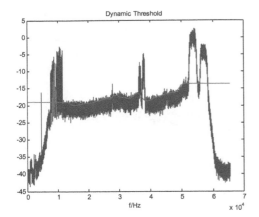

Figure 6.   Dynamic threshold.

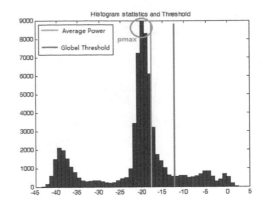

Figure 7.   Historgram statistics and threshold.

minimum) value on the left of $xSpec_{ave}$. The maxima is represented by p_{max} and the minima by p_{min}. The physical meaning of the minimum value p_{min} represents the edge portion of the power spectrum of noise or signal, and then we can estimate the global threshold $th_{overall}$:

If $xSpec_{ave} > p_{max}$, then $th_{overall} = p_{min}$;
If $xSpec_{ave} > p_{min}$, then $th_{overall} = xSpec_{ave}$;
b. The dynamic threshold estimation.
b.1. Power spectrum segmentation:

$$xSpec_{Len}^i = xSpec(iY + k), i = 0,1,2,......,N-1,$$
$$k = 1,2,3......,Len-1, \tag{2}$$

Len, N, Y is the integer. N_{FFT} is the FFT point, Len is the window length, N is the number of segments, and the overlap of each segment is $Len - Y$.

$$(N-1)Y + Len \leq N_{FFT} \tag{3}$$

b.2. Estimate the threshold of each segment $xSpec_{Len}^i$, which is represented by th_{Len}^i. The method is the same as (a), and compare th_{Len}^i with $th_{overall}$.
If $th_{Len}^i > th_{overall}$, $th_{Len}^i = th_{overall}$
Else, $th_{Len}^i = th_{Len}^{i-1}$
The dynamic threshold estimation algorithm is shown in Figure 8.

2.2.3 Background noise estimation

After segmentation threshold adaptive estimation, signal and noise separation can be carried out. As the wideband satellite noise is relatively smooth in a certain bandwidth, we can deal with the noise alone. The method is as follows:

a. Mark the noise. Comparing the power spectrum of each frequency energy with the dynamic threshold value.

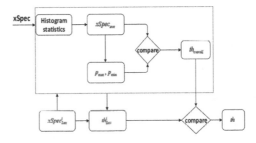

Figure 8.   Dynamic threshold estimation algorithm.

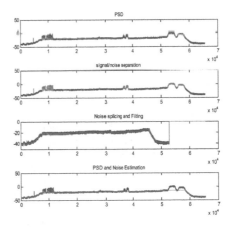

Figure 9.   Background noise estimation.

b. Noise extraction, splicing.
c. Low-pass filtering for stitching noise.
d. Noise position restoring and interpolation.
e. Low-pass filtering.

The signal power spectrum, signal noise separation diagram, noise splicing and filtering, noise restoring, and interpolation are shown in Figure 9, respectively.

491

(a)Signal intensive.

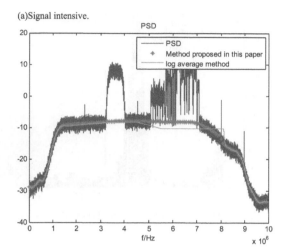

Figure 10. Background noise estimation (signal intensive).

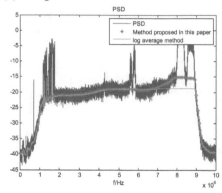

(b)Background noise not smooth.

Figure 11. Background noise estimation (noise uneven).

3 ALGORITHM TEST

In this section, we test the noise estimation algorithm by the actual satellite signal simulation, and then compare with the log average method. The log average method just uses the average power of the power spectrum as the threshold value to separate signal and noise. In the frequency range of wideband satellite signal, there may be dense signal, and the background noise is not smooth. This section shows the estimation results of background noise of different wideband satellite signal situations.

From the results, the traditional log average method may obtain accurate noise when the wideband signal is sparse and the background noise is smooth. However, when the signal is dense or background noise is not smooth, this method is invalid. This is because when we use the average method to estimate the threshold, when the background noise is not smooth, the threshold value may be higher than the partial signal, leading to that the signal error is considered as noise. On the other hand, signal in the frequency range can not completely obey uniform distribution. In the part of the band it will appear the phenomenon of signal intensity. Only for the entire spectrum of the threshold estimation, we cannot reflect the details of the part of the band signal distribution, resulting in that the band noise can not adaptive to the signal distribution characteristics, and can not reach the effect of tracking noise change. The method proposed in this paper can overcome the above problems, in the case of noise uneven and signal dense, adaptive tracking noise changes, and obtain a more accurate estimation result of the background noise.

4 SUMMARY

For satellite wideband signal may be intense and noise may be not smooth, we proposed a background noise estimation method. This method makes use of the distribution features of wideband spectrum energy statistics, estimates the dynamic threshold, and finally obtains the accurate estimation results of noise.

Compared with the previous method, on the one hand, this method does not need to do time-frequency analysis and signal frame detection, and on the other hand, this method can obtain more accurate noise estimation results when the band signal is dense and noise is not smooth.

REFERENCES

Emilia Nunzi, Paolo Carbone, Dario Petri. Spectral Estimation of Wideband Noise in Delta-Sigma Modulators (J). IEEE Transactions on Instrumentation and Measurement, 2006, 55(5):1691:1695.

Lalitha Venkataramanan, John L, Roman Kuc. Identification of Hidden Markov Models for Ion Channel Currents—Part I: Colored Background Noise (J). IEEE Transactions on Signal Processing, 1998, 46(7):1901–1915.

Martin R. Noise power spectral density estimation based on optimal smoothing and minimum statistics (J). IEEE. Trans. on Speech, and Audio Processing, 2001, 9(5): 504–512.

Miki Iwama. Estimation of Background Noise in HF-band (C). 2008, Asia-Pacific Symposium on Electromagnetic Compatib.

Xin J, Lu L, Bao X. Noise Energy Estimator Based on Sparseness of Time-Frequency Domain for Broadband Frequency-Hopping Signal (J). ACTA Electronica Sinica, 2014, 42(10):1932–1937.

Xu Qi-hua. The Signal Separation and Detection for Wide-band Reconnaissance in High Frequency (D). Xidian University.

Energy Science and Applied Technology ESAT 2016 – Fang (Ed.)
© 2016 Taylor & Francis Group, London, ISBN 978-1-138-02973-6

Research of a height measurement system based on information fusion

Yayang Cheng, Tongyue Gao & Shihao Zhu
School of Mechatronic Engineering and Automation, Shanghai University, Shanghai, China

ABSTRACT: The accuracy of positioning and navigation in the horizontal plane has been able to meet applications in various fields, but precision in the vertical direction is finite so that the navigation system is limited to certain areas. We developed a height positioning system combined with GPS and a barometer to demonstrate and evaluate the implementation of height information fusion. After finishing temperature compensation of the barometer, pressure height information with a specified precision can be obtained. Then, we establish the state equation and observation equation of the Kalman filtering algorithm, and fuse these data online through discretization of height information from GPS and the barometer to get the optimized height result. Finally, the high accuracy of the system is verified through several experiments.

Keywords: height, barometer, information fusion

1 INTRODUCTION

In present situation of high-precision navigation, the measurement precision of vertical direction is still limited. For instance, when cars drive down the viaduct, GPS signals are affected by multipath effect, so seriously that positioning signal is blocked and weakened (Di Liu, 2009). Navigation device is unable to reflect the change of height in time and recalculate the navigation path. These defects can cause a lot of trouble.

During the research, we find that: the positioning precision of latitude and longitude from GPS has been high. However, a well-known disadvantage of GPS systems is that they provide less accurate information about the vertical position. The standard deviation of the error of the estimated height is often more than 10m (J. Farrell, 1998). Therefore, the precision of traditional height measurement based on the GPS is difficult to satisfy the requirement of high-precision positioning navigation (Song Wang, 2012).

In this paper, we add a barometer in the original navigation system which is only based on GPS and obtain high-accuracy height data through the information fusion algorithm. This solution avoids the oneness of data sources and improves the precision of the height positioning system.

2 SYSTEM DESIGN

The architecture of the height measurement system we design is shown in Figure 1.

Figure 1. Height measurement system.

Considering the relationship between temperature, atmospheric pressure, and altitude, we must complete the temperature compensation of pressure sensor to obtain pressure information with a high precision. Then, we use the Kalman filtering algorithm to fuse height data from GPS and barometer information and obtain most excellent height result.

3 MULTI-SENSOR DATA FUSION ALGORITHM

The height information is provided by GPS and the barometer. Two modules have their own advantages and disadvantages. For example, the absolute precision of GPS is high; however, the short-term accuracy is low. In addition, height information based on barometer has opposite characteristics: the relative accuracy is high and the error will increase over time (Yi Yao, 2009). In order to maintain the high precision when it is used, we fuse different data from two kinds of sensor to get height in real time.

3.1 Mathematical model

1. Mathematical model of barometer

The standard formula of measuring the height of the barometer is:

$$H = \frac{T_b}{\beta}\left[\left(\frac{P_H}{P_b}\right)^{-\beta R/g_n} - 1\right] + H_b \qquad (1)$$

where H_b and T_b denotes the relevant aerosphere geopotential altitude and lower limit value of atmospheric temperature, β denotes the vertical change rate of temperature, g_n denotes the standard acceleration of free fall, P_b denotes the lower limit of atmospheric pressure, P_H denotes the atmospheric pressure at any altitude, and R denotes the constant of air special gas.

Mathematical model for measurement of barometer is defined as follows:

$$h_{bar} = h_{a-baro} + w_{baro} \qquad (2)$$

where h_{bar} denotes the true height, $h_{a\text{-baro}}$ denotes the measurement height, and w_{baro} denotes the barometric altitude noise.

Although data from barometer have the short-time high precision, sensor noise cannot be ignored. In order to eliminate this noise, we extract the difference in the adjacent barometer height data as altitude speeds further.

Therefore, the rate mathematical model of pressure is:

$$V_{bar} = \left(h_{bar-k2} - h_{bar-k1}\right)/dt \qquad (3)$$

where V_{bar} denotes the true altitude speed, h_{bar-k2} denotes the altitude from pressure measurement in k_2 moment, h_{bar-k1} denotes the altitude from pressure measurement in k_1 moment, and dt is the time difference between k_2 and k_1.

2. Mathematical model of GPS

According to the standard NEMA083 protocol of GPS, we can obtain the current height from GPS information. The mathematical model of height from GPS is:

$$h_{gps} = h_{a-gps} + w_{gps} \qquad (4)$$

where h_{gps} denotes the true height, $h_{a\text{-gps}}$ denotes the measurement height, and w_{gps} denotes the GPS altitude noise.

Owing to the short-term high precision of barometer information and the long-time high precision of GPS, the main ideas of data fusion in this paper are: predictive value about altitude is obtained by computing the integral for short-time

height difference from barometer, and measurements of altitude come from GPS. Optimizing the predictive result by eliminating the error is caused by noise in the process. Therefore, this system has the feature of both long-term and short-term accuracies of height information, which are very good.

3.2 Kalman data fusion algorithm

1. State equation

In order to use height information from barometer and GPS reasonably, the state equation of this system is

$$X = \begin{bmatrix} h & v \end{bmatrix} \qquad (5)$$

where h is the altitude and v is the rate of altitude change.

According to the state variable above, the state equation of the height data fusion system for continuous time is:

$$\dot{X} = AX + Q(t) \qquad (6)$$

$$\begin{bmatrix} \dot{h}(t) \\ \dot{v}(t) \end{bmatrix} = \begin{bmatrix} 0 & 1 \\ 0 & 0 \end{bmatrix}\begin{bmatrix} h(t) \\ v(t) \end{bmatrix} + \begin{bmatrix} 1 & 0 \end{bmatrix}\begin{bmatrix} Q(t)_h \\ Q(t)_v \end{bmatrix} \qquad (7)$$

where $h(t)$ is the altitude variable, $v(t)$ is the altitude rate variable, $Q(t)_h$ is the noise of system height state, and $Q(t)_v$ is the noise of system altitude rate state.

2. Measurement equation

As GPS signal and barometer data include altitude information with long-time and short-time high accuracies, respectively, we take both height data as the measured value to modify the predicted state vector.

System observation variable is:

$$Z = \begin{bmatrix} h_{gps} & v_{bar} \end{bmatrix} \qquad (8)$$

According to the observation variable above, measurement equation for continuous time of height data fusion system is:

$$Z = HX + R(t) \qquad (9)$$

$$\begin{bmatrix} h_{gps}(t) \\ v_{bar}(t) \end{bmatrix} = \begin{bmatrix} 1 & 0 \\ 0 & 1 \end{bmatrix}\begin{bmatrix} h(t) \\ v(t) \end{bmatrix} + \begin{bmatrix} 1 & 0 \end{bmatrix}\begin{bmatrix} R(t)_{gps} \\ R(t)_{vbar} \end{bmatrix} \qquad (10)$$

where $h_{gps}(t)$ denotes the altitude observation variable from GPS, $v_{bar}(t)$ denotes the barometer height rate variable, $R(t)_{gps}$ denotes the noise of system height measurement, and $R(t)_{vbar}$ denotes the noise of system altitude rate measurement.

3. Extended Kalman filter

The state equation and measurement equation of Kalman filtering algorithm are continuous, so they are used to calculate online after discretization and linearization (G. Welch, 2004). We need to construct the discrete extended Kalman filtering equation for the nonlinear system (J. Zhou, 2010).

Kalman filter cannot guarantee unbiased estimates originally, which means square error of the actual estimation is not smallest necessarily. However, with the increasing filtering steps, the influence of selection of filter for initial value X_0 and the mean square error P_0 to the filter will be weakened until they disappear. Therefore, the estimation tends to be unbiased gradually.

As a whole, we completed the design of value initialization for the model parameters, the state vector, and covariance matrix. Then, the optimal result can be obtained by fusing data through iterative operation of discrete Kalman filter algorithm.

4 EXPERIMENT

In order to evaluate the performance of integrated height measurement system, we did typical static and dynamic performance experiments. The static test shows that drift of height information from GPS is big, even when maximum error has reached around 6 m in the stationary process, as shown in Figure 2. The stability and accuracy of height after Kalman filter is better obviously. In addition, the optimal height value numerical precision is better than the information from GPS in the dynamic process, as shown in Figure 3.

5 SUMMARY

This article completed the design of height measurement system combined with GPS and a barometer. First of all, high precision data of pressure height after finishing temperature compensation of the air pressure sensor are obtained. Then,

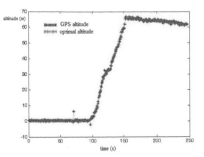

Figure 3.   The result of dynamic test.

mathematical models about the barometer and GPS are established. After obtaining accurate sensor information, the state equation and observation equation are designed according to the Kalman filtering algorithm. In order to get optimal height result, we fuse data online through discretization of the height from GPS and the barometer information. In the end, static and dynamic experiments show the combination height system has high precision on a wide variety of environments. The height system can be widely used in the field of outdoor positioning or environmental monitoring.

REFERENCES

Di Liu, Yuming Bo and Weijun Zou. GPS error modeling based on time series and research on the accuracy of point positioning(J). Acta Armamentarll, 2009, 30(6): 825–828(In Chinese).

Farrell J. and M. Barth, The Global Positioning System and Inertial Navigation, McGraw-Hill, 1998.

Maik Bevermeier, Oliver Walter, Sven Peschke and Reinhold Haeb-Umbach. Barometric Height Estimation Combined with Map-Matching in a Loosely-Coupled Kalman-Filter(J). 2010 7th Workshop on Positioning Navigation and Communication (WPNC), 2010: 128–134.

Nilsson, C. Hall Jr, S. Heinzen, et al. GPS Auto-Navigation Design for Unmanned Air Vehicles(R), AIAA-2002-0389, 2002

Song Wang, Juan Li and Sihong Wang. A Height Controller Design for Miniature UAVs Based on GPS and Barometer(J). 2012 2nd International Conference on Consumer Electronics, Communications and Networks (CECNet), 2012:31–34.

Welch G. and G. Bishop, "An introduction to the Kalman filter", Department of Computer Sciences, University of North Carolina, Chapper Hill, 2004: 1–16.

Yi Yao, Zhigang Huang and Rui Liu. Portable design and error modification of barometric altimeter(J). Journal of Telemetry, Tracking and Command, 2009,30(6): 48–51(In Chinese).

Zhou, J., S. Knedlik, E. Edwan and O. Loffeld, "Low-cost INS/GPS with nonlinear filtering methods," in Fusion 2010 (13th International Conference on Information Fusion). Edinburgh, UK. 2010.

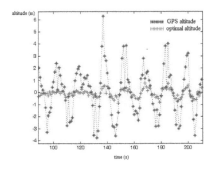

Figure 2.   The result of static test.

Energy Science and Applied Technology ESAT 2016 – Fang (Ed.)
© 2016 Taylor & Francis Group, London, ISBN 978-1-138-02973-6

Research on the synchronization clock of an electricity information acquisition terminal based on minimal commuting time difference algorithm

Q.Y. Shen, Y. Zhou & X.L. Gao
Jiangsu Electrical Power Company Research Institute, Nanjing, China
State Grid Key Laboratory Power Metering, Nanjing, China

Y.H. Yan & P. Li
Jiangsu Frontier Electric Technology Co. Ltd., Nanjing, China

ABSTRACT: Network time protocol is one of the most important methods to solve communication network time synchronization problems during the modern period. However, the current network structure is using wire or wireless net to realize information exchange, and the upper and bottom terminals are not kept in symmetrical shape. Owing to the high possibility of communication quality randomness, the info commute route could be largely different, which makes the traditional NTP synchronization calculation method hard to meet the requirement under current network structure. This paper now is raising a synchronization calculation method based on minimal commute time difference of NTP. By implementing this method in the synchronized structure of electricity information-collecting acquisition terminal clock, we can see that the new synchronization calculation method is largely improved in both accuracy and convergence compared with the older one.

Keywords: power-consuming information collecting, clock synchronization, Network Time Protocol (NTP), minimal commute time difference

1 INTRODUCTION

With the rapid growth of power grid development, the requirement of in-time data collection of power user is more and more strict (L.J. QIAN, 2013, G.D. GU, 2013). However, the delay of telecommunication transfer or the malfunction of data-collecting terminal will be ended up with big error in the timing system of power meter (G.C. Ren, 2009, Z.G. Liang. 2009). Network time protocol (NTP) is the right way to solve those problems both briefly and economically.

NTP is a standard network protocol for timing synchronism. NTP is a serious protocol that could pragmatically and effectively be used under various network situations in terms of size, speed, and connection level. NTP uses Client/Server structure and takes UTC as timing standard. Its theory is based on UDP/IP and has layer-stepping timing distribution model. Besides, it has high flexibility and it can be used under different network conditions. NTP not only could calibrate current time, but also can adjust time automatically as the change of time. Nowadays, under common working environment, the time accuracy that NTP provided could be calculated into millisecond. However, as

the tightening combination between wire and wireless networks, the asymmetry level will be higher and higher. This forces traditional NTP time comparing calculation method to collect relatively high time deviations error. This paper will raise a minimum round trip time difference calculation method, which could effectively reduce such error.

2 OVERVIEW

The way for NTP (D.L. Mills. 1991, D.L. Mills. 1998) to realize its accuracy is mainly due to its working theory. NTP calculation is based on commuting messages between servers and customer terminals, which can confirm the value differences between clocks of two different positions and transfer delay of message in network. We hereby identify the time differences between server and customer terminal as "θ"; and network transfer delay of time synchronization process will be identified as "δ".

In P1, T_1 and T_4 are clock values during terminal sending and receiving NTP message. T_2 and T_3 are clock values during server receiving and sending NTP messages. θ is the estimated time deviation

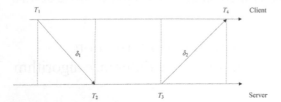

Figure 1. The synchronization process between server and customer terminal.

value between server and terminal. δ_1 stands for the trace from customer terminal to server while δ_2 stands for the trance from server to customer terminal. There are two formulas that could be introduced:

$$T_2 - T_1 = \theta + \delta_1 \tag{1}$$

$$T_4 - T_3 = \delta_2 - \theta \tag{2}$$

3 TRADITIONAL NTP SYNCHRONIZATION SYSTEM

The traditional NTP clock synchronization calculation method is to assume that the path delay from customer terminal to server equals the path delay from server to customer terminal. $\delta_1 = \delta_2 = \delta/2$

$$T_2 - T_1 = \theta + \delta/2 \tag{3}$$

$$T_4 - T_3 = \delta/2 - \theta \tag{4}$$

Clock time deviation between server and customer terminal:

$$\theta = \frac{(T_2 - T_1) \ (T_4 - T_3)}{2} \tag{5}$$

Total network transfer delay between customer terminal and server:

$$\delta = (T_2 - T_1) - (T_4 - T_3) \tag{6}$$

In customer server, we use θ to adjust clock time deviation between customer terminal and server.

In traditional NTP, the calculation of clock time deviation is to assume $\delta_1 = \delta_2$, but under the current network structure, messages are exchanged by wire or wireless network. In this network, the upper and downside message transfer paths are not in symmetrical structure. In addition, the randomness of message transfer is high. Those make very large differences to commuter path. The traditional NTP clock deviation could not solve this problem. Therefore, we raise NTP-MRTTD calculation method to reduce errors.

4 NTP-MRTTD TIME SYNCHRONIZATION SYSTEM

NTP-MRTTD (network time protocol-minimum round trip time difference) takes the path symmetrical character into account and based on (1) and (2), we get:

$$\delta_2 - \delta_1 = (T_4 - T_3) - (T_2 - T_1) + 2\theta \tag{7}$$

Formula (7) stands for the message transfer deviation in round trip. As the time deviation changes between server and customer terminal during consecutive synchronization, it could be ignored, and we can consider the θ of (7) as a constant value. During consecutive synchronization, we can get different $(\delta_2 - \delta_1)$

$$\begin{pmatrix} \delta_{21} - \delta_{11} \\ \delta_{22} - \delta_{12} \\ \vdots \\ \delta_{2n} - \delta_{1n} \\ \vdots \\ \delta_{2N} - \delta_{1N} \end{pmatrix} = \begin{pmatrix} 1 \\ -1 \\ -1 \\ 1 \end{pmatrix}^T \begin{pmatrix} T_{11} & T_{21} & T_{31} & T_{41} \\ T_{12} & T_{22} & T_{32} & T_{42} \\ \vdots & \vdots & \vdots & \vdots \\ T_{1n}. & T_{2n} & T_{3n} & T_{4n} \\ \vdots & \vdots & \vdots & \vdots \\ T_{1N} & T_{2N} & T_{3N} & T_{4N} \end{pmatrix}^T + 2\theta \tag{8}$$

$n = 1,2,3,\ldots N$. In traditional NTP calculation, $\delta_{2n} - \delta_{1n} = 0$, now we can get θ by finding the minimal value among $(\delta_{2n} - \delta_{1n})$.

$$\min\begin{pmatrix} \delta_{21} - \delta_{11} \\ \delta_{22} - \delta_{12} \\ \vdots \\ \delta_{2n} - \delta_{1n} \\ \vdots \\ \delta_{2N} - \delta_{1N} \end{pmatrix} = \min\begin{pmatrix} \begin{pmatrix} 1 \\ -1 \\ -1 \\ 1 \end{pmatrix}^T \begin{pmatrix} T_{11} & T_{21} & T_{31} & T_{41} \\ T_{12} & T_{22} & T_{32} & T_{42} \\ \vdots & \vdots & \vdots & \vdots \\ T_{1n}. & T_{2n} & T_{3n} & T_{4n} \\ \vdots & \vdots & \vdots & \vdots \\ T_{1N} & T_{2N} & T_{3N} & T_{4N} \end{pmatrix}^T \end{pmatrix} + 2\theta \tag{9}$$

θ in (9) is a constant value, and we can get minimal value in $\min[(1,-1,-1,1)(T_{1n}, T_{2n}, T_{3n}, T_{4n})^T]$ and $n = 1,2,3,\ldots,N$. If we get multi-group of data, we can choose the minimal value of $(\delta_{2n} + \delta_{1n}$, i.e., the minimal delay during transferring.

$$\min\begin{pmatrix} \delta_{21} + \delta_{11} \\ \delta_{22} + \delta_{12} \\ \vdots \\ \delta_{2n} + \delta_{1n} \\ \vdots \\ \delta_{2N} + \delta_{1N} \end{pmatrix} = \min\begin{pmatrix} \begin{pmatrix} -1 \\ 1 \\ -1 \\ 1 \end{pmatrix}^T \begin{pmatrix} T_{11} & T_{21} & T_{31} & T_{41} \\ T_{12} & T_{22} & T_{32} & T_{42} \\ \vdots & \vdots & \vdots & \vdots \\ T_{1n}. & T_{2n} & T_{3n} & T_{4n} \\ \vdots & \vdots & \vdots & \vdots \\ T_{1N} & T_{2N} & T_{3N} & T_{4N} \end{pmatrix}^T \end{pmatrix} \tag{10}$$

We can get θ through equation (5).

5 SIMULATION AND ANALYSIS

Build up synchronization structure between "master station and carrier". Master station holds

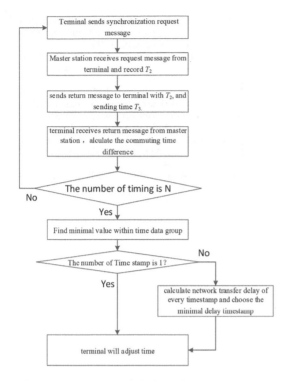

Figure 2. Flow chart of clock synchronization.

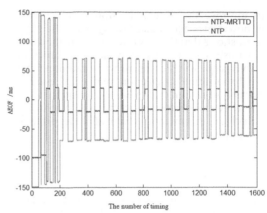

Figure 3. Synchronized error distribution map.

time server, exchanger, router, and firewall, the acquisition terminal realize message exchange with master station through GPRS. The flow shows as below:

1. Terminal sends synchronization request message that accords with NTP to master station based on synchronizing cycle.
2. Master station receives request message from terminal and record T2, then sends return message to terminal with T2, and sending time T3.
3. When terminal receives return message from master station it will record its receiving time T4 and it will analyze T2 and T3 according to NTP and calculate the commuting time difference based on formula (7).
4. Repeat the three points above for N times, and it will get a group data of commuting time differences.
5. Find minimal value within data group and record the value as timestamp (T1, T2, T3, and T4).
6. If there is only one timestamp, then we can calculate time deviation based on formula (5); if there are many timestamp existing, then it will need formula (10) to calculate network transfer delay of every timestamp and choose the minimal delay timestamp.
7. Acquisition terminal will adjust time simultaneously based on the time comes from timestamp.

8. Acquisition terminal will record time from power meter and compare the data with itself. If the result is beyond (more than) the threshold value, the acquisition terminal will form an abnormal instance and send such instance into master station.
9. Master station will perform clock synchronization by using NTP-MRTTD method to the power clock of abnormal instance. Synchronization method is the same as the way of master station to terminal; the operation of master station to power meter will through acquisition terminal, acquisition terminal will take transparent transfer method not to perform any handlings to clock synchronization message.

In one experience about commuting path, we will perform 1600 times test and set N = 8 to calculate average error of offset.

$$AFOF = \frac{\sum_{i=1}^{M}(T_i' - T_i'')}{M} \qquad (11)$$

In this formula, M is the synchronization time, value is 1~1600, T_i' is the customer terminal time of ith synchronization time during total M times. T_i'' is the server time of ith Synchronization time during total M times.

As Fig.3 shows, the synchronization accuracy of NTP-MRTTD is better than that of NTP. The accuracy is raised by 1 time approximately. In terms of convergence, NTP-MRTTD is faster than that of NTP. It began its convergence status when reaching 100 times, while the NTP would began its convergence status when reaching 200 times.

6 CONCLUSION

This paper raised a clock synchronization calculation method based on the NTP-MRTTD theory to improve the synchronization accuracy errors in power message acquisition system. The experiment shows that the new calculation method could obviously rise the synchronization accuracy and convergence levels compare with traditional NTP synchronization calculation method. Under the current situation that power acquisition system has more and stricter requirements to synchronization clock, this method is valuable to be promoted and applied.

REFERENCES

2008. IEEE Standard for a Precision Clock Synchronization Protocol for Networked Measurement and Control Systems. IEEE 3 Park Avenue.

Gu, G.D., Y. Zhou, L.J. Qian. 2013. Research on the Improvement of the Data Acquisition Success Rate of Power-consuming Information Acquisition System. Jiangsu Electrical Engineering 32(6): 29–31.

Liang, Z.G. 2009. Research and realization of clock synchronization technology for power system synchronization measurement. Chengdu UESTC.

Mills, D.L. 1991. Internet time synchronization: the network time protocol. IEEE Transactions on Communications 39(10): 1482–1493.

Mills, D.L. 1998. Adaptive hybrid clock discipline algorithm for the network time protocol networking[J]. IEEE ACM Transactions on Networking 6(5): 505–514.

Qian, L.J., X.J. Li. 2013. Power Usage Information Acquisition Data Accuracy Automatic Confirmation Technology and Its Application. Jiangsu Electrical Engineering 32(2): 64–65.

Ren, G.C., D.X. Wang. 2009. Reliability design and research. machine design and reform of remote meter reading system 12:46–48.

Energy Science and Applied Technology ESAT 2016 – Fang (Ed.)
© 2016 Taylor & Francis Group, London, ISBN 978-1-138-02973-6

Study on the fusion of index weight based on AHP and entropy

Xitong Sun, Qiusheng Liu & Hao Chen
Department of Ammunition Engineering, Ordnance Engineering College, Hebei, Shijiazhuang, China

ABSTRACT: Considering the problem of being unreasonable in quality assessment of electromagnetic induction setter, a new method of index weight fusion based on DSmT was proposed. First, the evaluation index system of object was established and the tested data of catenary system must be dealt with. Then, the method of information entropy was used to determine the index discrimination and the method of AHP was used to determine the index importance. Index discrimination and importance were as the basic belief assignment. DSmT theory was used to fuse belief assignment from different conflict sources to get more reasonable index weight. The practical example shows that the proposed method can overcome the fusion problem of the high conflict evidence. Fusion weight are given characteristics of importance and discrimination, so it has great value in engineering application.

Keywords: weight, quality assessment, induction setting, DSmT

1 INTRODUCTION

Quality assessment of weapon electronic system is process that is how to select index, divide weight and given result. Wherein, index weighting is a crucial work. Its theory and methods have been widely used in quality assessment process of ECG (Abdou et al, 2006), voice, electronics and aerospace. Index weighting is also the focus of quality assessment in various fields. Evaluation indexes' essential attribute determines the importance in quality assessment process (Cao F et al, 2014) (Deng J L, 2002). A reasonable degree of index empowerment directly affects the accuracy of the quality assessment process. Weight determination has been the focus of the study in field of decision-making experts (Dezert J. 2002) (Dezert J. 2003).

In modern quality evaluation, the index weights have been a difficult problem to solve (Gong Z W. 2006). More commonly used methods of determining the index weights contain principal component analysis, gray correlation analysis, rough set theory, cluster analysis, entropy method, analytic hierarchy process (Qiu W H et al, 2015) (Smarandache F, 2002). Regardless of the method, it is divided into two categories. One is to determine the subjective weights based on experts, another is to determine the objective weights based on source data (Yu X H et al, 2013). Even the subjective and objective methods are used to determine the weights, it is also the artificial leading. Mechanical allocation of subjective and objective weights is performed. These methods are a mechanical distribution of subjective and objective weights.

Considering the problem of subjectively or objectively serious bias in the existing method of determining index weights and highly conflict between different methods, a new method of index weight fusion based on DSmT (Zou Z H et al, 2006) was proposed. Data availability based entropy method and professional knowledge based expert judgment integrate efficiently. The index weights were considered highly conflicting evidence sources. According to reliability proportional relationship of source of evidence, the conflict reliability was reallocated and combined to build index weight. The practical example verified the feasibility and effectiveness of this method.

2 INDEX WEIGHT FUSION MODEL

The source of data is preprocessed then we calculate entropy value to obtain the relative size of the useful information between indexes. This value is used as index discrimination based on information entropy. Advantage of expert knowledge is fully used. We were able to get importance of index weight based on the analytic hierarchy process. This method usually obtained unanimous approval in the relevant field. There are different focusing in two sources. Generalized basic belief assignments are synthesized by DSmT fusion rule to get optimal index weights.

2.1 DSmT fusion theory

In the early 20th century, Florentin Smarandache in United States and Jean Dezert in France put forward the DSmT theory. DSmT is extension of the classical theory of DST. Two authors put paradoxical and contradictory reasoning methods in the field of information fusion. In order to make the reader clearly understand the DSmT theory, this paper presents the basic definition of DSmT theory by literature (Dezert J et al, 2004).

Definition 1 Hyper-power set D^Θ concept

In the framework of DSmT, Θ is a collection that contains n complete proposition $\{\theta_1,\cdots,\theta_n\}$ (also called a frame). Dedekind lattice model is called hyper-power set in DSmT frame. Symbol D^Θ indicates hyper-power set. It is a collection of proposition in Θ and all compound proposition by $\cup$ operation and $\cap$ operation. Its form is as follows.

1. $\varnothing, \theta_1, \cdots, \theta_n \in D^\Theta$;
2. If $A, B \in D^\Theta$, then $A \cap B \in D^\Theta, A \cup B \in D^\Theta$;
3. In addition to (1) and (2), no other propositions belong to K.

Definition 2 Generalized basic belief assignment

In a broad framework Θ, authors define a set of maps $m(\cdot): D^\Theta \to [0,1]$. It is associated with source of evidence. Satisfying the following formula:

$$m(\varnothing) = 0, \sum_{A \in D^\Theta} m(A) = 1 \qquad (1)$$

Wherein $m(A)$—Generalized basic belief assignment of A.

Definition 3 DSmT fusion rule

DSmT fusion rule contains DSmT classic rules, DSm hybrid rule and proportional conflict redistribution rule. In proportional conflict redistribution rule, it includes five PCR rule—PCR1 ~ PCR5.

When DSm model $\mu^f(\Theta)$ solves the integration problem, classic DSm combination rules of independent source of evidence S_1 and S_2 is $m_{\mu^f(\Theta)}(\cdot) \equiv m(\cdot) \triangleq [m_1 \oplus m_2](\cdot)$. Their generalized basic belief assignments are $m_1(\cdot)$ and $m_2(\cdot)$. This can be defined as follows:

$$\forall C \in D^\Theta, m_{\mu^f}(\Theta)(C) \equiv m(C) = \sum_{\substack{A,B \in D^\Theta \\ A \cap B = C}} m_1(A)m_2(B)$$

Hyper-power set D^Θ are closed under $\cup$ and $\cap$ operation. Its number increases sharply with the increase of Θ's potential. Hyper-power set D^Θ does not need to be classified finely in index weights fusion. Moreover, PCR5 is recognized as the highest precision in the distribution. Therefore, the paper chooses PCR5 as a principle of data fusion. This program is reasonable and accurate. It is defined as follows:

$$m_{PCR5}(A) = \sum_{\substack{X_1,X_2 \in D^\Theta \\ (X_1 \cap X_2)=A}} m_1(X_1)m_2(X_2) +$$
$$\sum_{\substack{X_1 \in D^\Theta \\ X_1 \cap A = \varnothing}} \left[\frac{m_1^2(A)m_2(X_1)}{m_1(A) + m_2(X_1)} + \frac{m_2^2(A)m_1(X_1)}{m_2(A) + m_1(X_1)} \right] \qquad (2)$$

X_1 of evidence 1 and X_2 of evidence 2 are in conflict. X_2 of evidence 1 and X_1 of evidence 2 are in conflict too. They are two types of conflict between the evidence. PCR5 distributes conflicting reliability to a combined reliability of X_1 and X_2 according to the proportion of X_1 and X_2's reliability.

2.2 Optimization of index discrimination based on entropy

The concept of entropy was originally proposed in thermodynamics. It is a physical quantity to show status of material system. It is an abstract physical quantity, which is used to illustrate irreversibility of the thermal process. We call it as information content in information theory. Entropy is a measure of uncertainty. The larger the entropy is, the more information the data can provide. Entropy can be expressed as follows:

$$H(p_1, p_2, \ldots, p_n) = -k \sum_{i=1}^{n} [p_i \ln(p_i)] \qquad (3)$$

where p_i is the probability.

In the evaluation of the setter, useful information can be used to determine how much weights are between various index. The larger the entropy, the worse the index directivity, and the lower the index availability. Therefore, the order degree of system information based on entropy is used to determine the index weight. In the process of calculating index weight, it can eliminate human disturbance.

In the problem of n's program and m's evaluation index, the original data matrix is shown below.

$$X = \begin{bmatrix} x_{11} & x_{12} & \cdots & x_{1m} \\ x_{21} & x_{22} & \cdots & x_{2m} \\ \vdots & \vdots & \ddots & \vdots \\ x_{n1} & x_{n2} & \cdots & x_{nm} \end{bmatrix} \qquad (4)$$

where x_{ij} – the j's evaluation of the i's program.

In order to eliminate the impact of indicators' dimension, the original data were normalized. The following is the normalization method in the paper.

The smaller the index, the better it is.

$$r_{ij} = \left| \frac{x_{ij} - x_{max}}{x_{max} - x_{min}} \right| \qquad (5)$$

Tolerance type.

$$r_{ij} = 1 - \left| \frac{x_{ij} - x_{mav}}{x_{max} - x_{min}} \right| \qquad (6)$$

where r_{ij} is the j's index normalized value of the i's program, $r_{ij} \in [0,1]$;

x_{max} is the upper limit or alert value of indicators;

x_{min} is the lower limit or alert value of indicators; and

$x_{m\,av}$ is the optimal value of indicators.

The original data are normalized to obtain matrix. It is $R = (r_{ij})n \times m$.

$$R = \begin{bmatrix} r_{11} & r_{12} & \cdots & r_{1m} \\ r_{21} & r_{22} & \cdots & r_{2m} \\ \vdots & \vdots & \ddots & \vdots \\ r_{n1} & r_{n2} & \cdots & r_{nm} \end{bmatrix} \qquad (7)$$

In the assessment, entropy of the j's evaluation is defined as follows by formula (3).

$$H_j = -k \sum_{i=1}^{n} f_{ij} \ln f_{ij}, j = 1,2,3,\cdots,m \qquad (8)$$

Wherein

$$f_{ij} = \frac{r_{ij}}{\sum_{j=1}^{m} r_{ij}}, k = \frac{1}{\ln n} \qquad (9)$$

If $f_{ij} = 0$, then $f_{ij} \ln f_{ij} = 0$.

After getting entropy of each index, we can calculate the index entropy weight value based on entropy. The first j's index entropy weight value is defined as follows:

$$\omega_j = \frac{1 - H_j}{\sum_{j=1}^{m}(1 - H_j)}, 0 \le \omega_j \le 1, \sum_{j=1}^{m} \omega_j = 1 \qquad (10)$$

2.3 Optimization of index importance based on analytic hierarchy process

In the 1970s, American operations research expert Saaty proposed an analytic hierarchy process. It is a multi-objective decision-making method. As the number of inductive setting index is more and it is difficult to distinguish, 0.5–0.9 scale method was adopted. Meaning of 0.5–0.9 scale is shown in Table 1.

According to Table 1-1, one compared with another in m's evaluation index to obtain judgment matrix $A = (a_{ij})_{m \times m}$.

Table 1. Meaning of judgement matrix 0.5–0.9 scale method.

Scale	Meaning
0.5	A compares with B, they are equally important
0.6	A compares with B, A is slightly more important than B
0.7	A compares with B, A is significantly more important than B
0.8	A compares with B, A is strongly more important than B
0.9	A compares with B, A is extremely more important than B
Complementary to the above scale	Conversely above meaning

$$A = \begin{bmatrix} a_{11} & a_{12} & \cdots & a_{1m} \\ a_{21} & a_{22} & \cdots & a_{2m} \\ \vdots & \vdots & \ddots & \vdots \\ a_{m1} & a_{m2} & \cdots & a_{mm} \end{bmatrix} \qquad (11)$$

where a_{ij} is the scale value which the i's index compares with the j's index.

$$AW = \lambda_{max} W \qquad (12)$$

According to the judgment matrix A, we calculated the largest eigenvalue λ_{max} and its corresponding eigenvector $W = (\omega_1, \omega_2, \cdots, \omega_m)^T$.

Eigenvectors that were obtained from the formula were as the evaluation index weight vector.

Consistency test of judgment matrix used K-factor test method. It was proposed by Herman in 1996. This approach eliminates the influence of random factors and is simple and effective. Calculate factor K was defined as follows:

$$K(A) = \frac{1}{2}\left[1 - \lambda_{max}(A) + \sqrt{(\lambda_{max}(A) - 1)^2 + 4m} \right] \qquad (13)$$

If $K(A) \ge K_0$, then $A = (a_{ij})_{m \times m}$ is consistent.

3 CASE ANALYSIS

3.1 Evaluation index system

Induction setter quality assessment model, which was constructed in this paper was used to determine the index weight. According to the characteristics of the system, signal characteristics and internal power characteristics were as the evaluation system. Then signal characteristics and power characteristics are subdivided into the specific parameters. It is shown in Figure 1.

Table 2. Normalized data of induction setter quality evaluation index.

	Signal characteristics						Power features
	u_{11}	u_{12}	u_{13}	u_{14}	u_{15}	u_{16}	u_{21}
1	0.8325	0.6421	0.9123	0.5048	0.2492	0.8866	0.7367
2	0.9188	0.9933	0.7963	0.5624	0.4756	0.9716	0.9567
3	0.7663	0.9398	0.8610	0.8044	0.6716	0.7302	0.6767
4	0.5300	0.6988	0.9407	0.8496	0.8844	0.5589	0.6067

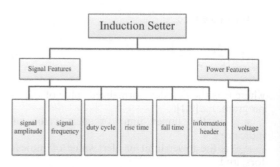

Figure 1. Induction setting quality evaluation system.

It is seen from the figure. Induction Setter quality evaluation factor set was as follows. $U = \{u_1, u_2\}$ = {signal characteristics, power characteristics }, this is the first assessment factors set. Each element contains n indicators. Mark as $u_i = \{u_{i1}, u_{i2}, \cdots, u_{in}\}$. This is the second assessment factor set.

3.2 Standardization of original data

In order to facilitate the integration of index and ensure the viability of quality assessment, original data need to be normalized by formulas (5) and (6). Data which were normalized are shown in the following Table 2.

3.3 Weight determination

From the perspective of induction setter comprehensive evaluation, signal characteristics and power characteristics were used commonly to determine the weight of each index. First, each index data was normalized to give information matrix F.

$$F = \begin{bmatrix} 0.1747 & 0.1348 & 0.1915 & 0.1060 & 0.0523 \\ 0.1861 & 0.1546 & & & \\ 0.1619 & 0.1750 & 0.1403 & 0.0991 & 0.0838 \\ 0.1712 & 0.1686 & & & \\ 0.1406 & 0.1724 & 0.1580 & 0.1476 & 0.1232 \\ 0.1340 & 0.1242 & & & \\ 0.1046 & 0.1379 & 0.1856 & 0.1676 & 0.1745 \\ 0.1103 & 0.1197 & & & \end{bmatrix}$$

We can get index entropy which was calculated by the formula (8), (9).

$$H_j = \begin{bmatrix} 0.8018 & 0.8306 & 0.8628 & 0.7565 & 0.6670 \\ 0.8133 & 0.7949 & & & \end{bmatrix}$$

According to equation (10), we obtained index entropy weight.

$$\omega_j = \begin{bmatrix} 0.1345 & 0.1150 & 0.0931 & 0.1656 & 0.2261 \\ 0.1267 & 0.1390 & & & \end{bmatrix}$$

Experts compared the importance of evaluation index to judgment matrix as shown in the formula (11).

$$A = \begin{bmatrix} 0.5 & 0.8 & 0.7 & 0.7 & 0.8 & 0.6 & 0.4 \\ 0.2 & 0.5 & 0.3 & 0.3 & 0.4 & 0.2 & 0.1 \\ 0.3 & 0.7 & 0.5 & 0.6 & 0.7 & 0.4 & 0.3 \\ 0.3 & 0.7 & 0.4 & 0.5 & 0.6 & 0.3 & 0.2 \\ 0.2 & 0.6 & 0.3 & 0.4 & 0.5 & 0.3 & 0.2 \\ 0.4 & 0.8 & 0.6 & 0.7 & 0.7 & 0.5 & 0.3 \\ 0.6 & 0.9 & 0.7 & 0.8 & 0.8 & 0.7 & 0.5 \end{bmatrix}$$

From the judgment matrix, we can draw that the important level of index is $u_{21} > u_{11} > u_{16} > u_{13} > u_{14} > u_{15} > u_{12}$.

According to the judgment matrix, the maximum eigenvalue was calculated by the formula (12). It is $\lambda_{max} = 3.1867$. Feature vector is $W = (0.4754, 0.1964, 0.3571, 0.2990, 0.2495, 0.4117, 0.5356)^T$. In the consistency check, $K(A)$ is 1.769. For the determination threshold value K_0, experts generally set 0.9. Because $K(A) \geq 0.9$, so the results meet the consistency requirements.

Feature vector W was normalized to obtain index weight.

$$\omega_j = \begin{pmatrix} 0.1883, 0.0778, 0.1414, 0.1184, 0.0988, \\ 0.1631, 0.2122 \end{pmatrix}^T$$

Two types of index weights are generalized basic belief assignment by the formula (1).

Induction setter index weights were fused by equation (2). The optimal weights were as follows:

$$\omega = \begin{pmatrix} 0.1688, 0.0856, 0.1049, 0.1394, 0.1663, \\ 0.1435, 0.1915 \end{pmatrix}^T$$

Index discrimination by the information entropy, index importance by experts and fusion weights by DSmT fusion rule were plotted in the following Figure 2.

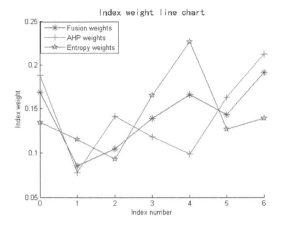

Figure 2. Line chart of induction setting system index weight.

As can be seen from the line graph, there is a high degree of conflict between index distinguish and the importance of index. They reflected the characteristics of index from their point of view. According to the ratio of source of evidence original reliability, conflict reliabilities were assigned to the combination of reliability by DSmT. There were the characteristics of expert judgment and index entropy in optimal weight. Fusion weights were more scientific and reasonable.

4 CONCLUSION

DSmT theory is a reasonable fusion method based on source of conflict evidence. Information entropy was as the index discrimination. Expert empowerment was as the importance of index. Weights are determined from the subjective and objective point of view. The weight is more scientific and reasonable. The demonstration showed that the method of determining the weights broken down the tendency of a single method and broken through inherent conflict of subjective and objective weights. Finally, we got both subjective and objective properties of the fusion weights. This is an effective conflict data fusion method.

REFERENCES

Abdou, Samir, Savoy, Jacques. Statistical and comparative evaluation of various indexing and search models [J]. Lecture Note in Computer Science, 2006:362–373.

Cao F, et al. Quality assessment weapons electronic systems [M]. Xidian University Press, 2014.1.

Deng J L. Grey theoretical basis [M]. Wuhan: Huazhong University of Science and Technology Press, 2002.

Dezert J., Foundations for a new theory of plausible and paradoxical reasoning, Inform. & Secur. J., Semerdjiev Ed., Bulg. Acad. of Scl., Vol.9, 2002.

Dezert J., Foundations pour une nouvelle theorie du raisonnement plausible et para-doxal, ONERA Tech. Rcp. RT 1/06769/DTIM, Jan.2003.

Dezert J, Smarandache F. Advances and Applications of DSmT for Information Fusion (collected works) [M]. America: University of Microfilm International, 2004.

Gong Z W. On the theories and methods of uncertainty Fuzzy judgment matrix[D].Nanjing: Nanjing University of Aeronautics and Astronautics, 2006,10–16.

Qiu W H, et al. Assessment method of health status based on DSmT and fuzzy comprehensive evaluation[J]. Computer Measurement & Control, 2015.23(11).

Smarandache F, (Editor), Proceedings of the First International Conference on Neutrosophics, Univ. of New Nexico Gallup Campus, NM, USA, 1–3 Dec.2001, Xiquan, Phoenix, 2002.

Yu X H, et al. Study on the adaptive integration algorithms with DST and DSmT[J]. Computer &. Digital Engineering, 2013,41(1):35–37.

Zou Z H, et al. Entropy method for determination of weight of evaluating indicators in fuzzy synthetic evaluation for water quality assessment[J]. Journal of Environmental Sciences, 2006,18(5):1020–1023.

Energy Science and Applied Technology ESAT 2016 – Fang (Ed.)
© 2016 Taylor & Francis Group, London, ISBN 978-1-138-02973-6

GPS/SINS integrated navigation system using an innovation-based adaptive Kalman Filter for land-vehicles

Shihao Zhu, Tongyue Gao & Yayang Cheng
School of Mechatronic Engineering and Automation, Shanghai University, Shanghai, China

ABSTRACT: Based on the complementary features of GPS and SINS, a low-cost, high precision and robust navigation system for land-vehicles can be designed by integrating them effectively. And Kalman Filter is often used to fusion the data of them. But, the adaptability of the traditional filtering algorithm is poor because its filtering parameters can't adjust with the change of GPS' measurement noise. So we designed an innovation-based adaptive Kalman Filter. It uses the innovation sequence to estimate the measurement noise covariance matrix of the system in real-time, and switches between IAE and EKF algorithm depend on the stability of GPS' signal. The test result showed that this method can improve the filtering accuracy and stability of the GPS/SINS integrated navigation system for land-vehicles effectively.

Keywords: GPS, SINS, integrated navigation, innovation sequence, AKF.

1 INTRODUCTION

Based on the complementary features of GPS (Global Position System) and SINS (Strapdown Inertial Navigation System) (L.R, 2011; A. Vydhyanathan, 2012), a low-cost, high precision and robust navigation system for land-vehicles can be designed by integrating them effectively. Nowadays, KF (Kalman Filter) is often used to fusion the data of GPS and SINS and realize this integrated navigation system. But, the filtering accuracy will be decline and diverge even when it is used in the environment in which the system's measurement noise is unstable (P. Davidson, 2009). In order to deal with the problem, many adaptive filtering algorithms have been proposed, including MMAE (Multi-Model Adaptive Estimation), IAE (Innovation-Based Adaptive Estimation) and so on. Among them, IAE can calculate and update the system noise matrix and measurement noise matrix in real-time during measuring (M. Narasimhappa, 2014). Thus, it can get better performance than the traditional filtering algorithm in high-precision and dynamic positioning applications. However, due to the limit of adaptive window length of IAE algorithm, it can't guarantee the system's accuracy and robust when the interference of the system is strong.

To solve the problem, we designed an innovation-based adaptive Kalman Filter in this paper. It uses the innovation sequence to estimate the measurement noise covariance matrix of the system in real-time, and switches between IAE and EKF algorithm depend on the stability of GPS' signal. On the basis of previous studies, we designed a MEMS sensors-based GPS/SINS integrated navigation system for land-vehicles. In the second part, the model of the system and the method to construct it will be introduced. In the third part, we will introduce the adaptive Kalman Filter proposed in this paper. And a test will be conducted to verify the navigation system designed in this paper at last.

2 THE MODEL OF GPS/SINS INTEGRATED NAVIGATION SYSTEM

We take the error of navigation parameters as the estimation objects and use the indirect method to filter. This method is convenient for the system to switch between integrated navigation mode and pure inertial navigation mode (M.S. Sheijani, 2013). Because every state parameters are error parameters and the magnitude of them is approximated, it will not affect the accuracy of the error's estimation and is more favorable for computer's numerical calculation. After the filtering process is completed, we use the estimated error to correct the original navigation parameters, and obtain the final navigation parameters. The model of the system is showed in Figure 1.

The method to construct it will be introduced in the next, including the selection of state vector

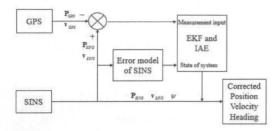

Figure 1. The model of GPS/SINS integrated navigation system.

and measurement vector, and the construction of system equation and measurement equation which describing the dynamic characteristics of the system and the relationship between the state vector and measurement vector respectively.

The State Vector and State Equation of the System. When we don't consider any errors, the ideal speed of SINS should be calculated by mechanics equation. But, there are a variety of errors in the actual system. After the installation angle error and the scale factor error of accelerometer are corrected, the real speed of SINS should be calculated by the equation below:

$$
\delta \dot{\mathbf{v}}_{en}^{n} = -\boldsymbol{\varphi}^{n} \times \mathbf{f}^{n} + \delta \mathbf{v}_{en}^{n} \times (2\boldsymbol{\omega}_{ie}^{n} + \boldsymbol{\omega}_{en}^{n}) \\ + \mathbf{v}_{en}^{n} \times (2\delta\boldsymbol{\omega}_{ie}^{n} + \delta\boldsymbol{\omega}_{en}^{n}) + \nabla^{i} \tag{1}
$$

The E-N-U geographic coordinate system is taken as the navigation system in this paper. In the equation above, superscript n indicates that the parameter is represented in the navigation system, and the same below. The attitude error matrix $(\boldsymbol{\varphi}^{n}\times)$ is a skew-symmetric matrix with elements defined as the elements of attitude error vector $\boldsymbol{\varphi}^{n}$.

After the installation angle error and the scale factor error of gyroscope is corrected, the attitude error equation of SINS is:

$$
\dot{\boldsymbol{\varphi}}^{n} = \boldsymbol{\varphi}^{n} \times (\boldsymbol{\omega}_{ie}^{n} + \boldsymbol{\omega}_{en}^{n}) + (\delta \hat{\boldsymbol{\omega}}_{ie}^{n} + \delta \hat{\boldsymbol{\omega}}_{en}^{n}) - \mathbf{C}_{b}^{n} \boldsymbol{\varepsilon}^{b} \tag{2}
$$

And the position error equation of SINS is:

$$
\begin{cases}
\delta \dot{L} = \dfrac{\delta v_{N}}{R_{M}+h} - \delta h \dfrac{v_{N}}{(R_{M}+h)^{2}} \\
\delta \dot{\lambda} = \dfrac{\delta v_{E}}{R_{N}+h}\sec L + \delta L \dfrac{v_{E}}{R_{N}+h}\tan L \sec L - \delta h \dfrac{v_{E}\sec L}{(R_{N}+h)^{2}} \\
\delta \dot{h} = \delta v_{U}
\end{cases} \tag{3}
$$

In the equation above, R_{M} and R_{N} are represent the curvature radius of the earth ellipsoid meridian and the earth prime vertical respectively. Take

the latitude error δL, longitude error $\delta \lambda$, height error δh; the velocity errors $\delta v_{E}, \delta v_{N}, \delta v_{U}$ along the three directions of the navigation system; the attitude angle errors of the digital platform $\varphi_{E}, \varphi_{N}, \varphi_{U}$; the bias of gyroscope along the three directions $\varepsilon_{x}, \varepsilon_{y}, \varepsilon_{z}$; the bias errors of accelerometer $\nabla_{x}, \nabla_{y}, \nabla_{z}$ to compose the state vector $\mathbf{X}_{SINS}$ of SINS:

$$
\mathbf{X}_{SINS} = [\delta L \quad \delta \lambda \quad \delta h \quad \delta v_{E} \quad \delta v_{N} \quad \delta v_{U} \quad \varphi_{E} \quad \varphi_{N} \quad \varphi_{U} \\ \varepsilon_{x} \quad \varepsilon_{y} \quad \varepsilon_{z} \quad \nabla_{x} \quad \nabla_{y} \quad \nabla_{z}]^{T}.
$$

Then the state equation of SINS is:

$$
\dot{\mathbf{X}}_{SINS} = \mathbf{F}_{SINS}\mathbf{X}_{SINS} + \mathbf{G}_{SINS}\mathbf{W}_{SINS} \tag{4}
$$

Take the rest term of SINS' error equations to compose matrix $\mathbf{F}_{N}$, then the state matrix is:

$$
\mathbf{F}_{SINS} = \begin{bmatrix} \mathbf{F}_{N} & \mathbf{F}_{S} \\ \mathbf{0}_{6\times9} & \mathbf{F}_{M} \end{bmatrix}_{15\times15} \tag{5}
$$

Where,

$$
\mathbf{F}_{S} = \begin{bmatrix} \mathbf{0}_{3\times3} & \mathbf{0}_{3\times3} \\ \mathbf{0}_{3\times3} & \mathbf{C}_{b}^{n} \\ -\mathbf{C}_{b}^{n} & \mathbf{0}_{3\times3} \end{bmatrix} \tag{6}
$$

The bias of gyroscope regarded as a first-order Markov process, then:

$$
\mathbf{F}_{M} = Diag\left[\dfrac{-1}{T_{rx}} \quad \dfrac{-1}{T_{ry}} \quad \dfrac{-1}{T_{rz}} \quad 0 \quad 0 \quad 0 \right]^{T} \tag{7}
$$

The Measurement Vector and Measurement Equation of the System. Take the position and velocity differences of SINS and GPS as the measurement vector, then we can get the measurement equation:

$$
\mathbf{Z}(t) = \begin{bmatrix} \mathbf{P}_{SINS} - \mathbf{P}_{GPS} \\ \mathbf{v}_{SINS} - \mathbf{v}_{GPS} \end{bmatrix} = \mathbf{H}(t)\mathbf{X}(t) + \mathbf{V}(t) \tag{8}
$$

Where $\mathbf{P}_{SINS}$ and $\mathbf{v}_{SINS}$ represent the position and velocity measurement value of SINS respectively, which including the error massages along the three directions of the navigation system; $\mathbf{P}_{GPS}$ and $\mathbf{v}_{GPS}$ represent the position and velocity measurement value of GPS respectively. The measurement output matrix $\mathbf{H}(t)$ is:

$$
\mathbf{H}(t) = \begin{bmatrix} Diag[R_{M} \quad R_{N}\cos L \quad 1] & \mathbf{0}_{3\times3} & \mathbf{0}_{6\times9} \\ \mathbf{0}_{3\times3} & Diag[1 \quad 1 \quad 1] & \end{bmatrix}_{6\times15} \tag{9}
$$

508

3 THE INNOVATION-BASED ADAPTIVE KALMAN FILTER

Innovation-Based Adaptive Kalman Filter. The filter model of the GPS/SINS integrated navigation system is introduced above, and the specific steps of this algorithm is:

1. One-step estimation of the system's state vector:

$$\delta\hat{\mathbf{X}}_{k/k-1} = \mathbf{\Phi}_{k,k-1}\delta\hat{\mathbf{X}}_{k-1} \qquad (10)$$

2. The calculation of mean square error of the one-step prediction:

$$\mathbf{P}_{k/k-1} = \mathbf{\Phi}_{k,k-1}\mathbf{P}_{k-1}\mathbf{\Phi}_{k,k-1}^{T} + \mathbf{Q}_{k-1} \qquad (11)$$

3. The calculation of adaptive matrix $\mathbf{R}_k$ based on the innovation sequence:

$$\mathbf{R}_k = \frac{1}{n}\sum_{i=0}^{n-1}\delta\mathbf{Z}_{k-i}\delta\mathbf{Z}_{k-i}^{T} - \mathbf{H}_k\mathbf{P}_{k/k-1}\mathbf{H}_k^{T} \qquad (12)$$

Where, $\delta\mathbf{Z}_{k-i}$ is the difference between the estimation value and the measurement value at each time the data updates, and is defined as the innovation sequence; $\sum_{i=0}^{n-1}\delta\mathbf{Z}_{k-i}\delta\mathbf{Z}_{k-i}^{T}$ is the sum of autocorrelation matrix of the innovation sequence which sample length is n.

4. Filtering gain updates:

$$\mathbf{K}_k = \mathbf{P}_{k/k-1}\mathbf{H}_k^{T}[\mathbf{H}_k\mathbf{P}_{k/k-1}\mathbf{H}_k^{T} + \mathbf{R}_k]^{-1} \qquad (13)$$

5. The estimation of state vector:

$$\delta\hat{\mathbf{X}}_k = \delta\hat{\mathbf{X}}_{k/k-1} + \mathbf{K}_k[\delta\mathbf{Z}_k - \mathbf{H}_k\delta\hat{\mathbf{X}}_{k/k-1}] \qquad (14)$$

6. The mean square error of the estimation:

$$\mathbf{P}_k = [\mathbf{I} - \mathbf{K}_k\mathbf{H}_k]\mathbf{P}_{k/k-1}[\mathbf{I} - \mathbf{K}_k\mathbf{H}_k]^{T} + \mathbf{K}_k\mathbf{R}_k\mathbf{K}_k^{T} \qquad (15)$$

When the measurement noise of GPS is small, the algorithm above can use the innovation sequence to estimate the measurement noise covariance matrix $\mathbf{R}_k$ effectively, and make the filtering parameters adjust with the change of GPS' measurement noise. So as to improve the filtering accuracy and suppress divergence. On the contrary, when the measurement noise of GPS is large, we remove the third step of the algorithm, and use the conventional EKF with the system saved and fixed measurement noise covariance matrix

to fusion the data of GPS and SINS. In order to avoid the filtering accuracy decline rapidly for the drastic fluctuations of $\mathbf{R}_k$. In the next, the stability detection of GPS signal will be introduced and the stability is the criterion for us to judge the GPS' measurement noise large or small.

The Stability Detection of GPS Signal. The stability is defined as the jump amplitude of GPS' output locus, and is obtained based on the statistical characteristics of the random error sequence of GPS signal in a certain time. The positioning accuracy of SINS is high in a short time, and the output locus of SINS in a short time can be seen as the ideal locus with no errors. So the difference of GPS' output locus and SINS' output locus in a short time can be seen as the error of GPS' output locus. Thereby, the position error sequence of GPS and the statistical characteristics of it in a short time can be obtained. In this way, we can realize the stability detection of GPS signal. The specific algorithm and steps are:

1. Record the position data of SINS and GPS respectively in a short time, and calculate the error sequence of GPS:

$$\mathbf{C}_m = \mathbf{I}_m - \mathbf{G}_m, \mathbf{I}_m = \begin{bmatrix} i_1 & i_2 & \cdots & i_m \end{bmatrix} \\ \mathbf{G}_m = \begin{bmatrix} g_1 & g_2 & \cdots & g_m \end{bmatrix}. \qquad (16)$$

The subscript m represents the number of position data.

2. Calculate the variance of the elements in $\mathbf{C}_m$:

$$\mathbf{C}_{var} = \frac{1}{m}\sum_{k=1}^{m}(\mathbf{C}_k - \mathbf{C})^2 \qquad (17)$$

3. When $\mathbf{C}_{var} > HT$, we think the measurement noise of GPS is large; And if $\mathbf{C}_{var} <= HT$, then we think the measurement noise of GPS is small. The threshold value HT is defined as 15 in this paper.

4 LAND-VEHICLE TESTS

The Test Equipment and Conditions. We use the hardware designed by ourselves to conduct the test. It is consist of a MEMS sensors-based IMU (Inertial Measurement Unit) module and a GPS module. The IMU module have three MEMS sensors, including a tri-axial accelerometer SCC1300 -D04, a tri-axial gyroscope L3G4200D and a tri-axial magnetometer HMC5883L. The antenna of GPS is mounted on the roof center of the vehicle and the SINS module is mounted on the middle point of the vehicle's chassis. The lever arm compensation between SINS and GPS has been treated properly.

The test is conducted in a viaduct environment. At first, the vehicle is travel on the straight viaduct with no shelter. Then, it drive into a curve road under the viaduct where the shelters are sparse. After drive out from the cure road, it entry another straight viaduct again where the shelters are intensive. At last, it drive along the same straight viaduct and into the environment with no shelter. The shelter mentioned in this paragraph is viaduct above the vehicle. The average speed of the vehicle is 40km/h on the straight road and 30km/h on the curve road.

The Test Result and Analysis. The traveling locus of the vehicle is showed in Figure 2. The red locus is generated by the GPS' data, and the blue locus is generated by the GPS/SINS integrated navigation system's data. And the length unit in this figure is meter.

At first, there is no shelter and the visibility of GPS satellite is good. At the moment, the measurement noise of GPS is small. As we can see in this figure, the two locus (the decumbent line in the figure) are overlapped and are consistent with the actual path. In the end, particularly, the measurement noise of GPS is small, and the two locus are overlapped almost and are consistent with the actual path too. This result illustrated that in the middle section in which the GPS signal is unstable and lost completely even, the integrated navigation system designed in this paper still can calculate the locus accurately. And in the middle section, the measurement noise of GPS is large. In summary, the innovation-based adaptive Kalman Filter designed in this paper can suppress divergence when the measurement noise of GPS is small and improve the filtering accuracy and stability when the measurement noise of GPS is large.

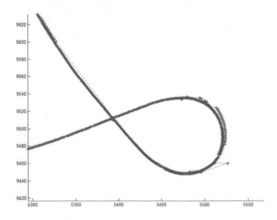

Figure 2. The traveling locus of vehicle in viaduct environment.

5 SUMMARY

Based on the complementary features of GPS and SINS, we designed a GPS/SINS integrated navigation system for land-vehicles. And using the innovation-based adaptive Kalman Filter designed in this paper to fusion the data of them. It uses the innovation sequence to estimate the measurement noise covariance matrix of the system in real-time, and switches between IAE and EKF algorithm depend on the stability of GPS' signal. The test result showed that this method can improve the filtering accuracy and stability of the system effectively.

REFERENCES

Davidson, P., J. Hautamäki, J. Collin and J. Takala, Improved vehicle positioning in urban environment through integration of GPS and low-cost inertial sensors, Proceedings of the European Navigation Conference, 2009.

Narasimhappa, M., S.L. Sabat, R. Peesapati and J. Nayak, An innovation based random weighting estimation mechanism for denoising fiber optic gyro drift signal, J. Optik, 2014, 125 (3) 1192–1198.

Sahawneh, L.R., M.A. al Jarrah, K. Assaleh and M.F. abdel Hafez, Real-time implementation of GPS aided low-cost strapdown inertial navigation system, Journal of Intelligent & Robotic Systems, 2011, 61 (1–4) 527–544.

Sheijani, M.S., A. Gholami, N. Davari and M. Emami, Implementation and Performance Comparison of Indirect Kalman Filtering Approaches for AUV Integrated Navigation System using Low Cost IMU, 21st Iranian Conference on Electrical Engineering, 2013, pp.1–6.

Vydhyanathan, A., H. Luinge, M.U. de Haag, and M. Braasch, Integrating GPS/MEMS-based-IMU with Single GPS Baseline for Improved Heading Performance, 2012 IEEE Aerospace Conference, 2012, pp.1–10.

Yuan, Q.Y. Inertial Navigation, second ed., Science Press, Bei Jing, 2014.

Energy Science and Applied Technology ESAT 2016 – Fang (Ed.)
© 2016 Taylor & Francis Group, London, ISBN 978-1-138-02973-6

Review on data visualization of the equipment support concept comprehensive evaluating system

YuQi Chen & Liu Zhang

Mechanical Engineering College, Shijiazhuang, Hebei, P.R. China

ABSTRACT: In order to use the Equipment Support Concept Comprehensive Evaluating System (ESCCES) well to evaluate equipment support concept, it is needed to study the efficient visualization of all kinds of data information in ESCCES, which are important evaluation bases. Specifically sorted out and summarized from the present situation of the research on data visualization, equipment support information visualization and relevant system data visualization laid the foundation for the next step in the research on visualization of ESCCES data information.

Keywords: equipment support concept, comprehensive evaluating system, data visualization

1 INTRODUCTION

The modern war is the comprehensive utilization of all kinds of weapons and equipment, which are based on the information system; it is the confrontation between systems of weapons and equipment. Among them, the quality of equipment support system operation is a key factor to ensure the effectiveness of weapon system playing. Equipment support concept is a whole description of equipment support system, and the pros and cons of its formulation are directly related to equipment support capability playing. Therefore, the focus of the work of equipment support at present is how to evaluate the equipment support concept scientifically and reasonably to form a quick, efficient, and accurate equipment integrated support capability.

Developed on the basis of previous research projects, Equipment Support Concept Comprehensive Evaluating System (ESCCES) is an intelligent platform including interaction, distribution, and coordination, which uses a mode of people-machine combination and group discussion, and adopts the method of combining quality and quantity, and maximize to integrate the existing equipment support program simulation evaluation tool, its input and output data information, mathematical evaluation model, and the expert resources, and provides some functions like information inquiry, model analysis, visualization, decision support, and so on. Large quantity of data information in ESCCES is an important basis on equipment support concept comprehensive evaluation. Current research focus on improving and optimizing ESCCES is how to make a scientific

and reasonable visualization of these data information to help experts evaluate the concept.

Data visualization is a technique of using the perception of the human eye to enhance cognition of visual expression of data interaction (Chen Xinggang, 2006) that can be used to express clearly, communicate and analyze data. Modern data visualization technology use computer graphics, image processing, human-computer interaction technology, and transform collected or simulated data into identifiable graphical symbols, images, videos, or animations to present valuable information to the users (Chen Wei. 2013). Therefore, in order to assist the evaluation personnel to accurately evaluate the equipment support concept, it is necessary to take reasonable data visualization processing method to the data information. This article will analyze the present situation of domestic and foreign research from three aspects: data visualization technology, equipment support information visualization, and relevant system data visualization.

2 DATA VISUALIZATION RESEARCH

2.1 *Foreign research status*

The origin of data visualization can be traced back to the early 1950s of computer graphics, with the popularization of personal computers, and people gradually began to use computer programming to design graphics to generate visualization. In October 1986, the National Science Foundation organized a workshop entitled "graphics, image processing and workstation workshop", applied computer graphics and image method into the subject of computational science to become "the visualization of scientific computing". In February 1987, the National Science Foundation held its first meeting of scientific visualization, the report of the meeting formally named and defined the scientific visualization (Chen Wei. 2013). Scientific visualization mainly presents and analyzes the scientific data and model of natural science, and its expression is usually in the three-dimensional and two-dimensional space, or includes the time dimension, roughly divided into the visualization of scalar field, vector field and tensor field. Card, MacKinlay and Robertson (Cai Zhuhua. 2014) in 1989 named information visualization, correspondingly extended and sublimated scientific visualization. Information visualization pay more attention to change some variables to improve the cognitive ability such as position, shape, orientation, color, texture, intensity, and size (the reference model is shown in Figure 1), and specifically deals with digital, non-geometrical, abstract data, such as visualization of financial transactions, social networks, text data which are multi-dimensional, time-varying, unstructured information including temporal and spatial data, hierarchical and network-structured data, text and cross-media data, and multivariate data.

Data visualization is a higher level technology in visualization, which was put forward following the scientific visualization and information visualization. It was produced and developed in the

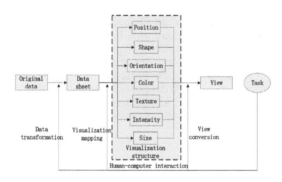

Figure 1. The reference model of information visualization.

Figure 2. Chart of the relationship between the three aspects.

pre-research on visualization, which included the scientific visualization and extended the research on information visualization to the field of knowledge visualization. The relationship between them is shown in Figure 2.

Up to now, the development of data visualization has improved in application and function. Many well-known universities in developed countries, national laboratories, and large companies research study and design some visualization software with actual application, such as medical data software VolView, 3D slicer, OsiriX, Amira, meteorological data processing software GrADS, Vis5D, and scientific engineering data-processing software OpenDX, AVS, IDL, and geographic data processing software World Wind, Google Earth, which are typically oriented to scientific visualization. Image-processing software includes GraphViz, Gephi, and CiteSpace, and high-dimensional multivariate data-processing software includes Xmdv Tool, infoScope, and ParSets, and text data-processing software includes Jigsaw and IN-SPIRE, which are typically oriented to information visualization. Business intelligence data-processing software includes Tableau, Spotfire, Splunk, Loggly, and Weave.

2.2 Domestic research status

At the beginning of the 1990s, Chinese scholars began to study scientific visualization, such as Zesheng Tang and his students at Tsinghua University, which is one of the first team to carry out scientific visualization research in China. In recent years, some colleges and research institutes in our country have made great progress in the research of visualization technology. Zhejiang University first studied on the concept of data visualization under the background of the era of big data (Chen Changyi, 2002). Based on this, they innovated semantic understanding visualization of high = dimensional variable data and medical image data visualization (Dai Zhongguo, 2013). Beijing University of Posts and Telecommunications, according to the characteristics and process of different data visualization tasks, developed a multidimensional data visualization tool based on Web, studied and realized part of the data graphics algorithm, and improved the ability of data analysis through scientific classification of data information. The National Institute of Metrology reasonably classified measurement data from the perspective of data visualization, established the measurement data visualization system with three characteristics of stereoscopic effect, immersion effect, and interactive function according to the basic process of data visualization, and realized the effective integration of measurement science and data visualization. In order to strengthen the link between visualization and data mining, Xiamen University established the visual model of hierarchical clustering, SOM clustering, and K-Means clustering algorithm based on MASI distance, realized the effective integration of visualization technology and data mining through the concrete application. East China University of Science and Technology, based on the concept and process of user experience, studied data visualization design for user satisfaction experience from the aspects of interaction design and visual design. Hebei University did a certain research on the visualization of the news data through the study of data visualization technology in the era of big data. National University of Defense Technology, aimed at defects of traditional multi variable network visualization method, proposed a multivariable network visualization framework MulNetVis and designed a visualization method for multivariable network. Shandong University established the data management system of scientific research realization in universities and presented the data scientifically by using a data visualization method. In addition, VINCT (Visual Information Communication and Interaction) conference held by Zhejiang University in 2012 and Tianjin University in 2013, respectively, guided the direction for the development of China's visual technology.

From the above research status, compared with China, data visualization research in foreign countries started earlier, the theory and technology is more mature, and there are more application areas and data visualization-processing software. However, no matter at home and abroad, there is a lack of research on specific information visualization for equipment support comprehensive evaluation and specific visualization software can be applied to the ESCCES. There is also a need to filter and improve current data visualization methods according to the working process and characteristics of equipment support comprehensive evaluation and help experts evaluate support concept scientifically by selecting or developing a visual software suitable for ESCCES to process all kinds of data information in the evaluation process.

3 EQUIPMENT SUPPORT INFORMATION VISUALIZATION RESEARCH

3.1 Foreign research status

In foreign countries, there are many related research on equipment support. In the field of equipment support information visualization, it is mainly focused on equipment support training information and equipment support material information.

In the field of equipment support training information, in recent years, the US army used advanced computer technology to combine the high-resolution satellite image with the battlefield digital map to obtain the detailed battlefield geographic data and took effectively visual processing to enable the troops to carry out a good equipment support training in the simulated battlefield environment during local wars such as the Gulf War and the Afghanistan war.

In the field of equipment support material information, the US army adapt a support way called "violence" that delivers logistics supplies at all costs in order to ensure the combat effectiveness in "Desert Shield" and "Desert Storm". However, in the delivery of more than 40000 containers in the Bay Area, US army had to carry out some work such as opening, registering, packaging, retransporting because all kinds of information of logistics supplies are not clear enough. At the end of the war, a large number of container materials did not really put into use and this support way greatly wasted time, manpower and material resources. Therefore, the US army proposed a concept of "joint asset visualization", formulated operational principles of master maneuver, precise engagement, full dimensional protection, focus support through 《Joint Vision 2010》 and 《Joint Vision 2020》 and adapt self recognition technology to realize real-time visual management of logistics supplies to improve the efficiency of logistics supply.

3.2 Domestic research status

The research on information visualization technology of domestic equipment support scholars has experienced the process from the introduction absorption to the independent innovation. The research results are mainly divided into equipment support command decision information visualization and equipment support material information visualization.

In the field of equipment support material information, scholars initially studied the overall framework of the support materials visualization system. Later, they gradually extended to study the overall framework and implementation technology of maintenance support material visualization and equipment logistics information visualization system.

In the field of equipment support command decision information, Mechanical Engineering College originally proposed the concept of equipment support management information visualization. The equipment established a framework of equipment support management information visualization based on equipment support command visualization, equipment material supply visualization, equipment technical support visualization, and equipment support task visualization. Key technologies such as 3S integration, data mining, intelligent decision support, and human-computer interaction were studied. Then, they constructed the visual model of equipment support progress and designed the system of equipment support progress visualization by using the multidimensional information visualization method. The Academy of Armored Forces Engineering constructed a visualization model based on the decision-making ability and analyzed the process for visualization processing by combining data acquisition with preprocessing.

From the above research status, domestic and foreign scholars have done some research on the visualization of equipment support information, accumulated experiences, and played a certain role in some specific work. However, for equipment support concept comprehensive evaluation, different evaluation focuses lead to different methods of data display requirements and data visualization processing methods. Therefore, it is necessary to make a reasonable distinction and efficient presentation among the relevant data information according to the specific requirements of equipment support concept comprehensive evaluation.

4 RELEVANT SYSTEM DATA VISUALIZATION RESEARCH

4.1 *Foreign research status*

Foreign literature related to information visualization of hall for workshop of meta-synthetic engineering are oppositely less. However, for visualization of other types of systems, foreign scholars have done a lot of research and realized remarkable results in the practical application. In New Brunswick University in Canada, Kavouras and Masry developed a 2D visualization of geographic information prototype system for mineral resources evaluation and exploitation in 1987. In 1994 in Holland, M.J. Kraak studied a 2D terrain visualization system based on vector, combined plane thematic maps with the corresponding area of two-dimensional stereoscopic graph and realized dynamic data roaming and spatial analysis. Rostock University and Stuttgart University in Germany established digital city simulation system in order to query, analyze, and display the urban infrastructure (including housing, roads, green spaces, etc.). "RAPID SCENE" software

EVENS & SUTHERLAND Company and digital photogrammetry workstation software system SET SOCET manufactured by Leica/Helava Company have data interface to produce interactively a two-dimensional landscape based on remote sensing image and real-timely and dynamically present the information. In addition, there are some other simulation systems, which focus on the combination of two-dimensional visualization, virtual reality, and computer network.

4.2 *Domestic research status*

A study on hall for workshop of meta-synthetic engineering of domestic scholars appeared earlier. Initially, they studied on "from qualitative to quantitative hall for workshop of meta-synthetic engineering" system proposed by Xuesen Qian. Then they realized hall for workshop of meta-synthetic engineering system software in many areas by designing architecture, organization and function. At present, the major universities and research institutes gradually tend to study on the optimization and improvement of the systems. In addition, data visualization is one of the aspects.

In the field of general information of hall for workshop of meta-synthetic engineering, Chinese Academy of Sciences applied Isomap algorithm to realize the dimension reduction of the expert evaluation opinions, visualized the results in the low-dimension space, and provided an ideal environment of the visualization for hall for workshop of meta-synthetic engineering, which reflects the clustering of experts thought. Nanjing University of Science and Technology divided discuss information visualization into natural visualization, autonomous visualization and intelligent visualization. Given the concept and visual object, every visualization technology was studied. It provided some reference for discussing information visualization. Northwestern Polytechnical University constructed information organization model of discuss system and designed a user driven discuss information visualization platform according to information requirement of users. Shanghai Jiao Tong University and Xi'an Jiao Tong University deeply studied and discussed information of hall for workshop of meta-synthetic engineering together. Xingxue Zhang and Pengzhu Zhang proposed the concept of discussion information-independent visualization, analyzed the information structure of group discussion, established a group discussion information-independent visualization model based on UML, and tried to construct the group discussion-independent visualization platform based on discussion tree and research network. Li Jia, Pengzhu zhang, and Yuzhu Jiang combined the discussion model of discussion system with discussion text feature, proposed feature vector selection method based on topic analysis, and designed an automatic clustering visualization tool based on the neural network theory of self-organizing maps. Yuzhu Jiang designed a human-machine combination of consensus evaluation, analysis, and prediction technology based on ECBAR and visualized it to add expert judgment easier with Internet technology, computer graphics, Java, and XML technology in order to realize the continuous process of discussion, analysis, evaluation, prediction, and further discussion.

In the field of specific domain information of hall for workshop of meta-synthetic engineering, Chinese Academy of Sciences took 3D visualization technology to solve the visualization of the large amount of data information including geological conditions, the characteristics of rock and soil, the logical relationship of the construction in underground engineering, and solved the problems in the amount of data of water conservancy and hydropower engineering to facilitate the discussion of engineering construction through 3D modeling visualization. Beijing University of Posts and Telecommunications studied the key technologies of data visualization and the monitoring system of Internet of things, and found a meeting point between the two. Then proposed an implementation plan for data visualization in the field of the Internet of things monitoring and designed a bridge safety remote monitoring system based on visualization. Academy of Armored Forces Engineering constructed an information visualization model of equipment support resources based on data visualization methods and realized intelligent and visualized management of equipment support resources by using Java, ActiveX, and ASP to design an equipment support resource visualization management system based on intelligent decision.

From the above research status, the domestic and foreign research on visualization of group decision support systems focused on the general information, not the specific domain information. Therefore, there is a need to seek common points and research ideas from related research in order to find an information visualization processing method suitable for ESCCES.

5 CONCLUSIONS

In view of the importance of data visualization for equipment support concept comprehensive evaluation, there is a need to analyze and model all kinds of data information in ESCCES by analyzing and summarizing the current status of research. In addition, it is necessary to seek different visualization methods to process different information, which are needed by comprehensive evaluation based on requirements of different processes and different roles. Finally, ESCCES can help experts evaluate equipment support concept reasonably and effectively by efficient presentation of data information.

REFERENCES

Cai Zhuhua. 2014. Research on Visualization Technology and Its Application Based on Clustering Analysis[D]. Xiamen University.

Chen Changyi, Gong Chuanxin, Wang Jixing. 2002. Research on the Framework of the Visualization System of Equipment Support Total Asset[J]. Journal of the Academy of Equipment Command & Technology. 13(3)14–17.

Chen Wei. 2013. Research Progress in Information Visualization and Visual Analysis[J]. International Academic Development. (5):18–20.

Chen Wei. 2013. Data Visualization[M].Publishing House of Electronics Industry.

Chen Xinggang. Liu Zhenhua. Baohua Guo. 2006. Anticipation of RFID and Bar Code Applying in the Fields of Military Logistics[J]. Packaging Engineering. 27(1):87–89.

Dai Zhongguo, Chen Wei, Hong Wenxue, etal. 2013. Information Visualization and Visual Analysis: Challenges and Opportunities-a summary report of the Beidaihe Information Visualization Strategy Conference[J]. Science in China:Information Science.

Fischer M M, Henk J S, Unwin D. 1996. Geographic Information systems, Spatial data analysis and Spatial Modelling: an introduction[A]. In: Fischer M M, Henk J S, Unwin D(Eds). Spatial Analytical Perspectives on GIS, GISDATA4[C]. Taylor and Francis.

General Herry. 2000. Chairman of the joint chiefs of staff Joint Vision 2020.

Gao Fei, Ge Qiang, Luo Lei. 2008. Study on Application of 2-dimensional Barcode Technology on General Equipment Support[J]. Packaging Engineering. 29(12):131–133.

John M, Shalikashvili. 1996. Chairman of the joint chiefs of staff Joint Vision 2010.

Jiao Yabing, Yuan Haibo. 2008. Study of Military Logistics Information System Based on RFID[J]. Packaging Engineering. 29(9):208–211.

Jiang Yuzhu, Zhang Pengzhu, Zhang Xingxue. 2009. Research on Intelligence Visualization in Group Argument Support System[J]. Journal of Management Science in China. 12(3):1–11.

Liu Yi. 2013. Analysis of Data Visualization Based on Good User Experience[D]. East China University of Science and Technology.

Li Xijuan. 2014. The Research of Data Visualization in the Big Data[D]. Hebei University.

Liu Chao. 2004. Information Visualization Technology and Its Application in Equipment Support [J]. Information Technology and Its Application. (2):26–28.

Li Rongqiang, Zhang Wenge, Du Jiaxing, Bi Mingguang. 2010. Research on Visualization of Equipment Support Decision Information [J]. Journal of Sichuan Ordnance. 31(8):25–28.

Li Chunmei, Dai Ruwei. 2004. Visualization of Experts Evaluation Opinions in the Hall for Workshop of Meta-synthetic Engineering[J]. PR&AI. 18(1):6–11.

Liu Mingfeng. 2009. Research on Visualization and opinion synthesis of Group Discussion in HWME[D]. Nanjing University of Science and Technology.

Li Jia, Zhang Pengzhu, Jiang Yuzhu. 2009. Research on Automatic Topic Visual Clustering in the Group Argument Support Systems[J].Journal of Systems & Management. 18(3):325–331.

Li Hui. 2013. Research and Application of Data Visualization on Internet of Things Monitoring[D]. Beijing University of Posts and Telecommunications.

Qiu Deqing, 2014. Research and Implementation of Multidimensional Data Visualization Tool Based on Web[D]. Beijing University of Posts and Telecommunications.

Robertson G, Card S K, Mackinlay J D. 1989. The cognitive coprocessor architecture for interactive user interfaces[C]//Proceedings of the 2nd Annual ACM SIGGRAPH Symposium on User interface Software and Technology. New York:10–18.

Shao Hong, Li Qinzhen. 2009. Application Research of Transportation Visualization System for Complicated Electromagnetic Environment[J]. Equipment Environmental Engineering. 6(2):89–92.

Sun Yang. 2010. Research on Multivariate Network Data Visualization[D]. National University of Defense Technology.

Su Yan, Xue Huifeng, Sun Jingle, Jiang Yuzhu. 2007. The Design and Realization of a User-center Information Visualization Platform[J]. Computer Engineering and Applications. 43(23):106–109.

Tang Zesheng. 1999. Field of The 3D Data Visualization[M]. Tsinghua University Press.

Tang Zesheng, Chen wei. 2011. Visualization Items[M]. China Computer Encyclopedia.

Tang Hua. 2006. Information Visualization and Intelligent Analysis and Management System to Safety Monitoring in Underground Powerhouse[D]. Chinese Academy of Sciences.

Wu Jianzhong, Liu Yile, Xu Zongchang. 2003. Visualized Management System of Equipment Support Resource Based on Intelligence Decision[J]. Computer Engineering. 29(16):165–169.

Xie Jiangning. 2014. Research on Key Technologies of Visualization Based on the University Scientific Research Data[D]. Shandong University.

Yang Zhenqian, Wang Xueyi. 2007. Design and Realization for Visualization System Equipment Guarantee Command[J]. Science Technology and Engineering. 7(16):4085–4087.

Zhang Yulin, Huang Jianqun. 2005. Design and Analysis on Equipment Maintenance Material Management Visualization System[J]. Journal of Ordnance Engineering College. 17(1):47–50.

Zeng You. 2014. The Concept study of Data Visualization Under the Background of Big Data[D]. Zhejiang University.

Zhou Xin. 2012. Research on Theories and Methods of Metrology Data Visualization[D]. National Institute of Metrology.

Zhang Haibing, Zhu Aihua. 2009. Visual Research of Weapon Equipment Maintenance Support Resource[J]. O I Automation. 28(12):49–50.

Zhang Wei, Wang Xueyi, Ma Weining. 2008. Visualization Framework of Information Under Equipment Support Management[J]. O I Automation. 27(10):46–48.

Zhang Xingxue, Zhang Pengzhu. 2003. The Elementary Research on the Independent Visualization of Argument Information in Group Decision Argumentation Web-based[J]. Computer Engineering and Applications. 32:42–44.

9   *Applied and computational mathematics*

Energy Science and Applied Technology ESAT 2016 – Fang (Ed.)
© *2016 Taylor & Francis Group, London, ISBN 978-1-138-02973-6*

A hybrid WENO/CPR method for the shallow water wave equations

S. Lin & S. Song
College of Science, National University of Defense Technology, Hunan, China

ABSTRACT: Correction Procedure via Reconstruction method (CPR) is a newly-proposed high-order efficient method. However, the shallow water wave equations may formulate shocks and rarefaction waves. CPR cannot calculate discontinuities directly since it is merely a linear method. On the other hand, Weighted Essential Non-Oscillatory (WENO) method can treat discontinuity well, but its calculating efficiency is relative low. Therefore, a novel hybrid CPR/WENO method is introduced here for solving the two-dimensional shallow water wave equations. The main idea is using smoothness indicators to find out the troubled cells that may contain discontinuity. Then smooth cells are calculating by CPR method while troubled cells are calculated by WENO method. High efficiency and accuracy of this hybrid method are verified by numerical experiments.

Keywords: shallow water waves equations, correction procedure via reconstruction, WENO method

1 INTRODUCTION

The shallow water wave equations are the basic equations describing fluid motion (Kinnmark, 2012). Ignoring frictions and slope rate of bottom, the two-dimension shallow water wave equations are

$$h_t + (hu)_x + (hv)_y = 0,$$

$$(hu)_t + (hu^2 + \frac{1}{2}gh^2)_x + (huv)_y = 0, \qquad (1)$$

$$(hv)_t + (huv)_x + (hv^2 + \frac{1}{2}gh^2)_y = 0,$$

where u, v denote velocities of fluid in x-direction and y-direction separately, g is gravitational constant, and h is the water height. Originally, shallow wave equations in one-dimensional case is proposed by de St Venant in 1871. With assistance of modern computers and development of numerical methods, shallow water wave equations can be solved well and have great applications in environmental fluid modeling (Pekeris, 1948).

Current numerical methods for solving shallow water wave equations are generally low-order (one-order or two-order) methods (Bermudez & Vazquez, 1994, Zoppou and Roberts, 2003). Low-order methods prevail in practical simulating mainly because of its simplicity and stability (Jha et al., 1995). On the other hand, high-order methods can provide more accurate solutions with a relative high efficiency. However, the current two types of high-order methods, WENO-type methods and Correction Procedure via Reconstruction (CPR) methods both have their own demerits and are not widely applied in practical problems. Therefore, it is of great importance to refine to performance of high-order methods by combining these two methods (Huynh, 2007).

The shallow water wave equations can be categorized into hyperbolic systems of conservation laws. And CPR method (also called CPR framework) is an effective method for solving conservation laws and can generalize substantial mainstream methods, e.g. Discontinuous Galerkin method (DG) (Cockburn et al., 2000), spectral volume methods (SV) (Wang, 2002), Staggered-Grid method (SG) (Huynh, 2007) and Spectral Difference methods (SD) (Huynh et al., 2014). It has several advantages as follows. The main advantage is high efficiency since it is a linear nodal numerical scheme. CPR framework just needs immediate cells to calculate the current cell thus it is compact.

Besides, CPR method can be written in conservative form therefore it is a conservative method. Nevertheless, CPR method is still a high-order linear numerical scheme, which is always unstable and cannot treat discontinuities. This is the principle difficulty for CPR method to solve the conservation laws as solutions may contain discontinuities. Therefore, it is necessary to adjust this method and improve its performance around discontinuities.

Another approach for solving shallow water wave equations is WENO method (Xing & Shu, 2005). Harten and Osher proposed the Essentially Non-Oscillatory method based on first-order Godunov method. Numerical experiments show that ENO method has high-order accuracy and at the same time ENO method can avoid nonphysical oscillations around discontinuities. To improve the accuracy in smooth regions, Weighted ENO (WENO) method was proposed by Liu and Osher. WENO method has been widely used in solving shallow water wave equations (Caleffi et al., 2006, Shi et al., 2002, Xu & Shu, 2006). A main drawback of WENO method is its relative high computation cost. In numerical experiments we can see that WENO method is much slower that the linear method, CPR method.

Therefore, we consider combining these two methods to maintain both high efficiency of CPR method and stability of WENO method. The main idea can be roughly stated as follows. First, we use smoothness detector to find out troubled cells. Then WENO method is adopted to update solution of trouble cells, while solution of other cells is updated by CPR method. In other words, computing domain for WENO method and CPR method is divided adaptively. To implement this method conveniently and efficiently, we adopt CPR method with equal-distance flux points and solution points. In numerical experiments, we test the accuracy and efficiency of our method. And we also solve several classical shallow water models that contain discontinuities successfully.

2 THE FORMULATION OF CPR METHOD

For simplicity, we consider solving one-dimension hyperbolic conservation law:

$$u_t + f(u)_x = 0,$$
$$u(x,0) = u_0(x),$$
(2)

First we divide the calculating domain into N cells and denote them as $I_j = [x_{j-1/2}, x_{j-1/2}], j = 1,2,...,N$. Inspired by finite element method, we consider to deal with the stand interval $I = [-1,1]$ (local element) instead of I_j (global element). And there is a simple linear function from the global element I_j to local element I:

$$x(\zeta) = x_j + \zeta h/2,$$

and its inverse is

$$\zeta(x) = 2(x - x_j)/h_j,$$

where h_j is the length of I_j. Then a function $r_j(x)$ in I_j can be transformed to $r_j(\zeta)$ in the local element. And the derivatives of these two functions can be related by the chain rule,

$$\frac{dr_j(x)}{dx} = \frac{2}{h_j} \frac{dr_j(\zeta)}{d\zeta}.$$
(3)

In I_j we use the Lagrange polynomial with K interpolating points $u_{j,k}, k = 1,...,K$, which are called solution points. Then the solution function is

$$u_j(\zeta) = \sum_{k=1}^{K} u_{j,k}\phi(\zeta),$$
(4)

where $\phi(\zeta)$ are the basis functions of Lagrange interpolation in the form of:

$$\phi_k(\zeta) = \prod_{i=1, i \neq k}^{K} \frac{\zeta - \zeta_i}{\zeta_k - \zeta_i}$$

And similarly, the flux function in the I_j is approximated as

$$f_j(\zeta) = \sum_{k=1}^{K} f_{j,k}\phi(\zeta),$$
(5)

where $f_{j,k} = f(u_{j,k})$. Generally, we have $f_j(1) \neq f_{j+1}(-1)$. Based on this fact, the flux polynomial that consists of $f_j(x), j = 0,1...$ is a piecewise smooth function and we may call it the discontinuous flux function.

To treat the discontinuity and include it into the calculation, we consider using a correction function to reconstruct a globally smooth flux function, which approximates the discontinuous function in some sense. To ensure the physical meaning of the reconstruction, the continuous flux function takes the upwind flux values on the boundaries of each element. We denote the values on the boundaries of I_j as $u_{j \pm 1/2} = u_j(\pm 1)$. And accordingly we compute the fluxes at boundaries, $f_j(\pm 1) = f(u_j(\pm 1))$. With the continuous flux function $F_j(x)$, (2) can be approximated by an ODE

$$\frac{du_{j,k}}{dt} = -\frac{2}{h_j}(F_j)_\zeta(\zeta_k).$$
(6)

There are several principles of reconstructing a continuous flux function $F(\zeta)$: (1) The continuous flux polynomial should take the Riemann flux at the boundaries of I_j, namely, $F_j(\pm 1) = f_{upwind}(\pm 1)$. (2) As a corrected function of original flux polynomial, $F_j(x)-f_j(x)$ needs to approximate the function 0 in some sense. (3) $F_j(x)$ should be one degree higher than solution polynomial to ensure that in (2), derivative of $F_j(x)$ has the same degree with $u(x)$.

Based on these principles, the continuous flux polynomials can be defined by

520

$$F_j(\zeta) = f_j(\zeta) + f_{\text{upwind}}(-1)g_L(\zeta) + f_{\text{upwind}}(1)g_R(\zeta), \quad (7)$$

where g_L is the correction function for the right boundary satisfying

$$g_L(-1) = 1, g_L(1) = 0, \quad (8)$$

G_R is the correction function for the left boundary function satisfying

$$g_R(-1) = 0, g_R(1) = 1. \quad (9)$$

Using the linear function from the global element I_j to local element I, we have

$$(F_j)_x(x_{j,k}) = \frac{2}{h_j}(F_j)_\zeta(\zeta_k). \quad (10)$$

Then we obtain a spatial discretion form:

$$\frac{u_{j,k}}{dt} + \frac{2}{h_j}(F_j)_\zeta(\zeta)_k = 0, \quad (11)$$

which can be solved by various method, e.g. Runge-Kutta method.

3 HYBRID WENO/CPR METHOD

3.1 The procedure of hybrid method

The procedure of hybrid WENO/CPR method is described as following:

i. As shown in Section 2, we first divide calculation domain into N small cells.

ii. We adopt a smoothness indicator to find out troubled cells that may contain discontinuities.

iii. For troubled cells and cells that are adjacent to troubled cells, we use WENO method to compute, while for other cells we adopt CPR method.

In practice, most cells are smooth and can be computed by CPR method which is highly efficient. Therefore, computing time could be largely saved.

3.2 Discontinuity indicators

The basis of implementing hybrid method is indicating cells that contain discontinuities. There are several types of discontinuity indicators.

Another indicator we adopt is the second-order difference indicator. For simplicity we denote $D^{(2)}$ $f_{j,k}$ as the finite difference operator of the second-order derivatives. It should be noted that when $k = 1$ or $k = K$, $D^{(2)}f_{j,k}$ stands for

$$D^{(2)}(f_{j,1}) = \frac{2[d_1 f_{j,2} + f_{j-1,K}d_2 - (d_1 + d_2)f_{j,1}]}{d_1 d_2 (d_1 + d_2)},$$

$$D^{(2)}(f_{j,K}) = \frac{2[d'_1 f_{j+1,1} + f_{j-1,K}d'_2 - (d'_1 + d'_2)f_{j,K}]}{d'_1 d'_2 (d'_1 + d'_2)},$$

where $d_1 = x_{j,1} - x_{j-1,K}$, $d_2 = x_{j,1} - x_{j-1,K}$, $d'_1 = x_{j,K} - x_{j,K-1}$, $d'_2 = x_{j+1,K} - x_{j,K}$.

If the second-order finite difference of any point is large than a given value, we can judge the cell as a troubled cell. To be more specific, cell I_j is judged as troubled cell if

$$\exists l, s.t. \mid D^{(2)}f_{j,l} \mid > M',$$

M' is a free parameter. Numerical tests show that this method can judge discontinuities correctly and efficiently.

3.3 Hybrid WENO method with CPR method

If the cell I_j is judged as a troubled cell, we calculate I_{j-1}, I_j and I_{j+1} using 5th order WENO finite difference method to avoid nonphysical oscillations. Note that we also compute two neighbor cells I_{j-1} and I_{j+1} with WENO method to ensure stability.

Then we give a brief outline of WENO method. For simplicity we denote $\boldsymbol{u}_j$ as $\{u_{j,1}, u_{j,2}, ..., u_{j,K}\}$. Then the values that need recalculating are $\{\boldsymbol{u}_{j-1}, \boldsymbol{u}_j, \boldsymbol{u}_{j+1}\}$. Since we need to use the WENO method, uniform cell with equidistant solution points are adopt here for convenience.

Let $\{u_m\}_{m=1,2,...,3K} = \{\boldsymbol{u}_{j-1}, \boldsymbol{u}_j, \boldsymbol{u}_{j+1}\}$. Standard WENO method used to calculate u_m are as follows.

We split the flux function $f(u)$ into positive flux $f^+(u) = f(u) + \alpha u$ and negative flux $f^-(u) = f(u) - \alpha u$, where $\alpha = \max(f'(u))$. For simplicity we only demonstrate the process of reconstructing $f^+(u)$. The procedure of reconstructing $f^-(u)$ is totally symmetry to that of $f^+(u)$.

$$f^+_{j+\frac{1}{2}} = \sum_{l=0}^{2} w_l^+ q_l^+,$$

where

$$\begin{cases} q_0^+(f^+_{j-2}, f^+_{j-1}, f^+_j) = \dfrac{1}{3}f^+_{j-2} - \dfrac{7}{6}f^+_{j-1} + \dfrac{11}{6}f^+_j, \\[2mm] q_1^+(f^+_{j-1}, f^+_j, f^+_{j+1}) = -\dfrac{1}{6}f^+_{j-1} + \dfrac{5}{6}f^+_j + \dfrac{1}{3}f^+_{j+1}, \\[2mm] q_2^+(f^+_j, f^+_{j+1}, f^+_{j+2}) = \dfrac{1}{3}f^+_j + \dfrac{5}{6}f^+_{j+1} + \dfrac{1}{3}f^+_{j+2}, \end{cases}$$

$$\omega_l^+ = \frac{\alpha_l^+}{\sum_{s=0}^{2}\alpha_s^+}, \quad \alpha_s^\pm = \frac{d_s^+}{(\epsilon + \beta_s^+)^2}, \quad l = 0,1,2.$$

521

In this article we take $\epsilon = 10^{-9}$ And the indicators of smoothness are defined as

$$\begin{cases} \beta_0 = \dfrac{13}{12}(f_{j-2}^+ - 2f_{j-1}^+ + f_j^+)^2 + \dfrac{1}{4}(f_{j-2}^+ - 4f_{j-1}^+ + 3f_j^+)^2, \\[2mm] \beta_1 = \dfrac{13}{12}(f_{j-1}^+ - 2f_j^+ + f_{j+1}^+)^2 + \dfrac{1}{4}(f_{j-1}^+ - f_{j+1}^+)^2, \\[2mm] \beta_2 = \dfrac{13}{12}(f_j^+ - 2f_{j+1}^+ + f_{j+2}^+)^2 + \dfrac{1}{4}(3f_j^+ - 4f_{j+1}^+ + f_{j+2}^-)^2, \end{cases}$$

And the combination coefficients are $d_0^+ = 1/10, d_1^+ = 3/5, d_2^+ = 3/10$.
Then we can compute $u_j^{(n+1)}$ by

$$\frac{u_j^{(n+1)} - u_j^{(n)}}{dt} = -\frac{f_{j+1/2} - f_{j-1/2}}{dx'}, \quad dx' = \frac{dx}{K}.$$

$$f_{j+1/2} = f_{j+1/2}^+ + f_{j+1/2}^-, \quad f_{j-1/2} = f_{j-1/2}^+ + f_{j-1/2}^-.$$

As described in section 3.3, after the troubled cells and their adjacent cells are computed by WENO5 method, other cells are computed by CPR method. The algorithm for two-dimension shallow water wave model can be easily extended from the one-dimension algorithm.

4 NUMERICAL RESULTS

In this section we implement hybrid WENO5/CPR5 method to solve shallow water wave equations. Third-order Runge-Kutta method are used in time discretization. As for the correction function of CPR method, we take the DG-type correction function described in (Huynh 2007).

4.1 *Two-dimension dam-break problem*

The computational domain is a 200 m × 200 m channel. The dam is located on the middle line of computational domain and its thickness is ignored. The breach is 75 m in length. One end of the breach is the middle point of dam. The initial upstream-water is 10 m in depth, while the downstream-water is 5 m. The cell length dx is 5 m and time step is 0.02 s. We adopt the original WENO method, the hybrid WENO/CPR method using finite-difference limiter to solve this problem. The results of water surface obtained by original WENO5 method and hybrid method are showed in Figure 1 and Figure 2. We can see that the results obtained by WENO5 method and hybrid method are similar. The hybrid WENO/CPR method can treat discontinuities well. On the other hand, Table 1 shows the CPU time of these two methods in solving this dam-break problem and we can that Hybrid CPR5/WENO5 method works

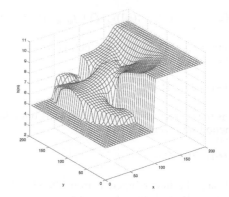

Figure 1. Water surface at $t = 7.2$ s using WENO method.

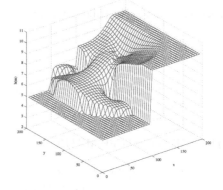

Figure 2. Water surface at $t = 7.2$ s using hybrid WENO/CPR method.

Table 1. Relative CPU time of different methods and ratio of cells that are computed by WENO method.

Method	Relative CPU time	Ratio
WENO	1	1
Hybrid method ($M = 1$)	0.133	0.86%
Hybrid method ($M = 0.1$)	0.291	1.35%

much more efficiently that WENO5 method. And it also indicates that smaller choice of M can cause large proportion of cells declared as troubled cells, which increases computing time. Nevertheless, if M is too large, some cells that contain discontinuities cannot be detected.

4.2 *Bores diffraction problem*

This subsection illustrate diffraction of a bores caused by dam breaking. Consider a 200 m × 400 m

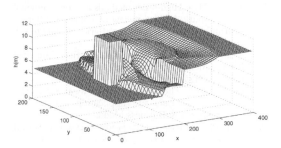

Figure 3. Water surface evaluation of bores diffraction problem at $t = 12$ s.

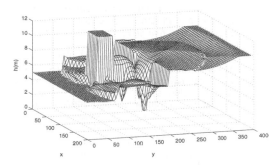

Figure 4. Water surface evaluation of bores diffraction problem at $t = 18$ s.

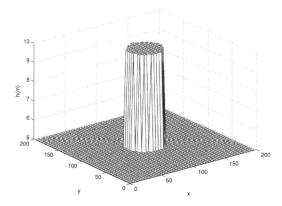

Figure 5. Initial water level of round-dam-break problem.

channel, with a dam locating on the middle line. The dam's thickness is ignored. Initially, a breach in length of 50 m is caused and water depth is 10 m and 5 m respectively. There is a 40 m × 40 m rectangular cylinder barrier, locating 100 m in front of the breach, 80 m from the left bank.

We adopt hybrid WENO/CPR method to solve this problem. The computation results of velocity field and 3D water surface elevation at $t = 12$ s and $t = 18$ s are presented in Figure 3 and Figure 4. It

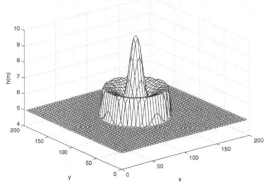

Figure 6. Water surface of round-dam-break problem at $t = 2$ s.

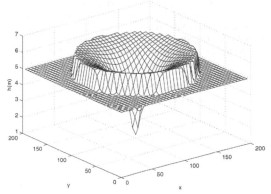

Figure 7. Water surface of round-dam-break problem at $t = 6$ s.

can be clearly seen that the discontinuities, location of vortices and structure of water surface change with time increasing.

4.3 *Round-dam-break problem*

Consider a 100 m × 100 m channel. A round dam with a radius of 25 m is located in the middle of the channel. The water depth in the dam is 10 m while the water outside the dam is 5 m in depth. After the dam is totally broken initially, the water surface begins to change, and shocks and vortices are generated. The initial water surface is showed in Figure 5. And the water surface at $t = 2$ s and $t = 6$ s are showed in Figure 6 and Figure 7.

5 CONCLUSION

The hybrid WENO/CPR method developed in this paper is an effective approach to solve the shallow water wave equations. This method can maintain the high efficiency of CPR method and can treat discontinuities well, which is also verified by numerical experiments. In numerical experiments, several problems including the dam-break problem, the bores diffraction problem and round-dam-break problem are simulated and the hybrid method performs well.

ACKNOWLEDGEMENT

This work was supported by Chinese National Natural Science Foundation no. 91530106.

REFERENCES

Bermudez & Vazquez, 1994. Upwind methods for hyperbolic conservation laws with source terms. *Computers & Fluids*, 23(8):1049–1071.

Caleffi et al. 2006. Fourth-order balanced source term treatment in central weno schemes for shallow water equations. *Journal of Computational Physics*, 218(1):228–245.

Cockburn et al. 2000. The development of discontinuous Galerkin methods. Springer.

Huynh. 2007. A flux reconstruction approach to high-order schemes including discontinuous Galerkin methods. *AIAA paper*, 4079: 2007.

Huynh et al. 2014. High-order methods for computational fluid dynamics: A brief review of compact differential formulations on unstructured grids. *Computers & Fluids*, 98:209–220.

Jha et al. 1995. First-and second-order flux difference splitting schemes for dam-break problem. *Journal of Hydraulic Engineering*, 121(12):877–884.

Kinnmark. 2012. The shallow water wave equations: formulation, analysis and application, volume 15. Springer Science & Business Media.

Pekeris, 1948. Theory of propagation of explosive sound in shallow water. Geological Society of America Memoirs, 27:1–116.

Shi et al. 2002. A technique of treating negative weights in weno schemes. *Journal of Computational Physics*, 175(1):108–127.

Wang, 2002. Spectral (finite) volume method for conservation laws on unstructured grids. basic formulation: Basic formulation. *Journal of Computational Physics*, 178(1):210–251.

Xing & Shu, 2005. High order finite difference weno schemes with the exact conservation property for the shallow water equations. *Journal of Computational Physics*, 208(1):206–227.

Xu & Shu. 2006. Anti-diffusive finite difference weno methods for shallow water with transport of pollutant. *Technical report, DTIC Document*.

Zoppou & Roberts. 2003. Explicit schemes for dam-break simulations. *Journal of Hydraulic Engineering*, 129(1):11–34.

Energy Science and Applied Technology ESAT 2016 – Fang (Ed.)
© *2016 Taylor & Francis Group, London, ISBN 978-1-138-02973-6*

Website usability comprehensive evaluation based on gray relational analysis

Hanping Zhang
Computer Department, Wuhan Polytechnic, Wuhan, China

ABSTRACT: Website construction has been paid high attention from all walks of life. To seek a scientific and reasonable website usability evaluation method is increasingly important. By using the basic theory of usability engineering, we have established the evaluation index system of website usability, and the website usability quantitative evaluation method is presented based on the grey relational analysis. This method can also solve the comprehensive evaluation problems about design quality of system, which concludes unknown or uncertain information.

Keywords: website construction, comprehensive evaluation, gray relational analysis, availability

1 PREFACE

With the popularity of Internet technology, all walks of life are setting up or have established their own website. Website usability is directly related to the user's experience level of the site, which affects the market competitiveness of website[1]. Website usability evaluation depends on the establishment of evaluation index system and scientific and quantitative evaluation method, to make quantitative analysis and evaluation about the degree of the website availability, providing reference basis of establishing website or selecting website.

Website usability connotation is quite rich, mainly referring to the following four aspects: (1) the content of the website contains the design of the interface and technical level of system; (2) the estimator is a person, who makes the subjective evaluation by finishing various operating tasks as a user; and (3) the website is targeted. For different users and domains, the evaluation index and parameter are different; (3) Website has the function of error correction. The user can complete the operation task under abnormal operation. Therefore, the website usability evaluation has certain subjectivity, which are partially known factors that interact with some unknown factors.

A gray system is known as a system containing unknown or uncertain information, and the relational analysis is called the gray relational analysis[2]. Website usability evaluation problem has the typical characteristics of gray system, and is suitable for comprehensive evaluation using the gray relational analysis method.

2 EVALUATION METHOD

The choice of website usability evaluation factors: selecting the main factors influencing the website usability to make evaluation. We select information content, interface design, navigation design, application conditions, and subjective satisfaction of the five elements as evaluation factors (Liu Zeng, 2009).

To determine the value of website usability evaluation factors: according to the user survey statistics, combined with the suggestions of experts to determine the weights p_k of evaluation factors.

Set up gray evaluation model of website usability: the basic principle of gray relational method is based on geometrical similarity of the contrast between a family of curves setting by sequence and curves in the composition of the reference sequence, to ensure and compare relational degree between sequence set and reference sequence. Comparing the curves of the sequence and the reference sequence, it can be observed that the more similar the geometry, the greater the degree of relational, and vice versa. It is equally applicable to the number of samples and there are no laws, and has a small amount of calculation, which is very convenient. There will be no situation that the quantitative results are not in conformity with the qualitative analysis results. The specific calculation method and the steps are as follows (Liu Huimin, 2009):

Step 1: identify the reference sequence and the compared sequence

The data series reflecting the characteristics of the system behavior is system evaluation standard, known as the reference sequence, denoted by $x_0 = \{x_0(k)|k = 1,2,\cdots,n\}$. The reference sequence should select the best performers from all of the design target, when the index belong to "efficiency", to select the maximum value of the index in all the programs; When the index belong to "cost", to select the multiplicative inverse of the minimum value of all the alternatives. The data sequence which combined with factors affecting the behavior of the system, that is to be evaluated in the system, is known as comparative sequence, denoted by

$$x_i = \{x_i(k)|k = 1,2,\cdots,n\}, i = 1,2,\cdots,m.$$

Step 2, the nondimensionalization of variable quantity

The different dimensions of data in all factors of the system may be different that affect the conclusions and even harder to obtain a correct conclusion, so in order to make the conclusion with sufficient reliability, when in the gray relational degree analysis, we generally need to perform dimensionless processing of data. $x_i(k)$ is still shown as $x_i(k)$ after the nondimensionalization. There many kinds of methods of nondimensionalization. We selected the following ways to carry on nondimensionalization:

$$x_i(k) = \frac{x_i(k)}{x_i(1)}, k = 1,2,\cdots,n; i = 1,2,\cdots,m. \tag{1}$$

Step 3, to calculate the relational coefficient of $x_0(k)$ and $x_i(k)$

$$\zeta_i(k) = \frac{\min_i \min_k |x_0(k) - x_i(k)| + \rho \max_i \max_k |x_0(k) - x_i(k)|}{|x_0(k) - x_i(k)| + \rho \max_i \max_k |x_0(k) - x_i(k)|}, \tag{2}$$

denoted by $\Delta_i(k) = |x_0(k) - x_i(k)|$ then

$$\zeta_i(k) = \frac{\min_i \min_k \Delta_i(k) + \rho \max_i \max_k \Delta_i(k)}{\Delta_i(k) + \rho \max_i \max_k \Delta_i(k)}.$$

$\rho \in (0,\infty)$ is named as distinguish coefficient. The smaller the value of ρ, the greater the resolution. Generally, the value of ρ is commonly between (0, 1), here $\rho = 0.5$.

$\zeta_i(k)$ is the correlation coefficient of the k element of the comparative series x_i and the k element of the reference sequence x_0.

As the correlation coefficient is a correlation degree of sequence and reference value in every moment, and there are many such values, it is not easy to compare the whole. We can use the weighted average method to centralized information as the

total number representation to compare relevance between comparison sequence and reference sequence. The correlation formula r_i is as follows:

$$r_i = \sum_{k=1}^{n} p_k \zeta_i(k), i = 1,2,\cdots,m. \tag{3}$$

Step 4, the repartition of the correlation degree ranking and grade

According to the size of the order sort, if $r_1 < r_2$, the correlation between reference number x_0 and the comparison sequence x_2 is greater than x_1. Accordingly, the website usability is stronger.

The correlation degree is divided into four sections to determine the correlation level. The specificity is given in Table 1 (Liu Zeng, 2009).

3 APPLICATION AND DISCUSSION

Assuming that there are three website designs, we need to evaluate their usability. We select information content, interface design, navigation design, application conditions, and subjective satisfaction as assessment factors. The weights of website assessment factors confirmed by experts are shown in Table 2.

To take system of 10 minutes for the evaluation factors, a questionnaire survey of the users and statistical data was made. The evaluation factors of the score are given in Table 3.

To evaluate the three website design programs.

The number list of three kinds of programs being compared are

$$x_1 = \{10,8,8,6,10\}$$

Table 1. Gray relational grade classification.

Grade	Relevancy	Evaluation description
Low correlation	0~0.3	Weak coupling between two indexes
Moderate correlation	0.3~0.6	The coupling effect between the two indexes is moderate
Higher correlation	0.6~0.8	Strong coupling effect between the two indicators
High correlation	0.8~1	The relative changes of the two indexes are almost the same, and the coupling effect is very strong.

Table 2. Weight of each evaluation factor.

Factor of evaluation (k)	Information content	UI	Navigation design	Application conditions	Subjective satisfaction
Evaluation weight p(k)	0.25	0.20	0.20	0.10	0.25

Table 3. Rating factor scores of three programs.

Factor of evaluation	Information content	Ui	Navigation design	Application conditions	Subjective satisfaction
Program A	10	8	8	6	10
Program B	8	8	6	10	6
Program C	10	6	8	8	8

$$x_2 = \{8,8,6,10,6\}$$

$$x_3 = \{10,6,8,8,8\}$$

The above sequence substituted into the formula (1), (2), (3), is calculated to

$$r_1 = 0.756, r_2 = 0.705, r_3 = 0.727.$$

According to the correlation degree shown in Table 1, these three programs are in a strong level of coupling with the standard program, and each has its own advantages. Comprehensive evaluation results show that $r_1 > r_3 > r_2$. On the whole, the coupling between program A and the standard scheme is the strongest, and is the most close to the standard program, so the program A website design is optimized.

4 CONCLUSION

Gray relational analysis method is applied in this article. Through the integration from qualitative analysis to ration analysis, we make comprehensive evaluation of the availability of the web site's design. Scientific and reasonable, simple and feasible, this method can also solve synthetic evaluation problems in which many factors interact with each other containing unknown or uncertain information system.

REFERENCES

Deng Julong. The grey control system. Wuhan: Huazhong University of science and technology press, 1990.

Liu Huimin. Urban residents' consumption structure evolution of the clustering analysis and grey forecasting j. statistics and decision, 2009,(3):93–95.

Liu Zeng, bing-fa Chen. User-centered website usability design and evaluation. China's manufacturing industry informationization, 2009,5(38):63–66.

Zhang Yiping, Xu Junliang, Zhao Xiping. Luoyang influence factors analysis of air quality based on grey relational. Journal of Henan university of science and technology (natural science edition), 2012,33(1): 100–104.

Energy Science and Applied Technology ESAT 2016 – Fang (Ed.)
© 2016 Taylor & Francis Group, London, ISBN 978-1-138-02973-6

Combination of big data analytics and traditional marketing analytics to predict sales of products

P.Y. Zhao & Y.M. Shi

School of Mathematics and Quantitative Economics, Shandong University of Finance and Economics, Shandong, P.R. China

ABSTRACT: Big data has brought about a lot of changes in our life and learning, for example, information extraction and information analysis methods have been fundamentally changed, which we usually classify into data mining. In order to find the differences between traditional market analytics and big data analytics, according to improve which to the predict sales of new products, this paper introduces the method of knowledge fusion. According to this paper, we find that to obtain its benefits the fusion to improve new product success is not automatic and sometimes needs strategic selections. To improve the practicability of this method, we extracted the numeric and textual data from customer reviews, and then the method of knowledge fusion is applied to sales of both existing products and new product forecast.

Keywords: big data; knowledge fusion; predicting

1 INTRODUCTION

Big data is a term involving large-scale database, for the Big Data are so large, unstructured, and complex that needs advanced and unique technologies to store, manage, analyze, and visualize (Ebach et al., 2016). For example, hosts of Facebook own over five hundred terabytes of knowledge every day. The knowledge covers uploaded files, hobbies, and invitations of users. Forrester defines big data as "techniques and technologies that make handling data at extreme scale affordable." Big data has brought about a lot of changes in our life and learning, for example, information extraction and information analysis methods have been fundamentally changed, which we usually classify into data mining.

An important source of marketing information for companies is user-generated online product comments. Companies can know the market through paying close attention to comments in order to gain a lot of data of interest. As customers depend on online product comments to make their buying deliberations, so customers comments of the content and price should contribute to forecast customers' behavior (Zhao et al., 2013, Chen & Xie, 2008). Collectively, customer comments contains a lot of helpful information which customers depend on to make their buying decisions. All this shows that data of the customers comments ought to contribute to prediction of new product sales.

In this paper, based on the Big Data Analytics (BDA) and Traditional Marketing Analytics (TMA), we design a method to forecast product sales. In order to find the differences between traditional market analytics and big data analytics, according to improve which to the predict sales of new products, this paper introduces the method of knowledge fusion. According to this paper, we find that to obtain its benefits the fusion to improve new product success is not automatic and sometimes needs strategic selections. To improve the practicability of this method, we extracted the numeric and textual data from customer reviews, and then the method of knowledge fusion is applied to sales of both existing products and new product forecast.

2 THEORETICAL FOUNDATIONS

In general, small-scale and big-scale analysis are spontaneously mapped to traditional data ware housing analysis. For example, Alberto et al. use OLAP (Abelló, A., & Romero, A, 2009) to get accurate understanding of data and build interested data sets firstly. Then, secondly through data mining or machine learning, they classify and predict trends from the data sets. The multidimensional method expresses knowledge be like put in a space of n-dimension, and makes it easy to understand and analyze knowledge in the areas of facts and dimensions forming the multidimensional space. In order to study the factual knowledge or measures, a dimension is formed by a concept hierarchy representing different granularities (or levels of detail). One piece of factual knowledge and a cluster of dimensions form a star schema. The structure is usually realized by keeping to the structure of star schema. Now, for its powerful foundations for data aggregation, the multi-dimensional method is the factual standard getting effective in data mining. More importantly, for the multi-dimensional method introduces the Roll-up operator (Abelló, A., & Romero, A, 2009), we can realize dynamic aggregation on measures along dimension hierarchies.

Today, more than 98% of information is digitally stored. Since nowaday market and new product development are the leading edge of the economic development, the destruction of the information is considered to grow up for unexpected threats of business. Firms decide something in real time to improve volumes and production using big data, of course, that use is just the beginning of BDA (Gustke, 2013). The point on many fronts of commercial functions exists in big data as the disruption moves forward. Further the degree of sensitivity in analyzing data is also important (Lamb, 2014).

2.1 Big data analytics

The commercial analytics for new product development contains estimating revenue cost, predicting sale, forecasting profit and assessing risks (Cravens & Piercy 2005, Srinivasan et al., 2005). In addition, the development of technologies, knowledge spread, sensitivity to market price, and fusing predicting for the products concerning new product development must be considered (Herbig et al., 1994).

Recently Polovets found that the better corporate performs the more progressive big data analytics technologies they adopt (Asay, 2014). The products that customers would rather buy are developed in method big data analytics provides in the web 3.0 era. However, because the marketing environment and consumer appetites change rapidly, understanding them is challenging. Companies can do many kinds of operations, such as searching the web, monitoring opponents, observing customers, delivering low cost questionnaires, testing models, and gaining reactivity by applying big data analytics. Although these repetitive things may have been done for decades, big data analytics can make it easy to promote their performance and lower the costs of the above operations. Enterprise can acquire from their opponents the main features, pricing strategy, and consumer reactivity in addition to designs. In a word, operational plans benefit from commercial wisdom. In addition, regarding peoples' sentiments and product evaluations, enterprise manager could draw real time knowledge, references, and product which is used to improve new product sales.

It is found that the revolution also has a significant effect on many aspects. The popular existing medium combines the ways of we gain knowledge, communicate with friends, support our preferred products, and select products with the consumer's perspective. Then, as important areas of BDA, text, mining, sentiment mining, and data mining become main tools for enterprises to make a rapid modification of marketing choices.

2.2 Knowledge fusion

An information fusion method of the big data analysis and traditional marketing analysis is proposed in this paper using both complexity theory and knowledge based theory (Anderson, 1999, Nickerson & Zenger, 2004). Thus the critical standard of big data analysis in market practical activities can be brought. Complexity can be changed suddenly. In a positive feedback loop, one or more tendencies can strengthen other tendencies until activities get out of hand and achieve a critical edge, over which activities transforms fundamentally (West, 2013).

Knowledge develops quickly because the data is easy to collect, the cost for analytics is significantly reduced, and the open information issues are shared on the web in the big data era. Prior scholars classify knowledge into two types: propositional knowledge and heuristic knowledge (Tsoukas & Vladimirou, 2001). Propositional knowledge is related and generalized with activities, whereas heuristic knowledge often emerges accidentally and informally.

Automated knowledge mainly refer to the disposing of natural language, databases, knowledge fusion, knowledge integration, and machine learning (Fan et al., 2012).

3 STRATEGIES AND EFFECTS

In this paper, the dataset titled "Market Dynamics and User Generated Content about Tablet Computers" is used to build and verify our strategies. (Wang, Mai, & Chiang, 2013). Many kinds of information about products and customer comments data is gathered from some big web sites.

3.1 Knowledge fusion strategies

Through the operation of knowledge fusion different traditional market analytics data and big data analytics data is fused to obtain a single effective knowledge. Consulting recent improvements of computer science and engineer design and then fusing information of both collecting from market and database, the term "knowledge fusion" (Jeong, 2012) is proposed to represent information. Four strategic options are presented which is combined with big data analytics and traditional marketing analytics. At the same time, how the data complex and which category the information belongs to aforementioned is discussed. By combing the features of information in years before and after the Internet, we classify the data into different categories. This classification of knowledge helps different firms classify the strategic choices.

3.2 *Pioneer, explorer and Bystander effects*

Knowledge fusion is defined as an operation. According to the operation different types of independent information is "merged" so the combinations are native, as that appear on the spot of information suppliers make a statement. New ideas and innovation may appear when the collaborative knowledge exploration is applied between applied subjects and other social subjects. And that is the argument. In the big data era, pioneer firms do not blind-trust automatic algorithms or models. Instead they increasingly combine IT expertise, marketing analytics, and customer knowledge to build knowledge fusion. In this process, the quality of knowledge state is a key notion by which collaborative fusion is measured. The insight is realized based on the dynamic information. To quickly develop prospective strategies in market, new product managers should adopt comprehensive method of information fusion extract from both the big data analytics and traditional marketing analytics. This study suggests the influence of TMA and BDA on both knowledge fusion and New Product Success (NPS). (a) Firms fuse the information the best when they adopt a highest big data analytics and traditional marketing analytics. (b) Firms obtain the most successful on new product when they adopt a highest big data analytics and traditional marketing analytics.

When TMA underperform, the explorer represents firms tend to employ BDA. Because the big data in real time is underused, traditional marketing analytics loses the uncertainty that brought by the things in real time which extract from multimedia files, blogs, and mobile telephone applications. Whereas the dynamic events may influence information fusion accuracy. Moreover, because of the excellent using of the big data analytics characteristics, the complex methods become more and more popular in the market. In addition, the above-mentioned algorithms have been used to catch automated knowledge and improve new product success as strong forecasting tools. However, when

the other limits are ignored, either method can be used to generate new knowledge in the short time and the long time. Our study proposes the following conclusions. (c) Firms fuse the information medially when they adopt big data analytics or traditional marketing analytics. (d) Firms obtain new product success medially when they adopt big data analytics or traditional marketing analytics.

Spectators represent firms resistant to analytics as the matrix defines them. For example, information from traditional marketing analytics, new product development, and the big data analytics can generate paradoxical conclusions. So firm administrators can use elicitation method to decide strategies. Certainly, new product success is almost impossible to occur if both TMA and BDA are underused. This study suggests the following conclusions. (e) Firms fuse the information least when they adopt little big data analytics or traditional marketing analytics. (f) Firms obtain new product success least when they adopt little big data analytics or traditional marketing analytics.

3.3 *Estimation strategy*

To predict the sale of new products, this study design the knowledge fusion method to recognize individual module in database. So we suggest marketing strategies to estimate the method which can be verified by existing products. Of course, using the information of marketing products to predict the sales of new products is harder than to predict the existing products. Then, we hope the knowledge fusion can act excellently on both the two types of data.

Errors are measured according to comparing the sales in real world and the sales in our prediction. Since our method is assessed in logarithmic space, so the results are expressed in same space. Power exponentiation can transform the results into sales of products. It is found that results may contain deviation. To correct the deviation the operation of multiplying the above results by $\exp(\frac{1}{2} \cdot MSE)$.

The following descriptions are the estimate strategies we adopt.

Firstly, the root mean squared error (Hyndman & Koehler, 2006) is used whose size is equal to the inputting data. It is important that the root mean squared error is more sensitive to outlier than existing methods.

Secondly, the mean absolute percentage error (Kolassa & Schütz, 2007, Hyndman & Koehler, 2006, Armstrong & Collopy, 1992) is used, whose size is not dependent on the input. It can be adopted to compare predicting results on different types of data. Thirdly, the correlation is used, which also is adopted to measure and identify the relationship of the predicted and factual results.

4 CONCLUSIONS

Collaborative knowledge fusion is presented by evaluating and merging different methods. A great deal of information from thousands of stakeholders is needed to refer to get new product success in current market.

Nowadays, as the burgeoning elements become critical to a particular product such as technology, regulation and inputs, the complexity and speed increase. As a result, the firms acquires and analyzes information rise. Aiming at the phenomenon, this study constructs a framework to help initiate the knowledge conceptually. Firms The knowledge fusion taxonomy suggests that Firms fuse the information the best when they adopt a highest big data analytics and traditional marketing analytics. Furthermore, this study provides insight to help managers decide when and which to use the different data for the cost of collecting and analyzing both types of data is not always be justified in all situations. In addition, this study identifies the knowledge fusion method to recognize individual module in database. So we suggest marketing strategies to estimate the method which can be verified by existing products.

In summary, in this paper we combine big data analytics and traditional marketing analytics that uses historical data of existing products and new products. According to improve which to the predict sales of new products, this paper introduces the method of knowledge fusion. According to this paper, we find that to obtain its benefits the fusion to improve new product success is not automatic and sometimes needs strategic selections. To improve the practicability of this method, we extracted the numeric and textual data from customer reviews, and then the method of knowledge fusion is applied to sales of both existing products and new product prediction.

ACKNOWLEDGMENT

This work was supported in part by the China National Nature Science Foundation (No.61305027) and the Academic Special Support Program for the Young Doctor from Shandong University of Finance and Economics.

REFERENCES

Abelló, A., Romero O., On-line analytical processing, in: Encyclopaedia of Database Systems, *Springer*, New York (USA), 2009: 1949–1954.

Anderson, P. (1999). Perspective: Complexity theory and organization science. *Organization Science*, 10(3): 216–232.

Armstrong, J.S., & Collopy, F. (1992). Error measures for generalizing about forecasting methods: Empirical comparisons. *International Journal of Forecasting*, 8(1): 69–80.

Asay, M. (2014). NoSQL databases are going mainstream-They actually have paying customers. *Read write*, November 21, http://readwrite.com/2014/09/23/nosql-databaseredis-labs-ofer-bengallast accessed Nov 21st, 2014.

Chen, H., Chang, R.H., & Storey, V.C. (2012). Business intelligence and analytics: From big data to big impact. *MIS Quarterly*, 36(4), 1165–1188.

Chen, Y., & Xie, J. (2008). Online consumer review: word-of-mouth as a new element of marketing communication mix. *Management Science*, 54(3), 477–491.

Cravens, D.W., & Piercy, N.F. (2005). In T. Emrich (Ed.), Strategic marketing Higher Education. *Veracity in big data reliability of routes*. Boston: McGraw-Hill Irwin (ISBN: 9780071263351). http://imsc.usc.edu/retreat2014/presentations/IMSC_retreat_2014_Tobias.pdf.

Fan, J., Kalyanpur, A., Gondek, D.C., & Ferrucci, D.A. (2012). Automatic knowledge extraction from documents. *IBM Journal of Research and Development*, 56(3–4): 5–1.

Forrester (2011). Expand your digital horizon with big data. *Forrester*, May 27 http://www.asterdata.com/newsletter-images/30-04-2012/resources/Forrester_Expand_Your_Digital_Horiz.pdf.

Goldberger, A.S. (1968). The interpretation and estimation of Cobb-Douglas functions. *Econometrica*, 35: 464–472.

Gustke, C. (2013). Big data takes turn as market darling. *CNBC*, http://www.cnbc.com/id/100638376 Access: Apr 21st, 2014.

Herbig, P., Milewicz, J., & Golden, J.E. (1994). Differences in forecasting behavior between industrial product firms and consumer product firms. *Journal of Business & Industrial Marketing*, 9(1): 60–69.

Hyndman, R.J., & Koehler, A.B. (2006). Another look at measures of forecast accuracy. *International Journal of Forecasting*, 22(4): 679–688.

Kolassa, S., & Schütz, W. (2007). Advantages of the MAD/MEAN ratio over the MAPE. *Foresight: The International Journal of Applied Forecasting*, 6: 40–43.

Lamb, J. (2014). Need for speed in data analysis. *Raconteur*, http://raconteur.net/ technology/need-for-speed-in-data-analysis last accessed Dec 21st, 2014.

Nickerson, J.A., & Zenger, T.R. (2004). A knowledge-based theory of the firm—The problem-solving perspective. *Organization Science*, 15(6): 617–632.

Provost, F., & Fawcett, T. (2013). Data science and its relationship to big data and data driven decision making. *Big Data*, 1(1), 51–59.

Srinivasan, S., Ramakrishnan, S., & Grasman, S.E. (2005).Identifying the effects of cannibalization on the product portfolio. Marketing Intelligence & Planning, 23(4): 359–371.

Tsoukas, H., & Vladimirou, E. (2001).What is organizational knowledge? *Journal of Management Studies*, 38(7): 973–993.

Wang, X., Mai, F., & Chiang, R.H.L. (2013). Market dynamics and user generated content about tablet computers. *Marketing Science*, 33(3): 449–458.

West, G. (2013).Wisdom in numbers. *Scientific American*, 308(5): 14–14.

Zhao, Y., Yang, S., Narayan, V., & Zhao, Y. (2013). Modeling consumer learning from online product reviews. *Marketing Science*, 32(1): 153–169.

Energy Science and Applied Technology ESAT 2016 – Fang (Ed.)
© 2016 Taylor & Francis Group, London, ISBN 978-1-138-02973-6

Coal logistics demand forecast based on an integrated prediction model

Yanfang Pan & Enyou Liu
School of Beijing, Jiaotong University, Beijing, China

ABSTRACT: Accurate demand forecasting coal logistics will facilitate railway logistic company to make effective decisions, including transportation planning, distribution yard, and pricing, so as to reduce the waste of transport capacity, improve customer satisfaction and competitiveness of enterprises. Logistics demand for coal as the research object analyzes the influencing factors, forming the corresponding index system, and establish an integrated forecasting model, combining time series analysis and multiple linear regression analysis. The model will be practiced using the real data, to verify the effectiveness of the method in coal logistics forecast.

Keywords: secondary exponential smoothing, multiple linear regression, coal logistics, adaptive smoothing factor, prediction

1 INTRODUCTION

Nowadays, rail freight is embarked on the management model of reform from the view of logistic. Railway logistics company, as an integrated logistics provider, generally transports bulk commodities such as coal, so an accurate prediction of coal logistics demand will increase the effectiveness of rail logistics company decision-making, so that production planning, route planning, construction of coal reserve base and so on are more reasonable.

At present, domestic and foreign scholars have done a lot of research about the forecasting of demand. Gardner describes the research status of exponential smoothing, emphasizing economic forecasting features of the method, and robust feature (Gardner J E S, 2006). Bates and Granger in the 1960s both proved that unbiased prediction combination is superior to every single individual prediction (Bates J N et al, 1964). Since then, research on combination forecasting abroad has made great progress. Ng-combined regression analysis and time series to predict tender price index, which showed that combination forecasting model predicted better than two separate prediction results (Ng S T et al, 2004). Zhou Zhuanshi proposed variable weight and Yongqing Liu proposed recombination prediction technique (Zhou Chuanshi et al, 1995), and Liao Xuezhen and Zhou proposed Logit Combined Forecasting technique (Zhou Hong et al, 2003), by adjusting the combination of weights to improve the prediction accuracy. In addition, domestic forecast of demand for coal was also analyzed in the same study. Jing Quanzhong, Zhang Jian studied GM (1, 1) model

in coal demand forecast in our country, and the total amount of short-term coal demand is predicted (Jing Quanzhong et al, 2004).

Through the above documents, it can be seen that the current prediction for coal is not tremendous. In addition, the most faced the national energy consumption, the less the decision-oriented rail logistics companies. This article includes intended railway logistics enterprises, using integrated forecasting model of multiple regression and time series to predict coal demand, facilitating rail logistics business decisions.

2 FACTORS OF COAL LOGISTICS DEMAND

Analyzing factors of coal logistics demand is based on coal logistics demand forecast. For a rail logistics business, try to analyze the factors area along its rail downstream demand for coal.

2.1 *Analyzation of effecting factors of coal demand*

2.1.1 *The level of economic development*
Logistics is an important part of social and economic activities, the development level of which can directly affect the level of demand for logistics development. A state or regional economic development in the region will promote the demand for raw materials, and low and high value-added products. With the development of the total size of the regional economic development, the level of the corresponding social energy consumption will

be raised, which will increase the demand of coal logistics.

2.1.2 *Industrial structure*

Industrial structure is an inherent link between the various industries in the economic area, and gross domestic products are relatively a percentage of the gross national product. A regional industrial structure, the primary industry, and secondary industry accounted for a larger amount and represented the rapid industrial development in the region, coal, and other raw materials for the corresponding logistics demand.

2.1.3 *Electric power consumption*

Power industry accounted for a pivotal position in coal consumption system, which includes the coal consumption by generating electricity and by heating. Coal consumption of generating electricity depends mainly on the amount of thermal power generation and coal consumption of heating depends on heat supply. There is approximately 80% of thermal power in the total power generation of China. Accordingly, coal consumption is affected by the consumption of electric power.

2.1.4 *Residents consumption*

When the level of consumption of an area is improved, it will generate more demand for material and services, and logistics in order to stimulate economic development in the region, and thus the introduction of foreign-invested enterprises, promoting production and consumption. Therefore, the level of residents' consumption of a region will affect the demand for coal in the area of logistics to a certain extent.

2.1.5 *Export levels*

In recent years, the trend of economic globalization led international logistics. The high export levels of a region will be easier to open up an international market for the coal logistics.

2.2 *Coal demand index system*

For a rail logistics business, analyze the impact of economic factors, industry structure, power consumption, etc., in the area along its rail downstream in which there is the enterprises of coal demand, and thus can build coal logistics demand index system, as shown in Table 1.

3 INTEGRATED PREDICTION MODEL

The demand for coal logistics will be influenced not only by the factors described above, but also by the demand value of the previous period.

Table 1. Coal logistics demand index system.

Index system	Index	Code	Unit
Coal logistics demand index system	Electric power consumption	X_1	TWh
	CPI	X_2	—
	GDP	X_3	100 million yuan
	Percentage of secondary industry	X_4	%
	Import and export value	X_5	100 million $
	Coal logistics demand	Y	10 kt

Therefore, this paper predict the demand for coal logistics by the secondary exponential smoothing with adaptive smoothing coefficient, and then consider the value of it as a factor of regression model to forecast.

3.1 *Time series method*

The secondary exponential smoothing is a kind of time series method, which is widely applied to predict owing to its convenient simplicity and of which the smoothing coefficient will influence the effect of prediction that always be selected by the experience or by testing many times, causing to an effect less accuracy. Besides, the smoothing coefficient does not have the adaptive capacity when the variation happens to change in the different period of time series, which will elicit a deviation to the effect. So this paper will use the secondary smoothing coefficient method with adaptive coefficient to predict the demand (Li Suoping et al, 2004).

3.2 *Integrated prediction model*

Combining the first part of the analysis of influencing factors on demand, select the appropriate factors to build a multiple linear regression forecast model including the time series method, then to develop an integrated model, as shown in Figure 1.

There should be a multiple linear regression forecast model based on the method of first part and the influencing factors for coal demand, as the following model:

$$Y_t = a_0 + \sum_{i=1}^{n} a_i X_{i,t} + a_{n+1} D_t + \varepsilon_t. \tag{1}$$

Y_t — the regression prediction value of period t
— the parameters to be estimated
— the i-th factors of period t
— the random factors, obeying the standard normal distribution

4 CASE ANALYSIS

It will take Shanxi as an example to test the integrated model.

4.1 Times series model to predict the demand for coal

The first stage of the integrated model will take the data from the year of 2000 to 2014, to obtain the predicted value D_t for each year, using JAVA language to choose a smooth coefficient initial value $a = 0.8$, the average model error is minimized.

4.2 Integrated model to predict the demand for coal

4.2.1 Factors test

Before an integrated prediction, you need to do a test of correlation of factors, the results of which are shown in Table 2.

Some factors should be removed because of that there is a strong correlation between them from the results of analysis, so this paper will remove the factor X_1 and the factor X_2, combining the prediction value of the first stage and getting the new index system as shown in Table 3.

4.2.2 Demand forecast

Select the actual consumption value of coal as the fitting target of integrated model, and put the value of new factor system from the year of 2000 to 2010 into the integrated model, getting the results as shown in Table 4.

Figure 1. The flow chart of integrated model.

Table 2. Results of correlation analysis.

Index	Y	X_1	X_2	X_3	X_4	X_5
Y	1					
X_1	0.980587	1				
X_2	0.301286	0.303025	1			
X_3	0.927321	0.977152	0.253765	1		
X_4	0.68226	0.577088	0.471735	0.453741	1	
X_5	0.949469	0.97381	0.452084	0.945145	0.60353	1

Then, predict the value of coal demand from the year 2012 to 2014 according to the above results, obtaining the results as shown in Table 5.

Compare these with the results of multiple linear regression forecast model and the results of times series model as shown in Table 6.

From the results of comparison, it is clear that the integrated model is better than the other two models.

Table 3. New index system and results of first stage.

year	Y	X_3	X_4	X_5	D_t
2000	12704	1845.72	45.9	17.6438	12704
2001	13271	2029.53	46.5	19.4098	13649
2002	16587	2324.8	48.1	23.114	18756.06
2003	18829	2855.22	50.5	30.8417	21125.28
2004	19112	3571.4	52.8	53.8173	20199.17
2005	22631	4180	54.8	55.4597	25097.27
2006	25514	4753	55.6	66.2779	28327.33
2007	27772	5733	56.4	115.7047	30268.76
2008	26879	7135	57.1	143.9004	27319.69
2009	26149	7358	54.1	85.5432	25751.68
2010	28180	9188.8	56.6	125.7839	29192.85
2011	30896	11214.2	58.6	147.5981	33027.87
2012	31085	12126.6	54.8	150.4325	32119.27
2013	33062	12665.3	52.2	157.9785	34584.19
2014	32056	12761.5	49.3	162.4852	32098.99

Table 4. Regression coefficients.

	Coefficients	P-value
Intercept	−27887.8	0.104283
X Variable 1	1.504094	0.006977
X Variable 2	816.159	0.06876
X Variable 3	−68.4525	0.026927
X Variable 4	0.254894	0.296295

Table 5. Results of prediction.

year	Consumption	Prediction	Relative error
2012	31085	35121.43	−0.1299
2013	33062	32966.75	0.0029
2014	32056	31766.74	0.0090

Table 6. Comparison of three models.

year	Integrated model	Regression model	Time series
2012	−12.99%	−12.39%	6.25%
2013	0.29%	4.13%	2.85%
2014	0.90%	9.75%	7.89%

5 SUMMARY

The integrated model is superior to a single model on prediction accuracy, which has been tested by actual data, making up a single prediction model for its univariate shortage, whereas the time series just make up for the shortage of regression model without considering the prediction value of next year. In addition, it has been tested whether the integrated model can also be adopted for the prediction of coal demand. Thus, it can be applied for the prediction of coal demand of rail logistics business.

REFERENCES

Bates J N, Granger C W J. Combination of forecasts [J]. Operations Research Quarterly. Vol. 4(1964), p. 451–468.

Gardner J E S. Exponential smoothing: The state of the art-Part II [J]. International Journal of Forecasting. Vol.22 (2006), p.637–666.

Jing Quanzhong, Zhang Jian. GM (1, 1) model on the application of forecast of coal demand [J]. China Coal. Vol. 30(2004), p.17–19.

Li Suoping, Liu Kunhui. Secondary smoothing coefficient method with adaptive coefficient and its application [J]. System Engineering Theory and Practice. Vol. 02(2004), p.95–99.

Ng S T, Cheung S O, Skitmore M, et al. An integrated regression analysis and time series model for construction tender price index forecasting [J]. Construction Management and Economics. Vol. 22(2004), p. 483–493.

Zhou Chuanshi, Liu Yongqing. Study on the variable weight combination of forecasting model [J]. Forecasting. Vol. 14(1995), p.47–48, 51.

Zhou Hong, Liao Xuezhen. Study on Logit combined forecasting of the market demand [J]. Systems Engineering: Theory &Practice. Vol. 23(2003), p.63–69.

Energy Science and Applied Technology ESAT 2016 – Fang (Ed.)

Research on initiator inversion data processing on the Bayes estimation

Y.D. Wang, Z.L. Xuan, W.B. Yu, K. Yao & T.P. Li
Ordnance Engineering College, Shijiazhuang, China

ABSTRACT: This paper will analyze the characteristics and the cause of initiator inversion data. They will be expressed as two distribution forms, which based on determining the reliability test data of initiator are the success or failure data. With the failure rate in different times being a criterion, the failure number and failure time in inversion data is modified by using Bayes estimation method. The results show that this method can effectively correct inversion data during life test of initiator.

Keywords: initiator; reliability; inversion; Bayes

1   INTRODUCTION

As the most sensitive initiation and ignition energy, initiator performance directly affects the weapon system's operational safety and reliability. Therefore, the practical need to carry out studies of initiator storage reliability is very urgent. To study in the failure mechanism and failure mode, it is a need for a large number of tests to obtain data validation. However, during doing actual life test to obtain the failure data, there is an "inversion" phenomenon that is taking initiators with short storage time and long storage time as test examples, when selecting the same sample capacity, the failure number of the former example is greater than the latter one. This situation obviously is contrary to products failure law and serious data inversion will make reliability estimation generate large deviations or even errors. Therefore, this paper will modify the failure number and failure time in inversion data by using the Bayes estimation method, which can provide an effective method for the preprocessing of initiator reliability data.

2   THE RELIABILITY OF INITIATOR

Initiator is a general term components, devices and systems disposable, which is after receiving ignition instructions, with smaller energy in order to inspire their inside reagents to produce the combustion or explosion, push them flame, blast, and high pressure gas, to achieve a predetermined function ignition, initiation, acting, etc. (Li, L. 2010).
There are mainly three initiator functions in the weapon system:

1. Used in weapon system of ignition, fire, extension, and its control system. Make weapons launch, carrying systems safety and reliability.
2. Used in weapon system of initiating explosive and control system, dominated the effect of the warhead, realize the damage of the targets' nearest enemy.
3. Used in weapons systems of push, pull, cutting, separation and scatters and posture control, work sequence and its control system, make the weapon system realize their own adjust or state transition and safety control.

As an important Component in ammunition, initiator was failed component easily in stress storage environment. Such as when a munitions depot that has been stored for a certain number of years of type terminal sensitive projectiles to shoot, which some section was lose efficacy, the situation partially warhead separation failure. According to the analysis and research, the reason is that initiator's detonating device, charge separation, and other section cannot be reliable.

Pyrotechnics failure is mainly affected by temperature, humidity, salt spray, and mold (Li, L. 2010). However, since the initiator is not an independent combat unit, but it is loaded into the shells together for storage and use. Therefore, the following are the characteristics of initiator storage:

1. some parts of initiator have their tightness,
2. project bodies having sealing,
3. elastomer packaging has some means of sealing and desiccant,
4. storage process adopted measures to prevent water erosion.

As the most sensitive initiation and ignition energy, initiator's performance directly affects the use of ammunition reliability and safety; therefore, we must be conducted for initiating explosive device reliability research.

3 INVERSION CAUSE ANALYSIS

The reasons causing the "inversion" of initiator failure data may be various. Through analysis of storage detection environment, product nature, and personnel factors, the likely main reasons will be summarized as follows (Li, M.L. 1997):

1. In actual work, constrained by time, fund, and others factors, the detection test cannot be carried out thoroughly and the number of sample is not big enough or its representativeness is not strong enough, so the test results are random.
2. Initiator belongs to products with high reliability. After storage in a long time, its failure rate is still very low, which means though the number of testing sample is very big, failure number is still small, and that increases the probability of the emergence of the "inversion".
3. Other reasons, including testing personnel differences, equipment status differences or calibration conditions differences, etc.

4 CHARACTERISTICS OF INITIATOR LIFE DATA

Initiator is generally believed that the life is characterized by lognormal distribution (Murtagh, F. et al, 2003), and life data generally has the following characteristics:

1. Binomial distribution form

Ammunition storage reliability test data can be written as the pass-fail data (Niu, Y.T. 2014). Therefore, whether it is irreproducible data of initiator components or performance test data of optical electronic components, it can be expressed as the binomial distribution form:

$$\left(t_i, n_i, f_i\right) \quad i = 1, 2, \cdots, k \qquad (1)$$

T_i is storage time, n_i is the test sample size. The failure number of this sample is f_i and k is the number of detection points.

In the life test, if we measured the invalid number f_i in the (t_{i-1}, t_i) time interval. The failure time t_{ij} of the j sample in the i test interval is:

$$t_{ij} = t_{i-1} + j\frac{t_i - t_{i-1}}{r_i + 1} \quad (j = 1, 2, \cdots, f_i) \qquad (2)$$

2. Incomplete data

Remember that the test sample amount is n_i, the reliable storage life of the j sample is X_{ij} ($j = 1, 2, \cdots, n_i$), if $X_{ij} > t_i$, it indicates that there is no failure; if $X_{ij} < t_i$, it indicates inefficiency. Order that:

$$Y_{ij} = \begin{cases} 1 & X_{ij} \leq t_i \\ 0 & X_{ij} > t_i \end{cases} \qquad (3)$$

So $f_i = \sum_{j=1}^{n_i} Y_{ij}$ is the failure number, which is the storage time t_i of n_i samples.

It can be seen that, in a specific sample, we can only know whether it is still reliable in time, but cannot get the specific failure time point.

3. The data "inversion" potentially

Explosive storage reliability test data are of the success or failure type (Sarhan et al, 2003). Data structure as shown in the formula (3). We mark the t_i moment cumulative failure rate is p_i, $p_i = f_i / n_i$. Due to the initiator storage, reliability is decreased with the increasing time. Therefore:

$$p_0 \leq p_1 \leq p_2 \leq \cdots \leq p_k \qquad (4)$$

P_0 is the failure rate for the factory time.

Set up the factory acceptance sampling plan is (n / c), We investigate all or no less than 10 batches of factory acceptance test data. The cumulative sample size is Σn, and the total number of unqualified samples is Σd. Therefore, the p_0 is satisfied:

$$\sum_{i=0}^{\Sigma n} \binom{\Sigma n}{i} p_0^i (1 - p_0)^{\Sigma n - i} = 1 - \gamma \qquad (5)$$

γ is the confidence level, and it is desirably 0.95. When determining whether the quality of the product is very good, it is desirably $p_0 = 0$.

Owing to the randomness of sampling, especially if the sample size is small, it may appear $p_i > p_j$ ($i < j$), the data "inversion". At this point, it is necessary to modify the data collected.

5 CORRECTION OF INVERSION DATA

Bayes estimation can modify the inversion data effectively, and the choice of prior distribution is important (Wang, K.M. 2014). Assuming that the failure rate $p(t_i)$ of the t_i moment is obeying the uniform distribution, the calculation is relatively simple. We select both ends and more reliable end data for failure of a period of time (t_i, t_j) on the "inversion" processing. If the assuming t_i moment data is more reliability, it would be standard for failure rate t_i and corrected after t_i moments data. If t_j is the end of the test, then take the p_{i+1} a priori distribution of the uniform distribution on $(p_i,1)$; if t_j is not the end of the test, then take p_{i+1} a priori distribution of (p_i,p_{i+1}) on the uniform distribution. Sign the $X=(X_1,X_2,\cdots,X_n)$ for $X=(X_1,X_2,\cdots,X_n)$ moment, which is the overall random sample test result, and the distribution is:

$$p\left(X|P_{i+1}\right)=P_{i+1}{}^{X}\left(1-P_{i+1}\right)^{1-X}, \quad X=0,1 \tag{6}$$

We can note that the density of the sample X under P_{i+1} is:

$$
\begin{aligned}
f\left(X|P_{i+1}\right)&=\prod_{i=1}^{n}P_{i+1}{}^{X_i}\left(1-P_{i+1}\right)^{1-X_i}\\
&=P_{i+1}{}^{\sum_{i=1}^{n}X_i}\left(1-P_{i+1}\right)^{n_{i+1}-\sum_{i=1}^{n}X_i}\\
&=P_{i+1}{}^{f_{i+1}}\left(1-P_{i+1}\right)^{n_{i+1}-f_{i+1}}
\end{aligned} \tag{7}
$$

In this formula, f_{i+1} is the t_{i+1} moments of the original number of failure, and n_{i+1} is the sample size. Therefore, when t_j is the end of the test, the p_{i+1} posterior distribution is:

$$\pi\left(P_{i+1}|X\right)=\frac{P_i{}^{f_{i+1}}\left(1-P_i\right)^{n_{i+1}-f_{i+1}}}{\int_{P_i}^{1}P_i{}^{f_{i+1}}\left(1-P_i\right)^{n_{i+1}-f_{i+1}}dP} \tag{8}$$

Thus, the Bayes estimation for p_{i+1} is:

$$\hat{p}_{i+1}=\frac{f_{i+1}+1}{n_{i+1}+2}+\frac{p_i{}^{f_{i+1}+1}\left(1-p_i\right)^{n_{i+1}-f_{i+1}+1}}{(n_{i+1}+2)\int_{P_i}^{1}p_i{}^{f_{i+1}}\left(1-p_i\right)^{n_{i+1}-f_{i+1}}dP} \tag{9}$$

Similarly, when t_j is not the end of test time, the Bayes p_{i+1} is estimated to be:

$$\hat{p}_{i+1}=\frac{\int_{p_i}^{p_{j+1}}p_i{}^{f_{i+1}+1}\left(1-p_i\right)^{n_{i+1}-f_{i+1}}dP}{\int_{p_i}^{p_{j+1}}p_i{}^{f_{i+1}}\left(1-p_i\right)^{n_{i+1}-f_{i+1}}dP} \tag{10}$$

$\hat{P}_{i+1}$ is the modified failure rate, and the correction number of failure is:

$$\hat{f}_{i+1}=n_{i+1}\cdot\hat{p}_{i+1} \tag{11}$$

Then, take the $\hat{p}_{i+1}$ as the benchmark, repeat the process until we obtain $\hat{P}_j$ and $\hat{f}_j$.

If the period of time (t_i,t_j) need to deal with "inversion" data, t_j time data are more reliable. Therefore, we can rely on the failure rate p_j, corrected before t_j moment data (Yan, N. 2006). If

$i=1$, then take the prior distribution of p_{j-1} for the uniform distribution on the $(0,p_j)$; if $i\neq 1$, then take the prior distribution of p_{j-1} for the uniform distribution on the (p_{i-1},p_j). Furthermore, we can use the p_{j-1} Bayes estimation, if $i=1$, we can modify by type (12); if $i\neq 1$, we can modify by type (13):

$$\hat{p}_{j-1}=\frac{f_{j-1}+1}{n_{j-1}+2}-\frac{p_j{}^{f_{j-1}+1}\left(1-p_j\right)^{n_{j-1}-f_{j-1}+1}}{(n_{j-1}+2)\int_0^{p_j}P^{f_{j-1}}\left(1-P\right)^{n_{j-1}-f_{j-1}}dP} \tag{12}$$

$$\hat{p}_{j-1}=\frac{\int_{p_{i-1}}^{p_j}P^{f_{j-1}+1}\left(1-P\right)^{n_{j-1}-f_{j-1}}dP}{\int_{p_{i-1}}^{p_j}P^{f_{j-1}}\left(1-P\right)^{n_{j-1}-f_{j-1}}dP} \tag{13}$$

The corrected failure number of t_{j-1} moment is:

$$\hat{f}_{j-1}=n_{j-1}\cdot\hat{p}_{j-1} \tag{14}$$

Then, we take $\hat{p}_{j-1}$ as the benchmark, repeated the process until we obtained $\hat{p}_i$ and $\hat{f}_i$. We can obtain monotone data sequence for the modified data, which is replaced by the original data sequence type (1).

In general, the failure number of samples will be changed from integer to non-integer after the "inversion" data processing. In order to facilitate, handle data, and avoid the error not to be too large, we can make the following agreement:
1) Order

$$f_i'=\begin{cases}\left[\hat{f}_i\right] & \left|\hat{f}_i-\left[\hat{f}_i\right]\right|<0.5\\[2mm]\left[\hat{f}_i\right]+1 & \left|\hat{f}_i-\left[\hat{f}_i\right]\right|\geq 0.5\end{cases} \tag{15}$$

$[\hat{f}_i]$ is an integral part of $\hat{f}$.
2) We modified the failure time, of the f_i' failure sample. If $\hat{f}_i=[\hat{f}_i]$, then the failure time should be revised forward; if $\hat{f}_i=[\hat{f}_i]+1$, then the failure time should be backward corrected. The calculation method is:

$$t'_{if_i'}=t_{if_i'}+\left(f_i'-\hat{f}_i\right)\frac{t_i-t_{i-1}}{r_i+1} \tag{16}$$

6 CASE ANALYSIS

The terminal-guided projectile was maintenance in the factory, and replacement of a number of electrical actuators. We randomly selected 80 electrical actuators for storage reliability test. There are 75 samples for primary experiment and 5 for standby experiment. We carried out the performance test for the electrical actuator, which is natural storage and after a certain accelerated life:

1.   Electrical actuator electrical parameter standard
Safety current test: give the product through the 1 A DC current, continuously for 1 min, and no ignition.

Ignition test: at –32°C, through the 4 A DC current, continuous 20 ms should be reliable ignition. At 60°C, through the 7 A DC current, continuous 10 ms should be reliable ignition.

2.   Testing of natural environment storage samples
The terminal-guided projectile has been stored for 20 years. We performed pulse signal test on 25 electrical actuators.

3.   Testing of the sample after accelerated life test
The other two groups were carried out for a period of 7.64 days and 17.82 days of accelerated life test. We can get the failure mechanism of explosive that did not change from the hypotheses of accelerated life testing and the initiator commonly used temperature stress accelerated life test. It can be considered that after the accelerated, the storage life of the two groups was achieved 23 years and 27 years. Take a pulse signal test for electrical actuators after accelerated life test.

Table 1.   Failure data statistics and correction of electrical actuator.

Storage time	Sample capacity	Failure data statistics	Failure data correction
20	25	1	1
23	25	0	1.43
27	25	2	3.171

Table 2.   Failure time correction of electrical actuator.

Failure number	1	2	3	4	5
Failure time	19.25	21.38	23.12	25.00	26.84

4.   Failure statistics and correction
According to the electrical parameter standard and test results, the failure condition of the test samples can be obtained from Table 1. Owing to the electrical actuator, there is the emergence of substandard products and replaced in factory, so it is believed that in the twentieth years of 1 failure is more reasonable. According to the formulas (12), (13), and (14), the failure data are corrected, as shown in Table 1.

According to the formulas (2), (15), and (16) for the failure time is corrected, as shown in Table 2.

7   CONCLUSION

This paper analyzes the characteristics and causes of initiator inversion data. The initiator reliability test data of success or failure type for binomial distribution form are considered. Take the failure rate in (t_i, t_j) time as the evolution criteria by using the Bayes estimation method. We fixed the failure number and failure time of inversion phenomenon, under different conditions. The results show that this method can correct the inversion data effectively, which provided a method for pre-processing the reliability data of the initiator.

REFERENCES

Li, L. 2010. Analysis on the Safety Influence Factors of Pyrotechnics, Journal of Naval Aeronautical and Astronautical University, 25(5): 545–548.
Li, M.L. 1997. Ammunition Storage Reliability, Beijing: National Defence Industry Press, 42–48.
Murtagh, F. & Starck, J.L. 2003. Quantization from Bayes factors with application to multilevel thresholding, Pattern Recognition Letters, 24(12): 2001–2007.
Niu, Y.T. 2014. Storage life assessment of an accelerometer under natural storage environment, Journal of Chinese Inertial Technology, 22(4): 552–556.
Sarhan & Ammar, M. 2003. Empirical Bayes estimates in exponential reliability model, Applied Mathematics and Computation, 135(2–3): 319–332.
Wang, K.M. 2014. Engineering of Initiators & Pyrotechnics, Beijing: National Defense Industry Press, 23–25.
Yan, N. 2006. Generality of Analysis on Initiating Explosive Device Failure, Failure Analysis and Prevention, 1(1): 10–14.

Energy Science and Applied Technology ESAT 2016 – Fang (Ed.)
© 2016 Taylor & Francis Group, London, ISBN 978-1-138-02973-6

Application and transition of the velocity gradient tensor from a rectangular coordinate to a spherical coordinate

Haishun Deng, Pan Xu & Ran Huang
School of Mechanical Engineering, Auhui University of Science and Technology, Huainan, China

ABSTRACT: The spherical coordinate and the cylindrical coordinate are often used in the computation fluid dynamic, so their differential equation tensor needs to be translated from the rectangular coordinate system, and the calculation of the process is more cumbersome. Depending on the equivalence relation of the differential equation in the different coordinate systems, the velocity tensor in the rectangular coordinate can be expressed by the spherical coordinate through derivation, and the first-order differential equation in the rectangular coordinate can be expressed by the spherical coordinate. It will lay the foundation for the transition of the velocity gradient tensor from the rectangular coordinate to the spherical coordinate. In addition, the velocity gradient tensor is applied to the fluid continuity differential equation, and the fluid continuity differential equation in the spherical coordinate system can be obtained. It is given to prove its feasibility through the example. This method can been applied to the transition of the second-order velocity gradient tensor in the spherical coordinate system, and the tensor transition of the cylindrical coordinate and the other orthogonal coordinate systems.

Keywords: velocity gradient; tensor structure; spherical coordinate system; rectangular coordinate system

1 INTRODUCTION

Tensor analysis becomes one of the important analysis tools of continuum mechanics, theoretical physics and other subjects, because of its successful application in the general theory of relativity (Braman K, 2010) (Kilmer M E et al, 2013). Its basic theories are derived on the conduction of continuum concept in fluid mechanics (Liu Bo-yun et al, 2013) (Li Zhan-hua et al, 2014). Under normal circumstances, the motion differential equations of fluid mechanics is expressed by the first- and second-order partial differential equations in the rectangular coordinate system or partial differential equations, where the first-order velocity gradient and the second-order velocity gradients, (i,j, k = 1,2,3) are the basic elements, namely the tensor structures in the rectangular coordinate. Despite the rectangular coordinate system being commonly used, it is not applicable to some special problems, when calculating the lubrication characteristics of port pair (Hu Xiao et al, 2012) and slipper pair (Ma J M et al, 2015) in axial piston pump. It is better to choose the cylindrical coordinate system, while calculating the lubrication characteristics of spherical port plate pair (Xiong Xiao-ping et al, 2009) and plunger head, the spherical coordinate system is better. Choosing appropriate coordinate system, such as cylindrical coordinate system and spherical coordinate system, can make the calculation and analysis process simple, but the transition of tensor structure from rectangular coordinate to other coordinates is difficult. The way to copy with this problem is converting the basic expression, the first-order and second-order velocity gradient, to the tensor structure in corresponding coordinate system. In this paper, the velocity gradient of the spherical tensor structure is discussed, which is prepared for the transition of differential equation from rectangular coordinate system to spherical coordinate system.

2 PREPARATORY CALCULATION AND SIGN CONTRACT

As shown in figure 1, the relationship between rectangular coordinates and spherical coordinates can be expressed as follows (Li Zhen et al, 2014):

$$\begin{cases} x = r\sin\theta\cos\varphi \\ y = r\sin\theta\cos\varphi \\ z = r\cos\theta \end{cases} \quad \begin{cases} r^2 = x^2 + y^2 + z^2 \\ \theta = \mathrm{atan}(\sqrt{x^2 + y^2}/z) \\ \varphi = \mathrm{atan}(y/x) \end{cases} \quad (1)$$

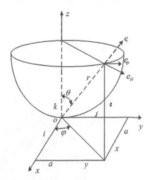

Figure 1. Rectangular-spherical coordinate system.

According to equation (1):

$$\begin{cases} \dfrac{\partial r}{\partial x} = \sin\theta\cos\varphi & \dfrac{\partial r}{\partial y} = \sin\theta\sin\varphi & \dfrac{\partial r}{\partial z} = \cos\theta \\[2mm] \dfrac{\partial \theta}{\partial x} = \dfrac{\cos\theta\cos\varphi}{r} & \dfrac{\partial \theta}{\partial y} = \dfrac{\cos\theta\sin\varphi}{r} & \dfrac{\partial \theta}{\partial z} = \dfrac{-\sin\theta}{r} \\[2mm] \dfrac{\partial \varphi}{\partial x} = \dfrac{-\sin\varphi}{r\sin\theta} & \dfrac{\partial \varphi}{\partial y} = \dfrac{\cos\varphi}{r\sin\theta} & \dfrac{\partial \varphi}{\partial z} = 0 \end{cases} \quad (2)$$

Arrange the relevant amount of equation (2) in vector form and introduce shorthand notation, equation (3) can be calculated:

$$\begin{cases} (r,\theta,\varphi)'_x = \left(\dfrac{\partial r}{\partial x} \quad \dfrac{\partial \theta}{\partial x} \quad \dfrac{\partial \varphi}{\partial x} \right) \\[3mm] (r,\theta,\varphi)'_y = \left(\dfrac{\partial r}{\partial y} \quad \dfrac{\partial \theta}{\partial y} \quad \dfrac{\partial \varphi}{\partial y} \right) \\[3mm] (r,\theta,\varphi)'_z = \left(\dfrac{\partial r}{\partial z} \quad \dfrac{\partial \theta}{\partial z} \quad \dfrac{\partial \kappa}{\partial z} \right) \end{cases} \quad (3)$$

The relationship of speed between rectangular coordinate system and spherical coordinate system can be calculated by the following equation:

$$\begin{cases} u_x = u_r \sin\theta\cos\varphi + u_\theta \cos\theta\cos\varphi - u_\varphi \sin\varphi \\ u_y = u_r \sin\theta\sin\varphi + u_\theta \cos\theta\sin\varphi + u_\varphi \cos\varphi \\ u_z = u_r \cos\theta - u_\theta \sin\theta \end{cases} \quad (4)$$

Record equation (4) into matrix form:

$$\begin{bmatrix} u_x \\ u_y \\ u_z \end{bmatrix} = \begin{bmatrix} \sin\theta\cos\varphi & \sin\theta\cos\varphi & -\sin\varphi \\ \sin\theta\sin\varphi & \cos\theta\sin\varphi & \cos\varphi \\ \cos\theta & -\sin\theta & 0 \end{bmatrix} \begin{bmatrix} u_r \\ u_\theta \\ u_\varphi \end{bmatrix} \quad (5)$$

Or further shorthand for:

$$\vec{u}_{x,y,z} = A\vec{u}_{r,\theta,\varphi} = \left[a_{ij} \right]\vec{u}_{r,\theta,\varphi} = \left[A_1 \ A_2 \ A_3 \right]^T \vec{u}_{r,\theta,\varphi} \quad (6)$$

where $\vec{u}_{x,y,z} = (u_x\, u_y\, u_z)^T$, $\vec{u}_{r,\theta,\varphi} = (u_r\, u_\theta\, u_\varphi)^T$, $A = [a_{ij}]_{3\times3}$.

$$\begin{cases} A_1 = (a_{11}, a_{12}, a_{13}) = (\sin\theta\cos\varphi, \cos\theta\sin\varphi, -\sin\varphi) \\ A_2 = (a_{21}, a_{22}, a_{23}) = (\sin\theta\sin\varphi, \cos\theta\sin\varphi, \cos\varphi) \\ A_3 = (a_{31}, a_{32}, a_{33}) = (\cos\theta, \sin\theta, 0) \end{cases} \quad (7)$$

3 MATHEMATICAL TRANSITION FROM VELOCITY GRADIENT $\partial u_i / \partial x_j$ TO SPHERICAL COORDINATES

According to the Euler method of speed, $u_i(i=1,2,3)$ is the function $u_i = u_i(x,y,z)$ of x, y, z. According to equation (1), $x = x(r,\theta,\varphi)$, $y = y(r,\theta,\varphi)$, $z = z(r,\theta,\varphi)$. In three-dimensional coordinates, velocity gradients of $u_1 = u_x$ are $\partial u_x / \partial x_x, \partial u_y / \partial x_y$ and $\partial u_z / \partial x_z$. Equation (8) can be calculated:

$$\begin{cases} \dfrac{\partial u_x}{\partial x} = \dfrac{\partial u_x}{\partial r}\dfrac{\partial r}{\partial x} + \dfrac{\partial u_x}{\partial \theta}\dfrac{\partial \theta}{\partial x} + \dfrac{\partial u_x}{\partial \varphi}\dfrac{\partial \varphi}{\partial x} = (r,\theta,\varphi)'_x \left[(u_x)'_{r,\theta,\varphi} \right]^T \\[3mm] \dfrac{\partial u_x}{\partial y} = \dfrac{\partial u_x}{\partial r}\dfrac{\partial r}{\partial y} + \dfrac{\partial u_x}{\partial \theta}\dfrac{\partial \theta}{\partial y} + \dfrac{\partial u_x}{\partial \varphi}\dfrac{\partial \varphi}{\partial y} = (r,\theta,\varphi)'_y \left[(u_x)'_{r,\theta,\varphi} \right]^T \\[3mm] \dfrac{\partial u_x}{\partial z} = \dfrac{\partial u_x}{\partial r}\dfrac{\partial r}{\partial z} + \dfrac{\partial u_x}{\partial \theta}\dfrac{\partial \theta}{\partial z} + \dfrac{\partial u_x}{\partial \varphi}\dfrac{\partial \varphi}{\partial z} = (r,\theta,\varphi)'_z \left[(u_x)'_{r,\theta,\varphi} \right]^T \end{cases} \quad (8)$$

where

$$\left[(u_x)'_{r,\theta,\varphi} \right]^T = \left(\dfrac{\partial u_x}{\partial r} \quad \dfrac{\partial u_x}{\partial \theta} \quad \dfrac{\partial u_x}{\partial \varphi} \right)^T \quad (9)$$

According to equation (8) and the method of coordinate rotation, $\partial u_i / \partial x_j$ and $\partial u_z / \partial x_j$ and can be calculated and shorthand for:

$$\dfrac{\partial u_i}{\partial x_j} = \left(\dfrac{\partial r}{\partial x_j} \quad \dfrac{\partial \theta}{\partial x_j} \quad \dfrac{\partial \varphi}{\partial x_j} \right) \left(\dfrac{\partial u_i}{\partial r} \quad \dfrac{\partial u_i}{\partial \theta} \quad \dfrac{\partial u_i}{\partial \varphi} \right)^T$$
$$= (r,\theta,\varphi)'_{x_j} \left[(u_i)'_{r,\theta,\varphi} \right]^T \quad (10)$$

The calculations of $\partial u_i / \partial r, \partial u_i / \partial \theta$ and $\partial u_i / \partial \varphi$ are involved in equation (10), for the convenience, taking $u_i = u_1 = u_x$.

According to equation (8), the speed u_x can be expressed as $u_x = A_1 u_{r,\theta,\varphi} = (\alpha_{11}\, \alpha_{12}\, \alpha_{13}) u_{r,\theta,\varphi}$, see u_r, u_θ, u_φ and A_1 as the function of r, θ and φ, then:

$$\begin{cases} \dfrac{\partial u_x}{\partial r} = \dfrac{\partial A_1}{\partial r}\vec{u}_{r,\theta,\varphi} + A_1 \dfrac{\partial}{\partial r}(\vec{u}r,\theta,\varphi)^T \\[3mm] \dfrac{\partial u_x}{\partial \theta} = \dfrac{\partial A_1}{\partial \theta}\vec{u}r,\theta,\varphi + A1 \dfrac{\partial}{\partial \theta}(\vec{u}r,\theta,\varphi)^T \\[3mm] \dfrac{\partial u_x}{\partial \varphi} = \dfrac{\partial A_1}{\partial \varphi}\vec{u}r,\theta,\varphi + A_1 \dfrac{\partial}{\partial \varphi}(\vec{u}r,\theta,\varphi)^T \end{cases} \quad (11)$$

Record equation (11) into matrix form

$$
\begin{bmatrix} \dfrac{\partial u_x}{\partial r} \\[2mm] \dfrac{\partial u_x}{\partial \theta} \\[2mm] \dfrac{\partial u_x}{\partial \varphi} \end{bmatrix} = \begin{bmatrix} \dfrac{\partial A_1}{\partial r} \\[2mm] \dfrac{\partial A_1}{\partial \theta} \\[2mm] \dfrac{\partial A_1}{\partial \varphi} \end{bmatrix} \vec{u}_{r,\theta,\varphi} + \begin{bmatrix} \dfrac{\partial}{\partial r}\left(\vec{u}_{r,\theta,\varphi}\right)^T \\[2mm] \dfrac{\partial}{\partial \theta}\left(\vec{u}_{r,\theta,\varphi}\right)^T \\[2mm] \dfrac{\partial}{\partial \varphi}\left(\vec{u}_{r,\theta,\varphi}\right)^T \end{bmatrix}
$$

$$
A_1^T = \begin{bmatrix} \dfrac{\partial a_{11}}{\partial r} & \dfrac{\partial a_{12}}{\partial r} & \dfrac{\partial a_{13}}{\partial r} \\[2mm] \dfrac{\partial a_{11}}{\partial \theta} & \dfrac{\partial a_{12}}{\partial \theta} & \dfrac{\partial a_{13}}{\partial \theta} \\[2mm] \dfrac{\partial a_{11}}{\partial \varphi} & \dfrac{\partial a_{12}}{\partial \varphi} & \dfrac{\partial a_{13}}{\partial \varphi} \end{bmatrix} \begin{bmatrix} u_r \\ u_\theta \\ u_\varphi \end{bmatrix}
$$

$$
+ \begin{bmatrix} \dfrac{\partial u_r}{\partial r} & \dfrac{\partial u_\theta}{\partial r} & \dfrac{\partial u_\varphi}{\partial r} \\[2mm] \dfrac{\partial u_r}{\partial \theta} & \dfrac{\partial u_\theta}{\partial \theta} & \dfrac{\partial u_\varphi}{\partial \theta} \\[2mm] \dfrac{\partial u_r}{\partial \varphi} & \dfrac{\partial u_\theta}{\partial \varphi} & \dfrac{\partial u_\varphi}{\partial \varphi} \end{bmatrix} \begin{bmatrix} a_{11} \\ a_{12} \\ a_{13} \end{bmatrix} \tag{12}
$$

According to the coordinate transition method, thus the following equation can be obtained:

$$
\begin{bmatrix} \dfrac{\partial u_i}{\partial r} \\[2mm] \dfrac{\partial u_i}{\partial \theta} \\[2mm] \dfrac{\partial u_i}{\partial \varphi} \end{bmatrix} = \begin{bmatrix} \dfrac{\partial a_{i1}}{\partial r} & \dfrac{\partial a_{i2}}{\partial r} & \dfrac{\partial a_{i3}}{\partial r} \\[2mm] \dfrac{\partial a_{i1}}{\partial \theta} & \dfrac{\partial a_{i2}}{\partial \theta} & \dfrac{\partial a_{i3}}{\partial \theta} \\[2mm] \dfrac{\partial a_{i1}}{\partial \varphi} & \dfrac{\partial a_{i2}}{\partial \varphi} & \dfrac{\partial a_{i3}}{\partial \varphi} \end{bmatrix} \begin{bmatrix} u_r \\ u_\theta \\ u_\varphi \end{bmatrix}
$$

$$
+ \begin{bmatrix} \dfrac{\partial u_r}{\partial r} & \dfrac{\partial u_\theta}{\partial r} & \dfrac{\partial u_\varphi}{\partial r} \\[2mm] \dfrac{\partial u_r}{\partial \theta} & \dfrac{\partial u_\theta}{\partial \theta} & \dfrac{\partial u_\varphi}{\partial \theta} \\[2mm] \dfrac{\partial u_r}{\partial \varphi} & \dfrac{\partial u_\theta}{\partial \varphi} & \dfrac{\partial u_\varphi}{\partial \varphi} \end{bmatrix} \begin{bmatrix} a_{i1} \\ a_{i2} \\ a_{i3} \end{bmatrix} \tag{13}
$$

For the general orthogonal curvilinear coordinate system, q_1, q_2, q_3, assume their corresponding speeds to be v_1, v_2, v_3, then:

$$
\begin{bmatrix} \dfrac{\partial u_i}{\partial q_1} \\[2mm] \dfrac{\partial u_i}{\partial q_2} \\[2mm] \dfrac{\partial u_i}{\partial q_3} \end{bmatrix} = \begin{bmatrix} \dfrac{\partial a_{i1}}{\partial q_1} & \dfrac{\partial a_{i2}}{\partial q_1} & \dfrac{\partial a_{i3}}{\partial q_1} \\[2mm] \dfrac{\partial a_{i1}}{\partial q_2} & \dfrac{\partial a_{i2}}{\partial q_2} & \dfrac{\partial a_{i3}}{\partial q_2} \\[2mm] \dfrac{\partial a_{i1}}{\partial q_3} & \dfrac{\partial a_{i2}}{\partial q_3} & \dfrac{\partial a_{i3}}{\partial q_3} \end{bmatrix} \begin{bmatrix} v_1 \\ v_2 \\ v_3 \end{bmatrix}
$$

$$
+ \begin{bmatrix} \dfrac{\partial v_1}{\partial q_1} & \dfrac{\partial v_2}{\partial q_1} & \dfrac{\partial v_3}{\partial q_1} \\[2mm] \dfrac{\partial v_1}{\partial q_2} & \dfrac{\partial v_2}{\partial q_2} & \dfrac{\partial v_3}{\partial q_2} \\[2mm] \dfrac{\partial v_1}{\partial q_3} & \dfrac{\partial v_2}{\partial q_3} & \dfrac{\partial v_3}{\partial q_3} \end{bmatrix} \begin{bmatrix} a_{i1} \\ a_{i2} \\ a_{i3} \end{bmatrix} \tag{14}
$$

According to equation (14), the structure form (tensor form) in the orthogonal curvilinear system of velocity gradient $\partial u_i/\partial x_j$ in the rectangular coordinate system can be expressed by the following equation.

$$
\frac{\partial u_i}{\partial x_j} = \begin{pmatrix} q_1 & q_2 & q_3 \end{pmatrix}'_{x_j} \left(\begin{bmatrix} \dfrac{\partial a_{i1}}{\partial q_1} & \dfrac{\partial a_{i2}}{\partial q_1} & \dfrac{\partial a_{i3}}{\partial q_1} \\[2mm] \dfrac{\partial a_{i1}}{\partial q_2} & \dfrac{\partial a_{i2}}{\partial q_2} & \dfrac{\partial a_{i3}}{\partial q_2} \\[2mm] \dfrac{\partial a_{i1}}{\partial q_3} & \dfrac{\partial a_{i2}}{\partial q_3} & \dfrac{\partial a_{i3}}{\partial q_3} \end{bmatrix} \begin{bmatrix} v_1 \\ v_2 \\ v_3 \end{bmatrix} \right.
$$

$$
\left. + \begin{bmatrix} \dfrac{\partial v_1}{\partial q_1} & \dfrac{\partial v_2}{\partial q_1} & \dfrac{\partial v_3}{\partial q_1} \\[2mm] \dfrac{\partial v_1}{\partial q_2} & \dfrac{\partial v_2}{\partial q_2} & \dfrac{\partial v_3}{\partial q_2} \\[2mm] \dfrac{\partial v_1}{\partial q_3} & \dfrac{\partial v_2}{\partial q_3} & \dfrac{\partial v_3}{\partial q_3} \end{bmatrix} \begin{bmatrix} a_{i1} \\ a_{i2} \\ a_{i3} \end{bmatrix} \right) \tag{15}
$$

4 NUMERICAL EXAMPLE

Fluid continuity theorem is one of the basic theories first introduced in computational fluid mechanics, also a basic differential equation of computational fluid dynamics simulation analysis. The fluid continuous differential equation in the rectangular coordinate system is:

$$
\frac{\partial}{\partial x}\left(\rho u_x\right) + \frac{\partial}{\partial y}\left(\rho u_y\right) + \frac{\partial}{\partial z}\left(\rho u_z\right) + \frac{\partial \rho}{\partial t} = 0 \tag{16}
$$

where $\rho = \rho(x,y,z,t)$ is the fluid density equation, while the density of steady flow does not change with time, namely $\partial \rho/\partial t = 0$. At the same time, assume the density $\rho(x,y,z) = \text{const}$, then equation (16) can be simplified to:

$$
\frac{\partial u_x}{\partial x} + \frac{\partial u_y}{\partial y} + \frac{\partial u_z}{\partial z} = 0 \tag{17}
$$

Equation (17) is one of the common forms of the fluid continuous differential equation, according to this, the form of $\partial u_i/\partial x$ in spherical coordinate system can be expressed as follows:

$$
\begin{cases}
\dfrac{\partial u_x}{\partial x} = \dfrac{\partial u_r}{\partial r}\sin^2\theta\cos^2\varphi + \dfrac{\partial u_\theta}{\partial r}\sin\theta\cos\theta\cos^2\varphi \\[2mm]
\qquad - \dfrac{\partial u_\varphi}{\partial r}\sin\theta\sin\varphi\cos\varphi + \dfrac{\partial u_r}{\partial\theta}\dfrac{\sin\theta\cos\theta\cos^2\varphi}{r} \\[2mm]
\qquad + \dfrac{\partial u_\theta}{\partial\theta}\dfrac{\cos^2\theta\cos^2\varphi}{r} - \dfrac{\partial u_\varphi}{\partial\theta}\dfrac{\cos\theta\sin\varphi\cos\varphi}{r} \\[2mm]
\qquad - \dfrac{\partial u_r}{\partial\varphi}\dfrac{\sin\varphi\cos\varphi}{r} - \dfrac{\partial u_\theta}{\partial\varphi}\dfrac{\cos\theta\sin\varphi\cos\varphi}{r\sin\theta} + \\[2mm]
\qquad \dfrac{\partial u_\varphi}{\partial\varphi}\dfrac{\sin^2\varphi}{r\sin\theta} + \dfrac{u_r}{r}(\cos^2\theta\cos^2\varphi + \sin^2\varphi) \\[2mm]
\qquad + \dfrac{u_\theta}{r}(\sin^2\varphi - \sin\theta\cos\theta\cos^2\varphi) + \dfrac{u_\varphi}{r}\dfrac{\sin\varphi\cos\varphi}{\sin\theta} \\[3mm]
\dfrac{\partial u_y}{\partial y} = \dfrac{\partial u_r}{\partial r}\sin^2\theta\sin^2\varphi + \dfrac{\partial u_\theta}{\partial r}\dfrac{\sin\theta\cos\theta\sin^2\varphi}{r} \\[2mm]
\qquad + \dfrac{\partial u_\varphi}{\partial r}\sin\theta\sin\varphi\cos\varphi + \dfrac{\partial u_r}{\partial\theta}\dfrac{\sin\theta\cos\theta\sin^2\varphi}{r} + \\[2mm]
\qquad \dfrac{\partial u_\theta}{\partial\theta}\dfrac{\cos^2\theta\sin^2\varphi}{r} + \dfrac{\partial u_\varphi}{\partial\theta}\dfrac{\cos\theta\sin\varphi\cos\varphi}{r} \\[2mm]
\qquad + \dfrac{\partial u_r}{\partial\varphi}\dfrac{\sin\theta\sin\varphi\cos\varphi}{r\sin\theta} + \dfrac{\partial u_\theta}{\partial\varphi}\dfrac{\cos\theta\sin\varphi\cos\varphi}{r\sin\theta} + \\[2mm]
\qquad \dfrac{\partial u_\varphi}{\partial\varphi}\dfrac{\cos^2\varphi}{r\sin\theta} + \dfrac{u_r}{r}(\cos^2\theta\sin^2\varphi + \cos^2\varphi) \\[2mm]
\qquad + \dfrac{u_\theta}{r}(\cos^2\varphi - \sin\theta\cos\theta\sin^2\varphi) - \dfrac{u_\varphi}{r}\dfrac{\sin\varphi\cos\varphi}{\sin\theta} \\[3mm]
\dfrac{\partial u_z}{\partial z} = \dfrac{\partial u_r}{\partial r}\cos^2\theta - \dfrac{\partial u_\theta}{\partial r}\sin\theta\cos\theta - \dfrac{\partial u_r}{\partial\theta}\dfrac{\sin\theta\cos\theta}{r} \\[2mm]
\qquad + \dfrac{\partial u_\theta}{\partial\theta}\dfrac{\sin^2\theta}{r} + \dfrac{u_r}{r}\sin^2\theta + \dfrac{u_\theta}{r}\sin\theta\cos\theta
\end{cases}
\tag{18}
$$

According to equation (18), the equivalent form of continuous differential equation in the spherical coordinate system can be calculated by the following equation:

$$
\begin{aligned}
&\dfrac{\partial u_x}{\partial x} + \dfrac{\partial u_y}{\partial y} + \dfrac{\partial u_z}{\partial z} = \dfrac{\partial u_r}{\partial r} + \dfrac{2u_r}{r} + \dfrac{1}{r}\dfrac{\partial u_\theta}{\partial\theta} \\[2mm]
&\quad + \dfrac{1}{r\sin\theta}\dfrac{\partial u_\varphi}{\partial\varphi} + \dfrac{u_\theta\cos\theta}{r\sin\theta} = \dfrac{1}{r^2}\left(\dfrac{\partial}{\partial r}(r^2 u_r)\right) \\[2mm]
&\quad + \dfrac{1}{r\sin\theta}\dfrac{\partial}{\partial\theta}(u_\theta\sin\theta) + \dfrac{1}{r\sin\theta}\dfrac{\partial u_\varphi}{\partial\varphi} = 0
\end{aligned}
\tag{19}
$$

If $\partial\rho/\partial t$ has nothing to do with the coordinates, at the same time, $\partial(\rho u_i)/\partial t$ can be regarded as a kind of velocity gradient, and equation (19) can be corrected in the form of equation (17):

$$
\begin{aligned}
&\dfrac{1}{r^2}\left(\dfrac{\partial}{\partial r}(\rho r^2 u_r)\right) + \dfrac{1}{r\sin\theta}\dfrac{\partial}{\partial\theta}(\rho u_\theta\sin\theta) \\[2mm]
&\quad + \dfrac{1}{r\sin\theta}\dfrac{\partial}{\partial\varphi}(\rho u_\varphi) + \dfrac{\partial\rho}{\partial t} = 0
\end{aligned}
\tag{20}
$$

Equation (20) is the expression of equation (16) in the spherical coordinate system. If selecting a cone curved hexahedron in the spherical coordinate system and establishing its physical model according to the law of conservation of mass, and then carrying on the mathematical derivation, we can get the above conclusion, obviously, the conversion process in this paper is simpler.

5 SUMMARY

The numerical example of fluid continuous differential equation proves that the spherical tensor structure analysis of velocity gradient $\partial u_i/\partial x_j$ is correct and it is established for general orthogonal curvilinear coordinate system. This method also opens up the analysis thought of the second-order velocity gradient $\partial^2 u_i/\partial x_j^2$, once $\partial^2 u_i/\partial x_j^2$ is confirmed, the basic differential equations in computational fluid dynamics can equivalently "translate" into the movement differential equations in the hope coordinate system (spherical or cylindrical). The analysis method and conclusion in this article are also established for partial differential equations (such as Laplace equation) in other disciplines.

ACKNOWLEDGMENTS

This work was financially supported by the National Natural Science Foundation of China (No. 51575002) and the Natural Science Foundation of Anhui Province (No.1508085ME80).

REFERENCES

Braman K. Third-Order Tensors as Linear Operators on a Space of Matrices[J]. Linear Algebra & Its Applications, 2010, 433(7):1241–1253.

Hu Xiao, Wang Shao-ping, Han Lei. Modeling and Simulation on Pressure Distribution of Plane Port Pair in Axial Piston Pump[J]. Hydraulics Pneumatics & Seals, 2012, 32(8): 68–71.

Kilmer M E, Braman K, Ning H, et al. Third-order tensors as operators on matrices: a theoretical and computational framework with applications in imaging[J]. Siam Journal on Matrix Analysis & Applications, 2013, 34(1):148–172.

Li Zhan-hua, Zheng Xu. The Problems and Progress in the Experimental Study of Micro/Nano-scale Flow[J]. Journal of Experiments in Fluid Mechanics, 2014 (3):1–11.

Li Zhen, Zhang Xi-wen, He Feng. Evaluation of Vortex Criteria by Virtue of the Quadruple Decomposition of Velocity Gradient Tensor[J]. Acta Physica Sinica, 2014, 05:259–265.

Liu Bo-yun, Pu Jin-yun, An Hong-rui, et al. Two-phase Flow Heat and Mass Transfer Model Construction and Experiment Study of the Fuel Evaporation Progress[J]. Journal of Wuhan University of Technology, 2013, 35(4): 59–63.

Ma J M, Li Q L, Ren C Y, et al. Influence Factors Analysis on Hydraulic Axial Piston Pump/Slipper Pair[J]. Journal of Beijing University of Aeronautics and Astronautics, 2015, (3):405–410.

Xiong Xiao-ping, Yu Lan-ying, Wang Guo-zhi, et al. The Adventure Analysis of Spherieal Valve Plate in Axial Piston Pump[J]. Fluid Power Transmission and Control, 2009 (4): 25–28.

Energy Science and Applied Technology ESAT 2016 – Fang (Ed.)
© *2016 Taylor & Francis Group, London, ISBN 978-1-138-02973-6*

Prediction of membrane protein types using an ensemble classifier

P.Y. Zhao & G.F. Yang
School of Mathematics and Quantitative Economics, Shandong University of Finance and Economics, Jinan, Shandong, P.R. China

ABSTRACT: Prediction of membrane protein types is an important project and is related to the function of a membrane protein. Based on the pseudo amino acid composition, which is presented by Chou originally to integrate the complementary information, the concept of approximate entropy is adopted to extract features of membrane proteins in this study. An ensemble classifier composed by multiple fuzzy K-nearest neighbor classifiers with individual parameters is built to predict the types of membrane proteins. The sequences digitized by pseudo amino acid composition is adopted to train the aforementioned each basic classifier. The jackknife dataset test is taken to verify the effectiveness. The high success rates indicating that we may use the ensemble classifier as a very helpful model to identify membrane proteins types.

Keywords: pseudo amino acid; K-nearest neighbor classifier; approximate entropy

1 INTRODUCTION

Membrane proteins implement the primary biological functions even the membranes are composed mainly by the lipid bilayer (B. Alberts et al. 1994, K.C. Chou & H.B. Shen 2007). As membrane proteins can mediate many interactions not only between cells and extracellular surroundings but also between the cytosol and organelles binding with membranes. As a result, membrane proteins are objectives of nearly half of the drugs in medicine.

Predicting membrane protein types is an important project and is related to the function of a membrane protein. Therefore, we can try to identify the types of membrane protein to analyze their functions and guide to production of drugs. Different types of membrane proteins carry out different functions biologically. Chou et al. classified membrane proteins to the following eight categories: type I, type II, type III, type IV, multipass, lipid-chain-anchored, GPI-anchored, and peripheral (K.C. Chou & H.B. Shen 2007), where the former four types are single-pass trans-membrane proteins (M. Spiess 1995).

Informative hints in understanding the functions of Golgi-resident proteins may be provided by correctly identifying the types of Golgi-resident proteins. However, the experimental approaches to this information are costly and time-consuming. Therefore, identifying protein types using only the protein sequences to present an automated method seems to be urgent, that is, transiting and using the proteins to carry our basic research and develop new drugs real-timely (K.C. Chou 2004, G. Lubec et al. 2005). Some investigators have done some work regarding this (S. Padhi et al. 2015, E.M. Rath et al. 2013, P. Du, Y. Li 2006, X. Pu et al. 2007, M. Hayat & A. Khan 2010, J.Y. Wang 2012).

Based on the pseudo amino acid composition, which is presented by Chou originally to integrate the complementary information, the concept of approximate entropy is adopted to extract features of membrane proteins in this study. An ensemble classifier composed by multiple fuzzy K-nearest neighbor classifiers with individual parameters is built to predict the types of membrane proteins. The sequences digitized by pseudo amino acid composition are adopted to train each aforementioned basic classifier.

2 DATASETS AND METHODS

2.1 Datasets

This study adopts the datasets built by Chou and Shen (K.C. Chou & H.B. Shen 2007). The Swiss-Prot database is an important source for them to collect protein sequences. In the training dataset, the numbers of membrane proteins belonging to type I, type II, type III, and type IV are 610, 312, 24, and 44, respectively. Moreover, based on the pseudo amino acid composition, another four types of membrane proteins is added to the training dataset. The numbers of membrane proteins belong to multi-pass transmembrane proteins, lipid chain-anchored membrane proteins, GPI-anchored membrane proteins, and peripheral membrane proteins are 1316, 151, 182, and 610, respectively. There are 3249 membrane protein sequences contained in the training dataset in total.

In the testing dataset the number of membrane proteins belong to type I, type II, type III and type IV is 444, 78, 6 and 12 respectively, whereas the number of membrane proteins belong to multi-pass transmembrane proteins, lipid chain-anchored membrane proteins, GPI-anchored membrane proteins and peripheral membrane proteins is 3265, 38, 46, and 444, respectively. There are 4333sequences contained in the testing dataset totally.

2.2 Methods

PseAAC

The Pseudo Amino Acid Composition (PseAAC) presented by Chou originally to integrate the complementary information is adopted (K.C. Chou 2007). Through the discrete model of pseudo amino acid composition, the membrane sequences can be formulated as

$$S = [s_1, s_2, ..., s_{20}, s_{21}, ... s_{20+\lambda}] \tag{1}$$

where,

$$
s_u = \begin{cases} \dfrac{g_u}{\sum_{i=1}^{20} g_i + a \sum_{j=1}^{\lambda} f_j} & (1 \le u \le 20) \\[4ex] \dfrac{af_{u-20}}{\sum_{i=1}^{20} g_i + a \sum_{j=1}^{\lambda} f_j} & (20+1 \le u \le 20+\lambda) \end{cases} \tag{2}
$$

where $g_i (1 \le i \le 20)$ is the frequency of occurrence, a is the effective weight element given by employing the efficient method (B. Hong et al. 1999), and $f_i (1 \le i \le \lambda)$ is the approximate entropy of a membrane protein while describing the else sequence characters. In short, each of the twenty types of amino acids contained in the numerical sequences is represented by combing frequency of occurrence and the else sequence characters. Where the above two parts are reflected in the former twenty dimensions components and the latter λ ones.

ApEn settings

ApEn (approximate entropy) has got successful application in some medical fields nowadays, which was first introduced by Pincus in 1991 (S.M. Pincus 1991).

ApEn is constant zero or a positive number. It presents the complexity of time series. A group of measurements of systematic complexity are correlated with ApEn in theory, which does well in curing Cardiovas. We improved the model characterized by Pincus and the new model is faster much than the primary one.

First, we must translate the original amino acid sequences into number forms. Hydrophobicity indexes of twenty sequences are used to calculate the ApEn values. For example, for a specific protein *Pro*, i.e. one of the digitized membrane proteins sequences S is described by $S = s_1 s_2 s_3 ... s_N$. We calculate the values of ApEn by the following procedures:

The first step: n-D vector $S(i)$ is fused by sequence $w(i)$ through the serial number.

$$S(i) = [w(i), w(i+1), ..., w(i+n-1)], \quad i = 1 \sim N-n+1 \tag{3}$$

The second step: Given a initial value p, if the values of distance d_{ij} and the distance $d_{(i+1)(j+1)}$ are smaller than p, then $C_i^n = C_i^n + 1$.

where $d_{ij} = w(i) - w(j)$ and $d_{(i+1)(j+1)} = w(i+1) - w(j+1)$.

$$C_i^n(p) = \sum_{j=1}^{N-n+1} d_{ij} \cap d_{(i+1)(j+1)} \tag{4}$$

$$C_i^{n+1}(p) = \sum_{j=1}^{N-n} d_{(i+1)(j+1)} \cap d_{(i+2)(j+2)} \tag{5}$$

The third step: Take the logarithm and compute the average of the of $C_i^n(p)$ and $C_i^{n+1}(p)$, the outcomes are expressed as $\Phi^n(p)$ and $\Phi^{n+1}(p)$.

$$\Phi^n(p) = \frac{1}{N-n+1} \sum_{i=1}^{N-n+1} \ln C_i^n(p) \tag{6}$$

$$\Phi^{n+1}(p) = \frac{1}{N-n} \sum_{i=1}^{N-n} \ln C_i^{n+1}(p) \tag{7}$$

The fourth step: Values of approximate entropy are the differences between $\Phi^n(p)$ and $\Phi^{n+1}(p)$.

$$ApEn(n,p) = \Phi^n(p) - \Phi^{n+1}(p) \tag{8}$$

In this paper, we preset $n = 2$ and $b = 0.10 / 0.15 / 0.20 / 0.25$ Where $p = b * SD(w)$ and $SD(w)$ indicates the standard difference of amino acid w. According to the steps, the approximate entropy of proteins is computed. Referring to the preceding experience, approximate entropy will catch a more rational feature statistically.

According to the description above, four different approximate entropy values corresponded to every value of b for a given membrane protein sequence, whereas the parameter p may have four different values.

Ensemble classifier

PseAA component combines twenty amino acid composition values and some else relevant information. In the pseudo amino acid composition, the parameter λ decides the amount of the pseudo amino acid compositions from the membrane proteins we described, while in the ApEn algorithm, value of parameter p determines the ApEn.

We integrate many individual basic classifiers, which is proposed above as $C(\lambda, P)$ to generate an ensemble classifier. Each basic classifier has different individual values of λ and b. The description above suggests that b catch one or more values from the dataset {0.10, 0.15, 0.20, 0.25}. At the same time the scale of the pseudo amino acid composition contained in P is λ. In another word, P is a subset that contains λ components. If we preset the λ to be 1, 2, 3, and 4, respectively, 15 FKNN classifiers can be formed through the values of approximate entropy.

In this paper, the FKNN classifier (J.M. Keller et al. 1985) is adopted as the basic classifier to generate the ensemble classifier. FKNN classifier fuses the fuzzy set theory with K-nearest neighbor classifiers which uses the pseudo amino acid compositions to train parameters. According to combination of the 15 individual classifiers, the ensemble classifier is formulated by

$$C = C_1(\lambda_1, P_1) \cup C_2(\lambda_2, P_2) \cup ... \cup C_{15}(\lambda_{15}, P_{15}) \quad (9)$$

where $\cup$ presents the combination operating, and C presents the above-mentioned classifier fused by the classifiers $C_1(\lambda_1, P_1)$, $C_2(\lambda_2, P_2)$, ..., and $C_{15}(\lambda_{15}, P_{15})$. The weighted fusion value of the exports calculated by the fifteen basic classifier exports is adopted to be the final export value of C. In order to identify the types of membrane proteins belong to, we propose a model. The protein Y belongs to the j-th type is defined by

$$Y_j = \sum_{i=1}^{15} a_i \Phi(C_i, T_j), (j = 1, 2, ..., 8) \quad (10)$$

Table 1. The 15 classifiers and the values of parameters.

Classifiers	Parameters		
	Λ	n	p
C_1	1	2	0.10
C_2	1	2	0.15
C_3	1	2	0.20
C_4	1	2	0.25
C_5	2	2	0.10, 0.15
C_6	2	2	0.10, 0.20
C_7	2	2	0.10, 0.25
C_8	2	2	0.15, 0.20
C_9	2	2	0.15, 0.25
C_{10}	2	2	0.20, 0.25
C_{11}	3	2	0.10, 0.15, 0.20
C_{12}	3	2	0.10, 0.15, 0.25
C_{13}	3	2	0.10, 0.20, 0.25
C_{14}	3	2	0.15, 0.20, 0.25
C_{15}	4	2	0.10, 0.15, 0.20, 0.25

where a_i is the weight, normally it is set at 1 to simplify the calculation, C_i is the value of predicting function of i-th individual classifier and T_j is the dataset of j-th type proteins sequences.

The function $\Phi(C_i, T_j)$ is defined as

$$\Phi(C_i, T_j) = \begin{cases} 1, & \text{if } C_i \in T_j \\ 0, & \text{otherwise} \end{cases} \quad (11)$$

As thus, we only need to find the subset with highest score of Equation (10), to which the query protein Y is predicted to belong.

3 RESULTS AND DISCUSSION

By conducting tests and comparing the ensemble classifier with pre-existing classifiers, our classifier is shown to have better performance in predicting accuracies by combining the fifteen different classifiers. In literatures, jackknife test and sub-sampling test are often carried out to verify the accuracies of the classifiers on the independent dataset test. Although both of the tests all cross validate the classifiers, the latter is usually regarded as the power of predictor, most strictly and objectively (K.C. Chou & C.T. Zhang 1995). Also it has been increasingly adopted by investigators (H.B. Shen & K.C. Chou 2005, H. Liu et al. 2005, S.Q. Wang et al. 2006, X.G. Yang et al. 2007, P.Y. Zhao & Y.S. Ding 2009, M. Hayat & A. Khan 2010, A. Majid & T.S. Choi 2010, B. Panda et al. 2013) to examine the results of different predicting models. As a result, the accuracies of the classifiers ought to be examined by the jackknife test and should have large values..

The accuracies of predicting membrane protein types by using the ensemble classifier are given in Table 2, where the effectiveness of fusion method is demonstrated.

4 CONCLUSIONS

In this work, we firstly propose a novel feature extraction method based on the pseudo amino acid compositions and approximate entropy. And then build a powerful classifier by fusing combines twenty amino acid composition such as the physicochemica property values and some else relevant information to predict membrane protein types. The outputs of jackknife test obtained by

Table 2. Prediction accuracy of jackknife tests.

Datasets	Accuracy of prediction								
	Type I	Type II	Type III	Type IV	Multi-pass	Lipid-chain-anchor	GPI-anchor	Peri pheral	Overall
D3249	86.8	79.6	55.8	77.3	95.2	63.3	77.7	78.2	89.4
D4333	88.4	77.9	43.3	75.0	95.3	49.5	79.3	84.9	92.3

547

our methods are 88.3% and 91.2% respectively on the training and testing datasets built by Chou and Shen. By integrating 15 effective basic FKNN classifiers, an ensemble classifier is presented. Tests results show that this method is more effective than the other individual feature extraction methods in view and combination of the individual classifier that significantly improves the accuracies of prediction compared with individual ones.

ACKNOWLEDGMENT

This work was supported in part by the China National Nature Science Foundation (No.61305027) and the Academic Special Support Program for the Young Doctor from Shandong University of Finance and Economics.

REFERENCES

Alberts, B., Bray, D., Lewis, J., Raff, M., Roberts, K. & Watson, J.D. 1994. *Molecular Biology of the Cell*. New York & London: Garland Publishing.

Cedano, J., Aloy, P., P'erez-Pons, J.A. & Querol, E. 1997. Relation between amino acid composition and cellular location of proteins. *Journal of Molecular Biology* 266(3): 594–600.

Chou, K.C. & Elrod, D.W. 1999. Prediction of membrane protein types and subcellular locations. *PROTEINS: Structure, Function, and Genetics* 34: 137–153.

Chou, K.C. & Shen, H.B. 2007. MemType-2 L: A Web server for predicting membrane proteins and their types by incorporating evolution information through Pse-PSSM. *Biochemical and Biophysical Research Communications* 360: 339–345.

Chou, K.C. & Zhang, C.T. 1995. Review: prediction of protein structural classes. *Critical Reviews in Biochemistry & Molecular Biology* 30(4):275–349.

Chou, K.C. 2001. Prediction of protein cellular attributes using pseudo amino acid composition, *PROTEINS: Structure, Function, and Genetics* 43: 246–255.

Chou, K.C. 2004. Review: Structural bioinformatics and its impact to biomedical science. *Current Medicinal Chemistry* 11: 2105–2134.

Du, P. & Li, Y. 2006. Prediction of protein submitochondria locations by hybridizing pseudo-amino acid composition with various physicochemical features of segmented sequence. *BMC Bioinformatics* 7(6):597–605.

Hayat, M. & Iqbal, N. 2014. Discriminating protein structure classes by incorporating Pseudo Average Chemical Shift to Chou's general PseAAC and Support Vector Machine. *Computer Methods and Programs in Biomedicine* 116(3): 184–192.

Hayat, M. & Khan, A. 2010.Membrane protein prediction using wavelet decomposition and pseudo amino acid based feature extraction. *Proceedings - 2010 6th International Conference on Emerging Technologies, ICET 2010* 1–6.

Hong, B., Tang, Q.Y. & Yang, F.S. 1999. ApEn and cross-ApEn: property, fast algorithm and preliminary application to the study of EEG and cognition. *Signal Process* 15: 100–108.

Keller, J.M., Gray, M.R. & Givens, J.A. 1985. A fuzzy k-nearest neighbor algorithm. *Systems Man & Cybernetics IEEE Transactions on* 15(4): 580–585.

Liu, H., Wang, M. & Chou, K.C. 2005. Low-frequency Fourier spectrum for predicting membrane protein types. *Biochemical & Biophysical Research Communications* 336: 737–739.

Lubec, G., Afjehi-Sadat, L., Yang, J.W. & John, J.P. 2005. Searching for hypothetical proteins: Theory and practice based upon original data and literature. *Progress in Neurobiology* 77: 90–127.

Majid, A. & Choi, T.S. 2010. A new ensemble scheme for predicting human proteins subcellular locations. *International Journal of Signal Processing, Image Processing and Pattern Recognition* 3(1): 1–8.

Nakashima, H., Nishikawa, K. & Ooi, T. 1986. The folding type of a protein is relevant to the amino acid composition. *Journal of Biochemistry* 99: 152–162.

Padhi, S., Ramakrishna, S. & Priyakumar, U.D. 2015. Prediction of the structures of helical membrane proteins based on a minimum unfavorable contacts approach. *Journal of Computational Chemistry* 36(8): 539–552.

Panda, B., Majhi, B., Mishra, A.P. & Rout, M. 2013. Performance evaluation of protein structural class prediction using artificial neural networks. *2013 International Conference on Human Computer Interactions, ICHCI 2013* 1–5.

Pincus, S.M. 1991. Approximate entropy as a measure of system complexity. *Proceedings of the National Academy of Sciences of the United States of America* 88: 2297–2301.

Rath, E.M., Tessier, D., Campbell, A.A., Lee, H.C., Werner, T., Salam, N.K., Lee, L.K. & Church, W. B. 2013. A benchmark server using high resolution protein structure data, and benchmark results for membrane helix predictions. *BMC Bioinformatics* 14(14): 1–11.

Shen, H.B. & Chou, K.C. 2006. Ensemble classifier for protein fold pattern recognition. *Bioinformatics* 22: 1717–1722.

Shen. H.B., Yang. J. & Chou. K.C. 2006. Fuzzy KNN for predicting membrane protein types from pseudo-amino acid composition. *Journal of Theoretical Biology* 240: 9–13.

Spiess, M. 1995. Heads or tails—what determines the orientation of proteins in the membrane. *FEBS Letters* 369: 76–79.

Wang, J.Y., Li Y.P., Wang, Q.Q., You, X.G., Man J.J., Wang, C. & Gao, X. 2012. ProClusEnsem: Predicting membrane protein types by fusing different modes of pseudo amino acid composition. *Computers in Biology and Medicine* 42(5): 564–574.

Wang, S.Q., Yang, J. & Chou, K.C. 2006. Using stacked generalization to predict membrane protein types based on pseudo amino acid composition. *Journal of Theoretical Biology* 242: 941–946.

Yang X.G., Luo, R.Y. & Feng, Z.P. 2007. Using amino acid and peptide composition to predict membrane protein types. *Biochemical & Biophysical Research Communications* 353: 164–169.

Zhao, P.Y. & Ding, Y.S. 2009. Using a fuzzy support vector machine classifier to predict interactions of membrane protein. *3rd International Conference on Bioinformatics and Biomedical Engineering, iCBBE 2009* 1–4.

Energy Science and Applied Technology ESAT 2016 – Fang (Ed.)
© *2016 Taylor & Francis Group, London, ISBN 978-1-138-02973-6*

Digital information classification based on singular value decomposition

C.S. Yuan, J.M. Wang, B.Z. Li, R. Lin & G.M. Lu
School of Electrical Information Engineering, Yunnan MinZu University, Kunming, China
Yunnan Province Colleges Minority Language Information Processing Engineering Research Center,
Yunnan MinZu University, Kunming, China

ABSTRACT: With the rapid development of information technology and Internet to promote the information technology revolution, allowing humans to enter the abundant data of the digital information age, a large number of structured data and unstructured data need to be processed by computer. In order to improve the processing speed of amount of digital information, it is very necessary to classify them. This paper presents a reconstruction algorithm, which is based on singular value decomposition of digital information. This new matrix, which is obtained by singular value decomposition and K-means is the classification of test data. In this method, we have modified the eigenvalues. The simulation results show that this new algorithm enhances the precision of classification and improves its generalization ability.

Keywords: digital information; K-means; singular value decomposition

1 INTRODUCTION

As the rise and popularization of Internet, cloud computing, Internet of things, and social networking technology, global data is growing faster than any other. The Times can be described as explosive growth (De-sheng Fu et al, 2011). Moreover, data storage unit also expands unceasingly by B, KB, MB, GB, TB to PB, EB, ZB, YB (Feng Li et al, 2013). People on the BBS (Feng Zhao et al, 2010), blog, microblog, WeChat (Jianping Zhu et al, 2014) (Yong-qing Chen et al, 2011), e-commerce transaction platform, search engine platform accumulate huge amounts of data, and it becomes the resources and wealth in the era of big data valuable information (Yu-lou Peng et al, 2012). With the wide application and the closely intertwined of network with the reality, it not only changed people's way of life, also impel the transformation of academic research (Yu-xia Lai et al, 2008). At present, a large number of digital information has been widely used in political elections, enterprises (especially e-commerce company) strategic layout, financial transactions, biological research, medical z, national security, public administration, public security, traffic management, weather monitoring, and many practice fields. On the one hand, vast amounts of Internet users use weibo, BBS, and other social media products and mobile Internet tools to record their life, break the human-computer interaction of traditional interpersonal of time and space constraints, and accumulate the unprecedented huge amounts of text, images, video information online. On the other hand, it is possible that mass human behavior data is efficient processing and analysis with the development of computer, information, and data mining technology. This technology is the foundation of huge amounts of data mining technology. Singular value decomposition, which is a kind of feature extraction of multivariate statistical analysis techniques, has been widely used in solving the optimization, the minimum power and multivariate statistical analysis. This paper puts forward an algorithm that bases on singular value decomposition of matrix reconstruction. Traditional idea is that larger eigenvalues reflect the main characteristic, while the smaller eigenvalues reflect information which is noise. We need to put the smaller eigenvalues of 0, in order to achieve the purpose of the filter out noise. However, any eigenvalues of matrix are reflected the characteristics of the matrix, they can increase or decrease eigenvectors. If all small eigenvalues transform into one big value, it will greatly increase the characteristics of the matrix, to make the digital information easier to handle. General idea is that test matrix is decomposed, using k-means algorithm to change the eigenvalues of matrix. Then we use Support Vector Machine (SVM) to verify the effect of reconstruction matrix with the original matrix.

2 K-MEANS

K-means is one of the most widely used clustering algorithm. All data samples of average in the every cluster sets represent the clustering. The main idea of the algorithm is divided into different categories through the iteration process, making the evaluation of clustering criterion function to achieve optimal performance, so that this method is compact within each cluster, independence between classes.

K-means steps:

1. choose k objects from n arbitrary data objects are as the initial clustering center;
2. according to the mean of every clustering objects, calculated the distance from the centers to every object; and corresponding objects are divided according to the minimum distance;
3. to calculate each the mean clustering of objects (center);
4. loops (2) to (3) until every cluster changes no longer occurs.

K-means has fast, simple, high efficiency and scalable. Therefore, the k-means may classify eigenvalues in this paper. Small eigenvalues are replaced, and then we can reconstruct matrix.

3 SINGULAR VALUE DECOMPOSITION

3.1 Singular value decomposition principle

Given a matrix $O^{M \times N}$. U is matrix of $M \times M$. U's row are orthogonal eigenvectors. V is matrix of $N \times N$. V's row are orthogonal eigenvectors. If t is the rank of matrix O, as follows,

$$O = U \times \Sigma \times V^T \tag{1}$$

where OO^T and O^TO are the same eigenvalues, U and V are orthogonal matrixes, and the superscript T denotes the transpose.

3.2 Improved algorithm of singular value decomposition

For testing matrix, we can obtain a new O by changing the eigenvalues of the matrix O. Specific algorithm is as follows,

1. Singular Value Decomposition decompose, as follows,

$$O = U \times \begin{bmatrix} \Sigma & 0 \\ 0 & 0 \end{bmatrix} \times V^T \tag{2}$$

where $\Sigma = diag(\sigma_1, \sigma_2, \cdots, \sigma_N)$.

2. The k-means is used to classify the eigenvalues $\lambda_1, \cdots, \lambda_N$ for two types which are A and B. Eigenvalues are large in class A, and class B is small. All the eigenvalues that are in class B replace into class A minimum of A value. Finally, we get new Eigenvalues.
3. Obtain new matrix $\overline{O^{M \times N}}$, as follows:

$$\overline{O^{M \times N}} = U \times \begin{bmatrix} \overline{\Sigma} & 0 \\ 0 & 0 \end{bmatrix} \times V^T \tag{3}$$

Therefore, new test matrix will have a higher accuracy than the original test matrix in the same situation. Based on the new test matrix to process digital information, we will get a better result.

4 SIMULATION

In this section, we carry out experiments to investigate and compare the performances of our different method: K-means and Singular Value Decomposition (K-SVD) and Direct Processing Method (DPM). Under the MATLAB, training matrixes are reconstructed matrixes, and testing matrixes are the original test matrix. Experimental data are divided into two parts, one are synthetic data sets, another is the real data sets. This method is verified.

4.1 Synthetic data sets

This experiment will be divided into two categories, K = 2.

1. Under the same hybrid synthetic data sets, the classification results are shown in the Figures 1 and 2,
2. We use the different methods to categorize the synthetic data sets. Experiment runs 5 times. Classification accuracy shows in Tables 1 and 2. Tables 1 and 2 show the different classification accuracy of the algorithm and data.

    Table 1 shows that K-SVD is better than DPM in classification accuracy. As K-SVD expands the secondary information, secondary information is also playing an important role rather than simply deleting all the secondary information. Thus, it improves the accuracy of classification.
3. Comparing two kinds of accuracies of training data and testing data under the same algorithm, specific data is given in Table 3. We can obviously find that K-SVD of training data and testing data of classification accuracy are similar, only 1.52%, but the DPM of testing data and training data classification accuracy is

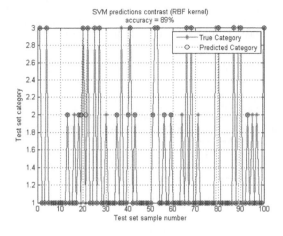

Figure 1. Classification accuracy of K-SVD.

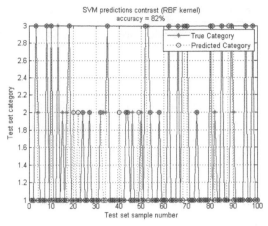

Figure 2. Classification accuracy of DPM.

Table 1. Classification accuracy of the different algorithm (training data)

	Training Data 1	Training Data 2	Training Data 3	Training Data 4	Training Data 5	Average
K-SVD	88.80%	82.00%	88.40%	83.20%	85.20%	85.52%
DPM	88.30%	83.20%	84.60%	82.00%	80.20%	83.66%

Table 2. Classification accuracy of the different algorithm (testing data).

	Testing Data 1	Testing Data 2	Testing Data 3	Testing Data 4	Testing Data 5	Average
K-SVM	80.00%	79.00%	86.00%	89.00%	88.00%	84.00%
DPM	76.00%	80.00%	82.00%	84.00%	80.00%	80.00%

Table 3. Minus between training data and test data classification accuracy.

	Training data	Testing data	minus
K-SVD	85.52%	84.00%	1.52%
PDM	83.66%	80.00%	3.66%

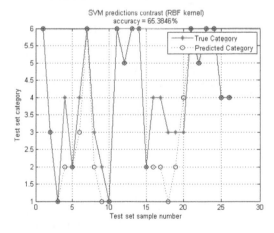

Figure 4. Classification accuracy of DPM.

3.66%. Therefore, K-SVD has good generalization ability, and good practical value.

4.2 Real data sets

This experiment will be divided into two categories, K = 2.

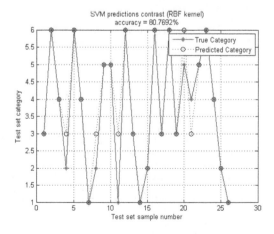

Figure 3. Classification accuracy of K-SVD.

Table 4. Classification accuracy of the different algorithm (training data)

	Training Data 1	Training Data 2	Training Data 3	Training Data 4	Training Data 5	Average
K-SVD	90%	80%	83.75%	92.5%	92.5%	88%
DPM	96%	95%	88.75%	97.5%	82.5%	92%

Table 5. Classification accuracy of the different algorithm (testing data)

	Testing Data 1	Testing Data 2	Testing Data 3	Testing Data 4	Testing Data 5	Average
K-SVM	84.62%	80.77%	68.54%	84.62%	73.08%	78.33%
DPM	80%	65.38%	65.38%	69.23%	65.38%	69%

Table 6. Minus between training data and test data classification accuracy.

	Training data	Testing data	minus
K-SVD	88.00%	78.33%	9.67%
PDM	92.00%	69%	23.00%

1. Under the same real data sets, the classification results are shown in the figures,
2. We use the different methods to categorize the real data sets. Experiment runs 5 times. Classification accuracy is shown in Tables 4 and 5. Tables 4 and 5 show the different classification accuracy of the algorithm and data.

Under the synthetic data sets and real data sets, we can clearly see that K-SVD has a great advantage in the classification, which greatly improves the accuracy of the classification.

3. Comparing two kinds of accuracies of training data and testing data under the same algorithm, specific data are given in Table 6. It can be obviously found that K-SVD of training data and testing data of classification accuracy is 9.67%, but DPM of testing data and training data classification accuracy reached 23%.

5 CONCLUSION

This paper using the improved algorithm of singular value decomposition improves the secondary useful information, increases the classification accuracy and improves the generalization ability. In synthetic data set, the classification accuracy rate increased by 4%. Moreover, in real data, the classification accuracy increased to 8%. Therefore, K-SVD is very important for large data processing. It provides a kind of effective method, which can be convenient for information management and positioning. And this way saves a large amount of information processing costs. However, the classification accuracy is not slightly enough. In the future, we will be the optimization of data and improve the classification accuracy.

ACKNOWLEDGMENTS

This work was supported by the regional project of the National Natural Science Fund (61363085); the commission project National Language Committee Research Fund (WT125-61); the major project of Yunnan Provincial Department of Education Research Fund (ZD2013013); the construction of high-level university for Nationalities project of Yunnan MinZu University Research Fund (ZZZC1501-JF12002); and the general project of Yunnan Minzu University Postgraduate Innovation Fund.

REFERENCES

Bouguila, N. Hybrid Generative/Discriminative Approaches for Proportional Data Modeling and Classification. IEEE Transactions on Knowledge and Data Engineering (Volume: 24, Issue: 12)

De-sheng Fu and Chen Zhou. Improved K-means algorithm and its implementation based on density[J]. Journal of Computer Applications, 2011, 02:432–434.

Feng Li, Zhixiang Zhu, and Shenghui Liu. The development status and the problems of large data[J]. Journal of Xian University of Posts and Telecommunications, 2013, 05:100–103.

Feng Zhao, Qingming Huang, and Wen Gao. An Image Matching Algorithm Based on Singular Value Decomposition[J]. Journal of Computer Research and Development, 2010, v.4701:23–32.

Jianping Zhu, Guijun Zhang, and Xiaowei Liu. Clarity of a Philosophy of Data Analysis During the Age of Big Data[J]. Statistical Research, 2014, 02:10–19.

Yong-qing Chen and Bin-bin Zhao. The application of both singular value decomposition and bi-dimensional empirical mode decomposition in extraction of gravity anomalies associated with gold mineralization[J]. Geological Bulletin of China, 2011, 05: 661–669.

Yu-lou Peng, Yi-gang He, Bin Lin. Noise signal recovery algorithm based on singular value decomposition in compressed sensing[J]. Chinese Journal of Scientific Instrument, 2012, 12:2655–2660.

Yu-xia Lai, Jian-ping Liu, and Guo-xing Yang. K-Means Clustering Analysis Based on Genetic Algorithm[J]. Computer Engineering, 2008, 20:200–202.

10 *Methods and algorithms optimization*

Energy Science and Applied Technology ESAT 2016 – Fang (Ed.)
© 2016 Taylor & Francis Group, London, ISBN 978-1-138-02973-6

Optimization of algorithm and performance simulation of electric air conditioning based on ADVISOR

H.L. Lin, J.R. Nan, W. Zhai & Z. Yang

Collaborative Innovation Center of Electric Vehicles in Beijing, Beijing Institute of Technology, Beijing, China

ABSTRACT: Taking an electric vehicle air conditioning system as the research object, the control system of the electric air conditioning system is studied and optimized by Fuzzy-PID composite control. The model of the electric air conditioning module is modeled and loaded to ADVISOR for secondary development. The simulation experiment is carried out about the effect of the air conditioning system on the performance of the vehicle under different driving conditions. The results show that the air conditioning system simulation model established in this paper can more accurately simulate the air conditioning system. Secondary development ADVISOR can improve the simulation accuracy of the vehicle system. Compared with the traditional switch control method, the Fuzzy-PID composite control method is significantly improved, the hysteresis is reduced, and the energy consumption is reduced by more than 7%, and the mileage of the whole vehicle is increased by more than 6%.

Keywords: electric vehicle; electric air conditioning; ADVISOR; fuzzy-PID

1 INTRODUCTION

Energy and environmental crisis have promoted new energy vehicles used widely (Cao Yunbo, 2009). Owing to the limited battery capacity, air conditioning is the second largest energy consumption unit on electric vehicle (Chen Qingquan, 2002). Therefore, its energy consumption will have important influence on the vehicle properties, such as limited driving distance (Khayyam H, 2009, Min Haitao, 2009). In order to reduce air conditioning energy consumption as much as possible, researchers at home and abroad have developed an electric air conditioning system using an electric compressor, which has effectively reduced the energy consumption through frequency conversion control. At present, ADVISOR is widely used in the simulation and analysis of the traditional vehicles and hybrid vehicles as a kind of backward simulation software (Y.P.B. Yeung, 2009). In the ADVISOR software loaded with Fuzzy-PID control algorithm to optimize the electric air conditioning model of secondary development, simulation experiment is carried out for the air conditioning system, and under different driving conditions of air conditioning system on the vehicle performance impact simulation experiment is carried out.

2 ELECTRIC AIR CONDITIONING SYSTEM

This paper adopts the electric air conditioning system for the pure electric city bus, which is top set, composition and arrangement of the program as shown in Fig. 1. The electric compressor, condenser, and evaporator are arranged at the top of the bus. The driving scheme of the electric air conditioning system is directly powered by the power battery to the electric compressor. When the cooling load of the air conditioning system varies, the speed of the electric compressor can be adjusted by the frequency converter, which can control the flow rate of the refrigerant. Therefore, it can reduce the energy consumption of air conditioning system by accurately controlling the speed of the compressor.

The compressor used in this paper is a scroll compressor, and the motor is used to choose a three-phase asynchronous motor. After calculating the heat load of a 10 meter bus, the component parameters are shown in Table 1.

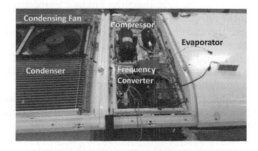

Figure 1. Layout scheme of top setair conditioning system for pure electric bus.

Table 1. Part parameters of electric air conditioning system.

Parameter	Value
Compressor rated speed ($r \cdot min^{-1}$)	3500
Maximum speed of compressor ($r \cdot min^{-1}$)	7000
Compressor displacement ($cm^3 \cdot r^{-1}$)	330
Voltage (V)	220/380
Input power (KW)	8
Refrigerant	R407C
Evaporator air discharge (m^3/h)	4059.2
Condenser air discharge (m^3/h)	4059.2

3 DEVELOPMENT OF SIMULATION MODULE OF ELECTRIC AIR CONDITIONER

This paper is based on the theory of thermal equilibrium in the car room. Air conditioning system is a typical nonlinear system. Owing to the hysteresis of the heat transfer, the temperature chge of the whole car room can be equivalent to a section of inertia link. Its transfer function can be expressed as:

$$t(\tau) = t_{final} + (t_0 - t_{final})e^{\frac{-\tau}{\tau_0}} \qquad (1)$$

Model of electric vehicle air conditioning system is based on MATLAB/Simulink simulation platform. Model building idea is shown in Fig. 2.

Firstly, the thermal load of the vehicle compartment is calculated according to the parameters of the input parameters in a step size of the car room heat load. At the same time, the control system calculated speed of electric air conditioning according to the temperature difference and rate of change of the setting temperature and the indoor temperature. Then result is delivered to the electric air conditioning module. The electric air conditioning module uses the speed, which is delivered to come over to calculate the refrigerating capacity and the power consumption in this step. Then, the temperature calculation module calculates the indoor temperature according to the thermal load of the car room, the real time outdoor temperature module, the current indoor temperature, and the refrigerating capacity of the air conditioner. Then, the current vehicle indoor temperature feedback to the control system, the car room heat load module, and the temperature calculation module for the next step of the calculation are analyzed. Finally, the simulation results are obtained.

In this paper, the model of the electric air-conditioning system includes the main parts of the air conditioning system, compressor, condenser, and evaporator. In the model of the electric compressor, the compressor speed converts its unit firstly. In order to prevent the compressor from exceeding the maximum speed of the motor, the speed limit is added. It

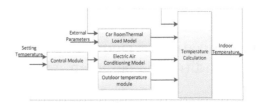

Figure 2. Electric air conditioning simulation system.

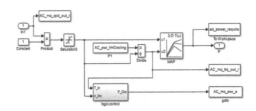

Figure 3. Electric compressor model.

checks the compressor efficiency table after limited speed input. The demand compressor torque which is calculated by the heat exchange module calculated the compressor power demand and the compressor speed dividing is also input to check the efficiency of the compressor. It calculates the power demand of the compressor motor According to the input data and refrigerating capacity can be calculated according to the speed of the compressor.

4 OPTIMIZATION OF CONTROL STRATEGY

The electric air conditioning system is a nonlinear and hysteretic system. Traditional switch control and PID control structure is simple, flexible, with a built steady state accuracy. However, the control parameters need to be built on the exact mathematical model of the controlled object, which is not suitable for the control of the nonlinear system. Fuzzy control is a nonlinear control theory based on fuzzy mathematics, knowledge representation of fuzzy linguistic form and fuzzy logic. In the control of the air conditioning system, it is very suitable to use fuzzy control to control the system due to the hysteresis and nonlinearity of the system. However, the fuzzy control has the problem of large error of steady state, which can cause the indoor temperature and the set value to have certain deviation. Therefore, combining the advantages of the two methods, fuzzy PID composite control method is adopted. The specific control logic is shown in Fig. 4. Fig. 5 is a Fuzzy-PID composite control model built in MATLAB/Simulink environment.

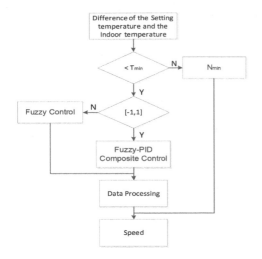

Figure 4. Fuzzy-PID composite control logic diagram.

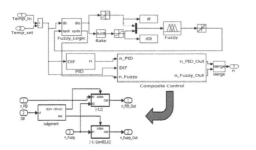

Figure 5. Fuzzy-PID composite control module.

5 ADVISOR SECONDARY DEVELOPMENT AND SIMULATION RESULTS ANALYSIS

5.1 Secondary development of electric vehicle model in ADVISOR

It uses ADVISOR to build the model for a pure electric medium-sized passenger car. The main technical parameters of the prototype are shown in Table 2.

The simulation model is built according to the actual power system layout. Vehicle model includes a cyclic conditions, vehicle, wheels, transmission, motor drive system, electronic auxiliary equipment sub-modules. Air conditioning is part of the electronic auxiliary equipment. Therefore, the replacement of the air conditioning parts in the electronic auxiliary equipment is found, and the new model is obtained, as shown in Fig. 6.

ADVISOR general software, air conditioning systems and other electrical accessories energy consumption is generally used to estimate the value of experience or rating and the vehicle performance calculation. As for the use of electric compressor to achieve automatic temperature control

of the electric vehicle air-conditioning system, different temperatures required for different power. The use of a single experience value calculation is bound to affect the accuracy of the vehicle simulation. Therefore, based on ADVISOR secondary development, the establishment of electric air-conditioning module is shown in Fig. 7.

5.2 Simulation results analysis

ADVISOR provides a variety of driving conditions, such as CYC_CONS, CYC_UDDS and so on. However, there is not driving conditions

Table 2. Main technical parameters of a pure electric bus.

Parameter	Value
Total vehicle mass (kg)	7600
Tire rolling radius (mm)	495
Motor voltage (V)	540
Maximum motor speed ($r \cdot min^{-1}$)	6000
The maximum torque of the motor ($N \cdot m$)	2100
Motor peak power (KW)	120
Wind resistance coefficient	0.56
Frontal area (m²)	6.2
Final drive gear ratio	11.2
Single battery voltage (V)	2.5~4.2
Battery sessions	750
Battery capacity ($A \cdot m$)	300

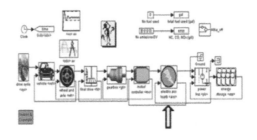

Figure 6. Electric vehicle simulation model.

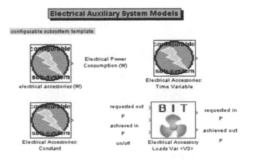

Figure 7. Secondary Development Auxiliary System.

557

for Chinese urban public transport. Therefore, it establishes Chinese urban bus condition data file according to the actual situation. According to the established vehicle model, the simulation experiment is carried out.

Figs. 8 and 9 compare the temperature change and energy consumption of the traditional switch mode control and Fuzzy-PID composite control strategy under the condition of Beijing in summer. Test time is 10 AM. Indoor and outdoor initial temperature is 32°C. The number of passengers is 20 people. The intensity solar radiation is 831 W/m^2 and the vertical solar radiation intensity is 831 W/m^2. The setting temperature is 26.5°C.

Traditional control can only achieve a simple switch function. When the indoor temperature is lower than the set temperature, the air conditioner is off. With the increase of temperature, when the temperature is higher than the setting temperature, the air conditioning is on. From 800 s to 1400s, the average temperature indoor is 27.0°C and the oscillation range is [26, 27.5]°C. The energy consumption is big and the response is lag affecting the comfort degree. Fuzzy-PID control operations in high power firstly. With the decrease of the indoor temperature, the stable power in about 4 KW. From 800 s to 1400s, the average temperature indoor is 26.3°C and the oscillation range is [26.1,

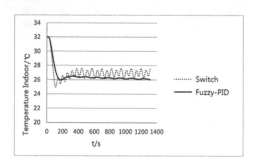

Figure 8. Temperature indoor.

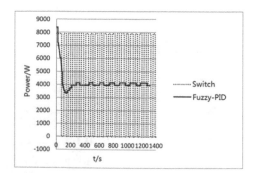

Figure 9. Power.

Table 3. Simulation results.

		OFF	Switch Control	Fuzzy-PID
Beijing	Consumption per 100 Km (KW·h)	69.23	78.95	73.41
	Mileage (Km)	232.67	204.15	217.73
CYC_UDDS	Consumption per 100 Km (KW·h)	66.44	75.62	70.02
	Mileage (Km)	240.81	211.58	228.51
CYC_CONS	Consumption per 100 Km (KW·h)	62.17	72.17	66.43
	Mileage (Km)	257.35	221.70	240.86

26.6]°C. The energy consumption is lower and the response is faster, which is more comfortable.

In different driving conditions, simulation results of the power consumption and mileage of the vehicle for two control strategies in the air conditioning system is shown in Table 3. The simulation results show that the power consumption of the vehicle with fuzzy PID composite control is reduced by 7.02%, and the driving range increased by 6.65%, compared with the switch control in the condition of Beijing. The power consumption of the vehicle with Fuzzy-PID composite control is reduced by 7.41%, and the driving range increased by 8.01%, compared with the switch control in the condition of CYC_UDDS. The power consumption of the vehicle with fuzzy PID composite control is reduced by 7.95%, and the driving range increased by 8.64%, compared with the switch control in the condition of CYC_UDDS. It shows that the use of Fuzzy-PID composite control of the electric air-conditioning system has a significant effect in energy saving.

6 CONCLUSIONS

Taking an electric vehicle air conditioning system as the research object, the control system of the electric air conditioning system is studied and optimized. The model of the electric air conditioning module is modeled and loaded with ADVISOR for secondary development. The simulation experiment is carried out on the effect of the air conditioning system on the performance of the vehicle under different driving conditions.

The results show that the air conditioning system simulation model established in this paper can be more accurately simulate the air conditioning system. Secondary development ADVISOR can improve the simulation accuracy of the vehicle system. Compared with the traditional switch control method, the Fuzzy-PID composite control method is significantly improved, the hysteresis is reduced, the energy consumption is reduced, and the mileage of the whole vehicle is increased.

REFERENCES

Cao Yunbo. Development and simulation analysis of electric vehicle air conditioning module based on ADVISOR [D]. Changchun: Jilin University, 2009.

Chen Qingquan, Sun Fengchun, Zhu Jiaguang. Modern electric vehicle technology [M]. Beijing: Beijing Institute of Technology press, 2002

Khayyam H, Kouzani A Z, Hu E J. Reducing energy consumption of vehicle air conditioning system by an energy management system [J]. Intelligent Vehicles Symposium, 2009:752–757.

Min Haitao, Cao Yunbo, Ceng Xiaohua et al. Modeling of electric vehicle air conditioning system and its influence on vehicle performance. Journal of Jilin University (Engineering Science Edition) [S]. 2009. 3, 39(1):53–57.

Popescu, M.C., A. Petrisor, M.A. Drighiciu. Modeling and Simulation of a Variable Speed Air Conditioning System. Automation [J], Quality and Testing, Robotics. 2008:115–120.

Wang Liangrno, Bai Weijun. Development and simulation of electric vehicle based on ADVISOR [J]. Journal of Southeast University (English Edition), 2006, 22(2):196–199.

Wang Yao, Pure electric vehicle air conditioning system simulation and control algorithm optimization [D], Beijing: Beijing Institute of Technology, 2012.

Yeung, Y.P.B., K.W.E. Cheng, W.W. Chan. Automobile Hybrid Air Conditioning Technology [C]. International Conference on Power Electronics Systems and Applications. 2009, 3.

Energy Science and Applied Technology ESAT 2016 – Fang (Ed.)
© 2016 Taylor & Francis Group, London, ISBN 978-1-138-02973-6

Improved Dijkstra algorithm for ship weather-routing

Z.H. Shen, H.B. Wang, X.M. Zhu & H.J. Lv
State Key Laboratory on Integrated Optoelectronics, College of Electronic Science and Engineering, Jilin University, Changchun, Jilin Province, China

ABSTRACT: Dijkstra algorithm is the academic foundation of engineering in the shortest path issue, which has also been widely applied to ship weather-routing. Although Dijkstra algorithm can find the global optimal route, it also has shortcomings. The main problem is lower search efficiency which is caused by traversing all weights at each node. The search direction is also blind and doesn't consider the position of the end node relative to the starting node. Therefore, when the ocean area to be navigated is divided by latitude-longitude line, this paper proposes a new improved Dijkstra algorithm which only searches five fixed directions at each node. Compared to the eight weights, the improved Dijkstra algorithm reduces three weights search direction at each node. This new method is more suitable for convex blockages. This paper also discusses the improved Dijkstra algorithm in the case that blockages are concave polygon. The simulation results of improved algorithm show that it can effectively increase the search speed and find the global optimal route.

Keywords: improved Dijkstra algorithm; ship weather-routing; optimal route

1 INTRODUCTION

Ship weather-routing is used to acquire the time-saving, minimum voyage distance or the most economical route from departure to destination by making use of meteorological information. Wind, waves and ocean currents have effects on ship velocity in open ocean area.

In order to serve the ship weather-routing better, many improved Dijkstra algorithms have been presented. (E.W. Dijkstra 1959) proposed the Dijkstra algorithm by a note on two problems in connection with graphs. Peter Hart, (Nils Nilsson and Bertram Raphael 1968) first described the A* (star) algorithm that it was an extension of Dijkstra algorithm. A* algorithm increased the speed of Dijkstra algorithm by using heuristics. (Zhan and Noon 1996) tested 15 of the 17 shortest path algorithms and identified a set of three shortest path algorithms that run fastest on real road networks. These three algorithms are: (1) TQQ (the graph growth algorithm implemented with two queues), (2) DKA (the Dijkstra algorithm implemented with approximate buckets), and (3) DKD (the Dijkstra algorithm implemented with double buckets). The TQQ algorithm is based on the theory of graph growth and use two FIFO queue to implement a deque structure to support the search process. The DKA algorithm and DKD algorithm are based on Dijkstra algorithm and use the bucket structure to improve search speed.

(Zheng et al. 2009) proposed sector Dijkstra algorithm to limit search area and reduce the nodes to be traversed. (Zhang et al. 2011) presented an improved new heap algorithm called Pairing heap to optimize Dijkstra algorithm. The improved new heap algorithm solves the problem that Dijkstra algorithm takes too much time to traverse the node table, delete the minimum node and update the node table.

To improve the performance of the Dijkstra algorithm and make few contributions to ship weather-routing, this paper proposes a new improved Dijkstra algorithm which only searches five fixed directions and makes search directions more reasonable at each node.

2 METEOROLOGICAL DATA

Routing of ships based on wave conditions was being practiced by seafarers and ship navigators since a long time. (R.W. James 1957) found that the most important parameter retarding a ship's progress was surface wave action. It is easy for navigators to select the optimum track to be navigated from departure to destination in a known wave climate. Therefore, this paper considers waves as weather condition. Wave prediction data are taken from ECMWF.

(Download link of meteorological data: http://apps.ecmwf.int/datasets/data/interim-full-daily/)

3 SHIP ROUTING OPTIMIZATION

Generally, the ship weather-routing optimization is related to optimization of several factors, namely: minimum voyage time linked to speed loss in seaway, minimum fuel cost related to added resistance, minimum structural damage, and so on. These factors are related to (i) the involuntary speed reduction and (ii) the voluntary speed.

3.1 Framework of Dijkstra algorithm

Dijkstra algorithm is essentially a kind of greedy algorithm. The route is defined as a series of connected nodes and the length of the path is arithmetic sum of the corresponding link lengths in the path. Consider a directed graph: $G = (V, E)$, with n vertices and e real valued weights sides. V is a collection of initial vertices. E is a collection of real valued weights sides. S is a collection of vertices which have found the shortest route from the starting node to themselves. V-S is a collection of vertices which have not found the shortest route from the starting node to themselves. The specific algorithm steps are as follows:

Step 1. Use the weighted adjacency matrix $arcs$ to represent the directed graph and $arcs(s,i)$ is the weight from node s to i. Suppose S equals $\{V_s\}$ and V_s is the starting node. Suppose $dist[i]$ equals the shortest distance from node V_s to V_i.

$$dist[i] = \begin{cases} 0 & i = s; \\ arcs(s,i) & i \neq s, \langle V_s, V_i \rangle \in E; \\ \infty & i \neq s, \langle V_s, V_i \rangle \notin E; \end{cases} \quad (1)$$

Step 2. V_j is the end of the next shortest path. Select the node V_j by:

$$dist[j] = Min\{dist[k] | V_k \in V - S\}, S = S \cup \{V_j\}.$$

Step 3. If $dist[j] + arcs(j,k) < dist[k]$, then:

$$dist[k] = dist[j] + arcs(j,k), (\forall V_k \in V - S).$$

Step 4. Repeat step 2 and step 3 until $S = V$.

3.2 Evaluation of weight

The weights can be expressed as: $w = \frac{L}{V}$. L is the distance between two neighboring nodes. V is calculated by the empirical formula which is recommended by Soviet Central Maritime Research Institute. The formula is expressed as follow:

$$V = V_0 - (0.745h - 0.257qh) \times (1.0 - 1.35 \times 10^{-6} D V_0) \quad (2)$$

where: V denotes the actual speed in the sea, V_0 represents the speed in the calm water, h means the wave height, q is the angle between ship heading and wave direction, D is the actual displacement of the ship (tons). This formula is applied to ships with displacement ranging from 5000 to 25000 tons, speeds between 9 and 20 knots, and the wave height is between 0 and 5 meters. According to the ship parameters table used in this paper and the above loaded wave data which meet the requirement of wave height ranging from 0 to 5 meters, the ship weather-routing simulation of this paper is suitable for above empirical formula. The ship parameters simulated in this paper are showed in Table 1.

3.3 Method

3.3.1 Traditional method

Firstly, download wave field data of the sea (latitude: from $0°N$ to $45°N$; longitude: from $120°E$ to $180°E$). The wave field data includes the significant wave height and means wave direction. Secondly, use above empirical formula to calculate weights. The sea to be navigated is divided by $1*1$ degrees latitude-longitude line. Each intersection of latitude-longitude line is a node. Apart from each other with $45°$ in a counterclockwise direction, each node has eight weights. The schematic diagram is shown below. Finally, the global optimal route can be acquired using the Dijkstra algorithm (Fig. 1).

3.3.2 Improved method

The main problem of traditional method is lower search efficiency which is caused by traversing all weights at each node. The search direction is also blind and doesn't consider the position of the end node relative to the starting node. Therefore, this paper proposes a new improved Dijkstra algo-

Table 1. Ship main parameters.

Type	Tanker
LOA	196 (m)
DWT	12000 (tons)
SPEED	18 (knots)

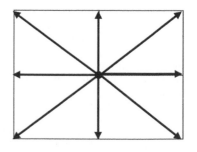

Figure 1. Eight weights distribution diagram of a node.

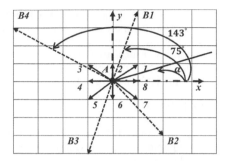

Figure 2. Demonstration of the improved method.

Table 2. Reserved weights and removed weights.

α	Reserved weights	Removed weights
$(0°, 22.5°] \cup (337.5°, 360°]$	2,1,8,7,6	3,4,5
$(22.5°, 67.5°]$	3,2,1,8,7	4,5,6
$(67.5°, 112.5°]$	4,3,2,1,8	5,6,7
$(112.5°, 157.5°]$	5,4,3,2,1	6,7,8
$(157.5°, 202.5°]$	6,5,4,3,2	7,8,1
$(202.5°, 247.5°]$	7,6,5,4,3	8,1,2
$(247.5°, 292.5°]$	8,7,6,5,4	1,2,3
$(292.5°, 337.5°]$	1,8,7,6,5	2,3,4

rithm which only searches five fixed directions at each node. Compared to the eight weights, the improved Dijkstra algorithm reduces three fixed search direction at each node.

The improved method is shown in Figure 2. In Figure 2, A is the starting node, $B_i(i=1,2,3,4)$ are the end points. The node A has eight weights. Use digitals (from 1 to 8) to distinguish among eight weights in a counterclockwise direction. Make the axis x point to the east. Make the axis y point to the north. Suppose that α is the angle from the east to the straight line which connects the starting node with the end node in a counterclockwise direction. Consider the counterclockwise direction as the positive direction.

The specific word explanations of the improved method are shown in Table 2.

Take an example from A to B_1, $\alpha = 75° \in (67.5°, 112.5°]$, so the fixed pointing directions of weights 4, 3, 2, 1 and 8 are reserved and the fixed pointing directions of weights 5, 6 and 7 are removed at each node. The other nodes also only search fixed five weights of reserved directions using Dijkstra algorithm.

4 RESULTS AND DISCUSSION

4.1 Convex blockages

The position of the starting node is $A(20°N,122°E)$. The position of the end node is $B(41°N,176°E)$. $\alpha = 21.25° \in (0°,22.5°]$. So the directions of

weights 2, 1, 8, 7 and 6 are reserved using the improved Dijkstra at each node. According to the distribution of waves, the position of the starting node and the position of the end node, consider wave field as convex blockages. The optimal track using Dijkstra algorithm is shown in Figure 3. The optimal track using improved Dijkstra algorithm is displayed in Figure 4. Both of these two tracks pass the same nodes. Therefore, these two tracks are the same track. It turns out that the improved algorithm also can find the global optimum track. These two tracks also show that the optimal track belongs to five reserved direction and doesn't go back. The improved Dijkstra algorithm only searches five reserved weights at each node. Programming operation time which aims at finding optimal track using two algorithms is shown in Table 3. The improvement of search efficiency is more obvious when there are a large number of nodes. So the improved Dijkstra algorithm increases search speed and makes search more reasonable.

4.2 Concave blockages

The position of the starting node is $A(44°N,161°E)$. The position of the end node is $B(41°N,176°E)$. $\alpha = 348.69° \in (337.5°,360°]$. So the directions of weights 2, 1, 8, 7 and 6 are reserved using the improved Dijkstra at each node. According to the distribution of waves, the position of the starting

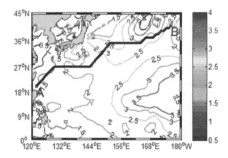

Figure 3. Optimal track using Dijkstra algorithm.

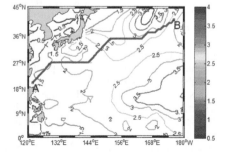

Figure 4. Optimal track using improved Dijkstra algorithm.

Table 3. Programming operation time using two algorithms.

Dijkstra algorithm	0.008 s
Improved Dijkstra algorithm	0.006 s

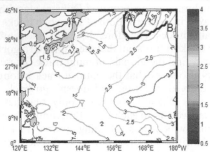

Figure 5. Optimal track using Dijkstra algorithm.

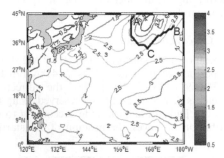

Figure 6. Optimal track using the second measure.

node and the position of the end node, consider wave field as concave blockages. The upper boundary and the upper wave field form concave blockages. The optimal track using Dijkstra algorithm is shown in Figure 5. This paper supposes that sea areas whose wave height is more than 3.2 meters are prohibited sailing areas. The improved Dijkstra algorithm means that the optimal track doesn't find back weights relative to the end node and the track can't go back. But the optimal track includes the node which needs to go back when the blockage is concave. So the optimal track using improved Dijkstra algorithm has no results. Therefore the improved method is more suitable for convex blockages.

There are two measures to solve the problem that the blockage is concave in this paper. One is increasing search weights from five to seven at each node. This measure means that the improved Dijkstra algorithms can find optimal track whether blockages are convex or concave. Although its efficiency is lower than the five weights search direction at each node, this improved algorithm with seven weights can find the optimal route in Figure 5. The second measure is adding an intermediate node. Consider the node $C(34°N, 166°E)$ as the intermediate node. Find firstly the optimal track

from A to C. Then, find the optimal track from C to B. In other words, the optimal track from A to B is made up of the optimal track from A to C and the optimal track from C to B using the improved Dijkstra algorithm. The optimal track using the second measure is shown in the Figure 6.

5 CONCLUSIONS

It is well known that the ship routing optimization for oceangoing vessels is widely applied, and a considerable number of studies have been carried out. In this paper, when the ocean area to be navigated is divided by latitude-longitude line, this paper proposes a new improved Dijkstra algorithm which only searches five fixed directions at each node. The improved Dijkstra algorithm considers the position of the end node relative to the starting node. The improvement of search efficiency is more obvious when there are a large number of nodes. While sailing, the blockages are divided into two kinds including convex blockages and concave blockages to discuss. The improved algorithm is more suitable for convex blockages. At the same time, this paper also provides two effective measures for solving the problem that blockage is concave. The simulation results of improved algorithm show that it can improve the search efficiency and find the global optimal route.

REFERENCES

Benjamin Zhan, F. Three Fastest Shortest Path Algorithms on Real Road Networks: Data Structures and Procedures. Journal of Geographic Information and Decision Analysis, Vol. 1, No. 1, Pp. 69–82.

Dijkstra, E.W. 1959. A Note on Two Problems in Connexion with Graphs. Numerische Mathematik 1, 269–271(1959).

Haltiner, G.J. & Hamilton, H.D. March 1962. Minimal-Time Ship Routing. Journal of Applied Meteorology. Vol. 1, No. 1, March 1962.

Masoud Nosrati*. 2012. Investigation of the * (Star) Search Algorithms: Characteristics, Methods and Approaches. World Applied Programming, Vol. 2, No. 4, April 2012. 251–256.

Panigrahi, J.K. May 2010. Application of Oceansat-1 MSMR analysed winds to marine navigation. International Journal of Remote Sensing, Vol. 31, No. 10, 20 May 2010, 2623–2627.

Tsou, M.C. & Cheng, H.C. 2013. An Ant Colony Algorithm for Efficient Ship Routing. Polish Maritime Researsh. Vol. 20, 2013, Pp. 28–38.

Zhang, W. & Jiang, C. 2011. Optimization Studies on an Improved Dijkstra Algorithm. Proceedings of the 2011 3rd International Conference on Information Technology and Scientific Management (ICITSM 2011).

Zheng, S.F. 2009. Sector Dijkstra Algorithm for Shortest Routes between Customers in Complex road networks. J Tsinghua Univ (Sci & Tech), Vol. 49, N0. 11, 2009.

Energy Science and Applied Technology ESAT 2016 – Fang (Ed.)
© *2016 Taylor & Francis Group, London, ISBN 978-1-138-02973-6*

Multiobjective route guidance systems based on real-time information

Zheng Zhou, Xuedan Zhang & Yuhan Dong
School of Tsinghua University, Beijing, China

ABSTRACT: The transportation network is always modeled as a static graph in the traditional route guidance systems. However, it ignores many changes in the real world obviously. At the same time, the single objective in driving is hard to reflect what the drivers want. What drivers think of is usually including many aspects, such as the travelling time, distance, and so on. Then, considering all of this, we propose a multiobjective route guidance algorithm based on real-time information. The local search process and studying process are mainly contained in the proposed algorithm. The local search process obtains local optimal route by improved heuristic algorithm, and the studying process mainly updates heuristic information of states by real-time information. At last, simulation results are compared with the traditional multiobjective search algorithm, and the search efficiency is improved obviously. The performance of our algorithms is evaluated.

Keywords: real-time information; heuristic search; multiobjective route guidance

1 INTRODUCTION

Route Guidance Systems (RGS) is an important branch of Intelligent Transport Systems (ITS) which has become a hot research area for many years. If the environment information of transportation network is all known and static, the optimal route can be proposed by many global planning algorithms, such as Dijkstra algorithm (Dijkstra E W, 1959). While the transportation environment changes all the time, the previous proposed route could be no longer optimal.

The traditional route guidance systems always consider the optimization of single objective, while it is obviously not fit for the actual transportation. A driver may want to spend shorter time and travel shorter distance at the same time. Therefore, we need consider more optimized objectives when planning optimal route, which belongs to the studying area of multiobjective route guidance systems. Research on multiobjective route guidance abstracts much attention (Harikumar S, 1996). Among much research, the heuristic search algorithms have played an important role. Heuristic search algorithms have been widely applied in the field of artificial intelligence, and also have attracted much attention of researchers of multiobjective route guidance. First, it has been proposed to solve problems of multiobjective route guidance by Stewart and White [Stewart B S, 1991]. Mandow proposed a method of extending route in search process, which improves the accuracy of proposed route (Machuca E, 2012).

Above, we know that transportation information changes at any time in dynamic environments.

Therefore, it is not wise to keep driving along with a so-called optimal route unchangeably. We need to plan our route and revise it by utilizing real-time information. At the same time, we must consider the varied demands of drivers, such as shorter time, shorter distance, and so on.

In the real world, the time of solving problem is always strictly limited for many engineering problems. Especially, when the scale of problems is very large, the efficiency of solving problems from global perspective is very low. Then, real-time search strategy has been proposed. It means that, during the whole driving, the local optimal path is planned many times and along which to move is based on the current transportation information and position, until we arrive at the destination. The real-time heuristic search algorithm-LRTA* was proposed by Korf, which only observes the forward local zone of depth of 1 (Korf R E, 1990). The LRTS algorithm was proposed by Bulitko, which observes the around zone of radius of d (Bulitko V, 2006). The RTAA algorithm, proposed by Koenig, adopts heuristic function achieving more easily, which improves studying process (Koenig S, 2006).

In this paper, we propose a multiobjective heuristic algorithm based on real-time information (RHMS). Unlike the traditional related algorithms, which plan an optimal route in the whole state space and travel along with the route unchangeably, the proposed algorithm performs the local search process and studying process in turn, then we can constantly revise the driving route by dynamic real-time information. The local search process plans the current optimal route in the local zone.

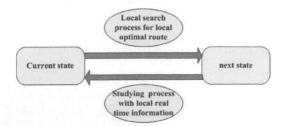

Figure 1. The basic procedure.

Then we travel along with it. The studying process is responsible for the update of heuristic information by utilizing real-time information. When we arrive at the next station, we can plan the new optimal route based on the updated heuristic information. The two processes are preformed constantly by turn until we arrive at the destination.

2 MATHEMATICAL FRAMEWORK

In this section, we develop the basic mathematical framework, which consists of model of transportation networks and some mathematical definitions.

We define the mathematical model of traffic networks as a space state graph $G = (N, L, C)$. The definitions of the symbols are given below:

N is the set of node states and the number of the states in the network is $|N| = n$, then $N = \{S_0, S_1,..., S_{n-1}\}$.

L is the set of links and $l_{i,j}$ denotes the link from state S_i to state S_j.

C is the set of q-dimensional link costs. $c_{m,n}$ denotes the cost of link $l_{i,j}$ when converts from state S_m to state S_n.

Problems of multiobjective route guidance can be described as the problem of a three elements vector $\langle G,S_0,\Gamma \rangle$, where S_0 denotes the initial state and Γ denotes the set of objective states. In this paper, we need to make some definitions, which are given below:

Suppose f_1, f_2 are k-dimensional vectors, if and only if $\forall i$, $1 \leq i \leq k$, it satisfies that $f_{1i} \leq f_{2i}$ and $f_1 \neq f_2$. We say that f_1 strongly controls f_2, recorded as $f_1 < f_2$, where f_{1i}, f_{2i} denote the i-th element of f_1, f_2, respectively.

Suppose f_1, f_2 are k-dimensional vectors, if and only if $\forall i$, $1 \leq i \leq k$, it satisfies that $f_{1i} \leq f_{2i}$. We say that f_1 weakly controls f_2, recorded as $f_1 \leq f_2$, where $f_1 i$ and $f_2 i$ denote the i-th element of f_1, f_2, respectively.

Suppose vector set X, if and only if $x \in X$, $\nexists\, y \in X$, $y < x$, we say vector set nondom(X) = $\{x \in X | \nexists\, y \in X, y < x\}$ is the non-domination vector set of X.

Suppose $P_{m\,n}$ is the set of all the routes from state S_M to state S_N, the route $p_1, p_2 \in P_{m\,n}$. If and only if $g_1 < g_2$, we say p_1 dominates p_2, recorded as $p_1 < p_2$, where g_1 and g_1 are the cost vector of p_1 and p_2, respectively.

Suppose $P_{m\,n}$ is the set of all the routes from state S_M to state S_N, the route $p_1 \in P_{m\,n}$. If and only if $\nexists\, p_2 \in P_{m\,n}$, $p_2 < p_1$, we say p_1 is the non-domination route, that is, there is no route better than p_1 in all objectives.

3 ALGORITHM DESCRIPTION

This section mainly contains three subsections, namely the basic process of proposed algorithm, the local search process and the studying process.

The Basic Process of Our Algorithm. We propose the Real-time Multiobjective Heuristic Search algorithm (RMHS), which contains two aspects, namely local search process and studying process. The basic process of RMHS is given below:

Step 1: At the current state S_{cur}, the depth of d local search process is performed, (LPATHS, CLOSED) = LMHS(S_{cur}, d);

Step 2: For every local optimal route p_m in LPATHS, the studying process LEARNING(S_m, CLOSED) is performed.

Step 3: Select a local optimal route p_u from LPATHS as the moving route, and arrive at the next state S_{next} along with the selected route. Then turn to step 1 until $S_{cur} \in \Gamma$.

Starting from the initial state S_0, the local search process search the local zone to get the local optimal route set. When arriving at the next state along with any route in the local optimal route set, the studying process updates all the extended heuristic information based on real-time information currently. Traditional related algorithms always calculate all the non-domination routes, which need much time because of the large amount of calculation.

However, RMHS divides the whole path planning problem into several local search problems. The scale of local state space is always small, and local state space is dependent compared with the whole state space to some degree, then local search process can be completed in a short time, which promises the optimization and real-time feature of the local route.

The Local Search Process. The Local Multiobjective Heuristic Search (LMHS) extends one certain route of some state every time. The search process is conducted by the heuristic information for improving the search efficiency. Some related definitions are given below:

1. DEF1: Suppose p_m is currently one of the routes from state S_{cur} to the next state S_m in local search process. The cost estimation of route p_m, f_m is given below:

$$f_m = g_m + h_m \qquad (1)$$

where g_m is the cost vector of route p_m, h_m is the heuristic information of route p_m which is used to estimate the minimum cost from state p_m to the objective state.

The set G_{cl}, G_{op} is used to record the extended and non-extending route of arriving state p_m in search process, respectively. The table CLOSED is used to record all extended state. The table OPEN records all the non-extending routes of all current states, and every non-extending route is recorded as a three-element vector (S_m, g_m, f_m).

2. DEF2: Suppose route p_m is one of table OPEN, f_m is the cost estimation of p_m. If and only if $\nexists$ $(S_n, g_n, f_n) \in$ OPEN, $f_n < f_m$, we say p_m is a non-domination route in table OPEN.

The local search process selects a non-domination route from table OPEN for extending, by using the relationship of vectors' domination every time. And the achieved extended route will inserted into table OPEN. When extending the objective state or the boundary of local zone, a local optimal route is achieved.

3. DEF3: Suppose PATHS is the set of all local routes from current state S_{cur} to the boundary of the local zone. If and only if $\nexists$ $p_n \in$ PATHS, $f_n < f_m$, we say p_m is a local non-domination path.

In addition, the specific procedures of LMHS is given below:

1. OPEN = $\{(S_{cur}, g_{cur}, f_{cur})\}$, CLOSED is empty.
2. Step 1: select a non-domination route p_m from table OPEN for extending, delete vector (S_m, g_m, f_m) corresponding to p_m from table OPEN and move g_m from $G_{op}(m)$ to $G_{cl}(m)$.
Step 2: if $S_m \in \Gamma$ or deep(S_m) = d, p_m is the local optimal route, and insert p_m into LPATHS, delete all the routes whose cost estimation is dominated by f_m from table OPEN.
Step 3: if deep(S_m)<d, insert S_m into table CLOSED. And for all forward state S_n of S_m, obtain the path p_n arriving at state S_n. We set $g_n = g_m + C(S_m, S_n)$. If g_n is not dominated by any route in $G_{op}(n) \cup G_{cl}(n)$, then:
Insert g_n into $G_{op}(n)$, and delete all routes dominated by g_n in $G_{op}(n) \cup G_{cl}(n)$.
Calculate the cost estimation f_n. If f_n is not dominated by any route in LPATHS, insert (S_n, g_n, f_n) into table OPEN and record related information.
3. Return local optimal route set LPATHS and CLOSED

The local search process constantly checks the domination relationships among all the routes by heuristic information, and discards the useless routes as early as possible. Then it efficiently reduces the amount of solving calculation. The more accurate heuristic information is, the more efficient the search process.

The Studying Process. After the local search process, we can get a set of local optimal routes. We perform the studying process for every route in the set, and update the heuristic information of every state in the local zone based on current real-time information. For obtaining more accurate heuristic information in the studying process, the local search process could perform well. The specific procedure of studying process STUDYING(S_w,CLOSED) is given below:

Step 1: Suppose Que = $\langle S_w \rangle$, S_x = Que.pop().
Step 2: For every forward state S_y of S_x, if $S_y \in$ CLOSED, then for any heuristic information of objective states, suppose $h_{yi} = \min(c_i(y,z) + h_{zi})$, where $1 \leq i \leq N$, N is the number of planning objectives.
Step 3: Que.push(S_y). Turn to step 2 until Que is empty.

In the local search process, the set CLOSED records all the expanded states the local search process, STUDYING (S_w, CLOSED) is called to revise heuristic information of all states in CLOSED.

4 SIMULATIONS

In this section, we test the proposed algorithm by using standard test of two-dimensional Grid World. The grid in the top left corner is the initial position, and the grid in the lower right corner is the destination, namely the objective position. Every dimensional element of cost vector is randomly generated from 1 to 10. First, with two objectives, we test the solution time and route cost under different size of local zone. In addition, the result is compared to the performance of traditional heuristic search NAMOA [Machuca L, 2012]. Secondly, under more than two objectives, we test the solution time and amount of path expanded between RMHS and NAMOA.

Fig. 2 shows the search time comparison between RHMS and NAMOA under different search deep in the 300*300 grid. It is obvious that the efficiency of RHMS is much better than that of NAMOA. In addition, with the increment in the search deeply, the efficiency of RHMS gets close to that of NAMOA gradually. The result is rational. When the search deep is low, the recorded information on table OPEN of RHMS is relatively little,

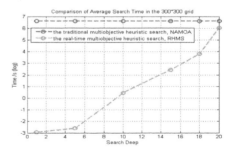

Figure 2. Comparison of average search time between two algorithms.

so the performance is better. In addition, when the search deep becomes larger, the advantage of search efficiency is losing little by little.

Fig. 3 shows the comparison of path cost at the same simulation environment of Fig. 2. For the whole performance, NAMOA is better than RHMS. Contacting with Fig. 2, we can see that when the search deep is low, the efficiency of RHMS is better, but the path cost is much than that of NAMOA. It means that we have to sacrifice the advantage of the path cost to some degree, to achieve much better search efficiency. In addition, this is what we will study during the next period.

In order to test the performance of RHMS under more planning objectives, more simulations have been made on different grid from 50*50 to 500*500. We select three, four and five objectives respectively. Figs. 4 and 5 show that whatever be the search time or the amount of path expanded, the

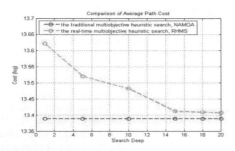

Figure 3. Comparison of average path cost between two algorithms.

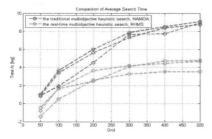

Figure 4. Comparison of average search time under different objective numbers.

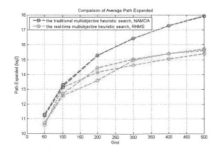

Figure 5. Comparison of path expanded under different objective numbers.

performance of RHMS is much better than that of NAMOA. Above all the results show that the real-time multiobjective heuristic search algorithm has two outstanding advantages. First, it searches on local zone, which can avoid much large calculation. Secondly, it improves the search efficiency as the local zone is dependent on the whole space. Then, in the situation of strictly limiting the solving time, RHMS is one of the best options to solve the multiobjective route guidance problems.

5 CONCLUSION

In this paper, we propose the real-time multiobjective heuristic search algorithm to solving the multiobjective route guidance problems. Unlike the traditional search algorithms, we utilize the real-time information fully by studying the process, and divide the whole planning space into several local search zones in order to improve the search efficiency. Moreover, the results indicate that we achieve the objective successfully. In the future, we try to reduce the path cost when keeping the search efficiency high.

REFERENCES

Bentley J L, Kung H T, Schkolnick M, et al. On the average number of maxima in a set of vectors and applications[J]. Journal of the ACM (JACM), 1978, 25(4): 536–543.

Bulitko V, Lee G. Learning in Real-Time Search: A Unifying Framework[J]. J. Artif. Intell. Res.(JAIR), 2006, 25: 119–157.

Dijkstra E W. A note on two problems in connexion with graphs[J]. Numerische mathematik, 1959, 1(1): 269–271.

Dasgupta P, Chakrabarti P P, DeSarkar S C. Multiobjective heuristic search in AND/OR graphs[J]. Journal of algorithms, 1996, 20(2): 282–311.

Harikumar S, Kumar S. Iterative deepening multiobjective A[J]. Information Processing Letters, 1996, 58(1): 11–15.

Korf R E. Real-time heuristic search[J]. Artificial intelligence, 1990, 42(2–3): 189–211.

Koenig S, Likhachev M. Real-time adaptive A*[C]//Proceedings of the fifth international joint conference on Autonomous agents and multiagent systems. ACM, 2006: 281–288.

Machuca E, Mandow L. Multiobjective heuristic search in road maps[J]. Expert Systems with Applications, 2012, 39(7): 6435–6445.

Stewart B S, White III C C. Multiobjective a*[J]. Journal of the ACM (JACM), 1991, 38(4): 775–814.

Rayner D C, Davison K, Bulitko V, et al. Real-Time Heuristic Search with a Priority Queue[C]//IJCAI. 2007, 382: 2372–2377.

Refanidis I, Vlahavas I. The MO-GRT system: Heuristic planning with multiple criteria[C]//AIPS–02. 2002: 46.

Refanidis I, Vlahavas I. Multiobjective heuristic state-space planning[J]. Artificial Intelligence, 2003, 145(1): 1–32.

Energy Science and Applied Technology ESAT 2016 – Fang (Ed.)
© *2016 Taylor & Francis Group, London, ISBN 978-1-138-02973-6*

The application of brain electrical signal filtering based on the matrix-coded intelligent algorithm

Zhang Jiao & Chen Feng
School of Nan Tong University, Nantong, China

ABSTRACT: Electroencephalogram (EEG) signal is a phenomenon of electrical activity in the brain cortex or on the scalp surface of brain cells. It is very important to study the EEG signal for clinical analysis. The EEG signal is a nonlinear and non-stationary signal that easily introduces noise in the acquisition process. An improved EEG signal filtering method can be achieved through a combination of intelligent algorithms such as matrix-coded genetic algorithm and particle swarm algorithm having less energy loss of the filter waveform. To analyze the performance of the algorithm, we obtained the actual brain electrical signal of a patient. The results indicated that the improved genetic algorithm has a better filtering effect in the GUI Matlab2010a platform.

Keywords: genetic algorithm, simulated annealing, EEG signal, filtering

1 INTRODUCTION

EEG signal contains a large number of physiological and pathological information. It plays a very important role in the research of clinical medicine and brain science. EEG signal is a kind of weak signal with complex non-stability, strong background noise and large individual difference. Thus, how to effectively extract useful signals from the EEG has become a hot topic in the research of brain science. In 1932, Dietch applied the Fu Liye transformation theory to study the EEG signal for the first time. With the rapid development of computer technology, the EEG signal has been analyzed using digital signal processing technology, modern methods of time-domain analysis and time-frequency analysis, nonlinear analysis and artificial neural network. For decades, researchers have proposed a large number of EEG signal noise processing methods. However, there is little breakthrough in its development. The patient's test environment can affect the characteristics of the EEG signal. With the development of advanced medical technology, research on the EEG signal filtering method has become increasingly practical.

Mental workload is also referred to as the psychological load or psychological pressure. Young and MS speculated that mental workload (mental stress) is the use of a wide range of concepts, one of the important research objects of human engineering and human, and also one of the most obscure concepts with many related definitions and measurement concepts. Jianqiao Liao studied mental workload by measuring the human information processing system and the degree of information processing system in information processing the resources being occupied from the EEG signal. The author found that the inactive information processing capacity is inversely proportional to the mental workload. A high mental workload can lead to rapid fatigue, reduced flexibility, stress response, and frustration. It also introduces errors in the analysis of information acquisition and decisions. In contrast, very low mental workload can result in waste of human resources and the sense of disgust. All these characteristics can be observed and determined from the waveform of EEG signals. Therefore, how to effectively use the most effective filtering method to extract EEG signals has become one of the important topics in the field of EEG signal.

The traditional filtering method can filter the noise of the EEG signal in the acquisition process. However, the energy loss of EEG waveform is a major concern, which cannot completely reflect the whole characteristics of the EEG signal. To achieve an improved EEG signal filtering method, this paper focuses on EEG signal feature extraction using filtering algorithm, intelligent algorithms and other brain electrical signal filtering algorithms enhance the performance of the algorithm.

This paper highlights the following innovations:

1. High-speed computing performance that adopts the matrix form encoding to facilitate the use of matrix calculation.
2. Combined characteristics of the powerful global search capability of the genetic algorithm. To search the approximate optimal solution, we

select their advantages and propose a hybrid algorithm.

2 COMPONENTS OF EEG SIGNALS

EEG signal mainly comprises a large number of neurons in the cortex, formed by the sum of the postsynaptic potentials, which is the result of the common activities of many neurons. EEG signal is divided into spontaneous EEG signals and evoked brain electrical signals. They are brain cells that perform electrophysiological activity in the brain cortex or on the scalp surface, exhibiting the phenomenon of electricity. Spontaneous EEG signal is a brain electrical signal that is produced when there is no specific external stimulation. It can also induce the brain electrical signal to stimulate the human's sense organ (including the light, sound or electricity). In this paper, we study the pattern classification of spontaneous EEG signals. The frequency of the EEG signal is generally between 0.5 and 30 Hz, including α (1–4 Hz), β (4–8 Hz), γ (9–13 Hz), δ (14–30 Hz), and some transient signals. However, the EEG signal is unstable and easily affected by factors such as the test environment. Therefore, it often contains a variety of EEG signal interference components. This makes the accurate analysis of the EEG signal difficult.

3 QUESTION PUT FORWARD

A series of experiments conducted by the US Air Force laboratory showed that the brain electrical activity is one of the sensitive indices that reflects the level of mental workload. Additionally, they demonstrated that it may be used in the classification of flight mental workload.

To achieve the purpose of the research on mental workload, the technical scheme of this paper is as follows:

1. EEG signal data were collected by the EEG signal analysis instrument (EEG data is P300 in ERP), and the characteristics of the EEG signal were analyzed;
2. EEG data were used through the genetic algorithm (Algorithm Genetic) EEG signal processing method for filtering;
3. Filtered EEG data were used to analyze the mental workload.

In the research process, Genetic Algorithm (GA) analysis of the EEG signal processing method includes two parts: preprocessing part and core algorithm part.

4 IMPROVED ALGORITHM DESIGN

Genetic algorithm solving process is essentially the stochastic optimization process. However, genetic algorithm involves a random optimization operation with respect to the probability. In general, the genetic algorithm can only have globally sub-optimal solution. It is not easy to find the optimum solution. This is because its search is random and has certain blindness.

The particle swarm optimization algorithm depends mainly on its own parameters. Moreover, the initial particle swarm is randomly generated, which usually has defects of a slow convergence rate, falling into local convergence easily.

In view of the shortcomings of the particle swarm algorithm and the genetic algorithm when applied individually, we combine the two algorithms to solve the optimal evacuation order set. The specific calculation steps are as follows:

Step 1: Randomly generate 10 particle positions. Get the corresponding 10 group evacuation instruction sequence as the initial population (particle swarm optimization) in accordance with the foregoing algorithm. Initialize the global and individual extremums to obtain the first individual (particle) evacuation fitness function values.

Step 2: Solve the evacuation fitness function value of instruction sequence in accordance with the aforementioned algorithm. The instruction sequence is described by the current population (particle swarm) of individuals (particles).

Step 3: According to the value of evacuation fitness function, sort the current population (particle swarm). Then, the individual extremum and global extremum are determined.

Step 4: Selection: 5 individuals (particles) were replaced by 5 new random individuals in the population (particles), which have a better evacuation fitness function value.

Step 5: Crossover: Make the variation of the part of the individual (particle) in the population (particle swarm) through the roulette gambling way.

Step 6: Variation: Update all individuals (particles) in the population (particle swarm) with the updating rule of the particle swarm algorithm.

Step 7: Observe whether it achieves the maximum number of iterations. If it achieves, then turn to Step 8, otherwise turn to Step 2.

Step 8: Output the final results.

The calculation process of the specific genetic particle swarm optimization algorithm is shown in Figure 1.

Matrix-coded Design

Particle swarm optimization, PSO, was first proposed by Eberhart and Kennedy in 1995. The basic idea is inspired by their previous study results of modeling and simulating on the group behavior

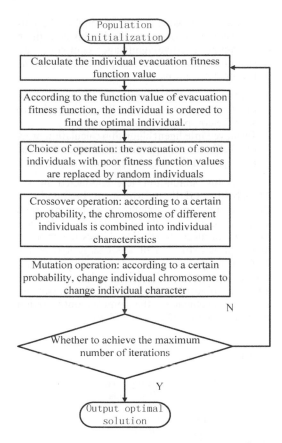

Figure 1. Process chart of the improved intelligent algorithm.

Table 1. Individual matrix coding table.

	1	2	3	4	5	6	7	8
1	1	1626	637	1082	339	232	538	1339
2	922	171	468	1870	312	1691	1102	2040
3	218	1969	9	1587	1673	1779	172	818
4	883	1865	372	540	298	278	1780	1187
5	177	1273	718	1051	822	155	491	252
6	854	101	1848	1934	1005	1001	691	1843
7	157	798	94	827	197	270	1929	1958
8	480	723	1681	31	88	346	1329	1498
9	110	606	1525	386	1406	375	754	1281
10	193	1588	996	892	915	627	1041	1046

Table 2. Matrix coding table of the individual variation component.

	1	2	3	4	5	6	7	8
1	36	−316	79	18	−38	112	96	12
2	−74	−50	−7	−68	88	−78	−70	−84
3	−48	108	−17	−80	103	122	−15	−99
4	24	−60	26	54	−71	−97	−52	−46
5	−16	−60	77	−120	109	58	−2	20
6	854	11	8	−68	−2	31	45	−26
7	157	98	105	75	−102	−60	−42	46
8	480	39	−1	71	55	103	100	−42
9	110	62	0	−5	103	28	30	92
10	193	−66	98	−10	−2	−85	122	54

of many birds. They used the model introduced by the biologist Hepper. Based on this, they presented a simplified model of the actual bird, using the particle swarm algorithm.

Genetic algorithm is a stochastic global search and optimization method that imitates the mechanism of natural biological evolution. It draws on Darwin's theory of evolution and Mendel's theory of heredity.

To use the intelligent algorithms particle swarm algorithm and genetic algorithm, we will encode sequences of instructions as particles. So, the computer can take advantage of matrix calculation to significantly improve the computational efficiency. A particle is composed of forms of code, such as random integers between 1 and 2048. The specific integer matrix encoding table of an individual is described in Table 1.

For the individual mutation operation, the coding of the individual variation component is necessary. Among them, the individual variation component selects a random integer between −128

and 128. A specific individual variation component matrix coding table is described in Table 2.

A particle swarm optimization algorithm and a genetic algorithm mutation operator are defined. The position of the particle matrix is updated by replacing the old individual coding matrix and individual variation component coding matrix with a new individual coding matrix elements of cross-border, with a new random integer replacement between 1 and 2048.

In this paper, the matrix calculation is included. for a quick calculation.

5 OPERATION RESULT

This paper uses the matlab2010a programming environment. To validate certain practicability and better convergence performance of the genetic particle swarm optimization, the genetic particle swarm algorithm and the basic PSO algorithm are tested based on the evacuation model under study. The test results are detailed as follows.

From Figures 2 and 3, it can be observed that the performance of the algorithm is iterative. Compared with the basic genetic algorithm, the combination of the genetic simulated annealing

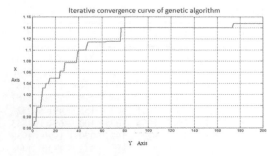

Figure 2. Sketch map of the waveform using the filtering method of the genetic algorithm.

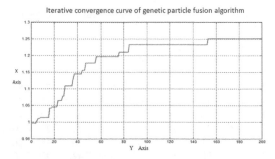

Figure 3. Diagram of filtering waveform using the improved genetic algorithm filtering method.

algorithm and the genetic algorithm proposed in this paper show better convergence performance. It can search to obtain a better approximation to the optimal solution. This is further evidence of the improved genetic algorithm proposed in this paper that is more suitable for EEG signal filtering processing.

6 SUMMARY

In this paper, a new EEG signal filtering method was proposed. This method can be used for the pretreatment of EEG feature extraction, which is helpful for the medical workers to diagnosis the disease conveniently by finding the EEG char-

acteristics of a patient. This highlights the current development of the medical industry with an extraordinary positive significance, compared with the traditional algorithm. This paper highlights the following innovations:

1. Matrix calculation is applied to the iterative calculation process of the intelligent algorithm to save the computation time greatly.
2. Improved intelligent algorithm calculation process facilitates the optimization of the results of an improved filtering waveform.

From the simulation test conducted in the Matlab2010a GUI platform, it can be found that the combination of intelligent algorithms such as matrix-coded genetic algorithm and Particle Swarm Optimization (PSO) achieves the approximate optimal solution of less energy loss of the EEG signal problem. Furthermore, the combination of the genetic algorithm and the simulated annealing algorithm proposed in this paper can solve the problem of the complex optimization problems such as EEG signal filtering.

REFERENCES

Eric A Pohlmeyer, Jun Wang, David C Jangraw, Bin Lou, Shih-Fu Chang, Paul Sajda. Closing the loop in cortically-coupled computer vision: a brain–computer interface for searching image databases [J]. Journal of Neural Engineering, (2011).

Holland, J.H. Adaptation in Natural Artificial Systems, MIT Press, 1–17, (1975).

Lim, M.H., Y. Yuan, and S. Omatu, Efficient Genetic Algorithms Using Simple Genes Exchange Local Search Policy for the Quadratic Assignment Problem, Computational Optimization and Applications (3): 249–268, (2000).

Nicol N. Schraudolph, Richard K. Belew. Dynamic Parameter Encoding for Genetic Algorithms, Machine Learning (1): 9–21, (1992).

Wang Y.N. The Research and Application of Genetic Algorithm, Jiangnan university, (2009).

Young M.S., Brookhuis K.A., Hancock C D W P A. State of science: mental workload in ergonomics [J]. Ergonomics, 58(1):1–17,(2014).

Energy Science and Applied Technology ESAT 2016 – Fang (Ed.)
© 2016 Taylor & Francis Group, London, ISBN 978-1-138-02973-6

Design and application of a three-dimensional direction recognition algorithm based on a hardware platform

Feng Lu
Key Laboratory of Fiber Optic Sensing Technology and Information Processing, Ministry of Education;
College of Information Engineering, Wuhan University of Technology, Wuhan, China

Wei Xie, Haoying Wu & Danfeng Wang
College of Information Engineering, Wuhan University of Technology, Wuhan, China

ABSTRACT: A robot with tactile sensors can perceive many physical features of the target object (e.g. weight, softness, texture, temperature, stiffness), which greatly improves situational awareness and real-time decision-making ability in unknown environments. Among them, the recognition of three-dimensional direction plays a very important role in human–computer interaction. The main aims of this paper are as follows: (1) a three-dimensional hardware platform was first built; (2) based on the hardware platform, the Sage–Husa adaptive Kalman filter algorithm was used to filter the data collected; (3) RBF (radial basis function) neural network algorithm was used to recognize the three-dimensional directions. The results indicated that in the three-dimensional direction recognition, the average error of nonlinear prediction based on the RBF neural network is 2.01×10^{-3}, the average accuracy is 94.6%, and the highest accuracy rate is 97.6%.

Keywords: intelligent robot; Sage–Husa adaptive Kalman filter; neural networks; three-dimensional direction recognition; algorithm design

1 INTRODUCTION

With the development of science and technology and the improvement in automation of industrialization processes (Boya Q et al, 2014), intelligent products (e.g. intelligent robot) have become the focus of academic research at home and abroad. Undoubtedly, tactile sensing technology plays an irreplaceable role in the perception ability of the intelligent robot (Chen J Y. 2013). It is also a supplement to the visual sensing technology. Therefore, tactile sensing can improve the level of human–robot interaction to make robots perform the autonomous operation in complex environments.

Currently, most tactile array systems have only uniaxial pressure sensing capability, meaning that it is easy to recognize the two-dimensional direction but difficult to recognize the three-dimensional direction that is very important to perceive human's intention during the human–robot interaction (Lu K et al. 2015). In this paper, a kind of single point tactile sensor is used to recognize the three-dimensional direction.

The rest of the paper is organized as follows: an intelligent robot hardware platform is designed in section 2. In section 3, the Sage–Husa adaptive Kalman filter algorithm is used to filter the col-lected data based on the intelligent robot hardware platform (Luo Y, Ren L. 2015). In section 4, the Radial Basis Function (RBF) network is used to process the filtered data, and the description of feature extraction from the tactile data is introduced (Wu H Y et al, 2013). Conclusions are provided in section 5.

2 DESIGN OF THE HARDWARE PLATFORM

2.1 *Selection of tactile sensor*

FSR tactile sensor is made of a kind of material that is sensitive to force. In fact, it is sensitive to the positive pressure exerted on the surface with no temperature drift. Due to its thin thickness and small size, it can be well attached to a three-dimen-

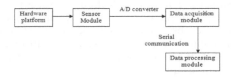

Figure 1. Flow chart of the data acquisition hardware system.

Figure 2. 3D print model and the location of sensor distribution.

sional surface. Therefore, FSR tactile sensor can meet the requirements.

This hardware platform comprises a 3D printing model, FSR tactile sensors, data acquisition software, and a data process system, as shown in Fig. 1.

2.2 *Experimental setup*

First, MK60 N512VLQ100 (K60) chip, FSR tactile sensors and Bluetooth module were used to design the hardware. Then, the SolidWorks software was used to design a three-dimensional model and the 3D printer was used to print it out. Subsequently, 20 pieces of FSR tactile sensors were pasted on the surface of the model. When holding the model by a hand, tactile sensors can sense the change of force. The 3D print model and the location of the sensor distribution are shown in Fig. 2.

3 DESIGN OF STATIC FORCE DATA FILTERING ALGORITHM BASED ON RBF NETWORK

In the process of transmission, signal is inevitably affected by internal or external interference. Thus, the measurement data should be filtered before the process. In this paper, the Sage–Husa adaptive Kalman filter algorithm is used to filter the data collected.

3.1 *Design of Sage–Husa adaptive kalman filter*

We assume that the introduction of noise is in conformity with the Gaussian distribution, and state noise and observation is not relevant. First, a discrete control process of the system is introduced. The system can be described by a linear stochastic differential equation:

$$x(k) = \varphi(k,k{-}1)x(k{-}1){+}w(k) \qquad (1)$$

The measured value of the system is given by:

$$z(k) = H(k)x(k){+}v(k) \qquad (2)$$

where $x(k)$ is the system state of the moment k, $\varphi(k, k{-}1)$ is the state transition matrix, $z(k)$ is the measured value of the moment k, and $H(k)$ is the observation matrix. $W(k)$ and $v(k)$ are, respectively, process and measurement noise, which are assumed as Gaussian white noise and their covariance are $Q(k)$ and $R(k)$. We assume that the expectations of $w(k)$ and $v(k)$ are, respectively, $q(k)$ and $r(k)$, which are unrelated. Some important formulas are given as follows:

$$X(k) = X(k,k{-}1){+}K(k)Z(k) \qquad (3)$$

$$X(k,k{-}1) = \varphi(k,k{-}1)X(k{-}1) {+}qs(k{-}1) \qquad (4)$$

$$Z(k) = z(k){-}H(k)X(k,k{-}1){-}rs(k) \qquad (5)$$

$$K(k) = P(k,k{-}1)H(k)T[H(k)P(k,k{-}1)\,H(k) \\ T{+}Rs(k)]{-}1 \qquad (6)$$

$$P(k,k{-}1) = \varphi(k,k{-}1)P(k{-}1)\,\varphi(k.k{-}1)T \\ {+}Qs(k{-}1) \qquad (7)$$

$$P(k) = [I{-}K(k)H(k)]P(k,k{-}1) \qquad (8)$$

where $X(k)$ and $Z(k)$ are, respectively, the state of the system and the estimation value after the kth iteration and filtering estimate of the noise parameter, $K(k)$ is the Kalman gain, I is the unit matrix, and $P(k)$ is the mean square error matrix. Among them, $qs(k)$, $rs(k)$, $rs(k)$, and $qs(k)$ can be calculated by the following formulas:

$$rs(k{+}1) = [1{-}d(k)]rs(k){+}d(k)[z(k{+}1){-}H(k{+}1) \\ X(k{+}1,k)] \qquad 9)$$

$$Rs(k{+}1) = [1{-}(k)]Rs(k){+}d(k)[Z(k{+}1)Z(k{+}1) \\ T{-}H(k{+}1)P(k{+}1,k)H(k{+}1)T] \qquad (10)$$

$$qs(k{+}1) = [1{-}d(k)]qs(k){+}d(k)[x(k{+}1){-} \\ \varphi(k{+}1,k)X(k)] \qquad (11)$$

$$Qs(k{+}1) = [1{-}d(k)]Qs(k){+}d(k)[K(k{+}1)Z(k{+}1) \\ Z(k{+}1)\,TK(k{+}1)T{+}P(k{+}1){-} \\ \varphi(k{+}1,k)P(k)\,\varphi(k{+}1,k)T] \qquad (12)$$

where $d(k) = (1{-}b)/(1{-}b^{k{+}1})$ and b is the forgetting factor with b.(0,1).

Using the Sage–Husa adaptive Kalman filter algorithm, we need to adjust only the forgetting factor. We can then estimate and correct the statistical characteristics of system noise and measurement noise to achieve the optimal filtering effect.

3.2 *Analysis and discussion after the filtering process*

A total of 20 pieces of FSR tactile sensor are attached to the experimental model in five direc-

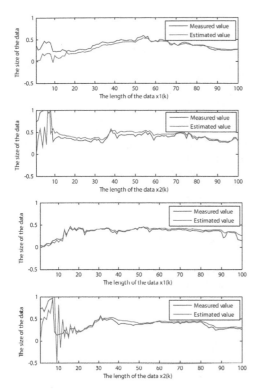

Figure 3.   (a) The filtering results when pressing down the model. (b) The filtering results when pushing back the model.

tions: the front, back, left, right and down. However, only a few sensors change obviously when the model is pressed. Thus, it is easy to analysis the data of sensors that change obviously. Sensors are similarly processed due to the data collected from all directions. In this paper, only two directions will be chosen for analysis and discussion.

When the model is pressed down or pushed back by a hand, the force on four slices of FSR tactile sensors changes obviously. Thus, we randomly select two groups of data, namely x1(k) and x2(k), for analysis. Each group of data contains 100 measurement values. We then use the Sage–Husa adaptive Kalman filtering algorithm to process the data x1(k) and x2(k). The results are shown in Fig. 3(a). and Fig. 3(b).

It can be seen from Fig. 3(a) and Fig. 3(b) that the measured value and the estimated value are very close, which means that the error is very small. Furthermore, although the real values are unknown, the data obtained after filtering can be close to the real value. The closer the measured value is, the smaller the error will be. This indicates that the Sage–Husa adaptive Kalman filter can achieve an ideal filtering effect.

4 THREE-DIMENSIONAL DIRECTION RECOGNITION USING THE RBF NEURAL NETWORK ALGORITHM

4.1 3D modeling and force analysis

To imitate the physical model, the MATLAB is used to model a sphere (radius 5 cm), and the sensor's position is replaced by a point. The sum of the force vectors of the 20 points is considered the resultant force of a set of data. In the experiment, due to several resultant forces by the size of the model, the three-dimensional space distribution is disordered. To observe the position of the resultant force, FY, FZ and FX will be scaled down to the sphere.

The direction of the resultant force exerted on the physical model is shown in Fig. 4. Then, the force direction of the physical model can be determined by the resultant force on the sphere.

If the tactile sensor is affected by the force F, it can be decomposed into three directions, X, Y, Z, as shown in Fig. 5. The vector of all directions can be solved by formula (13):

$$\begin{cases} F_z = F \bullet \cos\alpha \\ F_x = F \bullet \sin\alpha\cos\beta \\ F_Y = F \bullet \sin\alpha\sin\beta \end{cases} \tag{13}$$

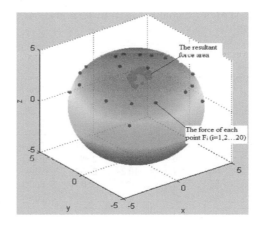

Figure 4.   Sensor stress on the model.

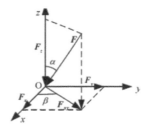

Figure 5.   Force analysis.

4.2 Extrcting features

Because of the small sample size of the original data and the large computation redundancy, we analyzed the effectiveness of the various features and selected the most representative characteristics, as shown in Table 1. For convenience, we selected one of the reference axes shown in Fig. 6.

4.3 Radial Basis Function network model

Radial Basis Function (RBF) network is an algorithm with three layers, which contains the input layer, the hidden layer and the output layer. The most commonly used basis functions is the Gauss function and the formula is as follows:

$$\phi_i(x) = \exp\left[-\frac{\|x - c_i\|^2}{2\sigma_i^2} \right], i = 1, 2 \ldots m. \quad (14)$$

where x is an n-dimensional input, c_i is the center of the ith basis function, and ci is a vector having the same dimension with x. σ_i is the ith perceptual variables and m is the number of units of perception. If there is a sufficient number of hidden layer neurons, we can approximate the nonlinear function by choosing appropriate center normalized parameters and output weights.

Table 1. Feature extraction.

Number of trials	Number of sets of data	Contact area	Contact time	Average force (average voltage, V)	Direction of the resultant force
1	86	4	10.25	4.8874	Down
2	86	6	10.28	2.9023	East
3	86	4	10.28	2.4985	West
4	86	5	10.28	3.9332	South
5	86	5	10.28	5.2434	North

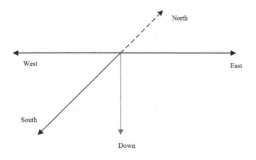

Figure 6. Reference axis.

4.4 Nonlinear prediction based on the RBF network

Radial Basis Function (RBF) network has a very good nonlinear mapping ability. Furthermore, RBF exhibits good generalization ability, and a fast convergence rate, making it more suitable for nonlinear network prediction.

In the experiments, the force is applied on the model in any direction, yielding 86 sets of data in 11.25 s. The data can be divided into two parts: the train data (69 groups) and test data (17 groups). In both data, the resultant force changes over time, as shown in Figure 7.

Then, the trained network is used to test the rest of the 16 sets of data. Thus, we get the residual error of the test data, as shown in Table 2. The average error is 2.01×10^{-3}, as shown in Table 2.

4.5 3D pattern recognition based on the RBF network

After the network has been completely trained, the new data is tested to verify the accuracy of the network. Then, the model is held in any direction to obtain 85 sets of data. The results of the RBF neural network classification in three-dimensions are shown in Figure 8.

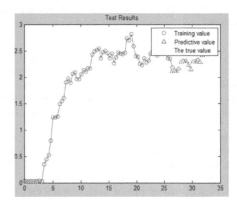

Figure 7. Resultant force varies with time.

Table 2. Average error of 17 groups of test data ($\times 10^{-3}$).

1	2	3	4	5	6	7
0.78	1.32	2.81	2.84	1.63	1.54	2.47
8	9	10	11	12	13	14
3.21	5.52	3.08	2.03	0.79	1.30	1.98
15	16	17	Average error			
2.22	0.43	0.21	2.01			

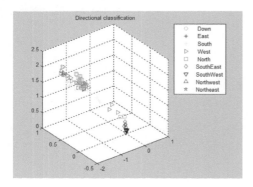

Figure 8.   Force classification.

Table 3.   Pattern recognition results.

Direction of the applied force	Number of measurement groups	Correct number of groups	Accuracy (%)
Down	85	81	95.2
East	86	80	93.0
West	86	80	93.0
South	85	83	97.6
North	86	81	94.2

The MATLAB statistical results are as follows: Number of Down: 16; Number of East: 15 Number of South: 5; Number of West: 18 Number of North: 8; Others: 23.

When the south force is applied on the sensors, the statistical results of MATLAB is as follows: the number of South is 83, the number of West is 2, and thus the correct rate can reach up to 97.6%.

To test the accuracy of the neural network pattern recognition, we randomly selected 25 people (16 men, 9 women) involved in the testing. They were instructed to press the model in any direction. We crawled the acquired data about 85 groups every time. Then, we calculated the correct number of crawl statistics group and the correct rate after crawling 25 times, averaging the correct rate. The results are summarized in Table 3.

5   CONCLUSION

In this study, a hardware platform was designed to recognize the three-dimensional direction using 3D printing technology, K60 chip, FSR tactile sensors and Bluetooth module. Furthermore, IAR and MATLAB software were successfully used to collect and process the data. Then, the Sage–Husa adaptive Kalman filter algorithm were used for data filtering. Finally, the RBF neural network algorithm was used to estimate the accuracy of three-dimensional direction recognition. The experimental results indicated that the average error of nonlinear prediction based on the RBF network was 2.01×10^{-3} and the average accuracy was 94.6% while the highest accuracy of three-dimensional direction recognition reached up to 97.6%.

ACKNOWLEDGMENTS

This research was supported by a grant from the National Natural Science Foundation of China. The authors thank Yali Feng and Shuaijun Wang for their valuable discussions.

REFERENCES

Boya Q, Cai Y. 2014. Champagne B, et al. Low-Complexity Variable Forgetting Factor Constrained Constant Modulus RLS Algorithm for Adaptive Beamforming. Signal Processing, 105(105).

Chen J Y. 2013. A 3D model classification and retrieval Graphics Technology based on integrated RBF neural network. 34 (2), 26–30.

Lu K et al. 2015. Vision Sensor-Based Road Detection for Field Robot Navigation. Sensors, 15(11): 29594–29617.

Luo Y, Ren L. 2015. Target Tracking Based on Amendatory Sage-Husa Adaptive Kalman Filtering. International Conference on Electronic Science and Automation Control. Atlantis Press.

Wu H Y et al, 2013. Two-Dimensional Direction Recognition Using Uniaxial Tactile Arrays, IEEE sensors journal, vol.13, no.12, pp. 4897–4903.

Energy Science and Applied Technology ESAT 2016 – Fang (Ed.)
© 2016 Taylor & Francis Group, London, ISBN 978-1-138-02973-6

Improved genetic algorithm to avoid premature convergence

X.G. Li, H.B. Wang, Q. Wu & Z.P. Sun
State Key Laboratory on Integrated Optoelectronics, College of Electronic Science and Engineering, Jilin University, Changchun, Jilin Province, China

ABSTRACT: Genetic algorithm is an efficient global random search algorithm, which has applications in many optimization problems, and achieved good results. But it also has shortcomings. One of the main problems is premature convergence. This paper proposes a new method to improve the performance of the genetic algorithm. In the genetic operators, the new individuals are mainly produced by crossover operator. This paper improves crossover operator by introducing new individuals in the process of crossing when the premature convergence occurred. Simultaneously, after each generation of genetic operations the reinsertion operation will be carried out. Improved genetic operators increased the diversity of the population and enhanced the accuracy of the solution. Experimental results show that the improved algorithm can effectively avoid the premature convergence and increase the probability to find the global optimal solution.

Keywords: genetic diversity; premature convergence; genetic operators; optimization

1 INTRODUCTION

Genetic Algorithm (GA) is proposed by J.H (1975). It is an optimization method, which search the optimal solution by simulating Darwinian natural evolution (survival of the fittest). Each solution for the problem is considered as one individual, which consists of by encoded characters. A certain number of individuals are combined together to form a population, the population is constantly evolving through a series of genetic operations (selection, crossover, mutation). At last, the solution that obtained by decoding the best individual of the final population can be regarded as the approximate optimal solution for the problem. GA has many advantages such as groups search, internal heuristic random search, parallel computing ability and scalability etc, it has been applied in many fields.

However, GA also has its limitations. Many researchers have found that the traditional GA's encoding is not standardized, constrained of the optimization problems can't be fully expressed, and prone to premature convergence. These problems reduce the efficiency of GA, how to prevent premature convergence is one of the research issues. Rocha & Neves (1999) proposed a random offspring generation method to prevent premature convergence and applied it to TSP (Traveling Salesman Problem). If the selected parent individuals are very similar then generate one or two offspring randomly. Liu et al. (2000) used Markov chain to analysis the degree of population diversity. He

denotes that GA can be improved by modification the size of the population, mutation probability and fitness function. The author used chaos operator to maintain the diversity of population, thus to prevent premature convergence. Simona (2009) proposed The Dynamic Application of Crossover and Mutation Operators and The Population Partial Reinitialization method. The author used two sets of operators instead of two operators. Pandey et al. (2014) reviewed and compared the methods to prevent premature convergence in GA, and analyzed the strengths and weaknesses of each method.

To enhance the efficiency and performance of the GA, this paper improves the genetic operators by increasing the diversity of the crossover individuals and using reinsertion operation to prevent the premature convergence.

2 THE PHENOMENON OF PREMATURE CONVERGENCE AND ITS ANALYSIS

This section describes the premature convergence phenomenon and by analyzing the function of genetic operators in the process of evolution, summarized the causes of this phenomenon.

2.1 *Premature convergence*

In the early stage of the evolution of the genetic algorithm, some "super individuals" occur within the population. These individuals' fitness is far greater than the average fitness of the current population. Under the action of selection operator these individuals quickly occupy the entire population, thereby the diversity of the population greatly reduced. Thus the entire population is no longer evolution limit to a local optimal solution.

2.2 *Analysis of genetic operators*

The theoretical basis for the effectiveness of the genetic algorithm is pattern theorem and building block hypothesis. Pattern theorem states that under the action of genetic operators, the pattern which has a low order, short length and high fitness will grow exponentially, these patterns are called building block. Building block hypothesis states that under the action of genetic operators, building blocks combined with each other can generate high order, long length and high fitness pattern, thus we can finally obtain the optimal solution.

For binary encoding, the population can be represented in the matrix form:

$$P = (P_1, P_2, \cdots, P_N)^T = \begin{bmatrix} p_{11} & p_{12} & \cdots & p_{1l} \\ p_{21} & p_{22} & \cdots & p_{2l} \\ \vdots & \vdots & \cdots & \vdots \\ p_{N1} & p_{N2} & \cdots & p_{Nl} \end{bmatrix}$$

The SGA (Simple Genetic Algorithm) can be represented as the following function:

$$P(j+1) = \{T_m^i[T_c^i(T_s^i\{P(j)\})]\}, i = 1, 2, \cdots N\}, j \geq 0$$

where T_s is the select operator, T_c is the one point crossover operator, T_m is the bit mutation operator, N is the population size.

As we can see genetic operators play a vital role in GA. Using the pattern theorem to analyze the genetic operators. Selection operator is under the criterion of "survival of the fittest" to choose individual, when the population appears with the "super individual", under the action of selection operator these individuals quickly occupy the entire population. However, some other individuals may contain superior patterns are eliminated, due to lack of new genes it can't jump out of local optima. Crossover is the main method for generating new individuals, which recombines the individuals in the population. Nevertheless, individuals that carry out crossover operation are selected from current population. When the population is occupied by "super individuals", the offspring obtained by the crossover operator will be very similar to the parent individual. Thus the population is stagnant unable to find the global optimal. The mutation operator is

able to produce new patterns, but mutation rate is small generally take 0.1 ~ 0.001. If the mutation rate chosen is too large, it may mutate a number of superior individuals, the algorithm will search with no purpose.

From the above analysis can be known, the essence of premature convergence is the individuals in the population are very similar, the population lack of genetic diversity and superior pattern. All of these lead to the genetic algorithm cannot converge to the global optimal solution.

3 IMPROVED GENETIC ALGORITHM

In order to increase the diversity of the population, so that the population can converge to the global optimal solution, this paper improved genetic operators. For selection operation, this paper uses the optimal individual retention strategy. The individuals who have a higher fitness in the parent's generation will be directly copied to the offspring. This ensures that the optimal individual is not eliminated and the convergence of the population.

For crossover operation, the selection of individuals to perform crossover operation is different from traditional genetic algorithm in this paper when the premature convergence risk reaches to a level that considered critical. This situation can be judged by the difference between the maximum fitness f_{max} and average fitness $\bar{f}$ of the current population. If $f_{max} - \bar{f} < k$, where k is the yardstick. By generating a new population randomly, this population will perform crossover operation with the current population. Select an individual from the new population and the current population respectively. These two individuals will perform several times crossover operation. Choose a highest fitness individual from the obtained individuals and add to the offspring. Select the next individual from the new population and the current population respectively. Carry out the same operation until the offspring is generated. The obtained offspring will perform reinsertion operation according to fitness value. The specific algorithm steps are as follows:

Step 1. Calculate the maximum fitness f_{max} and the average fitness $\bar{f}$ of the current population. If $f_{max} - \bar{f} < k$ then generate a new population randomly. Select the first individual from the new population and the current population respectively. These two individuals are used as the chromosomes to be cross.

Step 2. Generate a random number rand. If rand ≤Pc (crossover probability), these two individuals will perform several times crossover operation. Evaluate the fitness of the obtained individuals, choosing a highest fitness individual add to the offspring.

Step 3. If rand >Pc, copy the individual from the current population directly, and add it to the offspring without performing crossover operation.

Step 4. Select the next individual from the new population and the current population respectively as the chromosomes to be cross. Then go to step 2 and step3.

Step 5. Repeat step 4 until the offspring is generated.

Step 6. Calculate the fitness of the offspring. Choose a certain proportion individuals insert to the current population according to the fitness value. The obtained population will be considered as the new parent population to perform normal genetic operators.

For mutation operation, as a result of using binary code, all elements of each individual are mutated with a specific probability in this paper. The mutation probability is also varies according to k. This can increase the probability of generate a new individual. Likewise, the obtained individuals will perform reinsertion operation according to fitness value.

Improved genetic algorithm increases the diversity of the population, because it introduces new individuals in crossover operation when the premature convergence risk reaches to a level that considered critical. The increase of new genetic material is likely to make the algorithm out of local optimal solution, and find the global optimal solution. Simultaneously, the selection operation is embedded into crossover operation. It makes the algorithm faster to find the global optimal solution.

4 EXPERIMENTAL RESULTS

This paper selects two test functions, De Jong and Shubert functions to test the proposed algorithm.

4.1 F1: De jong function

The De Jong function expression is as follows:

$$f(x) = \sum_{i=1}^{n} x_i^2, \qquad -512 \le x_i \le 512$$

De Jong function is the sum of squares function. Only has one minimum point $(0, 0, \ldots, 0)$, theoretical minimum value is $f(0, 0, \ldots, 0) = 0$. This paper selects a twenty dimensional De Jong function, the two dimensional image is shown in Fig. 1.

4.2 F2: Shubert function

The Shubert function expression is as follows:

$$f(x_1, x_2) = \left\{ \sum_{i=1}^{5} i * \cos[(i+1) * x_1 + i] \right\}$$
$$* \left\{ \sum_{i=1}^{5} i * \cos[(i+1) * x_2 + i] \right\}$$
$$+ 0.5 * [(x_1 + 1.42513)^{\wedge} 2$$
$$+ (x_2 + 0.80032)^{\wedge} 2], -10 \le x_1, x_2 \le 10$$

This function is a multi-modal function. There are a total of 760 local minimum points in its domain. Only one point $(-1.42513, -0.80032)$ is the global minimum point, $f(\min) = -186.7309$. The function image is shown in Fig. 2.

The test results are listed in Table 1 and compared with simple genetic algorithm. Where M represents the population size, G represents the max iterations, Pc represents the crossover probability, Im GA represents improved GA. As can be seen from the table, the performance of the improved genetic algorithm is superior to the traditional genetic algorithm. Table 2 shows the 20 variables value of De Jong function when the objective function takes the optimal solution.

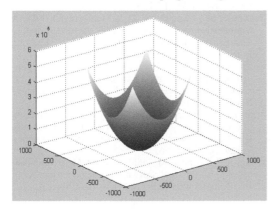

Figure 1.   Image of two-dimensional De Jong function.

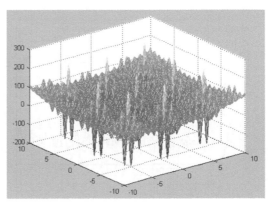

Figure 2.   Image of Shubert function.

Table 1. Test function instances.

F	M	G	Pc	Optimal solution	
				SGA	Im GA
F1	40	500	0.7	35.2739	0.0259
F2	40	200	0.7	−170.5308	−186.7309

Table 2. The value of 20 variables.

Variable	X_1	X_2	X_3	X_4	X_5	X_6	X_7
SGA	−0.37	1.02	−0.73	−1.03	0.67	0.81	2.08
Im GA	0.02	−0.00	−0.06	0.05	0.02	−0.01	−0.01

Variable	X_8	X_9	X_{10}	X_{11}	X_{12}	X_{13}	X_{14}
SGA	0.04	0.62	−0.73	−1.21	0.35	−0.08	−0.47
Im GA	0.04	0.01	−0.03	−0.07	−0.03	−0.07	−0.02

Variable	X_{15}	X_{16}	X_{17}	X_{18}	X_{19}	X_{20}
SGA	−4.01	−0.83	−1.91	1.58	0.42	1.11
Im GA	0.01	−0.07	−0.00	0.01	−0.03	−0.07

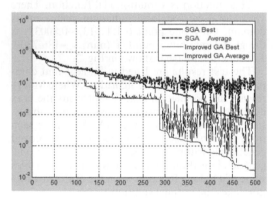

Figure 3. Objective function value curve (De Jong).

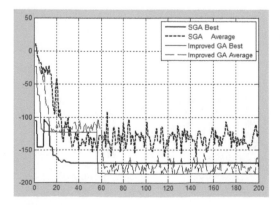

Figure 4. Objective function value curve (Shubert).

Figure 3 and Figure 4 show the average objective function value curve and the optimal objective function value variation curve of genetic algorithm. As can be seen from the figure, the improved genetic algorithm can find a better solution than simple genetic algorithm.

5 CONCLUSION

This paper briefly describes the genetic algorithm premature convergence and analyzes the causes of premature convergence according to the function of genetic operators in the process of evolution. This paper proposes a new method to improve genetic operators by generating new individuals randomly of the crossover individuals and using reinsertion operation to against the decrease of population diversity which lead to local optimal. Improved genetic algorithm has a higher probability to jump out of local optimal and better performance than simple genetic algorithm. At the same time, the complexity of the algorithm is increased, and it may take more time to deal with complex problems. This article will make further improvements to it in the future work in order to make the algorithm converge faster, and apply it to practical problems.

REFERENCES

Holland, J.H. 1975. Adaptation in natural and artificial systems. Ann Arbor, MI: University of Michigan Press.

Liu, J., Cai, Z.X. & Liu, J.Q. 2000. A novel genetic algorithm preventing premature convergence by chaos operator. Journal of central south university of technology. Vol.7. No.2. 100–103.

Nicoara, E.S. 2009. Mechanisms to Avoid the Premature Convergence of Genetic Algorithms. Bulletin of P.G. University of Ploiesti, Mathematics—Informatics-Physics Series. Vol.LXI. No.1/2009. 87–96.

Pandey, H.M., Chaudhary, A. & Mehrotra, D. 2014. A comparative review of approaches to prevent premature convergence in GA. Applied Soft Computing Journal, v 24. 1047–1077.

Rocha, M. & Neves. J. 1999. Preventing premature convergence to local optima in genetic algorithms via random offspring generation. Multiple Approaches to Intelligent Systems, Proceedings [C]. 127–136.

Srinivas, M. & Patnaik, L.M. 1994. Adaptive probabilities of crossover and mutation in genetic algorithms. IEEE Transactions on Systems, Man and Cybernetics. Vol.24. No.4. 656–667.

Wang, X.P. & Cao L.M. 2002. Genetic Algorithm-Theory, Application and Software Implementation. Xi'an Jiao Tong University Press. 51–55. (In Chinese)

Energy Science and Applied Technology ESAT 2016 – Fang (Ed.)
© 2016 Taylor & Francis Group, London, ISBN 978-1-138-02973-6

The methods to remove space debris

Anni Lin
School of North China Electric Power University, Baoding, China

ABSTRACT: Now space debris poses a serious threat to humanities efforts at space exploration as well as the expanding uses for earth-orbiting satellites. In light of these threats, certain measures have been taken to address the issue of space debris. However, countries have hesitated to develop space debris removal systems due to high costs, potential risks and other worries. The purpose of this paper is to determine the best alternative to clear out space debris. First, we predict the growth trend of space debris and obtain the distribution called spatial density respectively. Next, to evaluate the risk of collisions between satellites and space debris, we define the safety distance and the operation risk index in different altitudes. Finally, we formulate an optimized model to make the best alternative for the private firm considering various factors including costs, risks and benefits.

Keywords: space debris; safe distance; risk index; time-dependent; optimization

1 INTRODUCTION

From almost the first moment that man started traveling beyond Earth's atmosphere, we've been leaving behind all sorts of debris (non-functional and/or uncontrolled man-made objects and component parts thereof) in space, particularly in those orbits that are most used. The debris ranges in size and mass from paint flakes to abandoned satellites. Not only is it wasteful, but space junk can be dangerous as well—to satellites, to space stations, when some of it plummets back to Earth, to human life on the ground.

After many years of scientific research and exploration, in the field of space debris prevention, many methods have been proposed to remove the space debris. These methods include small, space-based water jets and high energy lasers used to target specific pieces of debris and large satellites designed to sweep up the debris, among others. But the debris' high velocity orbits make capture difficult.

2 SOLVING PROCESS

2.1 *The growth of the space debris*

Using the historical data from website, we draw the scattered point distribution of the space debris, and the results are as follows:

Based on Figure 1, we formulate a quadratic function: $y = ax^2 + bx + c$, applying the least square method to find the quadratic curve which makes the distance of scattered points the shortest.

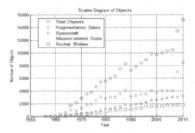

Figure 1. Scatter diagram of objects.

Simulated annealing algorithm to fit the non-linear curve process is to find the specific function: y, so that the residual square and the minimum of the original data y, is a dynamic estimation of the parameters of the process. So we take the parameter of the function as aim to obtain specific value of the parameter by adopting The Maximum Likelihood Criterion. In the process of figuring out parameters, we adopt simulated annealing algorithm to obtain the best fitting performance. By applying the method of fitting to calculate with MATLAB, we obtain the growth trend of various space debris for future years, and the formulations are shown as follows:

$$\begin{cases} Q_{Total} = 1.472t^2 + 459.2967t - 987.5186 \\ Q_{Fra} = 1.6448t^2 + 174.4339t - 194.8104 \\ Q_{Spa} = 1.6154t^2 + 84.0133t - 283.9498 \\ Q_{Spa} = 1.6154t^2 + 84.0133t - 283.9498 \\ Q_{Mis} = \begin{cases} -2.0035t^2 + 129.6994t - 267.5997(1950 \le t < 2040) \\ 1560(t \ge 2040) \end{cases} \\ Q_{Roc} = 0.2472t^2 + 70.1822t - 235.7015 \end{cases} \quad (1)$$

Where the Q_i is the amount of space debris.

2.2 The orbital distribution of space debris

At present, the society widely used the space debris environment model to predict the distribution of debris in orbits. And this model is built on the basis of observation data, which has many advantages such as convenience and accuracy. As mentioned above, space debris are divided into three categories (large debris, dangerous debris and small debris). And debris (larger than ten centimeters), can also incapacitate satellites but they are large enough to be tracked and thus potentially avoided. Debris smaller than one centimeter, in contrast, cannot be tracked or avoided, but can be protected against by using relatively simple shielding. The most dangerous pieces of space debris are those ranging in diameter from one to ten centimeters, of which there are roughly 300,000 in orbit. These are large enough to cause serious damage, yet current sensor networks cannot track them and there is no practical method for shielding spacecraft against them. Consequently, this class of orbital debris poses an invisible threat to operating satellites (Ansdell M, 2010). Considering these situations, we apply the probability theory to analysis the distribution of space debris since it's hard to formulate the space debris environment.

We extract the altitude and the number of space debris data from the data on NASA's website. Then we discard the data having limit value and draw the frequency distribution of space debris against altitude. The certain result is shown as follows:

2.3 Risk Assessment model of satellite operation

To evaluate the risks of collisions between space debris and satellites, we define the safety area of a satellite which can ensure its normal work to avoid collisions resulting from space debris. We take the satellite as the central and radius R to build a ball coordinate in Figure 2. When in a certain position, the distance between the satellite and space debris is greater than the safety radius R, in other words, once the space debris into the safety area, then there exists risks to the satellite. Otherwise, we consider space debris have no damage to the satellite.

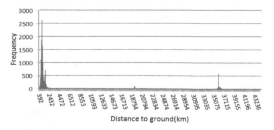

Figure 2. The frequency distribution histogram of total space debris.

Considering the distance between a satellite in a certain position of various orbits and space debris, we simulate various orbits by changing their inclination to obtain the minimal distance. In the process of simulation, if the distance between the satellite and space debris is less than R, then the frequency of collisions add 1 times. To quantify the risk of collision between a satellite and space debris, we define the operation risk index as follows:

$$Dan_h = \frac{n}{N} \qquad (2)$$

Where Dan_k is the operation risk index, N is the total number of space debris and n is times of collisions.

We adopt the method of probability distribution function (Cunha L. M. et al, 1998) to substitute for the origin curve of risk against the altitude. We fit the origin curve with the double Weibull distribution. Then we obtain the modified satellite risk in LEO (low earth orbit) against the altitude:

2.4 The optimization

In order to determine which time we are proposed to carry out space debris, we defined a new parameter A. The parameter is calculated by summing the risk index in one hundred orbits which are selected randomly in a certain altitude orbit and the formulation is shown as follows:

$$A = \sum_{i=1}^{100} Dan_{hi} \qquad (3)$$

Where Dan represents operation risk index in the orbit which altitude is hi. When A is larger than the threshold, we are proposed to carry out cleanup activities.

The certain steps are shown as follows:

Step 1. develop the object function

Taking the cost, risks, benefits and policy into consideration, we developed the following object function:

$$\min = \sum_{i=1}^{N} \sigma_i w_i h_i (f_{max} - f_i) \qquad (4)$$

Here, the σ_i presents the possibility that whether the space debris is selected to remove (when the space debris is selected to remove, the $\sigma_i = 1$; contrarily, the $\sigma_i = 0$); w_i presents the weight of the cost (when the debris is small debris, the $w_i = 1$; when the debris is dangerous debris, the $w_i = 10$). h_i represents the altitude of the orbit numbered I; f_i presents the risk of collision between spacecraft and a single space debris; N represents the total amount of the space debris.

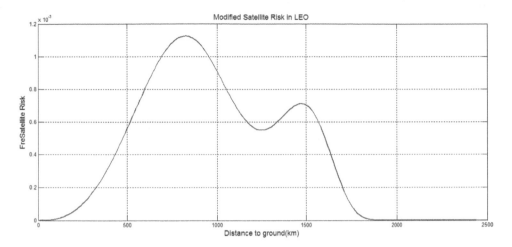

Figure 3. Modified satellite risk in LEO.

Step 2. Develop the constraint condition.

First, we are required to make sure the effect of cleaning well. So the condition is developed as follows:

$$\sum_{i=1}^{N} \sigma_i m_i \geq \Delta M \qquad (5)$$

Here, m_i presents the quality of the debris numbered i, ΔM presents the quality of the debris increased.

Then, the risks of the total debris that have been removed in this year should be larger than the risks of the total debris increased. So the condition is developed as follows:

$$\sum_{i=1}^{N} \sigma_i f_i \geq \Delta F \qquad (6)$$

Here, ΔF represents the risk of the total debris increased in this year.

Thus, we have developed the object function and the constraint condition. The final objective optimization model are shown as follows;

$$\min = \sum_{i=1}^{N} \sigma_i w_i h_i (f_{max} - f_i)$$

$$\begin{cases} \sigma_i = \begin{cases} 1 & \text{which is selected to remove} \\ 0 & \text{which is not selected to remove} \end{cases} \\ w_i = \begin{cases} 1 & \text{the cost of cleanup for small debris} \\ 10 & \text{the cost of cleanup for large debris} \end{cases} \\ \sum_{i=1}^{N} \sigma_i m_i \geq_\Delta M \\ \sum_{i=1}^{N} \sigma_i f_i \geq_\Delta F \end{cases} \qquad (7)$$

3 CONCLUSION

The paper is to determine the best alternative to clear out space debris. First, we predict the growth trend of space debris and obtain the distribution called spatial density respectively. Next, to evaluate the risk of collisions between satellites and space debris, we define the safety distance and the operation risk index in different altitudes. Finally, we formulate an optimized model to make the best alternative for the private firm considering various factors including costs, risks and benefits.

REFERENCES

Ansdell M. Active space debris removal: Needs, implications, and recommendations for today's geopolitical environment [J]. Journal of Public and International Affairs, 2010, 21: 7–22.

Cunha L. M., Oliveira F. A. R. & Oliveira J. C. Optimal experimental design for estimating the kinetic parameters of processes described by the Weibull probability distribution function[J]. Journal of Food Engineering, 1998, 37(2): 175–191.

Information on http://jyoder.com/

Energy Science and Applied Technology ESAT 2016 – Fang (Ed.)
© 2016 Taylor & Francis Group, London, ISBN 978-1-138-02973-6

Research on public key certificates

Changliang Cheng
School of Beijing University of Technology, Beijing, China

ABSTRACT: It is supposed that different random numbers are generated every time when public keys are selected differently. However, in real world, some public keys are not as secure as was supposed. We designed a method to choose high quality random numbers to generate more secure public keys. Our conclusion is that more secure public keys can be generated by our method that provides a new way to improve the quality of random number seeds.

Keywords: X.509 certificate, public keys, RSA, random number seed

1 INTRODUCTION

Many studies have been conducted on the public-key infrastructure, especially X.509 certificates (Stallings, 2013, Cooper, 2008). Moreover, many problems have been discussed about the certificates (Holz, R, 2011, Vratonjic, N, 2011). The key generation of the RSA algorithm (Rivest, R.L, 1978) has been discussed and found to be practical. A study (Loebenberger, 2011) has found that there is difficulty in factoring the output of RSA. However, Lenstra found that the conclusion can be correct only when the result is considered separately. For more results, the conclusion may sometimes be wrong (Lenstra, et al., 2012).

An important assumption of the security on the public key is that the public keys are generated by unrepeated random number seeds. However, RSA public keys and RSA moduli are not as secure as supposed. We focused on this issue and took some measure to improve the related problem.

Section 2 introduces public-key cryptography and RSA algorithm. Section 3 describes the problem and the solution. Finally, Section 4 summarizes our conclusion.

2 PUBLIC-KEY CRYPTOGRAPHY AND RSA

2.1 *Public-key cryptography*

It is a revolutionary matter to find the public-key cryptography. It solves a plenty of problems about sharing secret keys and digital signature. It is a system of encryption where cryptographic keys are paired, so that an encryption performed with one key can be a decryption by the other key, and possession of one key does not affect the practical application of the other. The public key can be opened and widely used, while the other is only known and used by the owner. This kind of cryptography system often relies on the cryptographic algorithm based on mathematical problems that currently has no efficient solution. In this paper, we will discuss the RSA algorithm based on integer factorization.

2.2 *RSA*

The principle of RSA is that it is difficult to find three large positive integers, e,d and n

s.t. $(m^e)^d = m(\bmod n)$

The RSA algorithm can be described as follows:

First, we select two prime pairs and compute $N = pq$. We then compute $\varphi(N) = (p-1)(q-1)$

Select $1 < e < N$ s.t. $\gcd(\varphi(N), e) = 1$. Then, we can use the Euclidean algorithm to calculate $1 \le d < N$ s.t. $ed \equiv 1(\bmod \varphi(N))$. After these steps, the user obtains (e,N) as the public key and d as the private key. It is difficult to find d *despite knowing n and e*. The (GNFS) is the most efficient algorithm known for factoring large integers and its complexity is:

$$e^{\left(\left(\frac{8}{3}\right)^{\frac{2}{3}} + o(1)\right)(\ln n)^{\frac{1}{3}}(\ln\ln n)^{\frac{2}{3}}}$$

The encryption of RSA is very simple. For example, using B's public key K, we compute the ciphertext C as follows:

$$C = K^e(\bmod N)$$

B sends C to A via a reliable, but not necessarily secret route. Then, A can compute K as follows:

$$K = C^d(\bmod N)$$

In the practical application, we often choose $e = 2^{16}+1 = 65537$. However, the smaller value of e (e.g. 3) has been shown to be less secure (L. Han, 2010, X. Wang, 2015).

3 PROBLEM AND SOLUTION

Because the public keys play an important role in cryptography, its security is always the concern of researchers. In his recent research, Lenstra found that there are two problems in some certificates based on the RSA algorithm, which are described below.

Certificates of different users use the same public key. Strictly speaking, because most users use e = 65537 as the public exponent, they will have the same N = pq. Thus, the users' certificates will not be effective because their private keys are the same.

1. Moduli have a shared prime factor. If the moduli of two users have a shared prime factor, then we can easily calculate the private key of the users.

To solve the above problem, we should ensure that public keys are generated after proper initialization of random number seed.

To ensure that different users have unrelated prime pairs, we use a parameter in the certificate, namely subject to construct a new RSA prime generation method. We need to use NIST symmetry encryption AES and hash function SHA3. Our random number generation pseudocodes are as follows:

Function GenerateRandNumber (S,id,k)
input: S random number seed
id user unique identification
k length of random number 128 bits
Output: R random number in 16 Kbytes length

$K \leftarrow SHA3 - 256(S \| SHA3 - 256(id))$
$C \leftarrow 1$
$R \leftarrow \lambda / / (\lambda is Null)$
$for(i = 1, \cdots, k) do$
$R \leftarrow R \| AES256_K(C)$
$C \leftarrow C + 1$
$return R$

Here the positive integer C is calculated as 128 bits string.

Using this random generation function, we can generate prime.

function GeneratePrime (n)
input: n prime length (multiples of 128)
S random number seed
id user unique identification
output: m

$m \leftarrow GenerateRandNumber \left(S, id, \dfrac{n}{128} \right)$
$while(Rabin - Miller(m) = 0) do$
$m \leftarrow GenerateRandNumber \left(S, id, \dfrac{n}{128} \right)$
$return m$

The Rabin–Miller (m) is an algorithm that tests whether the m is prime[10]. In the above algorithm, the prime pairs (p_A, q_A) and (p_B, q_B) of different users A and B cannot be the same because the hash function SHA3 has a strong collision $SHA3 - 256(id_A) \neq SHA3 - 256(id_B)$.
Then, $K_A \neq K_B$.

4 CONCLUSION

A new method was proposed in this paper to choose a proper random number seed to avoid the generation of repeated prime pairs. The properties of the hash function make it practical to avoid the generation of same random numbers by the proposed method. However, the RSA problem remains a complex issue. Therefore, we should ensure that the public keys are generated by means of selecting a proper random seed.

REFERENCES

Cooper, D., Santesson, S., Farrell, S. Boeyen, S., Housley, R., Polk, W.: Internet X. 509 Public Key Infrastructure Certificate and Certificate Revocation List (CRL) Profile. RFC 5280 (2008).

Holz, R., Braun, L., Kammenhuber, N., Carle, G.: The SSL landscape: a thorough analysis of the x.509PKI using active and passive measurements. In: Proceedings of the 2011 ACM SIGCOMM Conference on Internet Measurement Conference, IMC 2011, pp. 427–444. ACM (2011).

Loebenberger, D., Nusken, M.: Analyzing Standards for RSA Integers. In: Nitaj, A., Pointcheval, D. (eds.) AFRICACRYPT 2011, LNCS, vol. 6737 pp, 260–277, Springer, Heiddelberg (2011).

Lenstra, et al., Ron was wrong, Whit is right, Crypto 2012, Lecture Notes in Computer Science, Volume 7417, pp 626–642.

Han, L., X. Wang, and G. Xu, On an attack on RSA with small CRT-Exponents, Science China, [Ser. F], 53 (2010) 151–1518.

Wiener, M.J. Cryptanalysis of short RSA Secret exponents, IEEE Trans. On Info. Theory, 36 (1990), 553–558.

Rivest, R.L., Shamir, A., Adleman, L.: A method for obtaining digital signatures and public-key cryptosystems. Communications of the ACM 21, 120–126 (1978).

Stallings, W. Cryptography and Network Security. (6rd) United States PEARSON 2013.

Vratonjic, N., Freudiger, J., Bindschaedler, V., Hubaux, J.-P.: The inconvenient truth about web certificates. In: The Workshop on Economics of Information Security, WEIS (2011).

Wang, X., G. Xu, W. Wang, and X. Meng, Mathematical Foundations public key cryptography, CRC Press, 2015.

Energy Science and Applied Technology ESAT 2016 – Fang (Ed.)
© *2016 Taylor & Francis Group, London, ISBN 978-1-138-02973-6*

Optimal scheduling scheme selection model and algorithm based on the Bayesian game of cloud computing resources

Kan Niu, Hengwei Zhang, Jindong Wang, Na Wang & Tao Li
Zhengzhou Institute of Information Science and Technology, Zhengzhou, China

ABSTRACT: As a new business model, a deal is reached on ensuring interests based on cloud computing. Thus, it becomes necessary to consider the users' satisfaction degrees and the dispatch center of economic benefits in cloud computing resource scheduling. Game theory has been increasingly used in cloud computing resource scheduling in recent years. This paper performs the analysis of the cloud computing resource scheduling problem, and proposes a resource scheduling model based on the static Bayesian game of optimal scheduling scheme selection algorithm by using the game theory method. The algorithm improves the efficiency and economic benefit of resource scheduling. Finally, an experiment is conducted to verify the effectiveness and accuracy of the algorithm.

Keywords: cloud computing resource; Bayesian game; optimal selection; scheduling scheme

1   INTRODUCTION

Cloud computing is a new computing model developed on the basis of grid computing, parallel computing, and P2P technology. It adopts virtualization technology that links a large amount of computing resources, storage resources and software resources together to form a large-scale shared virtual resource pool, and provides cloud services for users or application systems (Buyya R. et al, 2009) (Armbrust M. et al, 2010).

The resource competition is similar to the free competition in market economics between the applications in the cloud platform interaction. Thus, many methods and strategies in economics theory are also applicable to solve the problem of cloud resources (K. Lee et al, 2009). Game theory is the most commonly used strategies in economics. It analyzes the interaction between the decision-making factors for playing the game. According to their respective interests, game theory considers their rational analysis and decision to obtain the optimal solution. The cloud resource scheduling problem is how to solve the competition among multiple users to meet the needs of all users. It just applies to the basic methods of game theory.

There are many related studies on the use of the game theory method to solve the problem of cloud computing resource scheduling. The non-cooperative game model was applied to the cloud resource management by Xin Jin et al. (J. Xin et al, 2013). The method obtained the optimal solution through solving the Nash equilibrium for the cloud multi-tenant resources. However, the main

research of this method was the cloud resource pricing problem. Carroll et al. put forward a kind of resource combination based on the game theory framework through maximizing the entire organization benefits to increase the income of the parties (T. E. Carroll et al, 2010). Nevertheless, each provider cannot be accurately calculated for each group of interest, so information was restricted to right decision-making. Guiyi Wei and others considered QoS constraints of resource allocation problems (G. Y. Wei et al, 2010). They used the game theory to solve the business layer parallel computing resource scheduling of tasks. The game algorithm considers two factors: optimization and fair. Mohammad et al. put forward a game method to solve the problem of resource management. Their algorithm was a fixed number of resource allocation problems that did not consider QoS factors. Therefore, this paper puts forward a reasonable and effective resource scheduling model and a method based on static Bayesian game in game theory to solve the competition among many users in cloud computing resource scheduling.

2   CLOUD COMPUTING RESOURCE SCHEDULING BASED ON THE STATIC BAYESIAN GAME MODEL

In the process of cloud computing resource scheduling, physical resources are virtualized to a virtual unit for running the program. The users' jobs often contain multiple child tasks. Each of the subtasks corresponds to a best virtual unit type and any abundant physical resources can create the corresponding virtual unit that can be assigned to the users to finish their homework task. However, a differently created solution can directly affect the user service quality, so the QoS is the key for the users.

Game theory studies the behavior of the participants with a direct interaction between decision-making and the balance of this decision-making problem. Static Bayesian game refers to the behavior of other participants and gains without complete information. Therefore, the Bayesian game is also known as incomplete information game. The participants who are the users submit the task requests according to each requirement in the cloud computing resource logic. They can be considered simultaneously in the game. Moreover, in the actual cloud computing resource scheduling process, different users can clearly know each other's task request and resource scheduling center feedback situation, namely income. It creates an incomplete information. Therefore, the process of the different user who has submitted a task can be as the incomplete information game process.

2.1 Model assumption

Assumption 1 for Rational Assumption. Assume that the user and resource scheduling center are fully rational and select strategies according to the principle of maximizing their income.

Assumption 2 for Type Hypothesis. Assume that each user' demand is different and their QoS are different. Dispatching center and strategic income uncertainty as to the user is not sure about the type of user. However, for each user type, the probability distribution can be decided.

Assumption 3 for Income Hypothesis. Assume that each user' goal is to request resources to complete tasks on the premise of meeting QoS. The goal of the resource scheduling center is to meet the needs of all users as far as possible and obtain income by feedback resources.

2.2 Model definition

Definition 1. The cloud computing resource scheduling model can use seven tuples (N, Q, Req, S, σ, P, U). Its specific meanings are as follows:

1. N is the set of game participants. It is the set $N = \{1, 2,..., i...\}$ of the resources of all requests of the users at a certain moment.
2. $Q = \{q_1, q_2, q_3,..., q_i,...\}$ means the type of space of the users. The type of the user determines the index weight of QoS, and the index of concrete can be divided into the response time, cost, reliability, stability and so on. For example, the users can be divided into notice to completion time and cost based on the type of the user and

the QoS of index weight. It needs to reduce the end-to-end delay to reduce the task completion time for the real-time service or online service users in cloud computing resource scheduling. However, some other users who pay more attention to costs are willing to reduce the amount of resources and sacrifice time on performance to reduce their own costs:

3. $Req = \begin{bmatrix} r_1(q_i) \\ r_2(q_i) \\ \vdots \\ r_i(q_i) \end{bmatrix}, i = 1, 2, \cdots,$ which indicates the resource request matrix of all users at a certain moment.

4. S means the strategy combination space of the resource scheduling in the game at a certain moment. The user i has two pure strategies: $S_i = \{s_{i1}, s_{i2}\}$, with Si = {Accept, Not Accept}. Resource scheduling center has K pure strategies, namely the scheduling scheme: $S_{center} = \{s^*_1, \cdots s^*_K\}$.

5. The probability distribution $\sigma = (\sigma_1, \cdots, \sigma_i, \cdots, \sigma_{center})$ means the mixed strategy of resource scheduling game at a certain moment. The probability distribution of $\sigma_i = (\sigma_{i1}, \sigma_{i2})$ is called a mixed strategy of user i. Here $\sigma_{i1} = \sigma(s_{i1})$ is the strategy s_{i1} probability which the user i chooses and $0 \le \sigma_{i1}, \sigma_{i2} \le 1, \Sigma_1^2 \sigma_{ik} = 1$. The probability distribution of $\sigma_{center} = (\sigma^*_1, \cdots, \sigma^*_K)$ is called a hybrid strategy of the center of resource scheduling, where $\sigma^*_k = \sigma^*(s^*_k)$ is the probability of the selection strategy s^*_k of the dispatching center and $0 \le \sigma_k \le 1, \Sigma_1^K \sigma_k = 1, k = 1, \cdots, K$.

6. P means the prior belief space of users. $P(N_i^{q_i})$ is the prior belief of the resource scheduling center deducing the user i for type of q_i. Among them, $P(N_i^{q_i}) = p_i, \Sigma_{i=1}^k p_i = 1$.

7. $U = \{u_1, u_2,..., u_i,..., u_{center}\}$ means the revenue function on users and resource scheduling. Earning reflects the profit and loss. Different results can be obtained by adopting a different strategy game, namely different earnings.

Definition 2. Harsanyi transformation refers that the information for the scheduling center is uncertain about the type of users is transformed into the natural choice of users' types. The notable ways are: (1) the introduction of "Nature" in the actual cloud resource scheduling and the type of user P is randomly decided for transforming not knowing the type of user P to not knowing the scheduling process. (2) Nature allows the user p to know the type of p, but does not allow the resource scheduling center to know the type of P. (3) After the Nature selection, the resource scheduling center selects the strategy from its own strategy space.

Definition 3. Cloud computing resource scheduling game tree is often used to represent a scheduling benefit for each strategy in the process of path expression form. It has a general tree structure with a triple system (V, E, U). V refers to all nodes, and different node represents different states of nodes in the process of resource scheduling. E is the collection of the edge in the game tree, based on the resource scheduling center strategy. U is a collection of scheduling center income, obtained based on a different strategy.

2.3 Quantitative calculation of the benefit

Cloud computing resource scheduling center of strategy benefits of quantitative calculation is the foundation of subsequent scheduling game analysis and directly affects the result of scheduling. Therefore, the strategies of the dispatching center are necessary to yield quantitative results reasonably. This paper shows that the benefit of applying for resources of user i can be divided into two kinds of circumstances, De_i and $De_i{}'$, respectively. De_i is the resource scheduling center's strategy that satisfies the QoS at a certain moment, and thus the user gets revenue. $De_i{}'$ is the resource scheduling center that cannot satisfy the QoS strategy at a certain moment, and thus the user gets loss. In the process of actual cloud computing resource scheduling, resource is similar to an infinite number, so total resources must satisfy all user application resources. Moreover, the type of user determines the quality of service, and the dispatching center meets all the user's QoS as far as possible and feedbacks to the user resources so as to reap the benefits. Instead, if dispatching center cannot satisfy the QoS of all users as much as possible, it will lose part of the users, resulting in loss.

For earnings expectations of user i:

1. Resource scheduling scheme satisfies the QoS of user i:

$$s.t. \begin{cases} Req < Resource \\ QoS_i \times S_{center} \in QoS \end{cases} \tag{2.1}$$

$$u_i = De_i \tag{2.2}$$

2. Resource scheduling scheme cannot meet the QoS of user k:

$$s.t. \begin{cases} Req < Resource \\ QoS_k \times S_{center} \notin QoS \end{cases} \tag{2.3}$$

$$u_k = De_k{}' \tag{2.4}$$

For dispatching center earnings expectations:

$$s.t. \begin{cases} Req < Resource \\ QoS_i \times S_{center} \cap QoS_k \times S_{center}, i \neq k \end{cases} \tag{2.5}$$

$$u_{center} = \prod_i^n P(N_i^{q_i}) \cdot De_i + \prod_k^n P(N_k^{q_k}) \cdot De_k{}', i \neq k \tag{2.6}$$

2.4 Nash equilibrium analysis

Any user for application resources wants feedback from the resource scheduling center to meet their QoS and to complete the work task. If the scheduling center cannot satisfy the users' QoS, the user can choose other cloud computing resources. Thus, the resource scheduling center will lose the users, resulting in loss. So, despite all the application resources of different QoS of the user at a certain moment, understanding how to allocate the resource selection strategy and as far as possible to meet all the needs of the users is the key to the cloud computing resources scheduling problem.

In the process of resource scheduling, the dispatching center hopes to satisfy the QoS of all users as much as possible in its prior information space to get the maximum gain U. Dispatching center and all users will eventually reach a balance under the principle, namely the Bayesian Nash equilibrium. Because the dispatching center does not want to lose any user so as to reduce its own profits, there is a strategy to satisfy the QoS of all users as much as possible under the Bayesian Nash equilibrium. The dispatching should choose the strategy and will not try to change the strategy.

Definition 4. In the cloud computing resource scheduling model (N, Q, Req, S, σ, P, U), a mixed strategy can satisfy $\max \sum_{i=1}^{n} \{u_i[\sigma_{ik}(S_{ik}); \sigma_k^*(S_k^*); q_i]P(N_i^{q_i})\}$ for the resource dispatching center and any user i of type $q_i \in Q$.

Strategy combination $\sigma = (\sigma_1, \cdots, \sigma_i, \sigma_{center})$ is a mixed strategy of the Bayesian Nash equilibrium for cloud computing resource scheduling in the game. Among them, σ_{ik} describes the probability of user i choosing the strategy s_{ik}. σ_k^* describes the resource scheduling center selection probability of the scheduling scheme s_k^* and meets $\sigma_{ik} \in \sigma_i, \sigma_k^* \in \sigma_{center}, \sigma_i, \sigma_{center} \subset \sigma$. S_{ik} describes that the user i chooses the first k strategy. S_{ik} describes the resource dispatching center that chooses the first k scheduling scheme and meets $S_{ik} \in S_i, S_k^* \in S_{center}, S_i, S_{center} \subset S$. Dispatching center belief $P(N_i^{q_i})$ describes that the user i is inferred for type q_i of uncertainty. By definition, the sum of the maximum is the sum of possible combinations of dispatching center and all users' summation and strategies of different types.

3 CLOUD COMPUTING RESOURCE SCHEDULING MODEL OF THE OPTIMAL SCHEDULING SCHEME SELECTION ALGORITHM

Algorithm: the optimal scheduling scheme selection algorithm is based on the cloud computing resource scheduling model.

Input: user application resource game tree (V, E, U).

Output: the optimal scheduling plan s_k^*.

BEGIN

1. Initialize the resource scheduling model (N, Q, Req, S, σ, P, U).
2. Build the set of user types the filing $Q\{q_1,q_2,q_3,\ldots,q_i,\ldots\}$.
3. Build users' strategies space collection S_i = {accept and not accept}.
4. Construct resource scheduling center strategy space collection $S_{center} = \{s_1^*,\cdots,s_K^*\}$.
5. $QoS_i \times S_{center} \in QoS$.
6. S_i_Move = accept.
7. For user i, income $u_i = De_i$.
8. $QoS_k \times S_{center} \notin QoS$.
9. S_2_Move = not accept.
10. For user k, income $u_k = De_k'$.
11. For dispatch center, income $u_{center} = \prod_i^n P(N_i^{q_i}) \cdot De_i + \prod_k^n P(N_k^{q_k}) \cdot De_k', i \neq k$.
12. For any user i and dispatch center.
13. The mixed strategy $\sigma = (\sigma_1,\cdots,\sigma_i,\cdots,\sigma_{center})$ meets $\max \sum_{i=1}^n \{u_i[\sigma_{ik}(S_{ik});\sigma_k^*(S_k^*);q_i]P(N_i^{q_i})\}$.
14. Return $S = \{s_{1k},s_{2k},\cdots,s_{ik},\cdots,s_k^*\}$.
15. END.

Based on the cloud computing resource scheduling model of the optimal scheduling scheme, the selection algorithm is the key of incomplete information static game model resource scheduling model (N, Q, Req, S, σ, P, U), including all probabilities of user types to ensure that the users and the establishment of dispatching center policy set scheduling scheme gains quantitative calculation and the final solution of the static Bayesian mixed strategy equilibrium. In the cloud computing resource scheduling process, the user and the game is incomplete information and the probability dis-

tribution selection strategy game between the users and dispatch center. The static Bayesian game model and mixed strategy Nash equilibrium are more operable and have stronger practicability.

4 APPLICATION EXAMPLES AND ANALYSIS

To further verify the cloud resource scheduling model for the effectiveness of the optimal scheduling scheme selection algorithm, this paper assumes that there are instances of resource scheduling, as shown in Figure 1. This example describes a Hypervisor virtualization cloud environment resource scheduling problem, with the dispatching center resource pool approaching infinity. The user's QoS indicators have response time, cost, reliability, availability, and fairness. Three users apply resources for dispatch center at a moment. Each user's type and QoS are different, namely the specific index weights difference. Dispatch center meets all the users' QoS principle to allocate resources according to the three users' resource request matrix Req, QoS index weights.

This game tree is shown in Figure 2. By running the optimal scheduling scheme selection algorithm, we can get the mixed strategy Bayesian Nash equilibrium of the game and the equilibrium result is $S = \{s_{11}(accept),s_{21}(accept),s_{31}(accept),s_2^*(scheme2)\}$.

For the mixed strategy, Nash equilibrium can be interpreted as follows: in the cloud computing resource scheduling process, the user types are under the premise of prior probability of knowing by the dispatch center. We give an optimal scheduling scheme according to the users' QoS weight and higher returns are produced when users accept resources than when they do not accept them, namely users choosing the accept strategy.

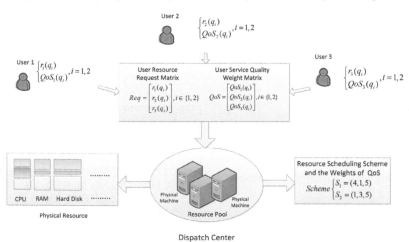

Figure 1.   Cloud computing resource scheduling instance.

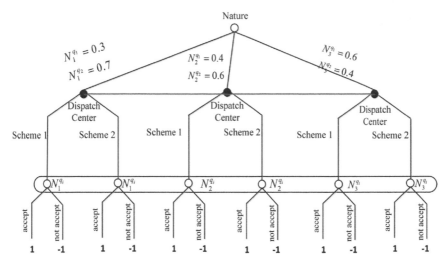

Figure 2. Cloud computing resource scheduling game tree.

The above results show that the proposed cloud computing resource scheduling model and the optimal scheduling scheme selection algorithm can be more reasonable and effective to reflect the costs and benefits of the impact on income of users and dispatching center. They can also effectively select the optimal scheduling scheme.

5 CONCLUSION

This paper establishes the static Bayesian game model and defines the model in detail. The application resource users are divided into various types and game equilibrium conditions are analyzed and proved in the model. From the user's perspective, the cloud computing user competition problems in resource scheduling and the resource allocation problem are analyzed by the Bayesian game theory model. The analysis shows that the resource scheduling scheme is more comprehensive, reasonable, and effective.

Currently, there are still some optimization factors in resource scheduling objectives such as energy consumption optimization problem of resource dispatch, load balancing, and so on. Our future study will focus on the multi-objective optimization of resource scheduling technology.

ACKNOWLEDGMENTS

This work was supported by the National Nature Science Foundation of China under Grant No. 61303074 and No. 61309013.

REFERENCES

Armbrust M., Fox A., Griffith R., et al. A view of cloud computing[J]. Communications of the ACM, 2010, 53(4): 50–58.

Buyya R., Yeo C.S., Venugopal S., et al. Cloud computing and emerging IT platforms: vision, hype, and reality for delivering computing as the 5th utility[J]. Future Generation Computer Systems, 2009, 25(6): 599–616.

Carroll, T.E., D. Grosu. Formation of virtual organizations in grids: a game-theoretic approach[J]. Concurrency and Computation: Practice and Experience. 2010, 22(14): 1972–1989.

Hassan, M., B. Song, E.N. Huh. Game-based distributed resource allocation in horizontal dynamic cloud federation platform[M]. Algorithms and Architectures for Parallel Processing. 2011. 194–205.

Lee, K., N.W. Paton, R. Sakellariou, et al. Utility driven adaptive workflow execution[A]. Proceedings of the 9th IEEE/ACM International Symposium on Cluster Computing and the Grid[C]. 2009. 220–227.

Wang Jindong, Yu dingkun, et al. Defense Strategies Selection Based on Incomplete Information Attack-Defense Game[J]. Journal of Chinese Computer Systems. 2015, 36(10).

Wei, G.Y., A.V. Vasilakos, Y. Zheng, et al. A game-theoretic method of fair resource allocation for cloud computing services[J]. Journal of Supercomputing. 2010, 54(2): 252–269.

Xin, J., K.Y. Kwong, Y. Yong. Competitive Cloud Resource Procurements via Cloud Brokerage[A]. Proceedings of the 5th IEEE International Conference on Cloud Computing Technology and Science[C]. 2013. 355–362.

Xu Xin. Game Theory Based Resource Scheduling Approaches of Cloud Computing[D]. Shanghai: East China University of Science and Technology. 2015.

Energy Science and Applied Technology ESAT 2016 – Fang (Ed.)
© *2016 Taylor & Francis Group, London, ISBN 978-1-138-02973-6*

Clustering based on skew-based boundary detection

Q. Han & B.Z. Qiu
School of Zhengzhou University, Zhengzhou, China

ABSTRACT: Nowadays, the clustering technique has been widely applied to digital mining across the board. Density-based clustering algorithm measures the density of digital points by calculating the number of points within neighborhood radius or determining with the shared relations between neighboring points. Apart from the existing density measurement, this paper, based on skew data distribution in margin areas, will propose a new skewness-based clustering method, which is able to distinguish core points from boundary points without the impact of neighborhood radius and data density. It can be seen that the clustering algorithm (CSBD) is on par with similar existing algorithms with respect to its higher accuracy and better-retrieved clustering boundary.

Keywords: skew distribution; boundary degree; clustering algorithm; data mining

1   INTRODUCTION

Clustering refers to the process of identifying data internal structures and potential models while partitioning data (Yang et al. 2015). Density-based clustering is applicable to the commercial, biological, geographical, and insurance sectors for its outstanding identification of clusters in different patterns, automatic discoveries of clustering number, and effective de-noising capability (Tekieh & Raahemi 2015) (Xu et al. 2014).

As a typical density-based clustering algorithm, DBSCAN (Ester et al. 1996) classifies data points into core points, boundary points, and noisy points, distributing from core points towards clustering boundaries. As the terminal conditions of scaling out, effective detection of boundary points holds the key to improve the algorithm performance. Another algorithm (Guo & Zang 2012) is designed to detect boundary points by substituting global density with local density to diminish the effect of data distribution density on the clustering algorithm. In contrast, IS-DBSCAN (Bakr et al. 2015) uses the shared relationship of neighboring points to compute density, further making the algorithm more accessible.

However, in the density-based clustering algorithm, the internal clustering is denser than the boundary areas, so that the data of lower-density areas are separated from that of higher-density region to develop into clustering. However, the method does not always win in the specific data integration. In some cases, for example, the data density in the boundary region is higher than that of the internal area, so the identified clustering boundary in the algorithm is not consistent with

actual boundaries. Therefore, the paper will focus on how to accurately determine cluster boundaries and effectively complete clustering.

2   ALGORITHM

2.1   *Motivation and boundary degree*

In the recent study on clustering, we find that the neighboring distribution of the boundary points tends to be skew, while that of the internal points within the cluster is not. As shown in Figure 1, points x_1 and x_2 are located at the core of the cluster and in the boundary, respectively. When mapping all points onto the X and Y axes, the corresponding distribution of x_2 is skew, while the mapped distribution of x_1 is non-skew.

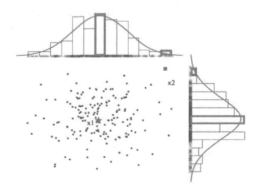

Figure 1.   Points x_1 and x_2 and their mapped coordinate system.

Statistically speaking, data distributions are commonly skew (Geng et al. 2015) (Georgiou & Voigt 2015). The gamma distribution is representative of them and has a wide range of applications as it can transform into distribution patterns, such as Normal, Inverted Gamma, Chi-square, Inverted Beta, and Beat. This paper sets the neighboring points distribution of boundary points to comply with that of gamma. In this paper, the maximum likelihood estimation is used to estimate the parameters α of the gamma distribution.

Definition 1. (*Skewness* (x_p)) It refers to a metric to evaluate the skew of data distribution. Define *skewness* (x_p) as:

$$skewness(x_p) = \frac{2}{\sqrt{\alpha}} \qquad (1)$$

Definition 2. (*Boundary Degree* (x_p)) It is used to measure the boundary of data points. A larger value of boundary degree indicates that a point is more likely to be in the boundary areas, and vice versa. The quotient of skewness degree and local density can be defined as:

$$Boundary\ \boldsymbol{degree}(x_p) = \frac{skewness(x_p)}{density(x_p)}$$
$$= \frac{2/\sqrt{\alpha}}{\frac{1}{k}\sum_{i=1}^{k} d_i} = \frac{2k}{\sqrt{\alpha}\sum_{i=1}^{k} d_i} \qquad (2)$$

2.2 Discussion on boundary degree

In this paper, a model is designed to assess the properties of the boundary degree in a way to visually and quantitatively investigate its changes. Assume x_1–x_8 are the neighboring points of p (as shown in Figure 2(a)) and p can move around. C_1 is at the core position of neighboring points x_1–x_8. C_1 expands out a unit distance each time to obtain four concentric circles, namely $C_{2,3,4,5}$, respectively. So, the characteristics of boundary degree can be detected by calculating its changes at different positions.

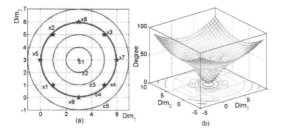

Figure 2. Test of boundary degree: (a) test model, (b) 3D image of boundary degree.

The minimum value of boundary degree at C_1 in the model is 1.54, while that of $C_{2,3,4,5}$ is 7.94, 19.41, 41.86, 44.05, respectively, with an increase of 7.94, 19.41, 41.86, 44.05, respectively, compared with the previous circumcircle. So, when C_1 to C_4 share the same neighboring points, their boundary degree rises with increasing increments. From C_4 to C_5, the boundary degree increases, but with decreasing increments, which is consistent with the changes in boundary degree extending from the center towards the outer (as shown in Figure 1(b)). In other words, internal points are separated from the external ones of the circumcircle C_4 by selecting threshold values.

Definition 3: (*Threshold Value* α) It can separate internal points from non-internal points. $\alpha = f$ (x_p) refers to the percentile of x_p after boundary degrees are in a descending order.

Definition 4: (*Connection Matrix Conn* (i,j)) It shows the connection of data points of x_i and x_j. If x_i is the neighboring point of the internal point x_j, *Conn* (i,j) value takes 1, otherwise 0. That means:

$$Conn(i,j) = \begin{cases} 1, x_i \in N_{k-dist}(x_j), f(x_j) < \alpha \\ 0, else \end{cases} \qquad (3)$$

where *Conn* is a matrix of n × n.

2.3 Algorithm introduction

The algorithm proposed in this paper first calculates the boundary degrees for all points, and then conducts subgraph searches based on the Conn constructed by the threshold value α, to obtain clustering results.

Name: CSBD

Input: dataset, neighbor amounts k, and threshold value α

Output: clustering results, boundary, noise

Step:

1. Compute the boundary degree of any point based on Definition 2.
2. Divide data points into internal and non-internal points by the threshold value α according to Definition 3.
3. Construct the connection matrix Conn by Definition 4.
4. Conduct subgraph searches within the Conn, where those subgraphs longer than k are clusters. The non-internal points refer to the single boundary of the cluster, while all single boundaries constitute the public boundaries of a dataset, while the rest of the points within the dataset are noise.

The paper selects synthetic datasets to display the clustering process by boundary degree. Synthetic dataset Syn1 (Figure 3(a)) has two nested

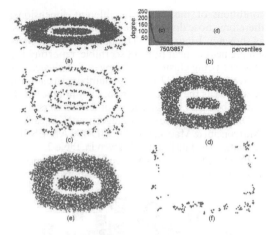

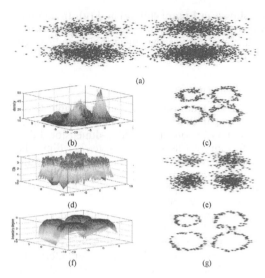

Figure 3. An example of synthetic dataset: (a) dataset, (b) boundary degree ranking, (c), (d) corresponding points of two sections, (e) clustering results, (f) noisy points.

Figure 4. Comparison on XOR: (a) original points, (b) density, (c) boundary of DBSCAN, (d) ISk, (e) boundary of IS-DBSCAN, (f) boundary degree, (g) boundary of CSBD.

target clustering and little amount of noise. Rank boundary degrees are shown in Figure 3(b); of which, the corresponding data points of the two sections (c) and (d) are shown in Figure 3(c) and (d). It can be seen that boundary degrees are able to separate boundary points and noisy points from the dataset. The most distinct difference between boundary points and noisy points is that boundary points are within clustering, while noisy points drift outside of clustering. We set $\alpha = 19.45\%$, build Conn, perform subgraph searching, and show subgraphs with the length larger than $k = 28$, as shown in Figure 3(e). Two clusters are found, whose boundary points are marked in red. The distribution of boundary points on the margins outlines the cluster.

3 EXPERIMENT AND ANALYSIS

3.1 *Experiment design and environment*

Based on the artificial dataset, image dataset, UCI datasets, the aim of the experiment is to test the algorithm performance in different dimensions, which is conducted on a MATLAB 2012 computer, equipped with Intel Core 2.93GHZ CPU, 4G RAM, and Window 7 OS.

3.2 *Artificial dataset XOR*

XOR, a two-dimensional dataset generated by MATLAB, consists of 5000 points of four normal distributed clusters (including 500, 1000, 1500 and 2000, respectively). As shown in Figure 4(a), the target cluster sizes and densities are different, and

scattered boundary points intertwine with each other. To make a comparison between DBSCAN, IS-DBSCAN, and the algorithm proposed herein, we show the corresponding *densities*, IS_k and *boundary degrees* of data points in Figure 4(b), (d) and (f), respectively.

It can be seen from Figure 4(b) that the 3D density view is in a cone shape. The value of *density* is associated with the amounts of points within neighbor radius, which is larger in the higher-density areas, while smaller in the lower-density regions. Since there is no significant density difference in the areas where boundary points should be separated, parameters cannot be easily distinguished. From Figure 4(d), it can be seen that the points in the IS-DBSCAN algorithm, where the IS_k value is smaller than 2k/3, are identified to be non-internal points. The super parameter 2k/3 can be used to reduce parameter numbers, but makes the algorithm less feasible, which may not work for some specific datasets. That means, only k = 4, and the algorithm is able to correctly complete the whole cluster analysis. Although it does not affect the results of clustering, the algorithm cannot obtain an effective boundary model. The value of *boundary degree* shown in Figure 4(f) is in the shape of mushrooms. In the non-boundary areas, boundary degree values are less influenced by density changes, with smooth variation. In the margin areas, the significant changes in boundary degree are conducive to the selection of the threshold value. The boundaries of the three algorithms are shown in Figure 4(c), (e) and (g). In comparison with the former two, the proposed algorithm accurately outlines the clustering boundary with less number of points.

3.3 Image dataset

The MNIST dataset is used as the graph dataset in this paper. Pixel points are taken as one-dimensional data to convert into high-dimensional data. So, the image (28 × 28) turns to 784 dims. The dataset is given in Table 1. Ranking the boundary degrees of data in data1, we select three typical sections, and display their corresponding digital graphs, as shown in Figure 5.

As shown in Figure 5, the boundary degree represents the levels of data point clustering on the boundary. A smaller boundary degree value indicates that points are within the cluster, referring to the overall status of the cluster; therefore, the digits are written neatly, as shown in Figure 5(d). A larger boundary degree value shows that the points are on the margin of the cluster, representing the limiting conditions of the cluster in different dimensions; therefore, those digits are illegible and in sloppiness, as shown in Figure 5(b). The numbers on the margin are neither neat nor indiscernible, reporting the phase of being clustered to being out of clusters, as shown in Figure 6. In reality, numbers should not be written as the pattern of boundary points, which cannot be easily recognized as other numbers.

On the datasets, the specific algorithm (K-means++ (Marconcini & Macucci 2015), IS-DBSCAN, CSBD) evaluation of data1–data4 is given in Table 2.

From Table 2, we can see that the proposed algorithm possesses advantages such as clustering

Table 1. Datasets data1–data4.

Set	No. of Cluster	Means
Data1	892,1028	5,7
Data2	892,6265	5,7
Data3	980,1010,1009	0,3, 9
Data4	1135,5842,1028	1,4,7

Figure 6. Results of digit 7 in data1.

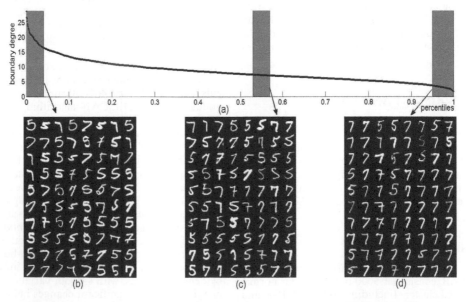

Figure 5. Boundary degree sections and the corresponding points: (a) order descending of boundary degree, (b)–(d) corresponding points of three sections.

Table 2. Comparison results of datasets data1–data4.

Set	Algorithm	Accuracy	Entropy	Purity
Data1	K-means++	87.45%	0.4828	0.8745
	ISDBSCAN	**99.84%**	**0.0170**	**0.9984**
	CSBD	99.64%	0.0299	0.9964
Data2	K-means++	55.06%	0.5242	0.8752
	ISDBSCAN	99.86%	0.0121	0.9986
	CSBD	**99.87%**	**0.0115**	**0.9987**
Data3	K-means++	93.50%	0.3872	0.9350
	ISDBSCAN	66.26%	0.7113	0.6626
	CSBD	**99.18%**	**0.0628**	**0.9918**
Data4	K-means++	83.22%	0.6768	0.8322
	ISDBSCAN	99.83%	0.0178	0.9983
	CSBD	**99.88%**	**0.0128**	**0.9988**

Table 3. UCI datasets.

Data	Dim	NO. of Cluster
Breast_cancer	9	458,241
seeds	7	70,70,70
Chart_T	60	100,100,100, 100,100,100

Table 4. Comparison results of UCI datasets.

Set	Algorithm	Accuracy	Entropy	Purity
Breast cancer	K-means++	65.52%	0.9076	0.6552
	IS-DBSCAN	71.14%	0.8644	0.7114
	CSBD	**94.64%**	**0.2994**	**0.9464**
seeds	K-means++	89.28%	0.4635	0.8905
	IS-DBSCAN	62.07%	0.8526	0.6355
	CSBD	**96.58%**	**0.1944**	**0.9658**
Chart T	K-means++	70.00%	0.7922	0.7000
	IS-DBSCAN	66.67%	0.6661	0.6678
	CSBD	**82.40%**	**0.3738**	**0.8240**

and non-internal points within the datasets. When the value of α is small, the recall rate of the algorithm results is higher, while the accuracy is lower. When the value of α is larger, the accuracy rate will increase, while the recall rate decreases. Users can select appropriate parameters depending on various requirements. With regard to time consumption, the time complexity of k neighborhood is O (nlogn), while that of clustering is O (kn); therefore, the time complexity of the algorithm will be O (nlogn).

accuracy, entropy evaluation, and purity degree. In the clustering process, the separation of internal points from non-internal points can improve the accuracy of the proposed algorithm. Determining the data points out of clusters to be noises can improve the entropy evaluation and purity degree, and remove those points that are easily recognized to be points in other clusters. Apart from the result of data1 being the second place and based on various evaluation indicators, the algorithm proposed in this paper has reached up to the current levels of existing algorithms, in terms of accuracy, entropy evaluation, and purity.

4 CONCLUSION

In this paper, we put forward a new method to measure the data distribution skewness, based on the skew distribution of boundary points. The method can effectively distinguish the boundary points in the dataset, and obtain clusters with higher accuracy and precision when searching within the clustering. The algorithm can identify clustering boundaries that tally with the actual situation, so as to play a significant role in the expanding boundary model.

3.4 UCI dataset

The UCI database includes a large amount of standard test datasets for data mining. We select three datasets, as shown in Table 3. In the UCI dataset, Algorithm, K-means++, IS-DBSCAN, and CSBD are considered, and the results are summarized in Table 4.

As shown in Table 4, the algorithm can achieve higher accuracy, entropy evaluation, and purity in datasets of different dimensions.

3.5 Algorithm analysis

Two parameters are in the algorithm: k and α. Parameter k indicates the number of neighboring points of data points, relating to the data density. Parameter α indicates the proportion of internal

REFERENCES

Bakr, A.M., Ghanem, N.M., and Ismail, M.A. (2015). Efficient incremental density-based algorithm for clustering large datasets. Alexandria Engineering Journal, 54(4):1147–1154.

Ester, M., Kriegel, H.-P., Sander, J., and Xu, X. (1996). A density-based algorithm for discovering clusters in large spatial databases with noise. In Kdd, volume 96, pages 226–231.

Geng, X., Meng, L., Li, L., Ji, L., and Sun, K. (2015). Momentum principal skewness analysis. IEEE Geoscience and RemoteSensing Letters, 12(11):2262–2266.

Georgiou, G. and Voigt, K. (2015). Stochastic computation of moments, mean, variance, skewness and kurtosis. Electronics Letters, 51(9):673–674.

Guo, C. and Zang, Y. (2012). Clustering algorithm based on density function and nichepso. J Syst Eng Electron, 23(3):445–452.

Marconcini, P. and Macucci, M. (2015). Effect of disorder on the local density of states of a graphene nanoribbon with a barrier. In Nanotechnology (IEEE-NANO), 2015 IEEE 15th International Conference on, pages 1126–1129.

Tekieh, M.H. and Raahemi, B. (2015). Importance of data mining in healthcare: A survey. In 2015 IEEE/ACM International Conference on Advances in Social Networks Analysis and Mining (ASONAM), pages 1057–1062.

Xu, L., Jiang, C., Wang, J., Yuan, J., and Ren, Y. (2014). Information security in big data: Privacy and data mining. IEEE Access, 2:1149–1176.

Yang, Y., Ma, Z., Yang, Y., Nie, F., and Shen, H.T. (2015). Multitask spectral clustering by exploring intertask correlation. IEEE Transactions on Cybernetics, 45(5):1083–1094.

Energy Science and Applied Technology ESAT 2016 – Fang (Ed.)
© *2016 Taylor & Francis Group, London, ISBN 978-1-138-02973-6*

Community detection in networks based on an immune genetic algorithm

Y. Yao & J.Q. Zhou
Nanjing University of Posts and Telecommunications, Nanjing, Jiangsu, China

ABSTRACT: In order to improve the optimization ability of the complex network community detection, this paper proposes a multi-objective community detection based on self-adaptive immune genetic algorithm. The algorithm applies an adaptive strategy to the genetic operators by using logistic function to set the corresponding crossover probability and mutation probability, and turns the multi-objective optimization problem into the minimal optimization of two objectives called Kernel K-Means (KKM) and Ratio Cut (RC). It can improve the diversity of the population. The algorithm also introduces an immune operator into the options of the genetic algorithm which prevents the degradation of the individuals in crossover and mutation. Experiments on both synthetic and real-world networks show that the proposed algorithm is highly efficient at discovering quality community structure.

Keywords: complex networks; community detection; self-adaptive; immune genetic; multi-objective

1 INTRODUCTION

In many real-world systems, complex networks constitute an effective formalism to represent the associations among objects, such as collaboration networks, biological networks, the Internet, social networks (Pizzuti, C. 2012), etc. Community structure is an important property in complex networks (Gong, M. G., et al. 2011). Communities are defined as groups of nodes which are densely interconnected but only sparely connected with the rest of the network. Nodes belonging to the same community have common properties (Girvan, M. et al., 2002). Therefore, the Community Detection (CD) is particularly useful in complex networks.

There have been many algorithms for CD. The most famous one is the GN algorithm proposed by Girvan and Newman (Girvan, M. et al., 2002) which is optimization based and relatively accurate but shows a high complexity. Newman made an improvement on GN and proposed the fast Newman algorithm (FN) (Newman, M. E. J. 2004). It regards the modularity (Q) as the objective function and is simple than GN. With the widespread use of Q function, researchers proposed a lot of optimization algorithms based on modularity optimization, such as Fast Modularity (FM) algorithm (Clauset, A. 2005). However, Fortunato and Barthlemy have found that the modularity optimization may fail to identify communities smaller than a scale which depends on the total size of the network. This is the reso-lution limit (Fortunato, S. et al, 2007). In order to overcome the problem, Li et al. introduced a quality function called modularity density (D) to analyze the network topology from different resolution by adjusting the parameters (Li, Z. P., et al. 2008). In recent years, the multi-objective optimization was introduced to CD. Because of the strong global optimization ability of the Genetic Algorithm (GA), some algorithms based on GA turned out to be efficient in CD, such as MOGA-net (Pizzuti, C. 2012), MOCD (Shi, C. et al. 2012), MOEA/D-net (Gong, M. G., et al. 2012). However, the parameters of basic GAs are often fixed, especially the crossover probability and mutation probability. The basic GAs cannot meet the dynamical requirements in genetic evolution; The genetic operators are random in iteration, so they not only provide opportunities for the individuals of population, but also produced a degradation possibility inevitably. These problems can lead to poor results.

In this paper, we propose a community detection method based on Self-adaptive Multi-objective Immune Genetic Algorithm (SMIGA). In accordance with the fitness values of individuals, the logistic function is used to set the corresponding crossover and mutation probability. The superior individuals have been reserved effectively and the mutation abilities of poor individuals are improved, so the convergence is accelerated; The immune theory is introduced to the selective operation of GA, which improves the diversity of the population and prevents the inherent degradation.

2 RELATED BACKGROUND

2.1 Community definition

A network N can be modeled as graph $G = (V, E)$, where V represents a set of objects, called nodes, and E represents a set of objects connect two elements of V, called edges. A community is a group of nodes in a network having dense connection within them, and relatively sparse connection between groups. A more formal definition has been introduced in (Radicchi, F, et al. 2004) by considering the degree k_i of a node i, defined as equation (1) where A is the adjacency matrix of G. If there is an edge between node i and node j, the position (i,j) is 1, otherwise 0. Suppose the node i belongs to a subgraph $S \subset G$, the degree of i concerning S can be split as follows.

$$k_i = \sum_j A_{ij} \tag{1}$$

$$k_i(S) = k_i^{in}(S) + k_i^{out}(S) \tag{2}$$

$$k_i^{in}(S) = \sum_{j \in S} A_{ij} \tag{3}$$

$$k_i^{out}(S) = \sum_{j \notin S} A_{ij} \tag{4}$$

2.2 Object functions

Li et al. introduced the modularity density (D) as the object function which is defined as follows (Li, Z. P., et al. 2008):

$$D = \sum_{i=1}^{k} \frac{L(V_i, V_i) - L(V_i, \overline{V_i})}{|V_i|} \tag{5}$$

$$L(V_1, V_2) = \sum_{i \in V_1, j \in V_2} A_{ij} \tag{6}$$

where V_i is the node set of the subgraph G_i, $|V_i|$ is the number of nodes in the subgraph. The greater the value of D is, the more accurate the community will be found. So as to detect the network topology structure in different resolutions, the expression of D is improved continuously and then decomposed to several parts to form a multi-objective optimization problem, one of the popular decompositions is KKM (Angelini, L, et al. 2007) and RC (Wei, Y. C. et al., 1991), which are defined as follows:

$$\min \begin{cases} KKM = (n-k)\sigma - \sum_{i=1}^{k} \frac{L(V_i, V_i)}{|V_i|} \\ RC = \sum_{i=1}^{k} \frac{L(V_i, \overline{V_i})}{|V_i|} \end{cases} \tag{7}$$

where $n = |V|$, σ is a real number. The smaller the value of KKM is, the closer the connections in the community will be. The smaller the value of RC

is, the sparser the connections between the internal nodes and external nodes will be.

3 DESCRIPTION OF PROPOSED ALGORITHM

3.1 Representation

SMIGA adopts the integer encoding to represent the network division, such as $x = \{x^1, x^2, ..., x^n\}$, where x denotes the chromosome which is the set of node labels in the network, n is the number of nodes, x^i is the label of node i, $x^i \in \{1, 2, ..., n\}$. Nodes with the same label belong to the same community. If a network is composed of n nodes, it can be divided into n communities most. In this encoding scheme, different codes may represent the same network partition. The benefits of using this direct coding is that the number of communities is not required in advance.

3.2 Initialization

The population initialization mechanism is based on label propagation. At the beginning, each node represents different communities, that is, $x^i = i$. Suppose the neighbor nodes set of node i is $S(i) = \{x_1, x_2, ..., x_k\}$, and $l(i)$ is the label of node i. In a label propagation mechanism, the label for each node in the network depends on the label which holds the maximum proportion in its neighbor set. If each label is not the same, then one is selected randomly and changed according to equation (8):

$$l(i) = \arg\max_k \sum_{j \in S(i)} \delta(l(j), k) \tag{8}$$

where $\delta(i, j)$ is 1 when node i and node j belong to the same community. This procedure is repeated 5 to 10 times. By this mechanism, the nodes connected closely will be set to the common label which enriches the diversity of population.

3.3 Genetic operators

Crossover. A one-point two-way crossover operation is used (Gong, M. G., et al. 2011). The procedure is defined as follows. In individuals we pick a node called v_i randomly, and mark its label as x_a^i. All of nodes belonging to this community in the x_a are also assigned to the same community in the x_b, that is $x_b^k \leftarrow x_a^i, \forall k \in \{k \mid x_a^k = x_a^i\}$. Simultaneously, the label x_b^i of node v_i in the x_b is found and all of nodes belonging to this community in the x_b are also assigned to the same community in the x_a, that is $x_a^k \leftarrow x_b^i, \forall k \in \{k \mid x_b^k = x_b^i\}$. This procedure will generate two new chromosomes x_c and x_d. Such crossover operation is exploratory and shows the ability of inheritance, so the produced offspring can carry features common to the parents and combine the features taken from their parents.

Mutation. A single-point mutation based on neighbors is used. We randomly pick a chromosome C to be mutated. A gene i is picked randomly on the chromosome C, and the possible values of it's allele are limited to the values of its neighbors. For each node in the offspring after mutation, it is associated only with one of its neighbors. Such mutation is able to prevent the invalid search, reduce the search space and improve the efficiency of the algorithm.

Logistic function is a common s-shaped function which is named by Pierre-François Verhulst in 1844 based on the study of the relationship between population growth and itself. Long term application shows that, the logistic function can describe some bounded growth phenomenon. In order to maintain the diversity of population in the evolutionary process, crossover probability should be decreased while mutation probability should be increased. This trend is similar with the logistic curve, so we modify the crossover probability p_c and mutation probability p_m based on the combination of parameters in GA and logistic function as follows:

$$p_c = \frac{1}{1+e^{-\frac{k_1}{\varphi}}} - 0.1 \tag{9}$$

$$p_m = \frac{k_2}{5(1+e^{\frac{1}{\varphi}})} \tag{10}$$

where k_1 and k_2 are two constants, $k_1 \in (0,\infty)$, $k_2 \in (0,1)$ and φ is the number of iterations.

3.4 *Selective operators*

In the process of evolution, selective operation is used to enlighten the evolutionary direction. In accordance with the immune theory, an adjustment factor based on the concentration is added in the mechanism of fitness selection, so as to maintain the diversity of population and prevent the premature convergence in the general GA. For a specific population (the number of individuals is m), the selective probability of the individual i ($i = 1,2,...,m$) is defined as follows:

$$p_i = \frac{F_i}{\sum_{j=1}^{m} F_j} \cdot e^{-\beta \cdot c_i} \tag{11}$$

where F_i is the fitness of i, $F = -f + f_{max} + \xi$, f is the weight sum of KKM and RC, $f = \omega_1 \cdot KKM + \omega_2 \cdot RC$, $\omega_1 = \omega_2 = 0.5$, $\xi = 0.1$; β is an adjustable parameter and $\beta \in (0,1)$; c_i is the concentration of individual i which represents the proportion of individuals

that are same or similar in the population and is defined as follows:

$$c_i = \frac{\sum_{j=1}^{m} S_{i,j}}{m} \tag{12}$$

$$S_{i,j} = \begin{cases} 1, & q_{i,j} \leq \varepsilon \\ 0, & otherwise \end{cases} \tag{13}$$

where $q_{i,j}$ is the index which represents the similarity of individual i and j; ε is the similarity threshold, $\varepsilon = 0.001$; $q_{i,j} = |F_i - F_j|/F_{max}$, where F_{max} is the bigger one between F_i and F_j.

The formula of the selective probability above indicate that the larger the individual fitness is, the greater the selective probability will be. The higher the concentration of the individual is, the lower the selective probability will be (suppression). This will not only retain the individuals with high fitness, but also reduce the proportion of the similar individuals. Such selected mechanism can accelerate the convergence of the algorithm and improve the diversity of the individuals in population effectively.

4 EXPERIMENTAL RESULTS

4.1 *Experimental evaluation criteria*

The Normalized Mutual Information (NMI), proposed by Leon Danon et al. (Danon, L, et al. 2005), is an evaluation metric to measure the similarity between the detected results of the algorithm and the real network partition. A larger value of NMI represents a greater similarity between two partitions. In this paper, the modularity (Q) and NMI are used as evaluation metrics, and the SMIGA is applied in both synthetic and real-world known networks. We make a comparison between our method and other algorithms based on GA to verify the effectiveness of our method.

4.2 *Computer-generated network*

In this section, the benchmark network proposed by Lancichinetti is used (Lancichinetti, A, et al. 2008). The network consists of 128 nodes divided into 4 communities of 32 nodes each. The average degree of each node is 16. Each node shares a fraction $1 - \mu$ of its links with the other nodes of its community and a fraction μ with the other nodes of the network. μ is called mixing parameter. When $\mu < 0.5$, the neighbors of a node inside its group are more than the neighbors belonging to other 3 groups. Therefore, a good algorithm should be able to detect these communities.

Figure 1 shows the comparison of the average NMI value among SMIGA, GA-net, MOGA-net,

MOGA-net and MOCD on the benchmark network. The mixing parameter μ is ranged from 0 to 0.5 to generate 11 different networks. The average NMI is computed over 20 independent runs.

As can be seen from Figure 1, when the mixing parameters $\mu < 0.35$ which means the fuzziness of the community is relatively low, SMIGA and MOEAD can discover the true partition correctly (NMI = 1) and are better than the results of GA-net, MOGA-net and MOCD. Along with the increase of μ, the community structure is becoming fuzzier, so finding the real structure of the community has become somewhat difficult. When $0.35 < \mu < 0.45$, it is difficult to discover the true partition, but SMIGA obtains higher NMI value than other four algorithms obtained and is the most close to the true one (NMI value remains at about 0.7 to 0.9). When μ continues to increase, finding the real structure has become very hard because the community structure is becoming extremely fuzzy, so it is very difficult for all algorithms to detect the real division of the network. This experiment shows that SMIGA has better performances than GA-net, MOGA-net, MOGA-net and MOCD on the benchmark network.

4.3 Real-world networks

SMIGA is applied to two real-world networks, namely Zachary Karate Club network (Zachary) (Zachary, W. W. 1997) and the American University Football Club network (Football) (Girvan, M. et al, 2002). We also make a comparison between our method and the Fast Modularity (FM) algorithm.

Zachary spent two years studying the relationships among members of a karate club and obtained the network of relationships (Karate). This network consists of 34 members. During the studying period, the club administrators and coaches hold different

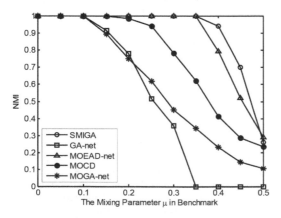

Figure 1. NMI obtained by SMIGA, GA-net, MOEAD-net, MOCD and MOGA-net on the GN benchmark.

opinions due to the costs of club, then the club split into two parts, forming two separate communities.

Football represents the American football games between Division IA colleges during the regular fall season in 2000, as compiled by Girvan and Newman. In this network, nodes represent teams, and edges represent the games between the two teams they connect. This network consists of 115 nodes, 616 edges, and all the teams are divided into 12 leagues.

Figure 2 shows the real result of Zachary network partition, and Figure 3 shows the detected result on Zachary based on SMIGA. As we can see from Figure 3(a), it is clear that SMIGA can discover the true partition correctly (corresponding NMI = 1), and Figure 3(b) also shows the community structure with the highest Q value which is also the same with the real network structure.

For comparison, Table 1 lists the results between SMIGA and FM algorithm on terms of the aver-

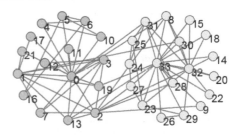

Figure 2. The real community structure of Zachary Karate Club.

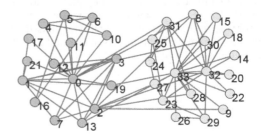

Figure 3(a). The detected community structure of Zachary Karate Club by SMIGA (NMI = 1).

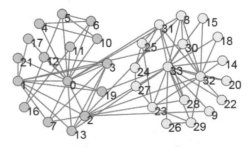

Figure 3(b). The detected community structure of Zachary Karate Club by SMIGA (Q = 0.453).

Table 1. The results of average NMI and Q values obtained by SMIGA and FM algorithm for the real-world datasets.

Network	SMIGA		FM algorithm	
	NMI_{avg}	Q_{avg}	NMI_{avg}	Q_{avg}
Zachary	1	0.429	0.693	0.380
Football	0.905	0.611	0.762	0.577

age NMI values and the average Q values. Both the algorithms have been executed 30 times independently on these two networks.

As is shown in Table 1, on Zachary network, the average NMI value obtained by SMIGA is 1. This means the true partition to Zachary network can be obtained at each run of SMIGA. However, the FM algorithm found a solution with the average NMI value of 0.693, respectively. When it comes to Football network, because of the complexity of the network structure, neither algorithm can discover the network structure correctly. While the highest average NMI value is obtained by employing SMIGA, as is known from Table 1. The FM algorithm found a solution with the average NMI value of 0.762, however, SMIGA found a solution with the average NMI value of 0.905 for 30 runs. The community structure detected by SMIGA is more accurate than by FM algorithm. Table 1 also shows that the average Q values obtained by SMIGA are larger than those obtained by FM algorithm on Zachary network and Football network. Therefore, SMIGA has better performances than FM algorithm on two real-world networks.

5 CONCLUSION

This paper proposes a new community detection algorithm called SMIGA. In order to find the pareto solutions, the algorithm applies the logistic function to set the adaptive crossover and mutation probabilities by optimizing two objective functions called Kernel K-Means (KKM) and Ratio Cut (RC). The immune principle is applied to the selective operation of the genetic algorithm, which prevents the degradation of the individuals in crossover and mutation and improves the diversity of the population efficiently. Experiments on synthetic and real-world networks show that, SMIGA can discover more accurate community structure compared to four traditional GA-based algorithms and the fast modularity algorithm. Therefore, complex network community detection mechanism based on SMIGA has certain advantages in solving complex network community detection problems.

ACKNOWLEDGMENTS

The authors thank the editors and anonymous reviewers for their valuable comments and helpful suggestions which greatly improved the quality of the paper. This work was supported by the Program for Normal College Graduate Innovation of Jiangsu Province (Grant No. SJLX_0377).

REFERENCES

Angelini, L., et al. 2007. Identification of network modules by optimization of ratio association. *Chaos* 17(2): 023114.

Clauset, A. 2005. Finding local community structure in networks. *Physical Review E* 72(7): 026132.

Danon, L., et al. 2005. Comparing community structure identification. *Journal of Statistical Mechanics* 78(09): P09008.

Fortunato, S. & Barthelemy, M. 2007. Resolution limit in community detection. *Proceedings of the National Academy of Sciences of the United States of America* 104(1): 36–41.

Girvan, M. & Newman, M. E. J. 2002. Community structure in social and biological networks. *Proceedings of the National Academy of Sciences of the United States of America* 99(12): 7821–7826.

Gong, M. G., et al. 2011. Memetic algorithm for community detection in networks. *Physical Review E* 84(5): 056101.

Gong, M. G., et al. 2012. Community detection in networks by using multiobjective evolutionary algorithm with decomposition. *Physical A-Statistical Mechanics and Its Applications* 391(6): 4050–4060.

Lancichinetti, A., et al. 2008. Benchmark graphs for testing community detection Algorithms. *Physical Review E* 78(4): 046110.

Li, Z. P., et al. 2008. Quantitative function for detection. *Physical Review E—Statistical, Nonlinear, and Soft Matter Physics* 77(3): 036109.

Newman, M.E.J. 2004. Fast algorithm for detecting community structure in networks. *Physical Review E* 69(6): 066133.

Pizzuti, C. 2012. A Multiobjective genetic algorithm to find communities in complex networks. *Ieee Transactions on Evolutionary Computation* 16(3): 418–430.

Radicchi, F., et al. 2004. Defining and identifying communities in networks. *Proceedings of the National Academy of Sciences of the United States of America* 101(9): 2658–2663.

Shi, C., et al. 2012. Multi-objective community detection in complex networks. *Applied Soft Computing Journal* 12(4): 850–859.

Wei, Y.C. & Cheng, C.K. 1991. Ratio cut partitioning for hierarchical designs. *Ieee Transactions on Computer-Aided Design of Integrated Circuits and Systems* 10(7): 911–921.

Zachary, W.W. 1997. An information flow model for confict and fission in small groups. *J. Anthropol. Res* 33(4): 452–473.

Energy Science and Applied Technology ESAT 2016 – Fang (Ed.)
© 2016 Taylor & Francis Group, London, ISBN 978-1-138-02973-6

Linear time approximation algorithms for the tree-sparse model

HuaiXiao Tou
Nankai University, Nankai, Tianjin, China

ABSTRACT: Compressive sensing is a technology for recording a signal $x \in R^n$ with the form of $y = Ax$, where A is an $m \times n$ matrix. It helps save a long signal with only a few measurements. Compressive sensing has applications in a wide variety of fields over the last decade. Thus, signal recovery problems attracted wide attentions. It is well known that it is possible to recover k-sparse signal with $m = O(k \log n/k)$ linear measurements. However, as the quantity m determines the compression rate, the logarithmic can worsen the rate tenfold for large k. To overcome the aforementioned problem, seminal researches propose a structured sparsity model M_k which makes it possible to reduce the number of measurements to $m = O(k)$. Moreover, there exists an efficient algorithm which solves the model-projection problem that returns a model-based sparse vector x', minimizing the tail error $\| x - x' \|_2$. In this paper, we limit our attention to the *tree-sparse model*. Assume that all entries of a signal $x \in R^n$ forms a tree, and all non-zero entries of x forms a subtree of k nodes containing the root node. We give two complementary approximation algorithms. One is for maximizing the total weight of the subtree, and the other one is for minimizing the left weight of the tree except the chosen subtree. Both algorithms run in time O(n). Then there exists an efficient algorithm that outputs a signal estimate $\hat{x}$ satisfies that $\| x - \hat{x} \|_p \le C \cdot \| e \|_p$.

Keywords: compressive sensing; mode-projection; tree sparsity; model-based compressive sensing

1 INTRODUCTION

1.1 *Compressive sensing*

Compressive sensing is a method for recording an n-dimensional vector (the signal) with a new linear measurement. For a k-sparse signal $x \in R^n$ (contains at most k non-zeros), the new linear measurement records the signal with the form of $y = Ax$, where A is an $m \times n$ matrix. The representation Ax takes little measurements for saving long signals, since m is much smaller than n. In reality, the measurements are often corrupted by a noise vector e (*i.e.* $y = Ax + e$). Compressive sensing has found applications in a wide variety of fields over the last decade, and the signal recovery problems have been well studied.

The problem of recovery is to find an estimate $\hat{X}$ which satisfies that $\| x - \hat{x} \|_2 \le C \cdot \| e \|_2$, where C is a constant approximation factor. Seminal researches show that it is possible to recover k-sparse signals with $m = O(k \log n/k)$ linear measurements. The recovery algorithms such as CoSaMP[6] run in polynomial time. The quantity m determines the compression rate, and k is a logarithmic factor in m. However, a large k will worsen the rate tenfold.

1.2 *Model projection*

Fortunately, decades of researches in signal processing have found that some signal supports have an additional structure, such as some time domain signals, the dominant coefficients of signal $x \in R^n$ tends to occur consecutively in cluster in practice. The family of signals which have additional structured supports are called structured sparsity model M_k. Based on the structured sparsity model, seminal results show that it is possible to reduce the number of measurements to $m = O(k)$, as long as the matrix A is chosen to satisfy the Model-based Restricted Isometry Property (RIP). Also, there exists an efficient algorithm that solves the model-projection problem. The model-projection problem is defined as follows: There exists an algorithm which returns a model-based sparse vector x' which minimizes the tail error $\| x - \hat{x} \|_2$.

While model-based compressive sensing algorithms decrease the number of measurements, the framework also raises a recovery problem of more expensive computation. The main obstacle lies in the availability of a model projection algorithm.

According to the theorem of [1], consider a structured sparsity model $M \in R^n$ and a norm parameter $p \in \{1,2\}$. Suppose that $x \in M$, we observe m noisy linear measurements $y = Ax + e$, where A satisfies the model-RIP. Instead of giving an exact model projection algorithm, the theorem requires two complementary approximation guarantees:

1. A tail approximation algorithm returns a support Ω_t in the model that the norm of the tail $\|x - x_{\Omega_t}\|_2$ is approximately minimized.
2. A head approximation algorithm returns a support Ω_t in the model such that the norm of the head $\|x_{\Omega_h}\|_2$ is approximately maximized. Formally,

$$\| x - x_{\Omega_t} \|_2 \leq C_T \cdot \min \| x - x_\Omega \|_2, \Omega \in M \quad (1.1)$$

$$\| x_{\Omega_h} \|_2 \geq C_H \cdot \max \| x_\Omega \|_2, \Omega \in \quad (1.2)$$

where $C_T \geq 1$, and $C_H \leq 1$

Then there exists an efficient algorithm which outputs a signal estimate $\hat{x}$ that satisfies $\| x - \hat{x} \|_p \leq C \cdot \| e \|_p$, constant $C > 0$.

2 OUR CONTRIBUTION

In this paper, we focus on the *tree-sparse model*. The coefficients of signal in this model occurred as a d-ary tree. The support of the signal x forms a subtree containing a root node. In fact, there exists a formal approach to capture this tree-sparse model in the wavelet-domain of natural images.

Let x be a signal in the tree-sparsity model with sparsity parameter k. The problem of projecting into the tree-sparsity model is to find the subtree which has the largest weight. Instead of providing a single exact model-projection algorithm, we give two novel approximate algorithms: one is for maximizing the weight of the subtree, and the other is for minimizing the left weight. In order to simplify the discussion, we give a symbol compare table.

Compared with the time complexity of the seminal researches $O(n\log n)$, our novel approximate algorithms performs the head approximation and the tail approximation algorithms in time $O(n)$. Together with the two approximate algorithms which satisfy the aforementioned two complementary approximation guarantees, we can obtain a quick recovery algorithm.

3 HEAD APPROXIMATION FOR THE TREE-SPARSITY MODEL

In this section, we propose a head approximation algorithm to give a maximization version of the problem. We want to find the subtree containing x_0 which has the largest weight. In other words, the maximization version of the problem is to

Table 1. Symbol compare table.

T	The tree defined by x
h	The height of the tree T
x_0	The root of the tree T
$\lvert x^i \rvert$	The i_{th} largest value among all $\lvert x \rvert$
M_k	The collection of all k-subtrees containing the root node
Ω_{opt}	$\Omega \in M_k$ with the largest weight
$\widehat{\Omega}_i$	The subtree of T rooted at x_i
S_i	The total weight of nodes in $\widehat{\Omega}_i$
S^i	The i_{th} largest value among all $\lvert S \rvert$
α	$\left\lvert \dfrac{k-1}{h-1} \right\rvert$

find $\left\| x_{\Omega_{opt}} \right\|_1$ which has the largest weight among all $\left\| x_\Omega \right\|_1, \Omega \in M_k$. This approximate algorithm returns Ω which satisfies that $\| x_\Omega \|_1 \geq \frac{1}{2h} \| x_{\Omega_{opt}} \|_1$. To simplify the analysis, we assume that $k \geq h$. Otherwise, we can always remove those nodes of depth greater than k on the tree, since those nodes will never be chosen. See Algorithm 1 for details.

Algorithm 1: (HeadApprox) Head approximation for the tree-sparsity model

Input: the d-ary tree T
Output: support Ω

1. $\alpha \leftarrow \frac{k-1}{h-1}$, $\Omega \leftarrow \varnothing$
2. Run BFS(T):
3. $\lvert x^\alpha \rvert \leftarrow$ the α_{th} largest value among all $\lvert x_i \rvert \ (1 \leq i \leq n)$
4. Run BFS(T):
5. if $\lvert x_i \rvert \geq \lvert x^\alpha \rvert$,
6. Add the nodes on the path from x_0 to x_i to Ω.
7. Return Ω

Both BFS (Breadth-Depth-First-Search) and finding the path from x_0 to x_i cost $O(n)$ time. So the HeadApprox algorithm also runs in time $O(n)$. The output Ω satisfies $\alpha \leq \lvert \Omega \rvert \leq k$.

Theorem 1. The output Ω from Algorithm 1 is a subtree of **T** rooted at x_0.

Proof: Note that the nodes satisfying $\lvert x_i \rvert \geq \lvert x^\alpha \rvert$ are added into © with their paths from x_0 to $x_i : \Omega \leftarrow x_i \cup root - path(x_i)$. It is obvious that Ω is a subtree of *T* rooted at x_0.

Theorem 2. Let $x \epsilon R^n$ be the coefficients corresponding to a d-ary tree rooted at x_0. Then HeadApprox returns Ω which satisfies that $\| x_{\Omega} \|_1 \geq \frac{1}{2h} \| x_{\Omega_{opt}} \|_1$.

Proof: Now consider $\| x_\Omega \|_1 = \sum_{i|\Omega} \lvert x_i \rvert$, Ω contains the x_i satisfying $\lvert x_i \rvert \geq \lvert x^\alpha \rvert$ so that:

$$\|x_{\Omega}\|_1 = \sum_{i \in \Omega} |x_i| \geq \sum_{|x_i| \geq |x^{\alpha}|} |x_i| \quad (3.1)$$

Since $k \geq h$, we consider the following situations:

1. If $2h \geq k \geq h$, then $\frac{h-1}{h-1} \leq \frac{k-1}{h-1} \leq \frac{2h-1}{h-1}$, we get $1 \leq \frac{k-1}{h-1} \leq 2$. It shows that $\alpha = 1$ or $\alpha = 2$, $\alpha * 2h \geq 2h \geq k$.
2. If $k \geq 2h$, then $2h * \frac{k-1}{h-1} \geq 2h * \frac{k-h}{h-1} > 2h * \frac{k-h}{h} \geq k + k - 2h \geq k$.

Combining the above two cases, we obtain that $2h * \frac{k-1}{h-1} \geq k, (k \geq h)$.

Moreover, $\sum_{|x_i| \geq |x^{\alpha}|} |x_i|$ have the largest $\frac{k-1}{h-1}$ nodes. Therefore,

$$\sum_{|x_i| \geq |x^{\alpha}|} |x_i| \geq \frac{1}{2h} \sum_{i \in opt} |x_i| \quad (3.2)$$

$$\sum_{i \in opt} |x_i| = \|x_{\Omega_{opt}}\|_1 \quad (3.3)$$

Combining (3.1)(3.2)(3.3), we have

$$\|x_{\Omega}\|_1 \geq \frac{1}{2h} \|x_{\Omega_{opt}}\|_1$$

Remark. Here is a simple example to prove that the approximation ratio of the algorithm is tight: $(\varepsilon \to 0)$

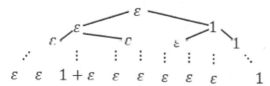

By Algorithm 1, we will get $\|x_{\Omega}\|_1 : 1 + h * \varepsilon$, while the $\|x_{\Omega_{opt}}\|_1$ is $h - 1$. Note that ε approaches to zero, we have that $\|x_{\Omega}\|_1 = \frac{1}{h-1} \|x_{\Omega_{opt}}\|_1$.

4 TAIL APPROXIMATION FOR THE TREE-SPARSITY MODEL

Next, we propose a tail approximation algorithm for giving a minimum version of the problem. We want to find the subtree containing x_0 which minimizes the left weight of T. In other words, the minimum version of the problem is to find $\|x - x_{\Omega_{opt}}\|_1$ which has the minimum weight among all $\|x - x_{\Omega}\|_1$, $\Omega \in M_k$. This approximate algorithm returns Ω which satisfies that $\|x - x_{\Omega}\|_1 \leq h \|x - x_{\Omega_{opt}}\|_1$. We denote $\hat{\Omega}_i$ as the subtree rooted at x_i, and s_i as the total weight of nodes in $\hat{\Omega}_i$. We still assume that $k \geq h$. See Algorithm 2 for a complete description.

Algorithm 2: (TailApprox) Tail approximation for the tree-sparsity model

Input: the d-ary tree T
Output: support Ω with $|\Omega| = k$

1. Run DFS(T), and compute $s_i \leftarrow \sum_{j \in \hat{\Omega}_i} x_j$. for each node x_i
2. $S \leftarrow$ Select the k_{th} large in s_i
3. Run BFS(T):
4. $\Omega \leftarrow \{x_i | s_i \geq S\}$
5. Return Ω

In TailApprox, we run DFS to calculate s_i. When the loop of $\hat{\Omega}_j$ is over, s_j have been calculated. Then we run recall move to find x_j's father x_i. In each recall move, we add s_j to x_i to calculate s_j.

Note that DFS(Depth-First-Search), BFS (Breadth- Depth-First-Search) and step 2 of the algorithm are all in time $O(n)$. So TailApprox runs in time $O(n)$.

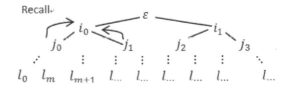

Theorem 1. The output Ω from Algorithm 2 is a subtree of T rooted at x_0.

Proof: Note that if node i is the father node of node j, then we have $s_i > s_j$. Assume that the output Ω is not a subtree of T rooted at x_0. Then Ω must contain some node x_i and not contain a node x_j which is on the path from x_0 to x_i. Note that $s_j > s$, so x_j must belong to Ω, which is a contradiction.

Theorem 2. Let $x \in R^n$ be the coefficients corresponding to a d-ary tree rooted at node x_0. Then TailApprox outputs Ω which satisfies that $\|x - x_{\Omega}\|_1 \leq h \|x - x_{\Omega_{opt}}\|_1$.

Proof: It is obviously that root x_0 belongs to Ω and Ω_{opt}. Then $h_{x-x\Omega} \leq h - 1$:

$$(h-1) * \|x - x_{\Omega_{opt}}\|_1 \geq \|s - s_{\Omega_{opt}}\|_1 \quad (4.1)$$

According to the TailApprox: $\|s_{\Omega}\|_1 \geq \|s_{\Omega_{opt}}\|_1$. Then,

$$\|s - s_{\Omega}\|_1 \leq \|s - s_{\Omega_{opt}}\|_1 \quad (4.2)$$

Combining (4.1)(4.2), we can conclude that $(h-1) * \|x - x_{\Omega_{opt}}\|_1 \geq \|s - s_{\Omega_{opt}}\|_1 \geq \|x - x_{\Omega}\|_1$

The height of tree T ranges from $\log_d n$ to n. In the following, we give two examples to show that the approximation ratio of the algorithm is tight.

1. T is a complete d-ary tree, $n = 15, h = 1, k = 8$:

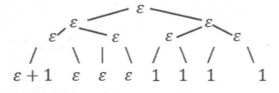

$\|x - x_{\Omega.}\|_1 = 3, \|x - x_{\Omega_{opt}}\|_1 = 1 + \varepsilon$

So, $3 * \|x - x_{\Omega_{opt}}\|_1 \geq \|x - x_{\Omega.}\|_1$.

2. More generally, $h_{x-x\Omega} \leq n - k$. There is another example:

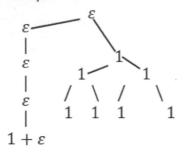

$n = 12, k = 8, n{-}k = 4$

$\|x - x_{\Omega.}\|_1 = 4, \|x - x_{\Omega_{opt}}\|_1 = 1 + \varepsilon$

$4 * \|x - x_{\Omega_{opt}}\|_1 \geq \|x - x_{\Omega.}\|_1$.

5 CONCLUSION

In this paper we seminal researches propose a structured sparsity model M_k which makes it possible to reduce the number of measurements to m = O(k). Moreover, there exists an efficient algorithm which solves the model-projection problem that returns a model-based sparse vector x^', minimizing the tail error $\| x{-}x^{\wedge\prime} \|\_2$. In this paper, we limit our attention to the tree-sparse model. Assume that all entries of a signal $x \in R^n$ forms a tree, and all non-zero entries of x forms a subtree of k nodes containing the root node. We give two complementary approximation algorithms. One is for maximizing the total weight of the subtree, and the other one is for minimizing the left weight of the tree except the chosen subtree. Both algorithms run in time O(n). Then there exists an efficient algorithm that outputs a signal estimate $\hat{x}$ satisfies that $\|x - \hat{x}\| - p \leq C \|e\| - p$.

REFERENCES

Baraniuk, R.G., Cevher, V., Duarte, M.F., & Hegde, C. 2015. Model-based compressive sensing. Information Theory IEEE Transactions on, 56(4), 1982–2001.

Eldar, Y.C., & Mishali, M. 2008. Robust recovery of signals from a structured union of subspaces. IEEE Transactions on Information Theory, 55(11), 5302–5316.

Hegde, C., Indyk, P., & Schmidt, L. 2014. Approximation-tolerant model-based compressive sensing. Proceedings of the Twenty-Fifth Annual ACM-SIAM Symposium on Discrete Algorithms:244–250.

Hegde, C., Indyk, P., & Schmidt, L. 2014. Nearly Linear-Time Model-Based Compressive Sensing. Automata, Languages, and Programming. Springer Berlin Heidelberg:588–599.

Hegde, C., Indyk, P., & Schmidt, L. 2014. Approximation al gorithms for model-based compressive sensing. Eprint Arxiv, 61(9), 244–250.

Needell, D., & Tropp, J.A. 2008. Cosamp: iterative signal recovery from incomplete and inaccurate samples ☆. Applied & Computational Harmonic Analysis, 26(3), 301–321.

11 *Network technology and application*

Energy Science and Applied Technology ESAT 2016 – Fang (Ed.)
© *2016 Taylor & Francis Group, London, ISBN 978-1-138-02973-6*

SDN-Specific DDoS/LDDoS and the machine learning based detection

XiaoFan Chen & ShunZheng Yu
School of Information Science and Technology, Sun Yat-Sen University, Guangzhou, China

ABSTRACT: We study two types of attacks particularly targeting SDN, i.e. controller-targeting DDoS and switch-targeting LDDoS (low-rate DDoS). Attackers of DDoS generate special packets to overload the bandwidth and computation resource of SDN controller. Attackers of LDDoS generate special packets periodically to overload the flow table of SDN switch. A new collaborative intrusion detection system is proposed in this paper to detect such attacks. It consists of cascaded two-level artificial neural networks which are distributed over the entire substrate of SDN. Each neural network disperses its computation power over the network that requires every participating switch to perform like a neuron. The first level neural networks discover the anomaly with a global view, while the second level one classifies the attacks based on their time pattern. Therefore, the system is robust without individual centralized targets and has a global view on the distributed attack without aggregating traffic over the network. Simulation results demonstrate its effectiveness.

Keywords: Distributed Denial-of-Service (DDoS), Low-rate DDoS (LDDoS), Software Defined Networking (SDN), Collaborative Intrusion Detection System (CIDS), Artificial Neural Network (ANN)

1 INTRODUCTION

SDN (Software Defined Networking) separates the network control plane and data plane. The attributes of centralized control and programmability associated with SDN brings great flexibility and benefit. However, these attributes also introduce network security challenges. An increased potential for DoS attacks due to flow table limitation in switch and the centralized controller are prime examples (S. Scott-Hayward, 2013). Attackers can launch such attacks by generating a large volume of traffic that keeps randomly changing flow attributes. It has two mainly damages (I. Alsmadi and D. Xu, 2015): (a) it will flood the switch's flow table and saturate it with illegitimate rules. (b) It will keep the controller busy from responding to legitimate flows. They both cause the failure of legitimate packets forwarding in SDN. During the past few years, some research works address such security problems. (D. Kreutz, F, 2013) proposes seven threat vectors that may enable the exploit of different SDN vulnerabilities. The vulnerabilities exist in the control & management plane, data plane, and the communication channel between controller and switch, etc. A DoS attack, which consumes resources of the SDN control plane and data plane, is demonstrated in (S. Shin and G. Gu, 2013). In (R. Kloti, 2013) the DoS attack witch overflows the flow table of openflow switches is studied. FloodGuard prevents the

SDN from data-to-control plane saturation attack. When saturation attack is detected, FloodGuard redirects the table-miss packets, i.e. the new arriving packets for which there are no matching rules installed in the switch, to the data plane cache. It generates proactive flow rules by reasoning the runtime logic of the controller and installs them into the switches. Then the cache slowly sends other table-miss packets as *packet-in* message, i.e. a type of openflow message that is sent from switch to controller when the arriving packet is table-miss or the action of the matching rule is sending the packet to controller, to protect the controller from being overloaded. The solution in first considers every source IP in the table-miss packet as DDoS source. It assigns short timeouts for their forwarding rules in the switches. It records the amount of connections of IPi as c_i. When $c_i \geq k$, where k is a preset threshold, it checks s_i, i.e. the average amount of packets of all connections of IPi. If $s_i \leq n$, where n is also a preset threshold, IPi is considered as a normal user. Long timeouts will be reassigned to the forwarding rules associating with IPi. When $c_i < k$, nothing happens. Packets of IPi will be blocked if and only if $c_i \geq k$, $s_i > n$. However, the solution is not effective when the destination IPs are not spoofed. It may cause $s_i \leq n$ for most source IPs and the system will do nothing. (N. Gde Dharma, 2015) proposes a time-based method to detect the controller-targeting DDoS. The controller checks the destination address of every packet.

If the arriving rate of the packets with non-valid destination addresses is over a threshold, it thinks that DDoS occurs.

Our work in this paper focus on two problems: (a) the DDoS that targets the centralized SDN controller; (b) the LDDoS that targets the flow table limitation of the SDN switches. We propose a cascaded two-level ANN-based CIDS to tackle with the problems. The first level ANN, i.e. ANN_1, is responsible for the space-based anomaly detection. It is similar to our previous work in (X.-F. Chen, 2016). (X.-F. Chen, 2016) focuses on traditional abnormality in legacy network and SDN, i.e. non SDN-specific attacks. But we focus on the SDN-specific DDoS and LDDoS in this paper. ANN_1 has a global view of the network status. Its output tells whether the network is normal or not. The second level ANN, i.e. ANN_2, is responsible for the time-based intrusion classification. It utilizes the time pattern of the ANN_1 output sequence to classify attacks and reduce the false positive rate of the CIDS. The workflow and working principle of the ANNs are similar to the solution in (X.-F. Chen, 2016), while the only differences are the implication of inputs and outputs of the ANNs. (X.-F. Chen, 2016) has validated the effectiveness of the first level ANN. The architecture of our CIDS for SDN is shown in Figure 1.

The main contributions of this paper are as follows: (a) we mention about the SDN-specific LDDoS attack for the first time. (b) Most existing schemes make the SDN controller acts as a centralized detection system. It is prone to suffer from single-point-failure problem, e.g. be attacked or overloaded. The distributed and collaborative architecture of our system not only eliminates the single-point-failure problem, but also keeps good detection accuracy with a global view. (c) Our approach performs well no matter the attackers launch DDoS/LDDoS with or without IP spoofing technology. The details are can be found in Section IV.

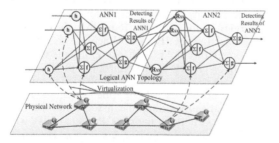

Figure 1. The architecture of our CIDS for SDN. h is the preprocess function of the input neuron of ANN1, ∑ the weighted sum up function, f the activation function of hidden neuron, g the activation function of output neuron, and Rxx the autocorrelation function of the input neuron of ANN2.

2 ADVERSARY MODEL

The features of the controller-targeting DDoS are as follows: the bots of DDoS incessantly generate a large volume of traffic that keep randomly changing their flow attributes. It ensures that most flows are new from each switch perspective. Hence they will be sent as table-miss *packet-in* messages to the controller for decision. This flood of abnormal flows keeps the controller busy from responding for the normal flows. Then the switches can not install rules for the new normal flows, which means that normal traffic will not be forwarded. When this attack occurs, the features of the flow tables of the switches are as follows: (a) the average usage of every flow is low. (b) The table-miss *packet-in* generating rate is fast all the time. (c) The flow tables of some switches may be full.

The features of the switch-targeting LDDoS are as follows: the switch will remove the useless flow entries periodically. The bots of LDDoS periodically generate relatively large volume of traffic that keeps changing flow attributes. Most flows are table-miss flows. It makes the victim switches periodically flood *packet-in* messages impulse to the controller. The flow tables of the victim switches are quickly occupied by the attack flows at the beginning of each period. As a result, the normal flow can not be forwarded. When this attack occurs, the features of the flow table of a switch are as follows: (a) the average usage of every flow is low. (b) The growth rate of new flow entry changes periodically, like the impulse feature of LDDoS in legacy TCP/IP network (C. Zhang, 2010). (c) The flow table is full or nearly full all the time.

The features of DDoS and LDDoS described above are quite different from how they are in legacy TCP/IP network. In the TCP/IP network, the packets of DDoS/LDDoS have similar attributes in the packet header, e.g. destination IP and transport layer protocol. The detection methods based on the similarity of packets attributes, which is useful in TCP/IP network, are useless for such SDN-specific attacks. The volume-based detection methods may not perform well when the flows of DDoS and LDDoS are small before they aggregate at the victim. Therefore we propose the new monitoring indices and collaborative intrusion detection system aiming at addressing such problems. The details are described in Section III.

Simulations results are presented to show the severity of the SDN-specific DDoS and LDDoS in a more intuitive way. We use OMNeT++ (D. Klein, 2013) to conduct simulations and BRITE (BRITE, 2014) to generate network topology. We use networks consists of 100 openflow switches, 1 SDN controller, and 100 hosts. Each switch connects

Table 1. The Simulation Results of the Damage of DDoS and LDDoS.

DDoS					
Attack rate	0	A	2A	4A	20A
Flow processing ratio/%s	100	90.84	58.38	42.69	15.99
Packets receiving ratio/%	100	93.09	66.88	51.31	21.53
LDDoS					
Attack rate	0	B	2B	4B	20B
Flow entry installing ratio/%	100	91.26	67.31	31.10	9.36

one host. We replay the normal traffic as what we did at (X.-F. Chen, 2016) and generate DDoS and LDDoS. The results are shown in Table 1. In the DDoS scenarios, we show the changes of the normal flow processing ratio of the controller and the average normal packets receiving ratio of hosts with different attack rate. A is assumed to be the processing rate of the SDN controller. It is about 3 Mbps. In the LDDoS scenarios, we show the changes of normal flow entry installing ratio of a switch with different attack rate. B is assumed to be normal packet generating rate of a host. It is set to be 10 thousand packets per second. The attack rate refers to the aggregated volume of DDoS/LDDoS traffic. As attacks become heavier, less normal packets can be forwarded in SDN. Table 1 illustrates that the SDN-specific DDoS and LDDoS mentioned in this paper have great impact on the regular running of the SDN.

3 SYSTEM ANALYSIS

Monitoring Indices & ANN1. The average utilization ratio of the flow entries:

$$u = \frac{1}{L}\sum_{i=1}^{L}\beta_i \cdot p_i \qquad (1)$$

where $p_i = n_i / \sum_{i=1}^{L} n_i$ is the utilization ratio of the ith entry, n_i the amount of packets of the ith flow in the current sampling period, and L the amount of flow entries. $\beta_i = \Delta\tau_i/t$ is the weight of the ith entry, where T is the sampling interval, and $\Delta\tau_i$ the duration of the ith entry in the current sampling period. The larger the $\Delta\tau_i$ is, the more effect p_i will have on u. Hence the entries, that expire soon after the beginning and are newly installed before the end of the sampling period, will have less effect on u. So u can represent the state of most flows in the switch during the sampling period.

- The table-miss *packet-in* ratio:

$$v = \frac{m_{packet-in}}{t} \qquad (2)$$

where $m_{packet-in}$ is the amount of table-miss *packet-in* messages during a sampling period. When the flow table is full, LDDoS stops. It restarts when most of the flow entries are going to be expired. Therefore, v varies periodically. DDoS keeps flooding table-miss packets into the network. Hence v is always high.
- The saturation of flow table:

$$\rho = \frac{L}{L_{max}} \qquad (3)$$

where L_{max} is maximum size of the flow table.

ANN$_1$ in this paper is similar to that in (X.-F. Chen, 2016). The difference is that we use three monitoring indices here instead of the six features. ANN$_1$ has a global view of the network state. It can detect abnormal from the space aspect of the network state. Its output only shows whether the network is normal or not. It sends its output to ANN$_2$ for further analysis so as to classify the attack types. We assume that all or most of the switches should act as input neurons of ANN$_1$. Then no matter which switches are chosen to be the neurons of other layer, our system still has a global sight of the whole network. Therefore, different choice of switches will have little or no effect on the detection ability of our system.

Autocorrelation & ANN2. We define the autocorrelation function as:

$$R_{xx}(m) = \frac{1}{N-m}\sum_{n=1}^{N-m} x(n) \cdot x(n+m), 1 \le m \le N-1 \qquad (4)$$

We find that the ANN$_1$ output sequences of normal traffic and different attack traffic have their own time patterns. The $Rxx(m)$ can reflect such time patterns.

Normal traffic with burst flows: the burst flows of normal traffic have the attribute of randomness. For most value of m, the corresponding $Rxx(m)$ is small.

DDoS traffic: the attribute of its flows is persistence. For many value of m, the corresponding $Rxx(m)$ is large.

LDDoS traffic: the attribute of its flows is periodicity. For some value of m, the corresponding $Rxx(m)$ is large.

Therefore, a set of properly selected values of m can be used to distinguish the normal, DDoS and

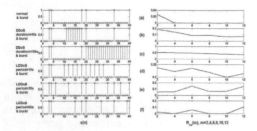

Figure 2. Examples of ANN1 output sequences of different traffic types and their autocorrelation curves.

LDDoS traffic. Assume that the sampling period of ANN_1 is T, and the timeout of each entry τ. Then m is related to T and τ. Take the simulations in this paper for example, $t = 5$ s and $\tau = 10$ s. But τ in some controllers or switches can be 30 s or 60 s. So we assume that $m=j \cdot \tau/t, j = 1, 2, \ldots, 6$, i.e. $m = 2, 4, 6, 8, 10, 12$. Their corresponding value of $Rxx(m)$ are respectively used to check whether the ANN_1 output sequence has the period of 10 s, 20 s, 30 s, 40 s, 50 s and 60 s.

We did some numeric simulation with MATLAB to test the effectiveness of $Rxx(m)$. The results are shown in Figure 2. The figures at the left side of Figure 2 are the ANN_1 output sequences, where 1 means abnormal and 0 normal. The sampling interval is 5 s. The figures at the right side are 6 values of $Rxx(m)$ of the sequences. There is only normal traffic in (a). The traffic is mixed with DDoS and normal traffic in (b) and (c). The DDoS in (b) and (c) has different duration. The traffic is mixed with LDDoS and normal traffic from (d) to (f). The LDDoS from (d) to (f) has different period. We find that the $Rxx(m)$ curves are quite different for different types of traffic, and similar for the same type of traffic. In Figure 2, we suppose that it is enough to detect abnormality with three values of $Rxx(m)$, where $m = 2,6,12$. But we preserve three more value of $Rxx(m)$, where $m = 4,8,10$, so as to classify the types of attacks better and decrease the false positive rate.

4 EVALUATION AND ANALYSIS

We conduct simulations with OMNeT++. We use BRITE to generate topology. The network size is 100-switch, which is large enough to validate the effectiveness and scalability of our CIDS. We use 10 different random topologies in our simulation. The normal traffic is a composition of the following protocols: 80% TCP, 15% UDP and 5% ICMP (P. Borgnat, 2000). The normal traffic generating rate of a host is about 3 Mpbs. We add burst normal flows at some random time interval to make

the accurate detection and classification harder. All hosts keep replaying normal traffic during the simulation. The sampling period of ANN_1 is $t = 5$ s. L_{max} is 1500 (S. Shin, 2013). The processing rate of the SDN controller is set to be 10 thousand packets per second. We train ANN_1 and ANN_2 offline. The connection weights of the trained ANNs are assigned to the corresponding neurons in the switches.

For the DDoS scenario, the duration of each simulation is 2000s. The attack lasts for 1000s. 20% of the hosts are randomly chosen to act as the bots to launch DDoS. They keep push table-miss packets into the network when DDoS starts. Two attack rates, i.e. high and low, are used. The volume of the aggregated traffic of the low rate attack is about twice that the controller can process per second. The volume of the high one is about four times. We compare the performance of our CIDS with the schemes proposed by (N.-N. Dao, 2015) and (N. Gde Dharma, 2015).

For the LDDoS scenario, three attack periods, i.e. 10 s, 30 s, 60 s, are used. They last for 1000s, 1800s and 3600s respectively. The duration of simulations are double compared with the LDDoS. Without loss of generality, the timeout of flow entry is set to be 10 s. 5% of the switches are randomly chosen to act as the victims. For each switch, 5 nearest hosts are selected to act as the bots to launch LDDoS aiming at overflowing its flow table. They keep push table-miss packets periodically when LDDoS starts. The destination IP addresses of LDDoS packets are not changed so that all the attack flows will be forwarded to the corresponding victim switches. There are also two attack rates in LDDoS scenario. The volume of the aggregated traffic of low rate attack is about twice of the normal packet generating rate of a host. The volume of the high one is about four times.

The results are shown in Figure 3 and 4. Every value is the average of the results from 10

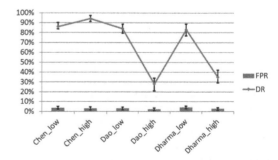

Figure 3. Simulation results of DDoS scenario.

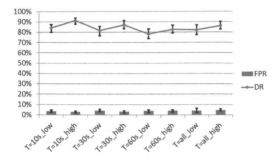

Figure 4. Simulation results of LDDoS scenario.

topologies. In Figure 3, DR is short for detection rate, and FPR for false positive rate. Chen, Dao and Dharma represent the schemes proposed in this paper, (N.-N. Dao, 2015) and (N. Gde Dharma, 2015), where low and high mean the attack rates. The error bars show the deviations. For the low attack rate scenario, the performance of our CIDS is a little better than others. However, Dao's and Dharma's schemes will not be effective if most destination IP addresses of the DDoS flows are valid. They may miss detecting much attack traffic. Hence, for the high attack rate scenario, the destination IP address in every LDDoS packet is randomly selected from a small range. The DR of our CIDS keeps high, while the DRs of Dao's and Dharma's schemes drop dramatically.

In Figure 4, T means the period of LDDoS. T = all means the scenario with mixture of periods. We can see that our CIDS performs well. The DR drops a little when T becomes larger. The reasons are as follows: (a) long attack interval makes the values of $Rxx(m)$ smaller than that of short attack interval. Smaller the values are, harder the attack can be detected. (b) If the LDDoS has long attack interval, there will be less attack samples at the beginning and end of the attack traffic. Less attack samples means it is harder to discover the periodical attribute. So these samples are more likely to be undetected.

The communication and computation overhead of our CIDS is similar to that of (X.-F. Chen, 2016). Its communication overhead comes from the neural messages. It is about 1.5% of the normal traffic rate of a switch, which is quite low. Its computation complexity is nearly $O(n\bar{L})$, where n is the size of the network, and $\bar{L}$ the average amount of flow entries in every switch. Its storage complexity is a small constant since the switch only needs to store the values of flow features, the weights of ANNs, and the ANN_1 output sequence. The of storage complexity of Dao's system is high since it has to store connection information for each source IP address.

5 SUMMARY

This paper proposes a cascaded two-level ANN-based collaborative intrusion detection system for SDN. It addresses two SDN-specific attacks: controller-targeting DDoS and switch-targeting LDDoS. The attacks, the architecture, working principle and implementation of the CIDS are described in detail in this paper. With a global view, the CIDS is able to detect distributed and stealthy intrusion behavior. Based on the time pattern analysis, the CIDS is able to classify attack types with high accuracy. The programmability and virtualization of SDN make it easy for our CIDS to be deployed and adjusted to fit to different security strategies. It is suitable to be deployed in any type of SDN-based networks, including SDN-based data center, campus and enterprise networks. The simulation results and analysis show that the ANN-based CIDS has high detection rate, low false positive rate and low communication and computation overhead.

ACKNOWLEDGEMENTS

XiaoFan Chen is the corresponding author. His e-mail is chxfanz@gmail.com. This work was supported by the Natural Science Foundation of Guangdong Province, China (Grant. No. 2014 A030313130), and the Natural Science Foundation of Guangdong Province, China (Grant. No. 2014 A030313637).

REFERENCES

Alsmadi and D. Xu, "Security of software defined networks: A survey," Elsevier Computers & Security, vol. 53, pp. 79–108, 2015.

Borgnat P., G. Dewaele, K. Fukuda, P. Abry, and K. Cho, "Seven years and one day: Sketching the evolution of internet traffic," in INFOCOM 2009, IEEE. IEEE, 2009, pp. 711–719.J. van der Geer, J.A.J. Hanraads, R.A. Lupton, The art of writing a scientific article, J. Sci. Commun. 163 (2000) 51–59.

BRITE, "BRITE: Boston university Representative Internet Topology Generator," http://www.cs.bu.edu/brite/, 2014. Wikipedia.

Chen X.F. and S.Z. Yu, "CIPA: a collaborative intrusion prevention architecture for programmable network and SDN," Elsevier Computers & Security, vol. 58, pp. 1–19, 2016.

Dao N.-N., J. Park, M. Park, and S. Cho, "A feasible method to combat against ddos attack in sdn network," in Information Networking (ICOIN), 2015 International Conference on. IEEE, 2015, pp. 309–311.

Gde Dharma N., M. F. Muthohar, J. Prayuda, K. Priagung, and D. Choi, "Time-based DDoS detection and mitigation for SDN controller," in Network Operations and Management Symposium (APNOMS), 2015 17th Asia-Pacific. IEEE, 2015, pp. 550–553.

Haykin S., Neural networks: a comprehensive foundation, 3rd ed. Upper Saddle River, NJ: Pearson Prentice Hall, 2008, pp. 230–263.

Klein D. and M. Jarschel, "An OpenFlow Extension for the OMNeT++ INET Framework," in 6th International Workshop on OMNeT++, Cannes, France, Mar. 2013. "Autocorrelation,"

Kloti R., V. Kotronis, and P. Smith, "Openflow: A security analysis," in Network Protocols (ICNP), 2013 21st IEEE International Conference on. IEEE, 2013, pp. 1–6.

Kreutz D., F. Ramos, and P. Verissimo, "Towards secure and dependable software-defined networks," in Proceedings of the second ACM SIGCOMM workshop on Hot topics in software defined networking (HotSDN). ACM, 2013, pp. 55–60.

OMNeT++, "Welcome to the OMNeT++ Community!" http://www.omnetpp.org/, 2014.

Shin S. and G. Gu, "Attacking software-defined networks: A first feasibility study," in Proceedings of the second ACM SIGCOMM workshop on Hot topics in software defined networking (HotSDN). ACM, 2013, pp. 165–166.

Scott-Hayward S., G. O'Callaghan, and S. Sezer, "Sdn security: A survey," in Future Networks and Services (SDN4FNS), 2013 IEEE SDN For. IEEE, 2013, pp. 1–7.

Wang H., L. Xu, and G. Gu, "FloodGuard: A DoS Attack Prevention Extension in Software-Defined Networks," in Dependable Systems and Networks (DSN), 2015 45th Annual IEEE/IFIP International Conference on. IEEE, 2015, pp. 239–250. https://en.wikipedia.org/wiki/Autocorrelation, 2014.

Zhang C., J. Yin, Z. Cai, and W. Chen, "RRED: robust RED algorithm to counter low-rate denial-of-service attacks," Communications Letters, IEEE, vol. 14, no. 5, pp. 489–491, 2010.

Energy Science and Applied Technology ESAT 2016 – Fang (Ed.)
© 2016 Taylor & Francis Group, London, ISBN 978-1-138-02973-6

Design and implementation of a staff training system based on JavaEE

Mei Bin
Wuhan University of Technology, Wuhan, China

ABSTRACT: The Web application framework that integrates the framework technology of Spring MVC, Spring, and Hibernate can improve the reusability of the project module. The aim of this paper is to design and implement the staff training system, which focuses on staff and has been applied in enterprises. The system uses the MVC programming model based on the aforementioned kinds of frameworks. In addition, it also discusses the functional modules of the system, as well as the analysis of the technical principles of the SSH framework.

Keywords: JavaEE; web application; SSH framework; MVC

1 INTRODUCTION

With the changing market, the competition between enterprises is becoming more and fiercer, and their competition is actually the competition of talents (Breivold, H.P. 2012). Therefore, enterprises pay more attention to the training of talents. The traditional way of staff training takes a lot of time and energy. In addition, it is bad to cultivate the self-learning ability and interest of employees, by face-to-face teaching and test papers (Bian Qingang et al, 2003). To solve this problem, the use of Online Autonomous Training is a good way.

Traditional JavaEE layered architecture is heavyweight and difficult to maintain because of the high cost and long duration (Wang Haitao et al, 2011). So, the developer cannot focus on business logic. Framework is a mature technology (Yang Gang. 2012) (Yuan Meileng. 2003), and the use of the framework to achieve the Web application has the advantages of high efficiency, stability, and easy maintenance. Therefore, this paper designs and implements the staff training system by using the SSH framework based on MVC.

2 INTEGRATION OF WEB APPLICATION

2.1 *MVC model*

MVC stands for "Model View Controller". It is a pattern to design and create Web application. Model is part of the application to process data and logic, whose Responsibility is to access data in the database. And View is the part of presenting data, which depends on the data of the model to be created. However, Controller handles user interaction, which means that it can get the data from the view and control the user's input and send the data to the model. MVC layered architecture can help manage complex applications. It is easy for the developer to develop different modules–development view, controller logic, and business logic without interferences.

This chapter describes how to design the staff training system by using the SSH framework. It is divided into three parts: Spring MVC–Web presentation layer, Spring–business logic layer, Hibernate–data access layer. The hierarchical structure separates the responsibility between different modules, according to the principle of "single responsibility" to reduce the degree of coupling between the layers, so that each framework fully demonstrated its role. This is easy to maintain.

2.2 *Spring MVC–presentation layer*

Spring MVC is the subsequent product of Spring. It separates the controller, the model object, dispatcher and the role of the handler object. This separation makes them easier to be customized. Spring MVC is mainly composed of Dispatcher Servlet, processor, processor (controller), view resolver, and view, as shown in Figure 1.

Dispatcher Servlet (the core controller of Spring MVC) receives the client HTTP request, and calls the Bean that implements the Handler Mapping interface, and then returns a handler (controller) for the request. In the next step, the controller will return a Model and View after processing the request by invoking corresponding methods. Thereafter, the Dispatcher Servlet resolves the Model and View by using View Resolver, and the view model will eventually be presented to the user. Spring MVC simplifies the use of native Servlet, so

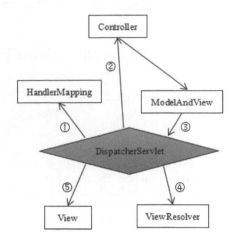

Figure 1. Http request is intercepted and processed by Spring MVC.

it is no longer necessary to give each Servlet declaration and configuration related information in the file named "web.xml". Besides, the use of filter and interceptor is also more simplified, which reduces the redundancy of the code.

2.3 Spring–business layer

Spring is an open-source framework developed in 2003, which is a lightweight Java development framework. Spring uses the basic JavaBean to accomplish things that could only be done by EJB. The core of Spring are Inversion of Control (IoC) and Aspect-Oriented Programming (AOP). They are described below.

IoC–The object is created by a control system, by which the external entities of all the objects in the control system pass on the reference to the object that it depends on. Dependency is also injected into the object. So, the inversion of control is about how to get the reference of the object that he relies on, which is inversion of the responsibility. The system uses IoC container to manage the lifecycle and dependencies of the object. So, the configuration and dependency specification of the application separate from the actual application code. It means that Configuration file is used to replace the code changes.

AOP–Spring provides rich support for aspect-oriented programming, allowing the development of cohesion by separating application business logic and system-level services. Application object implements only the business logic and is not responsible for other system-level services, such as logging or transaction support.

IoC container integrates presentation layer, business logic, and data access layer. On the other hand, AOP implements transaction control of the service layer. The implementation steps are as follows: when the Web container is started, the Dispatcher Servlet is instantiated, the Controller is injected into the Service class, the Service is injected into the Dao class, and the Hibernate configuration is loaded.

1. Configure the Bean. The element tag is as follows:

   <bean id = " " class = " ">......</bean>

   For example, to provide the data source,

   <bean id = " dataSource" class = "org. apache. commons. dbcp. Basic Data Source">
   ......
   </bean>

   Spring container manages Hibernate transactions,

   <bean id = "ts" class = "org. spring framework. orm. hibernate4. Hibernate Transaction Manager">
   ......
   </bean>

2. Transaction management
   The application does not need to be coupled with a specific transaction when using the Spring transaction management policy. Spring provides two types of transaction management. The first type is programming, where the program can directly obtain the transaction Manager Bean of the container, that is, the instance of Platform Transaction Manager. It involves three methods: the beginning of the transaction, the transaction, and the rollback transaction. The second type is declarative transaction, in which it is not necessary to write the code in the Java program, but in the XML file for configuration, and AOP is used for weaving. Weaving is done at the beginning of the transaction before the target method is executed, and the end of the transaction is finished after the target method is executed. Spring provides a consistent abstraction of things, regardless of the way chosen; therefore, a consistent programming model can be used in any environment. Of course, most developers choose declarative transaction management, in which the impact on the code is small, in accordance with the concept of non-intrusive lightweight containers.

3. Implementation of business logic
   Spring injects Session Factory instance into the Dao layer, and obtains session by session Factory. Get Current Session () method, so that the Dao layer can perform CRUD for the database.

By injecting the Dao Class into the Service layer, Service can achieve declarative transaction management with annotation @Transactional.

2.4 *Hibernate–persistence layer*

As the Data Persistence Framework, Hibernate is a tool of ORM. It provides lightweight object encapsulation for JDBC. The programmer can operate the database by an object-oriented programming task. The persistence layer focuses on a logical level of a particular system in the field of persistent application, and associates the data user with the data entity. The configuration is very complex if Hibernate is only used because of the hibernate.cfg.xml file. However, after combining with Spring, it becomes very simple: for each additional original entity Bean, it needs to be configured in "hibernate.cfg.xml"; now it is sufficient to use annotation @Entity. In general, the configuration of the Hibernate has included in Spring, including management of Hibernate transaction. The structure of Hibernate is shown in Figure 2.

A. Configuration: to read the Hibernate configuration file, and generate Session Factory object.
B. Session Factory: to produce Session instance.
C. Session: the core interface of Hibernate, which has some methods, such as get(), load(), save(), update(), and delete().
D. Query: to do query operation and get by Session. Create Query().
E. Transaction: to manage Hibernate transaction and get by Session. Begin Transaction().

3 SYSTEM DESIGN AND IMPLEMENTATION

3.1 *System function diagram*

As shown in Figure 3, a simple structure diagram, where the entry of the system is the user login, can be divided into two roles, namely staff and administrators. The employee can learn online video and test the knowledge that has been mastered. The administrator can view the staff's learning progress, check the examination records, and, according to the difficulty of problems, master the knowledge of the weak point of the staff to solve those problems. Indeed, there are also other functions.

1. Operations for employee
   - Show personal information
   - Upload personal picture
   - Make suggestions
   - Study from videos
   - Practice
   - Examination
   - Query personal exam records
2. Operations for administrator
   - CRUD for staff information
   - Deal suggestions
   - Manage question bank
   - Query all exam records
   - Set base information of exam
   - Statistics of question difficulty
   - Print reports

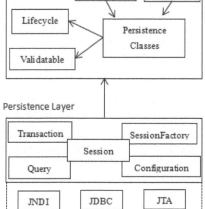

Figure 2.   The core of Hibernate to access the database.

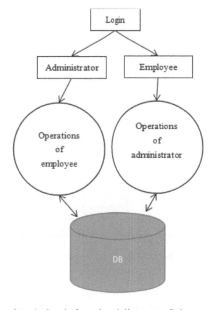

Figure 3.   A simple functional diagram of the system.

3.2 Client implementation

The client uses the B/S (Browser/Server) structure. With the popularity of Internet Technology, it is a kind of change or improvement of the structure of C/S. In this structure, the user interface is implemented through the WWW browser. Instead of in the front (browser), the main business logic is achieved in the server, which is a 3-tier structure. This greatly simplifies the client computer load, reduces the cost and workload of system maintenance and upgrade, and reduces the overall cost of the user.

3.3 Server-side deployment

In this paper, the system is designed using Tomcat as the server. Tomcat is the Apache Software Foundation (Apache Software Foundation Jakarta project, a core project) undertaken by the Apache, Sun and other companies and individuals for joint development. It is a free and open source web application server, which belongs to the lightweight application server. With the participation and support of Sun, the latest Servlet and JSP specifications can always be reflected in the Tomcat. Thus, Tomcat is the best server to run JSP and Servlet.

The Web application that has an integrated SSH framework can run on the Tomcat server if the runtime environment of Tomcat is configured. It provides access to the login page of the staff training system in the browser. In this system, Maven is used to deploy the Web application.

To date, the main body of the system has been realized, including how to build the Web application, the use of the client, and the server side of the deployment. The description of a complete "request-response" pattern is shown in Figure 4.

4 SUMMARY

In this paper, the staff training system was implemented based on Web application with an integrated SSH framework. The SSH framework in the system was described in detail. It meets the design idea of MVC, reduces the cost and complexity of the development, and reduces the development cycle of the application. The layered architecture design is conducive to the late maintenance, improving the reusability of the code, which follows the Open–Closed Principle. Obviously, there are some shortcomings of system performance, such as browser compatibility and concurrent access efficiency. These shortcomings of the system need to be solved.

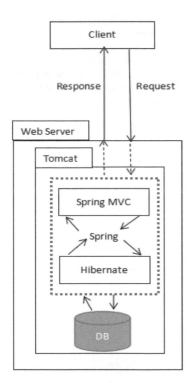

Figure 4. The process of the request-response pattern in the system.

REFERENCES

Bian Qingang & Pan Donghua. 2003. Discussion on integration of Tomcat and Apache to support JSP Technology. Application Research Of Computers.

Breivold, H.P. 2012. A systematic review of dynamic software architecture evolution research. Information and Software Technology.

Pubudu Gunawardena. 2012. Efficient Access to Hibernate Through JSP Powered by a New Tag Library. ICIAFS2012.

Wang Haitao & Jia Zongpu. 2011. Development of Web application based on Struts and Hibernate. Computer Engineering.

Yang Gang. 2012. Improvement of MVC mode view based on J2EE. Computer Technology and Development.

Yuan Meileng. 2003. Research and application of MVC software architecture in J2EE application model. Application Research of Computers.

Energy Science and Applied Technology ESAT 2016 – Fang (Ed.)
© *2016 Taylor & Francis Group, London, ISBN 978-1-138-02973-6*

Security technology of a wireless sensor network based on chaotic encryption

Yuxiang Hu

School of North China Electric Power University, Baoding, China

ABSTRACT: Wireless sensor network is a typical application of Ad-Hoc networks in the Internet of Things (IoT). The security of WSN influences the existence and uniqueness of real-time interactions in IoT services, and relates to the security metrics of the industrial real-time embedded control systems. According to the characteristics of WSN, a dual chaotic encryption algorithm was designed, which is a combination of Logistic map and Henon map. Moreover, it realized the dual encryption of the plaintext and effectively guaranteed the safety of the data. Finally, the experimental results indicated that the new algorithm is highly efficient and secure. Thus, it is suitable for WSN and can be used to design the security model and evaluate the performance of the industrial real-time embedded control systems.

Keywords: WSNs; information security; chaos encryption algorithm; data encryption

1 INTRODUCTION

With the rapid development of microelectromechanical systems, chip systems, wireless communication, and low power embedded technology, wireless sensor networks have emerged as the times require. WSN is a multi-hop *ad hoc* network that is composed of a large number of cheap microsensor nodes deployed in the monitoring area by means of wireless communication. With the characteristics of low power consumption (Zaibi G et al, 2014), low cost, distributed and self-organization, and wide application areas, such as military, aviation, medical, home, and business, it has caused a technological change in the field of Internet of Things (Yuan H E et al, 2013). However, due to the low computing power of WSN node, small storage capacity, limited energy, and other limitations, resulting in a more mature encryption algorithm on the PC, such as MD5 and RSA, they cannot be directly applied to the WSNs. This is because it needs to be used on the WSN security encryption algorithm (Zhang Z et al, 2013).

Chaos is a kind of non-periodic motion form, which is unique and widely exists in nonlinear systems. Due to its non-periodicity, sensitivity to initial value, and randomness, it has a natural relationship with cryptography (Kong D P et al, 2014) (Wang H Y, 2014). The chaotic system can be calculated by the random key, and the chaotic sequence can encrypt the information with only a few parameters. So, the amount of calculation is less and the basic requirement of WSNs is satisfied. At present, the chaotic encryption algorithm has been widely studied, but

most of them are based on the encryption and the authentication algorithm is based on single chaotic systems, such as the encryption algorithm proposed in the literature based on a one-dimensional logistic map. The algorithm proposed in this paper is based on a three-dimensional Lorenz system. Because the low-dimensional chaotic system can be easily broken, the security is insufficient, the high-dimensional chaotic system algorithm is complex (Jia W U et al, 2014), and the encryption speed is slow. Therefore, this paper proposes an encryption algorithm based on a double chaotic system, combined with Henon and Logistic mapping. Through the deepening of iterative generation of chaotic sequences and the text information encryption, it successfully overcomes the aforementioned shortcomings (Tong X J et al, 2015). Finally, the experimental analysis proves the effectiveness of the proposed algorithm.

2 DISCRETE CHAOTIC SYSTEM

2.1 *Logistic mapping*

One-dimensional Logistic mapping is a very simple chaotic map from the mathematical context, but the dynamic behavior of the system is very complex (Biswas, K et al, 2014). Moreover, it has a wide range of applications in the field of communication security. Its mathematical model is as follows:

$$x_{n+1} = x_n \times \mu \times (1 - x_n), \mu \in [0,4], x \in [0,1] \quad (1)$$

where μ is the control parameter and x_0 is the arbitrary initial value. When $x \in [0,1]$, $\mu = 4$, the logistic map is in a chaotic state. In other words, the generated sequence is random and ergodic, and has initial value sensitivity and other typical chaotic characteristics.

In view of the limitation of the WSN's special application requirement (Lu W et al, 2013), as well as the limitation of the distribution of the Logistic mapping sequence, the mathematical formula is improved. When the function of the fixed point of the slope is greater than or equal to 1, the Feigenbaum constant is not stable and the function in the fixed points cannot be ignored (Ye R et al, 2014) (Das A K, 2014). The greater the gradient value, the better the function of chaos. The improved formula is given as follows:

$$x_{n+1} = (1+\mu)\left(1+\frac{1}{\mu}\right)x_n(1-x_n)^\mu, \mu \in [1,4], x \in [0,1] \quad (2)$$

Meanwhile, the expansion of the system itself has an avalanche effect in the logistic system (Xue T et al, 2010) (Chen H B. 2010). The first iteration of the system is made at a certain number of times, and then the generated value is used, to better cover up the original information and thus obtain a better security.

2.2 Henon mapping

With the development of chaos technology, the encryption technology based on a one-dimensional chaotic mapping system is unable to meet the actual demand. Furthermore, the use of a multiple one-dimensional chaotic system or the high- and low-dimensional chaotic system becomes the trend of development. Henon mapping is a typical two-dimensional discrete nonlinear dynamic system, whose mathematical model is as follows:

$$\begin{cases} x_{i+1} = y_i + 1 - ax_i^2 \\ y_{i+1} = bx_i \end{cases} \quad (3)$$

When $a = 1.4, b = 0.3, x_0 = y_0 = 0.4$, the Henon mapping system is in a chaotic state. Its mapping diagram is shown in Figure 1.

Meanwhile, we can see that formula (3) has two control parameters a, b and two initial conditions x_0, y_0, which is more than two one-dimensional chaotic equations and more complex than one-dimensional chaotic equations, for the effective protection of the security of the information.

3 CHAOS ENCRYPTION ALGORITHM DESIGN

This algorithm is designed to fully consider the following characteristics of wireless sensor networks, in addition to its mobility: small node size, low cost, and limited computing power. The number of nodes

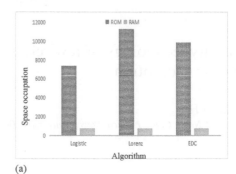

(a)

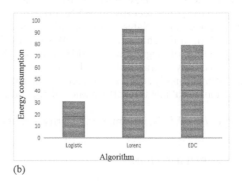

(b)

Figure 1. Comparative analysis of the three algorithms. (a) The histogram of space occupied for three algorithms. (b) The histogram of energy consumption for three algorithms.

is large, the failure is easy to occur, and the adaptability is strong; The communication radius is small, the bandwidth is low and the power supply is limited.

The core of the encryption algorithm based on the double chaotic system is the innovation of the algorithm, which is a breakthrough of the limitations of the simple one-dimensional chaotic system. The time complexity of the algorithm, the complexity of the space and the characteristics of wireless sensor networks, security, and other aspects are also considered. First, for simple one-dimensional chaotic systems (e.g. the typical Logistic mapping), the algorithm is relatively simple and has fast computing speed. Although it can meet the basic computing power of wireless sensor networks and solve the problem of node energy consumption, the key space is relatively small, security is poor, and the iterative sequence distribution is not uniform, showing two big middle small situations. However, for the high-dimensional chaotic systems (e.g. the typical Lorenz system), the algorithm is more complex, the sensor node itself has limited computing power conflict, and the computation speed is relatively slow, which leads to higher energy consumption. However, the key space is larger and the security is relatively better. Therefore, to integrate the high quality performance of chaotic systems, the characteristics of the wireless sensor networks are integrated, and the double chaotic system encryption algorithm (EDC) is proposed. Moreover, the encryption algorithm based on the combination of one-dimensional Logistic mapping and two-dimensional Henon mapping is proposed. The EDC algorithm uses the original computer operation, where the complexity is moderate, to evaluate the node computing power. While the algorithm is efficient, it is good to achieve the purpose of energy saving and calculate the required physical space, which is moderate, and security is greatly guaranteed.

The encryption process of the EDC algorithm is as follows:

1. Initialization key: select the initial value x_0, x'_0, y_0 of Logistic mapping and Henon mapping, and three control parameters (μ, a, b) as the initial key.
2. Initialization iteration: the key selects 15 iterations of the operation, this is done to better cover the original value of the situation, thereby expanding the avalanche effect of the Logistic system and improving the security.
3. Generating chaotic value: after obtaining the initial iteration of the secret key. First, the logistic map, x_0 as the initial parameters are substituted into formula (1) iteration a to x'_0, then x'_0 and y_0 product as Henon mapping the initial parameters are substituted into formula (2) iterative time to get x_n and y_n; the product of x_n and y_n is the final results of the iteration, denoted as P.
4. Residual operation: The 2, 4, and 6 effective bits of the iterative result P are performed on 256.

$$P' = P_i \bmod 256, i = 2, 4, 6 \qquad (4)$$

Get the chaotic key P'.

5. Termination judgment: check whether all the encryption of the text byte M is complete. If it reaches the end, then exit; otherwise, skip to step (3), the next round of encryption.

4 SIMULATION EXPERIMENT AND SECURITY ANALYSIS OF ALGORITHM

4.1 Simulation experiment

In this paper, the hardware platform of the simulation experiment is Dual-Core CPU i3 4200, the memory is 2 GB, and the network simulation software NS2 is used to establish the wireless sensor network scenarios. In the NS2 simulation environment, it sets the 100 m * 100 m monitoring area, randomly generates 150 sensor nodes, and the initial energy of nodes is 100 J, network topology for random deployment. The effectiveness of the EDC algorithm is verified by comparing the proposed algorithm with the one-dimensional Logistic mapping encryption

algorithm and the 3D Lorenz mapping encryption algorithm.

For the validation analysis of WSN encryption algorithm, the security of the algorithm, space occupation and energy consumption are in general considered. So, the simulation experiment used in this paper consists of two main aspects: space occupation and energy consumption. As can be seen from Figure 1(a), the ROM algorithm of the EDC space occupied by the number of bytes is in the middle, while the RAM space occupied by the number of bytes is equal. According to the literature, the RAM capacity of the standard sensor node is 4 ROM, the kB capacity is 128 kB, and the comprehensive analysis data can be known. Although the EDC algorithm needs to go through the two mapping iteration operation, the occupied space is still relatively ideal and better meets the WSN space occupied. Second, the energy consumption of the three encryption algorithms (Figure1 (b)) is given. As can be seen from Figure1 (b), the energy consumption of the EDC algorithm is between the Logistic algorithm and the Lorenz algorithm. Obviously, the more simple the algorithm is, the lower the energy consumption will be. The EDC algorithm is a class of two-dimensional Logistic mapping and two-dimensional Henon mapping, so there is energy consumption between the two.

Through the comprehensive analysis of the space occupation and energy consumption of the three algorithms, and according to the basic characteristics of the encryption algorithm, that is, the higher the security, the higher the complexity of the algorithm, the space occupation and energy consumption will be correspondingly increased. The logistic algorithm is safe and cannot meet the security requirements of WSN encryption algorithm. Furthermore, the Lorenz algorithm is large and cannot meet the needs of WSN encryption algorithm. The EDC encryption algorithm achieves performance tradeoff, security and energy consumption, thus reflecting its superiority in the WSN encryption algorithm.

4.2 Security analysis of algorithm

A good encryption algorithm focuses on the security, the need for a comprehensive analysis from the key space, key sensitivity and statistical and other aspects. An analysis of the security of the method is carried out in this paper.

4.2.1 Key space analysis

The encryption algorithm is most vulnerable to brute force attacks, so the encryption algorithm must have a key space demand. The EDC algorithm proposed in this paper uses Logistic mapping and Henon mapping to encrypt the text. Logistic map-

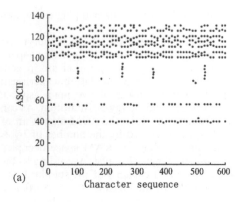

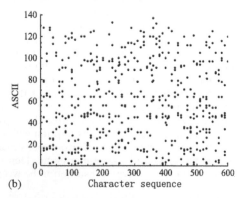

(a)

(b)

Figure 2. The scatter plot of the ASCII value for plaintext before and after encryption. (a) before encryption, (b) after encryption.

ping requires a control parameters and an initial value. The initial value of the initial value of 10^{-8}. Henon mapping requires two control parameters and two initial values. The accuracy of the two initial values is 10^{-16}. By the addition of three control parameters, the key space of the EDC algorithm is larger than that of 10^{32}, so it can effectively resist brute force attacks.

4.2.2 *Statistical analysis*

According to the Shannon theorem, any encryption password is based on the statistical analysis of the method for decryption. There are five groups based on the EDC algorithm before and after the encryption and encryption:

Group 1: The collection of all the clear text is called the clear text space;

Group 2: The collection of all the cipher text is called the cipher text space;

Group 3: The collection of all keys is called the key space;

Group 4: Wireless sensor network is a multi-hop *ad hoc* network;

Group 5: Wireless sensor network is a data centric information processing system;

Encryption is obtained through the 16 decimal number of the cipher text, which is expressed as follows.

Group 1: 72385E0731308B1EE63D23823438FD AA3315 A5582ED4CAE35034764DB3039F9C;

Group 2: 5F749FBCEF0D33F9011F9BE7 A403B FD 6BBA637953 A90EDAD608CA352 ACEDB 858;

Group 3: E4B7D9F4 ACB3B07D2783B45F44249 8085A71D1F9417018C8931DB5F1483EFEE5;

Group 4: 7F3264B320 A9B25F4595CAD58EC90 3 A3317B9298780FBE394988DC9586510217;

Group 5: 45C3CEC16E223DA6520C0C997E83D B821474ED878046E078DAA8D8C9DB6- C71519562DA098809640F8 AC;

By comparing before and after encryption, the EDC algorithm takes full advantage of the randomness and diffusion of chaotic maps, and there is a wide variation in the statistical characteristics of the encrypted text. Meanwhile, the EDC encryption algorithm is used in the text of the first test of the test encryption. For the summary before the character sequence ASCII code value distribution, Figure 2 shows the summary after the character sequence ASCII code value distribution. From the graph, we can see that there is a clear rule and statistical characteristics of the encrypted before, and the encrypted character sequence is disrupted, showing a chaotic random state. The ASCII code value distribution is more uniform, covering up the original information, and can effectively resist such as the only text attack, select text attack, select key attack, and other mathematical analysis. Moreover, it effectively improves the security of WSNs.

5 CONCLUSION

From the in-depth study of the characteristics of WSNs, this paper proposes a new encryption algorithm based on the combination of Logistic mapping and Henon mapping for the two chaotic systems. To make full use of the characteristics of the chaotic system, such as initial value sensitivity, scrambling, and randomness, it is applied to the generation of the key. Through the experimental analysis, it is found that the EDC algorithm not only ensures the high security of the encryption algorithm, but also overcomes the limitations of the WSN node storage space and weak computing ability. In the existence and uniqueness of real-time interactive behavior of Internet of Things, it is very feasible and practical, and can be used in the real-time embedded industrial control system.

REFERENCES

Biswas, K, Muthukkumarasamy, V, Singh, K. An Encryption Scheme Using Chaotic Map and Genetic Operations for Wireless Sensor Networks[J]. IEEE Sensors Journal, 2014, 15(5):2801–2809.

Chen H B. Reconstruction of Chaotic Signals in Wireless Sensor Networks Based on Threshold-Decision[J]. Journal of Southwest University, 2010.

Das A K. A secure and robust temporal credential-based three-factor user authentication scheme for wireless sensor networks[J]. Peer-to-Peer Networking and Applications, 2014:1–22.

Jia W U, Liu M, Yan J W, et al. Research of WSN Security Technology Based on Chaotic Ciphers and RC4[J]. Science & Technology Vision, 2014.

Kong D P, Zhu J D, He T, et al. The study of chaos encryption algorithm for wireless sensor networks based on the reconfigure technology of FPGA[J]. Journal of Chemical & Pharmaceutical Research, 2014.

Lu W, Liu Y, Wang D. A Distributed Secure Data Collection Scheme via Chaotic Compressed Sensing in Wireless Sensor Networks[J]. Circuits Systems & Signal Processing, 2013, 32(3):1363–1387.

Tong X J, Wang Z, Liu Y, et al. A novel compound chaotic block cipher for wireless sensor networks[J]. Communications in Nonlinear Science & Numerical Simulation, 2015, 22(s 1–3):120–133.

Wang H Y. Wireless Sensor Networks Based on a New Block Encryption Algorithm[J]. Applied Mechanics & Materials, 2014, 551:454–459.

Xue T, Yi Z, Yang G, et al. A novel encryption algorithm based on attractor computation for Wireless Sensor Network.[C]// Bio-Inspired Computing: Theories and Applications (BIC-TA), 2010 IEEE Fifth International Conference on. IEEE, 2010:693–697.

Ye R, Yu W. An Image Encryption Scheme Using 2D Generalized Sawtooth Maps[J]. International Journal of Computers & Technology, 2014, 12(6).

Yuan H E, Tian S. Block encryption algorithm based on chaotic S-box for wireless sensor network[J]. Journal of Computer Applications, 2013, 33(4):1081–1084.

Zaibi G, Peyrard F, Kachouri A, et al. A New Encryption Algorithm based on Chaotic Map for Wireless Sensor Network[J]. Architectures & Protocols for Secure Information Technology Infrastructures, 2014.

Zhang Z, Liu K, Niu X, et al. The research of hardware encryption card based on chaos[C]// Sensor Network Security Technology and Privacy Communication System (SNS & PCS), 2013 International Conference on. IEEE, 2013:116–119.

Energy Science and Applied Technology ESAT 2016 – Fang (Ed.)
© 2016 Taylor & Francis Group, London, ISBN 978-1-138-02973-6

A new architecture for a medical internet of things

Guoqiang Hu & Yanrong Yang
Network and Education Technology Center, Northwest A&F University, Yangling, Shaanxi, China

ABSTRACT: RFID technology has been widely applied into medical Internet of Things. To solve the problems brought by the independent co-existence of current traditional IP network and medical Internet of Things, new network technologies integrating WLAN and RFID are witnessing a rapid evolution. Based on an overall consideration of the technological advantages of multi-network integration, a new architecture for medical Internet of Things is proposed based on IoT AP architecture and hierarchical AC + AP architecture, which is designed for providing both the applications of WIFI and RFID sensor network. This architecture effectively integrates traditional IP network and sensor network, improving the liability of network transmission and reducing network maintenance costs.

Keywords: WLAN; RFID sensor network; integration; internet of things; architecture

1   INTRODUCTION

Using information sensors such as RFID, infrared sensor, GPS and laser scanner and relying on communication network, Internet of Things conducts transmission and interconnection according to agreement protocols, and carries out tasks of information processing and data mining with full use of hardware computing resources and software systems, realizing the information interactions between man and things and between different things, and making the liable transmission of information flow of production and living possible. The ultimate goal of Internet of Things is to realize the functions of smart recognition, positioning, tracking, monitoring and management (FAN Xuemei 2011a, b, ZHANG Xuefeng, 2010). Internet of Things is a new system for the real-time interaction between virtual network and the real world, but as its core and foundation is still the Internet, it can be interconnected with the public networks through 4G/5G. The concept of Internet of Things originated from RFID raised by the Auto DLabs which is set up by Massachusetts Institute of Technology in the US in 1999, with its core being connecting all the things to the Internet through information sensors such as RFID, thus realizing smart recognition and management (WANG Junping 2013). The medical area is a field where the technology of Internet of Things is mainly applied to. The technology of UHF RFID as one of the upstream technologies of Internet of Things is extensively applied in the medical Internet of Things with the characteristics of fast reading speed, convenience and long-distance (WU Wenjuan 2011). It is

regarded as one of the most promising information technologies. In practice, there are also some problems for the application of RFID technology in the medical Internet of Things:

1. The frequency bands of RFID vary from low frequency to high frequency, and as different frequency bands adopt different coding and modulation technologies, there are no unified standards for RFID hardware. RFID and the traditional networks do not have a unified standard interface because new businesses sometimes adopt different hardware. In many hospitals, different RFID systems are needed for different businesses, and the adoption of different RFID business results in complicated networks and great pressure on management.
2. When building Internet of things, hospitals also need to build a wireless access network, because RFID cannot provide user access, but large quantities of businesses need to be carried out by means of WLAN, such as returning the data obtained by PDA scanning, reading PACS data and browsing electronic medical records. As traditional hospital IP networks cannot co-exist with the wireless networks that support the Internet of Things, network administrators have to undertake a large amount of work.

To solve the above problems, an IoT architecture that can simultaneously support traditional IP network and RFID sensor network is designed in this paper by establishing the deep coupling of the traditional IP network and the wireless network that supports RFID through the same hardware. This architecture can support different RFID hardware equipment from different manufacturers through standard interfaces.

2 THE DESIGN OF IOT AP ARCHITECTURE

2.1 *RFID technology*

RFID is a contactless automatic identification technology that identifies objects and transfers data through radio frequency signals. Due to its large data capacity, fast reading and writing speed, high stability, and long lifespan, RFID technology is widely used, especially in the area of health care, where it shows unique advantages in terms of identity management, mobile prescriptions, entry of treatment and signs, mobile medicine management, mobile management of test specimens, storage and usage of mobile medical record, Infant stolen and baby care, receiving great attention from various countries (REN Shaojie et al. 2015). A typical RFID system is composed of an electronic tag, a reader and the system for data application management, as is shown in Figure 1.

2.1.1 *Electronic tag*
An electronic tag, also called a transponder or radio RF tag, is composed of circuit components, an electronic chip and a sensor antenna. Every electronic tag has a sole electronic code which is the data carrier of RFID and is usually used to store data in a fixed form.

2.1.2 *Reader*
A reader is also a radio frequency card mainly composed of RF receiver unit, video sender unit, control module and receiving antenna, and is usually classified as fixed reader and portable reader. Data communication is carried out between the reader and the electronic tag through the back-scattered electromagnetic signals of the electronic sensor module [5]. In addition, the reader communicates with the data management system by providing necessary information to the upper computer.

2.1.3 *The system for data application management*
A complete backstage application management system, which is used to store and process infor-

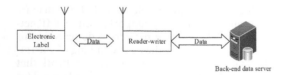

Figure 1. RFID system.

mation related to RFID system, mainly consists of middle ware, information processing system and database. As a significant component of the backstage data management system, the middle ware is an independent system software or a service program which can filter and process data, and has the functions such as coordinating and controlling the reader and reducing RF radiation.

RFID technology can realize wireless automatic recognition. The reader can automatically search for necessary information without needing to contact with the objects directly as long as it is within the reception range of RFID signals. Compared with the speed of entering the information through keyboard, that of RFID technology to collect data is extremely fast, and it can realize "instant data input" (ZHANU Miao 2014). As the RFID system is simple to operate, it has been widely applied into the fields such as military, industry and transportation.

2.2 *Hierarchical AC + AP wireless solution*

In the traditional AC + AP architecture, the ACs and APs are closely related, but ACs undertake most of the wireless functions because APs adopt the FIT AP. Therefore, traditional AC + AP architecture requires higher link quality between the ACs and APs, and it will be better for the communication link between the ACs and APs to have larger broadband and lower latency. Traditional AC + AP architecture can be applied to the environment where the ACs and APs are placed in the same area, but such a single-layered traditional AC and AP architecture will bring about problems such as low speed of user authentication and poor roaming performance for the environment where the head office of a hospital is linked with its branches through WAN. In order to meet the demand for the high-speed link between the ACs and APs, a hierarchical AC architecture is adopted. A single AC is replaced by the central AC and local AC. The central AC is of high performance while the local AC can be composed of switches that integrate wired and wireless networks and that can act as router and DPI (ZHANG Tao 2015). There is a specific protocol between the central AC and the local AC, which is only activated during data transmission. Most of the 802.11MAC, AP management and STA management are realized on the local AC, and such architecture not only keeps the traditional AC architecture, but adapts to the new environment. Although the hierarchical AC architecture is added with the local ACs, it is typical to modify network elements in users' current networks (routers are a must for the user exits). Thus, the hierarchical AC architecture has improved users' WLAN experience without increasing their costs. The network topology of the hierarchical AC architecture is as shown in Figure 2.

2.3 *RFID technology-based IoT AP architecture*

Information collection and object state sensing are solved by the traditional RFID sensor networks, but data transmission in the end is realized through LAN or WAN. The IoT AP can simultaneously provide applications through WIFI and RFID sensor network, deploying a set of APs that not only is available for hospital personnel to access the Inter-

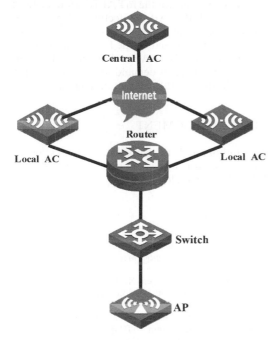

Figure 2. Hierarchical AC architecture.

net but also can carry the data on the RFID sensor networks. The IoT AP architecture can both monitor and record patients' position and state in real-time and deliver the data to backstage for processing through WLAN. In this paper, the IoT AP architecture will be divided into three layers from the top down: respectively sensing and interaction layer, network transmission layer and application services layer, with their roles being as follows.

1. Sensing and interaction layer: to identify and classify patients and hospital personnel, devices, medicines and equipment with the technologies of bar code and QR code; to obtain parameters such as signs and environment through various kinds of sensors; and to process data from the above sensors using readers and PDA.
2. Network transmission layer: to transfer the data collected by sensing and interaction layer to application services layer through WLAN so as to keep the data transmission liable and being real-time;
3. Application services layer: the particular scenarios and business into which the Internet of Things is applied, also the application system that uses the Internet of Things to realize certain functions.

According to the above roles, an IoT AP architecture is designed in this paper. The detailed design is as shown in Figure 3.

3 THE MEDICAL NETWORK SOLUTION BASED ON IOT AP ARCHITECTURE AND HIERARCHICAL AC + AP ARCHITECTURE

A RFID technology-based IoT AP architecture is applied to the wireless solution with hierarchical AC + AP in this paper, becoming a new solution for medical Internet of Things. It is composed of

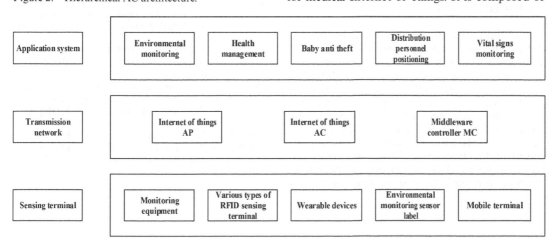

Figure 3. IoT AP architecture.

of different devices, and relieving the pressure on management. The new solution will effectively integrate wired networks, wireless networks, the Internet and sensor networks to help the application system platform collect information from physical entities, finally realizing smart control and interaction between things and man, and between different things, and thus ensuring better transmission.

5 CONCLUSION

A new architecture for the IoT AP is introduced and a hierarchical AC + AP architecture is deployed for the WLANs in this paper to design a solution for medical network based on the IoT AP architecture and hierarchical AC + AP architecture. This solution can meet the new demands of hospitals during the application of the Internet of Things.

ACKNOWLEDGMENT

The authors wish to thank Lilee, Thanks for his help. This work was supported in Network and Education Technology center.

REFERENCES

Fan Xuemei. 2011. A Survey on Development of Internet of Things. *Computer Measurement & Control* [J]. 05:1002–1004.

Ren Shaojie, Hao Yongshen, Xu Bohao. 2015. Review of radio frequency identification technology[J]. *Winged missile,* 01:70–73.

Wang Junping. 2013. Research on Key Technol of Service Delive and System IoT[D]. *China University of Posts and Telecommunications*

Wu Wenjuan, Zhuang Jiandong, Tao Yu, L1 Xin, Gao Huai. 2011. Novel Automatic Gain Adjust Circuit for RFID Readers[J]. *Information terminal & display,* 19:58–61.

Zhang Tao, Kang Huamin. 2015. H3C"cloud" business operators to enter the era of WiFi fusion[N]. *People's Posts and Telecommunications,* 09–09007.

Zhang Xuefeng. 2010. Design and Implementation of Sediment Vehicles Intelligent Monitoring System Based on The Internet of Things Technology[J]. *Software industry and Engineering,* 06:16–21+38.

Zhanu Miao. 2014. On Application of RFID Technology in School Physical Education[J]. J*ournal of Southwest China Normal University (Natural Science Edition),* 07:198–201.

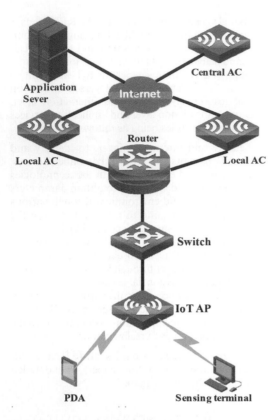

Figure 4. New IoT architecture.

five parts: application system, central ACs, local ACs, Internet of Things AP and terminals. The detailed architecture is as shown in Figure 4.

4 THE ADVANTAGES OF NEW IOT ARCHITECTURE

The medical network solution based on the IoT AP architecture and the hierarchical AC + AP architecture is Figure 4. New IoT architecture.

A new kind of medical network architecture, which couples the AP architecture based on the Internet of Things and the new hierarchical AC + AP architecture of WLANs. Through this set of network, the problems of hospital personnel accessing the Internet and the systematic application of the Internet of Things can be solved, reducing the workload of network administrators, lowering the hardware cost brought by the use

Energy Science and Applied Technology ESAT 2016 – Fang (Ed.)
© *2016 Taylor & Francis Group, London, ISBN 978-1-138-02973-6*

Design of the ZigBee gateway for an agricultural environmental parameter monitoring system

Z.Y. Zou & M. Zhou
College of Mechanical and Electronic Engineering, Sichuan Agricultural University, Ya'an, Sichuan Province, China
College of Water Resources and Hydropower, Sichuan Agricultural University, Ya'an, Sichuan Province, China

ABSTRACT: The ZigBee technology has provided effective technical support for precision agriculture. Therefore, this paper designed the ZigBee gateway by taking the high-performance 32-bit ARM controller as core. In hardware design, it realized power circuit, power-on reset circuit, real-time clock circuit, ZigBee RF circuit, LCD driver circuit, UART communication circuit, and Ethernet communication interface. In software design, it realized embedded WinCE6.0 operation system, embedded SQLite database, ZigBee protocol, and TCP/IP protocol communication program.

Keywords: ZigBee; gateway; ARM; agricultural environmental parameters; monitoring

1 INTRODUCTION

Parameters such as temperature, humidity, illumination intensity, precipitation, CO_2 concentration, soil temperature, soil moisture, soil pH value, and soil conductivity are vital to crop growth. Thus, realization of precision agriculture greatly depends on real-time, convenient, and effective acquisition of agricultural environmental parameters (Xiao et al. 2010).

Monitoring of agricultural environmental parameters is the key for supporting precision agriculture technology. The study on real-time monitoring technology for agricultural environmental parameters is one of the major tasks of international precision agriculture. At present, monitoring of agricultural environmental parameters focuses on fixed point monitoring or multi-sensor data acquisition, and data are sent to the monitoring center. As the sensors are not correlated, if any of the sensors fails, the data of corresponding area cannot be monitored. In addition, the data transfer is troublesome and the system function is simplex. Therefore, it is necessary to improve the monitoring efficiency, data transfer efficiency, and reliability of the system by correlation among sensors (Deng et al. 2010, Li et al. 2012). The emergence of the ZigBee network provides an approach for such a key technical problem in environmental parameter monitoring of fine agriculture.

2 DESIGN OF THE HARDWARE

On the one hand, the gateway is responsible for building and managing the ZigBee network, and on the other hand, it connects the ZigBee network and the monitoring center, and plays a role in data relay and transfer. With the high-performance 32-bit ARM controller as core, this paper designed a ZigBee gateway that can both dynamically configure the ZigBee network and monitor soil temperature, temperature moisture, soil pH value, and soil conductivity in remote areas.

The hardware structure of the ZigBee gateway is shown as Figure 1. To promote efficiency of design and facilitate debugging, the design idea of the core board combining baseboard is adopted. The core board consists of S3C6410 A embedded controller, memory, JTAG debugging interface, 64M NAND Flash formed by two pieces of K9F1208uom memory, and 64M SDRAM formed by two pieces of HY57V561620 memory. The base board circuit mainly consists of power circuit, power-on reset circuit, real-time clock circuit, ZigBee RF circuit, LCD driver circuit, UART communication circuit, and Ethernet communication interface (Sun et al. 2010, Zou et al. 2014).

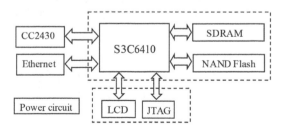

Figure 1. The hardware structure of the ZigBee gateway.

2.1 Power circuit

The circuit operating voltage of each unit of the ZigBee gateway varies. Thus, it designs transforming 12V DC into 5V and 3.3V DC voltage to power the circuit of each unit. The 12V DC voltage is supplied by an external power adapter or a storage battery. The circuit for transforming 12V DC voltage into 5V DC voltage is shown in Figure 2. The LM1117-3.3 chip realizes the transformation from 5V DC to 3.3V DC voltage. C1 and C2 in the circuit are used for front-end filtering. To reduce ripple voltage of the power source, the LCT filter formed by L1, C3 and C4 is used for rear-end filtering of the 5V DC voltage output. The real-time clock circuit is powered by a 3.3V lithium battery.

The M2596-5 chip is the core of the circuit that transforms 12V DC voltage into 5V DC voltage. This chip has the following functions and features: a) output 5V DC voltage and 3 A driving current. b) This part is integrated with frequency compensation and fixed frequency generator whose switching frequency is 150 KHz. Compared with low-frequency switching regulator, it can be used for filtering elements of smaller specification. c) Under a specific input voltage and output load, the error of output voltage can be guaranteed within ±4% and the error of oscillation frequency is within ±15%. d) Only an 80μA standby current is required and can realize external power off. e) There is self-protection circuit.

2.2 Power-on reset circuit

The state of the ARM controller is uncertain when it is powered on, while the reset circuit is an indispensable part for guaranteeing stability and reliability of circuits in the ARM controller system. The primary function of the reset circuit is power-on reset. As a result of low power consumption, high speed, and low operating voltage, the noise margin of the ARM chip is relatively low and has higher requirements such as transient response performance, power ripple wave, and stability of clock source. Therefore, a special micro-processor power source monitoring chip MAX811 is used on the power-on reset circuit to improve the stability

of the system. The reset signal is output by two pins of MAX811. Its circuit principle is shown in Figure 3.

2.3 ZigBee RF circuit

The ZigBee RF module adopts the CC2430 chip that can realize data receiving and transmitting with a few of peripheral components. Figure 4 shows the RF unit circuit in this design.

It takes the U4-CC2430 chip as core. DD and DC are download program signal lines. LED1 and LED2 are data receiving and transmitting signal indicators. R3 and C5 form the RC-type power-on reset circuit. ZTXD and ZRXD are used for communicating with the ARM controller. L2 and L3 form a folding dipole PCB antenna. The bias resistors are R7 and R8, of which R8 is used for setting precision bias current for the 32 MHz crystal oscillator. The external 32 MHz crystal oscillator X1, together with two load resistors (C20 and C21), are used for the 32 MHz crystal oscillator. Besides the 32 MHz crystal oscillator, it can also use X2, a 32.768 kHz crystal oscillator. Moreover, the ZigBee network can also function normally when it forms a mesh network. To realize excellent performance, power source decoupling is also required. The size and layout of the power source decoupling and wave filtering of power source play an important role in achieving the best performance in application. All of the other capacitances connecting the power source in circuit are for decoupling and filtering.

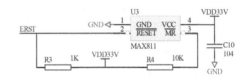

Figure 3. The reset circuit of turn-on power.

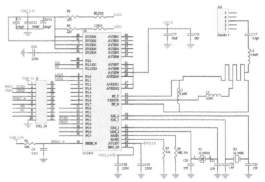

Figure 4. The radio of the frequency unit circuit.

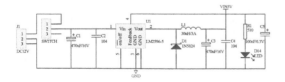

Figure 2. The circuit of 12VDC-5VDC.

2.4 Interface circuit

A 3.5" TFT LCD is adopted in this design, for which the control lines and data lines are drawn from the core board to the base board. To ensure the reliability of signal transfer, the 74 LVCH162245 A chip is used to drive the signals.

The Ethernet controller adopts the CS8900 A chip and communicates with the RJ45 interface. In this design, the UART1 interface of the ARM controller is used for communicating with CC2430. Its hardware is connected by means of three-wire connection.

3 DESIGN OF THE SOFTWARE

WinCE6.0 embedded OS is self-customized with Platform in the design of the system program for gateway node. To manage the data effectively, the SQlite embedded database is transplanted to WinCE6.0 embedded OS. SQlite has features such as small size, open sources, low RAM requirement, and convenient operation. Two library files, sqlite.lib and sqlite.dll, are added into the application project of WinCE 6.0 and the operation of SQlite embedded database can be realized by calling the API function of SQlite. Intelligent control of each parameter adopts fuzzy control to carry out a closed loop control (Zou et al. 2014).

CC2430 on the ZigBee gateway plays the role of a ZigBee network coordinator. It acquires data and forwards control commands, and then sends to the ARM controller on the ZigBee gateway via its serial port. The ARM controller sends data to the PC server through the TCP/IP protocol to realize mutual data transfer between the ZigBee protocol and the TCP/IP protocol. The protocol conversion flow chart is shown in Figure 5.

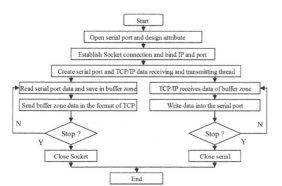

Figure 5. The flow chart of the ZigBee protocol to exchange the TCP/IP protocol.

4 SYSTEM TEST

The ZigBee gateway is tested in a 6.7 hectare farm. Five ZigBee network acquisition nodes, five ZigBee network control nodes, and several router nodes are arranged. The test includes both ZigBee network test and agricultural environmental parameter monitoring test.

4.1 ZigBee network test

To carry out the data transfer test, the ZigBee gateway sends 1,000 control commands to five ZigBee network acquisition nodes and five ZigBee network control nodes. The test shows that the maximum packet drop ratio of a single node is 0.23%, whereas the average packet drop ratio of the entire ZigBee network is 0.21%. These results indicate that the wireless transfer of data is highly stable.

4.2 Agricultural environmental parameter monitoring test

The soil temperature is acquired by the PT1000 temperature sensor, the soil moisture is acquired by the TDC220D module, the soil pH value is acquired by the E-201-C electrode, and the soil conductivity acquisition is designed with TDR-4 soil conductivity sensor.

To realize the detection of soil temperature parameters, the second port of the PT1000 temperature sensor is connected to the second port of the LM741 high-gain operational amplifier. The six port of the LM741 high-gain operational amplifier is connected to the third port of the OP07 precision operational amplifier. The six port of the OP07 precision operational amplifier is connected to the AIN1 port of the S3C6410 embedded controller.

To realize the detection of soil moisture parameters, the GPB1 port of the S3C6410 embedded controller is connected to the SDI-12 port of the TDC220D module.

To realize the detection of soil pH parameters, the first and second ports of the E-201-C electrode are, respectively, connected to the second and third port of the F011 low power operation amplifier. The sixth port of the F011 low power operation amplifier is connected to the third port of the INA826 AID instrument amplifier. Finally, the six port of the INA826 AID instrument amplifier is connected to the AIN2 port of the S3C6410 embedded controller.

To realize the detection of soil conductivity parameters, the second port of the TDR-4 soil conductivity sensor is connected to the third port of the F210 voltage follower. The sixth port of the F210 voltage follower is connected to the second port of the AD621 instrument amplifier. Finally,

Table 1. Test data of soil temperature, moisture, pH, conductivity.

Day	Temperature °C	Moisture %	pH	Conductivity mV
1	13.3	56	6.7	282
2	13.7	55	6.6	275
3	14.7	56	6.6	281
4	14.9	54	6.4	264
5	14.9	54	6.5	267
6	14.8	52	6.4	234
7	14.9	52	6.6	229
8	15.1	51	6.6	216
9	15.3	51	6.5	223
10	15.6	51	6.4	217
11	15.5	50	6.5	204
12	15.7	49	6.3	202
13	15.6	49	6.5	205
14	15.8	47	6.5	197
15	15.9	48	6.8	206
Max error	1.7	6	0.3	53

the six port of the AD621 instrument amplifier is connected to the AIN3 port of the S3C6410 embedded controller (Yue et al. 2013).

Regulation of soil temperature, soil moisture, soil pH value, and soil conductivity is realized by controlling the corresponding regulation equipment through ZigBee network control nodes. The soil temperature parameter is set at 15 °C. The soil moisture parameter is set to 50%. The soil pH parameter is set to 6.5. The soil conductivity parameter is set at 250 mV. The fuzzy controller is implemented on the ZigBee gateway. Each agricultural environmental parameter monitoring data is presented in Table 1.

5 CONCLUSION

1. Hardware circuit and software program for the ZigBee gateway are designed. The transmission of the ZigBee network is tested, and the maximum packet drop ratio of a single node is 0.23%.
2. The monitoring effect of agricultural environmental parameters including soil temperature, soil moisture, soil pH value, and soil conductivity is tested. The temperature error is controlled within ±1.7°C, the moisture error is controlled within ±6%, the pH error is controlled within ±0.3, and the soil conductivity error is controlled within ±53 mV, so the system fully meets the practical application.

ACKNOWLEDGMENTS

This research was supported Key Projects in Sichuan Agricultural University No. X2015040.

Zhiyong Zou is a Lecturer in College of Mechanical and Electrical Engineering, Sichuan Agriculture University, Ya'an, China. His research interests include agricultural engineering.

Man Zhou* (Corresponding author) is a Lecturer in College of Water Resources and Hydropower, Sichuan Agriculture University, Ya'an, China. Her research interests include agricultural engineering (Email: zouziyong111@163.com).

REFERENCES

Deng, X.L. Zheng, L.H. & Li, M.Z. 2010. Development of wireless sensor network of field information based on ZigBee and PDA, *Transactions of the Chinese Society of Agricultural Engineering (Transactions of the CSAE)*. 26(14):103–108.

Li, M.Z. Deng, X.L. & Zheng, L.H. 2012. Development of mobile soil moisture monitoring system integrated with GPRS, GPS and ZigBee, *Transactions of the Chinese Society of Agricultural Engineering (Transactions of the CSAE)*. 28(9):130–135.

Sun, Y.W. & Shen, M.X. 2010. Design of Embedded Agricultural Intelligence Services System Based on ZigBee Technology, *Transactions of the Chinese Society for Agricultural Machinery*. 41(5):148–151.

Xiao, K.H, Xiao, D.Q & L, X.W. 2010. Smart water-saving irrigation system in precision agriculture based on wireless sensor network, *Transactions of the Chinese Society of Agricultural Engineering (Transactions of the CSAE)*. 26(11):170–175.

Yue, X.J. Liu, Y.X & Chen, Z.L. 2013. Design and Experiment of Automatic Irrigation Control System Based on Soil Moisture Meter, *Transactions of the Chinese Society for Agricultural Machinery*. 44(S2):241–246,250.

Zou, Z.Y. Xu, L.J. & Chen, X.Y. 2014. Intelligent Facilities Fishpond Design of Breeding Schizothorax prenanti, *Transactions of the Chinese Society for Agricultural Machinery*. 45(S1):296–301.

Zou, Z.Y. Zhou, M. & Liu, M.D. 2014. Design of Methane Fermentation environmental factor detection system based on ZigBee, *Applied Mechanics and Materials*. (539) 567–571.

12 *System test, diagnosis, detection and monitoring*

Energy Science and Applied Technology ESAT 2016 – Fang (Ed.)
© *2016 Taylor & Francis Group, London, ISBN 978-1-138-02973-6*

Multi-lane vehicle detection method based on magneto-resistive sensors

Shimiao Yang, Congyuan Xu, Jizhong Shen, Wufeng Zhao & Wei Wang
College of Information Science and Electronic Engineering, Zhejiang University, Hangzhou, China

ABSTRACT: In traffic information acquisition system, vehicle detection method based on magneto-resistive sensors is more convenient and cheaper than traditional methods from the perspectives of installation and maintenance. This paper focuses on the magneto-resistive vehicle detection method for multi-lane scenery. A Single-Node Vehicle Detection algorithm (SNVD) based on multiple magnetic detection features is proposed to deal with dense traffic and low velocity vehicles. Moreover, a Multi-Lane Vehicle Detection method (MLVD) is proposed to solve the false detection by trans-lane and adjacent-lane vehicles, which is based on the combination of SNVD, vehicle position recognition and multi-node integration. With acquisition data on real road, SNVD is proved to be adaptable to different traffic situations. Test results also show the detection accuracy of MLVD is above 99%, the method features strong robustness and good scalability, which is applicable for urban multi-lane sceneries.

Keywords: magneto-resistive sensors; vehicle detection; multi-lane; vehicle position recognition; detection fusion

1 INTRODUCTION

With rapid development of intelligent transportation industry, traffic information acquisition applications become more and more common. However, traditional detection technologies such as video, induction coils, radar and ultrasonic have been applied in a limited scale, and most of these technologies are cost expensive, with complexity in installation and maintenance (A. Haoui et al, 2008). In recent years, the emerging magneto-resistive detection based on wireless sensor networks has attracted much attention due to its compact size, low costs, high reliability, ease of installation and maintenance (Z. Zhang et al, 2013).

Within a few kilometers range, the geomagnetic field can be seen as constant uniform. Vehicles containing ferromagnetic materials can lead to offset or disturbance on background geomagnetic field (A. Pascale et al, 2012), making it possible to detect the vehicle presence by sensing magnetic field changes. In normal method, one sensor node is deployed in the middle of each single lane, and for multiple lanes the total traffic flow equals the sum of flow on each lane (S. Yoo, 2013) (M. Bottero et al, 2013). Unfortunately, there exist two problems to be solved under multi-lane sceneries.

1. Adjacent-lane vehicles. Larger vehicles on adjacent lanes may induce large enough signals to result in false detection on current lane (E. Sifuentes et al, 2011).

2. Trans-lane vehicles. Vehicles travelling between two lanes may be detected by two nodes simultaneously and lead to redundant results of total flow. Meanwhile, vehicles also may be missed by both nodes.

This paper is organized as follows. Section II introduces the related work. Section III presents the framework of MLVD. Section IV describes the detection algorithms on distributed sensor nodes. Section V demonstrates the integration of multi-node detection. Section VI gives the experimental results and analysis. Conclusions are presented in Section VII.

2 RELATED WORK

At present, many researches about magneto-resistive vehicle detection have been carried out. Researchers from University of Berkeley proposed Adaptive Threshold Detection Algorithm (ATDA) (S. Y. Cheung et al, 2007). ATDA is based on threshold and Finite State Machine (FSM), realizing detection with high efficiency and real-time performance. Based on ATDA, improvements have been made on detection features selection, drift elimination, threshold setting and FSM design (J. Yoo et al, 2012) (B. Koszteczky et al, 2013) (S.G. Wei et al, 2009) (Z. Jian et al, 2011) (P. Kanathantip et al, 2010). Several algorithms introduced correlation similarity to achieve better signal-noise ratio at the expense of decreasing real-time perform-ance and increasing computational complexity (L. Zhang et al, 2010) (L. Zhang et al, 2011). Most algorithms were proved to be suitable for medium dense traffic in single-lane sceneries, yet the accu-racy for dense traffic had not been evaluated.

Single-lane detection algorithms can't be applied to multiple lanes directly due to the adjacent-lane and trans-lane vehicles (F. Ahdi et al, 2012). A fusion method for bidirectional two-lane scen-ery was introduced in (R. Wang et al, 2011). The method identifies adjacent-lane vehicles based on magnetic signal strength and driving direction, the detection accuracy rate is 93%. In (S. Taghvaeeyan et al, 2013), adjacent-lane vehicles are distinguished based on magnetic signal attenuation characteristics and Support Vector Machine (SVM) classification. The method can detect one of multiple lanes with accuracy rate of 99%. However, these methods can't deal with trans-lane vehicles. Besides, the methods must deploy sensor nodes at roadside, so they can't be expanded to sceneries of more than two lanes.

3 FRAMEWORK OF DETECTION SYSTEM

In this section, the framework of MLVD is intro-duced to achieve better detection performance in multi-lane sceneries. Nodes layout of MLVD is shown in Fig.1. Based on nodes layout of the normal methods, one added Sensor Node (SN) is deployed on the boundary line between every two lanes. When the number of lanes is N, the total nodes number is 2 N-1. The Access Point node (AP) provides time synchronization for SNs and integrates detection results from all SNs. Generally, the lane is 3~3.75 m wide and the vehicle width range from 1.5 m to 2.5 m. To avoid interference caused by vehicles in adjacent lanes, SNs are set at a small detection radius. It's ensured that a vehicle traveling across nodes array will be detected by at least one SN.

Fig. 2 shows the processing procedures of MLVD, which mainly consists of single-node vehi-cle detection algorithm SNVD, position recogni-tion and multi-node results integration.

SNs process magnetic sampling data, run SNVD algorithm to identify arriving and departure time of each vehicle. Position recognition algorithm is used to classify detected vehicles into two situa-tions, i.e. traveling over the SN or traveling from side of SN. By extracting features of signal magni-tude and peak-valleys, two traveling situations can be decided by a trained classifier. Both SNVD and position recognition algorithms are implemented on single SN to guarantee efficiency and real-time performance.

Vehicle detection results about time and position are packed into records and sent to AP for match-ing and fusion. Records from several SNs with approximate time information may indicate more than one vehicle passing. According to restrictions of vehicle position, nodes layout and vehicle width, a fusion algorithm will decide the actual number of vehicles traveling across nodes array.

4 DISTRIBUTED DETECTION

4.1 *Single-node vehicle detection*

Considering the limited computing ability and battery power of SNs, vehicle detection algorithm must be lightweight in complexity. SNVD algo-rithm is proposed based on the state machine of ATDA algorithm, which is given in Fig. 3. SNVD processes 3-axis sampling data once after sam-pling. The moving average filter is used to smooth

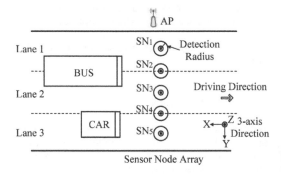

Figure 1. Nodes layout scheme of MLVD.

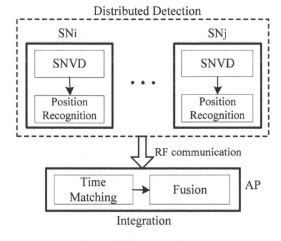

Figure 2. Procedures of multi-lane detection.

Figure 3. Single-node detection algorithm procedure.

the magnetic signal, and the baseline tracing is designed based on the moving median filter.

Detection feature reflects the magnetic signal changing caused by vehicles. In SNVD, more than one feature is extracted for detection. The net signal components of x, y, z are obtained by subtracting baseline from the raw signals. Component z decreases quickly when vehicle is leaving. Since x and y are always significant even when the vehicle is far away from the SN, thus, they are divided by a factor n to facilitate splitting the signals of successive close vehicles. Component dx, dy, dz are forward difference of 3-axis signals, which reflect the vehicle moving information with less signal drift. The threshold comparison conditions are given:

$$\max(z, x/n, y/n) \geq TH1 \tag{1}$$

$$\max(dx, dy, dz) \geq TH2 \tag{2}$$

In Eq. (1) and Eq. (2), TH1 and TH2 are both empirical thresholds upon analysis of actual signals. Either Eq. (1) or Eq. (2) is met, the input of state machine is true, otherwise, it is false.

4.2 Vehicle position recognition

As one or more SNs may detect vehicles at one point, the vehicle position information on each of these nodes can be gathered to help judging the real number of vehicles. So, the vehicle position recognition is introduced, and it will assist detecting the trans-lane and adjacent-lane vehicles.

As magnetic field points to the ground roughly in northern hemisphere, ferromagnetic materials gathering magnetic induction lines will enhance the vertical magnetic component. Consequently, when the vehicle travels over the SN, the signals of vertical Z-axis tend to be unidirectional. Besides, the signal has more significant magnitude and more peak-valleys since the vehicle is close to SN. However, when the vehicle travels from side of SN, the signal characteristics are opposite.

Two features of Z-axis signals are extracted for vehicle position recognition. One is the mean value of negative sample points (zneg); the other is the number of waveform peak-valleys (zpv), while small peak-valleys are removed. Upon feature extraction and normalization, Fisher Linear Discriminant Analysis (FLDA) is used to train a class boundary function for recognition (P. N. Belhumeur et al, 1999). In FLDA, samples in D-dimensional space (D ≥ 2) are projected onto 1-dimensional space, finding the best projection direction to get maximum between-class dispersion and minimum within-class dispersion.

In training, 138 vehicle samples are selected, which includes 51 samples travelling over SN and 87 samples travelling from side of SN. A training classification accuracy rate of 99% is achieved in results.

After detection and feature extraction in SNs, vehicle position can be recognized using the class boundary function trained off-line. Therefore, the position recognition algorithm is lightweight and real-time.

5 INTEGRATION OF MULTI-NODE DETECTION

Records from SNs should be integrated in AP to calculate traffic flow, especially when more than one SNs capture vehicle signals at the same time. First, time matching is implemented to find if there are any records reported almost simultaneously. Then, fusion algorithm extracts position information from these records and gets the actual vehicle number by an encoding mechanism.

5.1 Time matching

According to the record time of vehicle arrival and departure, two records from adjacent SNs are regarded as matching in time when following condition is satisfied:

$$\frac{L_{AB}}{\min(L_A, L_B)} \geq TH_{AB} \tag{3}$$

where L_A and L_B are vehicle signal durations on the two SNs respectively, and L_{AB} is the overlapping duration of two SNs, TH_{AB} is the matching degree threshold. Checking the records information of SNs one by one in sequence, all the matched records on adjacent SNs are found out for further processing.

5.2 Encoding and fusion

For ease of fusion, when a vehicle travels over the SN, the node is marked with 1, otherwise, it is marked with −1. Position marks are extracted from matched records, and these mark bits compose position code in the order of SNs layout. Regardless of the number of lanes, M is used to represent the number of vehicles travelling side by side across nodes array. The cases of M≤2 are discussed to design the fusion algorithm.

First, detection radius of SN is tuned to about 0.5 m by adjusting algorithm parameters. Consequently, a vehicle can be detected by 1~3 nodes simultaneously when travelling across nodes array. As Table. 1 shown, three bits are used to enumerate all position codes generated by case M = 1, when 0 is an invalid bit. Furthermore, these codes

Table 1. Position codes in case M = 1.

	Basic Code	
1	$\begin{pmatrix} 1 & 0 & 0 \\ 0 & 1 & 0 \\ 0 & 0 & 1 \end{pmatrix}$ →	(1)
2	$\begin{pmatrix} 1 & 1 & 0 \\ 0 & 1 & 1 \end{pmatrix}$ →	(1 1)
3	$\begin{pmatrix} -1 & -1 & 0 \\ 0 & -1 & -1 \end{pmatrix}$ →	(-1 -1)
4	$\begin{pmatrix} 1 & -1 & 0 \\ 0 & 1 & -1 \\ -1 & 1 & 0 \\ 0 & -1 & 1 \end{pmatrix}$ →	(1 -1) (-1 1)
5	(-1 1 -1) →	(-1 1 -1)

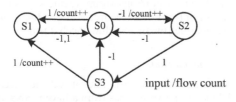

Figure 4. Code fusion state machine.

are summarized as five basic codes. Basic code 1 with only one valid bit may indicate a small car, and basic code 5 with three valid bits may indicate a large vehicle. What's important, these codes can uniquely represent one vehicle passing in actual.

In case M = 2, two vehicles traveling side by side may be detected by 3~6 nodes simultaneously. The codes for M = 2 must be the combination of two basic codes. As the situation shown in Fig.1, code $(-1\ 1\ -1)$ is acquired when the BUS travels over SN_2 and detected by SN_1 and SN_3 as well. Similarly, the CAR results in code $(1\ 1)$ on SN_4 and SN_5. As the BUS and the CAR travel side by side, a combination code $(-1\ 1\ -1\ 1\ 1)$ is acquired. All the codes for M = 2 can be enumerated when vehicles are of different size and travel across nodes array with variable positions.

The fusion aims to extract the number of basic codes in a complete position code. The position code consists of $\{-1, 1\}$ is processed bit by bit. If current processed bit with previous several bits can't compose one of the five basic codes, it is indicated that one more vehicle is detected on the basis of current flow count. The fusion state machine in Fig. 4 is designed to process the code so as to realize traffic flow counting.

The fusion state machine has four states:
S0: the first bit of next basic code is to be input;

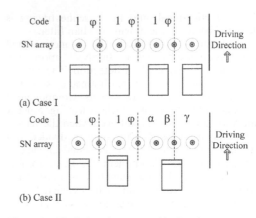

(a) Case I

(b) Case II

Figure 5. Typical code generated in four-lane scenery.

S1: the {1} has been input;
S2: the {−1} has been input;
S3: the {−1 1} has been input.

Each state corresponds to different prefixes of basic codes. As the first bits of code are entered, the state machine starts with state S0, and the flow count is incremented while state changing. Through exhaustive enumeration and state machine processing, it's verified that all codes for case M≤2 can be resolved correctly.

Under none-fusion methods a trans-lane vehicle may be detected by two lanes, which is confused with two vehicles travelling side by side. Within MLVD, the confusing situations can be distinguished by encoding and fusion.

5.3 Applicability to multi-lane

There are usually 1~4 lanes for unidirectional road. For general traffic the horizontal safety distance between vehicles is considered, so the value of M is regarded not to exceed the number of lanes N. The applicability of MLVD to 3 or 4 lanes is investigated as follows.

For sceneries with 3 or 4 lanes, if M = N, then the travelling vehicles are intensive horizontally; considering the driving safety, each vehicle should travel over the SN in the middle of each lane. Above situation in four-lane scenery is given in Fig. 5(a), and the position code can be expressed as $(1\ \varphi\ 1\ \varphi\ 1\ \varphi\ 1)$. The mark φ reported by SNs between every two lanes has the value of 1 or −1. Through enumeration, all valid codes will be resolved correctly by the designed state machine.

In four-lane sceneries, the situations of M = 3 are discussed. In general, trans-lane driving happens when two free lanes are available. When there is one trans-lane vehicle of the three, the position code can be expressed as $(1\ \varphi\ 1\ \varphi\ \alpha\ \beta\ \gamma)$ as Fig. 5(b) shown. The mark α must be −1, and β has the value of 1 or −1, γ is −1 or invalid. It is proved all codes will be resolved correctly by enumeration. Though the situations with two or three trans-lane vehicles are not common, most of the enumerated codes can also be resolved correctly.

According to the analysis above, MLVD method is applicable to sceneries of no more than four lanes. Meanwhile the software and hardware are compatible for nodes, the method can be easily expanded to accommodate different number of lanes. In addition, for motor way without isolation barrier at roadside, an extra SN can be deployed there to avoid missing vehicles.

6 EXPERIMENTAL RESULTS AND ANALYSIS

SNs are implemented to acquire magnetic signal data on road. SN consists of main modules such as MSP430 microcontroller and HMC5883 L magneto-resistive sensor. The sample rate is set to be 100 Hz, and the acquisition data is exported to PC. The proposed method is verified in Matlab via data acquisition and off-line algorithm simulation.

6.1 Single-lane detection experiment

Datasets are collected from the SN placed in the middle of single lane. Dataset DS1 is collected on urban road with medium dense traffic, which lasts about 10 min. DS2 is collected in the traffic congestion on urban road, which lasts about 40 min. And DS3 is collected on the toll-gate lane with extremely slow traffic.

Test results of the three datasets are shown in Tables 2–4. To compare the performance of different methods, the datasets are processed by another two methods as well. The method in (S. Y. Cheung et al, 2007) is based on z component, and the method in (J. Yoo et al, 2012) is based on the difference of total magnetic magnitude.

According to the results, all the algorithms achieve high accuracy (>95%) for medium dense traffic as shown in Table 2. The method in (S. Y. Cheung et al, 2007) and method in (J. Yoo et al, 2012) are prone to recognize one slow vehicle as several vehicles, both of them cause many false positives detections as shown in Table 3 and Table 4. The method in (J. Yoo et al, 2012) causes more false negative detections, for it's also prone to mix several close vehicles into one. SNVD performs better than other algorithms significantly for congestion and slow traffic. It's concluded that SNVD has excellent adaptability to different traffic density and velocity. Taking account of the total false detections for three datasets, the overall accuracy rate of proposed SNVD is 97.5%, while the methods in (S. Y. Cheung et al, 2007) and (J.

Table 2. Single-lane detection results for DS1 with medium dense traffic.

Type	Actual Flow	Detected Flow	False Negative	False Positive	Detection Accuracy
SNVD	125	126	0	1	99.2%
Ref[7]	125	130	0	5	96.0%
Ref[8]	125	128	0	3	97.6%

Table 3. Single-lane detection results for DS2 with congestion traffic.

Type	Actual Flow	Detected Flow	False Negative	False Positive	Detection Accuracy
SNVD	114	117	1	4	95.6%
Ref[7]	114	124	0	10	91.2%
Ref[8]	114	117	4	7	90.4%

Table 4. Single-lane detection results for DS3 with extremely slow traffic.

Type	Actual Flow	Detected Flow	False Negative	False Positive	Detection Accuracy
SNVD	37	38	0	1	97.3%
Ref[7]	37	44	0	7	81.1%
Ref[8]	37	49	0	8	78.4%

Yoo et al, 2012) achieve the same overall accuracy of 92.0%.

6.2 Multi-lane detection experiment

On the surface of urban road, three SNs are placed in a line with 1.8 m distance between every two nodes, and SN_B is on the boundary line of two lanes. According to video record, 219 vehicles passed in 20 min of the on-road data acquisition, which includes trans-lane vehicles as well as vehicles travelling side by side.

To evaluate the performance of MLVD method, several single-node detection algorithms without fusion are chosen as benchmark. These algorithms process the same acquisition data. Table 5 shows the results of experiment.

The detection accuracy of MLVD method with fusion is 99.6% as Table 5 shows, one vehicle causes false negative for it travels beyond the predefined two-lane boundary. For none-fusion methods the false negative II is in the majority of false detection. That's because trans-lane vehicles are missed by SN_A and SN_C with small detection radius. In addition, trans-lane or adjacent-lane vehicles may be detected by SN_A and SN_C. Consequently, some of these vehicles cause false negative II in the none-fusion methods. Overall, the accuracy of MLVD outperforms by about 16.0%~18.3% than none-fusion methods.

Table 5. Multi-lane experiment results of different methods.

Type	Fusion	Actual Flow	Detected Flow	SN$_A$ Flow	SN$_B$ Flow	SN$_C$ Flow	False Negative [1]	False Positive I [2]	False Positive II [3]	Detection Accuracy
MLVD	Yes	219	218	85	140	102	1	0	0	99.6%
SNVD	No	219	187	85	X	102	34	0	2	83.6%
Ref[7]	No	219	181	79	X	102	40	2	0	80.8%
Ref[8]	No	219	186	84	X	102	37	1	3	81.3%

1. False Negative: missed on single sensor node.
2. False Positive I: redundant detection on single sensor node
3. False Positive II: caused when calculating the total flow in AP

Obviously, it's impossible to reduce false positive and false negative cases within none-fusion methods. MLVD method solves this problem via adding more SNs, reducing detection radius and an effective fusion mechanism. Due to constraints of field environment, experiment is conducted on two lanes. Nevertheless, position recognition and fusion algorithms are validated in experiments, and MLVD can be applied to more lanes for its compatibility and scalability as discussed in section 5.

7 CONCLUSION

In this paper, the single-lane vehicle detection method SNVD is proposed. SNVD is based on multiple features, with better performance against exiting algorithms for low velocity and dense traffic situations. Then, the method of MLVD is proposed to expand single-lane vehicle detection for multiple lanes. Based on SNVD, MLVD improves nodes layout and implements a fusion mechanism using vehicle position recognition. MLVD achieves an accuracy of 99.6% in experiment, solving the problem of trans-lane and adjacent-lane vehicles in multi-lane sceneries. It's concluded that MLVD has good adaptability to unidirectional multi-lane road.

ACKNOWLEDGMENTS

This work was supported by National Natural Science Foundation of China under Gran No. 61471314 and Zhejiang Provincial Natural Science Foundation of China under Gran No. LY13F010001.

REFERENCES

Ahdi, F., M.K. Khandani, M. Hamedi, et al, "Traffic Data Collection and Anonymous Vehicle Detection Using Wireless Sensor Networks," *University of Maryland*, 2012.

Belhumeur, P.N., J.P. Hespanha, D. Kriegman, "Eigenfaces vs. fisherfaces: Recognition using class specific linear projection," *IEEE Transactions on Pattern Analysis and Machine Intelligence*, vol. 19, 1999, pp. 711–720.

Bottero, M., C.B. Dalla, F.P. Deflorio, "Wireless sensor networks for traffic monitoring in a logistic centre," *Transportation Research Part C: Emerging Technologies*, Vol. 26, 2013, pp. 99–124.

Cheung, S.Y., P. Varaiya, "Traffic Surveillance by Wireless Sensor Networks: Final Report," *University of California*, Berkeley, Jan.2007.

Haoui, A., R. Kavaler, P. Varaiya, "Wireless magnetic sensors for traffic surveillance," *Transportation Research Part C: Emerging Technologies*, Vol. 16, 2008, pp. 294–306.

Jian, Z., C. Hongbing, S. Jie, et al, "Data fusion for magnetic sensor based on fuzzy logic theory," *2011 IEEE International Conference on Intelligent Computation Technology and Automation(ICICTA)*, 2011, pp. 87–92.

Kanathantip, P., W. Kumwilaisak, J. Chinrungrueng, "Robust vehicle detection algorithm with magnetic sensor," *International Conference on Electrical Engineering/ Electronics Computer Telecommunications and Information Technology*, 2010, pp. 1060–1064.

Koszteczky, B., G. Simon, "Magnetic-based vehicle detection with sensor networks," *IEEE International Instrumentation and Measurement Technology Conference (I2MTC)*, 2013, pp. 265–270.

Pascale, A., M. Nicoli, F. Deflorio, et al, "Wireless sensor networks for traffic management and road safety," *IET Intelligent Transport Systems*, Vol. 6, 2012, pp. 67–77.

Sifuentes, E., O. Casas, R.A. Pallas, "Wireless Magnetic Sensor Node for Vehicle Detection With Optical Wake Up," *IEEE Sensors Journal*, Vol. 11, 2011, pp. 1669–1676.

Taghvaeeyan, S., R. Rajamani, "Portable Roadside Sensors for Vehicle Counting, Classification, and Speed Measurement," *IEEE Transactions on Intelligent Transportation Systems*, 2013.

Wang, R., L. Zhang, R. Sun, et al, "EasiTia: A Pervasive Traffic Information Acquisition System based on Wireless Sensor Networks," *IEEE Transactions on Intelligent Transportation Systems*, Vol. 12, 2011, pp. 615–621.

Wei, S.G., W. Jian, B.G. Cai, et al, "A Novel Vehicle Detection Method Based on Wireless Magneto-resistive Sensor," *Third International Symposium on Intelligent Information Technology Application(IITA)*, 2009, pp. 484–487.

Yoo, J., D.H. Kim, K.H. Kim, et al, "On-Road Wireless Sensor Networks for Traffic Surveillance," *The Second International Conference on Mobile Services*, 2012, pp. 131–135.

Yoo, S. "A Wireless Sensor Network-Based Portable Vehicle Detector Evaluation System," *Sensors*, Vol. 13, 2013, pp. 1160–1182.

Zhang, L., R. Wang, L. Cui, "Real-time Traffic Monitoring with Magnetic Sensor Networks," *Journal of Information Science & Engineering*, Vol. 27, 2011, pp. 1473–1486.

Zhang, L., R. Wang, Y. Wang, et al, "Accurate vehicle monitoring based on wireless sensor network," *IET International Conference on Wireless Sensor Network*, 2010, pp.13–16.

Zhang, Z., X. Li, H. Yuan, et al, "A Street Parking System Using Wireless Sensor Networks," *International Journal of Distributed Sensor Networks*, ID-107975, 2013. http://dx.doi.org/10.1155/2013/107975.

Energy Science and Applied Technology ESAT 2016 – Fang (Ed.)
© *2016 Taylor & Francis Group, London, ISBN 978-1-138-02973-6*

Influence of weather factors on weather-sensitive power load in the non-central city

Z.Y. Yin
Beijing Meteorological Service Center, Beijing, China

J.J. Fan
Huangshi Meteorological Bureau, Hubei Province, Huangshi, China

X. Liu
Public Meteorological Service Center of China Meteorological Administration, Beijing, China

J. Jiang
Wuhan Central Meteorological Observatory, Hubei Province, Wuhan, China

ABSTRACT: To analyze the trend of electric load during a long period and its response to the meteorological factors in the non-central city in China, the daily data of power load and weather factors from 2007 to 2013 in Huangshi City were selected and the sensible temperature and comfort index level were first calculated from the single meteorological factors. Based on the results, the power load influenced by financial crisis was divided. Then, the relationship between the sensible temperature and the sensitive load rate was calculated qualitatively and quantitatively. The results indicated that the meteorological factors have significant effects on the power load. The sensible temperature showed better results after analyzing its specific impact. The sensible temperature higher than 22.9°C (weekdays) or 21.5°C (holidays) would lead to sensitive load, while a 1°C increase of sensible temperature could reach up to 6% additional meteorological load rate during the study period.

Keywords: sensible temperature; sensitive load; 1°C effect

1 INTRODUCTION

With the rapid development of industry and national economy, the demand for electricity is increasing in China. However, the power capacity cannot completely meet the needs of various users. The power shortage brought great inconvenience to economic development and people's life.

The power load is greatly influenced by the weather factors, especially in summer (Selakov et al. 2014, Hu et al. 2014), as the air-conditioning load has been proved to be closely related to temperature, humidity, and other meteorological elements (Yang et al. 2010, Vaghefi et al. 2014). To understand how weather conditions influence the power load, current studies are more concerned about two aspects. The first aspect is meticulous analysis of the relationship between the power load and the external condition in which many factors are included such as big user load and specific events (Xia et al. 2010, Deihimi et al. 2013). In addition, other studies (Hernández et al. 2012, Fikru et al. 2015) have analyzed the trend of daily power load of long time series, as about 30% of the power load are influenced by weather factors (Liao et al. 2012), the effects of which have also been discussed. However, many problems need to be solved in this research. For example, the impact of financial crisis on power load is often neglected, so power load decomposition should be dealt more accurately.

In addition, it has been noted that the composite indicator characterized by multiple meteorological factors has a higher correlation with the power load (Zhang et al. 2013). However, the application of such indicator still has certain problems. For example, only a few studies on its application in power load decomposition have been reported, and the analysis of quantitative effects is still based on single weather factors such as temperature (He et al. 2011). Therefore, it is necessary to expand the application areas of composite indicators, and analyze their quantitative impact.

2 DATA COLLECTION AND METHODS

2.1 *Data collection*

Daily data of the maximum power load from 2007 to 2013 were obtained from State Grid Huangshi Power Supply Company, and 2477 samples were collected after quality control. The corresponding meteorological data were obtained from Huangshi Meteorological Bureau, including the data of mean temperature (T:°C), maximum temperature (T_H:°C), minimum temperature (T_L:°C), relative humidity (RH:%), precipitation (r;mm), and average wind speed (V:m/s).

According to the beginning of summer (May, 2) and autumn (September 12) of normal year in Huangshi City, the period between May and September was considered to be "Summer Time" in this study.

2.2 *Method*

Many studies (Vaghefi et al. 2014, Zhang et al. 2013) have shown that the composite index plays an important role in analyzing how weather factors influence the power load. So, the sensible temperature (T_g) and the comfort index (Zhang et al. 2014) were introduced in this study. Especially, the results of these studies were first used to select the appropriate samples to calculate the basic power load, and then the correlations between weather factors (including single factor and sensible temperature) and sensitive power load rate were obtained. Finally, the discussion on the quantitative effects of meteorological elements was provided.

According to the algorithms, the results of the daily sensible temperature and comfort index level from 2007 to 2013 in Huangshi City are shown in Figure 1.

3 DECOMPOSITE THE POWER LOAD

The daily power load (L) considered consists of three parts (Valor et al. 2001):

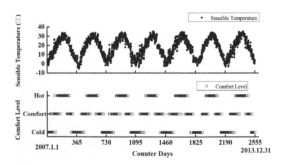

Figure 1. The daily sensitive temperature (up) and comfort index level (down) from 2007 to 2013 in Huangshi City.

$$L = L_t + L_m + \varepsilon \qquad (1)$$

where L_t, L_m and ε refer to the basic power load (depending on the economic situation), weather-sensitive power load, and the random components, including the holiday effect, price impact, and other random factors. In this study, only the holiday effect was considered due to data limitation.

Typically, the basic power load increases linearly with the economic development, that is:

$$L_t = a + bt \qquad (2)$$

where t refers to the sample sequence and the undetermined coefficients a, b can be obtained from the least squares method.

Furthermore, based on the results of L_m and L_t, the sensitive power load rate that was more concerned in power dispatching can be obtained (Liu et al. 2013), that is $R = L_m/L_t$.

The trend of daily power load from 2007 to 2013 is shown in Figure 2. It can be seen from Figure 2 that the daily power load showed a linear growth overall, but decreased between 2008 and 2009 due to the financial crisis. Such a phenomenon was also found in other underdeveloped regions (Liao et al. 2012), so the basic power load should be calculated section by section. In this study, the sensitive power load rate was obtained as follows.

First, all the samples were divided into two categories (weekdays and holidays) to reduce the impact of the holiday effect. Then, the specific period was determined when the daily power load was affected by the financial crisis. Here, we used the M-K test to examine the homogeneity of "hot days" power load data between 2008 and 2009. Since each sample would contain sensitive power load, the systemic changes were reduced. The homogeneity test results (Fig. 3) indicated that two discontinuous points were found at August 25, 2008 and June 13, 2009. So, the period between those two time nodes was considered to be the financial crisis period.

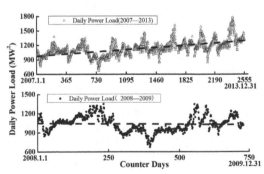

Figure 2. The trends of daily power load from 2007 to 2013 (up) and from 2008 to 2009 (down) in Huangshi City.

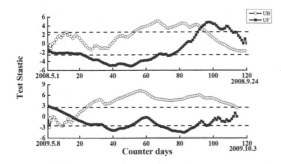

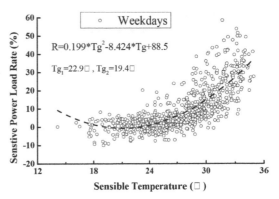

Figure 3. The M-K test of power load from 'hot days' in 2008 (up) and 2009 (down).

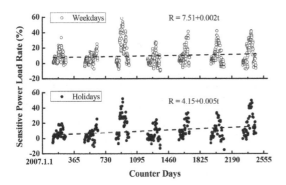

Figure 4. The trends of meteorological load rate from 2007 to 2013 in summer.

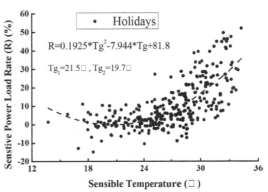

Figure 5. The scatter diagram of the sensible power load rate and the sensible temperature in weekdays and holidays.

Then, the basic power load of each section can be obtained from Eq. 2. Based on the comfort index level, only the power load data of comfort days were selected as statistical samples. So the results indicate how power load changes with economic development without the weather factor effect. As the samples were divided into the weekday category and the holiday category, the random factors ε were also considered. So, the daily sensible power load was calculated from Eq. 1. The daily weather-sensitive power load rate in summer is shown in Figure 4.

4 THE IMPACT OF SENSIBLE TEMPERATURE ON SENSIBLE POWER LOAD RATE

4.1 The threshold of sensible temperature

Generally, the sensible power load would emerge when the temperature was above 26°C. As the sensible temperature was obtained from several weather factors, the threshold would be different.

The scatter plots of the meteorological load rate and the sensible temperature in summer are shown in Figure 5. The nonlinear fitting method was used

to calculate the relationship between the sensible power load rate (R) and the sensible temperature (T_g). From Figure 5, it can be seen that the trends between R and T_g can be described as parabola curves. So, the thresholds of T_g can be obtained while $R = 0$, and the higher value was the threshold, above which would cause the sensible power load.

The thresholds of the sensible temperature between 2007 and 2013 were 22.9°C (weekdays) and 21.5°C (holidays), being somewhat less than 26°C. This difference is attributed to the fact that the maximum temperature (30.9°C) was 4.4°C higher than the mean temperature (26.5°C) during the study period in summer. Since the results of the sensible temperature were calculated based on the mean temperature, the maximum temperature would probably be 4°C higher, and the sensible power would have triggered. Therefore, the threshold value would be less than 26°C.

Furthermore, the air conditioners would continuously be opened during holidays, even at night, while the sensible temperature was not so hot. This is why the holiday's threshold value was higher.

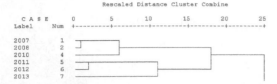

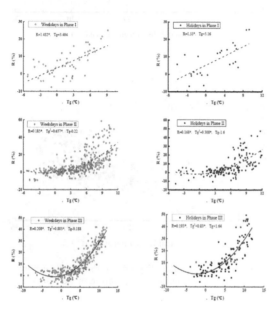

Figure 6. The results of cluster analysis of annual power load from 2007 to 2013 in Huangshi City.

Figure 7. The scatter diagram of the increments of the sensible temperature and sensitive power load rate during different periods.

4.2 The 1°C effect of sensible temperature

According to the results indicated in section 4.1, the sensible temperature above 22.9°C (weekdays) and 21.5°C (holidays) would cause the sensitive power load rate. So, it is more essential to understand the specific value caused by the 1°C sensible temperature variation, that is, the 1°C effect.

Furthermore, as the power load was influenced by the financial crisis, the 1°C sensible temperature variation would lead to different sensitive power load rates during different periods. So, the samples were classified using the cluster analysis method. According to the results shown in Figure 6 and the specific period of financial crisis, all the samples were divided into three categories. Phase I was considered to be the period from August 25, 2008 to June 12, 2009, which reflect the influence of the sensible temperature under the financial crisis. Phase II was considered to be the period from 2007 to 2013 without the financial crisis period, during

which the samples reflect the response of the sensible temperature while the electric load increased slowly. So, the rest of the samples in the period from 2011 to 2013 reflect the impact of the sensible temperature while the power load increases rapidly after the financial crisis.

Then, the fitting equation of R and ΔT_g was solved from the scatter plots shown in Figure 7. It can be inferred from Figure 7 that R increased linearly during the financial crisis period (phase I), while the relationships between R and ΔT_g in phase II and phase III could be described as quadratic parabolic functions. So, the 1°C effect could be obtained by the derivation of the equations.

According to the results, the sensitive power load rate in phase I would increase by 1.432% (weekdays) and 1.33% (holidays), while the sensible temperature increase by 1°C. However, for phase II and phase III, ΔR would increase linearly with the increase of ΔT_g, and change continuously between certain closed intervals. As the highest sensible temperature in this study was 34.9°C, the intervals were considered to be $\Delta R \in [0.644, 4.81]\%$ (weekdays) and $\Delta R \in [1.027, 5.097]\%$ (holidays) for phase II; $\Delta R \in [1.219, 5.795]\%$ (weekdays) and $\Delta R \in [1.226, 6.134]\%$ (holidays) for phase III.

5 CONCLUSION

In this study, we analyzed the trend of daily power load in Huangshi City from 2007 to 2013, and then the impact of weather factors on the sensitive power load rate was discussed. It was found that the daily power load increased linearly from 2007 to 2013 but decreased between 2008 and 2009 due to the financial crisis. So, the sensible temperature and the comfort index were used to decompose the electric load, and a good result was obtained.

Furthermore, the quantitative analysis of how the sensible temperature influenced the sensitive power load rate was carried out. The results indicated that the sensitive load rate would occur at a sensible temperature above 22.9°C (weekdays) or 21.5°C (holidays), and a 1°C variation of sensible temperature would lead to different sensitive power load rates during different periods. As to the financial crisis period, the variation of load rate was considered to be a constant value, but for the other periods, the load rate showed a quadratic change with sensible temperature and up to 6.134% sensitive load rate would be caused by only the 1°C change of sensible temperature.

ACKNOWLEDGMENTS

This work was funded by the Technology Plan Project of Beijing (Z151100005115045) and the Research

Project of Huangshi Meteorological Bureau (201507). The authors thank the staff of State Grid Huangshi Power Supply Company for their assistance.

REFERENCES

Deihimi, A., Orang, O. and Showkati, H. Short-term electric load and temperature forecasting using wavelet echo state networks with neural reconstruction. Energy, 2013, 57: 382–401.

Fikru, MG., Gautier, L. The impact of weather variation on energy consumption in residential houses. Applied Energy. 2015, 144: 19–30.

He, F.F., Jun, S. Impacts of Summer Temperature Variation on Power Load in Shanghai. Resources & Environment in the Yangtze Basin. 2011, 20(12): 1462–1467. (in Chinese).

Hernández, L., Baladrón, C., Aguiar, J. et al. A study of the relationship between weather variables and electric power demand inside a smart grid/smart world framework. Sensors, 2012, 12(9): 11571–11591.

Hu, Z., Bao, Y., Xiong, T. Comprehensive learning particle swarm optimization based memetic algorithm for model selection in short-term load forecasting using support vector regression. Applied Soft Computing. 2014, 25: 15–25.

Liao, F., Congying, X., Yao, J. Load Characteristics of Changde Region and Analysis on Its Influencing Factors. Power System Technology. 2012, 36(7):117–125. (in Chinese).

Liu, H.Y., Liang, C. The Relationship between Power Load and Meteorological Factors with Refined Power Load Forecast in Shanghai. Journal of Applied Meteorological Science. 2013, 24(4): 455–463. (in Chinese).

Selakov, A., Cvijetinović, D., Milović, L. et al. Hybrid PSO–SVM method for short-term load forecasting during periods with significant temperature variations in city of Burbank. Applied Soft Computing. 2014, 16: 80–88.

Vaghefi, A., Jafari, MA., Bisse, E. et al. Modeling and forecasting of cooling and electricity load demand. Applied Energy. 2014, 136: 186–196.

Valor, E., Meneu, V., Caselles, V. Daily air temperature and electricity load in Spain. Journal of applied Meteorology. 2001, 40(8): 1413–1421.

Xia, C., Wang, J. and Mcmenemy, K. Short, medium and long term load forecasting model and virtual load forecaster based on radial basis function neural networks. International Journal of Electrical Power & Energy Systems. 2010, 32(7): 743–750.

Yang, X., Yuan, J., Yuan, J. et al. An improved WM method based on PSO for electric load forecasting. Expert Systems with Applications. 2010, 37(12): 8036–8041.

Zhang, W., A distribution short-term load forecasting based on human comfort index. Power System Protection and Control. 2013, 9: 74–79. (in Chinese).

Zhang, Z.W., Hong, S., Wei, Jiang. The Relationship Between Human Comfort and Mortality from Circulatory System Disease in Nanjing, China. Advances in Climate Change Research. 2014, 10(1): 67–73.

Energy Science and Applied Technology ESAT 2016 – Fang (Ed.)
© 2016 Taylor & Francis Group, London, ISBN 978-1-138-02973-6

The method of long-time integration for radar echo on a shipborne platform

H.W. Cao & M.G. Gao
School of Information and Electronic, Beijing Institute of Technology, Beijing, P.R. China

ABSTRACT: With the development of micro-satellite and the increase of space debris, the detection, tracking and recognition of long-distance or micro space target become particularly important, especially the Specific applications of long-time integration technology on the shipborne motion platform. The piecewise fitting of phase disturbance compensation method based on the ship-swaying phase perturbation model is proposed. The piecewise third-order polynomial fitting results of the ship-swaying data to reduce the integration gain loss because of the phase disturbance; furthermore, the parameters search method can be utilized to compensate the remaining phase error.

Keywords: radar echo; long-time integration; phase compensation; piecewise fitting

1 INTRODUCTION

With the development of micro-satellite and the increase of space debris, the detection, tracking and recognition of long-distance or micro space target become particularly important[1]. The echoes of such space targets always suffer with low signal to noise ratios, and single echo usually cannot reach the radar detection threshold. Thus it is significance to study on long time integration technology[2] in space target detection for pulse instrumentation radar, which not only can support the real-time surveillance and the precise orbit determination, but also can achieve the feature extraction of the long-distance space target and micro space target.

The piecewise fitting of phase disturbance compensation method based on the ship-swaying phase perturbation model is presented to realize the long-time integration technology in space target detection for pulse instrumentation radar on the shipborne motion platform. Since the removal of harmful acceleration components in accelerometer output at the phase center of carrier platform, the accelerations are integrated to get the ship positions. After compensating the ship position information, the phase disturbance error is compensated with piecewise third-order polynomial fitting results of the ship-swaying data to reduce the integration gain loss because of the phase disturbance.

2 SHIPBORNE PLATFORM COORDINATES

The antenna of pulse measurement radar on the shipborne platform swing with the deck together.

Then radar measurement coordinate system and deck coordinate system is consistent, yet they are inconsistent with horizontal coordinate system. Usually the system console send orders according to horizon coordinates, and the Inertial Navigation System (INS) onboard provides the attitude angle based on horizontal coordinate system. The actual operation of shipborne antenna is in the deck coordinate system, so coordinates transformation from horizon coordinates to deck coordinates need to be considered. On the other hand, the actual antenna position detected by angle encoder is in deck coordinate system, and the terminal displays the pointing angle of antenna based on horizontal coordinate system. So transformation from deck coordinate system to horizontal coordinate system needs to be considered too.

Figure 1 shows the relationship of deck cartesian coordinate system (DCCS) $O - X_j Y_j Z_j$ and Horizon Cartesian Coordinate System (HCCS) $O - X_g Y_g Z_g$. Where, OX_g points to north, OY_g is upward along the plumb line, OZ_g is determined by right-hand rule. OX_j is the line between bow and stern of boat, and the bow points positive, OY_j is vertical to the deck plane, and the upward direction is positive, OZ_j is determined by right-hand rule.

The destabilization the ship suffered include that the sway, the surge and the heave of three position parameters, and similarly, the rolling, pitching and yawing of three attitude parameters. As shown in Fig. 1, the yawing angle K is the angle in XOZ plane between OX_j and OX_g, and the clockwise direction is positive, the pitching angle P is the angle in XOY plane between OX_j and OX_g, and the upward bow is positive, the rolling angle

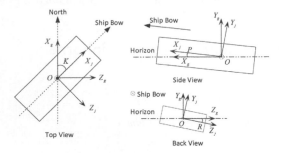

Figure 1. The relationship of DCCS and HCCS.

R is the angle on YOZ plane between OZ_j and OZ_g, the downwards starboard is positive.

3 CHARACTERISTICS OF SHIPBORNE MOTION PLATFORM

3.1 *Ship position information*

The ship position is constantly changing when the measurement devices onboard capture the tracking targets, the accelerometer installed on the three-axis center of platform can be used to obtain the ship position[3]. Two horizontal accelerometers are installed to measure the east acceleration and the north acceleration. The outputs of accelerometers contain harmful acceleration components, such as Coriolis acceleration and centrifugal acceleration. The harmful acceleration components can be compensated by the measurements of harmful accelerometer[4].

Suppose that the acceleration obtained by accelerometers include the East acceleration A_E and North acceleration A_N. The velocity component can be obtained by integrating A_E, A_N and adding the initial velocity V_{E0}, V_{N0}.

$$\begin{cases} V_E = \int_0^t A_E dt + V_{E0} \\ V_N = \int_0^t A_N dt + V_{N0} \end{cases} \quad (1)$$

Transform and integrate the velocity component V_N and V_E respectively, add to the initial latitude and longitude, the position of the carrier platform is

$$\begin{cases} L = 1/r_e \int_0^t V_E \sec B dt + L_0 \\ B = 1/r_e \int_0^t V_N dt + B_0 \end{cases} , \quad (2)$$

where, L and B are the latitude and longitude of the position of the moving platform, L_0 and B_0 are the latitude and longitude of the platform at the initial time, r_e is the radius of the earth.

The harmful acceleration components are not removed as the formula (1), it is discussed further below. For the measurement ship, the relationship of acceleration A_E, obtained by the accelerometer and the accelerations $\dot{V}_E, \dot{V}_N$ of the horizontal movement of platform relative to Earth's surface is

$$\begin{cases} \dot{V}_E = A_E - A_{BE} \\ \dot{V}_N = A_N - A_{BN} \end{cases} , \quad (3)$$

$$\begin{cases} A_{BE} = 2\omega V_N \sin B + V_E V_N / r_e \cdot \tan B \\ A_{BN} = -2\omega V_E \sin B - V_E^2 / r_e \cdot \tan B \end{cases} , \quad (4)$$

where, ω is the angular velocity of Earth rotation, A_{BE} and A_{BN} are harmful acceleration components, the first term is Coriolis acceleration, the second term is the centrifugal, acceleration. B, V_E and V_N are the true value, it is difficult to obtain, it can be approximated by the real time measurement value B_c, V_{Ec} and V_{Nc}.

3.2 *The ship-swaying data*

The offshore measured data is used as the ship-swaying data for analysis, the duration of 4500 seconds. Fig. 2 shows that the maximum amplitude of pitching, rolling, and yawing is about 0.5°, 2° and 2° respectively. The waveform of ship-swaying can be seen approximately as the superposition of sine waveforms, and the amplitudes, periods and initial phases of these sine waveforms vary continuously because of external disturbances.

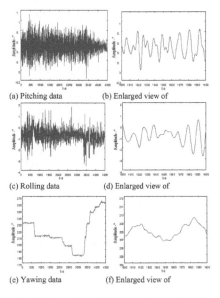

(a) Pitching data   (b) Enlarged view of

(c) Rolling data   (d) Enlarged view of

(e) Yawing data   (f) Enlarged view of

Figure 2. Schematics of ship-swaying period of measured data.

4 COMPENSATION METHOD OF PIECEWISE FITTING PHASE DISTURBANCE BASED ON SHIP-SWAYING DATA

4.1 The echo phase model with ship-swaying disturbance of space target

Figure 3 shows the INS Horizontal Coordinate System (IHCS), the origin O is the intersection of three-axis in the inertial navigation platform, the axis OX_g points to the north at the local level, the axis OY_g is along the local plumb line, the axis OZ_g is determined by the right-hand rule.

In INCS, the coordinates of phase center of radar is (x_r, y_r, z_r). Suppose that the hull rolling is R (radian), the hull pitching is P(radian) and the

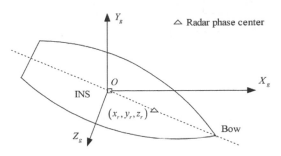

Figure 3. The schematic of phase center of radar and IHCS.

$$M_X(R) = \begin{bmatrix} 1 & 0 & 0 \\ 0 & \cos(R) & \sin(R) \\ 0 & -\sin(R) & \cos(R) \end{bmatrix} \quad (7)$$

$$
\begin{aligned}
M(R,P,K) &= M_X(R)M_Z(P)M_Y(-K) \\
&= \begin{bmatrix} \cos P\cos K & \sin P & \cos P\sin K \\ -\cos R\sin P\cos K - \sin R\sin K & \cos R\cos P & -\cos R\sin P\sin K + \sin R\cos K \\ \sin R\sin P\cos K - \cos R\sin K & -\sin R\cos P & \sin R\sin P\sin K + \cos R\cos K \end{bmatrix}
\end{aligned} \quad (8)
$$

hull yawing is K(radian). Suppose that DCCS and HCCS are coincident at the initial state, DCCS rotates due to ship-swaying, and DCCS rotates around the axis OY_g, OZ_g and OX_g by $-K$, P and R in order respectively. The rotation matrix is expressed in formula (11)[5][6].

After rotating the coordinates of phase center of radar (x_r, y_r, z_r) can be expressed as[7]

$$\begin{bmatrix} x_{r1} \\ y_{r1} \\ z_{r1} \end{bmatrix} = M(K,P,R)\begin{bmatrix} x_r \\ y_r \\ z_r \end{bmatrix}$$

$$
\begin{aligned}
&= \begin{bmatrix} \cos P\cos K & \sin P & \cos P\sin K \\ -\cos R\sin P\cos K - \sin R\sin K & \cos R\cos P & -\cos R\sin P\sin K + \sin R\cos K \\ \sin R\sin P\cos K - \cos R\sin K & -\sin R\cos P & \sin R\sin P\sin K + \cos R\cos K \end{bmatrix}\begin{bmatrix} x_r \\ y_r \\ z_r \end{bmatrix} \\
&= \begin{bmatrix} x_r\cos P\cos K + y_r\sin P + z_r\cos P\sin K \\ x_r(-\cos R\sin P\cos K - \sin R\sin K) + y_r(\cos R\cos P) + z_r(\sin R\cos K - \cos R\sin P\sin K) \\ x_r(\sin R\sin P\cos K - \cos R\sin K) - y_r\sin R\cos P + z_r(\sin R\sin P\sin K + \cos R\cos K) \end{bmatrix}
\end{aligned}
$$

$$M_Y(-K) = \begin{bmatrix} \cos(-K) & 0 & -\sin(-K) \\ 0 & 1 & 0 \\ \sin(-K) & 0 & \cos(-K) \end{bmatrix} \quad (5)$$

$$M_Z(P) = \begin{bmatrix} \cos(P) & \sin(P) & 0 \\ -\sin(P) & \cos(P) & 0 \\ 0 & 0 & 1 \end{bmatrix} \quad (6)$$

The disturbance of phase center of radar can be expressed as

$$\begin{cases} \delta x_r = x_{r1} - x_r \\ \delta y_r = y_{r1} - y_r \\ \delta z_r = z_{r1} - z_r \end{cases} \quad (9)$$

The disturbance vector of the phase center is $\mathbf{r_p} = (\delta x_r, \delta y_r, \delta z_r)$. Assume that the direction of the imaginary target is the unit vector

653

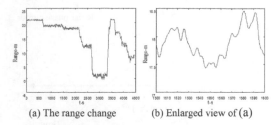

| (a) The range change | (b) Enlarged view of (a) |

Figure 4.   The range change of the measured data.

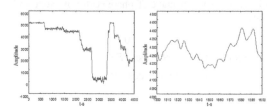

Figure 5.   The phase of the measured data.

$\mathbf{r_n} = (x_o, y_o, z_o)$, the slant range of ship-swaying disturbance is expressed as

$$\Delta r_c = \langle \mathbf{r_p}, \mathbf{r_n} \rangle = \delta x_r x_o + \delta y_r y_o + \delta z_r z_o \qquad (10)$$

The phase perturbations caused by ship-swaying is expressed as

$$\Delta \varphi_c = 4\pi f_{RF} \left(\delta x_r x_o + \delta y_r y_o + \delta z_r z_o \right) / c \qquad (11)$$

where, f_{RF} is the carrier frequency, c is speed of light.

Figures 4 and 5 shows the range change and the phase disturbances caused by ship-swaying of off-shore measured data.

From the figures above, the curve of phase disturbances can be seen as the superposition of sine waves. So polynomial fitting method can be used for phase fitting to compensate phase perturbations.

4.2   *Analysis on piecewise fitting for the phase perturbation*

The phase error will cause the integration losses when detecting targets. Therefore, the integration losses can be reduced by phase compensation, and the phase error is eliminated a lot by piecewise fitting. The influence of data length used to piecewise fitting and the fitting order on the fitting results are analyzed in turn in following.

The higher fitting order is, the smaller the integration gain loss is; the shorter data length for fitting is, the smaller the integration gain loss is. Fig. 6 shows the mean of integration gain loss by

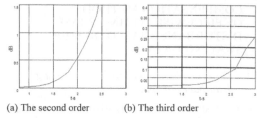

| (a) The second order | (b) The third order |

Figure 6.   The mean loss of long-time integration gain as different date length.

the second order fitting and the third order fitting using different data lengths. Based on the analysis above, the data with ship-swaying can meet the demand on integration detection of radar echo by phase compensation three times.

4.3   *Phase disturbance compensation under long-time integration*

For the ground-based pulse instrumentation radar with non-movement platform, phase compensation factor needs to be constructed for the long-time integration of space target echo. The phase compensation factor φ_n can be expressed as

$$\varphi_n = 4\pi f_{RF} \left(r_0 + v_n t_n + \frac{1}{2} a_n t_n^2 \right) \Big/ c \qquad (12)$$

According to the analysis above, for shipborne platform, the phase disturbance of echo caused by ship-swaying needs to be operated on phase compensation by piecewise fitting based on formula (16). From the conclusions above we can see that third order fitting can fulfil the requirements of echo integration detection. Assuming that each order coefficient of third order fitting are $\{b_{n0}, b_{n1}, b_{n2}, b_{n3}\}$ respectively during a certain integration time, then the fitted phase φ_c can be approximately expressed as

$$\varphi_c = 4\pi f_{RF} \left(b_0 + b_1 t_n + b_2 t_n^2 + b_3 t_n^3 \right) / c. \qquad (13)$$

In summary, for shipborne pulse instrumentation radar on moving platform, the technique of long-time integration is adopted, the total phase φ_A which needs to be compensated is

$$\begin{aligned}
\varphi_A &= \varphi_n + \varphi_c \\
&= 4\pi f_{RF} \left[(r_0 + b_0) + (v_n + b_1) t_n \right. \\
&\quad \left. + (a_n/2 + b_2) t_n^2 + b_3 t_n^3 \right] / c
\end{aligned} \qquad (14)$$

The error maybe appeared between φ_A and the echo phase because of the second order error of orbit reference model and the phase fitting error. The second order phase term will cause Doppler defocus and leads to the degradation on detection performance. The method of joint parameters search can be adopted to compensate the remaining first order and second order phase term, then the energy of echo signal can achieve self-focus.

5 CONCLUSION

The details in the technique of long-time integration for pulse instrumentation radar on shipborne movement platform are discussed in this paper. First, the transformation relationships between the deck coordinate system and the horizontal coordinate system used in target acquisition, tracking, and measuring are determined. The acquisition and compensation of ship position are analyzed based on the motion characteristics of the carrier platform. The characteristics of ship-swaying data is analyzed according to the measured data, then the echo phase with ship swaying disturbance is modeled. The phase disturbance can be compensated by INS data, but it cannot be fully compensated because of the error of INS data. Suppose that the phase disturbances caused by the ship rolling, pitching and yawing are the surplus term after compensation, the effect of the polynomial fitting on phase disturbance is further analyzed. Finally, the piecewise fitting compensation algorithm on phase disturbance is proposed based on ship swaying data to realize the space target detection by the technique of long-time integration for pulse instrumentation radar on shipborne motion platform.

REFERENCES

Chen, X. et al. 2013. Maneuvering Target Detection via Radon-fractional Fourier Transform based Long-time Coherent Integration[J]. IEEE Transactions on Signal Processing: 939–953.

Han, J.L. 2012. Engineering software design for shipborne ISAR detection signal processor [D]. Master's degree thesis, Harbin Institute of Technology.

Jian, S.L. et al. 2009. Introduction to maritime control technology for TT&C ship [M]. National Defense Industry Press.

Krepon, M. & Katz-Hyman, M. 2005. Space Weapons and Proliferation[J]. Nonproliferation Review 12(2): 323–341.

Li, X.Y. et al. 2005. Timing analysis and processing method of TT&C ship attitude data [J]. Manned Space Flight.

Pan, L. 2009. Measurement technology of TT&C ship attitude and position [M]. National Defense Industry Press.

Yang, L. 2007. Study on methods of the measurement data error separation of TT&C ship [D]. Master's degree thesis, National University of Defense Technology.

Energy Science and Applied Technology ESAT 2016 – Fang (Ed.)
© 2016 Taylor & Francis Group, London, ISBN 978-1-138-02973-6

Experimental study on the in-plane shear properties of Kevlar fabric reinforced epoxy composite material

Jian Zhao, Hai Wang & Longquan Liu
School of Aeronautics and Astronautics, Shanghai Jiao Tong University, Shanghai, China

ABSTRACT: In this paper, two kinds of shear test methods including ±45° tension test and Iosipescu V-Notched beam test are utilized to obtain the in-plane shear properties of a Kevlar fabric reinforced epoxy composite material. Static mechanical tests are performed according to the ASTM standards (ASTM D3518 and ASTM D5379) respectively. Test results from the two patterns of specimens according to different test methods are compared. The shear responses of the two patterns of specimens both present obvious nonlinear behavior under in-plane shear loading like carbon fiber reinforced composite material. The test data both present good uniformity, and the comparison results show that the average shear modulus from the Iosipescu V-Notched beam test is 11.7% higher than the average shear modulus from the ±45° tension test, the average shear strength from the V-Notched Iosipescu beam test is 8.2% higher than the average shear strength from the ±45° tension test. The results can be a reference and baseline for further study on the in-plane shear test methods for Kevlar fabric reinforced epoxy composite materials.

Keywords: shear property, Iosipescu specimen, Kevlar fabric

1 INTRODUCTION

Composite materials have been used widely in aerospace and other structures due to their various advantages such as high specific strength, high specific stiffness, better damage tolerance, etc. compared with traditional light alloys. Particularly, Kevlar aramid fabric reinforced composite materials are used in high velocity impact applications such as soft body armor due to the excellent energy absorption properties and high specific strength and specific modulus especially in aerospace applications like helicopter Anti-Prang apparatus (Subramani Sockalingam, 2014). Unlike metals, fiber reinforced composite laminates always present orthotropic mechanical properties. The mechanical properties such as tension modulus and tension strength, shear modulus and shear strength, etc. must be acquired before structure design start, which can be obtained through performing mechanical tests according to responding test methods. Among the typical mechanical properties of composite laminates, in-plane shear properties are quite important but difficult to obtain by test because that pure shear condition is very hard to realize experimentally (Morozov E V, 2013, G. Odegard, 2000). The requirement for determining the shear properties of composite laminates has motivated many researches on the test approaches.

Presently the most common test methods for in-plane shear behavior include tension test method using ±45° laminate specimens and V-notched beam method using Iosipescu specimen (H. Ho, 1993). Most of previous experimental studies are about carbon fiber reinforced composite materials, however, for Kevlar fabric reinforced composite materials, the difference of shear properties from different test methods is not quite clear [2–4]. In this paper, two kinds of shear methods including ±45° tension test and Iosipescu V-Notched beam test according to ASTM D3518 and ASTM D5379 standard respectively are utilized to obtain the in-plane shear properties of a Kevlar fabric reinforced epoxy composite material. And the in-plane shear properties acquired from the two test methods adopting different types of specimens are compared to check the difference of the two methods on determining the in-plane shear properties for Kevlar fabric reinforced composite materials.

2 TEST SPECIMENS AND DIMENSIONS

The ±45° tension specimen according to ASTM D3518 standard and the Iosipescu V-Notched beam specimen according to ASTM D5379 standard are shown in Fig. 1 and Fig. 2 respectively, and the responding geometry information is

250

25

2,15

Figure 1. Geometry of the +45° tension specimen with strain gage location.

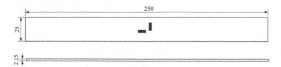

250

25

2,15

Figure 2. Geometry of Iosipescu V-Notched beam specimen with strain gage location.

indicated in the figures. Both kinds of the specimens used in this research are made from Kevlar fabric reinforced epoxy composite material. The stacking sequence of the ±45° tension specimen is $[(+45/-45)]_{6S}$, and the stacking sequence of the Iosipescu V-Notched beam specimen is $[(0/90)]_{10S}$. Strain gages are mounted on the specimen surface to record the strain data during the loading process. For the ±45° tension specimen, two strain gages are placed at the centroid of the plate at each side, with one gage parallel with the loading direction and the other gage perpendicular to the loading direction as shown in Fig. 1. For the Iosipescu V-Notched beam specimen, also two strain gages are mounted on the surface of the specimen at each side, with one strain gage mounted at +45° to the loading axis and the other one mounted at −45° to the loading axis, as shown in Fig. 2.

3 EXPERIMENTAL SETUP

Both of the static tests for the ±45° tension specimen and Iosipescu V-Notched specimen are performed on a MTS SANS 5105 electronic universal testing machine with a loading rate of 2 mm/min. For each pattern of specimen, Strain gages with a resistance of 120Ω and a length of 6.4 mm which is large enough compared with the Kevlar fabric unit cell are adopted. The test setups for the two test methods for measuring the in-plane shear properties of the responding specimens are shown in Fig. 3. For the Iosipescu V-Notched specimen, a test fixture produced by Wyoming Company is used to introduce shear loading into the specimen by applying compression load on the top head of the fixture, as shown in Fig. 3(a). For the ±45° tension specimen, shear load is directly introduced by the test machine through the two gripping heads.

(a) Setup for the Iosipescu V-Notched beam specimen
(b) Setup for the ±45° tension specimen.

Figure 3. Experimental setups for the measurement of in-plane shear properties.

The force and strain data is recorded automatically during the test process.

4 RESULTS AND DISCUSSION

The load versus displacement curves for both the ±45° tension specimen and Iosipescu V-Notched beam specimen can be seen in Fig. 4 and Fig. 5 respectively. As indicated in the figure, the load versus displacement curves present good consistency. A comparison of shear stress versus shear strain curves between the ±45° tension specimen and

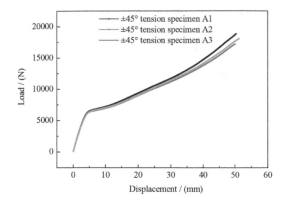

Figure 4. Load versus displacement curves for the ±45° tension specimen.

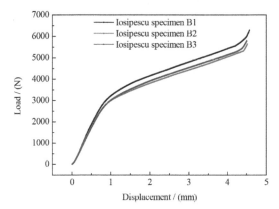

Figure 5. Load versus displacement curves for the Iosipescu V-Notched beam specimen.

Iosipescu V-Notched specimen is given in Fig. 6, both kinds of specimens present obvious nonlinear mechanical response under in-plane shear loading like carbon fiber reinforced composite laminates. As shown in Fig. 6, after the mechanical response process pass the linear stage, the shear stress versus strain curves of Iosipescu V-Notched beam specimen are obviously lower than the curves of ±45° tension specimen, indicating that the shear stress of Iosipescu V-Notched beam specimen increases slower than the ±45° tension specimen after the linear stage as the shear strain grow.

Shear properties including shear modulus, shear strength and maximum shear stress are computed from the test results for both kinds of specimens are given in Table 1. As can be read from the coefficient of variations of the data that all data present good consistency. It must be mentioned that the average shear modulus and average shear strength from the Iosipescu V-Notched beam specimens are higher than them from the ±45° tension specimens.

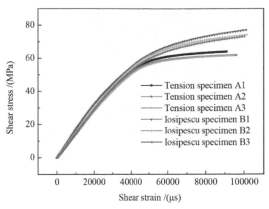

Figure 6. Shear stress versus shear strain curves for both kinds of specimens.

Table 1. Shear properties computed from the test results for both kinds of specimens.

Specimen	Shear modulus (GPa)	Shear strength (MPa)	Ultimate stress (MPa)
±45° Tension specimen	1.45	56.3	176.1
C.V. for ±45° tension specimen	1.1%	1.3%	4.7%
Iosipescu specimen	1.62	60.9	135.6
C.V. for Iosipescu specimen	6.7%	1.6%	5.3%

The average shear modulus computed from Iosipescu V-Notched beam specimen test result is 11.7% higher than that from the ±45° tension specimen test result, the average shear strength computed from Iosipescu V-Notched beam specimen test result is 8.2% higher than that from the ±45° tension specimen test result.

5 SUMMARY

In this paper, two kinds of in-plane shear test methods are utilized to measure the in-plane shear properties of a Kevlar fabric reinforced epoxy composite material. Two patterns of specimens including ±45° tension specimens according to ASTM D3518 standard standard and Iosipescu V-Notched beam specimen according to ASTM D5379 standard are adopted to measure the in-plane shear response. Comparisons are made to check the difference of the test results from the two patterns of specimens. Both of the two patterns of specimens made from Kevlar fabric reinforced composite material present obvious nonlinear shear behavior under shear loading like carbon fiber reinforced composite laminates. Comparison results show that the average shear modulus computed from the Iosipescu V-Notched beam specimen test results is 11.7% higher than that from the ±45° tension specimen test results, the average shear strength from the Iosipescu V-Notched beam specimen test results is 8.2% higher than that from the ±45° tension specimen test results. The test results and research can be a reference and baseline for further study on the in-plane shear test methods for Kevlar fabric reinforced epoxy composite materials.

REFERENCES

ASTM Standard D3518/D3518M, Standard test method for in-plane shear response of polymer matrix composite materials by tensile test of a ±45° laminate, Annual Book of ASTM Standards, 2001, 15.03.

ASTM Standard D5379/D5379M, Standard Test Method for Shear Properties of Composite Materials by the V-Notched Beam Method, Annual Book of ASTM Standards, 2000, 15.03.

Ho H., M.Y. Tsai, J. Morton and G.L. Farley. A comparison of three popular test methods for determining the shear modulus of composite laminates. Composite Engineering, 1993, 3: 69–81.

Morozov EV., Vasiliev VV. Determination of the shear modulus of orthotropic materials from off-axis tension tests. Composite Structures, 2003, 62: 379–382.

Odegard G., M. Kumosa. Determination of shear strength of unidirectional composite materials with the Iosipescu and 10° off-axis shear tests. Composite Science and Technology, 2000, 60: 2917–2943.

Subramani Sockalingam, John W. Gillespie and Jr. Michael Keefe. On the transverse compression response of Kevlar KM2 using fiber-level finite element model. International Journal of Solids and Structures, 2014, 51: 2504–2517.

Energy Science and Applied Technology ESAT 2016 – Fang (Ed.)
© 2016 Taylor & Francis Group, London, ISBN 978-1-138-02973-6

Design of the fault diagnosis terminal of an inertial navigation system based on ZigBee wireless communication technology

Zhilu Zhang, Guodong Li, Xiangjin Wang & Chaofeng Zhao
Wuhan Mechanical Technology College, Wuhan, Hubei, China

ABSTRACT: Aiming at the problems of high technological content, complicated structure and poor accessibility for detection and maintenance, we design the fault diagnosis terminal based on the ZigBee wireless communication technology to achieve rapid fault detection and diagnosis of the inertial navigation system. This terminal, according to the equipment workflow, employs the combination of passive and active tests as well as the reconfiguration technology of the built-in test mechanism, utilizes the established fault database for analysis of real-time dynamic information of the acquisition device, further completes the fault diagnosis of the inertial navigation system and realizes its intelligent and rapid testing.

Keywords: ZigBee; Inertial Navigation System (INS); fault diagnosis

1 INTRODUCTION

The Inertial Navigation System (INS) can provide information such as coordinate, azimuth and elevation of the carrier. Because of its fast and accurate navigation, good concealment and good environmental adaptability, the INS is widely used in large ships and aircraft etc. Since it has high technological content, complicated structure, poor accessibility for detection and maintenance, and too many faults in the system working process and invisible fault diagnosis, once the system is damaged, it is difficult to quickly determine the fault positions and therefore its normal functions in equipment are affected.

In this paper, due to the complicated structure and poor accessibility for detection and maintenance, we design a type of portable fault diagnosis terminal, which comprehensively employs the ZigBee wireless communication technology, passive and active testing technologies and intelligent fault diagnosis technology, controls the INS work based on wireless communication for real-time acquisition of internal dynamic data, uses the database for comparison and further completes the rapid diagnosis of the INS.

2 DESIGN OF FAULT DIAGNOSIS TERMINAL

2.1 Hardware system design

The fault diagnosis system is mainly made up of the transfer and test module, ZigBee wireless communication platform and portable fault diagnosis terminal. The fault diagnosis terminal employs the wireless communication and sends command to the INS through the transfer and test module to control the working state of the device. When the device is working, the transfer and test module can collect real-time data and uses the ZigBee wireless communication platform to upload the collected data to the fault diagnosis terminal. The internal software system processes the data and then completes the fault diagnosis of the INS.

2.1.1 Transfer and test module

The transfer and test module is the key part to achieve the data acquisition and device control with the following functions. First, it has the transfer function to ensure the normal original signal excitation and transfer. Second, reconfigure the built-in test mechanism of the tested device and with the combination of passive static and active dynamic tests, effectively prevent the reoccurrence of secondary faults. Third, it has the signal acquisition function and classifies and codifies the characteristic information of important signals, to achieve the preliminary processing f test information. Fourth, it

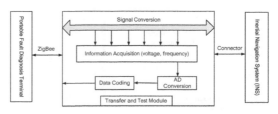

Figure 1. Schematic diagram of the transfer and test module.

has the wireless data transmission function, which can enable ZigBee wireless communication through wireless transceiver module and portable fault diagnosis terminal, and realize the digital diagnosis of the INS. The module comprises the A/D conversion circuit, acquisition circuit, main control unit circuit, ZigBee wireless communication circuit etc.

2.1.2 *ZigBee wireless communication platform*

The ZigBee wireless communication platform mainly provides communication service for the fault diagnosis terminal and the INS. To reduce cost and improve stability, we use the CC2530 wireless module to build the ZigBee wireless communication platform. Because the wireless sensor network built by the CC2350 chip needs only few external components and the compatible components are all low-cost and can support the rapid construction of wireless network, so it supports IEEE802.15.4 standard/ZigBee/ZigBee RF4 application (Li Zhirui, 2015).

Here, according to the diagnosis need of devices, we design the networking mode of 1 gateway + 3 nodes, as shown in Figure 2. The specific process is as follows. When the monitoring child node initiates for the first time after the deployment, it firstly sends the I-frame and then network connection request frame. The surrounding routing nodes respond with a corresponding beacon frame after receiving the network access request frame. The monitoring child node determines which routing node it belongs to on the basis of its own choice (the principle for selection is integrating the distance and strength of signals to the coordinator). If under the monitoring child node, it sends the acknowledgement frame to the routing node. If the monitoring child node does not receive a response from the routing node, after sleep, it regularly transmits the I-frame and network access request frame. Once it adds to some routing node, the network access request frame may be sent once for a longer time, but the I-frame will still be sent regularly. The monitoring child node records its parent routing node. If the monitoring child node receives the response information from many routing

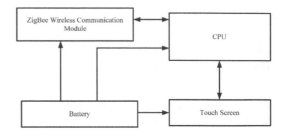

Figure 3. Architecture of portal fault diagnosis terminal.

nodes, one more routing node must be recorded as standby to avoid that there is no means to report data in case of parent node failure.

2.1.3 *Portable fault diagnosis terminal*

The portable fault diagnosis terminal as the core component of the INS fault diagnosis device, mainly achieves ZigBee networking control, wireless data communication, data analysis and processing, and Human-Computer Interaction (HCI), with the characteristics of light weight, small size and friendly HCI. It consists of the main control unit, the ZigBee wireless communication module, battery, and display screen etc., as shown in Fig. 3.

STM332F103 is the core part of the main control unit that achieves the communication data processing and works with the programming driver display screen and wireless module. ZigBee wireless communication module is used as a gateway for wireless communication with 3 nodes, and it also uses the CC2530 module produced by TI. The battery supplies +5V voltage to the main control unit, ZigBee wireless communication module and display screen, and generates 3.3V voltage through the voltage conversion circuit. Meanwhile, the battery charge management circuit is applied for the charge management.

2.2 *Software system design*

The fault diagnosis system software is the core part of the command and control of the whole system. The system fault detection and diagnosis need being controlled and processed by software. The entire software is composed of the test module software, ZigBee wireless communication software and portable fault diagnosis terminal software.

2.2.1 *Test module software*

According to the characteristics of the circuit under test, the test modules are divided into the passive and active tests by a certain time sequence. The specific process flow is shown in Fig. 4.

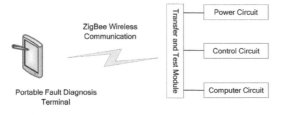

Figure 2. Network architecture of ZigBee wireless communication platform.

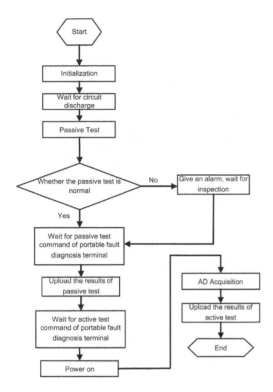

Figure 4. Flow chart of test module software.

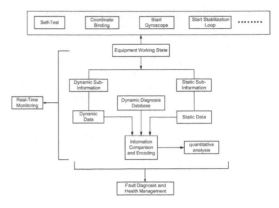

Figure 5. Flow chart of portable fault diagnosis terminal software.

2.2.2 ZigBee wireless communication software

The core chip of the wireless network node is CC2530 that has limited storage capacity, so we need to select IAR Embedded Workbench as the development environment for CC2530 chip-related software to achieve integrated download and debugging of related programs. ZigBee wireless communication software design must follow the Z-Stack software architecture. The Z-Stack protocol stack is based on a polling operating system, which may schedule the task through embedding an operating system to complete the functions at each layer and form the sequential scheduling mechanism of "event-task-operating system". In this way, the priority may be put in an important position. When the system is running, first make a judgment on which tasks will occur and implement them after automatically judging their priority. This ensures that the event with the highest priority in the task can be firstly executed (Sun Caiyun et al, 2010).

The system start needs to prepare all modules necessary for initializing the hardware platform and software architecture and be ready for running of the operating system. The operating system programs begin to execute after initialization. The operating system totally must accomplish 6 tasks, respectively at the MAC layer, network layer, hardware abstrac-

tion layer, application layer, ZigBee application layer and the application layer that can be completely processed by user. Their priority ranges from high to low, namely MAC layer has the highest priority and the user layer has the lowest priority.

2.2.3 Portable fault diagnosis terminal software

The portable fault diagnosis terminal software is programmed by C language. It can use the Zigbee wireless communication technology for sending commands to the INS to control its normal work, and real-time acquisition and processing of key signals output by the internal circuit board of the INS, so as to achieve the fault diagnosis.

The software adopts the equipment-oriented workflow and the fault phenomenon-based dynamic fault diagnosis strategy and uses the Zigbee wireless communication technology for real-time acquisition of the signal characteristic and law of change of the INS at each working state. The signals are divided into the static sub-information (no change to the characteristic) and the dynamic sub-information (signal characteristics change depending on operating conditions), which are coded and uploaded to the fault diagnosis software. The two types of information are compared with the database to determine whether or not the signal is normal, obtain the qualitative conclusion, and then complete the fault diagnosis and health management of the device.

3 FAULT DIAGNOSIS TERMINAL TESTING EFFECT ANALYSIS

Table 1 shows the testing result of some circuit in the INS by the fault diagnosis terminal. The interface includes the established database, detected value, working state and diagnosis result. It is 0.5h from the start of test to the display of result. The

Table 1. Testing result of azimuth holding instrument.

Circuit board	Pin	Definition	Nominal value	Finding, 1st position (detected value)	Working state	Result
Servo circuit board	B11	XLH	100 Hz, peak-peak value ±13V, PWM wave	00000	North finding abnormal	Fault of servo circuit board
	C11	XLL		00000	North finding abnormal	
	B12	YLH	100 Hz, peak-peak value ±13V, PWM wave	00000	North finding abnormal	Fault of servo circuit board
	C12	YLL		00000	North finding abnormal	

result is displayed as: north finding is abnormal, and the servo circuit board malfunctions, consistent with the actual situation. Test result shows that the fault diagnosis can locate the internal unit circuit, solve the problem of poor accessibility of INS detection and maintenance and realize rapid location of fault.

4 CONCLUSION

In this paper, due to the problems of complicated structure and poor accessibility for detection and maintenance, we design the INS fault diagnosis terminal based on the ZigBee wireless communication technology, which, according to the device workflow, adopts the combination of passive and active tests, uses the established fault database for acquisition and judgment of real-time dynamic data, and thus realizing the rapid diagnosis of the inertial navigation system.

REFERENCES

Li Zhirui. 2015. Design and Verification of Wireless MEMS Acceleration Sensor Based on CC2530. Internet of Things Technologies.

Sun Caiyun 2010. et al. Design of Devices Monitoring System Based on ZigBee Technology. Water Resources and Power.

Yu Jiangli 2014. et al. Research of Fire Fighting Equipment Power Monitoring System Based on ZigBee Technology. Chinese Journal of Power Sources.

Energy Science and Applied Technology ESAT 2016 – Fang (Ed.)
© 2016 Taylor & Francis Group, London, ISBN 978-1-138-02973-6

High-voltage circuit breaker online monitoring and fault diagnosis technology

Weihua Niu & Jing Zhang

School of North China Electric Power University, Baoding, China

ABSTRACT: High-voltage circuit breaker protection and control plays an important role in the normal operation of the grid. Its online monitoring and fault diagnosis has important significance. This paper introduces several common faults of high-voltage circuit breakers. Then, it introduces more advanced online monitoring techniques to provide technical support for fault diagnosis of high-voltage circuit breakers.

Keywords: high-voltage circuit breaker; online monitoring; fault diagnosis

1 INTRODUCTION

High-voltage circuit breaker as a grid of important equipment, protective and control action in the power grid, power stations, and substations are essential devices. It can not only cut off the load current and load current of the high-voltage circuit, but also, when the system fails, provide protection by the operation of the automatic device fitted, and the fault current can be quickly removed to reduce the power range and prevent the accident. At present, China's power system for online monitoring of key generators, transformers and other large electrical equipment, online monitoring of high-voltage circuit breaker fault diagnosis technology is still in the research and testing phase. Therefore, this paper analyzes the causes of high-voltage circuit breakers' common type of fault and cause. On this basis, the current fault diagnosis for online monitoring of high-voltage circuit breakers is used.

2 HIGH-VOLTAGE CIRCUIT BREAKERS' COMMON TYPE OF FAULT

According to statistics, the mechanical portion comprises about 64.8% and the operating mechanism of the total secondary circuit failure, followed by electrical faults, which accounts for the vast majority of the cause of the failure. Visible, monitoring, and diagnosis of mechanical failure accounted for a very important position in the high-voltage circuit breaker online monitoring. Common faults of high-voltage circuit breakers are described below.

2.1 *Refused action fault*

Breaker in the control loop has been given the opening (or closing) directive, but the opening or closing operation of the actuator is not carried out, causing the circuit breaker tripping to produce failure. Its causes include the following aspects (Bai Zhiyan, 2013):

1. Machinery. Hydraulic circuit includes a hydraulic mechanism with low oil pressure that leads to the electric control circuit lockout, hydraulic actuator control valve system failure, and opening and closing the solenoid valve stem. Although already started, a deformed valve is not opened or closed due to hydraulic actuator fault institutions, or pneumatic actuator air circuit and its component failure and spring mechanism of component failure.
2. Electrical. Maybe the operation control circuit voltage DC power declined sharply, though already make points (combined) brake solenoid coil power. However, on the low voltage terminal, we cannot start the electromagnet due to open fault in the control loop, disconnection of the open and close solenoid assembly coil, open circuit, poor contact or broken wiring fault removal, etc. in the control loop.
3. Circuit breaker body and the intermediate transmission fault. As interrupter dynamic and static contact are not adjusted properly, the movable contact insulated rod (or rod) and the transmission mechanism connecting shaft lockout, the movable contact insulated rod is damaged (broken).

2.2 *Malfunction fault*

Normal work (including standby) circuit breaker in the absence of its action instructions issued from time to time. Opening or closing operation occurs due to opening or closing error. The main reason for this failure is that actuator carve closing electromagnet voltage is low, when the operation control circuit in the DC system is grounded, circuit breaker has error in opening or closing, as well as sub-closing solenoid actuator latch into the buckle. There are outages during vibration, so that opening and closing are not operated on their own. Export operation is performed to protect the relay contact and automatic devices. The operating control loop is closed for some reason during the operation or standby.

2.3 *Insulation fault*

Insulation faults include internal insulation to ground arc-over breakdown fault, external insulation to ground arc-over breakdown fault, phase insulation arc-over breakdown fault, caused by lightning over-voltage arc-over breakdown, insulated rod arc-over, porcelain bushing, capacitor bushing, current transformer arc-over, breakdown, explosion, and arc-over. Inner insulation failure is mainly due to the presence of foreign material inside the circuit breaker. The foreign material may be introduced during the installation process, and it may be running for some time after the body and shield to wear off so that the metal particles produced inside the foreign body lead to the occurrence of discharge fault. External insulation and phase fault is due to the external insulation porcelain leakage ratio spacing, and dimensions do not meet the standard requirements.

2.4 *Carrier fault*

Breaker carrier failure was mainly due to poor contact contacts overheat or can lead to overheating. Grid run way increases sharply in the load. At a very large load, the contacts overheat. Newly installed circuit breaker contacts are mainly due to poor contact with the moving contact (no static contact). The contact carrier is due to poor insulation or accident.

2.5 *Other fault*

This type of circuit breaker operation malfunction caused only barrier without the accident, but it reflects the breaker potential accidents. The main reason is leakage failure and damage to components (Huang Xinbo, 2015).

1. Disclosure failures. For the hydraulic actuator may be due to the high pressure of the discharge valve during closing or poor sealing oil. Hydraulic oil pipeline leakage circuit connector has a high pressure relief valve and the cylinder and piston sliding seals or pump and check valve poor sealing accumulator barrel rod seals and poor hydraulic mechanism caused by oil spills. The pneumatic actuator may have compressed air circuit pipe joints that lead to oil spills, gas

tank drain valve closed lax, safety valve, and check valve seal bad pneumatic mechanism, leading to the leak.
2. Component damage. Damage to parts of the site is mainly a seal, transmission parts, rod, and valve body. The main cause of this type of failure is due to the high mechanical strength transmission components, poor quality seal, easy aging, short lifetime. In the assembly inspection process, improper installation or the seal position with very large deformation can affect their lifetime.

3 ONLINE MONITORING METHODS AND CHARACTERISTICS OF HIGH-VOLTAGE CIRCUIT BREAKERS

3.1 *Circuit breaker trip—Monitoring time characteristics*

During the high-voltage circuit breaker operation, one of the important characteristics is stroke–time characteristics. This characterization is an important factor in relation to the speed of opening and closing of the high-voltage circuit breakers, based on its calculations (Tang Jintao, 2012).

The current measurement of high-voltage circuit breaker trip–time characteristics uses photoelectric sensors and the corresponding displacement measuring circuit with the conduct, commonly used incremental rotary optical encoder or a linear optical encoder. The linear optical encoder is mounted on the circuit breaker linear motion components, or the rotary optical encoder is mounted on the spindle breaker operating mechanism, points (combined) measured by the sensor switch operation.

3.2 *Detecting electrical life of the circuit breaker*

Electrical circuit breaker contact wear is an important factor that affects the lifetime of the electrical circuit breakers in online monitoring. Usually, the breaker electrical lifetime refers to after breaking several short-circuit current due to contact and vents open burning until normal lifetime is not short-circuit breaking current of a new interrupter. IEC (International Electrotechnical Commission) and national standards for electrical lifetime are not regulated. Foreign manufacturers are generally breaking KA cumulative number of short-circuit current or accumulated rated breaking current breaking number of times to calibrate electrical lifetime. However, there are two obvious shortcomings. First, the cumulative breaking times because there is no distinction between each of the breaking current and arcing time, so the amount of contact wear is estimated to be rough. Second, the same external circuit breaker in the same breaking condition has twice the current value of the same size, which may have different degree burns, simply to cumulative breaking current is not accurate as the basis for judging the health status of the contacts (Liu Tao, 2015).

3.3 *Detection circuit breaker closing coil current*

Current coil will have some changes in the process of opening and closing the circuit breaker. It contains a wealth of information in the current change waveform. If we are in the right way of information processing and handling, we will get more useful information on breaker line monitoring and troubleshooting. Circuit breakers constantly in their opening during the operation will affect the current operation of the coil. In this process, through a proper analysis of overview breaker secondary control work and mechanical actuator status circuit, a secondary criterion is examined (Sun Guangpeng, 2012).

3.4 *Monitoring circuit breaker vibration signal*

An important element for monitoring the vibration signal is the circuit breaker monitoring. The advantages of monitoring vibration signals are collected vibration signal that does not involve electrical measurements, to a lesser extent by electromagnetic interference, and a higher signal acquisition authenticity, with high availability (Zhang Yanfei, 2014). In addition, the sensor that is mounted to the outside of the circuit breaker will not function and have an impact on the structure. Meanwhile, a vibration sensor is considered suitable for monitoring as it possess some advantages, such as small size, reliable, low cost, high sensitivity, and good anti-jamming features. According to the principle of similarity, for the same high-voltage circuit breakers, the process is repeated with the states under the external vibration signal within a certain range that is stable. The vibration waveform thus collected is similar.

4 SUMMARY

High-voltage circuit breaker plays a vital role in the safe and reliable operation of the power grid by their common failure causes and online monitoring of research and analysis of available signals. It provides a reliable basis for diagnosis, monitoring and making online fault diagnosis technology to achieve better development and application prospects.

REFERENCES

Bai Zhiyan. High Voltage Circuit Breaker Fault Diagnosis and Monitoring of the state[D]. Huazhong University of Science and Technology, 2013.13–15.
Huang Xinbo (eds). Intelligent Circuit Breaker-line monitoring and condition assessment technology[J]. High Voltage Electrical Apparatus, 2015, 03:129–134+139.
Liu Tao (eds). Online Monitoring of High Voltage Circuit Breakers[J]. Electronics and Software Engineering, 2015, 12:123.
Sun Guangpeng (eds).High voltage circuit breaker line monitoring and diagnosis of fault[J]. Silicon Valley, 2012, 13:123+92.
Tang Jintao (eds). On-line monitoring of high voltage circuit breaker technology[J]. China New Technology and New Products, 2012, 16:3–4.
Zhang Yanfei. High Voltage Circuit Breakers troubleshooting and status monitoring study[J]. Electronic World, 2014, 16:51–52.

Energy Science and Applied Technology ESAT 2016 – Fang (Ed.)
© 2016 Taylor & Francis Group, London, ISBN 978-1-138-02973-6

Present status and principle of heavy metal detection

Hongxia Li, Huanjun Chen & Hanqing Wang
Tianjin Capital Environmental Protection Group Company Limited, Tianjin, China
Tianjin Caring Technology Development Co. Ltd., Tianjin, China
Tianjin Municipal Wastewater Biochemical Treatment Technology Engineering Center, Tianjin, China

ABSTRACT: With the rapid development of the industry in recent years, more and more heavy metals are being used in our daily life and enriched in water and soil. Therefore, the study of heavy metal detection becomes increasingly important due to their features of causing carcinogenicity, teratogenicity, and mutagenicity in organisms. The present status and principles of heavy metal detection are summarized in this paper.

Keywords: heavy metal; detection method; principle

1 INTRODUCTION

Currently, more heavy metals are being used in our daily life and enriched in water and soil or other fields. Thus, the study of heavy metal detection becomes essential due to their features of causing carcinogenicity, teratogenicity and mutagenicity in organisms. It is important to grasp different applicable scopes, status, and principles of heavy metal detection. The details of these features are summarized in this paper.

2 DIFFERENT METHODS OF TESTING HEAVY METAL

2.1 Atomic absorption spectrometry

Atomic absorption spectrometry is a method of quantitatively measuring the absorption content of the element of specific gaseous atoms from light radiation with the absorption intensity of gaseous ground state atoms in outer electrons to ultraviolet and visible range.

2.2 Atomic fluorescence spectrometry

Atomic fluorescence spectrometry is a method of measuring the content of the element of fluorescence emission intensity generated from its atomic vapor by the laser radiation at a specific frequency. Although the method of atomic fluorescence spectrometry is based on the emission spectroscopy related closely to atomic absorption spectrometry, it combines the advantages of two methods of the atomic emission and the atomic absorption, and

overcomes their shortcomings (HOLAK W, 1969). The method of the atomic fluorescence spectrum has many advantages such as simple emission line and high sensitivity, compared with atomic absorption spectrometry, a wide range of linear and low interference. Additionally, it can perform multi-element simultaneous determination. The method of the atomic fluorescence spectrometer can analyze all kinds of elements such as mercury, arsenic, antimony, bismuth, selenium, tellurium, lead, tin, germanium, cadmium, and zinc. The method has been widely used in many fields, including environmental monitoring, medicine, geology, agriculture, and drinking water. Thus, the atomic fluorescence spectrometry has been legalized for the first method of the international standard methods of testing arsenic, mercury, and other elements in food.

2.3 X-ray fluorescence spectrometry

X-ray fluorescence spectrometry is a method of testing constituents qualitatively or quantitatively included in the sample by X-ray absorption changes in the composition of the sample. The method has a lot of features such as rapid analysis, simple sample preparation, a widely analysis range of elements, simple lines, less spectral interference, sample forms of diversity, and non-destructive sample morphology. X-ray fluorescence spectrometry can be used for the qualitative and quantitative analysis of major elements and trace elements whose detection limit can in fact reach to the level of PPB or ever higher if combined with other means of separation and enrichment. The method can measure a wide range of elements including all elements in the periodic table from F to U. Even the multi-channel analyzer can determine 20 elements within several minutes simultaneously.

2.4 *Anodic stripping voltammetry*

Anodic stripping voltammetry is an electrochemical analytical method that combines constant potential electrolysis enrichment and voltammetry determination. The method has high sensitivity and can determine a lot of metal ions continuously. The method has many advantages such as the relatively simple instrument use, convenient operation, and excellent analysis to trace elements. It has been promulgated that the national standards of the anodic stripping voltammetry are applied to determine metal impurities in chemical reagents in China. The method of anodic stripping voltammetry to determine the elements contains two steps. The first step is electrodeposition, which involves the electrolytic deposition of the measured ion accumulated on the working electrode and generation of amalgam at a constant potential. It is proportional to the amount of the electrodeposited metal and the measured concentration of metal ions to a given metal ion if the stirring speed and the time of pre-electrolysis are all constant. The second step is stripping, in which a reverse voltage is applied to the working electrode. After scanning from negative to positive, the metal in amalgam is re-oxidized to ion, and the oxidation current is generated and the voltage–current curve is recorded, and returns to the solution at the end of the enrichment or the general stationary of 30 seconds or 60 seconds. The peak current of voltammograms is proportional to the iron concentration measured and can be used as the basis for quantitative analysis. Peak potential can be used as a basis for qualitative analysis.

2.5 *Inductively coupled plasma mass spectrometry*

The high-frequency electromagnetic field of inductively coupled plasma is generated by the high-frequency current through the induction coil. The working gas can form plasma and show flaming discharge of high temperature over 7000 degree. The plasma is a spectrum light source of good performance in the process of evaporation/atomic/excitation/ionization. This is because the plasma torch of the cyclic structure is beneficial for injecting the sample from the central channel of the plasma and to maintain a stable flame. A lower flow rate of less than one liter per minute can penetrate the inductively coupled plasma and result in the residence time of two to three minutes of the completely evaporated atomization sample in the center channel. The temperature of the inductively coupled plasma

central channel of the clinical structure is more than any flame or arc-like spark temperatures. It is also the optimum excitation temperature of atoms and ions. The substance analyzed is heated indirectly in the central channel and has a little effect on the discharge properties of inductively coupled plasma. The light source of inductively coupled plasma is a thin light of electrodeless discharge, no electrode contamination and has a small self-absorption phenomenon. These features meet the requirements of an ideal analysis and make the inductively coupled plasma source have an excellent analytical performance. Due to poor tolerability to salt of the inductively coupled plasma mass spectrometry at the detection limit of solid concentration, some common light elements have serious interference in the inductively coupled plasma mass spectrometry in the detection process of elements, and the detection limit could be even worse.

2.6 *UV–visible spectrophotometry*

Complexation of heavy metal and chromogenic agent of organic compounds can lead to the formation of colored molecules. Moreover, the shade degree of the reaction solution is proportional to the concentration, and the solution can be detected by colorimetric detection at a specific wavelength (O'Haver TC et al, 1975) (O'Haver TC et al, 1976) (Cheng Shulin, 1992). The first spectrophotometric analysis method measures the uptake of the substance itself to absorb ultraviolet and visible light, and the other method tests the amount of colored compounds generated. The most widely used method is joining reagent in a measured solution and testing compound concentration absorption in the ultraviolet and visible region because of weak absorption of many inorganic ions in the ultraviolet and visible region and less directly for quantitative analysis. Reagent compromises inorganic and organic color chromogenic agents and more organic color-developing agent can be used. Most of the organic reagents is a kind of colored compound and usually generate stable chelate reaction with a metal ion compound. Color reaction has high selectivity and sensitivity. It easily dissolves in organic solvents, which can be colorimetric detected after extraction. In recent years, a major concern is the color system of multiple complexes containing three or more component forms. The multiple complexes formed can improve the sensitivity of spectrophotometric assay and analysis features. It is a very important research topic on the selection and use of reagent in pretreatment extraction and colorimetric detection in recent years compared with the method of spectrophotometry.

2.7 *Dipstick detection*

Dipstick detection is a new assay producing complex linear color reactions through fast semi-quantitative determination to heavy metal ions in water, which is carried out by basic dyes and heavy metal ions under acidic conditions reacting with potassium iodide.

3 CONCLUSION

Each method of heavy metal detection has advantages and disadvantages because of different applicable scopes, status, and principles. In real test work, we must select a suitable and efficient test method based on the actual detection situation. As environmentalists, we have an obligation to spread and promote better test methods in an increasingly polluted environment.

ACKNOWLEDGMENTS

The authors are grateful to Tianjin Municipal Science and Technology Commission for providing financial assistance for the program Tianjin Science and Technology Project (13RCGFSF14200, 14ZCZDSF00009, and 14ZCDGSF00032).

REFERENCES

Atomic Absorption Spectrometry, Geological Publishing House, 1979, 205–207.

Cheng Shulin, Analytical Chemistry, 1992.20(1):7.

HOLAK W: Gas-samp ling technique for arsenic determination by atomic absorption Spectrophotometry [J]. Anal Chem, 1969, 41(12): 1712–1713.

O'Haver TC, Green GL. Anal Chem, 1976, (48):312–318.

O'Haver TC, Green GL. Intern Lab, 1975, (5–6):11–17.

Energy Science and Applied Technology ESAT 2016 – Fang (Ed.)
© 2016 Taylor & Francis Group, London, ISBN 978-1-138-02973-6

Latin square test design and data analysis in SAS

Huameng Gao, Fan Yang & Qing Zhang
Academy of Equipment, Beijing, China

Yanli Lu
General Armament Department, Beijing, China

ABSTRACT: The Latin square design and data analysis in equipment tests were studied, and the influence and significance of factors on test results were defined. The Latin square and Latin square sampling were explained. The experiment of the Latin square design of an equipment was performed using SAS, and the test data were obtained. The test data were analyzed using variance analysis in the Latin square design. The main influential factors were determined, the interactions between factors were analyzed, and the influence difference in factors was also gained. The introduction of the experimental design and data analysis presented this study can be adopted in other fields.

Keywords: latin square design; variance analysis; main factors; difference in factors; significance test

1   INTRODUCTION

Latin square design is a class of constrained uniform sampling presented by M. D. McKay and R. J. Beckman in 1979. The test times of Latin square design can be controlled and the test times can sometimes be less than the factors. Latin square design has a better uniform character, and does not have the same test point aggregation phenomenon as orthogonal design. Latin square design is often used in test design. Liu Xiaolu researches the optimization and use of Latin square design (Liu Xiaolu et al, 2011). Ju Yarong researches the figure encrypt method based on logistic model and Latin square (Ju Yarong et al, 2008). The aforementioned papers pay attention to the theory research and do not present the detailed process of design, data processing, and software. This paper presents test design and variance analysis in detail based on test data and SAS. The mathematical resolving and function calling are presented. The experiment design and analysis presented in this paper can be adopted in other fields dealing with Latin square design and data processing.

2   LATIN SQUARE DESIGN

Disturbing factors should be system controlled in many tests, and the random complete block design is an effective way to make the test block to be homoscedascity. Latin square design has two orthogonal block factors and can control tow disturbing factors in tow dimensions.

2.1   Latin square

A $p \times p$ Latin square of p factors is an square of p rows and p columns. The test is denoted by Latin letters such as A, B, C, and D. Every unit of p^2 units has one of the p letters, and one letter appears one time in every row and column. If the letters in the first row and column have letter sequence, the Latin square is standard. A standard Latin square can be gained by allocating letters in the first row according to letter sequences, and left shifting letters allocated up rows one step in other rows (Yuan Zuoxiong, 2010) (Dong Xinfeng et al, 2014).

Latin square design has the following characteristics:

1. Control of two disturbing factors in two dimensions.
2. Numbers of row block and column block are equal.
3. Every test resolving appears one time in the row block and column block.

2.2   Data analysis

Using the Latin square test, the difference between factors can be analyzed, and the difference in factor levels can also be researched. Variance analysis is often used in Latin square data analysis.

The differences between factors and levels can be analyzed in the SAS ANOVA module.

3 SAS SOFTWARE

SAS is a professional scientific calculation software with many functions and great power. SAS can rapidly and accurately solve all kinds of test design, statistic, and data visualization. SAS can also process data mining and gene spectrum analysis.

3.1 *SAS products*

SAS has many products such as static and the use of each product is different. Module construction is used in SAS, and every module is one SAS product. Some products such as SAS/STAT have only the SAS procedure. Some products such as SAS/BASE have SAS language, SAS window, SAS macro, SAS SQL besides the SAS procedure. Some SAS products are software with serial functions such as SAS/ASSIST, SAS/ANALYST, and SAS/INSIGHT. Some SAS products are a tool of other SAS products such as SAS/AF.

3.2 *SAS procedures*

SAS procedure is a compiled program and the theory used and its way are recognized. So, the procedure is standardized, programmed, and systematized. The plan, sort, transpose, print, and anova procedure are used in this paper.

Procedure plan can construct many kinds of test design and realize randomization such as factorial design, random block design, nested design, cross design, and Latin square design. Procedure sort can sort the data according to some variables. Procedure transpose can transpose the dataset. Procedure print can print the data. Procedure anova can be used to carry out the analysis of variance. In anova, the response and the variables are tested in each condition.

4 DESCRIPTION OF A LATIN SQUARE DESIGN

A factory will test the explosive power of the explosion. A tester will research the effect of five kinds of mixer. One batch of raw material can only offer one sample of every five mixer. Five operators with different technical levels, respectively, prepare the five mixer.

4.1 *Test design*

The operator, batch of raw material and mixer are factors with five levels. Latin square design is realized using SAS.

```
proc plan seed = 20151119;
factors worker = 5 ordered batch = 5 ordered;
treatments mix = 5 cyclic; output out = a
  worker cvals = ('worker1' 'worker2' 'worker3'
'worker4' 'worker5') random
  batch cvals = ('batch1' 'batch2' 'batch3' 'batch4'
'batch5') random
  mix cvals = ('mix1' 'mix2' 'mix3' 'mix4' 'mix5')
random; run;
  proc sort data = a out = b; by worker batch; run;
  proc transpose
  data = b(rename = (batch = _name_))
  out = c(drop = _name_); by worker; var mix; run;
  ods html; proc print data = c noobs; run; ods
html close;
```

4.2 *Output of program*

The worker column is the operator factor with five levels (worker1, worker2, worker3, worker4, worker5). The batch of raw material has five levels (batch1, batch2, batch3, batch4, batch5). The content in the table is mixer. The test is carried out according to Table 1, and the data are summarized in Table 2.

5 ANOVA OF THE DATA

The anova is used to analyze the Latin square test data. In anova, the significance of square sums of error and random is compared and the effects of factors influencing the response are valued (Fang Kaitai et al, 2011).

The model of the Latin square test is:

$$y_{ijk} = \mu + \alpha_i + \tau_j + \beta_k + \varepsilon_{ijk} \quad (i,j,k = 1,2,\cdots,p)$$

Table 1. Latin square design table.

Worker	Batch1	Batch2	Batch3	Batch4	Batch5
Worker1	mix1	mix2	mix3	mix4	mix5
Worker2	mix2	mix3	mix4	mix5	mix1
Worker3	mix3	mix4	mix5	mix1	mix2
Worker4	mix4	mix5	mix1	mix2	mix3
Worker5	mix5	mix1	mix2	mix3	mix4

Table 2. Test result table.

Batches	Operators and date				
	1	2	3	4	5
1	A(24)	B(20)	C(19)	D(24)	E(24)
2	B(17)	C(24)	D(30)	E(27)	A(36)
3	C(18)	D(38)	E(26)	A(27)	B(21)
4	D(26)	E(31)	A(26)	B(23)	C(22)
5	E(22)	A(30)	B(20)	C(29)	D(31)

Here y_{ijk} is the test gaining of operation j in row i and column k, μ is the average of all, α_i is the effective of row i, τ_j is the effective of operation j, β_k is the effective of column k, and ε_{ijk} is the random error.

The square sum of the data observed is:

$$SS_T = \sum_i \sum_j \sum_k \frac{y_{ijk}^2}{p} - \frac{y_{...}^2}{N}$$

The square sum of row, column, operation, and error are:

$$SS_r = \sum_{j=1}^{p} \frac{y_{i..}^2}{p} - \frac{y_{...}^2}{N} \quad SS_c = \sum_{j=1}^{p} \frac{y_{..k}^2}{p} - \frac{y_{...}^2}{N}$$

$$SS_t = \sum_{j=1}^{p} \frac{y_{.j.}^2}{p} - \frac{y_{...}^2}{N} \quad SS_e = SS_T - SS_r - SS_c - SS_t$$

6 THE RESULT OF ANOVA

When the α level is 0.05, the differences in batches do not have static significance, but the differences in operators and mixers have static significance. The results of anova are summarized in Table 3.

7 SUMMARY

Latin square design is in order and uniform and can effectively reduce the samples, making easy analysis of the test data. Latin square design should be used when the tested equipment is costly and the test design is complex. The anova is the main method to analyze the Latin square test data. The effects of factors on the response can be defined using anova.

Table 3. Variance analysis table.

Source	DF	Anova SS	Mean square	F value	Pr > F
batch	4	68.0000000	17.0000000	1.59	0.2391
worker	4	150.0000000	37.5000000	3.52	0.0404
mixer	4	330.0000000	82.5000000	7.73	0.0025

REFERENCES

Dong Xinfeng, Zhou Yu, et. A Class of Latin Square with Higher Nonlinearity[J]. Communications Technology, 2014, 47(9):1058–1061.

Fang Kaitai, et. Test Design and Modeling [M]. Beijing: High Education Press. 2011:169 171.

Ju Yarong, Liu Xiaobing, et. An Encrypt Method Based on Logistic Model and Latin Square Transformation [J]. Journal of Chongqing University of Science and Technology: Natural Science Edition, 2008,10(4):143–146.

Liu Xiaolu, Chen Yingwu, et. Optimized Latin Hypercube Sampling Method and Its Application [J]. Journal of National University of Defense Technology, 2011, 33(5):73–77.

Yuan Zuoxiong. Programming of SAS Software to Analysis of Variance about the Data on Latin Square Design [J]. Medical Information, 2010, 23(12):453–454.

Energy Science and Applied Technology ESAT 2016 – Fang (Ed.)
© 2016 Taylor & Francis Group, London, ISBN 978-1-138-02973-6

Simulation analysis on factors affecting $PM_{2.5}$ monitoring

Tao Wei
Chinese Academy of Sciences, Shenzhen Institutes of Advanced Technology, Shenzhen, China
Shaanxi University of Science and Technology, Xi'an, China

Jingjing Ma
Testing and Technology Center for Industrial Products, Shenzhen, China

Zedong Nie
Shenzhen Institutes of Advanced Technology, Chinese Academy of Sciences, Shenzhen, China

Feng Zhang
Shaanxi University of Science and Technology, Xi'an, China

Lei Wang
Shenzhen Institutes of Advanced Technology, Chinese Academy of Sciences, Shenzhen, China

ABSTRACT: In order to study the influence of the factors on the accuracy of PM2.5 monitoring, including the sampling flow rate, particulate matter size distribution and number concentration, a simulated model for airborne particulates is proposed in this paper. A cylindrical mixing tank with a Mini-Vol Sampler is chosen, the diameter of the tank is 240 mm and the height is 369 mm, the Computational Fluid Dynamics (CFD) and orthogonal test methods are adopted, the air continuous phase is calculated and the moving trajectories of airborne particulates are simulated by adding discrete phase. The results show that the particulate matter number concentration has a relatively minor effect on the sampling efficiency, and the variances of both PM2.5 and PM10 sampling efficiency data are less than 0.3. The sampling efficiency of PM2.5 sampler is influenced both by sampling flow rate and particulate matter size distribution, and the measured accuracy can be improved when the sampling flow rate is 4 L/min.

Keywords: factor; sampling efficiency; PM2.5 sampler; CFD simulation

1 INTRODUCTION

Since the industrial revolution, human activities have significantly increased the concentrations of ozone and atmospheric particulate matters in both urban and rural regions, and these changes have been driven by direct changes in air pollutant emissions (Schuulz, M. 2006; Cheng, Z. 2013 & Parrish, D. 2014). Chinese Environmental Status Bulletin 2014 indicates that 88.8% of 161 monitored Chinese cities have not reached the standards for fine particulate matter (with aerodynamic diameter less than 2.5 μm, $PM_{2.5}$) (Information, 2014). Epidemiological studies have shown that $PM_{2.5}$ has a significant influence on human health, including premature mortality. Using simulation to separate the influences of past climate change on air quality and human health is studied, the simulation combines present-day emissions and preindustrial climate, and around 2.1 million deaths with anthropogenic $PM_{2.5}$-related cardiopulmonary diseases (93%) and lung cancer (7%) is estimated (Dagher, Z. 2006; Hoffmann, B. 2012 & Silva, R. A. 2013).

Monitoring $PM_{2.5}$ concentration precisely is important to obtain the data and study $PM_{2.5}$ pollution problem, and many research groups have investigated the factors for the accuracy of the sampling efficiency. On one hand, environmental condition near the monitoring sites can cause deviation of the monitoring data. A dispersion model is used to analyze the influence of the wind speed and wind direction on the dispersion of near-road air quality monitoring data, in particular, the standard deviation of the vertical fluctuation in wind (Venkatram, A. 2007). The influence of the turbulence level of wind-flow fields on the air quality monitoring data are studied in (Kalthoff, N. 2005). The environmental temperature and humidity exert a great influence on the movement and distribution of the particulate matter phase are proposed (Mészáros, E. 1999). On the other hand, different monitoring methods also can lead to the inaccuracy of the monitoring data. United States Environmental Protection Agency (U.S. EPA) uses the gravimetric measurement method to obtain the effective $PM_{2.5}$ monitoring data for the first time, and created a set of relevant international standards (Register, F. 1987; Watson, J. G. 1998 & W. H. Organization 2006). The differences between the 24 hours average particulate matter concentrations of the Wedding β-gauge monitor and Andersen or Wedding hi-vol sampler are studied (Chang, C. 2001). The TEOM data being generally lower than the manual methods is revealed (Allen, G. 1997 & Ayers, G.1999). by Comparing of 24 h mean $PM_{2.5}$ aerosol loadings determined by a TEOM and the manual gravimetric samplers. However, few works have studied the impact of the sampling flow rate, the particulate matter size distribution and number concentration in the aforementioned works.

In this paper, A simulated model for the mixing tank with a $PM_{2.5}$ sampler to imitate the actual situation is built, the Computational Fluid Dynamics (CFD) and orthogonal test methods are adopted, the trajectories of particles with different sizes (2.5μm, 5μm and 10μm) are simulated in different number concentrations (1024, 2048, 4096, 8192, 16384, 32768, 65536 and 131072 particles per cubic meter), and the movement of particles with different sizes (2.0 μm, 2.6 μm, 5.4 μm, 9.7 μm, 15.1 μm, and 20.0 μm) are simulated in different sampling flow rates (1 L/min, 2 L/min, 3 L/min, 4 L/min, 5 L/min). Meanwhile, we analyze and evaluate the influence degree of these factors on the sampling efficiency of $PM_{2.5}$ sampler.

The rest of the paper was organized into three sections. Section II presented the simulated methods and procedures. Section III introduced the simulation results. The final section concluded our work.

2 METHODS

2.1 Model

A cylindrical mixing tank with a Mini-Vol sampler is chosen as the simulated model, the tank is 240 mm in diameter and 369 mm in height. The tank have two air-inlets, a particle-inlet and an outlet, their diameters are 9.5 mm, 1.0 mm, 50 mm and the heights are 10 mm, 10 mm and 50 mm respectively.

The Mini-Vol sampler consists of a separator and a diaphragm pump. The schematic diagram of the model are illustrated in Figure 1 and the

working characteristic curve of diaphragm pump is given in Figure 2. Before the simulation begins, the model is meshed into 4629342 grids totally, and the pre-mesh quality is above 0.3, as Figure 3 and Figure 4 shown.

2.2 Boundary conditions and parameters

According to the hypothesis of fluid mechanics, particle motion equations are as follows:

$$\frac{du_p}{dt} = F_D(u - u_p) + \frac{g_x(\rho_p - \rho)}{\rho_p} + F_x \qquad (1)$$

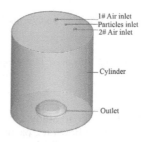

Figure 1. Schematic diagram of the model.

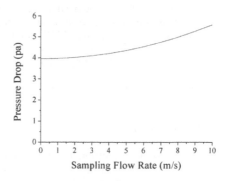

Figure 2. Working characteristic curve of diaphragm pump.

Figure 3. Mesh of the model.

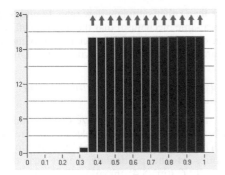

Figure 4. Pre-mesh quality of the model.

Table 1. Major boundary conditions.

Name	Continuous phase	Dispersed phase
Inlet	velocity-inlet	escape
Outlet	exhaust-fan	trap
Cylinder	wall	reflect

$$F_D = \frac{18\mu}{\rho_p d_p^2} \frac{C_D R_e}{24} \qquad (2)$$

$$R_e = \frac{\rho d_p |u_p - u|}{\mu} \qquad (3)$$

where u = gas phase velocity; u_p = particle velocity; μ = viscosity of fluid dynamics; ρ = gas density, ρ_p = particle density; d_p = particle diameter; and R_e = particle Reynolds number. And F_D (u-up) is the unit of grain quality drag resistance, g_x (ρ_p-ρ)/ρ_p is the resultant force of gravity and buoyancy for unit mass particle, and F_x is additional force, including the quality force, pressure gradient force, thermophoresis force, Basset force, brown, Saffman lift force, etc.

The Lagrangian model (also called tracking particle trajectory model) is adopted to solve these equations and calculate the unknowns in this simulation, the mayor boundary conditions are shown in Table 1, and some idealized assumptions are given as follows in order to better numerically simulate:

The shape of all simulated particles are regarded as spherical;

Only gravity, buoyancy, drag force are taken into account in the simulation (Li, A. 1992).

According to the conditions of the Lagrangian model, the interaction between particles is ignored, because the particle v volume fraction is far less than 10%;

3 RESULTS AND ANALYSES

3.1 Sampling flow rate and particulate matter size distribution

Both the sampling flow rate and particulate matter size distribution are chosen to be the variables, and the orthogonal test method is adopted in the simulation. 5 groups of sampling flow rate data (1 L/min, 2 L/min, 3 L/min, 4 L/min and 5 L/min) are selected to be one variable, the particles with different sizes (2.0 μm, 2.6 μm, 5.4 μm, 9.7 μm, 15.1 μm, and 20.0 μm) are regarded as another variable, and extra physical parameters of air and particle are given in Table 2.

The Rosin-Rammler distribution function is used to describe the distribution characteristic of particulate matter sizes (2.0 μm, 2.6 μm, 5.4 μm, 9.7 μm, 15.1 μm, and 20.0 μm) simultaneously, as Figure 5 shown, and the Rosin-Rammler distribution function is as follows:

$$Y_d = e^{-(d/D)^n} \qquad (4)$$

Table 2. Physical parameters of air and particle.

Name	Velocity m/s	Gauge pressure Pa	Temperature °C	Density kg/m³
Particle	0.05	–	25	1050
Air		101325	25	1.225

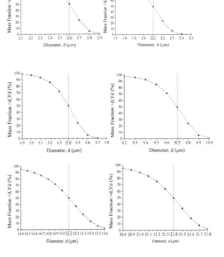

Figure 5. Rosin-Rammler size discribtions of the particles with different sizes (2.0 μm, 2.6 μm, 5.4 μm, 9.7 μm, 15 μm, and 20 μm).

where d = particle diameter; D = mean diameter, and D is obtained by noting the value of d at which Y_d = e − 1 ≈ 0.368, n = spread parameter, Y_d is the mass fraction of droplets with diameter greater than d.

Figure 6 shows the simulation results, 5 broken lines with different symbols represent the sampling efficiencies on the particles with different sizes (2.1 μm, 2.6 μm, 5.4 μm, 9.7 μm, 15 μm, and 20 μm), and the lines with spots, squares, triangles, pentagons and diamonds respectively represent that the sampling flow rates are 1 L/min, 2 L/min, 3 L/min, 4 L/min, and 5 L/min. Some characteristics can be obtained by analyzing the variation trends of these lines. The sampling efficiency data all tend to zero when the particle size is greater than 15 μm, and the reason is that these particles are too heavy to be brought into the PM$_{2.5}$ sampler by the airflow, instead of sinking in the bottom of the mixing tank. When the sampling flow rate is set to 4 L/min, the sampling efficiencies on PM$_{2.5}$, PM$_5$ and PM$_{10}$ particles are 87.68%, 73.88% and 22.07% respectively, and are

higher than other sampling flow rates. Figure 6 also shows a same mechanism that the sampling efficiencies of cies of PM$_{10}$ particles when the sampling flow rates are defined as 3 L/min, 4 L/min, and 5 L/min.

3.2 *Particulate matter size distribution and number concentration*

Both the particulate matter size distribution and number concentration are selected as the factors. The orthogonal test method is also adopted, 3 groups of the particulate matter size distributions (2.6 μm, 5.4 μm and 9.7 μm shown in Figure 5) are chosen to be one variable, the particles with different number concentrations (1024, 2048, 4096, 8192, 16384, 32768, 65536 and 131072 particles per cubic meter) are regard as another variable, and the extra physical parameters of air and particles are as the same as the Table 2.

Figure 7 shows the results that all the sampling efficiency data of PM$_{2.5}$ particles are distributed

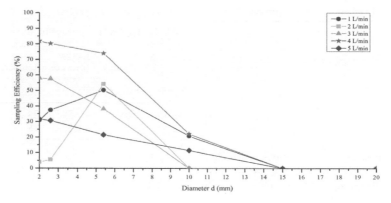

Figure 6.   Influencing characteristics of the trapping efficiencies on the different particle size discributions and sampling flow rates.

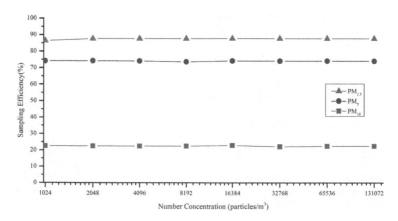

Figure 7.   Influencing characteristics of the trapping efficiencies on the different particle matter size distributions and number concemtrations.

between 86.29%~87.74%, and the variance is calculated to be about 0.2484. The sampling efficiency data of PM_5 particles are distributed between 73.59%~74.15%, and the variance is calculated to be about 0.0357. The sampling efficiency data of PM_{10} particles are distributed between 22.08%~22.79%, and the variance is calculated to be 0.0362. These variance data indicate that different particle number concentrations have little influence on the sampling efficiency of PM2.5 sampler.

4 CONCLUSIONS

In this paper, a simulated model for airborne particulates is proposed in order to study the influence of the factors on the accuracy of PM2.5 monitoring, including the sampling flow rate, particulate matter size distribution and number concentration, and few works have studied on these factors. A cylindrical mixing tank with a Mini-Vol sampler is chosen, the sampler is developed by AIRMETRICS Inc in America, the Computational Fluid Dynamics (CFD) and orthogonal test methods are adopted, the air continuous phase is calculated and the moving trajectory of airborne particulates are simulated by adding discrete phase. The results show that the variances of PM2.5 and PM10 sampling efficiency data are about 0.2484 and 0.0362 respectively. The sampling efficiency of PM2.5 sampler is influenced by sampling flow rate and particulate matter size distribution, and the measured accuracy on PM2.5 and PM10 particles could be improved when the sampling flow rate is setup to 4 L/min.

ACKNOWLEDGMENTS

This work is supported in part by the National Natural Science Foundation of China under Grant No. 61403366, No.U1505251 and No.71531004, Guangdong Science and Technology Planning Project under Grant No. 2015 A020214018, 2014B 010111008 Shenzhen Basic Research Project Fund under Grant No. JCYJ2014041713430695 and No. JCYJ2015040 1150223630, Shenzhen Technology Development Project Fund under Grant No. CXZZ 20150505093829778.

REFERENCES

Allen, G. & Sioutas, C. 1997. Evaluation of the TEOM® method for measurement of ambient particulate mass in urban areas. *Journal of the Air & Waste Management Association*, vol. 47, pp. 682–689.

Ayers, G. & Keywood, M. 1999. TEOM vs. manual gravimetric methods for determination of PM2. 5 aerosol mass concentrations. *Atmospheric Environment, vol. 33, pp. 3717–3721.*

Chang, C. & Tsai, C. 2001. Differences in PM 10 concentrations measured by β-gauge monitor and hivol sampler. *Atmospheric Environment, vol. 35, pp. 5741–5748.*

Cheng, Z. & Jiang, J. 2013. Characteristics and health impacts of particulate matter pollution in China (2001–2011). *Atmospheric Environment, vol. 65, pp. 186–194.*

Dagher, Z. & Garçon, G. 2006. Activation of different pathways of apoptosis by air pollution particulate matter (PM2. 5) in human epithelial lung cells (L132) in culture. *Toxicology, vol. 225, pp. 12–24.*

Hoffmann, B. & Luttmann-Gibson, H. 2012. Opposing effects of particle pollution, ozone, and ambient temperature on arterial blood pressure. *Environmental health perspectives, vol. 120, p. 241.*

Information on http://jcs.mep.gov.cn/hjzl/zkgb/2014zkgb /2015 06/t20150608303142.htm.

Kalthoff, N. & Bäumer, D. 2005. Vehicle-induced turbulence near a motorway. *Atmospheric Environment, vol. 39, pp. 5737–5749.*

Mészáros, E. 1999. Fundamentals of atmospheric aerosol chemistry. Acad. Kiado Budapest.

Li A. & Ahmadi, G. 1992. Dispersion and deposition of spherical particles from point sources in a turbulent channel flow. *Aerosol science and technology, vol. 16, pp. 209–226.*

Parrish, D. & Lamarque, J.F. 2014. Long-term changes in lower tropospheric baseline ozone concentrations: Comparing chemistry-climate models and observations at northern midlatitudes. Journal of Geophysical Research: Atmospheres, vol. 119, pp. 5719–5736.

Register, F. 1987. Reference method for the determination of particulate matter as PM10 in the atmosphere. Federal Register, vol. 52, p. 24664.

Schulz, M. & Textor, C. 2006. Radiative forcing by aerosols as derived from the AeroCom present-day and pre-industrial simulations. Atmospheric Chemistry and Physics, vol. 6, pp. 5225–5246.

Silva, R.A. & West, J.J. 2013. Global premature mortality due to anthropogenic outdoor air pollution and the contribution of past climate change. Environmental Research Letters, vol. 8, p. 034005.

Venkatram, A. & Isakov, V. 2007. Analysis of air quality data near roadways using a dispersion model. Atmospheric Environment, vol. 41, pp. 9481–9497.

W.H. Organization. 2006. WHO Air quality guidelines for particulate matter, ozone, nitrogen dioxide and sulfur dioxide: global update 2005: summary of risk assessment.

Watson, J.G. & Chow, J.C. 1998. Guidance for using continuous monitors in PM2. 5 monitoring networks. Nevada Univ. System, Desert Research Inst., Reno, NV (United States); National Oceanic and Atmospheric Administration, Las Vegas, NV (United States); Environmental Protection Agency, Office of Air Quality Planning and Standards, Research Triangle Park, NC (United States).

13 *Recognition, video and image processing*

Energy Science and Applied Technology ESAT 2016 – Fang (Ed.)
© *2016 Taylor & Francis Group, London, ISBN 978-1-138-02973-6*

Combination measurement method using laser tracker and 3D laser scanner

L. Yang, G.X. Li, B.Z. Wu & M. Zhang
College of Mechatronic Engineer and Automation, National University of Defense Technology, Changsha, China

ABSTRACT: To address the problems that measurement efficiency is low and the free curve and surface components cannot be measured in the assembly measuring process of large-scale products, this article puts forward a combination measurement method based on the 3D model using laser tracker and the 3D laser scanner. A new solution for the measurement method, unified coordinate system and surface fitting in this combination measurement method, is proposed. The proposed approach provides a new technical way for efficient and accurate assembly measurement for large-scale products.

Keywords: large scale; combination measurement; key point; 3D model

1 INTRODUCTION

With the development of modern industry, more and more large-scale products not only improve the accuracy of assembly, but also use a number of weak stiffness materials and free curve and surface in the design and manufacture of products. General large-scale measuring instruments (e.g. laser tracker), because of its contacted measuring, will cause great deformation when measuring weak stiffness materials, leading to an inaccurate result. Furthermore, it cannot accurately measure the free curve and surface because of its limited number of measuring points. Combination measurement can not only expand the measuring range, but also reduce data redundancy and improve the measurement of data reliability and accuracy. Finally, it makes the measurement of instruments' information complementary and the measuring tasks become optimal. To date, the combination measurement method has been used in aerospace and other fields.

At abroad, the American FARO company has introduced the portable 3D measurement system TrackArm that combines the FARO Laser Tracker's long distance and high accuracy with FaroArm's flexibility and consistency to expand the measuring scope of the FaroArm. Dr Lin in Chinese Changchun Science and Engineering University has linked the iGPS and laser radar with laser tracker to construct a measuring network, which avoids the limitation of a single measurement device on the plane measurement. The combination measurement based on the laser tracker and the 3D laser scanner, which is put forward by this article, makes the laser tracker's high accurate measurement and the 3D laser scanner's fast measurement complementary advantages, so as to solve the contradiction between assembly requirements and the measuring efficiency of large-scale products.

2 THE OVERALL DESIGN OF COMBINATION MEASUREMENT

The laser tracker is the most widely used measurement instrument in large-scale products' assembly measurement. It has high accuracy and wide measuring scope. However, its measuring efficiency is not high because it measures points slowly and its optical path would be blocked easily so that its measuring station transformation is complex during measurement. With the use of the 3D laser scanner in modern industry, it can quickly measure any surface prominently. However, its single instrument's measuring effective range is relatively small. Furthermore, its measuring accuracy has some disparity compared with the laser.

The laser tracker and the 3D laser scanner are used to perform combination measurement, as shown in Table 1. It combines the characteristics of single-point measurement and overall scanning measurement by improving the traditional measurement method, that is, unified coordinate system and surface fitting. It uses partitioned measurement, unifies the whole measuring points to the product coordinate, and uses key points to fit surface. All of these features make the combination measurement to have the advantage of both laser tracker's accurate measurement and 3D laser scanner's quick measurement.

Table 1. The comparison of measurement characteristics.

Measuring instruments	Measuring way	Measuring coordinate system	Surface fitting	Advantage
Laser tracker	Single-point measuring	Instrument coordinate system	Fitting by measuring points	Accurately
3D laser scanner	Overall scanning measuring	Tooling coordinate system	Fitting by scanning points	Quickly
Laser tracker and 3D laser scanner	Partitioned measuring	Product coordinate system	Fitting by key points	Accurately and quickly

3 PARTITIONED MEASURING WAY

To measure all of the assembly geometrical quantity, the large-scale product's assembly measurement needs to set many measuring stations around the product when only using the laser tracker in traditional measurement. One measuring station would have many problems such as blockage of the optical path, which is due to the product's shape, measuring environment, and other factors in the measuring process. So, it usually needs multiple stations, as shown in Figure 1. If only one laser tracker is used, it must move the measuring station repeatedly in the measuring process, which will severely affect the measuring efficiency. If many laser trackers are used to perform combination measurement, it will result in a substantial increase in the cost of measurement because of the laser tracker's high cost. Conversely, the laser tracker is a stable contact measurement by SMR. If only one SMR is used, it will take extra time to wait for the SMR to be stable when measuring every point. Nevertheless, the SMR moved between measuring points needs to complete the operation manually so that it cannot form the automation of the measuring process. If many SMR are used and arranging one SMR on each measuring point, it will substantially increase the cost of measurement because there are so many measuring points in some large-scale product's assembly measurement. Moreover, the SMR's cost is inexpensive. For laser tracker measurement, the measuring points are always very limited. These limited points cannot accurately fit the free curve and the surface. Finally, there is dead space for some complex shape products when the laser tracker is used.

If the 3D laser scanner is used to measure large-scale products, because of the 3D laser scanner's extremely limited single scanning area, it needs to measure the entire product or the whole assembly geometrical quantity that is discretely distributed in the entire product, and must scan the entire product. The final $pc(Point\ Cloud)$ is obtained by splicing the whole single scanning's single point cloud pc_i:

$$pc = \bigcup (pc_i) \tag{1}$$

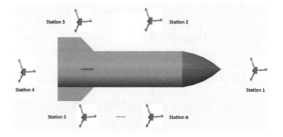

Figure 1. Multiple stations of laser tracker measurement.

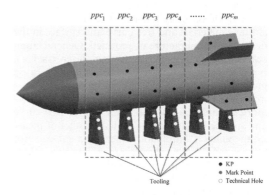

Figure 2. Combination measurement method.

However, splicing repeatedly will result in the accumulation of measurement errors for the 3D laser scanner, which cannot make the measurement results accurate because very large measurement errors are accumulated. On the other hand, to complete the scan's effective splicing, the traditional measurement method of the 3D laser scanner must paste a lot of mark points in the product, which is not allowed for some special products' final factory assembly test.

If combination measurement is adopted by using the laser tracker and the 3D laser scanner, shown in Figure 2, it needs to measure the entire product, could make the measuring area become discrete, and adopts partitioned measuring way to measure the product, dividing the product into m pieces of scanning area. It will install a tooling in each scanning area and paste the mark point on the tooling, which can avoid these problems. For example, some products may be repeatedly allow the pasting of mark points or do not allow the pasting of mark points. The measuring points of each scanning area are *ppc*(*piece of point cloud*). Each scanning area lays three more KP (key points). KP are completely laid in the design and manufacturing stages of the product. The laser tracker is used to measure the KP, and then KP are used to fit ppc_j directly onto the 3D model, thus avoiding the accumulation of measurement errors in the process so that splicing pc_i to be pc.

When the whole assembly geometrical quantities are discretely distributed in the entire product, it only needs to install tooling and lay the KP in the measuring area, and then begin to measure the product with the above method. This method can avoid the scanning of unnecessary areas and thus reduce the scanning workload.

4 UNIFIED COORDINATE SYSTEM

The descriptive information of space is different because each measuring instrument's spatial position is different. Measuring data for each measurement station are defined in the respective coordinate, and measurement instrument's coordinate definition is also different. Therefore, it must map the measuring data into the same reference space and unify the measuring coordinate to the global coordinate system.

In the measurement of large-scale products, most of the products would not be moved in the whole measuring process after being fixed by the tooling. It uses the product coordinate as the global coordinate and unifies the measuring data to the product coordinate because only the product coordinate is kept constant for each measuring product. The unified coordinate system of the laser tracker and the 3D laser scanner is shown in Figure 3. Because the laser tracker is using the spherical coordinate, which is the instrument coordinate based on itself, in actual measuring, it should use the laser tracker to measure the datum element of the product, and then fit the origin of coordinates, the X-axis, and other datum so as to construct the product coordinate. Then, it uses the coordinate transformation equation to transform the instrument coordinate of the laser tracker to the product coordinate:

$$F = \begin{bmatrix} R & T \\ 0 & 1 \end{bmatrix} M \tag{2}$$

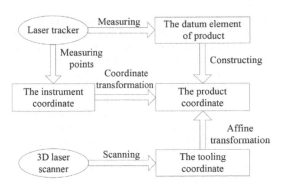

Figure 3. Unified coordinate system.

where $F = (x_p \ y_p \ z_p \ 1)^T$ is the measuring data in the product coordinate, $M = (x_i \ y_i \ z_i \ 1)^T$ is the measuring data in the instrument coordinate of the laser tracker, R is the rotation matrix, and T is the translation vector.

The measuring data's coordinate of the 3D laser scanner is defined by mark points. The above discussion shows that the mark points are pasted on the tooling. So, the 3D laser scanner's measuring data take in fact the tooling coordinate as datum. It will lay technical holes on the tooling and use the laser tracker to measure these technical holes. Then, KP and technical holes are used to solve the affine transformation matrix A, which can transform the tooling coordinate to the product coordinate. Finally, it uses affine transformation to transform the tooling coordinate system to the product coordinate system:

$$F = AN \tag{3}$$

where $N = (x_s \ y_s \ z_s \ 1)^T$ is the 3D laser scanner's measuring data in the tooling coordinate system.

5 SURFACE FITTING

Fitting is an operation that is based on specific criteria, making the ideal elements approach to the non-ideal elements. To fit the surface accurately, the 3D laser scanner's point cloud will fit the measuring point cloud onto the 3D model, and make the distance between the measuring points and the 3D model minimum. It is based on the least square method:

$$L = \sum_{k=1}^{N} d_k^2 \tag{4}$$

where N is the number of the points and d_k is the distance from the k-th point to the fitting surface.

For the large-scale product's measurement, the value of N may be millions and even billions, which makes the calculation for minimum L very large. It also makes the data processing time significantly longer. Therefore, the measurement efficiency will be affected seriously.

To increase the data processing speed and measuring efficiency, the combination measurement using the laser tracker and the 3D laser scanner could use the scanning area's KP to judge the fitting result, as stated earlier. The KP's measuring accuracy is higher than the scanning point cloud's because KP is laid in the design and manufacturing stages of product and measured with the laser tracker in the measuring stage. When the fitting surface that is based on the 3D model, it could only fit the measuring KP onto the KP of the 3D model. That is only performing the least square process for the KP fitted, which makes the sum of distance square between measuring KP and the model KP minimum:

$$S = \sum_{j=1}^{n} d_j^2 \qquad (5)$$

where d_j is the distance between the measuring KP and the fitted surface KP in each scanning area, n is the number of KP in a scanning area, and $n \ll N$. So, the calculation of minimum S is much lower than that of minimum L. Therefore, the data processing time is significantly decreased and the measuring efficiency is improved.

6 CONCLUSION

Several factors lead to the difficulties of rapid measurement for large-scale products, including large measuring size, high measuring accuracy, complex product structure, and high cost of measuring instrument. The traditional measuring method cannot consider the measuring efficiency and measuring accuracy in the same time. The combination measurement method using the laser tracker and the 3D laser scanner can improve the measurement efficiency of anti-aircraft missiles, reduce labor intensity, and increase the accuracy of measurement results, which is important for the development of digital measurement technology. This approach can also be used in other areas of aerospace such as the measurement of aircraft and satellites.

REFERENCES

Acerbi-Junior F.W. et al. 2006. The assessment of multi-sensor image fusion using wavelet transforms for mapping the Brazilian Savanna. *International Journal of Applied Earth Observation and Geo-information* 8: 278–288.

Cai Guo-zhu et al. 2013. A combined application of spatial analyzer in alignment and survey of Hirfl-CSR. *Science of Surveying and Mapping* 38(4): 162–163.

Jin Zhang-jun et al. 2015. Registration error analysis and evaluation in large-volume metrology system. *Journal of Zhejiang University (Engineering Science)* 49(4): 655–660.

Jing Xishuang et al. 2015. Digital combined measuring technology assisted quality inspection for aircraft assembly. *Journal of Beijing University of Aeronautics and Astronautics* 41(7): 1197–1200.

Liping Ding et al. 2013. Combined large-scale measurement system for large aircraft assembly. *Aeronautical Manufacturing Technology* 13: 76–80.

Ma Limin et al. 2005. Geometrical features based on the new Geometrical Product Specification and verification (GPS). *China Mechanical Engineering* 16(12): 1045–1049.

Shicheng Yu et al. 2015. Study of aircraft level-testing measurement technology based on laser tracker and iGPS. *Aeronautical Manufacturing Technology* 21: 119–121.

Xuezhu Lin et al. 2013. Aircraft digital measuring technology of multi-sensor fusion. *Aeronautical Manufacturing Technology* 7: 46–49.

Zhang Fu-min et al. 2008. Multiple sensor fusion in large scale measurement. *Optics and Precision Engineering* 16(7): 1236–1240.

Zhou Na et al. 2012. Design of multi-station network arrangement for aircraft digital measurement. *Optics and Precision Engineering* 20(7): 1487–1491.

Energy Science and Applied Technology ESAT 2016 – Fang (Ed.)

A popular keypoint-based method for effective image feature extraction

Yanfen Gan
School of Information Science and Technology, Guangdong University of Foreign Studies-South China Business College, Guangzhou, China

Junliu Zhong
School of Information Engineering, Guangdong Mechanical and Electrical College, Guangdong, China

Janson Young
School of Computer, Guangdong University of Technology, Guangdong, China

ABSTRACT: Image feature extraction is the important technique in image forensics. Image feature extraction technique contains block-based and keypoint-based methods. The block-based method usually divides the image into overlapping blocks first and then applies the feature extraction method to extract block features. The drawback of this method is low efficiency. The keypoint-based method computes and extracts features without any image subdivision, but only on image regions with high efficiency. Hence, this paper presents a popular keypoint-based method by applying Scale Invariant Feature Transform (SIFT) to extract image descriptors for Copy-Move Forgery Detection (CMFD). The experimental results indicate that the SIFT keypoint-based method can achieve superior performance in CMFD, especially in simple computation and with high efficiency.

Keywords: image feature extraction; image forensics; keypoint-based; Scale Invariant Feature Transform; high efficiency

1 INTRODUCTION

With the development of modern information technology, digital image has become easy to obtain and perform. Digital image as a main source of information; its reliability becomes more and more important. Image feature extraction is an indispensable technique in image forensics. Image feature extraction technique contains block-based and keypoint-based methods. The block-based method usually divides the image into overlapping and rectangular blocks first and then applies feature extraction method to extract relevant block features. Fridrich (J. Fridrich et al, 2003) presented 256 coefficients of the Discrete Cosine Transform (DCT) as the feature extraction method. Popescu (A.C. Popescu et al, 2005) presented Principal Components Analysis (PCA)-based method to reduce feature dimensions and increase efficiency. However, extraction of DCT and KPCA (V. Christlein, et al, 2012) features is most expensive in evaluated methods. In recent years, some moment invariant methods are presented to construct the invariant features and then extract them. Yap (P. Yap et al, 2010) presented Polar Harmonic Transforms (PHTs) to construct rotated invari-

ant. PHTs consist of Polar Complex Exponential Transform (PCET), Polar Sine Transform (PST), and Polar Cosine Transform (PCT). The kernels of PHTs are simple, but these moments only have rotation invariant features, missing the scaling invariant features. Zhong (J. Zhong et al, 2016) presented Discrete Analytical Fourier–Mellin Transform (DAFMT) and designed an auxiliary disk template to detect the features of the block. However, construction of the DAFMT method is complex and the time cost is also expensive. Compared with the block-based method, the keypoint-based method computes its features without any image subdivision, but only on image regions with high efficiency. A number of keypoint-based methods are available for feature extraction. Scale Invariant Feature Transform (SIFT) (J. Zheng et al, 2016) is a popular keypoint-based method applied for extracting feature descriptors. Unlike the block-based method where features are extracted from every block, The SIFT method needs to compute only a small number of descriptors and obtains a high efficiency. Debbarma (S. Debbarma et al, 2014) presented the SIFT keypoint-based method to extract 128 dimension descriptors from the whole image. For the lower time cost, high efficiency and good extracted performance, SIFT is a popular technique for feature extraction.

The rest of the paper is organized as follows. We provide a brief review on the SIFT method in section 2. A flowchart of feature extraction applied for CMFD is given in section 3. In section 4, extensive experiments are conducted to authenticate the superior performance compared with the block-based method. Conclusions are presented in section 5.

2 REVIEW OF THE SIFT METHOD

The template file B2ProcA4.dot (for A4 size paper) or B2ProcLe.dot (for Letter size paper) is copied to the template directory. The SIFT method based on scale space has the following four steps (S. Debbarma et al, 2010):

To detectthe extrema of scale space

In this step, it is aimed at constructing Difference of Gaussian (DOG) function to detect potential interest point for rotation and scale invariance. First Gaussian filter is applied to smooth the image. Then, each octave of Gaussian Pyramid is constructed by Gaussian smoothing and subsampling of the image resolution. Each DOG is subtraction of two adjacent Gaussian Pyramids. The structure of DOG is shown in Figure 1. Interest points are compared with adjacent points to select local extrema (minimum/maximum) in the scale space. The DoG function is given as follows:

$$D(x,y,\sigma) = (G(x,y,k\sigma) - G(x,y,\sigma)) \times I(x,y) \\ = L(x,y,k\sigma) - L(x,y,\sigma) \quad (1)$$

where $I(x,y)$ represents the original image, $D(x,y,\sigma)$ represents the DOG, $L(x,y,\sigma)$ is the Gaussian Pyramid function, which is the convolution of the image $I(x,y)$ and the Gaussian filter function $G(x,y,\sigma)$ at scale $k\sigma$.

To confirm the key point positioning. At the position of interest, to determine the location and scale of key points.

To determine the direction of the local image based on the gradient direction. To assign canonical orientations for each of the key point.

To generate the key point description. To measure gradient image part in each of the key points in the field. Finally, descriptors are generated and expressed as 128 dimension feature vectors.

3 FEATURE EXTRACTION APPLIED FOR CMFD

Copy move is a very common forgery by which the partial content of the image is copied from one area and then pasted into another area in the same image. Figure 2 shows a copy move forgery. In recent years, CMFD has become increasingly popular and attracted the attention of more researchers. The experimental results of (V. Christlein, et al, 2012) verify that SIFT is clearly the best choice when a copied area has suffered from large amounts of scaled or rotated distortion. The framework of CMFD is shown in Figure 3.

To input host image.

To apply Gaussian filter to perform pre-processing operation.

To detect the maxima points and compute the key point.

To extract feature vector of the keypoint.

To use the Best-Bin-First search method derived from the kd-tree method to get approximate nearest neighbors. Then, the Euclidean distance of two regions is used as a similarity measure.

(a)Original image

(b)Forgery image

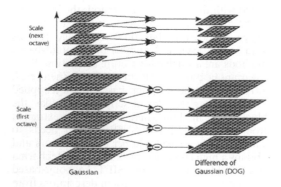

Figure 1.   Structure of DOG.

Figure 2.   Image of *Coccinella septempunctata*.

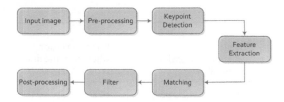

Figure 3. General framework of CMFD.

(a)Forgery image

(b)Extremal points

(c)Matched keypoints

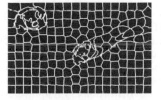

(d) SLIC segmentation

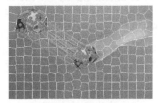

(e) Matched points in SLIC segmentation

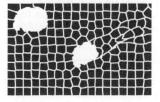

(f) Matting

Figure 4. Tested image of *Coccinella septempunctata*.

To use the correlation coefficient between two feature vectors to reduce the false matching.

To use morphological operations, such as Simple Linear Iterative Clustering (SLIC) to locate matched regions and fill them.

The feature extraction in step 4 is a narrow sense. The broader sense of feature extraction contains steps 2–6. The SIFT method can obtain the location, direction, and scaling information, and generate 128 dimension descriptors for CMFD.

4 EXPERIMENTAL RESULTS AND ANALYSIS

It is difficult to analyze and compare the overall performance, efficiency, and time cost of our proposed method and other relevant methods. In this paper, we proposed to use True Positive (TP) and False Positive (FP):

$$Precision = \frac{TP}{TP + FP} \qquad (2)$$

Precision in (2) should be used not only for evaluating the detected ratio of the image at the image level, but also for counting the matched ratio of the detected region and the ground-truth region at the pixel level. In this section, we adopt precision to analyze and evaluate the performance of our SIFT method and other relevant methods.

In this section, the tested whether datasets are downloaded from the MIC-F2000 dataset or from the Internet. The tested datasets do not conflict with copyrights. The tested dataset forms 50 high-resolution JPG color images whose size is 2048 × 1536, and 50 low-resolution JPG color images whose average size is 800 × 600.

4.1 *Performance evaluation*

The image of *Coccinella septempunctata* is taken as a test example, as shown in Figure 2. Figure 4 shows the CMFD results of the proposed SIFT method.

Figure 4(a) shows a copy move forgery. *Coccinella septempunctata* is copied and then performed scaling-up 50%, clockwise rotated at 30° distor-

Table 1. Precision comparison.

Precision	Average precision (pixel level)	Image precision when pixel precision is over 70%	Image precision when pixel precision is over 80%	Image precision when pixel precision is over 90%
SIFT	83%	96%	92%	84%
DAFMT	71%	96%	80%	62%

Table 2. Time of feature extraction.

Time of feature extraction	High-resolution	Low-resolution
Our SIFT method	20.5 s	3.2 s
DAFMT	59.1 s	8.1 s

tions. Figure 4(b) shows the extremal points of the detected image. Figure 4(c) shows the matched key points of two regions. Figure 4(d) shows the morphology segmentation by using SLIC. Figure 4(e) shows the matched keypoints in SLIC segmentation. Figure 4(f) shows the matched regions. From Figure 4, the proposed SIFT method can detect the duplicated region accurately.

4.2 Comparison with precision

In this section, we compare the precision of the proposed SIFT method and the DAFMT method [5]. The tested dataset is based on above 50 high-resolution and 50 low-resolution JPG color images. Each image is implemented by translated, 90° clockwise rotated, scaling-up 20% and added Gaussian noise distortions. First, average pixel precision of the suspicious region and the ground-truth region is counted. Based on the statistical pixel precision, the image precision is counted to describe the good or poor performance.

As shown in Table 1, the SIFT method has a higher precision at the pixel level. It means that the SIFT method can extract more useful key points. Table 1 also indicates that the level of average pixel precision increases from 70% to 90%, while the image precision of the SIFT method decreases only slightly, but the DAFMT method declines drastically. It also shows that the SIFT method can obtain a better image precision than the DAFMT method when pixel precision is over 90%.

4.3 Comparison with time costs

In this section, we compare efficiency and time costs of our SIFT method with DAFMT. The tested dataset is the same as shown in section 4.2. The tested result is summarized in Table 2.

From Table 2, it can be seen that our proposed SIFT method costs less time than the DAFMT method. It means that our proposed SIFT method has a higher efficiency in feature extraction.

5 CONCLUSION

Feature extraction is a significant step in CMFD. In this paper, we presented a popular key point-based method, namely SIFT, to extract image features for the next step of forgery detection. An extensive experiment proved that the proposed SIFT method has more precision and higher efficiency than the relevant method of CMFD, such as DAFMT.

ACKNOWLEDGMENTS

This work was supported by the 2014 Guangdong Province Young Innovative Talent (Natural Science) Class Project Fund (no. 2014 kqncx256). The authors are grateful for this support.

REFERENCES

Christlein, V., C. Riess, J. Jordan, C. Riess, E. Angelopoulou. An Evaluation of Popular Copy Move Forgery Detection Approaches. IEEE Transactions on information forgery and security, vol. 7, no. 6, pp.1841–1854, 2012.

Debbarma, S., A. Singh, K. Singh. Keypoints Based Copy-Move Forgery Detection of Digital Images. Electronics & Vision, 2014, pp.1–5.

Fridrich, J., D. Soukal, J. Lukás. Generation Detection of copy-move forgery in digital images. The Digital Forensic Research Workshop, pp.55–61, 2003.

Popescu, A.C., H. Farid. Exposing digital forgeries by detecting traces of resampling. IEEE Transactions on Signal Processing, vol.53, no.2, pp.758–767, 2005.

Yap, P., X. Jiang, A. Kot. Two-dimensional polar harmonic transforms for invariant image representation. IEEE Transactions on Pattern Analysis and Machine Intelligence, vol.32, no.7, pp.1260–1270, 2010.

Zheng, J., Y. Liu, J. Ren, T. Zhu, Y. Yan, H. Yang. Fusion of block and keypoints based approaches for effective copy-move image forgery detection. Multidimensional System Signal Process, April, 2016, DOI 10.1007/s11045-016-0416-1.

Zhong, J., Y. Gan. Detection of copy–move forgery using discrete analytical Fourier–Mellin transform. Nonlinear Dynamic, vol.84, pp.189–202, 2016.

Energy Science and Applied Technology ESAT 2016 – Fang (Ed.)
© 2016 Taylor & Francis Group, London, ISBN 978-1-138-02973-6

Digital pull image model based on smooth Bayes

Tingzhong Yu
School of Information Technology, Guizhou University of Engineering Science, Bijie, Guizhou, China

ABSTRACT: In the application of the standard Bayes algorithm in the digital image picking, it has the problems of calculating the speed delay and of picking out the image. In this paper, a digital image model based on Gauss hybrid smooth optimization Bayes algorithm is proposed. First, the standard algorithm is optimized by using the Gauss mixture model. In addition, the statistical mapping relationship between image visual features and segmentation points is established. Then, the improved Bayes algorithm is used to establish the color estimation model. Finally, a simple linear equation is simplified to be a simple linear equation. The results indicate that the Gauss hybrid smooth optimized Bayes algorithm has better accuracy.

Keywords: Bayes algorithm; smooth optimization; Gauss mixed model; digital image picking; color estimation

1 INTRODUCTION

In medical diagnosis, special effects, industrial test, and home entertainment, digital image has been widely used, with some digital images obtained directly from the natural scene. However, these images cannot satisfy people's aesthetic requirements. In the pursuit of aesthetic feeling or pure entertainment, people want to extract any part of the image they interested in, and combine other interesting part of the image to obtain a high quality synthetic image (Brinkman R. 2009). In digital image processing techniques, the cutout is a way to extract the interested part of the image technology (Guan Y et al, 2012).

In the early days of the technology of digital cutout, we usually placed the foreground image in the background of a single color, and then completely removed the background color. The background color is usually blue, so it is called the blue screen cutout (Rother C et al, 2014). Ben-Ezra Moshe put forward the use of infrared for the double technology cutout (Levin A. et al, 2013). Debevec Paul put forward the use of polarized light for the double technology cutout. The double technology put forward by Ben-Ezra Moshe (Sun J et al, 2013) can solve the problem of blue light shining, but it will cause the "according to the tooth" phenomenon at the edge of the object, and the application of this technology needs very expensive cameras, which greatly increases the difficulty in the popularization and application (McGuire M et al, 2014). In the current stage of digital image cutout algorithm development, most cutout algorithms adopt the way of artificial mark-

ers as *a priori* identification for the input image, whose marking methods are usually Trimap and Strokes (Gastal E.S.L et al, 2013). Trimap adopts the way of depicting, in which the border will be roughly divided into the image foreground region and the background region and the unknown area to calculate. Strokes uses the graffiti way to optionally tag on the image foreground region and the background region, and the unmarked rest is the unknown area to calculate (Chen X W et al, 2012). The unknown area to calculate in Trimap is usually small, and the known prospect region and the background region are larger. Therefore, the algorithm adopting the Trimap as the marker method requires relatively shorter time for cutout. In addition, Trimap is more orientative. People are more eager to know how to accurately divide Trimap as an auxiliary image cutout to obtain better results (Criminisi A et al, 2014). Accurate Trimap, however, requires more time, and for some special feature images, Trimap is idle. Strokes tag is relatively simple and convenient, and can more easily specify a special image of the foreground and background and the unknown area. However, the worker cannot clear where the tag is without guidance. Also, the cutout results of Strokes using the tag cutout algorithm are closely related to the location of the initial tag. So, the cutout effect is very poor due to the inappropriate mark in the location (Singaraju D et al, 2013). Meanwhile, for the cutout time, the tag on the image area is larger for the Strokes, and the cutout time is long.

Based on the defects of the standard Bayes algorithm in digital keyer application, this paper proposes a smooth optimal Bayes algorithm based on the Gaussian mixture model of digital keyer, and the validity of the model is verified by a simulation experiment.

2 THE DEFECT OF BAYES ALGORITHM ANALYSIS

Bayesian (Bayes theorem) algorithm is one of the numerous data analysis methods of the data mining classification method. The Bayes algorithm is based on the known data to calculate the unknown speculation, assuming that there are $C_1, C_2, ..., C_N$ sets in a data.

The following is a definite structure prediction formula:

$$P(x_{N+1} \mid D, S^h) = \int P(x_{N+1} \mid \theta_s, D, S^h) P(\theta_s \mid D, S^h) d\theta_s$$
$$= \prod_{i=1}^{n} \prod_{j=1}^{q_i} \frac{N'_{ijk} + N_{ijk}}{N'_{ij} + N_{ij}} \qquad (1)$$

The following is not a sure structure prediction formula:

$$P(x_{N+1} \mid D, S^h) = \sum_{S^h} P(x_{N+1} \mid D \mid, S^h) P(S^h \mid D) \quad (2)$$

The calculation with non-deterministic structure prediction is difficult because of all structural knowledge. So, in general, we use an average selective mode.

Given a training sample data, we can learn from the sample dataset, from which the new data can be classified into a category, rather than avoiding the new data and not knowing where to start. We can see that the Bayesian theory is consistent with this task:

$$s_{MAP} = \arg\max_{s_i \in S} P(s_i \mid x_1, x_2, ..., x_n) \qquad (3)$$

Using the Bayesian formula, this expression can be written as follows:

$$s_{MAP} = \arg\max_{s_i \in S} \frac{P(x_1, x_2, ..., x_n \mid s_i) P(s_i)}{P(x_1, x_2, ..., x_n)}$$
$$= \arg\max_{s_i \in S} P(x_1, x_2, ..., x_n \mid \qquad (4)$$

The Bayesian algorithm is based on a simple assumption, which is used to indicate that if the target value of examples is given, then the joint probability is the product of the probability of each property separately through the research:

$$P(x_1, x_2, ..., x_n \mid s_i) = \prod_k P(x_k \mid s_i) \qquad (5)$$

Substituting it into (4), the approach used in the Bayesian algorithm can be found:

$$s_{NB} = \arg\max_{s_i \in S} P(s_i) \prod_k P(x_k \mid s_i) \qquad (6)$$

Here s_{NB} is the target value of Bayes algorithm calculation. In Bayesian algorithm, estimates of the number of different $P(x_k \mid s_i)$ items from the training data are different in the different target number multiplied by the number of attribute values. This is much less than the estimated $P(x_1, x_2, ..., x_n \mid s_i)$ items.

However, in the Bayesian cutout method, there are clustering and statistical processes, so the computational speed is slow and the computing task is large.

3 SIMULATED ANNEALING ALGORITHM BASED ON GENETIC OPTIMIZATION

3.1 Smooth optimization based on Gaussian mixture

The distribution of the Gaussian mixture model can be seen as a multiple linear combination of the Gaussian distribution, in M Gaussian mixture D dimensions λ. Setting the observation vector to x, the distribution of x is:

$$p(x \mid \lambda) = \sum_{i=1}^{M} \omega_i b_i(x) \qquad (7)$$

Here ω_i is the mixed weights, satisfying $\sum_i \omega_i = 1$, and $b_i(x)$ is the D dimension Gaussian distribution:

$$b_i(x) = \frac{1}{(2\pi)^{D/2} \left| \sum_i \right|^{1/2}}$$
$$\exp\left\{ -\frac{1}{2} (x - \mu_i)^t \sum_i^{-1} (x - \mu_i) \right\} \qquad (8)$$

where μ_i is the mean vector and Σ_i is the covariance matrix.

In this paper, assuming visual image features $X : \{x_t\}, t \in [1, k]$, k is the number of visual features. In image segmentation points S_j, $j = 2$. The i joint probability distribution of the Bayes model is:

$$P_i(X_i, S_j, \theta_{S_j}) = \sum_{j=1}^{2} P_i(X_i \mid S_j, \theta_i) \omega_j \qquad (9)$$

Here $P_i(X_i, S_j, \theta_i)$ is the probability density function of j mixed component of i Bayes model and X_i is the visual feature vector of i Bayes model.

From formula (9), we set up statistical mapping of image visual feature and split points $Gap1$.

3.2 Digital cutout based on improved Bayes theorem

We must first define the rational Bayesian framework, and then establish a pixel color model to estimate the model and the calculation of the α value. Finally, we use the maximum *a posteriori* probability to get the optimal α. Given the unknown pixels C, the optimal estimate of a probability distribution is an optimization problem:

$$\arg\max_{F,B,\alpha} P(F,B,\alpha\,|\,C) = \arg\max L(C\,|\,F,B,\alpha) \\ + L(F) + L(B) - L(\alpha) \quad (10)$$

where $L(\cdot) = \log p(\cdot)$, F, B and α are unknown quantity, and $P(C)$ is a constant and can be rejected.

We define $L(C\,|\,F,B,\alpha)$ as a Gaussian distribution with standard deviation σ_c and center $\bar{C} = \alpha F + (1-\alpha)B$:

$$L(C\,|\,F,B,\alpha) = -\|C - \alpha F - (1-\alpha)B\|^2 / \sigma_c^2 \quad (11)$$

Clustering of sample points in RGB color space satisfies Gaussian distribution. The weighted mean and covariance are $\bar{F}$ and $\sum_F$, respectively:

$$\bar{F} = \frac{1}{W'}\sum_{i\in N} w_i F_i \quad (12)$$

$$\sum_F = \frac{1}{W}\sum_{i\in N} w_i (F_i - \bar{F})(F_i - \bar{F})^T \quad (13)$$

where $w_i = \alpha_i^2 g_i$, $W = \sum_{i\in N} w_i$, g_i is a Gaussian decay function with distance as the parameter. $L(F)$ also satisfies the Gaussian distribution:

$$L(F) = -(F - \bar{F})^T \sum\nolimits_F^{-1} (F - \bar{F})/2 \quad (14)$$

$L(B)$ satisfies the Gaussian distribution like $L(F)$, except that it only needs to take $1 - \alpha_i$ instead of α_i in w. Solving of the Gauss equation is essentially an iterative process. First, we assume a value, calculate the partial derivative of type (10) on the right side, and get a 6 yuan system of equations:

$$\begin{bmatrix} \sum_F^{-1} + I\alpha^2/\sigma_c^2 & I\alpha(1-\alpha)/\sigma_c^2 \\ I\alpha(1-\alpha)/\sigma_c^2 & \sum_B^{-1} + I(1-\alpha)^2/\sigma_c^2 \end{bmatrix} \begin{bmatrix} F \\ B \end{bmatrix} \\ = \begin{bmatrix} \sum_F^{-1}\bar{F} + C\alpha/\sigma_c^2 \\ \sum_F^{-1}\bar{B} + C(1-\alpha)/\sigma_c^2 \end{bmatrix} \quad (15)$$

So, we simplify the complex Gaussian equation for a simple system of linear equations to solve the problem.

We then use the results of type (15) using the projection method to calculate the new value, and then plug in type (16):

$$\alpha = \frac{(C-B)^T (F-B)}{\|F-B\|^2} \quad (16)$$

If the sample is divided into several clusters, we solve all the samples of the cluster, and take the solution to maximize the value of type (10).

4 ALGORITHM SIMULATION

To verify the effectiveness of the algorithm proposed in this paper, we perform a simulation. We dig out

Figure 1. Standard algorithm for digital pull like results.

Figure 2. Improved algorithm of digital pull like results.

the digital image with standard Bayes algorithm and Gaussian mixture smooth optimization Bayes algorithm in this paper. The results are presented below.

As can be seen from the comparison results shown in Fig. 1 and Fig. 2, the Gaussian mixture smooth optimization Bayes algorithm proposed in this paper has a better digital image picking performance.

5 CONCLUSION

Cutout technology is a kind of image processing technology. The image processing technique extracts the image that the users need from the whole image, which is a separation technology of the prospect of the image part and background part. Based on the defects of the standard Bayes algorithm in digital keyer application, this paper proposes a smooth optimal Bayes algorithm based on Gaussian mixture model of digital keyer, and the validity of the model is verified by conducting a simulation experiment.

ACKNOWLEDGMENTS

This work was supported by the Science and Technology Foundation of Guizhou Province, "China, Digital image matting and synthesis algorithm research based on OpenCV" (LH No. 2014–7539).

REFERENCES

Brinkman R. "The Art and Science of Digital Compositing". Morgan Kaufman, 2009, pp. 23–44.

Chen X.W., Zou D.Q. and Zhao Q.P., et al. "Manifold preserving edit propagation". ACM Transactions on Graphics, 2012, pp. 132–135.

Criminisi A., Sharp T. and Rother C, et al. "Geodesic image and video editing". ACM Transactions on Graphics, 2014, pp. 134–140.

Gastal E.S.L and Oliveira M.M. "Shared sampling for real-time alpha matting". Computer Graphics Forum, 2013, pp. 575–584.

Guan Y., Chen W. and Liang X, et al. "Easy matting: a stroke based approach for continuous image matting". Computer Graphics Forum, 2012, pp. 567–576.

Levin A. Acha R. and Lischinski D. "Spectral matting". IEEE Transaction on Pattern Analysis and Machine Intelligence, 2013, pp. 1699–1712.

McGuire M., Matusik W. and Pfister H, et al. "Defocus video matting". ACM Transactions on Graphics, 2014, pp. 567–576.

Rother C., Kolmogorov V. and Blake A. "'GrabCu': interactive foreground extraction using iterated graph cuts". ACM Transactions on Graphics, 2014, pp. 309–314.

Singaraju D. and Vidal R. "Estimation of Alpha Mattes for Multiple Image Layers". IEEE transactions on Pattern Analysis and Machine Intelligence, 2013, pp.1295–1309.

Sun J., Li Y. and Kang S B, et al. "Flash matting". ACM Transactions on Graphics, 2013, pp. 772–778.

Energy Science and Applied Technology ESAT 2016 – Fang (Ed.)
© *2016 Taylor & Francis Group, London, ISBN 978-1-138-02973-6*

A survey on deep neural networks for human action recognition in RGB image and depth image

Hongyu Wang

Department of Applied Mathematics, Northwestern Polytechnical University, China

ABSTRACT: Human action recognition has been a significant topic in computer vision and video surveillance. With the development of in-depth learning, the application of deep neural networks in related research is increasingly prevalent. This paper provides a survey on deep neural networks for human action recognition, which focus on the methods based on color images or depth images. The detailed description of each method is explained, and several related main datasets are briefly introduced. In this study, all papers published from 2013 to 2015 are reviewed, which provides an overview of the progress in this area.

Keywords: video surveillance, human action recognition, deep neural networks

1 INTRODUCTION

Human action recognition has been drawing more and more attention in computer vision, which is partially due to the widespread availability of cameras. The information on images and videos is so abundant and detailed that it is qualified to be analyzed for human action recognition (Poppe R, 2010). Human action recognition is a sophisticated but valuable issue in many fields for real-world applications Due to the difficulty in manual monitoring of large amounts of data from vision sensors, intelligent video surveillance is greatly worthy of study (Liu Y, 2012). So, more researchers have shifted their attention to human action recognition, and attempted to apply it to the human–computer interaction.

Nowadays, the study on human action recognition in general involves four aspects: gestures, actions, interactions, and group activities (Saad A, 2010). Gesture refers to a static state that is about a certain movement of body such as standing or bowing. Action generally consists of several sequential gestures such as running and waving. Interaction indicates a correlative action involving two persons or a person and an object such as brawling. Group activity always involves many persons such as meeting. Obviously, the above aspects range from simple to complex, and there are different-level approaches to deal with them. The overview process of handling these tasks consists of feature extraction, action representation learning, and classification (Tanaya G, 2012). Furthermore, the interested areas vary and need to be solved. So, the global and local features are extracted, respectively.

For example, background subtraction is a usual approach to extract global features, while local features are obtained by semantic segmentation.

Industrial digital camera is a novel camera that is capable of working robustly in a variety of complex scenes. The captured images contain RGB color. Each pixel in RGB images records a numerical value from 0 to 255, which indicates various colors in red, green and blue channels. However, RGB images only reflect two-dimensional information of the real world. To overcome this shortcoming, many techniques have been developed. Kinect is an outstanding equipment that acquires the distance between objects in view of camera. It is an intelligent combination of an infrared projector and an infrared camera to obtain depth information by utilizing Structured Light or ToF (Time of Flight) (Ganapathi V, 2010). The RGB image displays the color information, while the depth image shows the information of distance. There are multiple effective ways to obtain spatial information. Therefore, a large amount of useful images of objects encourage the development of human action recognition greatly.

In the recent years, many researchers have attempted to find many ways to solve the problems on human action recognition. Many methods are proposed for RGB images and depth images (Salas J, 2010) with the development of camera. Moreover, deep neural networks are especially and widely employed. There are several frequent and classical deep neural networks such as CNN (Convolutional Neural Network) (Salama M.A, 2010), DBN (Deep Belief Network), RNN (Recurrent Neural Network) (Gers F.A, 2003), and Denoising

Autoencoder (Memisevic R, 2010). CNN is a kind of feed-forward neural network inspired by network topology between neurons of visual cortex. The fundamental structure of CNN contains a convolutional layer and a sub-sampling layer, which aim to learn features and to sample from the data. The shared weights in CNN decrease the complexity of the network, so CNN is appropriate to be applied to large data. Likewise, Denoising Autoencoder simulates the animal's vision and is designed to cope with unlabeled data. It performs feature extraction in the form of unsupervised learning. DBN could be viewed as a probabilistic generative model. The component of DBN is RBM (Restricted Boltzmann Machines), which is a stochastic neural network to learn the probability distribution from the data. In brief, DBN consists of a number of RBMs in the stacked form. The structure of DBN makes it suitable for recognition and classification. Compared with forward propagation in CNN, the output of each neuron in RNN is not only an input for the next neuron, but also acts on itself. Based on its characteristics, RNN works on extraction of temporal features from the data effectively.

Further extending on these bases, many researchers have been continuously proposing their novel methods. For instance, Karpathy applied CNN to both high-resolution and low-resolution single frame images to obtain more information (A. Karpathy, 2014). Ji *et al.* exploited the conventional CNN to deal with the raw input video, which is able to extract spatio-temporal features directly (Shuiwang J, 2013). In the case of exploring temporal features, RNN with LSTM is adjusted to model the context between continuous frames, as done by Baccouche *et al.* (Baccouche M, 2010). Xie *et al.* proposed a pyramidal architecture to preprocess depth images and acquire the feature about human action (Xie L, 2014). In many papers, Denoising Autoencoder is always used as a significant component in their architecture. The model proposed by Memisevic is stable to learn valuable mapping with Denoising Autoencoder (Memisevic R, 2011).

This paper presents the state-of-the-art methods about human action recognition proposed in recent years. All papers reviewed were published between 2013 and 2015. Most of them are from CVPR, ECCV, PAMI, etc.; some latest papers are just submitted on arXiv. More precisely, the focus points in this paper are the methods based on RGB images and depth images. Through the review of these papers, a general framework of the methods is shown in Fig. 1. A detailed discussion on relevant methods is provided in the following sections.

The rest of this paper is organized as follows. In section 2, the datasets used in experiments are

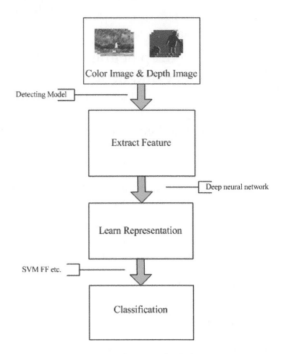

Figure 1.   The general framework of the methods.

reviewed briefly. Section 3 and section 4, respectively, describe the methods on RGB images and depth images in detail. The conclusion and future works are presented in section 5.

2   DATASET

In this section, the description of datasets that are regarded as a benchmark in the current research is provided (Table 1). Some datasets are classic and acknowledged such as KTH. In addition, there are also some new datasets and several datasets for specific problems.

2.1   *The KTH dataset*

The KTH dataset (Schuldt C, 2004) is a benchmark in the relevant field, which contains six actions including walking, jogging, running, boxing, hand waving, and hand clapping (see Fig. 2). Each action is done by 25 persons in four different scenes. There are a total of 600 videos in the dataset. The background of frames is relatively static, although the focal length changes slightly. All sequences are taken with a camera with a 25fps frame rate and are downsampled to 160×120 pixels.

Table 1. Description of datasets.

	Action types	Resolution	Scenes	Background	No. of videos	Format
KTH	6	160 × 120	4	static	600	AVI
Weizmann	10	180 × 144	N/A	static	90	AVI
HOHA2	12	N/A	10	cluttered	3669	AVI
TRECVID	many	N/A	many	cluttered	many	MPEG-4/H.264
UCF101	101	320 × 240	many	cluttered	13320	AVI

Figure 4. Twelve different scenes in Hollywood movies from the HOHA2 dataset.

Figure 2. Example of different classes from the KTH dataset.

Figure 3. Three classes of action in the Weizmann dataset.

2.2 The Weizmann dataset

This dataset (Lena G, 2007) consists of 90 videos. The videos show 10 types of action performed by nine persons including walking, running, jumping, gallop sideways, bending, one-hand waving, two-hand waving, jumping in place, jumping jack, and skipping. The frame is not quite complex because both the background and view angle are static. Moreover, it also contains videos labeled with foreground contour. Some examples are shown in Fig. 3.

2.3 HOHA (Hollywood Human Actions)-2 dataset

This dataset (Laptev I, 2008) provides 12 classes of human actions and 10 classes of scenes that are from 69 Hollywood movies. The 12 classes are about answering phone, getting out a car, hand shaking, hugging, etc. Then, the classes of scenes involved are leaving house, road and entering bedroom, car,

hotel, kitchen, living room, office, restaurant, and shop. There are twelve different scenes shown in Fig. 4. The whole action samples were collected by means of automatic script-to-video alignment in combination with text-based script classification.

2.4 TRECVID dataset

This dataset (Over P, 2015) is composed of many outdoor scenes in which actions occurring naturally. The video is continuously captured from the real world and about normal population instead of actors. The video of data includes Internet Archive videos, airport surveillance videos, video files from the BBC, as well as images and videos from Yahoo, which are mostly in the MPEG-4/H.264 format. Several examples extracted from the TRECVID dataset are shown in Fig. 5.

2.5 UCF101 dataset

UCF101 (Khurram Soomro, 2012) is an action recognition dataset of realistic action videos. It has 101 categories from 13320 videos, which provides the large diversity in terms of actions. It could be a challenging dataset due to the presence of large variations of camera motion. The videos can be grouped into 25 groups, where each group consists of 4 to 7 videos of action. There are five types of action categories including Human–Object Interaction, Body-Motion Only, Human–Human Interaction, Playing Musical Instruments and Sports. Fig. 6 shows some examples of UCF101.

Figure 5. Six different types of examples from the TRECVID dataset.

Figure 6. A lot of examples in the UCF101 dataset.

3 METHODS FOR RGB IMAGE

The structure, algorithm, and model of all reference articles will be presented below. Ji *et al.* proposed a novel 3D CNN to break through the limitation of 2D CNN (Lecun Y, 2004). 3D convolutional kernel is performed on STV (Spatio-Temporal Volume) with the feature maps in the convolutional layer connected with several continuous frames, which can obtain more spatiotemporal features. The structure of 3D CNN includes a hardwired layer, three convolution layers, two subsampling layers with max-pooling, and a full-connected layer. Seven continuous down-sampling frames are chosen as the cube of 3D convolution kernels. In the hardwired layer, the raw input is processed and concatenated in five channels including gray level, the gradient of x and y, and the optic flow of x and y. Then, two convolution kernels perform convolution operation on five channels; particularly, the

optic flow of x and y is obtained between two continuous frames. This paper takes the information in space and time into account simultaneously and extracts the features from multiple channels. The method proposed in this paper can process raw images directly, avoiding the noise of background image.

Zhang *et al.* proposed a deep network to model the attribute of the image. This paper introduced a method called poselet (L. Bourdev, 2011) to segment an image according to semantics. In this paper, an image is segmented into many patches. Then, a CNN is trained for each patch and concatenates the high-level activation in all CNN as a standard representation of a certain action. Finally, these representations are used to predict by using a linear SVM. The structure of CNN applied consists of four sets of convolution, max-pooling, normalization, and a fully connected layer. Each network branch outputs a fully connected layer to obtain an attribute for classification. To take the global region of an image, this paper also establishes another CNN with a similar structure of the above CNN. The contribution of this paper is that it decreases the scale of CNN and the number of training sample by segmenting images to semantic patches. By taking both global and local features into account, this paper achieves high performance.

The scheme of deep learning proposed by Moez Baccouche *et al.* is composed of two basic steps, namely an extension of CNN to 3D and a RNN. It could classify human action without any prior knowledge. It consists of 10 layers, including the input layer, two sets of alternating convolutional, rectification and sub-sampling layers, a third convolution layer, and two neuron layers. After trained on the KTH dataset, the above 3D CNN is capable of extracting features from raw inputs automatically. Based on the consideration of utilizing the temporal evolution of the features, these features are fed to a RNN to perform classification. Inspired by Gers, Felix A *et al.* (Gers F A, 2003) applied a RNN with LSTM, which is efficient to label sequences of descriptors about actions. The method is a combination of CNN and RNN, which extracts spatial and temporal information, respectively. Its main characteristic is that it learns representations and classifies sequences of human actions automatically without relying on *a priori* modeling.

Based on the achievement of (Hyvärinen A, 2009), Le *et al.* proposed an extension of the ISA (Independent Subspace Analysis) algorithm combined with deep learning techniques, including stacking and convolution. The method proposed in this paper aims to learn invariant spatio-temporal features and hierarchical representations from unlabeled video data. The ISA is capable of learning features from an unlabeled image without supervision. An ISA network consists of two layers. The ISA becomes a basic part in this method because of its robustness to local translation. For coping with large data, this paper introduces a CNN architecture where PCA and ISA learn features as subunits. After the ISA network trained on small input patches, the network is convolved with a larger region of the input image. Then, the responses of the step are inputted to the next ISA network with it whitened by PCA. The contribution of this paper is to expand the application of ISA in large data. Due to the translation invariance of the features learned, this method is competent to recognize human actions.

For capturing appearance and motion, Karen Simonyan and Andrew Zisserman devised a deep convolutional network architecture, which is divided into two streams. Generally, the spatial and temporal information in video are the components of interest. Therefore, each stream is actually a deep convolutional network that extracts spatial and temporal features, respectively. The configuration of two streams is identical, which is comprised of five convolutional layers, two fully-connected layers, and a softmax classifier. One of the streams is applied to analyze spatial information on a single frame; the other aims to describe the optical flow between continuous frames. Because the available datasets are small, multi-task learning is implemented, which utilizes two different datasets to train two softmax layers, respectively. As it can be seen, the method captures both spatial and temporal features throughout the two-stream architecture. The application of multi-task learning can effectively improve the performance, despite the data are limited.

The latest papers focus on learning data-driven features. The method proposed by F. Husain *et al.* can extract features from raw video data. It is based on the VGG network used in (Simonyan K, 2014), which is a 19-layer network. First, the first 16 layers of the VGG network are adopted for describing high-level spatial features. Afterwards, a 3D convolutional layer followed by three fully-connected layers is added to it, which composes a 20-layer network. There exists a fact that a deep network could learn different level features due to its characteristic of layer hierarchy. Therefore, for capturing local temporal changes, another network simpler than the first network is introduced, which contains eight 3D convolutional layers, five max-pooling, and two fully-connected layers. The sample videos are fed together into the above two networks. Throughout the two networks, a high-dimensional feature is processed by max-pooling to reduce the dimension. Finally, a softmax model

is built to perform classification. This method applies two different level networks to extract the spatial and temporal features, respectively. The advantage of the method lies on its high accuracy and less computational time without requiring any preprocessing or *a priori* knowledge on data capture.

4 METHODS FOR DEPTH IMAGE

A deep learning paradigm to recognition human actins is exploited by Pasquale Foggia *et al.* Inspired by (Megavannan V, 2012), this paper uses the method to capture three types of feature of raw data: ADI, MHI, and DDI. Particularly, ADI and MHI are encoded by Hu moment features, while DDI is encoded by R transform and Min-Max Depth Variations. Hu moment features estimate the distribution of pixels, while R transform and Min-Max Depth Variations capture the local attribute of images. Then, ADI, MHI, and DDI are concatenated as the first-level feature vector. The above step extracts the low-level features, which are as the input for DBN. For obtaining high-level features, the DBN used in this paper is trained by several RBMs that could gain hierarchical features from inputs. The DBN consists of six different scales of RBMs. These RBMs is stacked to a pyramidal structure. On the whole, the task is to model the joint distribution between the first-level feature vector and hidden units in RBMs. By defining an energy function, the coefficients linking every layer could be calculated. After extracting the whole features, the classification is implemented by a feedforward neural network, with an output layer whose size is equal to the number of the classes added. As it can be shown, the major advantage of this method is that the high-level representation is automatically extracted in an unsupervised way.

Liu *et al.* proposed a novel deep architecture to model the human action. The architecture consists of two crucial modules: local depth feature extraction and deep architecture representation learning. For extracting local features from the depth image, this paper introduces a creative descriptor abbreviated as CCD (Comparative Coding Descriptor). From the distance information used, the CCD depicts the structural relations between spatio-temporal points in action volumes and the related temporal transformations of the moving object. Specifically, the CCD feature is generated from an atomic cuboid. In an atomic cuboid, the depth values that record the absolute distances from the object to the depth sensor are encoded in a spiral order. Based on these local features, namely CCD, DA-CCD (Deep Architecture of CCD) could be constructed. The deep architecture proposed in this paper can learn multiple-level representation to obtain high-level semantic information automatically. Each layer in the architecture is able to generate a more high-level and more abstract representation. The basic component of this deep architecture is a denoising autoencoder (Vincent P, 2008). It learns data representation that is more robust by reconstructing raw data. For instance, X is a set containing N samples and several of them are corrupted by random element removal. After introducing an encoder mapping and a decoder mapping, the input X could be reconstructed. The encoder mapping and the decoder mapping can be calculated by using a non-iterative algorithm (Chen M, 2012). Finally, the DA-CCD feature is obtained by concatenating the hidden representation of all the layers and the original input histogram vector x of CCD. The deep architecture proposed could extract hidden factors from high-level semantic information. DA-CCD learns more high-level semantic information as the architecture becomes deeper, which is appropriate for low-dimensional local depth features.

Hu *et al.* proposed a deep learning architecture for human action recognition (Xie L, 2014). On account of the popularization of 3D vision sensor, the method is based on depth images. Instead of feeding raw data to the network, the depth images are needed to be preprocessed. In the process of pretreatment, the silhouette information is isolated and the contour images are extracted, which scales a raw depth image into 28×28. The hidden deep neural network is based on the de-noising auto-encoder. In particular, three de-noising auto-encoder layers are stacked to form a five-layer pyramidal architecture with the addition of the input layer and the output layer. As it is known, de-noising auto-encoder can reduce the noise of weight trained by combining the corrupted version with the raw data, which reserves the uncorrupted input for reconstructing. By adjusting the fraction of corruptions and the learning rate, the deep network has a good classification accuracy. Due to the characteristics of the de-noising auto-encoder and the deep learning architecture, the method could achieve robust results.

Wang *et al.* proposed a novel method composed of WHDMM (Weighted Hierarchical Depth Motion Maps) and three-channel deep CNN (P. Wang). The stages of the method include simulating different viewpoints, construction of WHD-MMs, and CNN initialization. There is a need to synthesize different viewpoints to capture a plenty of spatio-temporal information because the images from the RGB-D camera are not absolutely true 3D images. Specifically, a pixel in a depth map could be converted into a 3D point by implementing a virtual movement of the RGB-D camera, which involves matrix transformation about rotation and translation. Then, both the raw and synthesized depth images are used to extract features. As for construction of WHDMM, it maintains motion information by calculating the motion energy. For expanding speed invariance and decreasing noise, it generates hierarchical temporal scales via down sampling, which also increases training samples. Finally, it gives a larger weight to recent frames to discriminate motion direction. Moreover, pseudo-color coding is introduced to WHDMM which is capable of enhancing motion patterns of actions and improving the signal-to-noise ratio. After coding, three CNN are trained on the WHDMMs, which include five convolutional layers, three fully-connected layers, and a fusion layer, respectively. Given a test sample, the output of the network represents the final class score. The innovations of this method are that it obtains more spatial information by means of virtual rotation, and the introduction of pseudocolor coding improves the signal-to-noise ratio. Therefore, the method can achieve good performance in face of a dynamic or cluttered background.

5 CONCLUSION

As the application of human action recognition such as intelligent video surveillance becomes more and more widespread, it has not only been an active area, but also a challenging task in computer vision. In recent years, a trend of the research about human action recognition is popular, which addresses the issue by establishing deep neural networks. This is because deep learning neural networks can learn hierarchical nonlinear function relation to model the vision system. Most conventional methods construct handcrafted features for recognition depending on domain knowledge, which is time-consuming and inefficient. To tackle the deficiencies of the handcrafted feature, the methods with a deep learning architecture are driven by data, which are capable of extracting features from the original data automatically.

The typical datasets are generally divided into color and depth images. A deep network is devoted to extract the spatial and temporal features or local and global features. Similarly, the concerned aspects in most of the methods for color or depth images contain spatial and temporal features, as well as local and global features. As an example, CNN is usually applied to capture spatial features. Some of its extended methods are capable of obtaining spatio-temporal or strike a balance between the local and global features. Although much research is devoted to some datasets that provide a simple and static background, more and more research dedicates to real-world scenes. As the accuracy is improved, its significance in practice is boosted gradually. The area for human action recognition will be more and more active.

REFERENCES

Baccouche M., Mamalet F. & Wolf C. et al. Sequential Deep Learning for Human Action Recognition (M)// Human Behavior Understanding. Springer Berlin Heidelberg, 2011:29–39.

Bourdev L., S. Maji & J. Malik, "Describing people: A poselet-based approach to attribute classification," 2011 International Conference on Computer Vision, Barcelona, 2011, pp. 1543–1550.

Chen M., Xu Z. & Weinberger K., et al. Marginalized Denoising Autoencoders for Domain Adaptation (J). Computer Science—Learning, 2012.

Ganapathi V., Plagemann C. & Koller D., et al. Real Time Motion Capture Using a Single Time-Of-Flight Camera(C)// IEEE Conference on Computer Vision & Pattern Recognition. 2010:755–762.

Gers F.A., Schraudolph N.N. Schmidhuber, J.&#, et al. Learning precise timing with lstm recurrent networks (J). Journal of Machine Learning Research, 2003, 3(1):115–143.

Hyvärinen A., Hurri J. & Hoyer P.O. Natural Image Statistics (J). Computational Imaging & Vision, 2009.

Karpathy, A. G. Toderici, & S. Shetty, et al. Large-Scale Video Classification with Convolutional Neural Networks(C)// IEEE Conference on Computer Vision and Pattern Recognition. 2014:1725–1732.

Khurram Soomro, Amir Roshan Zamir and Mubarak Shah, UCF101: A Dataset of 101 Human Action Classes From Videos in The Wild, CRCV-TR-12-01, November, 2012.

Laptev I, Marszalek M, Schmid C, et al. Learning realistic human actions from movies(C)// Conference on Computer Vision and Pattern Recognition. IEEE Computer Society, 2008:1–8.

Lawrence S., Giles C.L. & Tsoi A.C., et al. Face recognition: a convolutional neural-network approach. (J). IEEE Transactions on Neural Networks, 1997, 8(1):98–113.

Lecun Y., Huang F.J. & Bottou L. Learning Methods for Generic Object Recognition with Invariance to Pose and Lighting(C)// IEEE Computer Society Conference on Computer Vision & Pattern Recognition. 2004:II-97-104 Vol.2.

Lena G., Moshe B. & Eli S., et al. Actions as space-time shapes.(J). IEEE Transactions on Pattern Analysis & Machine Intelligence, 2007, 29(12):2247–2253.

Liu Y., Fan J.L. & Wang D.W. Review of intelligent video surveillance with single camera(C)// International Conference on Machine Vision. 2012:834923-834923-5.

Megavannan V., Agarwal B. & Venkatesh Babu R. Human action recognition using depth maps(C)// Signal Processing and Communications (SPCOM), 2012 International Conference on. IEEE, 2012:1–5.

Memisevic R. Gradient-based learning of higher-order image features(C)// IEEE International Conference on Computer Vision. IEEE, 2011:1591–1598.

Memisevic R. Gradient-based learning of higher-order image features(C)// IEEE International Conference on Computer Vision. IEEE, 2011:1591–1598.

Over P., Fiscus J. & Sanders G., et al. TRECVID 2012 – An Overview of the Goals, Tasks, Data, Evaluation Mechanisms, and Metrics(J). Chinese Journal of Oceanology & Limnology, 2015, 33(2):1–8.

Poppe R. A survey on vision-based human action recognition (J). Image & Vision Computing, 2010, 28(6):976–990.

Saad A. & Mubarak S. Human action recognition in videos using kinematic features and multiple instance learning. (J). IEEE Transactions on Pattern Analysis & Machine Intelligence, 2010, 32(2):288–303.

Salama M.A., Ella Hassanien A. & Fahmy A.A. Deep Belief Network for clustering and classification of a continuous data(C)// The IEEE International Symposium on Signal Processing and Information Technology. IEEE Computer Society, 2010:473 477.

Salas J. & Tomasi C. People Detection Using Color and Depth Images(C)// Pattern Recognition—Third Mexican Conference, MCPR 2011, Cancun, Mexico, June 29–July 2, 2011. Proceedings. 2011:127–135.

Schuldt C., Laptev I. & Caputo B. Recognizing Human Actions: A Local SVM Approach(C)// Proceedings of the Pattern Recognition, 17th International Conference on (ICPR'04) Volume 3 - Volume 03. IEEE Computer Society, 2004:32--36.

Shuiwang J. Ming Y. & Kai Y. 3D convolutional neural networks for human action recognition.(J). IEEE Transactions on Pattern Analysis & Machine Intelligence, 2013, 35(1):221–31.

Simonyan K. & Zisserman A. Very Deep Convolutional Networks for Large-Scale Image Recognition (J). Eprint Arxiv, 2014.

Tanaya G. & Rabab Kreidieh W. Learning sparse representations for human action recognition (J). IEEE Transactions on Pattern Analysis & Machine Intelligence, 2012, 34(8):1576–88.

Vincent P., Larochelle H. & Bengio Y., et al. Extracting and composing robust features with denoising autoencoders(C)// International Conference, Helsinki, Finland, June. ACM, 2008:1096–1103.

Wang P.; W. Li; Z. Gao; J. Zhang; et al. C. Tang; P.O. Ogunbona, "Action Recognition From Depth Maps Using Deep Convolutional Neural Networks," in IEEE Transactions on Human-Machine Systems, vol. PP, no.99, pp.1–12.

Xie L., Pan W. & Tang C., et al. A pyramidal deep learning architecture for human action recognition (J). International Journal of Modelling Identification & Control, 2014, 21(2):139–146.

Xie L., Pan W. & Tang C., et al. A pyramidal deep learning architecture for human action recognition (J). International Journal of Modelling Identification & Control, 2014, 21(2):139–146.

Energy Science and Applied Technology ESAT 2016 – Fang (Ed.)
© 2016 Taylor & Francis Group, London, ISBN 978-1-138-02973-6

License plate recognition technology in the Chinese license plates

Tong Su & Shan Liu
Xiamen University, Fujian, China

ABSTRACT: In recent years, License plate recognition technology has become an important research topic in the field of intelligent transportation. Simply put, license plate recognition is using pattern recognition method to identify vehicles from their license plates. The crucial algorithm of license plate recognition is typically divided in three components, license plate localization, character segmentation, and character (numbers and Chinese characters) recognition. The paper presents morphological method to process the original image, and then use the traversal algorithm to segment the license plate in image. Moreover, the paper applies the method of Unicom Domain Detection to accomplish segmentation of the characters in the license plate. In the final step, the K-Nearest Neighbor algorithm is used to recognize the segmented characters. This paper also included the optimization method of image segmentation and slant correction of license plate in image. By adding these two processes, the rates of recognition can be improved.

Keywords: license plate recognition, segmentation, neural network, optimization method, slant correction

1 INTRODUCTION

1.1 Research situation

With the continuous economic and social development, the number of motor vehicles is also increasing. We must find out a solution to improve the management efficiency of the vehicle. Vehicle license plate location and recognition is one of the most important research topics in computer vision and pattern recognition technology. In the field of intelligent transportation applications, this technology also has quite a wide range of applications and great potential market value, which means the ability to crate tremendous social benefits.

Automatic License Plate Recognition research started earlier abroad. As early as the 1980s. After 1990s, systematic study of automatic license plate recognition started. Classic example is that A.S. Johnson came up with the vehicle license plate automatic identification system should be divided in Image Segment, Feature Extraction and the Template Formation, Character Recognition.

As Chinese license plate format and are quite different from other countries, we can only reference the study of foreign license plate recognition system, but not directly applied. In addition, the precious position of license plate is not unique.

1.2 The research of this paper

Learn more in the relevant research results, my research focused on the aspect of accurate locating license plate in the photos that taken under different conditions, and to maximize the recognition rate of license plate characters. For instance, identification of license plate photo that been taken in different angles, distance. Which to some degree, solved the difficulty, due to diverse conditions caused by image capture device to capture the image quality unevenly, in practical applications. Eventually improved the adaptability of the license plate recognition system.

2 METHOD

2.1 Overall design

As shown in Figure 1, the entire license plate recognition system consists of several parts. The first process procedure is to locate the license plate image, and then tilt correction, followed by detection and segmentation of the license plate. Processing continues for the segmented license plate, as shown in Figure 1, character segmentation and character recognition.

2.2 Plate localization

Most of the color images are based on the RGB color model. Therefore, when we are dealing with

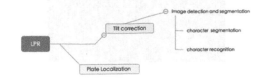

Figure 1. The overall design flow.

images, respectively, to the three RGB components of processing. But as a matter of fact, these three components do not reflect the morphological features of RGB image that we need for character segmentation and recognition. Thus, what we should make the image to a gray one at the beginning of whole processing to reduce the scale of data. This processing can help us shorten the processing time but without losing the character information required to identify the license plate.

Also reducing storage space and raise operation efficiency. And in China, the major background colors of license plates are blue. In order to extract and delimit character efficiently, we choose blue channel to accomplish shay processing. Next, in order to extend the dynamic range of the original image's pixel value, to change and improve the contrast of shades of gray, which is sharper and ease of handling, the method of histogram equalization is used in image.

After the equalization, we have to localize the edge of license in the image. The detection of thin license's edge is based on Robert Operator. But the image we acquired from this operator not only contains the edges of the plate or character, but also mixed with edge information of other objects. Since that, we have to apply morphological processing on the image contained useless information.

In this step, we should use etch operation to remove some thin contours, as shown in Figure 2. Then Closing Operation is applied to remove dark details in the image, while maintaining bright detail unaffected.

After that, we use open operation to remove small bright details and maintain a large bright areas unchanged, which means we remove the little size object, as shown in Figure 3.

2.3 Tilt correction

As a result of the shooting of license plates are mostly not quite right against the license plate, we need correct the tilt of image. Basically inclined plate can be approximated as a parallelogram. There are three kinds of tilts: horizontal tilt, vertical tilt, blend tilt.

Figure 2.   The etched image.

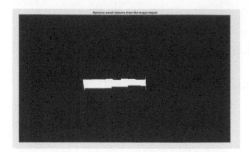

Figure 3.   Remove small objects from the target object.

Figure 4.   Sheared plate image.

The character does not occur wrong cut when the plate is horizontal tilt. There is a tilt angle α between the x-axis spindle and the horizontal axis x' of plate area in the image.

If we find the angle α and rotate the image to –α, we can attain the ideal image that we can use to recognition. When the car is on the road, broadside parallel to the horizontal axis, which also means vertical tilt angle is very small. Thus we can ignore vertical tilt direction.

In general, tilt of virtual plate consists of two parts: the tilt of license plate itself, there is a certain angle between plate and the horizontal axis, the second one is there is a tilt angle between the plane of image and the plane of the license plate. The total angle of tilt can be considered not more than thirty degrees.

We use the radon transform as tilt correction method. Radon transformation refers to do the projection on the matrix of given ray digital image.

Let f (x, y) be a two-dimensional image, the mathematical representation of the Radon transformation is as follows:

$$R_\theta(x') = \int_{-\infty}^{+\infty} f(x'\cos\theta - y'\sin\theta, x'\sin\theta - y'\cos\theta)dy'$$

Wherein

$$\begin{bmatrix} x' \\ y' \end{bmatrix} = \begin{bmatrix} \cos\theta & \sin\theta \\ -\sin\theta & \cos\theta \end{bmatrix}\begin{bmatrix} x \\ y \end{bmatrix}$$

The essence of radon transformation is an image projector in a different direction. For instance, f (x, y) in Figure 5 of could represent an image, and R (x') is the projection of the image to the bottom

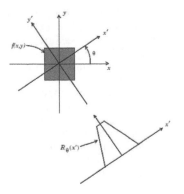

Figure 5. Radon transformation.

Figure 6. Radon transformation process.

right. And performance of this method, in mathematics, is the line integral projection direction, and that the image manifestation mode is reflected in the cumulative pixel along the projection direction.

The result of previous step can be cut out of the total image, and Ladon transformation can be applied to obtain the precise license plate after the correction, as shown in Figure 6.

2.4 *Character segmentation*

In character segmentation process, the mostly noticeable difference between the character and background is that the characters have special connectivity. Connected component analysis method is commonly used to accomplish the character segmentation and extraction processing. Connected Component generally refers to an image having an image region consisted of foreground pixel dots who have identical pixel values and neighboring positions.

Connected Component Analysis refers to the method that finding each connected region and marking them in the image. In other words, if we need to target and extract prospects objects for subsequent processing, the connected component analysis method is suitable for us. What we should

pay attention to is this method usually be utilized to process binarized image.

Before applying connected component analysis to target image, we need to put prior image into binarization process in order to extract the crucial information we need. And we can also remove noise, projecting the specific region of the characters in connectivity by applying mean filter operation and corrosion operation to image.

In the experiment, we can find that, the split plate also includes borders, and the borders are with feather similarity to the character. Therefore, we need to remove the upper and lower, left and right frames in order to exclude the influence of the borders when marking the connected region. I can set the appropriate threshold, traverse every image pixel to remove the outer border.

What we can learn from the definition of the connected component is that connected component is composed of a collection of adjacent pixels having the same pixel values. Therefore, we can find these connected regions in the image, which according to these requirements. For each connected region found, we can allocate it a unique label to distinguish from other connected components. We use the template, 8 neighborhoods, to detect peaks and valleys separately from the x direction and y direction to determine connected components.

Next, in order to exclude some small interference, we need to filter according to the area proportion of connected components. After debugging, I choose about 30% of the image area as a connected component filtering threshold, the domain consisting of characters can be better retained and exclude other interference, as shown in Figure 7.

Finally, according to the characteristics of the connected component, we can divide the license plate characters into single characters and normalization the size of pictures before saved as image, as shown in Figure 8. Then we can use these images to constitute a test sample prepared to identify.

Figure 7. Connected domain detection.

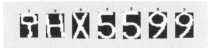

Figure 8. Character divided.

Figure 9. Training set and test set.

2.5 *Character recognition*

K-Nearest Neighbor (KNN) algorithm, a famous statistical method of pattern recognition, occupies a large position in machine learning classification algorithm.

Simply stated, the so-called K-nearest neighbor algorithm that is, given a training data set of new input instance, finding the nearest instance of K in the training data set. Most of this K instance belongs to a category, so the input instance will be classified into this class.

It is worth noting that the nearest neighbor matching is an unsupervised algorithm, so in order to achieve a good recognition results when training samples were collected pre-filter needs more good sample data to obtain a higher recognition rate.

Nearest neighbor matching is an unsupervised algorithm. In order to bring a good recognition effect, the previous collection of training samples needs more screening. Good sample would be helpful to a higher recognition rate.

As for character recognition, we can use the nearest neighbor matching method to achieve the goal. There are seven characters in license plate and the letters and numbers are targets we want to identify. Since there are few types of characters in the license plate, relatively simple, we can directly select pixel value as a stable characteristic. Using exhaustive method for nearest neighbor matching.

We have already made a test sample consisting of license plate characters and we can collect a variety of different fonts in digital images. So we can build training set and test set, take traversal method to find the nearest neighbor to obtain matching conclusions, as shown in Figure 9.

3 RESULT

Since the need of pattern classification and machine learning is always inconstant. Requirement also is variable when analyzing the classification capability of a classifier used for samples. For instance, there is only one positive case while the quantity of samples is 100. If you only take the accuracy (A) into consideration, the training of any model is not carried out, only need to directly test the samples of all the cases. The accuracy will

be able to reach 99%, but it cannot reflect the true ability of the model. Therefore, in order to identify the real effect of test procedures, 50 randomly selected pictures of license plate were examined in three parts of the algorithm: detection, segmentation and recognition.

We consider three test indicators: Accuracy (Precision), Recall, F-measure or balanced F-score.

Wherein, the accuracy $P = TP / (TP + FP)$, reflect the proportion of actually positive samples in the positive cases that determined by classifier. Recall rate $R = TP / (TP + FN)$, reflect the correct determination proportion of the total positive examples.

$$F = 2 \times recall \times accuracy\, rate / (recall + accuracy)$$

Detection-recall:80%, precision:90.9%,
F = 0.8510
Segment-recall:75.8%, precision:94.8%,
F = 0.8424
Recognition:F = 76.9%

4 CONCLUSION

In summary, the entire process includes plate location, morphology process, tilt correction, character segmentation, and character recognition. Final test results show that the recognition rate of a recognition system in more than 70 percent. However, in order to further improve the reliability and environmental adaptability of character recognition system, I suppose, the progress on Chinese character recognition system and choosing better tilt correction method, even turn to use neural network for character recognition, can further improve the effect of license plate recognition system.

REFERENCES

Ahmed M J, Sarfraz M, Zidouri A, et al. License plate recognition system[C]// IEEE International Conference on Electronics, Circuits and Systems. 2004:898–901, Vol.2.

Limin Yin,Yanying Liu, Lin He, et Enhancement [J] multi-scale image histogram-based electronics, 2006, 29 (2):431–433.

Rajput H, Som T, Kar S. An Automated Vehicle License Plate Recognition System [J]. Computer, 2015, 48(8):56–61.

Siqi Han, Lei Wang Summary threshold method image segmentation [J]. Systems Engineering and Electronics, 2002, 24 (6):91–94.

Xiaodan Jia, Wenju Li, Haijiao Wang, a new method of correcting the inclination [J] Based on Radon Transform plate Computer Engineering and Applications, 2008, 44 (3):245–248.

Energy Science and Applied Technology ESAT 2016 – Fang (Ed.)
© *2016 Taylor & Francis Group, London, ISBN 978-1-138-02973-6*

The research on color grading of green jade based on HSL chromaticity analysis

Lili Zhang & Xinqiang Yuan
Gemmological Institute, China University of Geosciences (GIC), Wuhan, P.R. China

ABSTRACT: This paper aims to find a simple and convenient method for the grading of jade color that is suitable for the market. In addition it uses the HSL color model as the foundation. Then, through the HSL analysis of green jade images acquired under the condition of camera flashlight, the values of images of the following six kinds of jade are obtained: melon green, emerald green, yellow green, bean green, light bean green, and light green. Based on this approach, the law and range of the values can be determined and respective range of HSL values and the threshold value of the seven green jades can be finally determined, laying an important foundation for the research of quick computer grading in the next step.

Keywords: green jade, color evaluation, image analysis method, HSL value

1 INTRODUCTION

This research method is image analysis, which is also known as the digital color camera colorimetric analysis method. Digital color camera adopts CCD (Zhang Hen, 2010) colored chip, which is full of light sensitive photosensitive diodes. Each small photosensitive diode constitutes the so-called pixel. Moreover, each pixel of the CCD colored chip is, respectively, covered with red, green, and blue filter, so that certain pixel can only sense red, green, or blue light. When the jade is imaged to the CCD colored chip by the lens of the digital camera, it will be sensed by the pixel of the chip and form electrical signals. The strength of the electrical signals will be expressed by 0–255 (Zhang Hui et al, 2006), and saved to the memory. These signals can be read on a computer, and color images can be displayed on the screen. If chromaticity analysis is made on the signals, quantitative data describing the image color such as R, G, and B data can be obtained (Mo Site et al, 2013). Although RGB data (also called tristimulus values) are the most basic colorimetric parameters, namely any color can be reproduced via the mixture of the three colors red (R), green (G), and blue (B), and RGB data are the quantity of the RGB contained in such color, such description is different from the daily color description habit of the author and thus inconvenient to use. Therefore, RGB values can be converted to be the more acceptable value of hue (H), saturation (S), and lightness (L).

2 HSL COLOR MODEL

This research is mainly based on the image acquisition research of the color of the common green jade in the market. The analysis method mainly adopted is the HSL color model. The HSL model is established based on human acceptance of color. It is a color system established by the scholar Smith, subject to artists' visual sense on color and luster, light and shade, and hue (Liu Xiao et al, 2002). In the HSL model, all colors are described with the three basic parameters: hue, saturation, and lightness. Thus, users without the knowledge of color mixture can also specify colors. The difference between the HSL color model and the common RGB color model is that the latter is a color system based on psychophysical color, while the former is a color system based on perception color.

3 GREEN JADE HSL COLORIMETRIC TEST

3.1 *Introduction of HSL colorimetric test for green jade*

To better understand the colorimetric characteristics of the green jade, the author has acquired a sample of 1590 green jade photos taken by cameras with flashlights, and selected 70 effective photos among the sample for this research. Next, the HSL chromatic value is obtained for each photo using the relevant software by setting to the appropriate value on the HSL color-block coordinate axis, as shown in Figure 1 (H = 120).

Although the green jade exemplifies a wide range of color changes, a few such series of changes can still be observed and distinguished via naked eyes. In the experiments, the jades are divided into various types, and a unanimous result of six distinc-

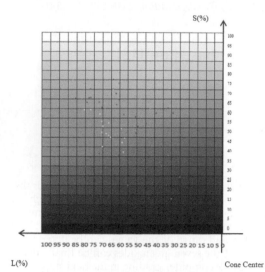

S(%)

L(%)   Cone Center

Figure 1.   HSL color analysis of color block coordinate of the green jade—average value.
*S (saturation) *L (lightness).
*Melon green is marked with brown points, emerald green is marked with bright-red points, yellow green is marked with yellow points, bean green is marked with purple points, light bean green is marked with reddish orange points, light green is marked with blue points, and pale green is marked with white points.

tive color types is obtained. According to the color depth sequence from the visual sense, the colors well-recognized in the jade industry are as follows: melon green, emerald green, yellow green, bean green, light bean green, light green, and pale green (Yuan Xinqiang, 2009). In this experiment, the HSL test was carried out on the six colors.

In addition, since jade has a complex color distribution, with some evenly distributed, while others are not, a color analysis solely dependent on the HSL values will without doubt lead to a bias. Therefore, for a more objective result, the author has, in particular, applied the software to get the average value of the jade image colors and selected the single point values for representative colors on the jade image. Besides, coordinate points indicating different colors for each kind of jade are used to mark on the color block coordinates (see Figure 1). Specific HSL analysis data are presented in Table 1, where data in the first row below the figure represent the average green values. For convenience, data recording is simplified, and only the data are kept here with their units omitted. For instance, in the table, the HSL values of the first photo are 128°, 25%, and 52%, which are simplified as 128, 25, and 52, respectively.

3.2   *Test data results and the distribution of the corresponding color block coordinate chart*

Table 1.   Images acquired and HSL parameters of the green jade.

	2	7	8	10	21	52	63
Melon green	128,25,52	135,28,47	124,32,41	131,55,25	128,31,48	126,43,47	141,30,18
	64	65	66	67			
	127,33,48	130,37,37	131,36,35	133,25,29			
	3	13	16	19	23	46	55
Emerald green	128,61,41	137,76,34	132,89,28	133,80,35	132,70,42	131,73,34	122,77,39
	56	60					
	138,72,42	129,90,37					

(Continued)

710

Table 1. (*Continued*)

	6	20	22	31	32	35	36
Yellow green	133,72,63	132,70,45	133,58,47	125,57,49	129,67,58	135,71,58	134,81,64
	39	40	43	54	57	61	62
	134,71,53	132,77,64	127,59,54	138,70,56	127,58,42	133,67,51	133,66,50
	11	14	15	18	24	25	26
Bean green	133,77,71	134,88,70	129,48,55	133,95,76	130,77,71	135,51,61	126,58,62
	42	44	45	48	49	50	51
	132,55,58	144,71,66	134,96,69	134,61,70	131,77,75	145,45,57	146,49,56
	53	58	59	68			
	130,63,67	141,79,67	141,73,68	131,51,53			
	1	4	5			27	28
Light bean green	136,88,80	137,31,61	123,29,57	126,28,56	128,34,59	131,43,65	129,48,70
	30	33	38	41	47	70	
	131,52,63	138,58,68	119,64,75	127,61,78	131,67,72	138,47,63	
	9	29	34	37	69		
Light green	137,63,72	124,64,80	121,33,65	141,39,62	146,29,63		

Table 2. HSL parameter distribution scope of all kinds of colors of the green jade.

Color grade	Color name	Hue (H)	Saturation (S)	Lightness (L)
1	Melon green	120–150	20–50	15–40
2	Emerald green	120–135	60–90	28–40
3	Yellow green	125–138	60–80	45–60
4	Bean green	130–140	50–80	60–70
5	Light bean green	125–135	20–60	50–70
6	Light green	120–140	20–60	60–80

4 TEST ANALYSIS AND DISCUSSION

The above HSL value test result indicates that the following: the six colors of the green jade are different, and based on such a difference, various colors can be differentiated. HSL values of each color have their own scope of law, which can be seen obviously from the distribution characteristics analysis of the HSL values of the green jade in the color block coordinates shown above. Jades with a bright color and uniform color are characterized by a high saturation of above 80%, moderate lightness is mostly of 20–30, and hue is of 120–149, which are green. Those whose hue is above 150 will have a blue tone, affecting the green beautiful degree of the jade. Those whose lightness is below 10 are too dark, which can be classified into black jadeite. Those whose lightness is above 80 and saturation is below 20 can be classified into white jade due to the light color, which is close to white (or colorless). The green beautiful degree of jade is mainly affected by saturation and lightness; definite relevance lies between the color saturation and lightness, while hue is secondary.

Problems emerged during the test: Since jade color is featured by hierarchy and complexity, the most common jades in the market often do not have the relatively single color change like diamond, lacking progressive color gradation.

Therefore, even if distinguished with naked eyes, under the condition where samples are uniform, such jade with a close color grade will have inconsistent color grading. When analyzing the HSL values, for green jades with a close color grade, it was found that the boundary overlap or confusion of the color values will occur.

5 CONCLUSION

Through the above test and analysis of the HSL values of the green jade, HSL values of seven kinds of green jade can be obtained. Through the analysis of these values, these seven kinds of green jade can be numeralized. Besides, the numerical range and the threshold interval of each color of the jade can be pinpointed (see Table 2). This research makes the fast digitization of jade color possible, which can be served as the colorimetric indicators for color grading of the green jade, thus creating a new way for the fast and effective grading of jade color. In other words, color grade of the jade can be obtained quickly and precisely by the image analysis of the HSL values of each jade. Meanwhile, it has laid an important foundation for the late stage research of the computer jade color automatic grading system.

REFERENCES

Liu Xiao, Jiang Gnagyi, Wu Xunwei., 2002, HSL Space-based Color Morphological Transformation. *Journal of Circuits and Systems*, 3 (7): 52–56.

Mo Site, Liu Tianqi, Li Bixiong., 2013, HSL Color Space-Based Automatic White Balance Algorithm. *Journal of Sichuan University (Engineering Science Edition)*, 6 (45): 95–99.

Yuan Xinqiang, 2009, *Application Jade Gemology*. Wuhan: China University of Geosciences Press Co., Ltd.:91.

Zhang Hen, 2010, Optimization of Green Jade color grading standard (MS., Kunming University of Science and Technology, China 2010):14–24.

Zhang Hui, Zhang Beili, Wang Manjun: Application of Method of Colour Measurement in Colour Appraisement of Jadeite Jade, *Journal of Gems and Gemmology*, Vol.8 (2006) No.3:16–20.

Energy Science and Applied Technology ESAT 2016 – Fang (Ed.)
© 2016 Taylor & Francis Group, London, ISBN 978-1-138-02973-6

Detection and location of copy-move forgery based on phase correlation

Jixiang Yang & Manman Xu

School of Information Science and Technology, Guangdong University of Foreign Studies-South China Business College, Guangzhou, P.R. China

ABSTRACT: To improve the efficiency of copy-move forgery detection, this paper presents an algorithm of detection and location based on phase correlation. The algorithm introduces overlapping window segmentation to detect suspicious images. Image blocks are selected according to the gray means, and then the similarity of image blocks is calculated on the basis of the phase correlation technology. Finally, according to the distance between candidate regions, mismatched blocks are erased and forgery regions are marked. The experiments indicate that the algorithm is of high detection efficiency and strong practicability.

Keywords: phase correlation; image detection; temper; copy-move

1 INTRODUCTION

Copy-move forgery is a technology that is a part of an image copied and pasted into another region of the image. It is a common image forgery technique and easily implemented. With the increasing popularization of image processing software and its powerful function, the phenomena of image forgeries are also becoming more common. Therefore, forgery images are hard to discern. In some special occasions such as presenting evidence in the court or news reports, it is required to provide image identification in advance. Thus, research on digital image detection technology is even more urgent. At present, digital image forgery detection technology is a hot research direction that has been widely focused at home and abroad. The digital image forgery authentication technology can be divided into active authentication and passive authentication. The latter is more widely used because it requires no pre-signing or pre-embedded information. There are many approaches to detect the authenticity of digital images with their advantages and disadvantages. The major factor that determines the advantages and disadvantages of an algorithm is its time complexity and the precision of detection results. Gu (Gu, 2011), in Anhui University, proposed an algorithm of digital image passive authentication based on Tchebichef moment, which extracts low frequency components of images, and then partitions them and extract their moment features. This algorithm is effectively anti-noise, anti-lossy compression, and anti-rotation. Wang (Wang 2010)

proposed a detection algorithm based on wavelet transformation and Zernike moment algorithm based on Gaussian pyramid decomposition and Hu moment. Zernike moment algorithm reduces dimensionality of detecting images using wavelet decomposition, to extract low-frequency components for research. In contrast, the core idea of Hu moment is to apply Gaussian pyramid decomposition for detecting images and replacing traditional rectangle blocks with round blocks, which resist the rotation attack during region coping. Liu (Liu et al, 2011) proposed a detection algorithm of fractal and statistical based on extracting feature vectors after the partition of images, which uses location information and the Euclidean distance of image blocks to locate the forged regions of the image. Du (Du et al, 2012) proposed a blind detection algorithm that uses the compound method of three image features to describe the image. Wu (Wu et al, 2009) proposed an image blind authentication algorithm based on zero-connectivity features and fuzzy membership. Wei (Wei, 2009) of Shanghai University proposed a blind detection algorithm to composite images by making use of the inconsistent approaches of JPEG blocks. Luo (Luo et al, 2007) put forward a feature extracting approach that uses 7seven-dimension image features as the feature vector of image blocks. The approach is capable of resisting post-processing such as smoothing filtering and compression, but without the anti-rotation ability. Gan (Gan et al, 2015) and Zhong (Zhong et al, 2016) used invariant moment to extract features of the image.

The method is more robust to resist post-processed operations, such as anti-translation, anti-rotation, anti-scaling, anti-mirror operations, and resisted Gaussian noise contamination.

One of the most concerned issues of existing approaches is that it is impossible to obtain satisfactory detection time and high detection rate at the same time. The amount of partitioned blocks is the major factor that determines the efficiency of the algorithm. If the number of partitioned blocks is too large, the space of algorithm searching and matching will be increased, and thus the algorithm time will be extended. This paper calculates the matching degree using phase correlation. Several pixels can be slided at the same time when the image is separated by slide window, which reduces the number of image blocks and increases the efficiency of detection.

2 THE ALGORITHM OF DETECTION AND LOCATION

2.1 The core idea of the algorithm

The most common technology of image forgery detection is block matching, which separates the image into small independent and overlapped image blocks, and then selects image features to describe image blocks, searches the feature matching image blocks in image block feature sets, and finally obtains the forged regions. The detection algorithm based on feature matching can only obtain the image block's similarity but not the relative location of image blocks, so the slide step-size in the image partition is merely a pixel. Thus, numerous image blocks after partition lead to long time in calculation. Using the phase correlation technique, this paper estimates the image block's matching degree and calculates the relative displacement between them. The overlapped region of image blocks can be obtained from their relative displacement even if the two image blocks are completely different. Based on this theory, this paper proposed sliding several pixels at same time when partitioning image blocks to reduce the number of image block matching, thus increasing the efficiency of program running.

Each image block is marked by the first pixel in the upper left corner of the image block, which can mark uniquely each one and calculate the distance of them, playing a decisive role in erasing mismatching blocks. It is not necessary to match all blocks, because the majority of blocks in the image are not matched. To reduce the space of matching search, the gray mean of image blocks is first compared. If the gray mean varies greatly, the matching probability of two image blocks is particularly small.

Two image blocks whose gray means are very close phase correlation are applied to match the image block. Before confirming the forged region, mismatched blocks are required to be erased. When detecting matching using phase correlation, the pulse function value of image blocks of many non-forgery regions are larger than the preset threshold, but they are not the marked forged regions. The corresponding image blocks of copy-move regions are concentrated and the distances between their matched blocks are the same, while the distances between mismatched blocks is scattered. Mismatched blocks can be erased through counting the distances between the image blocks.

2.2 Calculating the overlapping probability using phase correlation

Phase correlation algorithm is based on Fourier transforms. Because phase correlation makes use of the power spectral information of the image but not its contents, the technology is capable of resisting noise, and can be applied in the field of digital image detection, calculating the offset between the tiles to erase the mismatched block.

Fourier transforms and inverse transforms can transform signals between the time domain and the frequency domain. Signals can be analyzed in the time domain and the frequency domain, respectively but not simultaneously. This is because Fourier transforms and inverse transforms are global transforms.

Fourier transforms have various variant forms in different fields, such as discrete Fourier transform and continuous Fourier transform. Applying Fourier transforms in digital images can convert images from the time domain to the frequency domain, which uses Fourier spectral characteristics to process the image. Fast Fourier Transform (FFT) is the fast algorithm of discrete Fourier transform, which breaks down complex operation processes into countless simple add or subtraction operation, and retains all the characteristics of discrete Fourier transforms. Fast Fourier transform is widely used in image processing.

Assuming that $g(x,y)$ is the result of the translation of image $f(x,y)$, and the translation is (x_0, y_0). The relation between them can be represented as:

$$g(x,y) = f(x - x_0, y - y_0) \tag{1}$$

The corresponding Fourier transform is $G(u,v)$ and $F(u,v)$. According to Fourier transform's translation theorem:

$$G(u,v) = \exp(-j2\pi(ux_0 + vy_0))F(u,v) \tag{2}$$

The cross-power spectrum of image f(x,y) and image g(x,y) is defined as:

$$P(u,v) = F(u,v)G^*(u,v)/|F(u,v)G^*(u,v)|$$
$$= \exp(-j2\pi(ux_0 + vy_0)) \quad (3)$$

In the above expression, $G^*(u,v)$ is the conjugate complex numbers of $G(u,v)$. The phase correlation matrix $P(x, y)$, i.e. cross-power spectrum impulse function, of the two images can be obtained through applying Fourier inverse transform to cross-power spectrum $P(u, y)$. The peak of the matrix is related to their overlap degree, and the coordinate corresponding to the peak represents the translation of the two image blocks.

The maximum of cross-power spectrum impulse function represents the correlation between the two images. As the peak of the impulse function reduces, the correlation could decrease. The result of processing the 32 × 32 image block is shown in Figure 1. In Figure 1(a), the two images are completely the same, the peak of the impulse function at (1,1) is 1, and the other values are all 0; In Figure 1(b), the translation of the two images is (4,1), and the peak of the impulse function is 0.49; In Figure 1(c), the translation of the two images is (8,7), and the peak of the impulse function is 0.32; Images in Figure 1(d) have no apparent peak, and the maximum is only 0.12, so the two images do not correlate.

2.3 The steps of the algorithm

The size of the detecting image is A × B. The image can be partitioned into slide window. Each block is a square, and the sub-block's length of side is denoted as a and the slide step size as a_sli. When the rest of the image cannot be divided into an entire block, the sub-block should be zerofilled as an $a \times a$ pixel image block.

Each image block is marked in the location (xi, yi) of the first pixel in the upper left corner. Then, the gray mean of each image block is calculated and sorted in descending order.

Because the mean of the forged region gray mean is similar to that of the copied region, the image blocks that match each image block within the scope of [–L,L] are searched using the phase correlation. The specific steps are as follows:

First, the image blocks that overlap with them are excluded because the copy region and the paste region in the same image is disjoint. When |xi – xi'| ≤ a and |yi – yi'| ≤ a, the two image blocks are overlapped.

Phase correlation of non-overlapped image blocks is calculated. The result indicates the peak p and the peak's coordinate ($i1,j1$) of impulse function. p is compared with the preset threshold $T1$,

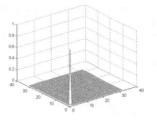

(a)The two images are completely the same

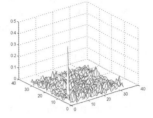

(b)The peak of the impulse when the translation is (4,1)

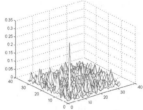

(c)The peak of the impulse when the translation is (8,7)

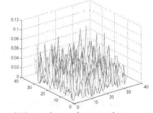

(d) The two images do not correlate

Figure 1. Two-dimensional diagram of impulse function.

and three conditions are obtained: (a) when the peak p is smaller than the preset threshold $T1$, the image blocks do not correlate; (b) when the peak is higher than the preset threshold $T1$ and translation's coordination is (1,1), the two image blocks completely match with each other, which are marked as Candidate Matched Block 1; (c) when the peak is higher than the preset threshold $T1$ and the translation's coordination is (m,n), the two image blocks are partially overlapped, which are marked as Candidate Matched Block 2.

In the Candidate Matched Blocks, every pair of blocks' offset is calculated. When the number of image blocks that have the same offsets is larger than the preset threshold $T2$, the image block pair is exactly the copy-move region, otherwise the mismatched block pair.

Finally, all the detected matched block pairs are marked with white pixel, and then the marked image is output.

3 EXPERIMENTAL RESULTS AND ANALYSIS

All the experiments were run in Windows 7, with CPU Intel Core (TM) 2.4 GHz, and internal storage 4GB. Matlab 7 was used to program the algorithm in this experiment.

The first image that we use in the experiment is a 256×256 gray image, as shown in Figure 2(a). The image in Figure 2(b) is a copy-move one, which is generated by copying the horse in the image and then pasting it in the other place of the grass with Photoshop.

To test the image, we set the sub-block size as 16×16, sliding step length as 8 pixels, $L = 10$, threshold $T1 = 0.28$, and $T2 = 9$. The detecting result is shown in Figure 3.

The size of the sub-block and the value of the slide step size are two significant factors that influence the precision of detection. In the second experiment, different sub-block sizes are imaged and slide step sizes are compared. Figure 4(a) shows the original image, while Figure 4(b) shows the forged image. Figure 5 shows the detected result when the sub-block $a = 16 \times 16$, the slide step size is 8 pixel, threshold $T1 = 0.28$, and $T2 = 20$; when the sub-block $a = 8 \times 8$, the slide step size is 4 pixel, threshold $T1 = 0.43$, and $T2 = 100$. The results are shown in Figure 5(b).

The experimental results indicate that different parameters have a significant impact on the location of forgery. When the size of the sub-block is 16, the program can precisely locate the copy-move region. Because the size of the block is relatively large, the marginal regions cannot be exactly marked. When the size of the sub-block is reduced to 8, as shown in Figure 5(b), the marked forged

Figure 3. Detected result.

(a) The original image   (b) The forged image

Figure 4. The original image and the temper image.

(a) a=16   (b) a=8

Figure 5. he detecting results at different parameters.

region becomes even exact; that is, the animal's feet in the image can also be marked. Thus, when the image is segmented, the smaller the sub-block and the slide step size are, the higher the precision probability of the algorithm's marking the copy-region will be. Meanwhile, the detecting time is also longer.

4 CONCLUSION

There are many problems in the existing approaches of image forgery detection, such as the low detect-

(a) The original image   (b) The forged image

Figure 2. The original and the forgery image.

ing efficiency due to too much partitioned blocks. Some approaches can only slide one pixel when partitioning images, which generates large numbers of image block pairs, thus increasing the latter's calculation and enlarging the searching scope. Within the same image, huge differences can be found in many image blocks, thus they cannot be matched. It is not necessary to calculate the phase correlation of one image block with that of all image block. The only thing to do is to calculate the phase correlation of the image blocks whose gray mean is in the same scope of the target block. The algorithm presented in this paper effectively solves low detecting efficiency when detecting the image forged region. It can slide several pixels at one time when partitioning sub-blocks, reducing the calculation of image blocks sharply. Gray means of image blocks are compared before matching using the phase correlation, and then similar image blocks are selected for further matching. In this way, the algorithm reduced the search scope to improve the detection efficiency of the program. The experimental results indicate that the algorithm can precisely locate the forged region and has a high detection efficiency.

ACKNOWLEDGMENT

This research was supported by the Special Funds for the Cultivation of Guangdong College Students' Scientific and Technological Innovation ("Climbing Program" Special Funds) Class Project Fund (No. pdjh 2016b0973) and the Student Innovation and Entrepreneurship Training Program Class Project Fund of Guangdong Province (No. 201512620013).

REFERENCES

Du Zhenlong, Yang Fan, Li Xiaoli, Shen Ganggang. Blind detection of copy-move forgery image based on feature combination. Computer Engineering and Design, 2012, Vol.33: 4264–4267.

Gu Zongyun. Research on passive digital image authentication with regional duplication forgery. Anhui University, 2011:24–33.

Gan Yanfen; Zhong Junliu. Research on 2D Composite Image Forensics Using Features of Discrete Polar Complex Exponential Transform. Source: International Journal of Bifurcation and Chaos, 1540018 [15 pages], Issue 14, 2015.

Luo Weiqi, Huang Jiwu, Qiu Guoping. Robust detection of region-duplication forgery in digital Image. Chinese Journal of Computers, 2007, Vol.30: 1998–2007.

Liu Meihong, Xu Weihong. Detection of copy-move forgery image based on fractal and statistics. Journal of Computer Applications, 2011, Vol.31:2236–2239.

Wu Qiong, Sun Shaojie, Zhu Wei, Li Guohui, Tu Dan, He Chaosheng. A Blind Forensic Algorithm for Detecting Doctored Image Region by Application of Exemplar-based Image Completion. Acta Automatica Sinica, 2009, Vol.35: 239–243.

Wei Weimin. Passive image forensics based on fourier spectrum analysis. Shanghai University, 2009:66–77.

Wang Junwen. Research on blind forensics for digital image content tampering. Nanjing University of Science and Technology, 2010:52–73.

Zhong Junliu; Gan Yanfen. Detection of copy move forgery using discrete analytical Fourier Mellin transform. Source: Nonlinear Dynamics April 2016, Volume 84, Issue 1, pp189–202.

Energy Science and Applied Technology ESAT 2016 – Fang (Ed.)
© 2016 Taylor & Francis Group, London, ISBN 978-1-138-02973-6

Detection of lane lines for vision-based micro-vehicles

Yongqiang Li & Yaping Zhu
School of Automation, Hangzhou Dianzi University, Hangzhou, Zhejiang, China

ABSTRACT: The micro-car lane mark detection algorithms have poor robust adaptation and real-time performance. Therefore, this paper presents an improved method combining mathematical morphology and least squares method to realize the micro-car lane mark detection. First, road images are preprocessed, and the influence of indoor lighting changes on lane images is removed by the improved morphological filtering algorithm. Then, the images are dividing by the OTSU threshold dividing algorithm. After dividing, Canny edge detection operators are used to realize edge extraction. Finally, the cumulative probability of the Hough transform combined with least squares method is used to realize the micro-car lane mark detection. The experimental results indicate that the proposed algorithm can detect the micro-car lane mark in strong light or lack of light and in other complex lighting conditions quickly and accurately. This helps solve the problem of poor robust adaptation and real-time performance.

Keywords: mathematical morphology, threshold segmentation, least squares method

1 INTRODUCTION

Intelligent Transport System (ITS) is a new generation of transport system that is comprehensively built with automatic control technology and computer technology. Autonomous driving technology of intelligent vehicles is the research focus of the intelligent transport system. The research platform of micro-intelligent transport of 1:10 is used (Figure 1) in this paper to carry out research on the autonomous driving technology of intelligent vehicles.

The detection of lane lines is the key to realize the autonomous driving of micro-vehicles. Because the micro-transport environment is created indoor, the indoor light forms the reflection region on the road for micro-vehicles, and this reflection region always has the same color and edge feature with lane lines, which causes a large noise interference for the detection of lane lines. To eliminate the interference of strong illumination, this paper combines the filtering of mathematical morphology and the least squares method to carry out the detection of lane lines. According to a previous study (Bi, Jian.quan. et al. 2013), first, preprocessing is carried out on road images, and then the illumination interference is filtered using the filtering algorithm of improved mathematical morphology. The images after filtering of illumination interference are segmented with OTSU threshold segmentation algorithm. Then, edge extraction of segmented images is carried out with Canny edge detection operator. Finally, Progressive Probability

Figure 1. Road environment of micro-transport.

Hough Transform (PPHT) and the least squares method are combined to carry out the detection of lane lines.

2 IMAGE PREPROCESSING

2.1 *Selection of image Region-of-Interest (ROI) and the gray scale*

As massive information on lane images is not useful, the region of the line lane is set as the ROI; the computation can be reduced only with processing of the ROI. Figure 2(1) shows the ROI image at strong illumination. and Figure 2(2) shows the ROI image at the insufficient light. Each pixel collected on the color image is composed of three basic components: red (R), green (G), and blue (B). The range of the component value is 0~255; thus, each pixel value is up to 16 million types (Chen, Long. et al. 2012). Large data size and long image processing time affect the real-time property of system processing. As the gray-scale image

(1)Image of high reflection

(2) Image of insufficient light

Figure 2. ROI image of the lane line.

(1)Image of high reflection

(2) Image of insufficient light

Figure 3. Gray-scale image of a single channel.

can also display the information on the road state, color images are generally converted to gray-scale images. In this paper, the weighted average method is used to convert the ROI color image to the gray-scale image, as shown in Figure 3. Figure 3(1) shows the image gray scale at reflection, and Figure 3(2) shows the gray-scale image at insufficient light.

2.2 *Image denoising*

The median filtering is carried out for the gray-scale image to eliminate the "salt-and-pepper" (Cheng, Huayao. et al. 2011) noise. The median filter is the nonlinear filter, which combines each pixel and pixels in its neighboring regions into a set of arrays. The median filter is the median calculating this group of numbers, which replaces the current pixel value for this median. The median filtering can well protect the edge of the lane line after removing the "salt-and-pepper" noise.

3 FILTERING PROCESSING

3.1 *Corrosion and expansion theory of morphology*

Corrosion and expansion are most basic operations of mathematical morphology (La, H.M. et al. 2011). The structural element is a basic tool in the mathematical morphology, which is defined as an element shape and an origin. When the origin of the structural element is aligned with a pixel of the image, the intersection part with the image defines a set of morphological operations. If the shape of the structural element is arbitrary, the circular with the radius of 4 is taken as the structural element in this paper.

Corrosion calculates the difference between each pixel and the gray scale of current pixels within the coverage of structural element, taking the gray scale of the pixel when the difference is minimal as the gray scale of current pixels. $B(x,y)$ indicates structural element and $f(x,y)$ indicates the input image. The corrosion operation can be recorded as $f \Theta B$, so

$$f \Theta B(m,n) = \min\{f(m+x,n+y) \\ -S(x,y) \,|\, (m+x,n+y) \in D_f, (x,y) \in D_B\} \quad (1)$$

Here D_f and D_B are, respectively, the domains of f and B. After the corrosion, the edges of the reflection region and the lane line shrink within the region of high gray scale. Proved by the width of lane lines and the simulation experiment, it is preferred to carry out four times of corrosion of the image continuously.

The expansion operation of the gray-scale image is recorded as $f \oplus B$, and it is defined as:

$$f \oplus B(m,n) = \max\{f(m-x,n-y) \\ +B(x,y) \,|\, m-x,n-y \in D_f, (x,y) \in D_B\} \quad (2)$$

Here D_f and D_B are, respectively, the domains of f and B. The maximum pixel value of the coverage image of the structural element B is calculated, to replace the current pixel value with this maximum pixel value. In this way, the edges of the reflection region and the lane line region shift from the region of high gray scale to the region of low gray scale.

3.2 *Effect of filtering of mathematical morphology on removal of illumination change*

According to corrosion and expansion theory, the corrosion operation is first carried out for the image after median filtering, and then the expansion is carried out to obtain the background image of lane lines, as shown in Figure 4. Figure 4(1) shows the background image at high reflection, and Figure 4(2) shows the background image at insufficient light. The differential operation of the gray-scale image and the background image after the median filtering is carried out, to obtain the image

(1) Image of high reflection   (2) Image of insufficient light

Figure 4. Background image of the lane line.

(1) Image of high reflection   (2) Image of insufficient light

Figure 5. Image after filtering of mathematical morphology.

of lane lines after the filtering of morphology, as shown in Figure 5. Figure 5(1) shows the image of the lane line of high reflection. Figure 5(2) shows the image of the lane line at insufficient light.

The experiment shows that the filtering of mathematical morphology can well process the image of lane lines in cases of strong illumination and insufficient light. Thus, it can extract the effective information on lane lines, and eliminate the effect of illumination on the detection of lane lines.

4 DETECTION OF LANE LINES

4.1 Segmentation of adaptive threshold

The image segmentation is the process of extracting the target from the image. The threshold (Chu, wei.dong. 2012) segmentation is a kind of image segmentation method. Setting threshold as T, the pixel with the gray scale smaller than T is considered as the background, and the pixel with the gray scale larger than T is considered as the target. The segmentation formula is as follows:

$$g(x,y) = \begin{cases} 0 & f(x,y) \geq T \\ 255 & f(x,y) < T \end{cases} \qquad (3)$$

Here $f(x, y)$ represents the gray-scale image. The selection of the threshold is divided into the fixed threshold and the adaptive threshold. The fixed threshold method cannot adapt to the complex scene. In this paper, the adaptive threshold segmentation method (Cheng, H.Y. et al. 2010) (OTSU) is used to take the maximum between-class variance as the threshold, and this algorithm has relatively strong robustness. We assume $f(i, j)$ as the gray scale of $M \times N$ image at the point (i, j). The image gray scale is divided into m classes and the probability of k-class gray scale, so $p(k)$ is:

$$p(k) = \frac{1}{M \times N} \sum_{f(i,j)=k} 1 \qquad (4)$$

The proportion of pixels of the segmented target image in the whole image is denoted as $w_1(t)$, the number of target pixels as $N_0(t)$, and the target

average gray scale as $u_0(t)$. The proportion of pixels of the background image in the whole image is denoted as $w_2(t)$, the number of background pixels as $N_1(t)$, and the average gray scale of the background as $u_1(t)$:

Proportion of the target pixel:

$$w_1(t) = \sum_{0 \leq i \leq t} P(i) \qquad (5)$$

Proportion of the background pixel:

$$w_2(t) = \sum_{0 \leq i \leq m-1} P(i) \qquad (6)$$

Average gray scale of the image:

$$u_0 = w_1(t)u_0(t) + w_2(t)u_1(t) \qquad (7)$$

The threshold of the OTSU threshold method:

$$T = Arg \max[w_1(t)(u_1 - u_0)^2 + w_2(t)(u_2 - u_0)^2] \qquad (8)$$

The flow chart of image segmentation after the filtering of mathematical morphology with the OTSU algorithm is shown in Figure 6.

4.2 Edge extraction

The image edge retains the important structural attribute of the target, which is the basis for the detection of lane lines (Canny Li, D.Y. et al. 2011). The operator is used in this paper to carry out edge detection of the image after the threshold segmentation. The importance of the Canny edge detection operator is to determine the edge of the image through finding the maximum value of the local gradient of the image, which has a strong stability and real-time property.

4.3 Probability Hough Transform

Hough Transform (Liu, Jin. long. 2014) maps the straight line in the image space to a point of the parameter space, and then the cumulative voting of this point is carried out, to obtain the peak in the parameter space. The linear equation is fit through inverse transform after the extraction of the peak. For the rectangular coordinate system, a linear expression is $y = ax + b$, which is converted to the polar coordinate system as $\rho = x\cos\theta + y\sin\theta$, in which ρ is the length in the vertical line section from the origin to the straight line, and $\rho \in (0, r)$, in which is the diagonal length of the image, and θ is included angle $\theta \in (0,180°)$ between this vertical and x-axis positive direction. The nature of Hough Transform is as follows:

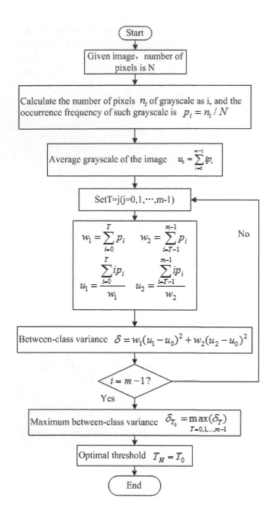

Figure 6. Flow chart of threshold processing of maximum between-class variance.

1. A point in the image space corresponds to a sine curve in the polar coordinate transform domain.
2. A point in the polar coordinate transform domain corresponds to a straight line in the image space.
3. N points on a straight line in the image space domain correspond to n curves through a public point in the transform domain.

Hough Transform needs to carry out coordinate transform of each pixel on the image, with large amounts of data. Second, in the transform process, the polar coordinate system should be quantified, to quantify the real-time property of excessive depth. The detection result would be poor if the amount is too low. In this paper, Probability Hough Transform (PPHT) is used to carry out the detection of lane lines. Its characteristic is

to use the method of random selection for fitting, and add it into counter. When the fitting times of a straight line reaches a certain threshold (*thr*), the straight line is fit.

Fitting steps for lane line of Probability Hough Transform:

1. Design the corresponding accumulator $ACC(\rho, \theta)$ for each parameter interval segmented by PPHT to set its initial value to 0, and place all the detected edge points at the edge point set S to be processed.
2. Detect whether the set S is null, and the algorithm ends if it is null. Otherwise, taking a pixel from the set S, mapped onto the parameter space, calculate the corresponding θ value under different values of ρ, and the corresponding accumulator $ACC(\rho, \theta)$ is plus with 1;
3. Determine whether the updated accumulator value is larger than the threshold *thr*; otherwise, return to the previous step.
4. When the value obtained from the accumulator is larger than *thr*, a straight line is determined, deleting the point located on such a straight line in the point set S to be processed, to clear the accumulator.
5. Return to Step 2, and the straight line determined by Probability Hough Transform is finally obtained.

4.4 Principle of the least squares method

Setting the functional relation between x and y, the formula $y = ax + b$ has two unknown parameters: a indicates the intercept and b indicates the slope. For N groups of measured accuracy observation data (x_i, y_i), $i = 1, 2......$, N. If the x_i value is accurate, all the errors are related to y_i. When estimating the parameters with the least squares method, the weighted sum of squares of the deviation of the observed value is y_i minimal. That is, the value of the following formula is minimal:

$$\sum_{i}^{N} (y_i - (a + bx_i))^2 \tag{9}$$

Best estimates of line parameters a and b can be obtained:

$$\hat{a} = \frac{(\sum x_i^2)(\sum y_i) - (\sum x_i)(\sum x_i y_i)}{N(\sum x_i^2) - (\sum x_i)^2} \tag{10}$$

$$\hat{b} = \frac{N(\sum x_i y_i) - (\sum x_i)(\sum x_i y_i)}{N(\sum x_i^2) - (\sum x_i)^2} \tag{11}$$

In addition to yielding a and b, the least squares method usually gives the correlation coefficient:

$$r = \frac{\sum(x_i - \bar{x})\sum(y_i - \bar{y})}{\sqrt{\sum(x_i - \bar{x})^2}\sqrt{\sum(y_i - \bar{y})^2}} \qquad (12)$$

in which $\bar{x} = \dfrac{\sum x_i}{n}$, $\bar{y} = \dfrac{\sum y_i}{n}$

where r indicates the functional relation on both sides, and linear degree, $r \in [-1,1]$. If $|r| \rightarrow 1$, the linear relation between x and y is good; if $|r| \rightarrow 0$, there is no linear relation between x and y. In this case, the fitting has no significance.

Probability Hough Transform and least squares method are combined to carry out the fitting of lane lines for the edge image. The lane line fitting technology combining Hough Transform and least squares method can solve the low fitting accuracy of lane lines directly with Hough Transform. The direct effect on the noise interference can also be solved using the least squares method.

5 EXPERIMENTAL RESULTS AND ANALYSIS

To validate the validity of the algorithm put forward in this paper, two groups of pictures of lane lines for micro-vehicles at high reflection and insufficient light are selected. The detection of lane lines is carried out for the above two groups of images without filtering processing with the traditional algorithm, to obtain the binary image, edge detection image, and fitting image of lane line, as shown in Figure 7. The processing of the algorithm put forward in this paper for the above two groups of images is carried out to obtain the binary image, edge detection image, and lane line image after illumination influence with mathematical morphology, as shown in Figure 8. The color of lane lines of lane line fitting on the gray-scale image is white.

The simulation experiment shows that the combination of the filtering algorithm of morphology and the least squares method put forward in this paper can well eliminate the interference of strong illumination for the detection of lane lines. The detection result of lane lines is relatively ideal. The adaptive threshold segmentation algorithm is used to carry out the image segmentation, and thus it well detects the lane lines under the condition of insufficient light. Meanwhile, the experimental results indicate that the time of an image processed by such algorithm is 20 ms, satisfying the requirement for the real-time property of identification of lane lines for autonomous driving of micro-vehicles.

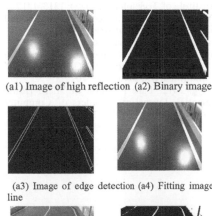

(a1) Image of high reflection (a2) Binary image

(a3) Image of edge detection (a4) Fitting image of the lane line

(b1) Image of insufficient light (b2) Binary image

(b3) Image of edge detection (b4) Fitting image of the lane line

Figure 8. Detection results of the algorithm put forward in this paper.

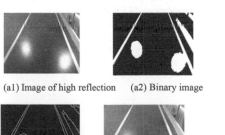

(a1) Image of high reflection (a2) Binary image

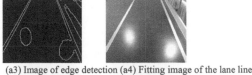

(a3) Image of edge detection (a4) Fitting image of the lane line

(b1) Image of insufficient light (b2) Binary image

(b3) Image of edge detection (b4) Fitting image of the lane line

Figure 7. Detection results of the traditional algorithm.

6 CONCLUSIONS

Aiming at the identification of lane mark of micro-vehicles, a kind of method of detecting lane lines under the environment of strong illumination interference and insufficient light is put forward in this paper. The results of the simulation experiment indicate that the algorithm has strong robustness. The algorithm considers requirements for the real-time property and robustness of micro-vehicles. The detection technology of lane lines for micro-vehicles provides a theoretical basis for the practical application of the intelligent transport system.

REFERENCES

Bai, X.Z. & Zhou, F.G. 2013. A unified form of multi-scale Top-Hat transform based algorithms for image processing[J]. Optik-International Journal for Light and Electron Optics, 124(13):1614–1619.

Bi, Jian.quan. et al. 2013. Micro intelligent software system design and implementation of computer engineering and application [J], 49 (01): 234–238.

Cheng, Hua.yao. et al. 2011. A lane detection method based on illumination invariant image [J]. Journal of Beijing Institute of Technology, 31 (11): 1313–1317.

Cheng, H.Y. et al. 2010. Environment classification and hierarchical lane detection for structured and unstructured roads[J]. IET Computer Vision, 4(1): 37–49.

Chen, Long. et al. 2012. Lane detection and tracking method based on imaging model [J]. China Journal of highway and transport, 24 (06): 96–102.

Chu, wei.dong. 2012. Lane Detetion in Micro-Traffic Environment[J]. Journal of Frontiers of Computer Science and Technology, 6(10):921–926.

Gechter, F. et al. 2012. Virtual intelligent vehicle urban simulator: Application to vehicle platoon evaluation[J]. Simulation Modelling Practice and Theory, 24(1): 103–114.

La, H.M. et al. 2011. A small-scale research platform for intelligent transportation systems[C] // Proceedings of the 2011IEEE International Conference on Robotics and Biomimetics Piscataway: IEEE Press: 1373–1378.

Li, D.Y. et al. 2011. Study on interaction behaviors of micro-autonomous vehicles.[C]// Proceedings of the 2011 10th International Symposium on Autonomous Decentralized Systems (ISADS). Los Alamitos: IEEE Computer Society Press:399–406.

Liu, Jin.long. 2014. A Novel Algorithm Research on Lane Recognition Based on Hough Space[J]. ISSN, 1(3):2331–9062.

Ma, Yu.lin. et al. 2010. Research on intelligent vehicle platoon driving simulation experiment system under the coordination between vehicle and highway[J]. Journal of Computers, 5(11):1767–1768.

Shen, Huan. et al. 2010. Structural road lane of monocular visual detection method [J]. Instruments of 1(2):397–403.

Sun, Wei. et al. 2011. Lane coordination Hough transform and least square fitting detection[J]. opto electronic engineering, 38(10): 13–19.

Sun, Yan.biao. et al. 2014. Design and implementation of DSP based lane detection system based on [J], 37 2(04):83–86.

Yu,Y. et al. 2013. Multi-agent based architecture for virtual reality intelligent simulation system of vehicles[C] // Proceedings of the 2013 10th IEEE International Conference on Networking, Sensing and Control (ICNSC). Piscataway: IEEE Press: 597–602.

Energy Science and Applied Technology ESAT 2016 – Fang (Ed.)
© 2016 Taylor & Francis Group, London, ISBN 978-1-138-02973-6

Applying polar sine transform for detecting copy-move forgery

Peiyu Lin
School of Information Engineering, Guangdong Mechanical and Electrical College, Guangzhou, China

Yanfen Gan
School of Information Science and Technology, Guangdong University of Foreign Studies-South China Business College, Guangzhou, China

Janson Young
School of Computer, Guangdong University of Technology, Guangdong, China

Jinhong Zhao & Huang Lian
School of Information Engineering, Guangdong Mechanical and Electrical College, Guangzhou, China

ABSTRACT: In recent years, constructing moment invariant and extracting invariant features of images has become popular in Copy-Move Forgery Detection (CMFD). Various moments have been proposed to detect simple forgery successfully. However, in most published literature, the rotational moment invariants have not been fully studied yet and the kernels of most moments are complex and inefficient. In this study, we present a novel method based on Polar Sine Transform (PST). The complexity of PST and its kernel are simpler than most of the moments. PST and its kernel also have orthogonal moments. The experiments show that PST has a good performance like Polar Complex Exponential Transform (PCET) in detecting the translated and rotated distortion. The computed cost of PST is lower than that of PCET. Therefore, PST serves as a powerful tool for CMFD.

Keywords: extracting invariant features; copy-move forgery detection; polar sine transform kernel

1 INTRODUCTION

With the development of image processing technology, image forensics has become an important concern for information forensics and security in today's society. Copy-move forgery is a very common type of image forgery (V. Christlein et al, 2012). Researchers have attempted to detect digital forged images accurately for decades. Fridrich (J. Fridrich et al, 2003), Popescu (A.C. Popescu et al, 2005) presented methods based on Discrete Cosine Transform (DCT) and Principal Components Analysis (PCA) to extract and analyze image features, respectively. However, their methods only perform the simple type of forgery, such as translational distortion. Morphology is considered a tool for the auxiliary analysis of the image. Geometric structure is an important image quality factor that mainly consists of shape and position. Geometric moment invariants are highly concentrated image features, including translational–rotational invariance. Hu (Hu MK 1962) first proposed the HU moments to analyze the image geometric features. However, classical HU moments are based only on

mathematical theory, not on the orthogonal function family. Therefore, it is quite sensitive to noise, contains a lot of redundant information, and has computational complexity that rapidly increases with the moment of the orders. In recent years, numerous effective moments have been proposed for CMFD. Representative orthogonal moments include Zernike Moments (ZMs) (S. Ryu et al, 2013) and Tchebichef Moments (TMs) (R. Mukundan et al, 2001). Furthermore, the Discrete Analytical Fourier–Mellin Moment (DAFMT) (J. Zhong et al, 2016) is an improved rotational moment invariant that better describes rotational invariant patterns. These orthogonal moment invariants that have rotational invariance features can be extracted to construct rotational moment invariants to describe the geometric structure of images, but these moments have prohibitive computation complexity. To overcome this defect, Yap (P. T. Yap et al, 2010) presented Polar Harmonic Transforms (PHTs) for an invariant image representation. PHTs consist of three different transforms, namely Polar Sine Transform (PST), Polar Cosine Transform (PCT), and Polar Complex Exponential

Transform (PCET). The orthogonal of PHTs and kernels are suited for representing the image features. Moreover, the complexities of moments are simpler than most of the moments. Yap confirmed that PST, PCT and PCET also show good performance in constructing rotational invariant and detecting rotational features. However, the kernel of PST is simpler than that of PCET. The formula is presented in section 2. The experiments conducted in (R. Mukundan, 2001) show that compared with ZMs, PHTs or PST is well suited for image matching applications owing to their ability to extract rotational invariances. PST is able to serve as a powerful tool for the extraction of image features. In this paper, PST is presented to construct the rotational moment invariant, and then extract the rotational invariance for CMFD.

The rest of the paper is organized as follows: the review of PST is described in section 2. The flow of PST is given in section 3. The experimental results as well as the evaluation and discussion are presented in section 4. Finally, conclusion is given in section 5.

2 REVIEW PST

PST is sine transform defined in a 2D polar coordinate system. PST of order k with repetition l is defined as follows:

$$M_{kl}^{S} = \frac{1}{\pi}\int_{0}^{2\pi}\int_{0}^{1}[H_{kl}^{S}(r,\theta)]^{*}f(r,\theta)drd\theta, \tag{1}$$

where $|k| = 0,1,...\infty, |l| = 0,1,...\infty$, M_{kl}^{S} is the PST function $f(r,\theta)$ and k is the order of PST, l is the repetition of PST, and $[\bullet]^{*}$ represents the complex conjugate, $H_{kl}^{S}(r,\theta)$ is the kernel of M_{kl}^{S}. The kernel function $H_{kl}^{S}(r,\theta)$ of M_{kl}^{S} is defined as follows:

$$H_{kl}^{S}(r,\theta) = R_{k}^{S}(r)e^{-il\theta} = \sin(\pi k r^2)e^{-il\theta} \tag{2}$$

where $R_{k}^{S}(r)$ is the radial kernel with complex exponential in the radial direction, $\theta = \arctan(y/x), i^2 = -1$, and r is the polar radius of PST defined on the unit circle.

The kernel of PCET is given as follows:

$$H_{kl}^{P}(r,\theta) = R_{k}^{P}(r)e^{-il\theta} \\ = \exp(i2\pi k r^2)\exp(-il\theta), \tag{3}$$

where $e^{iz} = \cos(z) + i\sin(z)$. From (3), it is easy to see that the time cost of PST is lower than that of PCET. From (1) and (2), it is supposed that pixels of the object or region are rotated β degree. The rotational moment invariant of PST is defined as follows:

$$M_{kl}^{S,R} = \frac{1}{\pi}\int_{0}^{2\pi}\int_{0}^{1}[H_{kl}^{S,R}(r,\theta)]^{*}f(r,\theta)drd\theta \\ = \frac{1}{\pi}\int_{0}^{2\pi}\int_{0}^{1}[R_{k}^{S,R}(r)e^{-il(\theta)}]^{*}f(r,\theta)drd\theta \\ = \frac{1}{\pi}\int_{0}^{2\pi}\int_{0}^{1}[R_{k}^{S,R}(r)e^{-il(\phi-\beta)}]^{*}f(r,\theta)drd\phi \\ = M_{kl}^{S}e^{-il\beta}, \tag{4}$$

where $\theta + \beta = \phi$. The rotational invariant moment of PST in (5) can be further defined as follows:

$$|M_{kl}^{S,R}| = |M_{kl}^{S}e^{-il\beta}| = |M_{kl}^{S,R}|. \tag{5}$$

From (5), moment invariant of PST can be constructed to extract the feature of each overlapping block.

3 FLOW OF PST IN CMFD

The flow of PST applied in CMFD is given as follows (V. Christlein et al, 2012):
Pre-processing
Block segmentation
Feature extraction
Sorting
Matching
Post-processing
The CMFD method is the block-based method that separates the image into overlapping and regular blocks. Our proposed PST-based method is applied for extracting the rotational invariance of each block.

4 EXPERIMENTAL RESULTS

In this section, the experimental images are downloaded from the MIC-F200 dataset or downloaded from the Internet. The sizes of images are from 300×300 to 2048×1536. The detected performance of PST is described in subsection 4.1. Then, precision comparison between PST and PCET, DAFMT are given in subsection 4.2 to discuss their performance in CMFD. The computed costs are given in subsection 4.3 to analyze the efficiency of compared moments.

4.1 *Performance in CMFD*

In this section, the moon and sunflower are taken as examples to show the good performance in CMFD.

The original image is shown in Figure 1(a) and 2(a). The copy region of the moon is implemented by clockwise rotated 45° distortion, as shown in

Figure 1(b). The matching and location result is then shown in Figure 1(c). In Figure 2(b), the sunflower is copied, rotated 45°, and pasted to other past of the image. In Figure 2(c), the ground-truth copied region and the forged-pasted region are successfully detected and marked in black pixels. The PST shows a good performance in detecting the rotated distortions of copy-move forgery.

4.2 *Evaluation of performance*

We further repeated our experiment by copying 100 templates from the "sunflower" image in different regions like Figure 2(b), rotating the template 30°, 60°, and 90° clockwise, and adding Gaussian noise by using Photoshop. The precision of DAFMT, PCET and our proposed PST are compared. and their experimental results are presented in Table 1.

As shown in Table 1, the precision of our PST and other relevant moments are consistently over 80%, thereby achieving our desired aims. These three moments show a good performance in CMFD.

(a) Original image

(b) Copy-move image

(a) Original image

(c) Detected image

Figure 2.   The sunflower example.

4.3 *Evaluation of efficiency*

Computed cost, namely efficiency, is also an important factor in CMFD. Table 2 presents the average time cost of the above moments for extracting the image features. This experiment is based on the dataset given in section 4.2.

As shown in Tables 1 and 2, PST obtains the similar performance in CMFD, but the proposed PST has a lower computed cost and a higher efficiency. From the above experiments, the proposed PST exhibits satisfactory detection results under geometric transforms.

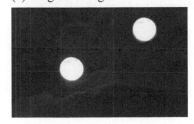

(b) Copy-move image

(c) Detected image

Figure 1.   The moon example.

Table 1. Precisions using PST and other relevant moments on 100 template images with different types of distortion.

Types of image distortion	DAFMT	PCET	PST
Rotation by 30°	90%	90%	91%
Rotation by 60°	85%	83%	83%
Rotation by 90°	87%	89%	88%
Additive Gaussian noise ($\mu = 0.01$, $\delta = 0.01$)	83%	85%	83%

Table 2. Computed cost of PST, PCET, and DAFMT for extracting the image features.

Moments	DAFMT	PCET	PST
Time cost	2.99 s	0.99 s	0.71 s

5 CONCLUSION

In this study, we presented PST to extract the image feature by constructing rotational moment invariants. Our experimental results show that the kernel of PST is simple. Moreover, PST not only accurately matches and locates original regions by extracting features, but also effectively resists noise and changes in rotation. Therefore, PST serves as a powerful tool for CMFD.

ACKNOWLEDGMENTS

This work was supported by the 2014 Guangdong Province Young Innovative Talent (Natural Science) Class Project Fund (no. 2014kqncx256). The authors are grateful for this support.

REFERENCES

Christlein, V., C. Riess, J. Jordan, C. Riess, E. Angelopoulou. An Evaluation of Popular Copy-Move Forgery Detection Approaches. IEEET Transactions on information forgery and security, vol. 7, no. 6, pp.1841–1854, 2012.

Fridrich, J., D. Soukal, J. Lukás. Detection of copy-move forgery in digital images. The Digital Forensic Research Workshop, pp.55–61, 2003.

Hu M.K. Visual pattern recognition by moment invariants. IEEE Transactions on Information Theory, no.8, pp.179–187, 1962.

Mukundan, R., S.H. Ong, P.A. Lee. Image analysis by Tchebichef moments. IEEE Transactions on Image Processing, vol.10, no.9, pp.1357–1364, 2001.

Popescu, A.C., H. Farid. Exposing digital forgeries by detecting traces of resampling. IEEE Transactions on Signal Processing, vol.53, no.2, pp.758–767, 2005.

Ryu, S., M. Kirchner, M. Lee, H. Lee. Rotation Invariant Localization of Duplicated Image Regions Based on Zernike Moments. IEEE Transactions on information forensics and security, vol.8, no.8, pp.1355–1370, 2013.

Yap, P.T., X.D. Jiang, C.H. Kot. Two-Dimensional Polar Harmonic Transforms for Invariant Image Representation. IEEE Transactions on Pattern Analysis and Machine Intelligence, vol. 32, no.7, p.1260–1270, 2010.

Zhong, J., Y. Gan. Detection of copy–move forgery using discrete analytical Fourier–Mellin transform. Nonlinear Dynamic, vol.84, pp.189–202, 2016.

Energy Science and Applied Technology ESAT 2016 – Fang (Ed.)
© *2016 Taylor & Francis Group, London, ISBN 978-1-138-02973-6*

HEVC intra-frame error concealment based on texture synthesis

Xiangqun Li & Jing Zhao
College of Electrical Engineering, Northwest University for Nationalities, Lanzhou, China

Xiaohai He
College of Electronics and Information Engineering, Sichuan University, Chengdu, China

Wenhui Jing
Xichang Satellite Launch Center, Xichang, China

ABSTRACT: In this paper, the packet loss of the analog channel transmission in HEVC test platform is conducted. The packet loss rate can be set at will and the packet loss location in each frame is selected randomly. In the repair process, the image repair technique of texture synthesis is adopted to recover the lost LCUs, to lead to better solution of the repair effects in high efficiency video coding, simultaneously, few influences on the coding time. At last, through experimental verification and the comparison with relevant algorithms, better repair effects are obtained by the method proposed in this paper based on the test results.

Keywords: HEVC, intra prediction, error concealment

1 INTRODUCTION

Spatial prediction technique is adopted for the intra-coding frame in the video coding. If the coding block is lost, the spatial correlation of each frame of information and the spatial error concealment algorithm are used to recover information. In general, this algorithm is applicable for the error concealment of intra-coding frame. In the inter-coding frame (Frame P) with complex movement, such method may also be applied when the temporal error concealment results are poor due to the reduction in temporal correlation proposed by Chen T (2002). The spatial error concealment algorithm is to recover the information in the frame of the lost block again by adjacent macro blocks, image blocks and pixel information. However, it should ensure that there is strong correlation between adjacent image blocks or adjacent pixel points. Therefore, if the lost image blocks and adjacent blocks are located in similar brightness area proposed by Zhang Y & Lin Y (2014), these methods will create excellent concealment results.

The spatial error concealment algorithm in video coding is similar to the repair algorithm in the study of still image. Both of them are to supplement and repair the lost or defective information by the spatial correlation of image information. The difference is that the error concealment technique in the video coding standard puts more focus on timeliness. Current main conditions are that common error concealment techniques consume less time, but their concealment results are not obvious enough; while the image repair technique consumes more time and its concealment results are better. In particular, for the image repair technique based on segmentation and based on texture synthesis, the image texture details are better recovered and the subjective quality is obviously improved by Patel H.M & Desai H.L (2014). The disadvantage is that the computational complexity is high, which is not good for real-time application. Currently, there are few related documents, in which the image repair technique is applied to the error concealment techniques of video coding.

2 PROPOSED

The basic process of error concealment is shown in Fig. 1. In the decoding process, each basic coding unit is tested and its decoding status is recorded. The unit with correct decoding is marked with location "a" and the lost unit is marked with location "c". For the concealed lost unit marked with location "b", after one frame is completely decoded, the lost data can be recovered with error concealment algorithm if there are lost image blocks in the frame.

This paper mainly studies the inter-frame error concealment (temporal error concealment) techniques in HEVC. For the error concealment under

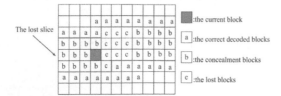

Figure 1. The Basic Process of the Concealment.

the new generation of coding standard HEVC, the error concealment mode applicable to H.264 shall be improved. But if the motion compensation method applicable to H.264 is still used under HEVC to conduct error concealment, its efficiency shows a slight shortcoming. According to idea of texture synthesis, Criminisi A. (2004) proposed the image repair algorithm based on the optimal sample block. This algorithm mainly refers to the method of texture generation, to seek the optimal sample matching area in the texture generation process, and then to copy the image information in the matching area to the pending-repair area. The core idea of this algorithm is the priority determination of the optimal repair block proposed by Zhu Feng & Zhang Wei ming (2011). Based on the information of images, it determines the filling priority of the pending-repair area. Then the repair is conducted based on the priority proposed by Zhang Y & Wang H (2013). In this algorithm, the image texture and structure information are kept to certain extent, and better repair effects on the lost or defective parts of the images with a large size are obtained. This paper simulates real environment to lose packet on data rate at any rate. In this paper, the output format of data rate is set as AnnexB and each LCU in the HEVC coding process is set as a film. In the packet loss simulation processing of AnnexB data rate, Picture Parameter Set (PPS) and Sequence Parameter Sets (SPS) information cannot be lost. These information is related to the normal decoding process of the whole video and the data started with the prefix "00000001" should not be lost, either. Once it is lost, the normal coding and decoding of the current frame will be affected. Through the analysis above, random overall rejection can be taken for the data with the format combination of prefix "000001"+NALU in this paper.

3 THE EXPERIMENTAL RESULTS AND ANALYSIS

At first, we select the sequences with different resolutions from the standard HEVC test sequences to conduct compressed coding according to different

Table 1. The experiment results of the proposed method.

Sequences	Loss rate (%)	PSNR(dB) HM10.0	After loss	Proposed
Party Scene	5	39.6	23.4	24.4
832 × 480	10	39.6	22.5	23.0
	20	39.6	20.6	21.5
Four People	5	42.2	25.5	27.0
1280 × 720	10	42.2	21.2	25.4
	20	42.2	20.1	22.8
Park Scene	5	40.1	27.5	32.2
1920 × 1080	10	40.1	27.1	30.5
	20	40.1	23.9	28.0
Traffic	5	41.4	28.8	30.6
2560 × 1600	10	41.4	25.8	28.3
	20	41.4	22.4	24.2

(a) original frame

(b) packet loss frame

(c) recovered by proposed method

Figure 2. The Basketball Drill experiment results by the proposed method.

sizes of CUs. The packet loss processing of analog channel is conducted for the compressed data rate according to the packet loss techniques of analog channel and different packet loss rates, to meet the conditions to realize error concealment algorithm. Then, we make use of the error concealment algorithm based on bilinear interpolation and the error concealment algorithm based on texture synthesis respectively to conduct information recovery processing for the processed data rate. The statistical results of experiments are shown in Table 1. In which, Fig. 2 reflects good visual quality obtained from the repair technique based on texture synthesis.

In Table 1, *Loss rate* refers to the loss percentage of the video sequence, which takes the coding unit as the unit. *HM10.0* refers to the objective quality (dB) of the video sequence after it is compressed on the standard HEVC test platform. *After loss* refers to the objective quality (dB) of the test video sequence, after which is compressed on the standard HEVC test platform and has part of coding units lost as per the packet loss rate, compared with the original test sequence. *Proposed* refers to the objective quality of the low-quality video sequence after packet loss, which is subject to the error concealment based on texture synthesis, compared with the original video sequence.

4 CONCLUSION

In this paper, the image repair technique based on texture synthesis is applied into the intra-frame error concealment algorithm of HEVC, to solve the problem that the repair results of the tradi-tional intra-frame error concealment algorithm based on bilinear interpolation are poor due to too large HEVC coding units.

ACKNOWLEDGEMENT

This work was supported by the Fundamental Research Funds for the Central Universities (No. 31920160070).

REFERENCES

Chen T. 2002. Refined boundary matching algorithm for temporal error concealment[C]. *Proc. Packet Video:* 875–887.

Criminisi A. 2004. Region filling and object removal by exemplar-based image inpainting[J]. *IEEE Transactions on Image Processing*, vol.13 (9): 1200–1212.

Patel H.M & Desai H.L. 2014. A Review on Design, Implementation and Performance analysis of the Image Inpainting Technique based on TV model[J]. *International Journal of Engineering Development and Research (IJEDR)*, vol.2 (1): 191–195.

Zhang Y & Lin Y. 2014. Improving HEVC intra prediction with PDE-based inpainting[C]. IEEE, Asia-Pacific Signal and Information Processing Association, 2014 Annual Summit and Conference (APSIPA): 1–5.

Zhu Feng & Zhang Weiming. 2011. Adaptive error resilient coding based on FMO in wireless video transmission[C]. *Proceeding of the 3rd International Conference on Multimedia Information Networking and Security:* 609–612.

Zhang Y & Wang H. 2013. Fast coding unit depth decision algorithm for interframe coding in HEVC[C]. *IEEE, Data Compression Conference (DCC):* 53–62.

14   *Civil engineering and geotechnical engineering*

Energy Science and Applied Technology ESAT 2016 – Fang (Ed.)
© *2016 Taylor & Francis Group, London, ISBN 978-1-138-02973-6*

Property research of high performance cement-based rapid repairing material

J. Yuan
School of Transportation Science and Engineering, Harbin Institute of Technology, Harbin, China

F. Liu
Jinan Quanjian Municipal Engineering Detection Ltd., Jinan, China

Z.C. Tan, Y. Liu & Y.Z. Wang
Harbin Institute of Technology, Harbin, China

ABSTRACT: The properties of high performance cement-based rapid repairing materials were investigated experimentally through the mixture of sulphoaluminate cement and Portland cement. Firstly, the statistical characteristics of testing materials were obtained by the orthogonal design. Secondly, the performance indexes like the early strength and the long-term strength of the mortar were analyzed and discussed through the comparison with the above characteristics. The results showed that through the best content of retarder, accelerator, mineral admixture, and the polypropylene fiber, the rapid repair mortar owned good performance like high early strength and enough long-term strength to resist the early cracking.

Keywords: rapid repair mortar, sulphoaluminate cement, orthogonal experiment, mineral admixtures, early cracking

1 INTRODUCTION

As a kind of high grade pavement, the cement concrete pavement has been widely applied to the airport road and municipal road (Mehta 1973). However, in recent years, cement concrete road has suffered a lot from overload and freeze-thaw cycles, causing numerous cracks endangering the traffic condition and passengers' safety (Berger et al. 2009). To restore the transportation system in time, a high performance rapid repairing material was in need (Bernardo et al. 2006).

Okamoto P.A. et al. (1994) used the $CaCl_2$ as an early strength agent in rapid hardening Portland concrete, so that the compressive strength in 4 hours could reach 14 MPa. In 1960, by adding 15% ~ 20% of mixed anhydrous calcium and calcium oxide in ordinary Portland cement, University of California developed a composite cement with expansion effect (Li & Victor 2003). The studies of Chatterjee A.K. (1991) showed that granulated blast furnace slag could improve the long-term strength of sulphoaluminate cement. The studies of Czerin W. (1980) showed that, in order to obtain the early strength effect, the content of ordinary Portland cement should not be higher than 20% of the total cement if mixed with sulphoaluminate cement. Kalogridis D. et al (2000) held the thought

that the sulphoaluminate cement swelling behavior occurred in the cement hydration process for the performance of micro expansion. In the view of Powers T.C. (1958) and Sshwiete H.E. et al (1966), surficial ettringite has formed up even when C_3A was still granular, and with hydration reaction continuing, growing ettringite crystals resulted in the expansion.

As discussed above, with the sulphoaluminate cement and Portland cement in a certain proportion as the basic composite cementitious system, properties of high performance cement-based rapid repairing materials were studied through the determination of early strength and long-term strength, and the best content of mineral admixtures.

To this end, the remainder of this paper is organized as follows. The details of the experimental material and procedure were introduced in next section. The experimental results were analyzed and discussed in Section 3. At last, conclusions are drawn.

2 MATERIALS AND EXPERIMENTS

The mixed based material consisted of 20% cement of P·O 42.5 and 80% sulphoaluminate cement, with coarse aggregate under 20 mm~40 mm single particle grade, and fine aggregate of quartz sand, which was a mixture of coarse sand (20–40 mesh), medium sand (40–70 mesh), fine sand (70–140 mesh) in the mass proportion of 2:2:1. In addition, class II fly ash, class S95 slag, silica fume and lime were used as mineral admixture for the repair material according to China's specifications. Plus, high efficiency water reducer and two kind of acid named HN1 and HN2 were used as the retarder to test the basic mix proportion of mortar, with 1% of accelerator. The densities of P·O 42.5, sulphoaluminate, coarse sand, medium sand, fine sand, and water reducer were respectively 200 kg/m³, 800 kg/m³, 440 kg/m³, 440 kg/m³, 220 kg/m³, and 15 kg/m³.

Orthogonal experiment design was a method for multiple factors study, picking a part of the representative points for test according to the orthogonality, with high effectiveness, fast speed, and economical efficiency. It could be used to arrange the multiple factor test plan scientifically and to analyze the test results with less times of testing. In order to study the best content of four mineral admixture, the 5×4 of L_{16} (4^5) orthogonal test table was applied.

According to China's 17671–1999 GB/T specification "Strength test method of cement mortar", the testing pieces were made as 40 mm × 440 mm × 160 mm to test the flexural strength and compressive strength of mortar. Also according to the content of China Civil Engineering Society (CCES) 01-2004 specification "Concrete crack resistance test", the early cracking test of the mortar was conducted with a self-made mold, and the size of the mold was 400 mm × 400 mm interior cavity with 16 mm thick outer walls.

3 RESULTS AND DISCUSSIONS

3.1 Retarder and accelerator

According to the test of two kinds of retarder, the initial setting time and the final setting time of mortar were determined by using the method of resistance method. The test results were shown in Figure 1.

According to the construction specification requirements, the initial setting time should be longer than 30 min and the final setting time should be between 45~50 min. Therefore, the optimal content of retarder would be 0.2% HN1 and 0.4% HN2.

Taking the retarder determined above, the optimal dosage of accelerator agent could be found according to the flexural strength and compressive strength of mortar under 2 hours. The results of the experiment were shown in Figures 2–3.

According to the construction specification requirements, the flexural strength should be more than 7 MPa in 2 hours and the compressive strength

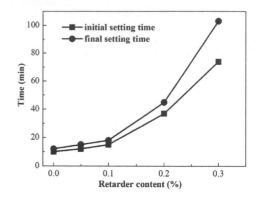

Figure 1. Effect of retarder on setting time of mortar.

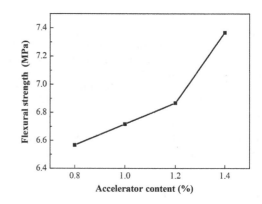

Figure 2. Effects of accelerator agent on flexural strength of mortar.

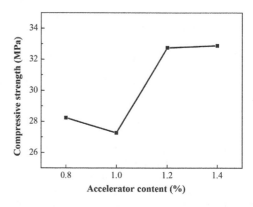

Figure 3. Effects of accelerator agent on compressive strength of mortar.

should be more than 30 MPa in 2 hours. When the content of the accelerator agent was 1.4%, both the flexural strength and the compressive strength could meet the demands.

3.2 *Mineral admixture*

According to the fluidity experimental results of orthogonal design of fresh mortar, silica fume, slag, lime and fly ash all had significant effects on the initial fluidity of mortar. Silica fume showed the highest influence on the initial flow rate, while fly ash showed the lowest. According to the influence of mineral admixture on mortar fluidity, the mineral synthetic material with the best content was 4% silica fume, 10% slag, 4% lime, and 6% fly ash.

Figure 4 showed the effects of four mineral admixtures on the strength of mortar in 2 hours were not significant. Moreover, silica fume had significant influences on the flexural strength of mortar in 1 day while lime in 28 days. According to the influence of mineral admixture on the flexural strength of mortar, the mineral material with the best content was 4% silica fume, 6% slag, 8% lime, and 2% fly ash.

Figure 5 showed lime had significant influences on the compressive strength of mortar at different ages. According to the influence of mineral admix-ture on the compressive strength of mortar, the best content of mineral material was 4% silica fume, 6% slag, 8% lime, and 4% fly ash. Table 1 showed the different content of mineral admixture according to different indexes.

Taking the flexural strength as the main reference, the comprehensively best content of mineral material was 4% silica fume, 6% slag, 8% lime, and 4% fly ash.

3.3 *Polypropylene fiber*

Four groups of experiments of early cracking of mortar were conducted with different content of polypropylene fiber, with Figure 6 showing the vivid influence results. With the increase of the content of polypropylene fiber, the early cracks of mortar gradually reduced, polypropylene fiber could effectively inhibit the early cracking of mortar.

Table 1. The best content of mineral admixture under different indexes.

	Mineral admixture content (%)			
Indexes	Silica fume	Slag	Lime	Fly ash
Fluidity	4	10	4	6
Flexural strength	4	6	8	2
Compressive strength	4	6	8	4

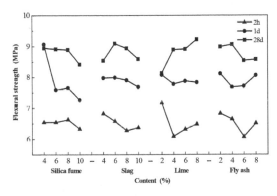

Figure 4. Influence of mineral admixtures on the flexural strength of mortar.

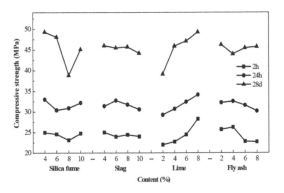

Figure 5. Influence of mineral admixtures on the compressive strength of mortar.

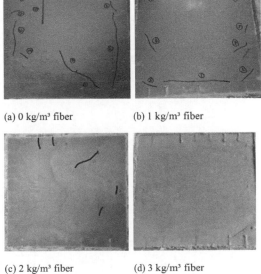

(a) 0 kg/m³ fiber

(b) 1 kg/m³ fiber

(c) 2 kg/m³ fiber

(d) 3 kg/m³ fiber

Figure 6. Influence of polypropylene fiber on the early cracks of mortar.

With the content of 1 kg/m^3 polypropylene fiber, the mortar cracking area got 47.7% reduction, and with the content of 3 kg/m^3 polypropylene fiber, the crack of the testing surface appeared to be only 0.1 mm of width and 46 mm of length, suppressing most of the early cracking phenomenon.

Four groups of experiments of early cracking of mortar were conducted with different content of polypropylene fiber, with Figure 5 showing the vivid influence results. With the increase of the content of polypropylene fiber, the early cracks of mortar gradually reduced, polypropylene fiber could effectively inhibit the early cracking of mortar. With the content of 1 kg/m^3 polypropylene fiber, the mortar cracking area got 47.7% reduction, and with the content of 3 kg/m^3 polypropylene fiber, the crack of the testing surface appeared to be only 0.1 mm of width and 46 mm of length, suppressing most of the early cracking phenomenon.

4 CONCLUSIONS

In this paper, the effects of four mineral admixture, silica fume, slag, lime, and fly ash, on the repair mortar were investigated, also the effects of polypropylene fiber on the early cracking phenomenon of the repair mortar, and the conclusions were drawn as follows:

1. Silica fume, slag, lime and fly ash all had significant effects on the initial fluidity of mortar, with the influence ranking: silica fume > lime > slag > fly ash. The addition of silica fume could reduce the fluidity of mortar while slag showed the opposite.
2. The effects of silica fume, slag, lime and fly ash on the early flexural strength of mortar were not significant. In the dosage of 8%, lime could significantly improve the later compressive strength and all-age flexural strength of mortar; in the dosage of 10%, the increase incorpora-

tion of silica fume could reduce the compressive strength of mortar in later period.
3. Polypropylene fiber could effectively inhibit the early cracking of mortar, with the 3 kg/m^3 incorporation of polypropylene fiber, the early cracking area of the mortar would decrease 98.6%.

REFERENCES

Berger, S. et al. 2009. Hydration of calcium sulfoaluminate cement by a ZnCl$_2$ solution: investigation at early age. *Cement and Concrete Research* 39(12): 1180–1187.

Bernardo, G. et al. 2006. A porosimetric study of calcium sulfoaluminate cement pastes cured at early ages. *Cement and Concrete Research* 6(36): 1042–1047.

Chatterjee, A.K. 1991. Special and new cements. India 9th International Congress on the Chemistry of Cement: 26–29.

Czernin, W. 1980. Cement chemistry and physics for civil engineers. *Bauverlag*: 16–18.

Kalogridis, D. et al. 2000. A quantitative study of the influence of non-expansive sulfoaluminate cement on the corrosion of steel reinforcement. *Cement and Concrete Research* 30(11): 1731–1740.

Li, C. & Victor, C. 2003. On engineered cementitious composites (ECC). *Journal of Advanced Concrete Technology* 1(3): 215–230.

Mehta, P.K. 1973. Mechanism of expansion associated with ettringite formation. *Cement and Concrete Research* 3(1): 1–6.

Mehta P.K. 1976. Scanning electron micrographic studies of ettringite formation. *Cement and Concrete Research* 6(2): 169–182.

Okamoto P.A. 1994. Use of maturity and pulse velocity techniques to predict strength gain of rapid concrete pavement repairs during curing period. *Transportation Research Record*: 1458.

Powers T.C. 1958. Structure and physical properties of hardened Portland cement paste. *Journal of the American Ceramic Society* 41(1): 1–6.

Schwiete H.E. et al. 1966. Symposium on structure of Portland cement paste and concrete. *Special Report 90, Highway Research Board* 1: 353.

Energy Science and Applied Technology ESAT 2016 – Fang (Ed.)
© 2016 Taylor & Francis Group, London, ISBN 978-1-138-02973-6

Analysis on the behavior of base-grouted bored piles embedded in various bearing layers

B.J. Li & H.M. Zang
Limited Company of Shandong Road and Bridge Group, Jinan, Shandong, China

X. Zheng
Research Center of Geotechnical and Structural Engineering, Shandong University, Jinan, China

ABSTRACT: Pile-bottom post grouting can increase both end bearing resistance and friction resistance. The histories about the analysis on case indicates that post grouting in gravel, pebble, silt, bedrock and clay can effectively increase single-pile bearing capacity and reduce pile-toe and pile-head settlements. It also shows that post grouting technique is most used in gravel and pebble stratum, followed by silty sand, and it can be used in silt and clay and bedrock as well. When used in bedrock, the main role of grouting is to solidify rock mass and sediment. For piles with the same diameter, the improvement of bearing capacity in gravel and pebble stratum is the biggest. By post grouting, pile-head and pile-toe settlements can be reduced by 24%~37% and 49%~78% respectively.

Keywords: bored cast-in-situ pile, sediment, mudcake, bearing layer, pile-bottom post grouting

1 INTRODUCTION

Pile foundation is presently the most common type of deep foundations in civil engineering constructions in China and around the world. Especially within the recent two decades, pile foundation technique has been developing very rapidly. However, there are two problems that unavoidably exist in the bored cast-in-situ piles, i.e., Mudcake and pile side soil stress relaxation, and pile end sediment and bearing layer disturbance.

Since the pile hole is usually bored below the water level, slurry is often used to balance the formation pressure, protect the hole-wall and prevent the hole collapse and shrinkage. As a result, the slurry drilled pile is usually surrounded by an unsubstantial layer as thick as several millimeters to a few centimeters, which is called mudcake (see Fig. 1). Mudcake with the characteristics of higher moisture content, higher compressibility, lower shear strength, will change the trait of pile-soil interaction. When the mudcake is too thick, the skin friction will be greatly reduced, because the physical and mechanical properties of soil around the pile will be difficult to be mobilized. Moreover, the problems caused by friction resistance degradation of thick-mudcake long piles will serious impair the bearing capacity of bored cast-in-situ pile.

It cannot be ignored that the weakening of pile side soil strength due to the stress relaxation of pile

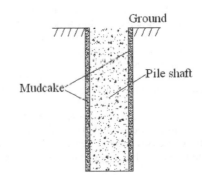

Figure 1. Mudcake of slurry drilled pile.

side soil in hole forming process, especially for large diameter bored pile in cohesionless soil, such as silt and gravel. Because of the stress relaxation of pile side soil in hole forming process, lateral relaxation deformation will occur, which will further weaken pile side soil strength and friction resistance. The decreases of soil strength and friction resistance are also related to soil properties, whether having hole-wall or not, pile diameter and so on.

As for dry-bored (dug) hole pile without slurry wall, radial displacement will occur because the pile side soil is free. The friction resistance still decreases ratherish, though the radial displacement recovers gradually after pile construction.

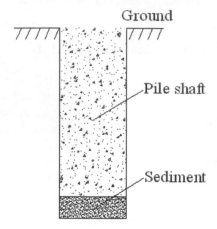

Figure 2. Sediment of slurry drilled pile.

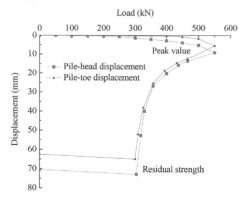

(a) Friction resistance degradation caused by mudcake

Another problem of slurry drilled pile is pile end sediment. During the construction process, it deposits a part of soil chipping in the bottom of hole. If the sediment has not been cleaned up, the end bearing resistance will be reduced. Figure 2 shows the sediment of slurry drilled pile. The existence of sediment makes the pile-toe rest on a weak layer, which will cause the decrease of end bearing, and the significant increase of pile-toe settlement.

Besides, for bored cast-in-situ pile, since the driller churns, cuts and compresses the pile tip soil in hole forming process, it also disturbs the bearing layer and decreases the end bearing. As a result, it may finally reduce the single-pile bearing capacity.

Besides, for bored cast-in-situ pile, since the driller churns, cuts and compresses the pile tip soil in hole forming process, it also disturbs the bearing layer and decreases the end bearing. As a result, it may finally reduce the single-pile bearing capacity.

2 TYPICAL LOAD DISPLACEMENT BEHAVIOR OF PILES WITH THICK SEDIMENT AND MUDCAKE

The existence of mudcake and sediment of bored cast-in-situ pile will reduce single-pile bearing capacity and increase pile settlement. This conclusion can be displayed by load displacement behavior from static load tests. For example, Fig. 3 (a) shows the friction resistance degradation of bored pile caused by the mudcake. The soil surrounding and at the tip of the pile are both soft clay. The failure mode of this pile is punching failure, and the load displacement curve can be divided into three parts. At small load, the load displacement curve presents as a linear elastic curve. And it turns to elasto-plastic curve while the load increases.

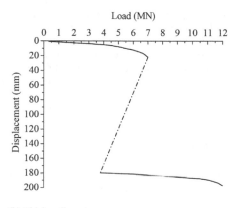

(b) Thick sediment

Figure 3. Load displacement behavior of piles with punching failure mode.

When the load reaches or exceeds pile's ultimate frictional resistance, it will bring on sharp increase of settlement, drop of applied load, and degradation of friction resistance. At the same time, the static resistance turns to the dynamic resistance. At last, the load is eventually maintained at residual strength. Typical failure curve of pile with thick sediment is shown in Fig. 3 (b). The sediment is too thick to bring end bearing resistance into play when friction resistance is already of full use, which causes 120 mm settlement and sharp fall of load. After the sediment is compacted, the load grows again as the end bearing resistance increases.

3 DISCRETENESS OF SINGLE-PILE BEARING CAPACITY

Because of the inherent defects in hole forming process, it is inevitable that bored cast-in-situ pile has those problems mentioned above. And the randomicity of those problems makes the single-pile bearing capacity have discrete distribution. For the same piles in the similar ground conditions, the bearing capacity varies greatly, which not only causes a serious resource waste, but also leads to uneven settlement.

Fig. 4 shows the load displacement curves of 9 test piles of a project located in Wenzhou City of Zhejiang province in China. In this project, bored cast-in-situ piles of the same type, diameter and bearing layer, vary greatly in bearing capacity with the construction conditions, especially with the thickness of sediment. And the variation coefficient of single-pile bearing capacity reaches 0.474.

In order to overcome these problems, post grouting technique of bored pile emerged as the time required. In the past decade, grouting technique for piles has experienced a fast development with a wide application.

A review of past and present base grouting methods used throughout the world has been given by Mullins (Mullins et al. 2000).

Chu Eu Ho (2003) clarified the mechanism of base grouting on weak granite and determined its effectiveness. In addition to the improvement of the pile toe bearing conditions, it was possible for grout to flow upwards along the pile shaft-soil interface during grouting at very high pressures, and thereby enhance the friction resistance as well. His review of base grouting records of 32 piles indicated that there was an approximate correlation between cement consumption, the number of grouting stages and the volume of pile shaft.

Castelli & Wilkins (2004) compared the results of Osterberg load tests at two test piles and discussed the improvement of end bearing resistance by applying base grouting. It showed that base grouting not only enhanced end bearing resistance but also effectively increased friction resistance along the lower section.

W.D. Wang (2006) compared the load displacement and distribution of axial loads in conventional bored piles with the behavior of similar piles that are base grouted after installation. Those conventional bored piles were constructed using bentonite slurry in thick sand soil. Base grouting technique was found to achieve a remarkable increase in friction resistance, end bearing and the basal stiffness of the pile, which significantly improved the overall performance of the pile foundation.

4 CASE STUDIES

4.1 Post grouting in gravel and pebble stratum

The bored piles with diameter of 1500 mm and length of 40 m to 44 m were rested on gravel. The single-pile bearing capacity should be no less than 16000 kN. So these piles were driven into the loose gravel of 8 m and the pile's concrete grade was C30. Two adjacent piles, S_1 (grouted pile) and S_2 (conventional pile), were tested. The results of the maintained load test are given in Fig. 5.

The measured results indicate that:

1. For test pile S_2 (conventional), the accumulated pile-head settlement is 34.87 mm and the incre-

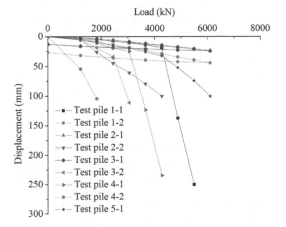

Figure 4. Load displacement behavior of piles in Wenzhou city.

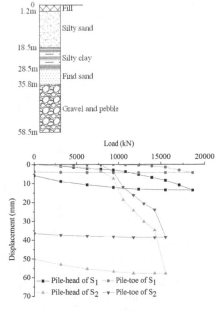

Figure 5. Soil profiles and load displacement behavior of piles with and without grouting.

mental pile-head settlement is 4.60 mm at the applied load of 14283 N. When the applied load reaches 15525 kN, the pile-head and pile-toe settlements increase sharply, and the incremental pile-head settlement is 23.06 mm, about 4 times greater than that at the previous load. These data suggest that the gravel has been gradually squeezed open and relative displacement occurred between pile and gravel. The single-pile ultimate bearing capacity can be defined as 14000 kN, and cannot reach the design value of 16000 kN.

2. For the adjacent test pile S_1 (grouted) of the same size, the accumulated pile-head and pile-toe settlements are only 13.57 mm and 4.6 mm respectively at the applied load of 18630 kN. So the single-pile ultimate bearing capacity of pile S_1 exceeds 18630 kN, which is at least 30 percent higher than that of pile S_2. Since the friction resistance of pile S_1 exceeds 10867 kN which is at least 40 percent greater than that 7762 kN of pile S_2, and end bearing resistance of pile S_1 is at least 19 percent greater than that of pile S_2.

The high friction resistance here is primarily due to two factors. One factor is, in gravel of high permeability, the grout under high pressure tends to compact, penetrate, and solidify the surrounding soil so as to increase the strength of surrounding soil and improve the friction resistance in gravel. The other factor is, the grout will also flows upwards along the pile shaft-soil interface, and enhances the friction resistance in other layers. In our succeeding construction practice, hardened cement grout with the thickness of 10–50 mm was found surrounding the pile shaft beneath the depth of 11 m when the foundation pit was excavated.

4.2 Post grouting in silt

The bored piles, S_1 (conventional pile) and S_2 (grouted pile), with diameter of 800 mm and length 58.33 m (from the ground) rested on silt layer. The pile's concrete grade was C25 and with the steel cage consisting of 12 bars diameter 16 mm. 60 cm thick gravel was placed beneath the pile tip before casting concrete. 10 days later, grout containing 1500 kg cement was injected into the pile end soil with the maximum grouting pressure of 3 MPa. In comparison, 30 cm thick gravel was placed beneath the pile tip while no grouting was carried out. The results of maintained load tests of piles S_1 and S_2 are shown in Fig. 6.

Fig. 6 shows that:

For the test pile S_1 (conventional pile), pile-toe settlement (0.06 mm) occurs when the applied load reaches 2898 kN, and the accumulated pile-head settlement is 3.92 mm then. When applied load reaches the maximum value (6210 kN), the accumulated pile-head and pile-toe settlements are

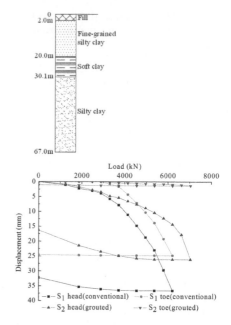

Figure 6. Soil profiles and load displacement behavior of piles resting on silt layer.

36.62 mm and 24.84 mm respectively. As the load is removed, the rebound of pile-head is 4.43 mm, and the residual displacements of pile-head and pile-toe are 32.79 mm and 24.54 mm respectively.

For the test pile S_2 (grouted pile with cement consumption of 1500 kg), pile-toe settlement (0.32 mm) occurs when applied load reaches 3312 kN, and the accumulated pile-head settlement is 1.36 mm then. At the applied load of 6210 kN, the accumulated settlements of pile-head and pile-toe are 14.35 mm and 0.97 mm respectively. When the load continues increasing to 7038 kN, the accumulated settlements of pile-head and pile-toe are 26.14 mm and 1.40 mm. As the load is removed, the rebound of pile-head is 9.87 mm, and the residual displacements of pile-head and pile-toe are 16.27 mm and 1.10 mm respectively.

It can be concluded that grouting in silt layer cuts down the settlement of piles and increases its bearing capacity by about 20 percent.

4.3 Post grouting in bedrock

The bored piles were 45.5 m in length and 800 mm in diameter, of which the concrete grade was C30. The piles were driven into moderate weathering porphyrite with a depth of 1.06 m. Maintained load tests were carried out on two piles, S_1 (conventional pile) and S_2 (grouted pile, cement consumption 700 kg, initial pressure 3~4 MPa, terminal

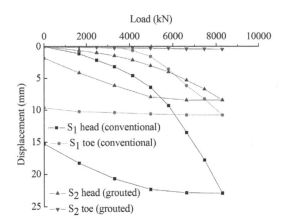

Figure 7. Load displacement behavior of piles resting on bedrock.

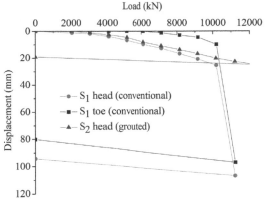

Figure 8. Load displacement behavior of piles resting on clay.

pressure 7~8 MPa, and grouting duration 30 min). The test results are shown in Fig. 7.

Fig. 7 shows that the maximum pile-toe settlement is 10.8 mm before grouting and it is only 0.1 mm after grouting. This indicates that grouting is an effective way to reduce settlement of piles on gravel. However, the single-pile bearing capacity of conventional pile has already satisfied the requirement in this project. Therefore it is not necessary to carry out grouting if the hole is clean and the bedrock is of good quality. Otherwise for the projects with special requirements or thick mudcake and sediment, base grouting can be used to guarantee the quality. The cement consumption in bedrock is usually less than 1 ton per pile.

4.4 Post grouting in clay

The bore piles with diameter of 1000 mm and length 68 m rested on clay. Maintained load tests were carried out on test piles S_1 (conventional pile) and S_2 (grouted pile with cement consumption of 1200 kg), and the results are shown in Fig. 8.

As shown in Fig. 8, the pile S_1 fails by the pile-tip punching into the bearing layer with a settlement of 113 mm when the applied load reaches 11000 kN. So its ultimate bearing capacity can be defined as 10000 kN. For grouted pile S_2, the accumulated pile-head settlement is only 33 mm at the applied load of 12000 kN, and its ultimate bearing capacity is about 20% larger than that of conventional pile.

The static load test results evince that failure mode of conventional pile at high load level is punching failure (the pile tip penetrates the sediment or the clay layer). Grouting in clay can reduce pile's settlement and increase single-pile bearing

capacity by about 15%. Thus whether carrying out grouting in clay depends on the design single-pile bearing capacity, convenience of construction, economical benefits and so on.

5 CONCLUSION

In order to overcome the defects of bored pile, such as mudcake, pile end sediment, and bearing layers disturbance, post grouting technique of bored pile emerged as the time required. It is proved that grouting can solidify the pile end soil and enhance the friction resistance. What's more, grouting can reduce the pile settlement and avoid the situation of uneven settlement. After grouting, the behavior of pile is better. End bearing resistance and friction resistance are enhanced and the settlements of pile-head and pile-toe together with the uneven settlement are lessened.

REFERENCES

Chu Eu Ho. 2003. Base Grouted Bored Pile on Weak Granite. In Proceeding of 3rd International Specialty Conference on Grouting and Ground Treatment: 716–727.

Mullins, G. & Dapp, S and Lai, P 2000. Pressure-Grouting Drilled Shaft Tips in Sand. Proceedings of Sessions of Geo-Denve: 1–17.

Raymond J. Castelli, Ed Wilkins. 2004. Osterberg Load Cell Test Results on Base Grouted Bored Piles in Bangladesh. GeoSupport Conference: 587–602.

Wang, W.D., J.B. Wu, and G.E. Di. 2006. Performance of Base Grouted Bored Piles in Specially Big Excavation Constructed Using Top-down Method. Proceedings of Sessions of GeoShanghai: 393–400.

Energy Science and Applied Technology ESAT 2016 – Fang (Ed.)
© *2016 Taylor & Francis Group, London, ISBN 978-1-138-02973-6*

Seismic behavior analysis of plan irregular reinforced concrete frame building structures

Donat Musonera & Jiejiang Zhu
Department of Civil Engineering, Shanghai University, Shanghai, China

ABSTRACT: While designing buildings, irregularity is inevitable due to the functional and architectural requirements of building as well as to the demands of the building owner. The purpose of this paper is to assess the behavior of four eight-storey reinforced concrete frame structures irregular in plan which presented in two different irregular shapes. The paper also compared their seismic vulnerability with the corresponding regular structure. The structures were modeled and designed using SAP2000 software in accordance with Chinese building codes, and each building structure is 8 floors high. Structural performance was assessed using response spectrum and time-history analysis methods.

Keywords: reinforced concrete frame structure; plan irregular structures; eccentricity

1   INTRODUCTION

It is widely acknowledged that the structural behavior of building structures during earthquakes depends on stiffness, mass, and strength distributions both horizontally and vertically. Irregular structures are characterized by having abrupt reductions in floor area and asymmetry. The building code provides some requirements of plan irregularity in structures, as shown in Fig. 1.

In this study, five different reinforced concrete structures are considered in three shapes (Regular, L-shaped and U-shaped). Their geometry layout is shown in Fig. 2. The buildings selected for study are modeled according to Shanghai seismic data, and the analysis was carried out with respect to the Chinese Code for seismic design of buildings (GB 50011-2010) in the seismic zone of design intensity 7, first seismic design group, and site-class IV. The column spacing is 6000 mm on the Y axis and 7200 m on the X axis. The section dimension equals 120 mm for the slab, 600 × 600 mm for columns, 600 × 300 mm for main beams, and 300 × 200 mm for secondary beams with reinforcement bar strength grade of HRB400 and C30 concrete strength. The storey height is 3.6 m.

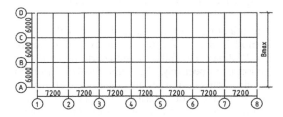

(a) – Regular model B=Bmax

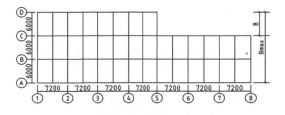

(b)– L-shaped model B=1/3Bmax

Figure 2.   (*Continued*)

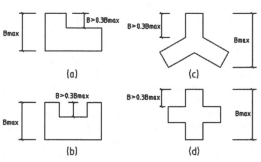

Figure 1.   Examples of concave–convex plane irregular structures.

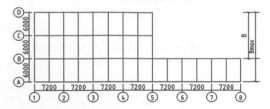

(c)– L-shaped model 01 B=2/3Bmax

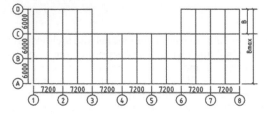

(d)– U-shaped model B=1/3Bmax

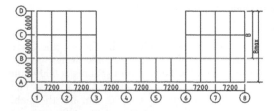

(c)– U-shaped 01 model B=2/3Bmax

Figure 2.   Structure in different shapes.

2   STRUCTURAL IRREGULARITY

The maximum entrant dimension of both U and L models is 6 m for B = 1/3Bmax and 12 m for B = 2/3Bmax. The full dimension corresponding projection is, respectively, 33.3% and 66.6% of 18 m in the given direction. They are greater than 30% given in the seismic code, which proves that both U and L models are re-entrant corner irregular structures. Considering accidental eccentricity seismic action, the maximum storey displacement computed, including accidental torsion, at one end of structure transverse to an axis is more than 1.2 times the average of the storey displacement at two ends of the structure, respectively, when diaphragms are not flexible. Therefore, the structures are torsional irregular.

3   SEISMIC PERFORMANCE FORTIFICATION TARGET

This study can be divided into two design stages: strength design under small earthquakes and deformation checking under strong earthquake to

implement three levels, namely no damage in minor earthquakes, repairable damage in moderate earthquakes, and no collapse in strong earthquakes", which are the basic seismic fortification goals. The main focus of the paper is performance-based fortification and irregularity, based on seismic design of building code appendix M, to achieve performance-based seismic design with respect to target design "performance 3". Particularly, the paper focuses on the structure's corner columns and horizontal transfer beams with respect to "performance 2" low-ductility construction.

4   STRUCTURAL COMPUTATIONAL ANALYSIS

4.1   *Response spectrum analysis under frequent earthquake*

Response spectrum analysis is useful for design decision-making because it relates structural type selection to dynamic performance. Structures with a shorter period experience greater acceleration, whereas those with a longer period experience greater displacement. The maximum transversal storey drift or inter-storey drift is greater than the average drift of the elastic transversal storey or inter-storey at the two ends of the structure, which is 1.2 times as expressed in formula (1). The inter-storey drift ratio for the reinforced concrete frame structure must be less than 1/550:

$$\mu_t = \frac{\Delta_{max}}{\Delta_\alpha} \leq 1.2 \qquad (1)$$

where Δ_{max} is the maximum drift of the storey and Δ_α is the average drift of the storey.

For the structure having obvious asymmetric mass and rigidity distribution, the torsion effects caused by horizontal earthquake actions in two directions are considered. The response spectrum analysis of elastic vibration mode decomposition was carried out under frequent earthquake with torsional coupling method under a bidirectional horizontal earthquake load.

The displacement ratio nearly increases gradually with the height of the building structure (Fig. 3). The maximum value of the displacement ratio occurs on the top floor of structures, while the inter-storey drift occurs on the third floor. The displacement ratio reflects the torsion of the structure layer, while the inter-storey drift reflects the torsional behavior of the overall building structure. From formula (1) for the displacement ratio and the results shown in figures, the inter-storey drift of all the structures is found to be less than 1/550, which satisfies the requirements of the Chinese seismic design of buildings code GB 50011-2010.

For the displacement ratio on the X and Y directions, the L-shaped 01 model presents more vulnerability to torsional effects (Fig. 3). However, the value is less than 1.2; therefore, it satisfies the building code's requirements. Regarding the inter-storey drift, the L-shaped 01 model has an overall greater value than the other structures (Fig. 4). From the period ratio shown in Table 1, except the regular structure model, we can see that the other models do not satisfy the code requirement that should be less than 90%. The non-conformity to this requirement means that torsional rigidity of the structure is slightly relatively bigger than lateral stiffness. Therefore, there must be an overall adjustment to strengthen the outer corner stiffness of the structure.

4.2 Time-history analysis under frequent earthquake

For irregular buildings in category A of tall buildings belonging to the height range listed in article 5.1.2-1 of code for seismic design of buildings, an additional calculation under frequent earthquake will be made by using the time-history analysis

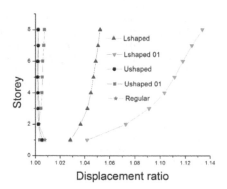

Figure 3. Displacement ratio.

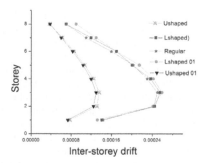

Figure 4. Inter-storey drift.

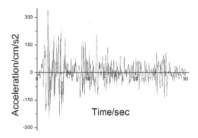

(a) El Centro wave

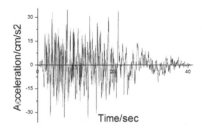

(b) Taft wave

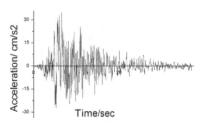

(c) Shanghai wave

Figure 5. Different seismic waves.

Table 1. Torsional period ratio.

Structural type	Period ratio
Regular structure	0.318
L-shaped structure	0.904
L-shaped structure 01	0.905
U-shaped structure	0.925
U-shaped structure 01	0.927

Table 2. Comparison between time-history and response spectrum analyses of a regular structure.

Wave	X(kN)	Vx/Vcx	Y(kN)	Vy/Vcy
El Centro	3871.75	0.95	3712.84	0.92
Taft	2023.19	0.50	2005.56	0.50
Shanghai2	4090.23	1.00	4049.47	1.00
Mean value	3328.39	0.82	3255.96	0.80
CQC method	4077.37	1	4046.45	1

Table 3. Comparison between time-history and response spectrum analyses of L-shaped structures.

Wave	L-shaped				L-shaped 01			
	X(kN)	Vx/Vcx	Y(kN)	Vy/Vcy	X(kN)	Vx/Vcx	Y(kN)	Vy/Vcy
El Centro	3558.96	1.03	3330.08	0.98	3025.58	1.09	2962.10	1.09
Taft	1838.67	0.53	1908.36	0.56	1388.12	0.50	1566.35	0.57
Shanghai2	3672.50	1.07	3584.36	1.06	3245.31	1.17	3104.29	1.14
Mean value	3023.38	0.88	2940.93	0.87	2553.00	0.92	2544.25	0.93
CQC method	3442.30	1	3387.06	1	2781.69	1	2724.94	1

Table 4. Comparison between time-history and response spectrum analyses of U-shaped structures.

Wave	U-shaped				U-shaped01			
	X(kN)	Vx/Vcx	Y(kN)	Vy/Vcy	X(kN)	Vx/Vcx	Y(kN)	Vy/Vcy
El Centro	2100.78	0.84	2127.99	0.85	1827.98	0.84	2043.80	0.94
Taft	1887.54	0.75	1926.19	0.77	1702.33	0.79	1741.56	0.80
Shanghai2	2282.20	0.91	2283.7	0.91	1982.66	0.92	1993.32	0.92
Mean value	2090.17	0.83	2112.63	0.84	1837.65	0.85	1926.23	0.89
CQC method	2508.17	1	2505.29	1	2165.09	1	2174.09	1

Vx/Vcx and Vy/Vcy are, respectively, shear ratios.

method. According to the location of the project site with frequency characteristics, we choose two actual earthquake records, namely El Centro wave and Taft wave, and one artificially simulated acceleration time history curve Shanghai2 wave. The main input values are the peak acceleration of 35 gal and the damping ratio of 0.05. The acceleration is adjusted to meet the site IV requirements.

From Table 2 to Table 4, we can see that the shear force effect of each wave on structures is greater than only 65% of the CQC method result. Also, the mean base sheer force of the three waves is greater than 80% of the CQC method result. This satisfies the GB 50011-2010 code article 5.1.2 requirements. The elastic time-history analysis reveals that the base shear force of the structure calculated with each time-history curve must not be less than 65% of the calculation results with the mode-decomposition response spectrum method. Moreover, the mean value of several time-history curves must not be less than 80% of the calculation results with the mode-decomposition response spectrum method. However, U-shaped structures present the smallest base shear force, while the regular structure has the greatest base shear force compared with the other different structures.

The maximum inter-storey drift is 1/571 for the L-shaped 01 model under the El Centro wave on the X and Y directions. Therefore, it is less than the 1/550 limit required in accordance with the GB 50011-2010 code in article 5.5.1, and meets the provisions of the seismic code. The above results show that the structural elastic analysis under frequent earthquake meets

the specification requirements. This guarantees that there will be no damage of the structures under frequent earthquake. The design will be carried out by considering the larger value between response spectrum and time-history analysis methods.

4.3 Elasto-plastic time history analysis under severe earthquake

Considering plan irregularity in structures, torsional response will obviously occur using pushover analysis. The horizontal loading should be distributed not only along the height of the structure, but also along its width. After entering the elasto-plastic phase, the structural effect is more apparent. SAP2000 software analysis results indicate the structural dynamic characteristics of the structure deformation pattern, component ductility and its damage, as well as the elasto-plastic behavior of the structures more accurately. Three seismic waves are selected under severe earthquake for elasto-plastic analysis, in which the main input values are the peak acceleration of 220 gal and the damping ratio of 0.05. Plastic hinges for the beam and column constitutive model are shown in Fig. 7. On both ends of the beam element, the M3 plastic hinge moment is used, while on both ends of the column element, the axial force with a bidirectional moment PMM coupling plastic hinge is used.

By comparing the effects of the three seismic wave functions on all the structures in the X and Y directions, the maximum inter-storey drift is found to occur on the third floor where the largest value is caused by the structural response of the El Centro wave (Fig. 8). The L-shaped 01 model shows the biggest inter-storey drift value equal to 1/92, which is less than 1/50. Therefore, the overall performance of the structure meets the fortification goal provided in the Chinese code for seismic design of buildings of

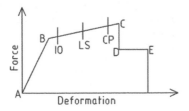

Figure 7.   Force load–deformation curve.

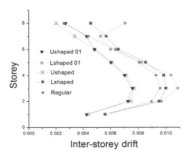

(a) Under El Centro wave

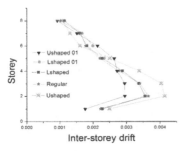

(b) Under Taft wave

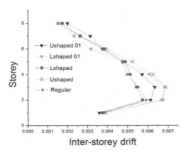

(c) Under Shanghai2 wave

Figure 8.   Inter-storey drift of structures under severe earthquake.

no collapse under severe earthquake. The maximum inter-storey drift in both X and Y directions is about 4 times the value of elastic time-history analysis, which satisfies the appendix M "performance 3" in code for seismic design of buildings GB 50011-2010. In the following analysis, the L-shaped 01structure will be chosen for further analysis to assess the behavior of plan irregularity in structures as a result of being vulnerable to the seismic effect.

The maximum average base shear of the L-shaped 01structure in the X direction is 4046 KN with a shear weight ratio of 12.8%. In the Y direction, it is 3874 KN with a shear weight ratio of 11.9%, which is about 3 times the value of elastic-time history analysis under frequent earthquake. The acceleration peak value of the selected under severe earthquake is about 6 times the value under frequent earthquake. However, the shear ratio is less than 6 times, which is due to certain structural damages under severe earthquake and concrete stiffness degradation, as shown in Fig. 9, by the hinge formation. By comparing the structural analysis results of the L-shaped 01 model, it is found that the structural response under the three seismic waves causes the most damage, whereas that under the El Centro wave is likely to cause the most serious damage, as shown in Fig. 9.

The analysis shows that at 1.82 s, the structure begins to have plastic hinges in the IO stage. At 5.10 s, the structure begins to have plastic hinges in the LS stage. At 8.04 s, the structure has plastic hinges in the CP and D stages on the second floor, which is due to beam yielding. The torsional displacement ratio is larger in the X direction, and the damage of side frame structural components is relatively serious. The above analysis shows that the structure satisfies the yield of the "strong column weak beam" mechanism, conforming to the expected damage mechanism of the whole framework. The worst damage of plastic hinge distribution mainly concentrates on the side frame, which is consistent with the structure torsion effect.

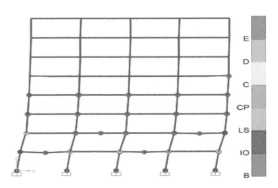

Figure 9.   Hinge formation.

4.4 *Floor stress analysis*

From the results of the SAP2000 structural analysis results, it can be found that the L-shaped 01 model has a smaller effective slab width in the corner. The floor stress under frequent and severe earthquakes showed that the structure response under the El Centro wave at second floor stress is the largest, as shown in Fig. 10. The floor area at the edge as well as the concave of the floor plan showed stress concentration, while a large tensile stress is produced on the slab with smaller effective width areas.

The maximum tensile stress of the floor under frequent and severe earthquakes appears at the narrow part of the slab. The maximum tensile stress produced under frequent earthquake is 0.48 MPa, which is not greater than 1.27 MPa. However, the maximum tensile stress produced under severe earthquake is 3.2 MPa, but most of the floor stress is not greater than 1.27 MPa. For slab positions that appear with larger stress during structural design, the slab reinforcement ratio must be improved and the slab should have two-way bidirectional reinforcement. For plan irregular structures, their center of mass and rigidity do not coincide; therefore, the torsional effect is generated. To control the plan structure's torsional behavior during earthquake, it is recommended to add an amount of concrete shear wall all around the structural plan corners as well as at the concave parts of irregular structure models (Lin Baoxin et al, 2012).

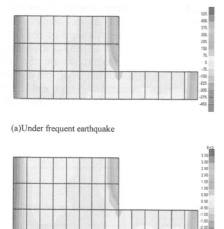

(a)Under frequent earthquake

(b)Under severe earthquake

Figure 10. Floor stress.

5 CONCLUSION

In this paper, the seismic behavior assessment of reinforced concrete frame structures with plan irregularities was carried out using SAP200 software. The results indicate that the U-shaped and U-shaped 01 models have a significant effect on torsion. Both elastic and elasto-plastic time-history analyses under frequent and severe earthquake were performed. The results indicate that the L-shaped 01 model shows more vulnerability than the other structures studied. The structural elastic analysis under small earthquakes meets all the properties and specification requirements, which ensures that there will be no damage of the structures under frequent earthquake. Furthermore, the overall performance of the structures studied meets the fortification goal provided in the Chinese code for seismic design of buildings of no collapse under severe earthquake. The fortification of plan irregular structures can guarantee the three basic rules of the seismic design of the building code. In accordance with the China seismological bureau files of structures, the fortification meets the appendix M of GB 50011-2010 for performance-based design goals and objectives.

REFERENCES

GB 50011-2010, Chinese code for seismic design of buildings.

Haijuan Duana, Mary Beth D. Huesteb. Seismic performance of a reinforced concrete frame building in China. 41 (2012) 77–89.

Herrera, R., Vielma, J., Ugel, R., Martínez, Y. and Barbat, A., Optimal design and earthquake-resistant design evaluation of low-rise framed RC structure. Natural Science, 4(8a), pp. 677–685 (2012).

Himanshu Gaur, R.K Goliya, Krishna Murari, Dr. A. K Mullick. A parametric study of multi-storey R.C building with horizontal irregularity. 2014, eISSN: 2319–1163 | pISSN: 2321–7308

Lin Baoxin, Tian Hao. The research on seismic performance of irregular frame structure with L-shaped plane. 1006–4540(2012) 06–023–05.

Raul Gonzalez Herrera and Consuelo Gomez Soberon, "Influence of Plan Irregularity of Buildings", 14th World Conference on Earthquake Engineering October 12–17, 2008, Beijing, China.

Ravikumar C M, Babu Narayan K S, Sujith B V, Venkat Reddy D. Effect of Irregular Configurations on Seismic Vulnerability of RC Buildings. 2012, 2(3): 20–26.

Wakchaure, M.R., Anantwad Shirish, Rohit Nikam, "Study of Plan Irregularity on High-Rise Structures", International Journal of Innovative Research & Development, Vol 1 Issue 8 October 2012.

Energy Science and Applied Technology ESAT 2016 – Fang (Ed.)
© *2016 Taylor & Francis Group, London, ISBN 978-1-138-02973-6*

Estimating mechanical parameters of rock mass relating to the coal-bearing formation

Zhiping Li
China Nerin Engineering Co. Ltd., Nanchang, Jiangxi, China

Jiaqing Du
Department of Civil Engineering, Shanghai Jiao Tong University, Shanghai, China

ABSTRACT: Estimation of rock mass mechanical parameters related to coal-bearing formation is important in the design and excavation of deep roadway. Some methods and their principles that are widely used to evaluate the mechanical parameters of rock mass are introduced. Based on the rock physico-mechanical parameters and the rock mass quality, a new method integrating the GSI index and the Hoek–Brown criterion is provided to evaluate the mechanical parameters of rock mass relating to coal-bearing formation. The results indicated that the new method is reasonable and the value of rock mass mechanical parameters relating to coal-bearing formation is closer to the field rock mass feature.

Keywords: Hoek–Brown method; rock mass mechanical parameters; coal-bearing formation

1 INTRODUCTION

The reliable estimation of mechanical rock mass parameters related to coal-bearing formation has always been a pivotal issue in the design of deep roadway. These rock characteristics are also influenced by discontinuities occurring in rocks, especially complex nonlinear mechanics characteristics. The field disturbance factor of deep surrounding rocks cannot be explained by the traditional linear theory. Thus, estimation of mechanical rock mass parameters related to coal-bearing formation has a great significance. Previous research has mostly focused on the Hoek–Brown criterion to estimate the parameters of rock mass (Hu Sheng-ming et al., 2011) (Zhuo Li et al., 2015) (Yang Xiao-li et al., 2006) (Behnam Yazdani Bejarbaneh et al., 2015) (LU Ping et al., 2010). In another study, laboratory test and rock mass quality evaluating the RMR value were combined to analyze the mechanical parameters of rock mass from the Yunxi Datun Tin Mine (Yang Ze et al., 2010). An intelligent displacement back analysis method has been proposed to estimate the mechanical parameters of rock mass relating to the Three Gorges Project permanent shiplock (Feng Xia-ting et al., 2000). This paper proposes the estimation of rock mass mechanical parameters by integrating the GSI index and the Hoek–Brown criterion, which provides the theory basis for numerical simulation research.

2 EVALUATION METHODS OF ROCK MASS PARAMETERS

2.1 *Hoek–Brown criterion (Hoek, E et al., 2008)*

The Hoek–Brown criterion was first introduced in 1980. Over the last 30 years, the criterion has been widely applied in the engineering field and accepted by the international rock mechanics community. The recent version of the Hoek–Brown criterion is described as follows:

$$\sigma_1' = \sigma_3' + \sigma_{ci}\left(m_b \frac{\sigma_3'}{\sigma_{ci}} + s\right)^a \tag{1}$$

where σ_1' is the maximum effective stress; σ_3' is the minimum effective stress; σ_{ci} is the uniaxial compressive strength of the intact face; m_b is a reduced value of the material constant m_i, which is given by:

$$m_b = m_i \exp\left(\frac{GSI - 100}{28 - 14D}\right) \tag{2}$$

s and a are given by the following relationships:

$$s = \exp\left(\frac{GSI - 100}{9 - 3D}\right) \tag{3}$$

$$a = \frac{1}{2} + \frac{1}{6}\left(e^{-GSI/15} - e^{-20/3}\right) \tag{4}$$

D is a factor that depends on the degree of disturbance and varies between 0 and 1.

2.2 Mohr–Coulomb criterion (S. D. Priest, 2005)

Equivalent angles of friction and cohesive strengths for each rock mass and stress range are found by fitting an average linear relationship to the curve generated by solving Eq. (1). The Mohr–Coulomb criterion is given by:

$$\phi' = \sin^{-1}\left[\frac{6am_b\left(s+m_b\sigma'_{3n}\right)^{a-1}}{2(1+a)(2+a)+6am_b\left(s+m_b\sigma'_{3n}\right)^{a-1}}\right] \quad (5)$$

$$c' = \frac{\sigma_{ci}\left[(1+2a)s+(1-a)m_b\sigma'_{3n}\right]\left(s+m_b\sigma'_{3n}\right)^{a-1}}{(1+a)(2+a)\sqrt{1+\left(6am_b\left(s+m_b\sigma'_{3n}\right)^{a-1}\right)/((1+a)(2+a))}} \quad (6)$$

where $\sigma_{3n} = \sigma'_{3\max}/\sigma_{ci}$.

The Mohr–Coulomb shear strength τ, for a given normal stress σ, is found by the substitution of the values of c' and ϕ' into the following equation:

$$\tau = c' + \sigma\tan\phi' \quad (7)$$

The equivalent plot, in terms of the major and minor principal stresses, is defined by:

$$\sigma'_1 = \frac{2c'\cos\phi'}{1-\sin\phi'} + \frac{1+\sin\phi'}{1-\sin\phi'}\sigma'_3 \quad (8)$$

3 ESTIMATION OF ROCK MASS PARAMETERS OF COAL-BEARING FORMATION

3.1 Rock mass quality evaluation

Recently, research on rock mass parameters of coal-bearing formation has been the key work. Rock deformation rate and strength characteristics of different coal-bearing formation represent different attenuation situations. The rock mechanics

properties of coal-bearing formation were determined directly from the engineering tests and previously monitored data (Table 1).

Considering the rock mass classification system by using the Hoek–Brown criterion, the value can be obtained from No. 9 Coal Mine of Hebi Coal Industry, China (Table 2).

3.2 Engineering transition of rock mechanical parameters

The generalized Hoek–Brown mechanics parameters of coal-bearing formation are computed using Roclab software. The depth H, the geological strength index GSI (0–100), the rock mass disturbance factor D (0–1), and the elastic modulus increase factor K are considered as input parameters. This paper integrated two sets of rock mass parameters calculated from the Hoek–Brown criterion and the Mohr–Coulomb criterion. The result of sandstone is shown in Fig. 1.

As shown in Fig. 1, the fitting effect meets the design requirements and the analysis method is feasible. The fitting results of the rock mass mechanical parameters related to coal-bearing formation are presented in Table 3.

After the investigation and understanding the rock mass characteristics, the estimation method for rock mass parameters of coal-bearing formation tended to be reasonable by correcting the parameter variables. Compared with the rock mass quality evaluation methods (e.g. BQ method, RMR classification method), the selected rock mass mechanical parameters of coal-bearing formation are closer to the overall understanding of the field rock mass feature.

Table 2. Rock mass variable value obtained from a case study (Li Zhiping, 2010).

Case	H/m	GSI	D	K
1	600	5	0.1	1.1

Table 1. Test results of the rock mechanics parameters of coal-bearing formation (Meng Zhao-ping et al., 2002).

Rock type	$\rho/$ $(g \cdot cm^{-3})$	$E/$ (GPa)	μ	$c/$ (MPa)	$\varphi/$ $(°)$	$\sigma_t/$ (MPa)	$\sigma_c/$ (MPa)
Sandstone	2.47~3.47	16.13~86.44	0.11~0.33	1.91~13.07	33.41~39.15	1.77~10.67	2.47~3.47
Siltstone	2.43~2.63	30.00~34.00	0.28~0.33	1.25~2.40	39.00~40.03	1.20~9.20	2.43~2.63
Sandy mudstone	2.64~2.98	7.60~44.00	0.10~0.30	4.00~11.90	31.90~38.39	0.70~8.70	2.64~2.98
Mudstone	2.05~2.97	2.01~19.71	0.15~0.34	0.14~8.40	31.8~41.52	0.30~7.29	2.05~2.97
Coal	1.30~1.46	0.70~4.74	0.11~0.38	0.04~4.13	30.20~33.42	0.19~0.55	1.30~1.46

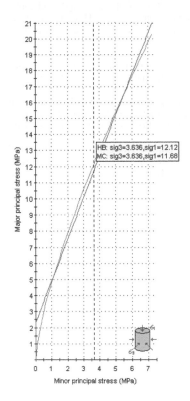

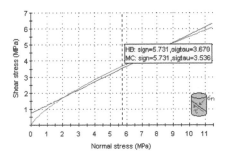

Hoek-Brown Classification
  intact uniaxial comp. strength (sigci) = 111.5 MPa
  GSI = 5   mi = 17   Disturbance factor (D) = 0.1
  intact modulus (Ei) = 59540 MPa

Hoek-Brown Criterion
  mb = 0.478   s = 1.81e-5   a = 0.619

Mohr-Coulomb Fit
  cohesion = 0.732 MPa   friction angle = 26.07 deg

Rock Mass Parameters
  tensile strength = -0.004 MPa
  uniaxial compressive strength = 0.129 MPa
  global strength = 5.377 MPa
  deformation modulus = 1521.39 MPa

Figure 1.   Analysis of sandstone set of rock mass parameters.

Table 3.   Fitting results of the rock mass mechanical parameters of coal-bearing formation.

Rock type	σ_0/MPa	K/GPa	G/GPa	c/MPa	φ/°
Sandstone	18	0.9	0.62	0.732	26
Siltstone	18	0.8	0.31	0.433	17.9
Sandy mudstone	18	0.38	0.24	0.342	14.1
Mudstone	18	0.184	0.103	0.252	10.9
Coal	18	0.047	0.028	0.103	9.0

4   CONCLUSION

1. The method used for estimating rock mass mechanical parameters of coal-bearing formation by integrating the GSI index and the Hoek–Brown criterion is more practical, and the parameters obtained can be effectively applied to the numerical calculation and simulation of rock mass in engineering.
2. This paper cannot accurately verify the selected rock mass parameters of coal-bearing formation due to lack of field experiments. However, it shows that the estimation method of rock mass parameters is reasonable as the results are

closer to the overall understanding of the field rock mass feature.

ACKNOWLEDGMENTS

This study was supported by the Key Technology Research and Development Program of Jiangxi Province of China (20111BBE50031). Corresponding author: Li Zhi-ping, Mine Construction Engineer. E-mail: lzp3143@163.com.

REFERENCES

Behnam Yazdani Bejarbaneh, Denial Jahed Armaghani, Mohd For Mohd Amin. Strength characterization of shale using Mohr-Coulomb and Hoek-Brown criteria [J]. Measurement, 2015, 63: 269–281.

Feng Xia-ting, Zhang Zhi-qiang, Qian Sheng. Estimating mechanical rock mass parameters relating to the Three Gorges Project permanent shiplock using an intelligent displacement back analysis method [J]. International journal of rock mechanics & mining sciences, 2000, 37: 1039–1054.

Hu Sheng-ming, Hu xiu-wen. Estimation of rock mass parameters based on quantitative GSI system and

Hoek-Brown criterion [J]. Rock and soil mechanics, 2011, 32(3): 861–865.

Kersten Lecture: Hoek, E., Carranza-Torres, C., Diederichs, M.S. et al. Integration of geotechnical and structural design in tunnelling. Proceedings University of Minnesota 56th Annual Geotechnical Engineering Conference. Minneapolis, 29 February 2008, 1–53.

Li Zhiping. Study on high convex strip and bolt-mesh-cable combined supporting effect for deep soft rock roadway [D]. Jiaozuo: Henan Polytechnical University, 2010.

Lu Ping, Wang Ning, Wu Yi. Research and application on mechanical strength parameters for deep rock mass based on Hoek-Brown failure criterion [J]. Journal of southwest university of science and technology, 2010, 25(4): 34–38.

Meng Zhao-ping, Peng Su-ping, Fu Ji-tong. Study on control factors of rock mechanics properties of coal-bearing formation [J]. Chinese journal of rock mechanics and engineering, 2002, 21(1): 102–106.

Priest S.D. Determination of shear strength and three-dimensional yield strength for the Hoek-Brown criterion [J]. Rock mechanics and rock engineering, 2005, 38(4): 299–327.

Yang Xiao-li, Yin Jian-hua. Linear Mohr-Coulomb strength parameters from the non-linear Hoek-Brown rock masses [J]. International journal of non-linear mechanics, 2006, 41: 1000–1005.

Yang Ze, Hou Ke-peng, Li Ke-gang, et al. Determination of mechanical parameters of rock mass from Yunxi Detun Tin Mine [J]. Rock and soil mechanics, 2010, 31(6): 1923–1927.

Zhuo Li, He Jiang-da, Xie Hong-qiang, et al. Study of new method to determine strength parameters of rock material based on Hoek-Brown criterion [J]. Chinese Journal of rock mechanics and engineering, 2015, 34: 2773–2783.

15   *Traffic and transportation engineering*

Energy Science and Applied Technology ESAT 2016 – Fang (Ed.)
© 2016 Taylor & Francis Group, London, ISBN 978-1-138-02973-6

Research on the text mining model of the database of the system in railways

Guimin Jia, Futian Wang & Lei Bai
State Key Laboratory of Rail Traffic Control and Safety, Beijing Jiaotong University, Beijing, China

Yunfeng Chen
Track Department of Lanzhou Railways Administration, Lanzhou, China

ABSTRACT: This paper proposes a model of text mining to solve the problem of long-text fields in the database table of the Safety Inspection Management Information System that are not subjected to in-depth analysis of security issues. The model comprises the Chinese word segmentation with the hidden Markov model algorithm. Then, it follows the word frequency statistics for the word after segmentation. Finally, it draws the picture of word cloud display. The model can be used to realize the deep mining of the long text data in the database table, analyze the security problems in the database of the railway system, and provide references for the safety management of railway system.

Keywords: Railway safety management; Safety Inspection Management Information System; weak equipment; disease types; text mining; segmentation; word cloud

1 INTRODUCTION

Railway operation safety refers to all production activities to maintain the normal operation of railway, and to ensure the safety of life and property of passengers and railway staff and the integrity of the equipment and cargo transport (Chen Liming, 2006) (Forney G D, 1973).

To improve the management level of railway transportation, the former Safety Supervision Department of the Ministry of Railways applied and popularized the "Safety Inspection Management Information System", which collects and stores all the data of accidents and failures in the process of transportation management, performs statistical analysis of the data to achieve the purpose of safety warning and problem prevention by correctly analyzing the situation of safety production.

The data of railway safety production problems in this system are related to the quality of equipment, safety management, fieldwork, and other categories of railway system. Some fields in the data table are presented in an unstructured form with long texts. It is difficult to analyze these texts directly, and it is necessary to carry out text mining on these data to get valuable information from the data.

Text mining refers to the process of extracting unknown, understandable, and available knowledge that can be used to better organize information for future reference from large amounts of text data (Forney G D, 1973). Since the introduction of the text mining concept by Feldman in 1995, text mining has been developing rapidly at home and abroad (Guo Jinlong et al, 2012). In China, the most popular text mining technologies, including text structure analysis, text summarization, text classification, text clustering, and association rules, have been widely used in many fields such as biology, medicine, information analysis, and humanities (Han Mailiang, 2008) (Guo Shulun, 2013) (Huang Xiao-bin et al, 2009) (Kang Dong, 2014).

In this paper, we first provide the data description of the "Safety Inspection Management Information System", pointing out its characteristics in the database table and the problems that need to be solved by text mining. Then, a text mining model is built appropriately based on the characteristics of the data table. Finally, the model is verified by R language.

2 DATA DESCRIPTION OF THE SYSTEM

2.1 Structure of data sheet for safety production problems

The main composition fields of safety production problems in the data sheet are presented in Table 1.

Among them, the msg_sort is divided into the sectors of engineering, electricity and signal,

Table 1. Main composition fields of safety production.

Fields	report_dept	report_intime	occur_dept	occur_time	occur_area	msg_code	msg_sort	msg_class
Length field	7 bytes	9 bytes	7 bytes	16 bytes	24 bytes	8 bytes	2 bytes	4 bytes
Fields	msg_content	report_desc	zr_man	zr_man_duty	zg_text	zg_dept	zgwc_date	
Length field	154 bytes	340 bytes	26 bytes	25 bytes	384 bytes	7 bytes	10 bytes	

traction power supply, maintenance, and vehicles. The msg_class is divided into quality of equipment, safety management, site operations, and quality of staff. As shown in Table 1, some fields are in the form of a relatively long text, such as the field of msg_content, report_desc, and zg_text.

2.2 Existing problems in data sheet

Most previous studies have not sufficiently considered the deep analysis and utilization of the long-text data in the database, which leads to the difficulty in obtaining the information of defects such as its categories and causes.

Sentences of long-text fields in the database provide abundant information. For example, the sentence "The X level is 6 mm in place of 20# turnout at Hekou South Railway Station" contains the type of defect, the extent of the defect, the location of the occurrence of the defect, and other information. The type of defect is X level, the extent of the defect is 6 mm, and the occurrence of the defect is the turnout. The information cannot be easily extracted and utilized directly, which requires the application of a text mining model to excavate long-text fields.

3 MODEL OF TEXT MINING

The author puts forward a model of text mining to analyze the problems that exist in the database table of the "Safety Inspection Management Information System". The idea of text mining is segmentation using the Hidden Markov Model (HMM) algorithm for the database table of long-text fields. Then, statistical results can be obtained by counting the word frequency and displaying word cloud. The model is shown in Figure 1.

3.1 Word segmentation

Chinese word segmentation is the key link to Chinese information processing. In the text analysis, a whole paragraph will be cut into entry information of the smallest semantic, which is called word segmentation (Ling Ming, 2014).

Figure 1. Text model of mining.

This paper uses the hidden Markov model algorithm based on the statistical algorithm for word segmentation. The statistical word segmentation algorithm based on probability theory has a good ability to identify ambiguity and unknown words. The hidden Markov model algorithm is a kind of statistical word segmentation algorithm, according to the transition probability of the front and rear units, to predict the final state through the Markov chain. The time and space overhead of HMM are smaller compared with other algorithms. Moreover, HMM is widely used and has a good effect and higher accuracy and coverage (Ronen Feldman et al, 1995).

Given a participle sequence of atoms $S = \{s_1, s_2, s_3, ..., s_n\}$, $W = \{w_1, w_2, w_3, ..., w_n\}$ is recorded as a word segmentation result of this segmentation sequence. The word sequence will be used as the observation value of HMM, and the corresponding category sequence of the word sequence is denoted as $C = \{c_1, c_2, c_3, ..., c_n\}$. Finally, as the state sequence of the HMM, we take the participle structure $W^{\#}$ with the highest probability as the final segmentation results, which is given by:

$$W^{\#} = \arg\max \prod_{i=1}^{n} p(w_i/c_i) p(c_i/c_{i-1}) \quad (1)$$

where C_0 is the category of the beginning of the sentence. For the convenience of calculation, a negative logarithm is often used, which is as follows:

$$W^{\#} = \arg\max \sum_{i=1}^{n} \left[-\ln p(w_i|c_i) - \ln p(c_i|c_{i-1}) \right] \quad (2)$$

$W^{\#}$ can be obtained by the Viterbi decoding algorithm (Viterbi A J, 2006) (Wang Min, 2014) (Xiao Ming, 2014).

Table 2. Frequency table of the first 10 words.

Words	Joint	Turnout	Surface	Gauge	Xlevel	Sleeper	Bolt	Switch rail	Train shaking	Rail fastening
Word frequency	8003	7881	7051	4118	3482	3412	3091	3082	3002	1790
Frequency	0.406	0.400	0.359	0.209	0.177	0.174	0.157	0.157	0.153	0.091

3.2 Word frequency statistics

Word frequency statistics is used to calculate the frequency of words, which is a basic project of the Chinese information process (Yuan Jinshi, 2004). In this paper, the frequency of each word is obtained through the statistics of the Chinese word segmentation. By counting the frequency of words accurately, we can find out which is the high frequency words and provide a basis for the display of the word cloud. By using word frequency statistics, the system's hidden information in the table of long text is mined, such as type of defect and weak equipment.

3.3 Word cloud

Word cloud is a technology of information text visualization that make up some words. It is a color graphics that is similar to the cloud. Each word has a weight, and the size of the text reflects the word frequency through the layout algorithm. Word cloud is supplemented by a variety of colors that directly reflects the importance of phrase and displays key information of the text. This paper will excavate mining information for visual display through the visualization technology of the word cloud.

4 MODEL TEST RESULTS

The author extracted 20,000 engineering data from the "Safety Inspection Management Information System's" database table that reflected the safety production problem about equipment quality in 2012–2015 of the Lanzhou Railway Administration. Those extracted data are investigated using the model of text mining shown in Fig. 1. The field "The description of problem details" is analyzed by R language. In this way, we can find the vulnerable equipment of engineering and the disease types of equipment quality. The statistics of word frequency is given in Table 2, and the corresponding word cloud is shown in Figure 2.

From the graph and the table, we can obtain the information on the equipment type and the quality of equipment with respect to the security problem. The types of equipment security problems consist

Figure 2. Picture of word cloud display.

of joint, turnout, sleeper, bolt, switch rail, and rail fastening. In contrast, examples of equipment quality problems are surface, gauge, Xlevel, and train shaking. The results show that:

1. The equipment that can be easily damaged are rail joint and turnout.

   From Figure 2, we reach the conclusion that the equipment type that is prone to safety problems is sorted by the number of security problems ranging from more to less as follows: joint, turnout, sleeper, bolt, switch rail, rail fastening, etc.

2. The disease types that can easily occur are surface, gauge, X level, etc.

   From the diagram, we can see that the most prominent disease type of equipment quality is surface, followed by rail gauge, X level, train shaking, track alignment, and twist warp.

We should not only focus on checking the problem of switches and joints, but also strengthen the level of surface, gauge, and X level management.

5 CONCLUSION

According to the characteristics of the railway safety production problems listed in the "Safety Inspection Management Information System", the author designs this text mining model, which has a reference value for the deep analysis and utilization of long-text fields of the data in the "Safety Inspection Management Information System".

The author extracts the four-year quality safety production data of permanent way equipment quality safety production of Lanzhou railway bureau, using R language software, to validate text mining. By extracting the key information of the text information, the author indeed researches on the type of equipment damages and on weak equipment that are prone to damages. The research results indicate that the model is helpful to strengthen the maintenance and protection management of vulnerable equipment, provide a reference for risk management and maintenance, and prevent the occurrence of railway operation accidents.

ACKNOWLEDGMENTS

This research was funded by Science and Technology Research and Development Program of China Railway Corporation (2015T001-B) and the National Natural Science Foundation (51578057).

REFERENCES

Chen Liming. 2006. Safety Risk Management Method of Railway Traffic, Shanghai Railway Science & Technology, 2006(02): 19–20.

Forney G D. 1973. The Viterbi Algorithm. Proceedings of the IEEE, Vol. 61, No. 3: 268–278.

Guo Jinlong, Xu Xin. 2012. Text Mining Research in Digital Humanities, Journal of Academic Libraries, 30(3): 11–18.

Guo Shulun. 2013. Chinese corpus application course, Shanghai: Publishing House of Shanghai Jiao Tong University.

Han Mailiang. 2008. Railway Traffic Safety Management, Beijing: China Railway Publishing House.

Huang Xiao-bin, Zhao Chao. 2009. Application of Text Mining Technology in Analysis of Net-Mediated Public Sentiment, Information Science, 2009(01): 94–99.

Kang Dong. 2014. Basic Theories and Application Techniques of Chinese Text Mining, Suzhou: Soochow University.

Ling Ming. 2014. R language and Website Analysis, Beijing, China Machine Press.

Ronen Feldman, Ido Dagan. 1995. Knowledge Discovery in Textual Databases (KDT), In proceedings of the First International Conference on Knowledge Discovery and Data Mining (KDD-95): 112–117.

Viterbi A J. 2006. A personal history of the Viterbi algorithm. IEEE Signal Processing Magazine, July 2006: 120–142.

Wang Min. 2014. Chinese Part-of-Speech Tagging Based on Ameliorated Hidden Makov Model, Taiyuan: Shanxi University, 2014.

Xiao Ming. 2014. Information Rate Statistic, Beijing: China Railway Publishing House.

Yuan Jinshi, 2004. Viterbi Algorithm: Analysis and Implement, Changsha: National University of Defense Technology.

Energy Science and Applied Technology ESAT 2016 – Fang (Ed.)
© *2016 Taylor & Francis Group, London, ISBN 978-1-138-02973-6*

Research on driving risk discrimination based on the Fisher analysis method

Xu Bao
Key Laboratory for Traffic and Transportation Security of Jiangsu Province, Huaiyin Institute of Technology, Huai'an, China

Lixing Yan
Intelligent Transport System Research Center, Wuhan University of Technology, Wuhan, China

Shengxue Zhu & Jun Zhou
Key Laboratory for Traffic and Transportation Security of Jiangsu Province, Huaiyin Institute of Technology, Huai'an, China

ABSTRACT: To prevent the occurrence of traffic accidents, early discrimination of driving risk and timely dissemination of warning to the driver have become a research hot spot. In this paper, driving simulator is used as the equipment for data collection. A total of 120 groups of data are collected. Of which, 100 groups are taken as learning samples and another 100 groups are taken as predicting samples. According to the data analyzed from driving simulation, six main factors are selected as the variables for driving risk discrimination. Therefore, the Fisher analysis method is used to build the driving risk discrimination model. The discrimination results indicate that the Fisher discrimination model can classify the driving risk effectively, and the accuracy prediction reaches up to 95% (a higher prediction precision compared with the logistic regression model, which is 85%). This paper can provide an accurate prediction of potential risk, and give early warning to motorists to improve the driving safety.

Keywords: driving risk, discrimination, Fisher analysis method

1 INTRODUCTION

Traffic accidents harm human seriously because about 1.25 million people die in road accidents every year worldwide. One of the most important reasons for road accidents is that drivers, important participants of road traffic, ignore or misestimate the driving risk easily (J.Q. Wang, 2015). Therefore, how to identify driving risk correctly and provide warning to drivers timely is becoming a key technique to improve driving safety.

As with the factors influencing traffic safety, drivers, and vehicles, the driving environment are also factors that have impacts on the accuracy of the driving risk discrimination model. However, a literature review has indicated that most driving risk discrimination models focus only on one of the aforementioned factors. Segovia-Gonzalez found the relationship between the driver's experience, sex, and the driving risk (M.M, 2009). Costa and McCrea studied the association between personality characteristics of drivers and driving risk (L.Q. Peng, 2014). Peng Liqun explored how the driver's behavior influences on the driving risk.

Cai Xiaonan developed a discrimination model to predict driving risk in rainy days (X.N. Cai, 2016).

Based on the single-factor analysis, many scholars have used different methods to develop the discrimination model. Guo Feng and Fang Youjia used the K-mean cluster method to discriminate different drivers according to three risk levels (F. Guo, 2013). Ye Xin developed a logistic regression model to research the injuries caused by vehicle crash characteristics (X. Ye, 2015). Khan set up the structural equation model to find the relationship between driving risk and driving skills (S.U.R. Khan, 2015). Choi applied a latent class analysis to classify old drivers into four risk groups (N.G. Choi, 2015).

In summary, most of the research conducted on driving risk prediction in China and abroad have only considered a single factor, and hardly considered the common effect of different factors. In contrast, discrimination accuracy is crucial to the model, while the accuracy of existing methods has room for further improvement. The Fisher analysis method is a traditional technique for discrimination, and has several advantages such as high accuracy and efficiency. This method has been applied in different areas.

Based on this method, a number of simulation experiments were conducted in this paper, and the statistic method was used to extract the impact factors of driving risk from the drive–vehicle–environment diverse perspective. Finally, the Fisher analysis method is adopted to estimate the driving risk state. The results of the study can be used in risk warning and auxiliary driving.

The rest of the paper is organized as follows. The estimation of driving risk using the Fisher analysis method is shown in Section 2. Experiments for data collection are presented in Section 3. Modeling by the Fisher analysis method and its effects are discussed in Section 4. Finally, conclusion is given in Section 5.

2 FISHER ANALYSIS METHOD

To overcome the problem of dimensionality, the Fisher analysis method maps the data on a high-dimensional space to a low-dimensional space. Suppose that there are two different totalities G_1, G_2, and two groups of samples including n_1, n_2 p-dimension samples that are extracted from G_1, G_2, respectively.

Suppose that the discrimination model is:

$$y_i^k = \sum_{l=1}^{p} c_l x_{il}^k \qquad (1)$$

Thus, the average value of two groups of samples, also known as center of gravity of totality, is given by:

$$\bar{y}^k = \sum_{l=1}^{p} c_l \bar{x}_l^k \qquad (2)$$

$$\bar{x}_l^k = \frac{1}{n_k} \sum_{i=1}^{nk} x_{il}^k \ (k=1,2) \qquad (3)$$

The best discrimination linear function must satisfy the following two conditions simultaneously: the distance between two different center of gravity should be as big as possible, and the sum of squares of deviations of G_1 and G_2, respectively, should be as small as possible. Thus, let I be the ratio of the distance and the sum of squares:

$$I = \frac{(\bar{y}^1 - \bar{y}^2)^2}{\sum_{i=1}^{n_1}(y_i^1 - \bar{y}^1)^2 + \sum_{i=1}^{n_2}(y_i^2 - \bar{y}^2)^2} \qquad (4)$$

The value of I should be as big as possible:
Let $P = P(c_1, c_2, ..., c_p) = (\bar{y}^1 - \bar{y}^2)^2$,

$$Q = Q(c_1, c_2, ..., c_p) = \sum_{i=1}^{n_1}(y_i^1 - \bar{y}^1)^2 + \sum_{i=1}^{n_2}(y_i^2 - \bar{y}^2)^2 \qquad (5)$$

So, the value of $c_1, c_2, ..., c_p$ can be calculated based on the extremum principle.

The discrimination critical value y_0 can be computed as:

$$\bar{y}^{(1)} = \sum_{l=1}^{p} c_k \bar{X}_l^{(1)} \ \bar{y}^{(2)} = \sum_{l=1}^{p} c_k \bar{X}_l^{(2)} \qquad (6)$$

Suppose that there is an observation data $X = (x_1, \cdots x_p)^T$. We can obtain the y value by the discriminant. When $\bar{y}^{(1)} > \bar{y}^{(2)}$, $y > y_0$, $X \in G_1$, else $X \in G_2$. When $\bar{y}^{(1)} < \bar{y}^{(2)}$, $y > y_0$, $X \in G_2$, else $X \in G_1$.

3 DATA COLLECTION FOR ANALYSIS OF INFLUENCING FACTORS OF DRIVING RISK

3.1 Experimental equipment

To increase the experiment's efficiency, several devices are selected in the experiments, as shown in Fig. 1. A single-channel driving simulator (Fig. 1 (A)) is applied to simulate vehicles, and the urban road environment (Fig. 1 (B)) including many signal intersections is selected as the experimental scene in the driving simulator. Meanwhile, the Biograph Infiniti system (Fig. 1 (C)), which can collect the driver's physiological indices timely, is used to compute the driver's state. Furthermore, the vehicle information collecting system that can timely collect the information of vehicle movement is also applied. Finally, data collecting interface (Fig. 1 (D)) is designed for the dialogue between human and computer.

3.2 Experimental process

A total of 25 volunteers (15 males and 10 females) participated in the driving experiment carried out in this research. All volunteers had a valid driver

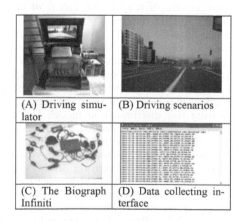

(A) Driving simulator	(B) Driving scenarios
(C) The Biograph Infiniti	(D) Data collecting interface

Figure 1. Experimental equipment.

Table 1. Influence of two states of variables on driving disk.

Variable	Discrimination	States	Comment
Driver's experience (x_1)	Experience	0-experienced 1-non-experienced	Obtained from the driver's experience and the driver's mileage
Driver's sex (x_2)	Sex	0-male 1-female	
Driver's state (x_3)	Driver's state	0-good 1-bad	Obtained from physical data
Vehicle's state (x_4)	Vehicle's state	0-good 1-bad	Obtained from vehicle data
Vehicle's speed (x_5)	Speed	0-< = 40km/h 1->40km/h	Vehicle's speed
Environment (x_6)	Weather and road	0-good 1-bad	Including visibility and slippery

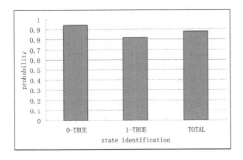

Figure 2. The accuracy of driving risk discrimination for training samples.

license, and their average age was 26.2 years. Their average driving experience was 2.5 years.

First, preliminary driving and training experiments were performed to make the volunteers familiar with vehicle driving simulator easily and to decrease the experiment time. Then, the volunteers filled out a questionnaire containing a number of basic driver attributes, such as driver's sex, driving license, and age. They accepted the task and requirements as drivers. Third, the formal experiments were conducted, in which the volunteers drove as usual and the experimental data were collected by the Biograph Infiniti system and the vehicle information collecting system.

The driving experiment comprised a 25 km-long road consisting of two expressways, three large-scale business districts, eight traffic lights, two tunnels, and many traffic scenarios such as the bad road environment and waiting for a red light. The scenarios covered all the scenes that the volunteers would meet in their daily driving.

Many variable data such as driver's state, vehicle's state, and environment were collected by the Biograph Infiniti system and the vehicle data collecting system. To implement the driving risk discrimination, a total of 120 groups of accident data and security data were extracted from the experimental data. Moreover, six main variables were selected by analyzing the data: driver's experience, driver's sex, driver's state, vehicle state, vehicle speed, and environment. These variables are divided into two states, as shown in Table 1.

The 120 groups of six factors and driving risk were obtained, as shown in Table 2, using the expert scoring and intelligent classification methods.

4 MODELING BASED ON THE FISHER ANALYSIS METHOD

The 100 samples, shown in Table 2, were regarded as the sample data to train the Fisher analysis model. Two different states of driving risk are considered to distinguish the driving risk level.

Suppose that two covariance matrices of different totality are equal. According to the Fisher discrimination method, the driving risk discrimination model is:

$$y = -0.349 + 1.038x_1 + 1.634x_2 + 0.971x_3 - 0.766x_4 - 1.593x_5 + 0.613x_6 \qquad (6)$$

To test the effect of the Fisher risk discrimination model, all the six indices of 100 training samples obtained in the above experiments were taken as the input of the model. Then, the outputs of the model were compared with the initial driving risk value. The results indicated that the accuracy of the discrimination reached up to 88%. The result is shown in Fig. 2. In terms of the accuracy of the discrimination, this model is qualified for driving risk discrimination.

The model was also validated by using another 20 samples that could not participate in the training and learning experiments. The accuracy of discrimination reached up to 95%, and only one secure sample showed an error. Meanwhile, the logistic regress model was also applied to conduct this research. However, the accuracy was only 85%, which was much lower than the Fisher analysis model. Thus, this model can be obviously used to discriminate the driving risk.

5 CONCLUSION

A total of six factors extracted based on the analysis, and 120 groups of data were achieved from driving simulator experiments. Expert scoring and experiencing method were selected to classify the entire variables into two states. Then, the Fisher analysis method was applied to develop the fit model between the driving risk and the six impact factors. Finally, 100 sample data were used to determine the model parameters, and another 20 samples were selected to measure the effect of the Fisher discrimination model. The discrimination results indicated that the Fisher discrimination model had a higher accuracy prediction of potential driving risk discrimination (up to 95%) than the logistic regression model. This model can also be used in auxiliary driving, as the accurate discrimination of the driving risk can increase the driving safety.

In addition, due to the limited data samples and experimental conditions, only six variables were chosen as the impact factors to build the model. In the future, the real vehicle experiment could be considered to get more actual traffic environment, which can make the model become more typical and practical. Meanwhile, because of the economic and time limitations, only two states of risk driving were considered in this model. This would be much insufficient if there is a demand for getting a more accurate discrimination.

ACKNOWLEDGMENT

This work was financially supported by the National Natural Science Foundation of China (Grant no. 51308246), the Social Science Foundation of Jiangsu Province (Grant no. 14EYD010), the peak of the six talents of Jiangsu Province (Grant no. WLW009), the University Natural Science major Basic Project of Jiangsu Province (Grant no. 15KJA580001), the Science and Technology Project of Ministry of Housing and Urban-Rural Development of China (Grant no. 2014-K5-013), and the QingLan Project of Jiangsu Province.

REFERENCES

Cai, X.N., C. Wang, S.D. Chen, J. Lu, Model Development for Risk Assessment of Driving on Freeway under Rainy Weather Conditions, Plos One. 11 (2016) 1–16.

Choi, N.G., D.M. Dinitto, C.N. Marti, Older adults who are at risk of driving under the influence: A latent class analysis, Psychology of Addictive Behaviors. 29 (2015) 725–732.

Costa, P.T., R.R. McCrea. Revised NEO personality inventory (NEO PI-R). The SAGE Handbook of Personality Theory and Assessment: personality Measurement and Testing, SAGE Publications Ltd, California. 1992.

Guo, F., Y.J. Fang. Individual driver risk assessment using naturalistic driving data, Accident Analysis & Prevention. 61 (2013) 3–9.

Khan, S.U.R., Z.B. Khalifah, Y. Munir, T. Islam, T. Nazir, H. Khan, Driving behaviours, traffic risk and road safety: comparative study between Malaysia and Singapore, International Journal of Injury Control and Safety Promotion, 22 (2015) 359–367.

Li, Z., Y.W. Pang, Y. Yuan, J. Pan, Relevance and irrelevance graph based marginal Fisher analysis for image search reranking, Signal Processing. 121 (2016) 139–152.

Machado-León, J.L., J. de Oña, R. de Oña, L. Eboli, G. Mazzulla, Socio-economic and driving experience factors affecting drivers' perceptions of traffic crash risk, Transportation Research Part F: Traffic Psychology and Behaviour. 37 (2016) 41–45.

Peng, L.Q., C.Z. Wu, Z. Huang, M. Zhong, Novel Vehicle Motion Model Considering Driver Behavior for Trajectory Prediction and Driving Risk Detection, Transportation Research Record: Journal of the Transportation Research Board. 2434 (2014) 123–134.

Segovia-Gonzalez, M.M., F.M. Guerrero, P. Herranz, Explaining functional principal component analysis to actuarial science with an example on vehicle insurance, Insurance: Mathematics and Economics. 45 (2009) 278–285.

Shi, H.T., J.C. Liu, Y.H. Wu, K. Zhang, L.X. Zhang, P. Xue. Rolison, A. Fault diagnosis of nonlinear and large-scale processes using novel modified kernel Fisher discriminant analysis approach, International Journal of Systems Science. 47 (2016) 1095–1109.

Sugiyama, M., Dimensionality reduction of multimodal labeled data by local fisher discriminant analysis, Journal of machine learning research. 8 (2007) 1027–1061.

WHO. Global status report onroad safety 2015, WHO press, Switzerland, 2015.

Wang, G.Q., N.F. Shi, Y.X. Shu, D.T. Liu, Embedded Manifold-Based Kernel Fisher Discriminant Analysis for Face Recognition, Neural Processing Letters. 43 (2016) 1–16.

Wang, J.Q., Y. Zheng, X.F. Li, C.F. Yu, K.J. Kodaka, K.Q. Li, Driving risk assessment using near-crash database through data mining of tree-based model, Accident Analysis & Prevention. 84 (2015) 54–56.

Ye, X., G. Poplin, D. Bose, A. Forbes, S. Hurwitz, G. Shaw, J. Crandall, Analysis of crash parameters and driver characteristics associated with lower limb injury, Accident Analysis & Prevention. 83 (2015) 37–46.

Yildirim-Yenie, Z., E. Vingilis, D.L. Wiesenthal, R.E. Mann, J. Seeley, A. High-Risk Driving Attitudes and Everyday Driving Violations of Car and Racing Enthusiasts in Ontario, Canada, Traffic Injury Prevention. 16 (2015) 545–551.

Energy Science and Applied Technology ESAT 2016 – Fang (Ed.)
© 2016 Taylor & Francis Group, London, ISBN 978-1-138-02973-6

Research of coordinated relation methods of land use and urban traffic for unit regulatory detailed planning

Jiadong Wang, Zhenzhou Yuan & Ye Zhang
MOE Key Laboratory for Urban Transportation Complex Systems Theory and Technology, Beijing Jiaotong University, Beijing, China

ABSTRACT: In this paper the basic theory is studied as well as quantitative methods of coordinated relations between land use and traffic for unit regulatory detailed planning. According to the selection principle of the evaluation index system, the land use and transport index system for unit regulatory detailed planning is constructed. Using principal component analysis algorithm to get principal components, a new DEA model with AHP restraint cone is proposed in which the land use system and traffic system are taken as input-output system mutually. A coordination degree membership function is constructed to determinate the degree of coordination between land use and traffic system. Finally the DEA model with AHP restraint cone has been proved rational by taking NH03 management unit land use planning in Handan city as an example.

Keywords: urban transport, land use, coordination evaluation, unit regulatory detailed planning, DEA

1 INTRODUCTION

Urban land use and traffic coordination is an important way to solve urban traffic problems and promote the urban sustainable development. Land use and urban transportation network coordination refers to that in the process of urban land development, land development intension should match with traffic placement scheme and road network scale (Zheng Lu, 2011).

In this paper, a DEA (Data Envelopment Analysis) model is established on the basis of index system of land use and traffic system. With AHP restraint cone, PCA (Principal Component Analysis) is used to obtain aggregative indicator of traffic system. Then making land use and traffic system as input-output system mutually, condition degree efficiency value can be calculated. According to the efficiency value we can decide whether a certain block in administrative unit is coordinative between land use and traffic. The DEA model with AHP restraint cone conquers both limitation of AHP method which relies too much on subjectivity and DEA model which ignores preference.

2 LITERATURE REVIEW

Zheng Lu proposed a coordinated evaluation model of land use and traffic system based on the relationship between floor area ratio and road network load. Hu Dong (Hu Dong, 2013) proposed a model based on job-housing balance and traffic capacity analysis. Qu Tao (Qu Tao, 2012) put forward a planning scheme unified urban land use and road traffic construction to shape a high efficient urban spatial structure. Hou Jinhua (Hou Quanhua, 2015) created a collaborative optimization of land use intensity and traffic capacity based on hierarchical regulatory detailed planning and traffic planning integration compilation. Luo Ming, Chen Yanyan (Luo Ming, 2008) proposed a coordinated evaluation model based on DEA method and membership function in fuzzy mathematics. J. D. Hunt, D. S. Kriger, E. J. Miler (J. D. Hunt, 2005) provides a detailed review of six frameworks for modeling urban land-use–transport interaction.

3 LAND USE AND TRAFFIC COORDINATION INDEX SYSTEM

The index of land use system must be able to comprehensively reflect how intensive the regulatory planning unit land is and to promote job-housing balance. Therefore, the indices of land use system are set as land mixedness, integrated floor area ratio, population density and local employment proportion. The evaluation indices of urban road network layout and network operation efficiency include road network density, artery network gap, road area ratio, road area per capita, network saturation, network accessibility, network reliability, level of service of external passageway and

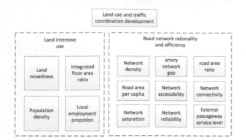

Figure 1. Hierarchy of appraisal system.

network connectivity. Index system of the coordinated development of land use and transportation is shown in Figure 1.

4 EVALUATION WITH A DEA MODEL WITH AHP RESTRAINT CONE

According to the concept of constraint cone in mathematical programming, a DEA model with AHP constraint cone is established which combines the objective analysis of DEA evaluation and AHP subjective judgment and is more suitable for practical problems.

Supposing n decision making units marked $DMU_i (i=1,2,...,n)$, there are m input indices and n output indices. The input index vector is assumed as $X = (x_1,x_2,...,x_m)^T$ and the output index vector is $Y = (y_1,y_2,...,y_s)^T$. For decision making unit DMU_j there is a corresponding efficiency evaluation index below. In formula (1), w and μ are weight coefficient.

$$h_j = \frac{\mu^T y_j}{w^T x_j} = \frac{\sum_{r=1}^{s} \mu_r y_{rj}}{\sum_{i=1}^{m} w_i x_{ij}}, \quad j=1,2,\cdots,n \quad (1)$$

$$\max P_i = \sum_{r=1}^{s} \mu_r y_{rj_0}$$

$$s.t. \sum_{r=1}^{s} \mu_r y_{rj} - \sum_{i=1}^{m} w_i x_{ij} \leq 0, \quad j=1,2,\cdots,n$$
$$\sum_{i=1}^{m} w_i x_{ij_0} = 1 \quad (2)$$
$$w \geq 0, \mu \geq 0$$

Making the efficiency index of decision making unit j_0 as the objective function and the efficiency indices of all decision making units as constraint conditions, a CCR model is constructed. With the help of Charnes-Cooper transform, the preceding model can be transformed into linear programming model shown in formula (2).

In order to reflect the preference of policy makers on the various indicators, AHP method is used to impose some restrictions on the weight choice, that is, increasing AHP restraint cone in the DEA evaluation model. AHP judgment matrices of land use and transportation system are marked respectively as $C_m' = (c_{ij})_{m \times m}$ and $B_s' = (b_{ij})_{s \times s}$, which meet the following demand $c_{ij} > 0, c_{ij} = c_{ji}^{-1}, c_{ii} = 1; b_{ij} > 0, b_{ij} = b_{ji}^{-1}, b_{ii} = 1$. Set the maximum eigenvalues of two matrices as λ_C and λ_B respectively and $C = C' - \lambda_C E_m$, $B = B' - \lambda_B E_s$. E_m and E_s are unit matrices with order of m and s respectively. And the DEA evaluation model with AHP constraint cone is shown in formula (3).

$$\max P_{j_0} = \mu^T Y_{j_0}$$
$$s.t. \ W^T X_j - \mu^T Y_j \geq 0, \quad j=1,2,\cdots,n$$
$$W^T X_{j_0} = 1 \quad (3)$$
$$W \in V, \mu \in U$$

In formula (3)

$$V = \{W | CW \geq 0, W \geq 0\}; U = \{|B\mu \geq 0, \mu \geq 0\}.$$

When serving land use system as input, the efficiency of land use planning under the existing traffic facilities can be worked out, being marked as value α, while serving traffic system as input, the efficiency of road transport infrastructure planning under the existing land planning scheme can be attained, being marked as value β. By constructing membership functions, both one-way and mutual coordination degrees between land use and transportation system can be worked out.

Because the DEA value is in the range of 0 to 1, the membership function can be just defined as $\mu(x) = x$. The coordination degree of land use system to traffic system marked $\mu_L = \alpha$ means that for land use system how approximately traffic planning scheme is close to the traffic demand of land development scheme. On the contrary, the corresponding degree marked $\mu_T = \beta$ means that for traffic system how approximately land use system is close to the land demand of traffic planning scheme.

The mutual coordination degree between land use system and traffic system in a block is defined as a valve marked μ_{TL}, which is shown as follow:

$$\mu_{TL} = \frac{\min(\mu_T, \mu_L)}{\max(\mu_T, \mu_L)} \quad (4)$$

In formula (4), the closer μ_T is to μ_L, the closer μ_{TL} is to 1, which is in accordance with higher coordination degree between traffic planning and land use planning scheme. With value μ_T and μ_L, we can analyze which indices result in the lower coordina-

Table 1. Judgment matrix of land intensification evaluation.

	Land mixedness	Integrated floor area ratio	Population density	Local employment proportion
Land mixedness	1	1/2	1/4	3
Integrated floor area ratio	2	1	1/3	5
Population density	4	3	1	6
Local employment proportion	1/3	1/5	1/6	1

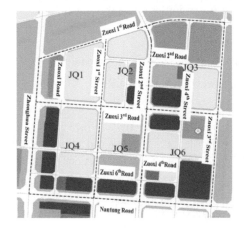

Figure 2. Block division in management unit.

Table 2. Land use index values of each block.

	Land mixedness	Integrated floor area ratio	Population density	Local employment proportion
	X_1	X_2	X_3	X_4
JQ_1	0.48	1.44	293.11	0.94
JQ_2	0.52	1.30	270.92	0.68
JQ_3	0.63	0.92	107.70	1.81
JQ_4	0.64	1.39	208.54	2.35
JQ_5	0.72	1.00	159.01	1.66
JQ_6	0.74	1.25	129.93	2.67

Table 3. Transport system index values of each block.

	y_1	y_2	y_3	y_4	y_5	y_6	y_7	y_8	y_9
JQ_1	10.53	0.25	0.50	8.64	0.12	2.29	0.42	1.00	0.44
JQ_2	13.20	0.30	0.35	11.02	0.09	2.60	0.41	1.00	0.50
JQ_3	10.79	0.18	0.55	16.83	0.12	2.73	0.41	1.00	0.62
JQ_4	11.31	0.28	0.37	13.51	0.10	2.83	0.39	0.88	0.53
JQ_5	11.07	0.25	0.35	15.72	0.12	2.30	0.38	0.88	0.53
JQ_6	10.31	0.25	0.55	19.38	0.11	2.77	0.40	0.88	0.50

tion in this block and then we can adjust the indicators of the uncoordinated block under the premise of total amount control in management unit.

5 CASE VERIFICATION

Based on the project Handan Urban Comprehensive Traffic Planning, the case to verify the DEA evaluation model with preference constraints is the administrative unit NH03 of south lake planning unit in Handan city. Firstly, the traffic flow data of road network in planning stage must be worked out through traffic demand forecasting. The administrative unit was divided into 68 traffic zones. Then according to the four-step method traffic planning model, the four steps including trip generation, trip distribution, model spilt and traffic assignment were performed successively.

In figure 2 the administrative unit NH03 is divided by main roads into six internal blocks named JQ1 to JQ6 serving as decision making units in the DEA model

In the index system set up in the preceding part of the text, the evaluation of land use intensification includes only four indicators between which the correlation is quite weak. Referring to the summary of relevant references (Hu Dong, 2013; Qu Tao, 2012), a nine-magnitude scale judgment matrix including four indicators is built as shown in table 1.

The consistency verification (λ_{Cm} = 4.096, CI_{Cm} = 0.032, CR_{Cm} = 0.036<0.1) shows that the judgment matrix can be accepted. Calculation of each block of land use and transportation indicators data is as shown in Table 2 and 3.

The principal component analysis of those traffic system indicators above with Statistic Package for Social Science (SPSS) is shown in Table 4. The accumulated variance contribution rates of the first three principal components are more than 85% while the other components have much lower rates. So the first three principal components are selected of which the eigenvectors are shown in Table 5.

Values form y_1 to y_9 successively represent: network density, road area ratio, artery network gap, road area per capita, network accessibility, network connectivity, network saturation, network reliability, level of service of external passageway.

The first three principal components (index Y_1, Y_2, Y_3) are linear representations of the original traffic indices. For example, the first principal component Y_1 can be represented as follow:

Table 4. Eigenvalue and contribution rate of the principal component.

Principle component	Eigenvalue	Variance contribution rates	Accumulated variance contribution rates
1	3.448	38.311%	38.311%
2	3.282	36.467%	74.778%
3	1.18	13.111%	87.889%

Table 5. Eigenvector of the principal component.

	Principle component		
	1	2	3
Network density	0.070	0.899	0.398
Road area ratio	0.130	0.863	−0.423
Artery network gap	0.582	0.355	−0.616
Road area per capita	0.819	−0.412	0.192
Network accessibility	0.323	0.915	0.028
Network connectivity	0.884	0.111	0.023
Network saturation	0.874	0.356	−0.283
Network reliability	0.711	0.076	0.568
Level of service of external passageway	−0.512	0.673	−0.153

Table 6. Judgment matrix of principal component.

B's	Y_1	Y_2	Y_3
Y_1	1	1	5
Y_2	1	1	5
Y_3	1/5	1/5	1

Table 7. Input & output index values.

	X_1	X_2	X_3	X_4	Y_1	Y_2	Y_3
JQ_1	0.48	1.44	293.11	0.94	15.62	16.84	4.15
JQ_2	0.52	1.30	270.92	0.68	19.14	20.91	5.37
JQ_3	0.63	0.92	107.70	1.81	23.19	14.02	5.63
JQ_4	0.64	1.39	208.54	2.35	20.94	17.17	5.15
JQ_5	0.72	1.00	159.01	1.66	22.06	14.85	5.35
JQ_6	0.74	1.25	129.93	2.67	24.43	14.01	5.67

$$Y_1 = 0.07y_1 + 0.13y_2 + 0.582y_3 + 0.819y_4 + 0.323y_5 + 0.884y_6 + 0.874y_7 + 0.711y_8 - 0.512y_9$$

After traffic system indicators are simplified, the judgment matrix B's involved traffic main components are built based on contribution rates, as shown in Table 6.

The consistency verification shows: $\lambda B's = 3$, $IB's = CRB's = 0 < 0.1$. Then the indices of both land use system and traffic system to be applied in DEA model are shown in Table 7.

Now the indices of land use system marked X and principal components of traffic system marked Y can be taken into the DEA model and the calculation can be implemented with Lingo or MATLAB. For example, for decision making unit JQ1 the DEA model is established as follows:

$$\text{Max } P_1 = 15.62\mu_1 + 16.84\mu_2 + 4.15\mu_3$$

$$\text{s.t. } 0.48w_1 + 1.44w_2 + 293.11w_3 + 0.94w_4 - 15.62\mu_1 - 16.84\mu_2 4.15\mu_3 \geq 0$$

$$0.74w_1 + 1.25w_2 + 129.93w_3 + 2.67w_4 - 24.43\mu_1 - 14.01\mu_2 - 5.67\mu_3 \geq 0$$

$$0.48w_1 + 1.44w_2 + 293.11w_3 + 0.94w_4 = 1$$

$$(w_1, w_2, w_3, w_4)^T \in V, (\mu_1, \mu_2, \mu_3)^T \in U$$

$$V = \left\{ \begin{bmatrix} w_1 \\ w_2 \\ w_3 \\ w_4 \end{bmatrix} \begin{bmatrix} -3.096 & 0.5 & 0.25 & 3 \\ 2 & -3.096 & 0.333 & 5 \\ 4 & 3 & -3.096 & 6 \\ 0.333 & 0.2 & 0.167 & -3.096 \end{bmatrix} \times \begin{bmatrix} w_1 \\ w_2 \\ w_3 \\ w_4 \end{bmatrix} \geq 0 \cap \begin{bmatrix} w_1 \\ w_2 \\ w_3 \\ w_4 \end{bmatrix} \geq 0 \right\}$$

$$U = \left\{ \begin{bmatrix} \mu_1 \\ \mu_2 \\ \mu_3 \end{bmatrix} \begin{bmatrix} -2 & 1 & 5 \\ 1 & -2 & 5 \\ 0.2 & 0.2 & -2 \end{bmatrix} \begin{bmatrix} \mu_1 \\ \mu_2 \\ \mu_3 \end{bmatrix} \geq 0 \cap \begin{bmatrix} \mu_1 \\ \mu_2 \\ \mu_3 \end{bmatrix} \geq 0 \right\}$$

The optimal solution is $(w_1, w_2, w_3, w_4)^* = (0.0009, 0.0016, 0.0034, 0.0004)$ with $(\mu_1, \mu_2, \mu_3)^* = (0.0096, 0.0096, 0.0019)$. The optimal value of objective function marked P* equals to 0.3204. In turn, when the indices of traffic system are served as input and the land use system as output, the optimal solution of traffic system is $(w_1, w_2, w_3, w_4)^* = (0.0009, 0.0016, 0.0034, 0.0004)$ with $(\mu_1, \mu_2, \mu_3)^* = (0.03, 0.03, 0.006)$ and the optimal value P* equals to 1.

In the same way other decision making units can be evaluated with the results of effective values and coordination degrees between land use system and traffic system, as shown in Table 8.

It can be inferred form Table 8 that the coordination degree of JQ1, JQ2, JQ3 and JQ6 are within the range between 0.3 and 0.6, which means the planning land use system and traffic system of these four blocks are in uncoordinated condition. The coordination degree of decision making units JQ4 and JQ5 are higher means that they are in coordinated condition. The main reason for higher μ_L is that the land development intensity is

Table 8. Effective values of each block.

	μT	μL	μTL
JQ$_1$	0.32	1	0.32
JQ$_2$	0.43	0.75	0.57
JQ$_3$	1	0.32	0.32
JQ$_4$	0.53	0.61	0.87
JQ$_5$	0.67	0.48	0.71
JQ$_6$	0.86	0.38	0.44

Table 9. Effective values of each block calculated by original DEA.

	μT	μL	μTL
JQ$_1$	0.88	1	0.88
JQ$_2$	1	0.87	0.87
JQ$_3$	1	0.98	0.98
JQ$_4$	0.89	1	0.89
JQ$_5$	0.96	1	0.96
JQ$_6$	0.90	1	0.90

so high that the traffic demand generated exceeds traffic capacity of traffic planning scheme. The land development of JQ3 and JQ6 is relatively reasonable so traffic planning scheme can effectively meet the current traffic demand. In turn the traffic capacity of JQ3 and JQ6 is much higher than the traffic demand and the utilization rate of planned traffic facility will be low, which accounts for the lower value $μ_L$. Keeping total planning amounts of administrative units, planning departments can appropriately increase the Floor Area Ratio (FAR) of residential and commercial land in JQ3 and JQ6, while they can reduce the intensity of land development of JQ1 and JQ2 especially FAR of residential land to ensure land and traffic system of the whole planning unit coordinate with each other.

By contrast, if the land use system serves as input and the traffic as output in the original DEA model without AHP constraint cone, the conclusion of coordination degree is shown in Table 9.

Comparing the results of the two models, Table 9 shows that most blocks are in high coordinated condition. While evaluation with the model with AHP constraint cone, the result shows that most blocks are in low coordinated condition. Through this example it can be illustrated that the

DEA model with AHP restraint cone can reflect the true coordination degree of land use and traffic system of regulatory detailed planning more accurately than the original model.

6 CONCLUSION

The evaluation index system and model are studied to evaluate coordinated relations between land use and traffic for unit regulatory detailed planning. Firstly, the land use and transport index system is constructed. Then using PCA to get three principal components of traffic system, a new DEA model with AHP restraint cone is proposed. A membership function is constructed to determinate the coordination degree between land use and traffic system before giving planning adjustment and suggestion under the total planning amount control. Through case verification it can be illustrated that DEA model with AHP restraint cone can reflect the true coordination degree of land use and traffic system more accurately than the original model.

REFERENCES

Hou Quanhua, Duan Yaqiong, Ma Rongguo. Collaborative optimization of land use intensity and traffic capacity under hierarchical control rules of regulatory detailed planning. Journal of Chang'an University (Natural Science Edition). 2015, 35(?)

Hu Dong, Analysis of Traffic Capacity in Regulatory Detailed Planning Based on Job-housing Balance. Annual National Planning Conference 2013.

Hunt, J.D., D.S. Kriger & E.J. Miler, Current operational urban land—use–transport modeling frameworks: A review. TRANSPORT REVIEWS 2005,25(3).

Luo Ming, Chen Yanyan, Liu Xiaoming, Study on Coordination Degree Model between Urban Transport and Land Use. Journal of Wuhan University of Technology (Transportation Science & Engineering) 2008,32(4).

Qu Tao. The Solution of City Development Problems Based on the Combination of Traffic and Land-use: Take Tianjin for an Example. Urban Studies, 2012,19(11).

Zheng Lu. Research of Coordination Relationship Between Traffic and Land Use in Regulatory Detailed Planning. Master Dissertation of Beijing Jiaotong University in 2011.

Energy Science and Applied Technology ESAT 2016 – Fang (Ed.)
© 2016 Taylor & Francis Group, London, ISBN 978-1-138-02973-6

Author index